Genomics of Crucifer's Host-Resistance

Govind Singh Saharan • Naresh K. Mehta •
Prabhu Dayal Meena

Genomics of Crucifer's Host-Resistance

Springer

Govind Singh Saharan
Department of Plant Pathology
CCS Haryana Agricultural University
Hisar, Haryana, India

Naresh K. Mehta
Department of Plant Pathology
CCS Haryana Agricultural University
Hisar, Haryana, India

Prabhu Dayal Meena
Crop Protection Unit
ICAR-Directorate of Rapeseed-Mustard Research
Bharatpur, Rajasthan, India

ISBN 978-981-16-0864-3 ISBN 978-981-16-0862-9 (eBook)
https://doi.org/10.1007/978-981-16-0862-9

This Springer imprint is published by the registered company Springer Nature Singapore Pte Ltd.
The registered company address is: 152 Beach Road, #21-01/04 Gateway East, Singapore 189721, Singapore

Foreword

Cruciferous plants are challenged by large number of biotic and abiotic stresses at global level. Most of these result in deterioration of crop quality and yield. In agriculture cropping system, plant protection is being practiced using different approaches such as chemical and biological control, alterations and improvement of various agronomic practices, integrated pest management (IPM), and cultivation of biotic and abiotic stress resistant cultivars. Among these approaches, use of resistance cultivars is the most economical, effective, and environmentally sound means to control different biotic stresses. Use of available resistance sources for specific diseases is being practiced extensively. The interaction between a host species and a pathogenic species is such that a host species often loses the race-specific resistance due to evolution of new races of the pathogen, and thus a new gene for resistance has to be incorporated in the host variety against this new race. So, identification of new specific genes for resistance is one of the crucial factors for the development of resistant varieties. The genes for resistance can now be transferred, combined, or pyramided using molecular tools and new techniques such as gene cloning and gene editing using CRISPER–Cas9 system which also can enhance understanding of resistance mechanism. The mapped disease resistance genetic loci and closely linked molecular markers to these loci can be used for marker-assisted selection in cruciferous resistance breeding of *Brassica* species.

The book "*Genomics of Crucifer's Host-Resistance*" authored by Prof. (Dr.) G. S. Saharan, Prof. (Dr.) Naresh K. Mehta, and Dr. Prabhu Dayal Meena is the seventh book in series on diseases of crucifers being published by Springer, Nature. Earlier books on *Albugo, Alternaria, Erysiphe*, *Hyaloperonospora*, *Plasmodiophora,* and *Sclerotinia* dealt with specific diseases in great details. The present book has been prepared after critical analysis of the world literature on all the major diseases of crucifers with detailed treatment of different aspects for better comprehension by the readers. The book contains the information on principles of host resistance, identification of sources of resistance, inheritance and transfer of disease resistance, host resistance signaling network system to multiple stresses, molecular mechanism of host resistance, management of disease resistance, and genomics of host resistance along with various techniques for the development of resistant cultivars through latest technology. The chapter on priorities areas of research on all major diseases of crucifers will motivate the researchers, teachers, and students to take further research work on these areas. Appropriate subheadings, photographs, figures, tables, and graphs in each chapter greatly enhanced the clarity and comprehension of the subject matter. I am sure that this book will provide much needed background and current information on host pathogen system of cruciferous crops for developing disease-resistant varieties.

My heartiest congratulations to the authors for bringing out their lifelong professional interest, and expertise in the preparation of this book. I am sure the book will be quite useful for the researchers, teachers, students, and all those who are concerned with research and development of cruciferous crops.

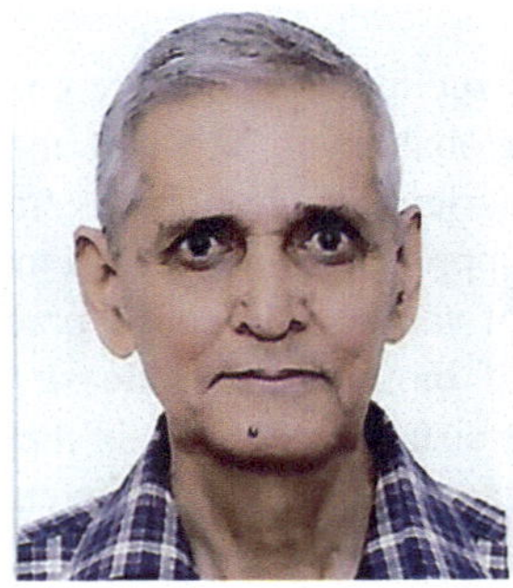

Former Dean, College of Agriculture,
CSKHPKVV
Palampur, India

B. M. Singh

Preface

Crucifers are challenged by large number of biotic and abiotic stresses but 16 pathogens are most important and widespread causing serious losses to these groups of crops. The variable losses in these crops have received attention of scientists to initiate the studies on genomics of crucifer's host resistance, host-pathogen interactions, and expression of molecular defense mechanisms actively operating in these crops. During the last three decades, substantial progress has been made on the investigations of mechanisms of host resistance using *Brassica* host-pathosystem as a model. The book "*Genomics of Crucifer's Host-Resistance*" has been attempted to present comprehensive information after scanning of large volume of available literature on fundamental and applied knowledge for developing varieties with resistance individually as well as to multiple major pathogens of crucifers like *Albugo, Alternaria, Erysiphe, Hyaloperonospora, Leptosphaeria, Plasmodiophora, Sclerotinia, Turnip mosaic virus, Verticillium,* and *Xanthomonas* of crucifers through the use of conventional as well as latest biotechnological approaches including transgenics with agronomically superior background. It is well known that wild plant species represent a gene pool that can potentially provide a rich source of durable, broad-spectrum disease resistance for use in breeding program. Species that are interfertile with a particular crop have been used for transferring monogenic resistance which is often genotype or race specific. Disease resistance can potentially be improved by altering the expression of defense regulation, through broad-spectrum resistance by overexpression of *NPR1,* regulator of acquired disease resistance. In plant disease resistance (R) genes, the nucleotide-binding site (NBS) plays an important role in offering resistance to pathogens. The complete genome sequences provide an important opportunity for researchers to identify and characterize NBS-encoding R-genes based on a comparative genomics approach. The evolutionary analysis of CNL-type NBS-encoding orthologous gene pairs suggested that orthologous genes in crucifers especially in *B. rapa* have undergone stronger negative selection than those in other *Brassica* species. Tandem duplication and whole-genome triplication (WGT) analyses revealed that after WGT of the *Brassica* ancestor, NBS-encoding homologous gene pairs on triplicated regions in *Brassica* ancestor which were deleted or lost quickly, but NBS-encoding genes in *Brassica* species experienced species-specific gene amplification by tandem duplication after divergence of *B. rapa* and *B. oleracea*. Molecular genetics investigations of disease resistance research of utilizing natural genetic variation (so-called *R*-genes) have revealed race-specific resistance to a wide assortment of pathogens now widely known by NB-LRR (**n**ucleotide-**b**inding site and **l**eucine-**r**ich **r**epeat domains). The use of DNA sequence from NB-LRR genes and other known *R*-gene are being used to develop more effective genetic markers for breeding of resistant varieties. The advances in the plant sciences indicated that signaling mechanisms governing genotype-specific resistance govern species-level resistance through *EDS1* (**e**nhanced **d**isease **s**usceptibility), restrictive host specialization, and molecular divergence. So, *EDS1* is an essential regulator of species-level resistance to some *Brassica* pathogens. The highly conserved protein, *SGT1* has also recently been reported as a key regulator for species-level resistance.

We have complied the information generated so far on all the major and minor diseases of crucifers in the form of published reports, research articles, popular scientific articles, books, bulletins, reviews, etc. in this book, which have been arranged in ten chapters, with appropriate headings and subheadings with the sections of references to consult original publications. It is a seventh book on the diseases of cruciferous crops series after *Sclerotinia*, *Albugo, Alternaria, Hyaloperonospora, Erysiphe,* and *Plasmodiophora* published by Springer Nature. The chapter-wise sections include the information, viz. principles of host resistance, identification of *R*-genes sources, inheritance of disease resistance, molecular mechanisms of disease resistance, host resistance signaling network system to multiple stresses, transfer of disease resistance, and management of disease resistance. A chapter on standardized, reproducible techniques has been included for the researchers of cruciferous crops for developing resistant cultivars. The last section deals with the gaps in understanding, knowledge of genomics, and offers suggestions for future research priorities in order to initiate the advance research programs on disease resistance. The subject matter has been vividly illustrated with photographs, graphs, figures, histogram, tables, and colored plates, which makes it stimulating, effective, and easy to comprehend by the readers. The headings and subheadings of each chapter have been arranged in numbered series to make the subject matter contiguous.

We believe that this book will be immensely useful to the researchers especially *Brassica* breeders, molecular biologists, pathologists, teachers, extension specialists, students, industrialists, farmers, and all others who are interested to grow healthy and profitable cruciferous crops all over the world. Any shortcomings, lacunae, and flaws in the book are responsibility of ours. Any suggestions by the readers are always a source of inspiration for the authors and suggestions for its improvement are most welcome.

Hisar, Haryana, India — Govind Singh Saharan
Hisar, Haryana, India — Naresh K. Mehta
Bharatpur, Rajasthan, India — Prabhu Dayal Meena

Acknowledgments

Authors are highly grateful to all the persons, scientists, publishers, societies, journals, institutes, websites, and all others whose valuable materials such as photographs (macroscopic, microscopic, electron micrographs, scanning electron micrographs), drawings, figures, histograms, graphs, tables, and flowcharts have been used through reproduction in the present document. The address of the author (s)/source (s) from where material was adopted whether institutes, websites, societies, etc. can be obtained from the reference which has been cited in the reference section of the book in each chapter. We have duly acknowledged all the authors for their excellent and valuable publications/articles in this manuscript by citing their names against each tables, figures, plates, dendrograms, etc. We have sought the necessary permission to reproduce their material in our document and majority of them have granted the permission to reuse them. The authors are sincerely thankful to all the scientists, publishers, journals, institutes, societies, and websites whose materials have been reproduced in one or the other form in the present manuscript but forgot to acknowledge their name(s) inadvertently. Their prompt help in providing permission, information, photographs, figures, and consents for reproduction in the present manuscript have been duly acknowledged and authors are grateful to all of them. There might be some errors, mistakes, and shortcomings in this manuscript. We shall greatly appreciate the healthy criticism and suggestions for the improvement of this publication in future.

Authors

Contents

1 Principles of Host Resistance . . . 1
1.1 Introduction . . . 2
1.2 The *Brassica* Crop Species and Their Wild Relatives . . . 9
1.3 *Brassica* Species Uses and Genetic Diversity . . . 9
1.4 The *Brassica* Wild Relatives: Coenospecies and Cytodemes . . . 11
1.5 Hybridization Between *Brassica* Species and Wild Relatives . . . 12
1.6 Interspecific Hybridization for Developing Disease Resistance *Brassica* Cultivars . . . 12
1.7 Identification and Use of R-Genes in *Brassica* Species . . . 13
1.8 R-Genes Recognition System . . . 15
1.9 Close Association of R-Genes to Confer Multiple Disease Resistance in *Brassica* . . . 15
1.10 Homoeologous Exchange: A Cause of R-Gene in *Brassica* . . . 18
1.11 Resistance Mechanisms Operating in *Brassica* . . . 19
1.12 Importance of *Arabidopsis* Model Host-Patho System in Genomics of Disease Resistance . . . 23
1.12.1 *Hyaloperonospora-Arabidopsis* Pathosystem . . . 23
1.12.2 *Albugo-Arabidopsis* Pathosystem . . . 24
1.12.3 Identification of Intracellular Receptors in *Arabidopsis* to Pathogens . . . 27
1.13 Phylogenetic Relationships Within and Among *Brassica* Species 30
1.14 *Brassica* Genome Complexity and R-Genes Identification . . . 32
1.15 Complexity of Genetic Variation in Hosts and Pathogens Variability . . . 34
1.16 High-Efficiency Integrated Breeding . . . 36
1.17 Deletion of R-Gene in *Brassica* and *Arabidopsis* . . . 36
1.18 Pathogen Perception and Induction of Host Resistance Genes . . 39
1.19 Challenges of *TuMV* Resistance Breeding . . . 40
1.20 Differential Genes Expression in *Brassica* Under Biotic Stresses 41
1.21 Significant Insight into Genomics of Host Resistance . . . 41
1.22 Approaches Used in the Genomics of Disease Resistance Breeding of Crucifers . . . 42
References . . . 46

2 Identification of *R*-Genes Sources . . . 65
2.1 Introduction . . . 66
2.2 *Brassica-Albugo* R-Genes . . . 67
2.2.1 Identification of R-Genes Genotypes . . . 67
2.2.2 Sources of Multiple Disease Resistance . . . 70
2.3 *Brassica-Alternaria* R-Genes . . . 73
2.3.1 Sources of Resistance . . . 73
2.3.2 Sources of Disease Resistance from Cruciferous Relatives . . . 74
2.3.3 Sources of Multiple Disease Resistance . . . 76
2.4 *Brassica-Colletotrichum* R-Genes . . . 78
2.4.1 Identification and Characterization of Resistance Genes . . . 78
2.5 *Brassica-Erysiphe* R-Genes . . . 79
2.5.1 Identification of R-Genes Genotypes Against Powdery Mildew . . . 79
2.5.2 Sources of Slow-Mildewing Resistance . . . 80
2.6 *Brassica-Fusarium* R-Genes . . . 86
2.6.1 Mapping of R-Genes of *Brassica* . . . 86
2.7 *Brassica-Hyaloperonospora* R-Genes . . . 87
2.7.1 Identification of R-Genes Genotypes . . . 87
2.7.2 Designation and Nomenclature of Downy Mildew Resistance Genes (R-genes) and Isolates (Races/Pathotypes) . . . 87
2.7.3 Mutational Approach to Identify Resistance Genes . . . 92
2.8 *Brassica-Leptosphaeria* R-Genes . . . 93
2.8.1 Identification of R-Genes Sources . . . 93
2.8.2 Novel Sources and Transfer of Genes . . . 94
2.8.3 Genomic Prediction to Identify Blackleg R-Genes of *Brassica* . . . 112
2.8.4 Identification of R-Genes in *Brassica* to *Leptosphaeria maculans* by Genome-Wide Association . . . 114
2.9 *Brassica-Plasmodiophora*: R-Genes . . . 116
2.9.1 Identification of R-Genes Sources . . . 116
2.9.2 Identification of Pathotype Specific R-Genes Sources . . 118
2.9.3 Identification of R-Genes by Genomic Approach . . . 121
2.9.4 Genetical Mechanism of R-Genes Sources . . . 124
2.9.5 Identification of R-Genes by Transcriptomic and Proteomic Approaches . . . 125
2.10 *Brassica-Sclerotinia* R-Genes . . . 127
2.10.1 Identification of Partial Resistance in *Brassica napus* to *Sclerotinia* . . . 127

2.10.2 Identification of Genomic Regions in Wild *Brassica oleracea* for Resistance to *Sclerotinia sclerotiorum* 127
2.10.3 Identification of R-Genes in *Brassica* Species to *Sclerotinia* 128
2.10.4 Identification of R-Genes Sources in *Brassica* Genotypes to *Sclerotinia* 129
2.10.5 Use of Oxalate Oxidase (OXO) to Identify R-Genes to *Sclerotinia* 132
2.11 *Brassica-TuMV* R-Genes 134
2.11.1 Identification of Pathotype Specific R-Genes Sources .. 134
2.12 *Brassica-Xanthomonas* R-Genes 134
2.12.1 Identification of R-Genes Sources in *Brassica* 134
2.12.2 Genetics of Sources of Resistance to *Xanthomonas* 138
2.12.3 Novel R-Genes Identification Through Molecular Approaches 139
2.13 Identification of Well-Characterized Sources of *R*-Genes for Major Pathogens of Crucifers 141
2.14 Identification of R-Genes Homologous DNA Fragments in *Arabidopsis thaliana* 148
2.15 Application of Omics Technologies in *Brassica* to Identify R-Genes 150
2.15.1 *Brassica* Species Genome Sequencing 150
2.15.2 Use of Pan-genomics in R-Genes Identification 151
2.16 Identification of QTLs and R-Genes Using NGS-Based SNP Methods 154
2.17 Identification of R-Gene Using In Silico *Methods* 156
2.18 NGS-Based Bulked Segregant Analysis (BSA) for R-Genes Identification 157
2.19 Resistance Gene Enrichment and Sequencing (RenSeq) 158
2.20 Effector-omics Approach to Detect R-Genes 158
References 159

3 Inheritance of Disease Resistance 195
3.1 Introduction 196
3.2 *Brassica-Albugo* Inheritance 197
3.2.1 Inheritance of Disease Resistance Genes (R-Genes) ... 197
3.2.2 Inheritance of Partial Resistance 201
3.3 *Brassica-Alternaria* Inheritance 206
3.3.1 Inheritance of Disease Resistance Genes (R-Genes) ... 206
3.3.2 Identification of Disease Tolerance Genotypes 209
3.3.3 Identification of Components of Horizontal Resistance .. 210
3.4 *Brassica-Erysiphe* Inheritance 211
3.4.1 Inheritance of Disease Resistance Genes (R-Genes) ... 211

3.4.2 Inheritance of Powdery Mildew Resistance in Progenies of *Brassica* . . . 214
3.4.3 Inheritance and Mapping of R-Genes in *Arabidopsis* to Powdery Mildew . . . 215
3.4.4 Inheritance of Enhanced *R*-Genes in *Arabidopsis* . . . 221
3.4.5 Inheritance and Mapping of *R*-Genes in *Arabidopsis* Mutants to Powdery Mildew . . . 221
3.5 *Brassica-Fusarium* Inheritance . . . 224
3.5.1 Inheritance and Mapping of R-Genes of *Brassica* . . . 224
3.6 *Brassica-Hyaloperonospora* Inheritance . . . 226
3.6.1 Inheritance of Disease Resistance Genes (R-Genes) . . . 226
3.6.2 Inheritance of Partial Resistance . . . 232
3.6.3 Inheritance of Avirulence Gene in *Arabidopsis–Hyaloperonospora* Pathosystem . . . 233
3.7 *Brassica-Plasmodiophora* Inheritance . . . 235
3.7.1 Inheritance of Disease Resistance Genes (R-Genes) . . . 235
3.7.2 Inheritance and Characterization of R-Genes in *Brassica rapa* . . . 238
3.7.3 Genetics of Clubroot Resistance Genes (R-Genes) . . . 239
3.7.4 Inheritance and QTL Mapping of Clubroot Resistance Genes of *Brassica rapa* subsp. *rapifera* (ECD02) . . . 240
3.7.5 Inheritance and Mapping of Clubroot Resistance Genes in *Brassica rapa* Using Bulk Segregant RNA Sequencing . . . 242
3.8 *Brassica-Sclerotinia* Inheritance . . . 247
3.8.1 Inheritance of Leaf and Stem Rot Resistance to *Sclerotinia* in *Brassica* . . . 247
3.8.2 Inheritance of Leaf Resistance to *Sclerotinia* in *Brassica* and Its Correlation with Cotyledon Resistance . . . 248
3.9 *Brassica-Turnip Mosaic Virus* Inheritance . . . 249
3.9.1 Inheritance of Disease Resistance to *TuMV* in *Brassica* . . . 249
3.10 *Brassica-Xanthomonas* Inheritance . . . 250
3.10.1 Inheritance of Disease Resistance to *Xanthomonas* in *Brassica* . . . 250
References . . . 251

4 Transfer of Disease Resistance . . . 265
4.1 Introduction . . . 266
4.2 Transfer of Disease Resistance in *Brassica* Through Introgression Breeding . . . 267
4.3 Transfer of Disease Resistance from Germplasm Sources . . . 267
4.4 Strategies for Developing Durable Alternaria Blight-Resistant *Brassica* Crops . . . 269

4.4.1 Inter- or Intraspecific Introgression of Candidate Genes from Non-host Plants to *Brassica* Crops 271
4.4.2 Microarray-Based Gene Identification and Transfer . . . 272
4.4.3 Gene Silencing 272
4.4.4 Identification and Use of Model Plants as Genetic Resources 274
4.4.5 Literature Mining and Targeted Functional Genomics . . 274
4.5 Transfer of an Endo-Chitinase Gene from *Trichoderma virens* Confers Enhanced Tolerance to *Alternaria* in Transgenic *Brassica* 275
4.6 Sources of Resistance 276
4.7 Sources of Disease Resistance from Cruciferous Relatives 276
4.8 Sources of Multiple Disease Resistance 277
4.8.1 Relationship Between Major Foliar Diseases 278
4.9 Development of Resistant Cultivars 279
4.9.1 Strategies and Methods of Screening for Resistance . . . 280
4.10 Bottlenecks in Resistance Breeding 281
4.11 Use of Biotechnological Approaches 284
4.11.1 In Vitro Embryo Rescue 284
4.11.2 Somatic Hybridization 285
4.11.3 Somaclonal Variations 285
4.11.4 Genetic Transformation 286
4.11.5 Molecular Markers 286
4.11.6 Induction of Systemic Resistance 287
4.11.7 Genetic Engineering 288
4.12 Transfer of R-Genes Approaches for Clubroot 289
4.12.1 Transferring R-Genes by Molecular Marker 289
4.12.2 Transfer of R-Genes by Interspecific Hybridization 289
4.12.3 Transfer of R-Genes by Distant Hybridization 290
4.12.4 Transfer of R-Genes from European Clubroot Differential (ECD) Set 291
4.12.5 Transferring of Powdery Mildew R-Genes Through Embryo Rescue 292
4.12.6 Mutagenic Resistance to Powdery Mildew 292
4.13 Clubroot Resistance Breeding in Cruciferous Crops 296
4.13.1 *Brassica rapa* 296
4.13.2 *Brassica oleracea* 297
4.13.3 *Brassica napus* 300
4.13.4 Radish (*Raphanus sativus* L.) 303
4.14 Development of Clubroot-Resistant Cultivars 304
4.14.1 Genetics and Breeding 304
4.14.2 Mechanism of Resistance 304
4.14.3 R-Genes Cultivars in the Field 305
4.14.4 Deployment of R-Genes Cultivars 305

4.15 Factors Affecting Transfer of Plant Disease Resistance 306
4.16 Accessing and Exploiting Genetic Diversity 306
4.17 Introgression and Pyramiding of Major R-QTL's into *Brassica napus* Against *Sclerotinia* 307
4.18 Introgression of Resistance from Wild *Brassica* Species into *Brassica juncea* to *Sclerotinia* 308
4.18.1 Marker-Assisted Introgression of *Sclerotinia* Resistance into *Brassica juncea* 309
4.19 Introgression of Black Rot Resistance from *Brassica carinata* to *Brassica oleracea* Through Embryo Rescue 310
4.20 Distant Hybridization for *Xcc* Resistance Through Somatic Hybridization and Embryo Rescue 313
4.21 Novel Sources and Transfer of R-Genes 314
4.22 Breeding Resistant Cultivars 315
4.23 Application of Molecular Markers in Breeding for Disease Resistance 316
4.23.1 Marker-Assisted Selection 317
4.23.2 Genome-Wide Marker-Assisted Selection 317
4.23.3 Genetic Diversity Analyses 318
4.23.4 Comparative Genomics 318
4.24 Development of Transgenic *Brassica* Crops Against Biotic Stresses 319
4.24.1 Improvement of *Brassica* Species by Transgenesis 319
4.24.2 Effective Transgenic Lines of *Brassica* Species 321
4.24.3 Transgenics Developed in *Brassica* Species Against Different Pathogens 322
4.24.4 Development of Transgenics in *Brassica* Species Against Pests 331
4.24.5 Use of Transgenics in Beneficial Biotic Interactions of *Brassica* Species 337
References 338

5 Host Resistance Signaling Network System to Multiple Stresses . . . 359
5.1 Introduction 360
5.2 Complex Network of Signaling Pathways in Host Resistance to Multiple Stresses 408
5.3 Molecular Mechanisms of Signaling System 411
5.3.1 Receptor Proteins as Frontline Defense Signals 411
5.3.2 Transcription Factors as Defense Signaling Pathways . . 413
5.3.3 Heat Shock Proteins and Heat Shock Transcription Factors as Signaling Pathways 415
5.3.4 MicroRNAs Analysis in Response to Defense Signaling 417
5.3.5 Epigenetics and Stress Response Signaling 418
5.3.6 Systemic Signaling Process 419

5.3.7 Signaling System in Response to Abiotic and Hormonal Stresses 420
5.4 Calcium as Key Signaling Pathways in Host Responses to Multiple Stresses 421
5.4.1 Role of CMLS in Multiple Stresses 423
5.4.2 CPKs as Signaling Molecules Mediating Plant Responses to Multiple Stresses 425
5.4.3 A Single Calcium Sensor in Specific Responses to Multiple Stresses 426
5.5 Crosstalk and Specificity Between Signaling Pathways 429
5.6 Crosstalk and Specificity in Abiotic Stress Signaling 430
5.7 Crosstalk and Specificity in Biotic Stress Signaling 430
5.8 Crosstalk Among Multiple Stresses 431
5.9 Abiotic Stress Diagnosing as Nutrient Deficiencies 432
5.9.1 Mobile and Immobile Nutrients 433
5.9.2 Identification Key 433
References 436

6 Glimpses of Molecular Mechanisms of Host Resistance 465
6.1 Introduction 467
6.2 Molecular Mechanisms of Disease Resistance 468
6.3 Molecular Mechanisms of Host Resistance to Biotrophs 471
6.4 Molecular Mechanisms of Host Resistance to Hemibiotrophs and Necrotrophs 474
6.5 Biometabolomics of Disease Resistance to Biotrophs 476
6.6 Biometabolomics of Disease Resistance to Hemibiotrophs and Necrotrophs 480
6.7 Techniques to Study Molecular Mechanisms of Host Resistance 483
6.8 Exploitation of Molecular Approaches for Breeding Disease Resistance 484
6.8.1 Through Introgression Breeding Approaches 484
6.8.2 In Vitro Embryo Rescue 484
6.8.3 Somatic Hybridization 485
6.8.4 Somaclonal Variations 485
6.8.5 Genetic Transformation 486
6.8.6 Molecular Marker-Assisted Breeding 486
6.8.7 Use of Systemic Resistance Induction 487
6.8.8 Use of Genetic Engineering 488
6.9 Transferring R-Genes by Molecular Marker 489
6.10 Transfer of R-Genes by Interspecific Hybridization 489
6.11 Transfer of R-Genes by Distant Hybridization 490
6.12 Transfer of R-Genes from European Clubroot Differential (ECD) Set 490

6.13 Transferring of Powdery Mildew R-Genes Through Embryo Rescue . . . 491
6.14 Use of Mutagenic Resistance to Powdery Mildew . . . 491
6.15 Strategies for Developing Durable Alternaria Blight-Resistant *Brassica* Crops . . . 492
6.15.1 Inter- or Intraspecific Introgression of Candidate Genes from Non-Host Plants to *Brassica* Crops . . . 493
6.15.2 Microarray-Based Gene Identification and Transfer . . . 493
6.15.3 Gene Silencing . . . 494
6.15.4 Identification and Use of Model Plants as Genetic Resources . . . 495
6.15.5 Targeted Functional Genomics Based on Literature Mining . . . 495
6.15.6 Transfer of an Endochitinase Gene from *Trichoderma virens* Confers Enhanced Tolerance to *Alternaria* in Transgenic *Brassica* . . . 496
References . . . 497

7 Management of Disease Resistance . . . 505
7.1 Introduction . . . 506
7.2 *Brassica-Albugo*-White Rust Management . . . 507
7.2.1 Use of Slow White Rusting Genotypes/Cultivars . . . 507
7.2.2 Use of Systemic Resistance Induction Approach . . . 507
7.3 *Brassica-Alternaria* Blight Management . . . 508
7.3.1 Exploring Non-host Disease Resistance Mechanisms . . . 508
7.3.2 Use of Host Resistance . . . 512
7.4 *Brassica-Erysiphe* Powdery Mildew Management . . . 513
7.4.1 Multilayered and Multicomponent Mechanisms to Manage Powdery Mildew Resistance . . . 513
7.5 *Brassica-Hyaloperonospora* Downy Mildew Management . . . 514
7.5.1 Disease Resistance Increases Competitive Ability of Host Plants . . . 514
7.5.2 Small RNA Inhibits Infection of *Hyaloperonospora arabidopsidis* . . . 516
7.6 *Brassica-Leptosphaeria* Blackleg Management . . . 517
7.6.1 Deployment of *Brassica R*-Genes Cultivars to *Leptosphaeria* . . . 517
7.6.2 Marker-Assisted Selection of Avirulence Gene in *Leptosphaeria* for Effective Deployment of *Brassica*-Resistant Cultivars . . . 520
7.7 *Brassica-Plasmodiophora* Clubroot Management . . . 520
7.7.1 R-Genes Cultivars of *Brassica* to Control Clubroot in the Field . . . 520
7.7.2 Deployment of R-Genes Cultivars of *Brassica* to Manage Pathogen Selection . . . 521

7.7.3 R-Genes Pyramiding in *Brassica* Cultivars 522
7.7.4 Non-host and Basal Resistance to Clubroot 523
7.7.5 Bacterial Biocontrol Agents to Manage Clubroot 524
7.7.6 Molecular Mechanism of Biocontrol Agent Against Clubroot 525
7.7.7 Role of ProCa in Modulation of Clubroot 526
7.7.8 Role of Arginase Gene in Modulation of Clubroot 533
7.7.9 Host Resistance Mechanisms of *Brassica* Cultivars . . . 536
7.7.10 Erosion of R-Genes Management Through Integrated Approaches 536
7.7.11 Clubroot QTL's Modulation by Nitrogenes 539
7.7.12 Management of R-Genes Resistance to Clubroot 540
7.7.13 Systemic Acquired Resistance in *Brassica* to Clubroot 543
7.7.14 Impact of Clubroot-Resistant Cultivars on Disease Management 544
7.8 *Brassica-Pseudocercosporella* White Leaf Spot Management 545
7.8.1 Deployment of Resistant Cultivars of *Brassica* Against *Pseudocercosporella* 545
7.9 *Brassica-Sclerotinia* Stem Rot Management 547
7.9.1 Reduction of Pathogen Virulence 547
7.10 *Brassica-Turnip Mosaic Virus* (*TuMV*) Management 549
7.10.1 *TuMV* Resistance Breeding in *Brassica* 549
7.10.2 Transformation of R-Genes into *Brassicas* 549
7.10.3 Molecular Marker-Assisted Breeding in *Brassica* 549
7.10.4 Pyramiding R-Genes for Durable Resistance to *Brassica* 550
7.10.5 Host-Induced Gene Silencing in *Brassica* 550
7.11 *Brassica-Xanthomonas* Black Rot Management 551
7.11.1 Distant Hybridization for *Xcc* Resistance Through Somatic Hybridization and Embryo Rescue 551
7.11.2 Use of Omics Approaches for Black Rot Resistance Management 552
7.12 *Brassica*-Cyst Nematode Management 554
7.12.1 Pyramiding of R-Genes in *Brassica* Against Beet Cyst Nematode 554
7.13 Multiple Pathogen Disease Resistance Management 556
7.13.1 Transgenic Approach for Biotic Stress Management . . . 556
7.13.2 Development of Transgenics for Multiple Pathogenic Resistances 560
7.13.3 Fungal Biocontrol Agents to Manage Multiple Pathogens 561

7.13.4 Bacterial Biocontrol Agents to Manage Multiple Pathogens . . . 563
7.13.5 Abiotic Conditions Modulate Diseases . . . 566
7.13.6 Use of Defense-Related Genes Other Than *R*-Genes Confers Resistance to Multiple Pathogens . . . 567
7.13.7 Fitness Cost to Multiple Disease Resistance in *Brassica* . . . 567
7.13.8 Molecular Bases for Assessment of Breakdown of R-Genes in *Brassica* . . . 568
7.13.9 Net Cost of R-Gene Resistance in the Diseased Crucifers . . . 570
References . . . 573

8 Genomics of Host Resistance at a Glance . . . 599
8.1 Introduction . . . 601
8.2 Functional Characterization of Genes for *Brassica* Crops Improvement . . . 602
8.3 Exploitation of R-Genes Sources . . . 607
8.4 Genetics of Disease Resistance . . . 610
8.5 Breeding of Disease Resistance *Brassica* Crops . . . 615
8.6 Signaling Network System of Host Resistance to Multiple Stress Responses . . . 617
8.7 Regulation of Genomics and Biometabolomics to Confer Molecular Mechanisms of Host Resistance . . . 620
8.8 Management of Disease Resistance Under Field Conditions . . . 622
8.9 Techniques to Develop Resistant Cultivars of *Brassica* . . . 623
8.10 Exploitation of Novel Protocols for Breeding Disease-Resistant Cultivars of Crucifers . . . 624
8.11 Use of Germplasm Sources to Transfer Disease Resistance . . . 625
8.12 Sources of Resistance Identified at Different Locations . . . 626
8.12.1 Use of Cruciferous Relatives as R-Genes Sources . . . 627
8.12.2 Multiple Disease Resistance Sources Identified from *Brassica* Species . . . 627
8.12.3 Assessing Relationship Between Major Foliar Diseases . . . 628
8.13 Use of R-Genes Sources for the Development of Clubroot-Resistant Cultivars . . . 629
8.13.1 Defense Mechanisms of Resistant Cultivars . . . 629
8.13.2 Performance of R-Genes Cultivars in the Field . . . 630
8.13.3 Field Deployment of R-Genes Cultivars . . . 630
8.14 Introgression of *Sclerotinia* Resistance into *Brassica juncea* Through Marker-Assisted Breeding . . . 631
8.14.1 *Sclerotinia* Resistance Introgression from Wild *Brassica* Species into *Brassica juncea* . . . 632
8.15 Use of Hexaploidy Approach for Pyramiding of Major R-QTLs into *Brassica napus* Against *Sclerotinia* . . . 632

8.16 Black Rot Resistance Introgression from *Brassica carinata* to *Brassica oleracea* Through Embryo Rescue 633
8.17 Black Rot Resistance Breeding Through Somatic Hybridization and Embryo Rescue 635
8.18 Accessing and Exploiting Genetic Diversity 637
8.19 Identification of Novel Sources and Transfer of R-Genes 637
8.20 Difficulties in Transfer of Plant Disease Resistance Trait 638
References 638

9 Development of Resistance Cultivars' Techniques 651
9.1 Introduction 652
9.2 *Brassica-Albugo* Techniques 652
9.2.1 Maintenance of *Albugo candida* Isolates and Inoculum Preparation 652
9.2.2 Pathogen Culture and Inoculation Method 653
9.2.3 Sporangial Viability Test 654
9.2.4 Sporangial Preservation 654
9.2.5 Inoculation Applicator 655
9.2.6 Components of Partial Resistance 655
9.2.7 Germplasm Screening 655
9.2.8 Growth Chamber and Greenhouse Screening 656
9.2.9 Induction of Stagheads 657
9.2.10 Identification of R-Genes in *Brassica* to *Albugo* 658
9.3 *Brassica-Alternaria* Techniques 660
9.3.1 Greenhouse Method for Testing Resistance 660
9.3.2 *Brassica* Germplasm Screening for Resistance Through AB-Toxin 660
9.3.3 Characterization of a Gene from *Alternaria* for Fungicidal Resistance 661
9.3.4 Biochemical and Molecular Analyses for Mechanism of Resistance in *Brassica* to *Alternaria* 665
9.3.5 Evaluation of Partial Resistance to *Alternaria brassicicola* 665
9.3.6 Assessment of Methods of Inoculation for Resistance to *Alternaria* 666
9.4 *Brassica-Erysiphe* Techniques 667
9.4.1 Collection, Preservation, and Cultivation of Crucifer's Powdery Mildew 667
9.4.2 Plant Material and Inoculation Methods 668
9.4.3 Primer Design and Test Specimen Methods 668
9.4.4 DNA Extraction and PCR 669
9.4.5 Use of qPCR, and Spore Count Assays to Quantify Powdery Mildew 670
9.4.6 Characterization of the Disease Reaction Phenotypes 672

9.4.7 Embryo Rescue Technique to Transfer Powdery Mildew Resistance 673
9.4.8 Identification of Molecular Markers Linked to Powdery Mildew R-Genes 673
9.4.9 Molecular Identification of Anamorphic Powdery Mildews (Erysiphales) 675
9.5 *Brassica-Hyaloperonospora* Techniques 675
9.5.1 Culturing of *Hyaloperonospora parasitica* 675
9.5.2 Maintenance of *Hyaloperonospora parasitica* Isolates and Production of Inoculum 676
9.5.3 Germplasm Screening and Evaluation 678
9.5.4 Preservation of *Hyaloperonospora parasitica* 679
9.5.5 Artificial Inoculation of Excised Cotyledons 680
9.5.6 Methods of Breeding for Multiple Disease Resistance . . 680
9.5.7 Identification of Heterothallism and Homothallism 682
9.5.8 Leaf Disc Test to Assess Resistance 683
9.5.9 Use of Rooted Leaves for Screening *Brassica* Germplasm 684
9.5.10 Artificial Inoculation Technique Under Growth Chamber (Williams 1985b) 685
9.5.11 Measuring Systemic Infection of the Downy Mildew Pathogen 687
9.6 *Brassica-Leptosphaeria* Techniques 689
9.6.1 Quantitative Genes Mediated Resistance to *Leptosphaeria* 689
9.6.2 Cloning and Transformation of *Leptosphaeria* Avirulence Gene 693
9.6.3 Identification of QTLs in *Brassica* to *Leptosphaeria* . . . 695
9.7 *Brassica-Plasmodiophora* Techniques 697
9.7.1 Pathogen Inoculum Preparation for Artificial Inoculation 697
9.7.2 *In Vitro* Culture of *Plasmodiophora brassicae* 698
9.7.3 Establishment of Callus Culture 699
9.7.4 Host Inoculation for Pathogenicity Test 699
9.7.5 Germplasm Evaluation Technique 700
9.7.6 Artificial Pathogen Inoculation Techniques 704
9.7.7 PCR Detection of Clubroot Pathogen 705
9.7.8 Isolation of Resting Spores 706
9.7.9 A Real-Time PCR Assay to Detect *Plasmodiophora brassicae* Under Field Conditions 708
9.7.10 Genetics and Molecular Mapping of R-Genes to *Plasmodiophora* Pathotypes 715
9.7.11 Identification of Parents and F2 Lines for R-QTLs in *Brassica* to *Plasmodiophora* 719
9.7.12 Transfer of Clubroot Resistance to *Brassica* 720

9.8 *Brassica-Sclerotinia* Techniques 722
9.8.1 A Rapid Screening Technique for Resistance 722
9.8.2 Germplasm Screening and Evaluation 722
9.8.3 A Green House Spray and Drop-Mycelium Inoculation Method 724
9.8.4 Detection of *Sclerotinia* by ELISA 724
9.8.5 Medium for Production of Oxalic Acid 725
9.8.6 Identification and Pyramiding of R-QTLs in *Brassica* to *Sclerotinia* 726
9.8.7 Identification of Genome-Wide Associated R-Loci in *Brassica* to *Sclerotinia* 728
9.8.8 Introgression of R-Genes from Wild Relatives in *Brassica* to *Sclerotinia* 730
9.8.9 Marker-Assisted Introgression of Resistance into *Brassica juncea* Against *Sclerotinia* 732
9.8.10 Germplasm Screening Against *Sclerotinia* Under Controlled Conditions 735
9.9 *Brassica-Xanthomonas* Techniques 736
9.9.1 Plant Culture and Pathogen Inoculation 736
9.9.2 Introgression of Black Rot Resistance Through Embryo rescue 739
9.9.3 Germplasm Screening for Black Rot Resistance 741
9.10 *Brassica-Turnip Mosaic Virus* Techniques 742
9.10.1 Collection of Samples in the Field 742
9.10.2 Virus Detection, Preservation, and Identification 743
9.10.3 Isolates of *TuMV* 744
9.10.4 Glasshouse Inoculations 744
9.10.5 Plants, Chemical Treatments, and Virus Inoculation Method 745
9.10.6 Serological Detection 746
9.10.7 Respiration in Systemic Infected Leaves 747
9.10.8 RNA Extraction and Transcript Level Estimation by Real-Time Quantitative PCR 747
9.11 *Brassica* Species Genomics Techniques 748
9.11.1 RNA Sequencing of *Brassica napus* 748
9.11.2 Identification of *Brassica oleracea* Genes That Encode NBS Domain and NBS-Associated Conserved Domains 750
References 752

10 Future Research Priorities 773
10.1 Introduction 773

Index 779

About the Authors

Govind Singh Saharan Former Professor and Head, Department of Plant Pathology, CCS HAU, Hisar. Dr. Saharan has conducted research in diverse fields of plant pathology and had 250 articles in national and international journals. He has been editor/author of several books, monographs, and Crop Production Compendium. He is on the panel of experts of SAU, ICAR, IARI, CSIR, UGC, and Department of Biotechnology. Dr. Saharan has been a visiting professor at University of Alberta, Edmonton, Canada; Agriculture and Agri-Food Canada, Saskatoon, Canada, and Rothamsted Research, IACR, Harpenden, UK. He has been president (NZ) of the Indian Phytopathological Society, editor-in-chief, Journal of Mycology and Plant Pathology, Journal of Oilseed Brassica, and president of the Indian Society of Mycology and Plant Pathology. He has been awarded with Y. L. Nene, Outstanding Plant Pathology Teacher by ISMPP, Udaipur, and Life Time Achievement Award by the Society for Rapeseed-Mustard Research, Bharatpur, India.

Naresh K. Mehta, Former Associate Dean, Professor and Consultant Faculty, Department of Plant Pathology, CCS HAU, Hisar. He has been teaching plant pathology courses to UG and PG students and conducted research in diverse fields of plant pathology especially on rapeseed-mustard. He has guided several M. Sc. and Ph. D students. He has published about 200 research articles and editor/author of several books, book chapters, review articles, and teaching manuals. Dr. Mehta has been president of INSOPP, Ludhiana, and editor-in-chief of ISMPP and has served as member of editorial board of various phytopathological societies in India. Dr. Mehta has been awarded Y. L. Nene,

Outstanding Plant Pathology Teacher by ISMPP, Udaipur, and Ms. Manju Utereja Memorial Gold Medal, HAU. He is on the panel of experts of SAU, ICAR, and UGC and member of various national and international committees. Dr. Mehta has been a visiting scientist to University of Alberta, Edmonton, Canada.

Prabhu Dayal Meena, Principal Scientist (Plant Pathology) at ICAR-Directorate of Rapeseed-Mustard Research, Bharatpur, Rajasthan, India. He has conducted research on different aspects of rapeseed-mustard diseases, including host resistance, management, epidemiology, and biology of crucifer's pathogen. He has published several research papers, reviews, and book chapters in reputed journals and number of books. Dr. Meena has been honored with Fellow of Indian Society of Mycology and Plant Pathology, Plant Protection Association of India, Indian Society of Oilseed Research, Society for Rapeseed-Mustard Research, Indian Phytopathological Society, and awarded Dr. P.R. Kumar Outstanding *Brassica* Scientist Award, Society for Rapeseed-Mustard Research. He is founder secretary and managing editor of Society for Rapeseed-Mustard Research. He visited the UK during 2007 under Indo-UK Collaborative Research on Oilseed Brassica crops. He has been guiding several M. Sc. and Ph.D students.

Abbreviations

AAFC	Agriculture and Agri-Food Canada
AAP	Acquisition access period
AB	Alternaria blight
ABA	Abscisic acid
ADI	Average disease on inflorescence
ADK2	Adenosine kinase 2
ADL	Average disease on leaf
AFLP	Amplified fragment length polymorphism
AICORPO	All India Coordinated Research Project on oilseeds
AICRP-RM	All India Coordinated Research Project on Rapeseed-Mustard
AMP	Antimicrobial peptides
ANOVA	Analysis of variance
APX	Ascorbate peroxidase
AUDPC	Area under disease progress curves
AUWPC	Area under wilt progress curve
avr	Avirulence
AVRDC	Asian Vegetable Research and Development Centre
BABA	β-amino butyric acid
BAP	Benzylaminopurine
BC	Backcross
BCN	Beet cyst nematode
bHLH	Basic helix-loop-helix
BLAST	Basic Local Alignment Search Tool
BLUPs	Best linear unbiased predictions
BnMPK4	*B. napus* mitogen-activated protein kinase
BSA	Bulked segregant analysis
BSMV	Barley stripe mosaic virus
BW	Burpee White
CAC	Clathrin adaptor complex
CaM	Calmodulin
CAMTA	Calmodulin-binding transcription activator
CaMV	Cauliflower mosaic virus

CAPs	Cationic antimicrobial peptides
CAPS	Cleaved amplified polymorphic sequence
CAT	Catalase
CBF	C-repeat-binding factor
CBLs	Calcineurin-B-like proteins
CC	Coiled coil
CCaMK	Calcium and calmodulin-dependent protein kinase
CCC	Chlormequat chloride
CD	Critical difference
CDPKs	Calcium-dependent protein kinases
CF	Culture filtrate
cM	centiMorgans
CMLs	Calmodulin-like protein
CNL	CC-NBS-LRR
CNV	Copy number variation
COIP	Co-immune precipitation
CP	Coat protein
CPKs	Calcium-dependent protein kinases
CR	Clubroot
CR	Clubroot resistance
CrGC	Crucifer Genetics Cooperation
CRISPR	Clustered regulatory interspaced short palindromic repeats
CRR	Cysteine-rich repeat
CRW	China Rose Winter
CTAB	Cetyltrimethyl ammonium bromide
DAI	Days after inoculation
DAS	Days after sowing
DEGs	Differentially expressed genes
DH	Doubled haploid
DI	Disease indices
DM	Downey mildew
DNA	Deoxyribonucleic acid
DOS	Date of sowing
DR	Disease reaction
DRMR	Directorate of Rapeseed-Mustard Research
DSI	Disease severity index
dsRNA	Double-stranded ribonucleic acids
DSTI	Disease stress tolerance index
ECD	European clubroot differential
EDS1	Enhanced disease susceptibility 1
EDTA	Ethylene diamine tetra-acetic acid
EF-Tu	Elongation factor Tu
eIF	Eukaryotic initiation factor
ELISAs	Enzyme-linked immunosorbent assays
ELWL	Excised leaf water loss

EMS	Ethyl methane-sulfonate
ENU	Ethyl nitrosourea
EPPO	European and Mediterranean Plant Protection Organization
ERK	Extracellular signal-regulated kinase
ESTs	Expressed sequence tags
ET	Ethylene
ETD	Ethylene thiuram disulfide
ETI	Effector-triggered immunity
FBA	Fructose bisphosphatealdolase
FDR	False discovery rate
FIL	Final intensity of rust on leaf
FIP	Final intensity of rust on plant
FPKM	Fragments Per Kilobase of transcript per Million mapped
$G \times E$	Genotype by environment
GBLUP	Genomic Best Linear Unbiased Prediction
GBS	Genotyping by sequencing
GCV	Genotypic coefficient of variance
GD	Geographical distribution
GG	Genetic gain
GISH	Genomic in situ hybridization
GLM	General linear model
GLM	Generalized linear models
GM	Genetically modified
GMP	Geometric mean productivity
GR	Genomics research
GRIN	Germplasm Resources Information Network
GSS	Genomic survey sequences
GWAS	Genome-wide association analysis
GWAS	Genome-Wide Association Study
HEs	Homoeologous exchanges
HGP	Human Genome Project
HIB	High-efficiency integrated breeding
HIGS	Host-induced gene silencing
HMM	Hidden Markov Model
HNRT	Homeologous non-reciprocal transposition
Hpa	*Hyaloperonospora arabidopsidis*
HR	Horizontal resistance
HR	Host range
HR	Hypersensitive response
HS	Highly susceptible
HSPs	Heat shock proteins
HTGs	High-throughput genome sequences
IAA	Indole-3-acetic acid
IAP	Inoculation access period
IBA	Indole-butyric acid

IBCN	International Blackleg of Crucifers Network
IBD	Isolation by distance
IC	Isochorismate
ICAR	Indian Council of Agricultural Research
ICIM	Inclusive composite interval mapping
iGS	Indole-glucosinolates
ILs	Introgression lines
INA	Isonicotinic acid
IP	Interaction phenotype
IP	Intron polymorphic
IPM	Integrated pest management
ISSRs	Inter-simple sequence repeats
ITS	Internal transcribed spacer regions
JA	Jasmonic acid
LB	Luria–Bertani
LC	Liquid chromatography
LCBs	Long chain bases
LD	Linkage disequilibrium
LDIA	Leaf disc inoculation assay
LGs	Linkage groups
Lm	*Leptosphaeria maculans*
LPs	Lipopeptides
LRR	Leucine-rich repeat
LRR-RLKs	Leucine-rich repeat receptor-like kinase
LRR-RLPs	Leucine-rich repeat receptors-like protein
LRRs	Leucine-rich repeats
LSD	Least significant difference
LZ	Leucine zipper
MAB	Marker-assisted backcross breeding
MAMPs	Microbe-associated molecular patterns
MAP	Mitogen-activated protein
MAPK	Mitogen-activated protein kinase
MAPKKK	MAP kinase kinase kinase
MAS	Marker-assisted selection
MDR	Multiple disease resistance
MET	Multi-environment trials
miRNAs	MicroRNAs
MKK	MAP kinase kinase
ML	Maximum likelihood
MP	Mean productivity
MPK	MAP kinase
MR	Moderately resistant
MS	Mass Spectrometry
MTAs	Marker trait associations
MTI	MAMPs-triggered immunity

MYA	Million years ago
NAA	Naphthalene acetic acid
NaOCl	Sodium hypochlorite
NBPGR	National Bureau of Plant Genetic Resources
NBS	Nucleotide-binding site
NBS-LRR	Nucleotide-binding site leucine-rich repeat
NCBI	National Centre for Biotechnology Information
NGS	Next-generation sequencing
NHR	Non-host resistance
NIa-Pro	Nuclear inclusion a-protease domain
NIRS	Near-infrared reflectance spectroscopy
NLRs	Nucleotide-binding site leucine-rich repeats
NMR	Nuclear magnetic resonance
NOA	Nicotinic acid
NPR1	Nonexpressor of PR genes 1
NWCVT	National winter canola variety trials
ONT	Oxford Nanopore Technologies
OXO	Oxalate oxidase
PAD4	Phytoalexin-deficient 4
PAL	Phenylalanine ammonia lyase
PAMP	Pathogen-associated molecular pattern
PAMP/MAPM	Pathogen/microbe associated molecular patterns
PAMPs	Pathogen-associated molecular patterns
PAV	Presence/absence variation
PBS	Phosphate buffer saline
PCA	Principal component analysis
PCD	Programmed cell death
PCR	Polymerase chain reaction
PCV	Phenotypic coefficient of variance
PDA	Potato dextrose agar
PDI	Per cent disease index
PDI	Per cent disease intensity
PEG	Glycerine polyethylene glycol
PGA	Polygalacturonase
PIC	Polymorphic information content
PM	Powdery mildew
POX	Peroxidase
PPO	Polyphenol oxidase
PPT	Phosphinothricin
PPV	Percentage of polymorphic variants
PR	Pathogenesis related
ProCa	Prohexadione-calcium
PRRs	Pattern recognition receptors
PSbMV	Pea seed-borne mosaic virus
PTI	PAMPs-triggered immunity

PTI	Pattern-triggered immunity
Pto	A Ser/Thr kinase protein
pv.	Pathovar
PVE	Phenotypic variation explained
QDR	Quantitative disease resistance
qPCR	Quantitative polymerase chain reaction
qRT-PCR	Quantitative reverse-transcription polymerase chain reaction
qRT-PCR	Real-time quantitative-PCR
QTL	Quantitative trait loci
R	Resistance
RAC	Recognition of *A. candida*
RAPD	Random Amplification of Polymorphic DNA
RBS	Round Black Spanish
RFLP	Restriction fragment length polymorphism
RGAs	Resistance gene analogs
RGL	Resistant gene like
RH	Relative humidity
RIP	Ribosome inactivating protein
RISC	RNA-induced silencing complex
RLCK	Receptor-like cytosolic kinase
RLKs	Receptor-like kinases
RLKs	Receptor-like protein kinases
RLPs	Receptor-like proteins
RNA	Ribonucleic acid
RNAi	RNA interference
ROS	Reactive oxygen species
RPP	Recognize *Peronospora parasitica*
RR-BLUP	Ridge Regression Best Linear Unbiased Prediction
RWC	Relative water content
SA	Salicylic acid
SAA	Systemic acquired acclimation
SAG	Salicylic acid glycoside
SAG101	Senescence-associated gene 101
SAR	Systemic acquired resistance
SCAR	Sequence characterized amplified region
SD	Standard deviation
SDW	Sterilized distilled water
Seq	Sequencing
SIGS	Spray-induced gene silencing
SMA	Single marker analysis
SNaP	Single Nucleotide Absence Polymorphism
SNP	Single nucleotide polymorphism
SOD	Superoxide dismutase
SRRs	Signal recognition receptors
SSH	Suppression Subtractive Hybridization

SSR	Simple sequence repeat
STK	Serine-threonine kinase
STS	Sequence-tagged sites
TALENs	Transcription activator-like effector nucleases
TEs	Transposable elements
TFs	Transcription factors
TIGS	Transient-induced gene silencing
TIR	Toll/Interleukin-1 receptor
TM	Trans-membrane
TNL	TIR-NBS-LRR
TOL	Disease tolerance
TuMV	Turnip mosaic virus
TuYV	Turnip yellows virus
UK	United Kingdom
ULVA	Ultra-low volume applicator
USA	United States of America
USDA	United States Department of Agriculture
UTR	Untranslated region
VIGS	Virus-induced gene silencing
Vol	Volume
VPg	Viral protein linked to the genome
WAK	Wall-associated kinase
WFP	Wisconsin Fast Plant
WGRS	Whole-genome resequencing
WGT	Whole-genome triplication
WR	White rust
WRR	White Rust Resistance
WUE	Water use efficiency
Xcc	Xanthomonas campestris pv. campestris
YDC	Yeast Dextrose Calcium Carbonate
YL	Yield losses
Yp	Potential yield under controlled environment
YS	Yellow sarson
Ys	Yield under disease-stress environment

Symbols

:	ratio
~	approximately
$\approx$	is approximately equal to
$<$	Less than
$>$	more than
$\leq$	Less than or equal to
$\geq$	more than or equal to
=	is equal to
$\pm$	plus-minus sign
e.g.	for example
%	per cent
h	hours
μmol/m^2/s	micromole per meter square per second
Subsp.	subspecies
var.	variety
Mb	megabyte
et al.	and others
M	million
S	susceptible
Mb	million bases
mm	millimeter
cm	centimeter
cv.	cultivar
viz.	videlicet which means namely
AU$	Australian dollar
g	gram
Gy	gray
dS/m	deciSiemens per metre
OC	degree Celsius
m^2	*Square Meter*
mL	milliliter
μL	microliter

min	minute
w/v	weight by volume
v/v	volume by volume
d	day
dpi	days-post-inoculation
μg	microgram
h^2	heritability
μmol	micromoles
sdH_2O	sterile de-ionized water
CFU	colony forming unit
£	pound
$	*dollar*
Σ	Sigma means "sum up"
N	nitrogen
P	potassium
S	sulfur
Ca	calcium
Mg	magnesium
B	boron

Principles of Host Resistance

1

Abstract

Crucifers include genus *Brassica* in the Brassicaceae family with 325 genera of 3740 species of 75 accepted species. Most of the species in the genus are cultivated for their edible roots, leaves, stems, buds, flower, and seeds for human consumption as raw, cooked, edible oil and as fodder for animals. Genetic diversity within *Brassica* species has been broadly studied with a special focus on six crop species widely cultivated for vegetables and oilseed production. Out of 44 pathogens known to infect crucifers, 16 pathogens have received attention for genomics investigations. The principles involved in the effective utilization of genomics technology largely depend on the several omics approaches for identification, inheritance, transfer, and management of R-genes for developing durable resistant cultivars of crucifers. Wild plant species represent a gene pool that can potentially provide a rich source of durable, broad-spectrum disease resistance for use in crops. Hybridization between *Brassica* species and wild relatives' cenospecies and cytodemes can be possible under controlled conditions with molecular technology. Interspecific hybridization for developing disease resistance in *Brassica* crops has potential since they share genomics in common. Genes containing a nucleotide-binding site (NBS) constitute one of the largest R-genes families which are co-located with resistance QTL's and facilitate multiple disease resistance breeding in *Brassica* crops. Elevated resistance against biotrophs is often related with increased susceptibility to necrotrophs and vice versa. Plants have evolved intracellular receptors with nucleotide-binding leucine-rich repeat domain (NLR's) to detect pathogens effectors or their activity to trigger immunity. Understanding the interspecific and intraspecific relationship among cultivated *Brassica* species will be useful for better parental selection to widen the genetic base for multiple disease resistance breeding programs of *Brassica* crops. Despite the genome complexity of *Brassica* species, homoeologous rearrangements/recombination occurs more frequently in

G. S. Saharan et al., *Genomics of Crucifer's Host-Resistance*,
https://doi.org/10.1007/978-981-16-0862-9_1

resynthesized species contributing towards genetic diversity to provide good sources of R-genes for breeding resistant cultivars. The complexity of R-genes resistance levels varies in *Brassica* species for major pathogens due to evolution of pathogenic variability. The genomic era is characterized by high-efficiency integrated breeding (HIB) in which multiple methods are combined including traditional ways such as microspore culture, backcrossing, and distant introgression with modern ways such as MAS, gene editing, and genome design. During HIB genome background analysis is helpful in eliminating undesirable linkage drags and rapidly identifying desirable individuals. Several numbers of genes have well documented that they are differentially expressed against different pathogens. Several omics technologies and approaches have provided tool to look at the differences in DNA, RNA, proteins, and other cellular molecules between species and among individuals of species to get better insight into the genomics of crucifers' disease resistance breeding programs. Using the various omics and multi-omics tools supplemented with further fine tuning of the bioinformatics methods will speed up the screening of favorable alleles in *Brassica* germplasm promoting resistance against major pathogens along with identification and cloning of favorable genes with increased precision.

Keywords

Principles · Resistance · Global ranking of crucifer's diseases · Brassica species · Wild relatives · Crucifers · Biotrophs · Hemibiotrophs · Necrotrophs · Genetic diversity · Wild relations · Hybridization · Interspecific hybridization · Phylogenetic relationships · Pathogens variability · Differential genes expression · Genomics of disease resistance · R-genes

1.1 Introduction

Cruciferous crop plants are always challenged by various biotic and abiotic stresses. In agriculture cropping system, plant protection is being delivered using different approaches such as chemical control, various agronomic practices, biological control, integrated pest management (IPM), and cultivation of resistance cultivars. Among these approaches, resistance cultivars are the most economical, environmentally sustainable solution to control different diseases in *Brassica* species. Previous studies suggested that the inheritance of disease resistance is either qualitative or quantitative in *Brassica* species. Recently mapped disease resistance genetic loci and closely linked molecular marker to these loci can be used for marker-assisted selection in cruciferous resistance breeding of *Brassica* species. Substantial use of recently developed resistant sources can be combined with molecular tools and new technologies such as gene/QTL mapping, fine mapping, gene cloning, comparative genomics and analysis of transcriptomic profiles through next-generation sequencing approaches, can enhance our understanding of resistance mechanism. Novel information can help controlling diseases in an effective way, so yield losses would be reduced and the quality of Brassica crop product would be improved. The principles involved in the effective utilization of genomics technology largely depend on the following approaches including molecular genetics.

In natural environ, cruciferous plants are challenged by 167 biotic (44 pathogen, 87 pests, and 36 weeds) and 19 abiotic stresses (environmental, soil salinity and alkalinity, nutrients deficiency, and hormonal changes) (Tables 5.1, 5.2, 5.3, and 5.4, Chap. 5) and some of them have successfully invaded crop plants and caused diseases which result in deterioration of crop quality and yield. The major pathogens of crucifers can be ranked at global level on the basis of host range, geographical distribution, yield losses, and emphasis given by the *Brassica* scientists on genomic investigations for improving the yield and quality of *Brassica* crops. On the basis of average rank at global level eight pathogens (1—*Plasmodiophora brassicae*; 2—*Alternaria brassicae*; 3—*Sclerotinia sclerotiorum*; 4— *Leptosphaeria maculans*; 5—*Erysiphe cruciferarum*; 6—*Albugo candida*; 7—*Turnip mosaic virus*; 8—*Xanthomonas campestris* pv. *campestris*) can be considered as of major consequences with 18.25 to 68.25 ranks (Table 1.1). In order to cope with disease attacks, the plants have developed multiple layers of defense mechanisms. Plant disease resistance (R) genes which specifically interact/ recognize with corresponding pathogen avirulence (avr) genes are considered as plant genetic factors of a major layer. The interactions of this gene-for-gene (or genes-for-genes) manner activate the signal transduction cascades that turn on complex defense responses against pathogen attack and this is called incompatible interaction. The interaction between a host species and a pathogenic species is dynamic where a host variety often lost the R-gene-dependent resistance due to its pathogen race evolution for a virulent gene and thus a new R-gene was selected against this new race. R-genes provide innate immunity whereas outcomes of defense responses lacking R-genes are partial resistance. Therefore, identification of R-genes is crucial for resistant variety development and relevant mechanism investigation.

More than one hundred R-genes, which are reported in PRGdb (http://prgdb.crg.eu/wiki), were functionally identified and comprise a super family in plants. Sequence composition analysis of R-genes indicates that they share high similarity and contain seven different conserved domains like NBS (nucleotide-binding site), LRR (leucine-rich repeat), TIR (Toll/Interleukin-1 receptor), CC (coiled coil), LZ (leucine zipper), TM (trans-membrane), and STK (serine-threonine kinase). Based on domain organization, R-gene products can be categorized into five major types: TNL (TIR-NBS-LRR), CNL (CC-NBS-LRR), RLK (Receptor like kinases), RLP (Receptor like proteins), and Pto (a Ser/Thr kinase protein). Most of the R-genes in plant kingdom are members of NBS-LRR (nucleotide-binding site-leucine-rich repeat) proteins. “NBS” and “LRR” domains play different roles in plant-microbe interaction, where the former have the ability to bind and hydrolyze ATP or GTP and the latter is involved in protein–protein interactions. NBS-LRR proteins in plants share sequence similarity with the mammalian NOD-LRR containing proteins which play a role in inflammatory and immune responses. On the basis of presence or absence of N-terminal domains (TOLL/ interleukin-1 receptor (TIR) and the coiled-coil (CC) motif), NBS-LRR class can be further divided into two major types, TNL

Table 1.1 Global ranking of crucifer's major diseases based on geographical distribution (GD), host range (HR), yield losses (YL), and Genomics Research (GR) (updated 2020)

Sr. no.	Pathogen	Disease	GD in number of countries	Host range number	Average/ (Range) yield loss (%)	GR No. of papers	Rank				Average (rank)	References
							GD	HR	YL	GR		
1.	*Albugo candida*	White rust	100	300	34	122	6	79	4	33	30.5 (VI)	Saharan et al. (2014)
2.	*Alternaria brassicae*	Alternaria blight	96	4000	45.1	183	12	114	11	46	45.75 (II)	Saharan et al. (2016)
3.	*Colletotrichum higginsianum*	Anthracnose	100	100	50	9	2	2	1	4	2.25 (–)	Cannon et al. (2012)
4.	*Erysiphe cruciferarum*	Powdery mildew	44	100	27.3	134	18	31	5	80	33.5 (V)	Saharan et al. (2019)
5.	*Fusarium oxysporum*	Wilt	100	100	15.9	20	1	0	1	18	5.0 (–)	Garibaldi et al. (2006), Matic et al. (2018), Lange et al. (2007)
6.	*Hyaloperonospora parasitica*	Downy mildew	102	99	37.3	15	3	4	1	7	3.75 (–)	Saharan et al. (2017)
7.	*Leptosphaeria maculans*	Blackleg	102	35	42	142	14	37	3	88	35.5 (IV)	Wang et al. (2020)
8.	*Plasmodiophora brassicae*	Clubroot	144	3700	28.5	273	21	127	17	108	68.25 (I)	Saharan et al. (2021)
9.	*Pseudocercosporella cepsillae*	White leaf spot	29	22	15	4	1	3	0	0	1.0 (–)	Gunasinghe et al. (2016)
10.	*Pyrenopeziza brassicae*	Light leaf spot	100	33	22	9	5	2	1	1	2.25 (–)	Fitt et al. (1997)
11.	*Rhizoctonia solani*	Seedling rot	70	200	30	5	0	1	0	4	1.25 (–)	Verma (1996)

12.	*Sclerotinia sclerotiorum*	Stem rot	102	600	40	166	22	70	3	71	41.5 (III)	Saharan and Mehta (2008)
13.	*Turnip mosaic virus*	TuMV	106	8		77	5	18	2	52	19.25 (VII)	Schwinghamer et al. (2014)
14.	*Verticillium longisporum*	Wilt	10	9		27	5	8	0	14	6.75 (-)	Tyvaert et al. (2014), Novakazi et al. (2015), Depotter et al. (2017)
15.	*Xanthomonas campestris pv. campestris*	Black rot	157	10		73	5	39	1	28	18.25 (VIII)	Singh et al. (2018)
16.	*Heterodera*	Nematode	91			3				1		Evans and Webb (1989)
						3138[a]						

[a]Total publications; Figure in the bracket are the ranking of the disease

(TIR-NBS-LRR) and CNL (CC-NBS-LRR). TNL type share homology with the Drosophila toll and human interleukin-1 receptor (TIR). The two types show divergence in their sequence and signaling pathways. Several partial NBS-LRR variants like TIR, TIR-NBS (TN), CC, CC-NBS (CN), and NBS (N) have also been identified in plant species (Wan et al. 2012; Inohara et al. 2005; Meyers et al. 2002, 2003).

The whole-genome sequence data enabled the genome-wide identification, mapping, and characterization of candidate NBS-containing R-genes in economically important plants. For example, the approximate arrays of 159 NBS-encoding R-genes in *A. thaliana* have been identified. Earlier genome-wide studies have demonstrated that TNL subfamily is abundant in dicots while absent in cereals (monocots). The presence of the full length of TNL and CNL types in the common ancestor (mosses) of both angiosperms and gymnosperms and exceptional presence of truncated domains of TN or TX type proteins in cereals indicate that the TNL class might have been lost in monocot plants. On the chromosomes, the NBS-LRR R-genes are arranged in clusters. The genes in the clusters could be homogenous (often tandem duplicated from single ancestor gene) or heterogenous (with different protein domains). However, the variation of the number and sequences of the R-genes presented in the *Brassica* lineage since split from the *Arabidopsis* lineage and their distributions in chromosomes are unknown.

The genera *Brassica* and *Arabidopsis*, both belong to the mustard family Brassicaceae (Cruciferae), are a model plant and a model crop, respectively. The two genera shared a latest and obviously detectable alpha genome duplication event before their divergence ~20 million years ago (MYA) and subsequently *Brassica* ancestor underwent a whole-genome triplication event (common to the tribe Brassicaceae) ~16 MYA. In *Brassica*, interspecific cytogenetic relationship between important crops (oilseed and vegetables) is well described by a "U" triangle where each two diploid species [*B. rapa* (AA, $2n = 20$), *B. oleracea* (CC, $2n = 18$), and *B. nigra* (BB, $2n = 16$)] formed a tetraploidy species [*B. napus* (AACC, $2n = 38$), *B. juncea* (AABB, $2n = 36$), or *B. carinata* (BBCC, $2n = 34$)]. This well-established phylogenetic relationship provides a chance to trace evolution of the R-genes between wild plants and their relative crops. Yu et al. (2014a, b) identified R-genes on genome-wide scale in *B. oleracea* and *B. rapa* and provided insights into their evolutionary history and disease resistance in crucifers.

Plant disease resistance (R) genes with the nucleotide-binding site (NBS) play an important role in offering resistance to pathogens. The availability of complete genome sequences of *Brassica oleracea* and *Brassica rapa* provides an important opportunity for researchers to identify and characterize NBS-encoding R-genes in *Brassica* species and to compare with analogs in *Arabidopsis thaliana* based on a comparative genomics approach. However, little is known about the evolutionary fate of NBS-encoding genes in the *Brassica* lineage after split from *A. thaliana*. Yu et al. (2014a, 2014b) presented genome-wide analysis of NBS-encoding genes in *B. oleracea, B. rapa,* and *A. thaliana*. Through the employment of HMM search and manual creation, they have identified 157, 206, and 167 NBS-encoding genes in *B. oleracea, B. rapa,* and *A. thaliana* genomes, respectively. Phylogenetic analysis among 3 species classified NBS-encoding genes into 6 subgroups. Tandem duplication and whole-genome triplication (WGT) analyses revealed that after WGT of the *Brassica* ancestor, NBS-encoding homologous gene pairs on triplicated regions in *Brassica* ancestor were deleted or lost quickly, but NBS-encoding genes in *Brassica* species experienced species-specific gene amplification by tandem duplication after divergence of *B. rapa* and *B. oleracea*. Expression profiling of NBS-encoding orthologous gene pairs indicated the differential expression pattern of retained orthologous gene copies in *B. oleracea* and *B. rapa*. Furthermore, evolutionary analysis of CNL type NBS-encoding orthologous gene pairs among 3 species suggested that orthologous genes in *B. rapa* species have undergone stronger negative selection than those in *B. oleracea* species. But for TNL type, there are no significant differences in the orthologous gene pairs between the two species. The identification and characterization of NBS-encoding genes in *B. rapa* and *B. oleracea* based on whole-genome sequences will be landmark in genomics research. Through tandem duplication and whole-genome triplication analysis in *B. oleracea*, *B. rapa*, and *A. thaliana* genomes, the study provides insight into the evolutionary history of NBS-encoding genes after divergence of *A. thaliana* and the *Brassica* lineage. These results together with expression pattern analysis of NBS-encoding orthologous genes provide useful resource for functional characterization of these genes and genetic improvement of relevant crops (Yu et al. 2014a, 2014b).

Molecular genetics investigations of disease resistance have been at the vanguard in *Arabidopsis* research of utilizing natural genetic variation. Numerous disease resistance genes (so-called *R*-genes) have been revealed that confer genotype- or race-specific resistance to a wide assortment of pathogens. A majority of these genes are of the same structural class, now widely known by the abbreviated name of NB-LRR (**n**ucleotide-**b**inding site and **l**eucine-**r**ich **r**epeat domains). Genes of this class have also been described in a wide diversity of crops such as maize, lettuce, sugar beet, tomato, and tobacco. An obvious spin-off, therefore, would be to use DNA sequence from NB-LRR genes and other known *R*-gene classes to develop more effective genetic markers for traditional breeding of resistance in crops. The prospect of using *Arabidopsis R*-genes themselves to confer resistance in crops has also been demonstrated. Most *Arabidopsis* research has been advanced using genetic variation that was derived artificially from mutagenesis induced chemically, by irradiation, or by transposon or T-DNA insertion. For investigation of disease resistance, numerous mutants classified broadly as ones that enhance susceptibility to avirulent pathogens or else enhance resistance to virulent pathogens have been molecularly characterized. A complex impression of defense signaling has emerged from these mutational analyses, and consequently redefines the frontline for *Arabidopsis* pathology as an effort to identify the intermediate components of defense regulation and dynamic relationships among distinct responses.

The underlying defense responses appear to be widely conserved among plant species. In some cases, genes have been conserved across Kingdoms, such as *SGT1* which has recently been identified as a major regulator of powdery mildew resistance in barley and downy mildew resistance in *Arabidopsis*, having first been described as a critical gene for cell cycling and proteolysis in yeast. Disease resistance can potentially be improved in crops by altering the expression of defense regulation, as suggested by broad-spectrum resistance that was achieved by over-expression in *Arabidopsis* of *NPR1,* a regulator of acquired disease resistance.

Wild plant species represent a gene pool that can potentially provide a rich source of durable, broad-spectrum disease resistance for use in crops. Species that are inter-fertile with a particular crop have often been used for transferring monogenic resistance that is often genotype or race specific. Unfortunately, such resistance derived from either crop or wild germplasm often lacks durability when deployed in field production because of selection it imposes that force the pathogen to adapt and overcome the resistance. Non-durable resistance is potentially still useful; however, plant breeders are usually challenged to continue the search for new sources of resistance.

Broad-spectrum resistance from wild germplasm, which may be polygenic and inherently more difficult to transfer and re-construct in a crop, will be increasingly more accessible as a result of recent advances in our knowledge of candidate resistance genes. The underlying basis for this resistance is that the pathogen as a taxonomic group (species or subspecies) is highly specialized to cause disease in the crop, but not in the wild crop relative. Broad-spectrum resistance that occurs universally in the wild relative is referred to below as *species-level ("non-host") resistance* to the pathogen.

Some of the advances demonstrate clearly that signaling mechanisms governing genotype-specific resistance can also govern species-level resistance. For example, mutations in one of the defense regulators discovered in *Arabidopsis*, designated *EDS1* (**e**nhanced disease **s**usceptibility), exhibit broad-spectrum susceptibility to several pathogens including isolates of *Albugo candida* (white rust) and *Hyaloperonospora parasitica* (downy mildew) from *Brassica oleracea*. Isolates of both pathogens collected from either *Brassica* or *Arabidopsis* represent distinct subspecies, based on their restrictive host specialization and molecular divergence. Therefore, *EDS1* is an essential regulator of species-level resistance to at least two *Brassica* pathogens. The highly conserved protein, *SGT1* has also recently been reported as a key regulator for species-level resistance.

Two bacterial *R*-genes, *RPS4* and *RPS5,* have been described in *Arabidopsis* which demonstrate that NBLRR genes can provide necessary components of species-level resistance. Both of these genes were characterized using corresponding avirulence determinants that were derived from legume isolates of *Pseudomonas syringae*. Arabidopsis exhibits species-level resistance, at least in symptomology, to these pathogens. The avirulence determinants, however, were introduced into an *Arabidopsis*-virulent isolate of *Ps. syringae* to facilitate molecular characterization of the resistance genes. Transient suppression of species-level resistance has also been observed in *Arabidopsis*. For example, *Bremia lactuca* (lettuce downy mildew) can reproduce asexually in cotyledons or true leaves following pre-infection with *A. candida* (Holub et al. 2002).

1.2 The *Brassica* Crop Species and Their Wild Relatives

The *Brassica* genus belongs to the tribe Brassiceae (family Brassicaceae). This family comprises 338 genera (assigned to 25 tribes) and 3709 species. The members of this family are mostly herbs with annual, biennial, or perennial growth habits. Initially this family was known as "Cruciferae" due to its characteristic flower conformation of four petals arranged in a cross-shape. Most of the member species are distributed in temperate regions, with the first center of diversification located in the Irano-Turanian region (~150 genera and ~900 species), followed by a second center of diversification in the Mediterranean region (>110 genera and ~630 species).

Brassica is the most prominent genus in the Brassicaceae family and includes 39 species. Many of the species in this genus are cultivated for their edible roots, leaves, stems, buds, flowers, mustard, and oilseeds. For 33 of the species the chromosome number has been determined, and ranges from $n = 7$ up to $n = 20$. During the 1930s, the chromosome number and genetic relationships between the cultivated *Brassica* species was established. The diploid species *B. rapa* (AA, $n = 10$), *B. nigra* (BB, $n = 8$), and *B. oleracea* (CC, $n = 9$) were determined to be the progenitors of the allopolyploid species *B. juncea* (AABB, $n = 18$), *B. napus* (AACC, $n = 19$), and *B. carinata* (BBCC, $n = 17$), in a relationship known as "U's Triangle." Based on chloroplast DNA data it was determined that *B. nigra* belongs to a different lineage (Nigra lineage) than *B. rapa* and *B. oleracea* (*Rapa*/*Oleracea* lineage), with the two lineages diverging approximately 7.9 Mya. The divergence between *B. rapa* and *B. oleracea* has been estimated to have occurred perhaps 3.75 Mya to about 5 Mya. Later on, approximately 7500 years ago or less, diploid species *B. rapa* and *B. oleracea* hybridized to produce *B. napus* L.

Genetic diversity within *Brassica* species has been broadly studied, with a special focus on the six crop species that form the U's triangle. Of these species, three are highly diverse: *B. oleracea, B. rapa,* and *B. juncea.* These species are quite morphologically variable, presenting different leaf types, numbers of branches per stem, inflorescence types, and stem thicknesses; these variations also lead to different end-product usage (e.g., oil or vegetable type). Genetic diversity observed in the *Brassica* allopolyploids can be due to (i) multiple hybridization events with diverse parents (or possibly subsequent backcrossing of the newly formed allotetraploids to the parent species), and (ii) genome changes occurring after polyploidization. Four *Brassica* species are mainly used as oilseed crops: *B. juncea*, *B. rapa, B. carinata,* and *B. napus* (Katche et al. 2019).

1.3 *Brassica* Species Uses and Genetic Diversity

Brassica napus (rapeseed, oilseed rape, swede) is the most economically important of the *Brassica* crop species, occupying the third position worldwide in the oil vegetable market, after soybean and palm oil. In Germany, a large proportion of the rapeseed oil produced is used to generate biodiesel. Rapeseed, as well as other members of the Brassicaceae, naturally contains 20–40% erucic acid and high glucosinolates in the seed meal. However, rapeseed has been extensively bred for low erucic acid and low glucosinolates to produce a type of rapeseed better known as canola. The main producers of rapeseed are Canada, China, and India, which together represent almost 60% of the total production worldwide. Winter-type rapeseed is mainly grown in Europe, and spring types are mostly grown in Canada, China, and Australia. *Brassica napus* (AACC, 2n = 4x = 38) is thought to have originated in the last 7500 years via at least two different hybridization events between *B. oleracea* and *B. rapa* in agricultural systems. Unfortunately, most of the genetic variation in oilseed rape has been reduced due to intensive selection for low erucic acid and low glucosinolate content traits. Rapeseed is not found in nature as a wild type, and most of the diversity existing nowadays comes from breeding programs or cultivars from different countries (Rahman et al. 2013).

Brassica juncea (AABB, 2n = 4x = 36) is also used as a vegetable, with leaf mustard or Indian mustard as the common name. A huge diversity of leaf morphotypes is present in this species that is thought to have been influenced by human selection, with two representative gene pools: East Europe and Indian. Mustard is mainly grown in India due to climate conditions, where the breeding objectives are mainly focused on improving seed yield. Although genetic resources available for *B. juncea* are not as comprehensive as those available for *B. napus* and its progenitor species, a reference *B. juncea* genome was published in the year 2016 (Yang et al. 2016).

Brassica rapa (AA, 2n = 2x = 20), initially named *B. campestris* and commonly known as turnip or Chinese cabbage, has its origins in the Mediterranean and Central Asia. The different subspecies of *B. rapa* can be used as a fodder (subsp. *rapifera*), vegetables (subsp. *chinensis* or *pekinensis*), or as an oilseed crop (subsp. *oleifera*). *Brassica rapa*, Chinese cabbage accession Chiifu-401-42, was the first *Brassica* species to get its genome sequenced. Of the estimated genome size of 485 Mb, 283.8 Mb was initially assembled. Later on, an improved assembly was released (v2.0) that increased the size of the genome assembly to 389.2 Mb. The *B. rapa* genome is rich in transposable elements, accounting for 32.3% (~54 Mb) of the assembled sequence, much more than the 10.0% observed in the related genome of *Arabidopsis thaliana* (Cai et al. 2017; Kaul et al. 2000).

Brassica oleracea (CC, 2n = 2x = 18) is mainly used as an edible vegetable. This species is composed of several varieties and morphotypes are usually referred to as coles. These vegetables are rich in vitamin C, folate, and calcium. Different varieties include Brussels sprouts (var. *gemmifera*), cabbage (var. *capitata*), cauliflower (var. *botrytis*), and Chinese kale (var. *alboglabra*). Some new vegetables have also been produced by crossing different varieties within this genus, such as broccoli. Two draft genome references for *B. oleracea* were published in 2014 (Liu et al. 2014).

Brassica carinata (BBCC, 2n = 4x = 34), also called Ethiopian mustard, possesses wide genetic variability and is also used as an oilseed crop. This crop has also been considered for use in biodiesel production and for other purposes including as a condiment, medicine, and vegetable (Kumar et al. 2015). *Brassica nigra* (BB, 2n = 2x = 16) was previously used as a condiment mustard but has now been mostly replaced by *B. juncea*. Compared to the major *Brassica* crops, *B. nigra* contains little variety in physical appearance, but it nevertheless possesses different agronomical traits of great value such as resistance to *Phoma lingam.* Although *B. nigra* is the least agriculturally significant of the six *Brassica* crop species, a scaffolded genome assembly (not yet assembled into pseudomolecules) was made available in 2016 alongside the *B. juncea* genome, and a new chromosome-level assembly was released in 2019 (Wang et al. 2019).

1.4 The *Brassica* Wild Relatives: Coenospecies and Cytodemes

In the 1970s, Harberd defined the term "coenospecies" for those species or genera that have sufficient relatedness to the six *Brassica* crops to be able to exchange genetic material with them. The coenospecies are composed of almost 100 wild species and genera that can potentially be used to increase diversity, and to introgress useful traits such as disease resistance or abiotic stress. Harberd also classified the *Brassica* coenospecies into biological units called "cytodemes." Each cytodeme can contain more than one genus or species, but all species within a cytodeme should have the same chromosome number, and readily cross with other species in the same cytodeme to produce fertile, vigorous hybrids. Based on these criteria, the *Brassica* coenospecies were initially classified into 38 cytodemes, covering nine genera from the sub-tribe Brassiceae (*Brassica, Coincya, Diplotaxis, Eruca, Erucastrum, Hirschfeldia, Sinapis, Sinapidendron,* and *Trachystoma*) and two genera from sub-tribe Raphaninae (*Enarthrocarpus* and *Raphanus*). This was later updated to 63, after the addition of three genera (*Moricandia*, *Pseuderucaria*, and *Rytidocarpus*) from the related sub-tribe Moricandiinae. The cross ability between cytodemes is low, but certain tools can be used to increase success rates. Cross ability can also be influenced by the direction of the cross, *i.e.*, which species is used as the maternal parent, which is referred to as "unilateral incompatibility" (Katche et al. 2019). An extended list of potentially useful agronomic traits for crop improvement present in wild allies of the *Brassica* species include resistance to white rust (*Albugo candida*) in *Brassica maurorum* and *Eruca versicaria* ssp. *sativa*, resistance to *Alternaria* blight in *Brassica fruticulosa* and *Trachystoma ballii*, resistance to beet cyst nematode in *Raphanus sativus* and *Sinapis alba*, and resistance to blackleg/Phoma disease (*Leptosphaeria maculans*) in *Sinapis arvensis*, *Sinapis alba*, *Thlaspi arvense*, and *B. tournefortii*. The *Brassica* crop species also contain unique, useful traits: examples include resistance to downy mildew (*Hyaloperonospora parasitica*) in *Brassica oleracea*, resistance to clubroot disease (*Plasmodiophora brassicae*) in *B. rapa*, *B. oleracea,* and *B. napus*, and pod shatter resistance and tolerance to heavy metals in *B. juncea*. More exotic traits of interest include a C3–C4 intermediate photosynthetic system in *Moricandia* and *Diplotaxis* species, and high erucic acid levels in *Crambe abyssinica*. Cytoplasmic male sterility in *Brassica* could also be conferred by hybridization with *Sinapis incana* and *Diplotaxis siifolia*, among other examples (Katche et al. 2019; Piao et al. 2009).

1.5 Hybridization Between *Brassica* Species and Wild Relatives

Direct wide hybridization has been attempted many times between *Brassica* and various wild relative species, with different levels of success. Originally such hybrids were produced to resolve chromosome homoeology (phylogenetic relationships) or simply out of curiosity. However, crossing with distant relatives is today attracting increasing recognition as a method with which to improve agronomic traits in high-end varieties. There are many examples of the successful introgression of new traits into *Brassica* crops. Initial attempts to create hybrids between *Brassica* species started in the early 1800s. At this time, some crosses were made between *B. napus* × *B. rapa* and *B. oleracea* × *B. rapa*. Different success rates were reported and the results were published in 1925. Later on, a compilation of cross ability between species in the *Brassica*, *Raphanus,* and *Sinapis* genera was published, showing that interspecific hybrids can be made between the *Brassica* crops and many closely related wild species (Prakash et al. 2009).

The occurrence of natural hybridization between distant relatives in natural conditions is low. It was found that hybridization between *Brassica napus*, *B. rapa,* and *B. juncea* and their two weedy relatives *B. nigra* and *Sinapis arvensis* does not occur under open pollination conditions in the field, although *B. rapa*, *B. juncea,* and *B. napus* all readily produce hybrid progeny with each other under the same conditions. The cross between *B. napus* ($2n = 38$) and *Raphanus raphanistrum* ($2n = 18$) has also been assessed under field conditions. In this case, just two allopolyploid hybrids ($2n = 56$) were obtained from more than 52 million *B. napus* seedlings when this species was used as a female, showing a hybridization frequency of $4 \times 10{-}8$ in field conditions. These results indicate that the likelihood of this cross in the wild is low, which shows the importance of conducting such hybridizations under controlled conditions (Rieger et al. 2001; Katche et al. 2019).

1.6 Interspecific Hybridization for Developing Disease Resistance *Brassica* Cultivars

Interspecific hybridization is widespread in nature, where it can lead to either the production of new species or to the introgression of useful adaptive traits between species. In agricultural systems, there is also great potential to take advantage of this process for targeted crop improvement. In the *Brassica* genus, several crop species share close relationships: rapeseed (*Brassica napus*) is an ancestral hybrid between turnip (*B. rapa*) and cabbage (*B. oleracea*), and mustard species *B. juncea*, *B. carinata,* and *B. nigra* share genomes in common. This close relationship, plus the abundance of wild relatives and minor crop species in the wider Brassiceae tribe which readily hybridize with the *Brassica* crop species, makes this genus an interesting example of the use of interspecific hybridization for crop improvement (Katche et al. 2019).

1.7 Identification and Use of R-Genes in *Brassica* Species

In order to cope with invader attacks, plants have evolved sophisticated immune mechanisms to protect themselves against their natural enemies. The best known are the vast numbers of resistance genes in plants, which play a central role in recognizing effectors from pathogens and in triggering downstream signaling during plant responses to pathogen invasions. Genes containing a nucleotide-binding site (NBS), namely NBS-encoding genes, constitute one of the largest plant resistance gene families (~80%). The NBS domain was found to bind and hydrolyze ATP or GTP, and primarily functions as a signal transduction switch following pathogen recognition. NBS-encoding genes typically comprise three principal domains, with the NBS domain in the middle region, flanked by a leucine-rich repeat (LRR) domain at the C-terminus, and by a toll/interleukin-1 receptor (TIR) or coiled-coil (CC) at the N-terminus. The central NBS domain encodes several motifs consisting of 10–30 amino acids (aa), and is typically highly conserved, whereas the C-terminal LRR domain exhibits high diversity and has been associated with pathogen recognition. According to the presence or absence of the N-terminal TIR domain, NBS-encoding genes are further classified into TIRNBS-LRR (TNL) or TIR-NBS (TN) genes and non-TIR-NBS-LRR (non-TNL) or (non-TN) genes. Based on the presence of the CC or other domains at the N-terminus, non-TNL and non-TN genes can be further divided into CC-NBS-LRR (CNL) or CC-NBS (CN) genes and X-NBS-LRR (XNL) or X-NBS (XN) genes (Dangl and Jones 2001). With the availability of genomic data for an increasing number of species, a set of NBS-encoding genes has been identified at the genome level in more than 30 angiosperms. Comparative studies of the evolutionary history of NBS encoding genes were further performed in a number of clades over recent years to illuminate the evolutionary characteristics of NBS-encoding genes.

Gene loss and retention patterns and a pattern of gene expansion followed by contraction were identified in various species in the Brassicaceae family. Oilseed rape (*Brassica napus*, AnAnCnCn, 2n = 38) is an allopolyploid that originated from spontaneous hybridization events between the two diploid *Brassica* species *B. rapa* (ArAr, 2n = 20) and *B. oleracea* (CoCo, 2n = 18) in the last 10,000 years. Despite the short domestication history of rapeseed (~300–400 years), it is today one of the most important oil crops worldwide. Hybridization and polyploidization often produce species with superior resistances or environmental tolerances relative to their progenitor species. The exact mechanisms responsible for this effect are unknown, but in rapeseed could result from hybridization, polyploidization, or domestication processes shaping genome evolution. The evolutionary impact of these processes on NBS encoding genes, which are major players in plant disease resistance, is hence of particular interest. With the availability of genomic data for *B. napus* and its two progenitor species *B. rapa* and *B. oleracea*, NBS-encoding genes can be systematically investigated to elucidate their role in contributing to differences in disease resistance between the three species, and help to decipher the mechanisms underlying disease resistance in *B. napus*. Previous studies have identified and compared numbers, types, and locations of NBS-encoding genes in

B. rapa, B. oleracea, and *B. napus*. Chalhoub et al. (2014) found 425 NBSLRRs in *B. napus* (245 in the C genome and 180 in the A genome) and similar numbers were reported for the *B. rapa* and *B. oleracea* genomes but with only 75% conservation of synteny. Alamery et al. (2017) identified 641 NBS-LRRs in *B. napus* in total, of which only 365 were intact and hence putatively functional. Yu et al. (2014a, 2014b) identified 239 NBS-LRRs in *B. oleracea*, which was updated to 556 NBS-LRRs in the *B. oleracea* pan-genome (Bayer et al. 2018).

A genome-wide characterization of NBS-encoding genes was performed in *B. rapa, B. oleracea,* and *B. napus*. Multiple approaches were utilized to assess the genome architecture and evolutionary characteristics of NBS genes, including genomic distribution, homologous genes, sequence similarities, selection signals, phylogenetic relationships, and expression patterns. Resistance genes were also co-localized with previously identified quantitative trait loci (QTL) conferring resistance against major oilseed rape pathogens blackleg (*Leptosphaeria maculans*), clubroot (*Plasmodiophora brassicae*), and Sclerotinia stem rot (*Sclerotinia sclerotiorum*), highlighting potential disease resistance candidate genes for future work. This analysis provided genome-level insights into the evolution of disease resistance genes in *B. napus*, shedding light on possible mechanisms of disease resistance for future disease resistance breeding in rapeseed (Fu et al. 2019). Genes containing nucleotide-binding sites (NBS) play an important role in pathogen resistance in plants. However, the evolutionary fate of NBS-encoding genes after formation of allotetraploid *Brassica napus* (AnAnCnCn, 2n = 38) is still unknown. Fu et al. (2019) performed a genome-wide comparison of putatively functional NBS-encoding genes in *B. napus* and its progenitor species *Brassica rapa* (ArAr, 2n = 20) and *Brassica oleracea* (CoCo, 2n = 18), identifying 464, 202, and 146 putatively functional NBS encoding genes, respectively, with genes unevenly distributed in several clusters. The An-subgenome of *B. napus* possessed similar numbers of NBS-encoding genes (191 genes) to the Ar genome of *B. rapa* (202 genes) and similar clustering patterns. However, the Cn genome of *B. napus* had many more genes (273) than the *B. oleracea* Co genome (146), with different clustering trends. Only 97 NBS-encoding genes (66.4%) in *B. oleracea* were homologous with NBS-encoding genes in *B. napus*, while 176 NBS-encoding genes (87.1%) were homologous between *B. rapa* and *B. napus*. These results suggest a greater diversification of NBS-encoding genes in the C genome may have occurred after formation of *B. napus*. Although most NBS encoding genes in *B. napus* appeared to derive from the progenitors, the birth and death of several NBS-encoding genes was also putatively mediated by non-homologous recombination. The Ka/Ks values of most homologous pairs between *B. napus* and the progenitor species were less than 1, suggesting purifying selection during *B. napus* evolution. The majority of NBS-encoding genes (60% in all species) showed higher expression levels in root tissue (out of root, leaf, stem, seed, and flower tissue types). Comparative analysis of NBS-encoding genes with mapped resistance QTL against three major diseases of *B. napus* (blackleg, clubroot, and Sclerotinia stem rot) found 204 NBS-encoding genes in *B. napus* located within 71 resistance QTL intervals. The majority of NBS-encoding genes were co-located with resistance QTLs against a single disease, while 47 genes were co-located with QTLs against two diseases and 3 genes were co-located with QTLs against all three. The results revealed significant variation as well as interesting evolutionary trajectories of NBS-encoding genes in the different *Brassica* sub-genomes, while co-localization of NBS-encoding genes and resistance QTL may facilitate resistance breeding in oilseed rape (Fu et al. 2019).

1.8 R-Genes Recognition System

In order to protect against biotic stresses, plants recognize invading pathogens via two different recognition systems: (1) pathogen-associated molecular pattern (PAMP)-triggered immunity (PTI), and (2) effector-triggered immunity (ETI), recognized by intracellular receptors (Jones and Dangl 2006). PAMPs are structural molecules that recognize the pathogen at the host surface cell level and induce PTI through pattern recognition receptors (PRR), which trigger a mitogen-activated protein kinase (MAPK) cascade as part of the defense response (Bigeard et al. 2015). The pathogen secretes a series of effectors within the host cell that inhibit recognition by the host, thereby interacting with resistant (R) proteins and initiating the second system of plant immunity, which includes leucine-rich repeats (LRR) or the LRR-like family, the toll/interleukin-1-receptor (TIR), and serine/threonine kinases (S/TK), which trigger a hypersensitive reaction in the plant, inhibiting the spread of the pathogen to adjacent healthy cell tissue and resulting in the plant acquiring systemic resistance to the pathogen (Thomma et al. 2011).

1.9 Close Association of R-Genes to Confer Multiple Disease Resistance in *Brassica*

The QTL mapping of blackleg, clubroot, and Sclerotinia stem rot, the three major diseases of *B. napus,* was analyzed by Fu et al. (2019). It was found that QTLs of these three diseases were unevenly distributed on chromosomes of *B. napus*, with the number of QTLs ranging from 4 to 14 (Table 1.2). This uneven distribution of QTLs across the chromosomes of *B. napus* was similar to the uneven distribution of NBS-encoding genes observed. The three chromosomes with the most QTLs detected for resistance against these pathogens were An09 with 14 QTLs, Cn3 with 14 QTLs, and Cn4 with 13 QTLs; this coincides with the finding that An09 and Cn3 were two of the three chromosomes with the most NBS-encoding genes (Fig. 1.1). The similar distribution of disease resistance QTLs and NBS-encoding genes supports the close associations of NBS encoding genes with disease resistance against blackleg, clubroot, and Sclerotinia stem rot in *B. napus*. Fu et al. (2019) also investigated overlapping and neighboring (within 1 Mb) QTL intervals to find repeatedly identified QTLs: approximately half (42/81) of resistance QTLs against blackleg were repeatedly identified. A total of 13 of these loci were identified for blackleg resistance (on An01, An02, An08, An09, An10, Cn1, Cn2, Cn4, Cn5, Cn6, and Cn8). One locus on chromosome An03 contained 4 of 19 resistance QTLs

Table 1.2 The summary of mapped QTL against blackleg, clubroot, and Sclerotinia stem rot in *Brassica napus* (Fu et al. 2019)

Chromosome	Total QTL number	Blackleg QTL number	Blackleg Repeatedly identified QTL[a]	Clubroot QTL number	Clubroot Repeatedly identified QTL[a]	*Sclerotinia* stem rot QTL number	*Sclerotinia* stem rot Repeatedly identified QTL[a]
A_n01	11	8	6, 2	1	0	2	0
A_n02	8	4	3	1	0	3	2
A_n03	12	4	0	5	4	3	0
A_n04	4	2	0	1	0	1	0
A_n05	4	2	0	0	0	2	0
A_n06	12	6	0	1	0	5	3
A_n07	6	3	0	0	0	3	0
A_n08	6	3	3	1	0	2	0
A_n09	14	9	6, 3	0	0	5	2, 2
A_n10	5	3	2	0	0	2	0
C_n1	7	5	3	0	0	2	0
C_n2	9	4	2	1	0	4	2
C_n3	14	2	0	3	0	9	0
C_n4	13	7	5	2	0	4	2
C_n5	8	3	2	1	0	4	0
$_n$6	10	5	3	0	0	5	4
C_n7	7	4	0	1	0	2	0
C_n8	10	3	2	0	0	7	0
C_n9	10	4	0	1	0	5	3
Total	170	81	42	19	4	70	20

[a]The overlapping or neighboring located QTL (less than 1 Mb) were considered as repeatedly identified QTL. The digit indicated the QTL number that repeatedly identified in a locus

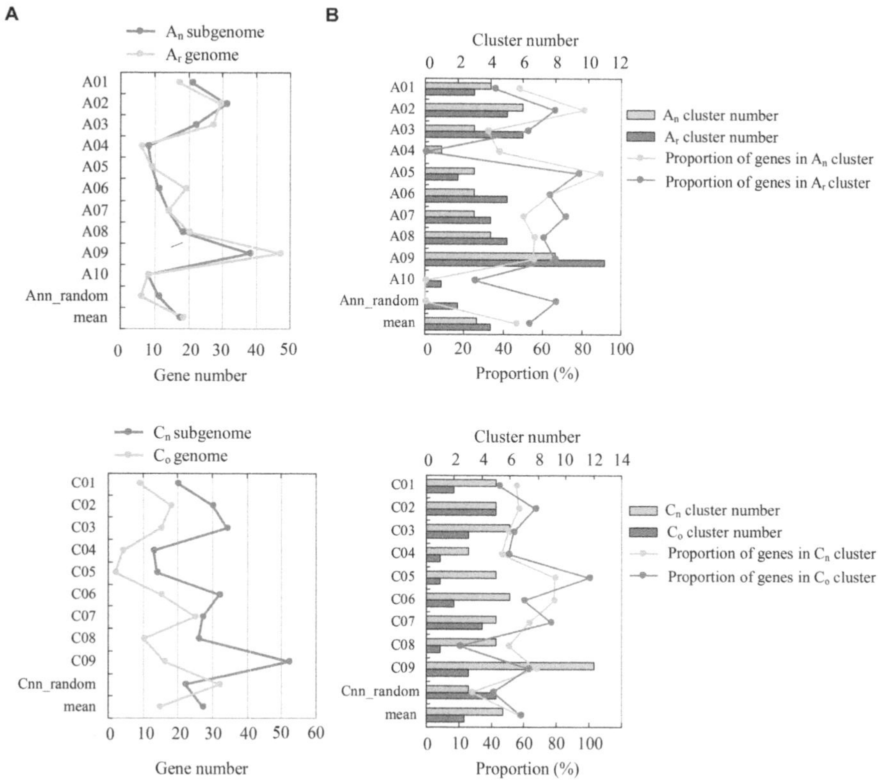

Fig. 1.1 Comparison of NBS-encoding genes between the sub-genomes of *Brassica napus* (AnAnCnCn) and the genomes of *Brassica rapa* (ArAr), and *Brassica oleracea* (CoCo). (**a**) The comparison of gene numbers between the genomes of Ar and An, and Co and Cn. The *y*-axis indicates the number of NBS-encoding genes located on each individual chromosome, and the *x*-axis represents the chromosomes. (**b**) The comparison of NBS-encoding gene clusters between the Ar and An, and Co and Cn genomes. The *y*-axis represents the chromosomes, while the *x*-axis represents the number of gene clusters and the proportion of genes located in each gene cluster, respectively (Fu et al. 2019)

against clubroot, most likely corresponding to the major resistance gene CRa previously identified in this region (Ueno et al. 2012). Of the Sclerotinia stem rot resistance QTLs, 20 of 70 were repeatedly identified in 8 loci on An02, An06, An09, Cn2, Cn4, Cn6, and Cn9 (Table 1.2). Co-localized QTL detected across multiple studies are highly credible candidates for resistance against these three diseases. Across all QTL, 41.76% of intervals were co-located with NBS-encoding genes, which verified the close associations between the NBS-encoding genes and QTLs against these pathogens. Three NBS-encoding genes, (GSBRNA2T00081287001, GSBRNA2T00021121001, and GSBRNA2T 000980 680 01) were located within overlapping QTLs against all three diseases, while 47 NBS-encoding genes were located within overlapping QTLs against two diseases. These results suggested that

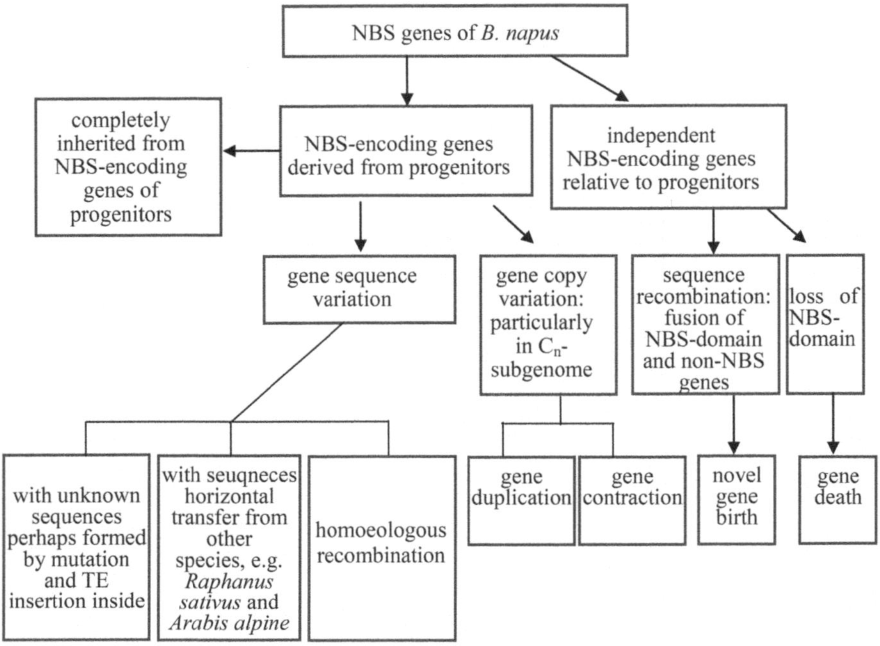

Fig. 1.2 The evolutionary routes of NBS-encoding genes in *Brassica napus* relative to its two progenitor species *Brassica rapa* and *Brassica oleracea* (Fu et al. 2019)

several common factors might exist for *B. napus* that confer resistance against clubroot, blackleg, and Sclerotinia stem rot, and that these key factors may comprise NBS-encoding genes (Fig. 1.2; Fu et al. 2019).

1.10 Homoeologous Exchange: A Cause of R-Gene in *Brassica*

The meiotic chromosome pairing that occurs between homoeologous chromosomes which share a high degree of sequence identity leads to increased homoeologous exchanges (HEs) and gene conversion events (Gaeta and Chris Pires 2010; Stein et al. 2017), and synthetic *B. napus* has been shown to exhibit a higher frequency of HEs than non-synthetic *B. napus* (Liu et al. 2014; Rousseau-Gueutin et al. 2017), making it an interesting model to study the impact of polyploidization on genome structure and agronomic traits (Rousseau-Gueutin et al. 2017; Schiessl et al. 2017; Stein et al. 2017; Zou et al. 2011). Two public reference genomes corresponding to the winter oilseed cultivars Darmor-bzh (Bayer et al. 2017; Chalhoub et al. 2014) and Tapidor (Bayer et al. 2017) are currently available; however, it was unknown how well these references represent the genetic diversity found in *B. napus*. The pan-genome represents the set of genes for a species, composed of core genes, which are present in all individuals, and variable genes, which are only present in some individuals. The concept of the pan-genome was introduced by Tettelin et al. (2005), who produced the first pan-genome for the bacterial species *Streptococcus agalactiae*. However, pan-genomics is increasingly being applied to higher organisms, including *B. rapa* (Lin et al. 2014) and *B. oleracea* (Golicz et al. 2016). Hurgobin et al. (2018) have analyzed data from a collection of 53 synthetic and non-synthetic accessions (Schmutzer et al. 2015a, 2015b; Snowdon et al. 2015) to produce the first estimate of the *B. napus* pan-genome and investigated the role of HEs in *B. napus* genomic diversity. These accessions come from diverse geographical locations, comprising a range of morphotypes including oilseeds, fodder, and vegetable types. They identified core and variable genes and predicted the size of the pan-genome and core genome for this species. They also assessed the variable gene content in relation to HEs and their potential association with agronomic traits and disease resistance. The results highlight the potential of using resynthesized *B. napus* accessions as a source of novel genetic structural variation for breeding improved varieties with resistance to diseases (Hurgobin et al. (2018). Synthetic *B. napus* has previously been shown to demonstrate greater genetic diversity than non-synthetic accessions (Golicz et al. 2016), and this difference has been attributed to the incorporation of novel alleles from diverse progenitor genomes. The diversity is amplified by PAV, with many of the variable genes due to HE events in the new synthetic accessions. These HE-related PAV events are useful to understand the association between genome structural rearrangement and phenotypic variation, particularly the role of genome duplications or deletions spanning genes with trait-related dosage effects. The observation that synthetic accessions experience HEs on a larger scale and more frequently than their non-synthetic counterparts suggests that they have the potential to increase the genetic diversity of *B. napus* accessions producing novel allele combinations and associated phenotypic variation beyond the addition of novel allelic variants (Chalhoub et al. 2014; Rousseau-Gueutin et al. 2017; Sharpe and Lydiate 2003). This also highlights niche exploitation and speciation (Gaeta and Chris Pires 2010).

The abundant PAV of R-genes in the synthetic accessions highlights their potential for the introgression of candidate disease resistance genes in *B. napus*, supporting adaptation of this important crop to diverse environments and pests. It also illustrates the value of the pan-genome in capturing additional information not contained within a single reference, with almost half of the candidate R-genes identified being in the variable genome. This information can be exploited to further characterize the relationships between candidate R-genes and resistance/susceptibility among accessions (Fig. 1.3; Hurgobin et al. 2018).

1.11 Resistance Mechanisms Operating in *Brassica*

Plants are very often exposed to a variety of biotic stresses, and thus have evolved multidimensional defense approaches to survive or retain their fitness (Roux et al. 2014). The plants display both preformed and inducible defense mechanisms to overcome pathogen challenges. However, much stronger and long lasting is inducible defense response such as systemic acquired resistance (SAR), which confers

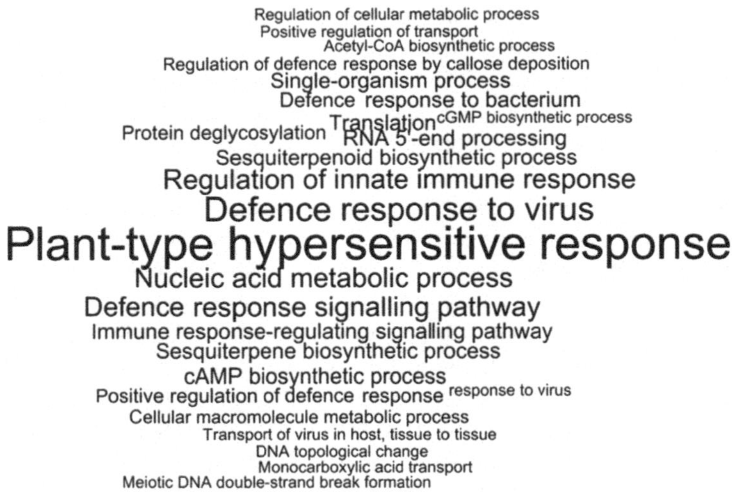

Fig. 1.3 GO enrichment analysis of variable genes. Significantly enriched GO terms among variable genes using all pan-genome genes as background. Font size is proportional to –log (P) (Hurgobin et al. 2018)

enhanced disease resistance to broad range of phytopathogens (Durrant and Dong 2004). In plants, SAR is generally activated by local infections and immunizes the whole plant to subsequent infectious diseases (Fu and Dong 2013; Shah and Zeier 2013). Realistic evidences have shown that activation of SAR is reliant on the higher levels of the endogenous salicylic acid (SA), and activation of a battery of pathogen-related (PR) genes. Most of these PR proteins such as glucanases, chitinases, thaumatins, and defensins possess antifungal activities and are known to play important role in disease resistance. Exogenous application of SA or its analogs have been also revealed to activate SAR pathway in plants (Durrant and Dong 2004; Makandar et al. 2006). Conversely, *Arabidopsis thaliana* plants expressing NahG transgene which codes for salicylate hydroxylase (SA-degrading enzyme) were deficient in accumulating SA, and hence failed to activate SAR (Delaney et al. 1995). In addition to SA, a group of heterogeneous proteins are crucial for the activation of SAR. Among them is the PR1 protein, a key regulator in the SA-mediated SAR signal transduction pathway. The quest to discover the SA receptor led to the discovery of a regulatory or transcription co-factor protein NPR1 (Cao et al. 1994). However, many studies have revealed that NPR1 is linked to SA signaling; however, its role as SA receptor remains largely unknown. In this context, Wu et al. (2012) have reported that NPR1 is the receptor for SA pathway in *Arabidopsis*. In addition, two NPR1 paralogs, namely NPR3 and NPR4, bind SA and control the proteasome-mediated degradation of NPR1 protein through their interaction with NPR1 (Fu et al. 2012). After pathogen assault, plants produce a variety of phytohormones; their composition, quantity, and timing significantly vary among plant species and depend mainly on the pathogens lifestyle and their mode of infection (De-Vos and Van-Oosten 2005). SA pathway generally provides resistance to biotrophic pathogens, whereas jasmonic acid/ethylene (JA/ET) pathways are commonly associated with resistance to necrotrophic pathogens, and to herbivorous pests (Glazebrook 2005; Bari and Jones 2009). Generally, SA and JA signaling pathways operate antagonistically, and thus elevated resistance against biotrophs is often related with increased susceptibility to necrotrophs and vice versa (Grant and Lamb 2006). Many regulatory components involved in SA/JA crosstalk have been identified; among them is NPR1 which plays a crucial role in regulating SA-mediated suppression of the JA pathway (Spoel et al. 2003; Pieterse et al. 2012; Thaler et al. 2012; Van der Does et al. 2013). Furthermore, SA/JA antagonism is commonly found in many plant species under various taxonomic groups, and therefore seems to be evolutionary conserved (Thaler et al. 2012).

As first discovered in *Arabidopsis*, various AtNPR1 homologs have been isolated thereafter in many agriculturally important crops (Chen et al. 2013; Zhong et al. 2015). *NPR1* is a multigene family in *Arabidopsis* with multifaceted functions. For example, *AtNPR1* and *AtNPR2* are notably considered as a key regulator of SAR (Cao et al. 1997, 1998; Zhang et al. 2003) while *AtNPR3* and *AtNPR4* are known as negative regulator of SAR (Fu et al. 2012). Moreover, another group of *AtNPR1* homologs are *AtBOP1*, and *AtBOP2*, which are related with lateral organ development (Hepworth et al. 2005). However, most of the studies were carried out on *Arabidopsis NPR1* (*AtNPR1*). Structurally, *AtNPR1* and its homologs contain an ankyrin repeat, N-terminal BTB/POZ broad-complex, Tramtrack, Bric-a-brac/pox-virus, and zinc finger domains, respectively (Cao et al. 1997; Aravind and Koonin 1999). In non-induced state, *NPR1* exists as an inactive oligomer form into cytosolic region. However, after SA accumulation, the redox status of the cell changes which leads to dissociation of the inactive oligomer *NPR1* to active monomers and their translocation to the nucleus where it binds to TGA factors, thereby inducing PR genes (Mou et al. 2003; Tada et al. 2008).

Transcriptional studies have shown that *NPR1* is expressed at low levels in mock plants but is induced significantly after microbial attack or treatment with SA or its biologically active analogs. Many studies have revealed *NPR1* mutant (*NPR1*) plants are more prone to diseases, and also show altered expression of defense marker PR genes compared to *NPR1* expressing plants (Glazebrook et al. 1996; Cao et al. 1997). Furthermore, *NPR1* also plays a role in cross talk of SA/JA signaling pathways and in antagonistic effect of SA on JA signaling (Spoel et al. 2003). Hence, *NPR1* is considered as the positive regulator of SA-mediated plant immune responses. To explore the defense role of *NPR1* against bacterial, viral, and fungal pathogens, various overexpression studies have been carried out in both model and crop plant systems. *NPR1* mediates the SA-induced expression of pathogenesis-related (PR) genes and SAR. Overexpression of *NPR1* results in the increase of transcript levels of antifungal genes like *PR1*, *PR2* (glucanase), and *PR5* (thaumatin) which are universally known to have antifungal activity. Many studies have revealed the potential antifungal activity of these PR genes against wide range of fungal pathogens. PR gene activity is regulated at the level of redox-dependent nuclear transport of *NPR1*. For example, overexpression of *NPR1* in *Arabidopsis* plants confers enhanced disease resistance to bacterial and fungal infections (Cao et al. 1998; Friedrich et al. 2001).

Transgenic carrot plants overexpressing AtNPR1 exhibit high disease resistance not only to biotrophs (*Erysiphe hordei*) but also to necrotrophic fungal pathogen (*Botrytis cinerea*, *Sclerotinia sclerotiorum*, and *Alternaria radicina*), respectively (Wally et al. 2009). Parkhi et al. (2010) also reported that cotton transgenic plants expressing *AtNPR1* exhibited broad spectrum of disease resistance not only to fungal pathogens but also to nematodes. Additionally, tobacco plants overexpressing *Malus hupehensis NPR1* confer resistance to *Botrytis cinerea* as well as induce battery of pathogen-related genes. Furthermore, studies have revealed that rice and wheat plants overexpressing *NPR1* gene confer broad spectrum of disease resistance against most disastrous pathogens *Magnaporthe oryzae, Fusarium verticillioides*, and *Fusarium oxysporum*, respectively (Makandar et al. 2006; Quilis et al. 2008). *NPR1* overexpression in *A. thaliana* was reported to result in an increase in the transcript levels of PR genes, and hence proves that NPR1-dependent PR gene mediated disease resistance (Cao et al. 1997; Friedrich et al. 2001). The overexpression of NPR1 was also revealed to confer disease resistance against broad range of pathogens in different crops (Dutt et al. 2015; Sundaresha et al. 2016). These results revealed that NPR1 is a potential candidate gene for developing disease resistant transgenic crops against multiple pathogens.

Studies on the mechanisms of resistance to Alternaria blight have implicated polygenes (Tripathi et al. 1980; Medhi 1985; Zhang et al. 1996; Krishnia et al. 2000), while some studies have attributed it to dominant nuclear genes (Tripathi et al. 1978; Subudhi and Raut 1994; Zhang et al. 1997). Thus far, different sources of resistance to *A. brassicae* have been identified from host germplasm, including wild and tolerant species of *Brassica*. The resistance in the host germplasm has different components, which mainly include structural components such as epicuticular waxes, as well as biochemical components like phenols and phytoalexins. Epicuticular waxes form a direct physical barrier in the plant, providing resistance to Alternaria (Conn 1986; Conn and Tewari 1989). High deposits of epicuticular wax, forming a protective hydrophobic coating on the leaf surface, reduce the adherence of inoculum, conidia germination, and germ tube formation in the plant (Saharan 1992). Some Brassicaceae members, including *B. napus, B. carinata*, and *Sinapis alba*, have been reported to have higher epicuticular wax compared to *B. rapa* and *B. juncea*, and therefore the former are less susceptible to Alternaria blight infection (Conn et al. 1984). High quantities of epicuticular wax have been observed in the progeny of interspecific crosses between *B. napus* and *B. juncea* (Singh et al. 1999).

Alternaria blight-resistant varieties accumulate high amounts of biochemical compounds such as phenols. After infection, blight-tolerant species of *Brassica* such as *B. carinata* and *B. napus* have been observed to accumulate a higher amount of total phenols compared to the susceptible species *B. juncea* and *B. rapa* (Gupta et al. 1990; Gupta et al. 1995; Gupta and Kaushik 2002). Meanwhile, the levels of soluble and reducing sugars and soluble nitrogen have been found to be lower in resistant species. In another study, resistance to the disease was reported to be associated with increased levels of antioxidant enzymes of the phenolic pathway, such as polyphenol oxidase, peroxidase, and catalase (Singh et al. 2009). Furthermore, phytoalexin accumulation in response to pathogen infection and its role in disease resistance have been well studied in *Brassica*.

Phytoalexins are low-molecular-weight antimicrobial compounds produced by plants after pathogen infection. Cultivars of both resistant (*B. napus, Camelina sativa, Eruca sativa*) and susceptible (*B. rapa*) species show the elicitation of phytoalexins (Tewari et al. 1987, 1988; Conn et al. 1988; Conn et al. 1991). The highly resistant *C. sativa* produces a large number of phytoalexins that are involved in regulating resistance to *A. brassicae*. Rapid accumulation of phytoalexins in *C. sativa* after pathogen infection has been found to inhibit fungal growth on the leaf surface (Jejelowo et al. 1991).

The *Alternaria* pathogen has also been found to secrete both host-specific and nonhost-specific toxic metabolites, enabling a wide range of infection symptoms (Nishimura and Kohmoto 1983; Kohmoto et al. 1995; Agrios 2005). These toxins have been shown to alter the permeability and functioning of the cell membrane and organelles, thereby inhibiting various physiological processes of the host plant (Mathur and Chand 1991). Destruxin B, a host-specific *Alternaria* phytotoxin, acts as a pathogenicity factor responsible for its aggressiveness and for the susceptibility of the host plant. The *Alternaria*-tolerant species *S. alba* detoxifies destruxin B, and this is followed by simultaneous phytoalexin formation and elicitation. Together, these processes constitute the resistance mechanism of *S. alba* against *A. brassicae* (Pedras and Smith 1997; Pedras et al. 2001). Some compounds related to camalexin and 6-methoxycamalex have also been found to confer toxicity to *A. brassicae* (Dzurilla et al. 1998; Fatima et al. 2019).

1.12 Importance of *Arabidopsis* Model Host-Patho System in Genomics of Disease Resistance

In nature, *Arabidopsis thaliana* is commonly infected by the downy mildew pathogen *Hyaloperonospora arabidopsidis* and two species of the white rust pathogen *Albugo* (*A. laibachii* and *A. candida*). These naturally occurring pathogens have imposed evolutionary pressures on *Arabidopsis* populations that resulted in a high level of intraspecific variation in resistance (Holub and Beynon 1996). Many of the underlying resistance genes have been cloned and their downstream signaling pathways have been characterized.

1.12.1 *Hyaloperonospora-Arabidopsis* Pathosystem

This species was the first eukaryotic pathogen of *Arabidopsis* to be documented (Koch and Slusarenko 1990a). It was initially described as *Peronospora parasitica* (Koch and Slusarenko 1990b), and was later re-named as *Hyaloperonospora parasitica* (Constantinescu and Fatehi 2002) and currently *Hyaloperonospora arabidopsidis* (Voglmayr et al. 2004; Goker et al. 2009). This species is a frequently occurring pathogen in natural *Arabidopsis* populations (Holub and Beynon 1996; Holub 2008). Moreover, the interactions are typified by abundant genetic polymorphism in the host and the pathogen (Holub et al. 1994b). For these reasons, *H. parasitica* was adopted as a reference pathogen during the early days of developing *Arabidopsis* as a system for molecular plant–microbe interactions (Dangl et al. 1992; Crute et al. 1994). These pioneering efforts substantially broadened the impact of *Arabidopsis* as model system. Like other phytopathogenic oomycetes, downy mildew species can quickly overcome host resistance genes and develop resistance to chemical control agents (Lucas et al. 1995). Thus, *H. parasitica* is a reference species for a group of pathogens with significant economic impact.

A significant advantage of *H. parasitica* is that it is a bona fide pathogen of *Arabidopsis* in the natural world, and has been co-evolving with its host (Holub 2001). The polymorphisms that have evolved in the host and pathogen were used to molecularly clone *Arabidopsis* disease resistance genes ("*R*-genes") and the corresponding effector/avirulence genes from the pathogen (Table 1.3). Recent studies have demonstrated additional natural variation in effector gene repertoires in *H. parasitica*, and in the responses of different *Arabidopsis* accessions to *Hpa* effectors, suggesting functional variability in the *Arabidopsis* proteins that are targeted by *Hpa* effectors (Fabro et al. 2011; Asai et al. 2014). Thus, natural genetic variation in the *Arabidopsis*/ *H. parasitica* interaction can be further exploited for mechanistic insight into plant-pathogen interaction and co-evolution. Furthermore, because of the experimental advantages of *Arabidopsis*, *H. parasitica* is also one of the best reference organisms for investigating obligate biotrophy (McDowell 2011). A final advantage of *Hpa* is the availability of a reference genome sequence (Baxter et al. 2010).

1.12.2 *Albugo-Arabidopsis* Pathosystem

White rust pathogens belonging to the genus *Albugo* are naturally occurring, obligate biotrophic pathogens of *Arabidopsis*, and their attributes will be summarized in comparison with *Hpa*. Like *Hpa*, *Albugo* is commonly found on *Arabidopsis* in the wild, and their interactions display natural variability, including gene-for-gene interactions (Holub et al. 1994a; Holub and Beynon 1996; Borhan et al. 2001; Holub 2008). In addition, *Albugo* species cause important crop diseases on cultivated *Brassica* species (Saharan et al. 2014).

Molecular phylogenies clearly illustrate that the *Albugo* and downy mildew lineages are distinct (Thines 2014; Ascunce et al. 2017). Thus, *Albugo* and *H. parasitica* lineages have independently evolved to an obligate lifestyle and to compatibility with *Arabidopsis* (Kemen and Jones 2012). Interestingly, *A. laibachii* and *A. candida* display different host ranges: the former, like *H. parasitica*, is restricted to *A. thaliana*, while host range of *A. candida* encompasses 63 genera and 241 species within the *Brassicaceae* (Thines et al. 2009; Saharan et al. 2014). Thus, these two species provide an opportunity for comparative studies to

Table 1.3 Features of cloned *Arabidopsis* resistance genes encoding NLRs mediating detection of the oomycetes *Hpa* (*RPP*) and *Albugo* (*RAC, WRR*) (Herlihy et al. 2019)

Gene	Accession	Col-0 gene/allele	Remark	Reference	Recognizing	References
RPP1-WsA	Ws	At3g44480[a] (*rpp1*)	Original *RPP1* locus	Botella et al. (1998)	Unknown	
RPP1-WsB	Ws	At3g44480[a] (*rpp1*)	Previous *RPP10* locus	Botella et al. (1998)	ATR1 alleles	Rehmany et al. (2005)
RPP1-WsC	Ws	At3g44480[a] (*rpp1*)	Previous *RPP14* locus	Botella et al. (1998)	Unknown	
RPP1-NdA	Nd	At3g44480[a] (*rpp1*)	Similar but different specificity as *RPP-WsB*	Botella et al. (1998)	ATR1 alleles	Rehmany et al. (2005)
RPP1-EstA	Est	At3g44480[a] (*rpp1*)	Similar but different specificity as *RPP-WsB*	Goritschnig et al. (2016)	WY motif 2 of ATR1 alleles	Goritschnig et al. (2016)
RPP1-ZdrA	Zdr	At3g44480[a] (*rpp1*)	Similar but different specificity as *RPP-WsB*	Goritschnig et al. (2016)	WY motif 2 of ATR1 alleles	Goritschnig et al. (2016)
RPP2A	Col-0	At4g19500	Requires *RPP2B*	Sinapidou et al. (2004)	Unknown	
RPP2B	Col-0	At4g19510	Requires *RPP2A*	Sinapidou et al. (2004)	Unknown	
RPP4	Col-0	At4g16860		van der Biezen et al. (2002)	ATR4	Asai et al. (2018)
RPP5	*Ler*	At4g16950, paralog of *RPP4*		Parker et al. (1997)	ATR5	Bailey et al. (2011)
RPP7	Col-0	At1g58602		Eulgem et al. (2007)	Unknown	
RPP8-Ler	*Ler*	At5g43470 (*rpp8*)	Alleles confer virus resistance	McDowell et al. (1998)	ATR8, not cloned	Gunn et al. (2002)
RPP13-Nd	Nd	At3g46530	Highly polymorphic in *A. thaliana* (Rose et al. 2004)	Bittner-Eddy and Beynon (2001)	ATR13	Allen et al. (2004)
RPP39	Wei-0	At1g61180 and At1g61190	CC-NB-LRR protein signaling through NDR1	Goritschnig et al. (2012)	ATR39	Goritschnig et al. (2012)
RAC1	Ksk-1	At1g31540	Resistance to *Albugo laibachii* Nc14	Borhan et al. (2004)	Unknown	

(continued)

Table 1.3 (continued)

Gene	Accession	Col-0 gene/allele	Remark	Reference	Recognizing	References
WRR4	Col-0	At1g56510	Confers resistance to multiple *A. candida* races	Borhan et al. (2008)	Unknown	
WRR4B	Ws-2	At1g56540	Resistance to *A. candida* isolate Ac2V	Cevik et al. (2019)	Unknown	
WRR8	Sf-2	At5g46270	Resistance to *A. candida* isolate Ac2V	Cevik et al. (2019)	Unknown	
WRR9	Hi-0	At1g63750	Resistance to *A. candida* isolate Ac2V	Cevik et al. (2019)	Unknown	
WRR12	Ler	At1g17600	Resistance to *A. candida* race AcBoT	Borhan et al. (2008)	Unknown	

RPP1 orthologs or family members in Col-0 are: At3g44000, At3g44480, At3g44630, and At3g44670
[a]Identified as a segregating locus in *H. parasitica* crosses

understand the factors that control host range (McMullan et al. 2015; Jouet et al. 2018). As with *H. parasitica* intraspecific variation for resistance/ susceptibility to *Albugo* is prevalent in *Arabidopsis* and has been used to clone several *Arabidopsis* genes for resistance to *Albugo* (Table 1.3). Perhaps most interestingly, *Albugo* species display a remarkable capacity to suppress defense in *Arabidopsis* (Cooper et al. 2008). Co-infection with *Albugo* can render *Arabidopsis* susceptible to otherwise incompatible pathogens, including the late blight pathogen *Phytophthora infestans* (Belhaj et al. 2017; Prince et al. 2017). The molecular mechanisms by which *Albugo* strongly suppresses plant immunity are beginning to be revealed (Prince et al. 2017). Further studies are expected to reveal *Albugo* effectors and their plant targets that are likely to be key components of the plant immune system (Kemen et al. 2011; Links et al. 2011). *Albugo-Arabidopsis* systems are also being developed to better understand how *Albugo* influences plant interactions with other microbes (Ruhe et al. 2016; Jouet et al. 2018), as well as the important phenomenon of non-host resistance (Cevik et al. 2019).

The experimental advantages and limitations of the *Albugo* species are similar to *H. parasitica*. Like *H. parasitica*, *Albugo* species cannot be cultured apart from their hosts but they are relatively straightforward to propagate on susceptible host genotypes (Crute et al. 1993). Long-term storage is possible and genetic crosses between isolates can be performed (Adhikari et al. 2003). Importantly, draft genome sequences are available from two different strains of *A. laibachii* and one of *A. candida* (Kemen et al. 2011; Links et al. 2011). As with *H. parasitica*, the genome data is having a major impact on research with this organism (Kemen and Jones 2012).

1.12.3 Identification of Intracellular Receptors in *Arabidopsis* to Pathogens

Plants have evolved intracellular receptors with nucleotide-binding, leucine-rich repeat domains (NLRs) to detect fungal effectors, or their activity, to trigger immunity (ETI). NLR proteins typically have an N-terminal variable region, either of the coiled-coil type (CC) or Toll/Interleukin-1 receptor-like (TIR), or an RPW8-like domain. The N-terminal domains are followed by a nucleotide-binding domain (NB) and C-terminal leucine-rich repeat region (LRR) (Cui et al. 2015). The genes encoding these intracellular receptors were first identified in *Arabidopsis* through genetic mapping of resistance genes. The *Arabidopsis* genome (Col-0) encodes a total of around 150 NLR proteins, of which only a limited number have an assigned function. The first cloned *NLR* gene was *RPS2* that confers resistance to *Pseudomonas* bacteria expressing the effector AvrRpt2 (Mindrinos et al. 1994). Soon after that also *R*-genes for resistance to downy mildew and *Albugo* were cloned (Table 1.3), based on natural variability of their functions in different accessions of *Arabidopsis* (Slusarenko and Schlaich 2003; Holub 2008; Coates and Beynon 2010). These included the first *R*-genes to be cloned against oomycetes (Parker et al. 1997; Botella et al. 1998; McDowell et al. 1998). More than 25 *R* loci conferring isolate-specific resistance to *H. parasitica* have been genetically identified in *Arabidopsis* and are named *RPP*, for RECOGNITION OF *PERONOSPORA PARASITICA*, the older name of this pathogen (Slusarenko and Schlaich 2003). So far, 8 *RPP* genes have been cloned: *RPP1, RPP2, RPP4, RPP5, RPP7, RPP8, RPP13*, and *RPP39* and all encode cytoplasmic NLRs. Alleles of *RPP1* confer different isolate-specific resistances (Botella et al. 1998), and also *RPP4* and *RPP5* can be considered allelic variants (Van der Biezen et al. 2002). Many of the other *RPP* loci map to the position of cloned *RPP* genes and could constitute allelic variants (Nemri et al. 2010). Broad resistance to downy mildew, observed in certain *Arabidopsis* accessions, appeared to be mediated by combinations of isolate-specific resistance loci (Lapin et al. 2012). The "helper" NLR proteins ADR1 and NRG1 were identified as signaling components for RPP proteins containing the TIR domain (Castel et al. 2019; Wu et al. 2019). It is becoming clear that ETI can be underpinned by networks of NLRs, and deconvolution of these networks is an emerging area of interest (Wu et al. 2018). At a broad scale, little is known about whether and how the activity of *RPP* genes is regulated (Lai and Eulgem 2018). In one case, feedback control involving salicylic acid and WRKY proteins has been implicated (Mohr et al. 2010). In another, gene activity is subject to post-transcriptional control involving alternative polyadenylation, which is regulated by histone marks (Tsuchiya and Eulgem 2013; Lai et al. 2019). This research has opened up a new avenue towards understanding the mechanisms and evolution of NLR gene regulation (McDowell and Meyers 2013). Another study identified eQTLs providing resistance against *Hpa*, emphasizing the potential of this pathosystem to understand epigenetic factors that are relevant to plant-pathogen interactions (Furci et al. 2019).

Several avirulence genes from *Hpa* have also been cloned (Table 1.3), (Coates and Beynon 2010; McDowell 2014). In fact, the first oomycete *Avr* gene to be cloned was *ATR13* from *Hpa* (Allen et al. 2004). This gene encodes a small, secreted protein with a signal peptide followed by the conserved protein motif Arg-X-Leu-Arg (Rehmany et al. 2005). Downy mildew pathogens and *Phytophthora* species contain hundreds of putative "RXLR genes" in their genomes, and it is clear that they play an important role in pathogen virulence. RXLR proteins are secreted from pathogen haustoria and enter the interior of plant cells to reprogram plant regulatory networks and thereby promote virulence. However, RXLR proteins can also be recognized inside plants cells by NLR surveillance proteins. Indeed, every *Avr* gene cloned from a downy mildew or *Phytophthora* species to date encodes an RXLR or RXLR-like protein, recognized by the corresponding plant NLR protein. To date, *H. parasitica* Avr proteins recognized by *RPP1*, *RPP13*, *RPP4*, *RPP5*, and *RPP39* have been molecularly identified, and the determinants of their virulence and avirulence functions are under investigation (Allen et al. 2004; Rose et al. 2004; Rehmany et al. 2005; Bailey et al. 2011; Chou et al. 2011; Krasileva et al. 2011; Leonelli et al. 2011; Goritschnig et al. 2012; Steinbrenner et al. 2015). Comparative and molecular studies of *RPP* genes have provided novel insights into the molecular mechanisms and selective forces that drive NLR gene evolution (McDowell and Simon 2006).

For example, *RPP5*, *RPP1*, and *RPP8* multigene families are physically linked in clusters and are subject to intra- and intergenic recombination, to produce new *NLR* gene variants in host-pathogen arms race (Botella et al. 1998; McDowell et al. 1998; Noel et al. 1999). RPP1 homologs have diversified through repeat duplication and sequence divergence, such that they can detect multiple surfaces of the corresponding effector (Goritschnig et al. 2016). ATR1 is a modular protein that can evolve to escape detection by mutations in any of several surfaces that mediate recognition by RPP1 (Chou et al. 2011). *RPP13* provides an example of a simple locus, in which a single copy gene displays substantial allelic polymorphism, driven by diversifying selection and intra-allelic recombination (Bittner-Eddy and Beynon 2001). Interestingly, the *H. parasitica ATR13* locus, encoding the RXLR effector protein recognized by RPP13, displays similar attributes, suggesting that co-evolution has been a major driver for this diversity, along with balancing selection to maintain repertoires of useful alleles in the plant and pathogen populations (Allen et al. 2004). The ATR13 protein displays a novel structure and polymorphisms that mediate recognition specificity by RPP13 map to a single, surface exposed region (Leonelli et al. 2011).

In the years following *RPP* gene cloning, only the RPP1 protein has been investigated in significant mechanistic detail. This protein binds directly to ATR1 via the LRRs. This interaction displays extensive allelic variability (Krasileva et al. 2010; Steinbrenner et al. 2015). In the absence of the ATR1 effector, RPP1 is likely maintained in an inactive state by intra-molecular interactions between the N-terminal TIR domain, the NB domain, and the LRRs. Binding of ATR1 via the LRRs disrupts these interactions and permits oligomerization of RPP1, thereby triggering cell death and other immune responses (Krasileva et al. 2010; Steinbrenner et al. 2015). This mode of inactivation/activation appears to be a general aspect of NLR proteins, although details vary between different NLR proteins. As with other NLR proteins, the molecular events between RPP1 activation and deployment of the ultimate cellular immune response remain to be identified.

Several *Arabidopsis* genes for resistance to *Albugo* have also been cloned (Table 1.3). The *RAC1* gene provides isolate-specific resistance to *Albugo laibachii* Nc14 (Borhan et al. 2004; Cevik et al. 2019). *WRR4* confers broad-spectrum resistance to many *Albugo candida* isolates and functions as a transgene in *Brassica juncea* (Borhan et al. 2008, 2010). These *R*-genes encode cytoplasmic NLRs that act similar to many of the RPP proteins, and helper NLRs have also been implicated for WRR-mediated resistance (Castel et al. 2019). Interestingly, the phenomenon of non-host resistance of *Arabidopsis* to *Brassica*-infecting races of *Albugo candida* is mechanistically underpinned by NLR receptors that can be identified by segregation analysis. This led to the molecular cloning of WRR4B, WRR8, WRR9, and WRR12. Some of these genes were shown to provide resistance as transgenes in *Brassica* species (Cevik et al. 2019). Some outstanding achievements of genomics of host resistance are given in Table 1.4 (Herlihy et al. 2019).

Table 1.4 Significant achievements of genomics of host resistance in crucifers (Herlihy et al. 2019)

Insight	Key references
Cloning of the first *R*-genes against oomycetes	Parker et al. (1997), Botella et al. (1998), McDowell et al. (1998)
First insights into evolutionary dynamics of *R*-genes against oomycetes	Botella et al. (1998), McDowell et al. (1998), Bittner-Eddy et al. (2000)
Definition of important immune system regulators	Parker et al. (1996)
Identification and map-based cloning of the first mutants exhibiting gain of resistance to oomycetes	van Damme et al. (2005, 2008)
Genetic definition of the complexity of the immune signaling network against oomycetes	Aarts et al. (1998), McDowell et al. (2000), Bittner-Eddy and Beynon (2001)
Definition of penetration resistance	Lipka et al. (2005)
Molecular cloning of the first oomycete *Avr* gene	Allen et al. (2004)
First insights into evolutionary dynamics of *Avr* gene loci	Allen et al. (2004), Armstrong et al. (2005), Rehmany et al. (2005)
Identification of the R x LR motif	Rehmany et al. (2005)
Test bed for new technologies to understand effector functions	Fabro et al. (2011), Caillaud et al. (2012)
Discovery that oomycete, fungal, and bacterial effectors can target the same plant signaling hubs	Mukhtar et al. (2011), Wessling et al. (2014)
First insights into genomic basis and evolution of obligate biotrophy	Baxter et al. (2010), Kemen et al. (2011)

1.13 Phylogenetic Relationships Within and Among *Brassica* Species

The gene pool of elite oilseed rape breeding material has been considerably reduced by emphasis on specific quality traits (Snowdon and Friedt 2004). This narrow genetic base has increased their vulnerability to diseases. Many important fungal pathogens reside in the soil and infect roots, leaves, stems, and fruits of crops, often causing significant yield losses. Fungal diseases such as white rust, black spot, clubroot, Sclerotinia, and blackleg (Ananga et al. 2006) and bacterial leaf spot, black rot, and soft rot (Rimmer and Buchwaldt 1995), as well as insect pests such as aphids, Japanese-beetle, and cabbage seedpod weevil (Salisbury et al. 1995) pose a major limitation to the cultivation of this important crop.

Despite being generally susceptible, a growing body of evidence show that some oilseed rape cultivars are resistant to blackleg disease (Ansan-Melayah et al. 1995, 1997; Ananga et al. 2006). Specifically, species that contain the B genome (*B. juncea*, *B. nigra* and *B. carinata*) can display total resistance to blackleg disease throughout the life of the plant (Roy 1984; Ananga et al. 2006). Resistant oilseed rape lines have also been obtained through interspecific crosses between *B. napus* and *B. nigra* (Rimmer et al. 1995; Chevre et al. 1997; Brun et al. 2001). Hence the use of resistant cultivars has been advocated as the most effective way of limiting yield losses due to blackleg disease (Rimmer et al. 1995).

Identification of resistant cultivars can involve either field trials (disease challenge and plant response data), or predictions based on genetic relationships. Field trials suffer practical drawbacks such as longer time to obtain results, are resource intensive, and the results may be unreliable when the trait is under a strong environmental influence, unlike molecular genetic analyses. Therefore, analysis of the genetic variation and relatedness in the *Brassica* germplasm is needed for genetic resource conservation and plant breeding programs. Resistance to plant pathogens has been shown to be under genetic control (Brun et al. 2001), and hence the genetic relationships among taxa can be used to predict susceptibility or resistance (Ochieng et al. 2007). Understanding the interspecific and intraspecific relationships among cultivated species of *Brassica* will inform better parental selection and to widen the genetic basis of resistance to blackleg through greater variations and the development of new genotypes in oilseed rape cultivars.

Studies on broader phylogenetic relationships among diploid species including *B. nigra, B. rapa,* and *B. oleracea*, showed that the B genome (*B. nigra*; B) was supported from a distinct clade separate from the other two (*B. rapa*; A, and *B. oleracea*; C) (Inaba and Nishio 2002). There is no reported phylogenetic placement of cultivated *B. napus* among these species; however, Inaba and Nishio (2002) suggested that *B. napus*, a tetraploid containing AC genome, is allied to the *rapa-oleracea* clade. Ananga et al. (2008) assessed the phylogenetic relationships within and among cultivated *B. napus, B. rapa*, and *B. oleracea* using RAPD markers associated with blackleg resistance in *B. nigra* (Ananga et al. 2006). Phylogenetic placement of oilseed rape cultivars from the NWCVT in relation to resistant or susceptible cultivars (Ananga et al. 2006) can help predict their potential response to blackleg. Because these RAPD markers are associated with resistance to a plant pathogen, the resultant phylogenies are predicted to reflect functional genetic variation and affinities. Since the diploid species *B. rapa* and *B. oleracea* each represent a parental genome of the amphidiploid species, the placement of *B. napus*, *B. rapa*, and *B. oleracea* cultivars within the *rapa-oleracea* clade may identify unique variants needed in breeding programs, expansion of the range of useful variation, and for future improvement of heterotic potential (Ananga et al. 2008).

Thirty accessions from USDA germplasm collection representing two diploid *Brassica* species (*Brassica rapa* and *Brassica oleracea* var. *virids*) and fifteen tetraploid cultivars (*Brassica napus*) from the national winter canola variety trials (NWCVT) were evaluated using 13 sets of random amplified polymorphic DNA (RAPD) associated with blackleg resistance in *Brassica nigra*. 126 highly polymorphic bands with an average of 10 per primer were detected. A UPGMA dendrogram showed *B. rapa* as highly diverse and was supported from three different basal branches, while *B. napus* accessions were generally monophyletic. Similarly, all of *B. oleracea* accessions were supported from the same basal node. Generally, the three species were reciprocally paraphyletic, suggesting that the RAPD markers showed both functional relationships and homology, possibly due to selection at the RAPD loci associated with blackleg resistance. Consequently, two potentially susceptible *B. napus* accessions were identified. The high polymorphic information content (PIC) and number of phylogenetically informative bands established RAPD

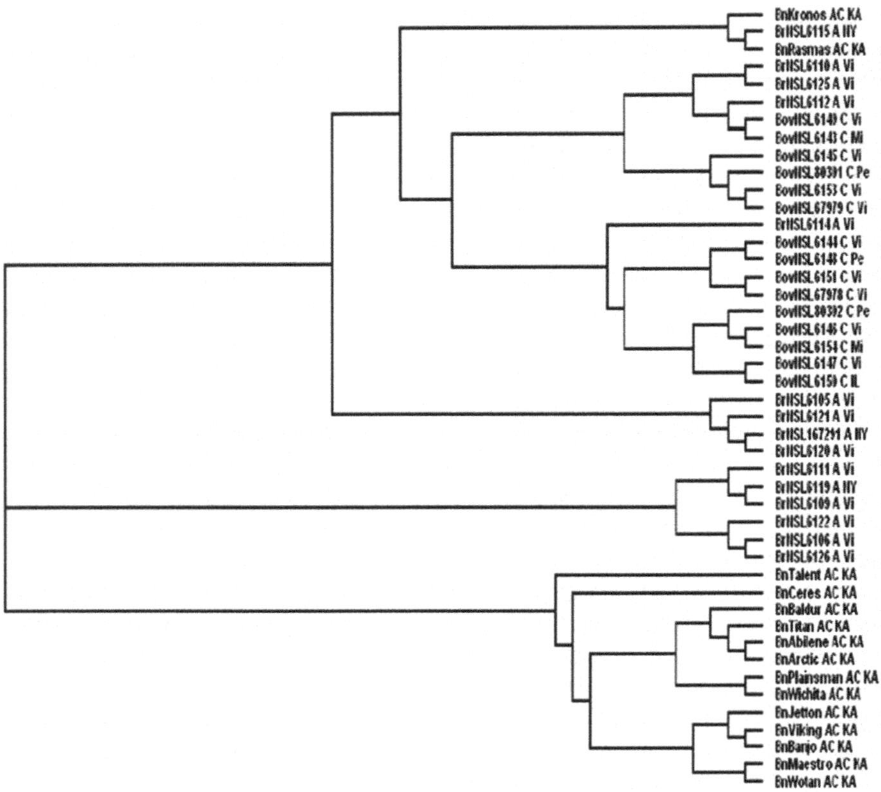

Fig. 1.4 UPGMA cluster showing the genetic relationships among three *Brassica* species (*B. rapa*, *B. oleracea*, *B. napus*) at 13 RAPD loci associated with blackleg resistance. The first two letters in the sample labels refer to the taxonomic abbreviations (Bo- *Brassica oleracea*; Br- *Brassica rapa*; Bn- *Brassica napus*), followed by the Accession number. The genomic grouping (A, C, or AC) precedes the abbreviation for sample (Ananga et al. 2008)

as a useful tool for phylogenetic reconstruction, quantification of genetic diversity for conservation, cultivar classification, and molecular breeding in *Brassica* (Fig. 1.4; Ananga et al. 2008).

1.14 *Brassica* Genome Complexity and R-Genes Identification

In cruciferous species, *Brassica napus* is a relatively young crop in evolutionary history that results from an accumulated series of polyploidization events (Mason and Snowdon 2016). Prior to hybridization, the whole-genome triplication in the diploids (*B. rapa* and *B. oleracea*) was preceded by three events of whole-genome duplication in the ancestor genome that have further contributed to complicated gene rearrangement, duplication, and loss (Town et al. 2006; Chen 2007). Despite the genome complexity, *Brassica* species have become models to study evolutionary events and how rich genetic resources can be exploited from the plant (Pires and Gaeta 2011; Liu et al. 2014). Domesticated *B. napus* has a lower genetic diversity compared to its diploid progenitors *B. rapa* and *B. oleracea*, evidenced by similar gene order and content between the three species due to strong genetic bottlenecks during cultivation (Cheung et al. 2009; Mason and Snowdon 2016). By contrast, homoeologous rearrangements/recombination occurs more frequently in resynthesized *B. napus*, thereby contributing greatly towards genetic diversity, which provides a good source of R-genes for breeding resistant *B. napus* cultivars (Gaeta and Chris Pires 2010). The efforts were made to produce introgressed *B. napus* in a novel way, from *B. juncea* (A genome source) and *B. carinata* (C genome source) (Chatterjee et al. 2016) (rather than from *B. oleracea* and *B. rapa*) from where more sources of R-genes can then be utilized and identified. While in *B. napus*, the homoeologous genes are often co-expressed (Chen 2007), there could also be bias in homoeology expression, as found in *B. juncea* (Yang et al. 2016) and *B. rapa* (Cheng et al. 2012). How the different levels of gene expression affect the stability of R-genes remains to be determined. The fate of homoeologous genes, whether they are silenced, expressed, or both copies diverge where the gene function may or may not change (sub- and neo-functionalization) is a result of the genetic and epigenetic interactions between redundant genes (Comai 2005), biased gene expression (Yang et al. 2016), gene fractionation (Subramaniam et al. 2013) or accelerated amino acid sequence evolution and positive selection (Liu and Adams 2010). These polyploidy events could potentially result in a selective advantage, for example higher seed yield in *B. napus* (Osborn et al. 2003).

Although race-specific R-genes are easier to manipulate for breeding resistant crops compared to polygenic resistance, which involves many genes and the effect of environmental factors, the identification of R-genes in *B. napus* still remains a challenge due to the complexity of its genome. A major QTL for resistance to *S. sclerotiorum* was duplicated through homoeologous non-reciprocal transposition (HNRT) in *B. napus* (Zhao et al. 2006). The quantitative resistance genes against *L. maculans* are affected by gene duplication in homoeologous regions in *B. napus* (Fomeju et al. 2014, 2015). *L. maculans* R-genes, *Lmr1* and *ClmR1* from cultivars Shiralee and Cresor, respectively, were mapped on regions that are highly duplicated within and between genomes in *B. napus*, and originated from the common ancestor before divergence of *B. oleracea* and *B. rapa* (Mayerhofer et al. 2005). Evolutionary information regarding resistance genes such as these is useful for breeding purposes, as it highlights specific genera/species likely to be most useful in developing new cultivars with highly effective and durable disease resistances. For *H. parasitica*, the single dominant gene Pp523 was mapped onto two regions on chromosome C8 and another on C5, reflecting triplicated regions in *B. oleracea* (Carlier et al. 2012).

The success stories of cloning *Crr1a* and *CRa* (Clubroot R-genes encoding TIR-NB-LRR) from *B. rapa* (Hatakeyama et al. 2013) and *LepR3/Rlm2* (Larkan et al. 2013) in *B. napus* revealed that the R-genes are organized as members of allelic variants or tandem repeats. The clustering effect of the R-genes for these two pathogens provides further evidence of complexity of the genome organization. Clustering of R-genes is also prevalent in association with Clubroot, Blackleg, and

Sclerotinia Stem Rot resistance on both *B. rapa* (Kato et al. 2013) and *B. napus* (Delourme et al. 2004; Kato et al. 2013; Li et al. 2015). Such gene clustering effects are common in plant genomes for quick adaptive response to pathogens via recombination events (Hulbert et al. 2001; Meyers et al. 2003). Clustering of R-genes may potentially enable recognition of different pathogen races carrying different avirulence genes, such as in Clubroot R-genes on chromosome A8 (Pang et al. 2014). Epistatic interaction was also observed for Clubroot in *B. napus* (Manzanares-Dauleux et al. 2000), where a single dominant resistance gene does not explain the resistance outcome in its entirety. It was also reported that the same genomic region in *B. napus* can act as both QTL and/or major gene based on different pathotypes of Clubroot (Manzanares-Dauleux et al. 2000; Rocherieux et al. 2004). The clustering and duplication of R-genes can also cause problems in R-gene identification as these genes tend to collapse in genome sequence assemblies. Reliance on the single reference genome sequence can therefore hamper efforts and future work should incorporate novel technologies such as isolated chromosome sequencing or targeted amplification of genomic regions of interest (Neik et al. 2017).

1.15 Complexity of Genetic Variation in Hosts and Pathogens Variability

The complexity of R-genes, resistance levels vary with different host species within *B. napus* for the key pathogens because of evolution of pathogenic variability. Often, intermediate/heterogeneous phenotypes were obtained when using the pathogen isolates either derived from differential sets or from host cultivars. This could be due to genetic variation within the host cultivars and/or even within the pathogen.

i. **Host species:** *Brassica napus* has a low genetic diversity due to the recent evolutionary history (Snowdon et al. 2007; Prakash et al. 2011). Although *B. napus* has a comparatively lower level of genetic variation (Wu et al. 2014), breeding programs have made extensive use of interspecific hybridization within the Brassicaceae family, including with *Sinapis arvensis*, to broaden the genetic diversity for Blackleg resistance (Winter et al. 2003; Snowdon et al. 2007). Depending on the genetic background of the *B. napus* host, for example winter, spring type cultivars or resynthesized lines, the R-gene diversity can be high or low (Rouxel et al. 2003). This can affect the assessment of disease on the plant where a wide range of scores is obtained (Larkan et al. 2016). Genetic mapping studies on the key pathogenic resistances showed that breeding materials or cultivars were mostly used to generate mapping populations (Diers and Osborn 1994; Plieske and Struss 2001). However, other genetic sources should be tapped into to increase the genetic pool for breeding resistant *B. napus*. For example, resynthesized lines can provide a good source of genetic diversity compared to cultivars (Becker et al. 1995). This is evidenced by the successful introgression of the resistance gene *LepR3* from *B. rapa* into *B. napus* (Larkan et al. 2013). Fodder and vegetable rape genetic material can also contribute significant amount of R-genes, because of high genetic variation in these species (Hasan et al. 2006). Genome-wide SNPs have been studied in natural accessions and synthetic *B. napus*, inferring sequence polymorphism in genes linked to agronomically important traits such as seed quality (Huang et al. 2013; Qian et al. 2014; Schmutzer et al. 2015a,b) and this could be expanded to disease resistance studies.

ii. **Pathogens:** A high level of genetic variation in *S. sclerotiorum* populations was found in *B. napus* fields in Australia (Sexton et al. 2006) and Canada (Kohn et al. 1991). In particular, the Australian *S. sclerotiorum*, with different isolates/pathotypes, caused variable resistance outcomes ranging from intermediate resistance to a HR to complete resistance in different *B. napus* genotypes (Garg et al. 2010; Ge et al. 2015). This poses important questions regarding this pathosystem: Does this host resistance come from specificity of the host species genotype, from the pathogen genotype, or from a combination of both? Is there a role for non-host resistance to adapted pathogens? Could there be specificity of the host cultivar? Can host specificity and non-host specificity be displayed in different genotypes of the same host? Until the genetic basis for the various resistance levels in this pathosystem is unveiled, we will not know if the *B. napus* genotypes are going through a process of losing or acquiring host status (marginal host), or are intermediate between host and nonhost status, much like the barley resistance to rust where the host-pathogen interaction is indeed very specialized (Niks and Marcel 2009).

In *L. maculans*, very high genetic diversity has been observed within populations. In France, evidence for this was found through an isolation by distance (IBD) study where there was high gene flow and high dispersal rates of spores among populations across large geographical distances (Travadon et al. 2011). These characteristics offer the pathogen a survival advantage to cope with the host resistance in the field, highlighting the role of co-evolutionary dynamics between pathogen and host in conferring resistance. Similarly, in Australia, *L. maculans* also displayed high genetic variation between geographic locations such that the isolates from Western Australia were shown to be genotypically different from those collected from the eastern states. Studies using SNPs (Zander et al. 2013; Patel et al. 2015), microsatellite and mini satellite markers (Hayden et al. 2007) indicated that some populations are panmictic. Gene loss in *L. maculans* has contributed to genetic variation between isolates (Golicz et al. 2015). The source of genetic diversity for *L. maculans* could also depend on the specific host morphotype from which the pathogen is adapted to. For instance *L. maculans* "brassicae" from cultivated *Brassica* and *L. maculans* "lepidii" from *Lepidium* sp. were identified in the International Blackleg of Crucifers Network (IBCN) collection as one of the seven subgroups of the species complex based on ITS-RFLPs (Mendes-Pereira et al. 2003). This species complex, differing in specificity and pathogenicity on *B. napus*, potentially complicates the cloning of R-genes in *B. napus* (Voigt et al. 2005).

High genetic diversity was reported for *P. brassicae,* in Japan, where the pathogen showed different pathogenicity levels on the cruciferous crops and on cruciferous weeds (Jones et al. 1982; Tanaka and Ito 2013). However, pure pathotype/genotype isolates of *P. brassicae* can still be obtained from single root hairs, which minimizes the variability of pathogenicity; thus a more reliable interpretation of the host pathogen relationship can be established (Diederichsen et al. 2016). For *H. parasitica*, a molecular study also revealed high genetic diversity within the species, probably due to the broad host range of this pathogen where it can infect, for example, various Brassica and Raphanus species (Choi et al. 2003; Neik et al. 2017).

1.16 High-Efficiency Integrated Breeding

The genomic era is also characterized by high-efficiency integrated breeding (HIB), in which multiple methods are combined, including traditional ways, such as microspore culture, backcrossing, and distant introgression, and modern ways, such as MAS, gene editing, and genome design (Fig.1.5). During HIB, genomic background analysis is helpful in eliminating undesirable linkage drags and rapidly identifying desirable individuals. In a study by Liu et al. (2018), resistance-specific markers and genome background markers were used to breed cabbage with resistance to Fusarium wilt. By combining these methods with microspore culture and backcrossing, the authors presented a rapid and effective approach for generating Fusarium wilt resistant ILs in the BC2 generation. Notably, the quickly emerging gene-editing technique helps realize accurate alteration of the target DNA sequence. Ma et al. (2019) applied CRISPR/Cas9-mediated multiple gene editing in cabbage, with the targets *BoPDS*, *BoSRK*, and *BoMS1*, and successfully generated albino, self-compatible, and male sterile lines, showing its great power in improving plant traits (Lv et al. 2020).

1.17 Deletion of R-Gene in *Brassica* and *Arabidopsis*

The outcome of many plant–pathogen interactions is determined by disease-resistance (*R*) genes that enable plants to recognize invading pathogens and activate inducible defenses. A typical *R*-gene allele encodes "race-specific" resistance to only one or a few strains of a single pathogen species. *R*-gene loci are functionally polymorphic within a plant species and encode alternate alleles that either recognize different strains of the same pathogen or do not recognize any tested pathogen. The molecular studies have revealed that *R*-genes often reside in complex loci consisting of the *R*-gene and tightly linked homologs. These complexes can exist in both disease-resistant and disease-susceptible plant genotypes. An *R*-gene specifies resistance only if the pathogen expresses a corresponding avirulence (*avr*) gene. If either component is nonfunctional, then the plant is unable to activate resistance responses. This model has successfully described a wide variety of plant–pathogen

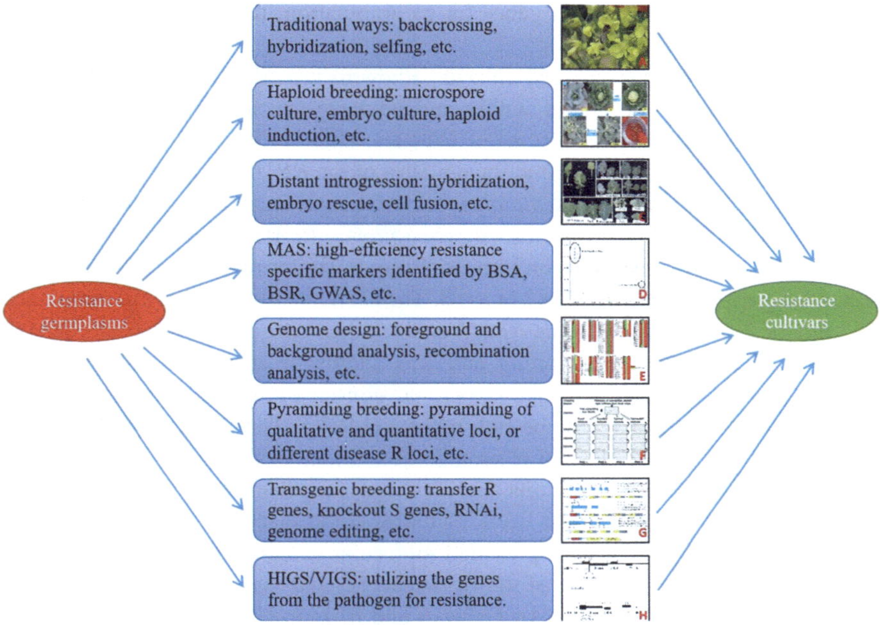

Fig. 1.5 Proposed high-efficiency integrated breeding (HIB) model in the genomic era. (**a**) Selfing using honeybees is one of the most traditional breeding methods. (**b**) A combined use of microspore culture and MAS helps promote the breeding cycle in *Brassica oleracea*. (**c**) BR resistance introgression from *Brassica carinata* to *Brassica oleracea* using distant hybridizing and embryo rescue. (**d**) SNP-based high-throughput KASP markers prove efficient and cost saving in genotyping during MAS in *Brassica rapa*. (**e**) Whole-genome background analysis helps eliminate the undesired linkage drags during MAS in *Brassica oleracea*. (**f**) Pyramiding both the qualitative and quantitative R loci generates durable BL resistance in *Brassica napus*. (**g**) CRISPR/Cas9-based gene editing helps knockout multiple target genes in *Brassica oleracea*. (**h**) Expressing the CP gene from TuMV confers high resistance in *Brassica napus*. (Lv et al. 2020)

associations; but little is known about the molecular basis and origin of *R*-gene polymorphism (Grant et al. 1998).

The most common molecular interpretation of the gene for-gene hypothesis is that *R*-genes encode specialized receptors that recognize the direct or indirect products (elicitors) of the corresponding *avr* genes. One implication of the receptor–elicitor model is that functional polymorphism at *R* loci could arise from gain-of-function mutations that enable the plant to recognize novel pathogen variants. Maintenance of multiple alleles or multiple linked genes with different recognition capabilities at an *R* locus would enable a host population to defend itself against the corresponding pathogens. *R*-gene polymorphism could also, in principle, arise from loss-of-function mutations in *R*-genes. Nonfunctional *R*-gene alleles could be maintained if there is a cost of resistance in the absence of pathogen selection. Nonfunctional alleles could also serve as a repository of divergent sequences that could be contributed to related genes by recombination, thereby, accelerating the evolution of novel *R*-genes.

The cloning of *R*-genes against diverse pathogens provides the tools for comparative analysis of *R*-gene alleles within and between species, which will provide insight into the evolutionary history of *R*-genes. Grant et al. (1998) described the structure of functional and nonfunctional alleles of the *RPM1* (resistance to *Pseudomonas syringae*, pathovar *maculicola*) bacterial resistance gene in *Arabidopsis thaliana* and a related crop species, *Brassica napus. RPM1* was initially identified in *A. thaliana* accession Col-0 through its ability to confer resistance to *P. syringae* isolates expressing either *avrRpm1* or *avrB*. The *avrB* and *avrRpm1* avirulence genes are sequence unrelated, and thus *RPM1* enables dual-specificity resistance. *RPM1* was isolated by map-based cloning and was shown to encode a protein with a putative amino-terminal leucine zipper, a consensus nucleotide-binding site (NBS), and 14 C-terminal leucine-rich repeats (LRRs). The RPM1 protein is thus a member of the largest class of *R* proteins functionally characterized to date, the so-called NBS-LRR class. Previous analyses suggested that *RPM1* functional polymorphism in *Arabidopsis* arose from an intraspecific insertion or deletion of the *RPM1* gene. Grant et al. (1998) provided molecular evidence that the evolution of *RPM1* predated the divergence of the Brassicaceae, and that independent deletions of *RPM1* have occurred in both *Arabidopsis* and *Brassica* (Grant et al. 1998).

Plant disease resistance (*R*) genes confer race-specific resistance to pathogens and are genetically defined on the basis of intraspecific functional polymorphism. Little is known about the evolutionary mechanisms that generate this polymorphism. Most *R* loci examined to date contain alternate alleles and/or linked homologs even in disease-susceptible plant genotypes. In contrast, the resistance to *Pseudomonas syringae* pathovar *maculicola* (*RPM1*) bacterial resistance gene is completely absent (*rpm1-null*) in 5/5 *Arabidopsis thaliana* accessions that lack *RPM1* function. The *rpm1-null* locus contains a 98-bp segment of unknown origin in place of the *RPM1* gene. Grant et al. (1998) undertook comparative mapping of *RPM1* and flanking genes in *Brassica napus* to determine the ancestral state of the *RPM1* locus. They cloned two *B. napus RPM1* homologs encoding hypothetical proteins with 81% amino acid identity to *Arabidopsis RPM1*. Co-linearity of genes flanking *RPM1* is conserved between *B. napus* and *Arabidopsis*. Surprisingly, four additional *B. napus* loci were found in which the flanking marker synteny is maintained but *RPM1* is absent. These *B. napus rpm1-null* loci have no detectable nucleotide similarity to the *Arabidopsis rpm1-null* allele. It was concluded that *RPM1* evolved before the divergence of the Brassicaceae and has been deleted independently in the *Brassica* and *Arabidopsis* lineages. These results suggest that functional polymorphism at *R*-gene loci can arise from gene deletions (Grant et al. 1998).

1.18 Pathogen Perception and Induction of Host Resistance Genes

Plants have evolved various ways to fend off pathogens. A suite of surface-exposed detectors, named pattern recognition receptors (PRRs, Fig. 1.6), can recognize conserved microbe-associated molecular patterns (MAMPs), such as bacterial flagellin or fungal chitin. Upon detection the PRRs activate a signaling cascade that leads to pattern-triggered immunity (PTI). Most PRRs require a co-receptor to initiate signaling. The receptor FLS2 interacts with the BAK1 co-receptor upon recognition of the MAMP flg22. BAK1 also interacts with other PRRs. PTI signaling via mitogen-activated protein kinase (MAPK) cascades and/or calcium-dependent protein kinases (CDPKs) (Bredow and Monaghan 2019) activates pathogen-nonspecific immune responses such as the production of reactive oxygen species and nitric oxide, cell wall reinforcement, and induction of defense genes (Hein et al. 2009; Nurnberger and Kemmerling 2009). Adapted pathogens are able to suppress PTI with effectors that have evolved to interact directly with host defense-associated proteins, resulting in effector-triggered susceptibility. Bacterial effectors can interfere directly with FLS2 or BAK1 (Toruno et al. 2016). Intracellular nucleotide-binding, leucine-rich repeat (NLR, Fig. 1.6) receptors that recognize these effectors or their activity have evolved in plants as a second layer of pathogen perception. NLR proteins are encoded by plant disease resistance genes and mediate so-called Effector-Triggered Immunity (ETI). This is the molecular basis of the genetic model of "gene-for-gene" resistance, in which pathogen "avirulence (*Avr*) genes" (now known to encode secreted effector proteins) are recognized inside plant cells by plant "resistance (*R*) genes" (now known to encode NLR proteins).

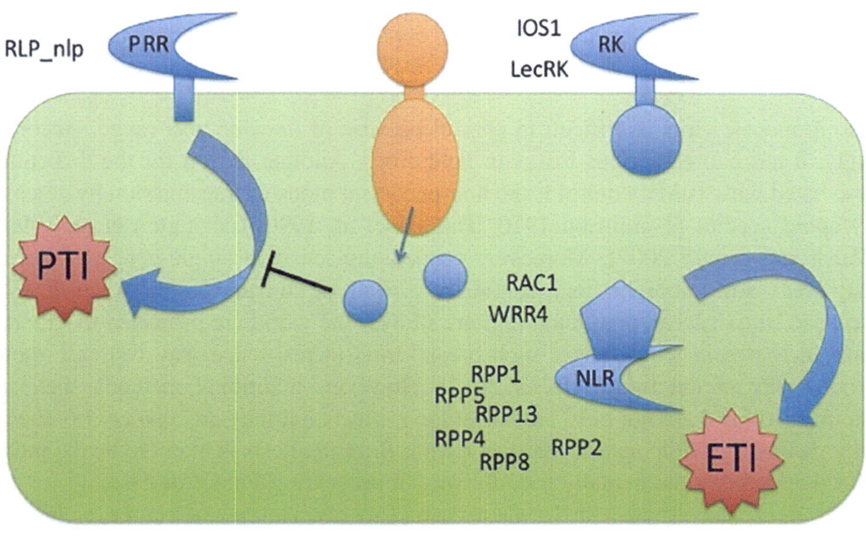

Fig. 1.6 Schematic of PTI and ETI against Oomycetes (Herlihy et al. 2019)

Resistance to adapted pathogen species is often rapidly broken by loss or mutation of avirulence effector genes or through suppression of ETI by different effectors (Woods-Tor et al. 2018).

The majority of plant PRRs identified so far belongs to the family of leucine-rich repeat receptor-like kinases (LRR-RLKs), which has more than 600 members in *Arabidopsis* (Shiu et al. 2004), or receptor-like protein (LRR-RLPs) that lack a C-terminal kinase domain and interact with RLKs for transducing signals (Gust and Felix 2014). The best-described plant PRR is FLS2 from *Arabidopsis* that recognizes bacterial flagellin, a major structural protein of bacterial flagella. An N-terminal 22-amino acid residue fragment of this protein (flg22) is sufficient to be recognized by FLS2 and to induce PTI (Gomez-Gomez and Boller 2000). Naito et al. (2008) showed that these residues are also essential for flagellum function and motility of bacteria, which makes it difficult for bacteria to circumvent this recognition by evolving different amino acid sequences in this region of the protein. A second well-described plant PRR that seems restricted to *Brassicaceae* is EFR, which recognizes the bacterial elongation factor Tu (EF-Tu) or its shorter N-terminal fragment elf18 (Kunze et al. 2004). When this PRR is expressed in the solanaceous plants *Nicotiana benthamiana* and tomato (*Solanum lycopersicum*), it confers resistance to a broad spectrum of bacteria (Lacombe et al. 2010), illustrating the potential use of PRRs as new sources for genetic resistance to crop diseases. Fungi can also be recognized by PRRs, e.g., through the *Arabidopsis* RLK CERK1 that is required for responses to fungal chitin and bacterial peptidoglycan (Miya et al. 2007; Willmann et al. 2011), and the RLP30 PRR that mediates recognition of necrotrophic fungi through a currently unknown proteinaceous MAMP (Zhang et al. 2013a, 2013b). However, only a few receptors have been identified so far that mediate the extracellular recognition of oomycete MAMPs (Herlihy et al. 2019).

1.19 Challenges of *TuMV* Resistance Breeding

Turnip mosaic virus is difficult to control because of its wide host range. Accordingly, it has caused serious losses in field crops. Another reason for the difficulty associated with TuMV control is the non-persistent mode of transmission by at least 89 aphid species (Tomlinson 1970; Thomas et al. 1990; Carrington et al. 1996; Walsh and Jenner 2002). Moreover, temperature-sensitive single-gene resistance may break down at higher temperatures as a result of climate change (Jones 2012). Because Brassicaceae-producing regions all over the world are projected to experience increases in temperature, such breakdowns in resistance may decrease their productivity in the future (Jones 2012). However, traditional methods such as chemical control are not only limited in their effects but they are also environmentally damaging, while natural resistance from *Brassica* hosts is likely to be the most effective and environmentally friendly way of preventing TuMV disease.

Progress has been made in TuMV resistance research in *Brassica* crops, including inheritance analysis, mapping and cloning of resistance genes, and molecular mechanism clarification, all of which have greatly facilitated *Brassica* resistance breeding.

Some genes for resistance to plant pathogens (R-genes) cloned in *Brassica* crops contain a nucleotide-binding site (NBS) domain and a leucine-rich repeat (LRR) domain (NBS-LRR, NLR), while others are genes from the eukaryotic initiation factor (eIF) family (Dangl and Jones 2001). However, NLR genes for resistance to TuMV have not been found in *Brassica* crops. Moreover, almost all of the recessive genes are eIF family genes that interact with the viral protein linked to the genome (VPg) of TuMV (Li et al. 2019).

1.20 Differential Genes Expression in *Brassica* Under Biotic Stresses

Biotic stress is a very devastating stress problem for cruciferous crops. Several numbers of genes have well documented that they are differentially expressed against different biotic stresses. However, when infection with *F. oxysporum* f. sp. *conglutinans*, six *BrWRKY* genes (*BrWRKY4*, *65*, *72*, *97*, *133*, and *141*) showed significantly higher expression and were about 8-, 6-, 6-, 3-, and 5-fold higher, respectively (Kayum et al. 2015a). *BrWRKY141* showed about 180-fold higher expression after *P. carotovorum* subsp. *carotovorum* infection (Kayum et al. 2015a). After infection with *F. oxysporum* f. sp. *conglutinans*, *BrAL2*, *3*, *4*, *7*, *9*, *10*, *13*, *14*, and *15* showed several fold higher expression, and *BrAL2*, *3*, *7*, *9*, *13*, *14*, and *15* were expressed in response to both biotic and abiotic stresses (Kayum et al. 2015b). Ahmed et al. (2012a) reported that three chitinase genes (*BrCLP1*, *BrCLP2*, and *BrCLP3*) response against *P. carotovorum* subsp. *carotovorum* infection. The defensin-like family protein (BrDLFP) expression significantly increased after infection with *P. carotovorum* subsp. *carotovorum* in Chinese cabbage (Ahmed et al. 2012b). On the other hand, Alfin-like two genes (BoAL8 and BoAL12) were induced after infection with *P. carotovorum* subsp. *carotovorum* in *B. oleracea* (Kayum et al. 2016). Among the 12 thaumatin-like proteins, three of which expressed differentially after *P. carotovorum* subsp. *carotovorum* infection in cabbage plants have been observed (Ahmed et al. 2013).

1.21 Significant Insight into Genomics of Host Resistance

The significant insight into genomics of host resistance has been given by Herlihy et al. (2019). Since the cloning of first R-gene against cruciferous pathogen significant contribution has been made in molecular mechanisms of host pathogen interaction and mechanism of R-genes function, expressions, and regulation of host resistance (Table 1.4).

1.22 Approaches Used in the Genomics of Disease Resistance Breeding of Crucifers

Technologies that measure some characteristic of a large family of cellular molecules, such as genes, proteins, or small metabolites, have been named by appending the suffix "**Omics**," as in "**genomics**." Omics refers to the collective technologies used to explore the roles, relationships, and actions of the various types of molecules that make up the cells of an organism. These technologies include: Genomics, "the study of genes and their function" (Human Genome Project (HGP) 2003); Proteomics, the study of proteins; Metabolomics, the study of molecules involved in cellular metabolism; Transcriptomics, the study of the mRNA; Glycomics, the study of cellular carbohydrates; Lipomics, the study of cellular lipids.

The development of genomic resources like whole-genome sequences, genomic survey sequences (GSS), large number of expressed sequence tags (ESTs), larger insert genomic libraries, high -density genomic maps, millions of molecular markers, and high-throughput genome sequences (HTGs) are essential to study the complex defense mechanism, metabolic pathways, and proteins/genes involved in plant pathogen resistance in *Brassica* crops. The genomic resources have been used for sequencing and a notation, genomic selection, mapping and cloning of genes or QTL's, association mapping, breeding by design, and MAS in crucifer's crops. The advent of next-generation sequencing (NGS) technologies like FLX-454, Illumina, SOLID/ion torrent, pacbio, HIC, bionano oxford nanopore, and helicase have enhanced the possibilities to generate genomic resources and their utilization in *Brassica* improvement programs. The molecular techniques and approaches used for comprehension of molecular mechanisms of crucifer's host resistance are briefly described as follows:

i. **Omics:** Technologies provide the tools needed to look at the differences in DNA, RNA, proteins, and other cellular molecules between species and among individuals of a species. These types of molecular profiles can vary with cell or tissue exposure to chemicals or drugs and thus have potential use in toxicological assessments. Omics experiments can often be conducted in high-throughput assays that produce tremendous amounts of data on the functional and/or structural alterations within the cell. "These new methods have already facilitated significant advances in understanding the molecular responses to cell and tissue damage, and perturbations in functional cellular systems" (Aardema and MacGregor 2003). The -omics technologies will continue to contribute to our understanding of toxicity mechanisms. Regulators are interested in these new technologies but are still sorting out how to incorporate the new information and technologies in regulatory decision-making. For example, the US Food and Drug Administration's Pharmaco-genomic Data submissions guidance document encourages the voluntary submission of genomics data but notes that the field of pharmaco-genomics is still in its early developmental stages.

ii. **Bioinformatics:** Bioinformatics is "the science of managing and analyzing biological data using advanced computing techniques" (HGP 2003). Bioinformatics tools include computational tools that mine information from large databases of biological data. These tools are most commonly used to analyze large sets of genomics data. However, bioinformatics tools are also being developed for other types of biological data, such as proteomics. The US National Centre for Biotechnology Information (NCBI) serves as an integrated source of genomics information and bioinformatics tools for researchers. An important bioinformatics tool available at NCBI for proteomics and genomics is the Basic Local Alignment Search Tool (BLAST), which compares gene or protein sequences against databases that contain many archived sequences, in order to find regions of local similarity. The statistical significance of the sequence matches is then calculated, and the results can be used to infer functional and evolutionary relationships.

iii. **Genomics:** The first of the -omics technologies to be developed, genomics has resulted in massive amounts of DNA sequence data requiring great amounts of computer capacity. Genomics has progressed beyond sequencing of organisms (structural genomics) to identifying the function of the encoded genes (functional genomics). The genome of each species is distinctive, but smaller genomic differences are also observed between each individual of a species. It was originally thought that obtaining the sequence of the human genome would immediately tell us the identity of the human genes. The genome has proved to be much more complex.

iv. **Epigenetics:** Epigenetics refers to mechanisms that persistently alter gene expression without actual changes to the gene/DNA sequence. DNA methylation is an example of an epigenetic mechanism. Scientists have shown that DNA methylation is an important component in a variety of chemical-induced toxicities, including carcinogenicity, and is a mechanism that should be assessed in the overall hazard assessment (Watson and Goodman 2002; Mogg et al. 2004).

v. **Transcriptomics:** Transcriptomics is the study of the transcriptome—the complete set of RNA transcripts that are produced by the genome, under specific circumstances or in a specific cell—using high-throughput methods, such as microarray analysis. Comparison of transcriptomes allows the identification of genes that are differentially expressed in distinct cell populations, or in response to different treatments. The term "transcriptome" is now widely understood to mean the complete set of all the ribonucleic acid (RNA) molecules expressed in some given entity, such as a cell, tissue, or organism.

vi. **Proteomics:** Proteins are the primary structural and functional molecules in the cell, and are made up of a linear arrangement of amino acids. The linear polypeptide chains are folded into secondary and tertiary structures to form the functional protein. Unlike the static nature of the cell's genes, proteins are constantly changing to meet the needs of the cell. Characterizing the identity, function, regulation, and interaction of all of the cellular proteins of an organism, the proteome, will be a major achievement. Studies of changes in the proteome of cells and tissues exposed to toxic materials, compared to normal cells, is being used to develop an understanding of the mechanisms of toxicity. As proteomics tools become more powerful and widely used, protein and proteome changes in response to exposures to toxic substances (fingerprints or response profiles) will be developed into databases that can be used to classify exposure responses at various levels of organization of the organism, thus providing a predictive in silico toxicology tool.

vii. **Metabolomics:** Metabolomics refers to the comprehensive evaluation of the metabolic state of a cell, organ, or organism, in order to identify biochemical changes that are characteristic of specific disease states or toxic insults. Typical metabolomics experiments involve the identification and quantization of large numbers of endogenous molecules in a biological sample (e.g., urine or blood) using chemical techniques such as chromatography and mass spectrometry. The output from these techniques is compared to computerized libraries of mass spectrometry tracings to facilitate identification of the compounds that are present. Environmental stresses such as exposure to chemicals or drugs alter the metabolic pathways in cells, and metabolite profiling can be used to assess toxic responses/exposures. The name metabolomics was coined in the late 1990s using the word metabolome (Oliver et al. 1998). Systematic functional analysis of the yeast genome has provided the first complete inventory of the working parts of a eukaryotic cell. Many of the bio-analytical methods used for metabolomics have been adapted (or in some cases simply adopted) from existing biochemical techniques. Metabolomics is the identification and quantification of all metabolites in a given biological situation.

viii. **Biometabolomics:** Biometabolomics is a term used to describe the destiny of metabolites in biometabolotic reactions (biochemical processes involving metabolites) and measuring of many small molecules in biological systems. These systems can be fluids within our bodies such as urine, blood, saliva as well as different tissues. Biometabolomics is a discipline of analytic biochemistry in living things. In Biometabolomics, nuclear magnetic resonance (NMR) or Mass Spectrometery (MS) are mostly used. In the field of obesity and metabolic disease, recent studies employing targeted metabolomics have identified a number of novel metabolites and pathways that may be involved in disease pathogenesis. In biometabolomics, the metabolome represents the collection of all metabolites in a biological cell, tissue, organ, or organism, which are the end products of cellular processes. Metabolites are the intermediates and products of metabolism. The metabolome forms a large network of metabolic reactions, where outputs from one enzymatic chemical reaction are inputs to other chemical reactions.

ix. **Toxicogenomics:** Toxicogenomics compares the genes expressed in organisms that have been exposed to a drug, chemical, or toxin to those of unexposed organisms (negative controls). The up- or down-regulation of certain genes or groups of genes may be linked to toxic responses occurring in the organism, and to particular organs or cell types in that organism. The goal of toxicogenomics is to identify patterns of gene expression related to specific chemicals or chemical classes so that these expression patterns can be used as endpoints for assessing toxicity. Thus far, toxicogenomics has been useful in refining animal experiments and identifying mechanisms of toxicity in lab animals where exposures can be controlled. There have also been experiments evaluating gene expression in cell cultures exposed to toxicants, which has been used in limited applications for prediction of in vivo toxicity.

x. **Pharmacogenetics:** Pharmacogenetics looks at the differences in response to a particular drug that are due to variations in the genetic makeup of individuals. For example, human genetic variation has been implicated in the variability of responses (effectiveness and/or toxicity) seen with some chemotherapeutic drugs (Crews 2006; Hahn et al. 2006).

xi. **Metagenomics:** Metagenomics is the study of a collection of genetic material (genomes) from a mixed community of organisms. Metagenomics usually refers to the study of microbial communities. "Metagenomics" is the two words "meta" and "genomics." So genomics is obtaining the DNA sequence, but meta implies that doing it for many organisms together and metagenomics is usually used when studying microbial communities where it can't separate one microbe from another. Like there may be two bacteria that grow together, and so when take the DNA sequence, getting the DNA sequence of two bacteria together.

xii. **Pathogenomics:** Pathogenomics is a field which uses high-throughput screening technology and bioinformatics to study encoded microbe resistance, as well as virulence factors (VFs), which enable a microorganism to infect a host and possibly cause disease. Pathogenomics is a discipline that seeks to mark out virulence factors and their contributions to overall pathogenesis by comparing gene repertoires of pathogenic and non-pathogenic strains/species

xiii. **Phylogenomics:** Phylogenomics is the intersection of the fields of evolution and genomics. The term has been used in multiple ways to refer to analysis that involves genome data and evolutionary reconstructions. It is a group of techniques within the larger fields of Phylogenetics and genomics. As more complete genomes are sequenced, phylogenetic analysis is entering a new era—that of phylogenomics. One branch of this expanding field aims to reconstruct the evolutionary history of organisms on the basis of the analysis of their genomes. The word "phylogenomics" was first introduced in the context of prediction of gene function for genome-scale data (Eisen 1998), and soon after in the context of phylogenetic inference (O'Brien and Stanyon 1999). The discipline of phylogenomics owes its existence to the advances made in DNA sequencing technology over the past two decades (Metzker 2010). It comprises several areas of research at the interface between molecular and evolutionary biology and has two major goals: (i) to infer phylogenetic relationships between taxa and gain insights into the mechanisms of molecular evolution; and (ii) to use multispecies phylogenetic comparisons to infer putative functions for DNA or protein sequences.

xiv. **Pan-genomics:** The pan-genome refers to a collection of genomic sequence found in the entire species or population rather than in a single individual; the sequence can be core, present in all individuals, or accessory (variable or dispensable), found in a subset of individuals only. The study of the pan-genome is called "pan-genomics." The concept of the pan-genome was introduced by Tettelin et al. (2005) who described the production of the first ever pan-genome, for a bacterial species *Streptococcus agalactiae*. It derives "pan" from the Greek word παν, meaning "whole" or "everything," while genome is a commonly used term to describe an organism's complete genetic material. The pan-genome can be open or closed (restricted). In the first instance, there appears to be no finite number of genes in the species, and with each new added individual, new gene is incorporated into the pan-genome. In the second instance, the gene pool is limited. After a certain number of individuals have been analyzed, the addition of new individuals to the analysis will not result in an expansion of the pan-genome.

References

Aardema MJ, MacGregor JT (2003) Toxicology and genetic toxicology in the new era of "Toxicogenomics": impact of "omics" technologies. In: Inoue T, Pennie WD (eds) Toxicogenomics. Springer, Tokyo. https://doi.org/10.1007/978-4-431-66999-9_22

Aarts N, Metz M, Holub E, Staskawicz BJ, Daniels MJ, Parker JE (1998) Different requirements for EDS1 and NDR1 by disease resistance genes define at least two R gene mediated signaling pathways in *Arabidopsis*. Proc Natl Acad Sci U S A 95:10306–10311

Adhikari TB, Liu JQ, Mathur S, Wu CX, Rismmer SR (2003) Genetic and molecular analyses in crosses of race 2 and race 7 of *Albugo candida*. Phytopathology 93:959–965

Agrios GN (2005) Plant pathology, 5th edn. Elsevier Academic Press, Burlington, 948 p

Ahmed NU, Park JI, Seo MS, Kumar TS, Lee IH, Park BS, Nou IS (2012a) Identification and expression analysis of chitinase genes related to biotic stress resistance in *Brassica*. Mol Biol Rep 39:3649–3657

Ahmed NU, Park JI, Jung HJ, Seo MS, Kumar TS, Lee IH, Nou IS (2012b) Identification and characterization of stress resistance related genes of *Brassica rapa*. Biotechnol Lett 34:979–987

Ahmed NU, Park JI, Jung HJ, Chung MY, Cho YG, Nou IS (2013) Characterization of thaumatin-like gene family and identification of *Pectobacterium carotovorum* subsp.*carotovorum* inducible genes in *Brassica oleracea*. Plant Breed Biotech 1:111–121

Alamery S, Tirnaz S, Bayer P, Tollenaere R, Chaloub B, Edwards D, Batley J (2017) Genome-wide identification and comparative analysis of NBS-LRR resistance genes in *Brassica napus*. Crop Pasture Sci 69:79–94. https://doi.org/10.1071/CP17214

Allen RL, Bittner-Eddy PD, Grenville-Briggs LJ, Meitz JC, Rehmany AP, Rose LE, Beynon JL (2004) Host–parasite co-evolutionary conflict between *Arabidopsis* and downy mildew. Science 306:1957–1960

Ananga A, Cebert E, Soliman K, Kantety R, Pacumbaba RP, Konan K (2006) RAPD markers associated with resistance to blackleg disease in *Brassica* species. African J Biotechnol 5 (22):2041–2048

Ananga AO, Cebert E, Soliman K, Kantety R, Konan K, Ochieng JW (2008) Phylogenetic relationships within and among *Brassica* species from RAPD loci associated with blackleg resistance. African J Biotechnol 7(9):1287–1293

Ansan-Melayah D, Balesdent HM, Buee M, Rouxel T (1995) Genetic characterization of AvrLml, the first avirulence gene of *Leptosphaeria maculans*. Phytopathology 85:1525–1529

Ansan-Melayah D, Balesdent HM, Delourme R, Pilet LM, Tanguy X, Renard M, Rouxel T (1997) Gene-for-gene interactions in the *Leptosphaeria maculans-Brassica napus* pathosystem. Acta Hortic 459:329–334

Aravind L, Koonin EV (1999) Fold prediction and evolutionary analysis of the POZ domain: structural and evolutionary relationship with the potassium channel tetramerization domain. J Mol Biol 285:1353–1361. https://doi.org/10.1006/jmbi.1998.2394

Armstrong MR, Whisson SC, Pritchard L, Bos JI, Venter E, Avrova AO, Rehmany AP, Bohme U, Brooks K, Cherevach I, Hamlin N, White B, Fraser A, Lord A, Quail MA, Churcher C, Hall N, Berriman M, Huang S, Kamoun S, Beynon JL, Birch PRJ (2005) An ancestral oomycete locus contains late blight avirulence gene Avr3a, encoding a protein that is recognized in the host cytoplasm. Proc Natl Acad Sci U S A 102:7766–7771

Asai S, Rallapalli G, Piquerez SJ, Caillaud MC, Furzer OJ, Ishaque N, Wirthmueller L, Fabro G, Shirasu K, Jones JD (2014) Expression profiling during *Arabidopsis*/downy mildew interaction reveals a highly-expressed effector that attenuates responses to salicylic acid. PLoS Pathog 10: e1004443

Asai S, Furzer OJ, Cevik V, Kim DS, Ishaque N, Goritschnig S, Staskawicz BJ, Shirasu K, Jones JDG (2018) A downy mildew effector evades recognition by polymorphism of expression and subcellular localization. Nat Commun 9:5192

Ascunce MS, Huguet-Tapia JC, Ortiz-Urquiza A, Keyhani NO, Braun EL, Goss EM (2017) Phylogenomic analysis supports multiple instances of polyphyly in the oomycete Peronosporalean lineage. Mol Phylogenet Evol 114:199–211

Bailey K, Cevik V, Holton N, Byrne-Richardson J, Sohn KH, Coates M, Woods-Tor A, Aksoy HM, Hughes L, Baxter L, Jones JD, Beynon J, Holub EB, Tor M (2011) Molecular cloning of ATR5 (Emoy2) from *Hyaloperonospora arabidopsidis*, an avirulence determinant that triggers RPP5-mediated defense in *Arabidopsis*. Mol Plant-Microbe Interact 24:827–838

Bari R, Jones JD (2009) Role of plant hormones in plant defense responses. Plant Mol Biol 69:473–488. https://doi.org/10.1007/s11103-008-9435-0

Baxter L, Tripathy S, Ishaque N, Boot N, Cabral A, Kemen E, Thines M, Ah-Fong A, Anderson R, Badejoko W, Bittner-Eddy P, Boore JL, Chibucos MC, Coates M, Dehal P, Dele-haunty K, Dong S, Downton P, Dumas B, Fabro G, Fronick C, Fuerstenberg SI, Fulton L, Gaulin E, Govers F, Hughes L, Humphray S, Jiang RHY, Judelson H, Kamoun S, Kyung K, Meijer H, Minx P, Morris P, Nelson J, Phuntumart V, Qutob D, Rehmany A, Rougon-Cardoso A, Ryden P, Torto-Alalibo T, Studholme D, Wang Y, Win J, Wood J, Clifton SW, Rogers J, Van den Ackerveken G, Jones JDG, McDowell JM, Beynon J, Tyler BM (2010) Signatures of adaptation to obligate biotrophy in the *Hyaloperonospora arabidopsidis* genome. Science 330:1549–1551

Bayer PE, Hurgobin B, Golicz AA, Chan CK, Yuan Y, Lee H, Renton M, Meng J, Li R, Long Y, Zou J, Bancroft I, Chalhoub B, King GJ, Batley J, Edwards D (2017) Assembly and comparison of two closely related *Brassica napus* genomes. Plant Biotechnol J 15:1602–1610

Bayer PE, Golicz AA, Tirnaz S, Chan CK, Edwards D, Batley J (2018) Variation in abundance of predicted resistance genes in the *Brassica oleracea* pangenome. Plant Biotechnol J 17 (4):789–800. https://doi.org/10.1111/pbi.13015

Becker HC, Engqvist GM, Karlsson B (1995) Comparison of rapeseed cultivars and resynthesized lines based on allozyme and RFLP markers. Theor Appl Genet 91:62–67

Belhaj K, Cano LM, Prince DC, Kemen A, Yoshida K, Dagdas YF, Etherington GJ, Schoonbeek HJ, van Esse HP, Jones JD, Kamoun S, Schornack S (2017) Arabidopsis late blight: infection of a nonhost plant by *Albugo laibachii* enables full colonization by *Phytophthora infestans*. Cell Microbiol 19

Bigeard J, Colcombet J, Hirt H (2015) Signaling mechanisms in pattern-triggered immunity (PTI). Mol Plant 8:521–539

Bittner-Eddy PD, Beynon JL (2001) The Arabidopsis downy mildew resistance gene, *RPP13-Nd*, functions independently of *NDR1* and *EDS1* and does not require the accumulation of salicylic acid. Mol Plant-Microbe Interact 14(3):416–421. https://doi.org/10.1094/MPMI.2001.14.3.416

Bittner-Eddy PD, Crute IR, Holub EB, Beynon JL (2000) RPP13 is a simple locus in *Arabidopsis thaliana* for alleles that specify downy mildew resistance to different avirulence determinants in *Peronospora parasitica*. Plant J 21:177–188

Block A, Alfano JR (2011) Plant targets for *Pseudomonas syringae* type III effectors: virulence targets or guarded decoys? Curr Opin Microbiol 14:39–46

Borhan MH, Brose E, Beynon JL, Holub EB (2001) White rust (*Albugo candida*) resistance loci on three *Arabidopsis* chromosomes are closely linked to downy mildew (*Peronospora parasitica*) resistance loci. Mol Plant Pathol 2:87–95

Borhan MH, Holub EB, Beynon JL, Rozwadowski K, Rimmer SR (2004) The *Arabidopsis* TIR-NB-LRR gene RAC1 confers resistance to *Albugo candida* (white rust) and is dependent on EDS1 but not PAD4. Mol Plant-Microbe Interact 17:711–719

Borhan MH, Gunn N, Cooper A, Gulden S, Tor M, Rimmer SR, Holub EB (2008) WRR4 encodes a TIR-NB-LRR protein that confers broad-spectrum white rust resistance in *Arabidopsis thaliana* to four physiological races of *Albugo candida*. Mol Plant-Microbe Interact 21:757–768

Borhan MH, Holub EB, Kindrachuk C, Omidi M, BozorgmaneshFrad G, Rimmer SR (2010) WRR4, a broad-spectrum TIR-NBLRR gene from *Arabidopsis thaliana* that confers white rust resistance in transgenic oilseed *Brassica* crops. Mol Plant Pathol 11:283–291

Botella MA, Parker JE, Frost LN, Bittner-Eddy PD, Beynon JL, Daniels MJ, Holub EB, Jones JD (1998) Three genes of the *Arabidopsis* RPP1 complex resistance locus recognize distinct *Peronospora parasitica* avirulence determinants. Plant Cell 10:1847–1860

Bredow M, Monaghan J (2019) Regulation of plant immune signaling by calcium-dependent protein kinases. Mol Plant-Microbe Interact 32:6–19

Brun H, Ruer D, Levivier S, Somda I, Renard M, Chevre MA (2001) Presence in *Leptosphaeria maculans* populations of isolates virulent on resistance introgressed into *Brassica napus* from the *B. nigra* B genome. Plant Pathol 50:69–74

Cai C, Wang X, Liu B, Wu J, Liang J, Cui Y, Cheng F, Wang X (2017) *Brassica rapa* genome 2.0: a reference upgrade through sequence re-assembly and gene re-annotation. Mol Plant 10:649–651

Caillaud MC, Piquerez SJ, Fabro G, Steinbrenner J, Ishaque N, Beynon J, Jones JD (2012) Subcellular localization of the *Hpa* RxLR effector repertoire identifies a tonoplast-associated protein HaR × L17 that confers enhanced plant susceptibility. Plant J 69:252–265

Cannon PF, Damm U, Johnston PR, Weir BS (2012) *Colletotrichum*–current status and future directions. Studies Mycol 72:181–213

Cao H, Bowling SA, Gordon AS, Dong X (1994) Characterization of an *Arabidopsis* mutant that is nonresponsive to inducers of systemic acquired resistance. Plant Cell 6:1583–1592. https://doi.org/10.1105/tpc.6.11.1583

Cao H, Glazebrook J, Clarke JD, Volko S, Dong X (1997) The *Arabidopsis* NPR1 gene that controls systemic acquired resistance encodes a novel protein containing ankyrin repeats. Cell 88:57–63. https://doi.org/10.1016/S0092-8674(00)81858

Cao H, Li X, Dong XN (1998) Generation of broad–spectrum disease resistance by over-expression of an essential regulatory gene in systemic acquired resistance. Proc Natl Acad Sci U S A 95:6531–6536. https://doi.org/10.1073/pnas.95.11.6531

Carlier JD, Alabaça CA, Coelho PS, Monteiro AA, Leitao JM (2012) The downy mildew resistance locus Pp523 is located on chromosome C8 of *Brassica oleracea* L. Plant Breed 131:170–175. https://doi.org/10.1111/j.1439-0523.2011.01904.x

Carrington JC, Kasschau KD, Mahajan SK, Schaad MC (1996) Cell-to-cell and long-distance transport of viruses in plants. Plant Cell 8:1669–1681

Castel B, Ngou PM, Cevik V, Redkar A, Kim DS, Yang Y, Ding P, Jones JDG (2019) Diverse NLR immune receptors activate defence via the RPW8-NLR NRG1. New Phytol 222(2):966–980

Cevik V, Boutrot F, Apel W, Robert-Seilaniantz A, Furzer OJ, Redkar A, Castel B, Kover PX, Prince DC, Holub EB, Jones JDG (2019) Transgressive segregation reveals mechanisms of *Arabidopsis* immunity to *Brassica*-infecting races of white rust *Albugo candida*. Proc Natl Acad Sci 116:2767

Chalhoub B, Denoeud F, Liu S, Parkin IA, Tang H, Wang X, Chiquet J, Belcram H, Tong C, Samans B, Corréa M, Da Silva C, Just J, Falentin C, Koh CS, Le Clainche I, Bernard M, Bento P, Noel B, Labadie K, Alberti A, Charles M, Arnaud D, Guo H, Daviaud C, Alamery S,

Jabbari K, Zhao M, Edger PP, Chelaifa H, Tack D, Lassalle G, Mestiri I, Schnel N, Le Paslier MC, Fan G, Renault V, Bayer PE, Golicz AA, Manoli S, Lee TH, Thi VH, Chalabi S, Hu Q, Fan C, Tollenaere R, Lu Y, Battail C, Shen J, Sidebottom CH, Wang X, Canaguier A, Chauveau A, Bérard A, Deniot G, Guan M, Liu Z, Sun F, Lim YP, Lyons E, Town CD, Bancroft I, Wang X, Meng J, Ma J, Pires JC, King GJ, Brunel D, Delourme R, Renard M, Aury JM, Adams KL, Batley J, Snowdon RJ, Tost J, Edwards D, Zhou Y, Hua W, Sharpe AG, Paterson AH, Guan C, Wincker P (2014) Early allopolyploid evolution in the post-Neolithic *Brassica napus* oilseed genome. Science 345(6199):950–953

Chatterjee D, Banga S, Gupta M, Bharti S, Salisbury PA, Banga SS (2016) Resynthesis of *Brassica napus* through hybridization between *B. juncea* and *B. carinata*. Theor Appl Genet 129:977–990. https://doi.org/10.1007/s00122-016-2677-3

Chen ZJ (2007) Genetic and epigenetic mechanisms for gene expression and phenotypic variation in plant polyploids. Ann Rev Plant Biol 58:377–406. https://doi.org/10.1146/annurev.arplant.58.032806.103835.

Chen X, Barnaby JY, Sreedharan A, Huang X, Orbović V, Grosser JW, Wang N, Dong X, Song WY (2013) Over–expression of the citrus gene CtNH1 confers resistance to bacterial canker disease. Physiol Mol Plant Pathol 84:115–122. https://doi.org/10.1016/j.pmpp.2013.07.002

Cheng F, Wu J, Fang L, Sun S, Liu B, Lin K, Bonnema G, Wang X (2012) Biased gene fractionation and dominant gene expression among the sub genomes of *Brassica rapa*. PLoS One 7:e36442. https://doi.org/10.1371/journal.pone.0036442

Cheung F, Trick M, Drou N, Lim YP, Park JY, Kwon SJ, Kim JA, Scott R, Pires JC, Paterson AH, Town C, Bancroft I (2009) Comparative analysis between homoeologous genome segments of *Brassica napus* and its progenitor species reveals extensive sequence-level divergence. Plant Cell 21:1912–1928. https://doi.org/10.1105/tpc.108.060376

Chevre AM, Barret P, Eber F, Dupuy P, Brun H, Tanguy X, Renard M (1997) Selection of stable *Brassica napus-B. juncea* recombinant lines resistant to blackleg (*Leptosphaeria maculans*). 1. Identification of molecular markers, chromosomal and genomic origin of the introgression. Theor Appl Genet 95:1104–1111

Choi YJ, Hong SB, Shin HD (2003) Diversity of the *Hyaloperonospora parasitica* complex from core brassicaceous hosts based on ITS rDNA sequences. Mycol Res 107:1314–1322. https://doi.org/10.1017/S0953756203008578

Chou S, Krasileva KV, Holton JM, Steinbrenner AD, Alber T, Staskawicz BJ (2011) *Hyaloperonospora arabidopsidis* ATR1 effector is a repeat protein with distributed recognition surfaces. Proc Natl Acad Sci U S A 108:13323–13328

Coates ME, Beynon JL (2010) *Hyaloperonospora arabidopsidis* as a pathogen model. Annu Rev Phytopathol 48:329–345

Comai L (2005) The advantages and disadvantages of being polyploid. Nat Rev Genet 6:836–846. https://doi.org/10.1038/nrg1711

Conn KL (1986) Leaf epicuticular wax of canola: ultrastructure, chemistry and interaction with *Alternaria brassicae*. M.Sc. thesis, p 159

Conn KL, Tewari JP (1989) Interactions of *Alternaria brassicae* conidia with leaf epicuticular wax of canola. Mycol Res 93:240–242. https://doi.org/10.1016/S0953-7562(89)80126-1

Conn KL, Tewari JP, Hadziyev D (1984) The role of epicuticular wax in canola in resistance to *Alternaria brassicae*. Phytopathology 74:851. (Abstr.)

Conn KL, Tewari JP, Dahiya JS (1988) Resistance to *Alternaria brassicae* and phytoalexin-elicitation in rapeseed and other crucifers. Plant Sci 55:21–25. https://doi.org/10.1016/0168-9452(88)90037-4

Conn KL, Jejelowa OA, Tewari JP, Bains PS (1991) *Camelina sativa* phytoalexins (Calmexin and methoxy camalexin) provide resistance against *Alternaria brassicae*. Can J Plant Pathol 13, Abstr 274.

Constantinescu O, Fatehi J (2002) *Peronospora*-like fungi (Chromista, Peronosporales) parasitic on Brassicaceae and related hosts. Nova Hedwigia 74:291–338

Cooper AJ, Latunde-Dada AO, Woods-Tor A, Lynn J, Lucas JA, Crute IR, Holub EB (2008) Basic compatibility of *Albugo candida* in *Arabidopsis thaliana* and *Brassica juncea* causes broad-spectrum suppression of innate immunity. Mol Plant-Microbe Interact 21:745–756

Crews KR (2006) Individualizing chemotherapeutic treatment of colorectal cancer. Am J Health Syst Pharm 63(Suppl 2):S12–S17. https://doi.org/10.2146/ajhp060113

Crute IR, Holub EB, Tor M, Brose E, Beynon JL (1993) The identification and mapping of loci in *Arabidopsis thaliana* for recognition of the fungal pathogens *Peronospora parasitica* (downy mildew) and *Albugo candida* (white blister). In: Nester EW, Verma DPS (eds) Advances in molecular genetics of plant-microbe interactions. Kluwer Academic, Dordrecht, pp 437–444

Crute IR, Beynon J, Dangl JL, Holub EB, Mauch-Mani B, Slusarenko A, Staskawicz BJ, Ausubel FM (1994) Microbial pathogenesis of *Arabidopsis*. In: Meyerowitz EM, Somerville CR (eds) *Arabidopsis*. Cold Spring Harbour Laboratory Press, Cold Spring Harbor, pp 705–748

Cui H, Tsuda K, Parker JE (2015) Effector-triggered immunity: from pathogen perception to robust defense. Annu Rev Plant Biol 66:487–511

Dangl JL, Jones JD (2001) Plant pathogens and integrated defence responses to infection. Nature 411:826–833. https://doi.org/10.1038/35081161

Dangl JL, Holub EB, Debener T, Lehnackers H, Ritter C, Crute IR (1992) Genetic definition of loci involved in *Arabidopsis*-pathogen interactions. In: Koncz C, Chua NH, Schell J (eds) Methods in *Arabidopsis* research. World Scientific Publishing Co., London, pp 393–418

Delaney TP, Friedrich L, Ryals JA (1995) *Arabidopsis* signal transduction mutant defective in chemically and biologically induced disease resistance. Proc Natl Acad Sci U S A 92:6602–6606. https://doi.org/10.1073/pnas.92.14.6602

Delourme R, Pilet-Nayel ML, Archipiano M, Horvais R, Tanguy X, Rouxel T, Brun H, Renard M, Balesdent MH (2004) A cluster of major specific resistance genes to *Leptosphaeria maculans* in *Brassica napus*. Phytopathology 94:578–583. https://doi.org/10.1094/PHYTO.2004.94.6.578

Depotter JRL, Rodriguez-Moreno L, Thomma BPHJ, Wood TA (2017) The emerging British *Verticillium longisporum* population consist of aggressive *Brassica* pathogens. Phytopathology 107:1399–1405

De-Vos M, Van-Oosten VR, van RMP P, van JA P, Pozo MJ, Mueller MJ, Buchala AJ, Métraux JP, van LC L, Dicke M, CMJ P (2005) Signal signature and transcriptome changes of *Arabidopsis* during pathogen and insect attack. Mol Plant-Microbe Interact 18:923–937. https://doi.org/10.1094/MPMI-18-0923

Diederichsen E, Wagenblatt B, Schallehn V (2016) Production of pure genotype isolates of *Plasmodiophora brassicae* Wor. comparison of inoculations with root hairs containing single sporangiosori or with single resting spores. Eur J Plant Pathol 145:621–627. https://doi.org/10.1007/s10658-016-0876-z

Diers BW, Osborn TC (1994) Genetic diversity of oilseed *Brassica napus* germplasm based on restriction fragment length polymorphisms. Theor Appl Genet 88:662–668. https://doi.org/10.1007/BF01253968

Durrant WE, Dong X (2004) Systemic acquired resistance. Annu Rev Phytopathol 42:185–209. https://doi.org/10.1146/annurev.phyto.42.040803.140421

Dutt M, Barthe G, Irey M, Grosser J (2015) Transgenic citrus expressing an *Arabidopsis* NPR1 gene exhibit enhanced resistance against huanglongbing (HLB; citrus greening). PLoS One 10: e0137134. https://doi.org/10.1371/journal.pone.0147657

Dzurilla M, Kutschy P, Tewari JP, Ruzinsky M, Senvicky S, Kovacik V (1998) Synthesis and antifungal activity of new indolypthiazinone derivatives. Coll Czechoslovakia Chem Comm 63:94–102. https://doi.org/10.1135/cccc19980094

Eisen JA (1998) Phylogenomics: improving functional predictions for uncharacterized genes by evolutionary analysis. Genome Res 8:163–167

Eulgem T, Tsuchiya T, Wang X, Beasley B, Cuzick A, Tor M, Zhu T, McDowell JM, Holub E, Dangl JL (2007) EDM2 is required for RPP7- dependent disease resistance in Arabidopsis and affects RPP7 transcript levels. Plant J 49:829–839

Evans K, Webb RM (1989) Nematode problems in oilseed rape. Production and protection of oilseed rape and other brassica crops. Asp Appl Biol 23:303–310

Fabro G, Steinbrenner J, Coates M, Ishaque N, Baxter L, Studholme DJ, Korner E, Allen RL, Piquerez SJ, Rougon-Cardoso A, Greenshields D, Lei R, Badel JL, Caillaud MC, Sohn KH, Van den Ackerveken G, Parker JE, Beynon J, Jones JD (2011) Multiple candidate effectors from the oomycete pathogen *Hyaloperonospora arabidopsidis* suppress host plant immunity. PLoS Pathog 7:e1002348

Fatima U, Bhorali P, Borah S, Senthil-Kumar M (2019) Perspectives on the utilization of resistance mechanisms from host and nonhost plants for durable protection of *Brassica* crops against Alternaria blight. Peer J 7:e7486. https://doi.org/10.7717/peerj.7486

Fitt BDL, Gladders P, Turner JA, Sutherland KG, Welham SJ, Davies JML (1997) Prospects for developing a forecasting scheme to optimize use of fungicides for disease control on winter oilseed rape in the UK. Aspects Appl Biol Optimizing Pesticide Appls 48:135–142

Fomeju BF, Falentin C, Lassalle G, Manzanares-Dauleux MJ, Delourme R (2014) Homoeologous duplicated regions are involved in quantitative resistance of *Brassica napus* to stem canker. BMC Genomics 15:498. https://doi.org/10.1186/1471-2164-15-498

Fomeju BF, Falentin C, Lassalle G, Manzanares-Dauleux MJ, Delourme R (2015) Comparative genomic analysis of duplicated homoeologous regions involved in the resistance of *Brassica napus* to stem canker. Front Plant Sci 6:772. https://doi.org/10.3389/fpls.2015.00772

Friedrich L, Lawton K, Dietrich R, Willits M, Cade R, Ryals J (2001) NIM1 overexpression in Arabidopsis potentiates plant disease resistance and results in enhanced effectiveness of fungicides. Mol Plant-Microbe Interact 14:1114–1124. https://doi.org/10.1094/MPMI.2001.14.9.1114

Fu ZQ, Dong X (2013) Systemic acquired resistance: turning local infection into global defense. Annu Rev Plant Biol 64:839–863. https://doi.org/10.1146/annurev-arplant-042811-105606

Fu ZQ, Yan S, Saleh A, Wang W, Ruble J, Oka N, Mohan R, Spoel SH, Tada Y, Zheng N, Dong X (2012) NPR3 and NPR4 are receptors for the immune signal salicylic acid in plants. Nature 486:228–232. https://doi.org/10.1038/nature11162

Fu Y, Zhang Y, Mason AS, Lin B, Zhang D, Yu H, Fu D (2019) NBS-encoding genes in *Brassica napus* evolved rapidly after allopolyploidization and co-localize with known disease resistance loci. Front Plant Sci 10:26. https://doi.org/10.3389/fpls.2019.00026

Furci L, Jain R, Stassen J, Berkowitz O, Whelan J, Roquis D, Baillet V, Colot V, Johannes F, Ton J (2019) Identification and characterization of hypo-methylated DNA loci controlling quantitative resistance in *Arabidopsis*. Elife 8

Gaeta RT, Chris Pires J (2010) Homoeologous recombination in allopolyploids: the polyploid ratchet. New Phytol 186:18–28

Garg H, Kohn LM, Andrew M, Li H, Sivasithamparam K, Barbetti MJ (2010) Pathogenicity of morphologicallydifferentisolatesof *Sclerotinia sclerotiorum* with *Brassica napus* and *B. juncea* genotypes. Eur J Plant Pathol 126:305–315. https://doi.org/10.1007/s10658-009-9547-7

Garibaldi A, Gilardi G, Gullino M L (2006). Evidence for an expanded host range of *Fusarium oxysporum* f. sp. *raphani*. Phytoparasitica34: 115–121.

Ge XT, You MP, Barbetti MJ (2015) Virulence differences among *Sclerotinia sclerotiorum* isolates determines host cotyledon resistance responses in Brassicaceae genotypes. Eur J Plant Pathol 143:527–541. https://doi.org/10.1007/s10658-015-0696-6

Glazebrook J (2005) Contrasting mechanisms of defense against biotrophic and necrotrophic pathogens. Annu Rev Phytopathol 43:205–227. https://doi.org/10.1146/annurev.phyto.43.040204.135923

Glazebrook J, Rogers EE, Ausubel FM (1996) Isolation of *Arabidopsis* mutants with enhanced disease susceptibility by direct screening. Genetics 143:973–982

Goker M, Voglmayr H, García-Blázquez G, Oberwinkler F (2009) Species delimitation in downy mildews: the case of *Hyaloperonospora* in the light of nuclear ribosomal ITS and LSU sequences. Mycol Res 113:308–325

Golicz AA, Martinez PA, Zander M, Patel DA, Van De Wouw AP, Visendi P, Fitzgerald T, Edwards D, Batley J (2015) Gene loss in the fungal canola pathogen *Leptosphaeria maculans*. Funct Integr Genomics 15:189–196. https://doi.org/10.1007/s10142-014-0412-1

Golicz AA, Bayer PE, Barker GC, Edger PP, Kim H, Martinez PA, Chan CK, Severn-Ellis A, McCombie WR, Parkin IA, Paterson AH, Pires JC, Sharpe AG, Tang H, Teakle GR, Town CD, Batley J, Edwards D (2016) The pangenome of an agronomically important crop plant *Brassica oleracea*. Nat Commun 7:13390

Gomez-Gomez L, Boller T (2000) FLS2: an LRR receptor-like kinase involved in the perception of the bacterial elicitor flagellin in *Arabidopsis*. Mol Cell 5:1003–1011

Goritschnig S, Krasileva KV, Dahlbeck D, Staskawicz BJ (2012) Computational prediction and molecular characterization of an oomycete effector and the cognate *Arabidopsis* resistance gene. PLoS Genet 8:e1002502

Goritschnig S, Steinbrenner AD, Grunwald DJ, Staskawicz BJ (2016) Structurally distinct *Arabidopsis thaliana* NLR immune receptors recognize tandem WY domains of an oomycete effector. New Phytol 210:984–996

Grant M, Lamb C (2006) Systemic immunity. Curr Opin Plant Biol 9:414–420. https://doi.org/10.1016/j.pbi.2006.05.013

Grant MR, McDowell JM, Sharpe AG, de Torres ZM, Lydiate DJ, Dangl JL (1998) Independent deletions of a pathogen-resistance gene in *Brassica* and *Arabidopsis*. Proc Natl Acad Sci U S A 95:15843–15848

Gunasinghe N, You MP, Cawthray GR, Barbetti MJ (2016) Cercosporin from *Pseudocercosporella capsellae* and its critical role in white leaf spot development. Plant Dis 100(8):1521–1531. https://doi.org/10.1094/PDIS-10-15-1192-RE

Gunn ND, Byrne J, Holub EB (2002) Outcrossing of two homo-thallic isolates of *Peronospora parasitica* and segregation of avirulence matching six resistance loci in *Arabidopsis thaliana*. In: Spencer-Phillips P, Gisi U, Lebeda A (eds) Advances in downy mildew research. Kluwer Academic Publishers, Dordrecht, pp 185–188

Gupta SK, Kaushik CD (2002) Metabolic changes in mustard leaf and silique wall due to the infection of Alternaria blight (*Alternaria brassicae*). Cruciferae Newslett 24:85–86

Gupta SK, Gupta PP, Yadav TP, Kaushik CD (1990) Metabolic changes in mustard due to *Alternaria* leaf blight. Indian Phytopath 43:64–69

Gupta SK, Gupta PP, Kaushik CD (1995) Changes in leaf peroxidase, polyphenols oxidase, catalase and total phenols due to Alternaria leaf blight in *Brassica* species. Indian J Mycol Plant Pathol 25:175–180

Gust AA, Felix G (2014) Receptor like proteins associate with SOBIR1-type of adaptors to form bimolecular receptor kinases. Curr Opin Plant Biol 21:104–111

Hahn KK, Wolff JJ, Kolesar JM (2006) Pharmacogenetics and irinotecan therapy. Am J Health Syst Pharm 63(22):2211–2217. https://doi.org/10.2146/ajhp060155

Hasan M, Seyis F, Badani A, Pons-Kuhnemann J, Friedt W, Luhs W, Snowdon R (2006) Analysis of genetic diversity in the *Brassica napus* L. gene pool using SSR markers. Genet Res Crop Evol 53:793–802

Hatakeyama K, Suwabe K, Tomita RN, Kato T, Nunome T, Fukuoka H, Matsumoto S (2013) Identification and characterization of Crr1a, a gene for resistance to clubroot disease (*Plasmodiophora brassicae* Woronin) in *Brassica rapa* L. PLoS One 8:e54745. https://doi.org/10.1371/journal.pone.0054745

Hayden HL, Cozijnsen AJ, Howlett BJ (2007) Microsatellite and mini-satellite analysis of *Leptosphaeria maculans* in Australia reveals regional genetic differentiation. Phytopathology 97:879–887. https://doi.org/10.1094/PHYTO-97-7-0879

Hein I, Gilroy EM, Armstrong MR, Birch PRJ (2009) The zig-zag-zig in oomycete-plant interactions. Mol Plant Pathol 10:547–562

Hepworth SR, Zhang Y, McKim S, Li X, Haughn GW (2005) BLADE-ON-PETIOLE–dependent signaling controls leaf and floral patterning in Arabidopsis. Plant Cell 17:1434–1448. https://doi.org/10.1105/tpc.104.030536

Herlihy J, Ludwig NR, van den Ackerveken G, McDowell JM (2019) Oomycetes used in *Arabidopsis* research. The *Arabidopsis* Book e0188. https://doi.org/10.1199/tab.0188
Holub EB (2001) The arms race is ancient history in *Arabidopsis*, the wildflower. Nat Rev Genet 2:516–527
Holub EB (2008) Natural history of *Arabidopsis thaliana* and oomycete symbioses. Eur J Plant Pathol 122:91–109
Holub EB, Beynon JL (1996) Symbiology of mouse ear cress (*Arabidopsis thaliana*) and oomycetes. Adv Bot Res 24:227–273
Holub EB, Brose E, Tor M, Clay C, Crute IR, Beynon JL (1994a) Phenotypic and genotypic variation in the interaction between *Arabidopsis thaliana* and *Albugo candida*. Mol Plant-Microbe Interact 8:916–928
Holub EB, Beynon JL, Crute IR (1994b) Phenotypic and genotypic characterization of interactions between isolates of *Peronospora parasitica* and accessions of *Arabidopsis thaliana*. Mol Plant-Microbe Interact 7:223–239
Holub EB, Tor M, Cooper A, Gordon P, Gunn N (2002) Genetics of disease resistance in Arabidopsis to crop pathogens. In: Proc 6th Conf EFPP 2002, Prague.Plant Prot Sci 38 (Spl issue 1): 151–154.
Huang S, Deng L, Guan M, Li J, Lu K, Wang H, Fu D, Mason AS, Liu S, Hua W (2013) Identification of genome-wide single nucleotide polymorphisms in allopolyploid crop *Brassica napus*. BMC Genomics 14:717. https://doi.org/10.1186/1471-2164-14-717
Hulbert SH, Webb CA, Smith SM, Sun Q (2001) Resistance gene complexes: evolution and utilization. Annu Rev Phytopathol 39:285–312. https://doi.org/10.1146/annurev.phyto.39.1.285
Human Genome Project (2003) Human Genome Project Information Archive 1990–2003. https://web.ornl.gov/sci/techresources/Human_Genome/redirect.shtml
Hurgobin B, Golicz AA, Bayer PE, Chan CKK, Tirnaz S, Dolatabadian A, Schiessl SV, Samans B, Montenegro JD, Parkin IAP, Pires JC, Chalhoub B, King GJ, Snowdon R, Batley J, Edwards D (2018) Homoeologous exchange is a major cause of gene presence/absence variation in the amphidiploid *Brassica napus*. Plant Biotech J 16(7):1265–1274
Inaba R, Nishio T (2002) Phylogenetic analysis of Brassicaceae based on the nucleotide sequences of the S-locus related gene, SLRI. Theor Appl Genet 105:1159–1165
Inohara N, Chamaillard M, McDonald C, Nuñez G (2005) NOD-LRR proteins: role in host-microbial interactions and inflammatory disease. Annu Rev Biochem 74:355–383
Jejelowo OA, Conn KL, Tewari JP (1991) Relationship between conidial concentration, germling growth, and phytoalexin production by *Camelina sativa* leaves inoculated with *Alternaria brassicae*. Mycol Res 95:928–934. https://doi.org/10.1016/S0953-7562(09)80089-0
Jones RAC (2012) Influence of climate change on plant disease infections and epidemics caused by viruses and bacteria. CAB Rev7: review no. 022.
Jones JD, Dangl JL (2006) The plant immune system. Nature 444:323–329
Jones DR, Ingram DS, Dixon GR (1982) Factors affecting tests for differential pathogenicity in populations of *Plasmodiophora brassicae*. Plant Pathol 31:229–238. https://doi.org/10.1111/j.1365-3059.1982.tb01273.x
Jouet A, Saunders DGO, McMullan M, Ward B, Furzer O, Jupe F, Cevik V, Hein I, Thilliez GJA, Holub E, van Oosterhout C, Jones JDG (2018) *Albugo candida* race diversity, ploidy and host-associated microbes revealed using DNA sequence capture on diseased plants in the field. New Phytol 221:1529–1543
Katche E, Quezada-Martinez D, Katche EI, Vasquez-Teuber P, Mason AS (2019) Interspecific hybridization for *Brassica* crop improvement crop breeding. Genet Genomics 1:e190007
Kato T, Hatakayema K, Fukino N, Matsumoto S (2013) Fine mapping of the clubroot resistance gene CRb and development of a useful selectable marker in *Brassica rapa*. Breed Sci 63:116–124. https://doi.org/10.1270/jsbbs.63.116
Kaul S, Koo HL, Jenkins J, Rizzo M, Rooney T, Tallon LJ, Feldblyum T, Nierman W, Benito MI, Lin X, Town CD, Venter JC, Fraser CM, Tabata S, Nakamura Y, Kaneko T, Sato S, Asamizu E, Kato T, Kotani H, Sasamoto S, Ecker JR, Theologis A, Federspiel NA, Palm CJ, Osborne BI,

Shinn P, Dewar K, Kim CJ, Buehler E, Dunn P, Chao Q, Chen H, Vysotskaia VS, Lenz CA, Hansen NF, Liu SX, Alta H, Sakano H, Lam B, Pham PK, Nguyen M, Yu G, Southwick A, Lee JM, Miranda M, Toriumi MJ, Davis RW, Conway AB, Conn L, Hootan A, Wambutt R, Murphy G, Dusterhöft A, Stiekema W, Pohl T, Entian KD, Terryn N, Volckaert G, Salanoubat M, Choisne N, Artiguenave F, Weissenbach J, Quetier F, Rieger M, Ansorge W, Unseld M, Fartmann B, Valle G, Wilson RK, Sekhon M, Pepin K, Murray J, Johnson D, Hillier L, de la Bastide M, Huang E, Spiegel L, Gnoj L, Habermann K, Dedhia N, Parnell L, Preston R, Marra M, McCombie WR, Chen E, Martienssen R, Mayer K, Lemcke K, Haas B, Haase D, Rudd S, Schoof H, Frishman D, Morgenstern B, Zaccaria P, Mewes HW, White O, Creasy TH, Bielke C (2000) Analysis of the genome sequence of the flowering plant *Arabidopsis thaliana*. Nature 408(6814):796–815

Kayum MA, Jung HJ, Park JI, Ahmed NU, Saha G, Yang TJ, NOU IS (2015a) Identification and expression analysis of WRKY family genes under biotic and abiotic stresses in *Brassica rapa*. Mol Gen Genomics 290:79–95

Kayum MA, Park JI, Ahmed NU, Jung HJ, Saha G, Kang JG, Nou IS (2015b) Characterization and stress-induced expression analysis of alfin-like transcription factors in *Brassica rapa*. Mol Gen Genomics 290:1299–1311

Kayum MA, Kim HT, Nath UK, Park JI, Kho KH, Cho YG, Nou IS (2016) Research on biotic and abiotic stress related genes exploration and prediction in *Brassica rapa* and *B. oleracea*: a review. Plant Breed Biotech 4(2):135–144 https://doi.org/10.9787/PBB.2016.4.2.135

Kemen E, Jones JD (2012) Obligate biotroph parasitism: can we link genomes to lifestyles? Trends Plant Sci 17:448–457

Kemen E, Gardiner A, Schultz-Larsen T, Kemen AC, Balmuth AL, Robert-Seilaniantz A, Bailey K, Holub E, Studholme DJ, Maclean D, Jones JD (2011) Gene gain and loss during evolution of obligate parasitism in the white rust pathogen of *Arabidopsis thaliana*. PLoS Biol 9:e1001094

Koch E, Slusarenko AJ (1990a) *Arabidopsis* is susceptible to infection by a downy mildew fungus. Plant Cell 2:437–445

Koch E, Slusarenko AJ (1990b) Fungal pathogens of *Arabidopsis thaliana* (L.) Heyhn. Bot Helvet 100:257–268

Kohmoto K, Otani H, Tsuge T (1995) *Alternaria alternata* pathogens. In: Kohmoto K, Singh US, Singh RP (eds) Turkey: Oxford Pergamon: 51–63. (Vol. 2 eukaryotes).

Kohn LM, Stasovski E, Carbone I, Royer J, Anderson JB (1991) Mycelial incompatibility and molecular markers identify genetic variability in field populations of *Sclerotinia sclerotiorum*. Phytopathology 81:480–485. https://doi.org/10.1094/Phyto-81-480

Krasileva KV, Dahlbeck D, Staskawicz BJ (2010) Activation of an *Arabidopsis* resistance protein is specified by the in planta association of its leucine-rich repeat domain with the cognate oomycete effector. Plant Cell 22:2444–2458

Krasileva KV, Zheng C, Leonelli L, Goritschnig S, Dahlbeck D, Staskawicz BJ (2011) Global analysis of *Arabidopsis*/downy mildew interactions reveals prevalence of incomplete resistance and rapid evolution of pathogen recognition. PLoS One 6:e28765

Krishnia SK, Saharan GS, Singh D (2000) Genetics of Alternaria blight resistance in inter and intraspecific crosses of *Brassica juncea* and *B. carinata*. Ann Biol 16:212–216

Kumar A, Banga SS, Meena PD, Kumar PR (2015) *Brassica* oilseeds: breeding and management. CABI Publishing, Oxfordshire (UK), p 261

Kunze G, Zipfel C, Robatzek S, Niehaus K, Boller T, Felix G (2004) The N terminus of bacterial elongation factor Tu elicits innate immunity in Arabidopsis plants. Plant Cell 16:3496–3507

Lacombe S, Rougon-Cardoso A, Sherwood E, Peeters N, Dahlbeck D, van Esse HP, Smoker M, Rallapalli G, Thomma BPHJ, Staskawicz B, Jones JDG, Zipfel C (2010) Interfamily transfer of a plant pattern-recognition receptor confers broad-spectrum bacterial resistance. Nat Biotech 28:365–U394

Lai Y, Eulgem T (2018) Transcript-level expression control of plant NLR genes. Mol Plant Pathol 19:1267–1281

Lai Y, Cuzick A, Lu XM, Wang J, Katiyar N, Tsuchiya T, Le Roch K, McDowell JM, Holub E, Eulgem T (2019) The *Arabidopsis* RRM domain protein EDM3 mediates race-specific disease resistance by controlling H3K9me2-dependent alternative polyadenylation of RPP7 immune receptor transcripts. Plant J 97:646–660

Lange RM, Gossmann M, Buttner C (2007) Yield loss in susceptible cultivars of spring rapeseed due to Fusarium wilt caused by *Fusarium oxysporum*. Commun Agril Appl Biol Sci 72 (4):723–734

Lapin D, Meyer RC, Takahashi H, Bechtold U, Van den Ackerveken G (2012) Broad-spectrum resistance of *Arabidopsis* C24 to downy mildew is mediated by different combinations of isolate-specific loci. New Phytol 196:1171–1181

Larkan NJ, Lydiate DJ, Parkin IA, Nelson MN, Epp DJ, Cowling WA, Rimmer SR, Borhan MH (2013) The *Brassica napus* blackleg resistance gene LepR3 encodes a receptor-like protein triggered by the *Leptosphaeria maculans* effector AVRLM1. New Phytol 197:595–605. https://doi.org/10.1111/nph.12043

Larkan NJ, Yu F, Lydiate DJ, Rimmer SR, Borhan MH (2016) Single R gene introgression lines for accurate dissection of the *Brassica* – *Leptosphaeria* pathosystem. Front Plant Sci 7:1771. https://doi.org/10.3389/fpls.2016.01771

Leonelli L, Pelton J, Schoeffler A, Dahlbeck D, Berger J, Wem-mer DE, Staskawicz B (2011) Structural elucidation and functional characterization of the *Hyaloperonospora arabidopsidis* effector protein ATR13. PLoS Pathog 7:e1002428

Li J, Zhao Z, Hayward A, Cheng H, Fu D (2015) Integration analysis of quantitative trait loci for resistance to *Sclerotinia sclerotiorum* in *Brassica napus*. Euphytica 205:483–489. https://doi.org/10.1007/s10681-015-1417-0.

Li LS, Luo C, Huang F, Liu ZL, An ZY, Dong L, He XH (2019) Identification and characterization of the mango eIF gene family reveals MieIF1A-a, which confers tolerance to salt stress in transgenic *Arabidopsis*. Sci Horticult 248:274–281

Lin K, Zhang N, Severing EI, Nijveen H, Cheng F, Visser RG, Wang X, de Ridder D, Bonnema G (2014) Beyond genomic variation – comparison and functional annotation of three *Brassica rapa* genomes: a turnip, a rapid cycling and a Chinese cabbage. BMC Genome 15:250

Links MG, Holub E, Jiang RH, Sharpe AG, Hegedus D, Beynon E, Sillito D, Clarke WE, Uzuhashi S, Borhan MH (2011) De novo sequence assembly of *Albugo candida* reveals a small genome relative to other biotrophic oomycetes. BMC Genomics 12:503

Lipka V, Dittgen J, Bednarek P, Bhat R, Wiermer M, Stein M, Landtag J, Brandt W, Rosahl S, Scheel D, Llorente F, Molina A, Parker J, Somerville S, Lefert PS (2005) Pre- and post-invasion defenses both contribute to nonhost resistance in *Arabidopsis*. Sci 310:1180–1183

Liu SL, Adams KL (2010) Dramatic change in function and expression pattern of a gene duplicated by polyploidy created a paternal effect gene in the Brassicaceae. Genome Biol Evol 27:2817–2828. https://doi.org/10.1093/molbev/msq169

Liu S, Liu Y, Yang X, Tong C, Edwards D, Parkin IA, Zhao M, Ma J, Yu J, Huang S, Wang X, Wang J, Lu K, Fang Z, Bancroft I, Yang TJ, Hu Q, Wang X, Yue Z, Li H, Yang L, Wu J, Zhou Q, Wang W, King GJ, Pires JC, Lu C, Wu Z, Sampath P, Wang Z, Guo H, Pan S, Yang L, Min J, Zhang D, Jin D, Li W, Belcram H, Tu J, Guan M, Qi C, Du D, Li J, Jiang L, Batley J, Sharpe AG, Park BS, Ruperao P, Cheng F, Waminal NE, Huang Y, Dong C, Wang L, Li J, Hu Z, Zhuang M, Huang Y, Huang J, Shi J, Mei D, Liu J, Lee TH, Wang J, Jin H, Li Z, Li X, Zhang J, Xiao L, Zhou Y, Liu Z, Liu X, Qin R, Tang X, Liu W, Wang Y, Zhang Y, Lee J, Kim HH, Denoeud F, Xu X, Liang X, Hua W, Wang X, Wang J, Chalhoub B, Paterson AH (2014) The *Brassica oleracea* genome reveals the asymmetrical evolution of polyploid genomes. Nat Commun 5:3930. https://doi.org/10.1038/ncomms4930

Liu Y, Xu A, Liang F, Yao X, Wang Y, Liu X, Zhang Y, Dalelhan J, Zhang B, Qin M, Huang Z, Shaolin L (2018) Screening of clubroot-resistant varieties and transfer of clubroot resistance genes to *Brassica napus* using distant hybridization. Breed Sci 68:258–267. https://doi.org/10.1270/jsbbs.17125

Lucas J, Hayter J, Crute I (1995) The downy mildews: host specificity and pathogenesis. In: Singh U, Singh R (eds) Pathogenesis and host specificity in plant diseases. Permagon, UK, pp 217–234

Lv H, Fang Z, Yang L, Zhang Y, Wang Y (2020) An update on the arsenal: mining resistance genes for disease management of *Brassica* crops in the genomic era. Hortic Res 7:34. https://doi.org/10.1038/s41438-020-0257-9

Ma C, Zhu C, Zheng M, Liu M, Zhang D, Liu B, Li Q, Si J, Ren X, Song H (2019) CRISPR/Cas9-mediated multiple gene editing in *Brassica oleracea* var. *capitata* using the endogenous tRNA-processing system. Hortic Res 6:20. https://doi.org/10.1038/s41438-018-0107-1

Makandar R, Essig JS, Schapaugh MA, Trick HN, Shah J (2006) Genetically engineered resistance to fusarium head blight in wheat by expression of *Arabidopsis* NPR1. Mol Plant-Microbe Interact 19:123–129. https://doi.org/10.1094/MPMI-19-0123

Manzanares-Dauleux MJ, Delourme R, Baron F, Thomas G (2000) Mapping of one major gene and of QTLs involved in resistance to clubroot in *Brassica napus*. Theor Appl Genet 101:885–891. https://doi.org/10.1007/s001220051557

Mason AS, Snowdon RJ (2016) Oilseed rape: learning about ancient and recent polyploid evolution from a recent crop species. Plant Biol 18:883–892. https://doi.org/10.1111/plb.12462

Mathur B, Chand L (1991) Effect of *Alternaria brassicae* toxin on photosynthesis and photorespiration in *Brassica* cultivars. Crop Res 4:146–151

Matic S, Gilardi G, Gullino ML, Garibaldi A (2018) Evidence for an expanded host range of *fusarium oxysporum* f. sp *chrysanthemi*. J Plant Pathol 100:97–104

Mayerhofer R, Wilde K, Mayerhofer M, Lydiate D, Bansal VK, Good AG, Parkin IA (2005) Complexities of chromosome landing in a highly duplicated genome: toward map-based cloning of a gene controlling blackleg resistance in *Brassica napus*. Genetics 171:1977–1988. https://doi.org/10.1534/genetics105.049098

McDowell JM (2011) Genomes of obligate plant pathogens reveal adaptations for obligate parasitism. Proc Natl Acad Sci U S A 108:8921–8922

McDowell JM (2014) *Hyaloperonospora arabidopsidis*: a model pathogen of *Arabidopsis*. In: Genomics of plant-associated fungi and oomycetes: Dicot pathogens, Dean RA, Lichens-Park A, Kole C (eds), (Springer Berlin Heidelberg), 209–234.

McDowell JM, Meyers BC (2013) A transposable element is domesticated for service in the plant immune system. Proc Natl Acad Sci U S A 110:14821–14822

McDowell JM, Simon S (2006) Recent insights into R gene evolution. Mol Plant Pathol 7:437–448

McDowell JM, Dhandaydham M, Long TA, Aarts MG, Goff S, Holub EB, Dangl JL (1998) Intragenic recombination and diversifying selection contribute to the evolution of downy mildew resistance at the RPP8 locus of *Arabidopsis*. Plant Cell 10:1861–1874

McDowell JM, Cuzick A, Can C, Beynon J, Dangl JL, Holub EB (2000) Downy mildew (*Peronospora parasitica*) resistance genes in Arabidopsis vary in functional requirements for *NDR1*, *EDS1*, *NPR1*, and salicylic acid accumulation. Plant J 22:523–530

McMullan M, Gardiner A, Bailey K, Kemen E, Ward BJ, Cevik V, Robert-Seilaniantz A, Schultz-Larsen T, Balmuth A, Holub E, van Oosterhout C, Jones JD (2015) Evidence for suppression of immunity as a driver for genomic introgressions and host range expansion in races of *Albugo candida*, a generalist parasite. elife 4

Medhi BN (1985) Inter-mating and mutagenesis for release of genetic variability in Indian mustard [*B. juncea* (L) Czern and Coss.]. Ph.D thesis.

Mendes-Pereira E, Balesdent MH, Brun H, Rouxel T (2003) Molecular phylogeny of the *Leptosphaeria maculans-L. biglobosa* species complex. Mycol Res 107:1287–1304. https://doi.org/10.1017/S0953756203008554

Metzker ML (2010) Sequencing technologies — the next generation. Nat Rev Genet 11:31–46

Meyers BC, Morgante M, Michelmore RW (2002) TIR-X and TIR-NB proteins: two new families related to disease resistance TIR-NB-LRR proteins encoded in *Arabidopsis* and other plant genomes. Plant J 32:77–92

Meyers BC, Kozik A, Griego A, Kuang H, Michelmore RW (2003) Genome-wide analysis of NB-LRR-encoding genes in *Arabidopsis*. Plant Cell 15:809–834

Mindrinos M, Katagiri F, Yu GL, Ausubel FM (1994) The *A. thaliana* disease resistance gene RPS2 encodes a protein containing a nucleotide-binding site and leucine-rich repeats. Cell 78:1089–1099

Miya A, Albert P, Shinya T, Desaki Y, Ichimura K, Shirasu K, Narusaka Y, Kawakami N, Kaku H, Shibuya N (2007) CERK1, a LysM receptor kinase, is essential for chitin elicitor signaling in *Arabidopsis*. Proc Natl Acad Sci U S A 104:19613–19618

Mogg K, Philippot P, Bradley BP (2004) Selective attention to angry faces in clinical social phobia. J Abnormal Psychol 113:160–165

Mohr TJ, Mammarella ND, Hoff T, Woffenden BJ, Jelesko JG, McDowell JM (2010) The *Arabidopsis* downy mildew resistance gene RPP8 is induced by pathogens and salicylic acid and is regulated by W box cis elements. Mol Plant-Microbe Interact 23:1303–1315

Mou Z, Fan W, Dong X (2003) Inducers of plant systemic acquired resistance regulate NPR function through redox changes. Cell 113:935–944. https://doi.org/10.1016/S0092-8674(03)00429-X

Mukhtar MS, Carvunis AR, Dreze M, Epple P, Steinbrenner J, Moore J, Tasan M, Galli M, Hao T, Nishimura MT, Pevzner SJ, Donovan SE, Ghamsari L, Santhanam B, Romero V, Poulin MM, Gebreab F, Gutierrez BJ, Tam S, Monachello D, Boxem M, Harbort CJ, McDonald N, Gai L, Chen H, He Y, Vandenhaute J, Roth FP, Hill DE, Ecker JR, Vidal M, Beynon J, Braun P, Dangl JL (2011) Independently evolved virulence effectors converge onto hubs in a plant immune system network. Sci 333:596–601

Naito K, Taguchi F, Suzuki T, Inagaki Y, Toyoda K, Shiraishi T, Ichinose Y (2008) Amino acid sequence of bacterial microbe-associated molecular pattern flg22 is required for virulence. Mol Plant-Microbe Inter 21:1165–1174

Neik TX, Barbetti MJ, Batley J (2017) Current status and challenges in identifying disease resistance genes in *Brassica napus*. Front Plant Sci 8:1788. https://doi.org/10.3389/fpls.2017.01788

Nemri A, Atwell S, Tarone AM, Huang YS, Zhao K, Studholme DJ, Nordborg M, Jones JD (2010) Genome-wide survey of *Arabidopsis* natural variation in downy mildew resistance using combined association and linkage mapping. Proc Natl Acad Sci U S A 107:10302–10307

Niks RE, Marcel TC (2009) Nonhost and basal resistance: how to explain specificity? New Phytol 182:817–828. https://doi.org/10.1111/j.1469-8137.2009.02849.x

Nishimura S, Kohmoto K (1983) Host-specific toxins and chemical structures from *Alternaria* species. Annu Rev Phytopathol 21:87–116. https://doi.org/10.1146/annurev.py.21.090183.000511

Noel L, Moores TL, van Der Biezen EA, Parniske M, Daniels MJ, Parker JE, Jones JD (1999) Pronounced intraspecific haplotype divergence at the RPP5 complex disease resistance locus of *Arabidopsis*. Plant Cell 11:2099–2112

Novakazi F, Inderbitzin P, Sandoya G, Hayes RJ, Tiedemann AV, Subbarao KV (2015) The three lineages of the diploid hybrid *Verticillium longisporum* differ in virulence and pathogenicity. Phytopathology 105:662–673

Nurnberger T, Kemmerling B (2009) PAMP-triggered basal immunity in plants. Adv Bot Res 51:1–38

O'Brien SJ, Stanyon R (1999) Phylogenomics: ancestral primate viewed. Nat 402:365–366

Ochieng JW, Muigai WT, Ude GN (2007) Phylogenetics in plant biotechnology: principles, obstacles and opportunities for the resource poor. Afr J Biotechnol 6:639–649

Oliver SG, Winson MK, Kell DB, Baganz F (1998) Systematic functional analysis of the yeast genome. Trends Biotechnol 16:373–378

Osborn TC, Butrulle DV, Sharpe AG, Pickering KJ, Parkin IAP, Parker JS (2003) Detection and effects of a homeologous reciprocal transposition in *Brassica napus*. Genetics 165:1569–1577

Pang W, Liang S, Li X, Li P, Yu S, Lim YP, Piao Z (2014) Genetic detection of clubroot resistance loci in a new population of *Brassica rapa*. Hortic Environ Biotechnol 55:540–547. https://doi.org/10.1007/s13580-014-0079-5

Parker JE, Holub EB, Frost LN, Falk A, Gunn ND, Daniels MJ (1996) Characterization of eds1, a mutation in Arabidopsis suppressing resistance to *Peronospora parasitica* specified by several different RPP genes. Plant Cell 8:2033–2046

Parker JE, Coleman MJ, Szabo V, Frost LN, Schmidt R, van der Biezen E, Moores T, Dean C, Daniels MJ, Jones JDG (1997) The *Arabidopsis* downy mildew resistance gene Rpp5 shares similarity to the toll and interleukin-1 receptors with N and L6. Plant Cell 9:879–894

Parkhi V, Kumar V, Campbell LM, Bell AA, Shah J, Rathore KS (2010) Resistance against various fungal pathogens and reniform nematode in transgenic cotton plants expressing *Arabidopsis* NPR1. Transgenic Res 19:959–975. https://doi.org/10.1007/s11248-010-9374-9

Patel D, Zander M, Van De Wouw A, Mason A, Edwards D, Batley J (2015) Population diversity of *Leptosphaeria maculans* in Australia. Intl J Biol 7:18–36. https://doi.org/10.5539/ijb.v7n3p18

Pedras MSC, Smith KC (1997) Sinalexin, a phytoalexin from white mustard elicited by destruxin B and *Alternaria brassicae*. Phytochemistry 46:833–837. https://doi.org/10.1016/S0031-9422(97)00362-2

Pedras MSC, Zaharia IL, Gai Y, Zhou Y, Ward DE (2001) In planta sequential hydroxylation and glycosylation of a fungal phytotoxin: avoiding cell death and overcoming the fungal invader. Proc Natl Acad Sci U S A 98:747–752. https://doi.org/10.1073/pnas.98.2.747

Piao Z, Ramchiary N, Lim YP (2009) Genetics of clubroot resistance in *Brassica* species. J Plant Growth Regul 28(3):252–264

Pieterse CM, Van der Does D, Zamioudis C, Leon-Reyes A, Van Wees SC (2012) Hormonal modulation of plant immunity. Annu Rev Cell Dev Biol 28:489–521. https://doi.org/10.1146/annurev-cellbio-092910-154055

Pires JC, Gaeta RT (2011) Structural and functional evolution of resynthesized polyploids. In: Schmidt R, Bancroft I (eds) Genetics and genomics of the Brassicaceae, Springer, New York, pp 195–214

Plieske J, Struss D (2001) Microsatellite markers for genome analysis in *Brassica*. I. Development in *Brassica napus* and abundance in *Brassicaceae* species. Theor Appl Genet 102:689–694. https://doi.org/10.1007/s001220051698

Prakash SR, Bhat SR, Quiros CF, Kirti PB, Chopra VL (2009) *Brassica* and its close allies: cytogenetics and evolution. Plant Breed Rev 31:21–187

Prakash S, Wu XM, Bhat SR (2011) History, evolution, and domestication of *Brassica* crops. In: Janick J (ed) Plant breeding reviews, vol 35. John Wiley and Sons, Inc., New Jersey, pp 19–84

Prince DC, Rallapalli G, Xu D, Schoonbeek HJ, Cevik V, Asai S, Kemen E, Cruz-Mireles N, Kemen A, Belhaj K, Schornack S, Kamoun S, Holub EB, Halkier BA, Jones JD (2017) Albugo-imposed changes to tryptophan-derived antimicrobial metabolite biosynthesis may contribute to suppression of non-host resistance to *Phytophthora infestans* in *Arabidopsis thaliana*. BMC Biol 15:20

Qian L, Qian W, Snowdon RJ (2014) Sub-genomic selection patterns as a signature of breeding in the allopolyploid *Brassica napus* genome. BMC Genomics 15:1170. https://doi.org/10.1186/1471-2164-15-1170

Quilis J, Penas G, Messeguer J, Brugidou C, Segundo BS (2008) The *Arabidopsis* AtNPR1 inversely modulates defense responses against fungal, bacterial, or viral pathogens while conferring hypersensitivity to abiotic stresses in transgenic rice. Mol Plant-Microbe Interact 21:1215–1231. https://doi.org/10.1094/MPMI-21-9-1215

Rahman H, Harwood J, Weselake R (2013) Increasing seed oil content in *Brassica* species through breeding and biotechnology. Lipid Technol 25(8):182–185

Rehmany AP, Gordon A, Rose LE, Allen RL, Armstrong MR, Whisson SC, Kamoun S, Tyler BM, Birch PR, Beynon JL (2005) Differential recognition of highly divergent downy mildew avirulence gene alleles by RPP1 resistance genes from two *Arabidopsis* lines. Plant Cell 17:1839–1850

Rieger MA, Potter TD, Preston C, Powles SB (2001) Hybridization between *Brassica napus* L. and *Raphanus raphanistrum* L. under agronomic field conditions. Theor Appl Genet 103 (4):555–560

Rimmer SR, Buchwaldt L (1995) Diseases. In: Kimber DS, Mcgregor DI (eds) *Brassica* oilseeds. CAB International, Wallingford, pp 65–92

Rimmer S, Scarth R, McVetty P, Woods D, Gugel R, Séguin-Swartz G (1995) Breeding for resistance to blackleg (stem canker) in Western Canada. Blackleg Newsl 5:11–15

Rocherieux J, Glory P, Giboulot A, Boury S, Barbeyron G, Thomas G, Manzanares-Dauleux MJ (2004) Isolate-specific and broad-spectrum QTLs are involved in the control of clubroot in *Brassica oleracea*. Theor Appl Genet 108:1555–1563. https://doi.org/10.1007/s00122-003-1580-x

Rose LE, Bittner-Eddy PD, Langley CH, Holub EB, Michelmore RW, Beynon JL (2004) The maintenance of extreme amino acid diversity at the disease resistance gene, RPP13, in *Arabidopsis thaliana*. Genetics 166:1517–1527

Rousseau-Gueutin M, Morice J, Coriton O, Huteau V, Trotoux G, Negre S, Falentin C, Deniot G, Gilet M, Eber F, Pelé A, Vautrin S, Fourment J, Lodé M, Bergès H, Chèvre A-M (2017) The impact of open pollination on the structural evolutionary dynamics, meiotic behavior, and fertility of resynthesized allotetraploid *Brassica napus* L. G3 (Bethesda) 7:705–717

Roux F, Voisin D, Badet T, Balague C, Barlet X, Huard-Chauveau C, Roby D, Raffaele S (2014) Resistance to phytopathogens e tutti quanti: placing plant quantitative disease resistance on the map. Mol Plant Pathol 15(5):427–432. https://doi.org/10.1111/mpp.12138

Rouxel T, Willner E, Coudard L, Balesdent MH (2003) Screening and identification of resistance to *Leptosphaeria maculans* (stem canker) in *Brassica napus* accessions. Euphytica 133:219–231. https://doi.org/10.1023/A:1025597622490

Roy NN (1984) Interspecific transfer of *Brassica juncea* – type high blackleg resistance to *Brassicanapus*. Euphytica 33:295–303

Ruhe J, Agler MT, Placzek A, Kramer K, Finkemeier I, Kemen EM (2016) Obligate biotroph pathogens of the genus *Albugo* are better adapted to active host defense compared to niche competitors. Front Plant Sci 7:820

Saharan GS (1992) Management of rapeseed and mustard diseases. In: Kumar D, Rai M (eds) Advances in oil seed research. Sci. Pub., Jodhpur.

Saharan GS, Mehta Naresh (2008) *Sclerotinia* diseases of crop plants: biology, ecology and disease management. Springer Science+Busines Media B.V. The Netherlands, 485 p.

Saharan GS, Verma PR, Meena PD, Kumar A (2014) White rust of crucifers: biology, ecology and disease management. Springer Verlag, Frankfurt, Germany, p 244. isbn:978-81-322-1791-6

Saharan GS, Naresh M, Meena PD (2016) Alternaria blight of crucifers: biology, ecology and disease management. Springer Verlag, Singapore, p 326. isbn:978-981-10-0019-5

Saharan GS, Naresh Mehta, Meena PD (2017) Downy mildew disease of crucifers: biology, ecology and disease management. Springer Nature, Singapore, LVI, p 357, ISBN 978-981-10-7499-8.

Saharan GS, Naresh M, Meena PD (2019) Powdery mildew disease of crucifers: biology, ecology and disease management. Springer Verlag, Singapore, p 362. isbn:978-981-13-9852-0

Saharan GS, Naresh M, Meena PD (2021) Clubroot disease of crucifers: biology, ecology and disease management. Springer Verlag, Singapore. (In press)

Salisbury PA, Ballinger JD, Wratten N, Plummer MK, Howlett JB (1995) Blackleg disease on oilseed *Brassicas* in Australia – a review. Austr J Exp Agric 3(5):665–674

Schiessl S, Huettel B, Kuehn D, Reinhardt R, Snowdon R (2017) Post-polyploidisation morphotype diversification associates with gene copy number variation. Sci Rep 7:41845

Schmutzer T, Samans B, Dyrszka E, Ulpinnis C, Weise S, Stengel D, Colmsee C, Lespinasse D, Micic Z, Abel S, Duchscherer P, Breuer F, Abbadi A, Leckband G, Snowdon R, Scholz U (2015a) Species-wide genome sequence and nucleotide polymorphisms from the model allo-polyploid plant *Brassica napus*. Sci Data 2:150072

Schmutzer T, Samans B, Dyrszka E, Ulpinnis C, Weise S, Stengel D, Colmsee C, Lespinasse D, Micic Z, Abel S, Duchscherer P, Breuer F, Abbadi A, Leckband G, Snowdon R, Scholz U (2015b) Species-wide genome sequence and nucleotide polymorphisms from the model allopolyploid plant *Brassica napus*. Sci Data 2:150072. https://doi.org/10.1038/sdata.2015.72

Schwinghamer MW, Schilg MA, Walsh JA, Bambach RW, Cossu RM, Bambridge JM, Hind-Lanoiselet TL, McCorkell BE, Cross P (2014) Turnip mosaic virus: potential for crop losses in the grain belt of New South Wales, Australia. Australasian Plant Pathol 43:663–678. https://doi.org/10.1007/s13313-014-0304-9

Sexton AC, Whitten AR, Howlett BJ (2006) Population structure of *Sclerotinia sclerotiorum* in an Australian canola field at flowering and stem-infection stages of the disease cycle. Genome 49:1408–1415. https://doi.org/10.1139/g06-101

Shah J, Zeier J (2013) Long–distance communication and signal amplification in systemic acquired resistance. Front Plant Sci 4:30. https://doi.org/10.3389/fpls.2013.00030

Sharpe AG, Lydiate DJ (2003) Mapping the mosaic of ancestral genotypes in a cultivar of oilseed rape (*Brassica napus*) selected via pedigree breeding. Genome 46:461–468

Shiu SH, Karlowski WM, Pan RS, Tzeng YH, Mayer KFX, Li WH (2004) Comparative analysis of the receptor-like kinase family in *Arabidopsis* and rice. Plant Cell 16:1220–1234

Sinapidou E, Williams K, Nott L, Bahkt S, Tor M, Crute I, Bittner-Eddy P, Beynon J (2004) Two TIR:NB:LRR genes are required to specify resistance to *Peronospora parasitica* isolate Cala2 in Arabidopsis. Plant J 38:898–909

Singh DN, Singh NK, Srivastava S (1999) Biochemical and morphological characters in relation to Alternaria blight resistance in rape-seed mustard. Ann Agric Res 20:472–477

Singh V, Kumar K, Bhajan R, Singh PK, Singh RB (2009) Identification of resistance sources against Alternaria blight and white rust in Indian mustard. J Oilseeds Res 26:435–436

Singh S, Dey SS, Bhatia R, Batley J, Kumar R (2018) Molecular breeding for resistance to black rot [Xanthomonas campestris pv. campestris (Pammel) Dowson] in Brassicas: recent advances. Euphytica 214:196. https://doi.org/10.1007/s10681-018-2275-3

Slusarenko AJ, Schlaich NL (2003) Downy mildew of *Arabidopsis thaliana* caused by *Hyaloperonospora parasitica* (formerly *Peronospora parasitica*). Mol Plant Pathol 4:159–170

Snowdon RJ, Friedt W (2004) Molecular markers in *Brassica* oilseed breeding: current status and future possibilities. Plant Breed 123:1–8

Snowdon RJ, Luhs W, Friedt W (2007) Oilseed rape. In: Kole C (ed) Genome mapping and molecular breeding in plants, 2: Oilseeds. Springer, Berlin Heidelberg, pp 55–114

Snowdon RJ, Abbadi A, Kox T, Schmutzer T, Leckband G (2015) Heterotic haplotype capture: precision breeding for hybrid performance. Trends Plant Sci 20:410–413

Spoel SH, Koornneef A, Claessens SM, Korzelius JP, Van-Pelt JA, Mueller MJ, Buchala AJ, Métraux JP, Brown R, Kazan K, Van Loon LC, Dong X, Pieterse CMJ (2003) NPR1 modulates cross–talk between salicylate–and jasmonate–dependent defense pathways through a novel function in the cytosol. Plant Cell 15:760–770. https://doi.org/10.1105/tpc.009159

Stein A, Coriton O, Rousseau Gueutin M, Samans B, Schiessl SV, Obermeier C, Parkin IAP, Chèvre AM, Snowdon RJ (2017) Mapping of homoeologous chromosome exchanges influencing quantitative trait variation in *Brassica napus*. Plant Biotechnol J 15:1478–1489

Steinbrenner AD, Goritschnig S, Staskawicz BJ (2015) Recognition and activation domains contribute to allele-specific responses of an *Arabidopsis* NLR receptor to an oomycete effector protein. PLoS Pathog 11:e1004665

Subramaniam S, Wang X, Freeling M, Pires JC (2013) The fate of *Arabidopsis thaliana* homeologous CNSs and their motifs in the paleohexaploid *Brassica rapa*. Genome Biol Evol 5:646–660. https://doi.org/10.1093/gbe/evt035

Subudhi PK, Raut RN (1994) White rust resistance and its association with parental species type and leaf waxiness in *B. juncea* L. Czern and Coss. X *B. napus* L. Crosses under the action of EDTA and Gamma ray. Euphytica 74:1–7. https://doi.org/10.1007/BF00033760

Sundaresha S, Rohini S, Appanna VK, Arthikala MK, Shanmugam NB, Shashibhushan NB, Hari Kishore CM, Pannerselvam R, Kirti PB, Udayakumar M (2016) Co-overexpression of *Brassica juncea* NPR1 (BjNPR1) and *Trigonella foenum*-graecum defensin (Tfgd) in transgenic peanut provides comprehensive but varied protection against *Aspergillus flavus* and *Cercospora arachidicola*. Plant Cell Rep 35:1189–1203. https://doi.org/10.1007/s00299-016-1945-7

Tada Y, Spoel SH, Pajerowska-Mukhtar K, Mou Z, Song J, Wang C, Zou J, Dong X (2008) Plant immunity requires conformational charges of NPR1 via S-nitrosylation and thioredoxins. Science 321:952–956. https://doi.org/10.1126/science.1156970

Tanaka S, Ito SI (2013) Pathogenic and genetic diversity in *Plasmodiophora brassicae* (clubroot) from Japan. J Gen Plant Pathol 79:297–306. https://doi.org/10.1007/s10327-013-0456-4

Tettelin H, Masignani V, Cieslewicz MJ, Donati C, Medini D, Ward NL, Angiuoli SV, Crabtree J, Jones AL, Durkin AS, Deboy RT, Davidsen TM, Mora M, Scarselli M, Ros IM, Peterson JD, Hauser CR, Sundaram JP, Nelson WC, Madupu R, Brinkac LM, Dodson RJ, Rosovitz MJ, Sullivan SA, Daugherty SC, Haft DH, Selengut J, Gwinn ML, Zhou L, Zafar N, Khouri H, Radune D, Dimitrov G, Watkins K, O'Connor KJB, Smith S, Utterback TR, White O, Rubens CE, Grandi G, Madoff LC, Kasper DL, Telford JL, Wessels MR, Rappuoli R, Fraser CM (2005) Genome analysis of multiple pathogenic isolates of *Streptococcus agalactiae*: implications for the microbial “pan-genome”. Proc Natl Acad Sci U S A 102:13950–13955

Tewari JP, Conn KL, Dahiya JS (1987) Resistance to *Alternaria brassicae* in crucifer. Proc. 7th IRC. Poznan 48:1085–1089

Tewari JP, Conn KL, Dahiya JS (1988) Dynamic response of rapeseed and other crucifers to *Alternaria brassicae* and *Rhizoctonia solani*. Proc. 5th Intl Cong plant pathol, Koyoto, p 242

Thaler JS, Humphrey PT, Whiteman NK (2012) Evolution of jasmonate and salicylate signal crosstalk. Trends Plant Sci 17:260–270. https://doi.org/10.1016/j.tplants.2012.02.010

Thines M (2014) Phylogeny and evolution of plant pathogenic oomycetes, a global overview. Eur J Plant Pathol:1–17

Thines M, Choi YJ, Kemen E, Ploch S, Holub EB, Shin HD, Jones JD (2009) A new species of *Albugo* parasitic to *Arabidopsis thaliana* reveals new evolutionary patterns in white blister rusts (Albuginaceae). Persoonia 22:123–128

Thomas PE, Evans DW, Fox L, Biever KD (1990) Resistance to beet western yellows virus among forage Brassicas. Plant Dis 74:327–330

Thomma BP, Nurnberger T, Joosten MH (2011) Of PAMPs and effectors: the blurred PTI-ETI dichotomy. Plant Cell 23(1):4–15

Tomlinson JA (1970) Turnip mosaic virus. Wallingford, UK: commonwealth mycological institute/ Association of Applied Biologists (CMI/AAB): descriptions of plant viruses, CMI/AAB publication no 8.

Toruno TY, Stergiopoulos I, Coaker G (2016) Plant-pathogen effectors: cellular probes interfering with plant defenses in spatial and temporal manners. Annu Rev Phytopathol54, 419-441.

Town CD, Cheung F, Maiti R, Crabtree J, Haas BJ, Wortman JR, Hine EE, Althoff R, Arbogast TS, Tallon LJ, Vigouroux M, Trick M, Bancroft I (2006) Comparative genomics of *Brassica oleracea* and *Arabidopsis thaliana* reveal gene loss, fragmentation, and dispersal after polyploidy. Plant Cell 18:1348–1359. https://doi.org/10.1105/tpc.106.041665

Travadon R, Sache I, Dutech C, Stachowiak A, Marquer B, Bousset L (2011) Absence of isolation by distance patterns at the regional scale in the fungal plant pathogen *Leptosphaeria maculans*. Fungal Biol 115:649–659. https://doi.org/10.1016/j.funbio.2011.03.009

Tripathi NN, Kaushik CD, Yadav TP, Yadav AK (1978) Studies on the inheritance of Alternaria blight resistance in raya. Indian Phytopathology 31:127

Tripathi NN, Kaushik CD, Yadav TP, Yadav AK (1980) Alternaria leaf spot resistance in raya. Haryana Agric Univ J Res 10:166–168

Tsuchiya T, Eulgem T (2013) An alternative polyadenylation mechanism coopted to the *Arabidopsis* RPP7 gene through intronic ret-rotransposon domestication. Proc Natl Acad Sci U S A 110:E3535–E3543

Tyvaert L, França SC, Debode J, Hofte M (2014) The endophyte *Verticillium* Vt305 protects cauliflower against *Verticillium* wilt. J Appl Microbiol 116:1563–1571

Ueno H, Matsumoto E, Aruga D, Kitagawa S, Matsumura H, Hayashida N (2012) Molecular characterization of the *CRa* gene conferring clubroot resistance in *Brassica rapa*. 80:621–629. https://doi.org/10.1007/s11103-012-9971-5

Van Damme M, Andel A, Huibers RP, Panstruga R, Weisbeek PJ, Van den Ackerveken G (2005) Identification of *Arabidopsis* loci required for susceptibility to the downy mildew pathogen *Hyaloperonospora parasitica*. Mol Plant-Microbe Interact 18:583–592

van Damme M, Huibers RP, Elberse J, Van den Ackerveken G (2008) *Arabidopsis* DMR6 encodes a putative 2OG-Fe(II) oxygenase that is defense-associated but required for susceptibility to downy mildew. Plant J 54:785–793

van der Biezen EA, Freddie CT, Kahn K, Parker JE, Jones JD (2002) *Arabidopsis* RPP4 is a member of the RPP5 multigene family of TIR-NB-LRR genes and confers downy mildew resistance through multiple signaling components. Plant J 29:439–451

Van der Does D, Leon-Reyes A, Koornneef A, Van Verk MC, Rodenburg N, Pauwels L, Goossens A, Korbes AP, Memelink J, Ritsema T, Van Wees SCM, Pieterse CMJ (2013) Salicylic acid suppresses jasmonic acid signaling downstream of SCFCOI1-JAZ by targeting GCC promoter motifs via transcription factor ORA59. Plant Cell 25:744–761. https://doi.org/10.1105/tpc.112.108548

Verma PR (1996) Biology and control of *Rhizoctonia solani* on rapeseed: a review. Phytoprotection 77(3):99–111

Voglmayr H, Riethmuller A, Goker M, Weiß M, Oberwinkler F (2004) Phylogenetic relationships of *Plasmopara*, Bremia and other genera of downy mildews with pyriform haustoria based on Bayesian analysis of partial LSU rDNA sequence data. Mycol Res 108:1011–1024

Voigt K, Cozijnsen AJ, Kroymann J, Pöggeler S, Howlett BJ (2005) Phylogenetic relationships between members of the crucifer pathogenic *Leptosphaeria maculans* species complex as shown by mating type (MAT1-2), actin, and β-tubulin sequences. Mol Phylogenet Evol 37:541–557. https://doi.org/10.1016/j.ympev.2005.07.006

Wally O, Jayaraj J, Punja ZK (2009) Comparative resistance to foliar fungal pathogens in transgenic carrot plants expressing genes encoding for chitinase, beta–1, 3–glucanase and peroxidase. Eur J Plant Pathol 123:331–342. https://doi.org/10.1007/s10658-008-9370-6

Walsh JA, Jenner CE (2002) Turnip mosaic virus and the quest for durable resistance. Mol Plant Pathol 3:289–300

Wan JR, Tanaka K, Zhang XC, Son GH, Brechenmacher L, Tran HNN, Stacey G (2012) LYK4, a lysin motif receptor-like kinase, is important for chitin signaling and plant innate immunity in *Arabidopsis*. Plant Physiol 160:396–406

Wang Z, Bao LL, Zhao FY, Tang MQ, Chen T, Li Y, Wang BX, Fu B, Fang H, Li GY, Cao J, Ding LN, Zhu KM, Liu SY, Tan XL (2019) BnaMPK3 is a key regulator of defense responses to the devastating plant pathogen *Sclerotinia sclerotiorum* in oilseed rape. Front Plant Sci 10:91

Wang Y, Strelkov SE, Hwang SF (2020) Yield losses in canola in response to blackleg disease. Can J Plant Sci 100(5). https://doi.org/10.1139/CJPS-2019-0259

Watson RE, Goodman JI (2002) Effects of phenobarbital on DNA methylation in GC-rich regions of hepatic DNA from mice that exhibit different levels of susceptibility to liver tumorigenesis. Toxicol Sci 68:51–58

Wessling R, Epple P, Altmann S, He Y, Yang L, Henz SR, McDonald N, Wiley K, Bader KC, Glasser C, Mukhtar MS, Haigis S, Ghamsari L, Stephens AE, Ecker JR, Vidal M, Jones JD, Mayer KF, Ver Loren van Themaat E, Weigel D, SchulzeLefert P, Dangl JL, Panstruga R, Braun P (2014) Convergent targeting of a common host protein-network by pathogen effectors from three kingdoms of life. Cell Host Microbe 16:364–375

Willmann R, Lajunen HM, Erbs G, Newman MA, Kolb D, Tsuda K, Katagiri F, Fliegmann J, Bono JJ, Cullimore JV, Jehle AK, Gotz F, Kulik A, Molinaro A, Lipka V, Gust AA, Nurnberger T (2011) Arabidopsis lysin-motif proteins LYM1 LYM3 CERK1 mediate bacterial peptidoglycan sensing and immunity to bacterial infection. Proc Natl Acad Sci U S A 108:19824–19829

Winter H, Snowdon RJ, Diestel A, Gartig S, Sacristan MD (2003) Transfer of new blackleg resistances into oilseed rape. In: Proc 11th Int rapeseed congress, Copenhagen, 19–21

Woods-Tor A, Studholme DJ, Cevik V, Telli O, Holub EB, Tor M (2018) A suppressor/avirulence gene combination in *Hyaloperonospora arabidopsidis* determines race specificity in *Arabidopsis thaliana*. Front Plant Sci 9:265

Wu Y, Zhang D, Chu JY, Boyle P, Wang Y, Brindle ID, Luca VD, Després C (2012) The *Arabidopsis* NPR1 protein is a receptor for the plant defense hormone salicylic acid. Cell Rep 1:639–647. https://doi.org/10.1016/j.celrep.2012.05.008

Wu J, Li F, Xu K, Gao G, Chen B, Yan G, Wang N, Qiao J, Li J, Li H, Zhang T, Song W, Wu X (2014) Assessing and broadening genetic diversity of a rapeseed germplasm collection. Breed Sci 64(4):321–330

Wu C-H, Derevnina L, Kamoun S (2018) Receptor networks underpin plant immunity. Science 360:1300

Wu Z, Li M, Dong OX, Xia S, Liang W, Bao Y, Wasteneys G, Li X (2019) Differential regulation of TNL-mediated immune signaling by redundant helper CNLs. New Phytol 222:938–953

Yang J, Liu D, Wang X, Ji C, Cheng F, Liu B, Hu Z, Chen S, Pental D, Ju Y, Yao P, Li X, Xie K, Zhang J, Wang J, Liu F, Ma W, Shopan J, Zheng H, Mackenzie SA, Zhang M (2016) The genome sequence of allopolyploid *Brassica juncea* and analysis of differential homoeology gene expression influencing selection. Nat Genet 48:1225–1232. https://doi.org/10.1038/ng.3657

Yu J, Tehrim S, Zhang F, Tong C, Huang J, Cheng X, Dong C, Zhou Y, Qin R, Hua W, Liu S (2014a) Genome-wide comparative analysis of NBS-encoding genes between *Brassica* species and *Arabidopsis thaliana*. Plant Pathol 56:317–323

Yu JY, Tehrim S, Zhang FQ, Tong CB, Huang JY, Cheng XH, Dong C, Zhou Y, Qin R, Hua W, Liu S (2014b) Genome-wide comparative analysis of NBS-encoding genes between *brassica* species and *Arabidopsis thaliana*. BMC Genomics 15:3. https://doi.org/10.1186/1471-2164-15-3

Zander M, Patel DA, Van De Wouw A, Lai K, Lorenc MT, Campbell E, Hayward A, Edwards D, Raman H, Batley J (2013) Identifying genetic diversity of avirulence genes in *Leptosphaeria maculans* using whole genome sequencing. Funct Integr Genomics 13:295–308. https://doi.org/10.1007/s10142-013-0324-5

Zhang FL, Xu JB, Takahata Y (1996) Inheritance of resistance to black leaf spot (*Alternaria brassicae*) in Chinese cabbage. Cruciferae Newslett 18:134–135

Zhang F, Xu J, Yan H, Li M (1997) Inheritance of resistance to black leaf spot in Chinese cabbage. Acta Agric Boreali-Sin 12:115–119

Zhang Y, Tessaro MJ, Lassner M, Li X (2003) Knockout analysis of *Arabidopsis* transcription factors TGA2, TGA5, and TGA6 reveals their redundant and essential roles in systemic acquired resistance. Plant Cell 15:2647–2653. https://doi.org/10.1105/tpc.014894

Zhang W, Fraiture M, Kolb D, Loffelhardt B, Desaki Y, Boutrot FF, Tor M, Zipfel C, Gust AA, Brunner F (2013a) *Arabidopsis* receptor-like protein30 and receptor-like kinase suppressor of BIR1-1/EVERSHED mediate innate immunity to necrotrophic fungi. Plant Cell 25:4227–4241

Zhang X, Zhou Y, Ding L, Wu Z, Liu R, Meyerowitz EM (2013b) Transcription repressor HANABA TARANU controls flowerdevelopment by integrating the actions of multiplehormones, floral organ specification genes, and GATA3family genes in *Arabidopsis*. Plant Cell 25:83–101

Zhao J, Udall JA, Quijada PA, Grau CR, Meng J, Osborn TC (2006) Quantitative trait loci for resistance to *Sclerotinia sclerotiorum* and its association with a homeologous non-reciprocal transposition in *Brassica napus* L. Theor Appl Genet 112:509–516. https://doi.org/10.1007/s00122-005-0154-5

Zhong X, Xi L, Lian Q, Luo X, Wu Z, Seng S, Yuan X, Yi M (2015) The NPR1 homolog GhNPR1 plays an important role in the defense response of *Gladiolus hybridus*. Plant Cell Rep 34:1063–1074. https://doi.org/10.1007/s00299-015-1765-1

Zou J, Fu D, Gong H, Qian W, Xia W, Pires JC, Li R, Long Y, Mason AS, Yang T-J, Lim YP, Park BS, Meng J (2011) *De novo* genetic variation associated with retrotransposon activation, genomic rearrangements and trait variation in a recombinant inbred line population of *Brassica napus* derived from interspecific hybridization with *Brassica rapa*. Plant J 68:212–224

Identification of *R*-Genes Sources

2

Abstract

Identification of effective and durable sources of R-genes with amicable introgression in the crucifers' genotypes using conventional as well as molecular/biotechnological approaches is essential to breed disease-resistant cultivars of *Brassica* to ameliorate production and productivity at global level. R-genes identification is the first and foremost step towards the disease-resistant hybrid/varieties development through appropriate breeding programs. R-genes conferring qualitative and quantitative (QTL's) loci have been identified from *Brassica* species, cruciferous wild relatives, land races, and *Brassica* coenospecies against major pathogens of crucifers. Several sources in *Brassica* species with R-genes and QTL's for inter- and intraspecific hybridization disease resistance cultivars development with multiple pathogens protection like *Albugo*, *Alternaria*, *Erysiphe* and *Hyaloperonospora* have been identified. R-genes governing seedling and adult plant resistance have been determined. Some R-genes show differential interaction on *Brassica* genotypes at different locations because of environmental and pathogenic variations. Multi R-genes genotypes have been identified from *Brassica* species against *Albugo candida*. R-genes in *Brassica* species to *Colletotrichum higginsianum* have been characterized using molecular and biochemical approaches. Several genotypes of *Brassica* species have slow mildewing attributes to provide protection against *Erysiphe cruciferarum*. In *B. oleracea*, a single R-genes and polygenic resistance to Fusarium wilt has been characterized. Several R-genes sources in *Arabidopsis thaliana* have been characterized as RPP genes against *H. arabidopsidis*.

In *Brassica* species, several *Leptosphaeria* qualitative race specific R-genes and quantitative (QTL's) have been identified. However, it is uncertain if some of these genes are same genes with different nomenclature or allelic variants of the same gene, since researchers have used different crosses, different isolates, and different markers system. R-genes effective against clubroot (*Plasmodiophora*

G. S. Saharan et al., *Genomics of Crucifer's Host-Resistance*,
https://doi.org/10.1007/978-981-16-0862-9_2

brassicae) have been identified from cruciferous specific as pathotypes specific as well as QTL's. GBS identified QTL's provide resistance to multiple pathotypes. Multiple wild *Brassica* species have shown high levels of resistance against stem rot pathogen (*Sclerotinia*). Isolate specific resistant sources to TuMV have been identified from *Brassica* species. Race specific sources of resistance to *Xanthomonas* have been identified from cultivated *Brassica* species and *A. thaliana*. Black rot (*Xanthomonas*) R-genes have been identified in cauliflower through molecular approaches. In general, the *Brassica* A genome is rich source for resistance to Turnip mosaic virus, Black rot, downy mildew, and clubroot, whereas B. genome possesses resistance against Black rot and Blackleg, and C genome harbors stem rot, Fusarium wilt, and Downy mildew resistance. Well-characterized R-genes/QTL's in *Brassica* crops have been identified and genetically mapped against clubroot, Sclerotinia stem rot, downy mildew, and *Leptosphaeria*. However, only limited R-genes/alleles have been cloned from *Brassica* species. Eight different bona fide R-genes like (RGL) PCR fragments have been obtained from *A. thaliana* accessions *Col* and L*er*. The application of omics technology in R-genes identification has widen the scope of *Brassica* species genome sequencing, use of pangenomics, use of NGS-based SNP methods, *In Silico* method, NGS-Based Bulked Segregant analysis (BSA), resistance gene enrichment and sequencing (Ren Seq), and effectromics approaches to detect R-genes.

Keywords

Identification · R-genes genotypes · Sources of resistance · Resistance · Multiple disease resistance · *Albugo* · *Alternaria* · *Colletotrichum* · *Fusarium* · *Hyaloperonospora* · *Erysiphe* · *Leptosphaeria* · *Plasmodiophora* · *Sclerotinia* · *Xanthomonas* · Slow-mildewing resistance · Designation and nomenclature of resistance genes · Mutational approach · Pathotype specific R-genes · Identification of QTL's by GBS · Genetical mechanism of R-genes sources · R-genes by transcriptomic, and proteomic approaches · TuMV R-genes · R-genes against *Xanthomonas* · Defense signaling pathways · Genomic approaches · R-genes homologus DNA fragments · Omics technologies to identify R-genes · NGS-based Bulked Segregant Analysis

2.1 Introduction

Identification of R-genes sources is the first most essential and crucial step towards breeding of disease resistance cultivars through conventional and molecular approaches. Crucifer's pathogens are of diverse nature from biotrophs and hemibiotrophs to necrotrophs with lot of pathogenic variability in the form of races, pathotypes, and strains. For introgression of durable resistance both qualitative and quantitative, R-genes sources are required to plan a breeding of cultivars with desired traits of yield and quality in *Brassica* crops. Considerable efforts have been made to take the R-genes from cultivated diploid and amphidiploid *Brassica* species in addition to *Arabidopsis thaliana*, *Brassica* wild relatives, and land races. With the use of omics approaches, several R-genes effective against major pathogens of crucifers have been identified, cloned, and mapped on chromosomes of *Brassica* with functional characterization using the various omics or multi-omics tools, supplemented with further fine tuning of the bioinformatics method, will not only speed up the screening of favorable alleles in *Brassica* germplasm promoting resistance against the major pathogens, but also expedite the identification and cloning of favorable genes with increased precision. Efforts have been made to incorporate and introgress desired R-genes in *Brassica* crops through different breeding approaches including molecular technology. Now, after genome sequencing of both host and pathogen, it has become easy to use the R-genes pyramiding in a cultivar with desirable gene editing and gene silencing approach.

2.2 *Brassica-Albugo* R-Genes

2.2.1 Identification of R-Genes Genotypes

The supply of gene(s) for disease resistance is the first concern in an ongoing resistance breeding program. Considerable efforts have been made to evaluate cultivar resistance to *Albugo* in oilseed *Brassica* crops. Cultivated *Brassicas* are represented by six interrelated species (Fig. 2.1), three of which are diploids: *B. nigra,* bb ($n = 8$), *B. oleracea,* cc ($n = 9$), and *B. rapa,* aa ($n = 10$), and the

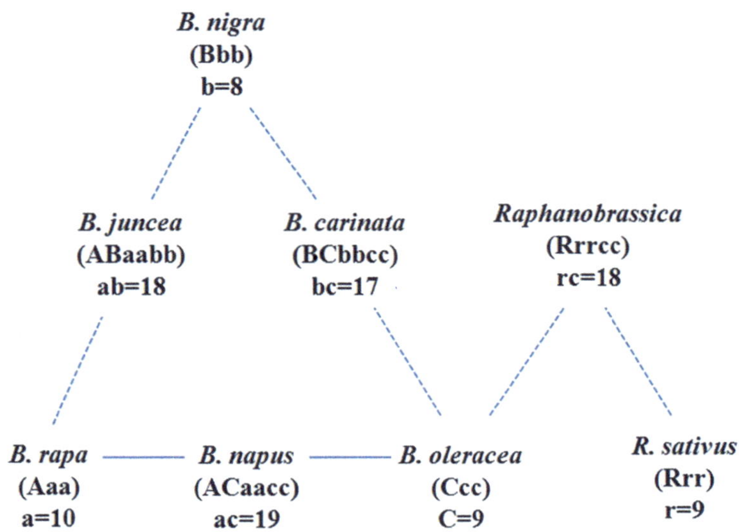

Fig. 2.1 Triangle of U is a theory about the evolution and relationship between cultivated Brassicas represented by six interrelated species

other three are the amphidiploid derivatives of the diploid species: *B. carinata,* bbcc ($n = 17$), *B. juncea,* aabb ($n = 18$), *B. napus,* aacc ($n = 19$), and *R. sativus,* rr ($r = 9$).

Many of the *Brassica* species consist of numerous subspecies or varieties representing a diverse range of morphotypes, and utilization from oils and condiments to vegetables and animal fodders. Brassica oil (rapeseed oil) ranks 5th in world commerce as a major edible and industrial oil; kales, rapes, turnips, and swedes are important sheep and cattle fodder in climates too cool for maize or soybeans, whereas the cole crops and oriental *Brassica* greens are a primary dietary vitamin source for over half of the world's population.

The cytogenetic interrelationships of the six *Brassica* spp. were first described by Morinaga (1934) and UN (1935), and since then numerous studies have been made on the interspecific transfer of genes among various species of Brassicas (Yarnell 1956; McNaughton and Ross 1976). Intergeneric relationships between various Brassicas and radish (*R. sativus*), rr ($n = 9$) have demonstrated the transfer of potentially useful characters including disease resistance and high dry matter content which have resulted in the development of the new crop genus *Raphanobrassica* (McNaughton and Ross 1976).

The three diploid species of *Brassica* are insect pollinated and strongly out breeding with self-incompatibility controlled by a multiple allelic series of genes at the locus and under saprophytic phenotypic expression. Occasionally, genetic-self-compatibility can be found and is predominant in cauliflower and sarson (yellow mustard). Selfing of incompatible plants can be accomplished by bud pollination (the placing of "self" pollen on the immature stigmas) 1–2 days prior to anthesis. Selfing in the diploid species normally results in inbreeding depression. Amphidiploid species are predominantly self-pollinating (75% in oil seed rape) though S-alleles do exist in some populations. Williams and Pound (1963) reported that resistance in radish cvs. China Rose Winter (CRW) and Round Black Spanish (RBS) are governed by a single dominant gene. Humaydan and Williams (1976) identified a single dominant gene, Ac-l, in radish cv. Caudatus. Bonnet (1981) reported that radish cv. Rubiso 2 contains a single dominant gene. Petrie (1985) reported that WR, *A. candida,* was not important due to widespread use of resistant cultivars, and later identified Chinese Rose Winter (CRW), Round Black Spanish (RBS), and Burpee White (BW) as resistant to WR among eleven cultivars of radish. Delwiche and Williams (1974) reported that all accessions of *B. napus* and most cultivars of *B. oleracea* are resistant to *A. candida.* Monteiro and Williams (1989) did not observe high resistance to *A. candida* in Portuguese cabbage and kale land races, although several accessions showed 20–30 per cent of plants with intermediate expression of resistance. Most accessions of *B. oleracea* including Portuguese cole land races, and two Portuguese cabbage were very susceptible, and no differential reaction to the 2 Portuguese *A. candida* isolates were found, although some expressed resistance to both isolates; Couve Algarvia (DI = 2.9), Couve Gloria de Portugal (Dl = 3.4), and Couve Portuguera (DI = 3.5) were the most resistant land races (Santos et al. 1996).

Fan et al. (1983) reported that the resistance in *B. napus* cv. Regent is governed by three genes, Ac-7-1, Ac-7-2, and Ac-7-3. In India, a number of *B. napus* genotypes resistant to *A. candida* have been reported: Gulivar (Gupta and Singh 1994a; Saharan et al. 1988), GSL-l (Saharan 1996, 1997), GLS-150 1 (Gupta and Singh 1994b; Saharan 1996, 1997), HNS-l (Bhardwaj and Sud 1993; Saharan et al. 1988), HNS-3, Tower (Bhardwaj and Sud 1993; Saharan 1996), HNS-4 (Jain et al. 1998; Saharan 1996, 1997), Midas (Jain et al. 1998; Saharan 1996, 1997; Saharan et al. 1988), Regent (Bhardwaj and Sud 1993; Saharan et al. 1988), GSB 7006, Norin-14 (Gupta and Singh 1994a), GSL Series (10 lines), H-715, HNS-1, Norin, Tower 1,2,3,4 (Saharan et al. 1988), EC 174243, GSB 101, GLS 706, HNS-8, Tower-GO (Jain et al. 1998), ABN, Altex, EC 129126, EC 129127, EC 131625, EC 131626, EC 132121, GSA, GSB, HSN-l, Karat, Mary, Niklas, VR-OLGA, VR-WW-1313 (Bhardwaj and Sud 1993). Whereas, all accessions of *B. napus* were reported to be resistant to *A. candida* (Bhardwaj and Sud 1993; Jain 1995; Lakra and Saharan 1989).

In India, a number of *B. juncea* genotypes resistant to existing races of *A. candida* have also been reported: CSR 142 (Kaushik and Saharan 1980; Saharan 1996), Domo-4 (Lakra and Saharan 1989; Yadav and Singh 1992), EC-126741 (Saharan et al. 1988; Yadav and Singh 1992), EC-126745 (Yadav and Singh 1992), EC 126746 (Saharan et al. 1988; Yadav and Singh 1992), EC 129126-1 (Jain 1995; Saharan 1996), KOS-l (Jat 1999), PHR-l (Jain 1995; Kolte 1987a, b), RC-781 (Kaushik and Saharan 1980; Kolte 1987a, b; Lakra and Saharan 1989), Scimitar (Wood and Petrie 1989), SSK-l (Gupta and Singh 1994b), T-4 (Parui and Bandyopadhyay 1973), YRT-1 (Kaushik and Saharan 1980; Kolte 1987a, b; Lakra and Saharan 1989), Zem-1 (Gupta and Singh 1994a; Jain, 1995), Domo, Lethbridge (Kolte 1987a, b; Saharan et al. 1988), Sikkim Sarson 1, YSIK 741 (Srivastava and Verma 1987), DIR 519, DIR 1507 (Jain et al. 1998), EC-126743-2 (Lakra and Saharan 1989; Yadav and Singh 1992), DIRA-313-7, GS 7027, RN 246 (Saharan 1996), EC 126126, EC 126747, RC 1401, RC 1499 (Saharan et al. 1988), C9b, DOMO, YS-7B, Zem-2 (Jain 1995), EC 126743, EC 126743-1, EC 129121-1, RH 8541-46, RW 81-59 (Lakra and Saharan 1989), Blaze, Metapolka, Newton, Purbiraya, Stoke (Kolte 1987a, b), RC-IOOl, RC-1405, RC-1408, RC-1424, RC-1425 (Kaushik and Saharan 1980; Saharan et al. 1988), RH 8545, WRR-3-1, SV7739035 × RH-30-12-15, SV7739035 × RH-30-10-3, SV7739035 × RH-30-16-6, SV 7739035 × RH-30-2-17 (Bhatia 1994), Chamba-l, CSR 741, RC 295, RC 398, RC 1424, RC 1499, (Kaushik and Saharan 1980), MLS-7, MLS-I0, MLS-11, MLS-13, MLS-16, MLS-17, MLS-18, MLS-29, MLS-30, MLS-31, MLS-32, MLS-35, MLS-39 (Velazhahan and Thiyagarajan 1994), DIRM 5, DIRM 11, Gonads 3, Gonads, IB 499-1, Kranti 43, R71-2, R75-2, RCI2, RC 43, RC 14-1, RH 861, RH 8121, RH 8176, RH 55, RLC 1015, RLM 39, RS 78, RW 15-6, RW 22, RW 33-2, RW 75-123-2, Trapal, NO 5422 (Bhardwaj and Sud 1993).

Brassica rapa cvs. Tobin (Kolte 1987a, b), NDYS-2, PI-303, YSK-8502 (Jain 1995), and BSH-l, and BS 15 of *B. rapa* var. Brown Sarson, and Type 6, Prain YST-6 of *B. rapa* var. Yellow Sarson were resistant to WR pathogen (Kolte and Tewari 1980; Lakra and Saharan 1989); *B. rapa* var. Toria IB-586 (Kolte and Tewari 1980), KIC (Jain et al. 1998) and PI 303, PT-30 (Kolte et al. 1985) were least susceptible, immune and tolerant to *A. candida,* respectively.

In *Brassica carinata* HC-l (Bhardwaj and Sud 1993; Jain 1995; Saharan 1996, 1997; Saharan et al. 1988), HC-2 (Bhardwaj and Sud 1993; Jain et al. 1998; Saharan et al. 1988), HC-3, HC-5 (Bhardwaj and Sud 1993; Saharan et al. 1988), HC-4 (Saharan et al. 1988), HC-7, PC-3 (Saharan 1996, 1997), PC-5 (Gupta and Singh 1994a), DIR 1510, DIR 1522, HC 9001 (Jain et al. 1998) were reported to be resistant to *A. candida.* Whereas, all accession of *B. carinata* were also reported to be resistant (Anand et al. 1985; Bhardwaj and Sud 1993; Lakra and Saharan 1989; Satyavir et al. 1994) to *A. candida. Eruca sativa* genotypes RTM 314, RTM 1263, RTM 1471 were resistant to *A. candida* (Jain et al. 1998). All accessions of *E. vesicaria* and *E. pinnatifida* were resistant to race 2; all wild and two cultivated *Eruca* species were resistant to race 7 (Bansal et al. 1997). Although, *B. chinensis,* cultivar Wong Bok (Sutton's) and its crosses were resistant to the foliar phase (Singh and Gangopadhyay 1976), all accessions were found to be free of stagheads infection of *A. candida* (Lakra and Saharan 1989). Although, all accessions of *B. alba* were resistant to *A. candida* (Lakra and Saharan 1989; Saharan et al. 1988), stag heads were observed in *B. pekinensis* (Lakra and Saharan 1989). *Brassica* species and its wild allies, including *B. spinescens, B. tenuifolia, B. incana,* and *Diplotaxis erucoides, D. siifolia, D. virgata, D. niuralis, D. vesicaria, D. desnottesi* were resistant to *A. candida* (Gupta et al. 1995) (Table 2.1).

2.2.2 Sources of Multiple Disease Resistance

Source of multiple disease resistance (MDR) in *Brassica* alongwith *A. candida* have been identified (Table 2.2). Although several genotypes are resistant to more than one disease, genotypes HC-1 and PCC-2 of *B. carinata* and GSL-1501 of *B. napus* consistently show resistance against WR, Alternaria blight, and powdery mildew diseases (Saharan and Krishnia 2001). Banga (1988) observed that some of the C genome chromosomes substitution lines of *B. juncea* and *B. napus* were practically free from WR infection. Canadian breeders and plant pathologists in 1980 licensed *B. rapa* cv. Tobin, which was specifically bred to be highly resistant to race 7 of *A. candida.* Tobin was also found to be resistant to WR in India. In Texas, USA, some resistance has also been reported in certain varieties of spinach against *A. occidentalis* (Dainello et al. 1981). The *B. napus* chromosome segment, carrying the WR resistance gene (*Ac2V1*), appeared to have recombined with the *B. juncea* DNA since recombinant individuals were identified (Somers et al. 2002). A single dominant gene controls avirulence in race Ac2 to *B. rapa* cv. Torch and provides further evidence for the gene-for-gene relationship in the *Albugo–Brassica* pathosystem (Adhikari et al. 2003).

The resistance of *B. napus* cv. EC 151964 was transferred to *B. juncea* cv. RLM-198 through interspecific hybridization following modified pedigree method; a derivative 'NRG-49', an advance generation of interspecific cross, was found as good as *B. napus* and superior to *B. juncea* cv. RLM-198 in agronomic characters including grain yield and earliness (Pal et al. 1999).

Table 2.1 Sources of R-genes in crucifers to *Albugo candida* (White rust) (Saharan 2010; Saharan et al. 2021 updated)

Crucifer's host	Sources	Country	References
Raphanus sativus	China Rose Winter (CRW), Round Black Spanish (RBS), Caudatus Rubiso 2 Burpee white	USA Canada	Williams and Pound (1963), Humaydan and Williams (1976), Bonnet (1981), Petrie (1986)
B. napus	Regent, Tobin GSL 1, S–II an S-IV from *B. napus* (Single dominant gene), Ac 2V1 (Single dominant gene)	Canada India	Fan et al. (1983), Bhardwaj and Sud (1993), Chauhan and Raut (2002), Somers et al. (2002)
B. juncea	EC 399300, EC 399301, EC 399299, EC 414299, IC 443623, IC 555891, BIOYSR, DRMR 2019, DRMR 2035, PWR 15-8, DRMRIJ 12-37, RH 1234; NDRE-08-14-01, BIO-YSR, DRMR 1-5, DRMR-12-39, DRMRJA-35, DRMRIJ-12-26, DRMRIJ-12-21, DRMRIJ-12-21, DRMRIJ-12-37, DRMR-5206, PDZ 2, PDZ 3, PDZ 5, PDZ 7, DRMRSJ-21, DRMRSJ-18, DRMRSJ-26, DRMRSJ-22, DRMRSJ-1, DRMRSJ-34, DRMRSJ-25, IC265495, IC313380, EC766091, EC766133, EC766134, EC766192, EC766230, EC766272; CBJ 001, CBJ 003, CBJ 004, JR 049, JM06011 RESJ-1004, RESJ-1005, RESJ-1033 and RESJ-1051	India China Australia UK	Saharan (2010), Meena et al. (2019a, b), Dahiya and Rashmi (2019), Gairola and Tewari (2017), Vignesh et al. (2009), AICRPRM (2020), Yadav et al. (2017) Li et al. (2008a, b), Li et al. (2007a, b, 2009a, b), Awasthi et al. (2012)
B. rapa var. Yellow Sarson	T 6, Prain, YST 6, Tobin, NDYS 2, YSK 8502	India	Kolte and Tewari (1980), Lakra and Saharan (1989)
B. rapa var. Brown Sarson	BSH 1, BS 15	India	Kolte and Tewari (1980), Lakra and Saharan (1989), Saharan (1988, 2005)
B. rapa var. toria	IB 586, KTC, PT 303, PT 30	India	Kolte et al. (1985)
B. carinata	PBC 9221, S-III derived from *B. carinata* (Single recessive gene	India	Chauhan and Raut (2002)
Eruca sativa	RTM 314, RTM 1263, RTM 1471	India	Jain et al. (1998)

(continued)

Table 2.1 (continued)

Crucifer's host	Sources	Country	References
B. chinensis	All accessions	India	Gupta et al. (1995), Lakra and Saharan (1989)
B. alba	All accessions	India	Saharan et al. (1988)
A. thaliana	RAC – 1, RAC-2, RAC – 3 from *Arabidopsis thaliana* (ksk-1 and ksk-2)	Columbia	Borhan et al. (2001, 2008)
B. spinescens	All accessions		
B. tenuifolia	All accessions		
B. incana	All accessions		
B. oleracea	Couve Algarvia, Couve Gloria, Couve Portuguera	Portugal	Santos et al. (1996)

Table 2.2 Sources of multiple disease resistance (MDR) in rapeseed-mustard (Saharan and Krishnia 2001; Saharan et al. 2021 updated).

Resistant to WR and AB	Resistant to WR and PM	Resistant to AB and PM	Resistant to WR, AB, and PM
B. juncea			
EC-322090	EC-129126-1	None	None
EC-322092	EC-129126-2		
EC-322093	PR-8805		
MCB-1,BIOYSR			
EC-399296			
B. rapa			
None	Candle	None	None
	Tobin, Torch		
B. carinata			
HC-1	DHC -1	HC-1	HC-1
PCC-2	DHC -4	PCC-1	PCC-2
	DHC-9601		
	HC -1		
	PCC -2		
B. napus			
GS-7027	GSL-1501	GSL-1501	GSL-1501
GSL-1	N-20-12-2		
GSL–1501	Wester		
Gulivar			
HNS -4			
Tower sel-1			

WR white rust, *PM* Powdery mildew, *AB* Alternaria blight

The inheritance of partial resistance to race 2 of *A. candida* was studied in *B. juncea* by crossing the partial resistant line with susceptible *B. juncea* cultivar

Commercial Brown; adult plants did not develop stagheads under greenhouse and field conditions (Bansal et al. 1999). *Brassica alba* [=*Sinapis alba*], *B. carinata* (HC-1), *B. juncea* (DIR-1507, DIR-1522), and *B. napus* (GS-7027, Midas, Tower)) had stable and multiple disease resistance (Dang et al. 2000). The segregation pattern of F_2 generation and test crosses indicated that WR resistance in *B. napus* cvs. Domo and Cutlass is under the control of a single dominant gene (Sachan et al. 2000).

The hybrids *tourn-CMS* × BJ-38 and *oxy-CMS* × PR-1108 exhibited significantly superior specific cross combinations for disease resistance (Sheikh and Singh 2001). A new *B. rapa* cultivar AC Sungold registered in Ontario, Canada, and derived from the cross between cultivars Tobin and SV8236580, has good resistance to WR (Woods and Falk 2001).

The F_2 population of Rajat × Shiva, Pusa Bahar × Domo, Varuna × EC 322092, and RH 30 × HC-1 fitted well in the ratio of 3R:1S, as well as 13R:3S, suggested that the resistance to *A. candida* was controlled by a single dominant gene, or two genes with dominant recessive epistasis. The F_2 population of cross, RH 30 × EC 322093 segregated in the ratio of 15R:1S which indicated the presence of two dominant resistant gene(s) in EC 322093 (Krishnia et al. 2000). Genotypes EC-129126-1, Shiva, RC 781, ZEM-1, and PR-8805 were found resistant to the WR disease (Gupta et al. 2002). Four Chinese genotypes (CBJ 001; CBJ 002; CBJ 003; CBJ 004) and two Australian genotypes (JR 049; JM06011) were highly resistant to *A. candida* pathotypes prevailing in Australia (Li et al. 2007b, 2008b).

2.3 *Brassica-Alternaria* R-Genes

2.3.1 Sources of Resistance

Digenomic *Brassica* species such as *B. napus* and *B. juncea* have better sources of resistance than the monogenomic (Tables 2.3 and 2.4) species like *B. rapa*. Some lines of *B. alba, B. carinata, B. spinescenes, B. maurorum, Eruca sativa, Camelina sativa, Capsella bursa-pastoris, Diplotaxis* sp., and *N. paniculata* have also shown resistance to *A. brassicae* in different areas of the world (Brun et al. 1987a, b; Conn et al. 1991; Grontoft 1986; Jejelowa et al. 1991; Saharan 1992a, b, 1997; Tewari and Conn 1993; Hansen and Earle 1997; Chrungu et al. 1999; Sharma et al. 2002). Sources of resistance to *A. brassicae* and *A. brassicicola* in different host species from different areas of the world are given in Table 2.4. In India, *Brassica* genotypes, viz. CSR 43, CSR I42, CSR-I42-2, CSR 343 , CSR 448, CSR 622, CSR 741, Gulivar, KRV-Tall, Midas, PHR 1, RC-781, TMV2, Tower, and YRT3 were consistently found to have field resistance proven after testing for several years at different locations under uniform disease nursery trials (Kolte 1985b, 1987a, b; Saharan 1984, 1992a, b, 1997; Saharan and Chand 1988; Saharan et al. 2003, 2005, 2016; Verma and Saharan 1994). *Brassica coenospecies* are rich reservoir for genetic resistance to *A. brassicae* (Sharma et al. 2002; Table 2.5). *Alternaria brassicae* pathotype specific resistance has been recorded in genotype, GS-05-1 of *B. napus* to pathotypes, Abr 1 and Abr 5 whereas genotypes, RH-8544,

Table 2.3 *Brassica* germplasm holdings at different organizations and research centers in the world

Sr. no.	Name of organization/research center	Germplasm accessions (Total number)	References
1.	The National Bureau of Plant Genetic Resources (NBPGR), New Delhi, India	10301	Radhamani et al. (2013)
2.	ICAR-All India Coordinated Research Project on Rapeseed-Mustard	12778	AICRP-RM (2011)
3.	ICAR-Directorate of Rapeseed-Mustard Research, Bharatpur (Raj) India	2452	Nanjundan et al. (2014)
4.	National Plant Germplasm System (USDA) Germplasm Resources Information Network (GRIN)	2958	http://www.ars-grin.gov/cgi-bin/npgs/html/taxon.pl?319659
5.	Biotechnology Department, Polytechnic University, Valencia, Caminode, Spain	150	Nuej et al. (1987)
6.	The Crucifer Genetics Cooperation (CrGC), Department of Plant Pathology, University of Wisconsin, Madison, USA	70 × 17 stocks in each *Brassica* crop	Williams (1988b)
7.	Wisconsin fast Plant (WFP), Department of Plant Pathology, University of Wisconsin, Madison, USA	Rapid cycling of *Brassica* seed stocks	Williams (1988a)
8.	The Asian Vegetable Research and Development Centre (A VRDC), Shanhua, Taiwan	822	Anonymous (1984), Opena and Lo (1981)

Pusa Swarnim, and HC-9605 of *B. juncea* showed moderately resistant reaction to three pathotypes (Kumar et al. 2014).

2.3.2 Sources of Disease Resistance from Cruciferous Relatives

Sources with high level of resistance against *A. brassicicola* and *A. brassicae* have not been identified among the cultivated *Brassica* species, although individual cabbage varieties differ considerably in their level of susceptibility to black spot (Otani et al. 2001). The highest level of *Alternaria* resistance among the oilseed *Brassica* crops is displayed by the Ethiopian mustard (*B. carinata*). Among the wild cruciferous plants closely related to the *Brassica* genus, the highest resistance against *A. brassicae* is found in white mustard (*Sinapus alba*), (Kolte 1985a; Brun et al. 1987a, b; Ripley et al. 1992; Sharma and Singh 1992; Verma and Saharan 1994; Hansen and Earle 1997). The highest overall *Alternaria* spp. resistance, however, has been identified in the crucifers more distant from the *Brassica*, such as *Camelina* (*Camelina sativa*; false flax), shepherd's purse (*Capsella bursa-pestoris*), taramira (*Eruca sativa*), and ball mustard (*Neslia paniculata*) (Conn and Tewari 1986; Conn et al. 1988, Tewari 1991). Resistance against *A. brassicae* and *A. brassicicola* has been reported among other wild members of the Brassicaceae

Table 2.4 Sources of R-genes in crucifer's to *Alternaria brassicae* and *Alternaria brassicicola* (Alternaria blight) (Saharan et al. 2015 updated)

Crucifer's host	Sources	References
A. brassicae		
Brassica rapa ssp. *oleifera* (Turnip rape/ Toria)	EC242660-61, EC242646, EC253287, EC253291	AICORPO (1989)
B. rapa var. Yellow Sarson	PYS6, BINA 1,2	Kolte (1987a, b), Rahman et al. (1987)
B. rapa rapifera (Turnip)	Edmonton ACC	Conn et al. (1988)
B. carinata (Ethiopian mustard)	PPSC1, HC-1, EC25381, EC253282, EC253284, EC253826, S67, PC3, PC5, CE9, HC-9605	AICORPO (1989), Bhowmik and Munde (1987), Kidane and Bekele (1986)
B. napus spp. *oleifera* (Oilseed rape)	Altex, Gulivar 1, Karat, Narde, Midas, Primer, Tower, WRGI5, Westar, Wei Bull-541, Wei Bull-521, Vuindsok, Regent 1, GS-1-1, GS 7027, Vestal, GSL-1501, GSL-1506, GSL-1508, GSL-1513, HNS1, HNS3, EC-338986-2, EC-338996-1, EC-338997, EC-339000, Surpass 404 CL, Hyalo 635 CC, Oscar, AG Outback, Rottenest, GS-05-1	AICORPO (1989), Conn et al. (1987), Conn and Tewari (1989), Kumar and Kumar (1989), Rozej (1974), Romero-Munoz and Jimenez (1979), Saharan (1992a, b, 1997), Stankova (1972), Kolte et al. (2008), AICRPRM (2011), Al-Lami et al. (2020), Kumar et al. (2014)
B. juncea (Indian mustard)	BECI07-109, BECI07-112, ECI29126-1, PR8805, PHR 1, KRV Tall, RC-781, Divya, Kranti, PR-8999, PR-9024; EC-399296, EC-399299, EC-399301, EC-399313, DRMR-2805, RH-8544, Pusa Swarnim, RSPN-29, CNH-13-2, AKGS-1, RSPN-28, CNH-13-1, Bina Sarisha-8, EC-399312m, PAB 9511	AICORPO (1989), Kolte (1985a), Saharan (1992a, b, 1997), Kolte et al. (2000, 2008), Kumar (2008), Pratap et al. (2014), Kumar et al. (2014), Das (2016), Hossain et al. (2018), Meena et al. (2020), Ghosh et al. (2019)
B. oleracea var. *Alboglabra* (Chinese kale)	All accessions, Local	Munde and Bhowmik (1985), Zaman and Biswas (1987)
B. oleracea var. *Botrytis* (Cauliflower)	Lines 1-6-1-2, 1-6-1-4, Pusa Shubhra	Singh et al. (1987)
B. oleracea ssp. *Gemmifera* (Brussels Sprouts)	Cambridge No.5	Berry and Lennard (1988), Williams and Pink (1987)
B. alba (White mustard)	All accessions, Local	Kolte (1987a), Saharan (1992a, b)

(continued)

Table 2.4 (continued)

Crucifer's host	Sources	References
B. hirta	All accessions, Local	Brun et al. (1987a, b)
Camelina sativa (False flax)	All accessions, Local	Tewari and Conn (1993)
Capsella bursa-pastoris (Shepherd's purse)	All accessions, Local	Conn et al. (1991)
Neslia paniculata (Ball mustard)	All accessions, Local	Jejelowa et al. (1991), Grontoft (1986)
Sinapis alba (White mustard)	S10001, S1004-10	Kolte (1985b), Rai et al. (1977)
A. brassicicola		
B. oleracea var. *Botrytis* (Cauliflower)	PI'S 231208-209, PI'S 217934, PI'S 231209, PI'S 267725, Pusa Shubhra	Braverman (1971), Singh et al. (1987)
B. oleracea var. *gemmifera* (Brussels sprouts)	Local accession	Braverman (1976, 1977)
B. rapa	Saori, Edononatsu	Doullah et al. (2006)
B. rapa var. *globra*	CR-Hagwang	Lee and Hong (2015)

family (Table 2.5) (Sharma et al. 2002; Tewari and Conn 1993; Warwick 2011), viz. *Alliaria petiolata*, *Barbarea vulgaris*, *B. elongate*, *B. desnottessi*, *B. fruticulosa*, *B. maurorum*, *B. nigra*, *B. souliei*, *B. spinescens*, *C. sativa*, *C. bursa-pestoris*; *Coincya* spp., *Diplotaxis catholica*, *D. berthautii*, *D. creacea*, *D. erucoides*, *D. tenuifolia*; *Erucastrum gallicum*, *E. vesicaria* subsp. *sativa*; *Hemicrambe fruticulosa*, *H. matronalis*, *N. paniculata*; *R. sativus*; *S. alba,* and *S. arvensis*. The completely immune plants remain symptom free both, under natural field infection as well as under controlled artificial inoculation (Sharma et al. 2002). Comparatively, broccoli and cauliflower varieties exhibit moderate *Alternaria* resistance, while the cabbage as susceptible.

2.3.3 Sources of Multiple Disease Resistance

Brassica alba species and genotypes of different *Brassica* species, viz. HC-1, PCC-2 (*B. carinata*), GSL-1501, GS-7027, Midas, Tower (*B. napus*), and DIR-1507, DIR-1522 (*B. juncea*) have been identified as sources of multiple disease resistance against white rust, Alternaria blight, downy mildew and powdery mildew diseases (Dang et al. 2000; Kumar and Saharan 2002). Earlier *B. napus* genotype, GSL-1501 was reported resistant to white rust and powdery mildew (Gupta and Singh 1994a).

Table 2.5 Classification of 38 *Brassica coenospecies* based on reaction to *Alternaria brassicae* under in vitro and in vivo inoculation conditions (Sharma et al. 2002)

Resistant	Moderately resistant	Susceptible	Highly susceptible
Brassica desnottessi	*Brassica oleracea* (broccoli)	*B. oleracea* (cabbage)	*B. juncea*
Camelina sativa	*B. oleracea* (cauliflower)	*B. napus*	*B. rapa*
Coincya pseuderucastrum	*B. nigra*	*B. fructiculosa*	*B. carinata*
Diplotaxis berthautii	*B. barrelieri*	*D. gomezcampoi*	*B. cossoniana*
Diplotaxis catholica	*B. oxyrrhina*	*D. assurgens*	*B. spinescens*
Diplotaxis cretacea	*B. gravinae*	*D. tenuisiliqua*	*Coincya longirostra*
Diplotaxis erucoides	*Coincya rupestris*	*Erucastrum abyssinicum*	*D. viminea*
Eruca gallicum	*D. harra*	*E. canarianse*	*Enarthocarpus lyratus*
	D. muralis	*Moricandia arvensis*	*Raphanus sativus*
	D. siifolia	*Moricandia moricandioides*	
	Sinapis alba	*S. flexuosa*	
	S. pubescens		

The variations in the incidence of powdery mildew scores among different Brassica genotypes differ considerably due to variations in date of sowing, stage of the crop at the time of observations and prevailing environmental conditions. Hence, early maturing genotypes or the genotypes planted early in the season would help in escaping the powdery mildew infection.

The genotypes, EC-322090, EC-322092, EC-322093, and RC-781 of *B. juncea*, HC-1 and PCC-2 of *B. carinata,* and GS-7027, GSL-1, GSL-1501, Gulivar 1, HNS-4, and Tower sel.1 of *B. napus* are resistant to *A. candida* and *A. brassicae*. Whereas genotypes, EC-129126-1, EC-129126-2, PR-8805 of *B. juncea*, Candle, Tobin, Torch of *B. rapa*, DHC-1, HDC-4, DHC-1960, HC-1, PCC-2 of *B. carinata*, and GSL-1501, N 20-12-2, and Wester of *B. napus* are resistant to *A. candida* and *Erysiphe cruciferarum*. However, HC-1, PCC-2 of *B. carinata,* and GSL-1501 (*B. napus*) are resistant to both *A. brassicae* and *E. cruciferarum*. GSL-1501 and Gulivar-1 of *B. napus* were found resistant to *A. candida* and *E. cruciferarum* by Gupta and Singh (1994a). The genotypes Wester (*B. napus*), Tobin, and Candle (*B. rapa*) are reported completely free from *A. candida* and *E. cruciferarum* (Shivpuri et al. 1997). The variations in the observations may be attributed to the occurrence of new races of the pathogen and congenial weather factors (Table 2.6; Kumar and Saharan 2002).

Table 2.6 Sources of multiple disease resistance in oilseed *Brassica* (Kumar and Saharan 2002)

Resistant to WR and AB	Resistant to WR and PM	Resistant to AB and PM	Resistant to WR, AB, PM, and DM
B. juncea			
EC-322090, EC-322092, EC-322093, RC-781	EC-129126-1, EC-29126-2, PR-8805	None	DIR 1507, DIR 1522, *Brassica alba*
B. rapa			
None	Candle, Tobin, Torch	None	None
B. carinata			
HC-1, PCC-2	DHC-1, DHC-4, DHC-9601, HC-1, PCC-2	HC-1, PCC-2	HC-1, PCC-2
B. napus			
GS-7027, GSL-1, GSL-1501, Gulivar-1, HNS-4, Towel Sel-1	GSL-1501, N-20-12-2, Wester	GSL-1501	GSL-1501, GS-7027, Midas, Tower

*WR*White rust, *AB* Alternaria blight, *PM* Powdery mildew, *DM* Downy mildew

2.4 *Brassica-Colletotrichum* R-Genes

2.4.1 Identification and Characterization of Resistance Genes

Molecular and biochemical bases of cultivar resistance to *Colletotrichum* spp. have been investigated using genetically diverse materials. Innate disease resistance responses in plants are triggered by a dual surveillance system composed of nucleotide-binding leucine-rich repeat (NB-LRR) proteins encoded by resistance genes and pattern recognition receptors (PRRs). The two layers are often called MAMPs-triggered immunity (MTI) and effector triggered immunity (ETI). By a combination of quantitative trait loci (QTL) and Mendelian mapping, a single putative R locus RCH1 was identified, at the tip of chromosome 4, in the resistant *A. thaliana* ecotype Eil-0 against *C. higginsianum*. By using map-based cloning and natural variation analysis of 19 *Arabidopsis* ecotypes, another dominant resistance locus RCH2 was identified against *C. higginsianum*. The locus RCH2 maps to an extensive cluster of disease-resistance loci known as MRC-J in the *Arabidopsis* ecotype Ws-0. These indicate that *Arabidopsis* resistance to *C. higginsianum* is controlled by a gene-for-gene interaction. In *A. thaliana*, NB-LRR-type resistance (R) genes to *Pseudomonas syringae* 4 (RPS4) and to *Ralstonia solanacearum* 1 (RRS1-R) were reported to also confer resistance to *C. higginsianum*. RRS1-R and RPS4 were also found as a complex that could help detect effectors which target WRKY proteins. Therefore, effectors in *C. higginsianum* that target WRKY proteins may be more likely to act as *Avr* genes (Birker et al. 2009; Sarris et al. 2015; Yan et al. 2018).

2.5 *Brassica-Erysiphe* R-Genes

2.5.1 Identification of R-Genes Genotypes Against Powdery Mildew

Major gene sources of resistance against powdery mildew of crucifers have been identified. These sources can be easily incorporated through conventional and biotechnological approaches to breed powdery mildew resistant cultivars of different crucifers. The number of resistant sources identified are belongs to five species of oil yielding crops (*Brassica alba* all available accessions, *B. carinata* 5, *B. juncea* 15, *B. napus* 5, *B. rapa* ssp. Yellow Sarson 2), one of fodder crops (*B. napus* ssp. *rapifera* 2), and a weed (*Arabidopsis thaliana* 7; Table 2.7). These sources are being used as model through powdery mildew–host pathosystem for molecular and genetical studies (Table 2.7).

In India, tolerant and resistant sources against powdery mildew of crucifers have been identified during 2000–2017 under AICRPRM program by different researchers and reactions of various *Brassica* species against powdery mildew with genotypes are presented in Table 2.8. However, registration of resistant source against powdery mildew of rapeseed-mustard in India is lacking so far.

Table 2.7 Sources of R-genes in crucifers to *Erysiphe cruciferarum* (Powdery mildew) (Saharan et al. 2021 updated)

Crucifer's hosts	Sources	References
Brassica napus ssp. *rapifera*	UG 3, UG 4	Shattuck (1993)
B. napus	GS-7027, Midas, Tower, GSL-1,MNS-9605, Trooper, Brove, TT, Summit, Tumby, Narender, Hyola 650TT	Dang et al. (2000), Mehta et al. (2008), Uloth et al. (2016)
B. alba, *Sinapis alba*	All accessions, Local	Mehta et al. (2008), Meena et al. (2018)
B. carinata	HC-1, HC-2, HC-9605, HC-9603, PI 360883	Dang et al. (2000), Mehta et al. (2008), Tonquc and Griffiths (2004a)
B. juncea	JMO 6014, JMO 6015, JMO 6009, JMO 6012 (from Australia), JM3, DIR-1507, DIR-1522, Kranti, DIR 621, IJWHJ 001, PCR-10, PCA 9201, RK 8602, RK 8615, RAUD 101, RDV 29	Dang et al. (2000), Singh et al. (2010), Singh and Singh (2003), Nanjundan et al. (2020)
Arabidopsis thaliana	Accessions Su-0, Ms-0, Wa-1, Kas-1, SI-0, Te-0, Stw-0	Adam et al. (1999), Adam and Somerville (1996)
B. rapa ssp. *Yellow Sarson*	YSPb-24, TH-68	Mehta et al. (2008)
B. rapa ssp. *Brown Sarson*	BSH-1	Mehta et al. (2008)

Table 2.8 Sources identified from crucifers for powdery mildew disease tolerance/resistance under AICRPRM (Anonymous 2000–2017; Meena et al. 2018)

Year	Powdery mildew tolerant/resistant sources[a]
2001	TM 18, RM 505 (Bj)
2004	EC -338997 (Bn), PBC-9221 (Bc), PBN-2001 (Bn) , PBN-2001 (Bn)
2005	NPC-14, JTC-55, PBC-2002-(Bc)
2006	OCN 3 (Bn)
2007	HNS-9605 (Bn), PT-303 (Brt)
2008	EC 338997 (Bn), ONK 1 (Bn)
2009	EC-414299 (Br), EC-339000 (Bn)
2010	NPJ-143 (Bj)
2011	DRMR 243 (Bc), DRMR 261 (Bc), DLSC 1 (Bc)
2012	DRMR 312 (Bc)
2014	NPC-16 (Bc), NPC-21 (Bc)
2015	PPBN-3 (Bn), PPBR-2 (Br), PT-2006-4 (Brt), RMT-10-7 (Brt)
2016	PRD 2013-3 (Bj), DRMR-316 (Bc), DRMR-100 (Bc)
2017	DRMR 1-5 (Bj)

[a]Bj (*Brassica juncea*); Bc (*B. carinata*), Bn (*B. napus*), Brt (*B. rapa* ssp Toria), Br (*B. rapa*)

2.5.2 Sources of Slow-Mildewing Resistance

To assess the nature of powdery mildew resistance in *Brassica* crops, seven cvs. of *Brassica juncea,* and one each of *Brassica napus,* and *Brassica carinata* were selected for evaluation of slow mildewing components (Singh 1994; Singh et al. 2008; Mehta et al. 2009; Meena et al. 2018). Various components of slow mildewing, viz. incubation period, latent period, no. of colony/speck per leaf, no. of conidia per colony/speck, progression of the disease, and disease intensity were recorded under field conditions. The incubation period of test cvs. ranged from 3 to 4 days. However, powdery mildew was not observed on variety HC-9603 even under artificial inoculation conditions. The maximum incubation period of 4 days was recorded on the varieties RH-9304 and RH-9801, whereas in rest of the varieties, it was 3 days. However, no significant differences in incubation period in case of *Brassica* cvs. was recorded (Table 2.9). The latent period of test cvs. ranged between 1 and 3 days. The variety HC-9603 did not contract powdery mildew. The latent period was 1–2 days in varieties belonging to the *Brassica juncea*. However, it was three days in case of variety GSL-1 (Table 2.9). The results also revealed that there was no significant difference in the latent period in the varieties belonging to *Brassica juncea*. However, it differed in case of *Brassica napus* where it was slightly higher. The number of powdery mildew specks/leaf was also recorded on all the nine varieties as a test of slow mildewing components. The variety GSL-1 showed minimum number of specks/leaf (5.52) whereas the variety RH-9801 contracted maximum number of the specks/leaf (39.40). It was followed

Table 2.9 Components of slow-mildewing resistance to powdery mildew on mustard cultivars/varieties (Singh 1994; Singh et al. 2008; Mehta et al. 2009)

Cultivars/ varieties	Disease intensity (%)	No. of specks/ leaf	No. of conidia/ microscopic field[a]	Incubation period (days)	Latent period (days)
RH-30	46.2 (42.8)	31.16	83.75	3	2
RH-8812	33.6 (35.4)	18.33	50.08	3	1
RH-9304	47.8 (43.7)	33.97	75.08	4	2
RH-9801	49.5 (44.7)	39.40	81.66	4	2
RH-9901	38.9 (38.6)	21.08	72.75	3	1
RC-781	43.4 (41.2)	27.30	41.58	3	1
Purple Mutant	36.7 (37.3)	21.67	62.58	3	1
GSL-1	4.1 (11.7)	5.52	10.00	3	3
HC-9603	–	–	–	–	–
C.D. at 5%	**8.45**	**7.85**	**17.77**		

Figures in parentheses are angular transformed values
[a]Five speck suspended in 1 ml of water; –Disease did not appear

by the variety RH-9304 (34.0). On the other varieties, viz. RH-8812, RH-9901, RC-781, and Purple Mutant, the number of specks ranged between 18 and 27 per leaf, which is moderate. The variety GSL-1 contracted less number of specks per leaf which significantly differed from the other varieties (less than ten specks /leaf) which may be considered as resistant whereas other varieties such as RH-9801, RH-9304, and RH-30 had higher number of specks/leaf (more than 30), and may be termed as susceptible. The other varieties such as RH-9901, RH-8812, RC-781, and Purple Mutant contracted the powdery mildew specks ranging between 10 and 30 specks/leaf and considered as moderately susceptible (Meena et al. 2018).

The disease intensity was recorded on all the varieties except HC-9603 on which disease did not appear till the end of the crop season to test the behavior of varieties against powdery mildew. The maximum disease (49.5%) was recorded on the variety RH-9801 followed by the variety RH-9304 (47.8%), and RH-30 (46.2%) though statistically at par. The minimum disease (4.1%) was observed on the variety GSL-1. Whereas, on other varieties including RH-8812, RH-9901, Purple Mutant, and RC-781, the disease intensity ranged from 33.6 to 43.4 per cent. There were no significant differences in all the varieties in relation to disease intensity except GSL-1 and HC-9603 which appeared as resistant to powdery mildew disease and others as susceptible (Table 2.9).

Number of conidia produced in each speck by *Erysiphe cruciferarum* on different cultivars/varieties of mustard was examined under the compound microscope ($10 \times 10\times$). The minimum number of conidia/speck was recorded on the variety

GSL-1 (10 conidia/speck). It was followed by the variety RC-781 where it was 41.58 conidia/speck. In other varieties such as RH-30, RH-9801, RH-9304, RH-8812, RH-9901, and Purple Mutant, the conidial production ranged between 50.08 and 83.75 conidia per speck being maximum on the variety RH-30 (83.8) and minimum on RH-8812 (50.1). It was revealed that the considerable amount of conidia per speck were produced in all the susceptible varieties belonging to *Brassica juncea* except GSL-1 and HC-9603 which belongs to *Brassica napus* and *Brassica carinata,* respectively (Table 2.9).

The progression of the powdery mildew on all the nine varieties was recorded from the appearance of the disease till the maturity of the leaves on ten randomly selected marked tagged leaves from each replication. The disease appeared on all the varieties in first week of March except that on HC-9603 (disease did not appear). The minimum speck size was recorded in the variety GSL-1 (1.98 mm) whereas maximum in the variety RH-30 (5.80 mm). It was followed by the variety RH-9901 (5.25 mm) and RH-9304 (4.98 mm), whereas on other varieties, the speck size ranged between 3.81 and 4.81 mm which is moderate (Table 2.10). The powdery mildew progression on different varieties/cultivars of mustard presented in Fig. 2.2 revealed that the progression of powdery mildew was maximum up to mid of March; after that the disease was slowed down. The minimum progression was recorded in the variety GSL-1 where it was almost static after initiation of the disease. Similarly, on Purple Mutant variety also, the disease progression was slow as compared to the other varieties.

The progression of the powdery mildew in relation to weather variables was evaluated which revealed the maximum R^2 value, i.e., 0.91 on variety RH-30 followed by RH-9901 (0.86), Purple Mutant (0.86), RH-9801 (0.83), and RH-8812 (0.83). The minimum value (R^2 0.47) was recorded on variety GSL-1 which indicated that in addition to weather variables included here other factors such as varietal resistance and some unknown factors have significant role in the disease development (Table 2.11). The variety GSL-1 and HC-9603 appeared as resistant to powdery mildew with the expression of slow-mildewing components whereas other varieties belonging to *B. juncea* group appeared as susceptible to the disease showing faster powdery mildew development under field conditions.

The correlation matrix for progression of the powdery mildew in relation to weather variables on all the test varieties/cultivars was analyzed. It was observed that T. Max. (X_1) had significant and positive role in the progression of powdery mildew on all the varieties/cultivars except GSL-1 where it was positive but non-significant. Similarly, Avp. M (X_5) also had significant and positive role in the disease progression on all the varieties/cultivars except HC-9603 (disease did not appear). The RHE (X_4) has negative and significant correlation in the disease development on all the varieties/cultivars except GSL-1 where it was negative but non-significant. Similarly, Sunshine (X_8) has negative and significant correlation in all the varieties/cultivars. Other weather variables such as T. Min.(X_2), RHM (X_3), and Avp. E (X_6) had positive but non-significant correlation in disease progression on all the varieties/cultivars (Table 2.12, Meena et al. 2018).

Table 2.10 Progression of powdery mildew (speck size, mm) on different cultivars/varieties of mustard in relation to slow-mildewing assessment (Singh 1994; Singh et al. 2008; Mehta et al. 2009)

Date of observations	Cultivars/verities								
	RH-30	RH-8812	RH-9304	RH-9801	RH-9901	RC-781	Purple Mutant	GSL-1	HC-9603
9-3-04	0.98[a]	0.94	1.06	1.03	1.11	0.98	0.98	1.15	–
11-3-04	1.81 (0.83)	1.83 (0.89)	1.88 (0.82)	1.90 (0.87)	2.11 (1.0)	1.90 (0.92)	1.56 (0.58)	1.58 (0.43)	–
13-3-04	3.16 (1.35)	3.03 (1.20)	3.11 (1.23)	3.23 (1.33)	3.31 (1.20)	3.13 (1.23)	2.60 (1.04)	1.98 (0.40)	–
15-3-04	4.75 (1.59)	4.06 (1.03)	4.55 (1.44)	4.55 (1.32)	4.65 (1.34)	4.33 (1.20)	3.36 (0.36)	1.98 (0.0)	–
17-3-04	5.70 (0.95)	4.35 (0.29)	4.93 (0.38)	4.81 (0.26)	5.21 (0.56)	4.46 (0.13)	3.81 (0.45)	1.98 (0.0)	–
19-3-04	5.80 (0.10)	4.43 (0.08)	4.98 (0.05)	4.81 (0.0)	5.25 (0.04)	4.48 (0.02)	3.81 (0.0)	1.98 (0.0)	–
21-3-04	5.80 (0.0)	4.43 (0.0)	4.98 (0.0)	4.81 (0.0)	5.25 (0.0)	4.48 (0.0)	3.81 (0.0)	1.98 (0.0)	–

Figures in parentheses are the periodical progression
[a]Size of specks (mm) –Disease did not appear

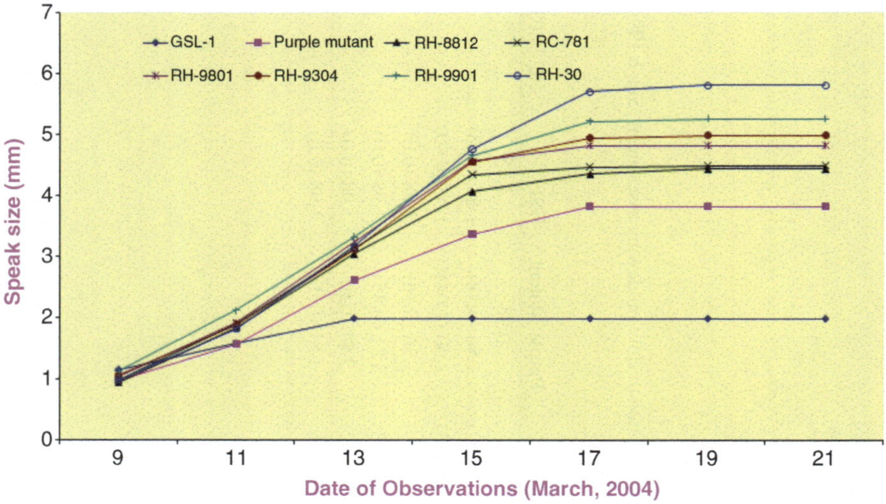

Fig. 2.2 Progression of powdery mildew on eight varieties of mustard (Singh 1994; Singh et al. 2008; Mehta et al. 2009)

Table 2.11 Regression equations for the progression of powdery mildew on different varieties of mustard in relation to weather parameters under different dates of sowing (Singh 1994; Singh et al. 2008; Mehta et al. 2009)

Varieties	Regression equations	R^2
	I[st] DOS*	
RH-30	$Y = -19.20 + 0.73X_1 - 0.08X_7$	0.94
RH-8113	$Y = -19.47 + 0.74X_1 - 0.09X_7$	0.92
RH-9304	$Y = -20.38 + 0.77X_1 - 0.09X_7$	0.94
RH-9801	$Y = -20.19 + 0.76X_1 - 0.08X_7$	0.93
	IInd DOS	
RH-30	$Y = -19.90 + 0.75X_1 - 0.10X_7$	0.94
RH-8113	$Y = -20.96 + 0.80X_1 - 0.09X_7$	0.90
RH-9304	$Y = -19.18 + 0.73X_1 - 0.11X_7$	0.94
RH-9801	$Y = -21.79 + 0.83X_1 - 0.10X_7$	0.91
	IIIrd DOS	
RH-30	$Y = 11.34 + 0.40X_1 - 2.26X_8$	0.92
RH-8113	$Y = 6.46 + 0.48X_1 - 2.01X_8$	0.94
RH-9304	$Y = 6.74 + 0.47X_1 - 2.03X_8$	0.94
RH-9801	$Y = 10.65 + 0.41X_1 - 2.20X_8$	0.92

DOS* – Date of sowing; X_1 = Temperature (Maximum); X_2 = Temperature (Minimum)

X_3 = Relative Humidity (Morning); X_4 = Relative Humidity (Evening)

X_5 = Average Evaporation (Morning); X_6 = Average Evaporation (Evening)

X_7 = Wind Speed; X_8 = Sunshine

Table 2.12 Correlation matrix between powdery mildew (speck size, mm) and weather parameters in mustard cultivars/varieties (Singh 1994; Singh et al. 2008; Mehta et al. 2009)

Weather variables	Cultivars/varieties								
	RH-30	RH-8812	RH-9304	RH-9801	RH-9901	RC-781	Purple Mutant	GSL-1	HC-9603
Temperature Maximum (X_1)	0.95*	0.90*	0.92*	0.90*	0.92*	0.88*	0.91*	0.68	–
Temperature Minimum(X_2)	0.41	0.46	0.42	0.43	0.45	0.45	0.43	0.62	–
Relative Humidity Morning (X_3)	0.56	0.59	0.59	0.61	0.57	0.61	0.60	0.54	–
Relative Humidity Evening (X_4)	–0.85*	–0.83*	–0.84*	–0.83*	–0.83*	–0.82*	–0.84*	–0.68	–
Average Evaporation Morning (X_5)	0.82*	0.84*	0.82*	0.82*	0.84*	0.83*	0.83*	0.88*	–
Average Evaporation Evening (X_6)	0.28	0.25	0.26	0.26	0.28	0.25	0.27	0.20	–
Wind Speed (X_7)	–0.63	–0.63	–0.64	–0.64	–0.62	–0.63	–0.64	–0.48	–
Sunshine (X_8)	–0.72	–0.79*	–0.78*	–0.79*	–0.77*	–0.81*	–0.75*	–0.80*	–

*Significant at 5% ($P = 0.05$)

2.6 *Brassica-Fusarium* R-Genes

2.6.1 Mapping of R-Genes of *Brassica*

Most resistance resources have been identified in *B. oleracea* (Table 2.13). Specifically, two types of resistance have been characterized, i.e., A and B. Type A resistance is stable under high or low temperature and follows a single dominant inheritance pattern; type B polygenic resistance is unstable under high temperatures above 24°C. The type A single dominant resistance gene for Foc race 1 has been explored extensively in the last several years. The Fusarium wilt R-gene *FocBo1* was first mapped to linkage group seven using both BSA and QTL analysis by Pu et al. (2012). Lv et al. (2013,2014) generated a genetic linkage map based on a cabbage DH population and mapped the R-gene *FOC1* to a 1.8 cM interval between two adjacent InDel markers. The authors further mapped the candidate gene *FOC1* to a repredicted *Bol037156*, which encodes a TIR-NBS-LRR, using an enlarged population. Shimizu et al. (2015) also mapped the resistance locus *FocBo1* by using 139recombinant F2 plants and identified a candidate gene, *Bra012688*. The two mapped candidates are homologous with high identity. However, the functions of these genes remain to be identified. Type A resistance to Foc race 1 conferred by a dominant single gene, *FOC1*, has been successfully mapped and molecular markers have been developed and applied to generate various resistant cultivars. In addition, MAS using these markers has been combined with other breeding methods to promote the breeding process. Lv et al. (2014) reported the use of isolated microspore cultures with MAS to rapidly obtain target DH lines with FW resistance, which could be used directly in resistance breeding, thereby shortening the breeding period by 2–3 years (Lv et al. 2020).

Table 2.13 Resistant sources identified in *Brassica* to *Fusarium* (Lv et al. 2020)

Species	Pathogen race/ Isolates	Techniques	Results
B. oleracea	Cong: 1-1 strain	SSR	A linked marker at 1.2 cM
B. oleracea	FGL3-6 race 1	InDel	FOC1 in an interval of 1.8 cM
B. oleracea	FGL3-6 race 1	InDel	The candidate is a repredicted *Bol037156*
B. rapa	Cong: 1-1strain	RNA-seq	Two candidate R-genes identified: *Bra012688* and *Bra012689*
B. oleracea	Cong: 1-1 strain	SSR	The candidate is *Bra012688*
B. oleracea	FGL3-6 race-1	SRR	A high-efficiency marker located 75 kb from the resistance gene

2.7 *Brassica-Hyaloperonospora* R-Genes

2.7.1 Identification of R-Genes Genotypes

Differential host resistance to isolates of *H. parasitica* has been identified in *B. campestris*, *B. juncea*, *B. napus*, *B. oleracea,* and *R. sativus* (Bonnet and Blancard 1987; Lucas et al. 1988; Nashaat and Rawlinson 1994; Nashaat et al. 1995a, b; Nashaat et al. 1997, 2004; Silue et al. 1996). Sources of resistance to *Hyaloperonospora parasitica* in different host species of crucifers identified from different countries of the world are given in Table 2.14 (Saharan et al. 2017).

2.7.2 Designation and Nomenclature of Downy Mildew Resistance Genes (R-genes) and Isolates (Races/Pathotypes)

Breeders have introgressed disease resistance (R) genes from both cultivated and wild cruciferous plants in their efforts to produce more resistant varieties. Even so, new races of downy mildew pathogen regularly evolve through sexual reproduction of the pathogen that can overcome individual R–genes. As per the term "gene–for–gene" hypothesis a plant to exhibit resistance (incompatibility) to a pathogen, a R-gene must be present in the plant, and a corresponding avirulence (AVR) gene must be present in the pathogen. An absence of either leads to disease (compatibility). This led to the hypothesis (elicitor/receptor model) that R-genes encode receptors that enable plant to detect the ingress of pathogens whose avirulence genes cause them to produce the corresponding legends. Thus, R-genes products might be expected to have two function: first, molecular recognition, and second, activation of plant defense upon recognition. Historically, and as per convention R-genes have been designated in different host-pathogen interactions on the basis of name of a disease, and /or a pathogen/host. To narrate the some of the R-genes, viz. *Sr* for stem rust, *Lr* for leaf rust, *Yr* for yellow rust, Pm for powdery mildew resistance of wheat; WRR for white rust resistance of crucifers, are based on the name of the diseases caused on respective hosts. R-genes Hm1 confers resistance to maize leaf blight (*Helminthosporium maydis*, *Cochliobolus carbonum*), Xa 21 confers resistance to *Xanthomonas* of rice, Cf9, Cf2 confers resistance to *Colletotrichum fulvum* of tomato, RPP1, RPP5 confers resistance to *Peronospora parasitica* (*H. parasitica*) of crucifers are based on the names of the pathogens. R-genes ATR1, ATR 13 conferring resistance to downy mildew pathogen (*H. arabidopsidis*) of *Arabidopsis thaliana* are based on the name of hosts generic, and specific names (AT). In the past specificity loci (R-genes) of *A. thaliana* were named as RPP loci (abbreviation of recognition of *P. parasitica* or else recognized by *P. parasitica* (AVR), and was numbered consecutively (i.e., RPP1, RPP 2). Specificity loci (AVR genes) of *P. parasitica* had been named ATR loci (abbreviation of *A. thaliana* recognized or else *A. thaliana* recognition), and were numbered the same as the corresponding RPP locus (R-genes). This nomenclature is descriptive of an interaction regardless of which partner is actively recognizing the other.

Table 2.14 Sources of R-genes in crucifers to *Hyaloperonospora parasitica* (Downy mildew) (Saharan et al. 2017 updated)

Crucifer's host	Sources	Country	References
Brassica alba (white mustard)	All Indian accessions	India	Saharan (1992a, b)
B. carinata (Ethiopian mustard)	All Indian accessions , HC1	India	Saharan (1992a, b), Saharan (1996), Dang et al. (2000)
B. campestris	Candle	India	Saharan (1992a, b)
B. campestris var. *toria* (Toria)	1B-586	India	Kolte and Tewari (1980)
B. campestris var. *Yellow Sarson* (Yellow Sarson)	YST-6	India	Kolte and Tewari (1980)
B. campestris var. *Brown Sarson* (Brown Sarson)	BS-16	India	Kolte and Tewari (1980)
B. juncea (Indian mustard)	IC 296685, IC 326253, IC 417020, DIR 1507, DIR 1522, PI340207, PI 340218, PI 347618 Domo, RC-781, EC 126743, Zem, YRT3, YRT 45, YRT 72, PR 8805, RN 248, EC 129126-1, PC-3 RESBJ-01, RESBJ-02, RESBJ-03, EC 399296, EC 399299,EC 399301, EC 399313, EC 414308 (NRCR-837), EC 414319 (NRCR-836)	India, UK	Bisht et al. (2015), Dang et al. (2000), Ebrahimi et al. (1976), Saharan (1992a, b), Saharan (1996), Nashaat and Awasthi (1995), Kolte et al. (2008), Ravi and Awasthi (2019)
B. oleracea var. *botrytis* (Cauliflower)	C300, Maudez, Igloo, Snowball y, Dok Elgon, RS-355, PI 181860, PI 188562, PI 189028 (MR), PI 204765, PI 204768, PI 204772, PI 204773, PI 204779, PI 241612, PI 264656, PI 291567, PI 373906, PI 462225 (MR), KPS-1, PI 231210, PI 189028, Aviso, Exale, B1704, Stanley, Adam's white head, IIHR-142, IIHR-217	USA, India	Ménard et al. (1999), Kontaxis et al. (1979), Thomas and Jourdain (1990), Sharma et al. (1991), Hoser-Krauze et al. (1991), Rooster et al. (1999), Pandey et al. (1995)
B. oleracea var. *capitata* (Cabbage)	January King, Balkan, Spitz Kool, PI 246063, PI 246077, PI 245013, Tromchuda cabbage "Algarvia" (ISA 207), PI 245015, Geneva 145-1	Canada, USA	Greenhalgh and Mitchell (1976), Elenkov (1979), Verma and Thakur (1989), Hoser-Krauze et al. (1991), Caravalho and Monteiro (1996), Sherf and Macnab (1986)

(continued)

Table 2.14 (continued)

Crucifer's host	Sources	Country	References
B. oleracea var. *accephala* gr. *ornamentalis*	Decorative cabbage	–	Vitanova (1996)
B. oleracea (Broccoli)	Ching-Long 45, Calabrese, Grand Central, PI 231210, Italian Green Sprouting, Hyb 1230 (Moran), Green surf (Moran), 2804 (Qualisal), GSV 82-4310 (Goldsmith), XPH 1117 (Asgrow), Hyb. 288 (Moran), AVX 7631 (Sun Seeds), PI 263056, PI 263057, PI 3573, PI 3574, PI 418984, PI 418985, PI 418986, PI 418987, PI 418988, OSU CR 2 to OSU CR 8, Citation, Excalibur, Nancy	Italy	Yang et al. (1998), Natti et al. (1956), Natti (1958), Laemmlen and Mayberry (1984), Hoser-Krauze et al. (1991), Baggett and Kean (1985), Sherf and Macnab (1986)
B. napus (Rape)	Hg Vestal, Eurora, Janetzki, Kubla, Lesira, Mogul, Primar, Rapot, Rapara, Sinus Cultivar 78-22, Cresor, PI 199949, PI 263056, Gulivar, Midas, Tower, GS 7027, RES 01-14, RES-02, RES-26, HNS3, HNS4, GSL-1, GSL 1501, EC338986-2, EC338996-1	UK, USA, India,	Jonsson (1966), Dixon (1975), Chang (1981), Kluczewski and Lucas (1983), Thomas and Jourdain (1992), Saharan (1992a,b), Dang et al. (2000), Nashaat and Awasthi (1995), Nashaat et al. (1996, 1997), Saharan (1996), Kolte et al. (2008)
B. chinensis (Chinese Cabbage)	Bau chin 26, PHW 64707, PHW 64710, PHW 64722, PHW 64620, Hyb. 77M (3)-27, Hyb.77M(3)-35, Hyb. 82-46, Hyb. 82-46R, Hyb. 82-156, Hyb. 82-157, Dwarf Resistant 5, Dwarf Resistant 6	India	Niu et al. (1983), Anonymous (1987a, b), Cao et al. (1998)
B. nigra (Black Mustard)	PI 19948		Thomas and Jourdain (1992)
B. rapa	PI 418984, PI 418988, PI 418987, PI 418988, *B. rapa* subsp. *Rapifera* Long Blanc de croissy, Stanis, Jaune Boule d'or	USA	Thomas and Jourdain (1992), Silue et al. (1996)
Raphanus sativus (Radish)	Okura, Tokinoshi (All season) Bamba, Noir Lon d'orloge, Rave a Forcer	Japan	Shiraishi et al. (1974), Bonnet and Blancard (1987), Silue et al. (1996)
Cheiranthus cheirii (Wall flower)	Convent Garden Blood Red	Canada	Greenhalgh and Dickinson (1975)

Ideally new loci (R-gene) should be named strictly on the basis of genetic recombination. Unfortunate consequence of change of pathogen name from *P. parasitica* to *H. parasitica* to *H. arabidopsidis* is no longer intuitively connected with the downy mildew pathogen from its current name to recognize *P. parasitica* (RPP) gene designation. Such changes in the names of host (*Sisymbrium thalianum* (L.) Grey to *Arabidopsis thaliana* (L.) Heynh), and pathogen (*P. arabidopsidis* Gaum. to *P. parasitica* (Pers. ex. Fr.) Fr. to *H. parasitica* to *H. arabidopsidis* (Gaum.) Goker) may be confusing, and irritating for students, and researchers but it is inevitable in this modern era of molecular genetics, and phylogenetic analysis of living beings. However, there is a need to develop, and adopt a standardized system, and procedure for the designation of R-genes. On the basis of host, and pathogen (both) recognition which can reflect both in their interaction phenotype, i.e., ATHA1, ATHA2 for recognition of R-genes by *H. arabidopsidis* from *A. thaliana*; ATAC 1, ATAC 2, etc. for recognition of R-genes by *A. candida* from *A. thaliana*; BJHP 1, BJHP 2 for recognition of R-genes by *H. parasitica* from *B. juncea*. The downy mildew resistance genes (R-genes) recognized by *H. parasitica* isolates (pathotypes) from *A. thaliana* accessions are given in Table 2.15. Similarly, for other crucifers a uniform system may be adopted, viz. BNHP for *B. napus-H. parasitica*, BOHP for *B. oleracea-H. parasitica*, and BRHP for *B. rapa–H. parasitica* interaction phenotypes (R-genes).

Designation and nomenclature of pathogenic isolates (races/pathotypes) have gone through evolutionary process, and methods. (1) Initially, physiologic races were generally designated as numbers or letters in an arbitrary manner in order of their discovery, i.e., *Puccinia* spp., *Melampsora lini*, *Albugo candida*, and *Peronospora parasitica*. (2) An improvement over the use of arbitrary numbers or letters was Black's nomenclature in which the races were designated on the basis of their virulence on particular genes for resistance, i.e., an isolate of *Phytophthora infestans* attacking a potato cv. carrying the R-genes, R1 was designated as race 1, the one attacking R4 as race 4, and an isolate attacking both R1 and R4 as race 1, 4. The race which was avirulent on all the genes for resistance was designated as race 0. (3) Virulence formulae were proposed to designate races of stem, and leaf rust of wheat virulent or avirulent on particular genes for resistance, e.g., the formula 6,7,10/5,8,9, 9a, 11 for a race of *P. graminis tritici* indicates that the race is virulent on Sr6, Sr7, Sr10 but avirulent on Sr5, Sr8, Sr9 a, and Sr11. (4) A very complicated method was proposed by Habgood using binary, and decanary values. (5) A virulence analysis method was suggested using mobile nurseries in case of powdery mildew of barley.

Like with any other host-pathosystem, the designation, and nomenclature of downy mildew of crucifers pathogenic isolates/races/pathotypes have not been standardized at International level. No standard method and procedure has been adopted. Each researcher has used his own vision, and system to name the pathogenic isolates collected from different locations/countries from cruciferous host species/varieties/accessions. However, a naming system for the isolates of *H. arabidopsidis* from *A. thaliana* was introduced by Dangl et al. (1992), Holub

Table 2.15 Resistance genes (R-genes) identified in crucifers (*Arabidopsis thaliana*) against downy mildew (*Hyaloperonospora arabidopsidis*) isolates (pathotypes) (Slusarenko and Schlaich 2003)

Arabidopsis accessions	R-genes	Downy mildew isolates (pathotypes)
Col-0 (Columbia)	RPP-4	EMOY 2, EMWA 1
Col-0	RPP 2	CALA-2
WS-0 (Wassilewskija)	RPP 1A,B	EMOY 2
Ler-0 (Landsberg erecta)	RPP 5, RPP 8	EMOY 2, NOCO 2, EMWA 1
Ws-0	RPP 1A	CALA 2
Ws-0	RPP 1A, B, C	NOCO 2
Nd-1 (Niederzenz)	RPP 13 (ATR 13Nd)	MAKS 9, ASWA, EDCO, EMCO, GOCO
Ws-0	RPP 1 (ATR 1 Ws B)	MAKS 9
Rld 2 (Reschew)	RPP 11	WELA 1
Col-0	RPP 6	WELA
Col-0	RPP 7	HIKS
Nd-0	RPP 25	AHCO
Ler-0	RPP 27	HIKS
Ws-0	RPP 10	NOCO, EMOY, MAKS, COLA
Ws-0	RPP 14	NOCO, EMOY, MAKS
Nd-1	RPP 26	WACO
Nd-1	RPP 16	ASWA
Nd-1	RPP 17	EMCO
Ws-0	RPP 12	WELA
Col-0	RPP 18	HIND
Ler-0	RPP 23	GOWA
Ler-0	RPP 21	MADI, MAKS
Ler-0	RPP 22	ASWA
Ler-0	RPP 24	EDCO
Oy-0 (Oystese)	RPP 3	CALA
Wei-0 (Weiningen)	RPP 9	HIKS
Cola-0	RPP 19	HIND 4
Cola-0	RPP 20	WAND
Cola-0	RPP 28	HIND 2
	Tightly linked genes	
	RPP 1-Ws A	CALA, EMOY, HIKS, MAKS, NOCO
	RPP 1-Ws B	
	RPP 1-Ws C	

et al. (1994a, b), and Slusarenko and Schlaich (2003) on the basis of geographical location and ecotypes infected.

As for example, an isolate collected from suburb of Zurich called Weiningen and virulent on (among others) the ecotype *La*nds bergerecta was named WELA using first two letters of the location where the isolate was found (WE), combined with first two letters of susceptible ecotype (LA). Thus, NOCO was found in Norwich, and is virulent on Columbia, EMWA at East Malling, and is virulent on Wassilewskija.

The isolate EM=East malling, UK; CA= Canterbery, UK; WE= Weiningen; CH, and NO Norwich, UK, and the susceptible host line used for the isolates, third and fourth letters OY=OY-0; LA=LA-er; ND=Nd-o, and CO= Col-0, etc. New isolates collected from the same location, and maintained on the same host genotypes were distinguished by a number (e.g., EMOY1 and EMOY 2). However, the system and procedure of naming of an isolate should reflect both host-pathogen interactions to recognize avirulence gene (AVR gene) of the pathogen along with R-gene of the host. i.e., HPBJ 1, HPBJ 2 etc. indicating H. parasitica isolate/pathotype recognized R-genes 1 and 2 from B. juncea after interaction of isolate (pathotype) HPBJ1 and HPBJ2. Like international code of Botanical Nomenclature for naming an organism a pattern or system of designation, and nomenclature of R-genes, and AVR genes should be developed at International level with code of conduct so that researchers can compare/confirm and validate each other's results. It will avoid unnecessary repetition and confusion among *Brassica* scientists (Saharan et al. 2017).

2.7.3 Mutational Approach to Identify Resistance Genes

A mutational approach has been adopted to identify genes that are necessary for resistance mediated by *RPP*-5 and *RPP*-14, and to attempt separation of RPP-14 from the other closely linked RPP gene specificities. Parker et al. (1996) screened mutagenized populations of *Ler*-0 and WsO for mutations that cause a change from N0CO2 resistance to susceptibility. They have described a recessive mutation of WsO called eds1 (for enhanced disease susceptibility), which abolishes the resistance mediated by *RPP*-14 as well as by other linked and unlinked *RPP* genes present in the WsO background. This mutation also partially suppresses resistance of WsO to five *Brassica oleracea*-infecting isolates of *H. parasitica* to which all *Arabidopsis* ecotypes so far tested exhibit resistance, implicating a possible common functional role for the EDS 1 protein in downy mildew resistance in *Arabidopsis* and *Brassica* plants.

The interaction between *Arabidopsis* and *H. parasitica* provides an attractive model pathosystem to identify molecular components of the host that are required for genotype-specific recognition of the parasite. These components are the so-called RPP genes (resistance to *H. parasitica*).

Mutational analysis of ecotype wassilwskija (Ws-O) revealed an RPP-nonspecific locus called EDS-1 (for enhanced disease susceptibility) that is required for the function of RPP genes on chromosomes 3 (*RPP1*/*RPP*-14 and *RPP*-10) and 4 (*RPP*-12). Genetic analysis demonstrated that the eds1 mutation is recessive, and is not a defective allele of any known *RPP* gene, mapping to the bottom arm of chromosome 3 (13 centimorgans below *RPP1*/*RPP14*). Phenotypically, the Ws-eds1 mutant seedlings supported heavy sporulation by *H. parasitica* isolates that are each diagnostic for one of the *RPP* genes in wild-type Ws-0; none of the isolates is capable of sporulating on wild-type Ws-0. Ws-eds1 seedlings exhibited enhanced susceptibility to some *H. parasitica* isolates when compared with a compatible wild-type ecotype, Columbia, and the eds1 parental ecotype, Ws-0. This was observed as earlier initiation of sporulation, and elevated production of conidia. Surprisingly, cotyledons of Ws-eds1 also supported low sporulation by five isolates of *H. parasitica* from *Brassica oleracea*. These isolates were unable to sporulate on >100 ecotype of *Arabidopsis*, including wild-type Ws-0. An isolate of *Albugo candida* from *B. oleracea* also sporulated on Ws-eds1, but the mutant exhibited no alteration in phenotype when inoculated with several oomycetes isolates from other host species. The bacterial resistance gene RPM1 conferring specific recognition of the avirulence gene *avrB* from *Pseudomonas syringae* pv. *glycinea* was not compromised in Ws-eds1 plants. The mutant also retained full responsiveness to the chemical inducer of systemic acquired resistance, 2, 6-dichloro-isonicotinic acid; Ws-eds1 seedlings treated with 2, 6 dichloro-isonicotinic acid became resistant to the ws-0 compatible and Ws-0 incompatible *H. parasitica* isolates, Emwa1 and NOCO2, respectively. The *EDS1* gene appears to be a necessary component of the resistance response specified by several *RPP* genes, and is likely to function upstream from the convergence of disease resistance pathways in *Arabidopsis* (Parker et al. 1996). While studying the *Arabidopsis* downy mildew resistance gene *RPP* 5, Parker et al. (1997) found that it encodes a protein that possesses a putative nucleotide-binding site, and leucine-rich repeats, and its product exhibits striking structural similarly to the plant resistance gene products N and L6. Like N and L6, the *RPP5* N-terminal domain resembles the cytoplasmic domains of the Drosophila Toll, and mammalian interleukin-1 trans-membrane receptors. In contrast to N, and L6, which produce predicted truncated products by alternative splicing. *RPP5* appears to express only a single transcript corresponding to full length protein. However, a truncated form structurally similar to those of N and L6 is encoded by one or more other members of the *RPP5* gene family that are tightly clustered on chromosome 4. The organization of repeated units within the leucine-rich repeats encoded by the wild-type *RPP5* gene and an *RPP5* mutant allele provides molecular evidence for the heightened capacity of this domain to evolve novel configuration, and potentially new disease resistance specificities. Nucleotide sequence data have been submitted to the Gen bank database under the accession number U97106.

2.8 *Brassica-Leptosphaeria* R-Genes

2.8.1 Identification of R-Genes Sources

The potential for *L. maculans* to cause severe damage was highlighted by the widespread total collapse of susceptible oilseed rape varieties in Western Australia in the early 1970s (Sivasithamparam et al. 2005) and subsequently in the early 2000s with the collapse of single dominant gene-based resistance from *B. rapa* ssp. *sylvestris* in various regions of Australia. *L. maculans* also causes severe damage to oilseed rape in the UK (Barnes et al. 2010), Canada, and mainland Europe. The less aggressive blackleg pathogen, *L. biglobosa,* is more prevalent in China and has been reported to be present in New Zealand. Although no severe *B. napus* yield loss has been reported from *L. biglobosa*, researchers have taken precautions to prevent the spread of *L. maculans* into China (Zhang et al. 2014a; Cai et al. 2017).

Blackleg resistance genes in *Brassica* species have been increasing over the years, with the focus more recently towards quantitative resistance. Although 12 Blackleg qualitative race specific *R*-genes have been genetically mapped and more have been recognized (Table 2.16; Fig. 2.3), it remains uncertain if some of these genes are actually the same gene with different nomenclature or allelic variants of the same gene. This uncertainty arises from researchers using different crosses, different isolates, and different marker systems. For example, the genes *LmR1*, *cRLMm*, *Rlm4*, *cRLMrb*, and *LEM1* were all mapped on chromosome A7 in *B. napus* using Australian cultivars Shiralee, Maluka, Skipton, and the French cultivar Major, respectively (Mayerhofer et al. 1997; Rimmer et al. 1999; Balesdent et al. 2001; Rimmer 2006), with different molecular markers and therefore, until one is cloned, it is unknown if they are the same gene. This uncertainty has implications in breeding as breeders do not know whether they are using the same or different sources of resistance. Furthermore, it is recommended in Australia to rotate the source of resistance gene grown annually so the pathogen population cannot build up to cause severe yield loss (Van De Wouw et al. 2016b). The lack of knowledge of the number of resistance genes makes this more difficult to achieve and often means a pool of fewer resistance genes is used. Another example relates to the *LepR3* gene in *B. napus* which has been shown to interact with *AvrLm1;* however, phenotypic studies suggest that *AvrLm1* may also be recognized by *Rlm1*, thus raising the question of the relationship between these two *R*-genes. The relationship between *Rlm4* (Raman et al. 2012b) and *Rlm7* (Balesdent et al. 2002) implicated to be responsible for the HR outcome on *B. napus* during attack by *L. maculans* carrying the allele *AvrLm4* or *AvrLm7,* respectively, is also unknown. Cloning studies on these effectors revealed that *AvrLm4* and *AvrLm7* are two distinct alleles of the same gene, termed *AvrLm4-7*, and can induce resistance in *B. napus* that harbors either *Rlm4* or *Rlm7*. However, the relationship between the two *R*-genes is yet to be determined (Parlange et al. 2009). It is also possible that *Rlm3*, *Rlm4,* and *Rlm7* may all be allelic variations of the same gene, based on evidence showing the location of these genes on a clustered region on chromosome A7 in *B. napus* and not being found together in the same *B. napus* pure lines, with a recent study showing an unusual interplay between *AvrLm4-7* and *AvrLm3* (Delourme et al. 2004; Larkan et al. 2016b; Plissonneau et al. 2017). Clearly there is a need to reconcile different nomenclatures for these genes and the relationship between them. The success story of clarifying *AvrLmJ1* and *AvrLm5* are indeed the same *Avr* gene (Plissonneau et al. 2017), is an example that should be followed by *Brassica* and blackleg researchers to clarify nomenclature for other genes (Table 2.16).

2.8.2 Novel Sources and Transfer of Genes

Novel sources of *R*-genes for Blackleg can be sourced from other A- and C-genome *Brassica* species or from the B genomes in *B. nigra* (Chevre et al. 1996), *B. juncea*

Table 2.16 Major resistance in *Brassica* crops and R-genes/QTL identified and genetically mapped or cloned against various diseases (Neik et al. 2017)

Resistance gene/QTL mapped	References	Resistance gene/ allele cloned
***1. Plasmodiophora brassicae*, an obligate biotroph chytrid (protista) causing clubroot disease**		
-*B. napus* (Canola lines "1CA0591.323" and "1CA0591.263" (Resistant, R) derived from cultivar "Mendel"× "A7-26NR" (Susceptible, S), using isolate SACAN-ss1, Williams pathotype 3): "Mendel" carries *CRa* and CRb^{Kato} on chromosome A3	Fredua-Agyeman and Rahman (2016)	- *Crr1a* (TIR-NB-LRR) from *B. rapa* derived from European fodder turnip cultivar "Siloga" resistant to isolate Ano-01 Williams race 3 (Hatakeyama et al. 2013)
CRa and CRb^{Kato} may be same allele also shown	Zhang et al. (2014a, b)	
-*B. napus* (Inbred line "12-3" (R) × inbred line "12-1" (S) using pathotype 3): New CR gene, act singly or combined with *CRa*	Zhang et al. (2016a, b)	
-*B. napus* (DH "263/11" (R) × Cultivar "Express" (S), 7 different isolates from Sweden, Germany, and France): 19 QTL mapped. Broad-spectrum resistance mapped on chromosome A2, A3, A8, A9, C3, C5, C6, and C9. QTL on A3 corresponds to the region for *CRk* and *Crr3*, some QTL have additive effects	Werner et al. (2007)	
-*B. napus* (Cultivar "Mendel"(R) × Breeding line (S), using isolate 1 that is highly virulent on *B. oleracea*): One single dominant locus from *B. rapa* ("ECD-04"), two recessive loci from *B. oleracea* ("ECD-15")	Diederichsen et al. (2006)	- *CRa* (DH lines "T136-8" and "Q5," Chinese cabbage cultivar "Ryutoku" and "CR Shinki," fodder turnip "Debra" and "Gelria R," using M85 clubroot isolate) (TIR-NB-LRR) from *B. rapa* (Ueno et al.2012)
-*B. napus* (DH progeny from Darmor-*bzh* (R) × Yudal (S), using single spore isolate pathotypes 4 and 7 according to: Monogenic or polygenic depending on isolate, major resistance dominant gene *Pb-Bn1* towards isolate Pb137-522 on LG	Some et al. (1996)	

(continued)

Table 2.16 (continued)

Resistance gene/QTL mapped	References	Resistance gene/ allele cloned
DY4, also show weaker partial resistance effect in association with other QTL. Additive and epistatic QTL were identified	Manzanares-Dauleux et al. (2000a, b)	
-*B. napus* (Natural lines *B. oleracea* "ECD-15" (R) × B. rapa" ECD-04" (R)), from which cultivar "Mendel" is produced, using virulent field isolates): Two dominant, unlinked genes	Diederichsen and Sacristan (1996)	
-*B. rapa* (Fine map *Rcr2* in Chinese cabbage cv. "Jazz" (R) × DH ACDC (S) using pathotype 3): *Rcr2* on 22 and 26 Mb of chromosome A3	Huang et al. (2017)	
-*B. rapa* (Breeding line "T19" originate from cultivar "Pluto" (R) × DH line "ACDC" (S), using Williams pathotype 2,3,5,6,8, 5x): GBS identified *Rcr4* on chromosome A3 and two novel QTL, *Rcr8* on A2 and *Rcr9* on A8	Yu et al. (2017)	
-*B. rapa* (Inbred lines Chinese cabbage "CCR13685" (R) × Pak choi "GHQ11021" (S), using isolate with unknown race/pathotype): single dominant gene	Chen et al. (2016)	
-*B. rapa* (Fine map *Rpb1* from Chu *et al.*2013): Single dominant allele *Rcr1* (or also known as *Rpb1*) on LG A3, close position to both *CRa* and *CRb*, against Williams pathotype 3. Rcr1 is predicted to be TIR-NBS-LRR)	Chu et al. (2014), Yu et al. (2017)	
-*B. rapa* (Pak choi "702-5" (S) × DH line "CR Shinki" (R), using Williams race 4): Mapped physical location of *CRb* on chromosome A3 on *B. rapa* with interval length of 83.5 kb with 15 candidate	Fredua-Agyeman and Rahman (2016), Zhang et al. (2014a, b)	

(continued)

Table 2.16 (continued)

Resistance gene/QTL mapped	References	Resistance gene/ allele cloned
genes including NBS-LRR genes. *CRb* is not an allele of *CRa*, but closely linked genes. *CRa and CRb*^{Kato} may be same allele		
- *B. rapa* (Inbred lines turnip "Siloga" (R) × Chinese cabbage "BJN3" (S), using isolate Williams race 4): One major locus *QS_B3.1* on chromosome A3 corresponding to *CRa* and *CRb*, and two minor loci *QS_B1.1* on chromosome A1 (homologous to *Arabidopsis* chromosome 3 independent of any other CR loci) and *QS_B8.1* on chromosome A8 (share same locus with *Crr1b* and *Crr1*). Additive effects and epistatic interactions were found in *B. rapa*	Pang et al. (2014)	
- *B. rapa* (Five Chinese cabbage cultivars (R) × *B. rapa* oilseed accession (S) using Canadian field isolate): linkage group N3 corresponding to chromosome A3	Gao et al.(2014)	
-*B. rapa* (Fine map *CRb* in DH line "CR Shinki" (R) × Chinese cabbage "702-5" (S) using pathotype 4): *CRb* gene was tightly linked to two other *CR* genes, *CRa* and *CRb*^{Kato}	Zhang et al. (2014a, b)	
-*B. rapa* (DH "G004" (R) × DH "A9709" (S), using field isolates Ano-01, Wakayama-01, and Nos. 5, 7, 9, and 14 based on *Two* genes, *Crr1a* with major effect and *Crr1b* with minor effect at *Crr1* locus	Hatakeyama et al. (2004, 2013)	
B. rapa (Inbred turnip line "ECD04" (R) × Inbred Chinese cabbage line "C59-1" (S) × using four different	Chen et al. (2013)	

(continued)

Table 2.16 (continued)

Resistance gene/QTL mapped	References	Resistance gene/ allele cloned
isolates Pb2, Pb4, Pb7, and Pb10): Partial resistance, *PbBa1.1 on A1, PbBa3.1, PbBa3.2 and PbBa3.3* on A3 (*PbBa3.1* and *PbBa3.3* on different region), *PbBa8.1* on A8		
- *B. rapa* (Cultivar Pak choi "FN" (R) × DH "ACDC" (S), using isolate Williams pathotype 2, 3, 5, 6, 8): Single dominant gene *Rpb1* located on LG A3, close to *CRa*	Chu et al. (2013)	
- *B. rapa* (Cultivars "Akiriso" and "CR Shinki," using isolate No. 14 or pathotype group 3 according to : Single dominant gene linked to *CRb*, or is *CRb* found in "Akiriso" *CRb* and *CRa* are one and the same clubroot-resistance gene	Hatakeyama et al. (2004, 2017), Kato et al. (2012)	
-*B. rapa* (Fine map *Crr1* in *Arabidopsis*): *Crr1* is likely consist of two genetic loci. The gene order is conserved except for one inversion in which insertion is found	Suwabe et al. (2012)	
***2. Sclerotinia sclerotiorum*, a necrotrophic fungal pathogen causing Sclerotinia Stem Rot disease**		
- *B. napus* (152 accessions from Canada, China, and Europe, as well as South Korea and Japan, using isolate #321 collected from oilseed rape fields in Alberta, Canada): 21 loci conferring resistance to *S. sclerotiorum* and 13 loci conferring susceptibility mapped to 12 of the 19 *B. napus* chromosomes	Gyawali et al. (2016)	None
- *B. napus* (347 Chinese accessions comprising spring, winter and semi-winter lines, using same isolate in: 17 SR QTL on chromosome A8 and chromosome C6 including five on A8, and 12 on C6. The C6 QTL corresponds to C6 in	Mei et al. (2013), Zhao et al. (2006), Wu et al. (2013), Li et al. (2015a, b), Wei et al. (2016a, b)	

(continued)

Table 2.16 (continued)

Resistance gene/QTL mapped	References	Resistance gene/ allele cloned
Candidate genes were predicted based on GWAS and transcriptome sequencing		
- *B. napus* (Physical map construction based on *B. napus* genome): 35 QTL mapped including eight LR and 27 SR. LR QTL distributed across chromosome A9 (corresponding to A9 region in; C1 and C9, SR QTL mapped to 12 chromosomes (A1, 2, 6, 8, 9, C2, 4, 5, 6, 7, 8, 9) where C9 corresponds to that in Some SR and LR QTL share same genomic region	Wu et al. (2013), Mei et al. (2013), Li et al. (2015a, b)	
B. napus (DH population from cultivar "Express" (female) × Chinese line "SWU7" (male), using same isolate in: Six field resistance (FR) QTL and five SR under controlled environment identified. Four FR and two SR mapped to chromosome C2 and one SR on chromosome A2, where C2 is homoeologous to A2 (corresponding to chromosome N2 and chromosome N12 in and both chromosomes syntenic with progenitor genomes	Mei et al. (2013)): Zhao et al. (2006), Wei et al. (2014)	
- *B. napus* (Pure line "J7005" (R) × Cultivar "Huashang 5" (S), using Chinese isolate): 13 QTL identified. Three LR on LG A3, A9, and C5, SR on LG A1, 2, 3, 6, 8, 9, C6, and C7, C8. Two Major QTL are *LRA9* and *SRC6* (LG A9 and LG C6 respectively); both quantitative traits with additive gene effects. *BnaC. IGMT5.a* is a candidate gene for QTL *SRC6*	Wu et al.(2013)	

(continued)

Table 2.16 (continued)

Resistance gene/QTL mapped	References	Resistance gene/ allele cloned
- *B. napus* (DH population from line "DH821" (R) × line "DHBa0604" (S), using Chinese isolate): 21 QTL, on chromosome N3 (or A3), chromosome N4 (or A4), chromosome N11 (or C1), chromosome N17 (or C7), chromosome N12 (or C2)	Yin et al.(2010)	
B. napus (Two DH populations: Chinese winter line "Hua dbl2" (R) × European spring line "P1804" and Cultivar "Major" (R) × Cultivar "Stellar" (S), using isolate 105HT derived from the US soybeans): Nine QTL on seven LGs including chromosome N2 (or A2) with its homeologous non-reciprocal transposition on chromosome N12 (or C2), chromosome N16 (or C6), chromosome N5 (or A5), chromosome N14 (or C4), chromosome N3 (or A3), chromosome N19 (or C9)	Zhao et al. (2006)	
B. napus (Breeding line "Ning RS-1" (R) × male sterility restorer line "H5200," using isolate HY-12): Six QTL including three LR QTL and three mature stage resistance with one major QTL *qLRS1* on LG 17 and one major QTL, *qSRM1* on LG 15 with additive effect	Zhao and Meng (2003)	
B. oleracea (Wild relative of *B. oleracea* that is *B. incana* (R) × Cultivated *B. oleracea* var. *alboglabra* (S), using same isolate in: Multiple epistatic interactions (polygenic genes) for LR and SR, control of these genes is different in both resistances	Mei et al. (2013) Disi et al. (2014)	

(continued)

Table 2.16 (continued)

Resistance gene/QTL mapped	References	Resistance gene/ allele cloned
B. oleracea (Wild relative of *B. oleracea* that is *B. incana* (R) × Cultivated *B. oleracea* var. *alboglabra* (S), using Chinese field isolate): 12 QTL for leaf resistance (LR) and 6 QTL for stem resistance (SR). Two major QTL on chromosome C9 for LR and SR with corresponding region on chromosome A9	Mei et al. (2013)	
***3. Hyaloperonospora parasitica* (syn. *Peronospora parasitica*), an obligate biotroph oomycete causing downy mildew disease**		
- *B. napus* (Resistant accessions “RES-26,” “RES-02,” and Susceptible cultivar “Callypso,” using isolates R1 and P003 provided by University of Nottingham): Single, partially dominant gene in RES-26 and two independent partially dominant genes in RES-02. These resistant genes could be closely linked, allelic or identical. Another single, incomplete dominant gene is found in RES-02	Nashaat et al.(1997)	None
B. napus (Cultivar “Victor,” “Jet Neuf,” and “Cresor,” using isolate R3 collected from winter oilseed rape crop in Leicestershire in 1982): Single dominant gene	Lucas et al. (1988)	
- *B. rapa* (Mapping population derived from (Yu *et al.*2009): Four major QTL including sBrDM8 (seedling resistance, identical to BraDM), yBrDM8 (young plant), rBrDM8 (rosette), and hBrDM8 on chromosome A8 and two minor QTL rBrDM6 on chromosome A6 and hBrDM4 on chromosome A4. Candidate gene for sBrDM8 is serine/threonine kinase (STK) family.	Yu et al. (2016b)	

(continued)

Table 2.16 (continued)

Resistance gene/QTL mapped	References	Resistance gene/ allele cloned
B. rapa (Inbred lines "RS1" (R) × "SS1" (S), using natural infection in Korea): Single dominant gene BrRHP1 on LG A1	Kim et al. (2011)	
B. rapa (DH population from line "T12-19" × line "91-112" (S), using Chinese isolate): Major QTL BraDM for seedling resistance on LG A8 and minor QTL on A6 . SSR markers for BraDM were developed	Yu et al. (2009, 2011)	
- *B. oleracea* (BAC libraries BoT01 and BoCig): Single dominant locus Pp523 on chromosome C8 and C5	Carlier et al. (2011)	
- *B. oleracea* (Mapping population of Portuguese genotypes "Couve Algarvia," "Penca de Chaves," and "Couve de Corte" and "DH-GK97362" (S), using Portuguese isolate P501): Two dominant genes at cotyledon stage and a single dominant gene at adult stage, inheritance is independent of stage	Monteiro et al. (2005)	
B. oleracea (52 entries including landraces, wild accessions and hybrid between wild and cultivars, using mixture of field isolates from *B. napus* in Sweden): Recessive resistance gene at cotyledon stage	Carlsson et al. (2004)	
- *B. oleracea* (S4 line derived from accession OL87125 (R) × rapid-cycling *B. oleracea* DH line "GK97362" (S), using (unknown isolate source): Pp523 locus on chromosome C8 confers adult resistance	Farinhó et al. (2004)	

(continued)

Table 2.16 (continued)

Resistance gene/QTL mapped	References	Resistance gene/ allele cloned
- *B. oleracea* (DH broccoli "USVL089" (R) × DH "USVL047" (S), using US field isolate): Single dominant gene at cotyledon stage	Farnham et al. (2002)	
- *B. oleracea* (DH broccoli "USVL012" (R) × DH "USVL047" (S), using the US field isolate): Two unlinked dominant genes at true leaf stage (seedling resistance)	Wang et al. (2001)	
B. oleracea (DH broccoli "USVL" series (R) and hybrid "Green Valiant" (S), using isolate SC1 that is race 2 isolate as described by (Thomas and Jourdain 1990) and CA1 from broccoli field in California with unknown race specificity): Polygenic	Wang et al. (2000)	
- *B. oleracea* (Cauliflower F1 hybrids 9304F1, 9305F1, 9306F1, 9311F1, and two open-pollinated cultivars "Perfection" and "Aberdeen," using Denmark isolate FYN93.cau and others from Europe, the UK and the US): Single dominant gene at cotyledon stage, associate with adult stage	Jensen et al. (1999a)	
B. oleracea (20 DH broccoli from "Corvet," "Shogun," "Skiff," "Atsumori," and "OSU" series, using three isolates, FYN82.cau, Lincs82.cau, and Moz82.cab obtained from cauliflower or broccoli host from Europe, UK, US and Mozambique): Partial resistance at cotyledon stage	Jensen et al.(1999b)	
B. oleracea (Cauliflower cc (R) × HR 5-4 (S), cc (R) × 244 (S), 3-5-1-1 (R) × 244 (S), cc (R) × 3-5-1-1 (R) and 244 (S) × 267-6-9	Mahajan et al. (1995)	

(continued)

Table 2.16 (continued)

Resistance gene/QTL mapped	References	Resistance gene/ allele cloned
(S) in India, using natural infection method): Single dominant gene PPA3 with recessive epistasis		
- *B. oleracea* (10 broccoli breeding lines, using isolates from the US): Single dominant gene and modifying genes at seven or more leaf stage)	Dickson and Petzoldt (1993)	
B. oleracea (Broccoli and cauliflower accessions from the US Plant Introduction collection, using US isolate): Single major gene against race 2)	Thomas and Jourdain (1990)	
- *B. juncea* (RESBJ accessions from Canada, Germany, and China, using 12 isolates from the UK and India derived from *B. juncea*, *B. rapa* and *B. napus*): Accessions RESBJ-200 and RESBJ-190 conditioned by single dominant genes, different in each accession but recognizing same isolate	Nashaat et al.(2004)	
B. juncea (31 spring type *B. juncea* accessions, susceptible control is winter type *B. napus* cv. Ariana, using four isolates from the UK and India derived from *B. napus* and *B. juncea*): mostly resistant with homozygous and heterozygous resistance.	Nashaat and Awasthi (1995)	
***Leptosphaeria maculans*, an facultative parasite causing Phoma leaf spot disease**		
- *B. napus* (GWAS panel of 179 accessions from DH population SAgS described in Raman et al. (2016), evaluated for resistance against 12 single spore isolates): Major R-gene for adult plant resistance Rlm12 on chromosome A1	Raman et al. (2016)	-*LepR3* (Larkan et al. 2013) -Allelic variant

(continued)

Table 2.16 (continued)

Resistance gene/QTL mapped	References	Resistance gene/ allele cloned
B. napus (DH populations from cultivar "AG-Castle" and "AV-Sapphire" (R) × "Topas" (S), field experiment in Australia): Three QTL for adult plant resistance on chromosome A1, A8, A9, and C6 where candidate genes include cysteine-rich receptor-like kinases on A1	Larkan et al. (2016a)	*Rlm2* (Larkan et al. 2015)
- *B. napus* (DH lines from BnaDYDH mapping population derived from "Darmor-bzh"(R) × "Yudal" (S) and developed in France, field experiments in the UK and France): 17 QTL for adult plant resistance across 13 LGs	Huang et al. (2016)	
- *B. napus* (Worldwide accessions from Germplasm Resources Information Network, using PG-4 isolate): one major QTL on chromosome A1	Raman et al. (2016)	
- *B. napus* ("DH12075" derived from cultivar "Cresor" that has R-gene *LmR1* × Westar, (S) using natural ascospores released from infected stubble): *LepR4* recessive on A genome	Yu et al. (2013)	
- *B. napus* (186 DH population SASDH, derived from Rlm4 cultivar "Skipton" and "Ag-Spectrum," using 11 single spore isolates from the national blackleg isolate collection in Australia): Single major gene *Rlm4* mapped on chromosome A7. Characterization of Rlm4 candidate genes in the same population	Raman et al. (2012b), Tollenaere et al. (2012)	
- *B. napus* (DH "Maxol" and "Columbus"): Mapped *Rlm1* on chromosome A7	Raman et al. (2012a)	

(continued)

Table 2.16 (continued)

Resistance gene/QTL mapped	References	Resistance gene/ allele cloned
- *B. napus* (SASDH population derived from "Skipton"/"Ag-Spectrum," using Australian isolates): *Rlm4* major qualitative locus mapped on chromosome A7	Raman et al. (2012b)	
B. napus (Mapping populations of cultivar "Surpass 400" (R) × "Westar" (S), using isolate 87-41): BLMR1 and BLMR2, single major gene on chromosome N10	Long et al. (2011)	
B. napus (Two different mapping populations,"DH12075" from cultivar "Cresor" (R) × re-synthesized line "PSA12" (S) and "Shiralee" (R) × "PSA12" (S), using unknown source of isolate): ClmR1 same genetic interval as *LmR1* on chromosome A7 ; gene control	Mayerhofer et al. (2005), Delourme et al. (2004)	
- *B. napus* (Cultivar "Surpass 400," using 31 isolates from Canada, Australia, Europe, Mexico and USA comprising PG2-4): *LepR3* single dominant allele, same linkage group as LepR2 on the A genome	Li and Cowling (2003), Yu et al. (2008)	
- *B. napus* (DH population, "DHP95" and "DHP96" with resistance introgressed from *B. rapa* subsp. sylvestris, using 30 isolates from Canada, Australia, Europe, and Mexico): LepR1 (complete, inhibit growth) and LepR2 (incomplete, reduced growth) on A genome chromosome A2 and A10, respectively	Yu et al.(2005)	
B. napus (Cultivars based on published differential set, using isolates from France,	Balesdent et al. (2002)	

(continued)

Table 2.16 (continued)

Resistance gene/QTL mapped	References	Resistance gene/ allele cloned
Australia, New Zealand, England and Portugal): *Rlm3*, *Rlm7* single gene control		
- *B. napus* (DH and F2 : 3 populations from "Darmor" (R) × "Samourai" (S), field experiment in France): 16 genomic regions for field resistance	Pilet et al. (1998, 2001)	
B. napus (Cultivar "Doublol," "Vivol," "Columbus," and "Capitol," "Jet Neuf," using isolate PG2-4): Rlm4 linked to *Rlm1*	Balesdent et al.(2001)	
B. napus (DH from cultivar "Maluka," "Cresor," and "RB87-62" × "Westar" (S), using isolate PG2): *cRLMm*, *cRLMrb* cited in single resistance gene at cotyledon stage and, aRLMc and aRLMrb adult stage linked to *cRLMm* and *cRLMrb*	Rimmer et al. (1999)	
- *B. napus* (Cultivar "Westar," "Quinta," and "Glacier," using isolate PG2, PG3, and PG4): *Rlm1* single dominant gene	Ansan-Melayah et al. (1995), Ansan-Melayah et al. (1998)	
B. napus (Cultivar "Westar," "Quinta," and "Glacier," using isolate PG2-4): *Rlm2* single dominant gene	Ansan-Melayah et al. (1998)	
B. napus (DH population from cultivar "Shiralee" and "Maluka" (R) × advanced breeding lines (S), using five single spore virulent isolates collected from provinces in Canada): LmR1 single major locus, could be linked/ identical	Mayerhofer et al. (1997)	
B. napus (DH population from cultivar "Major" (R) × "Stellar" (S), using isolate PHW1245): LEM1 single major locus	Ferreira et al. (1995)	

(continued)

Table 2.16 (continued)

Resistance gene/QTL mapped	References	Resistance gene/ allele cloned
B. napus (DH from cultivar "Cresor" (R) × "Westar" (S), using canola residues infected with virulent *L. maculans* and pycnidiospores of isolate Leroy): *LmFr1* single major gene	Dion et al. (1995)	
- *B. rapa* (Accession "02-159-4-1" (R) × DH "Z1" (S), and with "Darmor" and "Eurol," using 31 isolates from the IBCN and IMASCORE collections): Rlm11 single gene introgressed into *B. napus*	Balesdent et al. (2013)	
B. rapa (Line "156-2-1"): *Rlm8* single control	Balesdent et al. (2002)	
B. juncea (Cultivar "Aurea" and "Picra"): *Rlm5* and *Rlm6* epistatic interaction	Balesdent et al. (2002)	
- *B. juncea* (F2 population from F1 progeny of Cultivar "AC Vulcan" × Inbred line "UM3132," using PG2 isolate): Two independent genes, one dominant and one recessive	Christianson et al. (2006)	

(Christianson et al. 2006), and *B. carinata* (Plieske et al. 1998). Wild Brassicaceae species, such as *Sinapis arvensis*, also serve as a good source of *R*-genes for Blackleg (Snowdon et al. 2000; Chen et al. 2010). However *R*-genes from other wild species remain to be tapped. Besides identifying candidate *R*-genes through comparative genomic analysis, more screening work on wild and land races, including species from the Brassicaceae family, should be performed against the main pathogens to identify resistant plants. This would broaden the gene pool for introgression of novel *R*-genes or alleles into *Brassica* cultivars. Combining this information with genome sequencing will enable characterization of these novel resistance genes.

Although effector genes are usually unique and do not often share structural similarity but encode very diverse proteins, it is still possible to identify an *R*-gene that could recognize common or near identical effector gene(s) in different pathogen pathotypes/races/isolates, which already exists naturally in some plants/crops (Dangl et al. 1992). This type of *R*-gene is useful and can be transferred between hosts that are closely related or within the same genus or family. Some *R*-genes can also be transferred from host to non-host using GM technologies, providing broad-spectrum immunity against one or more pathogens. One such successful example is the pepper

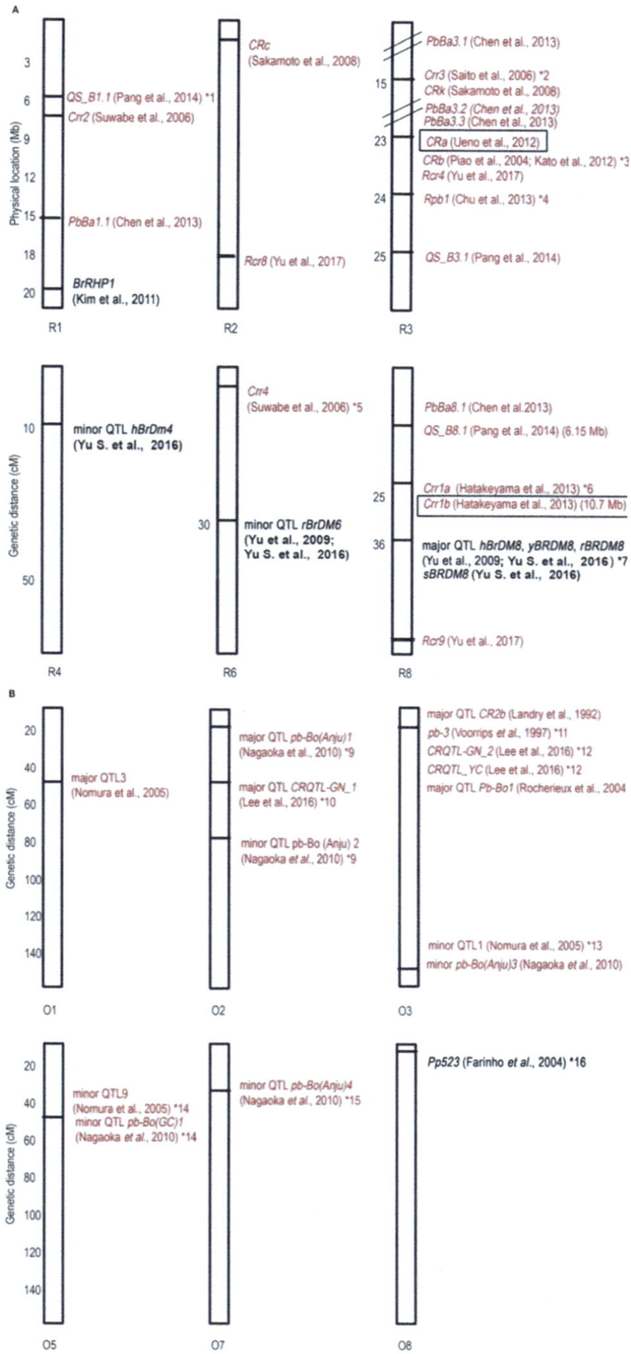

Fig. 2.3 *R*-genes/QTL mapped on *B. rapa* (**a**), *B. oleracea* (**b**), and *B. napus* (**c**). Each vertical rectangular is a chromosome, with the chromosome name followed by chromosome number at the bottom. Black font denotes *R*-genes/QTL mapped for *S. sclerotiorum*, blue font *L. maculans*, red font *P. brassicae* and bold black font *H. parasitica*. Boxed *R*-gene is cloned. Where genetic distance applies, the chromosome position is stated in parentheses at the end of the locus. *1, Independent minor loci, homologous to *A. thaliana* chromosome 3; *2, *Crr3* is independent of *Crr1* and *Crr2;* *3, Syntenic with *A. thaliana* chromosome 4. *Crr1* and *Crr2* also share the same syntenic region; *4, Close to *CRa* and *CRb*, also known as *Rcr1* (Chu et al. 2014), predicted as TIR-NBS-LRR (Yu et al. 2016a, b);

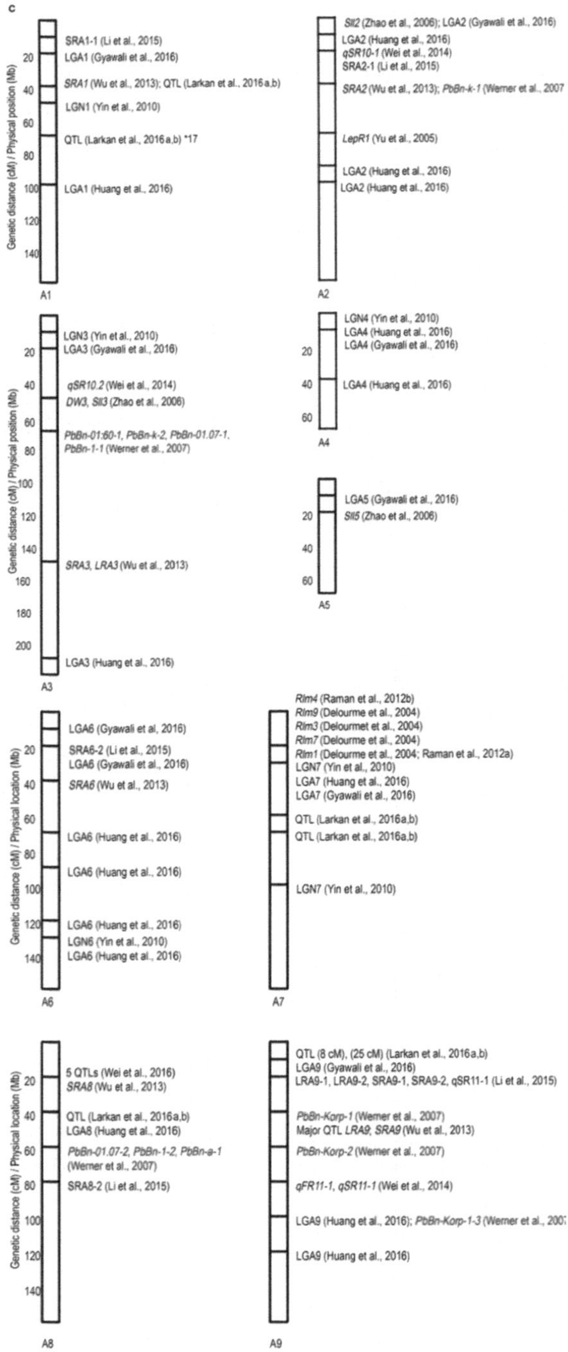

Fig. 2.3 (continued) *5, Independent of *Crr1*, *Crr2*, and *Crr3*; *6, Two genes, *Crr1a* and *Crr1b* were identified at originally single *Crr1* locus; *7, All at the same locus and are identical to major QTL *BraDM* (Yu et al. 2009); *8, Candidate gene, *Bra016457*, is serine/threonine (S/T) kinase. The major QTL in *7 and *8 lie between 17.6 and 17.8 Mb on chromosome R8; *9, Syntenic with *CRc* on chromosome R2; *10, Strong collinearity with chromosome A2 in *B. rapa* where *CRc* gene is located. Also overlap with *pb-Bo (Anju)1*; *11, Identified by Voorrips et al. (1997) but mapped on

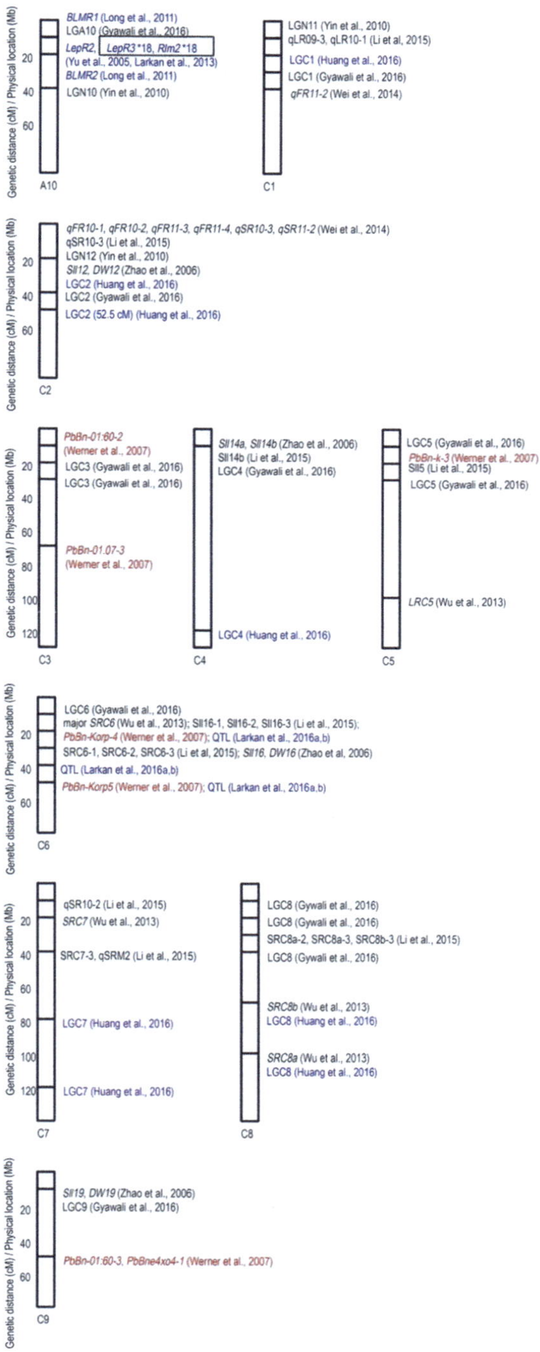

Fig. 2.3 (continued) chromosome O3 by Nagaoka et al. (2010); *12, *CRQTL-GN_2* (15.5–24.3 cM) and *CRQTL_YC* (8.5–38.1 cM) have similar location. Syntenic regions for these two QTL are on chromosome R3 (*Crr3*, *CRk*, *CRa*, and *CRb*), R8 (*Crr1*); *13, Same linkage group with QTL-LG1 (Moriguchi et al. 1999); *14, May be the same locus; *15, Closely linked to *CRb*, homologous to chromosome R3; *16, Adult resistance; *17, Field resistance for *L. maculans*. Candidate gene is cysteine-rich RLK; *18, *LepR3* and *Rlm2* are allelic variants (Neik et al. 2017)

Bs2 gene against bacterial spot disease transferred from its host pepper into non-host tomato (both members of Solanaceae family; Tai et al. 1999). The *Bs2* gene encodes the *R*-gene NBS-LRR class (Tai et al. 1999) while the gene *avrBs2* in *Xanthomonas campestris* pathovar *vesicatoria*, causal pathogen of bacterial spot disease, is essential for virulence activity and is highly conserved between and within the pathovars of *X. campestris* (Kearney and Staskawicz 1990). This has opened up new possibilities to protect economically valuable crops that lack *R*-gene resources, from disastrous pathogens.

2.8.3 Genomic Prediction to Identify Blackleg R-Genes of *Brassica*

Most resistant cultivars contain major genes conferring resistance to specific blackleg pathotypes, which is qualitative resistance. While very effective initially, the pathogen has proven adept at overcoming the different major gene resistances in areas where high disease pressure was created by the intense cultivation of varieties with the same major gene (*Rlm1* in France (Rouxel et al. 2003a, b) and *LepR3* in Australia (Sprague et al. 2006)). In contrast, quantitative resistance, which is due to many minor genes, offers incremental resistance improvements. Initial blackleg leaf lesions may appear similar to those on susceptible varieties, but later in the season no or less severe stem cankers are observed on varieties with improved quantitative resistance (Delourme et al. 2014). Phenotypic and pedigree selection can potentially exploit both types of resistance to breed varieties with durable field resistance to *L. maculans* (Brun et al. 2010; Delourme et al. 2014). Phenotyping for both qualitative and quantitative resistance can be done in field disease nurseries against a full fungal population (Kaur et al. 2009; Light et al. 2011) as well as in controlled environment conditions against specific single spore fungal isolates (Rouxel et al. 2003a, b; Marcroft et al. 2012a; Brun et al. 2010).

In the last 20 years, genomic data has been incorporated into breeding and quantitative trait loci (QTL) have been mapped that confer mainly qualitative resistance using a variety of statistical designs, such as F2-crosses (Ansan-Melayah et al. 1998), doubled haploid populations (Delourme et al. 2004), backcrosses (Long et al. 2011), and Winter and Spring diversity panels (Uzunova et al. 1995; Kaur et al. 2009; Jestin et al. 2011a; Raman et al. 2012a; Delourme et al. 2014; Van De Wouw et al. 2016a). These efforts have identified a growing list of *L. maculans* resistance genes (e.g., *LepR1, LepR2, LepR3, LepR4, BLMR2, Rlm1, Rlm2, Rlm3, Rlm4, Rlm5, Rlm6, Rlm7, Rlm8, Rlm9, Rlm10, Rlm11, Rlm12,* and *LmFr1*) in *Brassica* species (Yu et al. 2005; Delourme et al. 2006; Yu et al. 2008; Marcroft et al. 2012b; Raman et al. 2012b; Balesdent et al. 2013; Larkan et al. 2014; Fomeju et al. 2014; Raman et al. 2016a, b). Consequently, irrespective of some limitations, marker-assisted selection has been used for mapping and validating resistance QTL to *L. maculans* using progenies/populations derived from breeding programs. However, studies in canola and other species show that genetic variance explained by known QTL tends to be low (Yang et al. 2010; Huo et al. 2016).

Many plant genetic improvement schemes are transitioning from phenotype and pedigree based selection to genetic marker based genomic selection. Genomic selection uses a reference population that is genotyped and phenotyped to predict the genetic value for individuals that are only genotyped using genome-wide markers (Meuwissen et al. 2001). Several studies argue that genomic selection provides better accuracies than might be achieved on the basis of pedigree information alone (Crossa et al. 2010; Jia and Jannink 2012; Perez-Enciso et al. 2015; Tsai et al. 2015; Vela-Avitua et al. 2015). This paradigm shift is significantly improving genetic gain by reducing the breeding cycle, increasing selection accuracy in un-phenotyped germplasm, and increasing selection intensity. Furthermore, studies in a wide variety of plant species confirm the realized and potential benefits of using genomic selection (Heffner et al. 2011; He et al. 2016; Bentley et al. 2014; Grenier et al. 2015). Various genomic prediction models have been proposed such as Genomic Best Linear Unbiased Prediction (GBLUP) (Nejati-Javaremi et al. 1997; Habier et al. 2007; VanRaden 2008), Ridge Regression Best Linear Unbiased Prediction (RR-BLUP), Bayes A and B (Meuwissen et al. 2001), and BayesR (Erbe et al. 2012). BayesR has been extended to incorporate prior information obtained from variant, gene, or regulatory regions (BayesRC) (MacLeod et al. 2016). Genomic selection in canola was first highlighted by Wurschum et al. (2014), who reported prediction accuracies for agronomic traits using RR-BLUP and BayesB in a lower diversity set of 391 doubled haploid lines genotyped with 251 SNPs. A more diverse set of 475 spring canola lines was used by Jan et al. (2016) to examine the performance of testcross combinations using 24,403 SNPs markers based on the RR-BLUP model. Using three different genomic selection methods (GBLUP, BayesR, and BayesRC), Fikere et al. (2018) predicted R-genes in *B. napus* in a diverse set of 532 Winter and Spring canola lines genotyped for 98,054 SNPs and phenotyped for adult plant survival and stem internal infection in replicated field disease nurseries at two locations (Fikere et al. (2018).

Genomic prediction is becoming a popular plant breeding method to predict the genetic merit of lines. While some genomic prediction results have been reported in canola, none have been evaluated for blackleg disease. The genomic prediction for seedling emergence, survival rate, and internal infection using 532 Spring and Winter canola lines has been made. These lines were phenotyped in two replicated blackleg disease nurseries grown at Wickliffe and Green Lake, Victoria, Australia. A transcriptome genotyping-by-sequencing approach revealed 98,054 single nucleotide polymorphisms (SNPs) after quality control. Various genomic prediction scenarios based on Genomic Best Linear Unbiased Prediction (GBLUP), BayesR and BayesRC, which can make use of prior quantitative trait loci information, via cross-validation. Clustering based on genomic relationships showed that winter and spring lines were genetically distinct, indicating limited gene flow between sets. Genetic correlations within traits between Spring and Winter lines ranged from 0.68 to 0.90 (mean = 0.76). Based on GBLUP in the whole population, moderate to high genomic prediction accuracies were achieved within environments (0.35–0.74) and were reduced across environments (0.28–0.58). Prediction accuracy within the spring set ranged from 0.30 to 0.69, and from 0.19 to 0.71 within the winter set.

The BayesR model resulted in slightly lower accuracy to GBLUP. The proportion of genetic variance explained by known blackleg quantitative trait loci (QTL) was < 30%, indicating that there is a large reservoir of genetic variation in blackleg traits that remains to be discovered, but can be captured with genomic prediction. However, providing prior information of known QTL in the Bayes RC method resulted in increased prediction accuracy for survival and internal infection, particularly with spring lines. Overall, these promising results indicate that genomic prediction will be a valuable tool to make use of all genetic variation to improve blackleg resistance in canola (Fikere et al. 2018).

2.8.4 Identification of R-Genes in *Brassica* to *Leptosphaeria maculans* by Genome-Wide Association

Studies have shown that *L. maculans* populations evolve rapidly under the selection pressure from resistant canola/rapeseed cultivars carrying specific R-genes, and single R-genes can be overcome due to the high diversity of *L. maculans* and a shift of pathogen population towards virulence. Hence, it is critical for the canola industry to continue identifying novel alleles of resistance for sustainable management of blackleg. The GWAS allow us to rapidly identify and validate significant loci with associated markers. Twelve associated regions were identified from *B. napus*, including 93 Canadian and 150 Chinese canola/rapeseed accessions, against 22 *L. maculans* isolates, under controlled-environment and/or field experiments. Some race nonspecific QTLs were identified by Fu et al. (2020). The resistance-associated region on Chromosome A08 from Canadian varieties was effective against 15 of the 22 *L. maculans* isolates (Fig. 2.4). These significant regions with resistance to a broad range of *L. maculans* races are considered potential alleles with greater durability for blackleg resistance. Three most significant resistance-associated regions, identified among Canadian accessions against 22 Lm isolates based on cotyledon inoculation, were on chromosomes A03, A08,

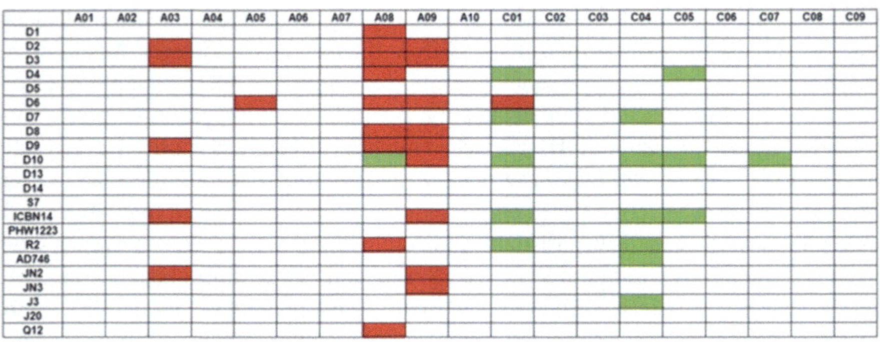

Fig. 2.4 Summary of the resistance loci in Canadian and Chinese accessions. Red box presents the loci in Canadian accessions. Green box presents the loci in Chinese accessions (Fu et al. 2020)

and A09. However, the known R-genes *Rlm1, Rlm2, Rlm3, Rlm4–7, Rlm6, LepR1, LepR2,* or *LepR3* were not detected in these significant SNP loci based on the *Avr* genes carried by the *L. maculans* isolates. The results show that the resistance alleles identified can be novel or they may interact with known R loci and produce new resistance specificity. Quantitative resistance has been identified on chromosomes A03, A08, and A09 previously. Three associated SNP loci were found in a 572-kb region on chromosome A03 (24,158, 622 to 24,701,214). One SNP and one QTL were identified also on chromosome A03 from studies of 179 Australia accessions and linkage mapping of a "Darmor-bzh" × "Yudal" DH population, respectively, but these regions appear far away from the resistance-associated regions identified. Significant SNP loci or QTLs associated with cotyledon or adult plant resistance have also been identified on chromosomes A08 and A09 using an association panel, DH, or F2/F3 Canola/rapeseed populations. However, these loci were detected using Australian or French materials based on genetic-linkage maps with SSRs or other markers, and it is difficult to determine physical positions of these SNPs/QTLs precisely against the *B. napus* reference genome. Fu et al. (2020) were able to narrow the range of resistance-associated SNP loci to an 11-kb region on chromosome A08 and 36 bp on chromosome A09, respectively, using GBS with *B. napus* reference genome for the resistance on cotyledons. This shows the advantage of using GWAS and GBS to identify resistance loci against blackleg disease of canola/ rapeseed.

Five resistance-associated regions were located on chromosomes A08, C01, C04, C05, and C07 in the Chinese rapeseed accessions against 12 Lm isolates on cotyledons. On chromosome A08, however, the physical position of the region appears different from that found with Canadian canola accessions. Although these regions have not been reported previously for Lm resistance at the seeding stage, the associated SNP loci were located in homologous regions, especially on chromosomes A01/C01, A04/C04/A05, and C07/A03/C03. These regions may either reside functionally redundant loci or be involved in increased allelic diversity of the genes controlling the resistance to blackleg (Fopa-Fomeju et al. 2014).

The results showed that there were less number of QTLs against L. maculans identified from the Chinese accessions than the Canadian accessions, and also SNPs associated with cotyledon resistance to *L. maculans* were distributed in both A and C genomes of *B. napus* from the Chinese accessions, but mainly in the A genome from the Canadian accessions. One of the reasons for these could be due to different breeding focuses for resistance to canola diseases between the two countries. The selection and accumulation of canola against *L. maculans* have never been performed in canola breeding programs in China. Chinese accessions are winter ecotype, and blackleg, caused by *L. maculans* has never been an issue in the canola growing areas in China. Hence, breeding for resistance to *L. maculans* has not been considered as one of the breeding objectives by Chinese canola breeders. On the other hand, *L. maculans* is an important disease on canola in western Canada. Great efforts have been made for breeding canola for resistance to blackleg in the past 3 years. Introgression of resistance genes derived from the A-genome species *B. rapa*, such as *LEM1*, *LmFr1*, *LmR1*/*CLmR1*, *Rlm1*, *Rlm3*, *Rlm4, Rlm7* and

Rlm9, and *LepR1* to *LepR4* have been performed. Therefore, it would not be surprising that more QTLs were identified in the Canadian accessions and the resistance loci were mainly distributed in A genome of *B. napus* (Fu et al. 2020).

The completely sequenced and annotated genome of *B. napus* provides a useful reference to identify blackleg resistance candidate genes in canola/rapeseed germplasm pools. Plant resistance genes, such as CC-LRRNBS, TIR-LRR-NBS, RLK, RLP and transmembrane proteins, can all be identified using the RGAugury pipeline. Several potentially new R-gene candidates were uncovered from the Canadian and Chinese canola/rapeseed accessions using genome-wide studies with the RGA pipeline, and these candidates have also been located close to the significant SNPs identified in GWAS, and can be potential blackleg-resistance genes supported by robust genetic and genomic analyses. They should be further explored for confirmation and applications. It is interesting to note that many of the well-known blackleg R-genes, including *Rlm1, Rlm3,* and *LepR3/Rlm2*, were not detected with GWAS in this study. This may be due to a lack of marker polymorphism in the mapping panel because of selected commercial varieties, such as the Canadian panel, has only moderate genome-wide coverage of makers and low frequencies of informative alleles (associated with blackleg resistance). Additionally, the Chinese panel was not selected originally for blackleg resistance, since the causal agent Lm had not been reported in China. The blackleg resistance loci identified seems novel, and provides insights into several potentially new regions for discovery of additional blackleg R-genes (Li et al. 2016a, b, c; Fu et al. 2020).

2.9 *Brassica-Plasmodiophora*: R-Genes

2.9.1 Identification of R-Genes Sources

Sources of resistance effective against clubroot pathogen, *P. brassicae* have been identified from cruciferous host species, viz. *B. napus, B. carinata, B. oleracea* var. *capitata*, *B. oleracea* var. *botrytis*, *B. oleracea* var. *gongylodes*, *Raphanus sativus*, *B. rapa* subsp. *pekinensis*, and *B. rapa* subsp. *rapifera* from Australia, Canada, China, Germany, India, Japan, and the UK (Table 2.17). Pathotype specific sources of resistance (Table 2.18) have been identified from *Arabidopsis thaliana* (Pathotype 2,5,6,e); *B. napus* (Pathotype 2,3,5,6,8); *B. oleracea* (Pathotypes 2,3,4,5); *B. nigra* (Pathotype 2,3,5,6,8), and *B. rapa* (Pathotype 2,3,5,6,8). The sources have been utilized to breed CR cvs. in different countries of the world which are providing protection to the crops from clubroot disease. Breeding disease resistance cvs. is a continuous process to replace breakdown of R-genes on the event of selection pressure of the pathogen and search of more stable sources of R-genes.

Table 2.17 Sources of R-genes in crucifers to *Plasmodiophora brassicae* (clubroot) (Saharan et al. 2021)

Crucifer's host	Sources	Country	References
B. napus	GSL-1, Culture -1, Culture -2, PCR -80, WW-1507, ISN-700, HNS -3	India	Chattopadhyay et al. (2001)
	Mendel, NPZ CR-21	Germany	Diederichsen et al. (2014)
	Wilhelmsburgar	Germany	Chiang and Crête (1983)
	Pioneer 45H29, D 3152, Dekalb 73-67, RR, 73-77 RR, Canterra 1960, Proven 9558C	Canada	Strelkov et al. (2011)
B. carinata	HC-1, HC-4, HC-5, HC-9001, PC-3, PCC-2, PPSC-1, PC-5, ACCBN-479	India	Chattopadhyay et al. (2001)
B. oleracea var. *capitata*	KK Cross, Drumhead	India	Chattopadhyay et al. (2001)
	Clapton	UK	Donald and Porter (2009)
	Highfield	Australia	Donald and Porter (2009)
B. oleracea var. *botrytis*	Danila, Snowball	India	Chattopadhyay et al. (2001)
	Tekila, Kilatan	UK	Donald and Porter (2009)
	Syngenta	Australia	Donald and Porter (2009)
	Bindsachsenar, Bohmerwaldkohl	Germany	Crute et al. (1983)
B. oleracea var. *gongylodes*	White viena	India	Chattopadhyay et al. (2001)
Raphanus sativus	Novitas, Poem van, Zwijndrechat, Flevo, Robijn, Saxafire, Scharo, Verano		Rowe (1980)
B. rapa subsp. *pekinensis*	Accession 1003, 1007, 1008	China	Liu et al. (2018)
B. rapa subsp. *rapifera*	ECD-04	Germany	Diederichsen et al. (2006, 2009)
	Milan white	Japan	Otani et al. (1982)
	Golria, Siloga, Debra, Milan white	Japan	Hirai (2006)

Table 2.18 Sources of pathotype specific R-genes in crucifers to *Plasmodiophora brassicae* (Clubroot) (Saharan et al. 2021)

Crucifers host	Sources	Pathotypes	References
A. thaliana	Liue ct-1	2	Sharma et al. (2013)
	Pu2-23	5	
	Ws-2, Sorbe	6	
	Tsu-0, Ze-0	e	Fuchs and Sacristan (1996)
B. napus	(Rapid cycling *Brassica* collection (RCBC)	All	Sharma et al. (2013)
	45H21, Invigor 5020 LL	6	Adhikari et al. (2012)
B. napus	Rutabaga	2, 3, 5, 6, 8	Hasan et al. (2012)
B. oleracea	RCBC lines	2,3,5	Sharma et al. (2013)
B. carinata	RCBC lines	2, 5	Sharma et al. (2013)
B. rapa	RCBC lines	2,5	Sharma et al. (2013)
B. nigra	BB	2,3,5,6,8	Hasan et al. (2012)
B. rapa subsp. *pekinensis* (Napa cabbage)	Deneko, Bilko, Yuki	6	Adhikari et al. (2012)
	Emiko, China Gold, Flower Nabana	6	Saude et al. (2012)
B. rapa var. *rapifera*	Turnip	2,3,5,6,8	Hasan et al. (2012)
B. rapa var. *oleifera*	Winter and spring rapeseed	2,3,5,6,8	Hasan et al. (2012)
B. oleracea var. *capitata*	Kilaton, Tekila, Kilaxy, Kilaherb	6	Saude et al. (2012)
	Xiangan 336, Verheul, Bindsachsener, Zhouyebai, 2358	4	Ning et al. (2018)
Pathotype-specific resistant species			
B. rapa (AA)			Hasan et al. (2012)
B. oleracea (CC)			
B. napus (AACC), Turnip, Rutabaga			

2.9.2 Identification of Pathotype Specific R-Genes Sources

This type of resistance is mediated by resistance (R) genes through the recognition of specific pathogen effectors. These effectors initially function as virulence/pathogenicity factors for the pathogen, but then become recognized by the host as signals of attempted infection (Bernoux et al. 2011). Most R-genes encode proteins carrying a nucleotide-binding site (NBS) in the central region and a leucine-rich repeat (NBS-LRR) domain at the C-terminus. These genes form a large multigene family in the plant genome and can be separated into two subclasses, the toll-interleukin-1 (TIR) class and the coiled-coil (CC) class (Rafiqi et al. 2009). The NBS–LRR proteins are thought to recognize effectors or effector-specific signals from the pathogen and activate the plant immune system to provide hypersensitive responses (Bernoux et al. 2011).

Although many clubroot resistance (CR) loci have been identified through genetic analysis and quantitative trait loci (QTL) mapping (Piao et al. 2009), the resistance mechanisms associated with these genes remained unknown until when two R-genes were cloned and characterized. The first gene is *CRa* (Ueno et al. 2012) from *B. rapa*, which confers specific resistance to *P. brassicae* isolate M85. Fine mapping of the *CRa* locus to the *A. thaliana* and *B. rapa* genomes revealed a candidate gene encoding a TIR-NBS-LRR protein. The cloned alleles of this gene in susceptible and resistant *B. rapa* lines have several structural differences, and *CRa* expression was observed only in the resistant line. Four mutant lines lacking clubroot resistance were obtained by the UV irradiation of pollen from a resistant line, and all of these mutant lines were found to carry independent mutations in this candidate gene. This genetic and molecular evidence strongly suggests that the identified gene is *CRa*. The second gene is *Crr1a* (Hatakeyama et al. 2013), which is also from *B. rapa*. The *Crr1a* is one of the two loci of *Crr1* that was previously considered as a single locus. This gene (named $Crr1a^{G004}$) was cloned from the resistant line G004 and encodes a TIR-NB-LRR protein. By comparison, the susceptible allele $Crr1a^{A9709}$, cloned from the same locus in the susceptible line A9709, encodes a truncated NB-LRR protein. The $Crr1a^{G004}$ protein is expressed in the stele and cortex of the hypocotyl and roots, but not in root hairs, suggesting that it controls resistance to secondary infection or later pathogen development. Gain-of-function analysis proved that $Crr1a^{G004}$ was sufficient to bestow resistance to isolate Ano-01 in susceptible *Arabidopsis* and *B. rapa* genotypes. Since resistance genes from different plant genomes often show similar structures because they are most likely to be derived from a common ancestral gene (Leister 2004), the information about *CRa* and *Crr1a* will facilitate the cloning and molecular characterization of other clubroot resistance genes (Feng et al. 2014).

A clubroot resistance gene, *Rcr1*, with efficacy against pathotype 3 of *P. brassicae*, was mapped to chromosome A03 of *B. rapa* in pak choy cultivar “Flower Nabana.” The resistance to pathotypes 2, 5, and 6 was shown to be associated with *Rcr1* region on chromosome A03. Bulked segregant RNA sequencing was performed and short read sequences were assembled into 10 chromosomes of the *B. rapa* reference genome v1.5. The resistant (R) bulks, a total of 351.8 million (M) sequences, 30,836.5 million bases (Mb) in length, produced 120-fold coverage of the reference genome. The susceptible (S) bulks, 322.9 M sequences, 28,216.6 Mb in length, produced 109-fold coverage. In total, 776.2 K single nucleotide polymorphisms (SNPs), and 122.2 K insertion/deletion (InDels) in R bulks, and 762.8 K SNPs and 118.7 K InDels in S bulks were identified, each chromosome had about 87% SNPs and 13% InDels, with 78% monomorphic and

22% polymorphic variants between the R and S bulks. Polymorphic variants on each chromosome were usually below 23%, but made up 34% of the variants on chromosome A03. There were 35 genes annotated in the *Rcr1* target region and variants were identified in 21 genes. The numbers of poly variants differed significantly among the genes. Four out of them encode Toll-Interleukin-1 receptor/nucleotide-binding site/leucine-rich-repeat proteins, *Bra019409* and *Bra019410* harbored the higher numbers of polymorphic variants, which indicates that they are more likely candidates of *Rcr1*. Fourteen SNP markers in the target region were genotyped using the Kompetitive Allele Specific PCR method and were confirmed to associate with *Rcr1*. Selected SNP markers were analyzed with 26 recombinants obtained from a segregating population consisting of 1587 plants, indicating that they were completely linked to *Rcr1*. Nine SNP markers were used for marker-assisted introgression of *Rcr1* into *B. napus* canola from *B. rapa*, with 100% accuracy (Yu et al. 2016a, b).

2.9.2.1 Identification of Pathotype Specific R-Genes Through Bulked Segregant RNA Sequencing

A clubroot resistance gene, *Rcr1*, with efficacy against pathotype 3 of *P. brassicae*, was mapped to chromosome A03 of *B. rapa* in pak choy cultivar "Flower Nabana." The resistance to pathotypes 2, 5, and 6 was shown to be associated with *Rcr1* region on chromosome A03. Bulked segregant RNA sequencing was performed and short read sequences were assembled into 10 chromosomes of the *B. rapa* reference genome v1.5. The resistant (R) bulks, a total of 351.8 million (M) sequences, 30,836.5 million bases (Mb) in length, produced 120-fold coverage of the reference genome. The susceptible (S) bulks, 322.9 M sequences, 28,216.6 Mb in length, produced 109-fold coverage. In total, 776.2 K single nucleotide polymorphisms (SNPs), and 122.2 K insertion/deletion (InDels) in R bulks, and 762.8 K SNPs and 118.7 K InDels in S bulks were identified, each chromosome had about 87% SNPs and 13% InDels, with 78% monomorphic and 22% polymorphic variants between the R and S bulks. Polymorphic variants on each chromosome were usually below 23%, but made up 34% of the variants on chromosome A03. There were 35 genes annotated in the *Rcr1* target region and variants were identified in 21 genes. The numbers of poly variants differed significantly among the genes. Four out of them encode Toll-Interleukin-1 receptor/nucleotide-binding site/leucine-rich-repeat proteins, *Bra019409* and *Bra019410* harbored the higher numbers of polymorphic variants, which indicates that they are more likely candidates of *Rcr1*. Fourteen SNP markers in the target region were genotyped using the Kompetitive Allele Specific PCR method and were confirmed to associate with *Rcr1*. Selected SNP markers were analyzed with 26 recombinants obtained from a segregating population consisting of 1587 plants, indicating that they were completely linked to *Rcr1*. Nine SNP markers were used for marker-assisted introgression of *Rcr1* into *B. napus* canola from *B. rapa*, with 100% accuracy (Yu et al. 2016a).

2.9.2.2 Identification of QTL's by GBS Conferring Resistance to Multiple Pathotypes

Lines with resistance to a broad range of pathotypes of *P. brassicae* have been identified in the canola progenitor species *B. rapa*. This species can be used to broaden the genetic base of clubroot resistance in canola. Introgression of traits from *B. rapa* into canola through interspecific crosses is possible, so resistance to clubroot from *B. rapa* could be transferred into canola through conventional breeding. The identification and genetic mapping of clubroot resistance genes has been carried out in *B. rapa*, *B.oleracea*, and *B. napus*. Two resistance genes, *CRa* and *Crr1*, have been isolated from Chinese cabbage lines of *B. rapa*. They encode Toll-Interleukin-1 receptor/nucleotide-binding site/leucine-rich-repeat (TIR-NBS-LRR, TNL) proteins. Genotyping-by-sequencing (GBS) offers a new tool to explore the genetic control of complex traits. GBS analysis was used in the current study to: (1) characterize genome-wide variants in *B. rapa*; (2) identify SNP sites that could be used for genetic mapping; and (3) detect QTL effectively resistant to multiple pathotypes of *P. brassicae* identified in western Canada; and (4) identify possible candidate genes for each QTL (Yu et al. 2017).

F_1 progeny from the *Brassica rapa* lines T19 (resistant) × ACDC (susceptible) were backcrossed with ACDC, then self-pollinated to produce BC1S1 lines. From genotyping-by-sequencing (GBS) of the parental lines and BC_1 plants, about 1.32 M sequences from T19 were aligned into the reference genome of *B. rapa* with 0.4-fold coverage, and 1.77 M sequences with 0.5-fold coverage in ACDC. The number of aligned short reads per plant in the BC1 ranged from 0.07 to 1.41 M sequences with 0.1-fold coverage. A total of 1584 high-quality SNP loci were obtained, distributed on 10 chromosomes. A single co-localized QTL, designated as *Rcr4* on chromosome A03, conferred resistance to pathotypes 2, 3, 5, 6, and 8. The peak was at SNP locus A03_23710236, where LOD values were 30.3 to 38.8, with phenotypic variation explained (PVE) of 85–95%. Two QTLs for resistance to a novel *P. brassicae* pathotype 5x, designated *Rcr8* on chromosome A02 and *Rcr9* on A08, were detected with 15.0 LOD and 15.8 LOD, and PVE of 36% and 39%, respectively. Bulked segregant analysis was performed to examine TIR-NBS-LRR proteins in the regions harboring the QTL (Yu et al. 2017).

2.9.3 Identification of R-Genes by Genomic Approach

P. brassicae infects all *Brassica* species and the model plant *Arabidopsis thaliana*. It is possible to apply currently available genomics tools and techniques for the cloning and characterization of CR genes in addition to comparative mapping, and identification of candidate genes through transcriptomic analysis. The whole-genome sequence information available for *Arabidopsis* has been used for comparative genome analysis of *Brassica* species. Comparative mapping between *Brassica* and *Arabidopsis* has revealed a conservation of gene order in small chromosomal blocks despite inversions and large-scale deletions (Cavell et al. 1998; O'Neill and Bancroft 2000). This information has been used to align linkage groups containing CR genes in *B. rapa* with those of *Arabidopsis* chromosomes. Genes *Crr1, Crr2*, and *CRb* are in synteny with the central region of chromosome 4 of *A. thaliana* (Suwabe et al. 2006; Fig. 3.7).The studies by Jubault et al. (2008) of *A. thaliana* have identified one QTL for clubroot resistance in this region, suggesting the presence of functionally active candidate gene(s). This region of an *Arabidopsis* chromosome has clusters of the CR genes such as leucine-rich repeats (LRRs) and nucleotide-binding sites (NBSs). Similarly, RPP for resistance to *Hyaloperonospora parasitica* (downy mildew), RPS for a resistance to *Pseudomonas syringae* (bacterial blight), and ACD, which accelerates cell death in response to pathogen infection (Suwabe et al. 2006) have been identified. It has been suggested that CR genes may be members of these clusters of resistance genes. The genes *Crr1, Crr2*, and *CRb,* however, are distributed on three different chromosomes in *B. rapa*, R8, R1, and R3, respectively (Saito et al. 2006; Suwabe et al. 2006). Based on this observation, it has been suggested that the evolution of CR genes occurs by one of two routes: First, clubroot resistance was originally controlled by a single major gene in the ancestral genome, which later differentiated and diverged as functionally duplicate genes during the course of evolution in the *Brassica* genome (Suwabe et al. 2006). Second, route might be that the resistance genes for clubroot were originally clustered in that region in the ancestral genome which was later distributed into different genomic regions following chromosomal rearrangement in *Brassica*. Current *Brassica* species, which diverged 17–18 million years ago from *Arabidopsis*, are evolutionarily believed to be derived from whole-genome triplication and rearrangement of one ancestral genome (Lagercrantz 1998; O'Neill and Bancroft 2000; Yang et al. 2006). This hypothesis explains why these CR genes are dispersed and located on different chromosomes in *B. rapa*.

Saito et al. (2006) suggested that the genomic region around *Crr3* exhibits homology to the top of the long arm of *Arabidopsis* chromosome 3, and possibly also to *CRk* (Sakamoto et al. 2008). It was concluded that *Crr3* has a different origin from that of *Crr1, Crr2*, and *CRb* (Fig. 2.5). Gene *CRk* is independent of the CR genes *Crr1, Crr2, CRa*, and *CRb* but has a similar QTL region with *Crr3* (Sakamoto et al. 2008). Another novel CR locus, *CRc*, which is independent of all other CR loci, is located on chromosome R2 (Sakamoto et al. 2008).

Fuchs and Sacristan (1996) mapped a CR locus (RPB1) in *Arabidopsis* chromosome 1. Their mapping for partial clubroot resistance identified two QTLs in chromosome 5 in F_2 and four QTLs in the RIL population, one each in chromosome 1 and 4 and two in the chromosome 5 of *A. thaliana* (Jubault et al. 2008). The CR QTL region in chromosome 5 co-localized with that containing several resistance gene clusters. These resistance genes could be candidates for clubroot disease resistance. Fine mapping and detailed analysis of the levels of expression possessed by these genes would help to identify specific ones capable of conferring clubroot resistance. The complete genome sequence of only one species of Brassicaceae family, *Arabidopsis thaliana*, is available. It has been reported that gene sequence identity of *Brassica* and *Arabidopsis* varies from 75% to 90% (Quiros et al. 2001). Therefore, sequence information from the model species will greatly enhance cloning and characterization of CR genes in other *Brassica* species because the

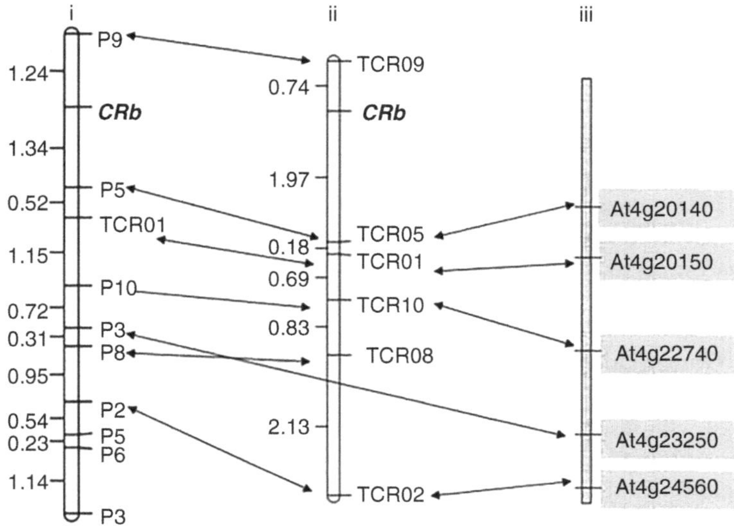

Fig. 2.5 Identification of the homologous region of *Brassica rapa* linkage group containing *CRb* locus with that of *Arabidopsis thaliana* chromosome 4 (Piao et al. 2004). (i) AFLP linkage group containing *CRb* locus, (ii) Linkage group converted to SCARs and CAPS markers, (iii) Homologous *Arabidopsis thaliana* chromosome 4 (Piao et al. 2009)

partial sequences of several accessions of *A. thaliana* showing degrees of response to the pathogen *P. brassicae* collected worldwide, are available now (Fuchs and Sacristan 1996; Siemens et al. 2002; Nordborg et al. 2005; Alix et al. 2007). Only a few studies in transcriptomic/expression levels have been completed with respect to disease development. These include investigation of involvement of several metabolic pathways in disease pathogenesis, such as hormonal regulation by auxins (Grsic et al. 1999; Neuhaus et al. 2000), cytokinins (Devos and Prinsen 2006; Siemens et al. 2006), and trehalose synthesis (Brodmann et al. 2002). Grsic et al. (1999) observed that de novoindole-3-acetic acid (IAA) biosynthesis plays a role in symptom development during later stages of disease development; it was suggested that jasmonic acid, which increased during club development, may be involved in the upregulation of enzymes involved in IAA synthesis. Neuhaus et al. (2000) further supported this hypothesis by showing a delay in clubroot development after transforming *A. thaliana* with an antisense construct of Nitrilase 1 and 2, enzymes required for auxin biosynthesis. Siemens et al. (2006) investigated host gene expression during clubroot development in *A. thaliana* using an ATH1 microarray at two points in time, an early initial stage of infection and a later stage at which 60 % of the host root cells were colonized.

More than 1000 genes were observed as being associated with the growth and cell cycle, sugar phosphate metabolism, and defense genes that were differentially expressed between infected versus control plants were observed. Upregulation of auxin biosynthesis genes such as nitrilases and members of the GH3 family and down-regulation of cytokinin homeostasis were observed. It was further observed that lines overexpressing cytokinin oxidase/hydrolases were resistant to clubroot thereby strongly suggesting cytokinin is a key factor in clubroot development. Of the 312 genes identified as defense- and disease-resistant-related genes, only 5 and 7% were up-regulated at the first and the second points in time. This functional analysis of gene expression in clubroot-resistant and susceptible lines provided initial preliminary information. To strengthen further the understanding of genes and genetic networks involved in the mechanisms of clubroot disease and host interaction, more detail analyses at the transcriptomic level are needed (Piao et al. 2009).

2.9.4 Genetical Mechanism of R-Genes Sources

Genotypes with resistance to one or more of the pathotypes of *P. brassicae* have been reported in all of the major *Brassica* crops, except *B. juncea* (L.) Czern. and *B. carinata* Braun (Diederichsen et al. 2009). Both qualitative (Wit and van de Weg 1964; Crute 1986) and quantitative (Chiang and Crête 1970; Figdore et al. 1993; Grandclément and Thomas 1996; Voorrips et al. 1997) types of resistance have been reported. Most of these sources of resistance, however, are pathotype or race specific. In *B. napus*, most studies have reported oligogenic control of resistance to *P. brassicae* (Crute 1986). This would make the pyramiding of resistance genes in *B. napus* genotypes more practical than in other species. Models based on three, four and five resistance genes have been proposed, and the most favored model was based on four genes (Gustafsson and Falt 1986).

A complex type of inheritance, with dominant genes from *B. rapa* and recessive genes from *B. oleracea*, was expected in a re-synthesized *B. napus*, with resistance from both ancestral species (Diederichsen and Sacristan 1996). Segregation analysis indicated that resistance in re-synthesized *B. napus* was controlled by at least two dominant and unlinked genes (Diederichsen and Sacristan 1996). One major gene (*Pb-Bn1*) for resistance against two *P. brassicae* isolates is located on chromosome N03 and additional minor QTL for each isolate on chromosomes N12 and N19 (Manzanares-Dauleux et al. 2000a). Previous reports indicate that *P. brassicae* resistance in canola is controlled by a combination of major genes and quantitative trait loci (Matsumoto et al. 1998; Suwabe et al. 2003, 2006; Hirai et al. 2004; Piao et al. 2009).

Accessions of *B. oleracea* with pathotype-independent resistance to *P. brassicae* have also been reported (Voorrips 1996). Most studies on the C genome indicate that *P. brassicae* resistance in *B. oleracea* is quantitative and under the polygenic control of one or two major QTLs and some QTLs with minor effects (Landry et al. 1992; Figdore et al. 1993; Grandclément and Thomas 1996; Voorrips et al. 1997; Moriguchi et al. 1999; Rocherieux et al. 2004; Nomura et al. 2005). A few studies, however, indicated that *P. brassicae* resistance in *B. oleracea* is qualitative and controlled by either dominant (Chiang and Crete 1983) or recessive (Yoshikawa 1993) genes. It is possible that both quantitative and qualitative resistance mechanisms may be at play in this species. Resistance genes from fodder turnip

(*B. rapa*) have been used in resistance breeding of various *Brassica* crops, including Chinese cabbage, oilseed rape, and *B. oleracea*. Although most turnip lines carry more than one resistance gene, cultivars of the other *Brassica* crops with resistance derived from turnip generally carry a single, dominant resistance gene that is pathotype-specific. While important to clubroot management, genetic resistance has generally been race- or pathotype-specific (Diederichsen et al. 2009) and can break down when virulent races increase in the pathogen population. Therefore, genetic resistance should be carefully managed in combination with other methods of clubroot control (Hwang et al. 2012a, b).

2.9.5 Identification of R-Genes by Transcriptomic and Proteomic Approaches

With the use of transcriptomic and proteomic approaches, the mechanisms of *P. brassicae* infection, pathogenesis, host specificity, and expression of host resistance has been elucidated. To date, the complete genome sequence of only one species of Brassicaceae, *A. thaliana* is available. The gene sequence identity of *Brassica* and *Arabidopsis* varies from 75% to 90% (Quiros et al. 2001). Using a full genome Affymetrix chip, Siemens et al. (2006) found that more than 1000 *Arabidopsis* genes are differentially expressed in *P. brassicae*-infected roots at either 10 or 23 days after inoculation versus non-inoculated roots. These included genes associated with growth and cell cycle, sugar phosphate metabolism, and defense. The involvement of plant hormones in club development further supported genes involved in auxin homeostasis, such as nitrilases and members of the GH3 family, are up-regulated, whereas genes involved in cytokinin homeostasis are strongly down-regulated at both time points. Cytokinin oxidase/dehydrogenase overexpressing lines are disease resistant, which clearly indicates the importance of cytokinin as a key factor in clubroot disease development. Most known defense- or resistance-related genes are either not differentially expressed or down-regulated at 23 vs. 10 days after inoculation. Feng et al. (2012a, b) observed the expression of several canola genes homologous to *Arabidopsis* genes found to be down-regulated at 35 days after inoculation, are highly expressed at early stages (7 days after inoculation) and down-regulated at later stages (42 days after inoculation) of infection. At the later stages of infection, the galls were forming and the resting spores were mature. Thus, the host genes involved in resistance/susceptibility may no longer be as active as during the early infection stages. Another explanation is that during the later stages, the pathogen already has successfully suppressed the expression of most resistance-related genes.

Proteome-level analysis, a powerful tool for high-throughput global protein expression analysis using two-dimensional gel electrophoresis (2-DE) coupled with mass spectrometry (MS) and bioinformatics, has been extensively used in the identification of proteins that are differentially expressed in plants in response to various abiotic or biotic stresses including those caused by plant pathogens. In order to investigate whether the pattern of specific host proteins is changed during clubroot formation, Hansen et al. (1994) employed 2-DE to compare the root protein profiles of *Brassica oleracea* with *B. oleracea* infected with *P. brassicae* four weeks after inoculation. They found that several proteins present in noninfected roots were absent or strongly reduced in the infected roots. In a proteomic study of a Chinese cabbage cultivar and two *P. brassicae* isolates, Ito et al. (1996) were able to resolve 133 differentially abundant protein spots, relative to noninfected controls, in infected roots 3 to 4 weeks after inoculation. The 133 protein spots were further classified into 6 groups, including spots that were enhanced in the susceptible response (59 spots), unique to the susceptible response (6 spots), repressed in the susceptible response (9 spots), enhanced in the resistant response (35 spots), unique to the resistant response (5 spots), and repressed in the resistant response (19 spots). N-terminal amino acid sequencing revealed that one protein (25 kDa, pI 7.0) that was enhanced in the susceptible response group showed high homology with pathogenesis-related protein group 5. These results suggest that the patterns of gene expression are different in the susceptible and the resistant responses.

In a differential protein analysis of infected versus noninfected roots and hypocotyls of *Arabidopsis* with 2-DE and MS, Devos et al. (2006) reported that 12% (46/390) of the visualized *Arabidopsis* proteins showed an altered abundance 4 days after inoculation with *P. brassicae* compared with the non-inoculated plants, including proteins involved in metabolism, cell defense, cell differentiation, and detoxification. The metabolism-related enzymes adenosine kinase 2 (ADK2) and fructose bisphosphatealdolase (FBA) were found to be down-regulated at 4 days after inoculation. Down-regulation of ADK2 during the early stages of *P. brassicae* infection may keep the elevated levels of active cytokinins high, which is consistent with the increased endogenous isopentenyl adenine content. Down-regulation of FBA upon *P. brassicae* infection suggests a flow towards the production of glucose, which correlates with the fact that galls of *P. brassicae* act as a carbon sink that may be induced by the high cytokine level present at the infection site (Devos et al. 2006). Myrosinase, an enzyme that can catalyze indole-3-methyl glucosinolate degradation to form indole-3-acetic acid, was also found to be six times more abundant in infected *Arabidopsis* at 4 days after inoculation. During *P. brassicae* infection, a number of detoxification-related proteins are found to be up-regulated, including catalase, glutathione S-transferase, thioredoxin, and ferredoxin-nitrate reductase. Out of the 46 differentially regulated proteins, 11 proteins including ADK2, FBA, peroxiredoxin, α-tubulin, and heat shock protein 70 are down-regulated, and 35 are up-regulated including myrosinase, glutathione S-transferase, ferredoxin-nitrite reductase, and pectin methyl-esterase (Devos et al. 2006).

The changes in the root protein profile were examined by 2-DE at 12, 24, 48, and 72 hours after inoculation of a susceptible canola (*B. napus*) genotype with *P. brassicae* (Cao et al. 2008). A total of 20 protein spots were identified as either up-regulated (13 spots) or down-regulated (7 spots). Decreased abundance of adenosine kinase, which is involved in cytokine homeostasis, supported the previous reports that cytokinins play a key role in the early phases of *P. brassicae* infection (Devos et al. 2006). An approximately sixfold reduction in caffeoyl-CoA O-methyltransferase abundance suggested a reduction in host lignin biosynthesis after inoculation, and is consistent with the compatible nature of the *B. napus/P. brassicae* interaction. Levels of enzymes involved in the metabolism of reactive oxygen species, such as copper/zinc superoxide dismutase and cytochrome c-oxidase, declined sharply at 12 hours after inoculation, but increased at 24–72 hours. Protein identification by MS analysis revealed that the 20 differentially regulated proteins can be classified into proteins involved in lignin biosynthesis, cytokinin metabolism, glycolysis, intracellular calcium homeostasis, and the detoxification of reactive oxygen species.

2.10 *Brassica-Sclerotinia* R-Genes

2.10.1 Identification of Partial Resistance in *Brassica napus* to *Sclerotinia*

B. napus accessions with partial stem resistance to a Canadian *S. sclerotiorum* isolate (#321) were identified using a stem test in which flowering plants were inoculated with mycelium plugs. The partial stem resistance of four of these accessions, PAK54, PAK93, DC21, and K22, were examined following inoculation with Australian isolates. Mycelial compatibility groups and intergenic spacer (IGS) region haplotypes were identified among 71 isolates from Australian oilseed rape and lupin fields. Eleven genetically diverse isolates showed differences in aggressiveness when inoculated onto nine oilseed rape varieties and one Chinese accession. Isolates CU8.24, CU10.17, and CU11.19 were selected based on genetic diversity, growth rate in vitro and high aggressiveness in the initial screen and subsequently inoculated onto the four *B. napus* accessions. These accessions developed significantly smaller lesions compared with the susceptible control varieties ("AV Garnet" and "Westar"), with the average frequency of soft and collapsed lesions being less than 20% in PAK54, DC21, and K22, 29% in PAK93 and greater than 88% in the susceptible controls. Microscopic examination revealed that hyphae were typically confined to the stem cortex in the smallest lesions, but could be found in the stem pith in larger lesions. These results show that *B. napus* accessions PAK54, PAK93, DC21, and K22 can be used in Australia for development of varieties with partial stem resistance to *S. sclerotiorum* (Denton-Giles et al. 2018).

2.10.2 Identification of Genomic Regions in Wild *Brassica oleracea* for Resistance to *Sclerotinia sclerotiorum*

Considering the wide genetic diversity in Brassicaceae, resources with high levels of resistance against *S. sclerotiorum* may exist in the relatives of *B. napus*. Some efforts have been made to identify resistance resources from wild crucifers such as *Erucastrum cardaminoides* (Garg et al. 2010), *E. abyssinicum* (Garg et al. 2010b), *E. gallicum* (Lefol et al. 1997b; Seguin-Swartz and Lefol 1999) and *Capsella bursa-pastoris* (Chen et al. 2007). Mei et al. (2011) have identified resources with high level of resistance against *S. sclerotiorum* from wild *B. oleracea* one of the parental species of rapeseed (UN 1935). This finding brings a new hope to improve Sclerotinia resistance of rapeseed. A resistant accession of wild *B. oleracea* (*B. incana*) was employed to cross with a susceptible cultivar, and an F2 population, with several clones of each F2 genotype, was developed from one individual F1 plant.

The lack of resistant source has greatly restrained resistance breeding of rapeseed (*Brassica napus*, AACC) against *Sclerotinia sclerotiorum* which causes severe yield losses in rapeseed production all over the world. Several wild *Brassica oleracea* accessions (CC) with high level of resistance have been identified (Mei et al. 2011), bringing a new hope to improve *Sclerotinia* resistance of rapeseed. To map quantitative trait loci (QTL) for *Sclerotinia* resistance from wild *B. oleracea*, an F2 population consisting of 149 genotypes, with several clones of each genotypes, was developed from one F1 individual derived from the cross between a resistant accession of wild *B. oleracea* (*B. incana*) and a susceptible accession of cultivated *B. oleracea* var. *alboglabra*. The F2 population was evaluated for Sclerotinia reaction in 2009 and 2010 under controlled condition. Significant differences among genotypes and high heritability for leaf and stem reaction indicated that genetic components accounted for a large portion of the phenotypic variance. A total of 12 QTL for leaf resistance and six QTL for stem resistance were identified in 2 years, each explaining 2.2–28.4% of the phenotypic variation. The combined effect of alleles from wild *B. oleracea* reduced the relative susceptibility by 22.5% in leaves and 15% in stems on average over 2 years. A 12.8-cM genetic region on chromosome C09 of *B. oleracea* consisting of two major QTL intervals for both leaf and stem resistance was assigned into a 2.7-Mb genomic region on chromosome A09 of *B. rapa*, harboring about 30 putative resistance-related genes. Significant negative corrections were found between flowering time and relative susceptibility of leaf and stem (Mei et al. 2012).

2.10.3 Identification of R-Genes in *Brassica* Species to *Sclerotinia*

There are number of promising candidates for introgression of disease resistances into commercial *Brassica* species. Multiple wild *Brassica* species have shown high levels of resistance against Stem rot (Seguin-Swartz and Lefol 1999; Chen et al. 2007; Uloth et al. 2013). A number of interspecific hybrids between wild and cultivated crucifers have shown an outstanding level of resistance to Stem rot, for example, introgression lines between *B. juncea* and *B. napus* from *Erucastrum cardaminoides*, *Diplotaxis tenuisiliqua* and *Erucastrum abyssinicum* (Garg et al. 2010a). Other examples of hybrids are between *Brassica rapa/Brassica oleracea* (Niu et al. 2005), *B. juncea/Brassica carinata*, *B. napus/B. carinata* (Barbetti et al. 2014), *B. napus/Sinapis alba* (Li et al. 2009a), and *E. cardaminoides/B. rapa* and *Brassica nigra* (Chandra et al. 2004), to name a few. The success of these intergeneric and/or interspecific crosses shows that there is enormous potential for incorporating high levels of resistance from wild crucifers into the germplasm of crop *Brassica* species. *Brassica oleracea*, usually grown as a leafy vegetable, is also a potential source for resistance in *B. oleracea* itself, and for introgression into other commercial vegetable and oilseed *Brassicas*. Partial resistance to *S. sclerotiorum* has been identified in *B. oleracea* var. *capitata* (cabbage) (Dickson and Petzoldt 1996; Cubeta et al. 1997). Mei et al. (2011) screened 68 accessions across six *Brassica* species, including 47 from *B. oleracea*, and found a large variation in resistance using excised leaves and stem inoculation in the field. The most resistant genotypes were from *B. oleracea*, and especially its wild relatives *Brassica rupestris, Brassica incana, Brassica insularis* and *Brassica villosa*. However, Uloth et al. (2013) found that mean lesion length for a group of *B. oleracea* genotypes tested against an aggressive strain of *S. sclerotiorum* ranged from 10 to 93 mm, so it is clear that there can be a wide range in resistance/susceptibility within selections of *B. oleracea*. The results from field screening genotypes of Chinese *B. oleracea* var. *capitata*, Indian *B. juncea* carrying wild weedy Brassicaceae introgression(s) or B-genome introgression, and Australian commercial *B. napus* varieties showed resistance to *Sclerotinia*.

Field resistances against Sclerotinia rot (*Sclerotinia sclerotiorum*) were determined in 52 Chinese genotypes of *Brassica oleracea* var. *capitata*, 14 Indian *Brassica juncea* genotypes carrying wild weedy Brassicaceae introgression(s) and four carrying B-genome introgression, 22 Australian commercial *Brassica napus* varieties, and 12 *B. napus* and *B. juncea* genotypes of known resistance. All plants were individually inoculated by securing an agar disc from a culture of *S. sclerotiorum* growing on a glucose-rich medium to the stem above the second internode with Parafilm tape. Mean stem lesion length across tested genotypes ranged from <1 to >68 mm. While there was considerable diversity within the germplasm sets from each country, overall, 65% of the *B. oleracea* var. *capitata* genotypes from China showed the highest levels of stem resistance, a level comparable with the highest resistance ever recorded for oilseed *B. napus* or *B. juncea* from China or Australia. One Indian *B. juncea* line carrying weedy introgression displayed a significant level of both stem and leaf resistance. However, the vast majority of commercial Australian oilseed *B. napus* varieties fell within the most susceptible 40% of genotypes tested for stem disease. There was no correlation between expressions of stem versus leaf resistance, suggesting their independent inheritance. A few Chinese *B. oleracea* var. *capitata* genotypes that expressed combined extremely high level stem (≤ 1 mm length) and leaf (≤ 0.5 mean number of infections/plant) resistance will be particularly significant for developing new stem rot resistant horticultural and oilseed *Brassica* varieties (You et al. 2016).

2.10.4 Identification of R-Genes Sources in *Brassica* Genotypes to *Sclerotinia*

Because of its very wide host range, breeding for resistance against *S. sclerotiorum* is very challenging. *Brassica napus* and *B. juncea* cv. Rugosa have been reported to possess resistance against white stem rot both in the field as well as in the green house conditions (Singh et al. 1994). Partial resistance has also been identified in some *B. napus* and, to a lesser extent, in *B. juncea* genotypes from China (Li et al. 1999, 2006, 2008a; Zhao et al. 2004), Australia (Li et al. 2006, 2008a), and India (Singh et al. 2008). Although a significant number of partially resistant genotypes have been identified, breeding to increase the levels of resistance against Sclerotinia disease in *B. napus* and/or *B. juncea* has been ineffective. This is mainly because resistance to *S. sclerotiorum* in existing cultivars of *Brassica* and in other cultivated germplasm appears to be of a complex nature, *i.e.*, it can either be monogenic and/or polygenic depending on the plant species and materials under investigation (Abawi et al. 1978; Baswana et al. 1991; Zhao and Meng 2003; Zhao et al. 2006). Complete resistance has not been identified in canola. Partial field resistance has been identified in the Chinese variety Zhongyou 821 (Li et al. 1999; Buchwaldt et al. 2003). A cultivar Zhongshuang No. 9, claimed to be better than Zhongyou 821, was reported in 2003 (Wang et al. 2003). The next most resistant *B. napus* genotypes that have been previously reported included 06-6-3792 (China), ZY004 (China), and RT 108 (Australia) with mean stem lesion lengths of <3.0 cm (Li et al. 2008a). The genotypes of *B. juncea*, namely, EC 597328 (Montara), EC 597329 (Berry), and EC 597331 (Ringot I) of Chinese origin were also found tolerant (Sharma et al. 2009). In addition, the levels of resistance previously reported in genotypes *B. juncea* JM06018 and JM 06006 with mean stem lesion lengths of 4.8 cm (Li et al. 2008b) was far lower as compared to the *B. napus* genotypes. However, the situation has begun to improve due to introduction of new sources of *Sclerotinia* resistance from Chinese native cultivars. For example, the double-low restorer line of the elite cv. Zhongshuang 9 was released as *Sclerotinia* resistant (Wang et al. 2004). Glucosinolates present in Brassicaceae plants have fungicidal and bactericidal properties which can also give resistance against fungal pathogens (Zukalová and Vasák 2002). Since significant differences in lesion lengths were detected, it is possible that isolates of *S. sclerotiorum* have a different pathogenic ability, as reported by Pratt and Rowe (1991) and Liu (1996). Li et al. (2001) found that Zhongyou 821 produced more indolic glucosinolates in the leaves after inoculation. Moderate *Sclerotinia* resistance has been reported in several related species, including *B. nigra*, *B. juncea* and *Sinapis alba* (Morrall and Dueck 1982; Kolte 1985a). Novel approaches to the introduction of increased *Sclerotinia* resistance into oilseed rape include the development of transgenic plants expressing oxalate oxidase activity (Thompson et al. 1995) capable of degrading oxalic acid, a patho-toxin of the fungus. Induced mutagenesis has also proved to be a successful technique for introducing disease resistance genes into crop varieties (Konzak 1956; Maluszynski et al. 1995); it is widely accepted that micro-mutations in minor genes, causing small quantitative changes in phenotypic characters, occur at much high frequencies (IAEA 1977). Induced mutants exhibiting increased partial (quantitative) disease resistance have been isolated from small populations (Varghese 1985; Worland and Law 1991; Kinane and Jones 1996). Against a non-specialized necrotrophic pathogen such as *S. sclerotiorum*, any resistance would probably be quantitative in nature. Successful isolation of *Sclerotinia*-resistant mutants from small M2 populations of oilseed rape could introduce potentially valuable genes into the *B. napus* gene pool.

In plant-pathogen interactions, development of new pathogenic races and break down of resistance are limiting factors in resistance deployment against plant diseases. The pathogen's life history characteristics and evolutionary potential are major factors leading to the pathogen overcoming host resistance (García-Arenal and McDonald 2003; Coletta-Filho and Machado 2002; McDonald and Linde 2002). Therefore, knowledge regarding the population genetic structure and evolutionary potential of the pathogen will provide insight into the most suitable breeding strategy for durable resistance (McDonald and Linde 2002). Nine genotypes, viz. Cutton, ZYR-6, PSM-169, PDM-169, Westar, PYM-7, Parkland, Tobin, and Candle showed resistance to stem rot in India (Shivpuri et al. 1997). Four genotypes, viz. PCR-10, RW-8410, RW-9401, and RGH-8006 had better resistance against *S. sclerotiorum* as compared to susceptible check (Pathak et al. 2002). Wang et al. (2009), isolated genes from *B. napus* known as *B. napus* mitogen-activated protein kinase, (BnMPK4), and found that *BnMPK4* along with *PDF* 1.2 are inducible in resistant line Zhongshuang 9, but both were consistently suppressed in susceptible line 84039 after inoculation with *S. sclerotiorum*. Transgenic oilseed rape over expressing *BnMPK4* markedly enhanced resistance to *S. sclerotiorum*. Further, it was observed that transgenic plants inhibited growth of *S. sclerotiorum* and constitutively activated *PDF1.2*, but decreased H_2O_2 production and constitutively suppressed *PR-1* expression. Treatment of roots of the transgenic plants with H_2O_2 solution resulted in enhanced susceptibility to the two pathogens. It was suggested that*MPK4* positively regulates jasmonic acid-mediated defense response, which might play an important role in resistance to *S. sclerotiorum* in oilseed rape. Sharma et al. (2009) evaluated indigenous and exotic genotypes of *B. napus* and *B. juncea* in pathogen-infested field and observed that genotype EC-597328 of *B. juncea* is tolerant to this disease. Sharma et al. (2012) evaluated seventy germplasm from India, Australia, and China by artificial stem inoculation and observed that none of the germplasm was resistant while partial tolerance was observed in lines BLN 3630 (EC597274) of *B. napus* and Berry (EC597329) of *B. juncea*. For evaluation of resistance to rape SR, Sang et al. (2013) transferred the *MSI-99m* gene (modified *MSI-99*) into Chinese rape variety Zhongyou 821 using *Agrobacterium*-mediated method. Nine transformed lines carrying a *MSI-99m* expression vector were detected by polymerase chain reaction (PCR), among which seven lines expressed *MSI-99m* gene according to qRT–PCR analysis. Disease resistance analysis consistently showed that the high-level expression of *MSI-99m* increased resistance to *S. sclerotiorum* in transgenic rape lines. It was demonstrated that *MSI-99m* gene might be applied as a resistant gene resource in rape for the improvement of rape varieties. Callogenic response to culture filtrate (CF) of the pathogen was evaluated and observed that MS medium supplemented with 1-naphthaleneacetic acid (NAA) and 1-benzylaminopurine (BAP) were suitable for callus induction; MS medium supplemented with CF of *S. sclerotiorum* @ 15 % was found inhibitory to callus production in *B. juncea* cv. RH-30. The activity of CF in callus production was lower in *B. carinata* cv. HC-9002 and *B. napus* cv. GSH-1 as compared to *B. juncea* (Watpade et al. 2012). The biochemical changes occurred during the development of callus in the presence of CF also suggests that the total soluble protein in *B. napus* and *B. carinata*, and activity of polyphenol oxidase and peroxidase increased in resistant calli; catalase activity decreased in resistant/tolerant calli as compared to susceptible calli (Watpade and Naresh 2012). Total soluble sugars and reducing sugars were higher after selection with *S. sclerotiorum* culture filtrate in calli of *B. napus* cv. GSH-1 and *B. carinata* cv. HC-9002, but other species showed lesser sugar contents. Generally, level of phenol, flavanol and total soluble protein were higher after selection in all tolerant calli (Watpade and Naresh 2013). Culture filtrates of 25 geographical *S. sclerotiorum* isolates reduced seed germination and length of radical and plumule in *B. juncea* cv. Rohini (Sharma et al. 2014). Lack of effective resistance to Sclerotinia disease in cultivated species has stimulated the interest of researchers towards exploitation of wild crucifer species to diversify the existing gene pool. Higher levels of resistance against Sclerotinia have already been reported in the secondary gene pool of bean (Abawi et al. 1978; Gilmore et al. 2002; Schwartz et al. 2006). Introgression of genomic segments responsible for resistance against *Sclerotinia* from wild to cultivated species of sunflower has also been attempted (Ronicke et al. 2004). Although, Brassicaceae family contains a wide array of different species, to date, only two wild crucifers, *Capsella bursa-pastoris* (Chen et al. 2007) and *Erucastrum gallicum* (Lefol et al. 1997a; Seguin-Swartz and Lefol 1999), have been reported to show high level of resistance against Sclerotinia. Although introgressive hybrids were successfully obtained between *B. rapa and B. napus* species and *Capsella bursa-pastoris* (Chen et al. 2007), it remains to be confirmed if the introgression of resistance against *S. sclerotiorum* from *E. Gallicum* into cultivated species has in fact been accomplished (Lefol et al. 1997a, b; Seguin-Swartz and Lefol 1999). Garg et al. (2010) developed Introgression lines following hybridization of three wild crucifers (viz. *Erucastrum cardaminoides, Diplotaxis tenuisiliqua* and *E. abyssinicum*) with *B. napus* or *B. juncea*; introgression lines had much higher levels of resistance. The various varieties which have been recognized to possessed resistance against *S. sclerotiorum* are presented in Table 2.19.

2.10.5 Use of Oxalate Oxidase (OXO) to Identify R-Genes to *Sclerotinia*

A common strategy for combating *Sclerotinia* is to degrade OA, the plant toxin and *Sclerotinia* pathogenicity factor secreted from the pathogen. There are three classes of known enzymes that can catabolize OA, namely oxalate oxidase (OXO) (EC 1.2.3.4) (Lane et al. 1991), oxalate decarboxylase (EC 4.1.1.2) (Mehta and Datta 1991), and oxalyl-CoA decarboxylase (EC 4.1.1.8) (Lung et al. 1994). The bacterial oxalyl-CoA decarboxylase gene could be used for oxalate degradation and engineering *Sclerotinia* resistance in plants (Dickman and Mitra 1992). However, both fungal and bacterial oxalate decarboxylases convert oxalate into CO2 and formic acid, which might have a toxic effect on plant cells. Therefore, scientists have been focusing on OXO, which releases CO_2 and H_2O_2 from $_{O2}$ and OA. This enzyme was first isolated and characterized from barley and wheat (Lane et al. 1993; Kotsira and Clonis 1997). The transgenics greatly reduced disease progression and

Table 2.19 Sources of R-genes in crucifer's to *Sclerotinia sclerotiorum* (Sclerotinia rot) (Saharan and Mehta 2008 updated)

Crucifer host	Sources	Country	References
Rapeseed-mustard	Omi nature, Isuzu (*Brassica napus*), Hallayucke, Norin-9, Doral, Librador, Lirama, Bor, BOH-1592, BOH-1693, MAH-1391, MAH-1592, BOH 2600, HH-1, OKEG-8, OKEG-94, POH-285, H-43/33, Jet Neuf, Koganenatane, Aburamasari, Kizakinonatane, Cutton, ZYR6, PSM 169, PDM-1969, Wester, PYM 7, Parkland, Tobin, PCR 10, Candle, Cutlass, Torch, RW 8410, RW 9401, RGN 8006, Hyola 401, PBN 9501, PWR 9541, Kiran, RH 9401, RH 492, PAB 9511, Bermuda, Capio, Mohican, Bor, Xiangyou 15, Zhongchuang 9, MAH 1996, Valesca, Passat, Libro, BKH 894, PNG 2170, MA 1615-1, MZL 236, BK 2466/93 and MA 1649-1, Rugosa, Montara, Berry, Ringot 1	India	Saharan and Naresh (2008), Singh et al. (1994), Sharma et al. (2009)
Brassica napus ssp. *rapifera*			
B. napus	06-6-3792 and ZY004 from China; RT108 from Australia,	China, Australia	Li et al. (2009)
	Mystic	Australia	Uloth et al. (2013, 2015)
	ZY006	Australia	Barbetti et al. (2014)
	169080, 286418, 436554, 458940, and 633119		Khot et al. (2011)
B. alba, Sinapis alba	All Accessions		
B. carinata	BRA 926/78, 054113, PI 193459	Australia	Uloth et al. (2013)
B. juncea	JM06018 and JM06006 from Australia	Australia	Li et al. (2009)
	B. juncea-B. fruticulosa introgression lines (ILs)	India	Atri et al. (2019a, b)
	DRMRSJ-22, DRMR 5206	India	AICRPRM-AR (2019)
Arabidopsis thaliana	All Accessions		
B. rapa ssp. *Yellow Sarson*			
R. raphanistrum	RRA 41, WARR 20	Australia	Uloth et al. (2013)
R. sativus	Krasnodar	Australia	Uloth et al. (2013)
B. cretica	Wild Brassica line	UK	Taylor et al. (2017)
Brassica species	*B. rupestris, B. incana, B. insularis,* and *B. villosa*	China	Mei et al. (2011)

lesion length following cotyledon and stem inoculation with *S. sclerotiorum*, indicating that OXO conferred resistance to the stem rot (Donaldson et al. 2001). Grison et al. (1996) introduced a hybrid endochitinase gene under a constitutive promoter by *Agrobacterium*-mediated transformation into a winter-type oilseed rape inbred line.

2.11 *Brassica-TuMV* R-Genes

2.11.1 Identification of Pathotype Specific R-Genes Sources

Sources of resistance to TuMV in *Brassicas* have been identified (Table 2.20). Walkey and Neely (1980) identified useful TuMV-resistance phenotypes in *Brassica* species during an evaluation of 88 cabbage lines. Using 19 major isolates of TuMV out of 5739 samples from 10 cities in China, Liu et al. (1996) screened 3000 Chinese cabbage germplasm accessions. They found that eight were immune or highly resistant to all TuMV isolates, resulting in a release of 28 new resistant cultivars. Breeders also made efforts to identify TuMV-resistance phenotypes in oilseed and forage *Brassica* species, including evaluation of 14 *B. rapa* and two *B. juncea* lines (using TuMV-C1, C2, C3, and C4), which revealed resistance to on any isolate of TuMV (Fjellstrom and Williams 1997). Hughes et al. (2002) reported that using isolates of TuMV pathotypes 1, 3, and 4 to test 42 *B. rapa* and *B. napus* lines, 14 lines were resistant to all three pathotypes, eight lines were resistant to two pathotypes, and nine were resistant to one pathotype. Coutts et al. (2007) tested 43 *B. napus* lines (TuMV pathotype 8) and Kehoe et al. (2010) assessed a collection of 44 *B. juncea* lines (TuMV isolates WA-Apl, NSW-1 and NSW-2); both of these studies revealed additional TuMV-resistance phenotypes to those named previously by Jenner and Walsh (1996). Nyalugwe et al. (2015) found TuMV-resistance phenotypes in 69 *B. juncea*, 11 *Camelina sativa*, 11 *B. oleracea*, five *B. rapa* and three *Raphanus sativus* lines using TuMV isolates WA-Apl, NSW-1, and NSW-2. In China, Qian et al. (2013) and Li et al. (2016a, b, c) reported five Chinese cabbage lines that were extremely TuMV-resistant (80122, 80124, BP058, 80186, and 2079) and three extremely susceptible lines (80403, 80461, and 80425) using TuMV isolate C4 (Li et al. 2019).

2.12 *Brassica-Xanthomonas* R-Genes

2.12.1 Identification of R-Genes Sources in *Brassica*

Resistance have been derived from *B. oleracea,* including the cabbage cultivar Early Fuji (Bain 1952), the inheritance of which was reported to be controlled by a single recessive gene with two modifiers (Williams et al. 1972). Resistance has also been documented in *B. oleracea* PIs, including the cabbage accession PI 436606 from China, also determined to be controlled by multiple genes (Dickson and Hunter

Table 2.20 Genes for resistance to TuMV, mapped or cloned from *Brassica* crops (Li et al. 2019)

Resistance gene	Species (genome)	Line	Mapping markers	Inheritance and mapping	Pathotype/isolate	References
TuRB01	*B. napus* (A)	N-o-9	RFLP	Single dominant, mapped to a 7.2 cM interval on N6	1	Walsh et al. (1999)
TuRB02	*B. napus* (C)	N-o-9	RFLP	Dominant, mapped to N14	CHN1, JPN1	Walsh et al. (1999)
TuRB03	*B. napus* (A)	22S	AFLP, SSR	Single dominant, mapped to a 7.9 cM interval on N6	CDN1	Hughes et al. (2003)
TuRB04	*B. napus* (A)	165	—	Single dominant, unmapped	1, 3	Jenner et al. (2002, 2003)
TuRB05	*B. napus* (A)	165	—	Single dominant, unmapped	1, 3	Jenner et al. (2002, 2003)
retr01	*B. rapa* (A)	RLR22	RFLP, SSR	Recessive, mapped to R4, and the candidate encoding eIF (iso)4E	1, 3, 4, 7, 8, 9, 12	Rusholme et al. (2007)
ConTR01	*B. rapa* (A)	RLR22	RFLP, SSR	Dominant, mapped to R8, and the candidate is an eIF(iso)4E-encoding gene	1, 3, 4, 7, 8, 9, 12	Rusholme et al. (2007)
—	*B. rapa* (A)	91-112	AFLP, RAPD, SSR, SCAR	Four QTLs	Tu1, Tu2, Tu3, Tu4	Zhang et al. (2008)
—	*B. rapa* (A)	Y195-93	AFLP, RAPD, SSR, SCAR	Three QTLs on R3, R4 and R6	C4	Zhang et al. (2009)
BcTuR3	*B. rapa* (A)	Duanbaigeng	—	Possibly related gene, TIR-NB-LRR type	—	Ma et al. (2010)
rnt1	*B. rapa* (A)	AS9	SSR	Dominant, mapped on R6	UK1	Fujiwara et al. (2011)
TuRBCH01	*B. rapa* (A)	Q048	AFLP, SSR	A single dominant gene on R6	C5	Wang et al. (2011)
retr02	*B. rapa* (A)	BP8407	SSR, indel	Recessive, mapped to A4, candidate is Bra035393, encoding eIF(iso)4E	C4	Qian et al. (2013)
trs	*B. rapa* (A)	SB18/SB22	CAPS, SCAR, SNP	Recessive, mapped to A4	CHN2, CHN3, CHN4, CHN5	Kim et al. (2013)

(continued)

Table 2.20 (continued)

Resistance gene	Species (genome)	Line	Mapping markers	Inheritance and mapping	Pathotype/isolate	References
TuMV-R	*B. rapa* (A)	VC40	SSR, indel, SNP	Mapped to a 0.34 Mb region on A6, with six candidates	—	Chung et al. (2014)
TuRB01b	*B. rapa* (A)	TD34-S1	RFLP	Single dominant, mapped to a 2.9 Mb region on A6	1	Lydiate et al. (2014)
TuRB07	*B. rapa* (A)	VC1/VC40	SSR	Single dominant, mapped to A6, candidate gene Bra018863 (CC-NB-LRR)	C4	Jin et al. (2014)
TuRBCS01	*B. rapa* (A)	8407	EST, SSR, indel	1.98 Mb region on A4	C4	Li et al. (2015a, b)
TuRBJU01	*B. juncea* (unknown)	Oasis CI	—	Dominant, unmapped	8 (WA-Ap1)	Nyalugwe et al. (2015, 2016)
retr03	*B. juncea* (A)	VC029	SNP	Recessive, encoding eIF2Bb	ZJ	Shopan et al. (2017)

AFLP amplified fragment length polymorphism, *CAPS* cleaved amplified polymorphic sequence, *EST* Expressed sequence tag, *Indel* insertion deletion, *QTL*, quantitative trait locus, *RAPD* random amplified polymorphic DNA, *RFLP* restriction fragment length polymorphism, *SCAR* sequence characterized amplified region, *SNP* single nucleotide polymorphism, *SSR* simple sequence repeat

1987; Hunter et al. 1987). These sources have been used in the development of black rot-resistant breeding lines, including Badger #16; NY4002; and Cornell 101, Cornell 102, and Cornell 103. However, the resistance is typically incomplete and difficult to incorporate into hybrid cultivars (Camargo et al. 1995). Resistance has been reported in the mustard species *B. nigra, B. juncea*, and *B. carinata* (Taylor et al. 2002; Tonguc and Griffiths 2004a; Westman et al. 1999), including *B. juncea* accessions PI 633077 and PI 633078 (previously A 19182 and A 19183), and *B. carinata* accessions PI 199947 and PI 199949 has been used in the creation of interspecific hybrids (Tonguc et al. 2003; Tonguc and Griffiths 2004b). Xcc has been characterized into at least six distinct races, the most important of these being races 1 and 4, which account for over 90% of black rot disease worldwide (Vicente et al. 2001). *B. oleracea* accessions have not been identified that exhibit complete resistance to races 1 and 4, but breeding lines have been developed that exhibit incomplete resistance to these races. Related crucifer species, including *B. carinata* and *B. juncea*, exhibit resistance to Xcc races 1 and 4, that appear to be controlled by a single gene (Tonguc et al. 2003; Tonguc and Griffiths 2004b). To effectively use germplasm resources for interspecific hybridization with *B. oleracea*, it will be important to determine the presence and frequency of Xcc resistance not just within cultivated *Brassica* vegetables, but also within related crucifer accessions.

Resistance may be exploited through interspecific crosses of *B. oleracea* with related species. Crucifer accessions could provide important resistance sources for the development of hybrid *Brassica* vegetables by contributing new resistance genes for Xcc or genes that can be incorporated into cole crops more effectively. Griffiths et al. (2009) evaluated crucifer species at the juvenile stage to quantify the presence, frequency, and potential use of accessions for introgression of Xcc resistance. To achieve this, the crucifer accessions from the NC-7 (Ames, IA) and NE-9 (Geneva, NY) USDA regional PI centers were inoculated and evaluated for resistance to race 1 and race 4 of Xcc using the wound inoculation technique (Griffiths and Roe 2005). The USDA NC-7 and NE-9 regional PI stations maintain vegetable, mustard, and oilseed crucifers, of which 4084 accessions were available for testing, representing 23 genera and 125 species. These accessions were evaluated for resistance to black rot after wound inoculation with race 1 of the pathogen. Accessions that were symptomless for race 1 of Xcc were replanted and inoculated with race 4 of the pathogen to identify accessions with resistance to both races. Symptomless responses were observed in 362 accessions of the mustard species (*Brassica juncea*, *Brassica carinata*, and *Brassica nigra*), in particular, *B. juncea* for which 353 of the 389 accessions tested were symptomless. Resistance was identified in five accessions of *B. carinata* out of 63 evaluated (PI 193460, PI 193959, PI 194254, PI 280230, PI 633077) determined by repeated symptomless responses after inoculation and four accessions of *B. nigra* from the 83 evaluated (PI 197401, A 25399, A 25401, PI 458981). Five accessions of *Brassica rapa* (PI 633154, A9285, PI 340208, PI 597831, PI 173847) were identified, which represent promising new sources of resistance to Xcc. Incomplete resistance was identified in an accession of *Eruca sativa* (PI 633207), an accession of *Lepidium* spp. (PI 633265), an accession of *Sinapis arvensis* (PI 296079), and two accessions of *B. napus* (PI 469733 and PI

469828). These identified accessions represent germplasm that can be used in breeding for resistance to Xcc in vegetable crucifers through interspecific crossing (Griffiths et al. 2009).

2.12.2 Genetics of Sources of Resistance to *Xanthomonas*

It is quite interesting that most of the *Arabidopsis thaliana* accessions, as wild relative to *Brassica* crops, harbor resistance to one or more races of *X. campestris* pv. *campestris*, and more than half of the accessions exhibit broad-spectrum resistance to all major races of the Xcc, indicating that *A. thaliana* could be used as a potential source of durable resistance to Xcc (Holub 2007). In different *Brassica* species a limited number of resistance sources have been identified (Table 2.21). It is broadly accepted that the *B. oleracea* with C genome are most susceptible to different races of Xcc (Dhar and Singh 2014; Saha et al. 2016). However, few sources of resistance have been identified in *B. oleracea* (Table 2.21). Most of the earlier reported sources of resistance (before 2014) in *B. oleracea* were race nonspecific. Therefore, use of the earlier identified resistant genotypes in studying the genetics of Xcc resistance and their molecular characterization is very limited and limited number of resistance sources have been identified (Table 2.21), including from the cabbage cultivar Early Fuji, the cabbage accession PI 436606 (cv. Heh Yeh da Ping Tou), cauliflower genotypes MGS 2-3, Pusa Kea, andSN445 (Sharma et al. 1977; Taylor et al. 2002; Vicente et al. 2002). QTLs for Xcc resistance have been genetically mapped in the inbred line Badger Inbred-16, which was derived from Early Fuji (Camargo et al. 1995). Jensen et al. (2005) reported field resistance to Xcc in hybrid cultivars T-689 F1, Gianty F1, No. 9690 F1, N 66 F1 and SWR-02 F1. Griffiths and Roe (2005), while testing different inoculation methods at different growing stages of the plant, identified resistance to Xcc in breeding lines Badger Inbred-16, Cornell 101, Cornell 102 and NY 4002 and in the Silver Dynasty commercial cultivar. Resistance sources to Xcc in cauliflower have been reported

Table 2.21 Sources of R-genes in crucifers to *Xanthomonas campestris* (Black rot) (Singh et al. 2010)

Crucifers host	Sources	Country	References
B. napus	Cobra, Giant English	Japan	Ignatov et al. (1998)
B. juncea	Florida Broad Leaf	Japan	Ignatov et al. (1998)
B. carinata	PI1 99947	Japan	Ignatov et al. (1998)
B. oleracea var. *botrytis*	BR-207	India	Saha et al. (2016)
	SN455	Japan	Ignatov et al. (1998)
B. oleracea var. *capitata*	Fujiwase, Singapura	Japan	Ignatov et al. (1998)
B. oleracea var. *alboglabra*	SI	Japan	Ignatov et al. (1998)
Raphanus sativus			
B. rapa subsp. *pekinensis*	RCBr	Japan	Ignatov et al. (1998)
B. rapa subsp. *rapifera*	Just Right	Japan	Ignatov et al. (1998)

by many researchers in India however, most of them are not found in the farmer's field due to their poor curd quality (Singh et al. 1987). Saha et al. (2014, 2016) identified resistance sources to race 1 and 4 in the C genome of *B. oleracea* accessions and reported only limited resistant sources specific to race 1 and 4 (Table 2.22). The lack of durable resistance sources in the C genome of *Brassica* to different Xcc races and quantitative inheritance of resistance complicates the complete control of disease (Singh et al. 2018).

The A and B genomes of *Brassica* species are a reservoir of the most efficient and durable genetic sources for resistance to Xcc, and a number of sources of resistance have been identified in the different species containing these genomes (Table 2.22) (Dey et al. 2015; Sharma et al. 2016). The race specific resistance to races 1 and 4 has been recorded in *B. nigra, B. rapa, B. carinata,* and *B. juncea* (Singh et al. 2011; Saha et al. 2014). The R1 genes conferring resistance to race 1 are present in the B genome of *B. carinata* (BBCC), *B. juncea* (AABB), and *B. nigra* (BB), while the R4 gene conferring resistance to race 4 is present in the A genome of *B. rapa* (AA), *B. napus* (AACC), and *B. juncea* (AABB) (Vicente et al. 2002; Taylor et al. 2002). In the C genome of *B. oleracea* (CC), specific resistance to races 1 and 4 is rarely expected. However, race-nonspecific resistance has been reported in the C genome of *Brassica* species (Sharma et al. 1995; Camargo et al. 1995; Taylor et al. 2002). A combined resistance to race 1, 4 and 6 of Xcc have been reported in alien *Brassica* species accessions, viz. Pusa Bold and Pusa Vijay (*B. juncea*), IC56072 (*B. nigra*) and NPC-9 (*B. carinata*) (Sharma et al. 2017; Dey et al. 2015).

2.12.3 Novel R-Genes Identification Through Molecular Approaches

Resistance (R-gene) genes identification is the first step towards the resistant hybrid/ variety development through appropriate breeding programs. Studying gene regulations in response to pathogens causing disease might give indications about the relevant defense genes. Suppression Subtractive Hybridization (SSH) is a powerful PCR-based technique to identify and isolate differentially expressed genes such as in the response to change in environmental conditions or plant pathogen interaction (Diatchenko et al. 1996; Jin et al. 2010; Luo et al. 2010; Jiang et al. 2011; Guo et al. 2013; Roohie and Umesha 2015). The application of the SSH approach for the identification of genes associated with black rot resistance in Brassicaceae crops has been documented (Jiang et al. 2011; Roohie and Umesha 2015). Jiang et al. (2011) employed SSH combined with RT-PCR to identify the black rot resistance genes in cauliflower (*B. oleracea* var. botrytis). They constructed a SSH cDNA library using resistant line C712 as tester and its susceptible near-isogenic line C731 as driver. Roohie and Umesha (2015) employed the SSH approach for the identification of resistance genes and to decipher the resistance mechanism in cabbage after infection with Xcc. In the resistant cultivar of cabbage "Pusa Mukta" they constructed a cDNA library enriched in differentially expressed transcripts. Out of 150 unigenes studied that were classified into five functional categories, 35% of the total unigenes accounted for the defense-related unigenes. Defense-specific representation of the

Table 2.22 Genetic resource and genetics of resistance to black rot in different *Brassica* species (Singh et al. 2018)

Genetic source/ accessions	*Brassica* genome	Species	Genetics	Pathogenic race(s)	Reference(s)
Sn 445	CC	*B. oleracea* var. *botrytis*	Single dominant gene	–	Sharma et al. (1972) and Jamwal and Sharma (1986)
MGS 2-3, Pusa Kea	CC	*B. oleracea* var. *botrytis*	Polygenic dominant		Sharma et al. (1972)
Sel-1-6-1-4	CC	*B. oleracea* var. *botrytis*	–	–	Sharma et al. (1977)
PI 436606 and DH line BOH85c	CC	*B. oleracea* var. *capitata*	Single dominant	Race 3	Vicente et al. (2002)
Early Fuji	CC	*B. oleracea* var. *capitata*	One or two dominant genes	–	Bain (1952) and Tonu et al. (2013)
BR 161	CC	*B. oleracea* var. *botrytis*	Monogenic dominant	Race 1	Saha et al. (2014)
BR 207	CC	*B. oleracea* var. *botrytis*	Monogenic dominant	Race 1	Saha et al. (2016)
NPC-9	BC	*B. carinata*	Monogenic dominant	Race 1	Sharma et al. (2016)
HCA-6	BC	*B. carinata*	Single dominant	Race 1 and race 4	Dey et al. (2015)
PI 199947	BC	*B. carinata*	Single dominant gene	Race 1 and race 4	Vicente et al. (2002)
Cobra, CrGC5	AC	*B. napus*	Single dominant gene	Race 4	Vicente et al. (2002)
PI 173847, PI 340208, PI 597831, PI 633154	AA	*B. rapa*	–	Race 1 and race 4	Griffiths et al. (2009)
A1982, A 1983	AB	*B. juncea*	–	–	Tonguc and Griffith (2004a, b)
IC56072	BB	*B. nigra*	–	–	Sharma et al. (2017)
DH line Reiho P01	CC	*B. oleracea*	Polygenic	Race 1	Doullah et al. (2011)
CY inbred line	CC	*B. oleracea*	Polygenic	–	Kifuji et al. (2013)
BI-16	CC	*B. oleracea* var. *capitata*	Quantitative recessive	Race 1	Vicente et al. (2002)

genes was confirmed through semi-quantitative RT-PCR and their elevated expression in the resistant cultivar was validated by qPCR. The resistant cultivar evoked a strong hypersensitive response upon attack by the *Xcc* pathogen (Singh et al. 2018).

2.13 Identification of Well-Characterized Sources of *R*-Genes for Major Pathogens of Crucifers

***Plasmodiophora*-Clubroot:** *P. brassicae*, causal agent for Clubroot, has a wide host range including non-cruciferous plants (Ren et al. 2016). Clubroot has caused severe cruciferous crop yield loss of up to 50% in Canada (Strelkov and Tewari 2005) and 20–30% in China (Chai et al. 2014). The pathogen physiology, progress in efforts to combat this disease, and an update on its epidemiology in Canada, Europe, and China have recently been the subject of various reviews (Chai et al. 2014; Diederichsen et al. 2014; Dixon 2014; Strelkov and Hwang 2014; Rícarova et al. 2016). The sources of *P. brassicae* R-genes and genetic mapping studies have been reviewed (Piao et al. 2004); Rahman et al. (2014). To date, at least 19 R-genes/QTL have been mapped for Clubroot resistance in *B. napus* by Werner et al. (2007). However, only two Clubroot resistance genes have been cloned, *Crr1a* and *CRa*, both in *B. rapa* (Table 2.16). Sources of *P. brassicae* R-genes in *B. rapa* have been particularly promising as there are many race-specific, single, dominant R-genes present, as compared with *B. oleracea* which has a more continuous resistance profile involving various combinations of both major and minor genes (Diederichsen et al. 2009; Rahman et al. 2011). Due to the different physiological resistance mechanisms towards *P. brassicae* in *B. rapa* and *B. oleracea*, complete resistance is often reported in the former but not in the latter (Tanaka et al. 2006; Donald et al. 2008; Hwang et al. 2012). Recent screening efforts in China have identified *B. rapa* to have good resistance against Canadian pathotype 3 (Zhang et al. 2015). Hence, it is not uncommon that breeding efforts make use of *B. rapa*, in particular using the European fodder turnip as the recurrent parent, to introduce resistance into *B. rapa, B. napus*, and *B. oleracea* (Diederichsen et al. 2009). The introgressed regions can be characterized using SNP genotyping and next-generation sequencing in order to identify and characterize these resistance genes and determine if they are the same gene or distinct. They could then be deployed in a more targeted manner.

In Canada, most oilseed *B. napus* shows little, or at best only moderate, resistance to Williams pathotypes 3 and 5, while *B. napus* rutabaga lines have better resistance (Jakir Hasan et al. 2012). However, in Australia, the majority of *B. napus* was found to be resistant (Donald et al. 2006), reflecting the influence of the local abiotic environments on the pathogenicity of the fungus, or the origin of host and/or pathogen influencing the resistance outcome (Donald et al. 2006; Donald and Porter 2014). Further QTL or Genome-Wide Association Study (GWAS) should be performed in *B. napus* in order to identify these resistance genes.

Traditionally, the AB genome amphidiploid *B. juncea* does not show high resistance. It often shows low, moderate (Peng et al. 2014) or intermediate (Sharma et al. 2012) resistance. However, more recently, one accession of *B. juncea* was found to be highly resistant to the local pathotype 4 under field conditions in China (Ren et al. 2016). Another good source of resistance from the B genome is diploid *B. nigra*, which has been reported to display high resistance against all local pathotypes in Canada (Jakir Hasan et al. 2012; Chu et al. 2013; Peng et al. 2014). There is therefore the potential for all A, B, and C genomes of the *Brassica* species to harbor R-genes against *P. brassicae.*

ii. ***Sclerotinia*—Stem Rot:** The Sclerotinia stem rot pathogen *Sclerotinia sclerotiorum* also has a wide host range, including members outside the crucifer family (Zhao et al. 2004). In the *Brassica* species, this pathogen can cause up to 80% yield losses in China (Mei et al. 2011) and has previously been reported to cause a loss of AU$ 23 million to the canola industry in Western Australia (Dafwa 2015). The resistance studies on Sclerotinia Stem Rot have mainly focused on the inheritance type at different developmental stages, namely seedling leaf and adult stem stages of the host plant. Thus far, no R-genes have been cloned for this pathogen (Wu et al. 2016a). However, candidate R-genes against *S. sclerotiorum* have been identified and physically mapped on the *B. napus* genome (Li et al. 2015a, b). These are NBSLRR genes with gene copies found on chromosomes A6 and C7 (Li et al. 2015b). Candidate R-genes against *S. sclerotiorum*, belonging to the CC-NBS-LRR class, were also identified on chromosome A9 in *B. napus* (Mei et al. 2013). Yet, the majority of the QTL for *S. sclerotiorum* have been identified in the C genome (C9 and C6) (Li et al. 2015a), indicating that *B. oleracea* is a good source of R-genes for S. sclerotiorum. This was recently confirmed for many *B. oleracea* var. capitata genotypes from China (You et al. 2016).

There can be strong correlation between leaf and stem resistance across some specific *Brassica* genotypes (Mei et al. 2011; Wu et al. 2013). For example, a few Chinese *B. oleracea* var. capitata genotypes were found to express extremely high level stem and leaf resistances (You et al. 2016). However, when diverse sets of *Brassica* genotypes from different parts of the world were evaluated, for most genotypes there was generally no correlation between leaf and stem resistances (Uloth et al. 2013; You et al. 2016). Where there is a strong correlation, such cases share common genes (Li et al. 2015a). This contrasting finding could reflect that high resistance against Sclerotinia Stem Rot in both stem and leaf is highly specific at the subpopulation level of the host species. The QTL region for stem resistance on chromosome C2 in *B. napus* was found to be homoeologous with that on chromosome A2, and falls within the syntenic region of their respective progenitor genomes (Wei et al. 2014), suggesting that the R-genes could be inherited from a common ancestral chromosome, that is largely conserved.

Although, monogenic control of single dominant genes against *S. sclerotiorum* has been found outside the *Brassica* genus, for example, in common bean

(*Phaseolus vulgaris*; Schwartz et al. 2006), it is highly likely that monogenic control is present in Brassicas where only a HR and/or lesions ≤2 mm diameter have been observed on stems (Uloth et al. 2013, 2015; You et al. 2016) or cotyledons (Garg et al. 2008, 2010b; Uloth et al. 2014; Ge et al. 2015). The genetic control of resistance genes is further complicated by the influence of inoculation methods (Zhao et al. 2004; Yin et al. 2010) and physiological traits, such as flowering time, as their QTL were found to be in the same genomic regions (Zhao et al. 2006; Mei et al. 2013; Wei et al. 2014). Despite this, Li et al. (2007a) both defined the importance of the type and time of inoculation and how expression of field stem resistance becomes independent of flowering time provided assessment is delayed until 3 weeks after inoculation. Identification of resistance genes against *S. sclerotiorum* has been hampered by inconsistencies in time and type of inoculation. Lines that have been shown to be highly resistant in one study can be susceptible in a different study. The use of common isolates and techniques will assist in moving this research forward. B genome *Brassicas* are commonly reported to have high levels of Sclerotinia Stem Rot resistance (Navabi et al. 2010). There is a range in resistance across the Australian *B. napus* rapeseed varieties carrying one or more *B. carinata* introgressions (Uloth et al. 2013). The Australian *B. juncea* varieties were shown to be generally more susceptible overall to Sclerotinia Stem Rot (Uloth et al. 2015), but the Indian (Barbetti et al. 2015) and Chinese (Li et al. 2006a, 2009) *B. juncea* varieties displayed high levels of host resistance. Wild weedy Brassicaceae species have shown particular promise for high-level resistance to Sclerotinia Stem Rot (Garg et al. 2010a; You et al. 2016), including genera such as *Raphanus* (Uloth et al. 2013). In particular, *B. napus* and *B. juncea* with introgression(s) from wild crucifers, including *Erucastrum cardaminoides*, *Diplotaxis tenuisiliqua*, and *E. abyssinicum*, showed high levels of stem resistance (Garg et al. 2010a). In addition, resistance sources from the wild *B. incana* and *B. insularis* have been identified (Mei et al. 2011, 2013). Characterization of these introgression regions using genomic technologies, and identification of candidate Sclerotinia R-genes, through sequencing the introgression regions, will be the first step towards understanding the interaction between the host and pathogen.

The characterization of R-genes for resistance against *S. sclerotiorum* in *B. napus* is particularly challenging. The interaction in this pathosystem is more complex than previously thought (Amselem et al. 2011). Host resistance studies in *B. napus* and other Brassicaceae genotypes have also demonstrated that not only did *S. sclerotiorum* cause pathotype specific HR in *B. napus* cotyledons (Garg et al. 2008, 2010c; Barbetti et al. 2014), it also caused non-pathotype-specific resistance response in some genotypes of *B. napus*, *Raphanus sativus*, *B. juncea*, and *B. nigra*, where the host was resistant against different pathotypes tested (Barbetti et al. 2014; Ge et al. 2015). These two resistance outcomes against *S. sclerotiorum* rely on the specific combination between host genotype-pathogen isolate. The question is how do we distinguish these two resistances using molecular tools so that we can exploit the R-genes? An enabler in the search for effective high-level resistance to Sclerotinia Stem Rot has been the ability to define the subspecific variation of *S. sclerotiorum* and to characterize pathotypes of this pathogen (Ge et al. 2012).

This has not only provided explanation for many conflicting findings in search for effective new resistances but, more importantly, allowed identification of the first Brassica resistances to Sclerotinia Stem Rot that are truly pathotype-independent (Barbetti et al. 2014). Utilization of this information and careful selection of pathotypes, inoculation methods and use of introgression lines, combined with technologies such as gene expression profiling and genome sequencing will be the first step to identification of resistance genes.

iii. ***Hyaloperonospora*—Downy Mildew:** The pathogen *Hyaloperonospora parasitica* is responsible for up to 90% yield loss of Chinese cabbage (*B. rapa*) in China (Yu et al. 2009). Although, five R-genes from *Arabidopsis* that encode both TIR- and CC-NBS-LRR have been cloned for resistance to *H. parasitica*; namely RPP2 (Sinapidou et al. 2004) and RPP5 (Parker et al. 1997) on chromosome 4, RPP8 (McDowell et al. 1998), RPP1 (Botella et al. 1998) and RPP13 (Bittner-Eddy et al. 2000) on chromosome 3 (Botella et al. 1998), no R-gene has been cloned from Brassica species. However, Ppa3, a single dominant R-gene against this pathogen was mapped in *B. oleracea* using molecular markers (Singh et al. 2012). The major locus Pp523 on chromosome C8 has also been identified to confer resistance at the adult stage in *B. oleracea* (Farinhó et al. 2004; Carlier et al. 2012), with a syntenic locus on chromosome 1 in *A. thaliana* (Farinhó et al. 2007). Another major gene *RPP31* for adult resistance against Downy Mildew was genetically mapped on chromosome 5 in *A. thaliana* (McDowell et al. 2005). Since many more R-genes have been cloned or mapped in *Arabidopsis*, the orthologous genes in *Brassica* can be investigated through genomics approaches, for example, comparative analysis of genomes (Yu et al. 2014) and pan-genome analysis (Golicz et al. 2016).

Genetic inheritance of specific resistance to *H. parasitica* has been reported in *B. oleracea* (Natti et al. 1967), *B. napus* (Lucas et al. 1988; Nashaat et al. 1997), and *B. juncea* (Nashaat and Awasthi 1995; Nashaat et al. 2004; Chattopadhyay and Séguin-Swartz 2005). Among these reports, Natti et al. (1967) was the only one that reported the existence of separate races (1 and 2) in broccoli (*B. oleracea* var. *italica*). Sherriff and Lucas (1990) used 33 isolates, collected from four different *Brassica* species (*B. napus*, *B. juncea*, *B. oleracea*, and *B. campestris*) and geographic locations to infect a set of *Brassica* accessions (including *B. napus*, *B. juncea*, *B. oleracea*, *B. carinata*, *B. nigra*, *B. campestris* and *Raphano brassica*). However, the study failed to delineate pathotypes of *H. parasitica*, which might explain variability in the resistance outcome. Mohammed et al. (2017) tested 131 Brassicaceae lines which highlighted excellent resistance to Downy Mildew, and was the first study to demonstrate the existence of very high levels of pathotype-independent resistance in Australian *B. napus* rapeseed varieties. However, no molecular analysis of these lines has been performed in order to identify resistance genes. The resistance loci expressed in the cotyledon does not necessarily translate to adult plant resistance in *B. oleracea* against *H. parasitica*, as the resistance outcome can be different between the growth stages (Dickson and Petzoldt 1993; Coelho et al.

1998). However, in some instances, there was association of resistance at cotyledon stage with that expressed at the adult stage (Jensen et al. 1999a; Wang et al. 2000). Similarly, screening tests on seedling cotyledons across *B. napus* cultivars were claimed to provide an accurate estimation of expression of field susceptibilities/ resistance on seedling and more mature leaves for Downy Mildew (Ge et al. 2008). These contrasting findings indicate that the resistance outcome in the host against Downy Mildew is likely to be QTL-specific where the QTL is associated with developmental stages of the plant. Identification of such QTL should be particularly useful for breeding purposes, especially where both seedling and adult plant resistances together are required.

The genetic mechanism of resistance in the *Brassica-H. parasitica* pathosystem needs further clarification, as a HR has been observed in both host and non-host species (Li et al. 2011). Further, there is no evidence of clear-cut differences between Brassica host species which displayed a hypersensitive, partially resistant or susceptible reaction, compared with non-host species, during the early stages of infection, at spore germination and host penetration (Li et al. 2011). This may simply mean that the expression of the important components of host resistance in *Brassicas* relates to resistance mechanisms occurring during later stages of infection. Besides, *H. parasitica* infects a wide host range including *Sinapis* and *Raphanus* species and even species outside of the Brassicaceae, with these latter hosts mostly resistant to *H. parasitica* (Dickinson and Greenhalgh 1977). Another intriguing aspect of the host resistance against *H. parasitica* is that the genetic control can be recessively inherited (Carlsson et al. 2004). Controlled infection studies using the wider host range, combined with molecular analysis such as gene expression and GWAS studies could be utilized as a first step towards understanding of the genetic mechanisms of resistance.

The sources of isolate, either geographic location from which the isolate was sourced or the host from which the isolate was found, influence the phenotypic outcome. Isolates sourced from different locations within the USA produced consistent disease scores within a test host genotype (Dickson and Petzoldt 1993). By contrast, in Australia, significant subspecific variation occurs and this needs to be considered in host-pathogen studies on resistance. Similarly, resistance was isolate specific whereby an English isolate was more virulent than a Danish isolate on *B. oleracea* cultivars (Jensen et al. 1999a). Unfortunately, none of these studies allows comparison of *H. parasitica* pathotype structure over time, or across different crops, different countries or continents. To date, it has not been possible to universally characterize *H. parasitica* pathotypes due to unavailability of host differentials used. Hence, there is clearly a need to develop and deploy a group of universal host differentials such that pathotypes of H. parasitica can be delineated, otherwise progress towards developing new varieties with resistance to the existing H. parasitica populations will be curtailed.

iv. ***Leptosphaeria* – Blackleg:** The potential for *L. maculans* to cause severe damage was highlighted by the widespread total collapse of susceptible oilseed rape varieties in Western Australia in the early 1970s (Sivasithamparam et al. 2005) and subsequently in the early 2000s with the collapse of single dominant gene-based resistance from *B. rapa* ssp. sylvestris in various regions of Australia. *L. maculans* also causes severe damage to oilseed rape in the UK (Barnes et al. 2010), Canada and mainland Europe (Hwang et al. 2016). The less aggressive blackleg pathogen, *L. biglobosa*, is more prevalent in China (Liu et al. 2014) and was recently reported to be present in New Zealand (Lob et al. 2013). Although no severe *B. napus* yield loss has been reported from *L. biglobosa*, researchers have taken precautions to prevent the spread of *L. maculans* into China (Zhang et al. 2014b; Cai et al. 2017). Published work on sources of Blackleg resistance genes in *Brassica* species have been increasing over the years, with the focus more recently towards quantitative resistance (Jestin et al. 2011a, b; Huang et al. 2016; Larkan et al. 2016a; Delourme et al. 2006, 2012; Rimmer 2006). Although 12 Blackleg qualitative race specific R-genes have been genetically mapped and more have been recognized (Table 2.16), it remains uncertain if some of these genes are actually the same gene with different nomenclature or allelic variants of the same gene. This uncertainty arises from researchers using different crosses, different isolates, and different marker systems. For example, the genes *LmR1, cRLMm, Rlm4, cRLMrb*, and *LEM1* were all mapped on chromosome A7 in *B. napus* using Australian cultivars Shiralee, Maluka, Skipton, and the French cultivar Major, respectively (Ferreira et al. 1995; Mayerhofer et al. 1997; Rimmer et al. 1999; Balesdent et al. 2001; Rimmer 2006), with different molecular markers and therefore, until one is cloned, it is unknown if they are the same gene. This uncertainty has implications in breeding as breeders do not know whether they are using the same or different sources of resistance. Furthermore, it is recommended in Australia to rotate the source of resistance gene grown annually so the pathogen population cannot build up to cause severe yield loss (Van De Wouw et al. 2016b). The lack of knowledge of the number of resistance genes makes this more difficult to achieve and often means a pool of fewer resistance genes is used.

Another example relates to the LepR3 gene in *B. napus* which has been shown to interact with AvrLm1 (Larkan et al. 2013); however phenotypic studies suggest that AvrLm1 may also be recognized by Rlm1, thus raising the question of the relationship between these two R-genes (Rouxel and Balesdent 2013). The relationship between Rlm4 (Raman et al. 2012b) and Rlm7 (Balesdent et al. 2002), implicated to be responsible for the HR outcome on *B. napus* during attack by *L. maculans* carrying the allele AvrLm4 or AvrLm7, respectively, is also unknown. Cloning studies on these effectors revealed that AvrLm4 and AvrLm7 are two distinct alleles of the same gene, termed AvrLm4-7, and can induce resistance in *B. napus* that harbors either Rlm4 or Rlm7. However, the relationship between the two R-genes, is yet to be determined (Parlange et al. 2009). It is also possible that Rlm3, Rlm4, and Rlm7 may all be allelic variations of the same gene, based on evidence showing the location of these genes on a clustered region on chromosome A7 in *B. napus* and not being found together in the same *B. napus* pure lines, with a study showing an unusual interplay between AvrLm4-7 and AvrLm3 (Delourme et al. 2004; Larkan et al. 2016b; Plissonneau et al. 2017). Clearly there is a need to reconcile different nomenclatures for these genes and the relationship between them. Towards this, a roundtable discussion among experts in the field was conducted during the Brassica2016 conference and a consensus reached to join research efforts to bridge this gap (Plissonneau et al. 2017). The success story of clarifying AvrLmJ1 and AvrLm5 are indeed the same Avr gene (Plissonneau et al. 2017) is an example that should be followed by *Brassica* and blackleg researchers to clarify nomenclature for other genes.

Compared with other *Brassica* pathogens, the molecular characterization of avirulence genes in L. maculans is more widely studied. The pathogen adapts to its new host via strict interaction at the molecular level. For example, the interaction between Rlm3 and AvrLm3 is suppressed by AvrLm4-7 when attacked by the *L. maculans* isolate that carries this allele (Plissonneau et al. 2016), although this isolate can be uncommon in the Canadian rapeseed fields (Kutcher et al. 2010; Zhang et al. 2016a). There is positive influence of the frequency of the virulent allele avrLm1 on avrLm6 due to genetic linkage between the two alleles in the pathogen genome (Van De Wouw et al. 2010). This observation is important when decision making is required for which variety to rotate in the field. For example, selecting a variety with Rlm1 directly after growing a variety containing Rlm6 would not be advised as there would be higher infection due to the presence of *avrLm1*.

The sources of Blackleg R-genes were more frequently found and more diversified in the winter type *B. napus* compared to the spring-type cultivars (Rouxel et al. 2003b), where Rlm4 is the most common R-gene in the latter (Rouxel et al. 2003b; Marcroft et al. 2012a). Both *B. rapa* and *B. napus* have proven to be good sources of Blackleg resistance genes on the A genome. Some genes from *B. rapa* have been introgressed into *B. napus*, such as the LepR-series of R-genes from *B. rapa* ssp. Sylvestris (Yu et al. 2005, 2008, 2013). Researchers have also attempted to exploit the B genome for resistance, and R-genes/QTL that are highly resistant to *L. maculans* have been introgressed from the B genome (Rimmer and van den Berg, 1992; Plieske et al. 1998), particularly into *B. napus* (Sacristan and Gerdemann 1986; Chevre et al. 1996; Fredua-Agyeman et al. 2014). The resistance level in the introgressed *B. napus* was the same as that in the donor B genome source, although expression of phenotype was shown to be influenced by temperature (Plieske et al. 1998). These novel sources of R-genes introgressed into *B. napus* will enable continued resistance against the pathogen, through expanding the number of resistance alleles available for breeding, particularly in times of climatic uncertainty. Both LepR1 and LepR2 are involved in cotyledon resistance and are also associated with resistance at the adult stage in *B. napus* (Yu et al. 2005). However, the mechanism of resistance at both stages is different because Yu et al. (2005) showed that presence of LepR1 provided complete resistance, while LepR2 only offered partial resistance at the cotyledon stage, yet both genes ensured complete resistance at the adult stage. A very different pattern of resistance exists for the Chinese *B. napus* lines derived from winter type and spring type parents, where R-gene-mediated resistance in these lines conferred by both Rlm3 and Rlm4 was effective at both seedling and adult plant stages against L. maculans (Zhang et al. 2016b).

Similarly, the Australian winter *B. napus* lines harboring either Rlm1, Rlm3 or a combination of both display stable resistance against L. maculans where resistance is effective at both seedling and adult plant stages (Light et al. 2011). Such examples, taken together, imply that while R-genes can be effective at both cotyledon and adult stages, they can also act interdependently or dependently at different plant developmental stages, and are likely affected by the genotype of host and pathogen, as well as by environmental conditions. The R-gene effective at seedling stage can also lie in the same QTL region as that of adult plant resistance (Raman et al. 2012b). Whether this adult plant resistance is mediated by the same R-gene or by quantitative nonspecific resistance genes is yet to be confirmed. An increased understanding of this molecular basis for resistance, for example, through further genome sequencing and gene cloning, will enable the genes to be deployed more effectively. Further, expression of R-gene-mediated resistance can also be expressed differently across other host plant components; for example, Rlm1 and Rlm4 exhibit major gene resistance on *B. napus* pods, not only on cotyledons (Elliott et al. 2016).

Besides qualitative resistance, quantitative resistance has also been assessed and reported in *B. napus* against L. maculans (Huang et al. 2009). Methods to determine quantitative resistance more efficiently such as inoculation at different parts of the leaves under controlled conditions have been evaluated (Huang et al. 2014). Several QTL responsible for quantitative resistance against L. maculans have been identified in French, European, and Australian *B. napus* cultivars (Jestin et al. 2011a, b; Raman et al. 2012b; Huang et al. 2016; Larkan et al. 2016a). However, the genetic control of quantitative resistance against Blackleg is poorly understood as it is very difficult to compare QTL, even with the same molecular markers and lines being utilized, due to environmental differences and a lack of knowledge of the pathogen presence in the field (Neik et al. 2017).

2.14 Identification of R-Genes Homologous DNA Fragments in *Arabidopsis thaliana*

Eight different bona fide resistant gene like (RGL) PCR fragments have been obtained from *A. thaliana* accessions Col and *Ler*, using degenerate PCR primers based on the NBS-LRR type of previously cloned disease resistance genes (Staskawicz et al. 1995). These fragments mapped to eight different loci on the *A. thaliana* genome. Six of these loci are closely linked to previously characterized disease resistance loci. Two fragments are actually derived from a functional R-gene or from a member of an R-gene family. The DNA sequences of all RGL fragments showed similarity to the NBS-LRR type of R-genes, which demonstrates that the primers used can amplify a range of RGL sequences. These primers were designed with a combination of inosines or multiple nucleotides at the third codon position, by which it is theoretically possible to amplify all genes coding for the consensus amino acid sequences. However, not all NBS-LRR type of R-genes present in Col or *Ler* was represented in the cloned RGL fragments. RGL fragments belonging to the *RPS2* or *RPM1* genes could not be cloned. An examination of the sequence of these

R-genes at the RG primer sites showed that especially the RG2 primer did not sufficiently match with the template (for both *RPM1* and *RPS2*, 23% mismatch). Slight alterations at the 3¢ end of the primers, selecting for different amino acid variants, will probably lead to the amplification of many other RGL fragments.

With the RG set of primers, a new type of RGL sequences represented by the L4, C4, and L3 fragments were observed. These are about 300 bp longer than what is expected based on the sequence of the R-genes used to design the RG1 and RG2 primers. Comparison of the DNA sequences of these RGL fragments showed the presence of unique sequences in addition to homologous sequences, which are flanked by *A. thaliana* consensus sequences for exon-intron splice junctions (Brown 1996). This suggests the presence of introns in the NBS region of the corresponding RGL genes, which is supported by the presence of a continuous open reading frame after artificial splicing, and the continuous alignment with the derived amino acid sequences of other RGL fragments and of cloned gene products. The next step following the isolation of RGL sequences was to establish a correlation between an RGL gene sequence and disease resistance. Accessions Col and Ler have been used for the RGL sequence PCRs, so the cloned fragments can only be part of resistance genes known to be active in these two accessions. Kunkel (1996) and Holub (1997) summarized the map positions of many disease resistance loci determined so far. Not all disease resistance loci have been characterized as yet, and only a portion of the R-genes residing at these loci will belong to the NBS-LRR gene class. Nevertheless, genetic linkage between an RGL fragment and a disease resistance locus has been found for nearly all isolated RGL fragments. The best correlation between RGL fragment and R-gene was found for the C3 fragment, which turned out to be part of the *RPS5* gene against *Pseudomonas syringae*. The other RGL fragments, the presence of an R-gene locus and a co-segregating RGL locus is often accession dependent. For instance, the *L4/C4* locus and the linked *RPP7* locus on chromosome 1 have been found in both Col and *Ler* (Crute et al. 1994), as well as the closely linked *L3A* and *EST177N18* loci. Also the *L2/C2* locus and the linked *RPP4* locus on chromosome 4 are present in both Col and *Ler* (Tor et al. 1994). Of the three segregating *L3* loci on chromosome 5, the *L3C* locus has an extra *Ler*-specific gene copy, which correlates with the linked *RPP8* locus from *Ler* (Holub et al. 1994a, b) and the *Ler*-specific *H. parasitica* recognitions against isolates Waco8, Waco11, and Boma1. This region of chromosome 5 contains more disease resistance loci, which are not specifically found in *Ler*, such as the *RPS4* locus against *Pseudomonas syringae* pv. *pisi* (in accession Ws-3) (Hinsch and Staskawicz 1996), the *HRT* locus against turnip crinkle virus (not in Col or Ler) (Dempsey et al. 1997), the *TTR1* locus against tobacco ring spot virus (in Col) (Lee et al. 1996), and the *RAC2* locus against *Albugo candida* (in Ksk-2).

With a correlation established between the chromosomal positions of R-gene loci and RGL loci, supportive evidence that RGL fragments are part of resistance genes was found in their genetic organization. Often, R-genes have been found as multi-copy, clustered sequences such as the *PTO, FEN,* and *PRF* complex or the *Cf-9/Cf-4* and *Cf-2/Cf-5* clusters in tomato (Dixon et al. 1996; Jones et al. 1994), or as dispersed multi-copy loci such as the *L6* and *M* loci in flax (Lawrence et al. 1995).

In contrast, the LRR receptor protein kinase genes such as *ERECTA, RLK1, RLK4, RLK5,* and *PR5K* are single or low-copy genes (Torii et al. 1996; Walker 1993; Wang et al. 1996). Nearly all RGL fragments detected several hybridizing bands upon DNA blot hybridization and they often revealed an RFLP between accessions *Col* and *Ler*. An exception to the rule is the C1 fragment, which was present as a single-copy sequence in Col and not at all in *Ler*. A similar case was observed for the *RPM1* gene (Grant et al. 1995), which is present as a single-copy gene in resistant and absent in susceptible accessions. Especially intriguing is the genome organization of the gene family to which fragments L3, C4, and L4 belong. This family accommodates clustered and dispersed family members that appear to be correlated with the presence of *RPP* loci on two chromosomes. The DNA sequences and DNA blot hybridization patterns of the L3, L4, C4, L3.1, and CK1 clones and the EST clones with RGL DNA sequence homology showed that all these fragments are part of genes coding for both NBS and LRR domains. The PCR fragments L3.1 and CK1 were produced after annealing of the L3, RG1, and STK PCR primers at unexpected positions. These annealing sites for the L3 and RG1 primers may be remnants of previous gene-shuffling events that are likely to have arisen during the generation of these multi-copy RGL genes.

Some of these genes are expressed, as demonstrated by the presence of the co-segregating EST sequences. For EST 108E20, this co-segregation was confirmed by partial DNA sequence analysis of BAC clone F24J19 containing part of the gene cluster at the *L4/C4* locus. Further analysis of the RGL copies present on this BAC clone may reveal clues about the organization and origin of this gene cluster and of the expression pattern of the individual genes in relation to disease resistance. The analysis of RGL sequences from the well-studied plant species *A. thaliana* may provide an insight into the evolution of gene families coding for NBS-LRR type of proteins towards a function in disease resistance, and offers starting points for direct cloning of these disease resistance genes (Aarts et al. 1998).

2.15 Application of Omics Technologies in *Brassica* to Identify R-Genes

The application of omics approaches like next-generation sequencing has allowed to sequencing *Brassica* species genomics widely activated for high-quality oilseeds. The use of another approach like pan-genomics has widened the scope for comprehensive gene pool analysis.

2.15.1 *Brassica* Species Genome Sequencing

With the introduction of next-generation sequencing (NGS) technology, five out of six *Brassica* crop genomes (*B. rapa, B. oleracea, B. nigra, B. napus,* and *B. juncea*) have now been sequenced, with some species having more than one genome from different individuals. The field of *Brassica* genomics has been "revolutionized" by the development of long-read sequencing technologies such as PacBio Single Molecule, Real-Time (PacBio) sequencing and Oxford Nanopore Technologies (ONT), along with high-throughput physical mapping technologies such as BioNano optical mapping and Chromosome Conformation Capture (Hi-C). Valuable genomics resources for interrogating the molecular aspects of *Brassica*–pathogen interactions have been provided via high-quality genome assemblies of *Brassica* species using PacBio sequencing, for example, *B. rapa* cultivar "Chiifu-401-42," *B. oleracea* cultivar "C-8," *B. napus* German winter cultivar "Express 617" and the recent improved *B. napus* "Darmor-bzh," along with the highly contiguous *B. nigra* assembly, both achieved via ONT technology (Table 2.23). These high-quality *Brassica* assemblies resolve the "difficult" genomic regions commonly found in polyploid crops, particularly highly repetitive DNA sequences related to transposable elements (TEs), copy number variation (CNV), presence–absence variation (PAV) and homoeologous exchange, many of which are associated with disease resistance genes.

Although high-quality *Brassica* genome assemblies are currently available, such reference genomes and other *Brassica* genomes that were previously sequenced represent only a fraction of the Brassica morphotypes. The reference *B. rapa* "Chiifu" is a heading type, while the reference "Z1" is a sarson type, the cauliflower *B. oleracea* "C-8" is an inflorescence type, and "TO1000" is a leafy type. Those genomes have not yet been assembled for other morphotypes in *Brassica* species, such as root or stem tubers in *B. rapa* (turnip), *B. oleracea* (kohlrabi), and *B. napus* (swede) (Li et al. 2017; Neik et al. 2020).

2.15.2 Use of Pan-genomics in R-Genes Identification

Pan-genomics has been developed to overcome the limitations of relying on a single reference genome, and allows more comprehensive genomic variations to be identified from the gene pools represented by many lines within a species. The pan-genome of *B. napus*, which was built with eight *B. napus* lines encompassing three ecotypes using a de novo approach, showed the PAV regions were enriched with genes associated with a defense-related response. Furthermore, in the *B. oleracea* and *B. napus* pan-genomes, it was found that a large proportion of the disease resistance genes were dispensable, meaning that these genes are not present in all lines (Table 2.23). These findings suggest that the R-genes in *Brassica* are highly variable, resulting from the strong selection pressure of arms-race evolution during host–pathogen interaction, superimposed by the frequent homoeologous exchanges between sub-genomes during the domestication process within the *Brassica* lineage. Hence, many candidate R-genes may have been missed from a single reference genome, thus hindering the speed of R-gene cloning in *Brassica* crops.

Pan-genomics has identified a large R-gene repertoire, collectively known as resistance gene analogs (RGAs), in *Brassica* species. The RGAs include nucleotide-binding site leucine-rich repeats (NLRs), mainly comprising the TIR-NBS-LRR (TNL) and CC-NBS-LRR (CNL) types, receptor-like protein kinases (RLKs),

Table 2.23 *Brassica* reference genomics useful for omics studies in the *Brassica* pathosystems (Neik et al. 2020)

Reference genome	Approach	Major findings relevant to R-gene study	References
Single genome			
B. napus winter cultivar "Express 617"	PacBio, ONT, Illumina HiSeq, Optical mapping	Resolved break-point sequence at homoeologous exchange regions	Lee et al. (2020)
		Cauliflower is the most recent var. to evolve within *Brassica* genus.	
B. oleracea cultivar "C-8"	PacBio, Illumina HiSeq, transcriptomics	It contains more repetitive sequences compared to other B. oleracea species	Sun et al. (2019)
		Hotspot of ALE-type retro element in the centromeric regions showed	
B. nigra accession CGN7651	ONT, Hi-C	that these retroelements play an important role in the divergence of *B. nigra* centromere V3.0, improved repeat reads,	Perumal et al. (2020)
B. rapa cultivar "Chiifu-401-42"	PacBio, Optical mapping, Hi-C	defined locations of centromeres and annotated more genes in these difficult regions. Annotated higher number of TEs	Zhang et al. (2018)
Pan-genome			
Eight *B. napus* accessions of three ecotypes	Alignment of de novo assembled genomes against "ZS11" Iterative mapping and assembly using improved "Darmor-bzh"(v8.1) from	PAV genes were highly represented by defense response gene	Song et al. (2020)
33 non-synthetic and 20 synthetic *B. napus* accessions	Bayer, Hurgobin, Golicz, Chan, Yuan, Lee, Renton, Meng, Li, Long, Zou, Bancroft, Chalhoub, King, Batley and Edwards (Bayer et al. 2017) as reference	Homoeologous exchange-related PAV genes highly represented by defense, stress and auxin pathways	Hurgobin et al. (2018)

(continued)

Table 2.23 (continued)

Reference genome	Approach	Major findings relevant to R-gene study	References
Nine *B. oleracea* subspecies and wild type comprising cabbage, kale, Brussels sprouts, kohlrabi, cauliflower, broccoli and *B. macrocarpa*	Iterative mapping and assembly using Chinese kale rapid cycling line (TO1000)as reference	18.7% of genes showed PAV with annotation of disease resistance genes	Golicz et al. (2016)
Two *B. rapa* subspecies: turnip and rapid cycling	Alignment of de novo assembled genomes against "Chiifu" reference	Peroxidase genes that are involved in phenylpropanoid biosynthesis response pathway during biotic stress are unique in turnip, with evidence of copy number variation	Lin et al. 2014)

receptor-like proteins (RLPs) and wall-associated kinases. Using pangenomics, 106 RGAs have been identified within the Blackleg QTL in the *B. napus* pan-genome while 59 RGAs were detected within the *Sclerotinia*, Fusarium wilt and Clubroot resistance QTLs in the *B. oleracea* pan-genome. These pan-genomics studies revealed that different classes of RGAs (RLKs, TNLs and others) show different percentages of variability across the lines, which leads to the following question: is there any association between R-gene variability and resistance outcomes in *Brassica* crops for the major diseases? It was also found in the *B. oleracea* pan-genome study that the wild relative *B. macrocarpa* harbors the most RGA candidates, suggesting that a large pool of genetic resources for R-genes can be found in wild *Brassicas*.

A super-pan-genome was reported that includes the genomes of wild relatives and/or different species within a genus, which adds an additional level of depth for investigating genomic variations within a crop genus. With a super-pan-genome developed in Brassica crops, not only can we identify many more novel candidate disease resistance genes from the wild genotypes, but we can also develop molecular markers to screen for resistant varieties in the field, therefore not only improving the speed and accuracy of crop breeding but also broadening the Brassica gene pool by using novel alleles from wild germplasm (Bayer et al. 2020; Golicz et al. 2016: Khan et al. 2020).

2.16 Identification of QTLs and R-Genes Using NGS-Based SNP Methods

In the past two decades, many genetic linkage maps of Brassica crops have been generated, using bi-parental crossing, selfing and backcrosses, with molecular markers such as Restriction Fragment Length Polymorphisms (RFLPs), Amplified Fragment Length Polymorphisms (AFLPs) and Randomly Amplified Polymorphic DNA (RAPDs) (Delourme et al. 2018). These molecular markers are often limited by low reproducibility and laborious techniques, thus limiting the quality of marker information in Brassica crops. The genomics era, driven by the innovation of high-throughput next-generation sequencing (NGS) technologies, has significantly increased the efficiency of the identification of QTL/candidate genes for disease resistance in *Brassica* crops through the development of genome-wide DNA-based molecular markers. This has brought great improvement in the resolution of genetic maps (Nguyen et al. 2019).

Among the DNA-based molecular markers, single nucleotide polymorphism (SNP) markers are most widely used for determining genotypic variation in a given species because they are uniformly distributed and highly abundant in the genome and are amenable in multiple genotyping platforms. With NGS, high-throughput genome-wide SNP marker development can be achieved rapidly and accurately in *Brassica* species through systems such as whole-genome resequencing (WGRS), genotyping-by-sequencing (GBS), and the *Brassica* 60K Illumina Infinium TM 60K SNP array. WGRS is an omics strategy for obtaining high-quality, high-density SNP markers at a whole-genome level by mapping sequence reads to the *Brassica* reference genome assemblies. In GBS, restriction enzymes are used to digest the genomic DNA and barcodes are used to ligate the fragmented DNA molecules before whole-genome sequencing is performed for SNP discovery. In this way, GBS is less complicated compared to WGRS because the sequencing reads cover only part of the genome instead of the whole genome, thus offering a cost-efficient approach to identifying SNPs yet achieving equally high-quality SNPs with wide applications in crop improvement studies. An extension of GBS called tGBS^{R} using oligonucleotides instead of adaptors has since been developed. The *Brassica* 60K SNP array offers a whole-genome SNP genotyping approach that is highly reproducible for genotyping hundreds of DNA samples in 48 h, making it an attractive option for the routine screening of *Brassica* germplasm. A *Brassica* 60K SNP array data repository called Crop SNPdb has been developed to enable users to access SNP information, currently containing genotypic information of 526 *Brassica* lines, permitting the easy retrieval of whole-genome SNP data for a wide range of downstream data analyses (Scheben et al. 2017; Chung et al. 2017; Clark et al. 2016).

These NGS-based SNP genotyping approaches have been applied widely for the QTL mapping of disease resistance traits and identification of candidate genes through genome-wide association studies (GWAS) in *Brassica* crops (Table 2.24). Highlights include the discovery of novel disease resistance QTLs/genes at an unprecedented speed, for example, in Blackleg, Sclerotinia, and Clubroot. The

Table 2.24 Application of next-generation sequencing (NGS)-based SNP genotyping approaches in resistance to *Brassica* diseases (Neik et al. 2020)

Approach	Disease type	Brassica sample	Main findings	References
WGRS, SNP genotyping	–	991 *B. napus* worldwide accessions	Selective-sweep regions enriched with genes related to stress response	Wu et al. (2019)
WGRS, SNP genotyping	–	588 *B. napus* worldwide accessions	A sub-genomic-specific selection contributes towards biotic stress response with several candidate genes identified	Lu et al. (2019)
WGRS, QTL mapping	Black rot	Mapping population of cabbage *B. oleracea* var. *capitata* inbred lines "C1234" (resistant) and "C1184" (susceptible)	21 candidate NBS-LRR genes associated with black rot resistance in *B. oleracea*	Lee et al. (2015)
GBS, GWAS	Blackleg	243 *B. napus* accessions from Canada and China	Significant SNPs were found on chromosome A08 with 25 RGAs identified consisting of NBS, RLK, RLP and TM-CC type R-genes	Fu et al. (2020)
GBS, GWAS	Sclerotinia	*B. juncea–B. fruticulosa* introgression lines	20 candidate genes mostly located on the A sub-genome of *B. juncea*	Atri et al. (2019a, b)
GBS, GWAS	Sclerotinia	*B. juncea–Erucastrum cardaminoides* introgression lines	QTL region on chromosomes A03 and B03 and candidate genes being LRR-RLK, LRR-PK and TIR-NBR-LRR	Rana et al. (2019)
		135 *B. oleracea* accessions including var. broccoli,	Resistant phenotype mostly found in kale. Candidate genes encoding	
tGBS[R] , GWAS	-	Brussels sprout, cabbage, cauliflower, Chinese kala, kale, kohlrabi and savoy cabbage	pathogenesis-related proteins were mainly found on chromosome C07	Farid et al. (2019)
Brassica 60K SNP array, GWAS	Clubroot	Mapping population of cabbage *B. oleracea* inbred lines "263" and "GZ87"	Significant QTL and novel loci found on C sub-genome	Peng et al. (2018)

(continued)

Table 2.24 (continued)

Approach	Disease type	Brassica sample	Main findings	References
Brassica 60K SNP array, Linkage disequilibrium (LD) analysis	–	327 *B. napus* worldwide accessions comprising three ecotypes	Selective-sweep regions enriched with Blackleg and Sclerotinia resistance QTLs	Wei et al. (2017)
Brassica 60K SNP array, GWAS	Clubroot	472 *B. napus* worldwide accessions	Most candidate genes were found on C sub-genome with novel QTLs and TIR-NBS gene clusters	Li et al. (2016a)
Brassica 60K SNP array, GWAS	Sclerotinia	448 worldwide *B. napus* accessions	Two novel loci with 39 candidate genes on C sub-genome	Wu et al. (2016b)
Brassica 60K SNP array	Blackleg	Seven *B. napus* seven donor parents for introgression lines	Genomic background of individual varieties and multiple defense-related gene interactions influence the resistance levels	Larkan et al. (2016)

other benefit of the application of NGS-based SNP genotyping is the successful breeding of *B. napus* varieties containing multiple improved traits. The breeding was achieved through the introgression of several major QTLs for Sclerotinia quantitative resistance from *B. oleracea* into *B. napus* with good seed yield and quality. In addition, bioinformatics pipelines are continuously being improved for whole-genome SNP data analysis. The Single Nucleotide Absence Polymorphism (SNaP) analysis has been successfully recovered numerous QTLs for Sclerotinia and Blackleg resistance in *B. napus* that were lost from the normal filtering of SNP data obtained from the *Brassica* 60K SNP array, with 3.2- and 2.2-fold increases in significant marker-trait associations for Sclerotinia and Blackleg resistance, respectively (Mei et al. 2020; Gabur et al. 2020).

2.17 Identification of R-Gene Using In Silico *Methods*

The *Brassica* genomics resources in public databases support the analysis and interpretation of many complex resistance mechanisms related to *Brassica*–pathogen interaction. It includes the in silico identification of the Blackleg resistance gene, LepR4, in the C genome of Korean cabbage (*B. oleracea* var. *capitata*) and the in silico exploration of 641 NBS-LRR-type disease resistance genes in *B. napus*, highlighting the genomic distribution and structural variation of these genes in *B. napus*. The in silico evolutionary study of NBS genes in *B. napus*, where comparative genomic analysis highlighted the NBS gene's distribution from its progenitors *B. rapa* and *B. oleracea* in relation to the three main *Brassica* diseases—Blackleg, Clubroot, and Sclerotinia is very significant achievement. Coupled with modern bioinformatics tools and the integration of multi-omics data sets, in silico methods are powerful tools that rapidly provide accurate and detailed models to answer various research questions ranging from candidate gene identification to evolutionary pathways of resistance mechanisms in both the *Brassica* host and the fungal pathogens. This is a time-, cost- and manpower-effective means of conducting higher-quality *Brassica* crop improvement research investigations (Ferdous et al. 2020; Fu et al. 2019a, b; Graham-Taylor et al. 2020).

In addition, database searches for protein motifs associated with disease resistance genes have enabled researchers to identify classes of R-genes in *B. napus* that are associated with Blackleg, Sclerotinia, and Clubroot resistance. Stotz et al. (2018) suggested most clubroot resistance genes are NLR type; Blackleg resistance genes are RLP type, while for *Sclerotinia*, neither NLR nor RLP were involved. Focusing on the NLR genes, comparative genomics and transcriptomics analyses supplemented with a database query on *B. napus* and its progenitors, *B. rapa* and *B. oleracea*, revealed many more NBS (also known as NLR) genes in the C sub-genome of *B. napus*. A number of these genes underling the QTL regions for resistance against Blackleg, Sclerotinia, and Clubroot, supporting the concept that the diversification of the R-genes likely happened after interspecific hybridization between *B. rapa* and *B. oleracea* (Fu et al. 2019b).

2.18 NGS-Based Bulked Segregant Analysis (BSA) for R-Genes Identification

NGS-based BSA is one of the more recent applications of omics in studying *Brassica*–pathogen interactions. This technique involves bulks/pools of DNA samples with representations of individuals with segregating phenotypes, where the pools are genotyped using NGS, either RNA sequencing (BSR-Seq) or whole-genome resequencing, followed by the detection of QTLs through SNP calling between the bulks (QTL-Seq). The traditional BSA technique allows the screening of many loci. With the NGS screening of BSA populations, the detection of sequence polymorphisms between the bulks is rapid and effective, as novel variation, such as PAVs or even novel QTLs/genes, can potentially be detected. Using BSR-Seq, an R-gene for resistance against Blackleg, Rlm1, was fine mapped in the *B. napus* cultivar "Quinta," and a candidate gene was identified, *BnA07G27460D*, that encodes a serine/threonine protein kinase. This gene is homologous to the protein kinase STN7 in *B. rapa, B.* oleracea and *A. thaliana*, which is involved in systemic plant immune responses by regulating reactive oxygen species (ROS)-induced cell signaling at the thylakoid membrane. BSR-Seq has also been applied in characterizing Clubroot resistance in some *Brassica* species. In *B. oleracea*, the first Clubroot major R-gene (*Rcr7*) in the *B. oleracea* cultivar "Tekila" was identified; *Bo7g108760* was the candidate TNL gene. In Chinese cabbage *B. rapa* var. pekinensis, the candidate gene Rcr2, which encodes a TIR-NBS-LRR, responsible for Clubroot resistance, has been identified on chromosome A03. In *B. nigra*, a novel Clubroot R-gene, *Rcr6*, was detected (*BniBo15819*, encoding a TNL gene), which is homologous to chromosome A08 of *B. rapa* and which provides a good source for gene introgression into *B. napus*. Using QTL-Seq, two novel QTL regions associated with Clubroot resistance on chromosomes A07 and A08 were detected in pak choi *B. campestris* var. *chinensis*. A single novel candidate R-gene, is also involved in Clubroot resistance, *CRd*, was identified on chromosome A03 of *B. rapa* (Zhu et al. 2019; Fu et al. 2019a, b; Chang et al. 2019; Pang et al. 2018).

2.19 Resistance Gene Enrichment and Sequencing (RenSeq)

RenSeq is a targeted resequencing method for identifying NLRs. RenSeq in combination with PacBio sequencing was applied in *A. thaliana* to study the R-gene sequence variants of the White Rust Resistance (WRR) gene against *Albugo candida*. This combinatorial approach was extended to 64 accessions of *A. thaliana* to study the evolution and variability of NLR genes in the model plant *Arabidopsis*, resulting in the construction of a species-wide pan-NLR-ome. Similar to the concept of pangenomics, the pan-NLR-ome is the collection of all the NLR genes and alleles contained within a species, distinguishing between the core and non-core NLRs in terms of the structural variations and diversity. New domain structures of NLRs were identified from the pan-NLR-ome study in *A. thaliana*. This implies that novel R-genes can be obtained from the diverse gene pool of NLRs within members of the Brassicaceae family, including *Brassica* crops. The discovery of a repertoire of NLRs is particularly important when pathogen-specific recognition can not only happen in the host species but also in the non-host species. The non-host A. thaliana displayed ETI-mediated defense against A. candida isolates derived from *B. juncea*, *B. rapa* and *B. oleracea*, implying that R-genes play a conserved role across members within the same family. This implies that screening for novel R-genes for Brassica crop improvement should also be applied across non-host species. Pan-NLR-ome type studies contribute significantly towards the discovery of R-gene diversity in crops (Cevik et al. 2019; Van de Weyer et al. 2019).

2.20 Effector-omics Approach to Detect R-Genes

Effectoromics, or effector-based screening, is an omics approach to detect R-genes in crop plants, although R-gene products may not necessarily interact with effectors directly. In this method, the target effector, as forecast from prediction tools, is transformed using Agrobacterium and infiltrated onto the host plants. The host genotypes that give a positive response to the target effector are then subjected to resistance gene mapping using molecular markers. Effectoromics allows screening for potential recognition targets in a particular crop species, immune receptors, and can be applied to screen different crop species with a selection of potential candidates that can be used in inter-crop species. This method was used for the identification of R-genes in the wild potato species *S. pinnatisectum* against the oomycete *Phytophthora infestans* and could be applied to *Brassica* pathosystems for the identification of R-genes (Gu et al. 2020).

References

Aarts N, Metz M, Holub E, Staskawicz BJ, Daniels MJ, Parker JE (1998) Different requirements for EDS1 and NDR1 by disease resistance genes define at least two R gene-mediated signaling pathways in *Arabidopsis*. Proc Natl Acad Sci U S A 95:10306–10311

Abawi GS, Provvidenti R, Crosier DC, Hunter JE (1978) Inheritance of resistance to white mold disease in *Phaseolus coccineus*. J Heredity 69:200–202

Adam L, Somerville SC (1996) Genetic characterization of five powdery mildew disease resistance loci in *Arabidopsis thaliana*. Plant J 9:341–356

Adam L, Ellwood S, Wilson I, Xiao S, Saenz G, Oliver R, Turner JG, Somerville S (1999) Comparison of *Erysiphe cichoracearum* and *E. cruciferarum* and a survey of 360 *Arabidopsis thaliana* accessions for their resistance to these two powdery mildew pathogens. Mol Plant Microbiol Interac 12:1031–1043

Adhikari TB, Liu JQ, Mathur S, Wu CX, Rimmer SR (2003) Genetic and molecular analyses in crosses of race 2 and race 7 of *A. candida*. Phytopathology 93:959–965

Adhikari KC, Mcdonald MR, Gossen BD (2012) Reaction to *Plasmodiophora brassicae* pathotype 6 in lines of *Brassica* vegetables, Wisconsin Fast Plants, and Canola. Hort Sci 47(3):374–377

AICORPO (1989) All India Co-ordinated Research Project on Rapeseed-Mustard, Hisar.

AICRPRM (All India Coordinated Research Project on Rapeseed-Mustard) (2011) Annual Progress Report. ICAR-Directorate of Rapeseed-Mustard Research, Bharatpur (Raj) India

AICRPRM-AR (2019) Annual Progress Report.All India Co-ordinated Research Project on Rapeseed-Mustard. ICAR-Diretorate of Rapeseed-Mustard Research, Bharatpur, India

AICRPRM (All Indian Coordinated Research Project on Rapeseed-Mustard) (2020) Annual Progress Report. ICAR-Directorate of Rapeseed-Mustard Research, Bharatpur, India

Al-Lami HFD, You MP, Barbetti MJ (2020) Relative host resistance to Alternaria leaf spot in canola and mustard varieties is defined by *Alternaria* species. Crop Pasture Sci 71(7):689–699. https://doi.org/10.1071/CP20024

Alix K, Lariagon C, Delourme R, Manzanares-Dauleux MJ (2007) Exploiting natural genetic diversity and mutant resources of *Arabidopsis thaliana* to study the *A. thaliana-Plasmodiophora brassicae* interaction. Plant Breed 126:218–221

Amselem J, Cuomo CA, van Kan JA, Viaud M, Benito EP, Couloux A, Coutinho PM, de Vries RP, Dyer PS, Fillinger S, Fournier E, Gout L, Hahn M, Kohn L, Lapalu N, Plummer KM, Pradier JM, Quévillon E, Sharon A, Simon A, ten Have A, Tudzynski B, Tudzynski P, Wincker P, Andrew M, Anthouard V, Beever RE, Beffa R, Benoit I, Bouzid O, Brault B, Chen Z, Choquer M, Collémare J, Cotton P, Danchin EG, Da Silva C, Gautier A, Giraud C, Giraud T, Gonzalez C, Grossetete S, Güldener U, Henrissat B, Howlett BJ, Kodira C, Kretschmer M, Lappartient A, Leroch M, Levis C, Mauceli E, Neuvéglise C, Oeser B, Pearson M, Poulain J, Poussereau N, Quesneville H, Rascle C, Schumacher J, Ségurens B, Sexton A, Silva E, Sirven C, Soanes DM, Talbot NJ, Templeton M, Yandava C, Yarden O, Zeng Q, Rollins JA, Lebrun MH, Dickman M (2011) Genomic analysis of the necrotrophic fungal pathogens *Sclerotinia sclerotiorum* and *Botrytis cinerea*. PLoS Genet 7:e1002230

Anand IJ, Reddy WR, Rawat DS (1985) Inheritance of seed coat colour in mustard. Indian J Genet 45:34–37

Anonymous (1984) Annual Report. Asian Vegetable Research and Development Centre (AVRDC), Shanhua, Taihnan, Taiwan

Anonymous (1987a) In vitro studies in downy mildew resistance in Chinese cabbage. Progress report 1985. Asian Vegetable Research and Development Centre, Shanhua, Taiwan, pp 39-42.

Anonymous (1987b).Resistance of Chinese cabbage hybrids to downy mildew under field conditions. Progress Report 1985. Asian Vegetable Research and Development Centre. Shanhua, Taiwan: 37-39.

Anonymous (AICRPRM) All India Co-ordinated Research Project on Rapeseed-Mustard (2000–2017) Annual Progress Report. ICAR-Directorate of Rapeseed-Mustard Research, Bharatpur, India

Ansan-Melayah D, Balesdent MH, Buee M, Rouxel T (1995) Genetic characterization of *AvrLm1*, the first avirulence gene of *Leptosphaeria maculans*. Phytopathology 85:1525–1529

Ansan-Melayah D, Balesdent MH, Delourme R, Pilet ML, Tanguy X, Renard M, Roulx T (1998) Genes for race-specific resistance against blackleg disease in *Brassica napus* L. Plant Breed 117:373–378. https://doi.org/10.1111/j.1439-05231998.tb01956.x

Atri C, Akhatar J, Gupta M, Gupta N, Goyal A, Rana K, Kaur R, Mittal M, Sharma A, Singh MP, Sandhu PS, Barbetti MJ, Banga SS (2019a) Molecular and genetic analysis of defensive responses of *Brassica juncea–B. fruticulosa* introgression lines to Sclerotinia infection. Sci Report 9:17089

Atri C, Akhatar J, Gupta M, Gupta N, Goyal A, Rana K, Kaur R, Mittal M, Sharma A, Singh MP et al (2019b) Molecular and genetic analysis of defensive responses of *Brassica juncea—B. fruticulosa* introgression lines to *Sclerotinia* infection. Sci Report 9:17089

Awasthi RP, Nashaat NI, Kolte SJ, Tewari AK, Meena PD, Renu B (2012) Screening of putative resistant sources against Indian and exotic isolates of *Albugo candida* inciting white rust in rapeseed-mustard. J Oilseed Brassica 3:27–37

Baggett JR, Kean D (1985) Clubroot resistant broccoli breeding lines OSU CR-2 to OSC CR-8. Hort Sci 20:784–785

Bain DC (1952) Reaction of *Brassica* seedlings to black rot. Phytopathology 42:497–500

Balesdent MH, Attard A, Ansan-Melayah D, Delourme R, Renard M, Rouxel T (2001) Genetic control and host range of avirulence toward *Brassica napus* cultivars Quinta and Jet Neuf in *Leptosphaeria maculans*. Phytopathology 91:70–76. https://doi.org/10.1094/PHYTO.2001.91.1.70

Balesdent MH, Attard A, Kuhn ML, Rouxel T (2002) New avirulence genes in the phytopathogenic fungus *Leptosphaeria maculans*. Phytopathology 92:1122–1133. https://doi.org/10.1094/PHYTO.2002.92.10.1122.

Balesdent MH, Fudal I, Ollivier B, Bally P, Grandaubert J, Eber F, Chèvre AM, Leflon M, Rouxel T (2013) The dispensable chromosome of *Leptosphaeria maculans* shelters an effector gene conferring avirulence towards *Brassica rapa*. New Phytol 198:887–898. https://doi.org/10.1111/nph.12178

Banga SS (1988) C-genome chromosome substitution lines in *Brassica juncea* (L.)Coss. Genetica 77:81–84

Bansal VK, Tewari JP, Tewari I, Gomez-Campo C, Stringam GR (1997) Genus Eruca: a potential source of white rust resistance in cultivated Brassicas. Plant Genet Res Newslett 109:25–26

Bansal VK, Thiagarajah MR, Stringam GR, Tewari JP (1999) Inheritance of partial resistance to race 2 of *Albugo candida* in canola-quality mustard (*Brassica juncea*) and its role in resistance breeding. Plant Pathol 48:817–822

Barbetti MJ, Banga SK, Fu TD, Li YC, Singh D, Liu SY, Ge XT, Banga SS (2014) Comparative genotype reactions to *Sclerotinia sclerotiorum* within breeding populations of *Brassica napus* and *B. juncea* from India and China. Euphytica 197:47–59

Barbetti MJ, Li CX, Banga SS, Banga SK, Singh D, Singh PS, Singh R, Liu SY, You MP (2015) New host resistance in *Brassica napus* and *Brassica juncea* from Australia, China and India: Key to managing Sclerotinia stem rot (*Sclerotinia sclerotiorum*) without fungicides. Crop Protect 78:127–130

Barnes AP, Wreford A, Butterworth MH, Semonov MA, Moran D, Evans N, Fitt BDL (2010) Adaptation to increasing severity of Phoma stem canker on winter oilseed rape in the UK under climate change. J Agric Sci 148:683–694

Baswana KS, Rastogi KB, Sharma PP (1991) Inheritance of stalk rot resistance in cauliflower (*Brassica oleracea* var. *botrytis* L.). Euphytica 57:93–96

Bayer PE, Hurgobin B, Golicz AA, Chan CK, Yuan Y, Lee H, Renton M, Meng J, Li R, Long Y, Zou J, Bancroft I, Chalhoub B, King GJ, Batley J, Edwards D (2017) Assembly and comparison of two closely related *Brassica napus* genomes. Plant Biotechnol J 15:1602–1610

Bayer PE, Golicz AA, Scheben A, Batley J, Edwards D (2020) Plant pan-genomes are the new reference. Nat Plants 6:914–920

Bentley A, Scutari M, Gosman N, Faure S, Bedford F, Howell P, Cockram J, Rose GA, Barber T, Irigoyen J, Horsnell R, Pumfrey C, Winnie E, Schacht J, Beauchêne K, Praud S, Greenland A, Balding D, Mackay IJ (2014) Applying association mapping and genomic selection to the dissection of key traits in elite European wheat. Theor Appl Genet 127(12):2619–2633. https://doi.org/10.1007/s00122.014.2403.y

Bernoux M, Ve T, Williams S, Warren C, Hatters D, Valkov E, Zhang X, Ellis JG, Kobe B, Dodds PN (2011) Structural and functional analysis of a plant resistance protein TIR domain reveals interfaces for self-association, signaling, and autoregulation. Cell Host Microbe 9:200–211. https://doi.org/10.1016/j.chom.2011.02.009

Berry LA, Lennard JH (1988) Foliar infection of plants by *Alternaria brassicae* in relation to leaf surface characteristics. Cruciferae Newslett 8:104–105

Bhardwaj CL, Sud AK (1993) Reaction of *Brassica* cultivars against *Albugo candida* isolates from Kangra valley. Indian Phytopath 46:258–260

Bhatia JN (1994) Field evaluation of mustard germplasm resistant to white rust. Cruciferae Newslett 16:119

Bhowmik TP, Munde PN (1987) Identification of resistance in rapeseed and mustard against *Alternaria brassicae* (Berk) Sacc and some resistant sources. Beitr Trop Landwirtsch Veterinarmed 25:49–53

Birker D, Heidrich K, Takahara H, Narusaka M, Deslandes L, Narusaka Y, Reymond M, Parker JE, O'Connell R (2009) A locus conferring resistance to *Colletotrichum higginsianum* is shared by four geographically distinct *Arabidopsis* accessions. Plant J 60:602–613

Bisht KS, Meenakshi R, Kalpana G, Sharma BC, Tewari AK, Awasthi RP (2015) Screening of *Brassica* germplasm for resistance to major diseases of rapeseed- mustard. The Bioscan 10:2111–2119

Bittner-Eddy PD, Crute IR, Holub EB, Beynon JL (2000) *RPP13* is a simple locus in *Arabidopsis thaliana* for alleles that specify downy mildew resistance to different avirulence determinants in *Peronospora parasitica*. Plant J 21:177–188

Bonnet A (1981) Resistance to white rust in radish (*Raphanus sativus* L.). Cruciferae NewsLett 6:60.

Bonnet A, Blancard D (1987) Resistance of radish (*Raphanus sativus* L.) to downy mildew, *Peronospora parasitica*. Cruciferae NewsLett 12:98–99

Borhan MH, Brose E, Beynon JL, Holub EB (2001) White rust (*Albugo candida*) resistance loci on three *Arabidopsis* chromosomes are closely linked to downy mildew (*Peronospora parasitica*) resistance loci. Mol Plant Pathol 2:87–95

Borhan MH, Gunn N, Cooper A, Gulden S, Tor M, Rimmer SR, Holub EB (2008) WRR4 encodes a TIR-NB-LRR protein that confers broad-spectrum white rust resistance in *Arabidopsis thaliana* to four physiological races of *Albugo candida*. Mol Plant-Microbe Interact 21:757–768

Botella MA, Parker JE, Frost LN, Bittner-Eddy PD, Beynon JL, Daniels MJ, Holub EB, Jones JDJ (1998) Three genes of the *Arabidopsis RPP1* complex resistance locus recognize distinct *Peronospora parasitica* avirulence determinants. Plant Cell 10:1847–1860

Braverman SW (1971) Reaction of broccoli and cauliflower introductions to *Alternaria brassicicola*. Plant Dis Repor 55:454–457

Braverman SW (1976) Resistance to *Alternaria brassicicola* in brusssels sprouts introductions. Proc Amer Phytopathol Soc 3:302

Braverman SW (1977) Reaction of brussels sprouts introductions to artificial inoculation with *Alternaria brassicicola*. Plant Dis Reptr 61:360–361

Brodmann A, Schuller A, Ludwig-Muller J, Aeschbacher RA, Wiemken A, Boller T, Wingler A (2002) Induction of trehalase in *Arabidopsis* plants infected with the trehalose producing pathogen *Plasmodiophora brassicae*. Mol Plant-Microbe Interact 15:693–700

Brown JW (1996) *Arabidopsis* intron mutations and pre-mRNA splicing. Plant J 10:771–780

Brun H, Plessis J, Renard M (1987a) Resistance of some crucifers to *Alternaria brassicae* (Berk) Sacc. In: Proc 7th Rapeseed Conf, Pozanan, Poland, 1222–1227

Brun H, Renard M, Plessis J (1987b) Resistance of some crucifers to *Alternaria brassicae* (Berk.) Sacc. In: Proc of 7th Intl Rapeseed Cong, Pozanan, Poland, p 247

Brun H, Chèvre AM, Fitt BD, Powers S, Besnard AL, Ermel M, Huteau V, Marquer B, Eber F, Renard M, Andrivon D (2010) Quantitative resistance increases the durability of qualitative resistance to *Leptosphaeria maculans* in *Brassica napus*. New Phytol 185:285–299. https://doi.org/10.1111/j.1469-8137.2009.03049.x

Buchwaldt L, Yu FQ, Rimmer SR, Hegedus DD (2003) Resistance to *Sclerotinia sclerotiorum* in a Chinese *Brassica napus* cultivar. 8th. Intl Cong of Plant Pathol, Christchurch, New Zealand, p 289

Cai X, Huang Y, Jiang D, Fitt BDL, Li G, Yang L (2017) Evaluation of oilseed rapeseed yield losses caused by *Leptosphaeria biglobosa* in central China. Eur J Plant Pathol. https://doi.org/10.1007/s10658-017-1266-x

Camargo LEA, Williams PH, Osborn TC (1995) Mapping of quantitative loci controlling resistance to *Brassica oleracea* to *Xanthomonas campestris* pv. *campestris* in the field and green house. Phytopathology 85:1296–1300

Cao H, Li X, Dong X (1998) Generation of broad-spectrum disease resistance by over expression of an essential regulatory gene in systemic acquired resistance. Proc Natl Acad Sci U S A 95:6531–6536

Cao T, Srivastava S, Rahman MH, Kav NNV, Hotte N, Deyholos MK, Strelkov SE (2008) Proteome-level changes in the roots of Brassica napus as a result of Plasmodiophora brassicae infection. Plant Sci 174:97–115

Caravalho T, Monteiro A (1996) Preliminary study on the inheritance of resistance to downy mildew (*Peronospora parasitica* (Pers. Ex. Fr.) at cotyledon stage in Tranchuda cabbage "Algarvia". Cruciferae NewsLett 18:104

Carlier JD, Alabaça CS, Sousa NH, Coelho PS, Monteiro AA, Paterson AH, Leitao JM (2011) Physical mapping in a triplicated genome: mapping the downy mildew resistance locus Pp523 in *Brassica oleracea* L. G3 (Bethesda) 1:593–601. https://doi.org/10.1534/g3.111.001099

Carlier JD, Alabaça CA, Coelho PS, Monteiro AA, Leitao JM (2012) The downy mildew resistance locus Pp523 is located on chromosome C8 of *Brassica oleracea* L. Plant Breed 131:170–175. https://doi.org/10.1111/j.1439-0523.2011.01904.x

Carlsson M, Bothmer RV, Merker A (2004) Screening and evaluation of resistance to downy mildew (*Peronospora parasitica*) and clubroot (*Plasmodiophora brassicae*) in genetic resources of *Brassica oleracea*. Hereditas 141:293–300. https://doi.org/10.1111/j.1601-5223.2004.01818.x

Cavell AC, Lydiate DC, Parkin IAP, Dean C, Trick M (1998) Co-linearity between a 30-centimorgan segment of *Arabidopsis thaliana* chromosome 4 and duplicated regions within the *Brassica napus* genome. Genome 41:62–69

Cevik V, Boutrot F, Apel W, Robert-Seilaniantz A, Furzer OJ, Redkar A, Castel B, Kover PX, Prince DC, Holub EB et al (2019) Transgressive segregation reveals mechanisms of *Arabidopsis* immunity to *Brassica*-infecting races of white rust (*Albugo candida*). Proc Natl Acad Sci U S A 116:2767

Chai AL, Xie XW, Shi YX, Li BJ (2014) Research status of clubroot (*Plasmodiophora brassicae*) on cruciferous crops in China. Can J Plant Pathol 36:142–153. https://doi.org/10.1080/07060661.2013.868829

Chandra A, Gupta ML, Ahuja I, Kaur G, Banga SS (2004) Intergeneric hybridization between *Erucastrum cardaminoides* and two diploid crop *Brassica* species. Theor Appl Genet 108:1620–1626

Chang ZM (1981) Evaluation of rape downy mildew and white rust. Fukien Agric Sci Technol 5:8–10

Chang A, Lamara M, Wei Y, Hu H, Parkin IAP, Gossen BD, Peng G, Yu F (2019) Clubroot resistance gene *Rcr6* in *Brassica nigra* resides in a genomic region homologous to chromosome A08 in *B. rapa*. BMC Plant Biol 19:224

Chattopadhyay C, Séguin-Swartz G (2005) Breeding for disease resistance in oilseed crops in India. Annu Rev Plant Pathol 3:101–142

Chattopadhyay AK, Moitra AK, Bhunia CK (2001) Evaluation of *Brassica* species for resistance to *Plasmodiophora brassicae* causing club root of rapeseed mustard. Indian Phytopath 54 (1):131–132

Chauhan SK, Raut RN (2002) A source of white rust resistance in Indian mustard derived from *Brassica carinata*. Cruciferae Newslett 24:79–80

Chen HF, Wang H, Li ZY (2007) Production and genetic analysis of partial hybrids in intertribal crosses between *Brassica* species (*B. rapa, B. napus*) and *Capsella bursa-pastoris*. Plant Cell Reptr 26:1791–1800

Chen S, Zou J, Cowling WA, Meng J (2010) Allelic diversity in a novel gene pool of canola-quality *Brassica napus* enriched with alleles from *B. rapa* and *B. carinata*. Crop Pasture Sci 61:483–492. https://doi.org/10.1071/CP09327

Chen J, Jing J, Zhan Z, Zhang T, Zhang C, Piao Z (2013) Identification of novel QTLs for isolate-specific partial resistance to *Plasmodiophora brassicae* in *Brassica rapa*. PLoS One 8:e85307. https://doi.org/10.1371/journal.pone.0085307

Chen L, Zhang X, Xu H, Song B, Fan X (2016) Introgression of clubroot resistance into an elite pak choi inbred line through marker-assisted introgression breeding. Plant Breed 135:471–475. https://doi.org/10.1111/pbr.12379

Chevre AM, Eber F, This P, Barret P, Tanguy X, Brun H, Dlseny M, Renard M (1996) Characterization of *Brassica nigra* chromosomes and of blackleg resistance in *B. napus–B. nigra* addition lines. Plant Breed 115:113–118. https://doi.org/10.1111/j.1439-05231996.tb00884.x

Chiang MS, Crête R (1970) Inheritance of clubroot resistance in cabbage (*Brassica oleracea* L. var. *capitata* L.). Can J Genet Cytol 12:253–256

Chiang MS, Crete R (1983) Transfer of resistance to race 2 of *Plasmodiophora brassicae* from *Brassica napus* to cabbage (*B. oleracea* ssp. *capitata*). V. The inheritance of resistance. Euphytica 32:479–483

Christianson JA, Rimmer SR, Good AG, Lydiate DJ (2006) Mapping genes for resistance to *Leptosphaeria maculans* in *Brassica juncea*. Genome 49(1):30–41. https://doi.org/10.1139/g05-08

Chrungu B, Verma N, Mohanty A, Pradhan A, Shivanna KR (1999) Production and characterization of interspecific hybrids between *Brassica maurorum* and crop *Brassicas*. Theor Appl Genet 98(3-4):608–613

Chu M, Yu F, Falk KC, Liu X, Zhang X, Chang A, Peng G (2013) Identification of the clubroot resistance gene Rpb1 and introgression of the resistance gene into canola breeding lines using a marker-assisted selection approach. Acta Hortic 1005:599–605. https://doi.org/10.17660/ActaHortic.2013.1005.74.

Chu M, Song T, Falk KC, Zhang X, Liu X, Chang A, Lahlali R, McGregor L, Gossen BD, Peng G, Yu F (2014) Fine mapping of Rcr1 and analyses of its effect on transcriptome patterns during infection by *Plasmodiophora brassicae*. BMC Genomics 15:1166. https://doi.org/10.1186/1471-2164-15-1166

Chung H, Jeong YM, Mun JH, Lee SS, Chung WH, Yu HJ (2014) Construction of a genetic map based on high-throughput SNP genotyping and genetic mapping of a TuMV resistance locus in *Brassica rapa*. Mol Genet Genomics 289:149–160

Chung YS, Choi SC, Jun TH, Kim C (2017) Genotyping-by-sequencing: A promising tool for plant genetics research and breeding. Hortic Environ Biotechnol 58:425–431

Clarke WE, Higgins EE, Plieske J, Wieseke R, Sidebottom C, Khedikar Y, Batley J, Edwards D, Meng J, Li R et al (2016) A high-density SNP genotyping array for *Brassica napus* and its ancestral diploid species based on optimised selection of single-locus markers in the allotetraploid genome. Theor Appl Genet 129:1887–1899

Coelho P, Bahcevandziev K, Valério L, Monteiro A, Leckie D, Astley D, Crute IR, Boukema I (1998) The relationship between cotyledon and adult plant resistance to downy mildew (*Peronospora parasitica*) in *Brassica oleracea*. Acta Hortic 459:335–342. https://doi.org/10.17660/ActaHortic.1998.459.39.

Coletta-Filho HD, Machado MA (2002) Evaluation of the genetic structure of *Xylella fastidiosa* populations from different *Citrus sinensis* varieties. App Environ Microbiol 68:3731–3736. https://doi.org/10.1128/AEM.68.8.3731-3736.2002

Conn KL, Tewari JP (1986) Hypersensitive reaction induced by *Alternaria brassicae* in *Eruca sativa*, an oil yielding crucifer. Can J Plant Pathol 8:348

Conn KL, Tewari JP (1989) Interactions of *Alternaria brassicae* conidia with leaf epicuticular wax of canola. Mycol Res 93:240–242. https://doi.org/10.1016/S0953-7562(89)80126-1

Conn KL, Tewari JP, Dahiya JS (1987) Resistance to *Alternaria brassicae* and phytoalexin-elicitation in rapeseed and some other crucifers. Phytopathology 77:1703. (Abstr)

Conn KL, Tewari JP, Dahiya JS (1988) Resistance to *Alternaria brassicae* and phytoalexins-elicitation in rapeseed and other crucifers. Plant Sci 56:21–25. https://doi.org/10.1016/0168-9452(88)90180-X

Conn KL, Jejelowa OA, Tewari JP, Bains PS (1991) *Camelina sativa* phytoalexins (Calmexin and methoxy camalexin) provide resistance against *Alternaria brassicae*. Abstract 274 Can J Plant Pathol 13

Coutts BA, Walsh JA, Jones RAC (2007) Evaluation of resistance to Turnip mosaic virusin Australian *Brassica napus* genotypes. Aust J Agric Res 58:67–74

Crossa J, de Los CG, Pérez P, Gianola D, Burgueno J, Araus JL, Makumbi D, Singh RP, Dreisigacker S, Yan J, Arief V, Banziger M, Braun HJ (2010) Prediction of genetic values of quantitative traits in plant breeding using pedigree and molecular markers. Genetics 186 (2):713–724. https://doi.org/10.1534/genetics.110.118521

Crute IR (1986) The relationship between *Plasmodiophora brassicae* and its hosts: the application of concepts relating to variation in inter-organism associations. Adv Plant Pathol 5:1–52

Crute IR, Phelps K, Barnes A, Buczacki ST, Crips P (1983) The relationship between genotypes of three *Brassica* species and collections of *Plasmodiophora brassicae*. Plant Pathol 32:405–420

Crute I, Beynon J, Dangl J, Holub E, Mauch-Mani B, Slusarenko A, Staskawicz B, Ausubel F (1994) Microbial pathogenesis of *Arabidopsis*. In: Meyerowitz EM, Somerville CR (eds) Arabidopsis. Cold Spring Harbor Laboratory, Cold Spring Harbor, NY, pp 705–747

Cubeta MA, Cody BR, Kohli Y, Kohn LM (1997) Clonality in *Sclerotinia sclerotiorum* on infected cabbage in Eastern North Carolina. Phytopathology 87:1000–1004

Dafwa WA (2015) Agribusiness Crop Updates: Research Delivers Opportunities to Cut Canola Production Costs. Department of Agriculture and Food Western Australia, Western Australia

Dahiya N, Rashmi PU (2019) Inheritance of white rust in newly developed resistant line of *Brassica juncea* from local germplasm. Intl J Chem Stud 7(2):677–679

Dainello FJ, Heineman RR, Jones RK (1981) Relative white rust (*Albugo occidentalis*) resistance and adaptability of spinach (*Spinacia oleracea*) varieties in south west Texas. USA Tex Agric Expt Stn Prog Reptr 3878:1–7

Dang JK, Sangwan MS, Naresh M, Kaushik CD (2000) Multiple disease resistance against four fungal foliar diseases of rapeseed-mustard. Indian Phytopath 53:455–458

Dangl JL, Holub EB, Debener T, Lehnackers H, Ritter C, Crute IR (1992) Genetic definition of loci involved in Arabidopsis-pathogen interaction. In: Chua NH, Schell J (eds) Methods in *Arabidopsis* research Koncz C. World Sci Press, Singapore, pp 393–418

Das R (2016) Identification of resistant source against Alternaria blight disease of rapeseed-mustard. J Crop Weed 12(2):134–136

Delourme R, Pilet-Nayel ML, Archipiano M, Horvais R, Tanguy X, Rouxel T, Brun H, Renard M, Balesdent MH (2004) A cluster of major specific resistance genes to *Leptosphaeria maculans* in *Brassica napus*. Phytopathology 94:578–583. https://doi.org/10.1094/PHYTO.2004.94.6.578

Delourme R, Chèvre AM, Brun H, Rouxel T, Balesdent MH, Dias JS, Salisbury P, Renard M, Rimmer SR (2006) Major gene and polygenic resistance to *Leptosphaeria maculans* in oilseed rape (*Brassica napus*). Eur J Plant Pathol 114:41–52. https://doi.org/10.1007/s10658-005-2108-9

Delourme R, Barbetti MJ, Snowdon R, Zhao J, Manzanares-Dauleux M (2012) genetics and genomics of resistance. In: Edwards D, Batley J, Parkin IAP, Kole C (eds) Genetics, genomics and breeding of oilseed *Brassicas*. Science Publishers, CRC Press, pp 276–318

Delourme R, Bousset L, Ermel M, Duffé P, Besnard AL, Marquer B, Fudal I, Linglin J, Chadœuf J, Brun H (2014) Quantitative resistance affects the speed of frequency increase but not the diversity of the virulence alleles overcoming a major resistance gene to *Leptosphaeria maculans* in oilseed rape. Infect Genet Evol 27:490–499. https://doi.org/10.1016/j.meegid.2013.12.019

Delourme R, Laperche A, Bouchet AS, Jubault M, Paillard S, Manzanares-Dauleux MJ, Nesi N (2018) Genes and quantitative trait loci mapping for major agronomic traits in *Brassica napus* L. In: Liu S, Snowdon R, Chalhoub B (eds) The *Brassica napus* genome. Springer International Publishing, Cham, Switzerland, pp 41–85

Delwiche PA, Williams PH (1974) Resistance to *Albugo candida* race 2 in *Brassica* sp. Proc Am Phytopathol Soc 1:66. (Abstr)

Dempsey DA, Pathirana MS, Wobbe KK, Klessig DF (1997) Identification of an *Arabidopsis* locus required for resistance to turnip crinkle virus. Plant J 11:301–311

Denton-Giles M, Derbyshire MC, Khentry Y, Buchwaldt L, Kamphuis LG (2018) Partial stem resistance in *Brassica napus* to highly aggressive and genetically diverse *Sclerotinia sclerotiorum* isolates from Australia. Can J Plant Pathol 40(4):551–561. https://doi.org/10.1080/07060661.2018.1516699

Devos S, Prinsen E (2006) Plant hormones: a key in clubroot development. Commun Agric Appl Biol Sci 71:869–872

Devos S, Laukens K, Deckers P, Van Der Straeten D, Beeckman T, Inze D, van Onckelen H, Witters E, Prinsen E (2006) A hormone and proteome approach to picturing the initial metabolic events during *Plasmodiophora brassicae* infection on Arabidopsis. Mol Plant-Microbe Interact 19:1431–1433

Dey SS, Sharma K, Dey RB, Kumar SGM, Singh D, Kumar R, Parkash C (2015) Inter specific hybridization (*Brassica carinata* × *Brassica oleracea*) for introgression of black rot resistance genes into Indian cauliflower (*B. oleracea* var. *botrytis* L.). Euphytica 204:149–162

Dhar S, Singh D (2014) Performance of cauliflower genotypes for yield and resistance against black rot (*Xanthomonas campestris* pv. *campestris*). Indian J Hort 71:197–201

Diatchenko L, Lau YFC, Campbell AP, Chenchik A, Moqadam F, Huang B, Lukyanov K, Gurskaya K, Sverdlov E, Siebert PD (1996) Suppression subtractive hybridization: a method for generating differentially regulated or tissue-specific cDNA probes and libraries. Proc Natl Acad Sci U S A 93:6025–6030

Dickinson CH, Greenhalgh JR (1977) Host range and taxonomy of *Peronospora* on crucifers. Trans Br Mycol Soc 69:111–116. https://doi.org/10.1016/S0007-1536(77)80121-6

Dickman MB, Mitra A (1992) *Arabidopsis thaliana* as model for studying *Sclerotinia sclerotiorum* pathogenesis. Physiol Mol Pl Pathol 41:255-263.

Dickson MD, Hunter JE (1987) Sources of resistance to black rot of cabbage expressed in seedlings and adult plants. Plant Dis 71:263–266

Dickson MH, Petzoldt R (1993) Plant age and isolate source affect expression of downy mildew resistance in Broccoli. Hort Sci 28:730–731

Dickson MH, Petzoldt R (1996) Breeding for resistance to *Sclerotinia sclerotiorum* in *Brassica oleracea*. Acta Hort 407:103–108

Diederichsen E, Sacristan MD (1996) Disease response of resynthesized *Brassica napus* L. lines carrying different combinations of resistance to *Plasmodiophora brassicae* Worn. Plant Breed 115:5–10. https://doi.org/10.1111/j.1439-0523.1996.tb00862.x

Diederichsen E, Beckmann J, Schondelmeier J, Dreyer F (2006) Genetics of clubroot resistance in *Brassica napus* 'Mendel'. Acta Hortic 706:307–312. https://doi.org/10.17660/ActaHortic2006.706.35.

Diederichsen E, Frauen M, Linders EGA, Hatakayema K, Hirai M (2009) Status and perspectives of clubroot resistance breeding in crucifer crops. J Plant Growth Regul 28:265–281

Diederichsen E, Frauen M, Ludwig-Muller J (2014) Clubroot disease management challenges from a German perspective. Can J Plant Pathol 36:85–98

Dion Y, Gugel RK, Rakow GFW, Seguin-Swartz G, Landry BS (1995) RFLP mapping of resistance to the blackleg disease [causal agent, *Leptosphaeria maculans* (Desm.) Ces.et de Not.] in canola (*Brassica napus* L.). Theor Appl Genet 91:1190–1194

Disi JO, Mei J, Wei D, Ding Y, Qian W (2014) Inheritance of leaf and stem resistance to *Sclerotinia sclerotiorum* in a cross between *Brassica incana* and *Brassica oleracea* var. *alboglabra*. J Agric Sci 152:146–152. https://doi.org/10.1017/S0021859613000087

Dixon GR (1975) The reaction of some oil rape cultivars to some fungal pathogens. Proc 8th British Insecticide and Fungicide Conf 2:503–506

Dixon GR (2014) Clubroot (*Plasmodiophora brassicae* Woronin)–an agricultural and biological challenge worldwide. Can J Plant Pathol 36:5–18. https://doi.org/10.1080/07060661.2013.875487

Dixon MS, Jones DA, Keddie JS, Thomas CM, Harrison K, Jones JDG (1996) The tomato Cf-2 disease resistance locus comprises two functional genes encoding leucine-rich repeat proteins. Cell 84:451–459

Donald C, Porter I (2009) Integrated control of clubroot. J Plant Growth Regul 28:194–202

Donald EC, Porter IJ (2014) Clubroot in Australia: the history and impact of *Plasmodiophora brassicae* in *Brassica* crops and research efforts directed towards its control. Can J Plant Pathol 36:66–84. https://doi.org/10.1080/07060661.2013.873482

Donald EC, Cross SJ, Lawrence JM, Porter IJ (2006) Pathotypes of *Plasmodiophora brassicae*, the cause of clubroot, in Australia. Ann Appl Biol 148:239–244. https://doi.org/10.1111/j.1744-7348.2006.00062.x

Donald EC, Jaudzems G, Porter IJ (2008) Pathology of cortical invasion by *Plasmodiophora brassicae* in clubroot resistant and susceptible *Brassica oleracea* hosts. Plant Pathol 57:201–209. https://doi.org/10.1111/j.1365-3059.2007.01765.x

Donaldson PA, Andersonb T, Lanec BG, Davidsona AL, Simmondsa DH (2001) Soybean plants expressing an active oligomeric oxalate oxidase from the wheat gf-2.8 (germin) gene are resistant to the oxalate-secreting pathogen *Sclerotinia sclerotiorum*. Physiol Mol Pl Pathol 59:297–307

Doullah M, Meah M, Okazaki K (2006) Development of an effective screening method for partial resistance to *Alternaria brassicicola* (dark leaf spot) in *Brassica rapa*. Eur J Plant Pathol 116:33–43

Doullah MAU, Mohsin GM, Ishikawa K, Hori H, Okazaki K (2011) Construction of a linkage map and QTL analysis for black rot resistance in *Brassica oleracea* L. Int J Nat Sci 1:1–6

Ebrahimi AG, Delwiche PA, Williams PH (1976) Resistance in *Brassica juncea* to *Peronospora parasitica* and *Albugo candida*, Race 2. Proc Am Phytopathol Soc 3:273. (Abstr.)

Elenkov E (1979) Breeding disease resistant varieties of vegetable crops. Rastitelna Zashchita 27:13–16

Elliott VL, Marcroft SJ, Howlett BJ, Van De Wouw AP (2016) Gene-for-gene resistance is expressed in cotyledons, leaves and pods, but not during late stages of stem colonization in the *Leptosphaeria maculans–Brassica napus* pathosystem. Plant Breed 135:200–207. https://doi.org/10.1111/pbr.12343/.

Erbe M, Hayes BJ, Matukumalli LK, Goswami S, Bowman PJ, Reich CM (2012) Improving accuracy of genomic predictions within and between dairy cattle breeds with imputed high-density single nucleotide polymorphism panels. J Dairy Sci 95(7):4114–4129. https://doi.org/10.3168/jds.2011-5019

Fan Z, Rimmer S, Stefansson B (1983) Inheritance of resistance to *Albugo candida* in rape (*Brassica napus* L.). Can J Genet Cytol 25:420–424

Farid M, Yang RC, Kebede B, Rahman H (2019) Evaluation of *Brassica oleracea* accessions for resistance to *Plasmodiophora brassicae* and identification of genomic regions associated with resistance. Genome 63:91–101

Farinhó M, Coelho P, Carlier J, Svetleva D, Monteiro A, Leitão J (2004) Mapping of a locus for adult plant resistance to downy mildew in broccoli (*Brassica oleracea* convar. *italica*). Theor Appl Genet 109:1392–1398. https://doi.org/10.1007/s00122-004-1747-0

Farinhó M, Coelho P, Monteiro A, Leitao J (2007) SCAR and CAPS markers flanking the *Brassica oleracea* L. Pp523 downy mildew resistance locus demarcate a genomic region syntenic to the top arm end of *Arabidopsis thaliana* L. chromosome 1. Euphytica 157:215–221. https://doi.org/10.1007/s10681-007-9414-6

Farnham M, Wang M, Thomas C (2002) A single dominant gene for downy mildew resistance in broccoli. Euphytica 128:405–407. https://doi.org/10.1023/A:1021279426050

Feng J, Hwang SF, Strelkov SE (2012a) Studies into primary and secondary infection processes by *Plasmodiophora brassicae* on canola. Plant Pathol. https://doi.org/10.1111/j.1365–3059.2012.02612.x

Feng J, Xiao Q, Hwang SF, Strelkov SE, Gossen BD (2012b) Infection of canola by secondary zoospores of *Plasmodiophora brassicae* produced on a nonhost. Eur J Plant Pathol 132: 309–315

Feng J, Cao T, Rennie DC, Strelkov SE, Hwang SF (2014) Host–parasite interactions in clubroot of crucifers. Can J Plant Pathol 36 (Sup1): 113-121.

Ferdous MJ, Hossain MR, Park JI, Robin AHK, Natarajan S, Jesse DMI, Jung HJ, Kim HT, Nou IS (2020) In-silico identification and differential expressions of LepR4-syntenic disease resistance related domain containing genes against Blackleg causal fungus *Leptosphaeria maculans* in *Brassica oleracea*. Gene Rep 19:100598

Ferreira ME, Rimmer SR, Williams PH, Osborn TC (1995) Mapping loci controlling *Brassica napus* resistance to *Leptosphaeria maculans* under different screening conditions. Phytopathology 85:213–217

Figdore SS, Ferreira ME, Slocum MK, Williams PH (1993) Association of RFLP markers with trait loci affecting clubroot resistance and morphological characters in *Brassica oleracea* L. Euphytica 69(1/2):33–44

Fikere M, Barbulescu DM, Malmberg MM, Shi F, Koh JCO, Slater AT, MacLeod IM, Bowman PJ, Salisbury PA (2018) Genomic prediction using prior quantitative trait loci information reveals a large reservoir of underutilised blackleg resistance in diverse canola (*Brassica napus* L.) lines. Plant Genome 11:1–16. https://doi.org/10.3835/plantgenome2017.11.0100

Fjellstrom RG, Williams PH (1997) Fusarium yellows and Turnip mosaic virus resistance in *Brassicarapa* and *B. juncea*. Hort Sci 32:927–930

Fomeju BF, Falentin C, Lassalle G, Manzanares-Dauleux MJ, Delourme R (2014) Homoeologous duplicated regions are involved in quantitative resistance of *Brassica napus* to stem canker. BMC Genomics 15:498. https://doi.org/10.1186/1471-2164-15-498

Fredua-Agyeman R, Rahman H (2016) Mapping of the clubroot disease resistance in spring *Brassica napus* canola introgressed from European winter canola cv. 'Mendel'. Euphytica 211:201–213. https://doi.org/10.1007/s10681-016-1730-2

Fredua-Agyeman R, Coriton O, Huteau V, Parkin IA, Chèvre AM, Rahman H (2014) Molecular cytogenetic identification of B genome chromosomes linked to blackleg disease resistance in *Brassica napus* × *B. carinata* interspecific hybrids. Theor Appl Genet 127:1305–1318. https://doi.org/10.1007/s00122-014-2298-7

Fu F, Liu X, Wang R, Zhai C, Peng G, Yu F, Fernando WGD (2019a) Fine mapping of *Brassica napus* Blackleg resistance gene *Rlm1* through bulked segregant RNA sequencing. Sci Rep 9:14600

Fu Y, Zhang Y, Mason AS, Lin B, Zhang D, Yu H, Fu D (2019b) NBS-Encoding Genes in *Brassica napus* evolved rapidly after allopolyploidization and co-localize with known disease resistance loci. Front Plant Sci 10:26

Fu F, Zhang X, Liu F, Peng G, Yu F, Fernando D (2020) Identification of resistance loci in Chinese and Canadian canola/rapeseed varieties against *Leptosphaeria maculans* based on genome-wide association studies. BMC Genomics 21:501

Fuchs H, Sacristan MD (1996) Identification of a gene in *Arabidopsis thaliana* controlling resistance to clubroot and characterization of the resistance response. Mol Plant-Microbe Interact 9:91–97

Fujiwara A, Inukai T, Kim BM, Masuta C (2011) Combinations of a host resistance gene and the CI gene of Turnip mosaic virus differentially regulate symptom expression in *Brassica rapa* cultivars. Archiv Virol 156:1575–1581

Gabur I, Chawla HS, Lopisso DT, von Tiedemann A, Snowdon RJ, Obermeier C (2020) Gene presence-absence variation associates with quantitative *Verticillium longisporum* disease resistance in *Brassica napus*. Sci Rep 10:4131

Gairola K, Tewari AK (2017) Evaluation of *Brassica* germplasm for resistance sources against white rust. Intl J Env Agric Biotech (IJEAB) 2(3):1215–1226

Gao F, Hirani AH, Liu J, Liu Z, Fu G, Wu C, McVetty PBE, Li G (2014) Fine mapping a clubroot resistance locus in Chinese cabbage. J Am Soc Hortic Sci 139:247–252

García-Arenal F, McDonald BA (2003) An analysis of the durability of resistance to plant viruses. Phytopathology 93:941–952. https://doi.org/10.1094/PHYTO.2003.93.8.941

Garg H, Sivasithamparam K, Banga SS, Barbetti MJ (2008) Cotyledon assay as a rapid and reliable method of screening for resistance against *Sclerotinia sclerotiorum* in *Brassica napus* genotypes. Aus Plant Pathol 37:106–111

Garg H, Atri C, Sandhu PS, Kaur B, Renton M, Banga SK, Singh H, Singh C, Barbetti MJ, Banga SS (2010) High level of resistance to *Sclerotinia sclerotiorum* in introgression lines derived from hybridization between wild crucifers and the crop *Brassica* species *B. napus and B. juncea*. Field Crops Res 117:51–58. https://doi.org/10.1016/j.fcr.2010.01.013

Garg H, Kohn LM, Andrew M, Li H, Sivasithamparam K, Barbetti MJ (2010a) Pathogenicity of morphologically different isolates of *Sclerotinia sclerotiorum* with *Brassica napus* and *B. juncea* genotypes. Eur J Plant Pathol 126:305–315

Garg H, Sivasithamparam K, Barbetti MJ (2010b) Scarification and environmental factors that enhance carpogenic germination of sclerotia of *Sclerotinia sclerotiorum*. Plant Dis 94:1041–1047

Garg H, Li H, Sivasithamparam K, Kuo J, Barbetti MJ (2010c) The infection processes of *Sclerotinia sclerotiorum* in cotyledon tissue of a resistant and a susceptible genotype of *Brassica napus*. Ann Bot 106:897–908. https://doi.org/10.1093/aob/mcq196

Ge XT, Li H, Han S, Sivasithamparam K, Barbetti MJ (2008) Evaluation of Australian *Brassica napus* genotypes for resistance to the downy mildew pathogen, *Hyaloperonospora parasitica*. Aust J Agric Res 59:1030–1034. https://doi.org/10.1071/AR08032

Ge X, Li YP, Wan ZJ, You MP, Finnegan PM, Banga SS, Sandhu PS, Garg H, Salisbury PA, Barbetti MJ (2012) Delineation of *Sclerotiniasclerotiorum* pathotypes using differential resistance responses on *Brassica napus* and *B. juncea* genotypes enables identification of resistance to prevailing pathotypes. Field Crops Res 127:248–258. https://doi.org/10.1016/j.fcr.2011.11.022

Ge XT, You MP, Barbetti MJ (2015) Virulence differences among *Sclerotinia sclerotiorum* isolates determines host cotyledon resistance responses in Brassicaceae genotypes. Eur J Plant Pathol 143:527–541. https://doi.org/10.1007/s10658-015-0696-6

Ghosh A, Mondal D, Bandopadhyay P, Ghosh R (2019) Rapeseed yield loss estimates through selected biotic pressures. J Entomol Zool Stud 7(3):1101–1105

Gilmore B, Myers JR, Kean D (2002) Completion of testing of *Phaseolus coccineus* plant introductions (Pls) for white mold, *Sclerotinia sclerotiorum*, resistance. Ann Rep Bean Improv Coop 45

Golicz AA, Bayer PE, Barker GC, Edger PP, Kim H, Martinez PA, Chan CK, Severn-Ellis A, McCombie WR, Parkin IA, Paterson AH, Pires JC, Sharpe AG, Tang H, Teakle GR, Town CD, Batley J, Edwards D (2016) The pangenome of an agronomically important crop plant *Brassica oleracea*. Nat Commun 7:13390. https://doi.org/10.1038/ncomms13390

Graham-Taylor C, Kamphuis LG, Derbyshire MC (2020) A detailed in silico analysis of secondary metabolite biosynthesis clusters in the genome of the broad host range plant pathogenic fungus *Sclerotinia sclerotiorum*. BMC Genomics 21:7

Grandclément C, Thomas G (1996) Detection and analysis of QTL based on RAPD markers for polygenic resistance to *Plasmodiophora brassicae* Wor. in *Brassica oleracea* L. Theor Appl Genet 93:86–90

Grant MR, Godiard L, Straube E, Ashfield T, Lewald J, Sattler A, Innes RW, Dangl JL (1995) Structure of the *Arabidopsis RPM1* gene enabling dual specificity disease resistance. Science 269:843–846

Greenhalgh JR, Dickinson CH (1975) Differential reactions of three crucifers to infection by *Peronospora parasitica* Pers. Ex Fr. Fr. Phytopathology Z 84:131–141

Greenhalgh JR, Mitchell ND (1976) The involvement of flavour volatiles in the resistance to downy mildew of wild and cultivated forms of *Brassica oleracea*. New Phytol 77:391–398

Grenier CCT, Ospina Y, Quintero C, Chatel MH, Tohme J, Courtois B, Ahmadi N (2015) Accuracy of genomic selection in a rice synthetic population developed for recurrent selection Breeding. PLoS One 10(8):e0136594. https://doi.org/10.1371/journal.pone.0136594

Griffiths PD, Roe C (2005) Response of *Brassica oleracea* var. *capitata* to wound and spray inoculations with *Xanthomonas campestris* pv. *campestris*. Hort Sci 40:47–49

Griffiths PD, Marek LF, Robertson LD (2009) Identification of crucifer accessions from the NC-7 and NE-9 plant introduction collections that are resistant to black rot (*Xanthomonas campestris* pv. *campestris*) races 1 and 4. Hort Sci 44:284–288

Grison R, Grezes BB, Schneider M, Lucante N, Olsen L, Leguay JJ, Toppan A (1996) Field tolerance to fungal pathogens of *Brassica napus* constitutively expressing a chimeric chitinase gene. Nat Biotech 14:643–646

Grontoft M (1986) Resistance to *Alternaria* spp. in oilseed crops. Sveriges-Utsadesforenings Tidskrift (Sweden) 96:293

Grsic S, Kirchheim B, Pieper K, Fritsch M, Hilgenberg W, Ludwig-Muller J (1999) Induction of auxin biosynthetic enzymes by jasmonic acid and in clubroot diseased Chinese cabbage plants. Physiol Plant 105:521–531. https://doi.org/10.1034/j.1399-3054.1999.105318.x

Gu B, Cao X, Zhou X, Chen Z, Wang Q, Liu W, Chen Q, Zhao H (2020) The histological ectoromic and transcriptomic analyses of *Solanum pinnatisectum* reveal an upregulation of multiple NBS-LRR genes suppressing *Phytophthora infestans* infection. Int J Mol Sci 21:3211

Guo WL, Chen R-G, Gong ZH, Yin YX, Li DW (2013) Suppression subtractive hybridization analysis of genes regulated by application of exogenous abscisic acid in pepper plant (*Capsicum annuum* L.) leaves under chilling stress. PLoS ONE 8:e66667. https://doi.org/10.1371/journal.pone.0066667

Gupta RBL, Singh M (1994a) Source of resistance to white rust and powdery mildew of mustard. Int J Trop Plant Dis 12:225–227

Gupta RBL, Singh M (1994b) Source of resistance to white rust and *powdery mildew* of mustard. Intl J Tropical Pl Dis 21(2):225–227

Gupta S, Sharma TR, Chib HS (1995) Evaluation of wild allies of *Brassica* under natural conditions. Cruciferae Newslett 17:10–11

Gupta K, Saharan GS, Singh D (2002) Assessment of white rust severity on *Brassica juncea* under genotype-environment interactions. J Mycol Plant Pathol 32:106–109

Gustafsson M, Falt AS (1986) Genetic studies on resistance to clubroot in *Brassica napus*. Ann Appl Biol 108:409–415

Gyawali S, Harrington M, Durkin J, Horner K, Parkin IA, Hegedus DD, Bekkaoui D, Buchwaldt L (2016) Microsatellite markers used for genome-wide association mapping of partial resistance to

Sclerotinia sclerotiorum in a world collection of *Brassica napus*. Mol Breed 36:72. https://doi.org/10.1007/s11032-016-0496-5

Habier D, Fernando RL, Dekkers JCM (2007) The impact of genetic relationship information on genome-assisted breeding values. Genet 177(4):2389–2397. https://doi.org/10.1534/genetics.107.081190

Hansen LH, Earle ED (1997) Somatic hybrids between *Brassica oleracea* (L.) and *Sinapis alba* (L.) with resistance to *Alternaria brassicae* (Berk) Sacc. Theor Appl Genet 94:1078–1085

Hansen AK, Andersen HV, Svendsen O (1994) Studies on the diagnosis of Tyzzer's disease in laboratory rat colonies with antibodies against *Bacillus piliforrnis* (*Clostridiurn piliforrne*). Lab Anirn Sci 44:424–429

Hasan M, Strelkov SE, Howard RJ, Rahman H (2012) Screening of *Brassica* germplasm for resistance to *Plasmodiophora brassicae* pathotypes prevalent in Canada for broadening diversity in clubroot resistance. Can J Plant Sci 92:501–515

Hatakeyama K, Fujimura M, Ishida M, Suzuki T (2004) New classification method for *Plasmodiophora brassicae* field isolates in Japan based on resistance of F1 cultivars of Chinese cabbage (*Brassica rapa* L.) to clubroot. Breed Sci 54:197–201

Hatakeyama K, Suwabe K, Tomita RN, Kato T, Nunome T, Fukuoka H, Matsumoto S (2013) Identification and characterization of Crr1a, a gene for resistance to clubroot disease (*Plasmodiophora brassicae* Worn) in *Brassica rapa* L. PLoS One 8:e54745. https://doi.org/10.1371/journal.pone.0054745

Hatakeyama K, Niwa T, Kato T, Ohara T, Kakizaki T, Matsumoto S (2017) The tandem repeated organization of NB-LRR genes in the clubroot-resistant CRb locus in *Brassica rapa* L. Mol Gen Genomics 292:397–405. https://doi.org/10.1007/s00438-016-1281-1

He S, Schulthess AW, Mirdita V, Zhao YS, Korzun V, Bothe R, Ebmeyer E, Reif JC, Jiang Y (2016) Genomic selection in a commercial winter wheat population. Theor Appl Genet 129 (3):641–651. https://doi.org/10.1007/s00122-015-2655-1

Heffner EL, Jannink JL, Sorrells ME (2011) Genomic selection accuracy using multi-family prediction models in a wheat breeding program. Plant Genome 4:65–75

Hinsch M, Staskawicz B (1996) Identification of a new *Arabidopsis* disease resistance locus, RPS4, and cloning of the corresponding avirulence gene, avrRps4, from *Pseudomonas syringae* pv. *pisi*. Mol Plant-Microbe Interact 9:55–61

Hirai M (2006) Genetic analysis of clubroot resistance in *Brassica* crops. Breed Sci 56:223–229

Hirai M, Harada T, Kubo N, Tsukada M, Suwabe K, Matsumoto S (2004) A novel locus for clubroot resistance in *Brassica rapa* and its linkage markers. Theor Appl Genet 108:639–643

Holub EB (1997) Organization of resistance genes in *Arabidopsis*. In: Crute IR, Holub EB, Burdon JJ (eds) The gene-for-gene relationship in plant-parasite interaction. CAB Int, Wallingford, UK, pp 5–26

Holub EB (2007) Natural variation in innate immunity of a pioneer species. Curr Opin Plant Biol 10:415–424

Holub EB, Beynon JL, Crute IR (1994a) Phenotypic and genotypic characterization of interactions between isolates of *Peronospora parasitica* and accessions of *Arabidopsis thaliana*. Mol Plant-Microbe Interact 7:223–239

Holub EB, Beynon JL, Crute IR (1994b) Phenotypic and genotypic characterization of interactions between isolates of *Peronospora parasitica* and accessions of *Arabidopsis thaliana*. Mol Plant-Microbe Interact 7:223–239

Hoser-Krauze J, Lakowska-Ryk E, Antosik J (1991) Resistance to some *Brassica oleracea* L. plant introductions to downy mildew (*Peronospora parasitica*). Cruciferae NewsLett 14 (15):144–145

Hossain MA, Habib A, Islam MS, Tuz-Zohura F, Khokon MAR (2018) Validation of inoculation test and screening for resistance sources of mustard-rapeseed cultivars in Bangladesh against *Alternaria brassicicola*. Curr Agri Res 6(1). doi.org/10.12944/CARJ.6.1.02.

Huang YJ, Pirie EJ, Evans N, Delourme R, King GJ, Fitt BDL (2009) Quantitative resistance to symptomless growth of *Leptosphaeria maculans* (phoma stem canker) in *Brassica napus* (oilseed rape). Plant Pathol 58:314–323. https://doi.org/10.1111/j.1365-3059.2008.01957.x

Huang YJ, Qi A, King GJ, Fitt BDL (2014) Assessing quantitative resistance against *Leptosphaeria maculans* (phoma stem canker) in *Brassica napus* (oilseed rape) in young plants. PLoS One 9: e84924. https://doi.org/10.1371/journal.pone.0084924

Huang YJ, Jestin C, Welham SJ, King GJ, Manzanares-Dauleux MJ, Fitt BD, Delourme R (2016) Identification of environmentally stable QTL for resistance against *Leptosphaeria maculans* in oilseed rape (*Brassica napus*). Theor Appl Genet 129:169–180. https://doi.org/10.1007/s00122-015-2620-z

Huang Z, Peng G, Liu X, Deora A, Falk KC, Gossen BD, McDonald MR, Yu F (2017) Fine mapping of a clubroot resistance gene in Chinese cabbage using SNP markers identified from bulked segregant RNA sequencing. Front Plant Sci 8:1448. https://doi.org/10.3389/fpls.2017.01448

Hughes SL, Green SK, Lydiate DJ, Walsh JA (2002) Resistance to Turnip mosaic virus, in *B. rapa*, and *B. napus*, and the analysis of genetic inheritance in selected lines. Plant Pathol 51:567–573

Hughes SL, Hunter PJ, Sharpe AG, Kearsey MJ, Lydiate DJ, Walsh JA (2003) Genetic mapping of the novel Turnip mosaic virus resistance gene TuRB03 in *Brassica napus*. Theor Appl Genet 107:1169–1173

Humaydan HS, Williams PH (1976) Inheritance of seven characters in *Raphanus sativus*. Hort Sci 11:146–147

Hunter JE, Dickson MH, Ludwig JW (1987) Sources of resistance to black rot of cabbage expressed in seedlings and adult plants. Plant Dis 71:263–266

Huo D, Ning Q, Shen X, Liu L, Zhang Z (2016) QTL mapping of kernel number-related traits and validation of one major QTL for ear length in maize. PLoS One 11(5):e0155506. https://doi.org/10.1371/journal.pone.0155506

Hurgobin B, Golicz AA, Bayer PE, Chan CKK, Tirnaz S, Dolatabadian A, Schiessl SV, Samans B, Montenegro JD, Parkin IAP et al (2018) Homoeologous exchange is a major cause of gene presence/absence variation in the amphidiploid *Brassica napus*. Plant Biotechnol J 16:1265–1274

Hwang SF, Strelkov SE, Feng J, Gossen BD, Howard RJ (2012a) *Plasmodiophora brassicae*: a review of an emerging pathogen of the Canadian canola (*Brassica napus*) crop. Mol Plant Pathol 13:105–113. https://doi.org/10.1111/j.1364-3703.2011.00729.x

Hwang SF, Ahmed HU, Zhou Q, Rashid A, Strelkov SE, Gossen BD, Peng G, Turnbull GD (2012b) Assessment of the impact of resistant and susceptible canola on *Plasmodiophora brassicae* inoculum potential. Plant Pathol 61(945):952

Hwang SF, Strelkov SE, Peng G, Ahmed H, Zhou Q, Turnbull G (2016) Blackleg (*Leptosphaeria maculans*) severity and yield loss in canola in Alberta, Canada. Plants 5:31. https://doi.org/10.3390/plants5030031.

Ignatov A, Hida K, Kuginuki Y (1998) Black rot of crucifers and sources of resistance in Brassica crops. JARQ 32:167–172

Ito S, Tanaka S, Miyanaga C, Takabayashi S, Yano S, Kameya-Iwaki M (1996) Constitutive and inducible proteins in the root of Chinese cabbage infected with *Plasmodiophora brassicae*. J Phytopathol 144:89–95

Jain JP (1995) Management of white rust and Alternaria blight diseases of mustard. Proc Global Conf on Advances in Research on Plant Diseases and their management. Rajasthan College of Agriculture, Udaipur, Rajasthan, India: 80 (Abstr.).

Jain KL, Gupta AK, Trivedi A (1998) Reaction of rapeseed-mustard lines against white rust pathogen *Albugo candida*. J Mycol Plant Pathol 28:72–73

Jakir Hasan M, Strelkov SE, Howard RJ, Rahman H (2012) Screening of Brassica germplasm for resistance to *Plasmodiophora brassicae* pathotypes prevalent in Canada for broadening diversity in clubroot resistance. Can J Plant Sci 92:501–515. https://doi.org/10.4141/cjps2010-006

Jamwal RS, Sharma PP (1986) Inheritance of resistance to black rot (*Xanthomonas campestris* pv. *campestris*) in cauliflower (*Brassica oleracea* var. *botrytis*). Euphytica 35:941–943

Jan H, Abbadi A, Lucke S, Nichols R, Snowdon R (2016) Genomic prediction of test cross performance in Canola (*Brassica napus*). PLoS One 11(1):e0147769. https://doi.org/10.1371/journal.pone.0147769

Jat RR (1999) Pathogenic variability and inheritance of resistance to *Albugo candida* in oilseed *Brassica*. PhD Thesis, CCSHAU, Hisar: pp 129.

Jejelowo OA, Conn KL, Tewari JP (1991) Relationship between conidial concentration, germling growth, and phytoalexin production by *Camelina sativa* leaves inoculated with *Alternaria brassicae*. Mycol Res 95:928–934. https://doi.org/10.1016/S0953-7562(09)80089-0

Jenner CE, Walsh JA (1996) Pathotypic variation in Turnip mosaic virus with special reference to European isolates. Plant Pathol 45:848–856

Jenner CE, Tomimura K, Ohshima K, Hughes SL, Walsh JA (2002) Mutations in Turnip mosaic virus P3 and cylindrical inclusion proteins are separately required to overcome two *Brassica napus* resistance genes. Virology 300:50–59

Jenner CE, Wang X, Tomimura K, Ohshima K, Ponz F, Walsh JA (2003) The dual role of the potyvirus P3 protein of Turnip mosaic virus as a symptom and a virulence determinant in Brassicas. Mol Plant-Microbe Interact 16:777–784

Jensen BD, Hockenhull J, Munk L (1999a) Seedling and adult plant resistance to downy mildew (*Peronospora parasitica*) in cauliflower (*Brassica oleracea* convar. *botrytis* var. *botrytis*). Plant Pathol 48:604–612. https://doi.org/10.1046/j.1365-3059.1999.00388.x

Jensen BD, Værbak S, Munk L, Andersen SB (1999b) Characterization and inheritance of partial resistance to downy mildew, *Peronospora parasitica*, in breeding material of broccoli, *Brassica oleracea* convar. *botrytis* var. *italica*. Plant Breed 118:549–554. https://doi.org/10.1046/j.1439-0523.1999.00409.x

Jensen BD, Massomo SMS, Swai IS, Anderson SB (2005) Field evaluation for resistance to the black rot pathogen *Xanthomonas campestris* pv. *campestris* in cabbage (*Brassica oleracea*). Eur J Plant Pathol 113:297–308

Jestin C, Lodé M, Vallée P, Domin C, Falentin C, Horvais R, Coedel S, Manzanares B, Dauleux M, Delourme R (2011a) Association mapping of quantitative resistance for *Leptosphaeria maculans* in oilseed rape (*Brassica napus* L.). Mol Breed 27:271–287

Jestin C, Lodé M, Vallée P, Domin C, Falentin C, Horvais R, Coedel S, Manzanares-Dauleux MJ, Delourme R (2011b) Association mapping of quantitative resistance for *Leptosphaeria maculans* in oilseed rape (*Brassica napus* L.). Mol. Breed 27:271–287. https://doi.org/10.1007/s11032-010-9429-x

Jia Y, Jannink JL (2012) Multiple-trait genomic selection methods increase genetic value prediction accuracy. Genet 192(4):1513–1522. https://doi.org/10.1534/genetics.112.144246

Jiang H, Song W, Li A, Yang X, Sun D (2011) Identification of genes differentially expressed in cauliflower associated with resistance to *Xanthomonas campestris* pv. *campestris*. Mol Biol Rep 38:621–629

Jin H, Sun Y, Yang Q, Chao Y, Kang J, Jin H, Li Y, Margaret G (2010) Screening of genes induced by salt stress from Alfalfa. Mol Biol Rep 37:745–753

Jin M, Lee SS, Ke L, Kim JS, Seo MS, Sohn SH, Park BS, Bonnema G (2014) Identification and mapping of a novel dominant resistance gene, TuRB07 to Turnip mosaic virus in *Brassica rapa*. Theor Appl Genet 127:509–519

Jones DA, Thomas CM, Hammond-Kosack KE, Balint-Kurti PJ, Jones JD (1994) Isolation of the tomato Cf-9 gene for resistance to *Cladosporium fulvum* by transposon tagging. Science 266:789–793. https://doi.org/10.1126/science.7973631

Jonsson R (1966) *Peronospora* on oil yielding *Brassicas*. Methods for testing resistance in winter rape and their results. Sver Utsadestor Tidskr 76:54–62

Jubault M, Hamon C, Gravot A, Lariagon C, Delourme R, Bouchereau A, Manzanares-Dauleux MJ (2008) Differential regulation of root arginine catabolism and polyamine metabolism in club-root-susceptible and partially resistant *Arabidopsis* genotypes. Plant Physiol 146:2008–2019

Kato T, Hatakayema K, Fukino N, Matsumoto S (2012) Identification of a clubroot resistance locus conferring resistance to a *Plasmodiophora brassicae* classified into pathotype group 3 in Chinese cabbage (*Brassica rapa* L.). Breed Sci 62:282–287. https://doi.org/10.1270/jsbbs.62.282

Kaur S, Cogan NO, Ye G, Baillie RC, Hand ML, Ling AE, McGearey AK, Kaur J, Hopkins CJ, Todorovic M, Mountford H, Edwards D, Batley J, Burton W, Salisbury P, Gororo N, Marcroft S, Kearney G, Smith KF, Forster JW, Spangenberg GC (2009) Genetic map construction and QTL mapping of resistance to blackleg (*Leptosphaeria maculans*) disease in Australian canola (*Brassica napus* L.) cultivars. Theor Appl Genet 120:71–83. https://doi.org/10.1007/s00122-009-1160-9

Kaushik CD, Saharan GS (1980) Screening of raya germplasm against foliar diseases. Haryana Agric University, Hisar, India, Ann Prog Reptr Rapeseed-Musatrd

Kearney B, Staskawicz BJ (1990) Wide spread distribution and fitness contribution of *Xanthomonas campestris* avirulence gene avrBs2. Nat 346:385–386. https://doi.org/10.1038/346385a0

Kehoe MA, Coutts BA, Jones RAC (2010) Resistance phenotypes in diverse accessions, breeding lines, and cultivars of three mustard species inoculated with Turnip mosaic virus. Plant Dis 94:1290–1298

Khan AW, Garg V, Roorkiwal M, Golicz AA, Edwards D, Varshney RK (2020) Super-pangenome by integrating the wild side of a species for accelerated crop improvement. Trends Plant Sci 25:148–158

Khot SD, Bilgi VN, del Rio LE, Bradley CA (2011) Identification of *Brassica napus* lines with partial resistance to *Sclerotinia sclerotiorum*. Plant Health Progr. https://doi.org/10.1094/PHP-2010-0422-01-RS

Kidane A, Bekele E (1986) Major diseases of *Brassica* and preliminary observations on yield loss due to Alternaria leaf spot. In: Proc IDRC (Canada) 3rd Oil Crops Network held in Addis Ababa, 6-10 October 1986: 53-57.

Kifuji Y, Hanzaea H, Terasawa Y, Nishio T (2013) QTL analysis of black rot resistance in cabbage using newly developed EST-SNP markers. Euphytica 190:289–295

Kim S, Song YH, Lee JY, Choi SR, Dhandapani V, Jang CS, Lim YP, Han T (2011) Identification of the BrRHP1 locus that confers resistance to downy mildew in Chinese cabbage (*Brassica rapa* ssp. *pekinensis*) and development of linked molecular markers. Theor Appl Genet 123:1183–1192. https://doi.org/10.1007/s00122-011-1658-9

Kim J, Kang WH, Yang HB, Park S, Jang C, Yu HJ, Kang BC (2013) Identification of a broad spectrum recessive gene in *Brassica rapa* and molecular analysis of the eIF4E gene family to develop molecular markers. Mol Breed 32:385–398

Kinane SJ, Jones P (1996) Isolation and characterization of induced wheat mutants exhibiting partial resistance to powdery mildew. Cereal Rusts Powdery Mildew Bull 24:214–217

Kluczewski SM, Lucas JA (1983) Host infection and oospore formation by *Peronospora parasitica* in agricultural and horticultural *Brassica* species. Trans Br Mycol Soc 81:591–596

Kolte SJ (1985a) Disease management strategies for rapeseed-mustard crops in India. Agric Rev 6:81–88

Kolte SJ (1985b) Diseases of annual edible oilseed crops, Vol II.Rapeseed-Mustard and sesame diseases.CRC Press, Boca Raton, Florida, USA, 135.

Kolte SJ (1987a) Important diseases of rapeseed and mustard in India: Present research progress and future research needs. Omran A (ed), Proc. 3rd Oil Crop Network Workshop, Ethiopia, 6-10 Oct 1986. International Development Research Centre, Ottawa, Ontario, Canada: IDRC-MR 153e, pp 91–106

Kolte SJ (1987b) Screening for resistance to diseases. Oil Crops: Niger, rapeseed-mustard. Proc. 3rd Oil Crop Network Workshop, Ethiopia, 6-10 Oct 1986. International Development Research Centre, Ottawa, Ontario, Canada: IDRC-MR 153e, pp 225–229

Kolte SJ, Tewari AN (1980) Note on the susceptibility of certain oleiferous Brassicaceae to downy mildew and white blister diseases. Indian J Mycol Plant Pathol 10:191–192

Kolte SJ, Awasthi RP, Vishwanath (1985) Field performance of improved toria varieties against Alternaria blight, downy mildew and white rust diseases. Indian J Mycol Plant Pathol 15:211–213

Kolte SJ, Awasthi RP, Vishwanath (2000) Divya mustard: a useful source to create Alternaria black spot tolerant dwarf varieties of oilseed Brassicas. Plant Varieties Seeds (UK) 13:107–111

Kolte SJ, Nashaat NI, Kumar A, Awasthi RP, Chauhan JS (2008) Towards improving the genetic base of rapeseed-mustard through an Indo-UK research collaboration. Indian J Plant Genet Resour 21:132–137

Kontaxis DG, Mayberry KS, Rubatzky VE (1979) Reaction of cauliflower cultivars to downy mildew in Imperial Valley. California Agric 33:19

Konzak CF (1956) A note on the use of radiation for the production of mutations for victoria blight resistance in oats. Phytopathology 46:177–178

Kotsira VP, Clonis YD (1997) Oxalate oxidase from barley roots: purification to homogeneity and study of some molecular, catalytic, and binding properties. Arch Biochem Biophy 340:239–249

Krishnia SK, Saharan GS, Singh D (2000) Genetic variation for multiple disease resistance in the families of interspecific cross of *Brassica juncea* × *B. carinata*. Cruciferae NewsLett 22:51–53

Kumar B (2008) Assessment of slow blighting (*Alternaria brassicae*) resistance and yield in Indian mustard (*Brassica juncea* (L.)Czern. & Coss.) Indian Phytopath 61:171–183.

Kumar PR, Kumar P (1989) Genetic improvement of rapeseed-mustard, achievements critical gaps and future priorities. J Oilseed Res 6:211–219

Kumar S, Saharan GS (2002) Sources of multiple disease resistance in *Brassica* spp. J Mycol Plant Pathol 32:184–188

Kumar D, Maurya N, Bharati YK, Kumar A, Kumar K, Srivastava K, Chand G, Singh SK, Mishra RK, Kumar A (2014) Alternaria blight of oilseed *Brassicas*: a comprehensive review. African J Microbiol Res 8:2816–2829. https://doi.org/10.5897/AJMR2013.6434

Kunkel BN (1996) A useful weed put to work: Genetic analysis of disease resistance in *Arabidopsis thaliana*. Trends Genet 12:63–69

Kutcher HR, Balesdent MH, Rimmer SR, Rouxel T, Chèvre AM, Delourme R, Brun H (2010) Frequency of avirulence genes in *Leptosphaeria maculans* in western Canada. Can J Plant Pathol 32:77–85. https://doi.org/10.1080/07060661003594109

Laemmlen FF, Mayberry KS (1984) Broccoli resistance to downy mildew. California Agric 38:17

Lagercrantz U (1998) Comparative mapping between *Arabidopsis thaliana* and *Brassica nigra* indicates that *Brassica* genomes have evolved through extensive genome replication accompanied by chromosome fusions and frequent rearrangements. Genetics 150:1217–1228

Lakra BS, Saharan GS (1989) Sources of resistance and effective screening techniques in *Brassica-Albugo* system. Indian Phytopath 42:293. (Abstr)

Landry BS, Hubert N, Crete R, Chiang MS, Lincoln SE, Etoh T (1992) A genetic map for *Brassica oleracea* based on RFLP markers detected with expressed DNA sequences and mapping of resistance gene to race 2 of *Plasmodiophora brassicae*. Genome 35:409–420

Lane BG, Bernier F, Dratewka-Kos E, Shafai R, Kennedy TD, Pyne C, Munro JR, Vaughan T, Walters D, Altomare FE (1991) Homologies between members of the germin gene family in hexaploid wheat and similarities between these wheat germins and certain *Physarum spherulins*. J Biol Chem 266:10461–10469

Lane BG, Dunwell JM, Ray JA, Schmitt MR, Cuming AC (1993) Germin, a protein marker of early plant development, is an oxalate oxidase. J Biol Chem 268:12239–12242

Larkan NJ, Lydiate DJ, Parkin IA, Nelson MN, Epp DJ, Cowling WA, Rimmer SR, Borhan MH (2013) The *Brassica napus* blackleg resistance gene LepR3 encodes a receptor-like protein triggered by the *Leptosphaeria maculans* effector AVRLM1. New Phytol 197:595–605. https://doi.org/10.1111/nph.12043.

Larkan NJ, Lydiate DJ, Yu F, Rimmer SR, Borhan MH (2014) Co-localization of the blackleg resistance genes *Rlm2* and *LepR3* on *Brassica napus* chromosome A10. BMC Plant Biol 14:387. https://doi.org/10.1186/s12870-014-0387-z

Larkan NJ, Ma L, Borhan MH (2015) The *Brassica napus* receptor-like protein RLM2 is encoded by a second allele of the LepR3/Rlm2 blackleg resistance locus. Plant Biotechnol J 13:983–992. https://doi.org/10.1111/pbi.12341

Larkan NJ, Yu F, Lydiate DJ, Rimmer SR, Borhan MH (2016) Single R gene introgression lines for accurate dissection of the *Brassica-Leptosphaeria* pathosystem. Front Plant Sci 7:1771

Larkan NJ, Raman H, Lydiate DJ, Robinson SJ, Yu F, Barbulescu DM, Raman R, Luckett DJ, Burton W, Wratten N, Salisbury PA, Rimmer SR, Borhan MH (2016a) Multi-environment QTL studies suggest a role for cysteine-rich protein kinase genes in quantitative resistance to blackleg disease in *Brassica napus*. BMC Plant Biol 16:183. https://doi.org/10.1186/s12870-016-0877-2

Larkan NJ, Yu F, Lydiate DJ, Rimmer SR, Borhan MH (2016b) Single R gene introgression lines for accurate dissection of the *Brassica-Leptosphaeria* pathosystem. Front Plant Sci 7:1771. https://doi.org/10.3389/fpls.2016.01771

Lawrence GJ, Finnegan EJ, Ayliffe MA, Ellis JG (1995) The L6 gene for flax rust resistance is related to the *Arabidopsis* bacterial resistance gene *RPS2* and the tobacco viral resistance gene N. Plant Cell 7:1195–1206

Lee YH, Hong JK (2015) Differential defense responses of susceptible and resistant kimchi cabbage cultivars to anthracnose, black spot and black rot diseases. Plant Pathol 64:406–415. https://doi.org/10.1111/ppa.12262

Lee JM, Hartman GL, Domier LL, Bent AF (1996) Identification and map location of TTR1, a single locus in *Arabidopsis thaliana* that confers tolerance to tobacco ring spot nepovirus. Mol. Plant-Microbe Interact 9:729–735

Lee J, Izzah NK, Jayakodi M, Perumal S, Joh HJ, Lee HJ, Lee SC, Park JY, Yang KW, Nou IS et al (2015) Genome-wide SNP identification and QTL mapping for black rot resistance in cabbage. BMC Plant Biol 15:32

Lee H, Chawla HS, Obermeier C, Dreyer F, Abbadi A, Snowdon R (2020) Chromosome-scale assembly of winter oilseed rape *Brassica napus*. Front Plant Sci 11:496

Lefol C, Seguin-Swartz G, Morrall RAA (1997a) Resistance to *Sclerotinia sclerotiorum* in a weed related to canola. Can J Plant Pathol 19:113.

Lefol C, Seguin-Swartz G, Downey RK (1997b) Sexual hybridization in crosses of cultivated *Brassica* species with the crucifers *Erucastrum gallicum* and *Raphanus aphanistrum*: potential for gene introgression. Euphytica 95:127–139

Leister D (2004) Tandem and segmental gene duplication and recombination in the evolution of plant disease resistance gene. Trends Genet 20:116–122

Li CX, Cowling WA (2003) Identification of a single dominant allele for resistance to blackleg in *Brassica napus* 'Surpass 400'. Plant Breed 122:485–488. https://doi.org/10.1111/j.1439-0523.2003.00863.x

Li Y, Chen J, Bennett R, Kiddle G, Wallsgrove R, Huang Y, He Y (1999) Breeding, inheritance, and biochemical studies on *Brassica napus* cv. Zhongyou 821: tolerance to *Sclerotinia sclerotiorum* (stem rot).), Proc 10th. Intl Rapeseed Cong, Wratten N, Salisbury PA, (eds) Canberra, Australia, September 26–29, p. 61

Li Y, Mei D, Huang Y, Xu Y, Roger W, Guy K, Richard B (2001) Biochemical changes of Zhongyou 821 after inoculation with mycelia of *Sclerotinia sclerotiorum*. Chinese J Oil Crop Sci 23:63–65

Li CX, Li H, Sivasithamparam K, Fu TD, Li YC, Liu SY, Barbetti MJ (2006) Expression of field resistance under Western Australian conditions to *Sclerotinia sclerotiorum* in Chinese and Australian *Brassica napus* and *Brassica juncea* germplasm and its relation with stem diameter. Australian J Agril Res 57:1131–1135

Li H, Barbetti MJ, Sivasithamparam K (2006a) Concomitant inoculation of an avirulent strain of *Leptosphaeria maculans* prevents break-down of a single dominant gene-based resistance in *Brassica napus* cv. Surpass 400 by a virulent strain. Field Crops Res 95:206–211. https://doi.org/10.1016/j.fcr.2005.02.010

Li CX, Li H, Siddique AB, Sivasithamparam K, Salisbury P, Banga SS, Banga S, Chattopadhyay C, Kumar A, Singh R, Singh D, Agnihotri A, Liu SY, Li YC, Tu J, Fu TD, Wang YF, Barbetti MJ

(2007a) The importance of the type and time of inoculation and assessment in the determination of resistance in *Brassica napus* and *B. juncea* to *Sclerotinia sclerotiorum*. Aust J Agric Res 58:1198–1203. https://doi.org/10.1071/AR07094

Li CX, Sivasithamparam K, Walton G, Salisbury P, Burton W, Banga SS, Banga S, Chattopadhyay C, Kumar A, Singh R, Singh D, Agnihotri A, Liu SY, Li YC, Fu TD, Wang YF, Barbetti MJ (2007b) Expression and relationships of resistance to white rust (*Albugo candida*) at cotyledonary, seedling, and flowering stages in *Brassica juncea* germplasm from Australia, China, and India. Aus J Agril Res 58:259–264

Li CX, Liu SY, Sivasithamparam K, Barbetti MJ (2008a) New sources of resistance to Sclerotinia stem rot caused by *Sclerotinia sclerotiorum* in Chinese and Australian *Brassica napus* and *Brassica juncea* germplasm screened under Western Australian conditions. Australasian Plant Pathol 38:149–152

Li CX, Sivasithamparam K, Walton G, Fels P, Barbetti MJ (2008b) Both incidence and severity of white rust disease reflect host resistance in *Brassica juncea* germplasm from Australia, China and India. Field Crops Res 106:1–8

Li QY, Zhang ZG, Cheng WH, Zhao ZZ (2009) Determination of TuMV resistance of Chinese cabbage by ELISA method. Sci Technol Rev 27:42–45. (in Chinese)

Li A, Wei C, Jiang J, Zhang Y, Snowdon RJ, Wang Y (2009a) Phenotypic variation in progenies from somatic hybrids between *Brassica napus* and *Sinapis alba*. Euphytica 170:289–296. https://doi.org/10.1007/s10681-009-9979-3

Li CX, Sivasithamparam K, Barbetti MJ (2009b) Complete resistance to leaf and staghead disease in Australian *Brassica juncea* germplasm exposed to infection by *Albugo candida* (white rust). Aus Plant Pathol 38:63–66

Li QY, Tong HS, Zhang ZG, Zhao ZZ, Song XY (2011) Inheritance and development of EST-SSR marker associated with Turnip mosaic virus resistance in Chinese cabbage. Can J Plant Sci 91:707–715

Li J, Zhao Z, Hayward A, Cheng H, Fu D (2015a) Integration analysis of quantitative trait loci for resistance to *Sclerotinia sclerotiorum* in *Brassica napus*. Euphytica 205:483–489. https://doi.org/10.1007/s10681-015-1417-0.

Li Q, Zhang X, Zeng Q, Zhang Z, Liu S, Pei Y, Wang S, Liu X, Xu W, Fu W, Zhao Z, Song X (2015b) Identification and mapping of an ovel Turnip mosaic virus resistance gene TuRBCS01 in Chinese cabbage (*Brassica rapa* L.). Plant Breed 134:221–225

Li GL, Qian W, Zhang SJ, Zhang SF, Li F, Zhang H, Wu J, Wang ZW, Sun RF (2016a) Development of gene-based markers for the Turnip mosaic virus resistance gene retr02 in *Brassica rapa*. Plant Breed 135:466–470

Li L, Luo Y, Chen B, Xu K, Zhang F, Li H, Huang Q, Xiao X, Zhang T, Hu J et al (2016b) A genome-wide association study reveals new loci for resistance to clubroot disease in *Brassica napus*. Front Plant Sci 7:1483

Li P, Quan X, Jia G, Xiao J, Cloutier S, You FM (2016c) RGAugury: a pipeline for genome-wide prediction of resistance gene analogs (RGAs) in plants. BMC Genomics 17:852. https://doi.org/10.1186/s12864-016-3197-x.

Li C, Lin F, An D, Wang W, Huang R (2017) Genome sequencing and assembly by long reads in plants. Genes 9:6

Li G, Lv H, Zhang S, Zhang S, Li F, Zhang H, Qian W, Fang Z, Sun R (2019) TuMV management for *Brassica* crops through host resistance: retrospect and prospects. Plant Pathol:1–10. https://doi.org/10.1111/ppa.13016

Light KA, Gororo NN, Salisbury PA (2011) Usefulness of winter canola (*Brassica napus*) race-specific resistance genes against blackleg (causal agent *Leptosphaeria maculans*) in southern Australian growing conditions. Crop Pasture Sci 62:162–168. https://doi.org/10.1071/CP10187

Lin K, Zhang N, Severing EI, Nijveen H, Cheng F, Visser RGF, Wang X, de Ridder D, Bonnema G (2014) Beyond genomic variation—Comparison and functional annotation of three *Brassica rapa* genomes: a turnip a rapid cycling and a Chinese cabbage. BMC Genomics 15:250

Liu XP, Lu WC, Liu YK, Wei SQ, Xu JB, Liu ZR, Zhang HJ, Li JL, Ke GL, Yao WY, Cai YS, Wu FY, Cao SC, Li YH, Xie SD, Lin BX, Zhang CL (1996) Occurrence and strain differentiation on Turnip mosaic potyvirus and sources of resistance in Chinese cabbage in China. Acta Hortic 407:431–440

Liu Z, Latunde-Dada AO, Hall AM, Fitt BDL (2014) Phoma stem canker disease on oilseed rape (*Brassica napus*) in China is caused by *Leptosphaeria biglobosa* '*brassicae*.'. Eur J Plant Pathol 140:841–857. https://doi.org/10.1007/s10658-014-0513-7

Liu C, Yang Z, He P, Munir S, Wu Y, Ho H, He Y (2018) Deciphering the bacterial and fungal communities in clubroot-affected cabbage rhizosphere treated with *Bacillus subtilis* XF-1. Agric Ecos Environ 256:12–22

Lob S, Jaspers MV, Ridgway HJ, Jones EE (2013) First Report of *Leptosphaeria biglobosa* as a stem canker pathogen of *Brassicas* in New Zealand. Plant Dis 97:1113–1113. https://doi.org/10.1094/PDIS-12-12-1122-PDN

Long Y, Wang Z, Sun Z, Fernando DW, Mcvetty PB, Li G (2011) Identification of two blackleg resistance genes and fine mapping of one of these two genes in a *Brassica napus* canola cultivar 'Surpass 400.'. Theor Appl Genet 122:1223–1231. https://doi.org/10.1007/s00122-010-1526-z

Lu K, Wei L, Li X, Wang Y, Wu J, Liu M, Zhang C, Chen Z, Xiao Z, Jian H et al (2019) Whole-genome resequencing reveals *Brassica napus* origin and genetic loci involved in its improvement. Nat Commun 10:1154

Lucas JA, Crute IR, Sherriff C, Gordon PL (1988) The identification of a gene for race-specific resistance to *Peronospora parasitica* (downy mildew) in *Brassica napus* var. *oleifera* (oilseed rape). Plant Pathol 37:538–545. https://doi.org/10.1111/j.1365-3059.1988.tb02112.x

Lung HY, Baetz AL, Peck AB (1994) Molecular cloning, DNA sequence, and gene expression of the oxalyl-coenzyme A decarboxylase gene, oxc, from the bacterium *Oxalobacter formigenes*. J Bacteriol 176:2468–2472

Luo Y, Lv GL, Wu WT, Chen SN, Cheng ZQ (2010) Analysis of genome expression in the response of *Oryza granulata* to *Xanthomonas oryzae* pv *oryzae*. Mol Biol Rep 37(2):875–892. https://doi.org/10.1007/s11033-009-9694-0

Lv H, Yang L, Kang J, Wang Q, Wang X, Fang Z, Liu Y, Zhuang M, Zhang Y, Lin Y, Yang Y, Xie B, Liu B, Liu JS (2013) Development of InDel markers linked to Fusarium wilt resistance in cabbage. Mol Breed 32:961–967. https://doi.org/10.1007/s11032-013-9925-x

Lv H, Fang Z, Yang L, Zhang Y, Wang Q, Liu Y, Zhuang M, Yang Y, Xie B, Liu B, Liu J, Kang J, Wang X (2014) Mapping and analysis of a novel candidate Fusarium wilt resistance gene FOC1 in *Brassica oleracea*. BMC Genomics 15:1094. https://doi.org/10.1186/1471-2164-15-1094

Lv H, Fang Z, Yang L, Zhang Y, Wang Y (2020) An update on the arsenal: mining resistance genes for disease management of *Brassica* crops in the genomic era. Hortic Res 7:34. https://doi.org/10.1038/s41438-020-0257-9

Lydiate DJ, Pilcher RL, Higgins EE, Walsh JA (2014) Genetic control of immunity to Turnip mosaic virus (TuMV) pathotype 1 in *Brassica rapa* (Chinese cabbage). Genome 57:419–425

Ma JF, Hou XL, Xiao D, Qi L, Wang F, Sun F, Wang Q (2010) Cloning and characterization of the BcTUR3, gene related to resistance to Turnip mosaic virus (TuMV) from non-heading Chinese cabbage. Plant Mol Biol Reptr 28:588–596

MacLeod IM, Bowman PJ, van der Jagt CJ, Haile-Mariam M, Kemper KE, Chamberlain AJ, Schrooten C, Hayes BJ, Goddard ME (2016) Exploiting biological priors and sequence variants enhances QTL discovery and genomic prediction of complex traits. BMC Genomics 17:144. https://doi.org/10.1186/s12864-016-2443-6

Mahajan V, Gill HS, More TA (1995) Inheritance of downy mildew resistance in Indian cauliflower (group III). Euphytica 86:1–3. https://doi.org/10.1007/BF00035932

Maluszynski M, Ahloowalia BS, Sigurbjonsson B (1995) Application of in vivo and in vitro mutation techniques for crop improvement. Euphytica 85:303–315

Manzanares-Dauleux MJ, Delourme R, Baron F, Thomas G (2000a) Mapping of one major gene and of QTLs involved in resistance to clubroot in *Brassica napus*. Theor Appl Genet 101:885–891. https://doi.org/10.1007/s001220051557

Manzanares-Dauleux MJ, Divaret I, Baron F, Thomas G (2000b) Evaluation of French *Brassica oleracea* landraces for resistance to *Plasmodiophora brassicae*. Euphytica 113:211–218

Marcroft SJ, Elliott VL, Cozijnsen AJ, Salisbury PA, Howlett BJ, Van De Wouw AP (2012a) Identifying resistance genes to *Leptosphaeria maculans* in Australian *Brassica napus* cultivars based on reactions to isolates with known avirulence genotypes. Crop Pasture Sci 63:338–350. https://doi.org/10.1071/CP11341

Marcroft SJ, Van De Wouw AP, Salisbury PA, Potter TD, Howlett BJ (2012b) Effect of rotation of canola (*Brassica napus*) cultivars with different complements of blackleg resistance genes on disease severity. Plant Pathol 61:934–944. https://doi.org/10.1111/j.1365-3059.2011.02580.x

Matsumoto E, Yasui C, Ohi M, Tsukada M (1998) Linkage analysis of RFLP markers for clubroot resistance and pigmentation in Chinese cabbage (*Brassica rapa* ssp. *pekinensis*). Euphytica 104:79–86

Mayerhofer R, Good AG, Bansal VK, Thiagarajah MR, Stringam GR (1997) Molecular mapping of resistance to *Leptosphaeria maculans* in Australian cultivars of *Brassica napus*. Genome 40:294–301. https://doi.org/10.1139/g97-041

Mayerhofer R, Wilde K, Mayerhofer M, Lydiate D, Bansal VK, Good AG, Parkin IAP (2005) Complexities of chromosome landing in a highly duplicated genome: toward map-based cloning of a gene controlling blackleg resistance in *Brassica napus*. Genetics 171:1977–1988. https://doi.org/10.1534/genetics.105.049098

McDonald BA, Linde C (2002) The population genetics of plant pathogens and breeding strategies for durable resistance. Euphytica 124:163–180. https://doi.org/10.1023/A:1015678432355

McDowell JM, Dhandaydham M, Long TA, Aarts MGM, Goff S, Holub EB, Dangl JL (1998) Intragenic recombination and diversifying selection contribute to the evolution of downy mildew resistance at the *RPP8* locus of *Arabidopsis*. Plant Cell 10:1861–1874

McDowell JM, Williams SG, Funderburg NT, Eulgem T, Dangl JL (2005) Genetic analysis of developmentally regulated resistance to downy mildew (*Hyaloperonospora parasitica*) in *Arabidopsis thaliana*. Mol Plant-Microbe Interact 18:1226–1234. https://doi.org/10.1094/MPMI-18-1226

McNaughton IH, Ross CL (1976) Inter-specific and inter-generic hybridization in the Brassicae with special emphasis on the improvement of forage crops. Scottish Plant Breed Stat Ann Rep 57:75–110

Meena PD, Mehta N, Rai PK, Saharan GS (2018) Geographical distribution of rapeseed-mustard powdery mildew disease in India. J Mycol Plant Pathol 48:60–79

Meena SS, Meena PD, Singh VV, Meena HS, Singh D, Yadav R, Singh KH, Sharma P, Nanjundan J, Singh BK, Singh YP, Ram B (2019a) DRMR-2019 (IC0598622; INGR17077), an Indian mustard (*Brassica juncea*) germplasm with white rust resistance. Indian J Plant Genet Resour 32(2):281–281

Meena SS, Meena PD, Singh VV, Meena HS, Singh D, Yadav R, Singh KH, Sharma P, Nanjundan J, Singh BK, Singh YP, Ram B (2019b) DRMR-2035 (IC0598623; INGR17078), an Indian mustard (*Brassica juncea*) germplasm with white rust resistance. Indian J Pl Genetic Resour 32(2):281–281

Meena PD, Sujith Kumar MS, Meena HS, Jambhulkar S, Gohartaj, Pathak D, Srivastava S, Gupta R, Singh D, Gurung B, Rai PK (2020) Confirmation of induced tolerance to Alternaria blight disease in Indian mustard (*Brassica juncea* L.). Appl Biochem Biotechnol. https://doi.org/10.1007/s12010-020-03362-2

Mehta A, Datta A (1991) Oxalate decarboxylase from *Collybia velutipes*: purification, characterization, and cDNA cloning. J Biol Chem 266:23548–23553

Mehta N, Singh K, Sangwan MS (2008) Assessment of yield losses and evaluation of different varieties/ genotypes of mustard against powdery mildew in Haryana. Plant Dis Res 23:55–59

Mehta N, Karn S, Sangwan MS (2009) Resistance to powdery mildew (*Erysiphe cruciferarum*) in mustard in relation to rate reducing factors. Pl Dis Res 24:114–119

Mei J, Qian L, Disi JO, Yang X, Li Q, Li J, Frauen M, Cai D, Qian W (2011) Identification of resistant sources against *Sclerotinia sclerotiorum* in *Brassica* species with emphasis on *B. oleracea*. Euphytica 177:393–399

Mei J, Ding Y, Lu K, Wei D, Liu Y, Disi JO, Li J, Liu L, Liu S, McKay J, Qian W (2012) Identification of genomic regions involved in resistance against *Sclerotinia sclerotiorum* from wild *Brassica oleracea*. Theor Appl Genet DOI. https://doi.org/10.1007/s00122-012-2000-x

Mei J, Ding Y, Lu K, Wei D, Liu Y, Disi JO, Li J, Liu L, Liu S, McKay J, Qian W (2013) Identification of genomic regions involved in resistance against *Sclerotinia sclerotiorum* from wild *Brassica oleracea*. Theor Appl Genet 126:549–556. https://doi.org/10.1007/s00122-012-2000-x

Mei J, Shao C, Yang R, Feng Y, Gao Y, Ding Y, Li J, Qian W (2020) Introgression and pyramiding of genetic loci from wild *Brassica oleracea* into *B. napus* for improving *Sclerotinia* resistance of rapeseed. Theor Appl Genet 133:1313–1319

Ménard R, Larue JP, Silue D, Thouvenot D (1999) Glucosinolates in cauliflower as biochemical markers for resistance against downy mildew. Phytochemistry 52:29–35

Meuwissen TE, Hayes B, Goddard M (2001) Prediction of total genetic value using genome-wide dense marker maps. Genet 157(4):1819–1829

Mohammed AE, You MP, Barbetti M (2017) New resistances offer opportunity for effective management of the downy mildew (*Hyaloperonospora parasitica*) threat to canola. Crop Pasture Sci 68:234–242. https://doi.org/10.1071/CP16363

Monteiro AA, Williams PH (1989) The exploration of genetic resources of Portuguese cabbage and kale for resistance of several Brassica diseases. Euphytica 41:215–225

Monteiro AA, Coelho PS, Bahcevandziev K, Valério L (2005) Inheritance of downy mildew resistance at cotyledon and adult-plant stages in 'Couve Algarvia' (*Brassica oleracea* var. *tronchuda*). Euphytica 141:85–92. https://doi.org/10.1007/s10681-005-5696-8

Moriguchi K, Kimizuka-Takagi C, Ishii K, Nomura K (1999) A genetic map based on RAPD, RFLP, isozyme, morphological markers and QTL analysis for clubroot resistance in *Brassica oleracea*. Breed Sci 49:257–265. https://doi.org/10.1270/jsbbs.49.257.

Morinaga T (1934) Interspecifichybridization in *Brassica* VI. The cytology of F1 hybrids of *B. juncea* and *B. nigra*. Cytologia 6:62–67

Morrall RAA, Dueck J (1982) Epidemiology of Sclerotinia stem rot of rapeseed in Saskatchewan. Can J Plant Pathol 4:161–168

Munde PN, Bhowmik TP (1985) A source of morphological resistance to leaf blight disease of rapeseed and mustard caused by *Alternaria brassicae* (Berk) Sacc. Curr Sci (India) 54:514–515

Nagaoka T, Doullah MA, Matsumoto S, Kawasaki S, Ishikawa T, Hori H, Okazaki K (2010) Identification of QTLs that control clubroot resistance in *Brassica oleracea* and comparative analysis of clubroot resistance genes between *B. rapa* and *B. oleracea*. Theor Appl Genet 120:1335–1346. https://doi.org/10.1007/s00122-010-1259-z

Nanjundan J, Singh K, Singh KH, Singh D (2014) Catalogue on rapeseed-mustard germplasm. ICAR-Directorate of Rapeseed-Mustard Research, Bharatpur (Raj) India, ` 180

Nanjundan J, Manjunatha C, Radhamani J, Thakur AK, Yadav R, Kumar A, Meena ML, Tyagi RK, Yadava DK, Singh D (2020) Identification of new source of resistance to powdery mildew of Indian mustard and studying its inheritance. Plant Pathol J 36(2):111–120. https://doi.org/10.5423/PPJ.OA.07.2019.0205

Nashaat NI, Awasthi RP (1995) Evidence for differential resistance to *Peronospora parasitica* (downy mildew) in accessions of *Brassica juncea* (mustard) at the cotyledon stage. J Phytopathol 143:157–159. https://doi.org/10.1111/j.1439-0434.1995.tb00250.x

Nashaat NI, Rawlinson CJ (1994) The response of oilseed rape (*Brassica napus* ssp. *oleifera*) accessions with different glucosinolate and erucic acid contents to four isolates of *Peronospora parasitica* (downy mildew) and the identification of new sources of resistance. Plant Pathol 43:278–285

Nashaat NI, Mitchell S, Awasthi RP (1995a) Genetic analysis of a noval factor for resistance to *Peronospora parasitica* in *Brassica napus* spp. *oleifera*. Indian J Mycol Plant Pathol 25:55

Nashaat NI, Mitchell SE, Awasthi RP (1995b) Inheritance of two new factors for resistance to downy mildew (*Peronospora parasitica*) in oilseed rape (*Brassica napus* spp. *oleifera*). Proc. 9th Intl Cong, Cambridge, July 4-7, 1995, 4: pp 1283–1285

Nashaat NI, Heran A, Mitchell S (1996) Resistance to downy mildew and its interaction with white rust in rapeseed-mustard. In: 2nd Intl Crop Sci Cong, New Delhi, India, November 17–24, 1996, p 199

Nashaat NI, Heran A, Mitchell SE, Awasthi RP (1997) New genes for resistance to downy mildew (*Peronospora parasitica*) in oilseed rape (*Brassica napus* ssp. *oleifera*). Plant Pathol 46:964–968. https://doi.org/10.1046/j.1365-3059.1997.d01-76.x

Nashaat NI, Heran A, Awasthi RP, Kolte SJ (2004) Differential response and genes for resistance to *Peronospora parasitica* (downy mildew) in *Brassica juncea* (mustard). Plant Breed 123:512–515. https://doi.org/10.1111/j.1439-0523.2004.01037.x

Natti JJ (1958) Resistance of Broccoli and other crucifers to downy mildew. Plant Dis Reptr 42:656–662

Natti JJ, Harvey GER, Sayre CB (1956) Factors contributing to the increase of downy mildew of Broccoli in New York State and its control with fungicides and agrimycin. Plant Dis Reptr 40:118–124

Natti JJ, Dickson MH, Aktin JD (1967) Resistance of *Brassica oleracea* cultivars to downy mildew. Phytopathology 57:144–147

Navabi ZK, Strelkov SE, Good AG, Thiagarajah MR, Rahman MH (2010) *Brassica* B-genome resistance to stem rot (*Sclerotinia sclerotiorum*) in a doubled haploid population of *Brassica napus* × *Brassica carinata*. Can J Plant Pathol 32:237–246. https://doi.org/10.1080/07060661.2010.484229

Neik TX, Barbetti MJ, Batley J (2017) Current status and challenges in identifying disease resistance genes in *Brassica napus*. Front Plant Sci 8:1788. https://doi.org/10.3389/fpls.2017.01788

Neik TX, Amas J, Barbetti M, Edwards D, Batley J (2020) Understanding host-pathogen interactions in *Brassica napus* in the omics era. Plan Theory 9:1336. https://doi.org/10.3390/plants9101336

Nejati-Javaremi A, Smith CW, Gibson JP (1997) Effect of total allelic relationship on accuracy of evaluation and response to selection. J Anim Sci 75(7): 1738–1745. doi: https://doi.org/10.2527/1997.7571738x.

Neuhaus K, Grsic-Rausch S, Sauerteig S, Ludwig-Muller J (2000) Arabidopsis plants transformed with nitrilase 1 or 2 in antisense direction are delayed in clubroot development. J Plant Physiol 156:756–761

Nguyen KL, Grondin A, Courtois B, Gantet P (2019) Next-generation sequencing accelerates crop gene discovery. Trends Plant Sci 24:263–274

Ning LY, Lin ZW, Gu JW, Gan L, Li YH, Wang H, Miao L, Zhang L, Wang B, Li M (2018) The initial deficiency of protein processing and flavonoids biosynthesis were the main mechanisms for the male sterility induced by SX-1 in *Brassica napus*. BMC Genomics 19:806

Niu XK, Leung H, Williams PH (1983) Sources and nature of resistance to downy mildew and turnip mosaic in Chinese cabbage. J Am Soc Hortic Sci 5:775–778

Niu YZ, Guo SX, Fu SH (2005) Development of a specially long pod variant line from resynthesized *Brassica napus* L. J Plant Genet Res 6:151–155

Nomura K, Minegishi Y, Kimizuka-Takagi C, Fujioka T, Moriguchi K, Shishido R, Ikehashi H (2005) Evaluation of F2 and F3 plants introgressed with QTL for clubroot resistance in cabbage developed by using SCAR markers. Plant Breed 124:371–375

Nordborg M, Hu TT, Ishino Y, Jhaveri J, Toomajian C, Zheng HG, Bakker E, Calabrese P, Gladstone J, Goyal R, Jakobsson M, Kim S, Morozov Y, Padhukasahasram B, Plagnol V, Rosenberg NA, Shah C, Wall JD, Wang J, Zhao KY, Kalbfleisch T, Schulz V, Kreitman M, Bergelson J (2005) The pattern of polymorphism in *Arabidopsis thaliana*. PLoS Biol 3:e196

Nuej F, Diez MJ, Ferrando C, Cuartero J, Costa J (1987) Germplasm resources of *Brassica* and *Raphanus* from Spain. Cruciferae NewsLett 12:4

Nyalugwe EP, Barbetti MJ, Jones RAC (2015) Studies on resistance phenotypes to Turnip mosaic virus, in five species of Brassicaceae, and identification of a virus resistance gene in *Brassica juncea*. European J Plant Pathol 141:647–666

Nyalugwe EP, Barbetti MJ, Clode PL, Jones RAC (2016) Programmed cell death pathways induced by early plant-virus infection are determined by isolate virulence and stage of infection. Plant Pathol 65:1518–1528

O'Neill CM, Bancroft I (2000) Comparative physical mapping of segments of the genome of *B rassica oleracea* var. *alboglabra* that are homoeologous to sequenced regions of chromosomes 4 and 5 of *Arabidopsis thaliana*. The. Plant J 33(2):233–243

Opena RT, Lo SE (1981) Breeding for heat tolerance in heading Chinese cabbage. Proc first Intl Sym Asian Veg Res Dev Centre, Shanhua, Taihnan, Taiwan, pp 431–441

Otani F, Maruyama S, Tsukada M, Nagase Y (1982) Studies on the breeding of clubroot resistance in Chinese cabbage and Chinese mustard II. Process of breeding the clubroot resistant line of Nozawana, a variety of Chinese mustard, and its characteristics (in Japanese with English summary). Bull Nagano Veg Ornam Crop Exp Sta Jpn 2:1–8

Otani H, Kohnobe A, Narita M, Shiomi H, Kodama M, Kohmoto K (2001) A new type of host-selective toxin, a protein from *Alternaria brassicicola*. In: Keen NT, Mayama S, Leach JE, Tsuyumu S (eds) Delivery and perception of pathogen signals in plants. APS Press, St. Paul, MN, pp 68–76

Pal SS, Kumar V, Singh I, Kumar V, Singh I (1999) Inheritance of white rust resistance in mustard. Plant Dis Res 14:195–197

Pandey SC, Naik G, Ram K, Sridhar TS (1995) Seedling resistant varieties in cauliflower and cabbage. In: Mandal NC, Dasgupta MK, Ghosh DC, Mukhopadhyay SK, Gupta DD, Majumdar DK, Sriniketan (eds) Agro-ecosystems management. India Palli-Siksha Bhawana, pp 144–145

Pang W, Liang S, Li X, Li P, Yu S, Lim YP, Piao Z (2014) Genetic detection of clubroot resistance loci in a new population of *Brassica rapa*. Hortic Environ Biotechnol 55:540–547. https://doi.org/10.1007/s13580-014-0079-5

Pang W, Fu P, Li X, Zhan Z, Yu S, Piao Z (2018) Identification and mapping of the Clubroot resistance gene CRd in Chinese Cabbage (*Brassica rapa* ssp. *pekinensis*). Front. Plant Sci 9:653

Parker JE, Holub EB, Frost LN, Falk A, Gunn ND, Daniels MJ (1996) Characterization of eds1, a mutation in *Arabidopsis* suppressing resistance to *Peronospora parasitica* specified by several different RPP genes. Plant Cell 8:2033–2046

Parker JE, Coleman MJ, Szabo V, Frost LN, Schmidt R, Van der Biezen EA, Moores T, Dean C, Daniels MJ, Jones JDJ (1997) The *Arabidopsis* downy mildew resistance gene *RPP5* shares similarity to the toll and interleukin-1 receptors with *N* and *L6*. Plant Cell 9:879–894

Parlange F, Daverdin G, Fudal I, Kuhn ML, Balesdent MH, Blaise F, Grezes-Besset B, Rouxel T (2009) *Leptosphaeria maculans* avirulence gene AvrLm4-7 confers a dual recognition specificity by the *Rlm4* and *Rlm7* resistance genes of oilseed rape, and circumvents *Rlm4*-mediated recognition through a single amino acid change. Mol Microbiol 71:851–863. https://doi.org/10.1111/j.1365-2958.2008.06547.x

Parui NR, Bandyopadhyay DC (1973) A note on the screening of rai (*Brassica juncea* L.) against white rust (*Albugo candida*) (Pers.) Kuntz. Curr Sci 42:198–799

Pathak AK, Godika S, Jain JP, Muralia S (2002) Screening of *Brassica* genotypes against stem rot disease of mustard caused by *Sclerotiniasclerotiorum* (Lib.) de Bary. J Mycol Pl Pathol 32:111–112

Peng G, Falk KC, Gugel RK, Franke C, Yu F, James B, Strelkov SE, Hwang SF, McGregor L (2014) Sources of resistance to *Plasmodiophora brassicae* (clubroot) pathotypes virulent on canola. Can J Plant Pathol 36:89–99. https://doi.org/10.1080/07060661.2013.863805

Peng L, Zhou L, Li Q, Wei D, Ren X, Song H, Mei J, Si J, Qian W (2018) Identification of quantitative trait loci for Clubroot resistance in *Brassica oleracea* with the use of *Brassica* SNP microarray. Front Plant Sci 9:822

Perez-Enciso M, Rincon, JC, Legarra A (2015) Sequence- vs. chip-assisted genomic selection: accurate biological information is advised. Genet Sel Evol 47:43. https://doi.org/10.1186/s12711-015-0117-5.

Perumal S, Koh CS, Jin L, Buchwaldt M, Higgins EE, Zheng C, Sanko D, Robinson SJ, Kagale S, Navabi ZK et al (2020) A high-contiguity *Brassica nigra* genome localizes active centromeres and defines the ancestral *Brassica* genome. Nat Plants 6:929–941

Petrie GA (1985) Yield losses in Saskatchewan rapeseed/canola disease survey, 1983. Can Plant Dis Surv 65:47–49

Petrie GA (1986) *Albugo candida* on *Raphanus sativus* in Saskatchewan. Can Plant Dis Surv 66:43–47

Piao ZY, Deng YQ, Choi SR, Park YJ, Lim YP (2004) SCAR and CAPS mapping of CRb, a gene conferring resistance to *Plasmodiophora brassicae* in Chinese cabbage (*Brassica rapa* ssp. *pekinensis*). Theor Appl Genet 108:1458–1465. https://doi.org/10.1007/s00122-003-1577-5

Piao Z, Ramchiary N, Lim YP (2009) Genetics of clubroot resistance in *Brassica* species. J Plant Growth Reg 28(3):252–264

Pilet ML, Delourme R, Foisset N, Renard M (1998) Identification of loci contributing to quantitative field resistance to blackleg disease, causal agent *Leptosphaeria maculans* (Desm.) Ces.Et de Not., in winter rapeseed (*Brassica napus* L.). Theor Appl Genet 96:23–30

Pilet ML, Duplan G, Archipiano M, Barret P, Baron C, Horvais R, Tanguy X, Lucas MO, Renard M, Delourme R (2001) Stability of QTL for field resistance to blackleg across two genetic backgrounds in oilseed rape. Crop Sci 41:197–205

Plieske J, Struss D, Robbelen G (1998) Inheritance of resistance derived from the B-genome of Brassica against *Phoma lingam* in rapeseed and the development of molecular markers. Theor Appl Genet 97:929–936. https://doi.org/10.1007/s001220050973

Plissonneau C, Daverdin G, Ollivier B, Blaise F, Degrave A, Fudal I, Rouxel T, Balesdent MH (2016) A game of hide and seek between avirulence genes AvrLm4-7 and AvrLm3 in *Leptosphaeria maculans*. New Phytol 209:1613–1624. https://doi.org/10.1111/nph.13736

Plissonneau C, Blaise F, Ollivier B, Leflon M, Carpezat J, Rouxel T, Balesdent MH (2017) Unusual evolutionary mechanisms to escape effector-triggered immunity in the fungal phytopathogen *Leptosphaeria maculans*. Mol Ecol 26:2183–2198. https://doi.org/10.1111/mec.14046

Pratap P, Meena PD, Singh BK, Meena HS, Meena SS, Sharma P, Majumdar R, Singh D (2014) Development and evaluation of Alternaria blight tolerant lines in Indian mustard (*Brassica juncea*). J Oilseed Brassica 5:141–148

Pratt RG, Rowe DE (1991) Differential responses of alfalfa genotypes to stem inoculations with *Sclerotinia sclerotiorum* and *S. trifoliorum*. Plant Dis 75:188–191

Pu Z, Shimizu M, Zhang Y, Nagaoka T, Hayashi T, Hori H, Matsumoto S, Fujimoto R, Okazaki K (2012) Genetic mapping of a Fusarium wilt resistance gene in *Brassica oleracea*. Mol Breed 30:809–818. https://doi.org/10.1007/s11032-011-9665-8

Qian W, Zhang SJ, Zhang SF, Li F, Zhang H, Wu J, Wang X, Walsh JA, Sun R (2013) Mapping and candidate-gene screening of the novel Turnip mosaic virus resistance gene retr02 in Chinese cabbage (*Brassica rapa* L.). Theor Appl Genet 126:179–188

Quiros CF, Grellet F, Sadowski J, Suzuki T, Li G, Wroblewski T (2001) *Arabidopsis* and *Brassica* comparative genomics: sequence, structure and gene content in the ABI1-Rps2-Ck1 chromosomal segment and related regions. Genet 157:1321–1330

Radhamani J, Ranbir S, Srinivasan K, Tyagi RK (2013) Conservation of trait-specific germplasm of *Brassica* in National Gene bank. National Bureau of Plant Genetic Resources, New Delhi, p 114

Rafiqi M, Bernoux M, Ellis JG, Dodds PN (2009) In the trenches of plant pathogen recognition: role of NB-LRR proteins. Semin Cell Dev Biol 20:1017–1024

Rahman A, Das ML, Howlider MAR, Mansur MA (1987) Promising mutants in *Brassica campestris*. Mutat Breed Newsl 29:14–15

Rahman H, Shakir A, Jakir Hasan M (2011) Breeding for clubroot resistant spring canola (*Brassica napus* L.) for the Canadian prairies: can the European winter canola cv. Mendel be used as a source of resistance? Can J Plant Sci 91:447–458. https://doi.org/10.4141/cjps10073

Rahman H, Peng G, Yu F, Falk KC, Kulkarni M, Selvaraj G (2014) Genetics and breeding for clubroot resistance in Canadian spring canola (*Brassica napus* L.). Can J Plant Pathol 36:122–134. https://doi.org/10.1080/07060661.2013.862571

Rahman M, Mamidi S, Del Rio L, Ross A, Kadir MM, Rahaman MM, Arifuzzaman (2016) Association mapping in *Brassica napus* (L.) accessions identifies a major QTL for blackleg disease resistance on chromosome A01. Mol Breed 36:90 https://doi.org/10.1007/s11032-016-0513-8.

Rai B, Kolte SJ, Tewari AN (1977) Evaluation of oleiferous *Brassica* germplasm for resistance to Alternaria leaf blight. Indian J Mycol Plant Pathol 6:76–77

Raman R, Taylor B, Lindbeck K, Coombes N, Barbulescu D, Salisbury P, Raman H (2012a) Molecular mapping and validation of *Rlm1* genes for resistance to *Leptosphaeria maculans* in canola (*Brassica napus* L). Crop Pasture Sci 63:1007–1017. https://doi.org/10.1071/CP12255

Raman R, Taylor B, Marcroft S, Stiller J, Eckermann P, Coombes N, Rehman A, Lindbeck K, Luckett D, Wratten N, Batley J, Edwards D, Wang X, Raman H (2012b) Molecular mapping of qualitative and quantitative loci for resistance to *Leptosphaeria maculans* causing blackleg disease in canola (*Brassica napus* L.). Theor Appl Genet 125:405–418. https://doi.org/10.1007/s00122-012-1842-6

Raman H, Raman R, Coombes N, Song J, Diffey S, Kilian A, Lindbeck K, Barbulescu DM, Batley J, Edwards D, Salisbury PA, Marcroft S (2016a) Genome-wide association study identifies new loci for resistance to *Leptosphaeria maculans* in canola. Front Plant Sci 7:1513. https://doi.org/10.3389/fpls.2016.01513

Raman R, Diffey S, Carling J, Cowley RB, Kilian A, Luckett DJ, Raman H (2016b) Quantitative genetic analysis of grain yield in an Australian *Brassica napus* doubled-haploid population. Crop Pasture Sci 67:298–307. https://doi.org/10.1071/CP15283

Rana K, Atri C, Akhatar J, Kaur R, Goyal A, Singh Mohini P, Kumar N, Sharma A, Sandhu Prabhjodh S, Kaur G et al (2019) Detection of first marker trait associations for resistance against *Sclerotinia sclerotiorum* in *Brassica juncea–Erucastrum cardaminoides* introgression lines. Front Plant Sci 10:1015

Ravi S, Awasthi RP (2019) Screening of *Brassica juncea* (L.) against downy mildew and white rust in tarai region of Uttarakhand. Intl J Pure App Biosci 7(1):327–332

Ren L, Xu L, Liu F, Chen K, Sun C, Li J, Fang X (2016) Host range of *Plasmodiophora brassicae* on cruciferous crops and weeds in China. Plant Dis 100:933–939. https://doi.org/10.1094/PDIS-09-15-1082-RE

Rícarova V, Kazda J, Singh K, Rysanek P (2016) Clubroot caused by *Plasmodiophora brassicae* Wor.: a review of emerging serious disease of oilseed rape in the Czech Republic. Plant Protect Sci 52: 71–86. https://doi.org/10.17221/87/2015-PPS.

Rimmer SR (2006) Resistance genes to *Leptosphaeria maculans* in *Brassica napus*. Can J Plant Pathol 28:S288–S297. https://doi.org/10.1080/07060660609507386

Rimmer SR, van den Berg CGJ (1992) Resistance of oilseed *Brassica* spp. to blackleg caused by *Leptosphaeria maculans*. Can J Plant Pathol 14:56–66. https://doi.org/10.1080/07060669209500906

Rimmer SR, Borhan MH, Zhu B, Somers D (1999) Mapping resistance genes in *Brassica napus* to *Leptophaeria maculans*. In: 10th Intl Rapeseed Cong. The Regional Institute, Canberra, Australia, p 155

Ripley V, Thorpe M, Iler S, Mizier K, Beversdorf WD (1992) Isozyme analysis as a tool for introgression of *Sinapis alba* germplasm into *Brassica napus*. Theor Appl Genet 84:403–410. https://doi.org/10.1007/BF00229500

Rocherieux J, Glory P, Giboulot A, Boury S, Barbeyron G, Thomas G, Manzanares-Dauleux MJ (2004) Isolate-specific and broad-spectrum QTLs are involved in the control of clubroot in *Brassica oleracea*. Theor Appl Genet 108:1555–1563

Romero-Munoz F, Jimenez DRM (1979) The black spot: a rape disease recently registered in Spain. Annales del Institute Nacional de investigaciones Agrarias, Proccion Vegetal 9:11–31

Ronicke S, Hahn V, Horn R, Grone I, Brahm L, Schnab H, Friedt W (2004) Inter-specific hybrids of sunflower as a source of *Sclerotinia* resistance. Pl Breed 123:152–157

Roohie RK, Umesha S (2015) Identification of genes associated with black rot resistance in cabbage through suppression subtractive hybridization. eBiotech 5:1089–1100

Rooster LDE, Spiessens K, Reycke LDE, Janssens M (1999) Late autumn cultivation of cauliflowers: difference in resistance to downy mildew. PRO 9:12–13

Rouxel T, Balesdent MH (2013) From model to crop plant–pathogen interactions: cloning of the first resistance gene to *Leptosphaeria maculans* in *Brassica napus*. New Phytol 197:356–358. https://doi.org/10.1111/nph.12099

Rouxel T, Penaud A, Pinochet X, Brun H, Gout L, Delourme R, Schmit J, Balesdent MH (2003a) A 10-year survey of populations of *Leptosphaeria maculans* in France indicates a rapid adaptation towards the *Rlm1* resistance gene of oilseed rape. Eur J Plant Pathol 109:871–881. https://doi.org/10.1023/A:1026189225466

Rouxel T, Willner E, Coudard L, Balesdent MH (2003b) Screening and identification of resistance to *Leptosphaeria maculans* (stem canker) in *Brassica napus* accessions. Euphytica 133:219–231. https://doi.org/10.1023/A:1025597622490

Rowe RC (1980) Evaluation of radish cultivars for resistance to clubroot (*Plasmodiophora brassicae*) race 6 for mid-western United States. Plant Dis 64:462–464

Rozej A (1974) Susceptibility of different varieties of winter rape (*Brassica napus* L. var *oleifera*) to infection with *Alternaria brassicae* (Berk) Sacc Roczniki Nauk Rolniczch A 100:67–71

Rusholme RE, Higgins E, Walsh JA, Lydiate DJ (2007) Genetic control of broad-spectrum resistance to Turnip mosaic virus in *Brassica rapa* (Chinese cabbage). J Gen Virol 88:3177–3186

Sachan JN, Kolte SJ, Basudeo S, Singh B (2000) Inheritance of resistance of white rust (*Albugo candida*race 2) in *Brassica juncea*. Indian Phytopath 53:206–209

Sacristan MD, Gerdemann M (1986) Different behavior of *Brassica juncea* and *B. carinata* as sources of *Phoma lingam* resistance in experiments of interspecific transfer to *B. napus*. Plant Breed 97:304–314. https://doi.org/10.1111/j.1439-0523.1986.tb01071.x

Saha P, Kalia P, Sharma P, Sharma TR (2014) Race-specific genetics of resistance to black rot disease [*Xanthomonas campestris* pv. *campestris* (*Xcc*) (Pammel) Dowson] and the development of three random amplified polymorphic DNA markers in cauliflower. J Hort Sci Biotechnol 89:480–486

Saha P, Kalia P, Sharma M, Singh D (2016) New source of black rot disease resistance in *Brassica oleracea* and genetic analysis of resistance. Euphytica 207:35–48

Saharan GS (1984) A review of research on rapeseed-mustard pathology in India. Ann Workshop AICORPO, ICAR, Jaipur, 6–10 August, 1984

Saharan GS (1992a) Management of rapeseed and mustard diseases. In: Kumar D, Rai M (eds) Advances in oilseeds research. Sci Pub Jodhpur, India 1(7):152–188

Saharan GS (1992b) Disease resistance. In: Labana KS, Banga SS, Banga SK (eds) Breeding oilseed Brassicas, vol 12. Narosa Pub House, New Delhi, India, pp 181–205

Saharan GS (1996) Studies on physiologic specialization, host resistance and epidemiology of white rust and downy mildew disease complex in rapeseed-mustard. Final report of adhoc research project ICAR, Dept of Plant Pathology, CCSHAU, Hisar, India, pp 1083

Saharan GS (1997) Disease resistance. In: Kalia HR, Gupta SC (eds) Recent advances in oilseed Brassicas. Kalyani Pub, Ludhiana, Chap 12. pp 233–259

Saharan GS (2010) Analysis of genetic diversity in *Albugo*-Crucifer system. J Mycol Plant Pathol 40:1–13

Saharan GS, Chand JN (1988) Diseases of oilseed crops (Hindi). Min. Human Resources, Govt of India.HAU Press, Hisar, India. Ch 3:84–91

Saharan GS, Krishnia SK (2001) Multiple disease resistance in rapeseed and mustard. In: Nagarajan S, Singh DP (eds) Role of resistance in intensive agriculture. Kalyani Pub, New Delhi, pp 98–108

Saharan GS, Naresh M (2008) Sclerotinia diseases of crop plants: Biology, ecology and disease management. Springer Science+Busines Media B.V, The Netherlands, 485 p

Saharan GS, Kaushik CD, Kaushik JC (1988) Sources of resistance and epidemiology of white rust of mustard. Indian Phytopath 41:96–99

Saharan GS, Naresh M, Sangwan MS (2003) Nature and mechanism of resistance to Alternaria blight in rapeseed-mustard system. Ann Rev Plant Pathol 2:85–128

Saharan GS, Naresh M, Sangwan MS (2005) Development of disease resistance in rapeseed-mustard. In: Saharan GS, Naresh M, Sangwan MS (eds) Diseases of oilseed crops. Indus Pub Co, New Delhi, pp 561–617

Saharan GS, Mehta Naresh, Verma PR, Meena PD (2015) Analysis of variability determinants in *Alternaria-Brassica*-interactions. J Oilseed Brassica 6 (1):126–141

Saharan GS, Mehta Naresh, Meena PD (2016) Alternaria blight of crucifers: Biology, ecology and management. Springer Verlag, Singapore: 326, ISBN 978-981-10-0019-5.

Saharan GS, Mehta Naresh, Meena PD (2017) Downy mildew disease of crucifers: Biology, ecology and disease management. Springer Verlag, Singapore, LVI, p 357. ISBN 978-981-10-7499-8.

Saharan GS, Mehta Naresh, Meena PD (2021) Clubroot of crucifers: biology, ecology and management. Springer Verlag, Singapore, (Inpress).

Saito M, Kubo N, Matsumoto S, Suwabe K, Tsukada M, Hirai M (2006) Fine mapping of the clubroot resistance gene, Crr3, in *B rassica rapa*. Theor Appl Genet 114:81–91. https://doi.org/10.1007/s00122-006-0412-1

Sakamoto K, Saito A, Hayashida N, Taguchi G, Matsumoto E (2008) Mapping of isolate-specific QTLs for clubroot resistance in Chinese cabbage (*Brassica rapa* L. ssp. *pekinensis*). Theor Appl Genet 117:759–767. https://doi.org/10.1007/s00122-008-0817-0

Sang X, Jue D, Yang L, Bai X, Chen M, Yang Q (2013) Genetic transformation of *Brassica napus* with MSI-99m gene increases resistance in transgenic plants to *Sclerotinia sclerotiorum*. Mol Plant Breed 4(30):247–253

Santos MP, Dias JS, Monteiro AA (1996) Screening portuguese cole landraces for resistance to white rust (*Albugo candida* (Pers.) Kuntze). Acta Horticul 407:453–459

Sarris PF, Duxbury Z, Huh SU, Ma Y, Segonzac C, Sklenar J, Derbyshire P, Cevik V, Rallapalli G, Saucet SB, Wirthmueller L, Menke FLH, Sohn KH, Jones JDG (2015) A plant immune receptor detects pathogen effectors that target WRKY transcription factors. Cell 161:1089–1100

Satyavir RN, Saharan GS, Ram B, Gandhi SK (1994) Plant Pathology at a glance, (1970-1994) A Silver Jubilee Publication. CCS Haryana Agric Univ, Hisar, India, Dept Plant Pathol, pp 48–55

Saude C, Mckeown A, Gossen BD, McDonald MR (2012) Effect of host resistance and fungicide application on clubroot pathotype 6 in green cabbage and rapa cabbage. Hort Technol 22 (311):319

Scheben A, Batley J, Edwards D (2017) Genotyping-by-sequencing approaches to characterize crop genomes: choosing the right tool for the right application. Plant Biotechnol J 15:149–161

Schwartz HF, Otto K, Teran H, Lema M, Singh SP (2006) Inheritance of white mold resistance in *Phaseolus vulgaris* × *P. coccineus* crosses. Plant Dis 90:1167–1170

Seguin-Swartz G, Lefol C (1999) Sclerotinia stem rot resistance in crucifers. In: Proceedings of the 10th Intl Rapeseed Cong. Canberra, Australia, p 153

Sharma TR, Singh BM (1992) Transfer of resistance to *Alternaria brassicae* in *Brassica juncea* through interspecific hybridization among *Brassica*. J Genet Breed 46:373–378

Sharma BR, Swarup V, Chatterjee SS (1972) Inheritance of resistance to black rot in cauliflower. Can J Genet Cytol 14:363–370

Sharma BR, Swarup V, Chatterjee SS (1977) Resistance to black rot disease in cauliflower. Sci Hortic 7:1–7

Sharma BR, Dhiman JS, Thakur JC, Singh A, Bajaj KL (1991) Multiple disease resistance in cauliflower. Adv Hort Sci 5:30–34

Sharma SR, Kapoor KS, Gill HS (1995) Screening against Sclerotinia rot (*Sclerotinia sclerotiorum*), downy mildew (*Peronospora parasitica*) and black rot (*Xanthomonas campestris)* in cauliflower *Brassica oleracea* var. *botrytis* sub var. *cauliflora*. Indian J Agric Sci 65:916–918

Sharma G, Kumar VD, Haque A, Bhat SR, Prakash S, Chopra VL (2002) *Brassica* coenospecies: a rich reservoir for genetic resistance to leaf spot caused by *Alternaria brassicae*. Euphytica 125:411–417

Sharma P, Kumar A, Meena PD, Goyal P, Salisbury P, Gurung A, Fu TD, Wang YF, Barbetti MJ, Chattopadhyay C (2009) Search for resistance to *Sclerotinia sclerotiorum* in exotic and indigenous *Brassica* germplasm. Proc 16th Australian Research Assembly on Brassicas. Ballarat, Australia, September 14–16, pp 169–173

Sharma P, Kumar A, Chauhan JS (2012) Evaluation of Indian and exotic *Brassica* germplasm for tolerance to stem rot caused by *Sclerotinia sclerotiorum*. J Mycol Plant Pathol 42:297–302

Sharma K, Gossen BD, Greenshields D, Selvaraj G, Strelkov SE, McDonald MR (2013) Reaction of lines of the rapid cycling *Brassica* collection and *Arabidopsis thaliana* to four pathotypes of *Plasmodiophora brassicae*. Plant Dis 97:720–727

Sharma P, Deep S, Bhati DS, Sharma M, Chowdappa P (2014) Penetration and infection processes of *Alternaria brassicicola* on cauliflower leaf and *Alternaria brassicae* on mustard leaf: a histopathological study. Plant Pathol J 13:100

Sharma BB, Kalia P, Yadava DK, Dingh D, Sharma TR (2016) Genetics and molecular mapping of black rot resistance locus *Xca1bc* on chromosome B-7 in Ethiopian mustard (*Brassica carinata* A. Braun). PLoS One 11:e0152290. https://doi.org/10.1371/journal.pone.0152290

Sharma BB, Kalia P, Dingh D, Sharma TR (2017) Introgression of black rot resistance from *Brasssica carinata* to cauliflower (*Brassica oleracea botrytis* group) through embryo rescue. Front Plant Sci 8:1255

Shattuck VI (1993) Glucosinolates and glucosinolate degradation in seeds from turnip mosaic virus-infected rapid cycle *Brassica campestris* L. plants. J Exp Bot 262:963–970

Sheikh IA, Singh JN (2001) Genetics of resistance to white rust in Indian mustard [*Brassica juncea* (L) Czern & Coss]. Crop Res (Hisar) 21:341–344

Sherf AF, Macnab AA (1986) Downy mildew. In: Vegetable diseases and their control. Wiley, New York, pp 269–273

Sherriff C, Lucas JA (1990) The host range of isolates of downy mildew, *Peronospora parasitica*, from *Brassica* crop species. Plant Pathol 39:77–91

Shimizu M, Pu ZJ, Kawanabe T, Kitashiba H, Matsumoto S, Ebe Y, Sano M, Funaki T, Fukai E, Fujimoto R, Okazaki K (2015) Map–based cloning of a candidate gene conferring *Fusarium* yellows resistance in *Brassica oleracea*. Theor Appl Genet 128:119–130. https://doi.org/10.1007/s00122-014-2416-6

Shiraishi M, Sakamoto K, Yoshida S, Asada Y (1974) An electron microscope study of Japanese radish leaves infected with *Peronospora parasitica*. Memoirs College Agric, Ehime Univ 19:137–168

Shivpuri A, Chhipa HP, Gupta RBL, Sharma KN (1997) Field evaluation of mustard genotypes against white rust, powdery mildew and stem rot. Annals Arid Zone 36:387–389

Shopan J, Mou H, Zhang L, Zhang C, Ma W, Walsh JA, Hu Z, Yang J, Zhang M (2017) Eukaryotic translation initiation factor 2B-beta (eIF2Bb), a new class of plant virus resistance gene. The Plant J 90:929–940

Siemens J, Nagel W, Ludwig-Muller J, Sarcristán MD (2002) The interaction of *Plasmodiophora brassicae* and *Arabidopsis thaliana*: parameters for disease quantification and screening mutant lines. J Phytopathol 150:592–605

Siemens J, Keller I, Sarx J, Kunz S, Schuller A, Nagel W, Schmulling T, Parniske M, Ludwig-Muller J (2006) Transcriptome analysis of *A rabidopsis* clubroots indicate a key role for cytokinins in disease development. Mol Plant-Microbe Interact 19:480–494. https://doi.org/10.1094/MPMI-19-0480

Silue D, Nashaat NI, Tirilly Y (1996) Differential responses of *Brassica oleracea* and *B. rapa* accessions to seven isolates of *Peronospora parasitica* at the cotyledon stage. Plant Dis 80:142–144

Sinapidou E, Williams K, Nott L, Bahkt S, Tor M, Crute I, Bittner-Eddy P, Beynon J (2004) Two TIR:NB:LRR genes are required to specify resistance to *Peronospora parasitica* isolate Cala2 in *Arabidopsis*. Plant J 38:898–909

Singh K (1994) Studies on the ecofriendly management of powdery mildew (*Erysiphe cruciferarum* Opiz ex. Junell) of mustard [*Brassica juncea* (Linn.)Czern & Coss.]. MSc Thesis, Department of Plant Pathology, CCS HAU, Hisar. Pages 106+ xvii.

Singh N, Gangopadhyay S (1976) Reaction of Chinese cabbage lines to white rust. Veg Sci 3:71–73

Singh RB, Singh RN (2003) Management of mildew of mustard. Indian Phytopathol 56:147–150

Singh R, Trivedi BM, Gill HS, Sen B (1987) Breeding for resistance for black rot, downy mildew and curd blight in Indian cauliflower. Cruciferae Newslett 12:96–97

Singh R, Tripathi NN, Kaushik CD, Singh R (1994) Management of Sclerotinia rot of Indian mustard [*Brassica juncea* (L.)Czern and Coss] by fungicides. Crop Res 7:276–281

Singh R, Singh D, Li H, Sivasithamparam S, Yadav NR, Salisbury P, Barbetti MJ (2008) Management of Sclerotinia rot of oilseed *Brassicas*—a focus on India. J OilseedRes10: 1–27.

Singh R, Singh D, Salisbury P, Barbetti MJ (2010) Field evaluation of indigenous and exotic *Brassica juncea* genotypes against blight, white rust, downy mildew and powdery mildew diseases in India. Indian J Agril Sci 80:155–159

Singh D, Dhar S, Yadava DK (2011) Genetic and pathogenic variability of Indian strains of *Xanthomonas campestris* pv. *campestris* causing black rot disease in crucifers. Curr Microbiol 63:551–560

Singh S, Sharma SR, Kalia P, Deshmukh R, Kumar V, Sharma P, Sharma TR (2012) Molecular mapping of the downy mildew resistance gene Ppa3 in cauliflower (*Brassica oleracea* var. *botrytis* L.). J Hortic Sci Biotechnol 87:137–143. https://doi.org/10.1080/14620316.2012.11512844

Singh S, Dey SS, Bhatia R, Batley J, Kumar R (2018) Molecular breeding for resistance to black rot [*Xanthomonas campestris* pv. *campestris* (Pammel) Dowson] in *Brassicas*: recent advances. Euphytica 214:196

Sivasithamparam K, Barbetti MJ, Li HUA (2005) Recurring challenges from a necrotrophic fungal plant pathogen: a case study with *Leptosphaeria maculans* (causal agent of Blackleg disease in *Brassicas*) in Western Australia. Ann Bot 96:363–377. https://doi.org/10.1093/aob/mci194

Slusarenko AJ, Schlaich NL (2003) Downy mildew of *Arabidopsis thaliana* caused by *Hyaloperonospora parasitica* (formerly *Peronospora parasitica*). Mol Plant Pathol 4:159–170

Snowdon R, Winter H, Diestel A, Sacristán MD (2000) Development and characterization of *Brassica napus-Sinapis arvensis*addition lines exhibiting resistance to *Leptosphaeria maculans*. Theor Appl Genet 101(7):1008–1014

Somé A, Manzanares MJ, Laurens F, Baron F, Thomas G, Rouxel F (1996) Variation for virulence on *Brassica napus* L. amongst *Plasmodiophora brassicae* collections from France and derived single-spore isolates. Plant Pathol 45:432–439. https://doi.org/10.1046/j.1365-3059.1996.d01-155.x

Somers DJ, Rakow G, Rimmer SR (2002) *Brassica napus* DNA markers linked to white rust resistance in *Brassica juncea*. Theor Appl Genet 104:1121–1124

Song J-M, Guan Z, Hu J, Guo C, Yang Z, Wang S, Liu D, Wang B, Lu S, Zhou R et al (2020) Eight high-quality genomes reveal pan-genome architecture and ecotype differentiation of *Brassicanapus*. Nat Plants 6:34–45

Sprague SJ, Marcroft SJ, Hayden HL, Howlett BJ (2006) Major gene resistance to blackleg in *Brassica napus* overcome within three years of commercial production in Southeastern Australia. Plant Dis 90:190–198. https://doi.org/10.1094/PD-90-0190

Srivastava LS, Verma RN (1987) Field screening of mustard cultivars against white rust in Sikkim. Indian J Mycol Plant Pathol 17:46

Stankova J (1972) Varietal variability of winter rape in its susceptibility to dark leaf spot and the factors influencing the development of the disease. Rostlinna vyroba 18:625–630

Staskawicz BJ, Ausubel FM, Baker BJ, Ellis JG, Jones JDG (1995) Molecular genetics of plant disease resistance. Science 268:661–667

Stotz HU, Harvey PJ, Haddadi P, Mashanova A, Kukol A, Larkan NJ, Borhan MH, Fitt BDL (2018) Genomic evidence for genes encoding leucine-rich repeat receptors linked to resistance against the eukaryotic extra- and intracellular *Brassica napus* pathogens *Leptosphaeria maculans* and *Plasmodiophora brassicae*. PLoS One 13:e0198201

Strelkov SE, Hwang SF (2014) Clubroot in the *Canadian canola* crop: 10 years into the outbreak. Can J Plant Pathol 36:27–36. https://doi.org/10.1080/07060661.2013.863807

Strelkov SE, Tewari JP (2005) Clubroot on canola in Alberta in 2003 and 2004. Can Plant Dis Surv 85:72–73

Strelkov SE, Hwang SF, Howard RJ, Hartman M, Turkington TK (2011) Progress towards the sustainable management of clubroot [*Plasmodiophora brassicae*] of canola on the Canadian Prairies. Prairie Soils Crops J 4:114–121

Sun D, Wang C, Zhang X, Zhang W, Jiang H, Yao X, Liu L, Wen Z, Niu G, Shan X (2019) Draft genome sequence of cauliflower (*Brassica oleracea* L. var. *botrytis*) provides new insights into the C genome in *Brassica* species. Hortic. Res 6:82

Suwabe K, Tsukazaki H, Iketani H, Hatakeyama K, Fujimura M, Nunome T, Fukuoka H, Matsumoto S, Hirai M (2003) Identification of two loci for resistance to clubroot (*Plasmodiophora brassicae* Woronin) in *Brassica rapa* L. Theor Appl Genet 107:997–1002

Suwabe K, Tsukazaki H, Iketani H, Hatakeyama K, Kondo M, Fujimura M, Nunome T, Fukuoka H, Hirai M, Matsumoto S (2006) Simple sequence repeat-based comparative genomics between *Brassica rapa* and *Arabidopsis thaliana*: the genetic origin of clubroot resistance. Genetics 173:309–319

Suwabe K, Suzuki G, Nunome T, Hatakeyama K, Mukai Y, Fukuoka H, Matsumoto S (2012) Microstructure of a *Brassica rapa* genome segment homoeologous to the resistance gene cluster on *Arabidopsis* chromosome 4. Breed Sci 62:170–177. https://doi.org/10.1270/jsbbs.62.170

Tai TH, Dahlbeck D, Clark ET, Gajiwala P, Pasion R, Whalen MC, Stall RE, Staskawicz BJ (1999) Expression of the Bs2 pepper gene confers resistance to bacterial spot disease in tomato. Proc Natl Acad Sci U S A 96:14153–14158. https://doi.org/10.1073/pnas.96.24.14153

Tanaka S, Mido H, Ito SI (2006) Colonization by two isolates of *Plasmodiophora brassicae* with differing pathogenicity on a clubroot-resistant cultivar of Chinese cabbage (*Brassica rapa* L. subsp. *pekinensis*). J Gen Plant Pathol 72:205–209. https://doi.org/10.1007/s10327-006-0276-x

Taylor JD, Conway J, Roberts SJ, Astley D, Vicente JG (2002) Source and origin of resistance to *Xanthomonas campestris* pv. *campestris* in *Brassica* genomes. Phytopathology 92:105–111

Taylor A, Rana K, Handy C, Clarkson JP (2017) Resistance to *Sclerotinia sclerotiorum* in wild *Brassica* species and the importance of *Sclerotinia subarctica* as a *Brassica* pathogen. Plant Pathol 67:433–444

Tewari JP (1991) Resistance to *Alternaria brassicae* in crucifers. Bulletin SROP 14(6):154–161

Tewari JP, Conn KL (1993) Reactions of some wild crucifers to *Alternaria brassicae*. Bulletin-OILS-CROP 16:53–58

Thomas CE, Jourdain EL (1990) Evaluation of broccoli and cauliflower germplasm for resistance to Race 2 of *Peronospora parasitica*. Hort Sci 25:1429–1431

Thomas CE, Jourdain EL (1992) Resistance to race 2 of *Peronospora parasitica* in U.S. Plant Introductions of *Brassica oleracea* var. *capitata*. Hort Sci 27:1120–1122

Thompson C, Dunwell JM, Johnstone CE, Lay V, Ray J, Schmitt M, Watson H, Nisbet G (1995) Degradation of oxalic acid by transgenic oilseed rape plants expressing oxalate oxidase. Euphytica 85:169–172

Tollenaere R, Hayward A, Dalton-Morgan J, Campbell E, Lee JR, Lorenc MT, Manoli S, Stiller J, Raman R, Raman H, Edwards D, Batley J (2012) Identification and characterization of candidate Rlm4 blackleg resistance genes in *Brassica napus* using next-generation sequencing. Plant Biotechnol J 10:709–715. https://doi.org/10.1111/j.1467-7652.2012.00716.x

Tonguc M, Griffiths PD (2004a) Transfer of powdery mildew resistance from *Brassica carinata* to *Brassica oleracea* through embryo rescue. Plant Breed 123:587–589. https://doi.org/10.1111/j.1439-0523.2004.00987.x

Tonguc M, Griffiths PD (2004b) Development of black rot resistant interspecific hybrids between *Brassica oleracea* L. cultivars and *Brassica* accession A 19182, using embryo rescue. Euphytica 136:313–318

Tonguc M, Earle ED, Griffiths PD (2003) Segregation distortion of *Brassica carinata* derived black rot resistance in *Brassica oleracea*. Euphytica 134:269–276

Tonu NN, Doullah MA, Shimizu M, Karim MM, Kawanabe T, Fujimoto R, Okazaki K (2013) Comparison of positions of QTLs conferring resistance to *Xanthomonas campestris* pv. *campestris* in *Brassica oleracea*. Am J Plant Sci 4:11–20

Tor M, Holub EB, Brose E, Musker R, Gunn N, Can C, Crute IR, Beynon JL (1994) Map positions of three loci in *Arabidopsis thaliana* associated with isolate-specific recognition of *Peronospora parasitica* (downy mildew). Mol Plant-Microbe Interact 7:214–222

Torii KU, Mitsukawa N, Oosumi T, Matsuura Y, Yokoyama R, Whittier RF, Komeda Y (1996) The *Arabidopsis* ERECTA gene encodes a putative receptor protein kinase with extracellular leucinerich repeats. Plant Cell 8:735–746

Tsai HY, Hamilton A, Tinch AE, Guy DR, Gharbi K, Stear MJ, Matika O, Bishop SC, Houston RD (2015) Genome wide association and genomic prediction for growth traits in juvenile farmed Atlantic salmon using a high density SNP array. BMC Genomics 16:969. https://doi.org/10.1186/s12864-015-2117-9

Ueno H, Matsumoto E, Aruga D, Kitagawa S, Matsumura H, Hayashida N (2012) Molecular characterization of the CRa gene conferring clubroot resistance in *Brassica rapa*. Plant Mol Biol 80:621–629. https://doi.org/10.1007/s11103-012-9971-5

Uloth MB, Ping PY, Finnegan PM, Banga SS, Banga SK, Sandhu PS, Phillip HY, Salisbury A, Brbetti MJ (2013) New sources of resistance to *Sclerotinia sclerotiorum* for crucifer crops. Field Crops Res 154:40–52. https://doi.org/10.1016/j.fcr.2013.07.013

Uloth M, Ming PY, Finnegan PM, Banga SS, Yi H, Barbetti MJ (2014) Seedling resistance to *Sclerotinia sclerotiorum* as expressed across diverse cruciferous species. Plant Dis 98:184–190. https://doi.org/10.1094/PDIS-06-13-0612-RE

Uloth MB, Ming PY, Barbetti MJ (2015) Host resistance to Sclerotinia stem rot in historic and current *Brassica napus* and *B. juncea* varieties: critical management implications. Crop Pasture Sci 66:841–848. https://doi.org/10.1071/CP15064

Uloth MB, Ming PY, Barbetti MJ (2016) Cultivar resistance offers the first opportunity for effective management of the emerging powdery mildew (*Erysiphe cruciferarum*) threat to oilseed *Brassicas* in Australia. Crop Pasture Sci 67:1179–1187

UN (1935) Genomic analysis in Brassica with special reference to the experimental formation of *B. napus* and peculiar mode of fertilization. Jpn J Bot 7:389–452

Uzunova M, Ecke W, Weissleder K, Robbelen G (1995) Mapping the genome of rapeseed (*Brassica napus* L.). I. Construction of an RFLP linkage map and localization of QTLs for seed glucosinolate content. Theor Appl Genet 90:194–204. https://doi.org/10.1007/BF00222202

Van de Weyer A-L, Monteiro F, Furzer OJ, Nishimura MT, Cevik V, Witek K, Jones JDG, Dangl JL, Weigel D, Bemm F (2019) A species-wide inventory of NLR genes and alleles in *Arabidopsisthaliana*. Cell 178:1260–1272.e14

Van de Wouw AP, Cozijnsen AJ, Hane JK, Brunner PC, McDonald BA, Oliver RP, Howlett BJ (2010) Evolution of linked avirulence effectors in *Leptosphaeria maculans* is affected by genomic environment and exposure to resistance genes in host plants. PLoS Pathol 6:e1001180

Van de Wouw AP, Idnurm A, Davidson JA, Sprague SJ, Khangura RK, Ware AH, Lindbeck KD, Marcroft SJ (2016a) Fungal diseases of canola in Australia: identification of trends, threats and potential therapies. Australas Plant Pathol 45:415–423. https://doi.org/10.1007/s13313-016-0428-1

Van De Wouw AP, Marcroft SJ, Howlett BJ (2016b) Blackleg disease of canola in Australia. Crop Pasture Sci 67:273–283. https://doi.org/10.1071/CP15221

VanRaden PM (2008) efficient methods to compute genomic predictions. J Dairy Sci 91 (11):4414–4423. https://doi.org/10.3168/jds.2007-0980

Varghese YA (1985) Selection of mutants showing partial resistance to powdery mildew (*Erysiphe graminis* f. sp. *hordei*) in barley after sodium azide mutagenesis. Indian J Genet Pl Breed 45:57–66

Vela-Avitua S, Meuwissen TH, Luan T, Odegard J (2015) Accuracy of genomic selection for a sib-evaluated trait using identity-by-state and identity-by-descent relationships. Genet Sel Evol 47:9. https://doi.org/10.1186/s12711-014-0084-2

Velazhahan R, Thiyagarajan K (1994) Mustard genotypes resistance to white rust. Madras Agric J 81:459–460

Verma PR, Saharan GS (1994) Monograph on Alternaria diseases of crucifers. In: Saskatoon Research Centre Technical Bulletin 1994–6E. Canada, Agriculture and Agri-Food Canada, Saskatoon, SK

Verma TS, Thakur PC (1989) Comparative field resistance of cabbage collection to downy mildew at seedling stage. Indian J Plant Prot 17:79–80

Vicente JG, Conway J, Roberts SJ, Taylor JD (2001) Identification and origin of *Xanthomonas campestris* pv. *campestris* races and related pathovars. Phytopathology 91:492–499

Vicente JG, Taylor JD, Sharpe AG, Parkin IAP, Lydiate DJ, King GJ (2002) Inheritance of race-specific resistance to *Xanthomonas campestris* pv. *campestris* in *Brassica* genomes. Phytopathology 92:1134–1141

Vignesh M, Yadava DK, Sujata V, Mohapatra T, Jain N, Yadav AK, Malik D, Yadav MS, Prabhu KV (2009) Genetics of white rust resistance in [*Brassica juncea* (L.) Czern. & Coss.]and allelic relationship between interspecific sources of resistance. Indian J Genet 69(3):205–208

Vitanova G (1996) Sensitivity of the decorative species from the family Brassicaceae to downy mildew (*Peronospora parasitica*). Cruciferae NewsLett 18:132

Voorrips RE (1996) Production, characterization and interaction of single-spore isolates of *Plasmodiophora brassicae*. Eur J Plant Pathol 102:377–383

Voorrips RE, Jongerius MC, Kanne HJ (1997) Mapping of two genes for resistance to clubroot (*Plasmodiophora brassicae*) in a population of doubled haploid lines of *Brassica oleracea* by means of RFLP and AFLP markers. Theor Appl Genet 94:75–82. https://doi.org/10.1007/s001220050384

Walker JC (1993) Receptor-like protein kinase genes of *Arabidopsis thaliana*. Plant J 3:451–456

Walkey DGA, Neely HA (1980) Resistance in white cabbage to necrosis caused by turnip and cauliflower mosaic viruses and pepper-spot. J Agril Sci 95:703–713

Walsh JA, Sharpe AG, Jenner EC, Lydiate DJ (1999) Characterization of resistance of turnip mosaic virus in oilseed rape (*Brassica napus*) and genetic mapping of TuRB01. Theor Appl Genet 99:1149–1154

Wang X, Zafian P, Choudhary M, Lawton M (1996) The PR5K receptor protein kinase from *Arabidopsis thaliana* is structurally related to a family of plant defense proteins. Proc Natl Acad Sci U S A 93:2598–2602

Wang J, Yang G, Fu T (2000) Meng J. Development of PCR-based markers linked to the fertility restore gene for the Polima cytoplasmic male sterility in rapeseed (*Brassica napus* L.) Acta Genet Sin 27:1012–1017

Wang M, Farnham MW, Thomas CE (2001) Inheritance of true leaf stage downy mildew resistance in broccoli. J Am Soc Hortic Sci 126:727–729

Wang HZ, Liu GH, Zheng YB, Wang XF, Yang Q (2003) Breeding of a *Brassica napus* cultivar Zhongshuang No. 9 with high-resistance to *Sclerotinia sclerotiorum* and dynamics of its important defense enzyme activity.11th Intl Rapeseed Cong, Copenhagen, Denmark

Wang HZ, Liu GH, Zheng YB, Wang XF, Yang Q (2004) Breeding of the *Brassica napus* cultivar Zhongshuang No 9 with high-resistance to *Sclerotinia sclerotiorum* and dynamics of its important defense enzyme activity. Sci Agril Sin 37:23–28

Wang Z, Mao H, Dong C, Ji R, Cai L, Fu H, Liu S (2009) Over expression of *Brassica napus MPK4* enhances resistance to *Sclerotinia sclerotiorum* in oilseed rape. Mol Plant Microbiol Interter 22:235–244

Wang X, Wang H, Wang J, Sun R, Wu J, Liu S, Bai Y, Mun JH, Bancroft I, Cheng F, Huang S, Li X, Hua W, Wang J, Wang X, Freeling M, Pires JC, Paterson AH, Chalhoub B, Wang B, Hayward A, Sharpe AG, Park BS, Weisshaar B, Liu B, Li B, Liu B, Tong C, Song C, Duran C,

Peng C, Geng C, Koh C, Lin C, Edwards D, Mu D, Shen D, Soumpourou E, Li F, Fraser F, Conant G, Lassalle G, King GJ, Bonnema G, Tang H, Wang H, Belcram H, Zhou H, Hirakawa H, Abe H, Guo H, Wang H, Jin H, Parkin IA, Batley J, Kim JS, Just J, Li J, Xu J, Deng J, Kim JA, Li J, Yu J, Meng J, Wang J, Min J, Poulain J, Wang J, Hatakeyama K, Wu K, Wang L, Fang L, Trick M, Links MG, Zhao M, Jin M, Ramchiary N, Drou N, Berkman PJ, Cai Q, Huang Q, Li R, Tabata S, Cheng S, Zhang S, Zhang S, Huang S, Sato S, Sun S, Kwon SJ, Choi SR, Lee TH, Fan W, Zhao X, Tan X, Xu X, Wang Y, Qiu Y, Yin Y, Li Y, Du Y, Liao Y, Lim Y, Narusaka Y, Wang Y, Wang Z, Li Z, Wang Z, Xiong Z, Zhang Z (2011) *Brassica rapa* genome sequencing project consortium. The genome of the mesopolyploid crop species *Brassica rapa*. Nat Genet 43:1035–1039. https://doi.org/10.1038/ng.919

Warwick SI (2011) Brassicaceae in agriculture. In: Schmidt R, Bancroft I (eds) Genetics and genomics of the Brassicaceae. Plant genetics and genomics: Crop models 9. Springer Science +Business Media, LLC, pp 33–65. https://doi.org/10.1007/978-1-4419-7118-0_2

Watpade S, Naresh M (2012) Effect of culture filtrate of *Sclerotinia sclerotiorum* (Lib.) de Bary on the activities of oxidative enzymes in calli of *Brassica* species. Plant Dis Res 27:182–185

Watpade S, Naresh M (2013) Biochemical characterization of rapeseed-mustard calli tolerant to *Sclerotinia sclerotiorum* culture filtrate. Plant Dis Res 28:1–4

Watpade S, Naresh M, Sangwan MS (2012) Evaluation of relative resistance of *Brassicas* to white stem rot incited by *Sclerotinia sclerotiorum* through tissue culture technique. J Mycol Plant Pathol 42:238–243

Wei D, Mei J, Fu Y, Disi JO, Li J, Qian W (2014) Quantitative trait loci analyses for resistance to *Sclerotinia sclerotiorum* and flowering time in *Brassica napus*. Mol Breed 34:1797–1804. https://doi.org/10.1007/s11032-014-0139-7

Wei D, Cui Y, He Y, Xiong Q, Qian L, Tong C, Lu G, Ding Y, Li J, Jung C et al (2017) A genome-wide survey with different rapeseed ecotypes uncovers footprints of domestication and breeding. J Exp Bot 68:4791–4801

Werner S, Diederichsen E, Frauen M, Schondelmaier J, Jung C (2007) Genetic mapping of clubroot resistance genes in oilseed rape. Theor Appl Genet 116:363–372. https://doi.org/10.1007/s00122-007-0674-2

Westman AL, Kresovich S, Dickson MH (1999) Regional variation in *Brassica nigra* and other weedy crucifers for disease reaction to *Alternaria brassicicola* and *Xanthomonas campestris* pv. *campestris*. Euphytica 106:253–259. https://doi.org/10.1023/A:1003544025146

Williams PH (1988a) Wisconsin fast plants.Cruciferae Newsl ett 13:7

Williams PH (1988b) Activities of the crucifer genetics cooperative in 1988. Cruciferae Newslett 13:91

Williams J, Pink DAC (1987) Development of an in vitro screen in Brussels sprout for resistance to *Alternaria brassicae*. Cruciferae News Lett 12:91

Williams PH, Pound GS (1963) Nature and inheritance of resistance to *Albugo candida* in radish. Phytopathology 53:1150–1154

Williams PH, Staub T, Sutton JC (1972) Inheritance of resistance in cabbage to black rot. Phytopathology 62:247–252

Wit F, van de Weg M (1964) Clubroot resistance in turnips (*Brassica campestris* L.). I. Physiological races of the parasite and their identification in mixtures. Euphytica 13:9–18

Woods DL, Falk KC (2001) AC Sungold summer turnip rape. Can J Plant Sci 81(2):285–287

Woods DL, Petrie GA (1989) Scimitar brown mustard. Can J Plant Sci 69:247–248

Worland AJ, Law CN (1991) Improving disease resistance in wheat by inactivating genes promoting disease susceptibility. Mutation Breed Newslett 38:2–5

Wu J, Cai G, Tu J, Li L, Liu S, Luo X, Zhou L, Fan C, Zhou Y (2013) Identification of QTLs for resistance to Sclerotinia stem rot and BnaC.IGMT5.a as a candidate gene of the major resistant QTL SRC6 in *Brassica napus*. PLoS One 8:e67740. https://doi.org/10.1371/journal.pone.0067740

Wu J, Zhao Q, Liu S, Shahid M, Lan L, Cai G, Zhang C, Fan C, Wang Y, Zhou Y (2016a) Genome-wide association study identifies new loci for resistance to Sclerotinia Stem Rot in *Brassica napus*. Front Plant Sci 7:1418

Wu J, Zhao Q, Yang Q, Liu H, Li Q, Yi X, Cheng Y, Guo L, Fan C, Zhou Y (2016b) Comparative transcriptomic analysis uncovers the complex genetic network for resistance to *Sclerotinia sclerotiorum* in *Brassica napus*. Sci Rep 6:19007. https://doi.org/10.1038/srep19007

Wu D, Liang Z, Yan T, Xu Y, Xuan L, Tang J, Zhou G, Lohwasser U, Hua S, Wang H et al (2019) Whole-genome resequencing of a worldwide collection of rapeseed accessions reveals the genetic basis of ecotype divergence. Mol Plant 12:30–43

Wurschum T, Abel S, Zhao YS (2014) Potential of genomic selection in rapeseed (*Brassica napus* L.) breeding. Plant Breed 133(1):45–51. https://doi.org/10.1111/pbr.12137

Yadav YP, Singh H (1992) The potential exotic sources for white rust resistance in India mustard [*Brassica juncea* (L.) Czern & Coss]. IDRC, Ottawa, Ontario, Canada (Pub), A Omran (ed). Oil Crops NewsLett 9:18.

Yadav R, Prasad L, Nanjundan J, Tewari AK, Singh P, Sandhu SS, Pant U, Avtar R, Jalli R, Kumar S, Rao M, Rana JS (2017) Identification and evaluation of Indian mustard genotypes for white rust resistance and agronomic performance. Indian J Genet Plant Breed 78(1):81

Yan D, Scott RL, Biederman JA, Barron-Gafford G, Yang X, Moore DJP, Dannenberg MP, Wang X, Knowles JF, Yang J, Smith WK (2018) Quantifying plant physiological response to water stress with high-frequency, near-surface observations of chlorophyll fluorescence and photochemical reflectivity. Abstract B31N-2694 presented at 2018 AGU Fall Meeting, Washington, DC, 10-14 Dec 2018

Yang YW, Change TC, Wang TT (1998) A heat tolerant broccoli F1 hybrid, 'Ching- Long 45'. Hort Sci 33:1090–1091

Yang TJ, Kim JS, Kwon SJ, Lim KB, Choi BS, Kim JA, Jin M, Park JY, Lim MH, Kim HI, Lim YP, Kang JJ, Hong JH, Kim CB, Bhak J, Bancroft I, Park BS (2006) Sequence-level analysis of the diploidization process in the triplicated FLOWERING LOCUS C region of *Brassica rapa*. Plant Cell 18:1339–1347

Yang M, Ding G, Shi L, Feng J, Xu F, Meng J (2010) Quantitative trait loci for root morphology in response to low phosphorus stress in *Brassica napus*. Theor Appl Genet 121(1):181–193

Yarnell SH (1956) Cytogenetics of the vegetable crops. II. Crucifers. Bot Rev 22:81–166

Yin X, Yi B, Chen W, Zhang W, Tu J, Fernando WGD, Fu T (2010) Mapping of QTLs detected in a *Brassica napus* DH population for resistance to *Sclerotinia sclerotiorum* in multiple environments. Euphytica 173:25–35. https://doi.org/10.1007/s10681-009-0095-1

Yoshikawa H (1993) Studies on breeding of clubroot resistance in cole crops. Bull Natl Res Inst Veg Ornam Plants Tea Jpn Ser A 7:1–165

You MP, Uloth MB, Li XX, Banga SS, Banga SK, Barbetti MJ (2016) Valuable new resistances ensure improved management of Sclerotinia stem rot (*Sclerotinia sclerotiorum*) in horticultural and oilseed *Brassica* species. J Phytopathol 164:291–299. https://doi.org/10.1111/jph.12456

Yu F, Lydiate DJ, Rimmer SR (2005) Identification of two novel genes for blackleg resistance in *Brassica napus*. Theor Appl Genet 110:969–979. https://doi.org/10.1007/s00122-004-1919-y

Yu F, Lydiate DJ, Rimmer SR (2008) Identification and mapping of a third blackleg resistance locus in *Brassica napus* derived from *B. rapa* subsp. *sylvestris*. Genome 51:64–72. https://doi.org/10.1139/G07-103

Yu S, Zhang F, Yu R, Zou Y, Qi J, Zhao X, Yu Y, Zhang D, Li L (2009) Genetic mapping and localization of a major QTL for seedling resistance to downy mildew in Chinese cabbage (*Brassica rapa* ssp. *pekinensis*). Mol Breed 23:573–590. https://doi.org/10.1007/s11032-009-9257-z

Yu S, Zhang F, Zhao X, Yu Y, Zhang D (2011) Sequence-characterized amplified region and simple sequence repeat markers for identifying the major quantitative trait locus responsible for seedling resistance to downy mildew in Chinese cabbage (*Brassica rapa* ssp. *pekinensis*). Plant Breed 130:580–583. https://doi.org/10.1111/j.1439-0523.2011.01874.x

Yu F, Gugel RK, Kutcher HR, Peng G, Rimmer SR (2013) Identification and mapping of a novel blackleg resistance locus LepR4 in the progenies from *Brassica napus* × *B. rapa* subsp. *sylvestris*. Theor Appl Genet 126:307–315. https://doi.org/10.1007/s00122-012-1919-2

Yu J, Tehrim S, Zhang F, Tong C, Huang J, Cheng X, Dong C, Zhou Y, Qin R, Hua W, Liu S (2014) Genome-wide comparative analysis of NBS-encoding genes between *Brassica* species and *Arabidopsis thaliana*. BMC Genomics 15:3. https://doi.org/10.1186/1471-2164-15-3

Yu F, Zhang X, Huang Z, Chu M, Song T, Falk KC, Deora A, Chen Q, Zhang Y, McGregor L, Gossen BD, McDonald MR, Peng G (2016a) Identification of genome-wide variants and discovery of variants associated with *Brassica rapa* clubroot resistance gene Rcr1 through bulked segregant RNA sequencing. PLoS One 11:e0153218. https://doi.org/10.1371/journal.pone.0153218

Yu S, Su T, Zhi S, Zhang F, Wang W, Zhang D, Zhao Z, Yu Y (2016b) Construction of a sequence-based bin map and mapping of QTLs for downy mildew resistance at four developmental stages in Chinese cabbage (*Brassica rapa* L. ssp. *pekinensis*). Mol Breed 36:1–12. https://doi.org/10.1007/s11032-016-0467-x

Yu F, Zhang X, Peng G, Falk KC, Strelkov SE, Gossen BD (2017) Genotyping-by-sequencing reveals three QTL for clubroot resistance to six pathotypes of *Plasmodiophora brassicae* in *Brassica rapa*. Sci Rep 7:4516. https://doi.org/10.1038/s41598-017-04903-2

Zaman MW, *Biswas* KP *(*1987*) Brassica napus*: a potential new oilseed crop for Bangladesh. Cruciferae Newslett 12*:*36–37

Zhang FL, Wang M, Liu XC, Zhao XY, Yang JP (2008) Quantitative trait loci analysis for resistance against Turnip mosaic virus based on a doubled-haploid population in Chinese cabbage. Plant Breed 127:82–86

Zhang XW, Yuan YX, Wang XW (2009) QTL mapping for TuMV resistance in Chinese cabbage [*Brassica campestris* L. ssp. *pekinensis* (Lour.)Olssom]. Acta Horticult Sin 36:731–736

Zhang T, Zhao Z, Zhang C, Pang W, Choi SR, Lim YP, Piao Z (2014a) Fine genetic and physical mapping of the CRb gene conferring resistance to clubroot disease in *Brassica rapa*. Mol Breed 34:1173–1183. https://doi.org/10.1007/s11032-014-0108-1

Zhang X, White RP, Demir E, Jedryczka M, Lange RM, Islam M, Li ZQ, Huang YJ, Hall AM, Zhou G, Wang Z, Cai X, Skelsey P, Fitt BDL (2014b) *Leptosphaeria* spp. Phoma stem canker and potential spread of *L. maculans* on oilseed rape crops in China. Plant Pathol 63:598–612

Zhang H, Feng J, Zhang S, Zhang S, Li F, Strelkov SE, Sun R, Hwang SF (2015) Resistance to *Plasmodiophora brassicae* in *Brassica rapa* and *Brassica juncea* genotypes from China. Plant Dis 99:776–779. https://doi.org/10.1094/PDIS-08-14-0863-RE

Zhang H, Feng J, Hwang SF, Strelkov SE, Falak I, Huang X, Sun R (2016a) Mapping of clubroot (*Plasmodiophora brassicae*) resistance in canola (*Brassica napus*). Plant Pathol 65:435–440. https://doi.org/10.1111/ppa.12422

Zhang X, Peng G, Parks P, Hu B, Li Q, Jiang L, Niu Y, Huang S, Fernando WGD (2016b) Identifying seedling and adult plant resistance of Chinese *Brassica napus* germplasm to *Leptosphaeria maculans*. Plant Pathol 66:752–762. https://doi.org/10.1111/ppa.12626

Zhang L, Cai X, Wu J, Liu M, Grob S, Cheng F, Liang J, Cai C, Liu Z, Liu B et al (2018) Improved *Brassica rapa* reference genome by single-molecule sequencing and chromosome conformation capture technologies. Hortic Res 5:50

Zhao J, Meng J (2003) Genetic analysis of loci associated with partial resistance to *Sclerotinia sclerotiorum* in rapeseed (*Brassica napus* L.). Theo Appl Genet 106:759–764

Zhao J, Peltier AJ, Meng J, Osborn TC, Grau CR (2004) Evaluation of Scleroinia stem rot resistance in oilseed *Brassica napus* using an inoculation technique under green house conditions. Plant Dis 88:1033–1039

Zhao J, Udall JA, Quijada PA, Grau CR, Meng J, Osborn TC (2006) Quantitative trait loci for resistance to *Sclerotinia sclerotiorum* and its association with a homeologous non-reciprocal transposition in *Brassica napus* L. Theo Appl Genet 112:509–516

Zhu H, Zhai W, Li X, Zhu Y (2019) Two QTLs controlling Clubroot resistance identified from Bulked Segregant Sequencing in Pakchoi (*Brassica campestris* ssp. *chinensis* Makino). Sci Rep 9:9228.

Zukalová H, Vasák J (2002) The role and effects of glucosinolates of *Brassica* species – a review. Rostliniá Viroba 48:175–180

Inheritance of Disease Resistance

3

Abstract

The inheritance of disease resistance in *Brassica* crops is by both qualitative one gene and polygenes as well as quantitative conferred by QTLs. The resistance to white rust (*Albugo*) in different *Brassica* species is governed by single dominant gene, one or two dominant genes with epistatic effects, additive dominant with epistatic effects, and partial resistant single recessive genes *wpr*. In *Arabidopsis thaliana*, *WRR* 4 genes confer broad-spectrum resistance to races AC-2, 4, 7, and 9. In Brassica genotypes, significant additive gene effects and dominant gene effects govern resistance to Alternaria blight disease. In some crosses, additive × dominance epistatic effects are significant. In *Brassica juncea* genotypes, Alternaria tolerance with components of horizontal resistance attributes has been identified. In the interspecific crosses of *B. juncea* × *B. carinata*, single resistance gene with complete dominance confers resistance to *Erysiphe*. In *A. thaliana*, accessions resistance to powdery mildew is either polygenic based on atypical R-gene RPW8 or combination thereof with RPW8 representing major QTL. The inheritance of enhanced R-genes in *edr* 1 mutant is conferred by recessive genes to powdery mildew. In *Brassica,* resistance to *Fusarium* is conferred by a dominant gene. Inheritance of resistance in crucifers to downy mildew has been observed to be governed by one to four dominant complementary genes, additive genes, one to two partial dominant genes, partial dominant epistatic and gene dosage affects along with quantitative genes depending on host species and pathotype interactions. In *Brassica*, most resistant cultivars contain major genes (QTLs) conferring resistance to specific blackleg pathotypes (*Leptosphaeria maculans*). Resistant loci to *L. maculans* are distributed in A and B genome of *B. napus*. Clubroot (CR) resistance in *B. oleracea* is governed by 2 to 4 dominant genes, additive and recessive genes and more than 22 QTLs. In *B. napus*, clubroot resistance is governed by one to two dominant genes and more than 19 QTLs, whereas in *B. rapa*, two dominant genes interact in a non-additive manner to

G. S. Saharan et al., *Genomics of Crucifer's Host-Resistance*,
https://doi.org/10.1007/978-981-16-0862-9_3

confer resistance to different *Plasmodiophora brassicae* pathotypes. Across all *Brassica* spp. at least 15 CR loci from *B. rapa*, 22 QTL's from *B. oleracea*, and 16 QTL's from *B. napus* have been identified. In *B. rapa*, chromosomes AO3 carry the highest number of loci. *P. brassicae* pathotype 5 × identified two independent QTL's, *Rcr* 8 and *Lc9*, in the A genome on chromosomes AO2 and AO8. A single dominant gene, *Rcr* 7 has been identified to provide resistance to pathotypes 3H and 5X in the C genome on chromosomes CO7. A modifier gene has been identified to increase efficiency of a dominant CR gene against virulent pathotypes. The inheritance of resistance to *Sclerotinia* is complex since no single consistent mode exists to provide resistance to *Brassica* crops. In *Brassica*, partial dominant, polygenic, additive epistatic interactions confer resistance to *Sclerotinia*. Quantitative resistance in both leaf and stem by dominant and additive × additive epistatic interactions controls the genetic effects. In *B. rapa*, resistance to Turnip mosaic virus (TuMV) is controlled by dominant and recessive genes. In *B. oleracea*, resistance to TuMV has been observed to be controlled by both dominant genes and multiple loci. The inheritance of resistance in *Brassica* species to black rot (*Xanthomonas*) is very complex and depends on genotype-race interactions. In *B. oleracea*, resistance in controlled by single dominant gene to race 1 and 4. It is polygenic recessive to races 1 and 3 in *B. oleracea* line Badger Inbred-16 and high level of resistance to races 1 and 4 in *B. carinata* controlled by a single dominant locus (Xca1). Resistance to race 4 in three *B. napus* lines is controlled by a single dominant locus (Xca4).

Keywords

Inheritance of disease resistance · Partial resistance · Tolerance genotypes · Horizontal resistance · Characterization of R-genes · *Brassica* pathogens · Mapping of R-genes · *Arabidopsis* · Avirulence genes · Seedling and adult plant resistance

3.1 Introduction

Identification of effective R-genes sources along with the nature of inheritance in *Brassica* genotypes is the basic impact in breeding of disease resistance cultivars of crucifers against different range of pathogens. There are two different genetic mechanisms for disease resistance traits being inherited in generations. One is monogenic resistance which is based on single gene and second is quantitative which is governed by two or more genes. In most cases, single R-genes confer complete resistance but are active against few races of the pathogen. R-genes are clustered at some loci in the genome or exist as different alleles conferring resistance towards specific pathogen races. Quantitative resistance shows efficacy against more number of pathogenic races. Resistance breeding in crucifers depends on both types of resistance. Monogenic resistance is easy to incorporate but it is not durable. Mostly quantitative resistance is preferred. The use of molecular markers has allowed the isolation, cloning, and gene pyramiding to develop more durable resistance cvs. of *Brassica* crops. Qualitative resistance mostly provides defense against biotrophs, whereas quantitative resistance is involved in defense response to all plant pathogens from biotrophs, hemibiotrophs to necrotrophs pathogenic to crucifers.

3.2 *Brassica-Albugo* Inheritance

Studies on the genetics of host–parasite interactions in WR diseases has focused on the level of specificity among races of pathogens and genotypes of related host species. Even within the confines of race-cultivar-specificity, the studies have been one-sided in that no genetic information has been generated on *Albugo*, the causal organism. Interest in such studies was stimulated by Hougas et al. (1952), who studied the genetic control of resistance in WR of horseradish.

3.2.1 Inheritance of Disease Resistance Genes (R-Genes)

The exhaustive work of Williams and Pound (1963) clearly demonstrated that resistance to WR in radish (*Raphanus sativus* var. *Caudatus*) cultivar China Rose Winter (CRW) and Round Black Spanish (RBS) was controlled by a single dominant gene. Histological studies revealed that resistance in CRW is manifested as a hypersensitive reaction, which may be modified to a sporulating tolerant reaction by environmentally controlled minor genes. Humaydan and Williams (1976), while studying the inheritance of resistance in *R. sativus* var. *Caudatus* to *A. candida* race 1, changed the gene designation "R" into the more descriptive symbol "AC-1," derived from the initials and race number of *A. candida*. The resistance to *A. candida* race 1 in radish is controlled by a single dominant gene, AC-1. The resistance gene AC-1 and the gene Pi, controlling pink pigmentation, was found to be linked with a recombination value of 3.28%. Bonnet (1981) found that WR resistance in radish variety "Rubiso 2" is also controlled by one dominant gene. Among *Brassica* species, monogenic dominant resistance to *A. candida* race 2 has been found in *B. nigra, B. rapa, B. carinata,* and *B. juncea* (Ebrahimi et al. 1976; Delwiche and Williams 1976, 1977; Thukral and Singh 1986). A single dominant gene, AC-2, controlling resistance to *A. candida* race 2 in *B. nigra* was identified by Delwiche and Williams (1981). In a study determining quantitatively inherited resistance to *A. candida* race 2 in *B. rapa* cv. CGS-1, Edwards and Williams (1982, 1987) found that variability in reaction among susceptible *B. rapa* strain PHW-Aaa-1 was due to quantitative genetic regulation, and suggested that rapid progress in resistance breeding could be made via mass selection when starting with a susceptible base population.

Canadian cultivars of *B. napus* are resistant to *A. candida*, but many cultivars of this species grown in China are susceptible (Fan *et al.* 1983). The inheritance of WR resistance in *B. napus* cv. Regent is conditioned by independent dominant genes at three loci, designated as AC-7-1, AC-7-2, and AC-7-3. Resistance is conferred by dominance at any one of these loci, while plants with recessive alleles at all loci are susceptible. Results of a study by Verma and Bhowmik (1989) are in part agreement with those of Fan et al. (1983) and suggest that resistance of *B. napus* cv. BN-Sel to the *B. juncea* pathotype of *A. candida* in India is conditioned by dominant duplicate genes. In 1980, Canadian researcher at the Agriculture Canada Research Station in Saskatoon registered the world's first highly resistant *B. rapa* variety Tobin (canola) against *A. candida* race 7. In their inheritance of resistance studies, in crosses involving resistant (*B. rapa* accession from Maxico), and an adapted susceptible *B. rapa* cv. Torch (rapeseed), they showed that the reaction of the F_1 was like the resistant parent, indicating that resistance is dominant and controlled by nuclear genes. Backcrosses of F_1 plants to the resistant parent showed the same reactions as that of the resistant parent. Backcrosses of F_1 to the susceptible parent segregated in a 1:1 ratio of resistant and susceptible. The F_2 segregation of resistant and susceptible plants gave a good fit to 3:1 ratio. From the same Agriculture Canada Research Station, study of inheritance of resistance to *A. candida* race 2 in Indian mustard, Tiwari et al. (1988) also found that resistance is dominant, monogenic, controlled by nuclear genes and is easily transferred to adapted susceptible genotypes via backcrossing. In a study evaluating performance of fifteen advanced generations (F6) progenies of 2 interspecific crosses of *B. juncea* and *B. carinata* against white rust (WR), Singh et al. (1988) showed significant differences among the progenies, and that all the hybrid progenies gave resistant reaction; five interspecific crosses between *B. juncea* and *B. carinata* revealed that the dominant gene which conferred resistance to WR is located in "C" genome of *B. oleracea*, a progenitor of *B. carinata*. Williams and Hill (1986) and Edwards and Williams (1987) have opened unusual potential for resolving many problems relating to host–parasite interactions and breeding for disease resistance through development of rapid-cycling *Brassica* populations. Their preliminary studies demonstrated considerable isozyme variations among individuals in populations which, when inoculated with several pathogens, showed a wide range of plant-to-plant variation in the levels of resistance and susceptibility. This will assist plant breeders in developing cultivars with genetic resistance to plant diseases. These authors and Hill et al. (1988) constructed gene pools of both major and minor genes for resistance to various crucifer pathogens. The occurrence of systemic protection and changes detected in phenylalanine ammonia lyase and total soluble peroxidase activities in inoculated cotyledons, particularly after the induced inoculation, suggested that host-mediated factors may also be involved in the interaction between the two isolates of *A. candida* that were compatible (CO) and incompatible (IN) on a *B. juncea* accession (Singh et al. 1999). Currently, markers are available for the selection of *B. juncea* plants carrying the resistance gene (*Ac2A1*) to *A. candida* race 2A (Prabhu et al. 1998).

Singh and Singh (1987) studied that when *A. candida* resistant Ethiopian mustard (*B. carinata*) was crossed with *B. juncea*, the interspecific hybrids showed tolerance to *A. candida*. In a study on the inheritance of resistant to *A. candida* race 7 in *B. napus*, Liu et al. (1987) found that, in a digenic type, dominant resistance is conferred by R_1 and R_2 genes. Presence of a dominant allele at either of the two loci will confer resistance to a plant, whereas homozygous recessive at both loci will result in a susceptible phenotype expression. Liu and Rimmer (1992) suggested that resistance to an Ethiopian *A. candida* isolate from *B. carinata* was conditioned by a single dominant resistant gene.

Pal et al. (1991) evaluated the genetic component of variation for WR resistance through a 12 × 12 diallel crosses involving resistant and susceptible parents of both Indian and exotic-origin *B. juncea* under four sets of environments, viz. normal-sown in natural conditions, normal-sown in artificially created epiphytotic conditions, late-sown in natural conditions, and late-sown in artificially epiphytotic conditions. Based on these results, they suggested that both additive and non-additive components of variation are significant for WR resistance in all four sets of environments, but an overdominance under late-sown environment. Gadewadikar et al. (1993) in the study of inheritance of resistance to *A. candida* in crosses involving exotic and national promising varieties suggested that resistance was governed by a single dominant nuclear gene pair, and as such, resistance can easily be transferred via backcross breeding. Paladhi et al. (1993) while conducting the study on inheritance of field reaction to WR in Indian mustard concluded that gene PI-15, a resistance source in *B. juncea,* can easily be transferred to susceptible type via backcrossing as resistance was controlled by a single gene.

Bains (1993) reported that resistance in the leaves differed from that of the young flowers; resistance in the leaves was due to the CC genome transferred from sarson, which can be utilized for breeding purposes. Rao and Raut (1994) observed that the susceptibility of *B. juncea* cv. Varuna to the local Delhi pathotype of *A. candida* was conditioned by two genes, with dominant and recessive gene interaction. Interspecific crosses between *B. juncea* and *B. napus* suggested that resistance in genotypes WW 1507 and ISN 114 to *A. candida* was controlled by a single dominant gene (Jat 1999). In the study of three interspecific crosses between *B. juncea* and *B. napus,* Subudhi and Raut (1994) revealed digenic control with epistatic interaction, and a close association of parental species and different grades of leaf waxiness for WR resistance. In a diallel fashion, Sachan et al. (1995) crossed two WR resistant mustard cultivars Domo and Cutlass with two susceptible cultivars Kranti and Varuna and reported that F_1 hybrids, except susceptible × susceptible, were resistant. Segregation pattern for resistance to WR in F_2 and test crosses were under the control of a single dominant gene present in Domo and Cutlass, while a recessive gene for susceptibility was observed to be present in Kranti and Varuna. Liu et al. (1996) in Canada developed monogenic lines for resistance to *A. candida* from a Canadian *B. napus* cultivar Reagent and suggested that these monogenic lines can be used to study the mechanism of resistance response conditioned by the individual genes. These lines also facilitate molecular mapping of the loci in *B. napus* for resistance to *A. candida* race 7. In intervarietal crosses between susceptible Indian mustard cvs. Pusa Bold and resistant genotype DIRA 313, Mani et al. (1996) suggested that final intensity of rust on plant (FIP), final intensity of rust on leaf (FIL), and area under disease progress curves (AUDPC) showed significant additive × additive interaction (i) along with the association of complimentary epistatic interactions indicating close association between the nature of inheritance for AUDPC on one hand, and FIP and FIL on the other. This was also substantiated by a significant correlation of FIP and FIL with AUDPC suggesting ease in selection for lower AUDPC (slow rusting) through FIP or FIL. Sridhar and Raut (1998) reported presence of a monogenic complete dominance in four crosses, and absence of monogenic dominance in seven crosses between *B. juncea* and resistance sources from different *Brassica* species. According to Jat (1999), resistance was dominant in all the crosses except susceptible × susceptible where it was recessive. In intraspecific crosses, inheritance of resistance to *A. candida* was governed by one dominant gene, or two genes with either dominant-recessive epistatic interaction, or complete dominance at both gene pairs where either gene was dominant epistatic to other. The inoculation studies in F_2 population of crosses between R × R showed that resistant genes may be located on the same locus, or on different loci. In different intraspecific crosses of *B. juncea,* and interspecific crosses of *B. juncea* × *B. carinata, A. candida,* resistance was dominant in all the crosses; resistance was governed by one dominant gene, or two genes with either as dominant, recessive or epistatic interaction, or complete dominance at both gene pairs (Saharan and Krishnia 2001). Partial resistance in *B. napus* was controlled by a single recessive gene designated a *wpr* with a variable expression (Bansal et al. 2005). Dominant alleles at three unlinked loci (Ac 7h, AC 7z, and AC 7_3) conferred resistance in *B. napus* cv. Regent to *A. candida* race AC 7 (Fan et al. 1983; Liu et al. 1996). Two loci also controlled resistance in *B. napus* to *A. candida* race AC 2 collected from *B. juncea* (Verma and Bhowmik 1989). The Chinese *B. napus* accession 2282-9, which is susceptible to AC 7, has only one locus controlling resistance to *A. candida* isolate from *B. carinata* (Liu and Rimmer 1992). These studies indicated that only one allele for resistance was sufficient to condition an incompatible reaction in this pathosystem (Ferreira et al. 1995). In addition, a single locus controlling resistance to AC 2 in *B. napus* and *B. rapa* was mapped using restriction fragment length polymorphism (RFLP) marker (Ferreira et al. 1995). Three host genes for recognition of *A. candida* (*RAC*) each associated with different interaction phenotypes were identified through inheritance studies with three accessions of *Arabidopsis thaliana*. One of these genes at locus *RAC1* appeared to be completely dominant, whereas the other two genes were only partially dominant or recessive under certain conditions, possibly including the effect of genetic background. One of the latter two genes defined a second locus *RAC2*. *RAC1* was mapped to the top arm of chromosome 1 in the 1 cM interval between RFLP markers M254 and M253 (Holub et al. 1995; Fig. 3.1).

A dominant allele at a single locus or two tightly linked loci confer resistance to both *A. candida* races AC 2 and AC 7 (Kole et al. 2002). According to Borhan et al. (2008), a dominant WR resistant gene, *WRR* 4 encodes a TIR-NB-LRR protein that confers broad-spectrum resistance in *A. thaliana* to *A. candida* races AC 2, AC 4, AC 7, and AC 9 (Tables 3.1 and 3.2).

Genetic variation and the possibilities of increasing the level of WR resistance was evaluated in seven families of interspecific cross between *B. juncea* cv. Varuna × *B. carinata* cv PCC-2 (Krishnia et al. 2000). Results revealed that within family, variation in cross Varuna × PCC-2 on leaves were significant. The

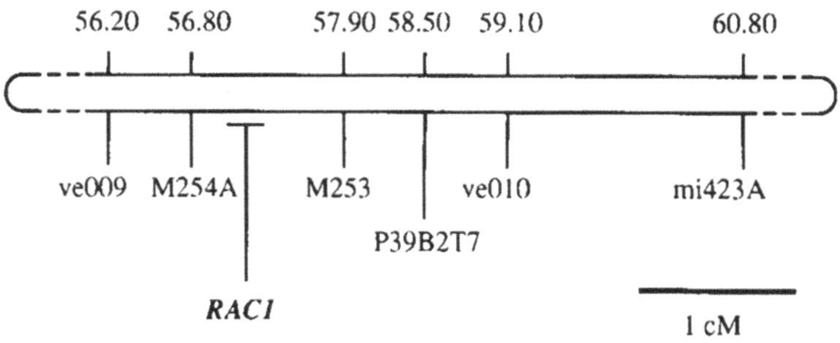

Fig. 3.1 Location of *RAC1* on the upper arm of chromosome 1 in *Arabidopsis thaliana* mapped relative to loci defined by restriction fragment length polymorphisms (Holub et al. 1995)

minimum disease severity index (DSI) was recorded in progenies of R × R (0.96) which was statistically at par with progenies of R-self (1.18), and significantly different from R-open (1.34). The maximum DSI was observed in progenies of S-open (3.51) which was statistically at par with progenies of S × S, and self families. The 18.64% reduction in disease score was observed in progenies of R × R over R-self. The minimum limit of disease score range was observed in R × R family, whereas it was maximum in S-self. The highest genotypic coefficient of variance (GCV) (0.210), h^2 heritability broad sense (75.31), and genetic gain per cent as mean (GG) (55.21) were recorded in progenies of R × S, R × R, and R-self, respectively; these indices were non-significant in the progenies of S × S. The progenies of all seven families, except R × R and S × S, showed significant variation for WR-SP; between families, variations were significant. The lowest and highest DSI were recorded in progenies of S-self and R × R, respectively. The GCV (0.012) and GG (12.63) were maximum in R-open, and minimum in progenies of R × R and S × S families (Table 3.3).

3.2.2 Inheritance of Partial Resistance

The partial resistance in *B. juncea* lines against race 2 is controlled by a single dominant gene that has variable expressions (Bansal et al. 1999). It is similar to that reported in *B. napus, B. rapa* and *R. sativus* (Williams and Pound 1963; Edwards and Williams 1987; Kole et al. 1996). Traditional and doubled haploid (DH) genotypes of oilseed Brassica spp. resistant, partially resistant, moderately susceptible, and susceptible to *A. candida* were compared for both phenotypic development and histology of host–pathogen interaction. The partially resistant genotype showed pinhead-size pustules, mainly on the upper surface of cotyledonary leaves. Relatively, less mycelium was observed in the partially resistant than in the susceptible genotypes. In resistant *B. napus* genotypes, there was neither pustule development nor any mycelial growth. In the moderately susceptible genotype, the pustules were similar to those in the partially resistant genotype except that the pinhead-size pustules were occasionally coalescing in the former. In the moderately

Table 3.1 Inheritance of resistance in crucifers to *Albugo candida* causing white rust (Saharan et al. 2014)

Crucifers genotypes	Resistance gene	Reference (s)
R. sativus cv. China Rose winter, round black Spanish	Single dominant	Pound and Williams (1963)
R. sativus cv. Caudatus	AC 1	Hu Humaydan and Williams (1976)
R. sativus cv. Rubiso 2		Bonnet (1981)
B. nigra	AC 2	De Delwiche and Williams (1981)
B. rapa		Delwiche and Williams (1974)
B. juncea		Ebrahimi et al. (1976)
B. carinata		Thukral and Singh (1986)
B. napus cv. Regent	Dominant geneAC-7-1, AC7-2, AC-7-3 to race AC-7	Fan et al. (1983), Liu et al. (1996)
B. napus cv. BN-Sel.	Dominant duplicate gene to race AC 2	Verma and Bhowmik (1989)
B. napus	Partial resistant single recessive gene *WPr*	Bansal et al. (2005)
B. napus	Single dominant gene	Liu and Rimmer (1992)
B. juncea cv. EC 12749	Additive dominant gene with epistatic effects	Thukral and Singh (1986)
B. juncea cv. Domo and cutlass	Single dominant gene	Sachan et al. (1995)
B. juncea cv. EC 333590	One/two dominant genes with epistatic effects	Jat (1999)
B. juncea × *B. carinata*	Monogenic dominant	Tiwari et al. (1988)
B. juncea × *B. carinata*	1/2 dominant gene dominant recessive epistatic or complete dominant at both gene pairs	Saharan and Krishnia (2001)
B. juncea × *B. napus* cv. ISN 733, WW 1507	Single dominant gene	Jat (1999)
B. juncea × *B. napus*	Digenic with epistatic interaction	Subudhi and Raut (1994)
A. thaliana	WRR4 gene confers broad-spectrum resistance to races AC 2, 4, 7 and 9	Borhan et al. (2008)

Table 3.2 Mode of segregation for *Albugo candida* reaction to *Brassica juncea* × *Brassica juncea* and *Brassica juncea* × *Brassica carinata* F_2 progenies, BC_1 and BC_2 (Saharan and Krishnia 2001)

Crosses	Host reaction		Total no. of plants	Expected ratio (R: S)	χ2	*P* value
	Resistant (R)	Susceptible (S)				
F_2						
PCR 3 × Shiva	195	75	270	3:1	1.1	0.2–0.3
Pusa Basant × Shiva	172	66	238	3:1	0.9	0.3–0.5
Rajat × Shiva	204	58	262	3:1	1.1	0.2–0.3
				13:3	2.0	0.1–0.2
Pusa Bahar × Domo	156	42	198	3:1	1.5	0.2–0.3
				13:3	0.8	0.3–0.5
Pusa Basant × EC 322092	174	64	240	3:1	0.4	0.5–0.7
Kranti × EC 322092	142	38	180	3:1	1.5	0.2–0.3
Varuna × EC 322092	181	53	234	3:1	0.7	0.3–0.5
				13:3	2.3	0.1–0.2
RH 30 × EC 322093	246	12	258	15:1	1.1	0.2–0.3
RH 30 × HC-1	186	48	234	3:1	2.5	0.1–0.2
				13:3	0.5	0.3–0.5
Varuna × PCC-2	154	58	212	3:1	0.6	0.3–0.5
BC_1 (F_1 × S)						
F_1 (PCR 3 × Shiva) × PCR 3	43	55	98	1:1	1.5	0.2–0.3
F_1 (Pusa Basant × Shiva) × Pusa Basant	45	36	81	1:1	1.0	0.3–0.5
F_1 (Rajat × Shiva) × Rajat	52	44	96	1:1	0.7	0.3–0.5
F_1 (Pusa Bahar × Domo) × Pusa Bahar	38	32	70	1:1	0.5	0.3–0.5
F_1 (Pusa Basant × EC 322092) × Pusa Basant	46	35	81	1:1	0.5	0.2–0.3
F_1 (Kranti × EC 322092) × Varuna	28	36	64	1:1	1.0	0.3–0.5
F_1 (RH 30 × EC 322093) × RH 30	55	15	70	3:1	0.5	0.3–0.5
F_1 (RH 30 × HC-1) × RH 30	34	43	77	1:1	1.1	0.3–0.5
F_1 (Varuna × PCC-2) × Varuna	32	36	68	1:1	0.2	0.5–0.7

BC2 (F1 × R) in all the 10 crosses no segregation; R = resistant; S = susceptible

Table 3.3 Disease score and selection parameters for WR infection in the progenies of seven families of interspecific cross between *Brassica juncea* cv. Varuna × *Brassica carinata* cv. PCC-2 (Krishnia et al. 2000)

Families	Disease score		CD	GCV	Selection parameter		
	Mean	Range			PCV	h^2	GG
White rust (leaf phase)							
R open	1.3	0.4–2.2	0.5	0.121	0.205	59.0	41.1
R × R	1.0	0.5–1.5	0.3	0.061	0.081	75.3	46.1
R self	1.2	0.3–1.8	0.4	0.138	0.191	72.3	55.2
R × S	2.2	1.4–3.0	0.5	0.210	0.283	74.2	37.7
S self	3.5	2.9–4.1	0.3	0.108	0.155	69.7	16.3
S × S	3.3	3.1–3.5	NS[a]	–	–	–	–
S open	3.5	2.8-4.1	0.3	0.084	0.114	73.7	14.6
Mean	2.3						
CD	0.3						
White rust (Staghead phase)							
R open	0.5 (1.2)	0.1–0.9 (1.1–1.4)	0.2	0.012	0.025	48.0	12.6
R × R	0.7(1.3)	0.6–0.8 (1.3–1.4)	NS[a]	–	–	–	–
R self	0.5 -(1.2)	0.2–0.7 (1.1–1.3)	0.2	0.010	0.022	45.5	11.4
R × S	0.5 (1.2)	0.3–0.7 (1.2–1.3)	0.2	0.009	0.018	50.0	11.3
S self	0.4 (1.2)	0.2–0.7 (1.1–1.3)	0.2	0.006	0.015	40.0	8.6
S × S	0.5 (1.2)	0.3–0.7 (1.13–1.30)	NS[a]	–	–	–	–
S open	0.5 (1.2)	0.3–0.8 (1.1–1.3)	0.2	0.007	0.016	43.8	9.4
Mean	0.5 (1.2)						
CD	0.1						

R = Resistant; S = Susceptible; GCV = Genotypic coefficient of variance; PCV = Phenotypic coefficient of variance; h^2 = Heritability (broad sense); GG = Genetic gain (per cent of mean); Figures in parenthesis are transformed values
[a]NS therefore no need to calculate selection parameters

susceptible genotype, ample mycelial growth in the mesophyll tissue was similar to that in the susceptible control *B. rapa* cv. Torch. In the resistant *B. juncea* cv. Commercial Brown, no pustules were formed although some mycelial growth was observed beneath the epidermal cell layer, and in the mesophyll cell layer of the cotyledonary leaf tissue. For inheritance studies, two partially resistant *B. napus* genotypes were crossed with a resistant *B. napus* genotype, and plants of F_1, F_1 (reciprocal), F_2, and DHs were inoculated with a zoospore suspension of *A. candida* race 7v (Bansal et al. 2005). Results showed that the partially resistant phenotype appeared to be controlled by a single recessive gene designated as *wpr* with variable expression (Table 3.4). The simple inheritance of partial resistance against WR can very easily be incorporated into elite breeding lines through conventional and DH breeding methods (Bansal et al. 2005).

Table 3.4 Segregation for resistance and susceptibility to *Albugo candida* race 7v in parents, F_1, F_2 and doubled haploid (DH) populations in *Brassica napus* (Bansal et al. 2005)

Pedigree	Generation and putative genotype	Observed reaction				Expected R: PR ratio	χ2 calculated	*P*-value
		R	PR	S	Total			
Parent A00-63N (PR)	P_1 *wpr wpr*	0	68	0	68	–	–	–
Parent A00-66N (PR)	P_2 *wpr wpr*	0	63	0	63	–	–	–
Parent A00-65N (R)	P_3 *Wpr wpr*	60	0	0	60	–	–	–
PR (P_1) × R (P_3)	F_1 *Wpr wpr*	61	0	0	61	–	–	–
R (P_3) × PR (P_1)	F_1(rec.)*Wpr wpr*	155	0	0	155	–	–	–
PR (P_2) × R (P_3)	F_1 *Wpr wpr*	58	0	0	58	–	–	–
R (P_3) × PR (P_2)	F_1(rec.)*Wpr wpr*	145	0	0	145	–	–	–
From F_1 of P_1 and P_3	F_2	108	36	0	144	3:1	0.00	1.00
From F_1 (rec.)	F_2	105	39	0	144	3:1	0.33	0.56
From F_1 of P_2 and P_3	F_2	109	35	0	144	3:1	0.04	0.85
From F_1 (rec.)	F_2	106	38	0	144	3:1	0.15	0.70
From F_1 of P_1 and P_3	DH	29	27	0	56	1:1	0.07	0.79
From F_1 (rec.)	DH	25	46	0	71	1:1	6.21	0.01

Rec. = reciprocal; R = resistant; PR = partial resistant; S = Susceptible; *wpr* = recessive partial resistant; *Wpr* = dominant resistant

3.3 *Brassica-Alternaria* Inheritance

3.3.1 Inheritance of Disease Resistance Genes (R-Genes)

Genetics of Alternaria blight resistance in inter- and intraspecific crosses of *B. juncea* and *B. carinata* have been determined by Krishnia et al. (2000). Population of six generations, viz. P_1, P_2, F_1, F_2, BC_1, and BC_2 of six crosses, viz. Pusa Basant × EC 322092, Kranti × EC-322092, Varuna × EC 322092, RH 30 × EC 322093, RH 30 × HC-1, and Varuna × PCC-2, were evaluated for the inheritance of resistance to *A. brassicae*. The F_1 hybrids showed intermediate disease intensity compared to their parents. The F_2 populations of all six crosses exhibited more susceptibility to *A. brassicae* than their respective F_1s. Performance of backcross progenies tends to deviate towards their respective recurrent parents. The χ^2 values in joint scaling test were highly significant in all the crosses indicating simple additive dominance model to be inadequate for all the crosses. The six-parameter model revealed significant additive gene effects in all six crosses, whereas dominance gene effects only in the three cross, viz. Varuna × EC 322092, RH 30 × EC 322092, and Varuna × PCC-2. However, there was preponderance of additive gene than dominance gene effects in all the six crosses. Non-allelic interaction and additive × additive (i) were significant in all the crosses except Pusa Basant × EC322092 and Kranti × EC 322092, whereas additive × dominance epistasis was significant only in Pusa Basant × EC 322092, Kranti × EC 322092, and RH 30 × HC-1 crosses. The crosses Kranti × EC 322092, Varuna × EC 322092, and RH 30 × EC 322093 exhibited significant dominance × dominance (I) type epistasis. Based upon these results, it is advocated that such breeding methods should be employed which exploit both kinds of gene effects. Therefore, reciprocal recurrent selection or diallel selective matings can be successfully used to develop genotypes resistant to Alternaria blight (Tables 3.5, 3.6, and 3.7; Saharan and Krishnia 2001).

The per cent disease intensity (PDI) of Alternaria blight at leaf stage is less in the progenies of R × R cross families as compared to R self and R open families in each cross indicating that inter-mating between resistant plants helps in increasing the level of resistance. A similar improvement has also been reported for multiple disease resistance in other crops where the frequency of favorable genes increases in the population enhancing the probability of obtaining multiple disease resistance (Saharan and Krishnia 2001). The utility of bi-parental techniques can be more pronounced if the additive or additive × additive types of genetic variations are coupled with repulsion phase linkage between genes (Singh et al. 1986). The selection of resistance to Alternaria blight in the inter-mated populations rather than in F_2, and F_3 populations had been advocated by Singh and Singh (1989), since the generations of intervarietal hybrids would obviate the harmful effects of linkages and linkage disequilibrium, and shuffle the desirable genes in one recombinant (Comstock and Robinson 1952; Matzinger and Cockerham 1963; Gates et al. 1957).

The reduction in Alternaria blight at leaf stage is less in four crosses, viz. PCR 3 × Shiva, Pusa Basant × Shiva, Rajat × Shiva, Pusa Bahar × Domo as compared to

Table 3.5 Per cent disease intensity (PDI) of Alternaria leaf blight (*Alternaria brassicae*) on different generations of *Brassica* crosses (Krishnia et al. 2000; Saharan and Krishnia 2001)

Cross	P_1	P_2	F_1	F_2	BC_1	BC_2
B. juncea × *B. juncea*						
Pusa Basant × EC-322092	43.42	5.75	23.55	26.65	34.35	17.52
	(41.21)	(13.81)	(29.06)	(31.05)	(35.91)	(24.73)
	±0.314	±0.184	±0.132	±0.088	±0.372	±0.190
Kranti × EC-322902	36.35	4.54	19.68	22.85	30.15	15.12
	(37.11)	(12.11)	(26.35)	(28.52)	(33.27)	(22.87)
	±0.290	±0.121	±0.143	±0.128	±0.218	±0.160
Varuna × EC 322092	46.95	6.32	28.35	33.33	38.85	16.98
	(43.72)	(14.54)	(32.20)	(35.24)	(38.59)	(24.27)
	±0.234	±0.211	±0.122	±0.086	±0.226	±0.134
RH 30 × EC-322093	38.25	5.58	20.75	22.26	34.12	16.45
	(38.23)	(13.69)	(27.06)	(28.39)	(35.73)	(24.04)
	±0.340	±0.290	±0.208	±0.090	±0.386	±0.342
B. juncea × *B. carinata*						
RH 30 × HC-1	41.41	6.58	27.35	30.45	36.35	20.75
	(40.05)	(15.00)	(31.56)	(33.52)	(37.11)	(27.06)
	±0.284	±0.202	±0.192	±0.100	±0.252	±0.222
Varuna × PCC-1	45.52	7.42	31.12	34.88	40.92	19.76
	(42.48)	(15.79)	(33.90)	(36.15)	(39.76)	(26.42)
	±0.384	±0.282	±0.204	±0.114	±0.306	±0.238

Figures in parentheses are angular transformed values

others indicating that the parents involved in these crosses have similar degree of susceptibility; only slight improvement is made through selection in segregating generations. In general, the progenies of R × S families record greater variation in disease score. The numerical reduction in Alternaria blight (leaf phase) disease is observed in the progenies of S × S families in comparison to S-self-plant progenies in each cross indicating that even susceptible plants possess resistant genes at different loci with minor effects.

The significant "j" type epistasis is observed in crosses Pusa Basant × EC 322092, Kranti × EC 322092, and RH 30 × HC-1, whereas "I" type epistasis is significant in crosses Kranti × EC 322092, Varuna × EC 322092, and RH 30 × EC 322093. The selection of resistant plants should be therefore done in advanced segregating generation rather than F_2. Therefore, the crosses possessing predominantly significant additive gene effect with additive × additive kinds of epistasis must be exploited to improve through simple selections. The highest GCV, PCV, h^2, and GG are recorded in progenies of R × S cross families in most of these crosses.

Table 3.6 Estimates of components of generation means on three-parameter model for *Alternaria brassicae* on different crosses of oilseed *Brassica* (Krishnia et al. 2000)

Cross	m	d	h	i	j	k
B. juncea × *B. juncea*						
Pusa Basant × EC-322092	31.05	11.18**	−1.37	−2.92	−2.52**	−5.22
	±0.30	±0.75	±1.97	±1.91	±0.83	±3.38
Kranti × EC-322902	28.52	10.40**	−0.06	−1.80	−2.10**	−8.56**
	±0.36	±0.61	±1.95	±1.69	±0.69	±3.01
Varuna × EC 322092	35.24	14.32**	−12.17**	−15.24**	−0.27	12.18**
	±0.29	±0.60	±1.75	±1.68	±0.69	±2.84
RH 30 × EC-322093	28.39	11.69**	7.08**	5.98**	−0.58	−19.48**
	±0.30	±0.85	±2.17	±2.09	±0.94	±3.81
B. juncea × *B. carinata*						
RH 30 × HC-1	33.52	10.05**	1.71	−5.74**	−2.48**	−4.43
	±0.32	±0.69	±1.95	±1.87	±0.77	±3.23
Varuna × PCC-1	36.15	13.34**	−7.48**	−12.24**	−0.01	5.95
	±0.34	±0.74	±2.09	±2.00	±0.84	±3.47

*, ** Significant at $P = 0.05$ and $P = 0.01$ level respectively
m = general mean, d = sum of additive effects of genes, h = sum of dominance effects of genes

Table 3.7 Estimates of gene effects under six generation mean analysis for *Alternaria brassicae* in different crosses of oilseed *Brassica* (Krishnia et al. 2000; Saharan and Krishnia 2001)

Cross	*m*	*d*	*h*	χ^2
B. juncea × *B. juncea*				
Pusa Basant × EC-322092	28.79	13.42**	1.81**	71.93**
	±0.31	±0.32	±0.50	
Kranti × EC-322902	25.68	12.39**	2.53**	86.42**
	±0.28	±0.28	±0.49	
Varuna × EC 322092	30.60	14.68**	3.21**	154.46**
	±0.28	±0.29	±0.48	
RH 30 × EC-322093	27.19	12.21**	1.51*	64.35**
	±0.34	±0.36	±1.51	
B. juncea × *B. carinata*				
RH 30 × HC-1	28.99	12.21**	5.04**	105.87**
	±0.31	±0.31	±0.55	
Varuna × PCC-1	30.87	13.57**	5.16**	105.26**
	±0.35	±0.36	±0.61	

*, **Significant at $P = 0.05$ and $P = 0.01$ level, m = general mean, d = sum of additive effects of genes, h = sum of dominance effects of genes, i = sum of additive × additive epistatic effects of genes, j = sum of additive × dominance epistatic effects of genes, k = sum of dominance × dominance epistatic effects of genes

The values of heritability and genetic gain are low in susceptible plant progenies as compared to resistant progenies. Almost a similar trend in all the crosses for Alternaria blight PDI at siliquae phase indicates a high-order correlation in Alternaria blight resistance at both leaf and siliquae phases (Saharan and Krishnia 2001).

3.3.2 Identification of Disease Tolerance Genotypes

Disease attributes in Indian mustard genotypes have been identified by Gupta et al. (2002). Disease stress tolerance index (DSTI) was found effective selection criteria for assessing the genotypes for their disease stress tolerance, and yield potential. The genotypes Rajat, Kranti, and RH-781 under normal sown, and Rajat, RL- 1359, and Kranti under late sown conditions, performed with uniform superiority under both non-disease-stress and disease-stress environments. Potential yield under controlled environment (Yp) was significantly and positively correlated with yield under disease-stress environment (Ys). Potential yield, mean productivity (MP), disease tolerance (TOL), geometric mean productivity (GMP) and disease stress tolerance index (DSTI) under normal sown and non-disease stress conditions (Yp) had

Table 3.8 Estimates of disease stress tolerance attributes from the potential yield and yield under disease stress environment (DSI = 0.195) in Indian mustard (*Brassica juncea*) under normal date of sowing (Gupta et al. 2002)

Genotypes	Yp	Ys	MP	TOL	DSSI	GMP	DSTI
EC-129126-1	2508	2036	2272	472	0.96	2260	0.87
RH-781	2635	2063	2349	572	1.11	2332	0.93
PHR-1	2034	1753	1894	281	0.71	1888	0.61
Varuna	2586	2087	2337	499	0.99	2323	0.92
Kranti	2738	2217	2478	521	0.98	2464	1.04
RH 8113	2318	1886	2102	432	0.96	2091	0.75
RL 1359	2680	2025	2353	655	1.25	2330	0.93
Rajat	2807	2234	2524	573	1.05	2504	1.07
RH 30	2769	1897	2333	872	1.61	2292	0.90
Pusa Bold	2463	1856	2160	607	1.26	2138	0.78
PR- 8805	2107	1826	1967	281	0.68	1961	0.66
RC- 781	1838	1781	1810	57	0.16	1757	0.56
ZEM -1	1973	1564	1769	409	1.06	1757	0.53
Shiva	2413	2027	2220	386	0.82	2212	0.84
Mean	2419	1947	2183	473	1.00	2169	1.00
CD ($P = 0.05$)	183.07	106.87	139.06	111.92	0.19	136.04	0.09

Significant at $P = 0.05$; Yp = yield under disease-controlled environment (kg/ha); Ys = Yield under artificially pathogen inoculated environment; DSI = Disease stress intensity; MP = Mean productivity; TOL = Disease tolerance; DSSI = Disease stress susceptibility index; GMP = Geometric mean productivity; DSTI = Disease stress tolerance index

significant positive association with yield under disease stress environment (Ys), under late sown conditions (Gupta et al. 2002) (Tables 3.8 and 3.9).

3.3.3 Identification of Components of Horizontal Resistance

While analyzing different components of horizontal resistance (HR) in *Brassica* genotypes against *A. brassicae,* Saharan and Kadian (1983) reported large differences in the number and size of lesions, latent period, rate of sporulation, and infection rate in *B. napus* cv. Tower and *B. juncea* cv. RC-781; *B. juncea* genotypes CSR 142 and CSR 448 showed moderate level of HR. The latent periods of Tower and RC-781, respectively, were 18 and 22 days compared to only 3 days in Prakash. The amount of conidial production was also considerably higher in Prakash (265 conidia per lesion) compared to Tower (92 conidia per lesion). Kolte (1987) suggested that size of lesions and amount of sporulation are also important factors in evaluating resistance against *A. brassicae*. In genotypes PR-8988 and PR-9024, *A. brassicae* produced significantly reduced number of spots (4.36–15.89), smaller size of spots (2.12–6.17 mm), lower sporulation ($0.30–1.84 \times 10^3$conidia), lower disease index (36.51–42.2%), reduced apparent infection rate ($r = 0.047–0.080$), and lesser values of AUDPC (45.35–126.70) on leaf and pod, respectively, along with reduced leaf defoliation (38.40–44.40%) in comparison to on national

Table 3.9 Estimates of disease stress tolerance attributes from the potential yield and yield under disease stress environment (DSI = 0.209) in Indian mustard (*Brassica juncea*) under late date of sowing (Gupta et al. 2002)

Genotypes	Yp	Ys	MP	TOL	DSSI	GMP	DSTI
EC-129126-1	1691	1149	1420	542	1.53	1393	0.70
RH-781	1656	1281	1469	375	1.08	1456	0.76
PHR-1	1576	1294	1435	282	0.86	1428	0.73
Varuna	1865	1483	1674	382	0.98	1663	0.99
Kranti	1768	1463	1616	305	0.83	1608	0.93
RH 8113	1625	1493	1559	132	0.39	1557	0.87
RL 1359	1865	1481	1673	384	0.98	1662	0.99
Rajat	1849	1160	1727	244	0.63	1723	1.07
RH 30	1796	1452	1624	344	0.92	1615	0.94
Pusa Bold	1699	1358	1529	341	0.96	1519	0.83
PR- 8805	1598	1005	1302	593	1.77	1267	0.58
RC- 781	1575	1201	1388	374	1.14	1375	0.68
ZEM -1	1440	1125	1283	315	1.05	1272	0.58
Shiva	1357	1094	1226	263	0.93	1218	0.53
Mean	1669	1320	1494	348	1.00	1482	0.80
CD (*P=0.05*)	88.97	105.34	91.62	66.40	0.19	93.72	0.10

Significant at $P = 0.05$; Yp = yield under disease-controlled environment (kg/ha); Ys = Yield under artificially pathogen inoculated environment; DSI = Disease stress intensity; MP = Mean productivity; TOL = Disease tolerance; DSSI = Disease stress susceptibility index; GMP = Geometric mean productivity; DSTI = Disease stress tolerance index

susceptible genotype Varuna (Tables 3.10 and 3.11). Therefore, these genotypes showed higher degree of partial resistance or slow blighting. The development and progression of Alternaria blight was slower in these two genotypes in comparison to the others. The yield (q ha^{-1}) of the genotype Kranti was highest (14.01) followed by genotype PR-9024 (11.14); yield day^{-1} was highest in genotype Divya (11.12 kg ha^{-1} day^{-1}) in comparison to the others (Kumar and Kolte 2001; Kumar and Singh 2006). Genotypes PR-8988 and PR-9024 exhibit slow blighting and have lowest number of lesions/10 cm^2 leaf area, reduced spot size, least number of conidia per spot, and lowest disease index on leaf and pod (Kumar and Kolte 2006).

3.4 *Brassica-Erysiphe* Inheritance

3.4.1 Inheritance of Disease Resistance Genes (R-Genes)

Inheritance of powdery mildew resistance in crucifers is the result of interaction of the genetic system of two distinct organisms, a host and powdery mildew pathogen. In cabbage, resistance to powdery mildew is controlled by a single dominant gene with modifiers (Walker and Williams 1965; Williams et al. 1968).

Table 3.10 Components of Alternaria blight disease resistance and yield of mustard (*Brassica juncea*) genotypes (Kumar and Kolte 2001)

Genotypes	No. of spots per 10 cm^2		Size of spot (mm)		Sporulation/ spots (No. of conidia × 10^3)		Disease Index (%)		Apparent infection rate (r)		AUDPC (mm^2)		Leaf defoliation (%)	Yield (q/ha)	Yield (kg ha^{-1} day^{-1})
	Leaf	Pod	Leaf	Pod	Leaf	Pod	Leaf	Pod	Leaf	Pod	Leaf	Pod			
Kranti	4.48	17.78	6.25	2.38	2.14	0.33	46.4	38.7	0.054	0.065	124.8	52.5	38.8	14.01	10.51
Varuna	5.10	25.50	7.71	5.76	3.04	0.63	53.9	52.5	0.063	0.115	153.0	58.5	57.6	10.05	8.21
PR-9650	5.56	23.47	7.74	3.27	3.48	0.63	56.1	53.8	0.065	0.140	155.5	65.0	64.8	7.73	6.64
Krishna	4.75	19.56	6.93	2.46	2.74	0.40	48.2	47.0	0.055	0.090	139.1	55.6	50.6	10.18	8.39
PR-8988	4.42	15.89	6.11	2.12	1.94	0.30	44.4	36.6	0.047	0.075	122.3	45.3	44.4	10.07	8.47
PR-9024	4.36	14.38	6.17	2.13	1.81	0.36	45.2	39.2	0.057	0.080	126.7	45.7	38.4	11.14	8.70
PR-8943	4.60	17.82	6.34	2.38	2.40	0.33	46.7	45.0	0.056	0.090	128.4	52.7	40.0	10.78	8.32
Divya	4.50	21.13	6.60	2.70	2.59	0.36	47.3	49.0	0.059	0.100	131.1	60.1	55.3	10.12	11.12
PR-9301	5.97	29.91	8.58	3.24	3.75	0.66	59.5	57.0	0.073	0.140	172.7	70.5	68.6	7.70	6.72
CD at 5%	0.284	1.685	0.506	0.223	0.404	0.001	2.934	1.839	NS	NS	NS	NS	3.354	1.710	1.270

Table 3.11 Correlation coefficients (R) among different components of Alternaria blight disease resistance and yield of mustard (*Brassica juncea*) genotypes (Kumar and Kolte 2001)

Disease components and yield	Size of spot	Disease Index	Leaf defoliation	Sporulation	AUDPC	Infection rate	Yield
No. of spots	0.883**	0.897**	0.812**	0.923**	0.956**	0.864**	–0.788*
Size of spots	–	0.985**	0.934**	0.982**	0.956**	0.893**	–0.668*
Disease index	–	–	0.905**	0.974**	0.949**	0.876**	–0.684*
Leaf defoliation	–	–	–	0.952**	0.884**	0.743*	–0.625 NS
Sporulation	–	–	–	–	0.973**	0.851**	–0.679*
AUDPC	–	–	–	–	–	0.903**	–0.790*
Infection rate	–	–	–	–	–	–	–0.770*

*Significant at 5%; **Significant at 1%; NS = Non-significant

The six generations of two interspecific crosses *B. juncea* × *B. carinata* (RH 30 × HC-1; Varuna × PCC 2) were evaluated to understand the nature of resistance against powdery mildew. The F_1 hybrids of both crosses RH 30 × HC-1 and Varuna × PCC 2 recorded 1.54 and 0.62 disease score, respectively, indicating the complete dominance of resistance (Table 3.12). F_2 population of both the crosses segregated in the ratio of 3R:1S to *E. cruciferarum*, indicating that the single resistance gene is present in HC-1 and PCC-2, and controlled the resistance with complete dominance. The backcross progenies with susceptible parents segregated in the ration of 1R:1S, well supporting the segregation pattern of F_2 population. The backcross progenies with resistant parents had no segregation (Table 3.13; Krishnia et al. 2000; Saharan and Krishnia 2001; Kumar et al. 2002).

3.4.2 Inheritance of Powdery Mildew Resistance in Progenies of *Brassica*

In all the crosses, the R × R plant progenies possessed comparatively less disease score for powdery mildew infection as compared to resistant self, and resistant open plant progenies, except PCR 3 × Shiva, Pusa Basant × Shiva, and Pusa Bahar × Domo. The level of susceptibility in progenies of S × S families as compared to S self families was less in all the crosses, except PCR 3 × Shiva, Pusa Basant × EC 322092, Varuna × EC322092, and RH 30 × EC 322093. The maximum GCV and PCV were observed in progenies of R × S for powdery mildew in most of the crosses while in other crosses the R × R family (Pusa Basant × Shiva), R open family (Pusa Basant × EC 322092, RH 30 × EC 322092), and S × S families (Rajat × Shiva; RH 30 × HC-1) had highest GCV and PCV. The h^2 was observed highest in R × S families of most of the crosses. However, the R × R family of Pusa Basant × Shiva, Pusa Bahar × Domo, and RH-30 × HC-1 recorded higher h^2. The genetic gain was recorded highest in progenies of R × S families of most of the crosses. However, in other crosses, the progenies of R × R (Pusa

Table 3.12 Reaction of *Brassica juncea* and *Brassica carinata* parents, and their F1 hybrids (*B. juncea* × *B. carinata*) to powdery mildew (*Erysiphe cruciferarum*) (Saharan and Krishnia 2001)

	Host Reaction				
Parents/F_1 hybrids	Resistant	Susceptible	Total no. of plants	DSI	DR
Parents					
RH 30	0	24	24	3.78	S
Varuna	0	28	28	3.25	S
HC-1	23	0	23	0.65	R
PCC 2	20	0	20	0.25	R
F_2 hybrids					
RH 30 × HC-1	18	0	18	1.54	R
Varuna × PCC 2	15	0	15	0.62	R

DSI = Disease severity index, DR = Disease reaction, R = Resistant, S = Susceptible

Table 3.13 Mode of segregation for *Erysiphe cruciferarum* reaction to *Brassica juncea* × *Brassica carinata* F_2 progenies, BC_1, and BC_2 (Saharan and Krishnia 2001)

Crosses	Host Reaction R	Host Reaction S	Total no. of plants	Expected ratio (R:S)	χ^2	*P* value
F_2						
RH 30 × HC 1	177	57	234	3:1	0.05	0.80–0.90
Varuna × PCC2	155	57	212	3:1	0.40	0.50–0.70
BC_1 (F1 × S)						
F_1 (RH 30 × HC 1) × RH 30	45	32	77	1:1	2.19	0.10–0.20
F_1 (Varuna × PCC2) × Varuna	38	30	68	1:1	0.94	0.30–0.50
BC_2 (F_1 × R)						
F_1 (RH 30 × HC 1) × HC 1	62	0	62		No segregation	
F_1(Varuna × PCC2) × PCC 2	58	0	58		No segregation	

Basant × Shiva, RH-30 × HC-1), R self families (Pusa Bahar × Domo; Pusa Basant × EC 322092), and R open families (RH 30 × EC 322093) recorded highest genetic gain. R × R families of each cross had the lowest disease score for powdery mildew; therefore through selective inter-mating of resistant plants in advance generations, the level of resistance can be increased. However, the degree or magnitude of increase will depend on parents involved in the crosses, and host–pathogen–environment interaction. Two crosses, i.e., RH-30 × HC-1 and Varuna × PCC 2, showed multiple resistances to all three diseases of rapeseed-mustard (white rust, Alternaria blight, and powdery mildew). Therefore, these parents are the best donors for developing the multiple disease resistance cultivars in oilseed *Brassicas*. The positive genetic correlations obtained here indicate that selection for resistance to a single pathogen will also bring increasing resistance to other pathogens (Krishnia et al. 2000; Saharan and Krishnia 2001; Kumar et al. 2002). Similar results were reported by Mitchell-Olds et al. (1995) in *B. rapa*.

3.4.3 Inheritance and Mapping of R-Genes in *Arabidopsis* to Powdery Mildew

In most of the *Arabidopsis* accessions resistance is either polygenic, based on the atypical resistance (R) gene [Resistance to powdery mildew 8 (RPW)] or and combinations thereof, with RPW 8 representing the major Qualitative Trait Locus (Adam and Somerville 1996; Xiao et al. 1997; Adam et al. 1999; Schiff et al. 2001; Wilson et al. 2001; Orgil et al. 2007; Gollner et al. 2008). In general, more accessions are resistant to *E. cruciferarum* than to *G. cichoracearum,* and *G. orontii* (Adam et al. 1999; Gollner et al. 2008). RPW 1, RPW2, RPW 4, and RPW 5 were described as semi-dominant R loci, and localized to chromosomes II, III, IV, and V, respectively (Fig. 3.2; Adam and Somerville 1996).

To determine the genetic basis of powdery mildew resistance in *Arabidopsis,* segregation analysis of progeny from crosses between each of the R–accessions, and

Columbia (gl1), which is S to powdery mildew pathogen, were performed by Adam and Somerville (1996). For all accessions except S1-0, resistance was conferred by a single locus. S1-0 was unique in that two unlinked loci controlled the disease reaction phenotype. In accessions Wa-1, Kas-1, Stw-0, and su-0, powdery mildew resistance was encoded by a semi-dominant allele. Mapping studies revealed that powdery mildew resistance in Kas -1, Wa-1, Te -0, Su-0, and Stw-0 were controlled by five independent loci (Fig. 3.2).

(a)

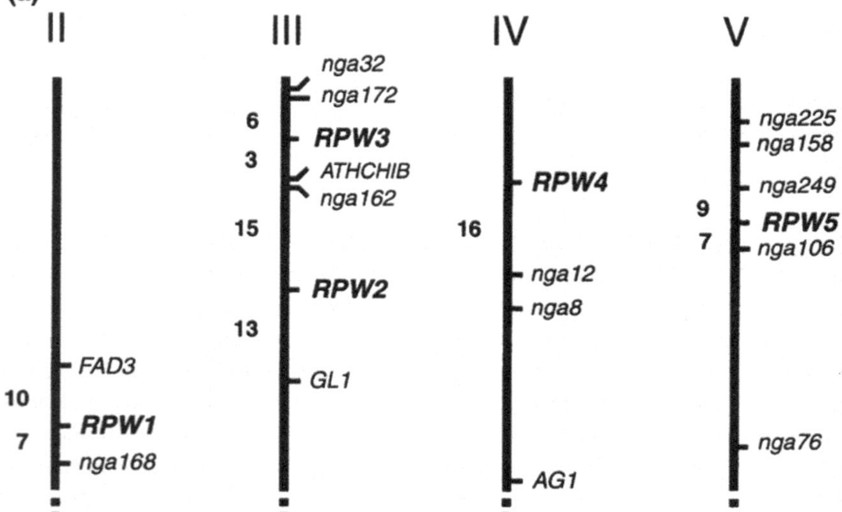

(b)

	I					II		III						
	nga59 1.9[a]	nga63 14.2	nga280 114.6	nga111 149.2	ATHATPase 153.6	FAD3 60	nga168 78.5	nga32 1.2	nga172 2.2	GAPC 3.8	nga126 13.1	ATHCHIB 19.4	nga162 21.2
RPW1	-[b]	U[c]	U	-	U	10[d]	7	U	-	-	-	-	U
RPW2	-	U	-	#[e]	-	-	U	U	-	U	-	19	17
RPW3	U	U	-	U	-	-	U	10	7	-	5	4	8
RPW4	U	-	U	-	U	-	U	-	U	-	-	-	U
RPW5	-	U	U	-	U	-	U	-	-	-	U	-	U

	III			IV				V						
	GL1 64.6	nga6 123.3	nga112 125.3	nga12 29.8	nga8 35.0	AG1 72.6	SC5 80	nga225 11.0	nga158 13.7	nga249 22.3	nga106 33.7	nga139 61.0	nga76 75.8	nga129 118.9
RPW1	U	U	-	-	U	-	U	U	-	U	-	U	-	U
RPW2	14	-	#	-	U	-	U	#	-	#	-	U	-	U
RPW3	U	-	U	-	U	-	U	U	-	-	-	U	-	U
RPW4	U	-	U	19	20	U	-	-	-	U	-	U	-	U
RPW5	U	-	U	-	-	-	U	13	14	9	7	22	U	U

[a] Position on map generated from recombinant inbred lines developed by Lister and Dean (version released on network June 29, 1995).
[b] -, linkage not determined
[c] U, unlinked
[d] Genetic distance in cM as determined by two-point linkage analysis.
[e] #, abnormal segregation ration ($P<0.01$)

Fig. 3.2 Map locations of five RPW loci. (**a**) Graphical representation of the map positions of the RPW loci. The most likely map order and the genetic distances (in cM) were derived via multipoint analysis of data from about 54 F_3 families for each cross. (**b**) Two point linkage data for the RPW loci, and various SSLP and CAPS markers. Genetic distances are given in cM (Adam and Somerville 1996)

Powdery mildew resistance in *Arabidopsis* accessions Kas-1, Wa 1, Te-0, Su-0, and Stw-0 to *E. cichoracearum* was controlled by five independent loci/genes (Table 3.14). In Kas-1, resistance to *E. cruciferarum* was conferred by RPW1, which mapped to chromosome 2, between nga 168 and FAD 3 (Fig. 3.2a). Resistance in Wa-1 (RPW2) and in Te-0 (RPW3) was encoded by loci on the same arm of chromosome 3. RPW 2 was placed between markers nga 162 and GL1, while RPW3 mapped to a 9 cM interval defined by nga 172 and ATHCHIB. Although SSLP marker nga 126 also maps in this interval, it is a dominant marker in the Col-5 by Te-0 cross and thus is of limited usefulness.

The resistance of *A. thaliana* to the powdery mildew pathogen *E. cruciferarum* UEA1 is conferred by two dominant alleles, RPW6 on chromosome 5 and RPW7 on chromosome 3. Neither RPW6 nor RPW7 map in the close vicinity of the other resistance genes that have been located on *A. thaliana* chromosomes (Kunkel 1996). RPW7 lies more than 10 cM below the bottom of a 25 cM interval on chromosome 3 containing four loci for resistance to *Peronospora parasitica*, and RPW6 maps more than 20 cM below RPW5 for resistance to *E. cichoracearum*, on chromosome 5 (Fig. 3.3).The resistance of Ms-0 to *E. cruciferarum* was associated with the early arrest of the pathogen, collapse of epidermal cells underlying appressoria, and occasionally with necrotic palisade mesophyll cells at the infection site. Resistance of Ms-0 to *E. cichoracearum* was consistently associated with localized necrosis visible to the naked eye. These host responses are comparable to the hypersensitive reaction to cereal powdery mildews (Jorgensen 1994), and therefore distinguish the resistance of Ms-0 to *E. cruciferarum* UEA1 and *E. cichoracearum*, from the resistance of other *A. thaliana* accessions to *E. cichoracearum*, which was not consistently associated with host cell collapse (Adam and Somerville 1996).

Papillae are formed in *A. thaliana* at the site of penetration pegs of *E. cichoracearum* UCSC1 (Adam and Somerville 1996) and *E. cruciferarum* (Hall 1994), but could not be associated with the expression of resistance. Xiao et al. (1997) have defined resistant plants as those which do not support vegetative reproduction of the powdery mildew pathogen. Thus, RPW6 and RPW7 were both required for disease resistance to *E. cruciferarum* UEAI. However, plants with dominant alleles at one of these loci, and recessive alleles at the other, exhibited a weak resistance reaction characterized by reduced fungal growth, and conidiospores production, which were referred to as an intermediate reaction. Other features of the intermediate reaction were its dependence upon high humidity, and the extensive collapse of host epidermal cells (Plate 3.1d), and necrosis of mesophyll cells, in infected tissues (Plate 3.1d). These characteristics bear some similarity to the phenomenon of partial resistance, which is well known in barley (Jorgensen 1994). Some forms of partial resistance are associated with delayed necrosis (Carver 1986), and the pathogen grows, and sporulate in the infected tissues. Like the intermediate resistance reported in *A. thaliana*, partial resistance is strongly influenced by the environment, and is notoriously difficult to score, usually requiring laborious laboratory methods (Newton 1993), and individual genes have not been traced in barley. However, unlike intermediate resistance conferred by either RPW6 or RPW7, partial resistance in barley is thought to be controlled by many genes

Table 3.14 F_2 segregation and test cross data illustrating the inheritance of powdery mildew resistance in six *Arabidopsis* accessions (Adam and Somerville 1996)

Accessions (Locus)	DR score[c]		F_2 Segregation Analysis[a]										Test cross[b]	
			# F_2 plants in each DR group								Ratio[d]		Ratio	
	Parent	F_1	3.1	3.0	2.1	2.0	1.1	1.0	0.1	0.0	(R:1:S)	χ^2	(R:S)	χ^2
Kas -1 (RPW1)	0,0–0,1	1,0–2,1	2	52	10	25	38	41	37	9	46:114:54 (1:2:1)	1.51 ($0.50 < p < 0.25$)	19. 14 (1:1)	0.76 ($0.50 < p < 0.25$)
Wa-1 (RPW2)	0,1	1,0–2,1	19	118	94	34	113	22	125	3	128: 263:137 (1:2:1)	0.31 ($0.90 < p < 0.75$)	34.29 (1:1)	0.40 ($p = 0.50$)
Te-0 (RPW3)	0,0–1,1	2,0–3,1	12	145	7	49	29	11	17	4	61:213 (1:3)	1.10 ($0.75 < p < 0.50$)	17.16 (1:1)	0.03 ($p = 0.90$)
Su-0 (RPW4)	0,0–1,1	2,0–2,1	10	136	108	158	46	31	40	30	147: 266:146 (1:2:1)	1.31 ($p = 0.50$)	18.17 (1:1)	0.03 ($p = 0.90$)
Stw-0 (RPW5)	0,1	1,0–2,1	11	62	45	12	73	9	81	0	81:139:73 (1:2:1)	1.20 ($0.75 < p < 0.50$)	27.20 (1:1)	1.04 ($p = 0.75$)
SI-0	0,0	3,0–3,1	46	210	48	61	39	12	64	31	31:480 (1:15)	0.03 ($0.90 < p < 0.75$)	20.57 (1:3)	0.04 ($0.90 < p < 0.75$)

[a]Data from F_2 of crosses between the susceptible accession Col-5, and each resistant accession

[b]Data from test crosses between the F_1 Col-5 × resistant accession, and the recessive parent

[c]Criteria for assessing the disease reaction phenotypes are given in the experimental procedures

[d]For F_2 progeny from crosses with Kas-1, Wa-1, Su-0, and Stw-0, plants were placed into groups consisting of those with phenotypes like the resistant parent (R), those with fully susceptible phenotypes like Col-5 (DR 3,0 or 3,1) (S), and those with intermediate phenotypes distinct from either parent (I). F_2 progeny of crosses with Te-0 and Sl-0 were grouped into two groups with one group consisting of progeny that resembled the resistant parent (R) and the second group (S) consisting of all other progeny. See text for a complete description. Resistances in Stw-0 and Su-0 were conferred by RPW4 (chromosome 4) and RPW 5 (chromosome 5), respectively. RPW4 was located about 16 cM telomeric from marker nga 12. No polymorphic CAPS or SSLP marker telomeric to RPW 4 was found for the Col-5 by Su-0 cross. However, the absence of linkage to AG-1 and slightly closer linkage to nga 12 than nga 8 place RPW 4 to the telomeric side rather than the centromeric side of nga 12. Multipoint linkage analysis placed RPW5 to the nga 249-nga 106 interval on chromosome 5 (Adam and Somerville 1996)

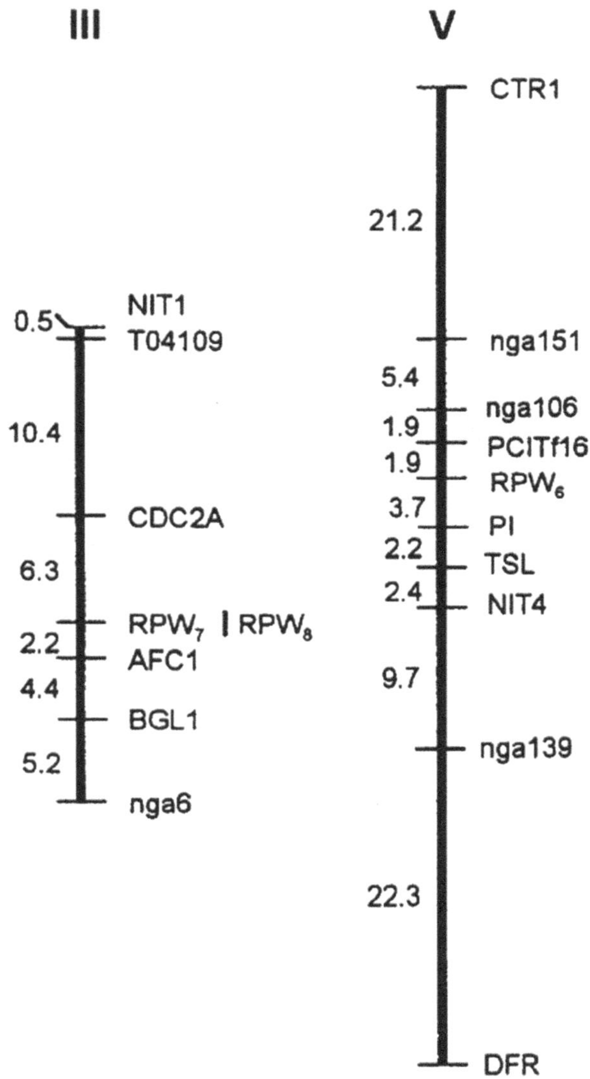

Fig. 3.3 Genetic map positions of powdery mildew resistance loci RPW6, RPWT, and RPW8, from *A. thaliana* accession Ms-0. RPW6 and RPW7 were both required for resistance of *A. thaliana* to *E. cruciferarum*, and were mapped simultaneously on 224 F 3 families. RPW8 conferred resistance to *E. cichoracearum* UCSC1, and was mapped on 81 F_3 families from a cross between accessions La-*er* (susceptible) × Ms-0 (resistant). Horizontal bars indicate position of locus; numbers are centimorgans (Xiao et al. 1997)

(Geiger and Heun 1989). RPW6 and RPW7 actin dependently as weak resistance genes, but in combination, they cooperate synergistically to elicit effective disease resistance against *E. cruciferarum* UEA1, apparently involving a rapid hypersensitive reaction. The genetics of resistance of *A. thaliana* to *E. cruciferarum* might therefore be similar to the resistance of mungbean to *Erysiphe polygoni* DC, which segregates as two unlinked genes with additive effects (Reddy et al. 1994). However, the independent action of the mungbean powdery mildew resistance loci was indicated by a distinctive resistance reaction which could be scored in F_2 populations. Resistance of *A. thaliana* accession Ms-0 to *E. Cichoracearum*

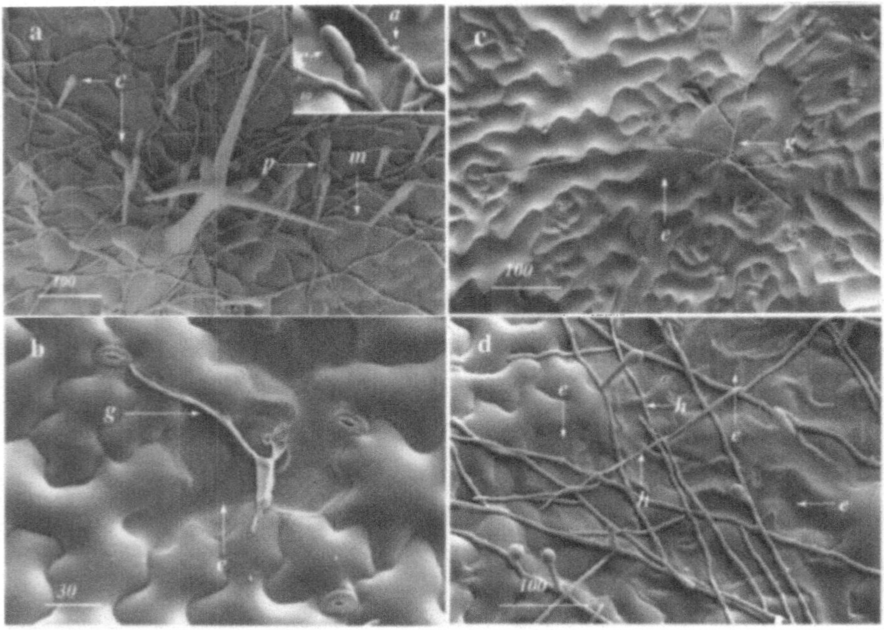

Plate 3.1 Scanning electro-micrographs of *Erysiphe cruciferarum* on *Arabidopsis thaliana* leaves, 8 days after inoculation. (**a**) Leaves of the susceptible *A. thaliana* accession, La-*er*, supported an extensive mycelial (m) network bearing conidiophores (p), and conidia (c), shown here surrounding a branched leaf trichome. Insert: Lobed appressoria (a) attached to epidermal cells of La-*er* leaves, and large terminal conidium (c) was characteristic features of *E. cruciferarum*. (**b**) Resistance of *A. thaliana* accession Ms-0 was associated with collapse of germ tubes (g), shown here arising from a conidium. The underlying host epidermal cells (e) have also collapsed. (**c**) Resistance of F_1 plants from a cross La-*err* × Ms-0 was associated with partial collapse of hyphal germ tubes (g) from germinating conidium, and collapse of several host epidermal cells (e) around the germinated conidium. (**d**) The intermediate reaction on plants from a F_3 line was characterized by a mycelial network, with conidiophores, and conidia, similar to the susceptible reaction shown in plate (a). However, in the intermediate reaction many host epidermal cells underlying hyphae had collapsed (e). Some hyphae had also collapsed (h). Bars represent indicated length in 1 μm (Xiao et al. 1997).

UCSC1 was controlled by a single locus, RPW8, which mapped to the same position as RPW7. Possibly, RPW7 and RPW8 are linked genes in a complex resistance locus similar to Mlo in barley (Jorgensen 1994) and Pm3 in wheat (Zeiler et al. 1993). Alternatively, RPW7 and RPW8 may define a single resistance gene, which recognizes products of both *E. cruciferarum* UEA1 and *E. cichoracearum* UCSC1. A precedent for a dual-specificity resistance gene is RPM1, which interacts with two avirulence genes from different bacterial pathogens of *A. thaliana* (Bisgrove et al. 1994; Grant et al. 1995).

3.4.4 Inheritance of Enhanced *R*-Genes in *Arabidopsis*

Genetic analysis of edr1 mutant was made by Frye and Innes (1998) to determine the inheritance of the enhanced resistance phenotype. The *edr1* mutant was crossed with *Arabidopsis* ecotype Landsberg *erecta* (L*err*), which is susceptible to *E. cichoracearum*. The F_2 progeny were inoculated with *E. cichoracearum* conidia, and scored 7 to 9 days later for development of necrotic lesions and lack of visible powdery mildew. These two traits co-segregated and behaved as a recessive mutation, producing approximately 1:3 ratios of resistant-to-susceptible plants (85:266; 250.115; P.0.1). To obtain a chromosomal map position for the mutation in *edr1* plants, a total of 1223 F_2 plants from the L*er* cross were scored for *E. cichoracearum* resistance, and 235 plants displaying resistance to *E. cichoracearum* were selected for mapping. DNA was isolated from the resistant F_2 plants and analyzed for linkage to simple sequence length polymorphism (SSLP) and co-dominant amplified polymorphic sequence (CAPS) markers (Konieczny and Ausubel 1993; Bell and Ecker 1994). The *edr1* mutation mapped 3.2 centimorgans centromeric from the SSLP marker *ATEAT1* (15 recombinant chromosomes), and 0.85 centimorgans telomeric from the CAPS marker *NCC1* (four recombinant chromosomes) on chromosome 1. Multiple defenses are induced more rapidly in edr1 plants than in wild-type plants when infected with a virulent strain of powdery mildew.

3.4.5 Inheritance and Mapping of *R*-Genes in *Arabidopsis* Mutants to Powdery Mildew

From a screen of approximately 26,000 M2 *Arabidopsis* plants, 32 confirmed powdery mildew-resistant mutants that did not exhibit spontaneous macroscopic lesions or constitutively express *PR1* were identified. Twelve of these mutants had obvious lesions after inoculation, reminiscent of *edr1*, and were set aside. The remaining 20 mutants showed no macroscopic lesions after inoculation, and they were crossed for inheritance pattern. Crosses between the 20 mutants defined 4 complementation groups, designated powdery mildew resistant 1–4(*pmr1–4*) (Plate 3.2). Multiple, independent alleles were identified for *pmr1* (7), *pmr2* (6), and *pmr4* (2). No obvious phenotypic differences were noted among the alleles at each locus. The number of alleles is less than 20 because multiple mutants from the same lot of mutagenized seed were counted as one allele. The mutants were backcrossed to the wild type, Col, two times before characterization, excluding mapping. Segregation of powdery mildew resistance in the F_1 and F_2 generations indicated that all mutations are recessive, and that *pmr2*, *pmr3*, and *pmr4* segregate in a simple 3:1 Mendelian fashion (Table 3.15). By contrast, all seven alleles of *pmr1* had a more complex segregation pattern (Table 3.15). Reciprocal crosses between plants heterozygous and homozygous for *pmr1* indicated that transmission of *pmr1* through the pollen is reduced (Table 3.15). To further characterize the decreased transmission of *pmr1*, a pollen competition experiment was conducted. Pollen from heterozygous plants was placed on stigmas of homozygous *pmr1* plants. When the

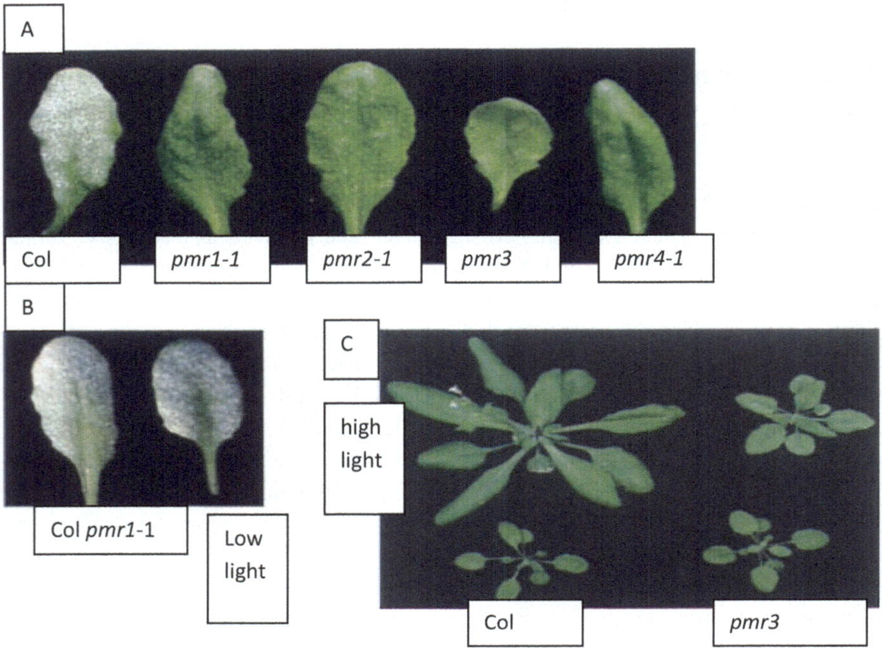

Plate 3.2 Phenotype of *pmr* mutants. (**a**) Plants were inoculated with *Erysiphe cichoracearum* 8 days before being photographed. Note the extensive fungal growth on Col. (**b**) Plants were inoculated with *Erysiphe orontii* 10 days before being photographed. (**c**) Effect of light levels on *pmr3*. Twenty-two-day-old plants grown under high, 150 mEym^2 per sec, or low, 45 mEym^2 per sec, light conditions (Vogel and Somerville 2000)

Table 3.15 Genetic analysis of powdery mildew-resistant mutants (Vogel and Somerville 2000)

Cross (female × male)	Type	Total	Disease response: Susceptible	Disease response: Resistant	χ^2
PMR1/PMR1 Xpmr1/pmr1	F_1	28	28	0	
PMR1/PMR1 Xpmr1/pmr1	F_2	1841	1742	99	378[a]; $p < 0.05$
PMR1/pmr1 Xpmr1/pmr1	Testcross	124	66	58	0.47[b]; $p < 0.05$
PMR1/pmr1 XPMR1/pmr1	Testcross	199	182	17	137[b]; $p < 0.05$
PMR2/PMR2 Xpmr2/pmr2	F_1	71	71	0	
PMR2/PMR2 Xpmr2/pmr2	F_2	193	140	53	0.43[a]; $p < 0.05$
PMR3/PMR3 Xpmr3/pmr3	F_1	11	11	0	
PMR3/PMR3 Xpmr3/pmr3	F_2	98	71	27	0.56[a]; $p < 0.05$
PMR4/PMR4 Xpmr4/pmr4	F_1	22	22	0	
PMR4/PMR4 Xpmr4/pmr4	F_2	443	340	103	0.4[a]; $p < 0.05$

[a]χ^2 = calculated for an expected 3:1, wild type to mutant ratio
[b]χ^2 = calculated for an expected 1:1, wild type to mutant ratio

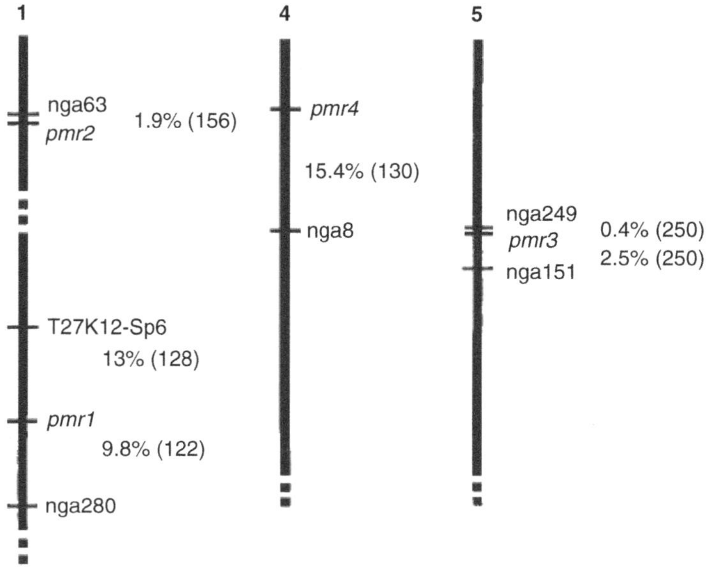

Fig. 3.4 Map positions of *pmr* mutants. Vertical bars represent chromosomes. Percent recombination between markers and the *pmr* mutants is indicated. The number of chromosomes scored is in parentheses (Vogel and Somerville 2000)

siliques were fully ripe, but not yet brittle, they were cut into three sections: top, middle, and bottom. The segregation of powdery mildew resistance among the progeny, and, by inference, the genotype of the pollen fertilizing the ovules derived from the different sections was determined. Plants originating from the top sections segregated in a 2.9:1 ratio (55 susceptible; 19 resistant), plants from middle sections segregated in a 9.3:1 ratio (93 susceptible; 10 resistant), and plants from the bottom sections segregated 52:1 (103 susceptible; 2 resistant).

Thus, *pmr1* pollen fertilizes ovules close to the stigmatic surface more efficiently than distal ovules. Slower pollen tube growth or an inability of the pollen tubes to sense and grow towards unfertilized ovules could explain these results. It is important to note that homozygous *pmr1* plants are fertile, indicating that *pmr1* pollen can fertilize all ovules; it is just less efficient than wild-type pollen. *pmr1* and *pmr2* mapped to chromosome 1, and *pmr4* mapped to chromosome 4 (Fig. 3.4). *pmr3* mapped to chromosome 5, close to the position of the dwarf mutant *lu-1* (lutescens). Because the description of the *lu-1* phenotype was similar to *pmr3*, the mutants were crossed. The F_1 progeny were not dwarf, indicating that *pmr3* and *lu-1* complement one another and presumably affect different genes (Vogel and Somerville 2000). In addition to powdery mildew resistance, *pmr3* plants are much smaller and paler than wild type at light intensities 150 mEym^{2}per sec. However, at light intensities 75 mEym2 per sec, *pmr3* plants are nearly wild type in stature and are susceptible to powdery mildew. A careful analysis reveals that the size of *pmr3* plants under high- and low-light conditions is similar, but wild-type plants are much larger under high light (Plate 3.2c).

The dwarf phenotype co-segregated 100% with powdery mildew resistance through two backcrosses, and among 598 F_2 plants. In addition, the fact that both the dwarf phenotype and the resistance phenotype are dependent on light intensity strongly suggests that the same mutation is responsible for both phenotypes. Leaves from *pmr4* plants are epinastic, especially when grown under short-day conditions. Both alleles of *pmr4* display this phenotype, indicating that the same mutations are responsible for both the powdery mildew resistance and the epinasty. Adult *pmr1* and *pmr2* plants are indistinguishable from wild type.

To determine whether fungal growth on the mutants was blocked at a defined step in the infection cycle, the growth of hyphae and level of conidiation were measured over time. The same general trend was observed for all mutants. Conidia germinate and begin growth in a normal fashion, but colonies on mutant plants grow more slowly than colonies on wild-type plants. By 3 dpi, hyphal length was significantly (ANOVA, *P*, 0.05) less on all the mutants than on wild-type plants (Fig. 3.5a). Eventually, most colonies on the mutants appeared to be dead or dying and produced few, if any, conidiophores, the aerial stalks bearing asexual conidia. In contrast, colonies on wild-type plants appeared healthy and produced many conidiophores (Fig.3.5b). However, on all mutants a small subset of colonies does produce some conidiophores, indicating that none of the mutations result in a complete block at any stage of fungal development (Vogel and Somerville 2000).

3.5 *Brassica-Fusarium* Inheritance

3.5.1 Inheritance and Mapping of R-Genes of *Brassica*

Fusarium wilt, caused by the fungus *Fusarium oxysporum* f. sp. *conglutinans* (Foc), is posing a threat to *Brassica* production worldwide, especially for cole crops. Fusarium wilt was first observed on cabbage by Smith (1899) in the United States in 1895. Since 1910, Fusarium wilt has spread quickly from the United States to almost the whole world (Liu et al. 2019). So far, two Foc races have been reported, but only race 1 is found worldwide. Fusarium wilt is a soil-borne disease and Foc can survive for more than 10 years, even without a host. Type A resistance conferred by a dominant monogene has been identified and applied successfully. However, race 2 can overcome type A resistance, which indicates that single resistance application is at high risk (Lv et al. 2020).

The mapped candidate gene *FOC1* is repredicted as *Bol037156*, which encodes a TIR-NBS-LRR, using an enlarged population. Shimizu et al. (2015) also mapped the resistance locus *FocBo1* by using 139 recombinant F2 plants and identified a candidate gene, *Bra012688*. The two mapped candidates are homologous with high identity. However, the functions of these genes remain to be identified. Type A resistance to Foc race 1 conferred by a dominant single gene, *FOC1*, has been successfully mapped and molecular markers have been developed and applied to

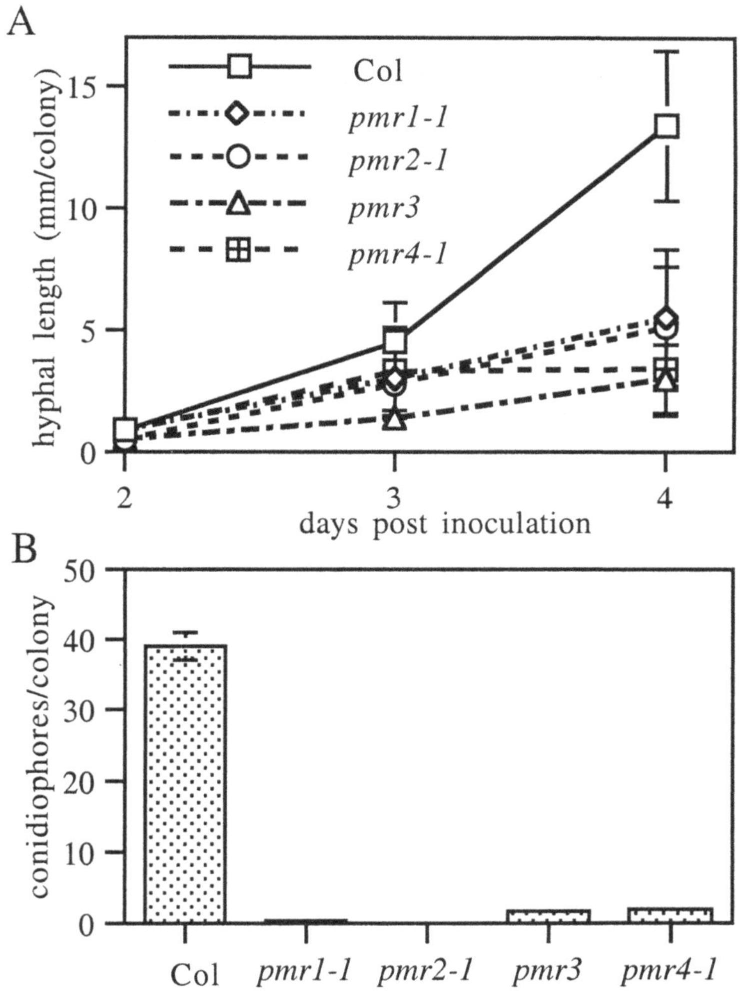

Fig. 3.5 Quantification of *Erysiphe cichoracearum* growth on *pmr* mutants. (**a**) Hyphal length per colony. (**b**) Conidiophores per colony at 6 dpi. The means 6 SD based on at least 15 replicates are plotted in *A* and *B*. The entire experiment was repeated once with similar results (Vogel and Somerville 2000)

generate various resistant cultivars. In addition, MAS using these markers has been combined with other breeding methods to promote the breeding process. Lv et al. (2014) reported the use of isolated microspore cultures with MAS to rapidly obtain target DH lines with FW resistance, which could be used directly in resistance breeding, thereby shortening the breeding period by 2–3 years (Lv et al. 2020).

3.6 *Brassica-Hyaloperonospora* Inheritance

3.6.1 Inheritance of Disease Resistance Genes (R-Genes)

Resistance derived from the Broccoli Introduction No. PI 189028 to *H. parasitica* race 1 is found to be governed by one dominant gene. The distribution of resistant plants in populations segregating for both downy mildew resistance and wax less foliage indicates that resistance is independent of foliage wax. Resistance to race 1 and race 2 obtained from the cabbage introduction PI 245015 is found to be inherited independently. Resistance is governed by one dominant gene for each race (Natti et al. 1967). However, in a later study, Hoser-Krauze et al. (1991) found that in broccoli, (*B. oleracea* var. *botrytis*) resistance to a Polish isolate of *H. parasitica* at the 4–5 leaf stage is determined by a single recessive gene different from genes determining resistance at the cotyledon stage. Subsequently they (Hoser-Krauze et al. 1995) found that in broccoli resistance to downy mildew is governed by 3 or 4 dominant complimentary genes. Both seedling and mature plant resistance have been reported in *B. oleracea* with the latter being quantitative (Dickson and Petzoldt 1996). Caravalho and Monterio (1996) crossed "R" *B. oleracea* cv. Algarvia with "S" cv. Penca de chaves and observed that the resistance to downy mildew is dominant and controlled by two independent loci with Algarvia being heterozygous at both loci (R1 r1, R2 r2) and Penca de chaves being homozygous recessive r1 r1 r2 r2 . Resistance to downy mildew in Broccoli and cauliflower is controlled by a single dominant gene. No susceptibility was observed among F_2 seedlings derived from intercrossing the R lines indicating that they all are the same or closely linked broad-spectrum R-genes (Vicente et al. 2012).

In Chinese cabbage resistance to downy mildew at the cotyledon stage expressed as a reduction in the sporulation capacity of *H. parasitica* was found to be under dominant monogenic control (Niu et al. 1983). However, Yuen (1991) while analyzing Chinese cabbage lines with a reduced rate of mildew development found additive effects with involvement of several resistant genes. The resistance to *H. parasitica* in radish cultivars Tokinaski (All Season) and Okura was found to be controlled by two dominant and independent genes (Bonnet and Blancard 1987). Cytoplasmic male sterile (*B. campestris*) breeding lines with resistance to downy mildew have been identified by Leung and Williams (1983). Downy mildew resistance was expressed in cotyledons as a reduction in the sporulation capacity of *H. parasitica*. In high and partially resistant hosts, spore production ranged from 7 to 30 spores/g host tissue as compared to 260 spores/g host tissue in the fully susceptible hosts. In the chinensis lines, 80% of the plants showed high to partial resistance to *H. parasitica* whereas in the other lines resistance ranged between 10% and 50%.

Differential host resistance to homologous isolates of *H. parasitica* has been identified in *B. rapa* (*B. campestris*), *B. napus*, *B. juncea,* and *B. oleracea*. In *B. napus*, resistance in the oilseed rape cultivar Cresor is controlled by a single dominant allele (Lucas et al. 1988). In *B. oleracea*, differential resistance has been located in a land race cauliflower "Palermo Green." A model based on two or possible three major genes has been proposed by Moss et al. (1988) to account for

Table 3.16 Inheritance of resistance in cauliflower to *Hyaloperonospora parasitica* using Palermo green model (Moss et al. 1988)

	Virulence		Resistance phenotype				
			–	R1	–	R1	Approx. resistant seedlings
H. parasitica isolates	A1	A2	–	–	R2	R2	observed
P005	1	2	+	–	-	–	95
P015	1	–	+	-	+	–	80
P018	–	2	+	+	–	–	80
P006	–	–	+	+	+	+	0
% phenotype in seed stock			5	15	15	65	

A = avirulence gene, R = Resistance gene, + + Susceptibility, – = Resistance

Table 3.17 Response of groups A, B, C, D, and E of *Brassica juncea* accessions and of one accession of *Brassica napus* at the cotyledon stage to infection with four isolates of *Hyaloperonospora parasitica* (Nashaat and Awasthi 1995)

		Disease Index		
		Isolates of *H. parasitica*		
Brassica species	Host category[a]	IP01/IP02	P003	R1
B. juncea	A (4)	1	1	1
	B (2)	2–3	1	1
	C (19)	6–8	1	1
	D (5)	6–8	2–3	1
	E (1)	6–7	5–6	1
B. napus	Ariana	1	7–8	7–8

[a]Number of accessions of each group in parentheses

the reaction patterns of individual plants to select fungal isolates within a host population (Table 3.16). In *B. campestris*, both rapid cycling and commercial genotypes have been identified for differential resistance (Table 3.16). Four homologous isolates have identified a range of differential responses. According to Nashaat et al. (1995a, b, 1996) resistance of the RES-01-1-4 and RES-26 lines of *B. napus* to isolate P003 of *H. parasitica* is conditioned by a single dominant resistant gene, whereas resistance of RES-02 is conditioned by two independent dominant resistance genes. Later, Nashaat and Awasthi (1995) selected differential putative homozygous resistant lines from seedling populations of accessions that exhibited a heterogeneous reaction to the isolates from *B. juncea* (Tables 3.17 and 3.18).

In inoculation tests on *Arabidopsis thaliana* with seven pathogen isolates and eleven host accessions, a range of interaction phenotypes were observed including localized necrosis (flecking), more extensive cell collapse (pitting), delayed sporulation, or complete susceptibility (Holub et al. 1994). Segregation for these phenotypes among F_2 progeny from a half-diallel cross between nine

Table 3.18 Examples of a successful selection for putative homozygous resistance response to *Hyaloperonospora parasitica* from a heterogeneous starting population of *Brassica juncea* at the cotyledon stage (Nashaat and Awasthi 1995)

	Isolates of *H. parasitica*			
	Disease Index (Standard deviation, $n = 24$)			
Brassica juncea[a]	IP01	IP02	P003	R1
Kranti	3.0 (3.0)	4.1 (3.6)	1.0(0.0)	1.0(0.0)
RES-BJ 01	1.0 (0.0)	1.0 (0.0)	1.0(0.0)	1.0(0.0)
Krishna	5.4 (2.5)	5.7 (2.5)	1.0(0.0)	1.0(0.0)
RES-BJ02	1.0 (0.0)	1.0 (0.0)	1.0(0.0)	1.0(0.0)
Varuna	6.1 (1.8)	7.1 (2.2)	1.0(0.0)	1.0(0.0)
RES-BJ03	1.0 (0.0)	1.0 (0.0)	1.0(0.0)	1.0(0.0)

[a]RES-BJ01, RES-BJ02 and RES-BJ03 lines selected from seedling populations of Kranti, Krishna and Varuna, respectively

A. thaliana accessions, and ten host loci (termed RPP, recognition of *H. parasitica* loci) has been observed. Four of these loci (RPP1, RPP2, RPP4, and RPP7) were mapped (Tor et al. 1994), along with a further locus RPP5 (Parker et al. 1993). Three loci were apparently clustered together on chromosome 4. There was also evidence for the existence of different alleles at a single locus, although corresponding crosses between pathogen isolates differing in specificity are required to confirm whether different alleles at closely linked loci might explain such results. A genetic model postulating the existence of complementary recognition loci in the fungus designated ATR-*A. thaliana* recognition has been proposed (Holub et al. 1994). These results provide support for the gene-for-gene model of specificity in crucifers' downy mildew. The interaction phenotypes between different host lines and pathogen isolates reveal a degree of complexity in the system, with partial dominance, epistasis, and gene dosage effects.

The resistance of RES-26 to isolate P003 seemed to be conditioned by a single, partially dominant gene, and the resistance of RES-02 which belong to group "A" (resistant to R1 and P003) by 2 independent partially dominant genes. The genes for resistance to P003 in RES-26 were closely linked, allelic or identical to one of the 2 genes for resistance in RES 2. Resistance of RES-02 to R1 was conditioned by a single incompletely dominant gene. The genes for resistance to isolates R1 and P003 in RES-02 were either closely linked allelic or identical (Table 3.19). The cotyledonary leaves of each seedling responded independently when inoculated simultaneously each with a different isolate of the pathogen (Nashaat et al. 1997).

Differential response and genes for resistance to *Hyaloperonospora parasitica* in *Brassica juncea* have been identified by Nashaat et al. (2004). The response of a wide range of *Brassica juncea* accessions to 14 isolates of *Hyaloperonospora parasitica*, 12 from India (IP00A, IP02, IP03, IP04, IP04A, IP05, IP05B, IP33, and IP33A) were derived from *B. juncea*; (IP09, IP14, and IP13A) from *B. rapa*, and two from *B. napus* in the UK (R1 and P003), was screened. Sixteen differential host response groups to these isolates (classified as groups A–P) were identified. Groups "A" and "B" expressed the widest resistance profiles to these isolates. Group "A"

Table 3.19 Reaction to *Hyaloperonospora parasitica* isolate P003 of F_1, F_2 and backcross F_1 (BC_1F_1) progeny from crosses involving spring *Brassica napus* accessions RES-26 and Callypso (Nashaat et al. 1997)

	Number of plants with a disease reaction rating of				
Generation and cross	1–5[a]	6–7	Ratio	χ^2	*P*
F_1 RES-26/Callypso and reciprocal	240 (2–3)	0			
F_2 RES-26/Callypso and reciprocal	1581	520	1:1	0.07	0.79
BC_1F_1RES-26 Callypso and reciprocal/Callypso	920	876	1:1	1.08	0.30
F_1 RES-02/Callypso and reciprocal	240 (2–3)	0			
F_2 RES-02/Callypso and reciprocal	1989	139	15:1	0.29	0.59
BC_1F_1RES-02/Callypso and reciprocal/Callypso	1532	482	3:1	1.22	0.27
F_1RES-26/RES-02 and reciprocal	48 (1.0)	0			
F_2 RES-26/RES-02 and reciprocal	5000 (1.0)	0			

[a]Except where indicated in parenthesis

was susceptible to isolates IP05 and IP05B, moderately resistant to isolate IP33, and resistant to all other isolates. Group "B" was susceptible to isolates IP03, IP04, and IP04A, and resistant to the other isolates. Putative homozygous lines resistant to all 14 isolates were selected from the F_4 progeny of crosses involving lines RESBJ-200 from group "A" (selection from cv. Kranti) and RESBJ-190 from group "B" (selection from cv. Krishna). Both selections were selfed and tested for uniformity of reactions to all isolates for three generations. The resistance of RESBJ-200 to isolates IP00A, IP04A, and IP33A seems to be conditioned by single dominant genes. The resistance of RESBJ-190 to isolates IP00A, IP05B, and IP33A was also conditioned by single dominant genes. The gene for resistance to IP00A and IP33A in RESBJ-200 seems to be independent of the genes for resistance to the same isolates in RESBJ-190 (Table 3.20). The new genes for differential resistance to *H. parasitica* will be of value in future studies of the genetics of the host–pathogen interaction and for breeding for disease resistance.

In cabbage, resistance to downy mildew was dominant controlled by 2 independent loci with *Algarvia* being heterozygous at both loci (r1r1 R2r2) and Penca de Chaves being homozygous recessive (r1r1 r2r2) (Caravalho and Moteiro 1996). Downy mildew resistance in Indian cauliflower crosses CC × HRS-4 and 3-5-1-1 × 244 was governed by single dominant gene PPA3 but in cross CC × 244 recessive epistasis was observed (Mahajan et al. 1995). According to Hoser-Krauze et al. (1995) in a cross P1246077 × P123210 to *H. parasitica* polish isolate proved to be dominant character. At the cotyledon stage it was determined by four (in the broccoli-cauliflower line) or three (in the P/S) dominant complementary genes, while at the stage of 4–5 leaves by one (inP123210), two (in P1246077), and three (in the broccoli-cauliflower line) additive dominant genes. In *A. thaliana*, alleles controlling the genotype-specific variation were identified at 19RPP (resistance to *H. parasitica*) loci and most were mapped with varying degrees of resolution. RPP loci occur in several clusters and at least 2 clusters also contain loci identified by involvement in recognition of other pathogens (Crute et al. 1994). Genetic crosses between

Table 3.20 Reaction to *Hyaloperonospora parasitica* isolates of F_1, F_2 and backcross $F_1(BC_1F_1)$ progeny from crosses involving *Brassica juncea* accessions RESBJ-190 and RESBJ-200 (Nashaat et al. 2004)

Generation and cross	IsolateIP04A					Isolate IP05B					Isolate IP00A					IsolateIP33A				
	No. of plants					No. of Plants					No. of plants					No. of plants				
	R	S	Ratio	χ^2	*P*	R	S	Ratio	χ^2	*P*	R	S	Ratio	χ^2	*P*	R	S	Ratio	χ^2	*P*
F_1 RESBJ-200/ RESBJ-190 and reciprocal	184	0				184	0				120	0				120	0			
F_1 RESBJ-200/ RESBJ-190 and reciprocal	1521	520	3:1	0.25	0.62	1803	604	3:1	0.01	0.92	1006	65	15:1	0.06	0.81	903	60	15:1	0.0	1.0
BC_1F_1 RESBJ-200/RESBJ-190//RESBJ-200	355	0				126	111	1:1	0.95	0.33										
BC_1F_1 RESBJ-190/RESBJ-200//RESBJ-190	199	173	1:1	1.82	0.18	349	0													

Table 3.21 Inheritance of resistance in crucifers to *Hyaloperonospora parasitica*

Crucifer genotype	Resistance gene	References
Broccoli PI 189028	One dominant gene	Natti et al. (1967), Vicente et al. (2012)
Cabbage PI245015	One dominant gene	Natti et al. (1967), Vicente et al. (2012)
Broccoli	One recessive gene	Hoser-Krauze et al. (1991)
Broccoli	3 or 4 dominant complementary gene	Hoser-Krauze et al. (1995)
B. oleracea	Quantitative genes	Dickson and Petzoldt (1996)
Chinese cabbage	One dominant gene	Niu et al. (1983)
Chinese cabbage	Additive genes	Yuen (1991)
Radish cvs. Tokinaski and Okura	Two dominant gene	Bonnet and Blancard (1987)
B. rapa	Partial resistance	Leung and Williams (1983)
B. napus cv. Cresor	One dominant gene	Lucas et al. (1988)
B. oleracea	2 or 3 dominant gene	Moss et al. (1988)
B. napus lines	One dominant gene	Nashaat et al. (1995a, 1995b, 1996)
RES 01-1-4 and RES-26	One partial dominant gene	Nashaat et al. (1997)
RES-02	Two partial dominant gene	Nashaat et al. (1997)
Arabidopsis thaliana	Partial dominant epistasis and gene dosage effects	Holub et al. (1994)
B. oleracea	2 dominant gene	Caravalho and Monteiro (1996)
	One dominant gene	Vicente et al. (2012)
B. juncea	One dominant gene	Nashaat et al. (2004)
RESBJ-200, RESBJ-190	One dominant gene	Nashaat et al. (2004)
Indian cauliflower	One dominant gene	Mahajan et al. (1996)
A. thaliana	One dominant gene	Reignault et al. (1996), Joos et al. (1996)
Broccoli	Partial resistance	Jensen et al. (1999a)

A. thaliana genotypes RLD R and Wei-0(s) showed that resistance to *H. parasitica* was inherited through a monogenic dominant trait. The RPP 11 resistance gene was mapped by following the co-segregation of the resistance phenotype with RFLP markers in a mapping population of 254 F_3 families derived from RLD × Wei-0 F_2 individuals. Linkage analysis placed the RPP 11 resistance locus on chromosome III between marker m 249 (two recombinants) and marker g 2534 (six recombinants) (Joos et al. 1996). It is clear from Table 3.21 that resistance in different cruciferous crops against *H. parasitica* is governed by one to four dominant genes, additive genes, partial or partial dominant genes with epistasis and gene dosages effects.

Twelve host loci associated with different interaction phenotypes were postulated in part by studying segregations among F_2 progeny from crosses of half-diallel among nine *A. thaliana* accessions. Advanced generations including recombinant inbred lines were used to confirm the identity of loci. It was suggested that a single cross between two host accessions could be used to characterize and map several loci associated with isolate specific recognition of *H. parasitica* and distinct interaction phenotypes. Partial dominance and genetic epistasis are postulated as being common features of these parasitic symbols (Holub et al. 1994). As a prelude to cloning, the positions in *A. thaliana* genome were investigated for four loci (RPP1, RPP2, RPP4, and RPP7) of the 10 that have been identified as associated with the genotype-specific recognition of the downy mildew pathogen *H. parasitica* (Tor et al. 1994). Segregating populations of *A. thaliana* generated from crosses between the susceptible accessions Col-0 and the resistant accessions WS-0, Pr-0, OY-0, P0-1, gc-1, Di-1, Ji-1, and Te-0 were screened with *H. parasitica* isolate N0C0. The genetic data were consistent with the presence of single resistance (RPP) loci in all of these accessions except Oy-0, in which resistance was inherited as a digenic trait. As a first step to molecular cloning, the map positions of 4 resistant loci were determined. These have been designated RPP 14.1 from Ws-0, RPP 14.2 from Pr-0, and RPP 14.3, and RPP 5.2 from Oy-0. RPP 14.1 was mapped to a 3.2 CM interval on chromosome 3 that is linked to a region between the markers Gl-1 and M 249 known to contain other *H. parasitica* specifications. RPP 14.2 from Pr-0 and RPP 14.3 from Oy-0 were also positioned in this interval. RPP 14.1 and RPP 14.2 showed linkage of <0.05 CM suggesting possible allelism. The second RPP locus from Oy-0, RPP 5.2 was located on chromosomes 4 and exhibited strong linkage (<2 CM) to RPP 5.1, a locus previously identified in the *Arabidopsis* accession Landsberg-crecta (Reignault et al. 1996).

According to Botella et al. (1998) in the *Arabidopsis* accessions Wasilewskija, the RPP 1 region on chromosome 3 contain four genetically linked recognition specificities conditioning resistance to different isolates of *H. parasitica*. Three of the four tightly linked genes in this region, designated RPP-1- WSA, RPA-1-WsB, and RPP-1-WsC, encode functional products of the NBS-LRR (Nucleotide binding site-leucine-rich repeat) R protein class. They possess a TIR (Toll, interlcukin-1, resistance) domain that is characteristic of certain other NBS-LRR- type R protein but, in addition, they have unique hydrophilic or hydrophobic N termini. Together, the three RPP-1 genes account for the spectrum of resistance previously assigned to the RPP-1 region and thus comprise a complex R locus. The distinct but partially overlapping resistance capabilities conferred by these genes are best explained by the hypothesis that each recognizes a different pathogen avirulence determinant.

3.6.2 Inheritance of Partial Resistance

Latent period and sporulation capacity are two major components of partial resistance, and governed by the same mechanism in the interaction between Broccoli (*B. oleracea*) and *H. parasitica*. Twenty doubled haploid broccoli lines from breeding material were evaluated for resistance to *H. parasitica* at the seedling stage. All lines supported sporulation of the pathogen, but to varying extents and intensities.

Partial resistance of the more resistant lines "br8" and "br9" reduced conidia production on cotyledons by 50-70% compared with the most susceptible lines. Inoculation of the two most resistant lines with 13 isolates of different geographical origins revealed that "br9" showed a rather uniform level of resistance to all isolates while "br8" showed some isolate specificity. Partial resistance was evaluated in six of the broccoli lines in a half-diallel set of crosses. Disease assessment of seedlings showed that additive genetic effects explained 45.8 and 31.8% of the total variation of sporulation score and conidia production, respectively. This suggests that recurrent selection for partial resistance to *H. parasitica* in early generation inbreds or in populations of broccoli will be efficient to obtain cotyledon resistance (Jensen et al. 1999b). Seedling resistance conditioned by a single major gene is isolate specific, and may lose the effect through selection of virulent types of the pathogen, if resistant cultivars are produced on large areas (Lucas et al. 1988). Partial quantitative disease resistance by definition provides reduced host invasion without complete protection (Parlevliet 1992), and is generally anticipated to be based on several genes. It may be a source of durable resistance with less isolate specificity than resistance based on single genes (Parlevliet 1992). In *Brassica oleracea* partial resistance to *H. parasitica* at the seedling stage has been reported in Portuguese coles (Monteiro and Williams 1989; Dias et al. 1993) and in gene bank accessions of cabbage (Thomas and Jourdain 1992) and broccoli (Thomas and Jourdain 1990). Jensen et al. (1999b) characterize partial resistance to *H. parasitica* at the cotyledon stage within 20 doubled haploid broccoli lines originating from a wide range of breeding material. Inheritance of the resistance was analyzed in progeny from a half-diallel set of crosses (Jensen et al. 1999b).

3.6.3 Inheritance of Avirulence Gene in *Arabidopsis–Hyaloperonospora* Pathosystem

The biotrophic *Hyaloperonospora arabidopsidis* (*Hpa*; *H. parasitica*) has co-evolved as a downy mildew pathogen in wild populations of *Arabidopsis thaliana* and has been used for more than 25 years as an experimental model for investigating the molecular basis of the gene-for-gene theory evidence that for each gene controlling resistance in the host there is a corresponding gene controlling pathogenicity in the pathogen. At least seven downy mildew R-genes have been characterized in this pathosystem, and all encode cytoplasmic NLR proteins that contain a conserved nucleotide-binding site (NBS) with a variable leucine-rich repeat (LRR) domain. Three of the seven predicted AVR elicitors have also been characterized and all encode secreted effector-like proteins. Plant pathogens deliver effectors to establish infection by suppressing the host immune system, altering plant physiology and colonizing the host cell. Current consensus suggests that effectors target apoplastic and cytoplasmic components. Fungal pathogens secrete effectors with the involvement of an N-terminal signal peptide (Kamoun 2006; Song et al. 2009), and some studies suggest that R $\times$ LR and dEER motifs may be involved in the translocation of these effectors into the host cell (Whisson et al. 2007; Tyler et al.

2013). Effectors that elicit a defense response due to detection by an R-protein are referred to as avirulence elicitor proteins (Jones and Dangl 2006).

While genetic studies into *Arabidopsis-Hpa* interactions have shown the existence of a gene-for-gene relationship, molecular, and bioinformatics investigations have helped the cloning of several avirulent determinants in the pathogen and the corresponding R-genes in the host (Holub et al. 1994; Holub 2006; Baxter et al. 2010; Bailey et al. 2011). Some of the *Hpa* RXLR-type effector proteins have been molecularly characterized. ATR1 (Rehmany et al. 2005) is recognized by RPP1 by direct association via C-terminal leucine-rich repeats (Steinbrenner et al. 2015). ATR13 (Allen et al. 2004) is a highly polymorphic and dynamic protein with two surface-exposed patches of polymorphism, only one of which is involved in specific recognition of RPP13-Nd (Leonelli et al. 2011). ATR39-1 has been identified by computational methods and is recognized by RPP39 in *Arabidopsis* accession Weiningen (Wei-0) (Goritschnig et al. 2012). Woods-Tor et al. (2018) cloned ATR5 Emoy2, which does not have the R × LR motif but contains a dEER motif (Bailey et al. 2011). All of these *Hpa* effectors are avirulence determinants and, until now, no suppressor has been identified for any of them. They reported the genetic identification and map-based cloning of a *H. arabidopsidis* cryptic (*HAC1*^{Cala2}) avirulence determinant and a predicted matching suppressor (*S-HAC1*$^{\mathrm{Cala2}}$) from the same interval using asexual cross between two virulent isolates of *H. arabidopsidis*. HAC1$^{\mathrm{Cala2}}$ is an R × LR-dEER type putative effector protein inherited in progeny from the parent isolate *Hpa*-Cala2 and can trigger an immune response in the *A. thaliana* accession RMX-A02. *S-HAC1*$^{\mathrm{Cala2}}$ is in the same parent isolate within the genetic interval next to the *HAC1*$^{\mathrm{Cala2}}$ gene and encodes a putative effector protein with a dKEE motif *S-HAC1*$^{\mathrm{Cala2}}$ is epistatic over *HAC1*$^{Cala2.}$

The pathosystem of *Arabidopsis thaliana* and diploid biotrophic oomycete *Hyaloperonospora arabidopsidis* has been a model for investigating the molecular basis of Flor's gene-for-gene hypothesis. The isolates *Hpa-Noks1* and *Hpa-Cala2* are virulent on *Arabidopsis* accession RMX-A02 while an F1 generated from a cross between these two isolates was avirulent. The F2 progeny segregated into 3:1 (avirulent: virulent), indicating a single major effect AVR locus in this pathogen. SNP-based linkage mapping confirmed a single AVR locus within a 14 kb map interval containing two genes encoding putative effectors. The *Hpa*-Cala2 allele of one gene, designated *H. arabidopsidis* cryptic1 (HAC1), encodes a protein with a signal peptide and an R × LR/deer motif, and triggers a defense response in RMX-A02. The second gene is heterozygous in *Hpa*-Cala2. One allele, designated Suppressor of *HAC1*$^{\mathrm{Cala2}}$(*S-HAC1*$^{\mathrm{Cala2}}$), encodes a protein with a signal peptide and a dike motif with no R × LR motif; the other allele (*s-hac1*$^{\mathrm{Cala2}}$) encodes a protein with a signal peptide, a dEEE motif and is divergent in sequence from the S-HAC1Cala2 allele. In selfed progeny from *Hpa*-Cala2, dominant *S-HAC1*$^{\mathrm{Cala2}}$ allele carrying progeny correlates with virulence in RMX-A02, whereas homozygous recessive *s-hac1*$^{\mathrm{Cala2}}$ carrying progeny were avirulent. Genetic investigations suggested other heterozygous suppressor loci might exist in the *Hpa*-Cala2 genome (Woods-Tor et al. 2018).

3.7 *Brassica-Plasmodiophora* Inheritance

3.7.1 Inheritance of Disease Resistance Genes (R-Genes)

The inheritance of resistance to clubroot was found to be dominant in crosses between resistant and susceptible cabbage cultivars and disease response appeared to be governed by a polygenic type of inheritance (Walker and Larson 1951). Strandberg and Williams (1967) reported that Chinese cabbage (*B. pekinensis* Rupr.) cvs. Michihili possessed one dominant gene for resistance to races 6 and 7 of the clubroot pathogen. Chiang and Crête (1970) crossed resistant and susceptible cabbage lines, F_1 data showed that susceptibility was dominant over resistance and data from F_2 and backcross progenies supported a duplicate gene hypothesis. Lammerink (1967) crossed New Zealand clubroot resistant rape and New Zealand Wilhelms burger rutabaga and found that resistance to races B and C was determined by single dominant non-allelic genes. Johnston (1970), working with a new racial isolate, crossed susceptible Giant Rape and resistant New Zealand rape and found that resistance was dominant and controlled by a single gene; similar findings were reported when susceptible Giant Rape was crossed with resistant Wilhelms burger rutabaga.

The York and Wilhelms burger rutabaga cultivars, both resistant to races 2 and 3 of the clubroot pathogen, were crossed with the race 2 and 3 susceptible Laurentian cultivar. Race 2-resistant Ditmars S2 rutabaga selection was also crossed with susceptible Laurentian. Race exposure of F_1 seedlings of resistant × susceptible crosses showed that resistance to races 2 and 3 was dominant. Susceptibility studies of F_2 and backcross progenies showed that resistance to race 2 in York was controlled by a single major gene, whereas resistance to this race in Wilhelms burger was controlled by two major genes. Single major gene resistance to race 2 was indicated for the Ditmars S2 selection. This crucifer line is susceptible to race 3. The F_2 progeny exposures to race 3 showed that resistance was controlled by one major gene in both the York and Wilhelms burger rutabaga cultivars (Ayers and Lelacheur 1972). In case of *A. thaliana,* resistance to *P. brassicae* isolate is conferred by a dominant allele of a single nuclear gene designated as RPB1. With the aid of morphological marker, this locus is mapped near the markers disease 2 to chromosomes 1 (Fuchs and Sacristan 1996).

In the major cultivated cruciferous species, *B. rapa*, *B. napus*, *B. oleracea*, *R. sativus,* and model crucifer species *A. thaliana,* clubroot disease has become a cause of serious concern because of rising crop failures. To breed clubroot-resistant cultivars of these species, a number of germplasms have been evaluated. This allowed sources of resistance to be identified in crucifer's species. Inheritance of resistance to clubroot varies between these species up to molecular level.

3.7.1.1 *Brassica rapa*

The breeding of clubroot-resistant cultivars of Chinese cabbage (*B. rapa* ssp. *pekinensis*) has been impeded because most cultivars are highly susceptible to clubroot disease (Yoshikawa 1981; Cho et al. 2002). The identification of resistant sources in European fodder turnips (*B. rapa* ssp. *rapifera*) (Crute et al. 1983; Crisp et al. 1989) allowed the transfer of CR genes from European fodder turnip to Chinese cabbage (Yoshikawa 1981). The introgression of CR genes from European fodder turnip cultivars, including Gelria R, Siloga, Debra, and Milan White, has provided and broadened the genetic diversity of Chinese cabbage. The European Clubroot Differential (ECD) hosts 01–04 represent the spectrum of resistance to virulent pathotypes of *P. brassicae* (Toxopeus and Janssen 1975; Toxopeus et al. 1986; Buczacki et al. 1975). These have been used as resistant sources inbreeding clubroot-resistant cultivars either in Chinese cabbage or in *B. napus*. In these studies, at least three independent dominant genes, which conferred differential (race-specific or vertical) resistance to particular pathotypes of *P. brassicae*, are present in turnip genotypes (James et al. 1978; Crute et al. 1980). Yoshikawa (1993) demonstrated that clubroot resistance of European fodder turnips, including cv. Siloga, was controlled mainly by a major gene and a few minor genes. James et al. (1978) identified three independent dominant genes that conferred resistance in three *B. rapa* genotypes to race 6 of *P. brassicae*. Crute et al. (1980) also demonstrated that three genes controlled resistance in turnip. This suggests that clubroot resistance in *B. rapa* is controlled by several genes independently. Because clubroot-resistant cultivars of Chinese cabbage were released by introducing CR genes from clubroot resistant European fodder turnip, it is believed that clubroot-resistant cultivars of Chinese cabbage have the potential for introducing one to several genes from turnip. The data obtained by marker trait analysis confirmed that at least eight CR loci are present in *B. rapa*. Two loci crr1 and crr2 have been identified (Hirai et al. 2004; Piao et al. 2004; Sakamoto et al. 2008).

3.7.1.2 *Brassica oleracea*

Various sources of clubroot resistance through the screening of germplasm have been identified (Crisp et al. 1989; Dias et al. 1993; Voorrips and Kanne 1997a, b; Voorrips et al. 1997; Manzanares-Dauleux et al. 2000b; Carlsson et al. 2004). In contrast to *B. rapa*, completely resistant accessions have only rarely been identified in *B. oleracea*. Crisp et al. (1989) evaluated about 1000 *B. oleracea* accessions, resistant sources were confirmed in some kales, including the curly marrow stem and 1000-head types, and in cabbage, including Bohmerwaldkohl, Bindsachsener cabbage, Badger Shipper, and Ladoszkaya cabbage. Some open pollinated Brussels sprouts, including cvs. Cambridge, Continuity, Rubine, and Catskill, and forms of southern European cabbage, cauliflower, and broccoli exhibited lower levels of susceptibility. Among the kale, cabbage, and winter cauliflower *otrytis* accessions evaluated, only some kale accessions, mainly the leafy and short leafy kale groups, exhibited high levels of resistance to clubroot, whereas all cabbage and cauliflower accessions were susceptible (Manzanares-Dauleux et al. 2000b). Some of these resistant sources are widely used in breeding programs for *B. oleracea*.

The detailed genetics of clubroot resistance were studied in *B. oleracea* using either diallel crossing methods or segregating population. Most of these studies concluded that inheritance of this trait in *B. oleracea* is polygenic (Table 3.22). Yoshikawa (1993) observed that progenies of crosses between cv. Bohmerwaldkohl

Table 3.22 Inheritance of clubroot resistance in *Brassica oleracea* (Piao et al. 2009)

Resistant source	Population	Test (pathotype)	Genes involved	References
Bohmerwaldkohl	F2, BC	Glasshouse (field isolate)	Four genes	Yoshikawa (1993)
Bohmerwaldkohl	F2, BC	Glasshouse (field isolate)	Additive and recessive	Crute and Pink (1989)
Kale accessions	Diallel	Glasshouse (field isolate)	Many dominant genes	Laurens and Thomas (1993)
Bohmerwaldkohl, Badger Shipper, Larson 8353 T, Resistant Detroitand Curly kale	F1	Glasshouse (field isolate)	Recessive	Voorrips and Visser (1993)
Cabbage	F1, F2, BC1	Glasshouse (field isolate, single spore isolate)	Two complementary genes	Voorrips and Kanne (1997a, b)
Resistant Detroit			Two genes	
Bindsachsener			Two genes	
Curly kale			More than two genes	

and a susceptible cabbage cv. have at least one to four genes to control this trait. Voorrips and Visser (1993) found that the inheritance of resistance to clubroot was recessive based on the genetic analysis of the 11 F_1 populations derived from crosses between resistant sources, including three accessions of Bohmerwaldkohl, cv. Badger Shipper, two lines derived from "Larson 8353 T," cv. Resistant Detroit, and four accessions of curly kale and susceptible cabbage, respectively. In kale (Laurens and Thomas (1993), the inheritance of resistance is controlled by many dominant alleles with a predominance of additive genetic effects with incomplete dominance. Based on qualitative and quantitative analyses of the F_1, F_2, and backcross progenies of four crosses derived from four sources of resistance and one common susceptible doubled haploid line, Voorrips and Kanne (1997b) suggested that one was controlled by two complementary genes. Chiang and Crête (1970) agreed that resistance was controlled by two loci.

3.7.1.3 *Brassica napus*

Clubroot resistance in swedes (*B. napus*) genotypes, including cvs. Wilhelms burger and Studsgaard Bangholm, exhibited resistance under certain environmental conditions (Karling 1968; Dixon et al. 1972). These sources are used to breed clubroot resistant cultivars of *B. napus*. During genetic studies of resistance to clubroot in swedes, different sources of resistance have been used including Wilhelms burger, clubroot-resistant rape, Wye swede group, Yorkswede, and Ditmar S2 swede (Ayers 1972; Ayers and Lelacheur 1972). The inheritance of resistance in *B. napus* is controlled by one or two single independent dominant genes. At least one of these genes is shared by several sources. Two recessive genes with additive effect were also identified in Danish Giant swede. Gustafsson and Falt (1986) suggested that four resistance genes are present in the *B. napus* ECD hosts. Based on genetic analysis using partial or genome-wide marker surveys in several *B. napus*, over 20 quantitative trait loci for CR were identified (Manzanares-Dauleux et al. 2000a; Werner et al. 2008). Additive effects in genetic control of CR in *B. napus* loci were found by Li et al. (2016) who predicted the role of TIR-NBS gene family. Because *B. napus* is a natural amphidiploid between *B. oleracea* and *B. rapa*, however, it is not clear where these resistance genes are located and their origins. It is suggested that genes for resistance in *B. napus* are located in the A genome from *B. rapa* (Chiang et al. 1977).

3.7.2 Inheritance and Characterization of R-Genes in *Brassica rapa*

European fodder turnips (*Brassica rapa*) were identified as sources of resistance and have been used to transfer their clubroot resistance (CR) genes into Chinese cabbage, oilseed rape, and *Brassica oleracea*. The CR trait of European turnip cultivars such as Siloga, Gelria R, and Debra was considered at first to be controlled by a single dominant gene. Yet after the release of more than 50 CR F_1 cultivars of Chinese cabbage so far, the breakdown of resistance, caused by variation in the pathogenicity of *P. brassicae*, has been reported (Yoshikawa 1981; Kuginuki et al. 1999). Genetic analysis and quantitative trait locus (QTL) mapping studies have identified at least 8 CR loci in *B. rapa*, 22 QTLs in *B. oleracea*, and 16 QTLs in *Brassica napus*. In *B. rapa*, *Crr1* and *Crr2* were identified on chromosomes A08 and A01, respectively. These two loci were detected by using two *P. brassicae* isolates, the mild Ano-01 and the more virulent Wakayama-01. *Crr1* was necessary for the resistance to both isolates, but plants having *Crr1* alone were susceptible to Wakayama-01. *Crr2*, which by itself does not show any effect against either isolate, was necessary for resistance to Wakayama-01 in combination with *Crr1*. Therefore, *Crr1* may play a role in a common pathway of resistance, and *Crr2* may be a modifier locus for the resistance expressed by *Crr1*. Four CR loci–*CRa*, *CRb*, *Crr3*, and *CRk* were identified on chromosome A03.The relationships between common markers and the location of the CR loci suggest that *CRa* and *Crr3* are identical, allelic, or closely linked to *CRb* and *CRk*, respectively. *CRc* was mapped on A02, and a weak QTL, *Crr4*, on A06. Comparative mapping of CR loci between *B. rapa* and *Arabidopsis* revealed that *Crr1, Crr2*, and *CRb* are syntenic with the central region of *Arabidopsis* chromosome 4. Because this region is located within a disease resistance gene cluster, it has been suggested that CR genes are members of these clusters.

Suwabe et al. (2011) fine-mapped the *Crr1* locus and found that it was likely to comprise two gene loci, a major locus for clubroot resistance and another locus with minor effect named *Crr1a* and *Crr1b*. The map-based cloning and characterization of *Crr1a*, the major locus for resistance to Ano-01, derived from European fodder turnip "Siloga." *Crr1a*^{G004} encodes a TIR-NB-LRR disease resistance protein and is expressed in the stele or cortex of hypocotyl and roots, where the secondary infection phase occurs. Transgenic *Arabidopsis* and susceptible *B. rapa* harboring *Crr1*^{G004} showed resistance to *P. brassicae* isolates similar to that of the resistant *B. rapa* (Hatakeyama et al. 2013). Fine mapping of the *Crr1* locus, which was originally identified as a single locus, revealed that it comprises two gene loci, *Crr1a* and *Crr1b*. The map-based cloning and characterization of *Crr1a*, which confers resistance to clubroot in *Brassica rapa*. *Crr1a*^{G004}, cloned from the resistant line G004, encodes a Toll-Interleukin-1 receptor/nucleotide-binding site/leucine-rich repeat (TIR-NB-LRR) protein expressed in the stele and cortex of hypocotyl and roots, where secondary infection of the pathogen occurs, but not in root hairs, where primary infection occurs (Fig. 3.6). Gain-of-function analysis proved that *Crr1a*^{G004} alone conferred resistance to isolate Ano-01 in susceptible *Arabidopsis* and *B. rapa*. In comparison, the susceptible allele *Crr1a*^{A9709} encodes a truncated NB-LRR protein, which lacked more than half of the TIR domain on account of the insertion of a solo-long terminal repeat (LTR) in exon 1 and included several substitutions and insertion-deletions in the LRR domain.

3.7.3 Genetics of Clubroot Resistance Genes (R-Genes)

The genetic basis of clubroot resistance (CR) has been determined in economically important cruciferous species: *Brassica rapa, B. oleracea*, *B. napus*, *R. sativus*, and *Arabidopsis thaliana*. In *B. rapa*, mainly in Chinese cabbage, one minor and seven major CR genes introduced from European fodder turnips have been identified. Mapping of these CR genes localized *Crr1* on R8, *Crr2* on R1, *CRc* on R2, and *Crr4* on R6 linkage groups of Chinese cabbage. Genes *Crr3, CRa, CRb*, and *CRk* mapped to R3, but at two separate loci, *CRa* and *CRb* are independent of *Crr3* and *CRk*, which are closely linked. Further, analysis suggested that *Crr1, Crr2*, and *CRb* have similar origins in the ancestral genome as in chromosome 4 of *Arabidopsis thaliana*. Genetic analysis of clubroot resistance genes in *B. oleracea* suggests that these are quantitative traits. Twenty-two quantitative trait loci (QTLs) have been mapped in different linkage groups of *B. oleracea*. In *B. napus*, genetic analysis of clubroot resistance was found to be governed by one or two dominant genes. The quantitative analysis approach, however, proved that these are polygenic. In total, at least 19 QTLs have been detected on eight chromosomes of *B. napus*, N02, N03, N08, N09, N13, N15, N16, and N19. The chromosomal location of the other six QTLs is not clear. Cloning of these QTLs or resistance loci is possible. Progress in genomics, particularly the techniques of comparative mapping and genome sequencing, supplements cloning, and allows improved characterization of CR genes. Further, the development of DNA markers linked to CR genes will in turn hasten the breeding of clubroot-resistant crucifer's cultivars.

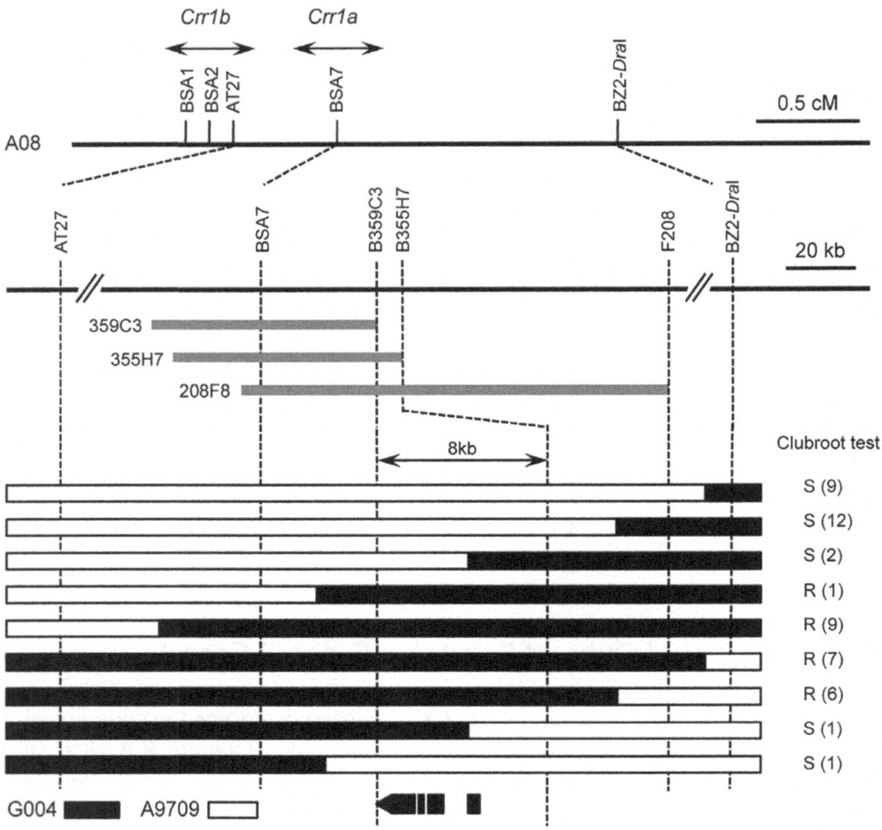

Fig. 3.6 Map-based cloning of *Crr1a*: a genetic map around the *Crr1* locus, and graphical genotypes of F_3 populations in which recombination occurred between AT27 and BZ2–DraI. The major locus for resistance to Ano-01 was predicted around BSA7, and another locus with minor effect was predicted around BSA2 (Suwabe et al. 2012). Named the former locus *Crr1a* and the latter *Crr1b*. Gray bars indicate BAC clones isolated by using BSA7 as a probe. Black bars, homozygous for resistant G004 allele; white bars, homozygous for susceptible A9709 allele. Phenotypes of resistance (R), and susceptibility (S) of F_3 populations to Ano-01 are indicated to the right of the graphical genotype with the number of F_3 populations of each in parentheses. The 3 F_3 populations in which recombination occurred between B359C3 and B355H7 were susceptible to Ano-01. The position of *Crr1a* was delimited to an 8-kb region between B355H7 and B359C3. The predicted ORFs of the *Crr1a* are presented (Hatakeyama et al. 2013)

3.7.4 Inheritance and QTL Mapping of Clubroot Resistance Genes of *Brassica rapa* subsp. *rapifera* (ECD02)

The identification of effective resistance is the first step in breeding for this trait, with *Brassica rapa* (2n = 20, AA) considered a superior source of dominant major clubroot resistance genes than *Brassica oleracea* (2n = 18, CC) (Toxopeus et al. 1986; Hirai 2006; Piao et al. 2009). Over the past 20 years, at least 15 CR genes have been identified in *B. rapa*, including *CRa* (Matsumoto et al. 1998), *CRb* (Piao et al. 2004), CRb^{Kato} (Kato et al. 2012, 2013), *CRk* (Matsumoto et al. 2012), *Crr3* (Hirai et al. 2004; Saito et al. 2006), *Rcr1* (Chu et al. 2014; Yu et al. 2016), *CRc* (Sakamoto et al. 2008; Matsumoto et al. 2012), *Crr1,Crr2, Crr4* (Suwabe et al. 2003, 2006), *CRd* (Pang et al. 2018), *BraA.Cr.a, BraA.Cr.b, BraA.Cr.c* (Hirani et al. 2018), and *CrrA5* (Nguyen et al. 2018). Clubroot resistance genes from *B. rapa* have been introgressed into several European *B. napus* oilseed cultivars, including "Mendel" and "Tosca" (Frauen 1999). In Canada, different *Brassica* genotypes have been used as resistance donors in the breeding of CR canola/oilseed rape (*B. napus*). Fredua-Agyeman and Rahman (2016) found that the *B. napus* cultivar "Mendel" contained one dominant CR gene effective against pathotype 3H of *P. brassicae*, as defined on the Canadian Clubroot Differential Set (Strelkov et al. 2018). This was the dominant pathotype in Alberta, Canada (Strelkov et al. 2006, 2007), at least prior to the introduction of CR canola cultivars beginning in 2009 (Strelkov et al. 2018). However, the planting of CR canola in short rotations over large acreages led to the rapid development of new pathotypes of *P. brassicae* (Strelkov et al. 2016, 2018), and "Mendel" resistance was eroded and is no longer a good choice for breeding cultivars resistant to the new pathotypes. Rutabaga (*Brassica napus* ssp. *napobrassica*) is another potential donor of clubroot resistance genes. The rutabaga cultivars "Brookfield" and "Polycross" possessed excellent resistance to Canadian *P. brassicae* pathotypes 2, 3, 5, 6, and 8 (Hasan and Rahman 2011, 2013, 2016). The ratio of resistant (R) to susceptible (S) in the F2 generation derived from crossing both Brookfield and Polycross with susceptible *B. napus* lines was 3R:1S while segregation in the test cross family of the latter deviated from a 1R:1S ratio. This suggested that CR in Brookfield was controlled by a single dominant gene while resistance in Polycross was more complex (Hasan and Rahman 2011).

With the recent emergence of new pathotypes of *P. brassicae* and their ability to overcome clubroot resistance in most CR canola cultivars, additional sources of resistance are needed. The *CRa* resistance gene was first detected in the European Clubroot Differential (ECD) 02 (*B. rapa* ssp. *rapifera* line AAbbCC) (Buczacki et al. 1975; Matsumoto et al. 1998). Zhang et al. (2016) reported one marker (i3e4) that was tightly linked to *CRa*. While ECD 02 appears to be resistant to all *P. brassicae* pathotypes identified in Canada (Strelkov et al. 2006, 2018; Leboldus et al. 2012), and was used as a resistance source in studies from Japan (Sakamoto et al. 2008), its application in clubroot resistance breeding in Canada has not yet been reported. The introgress clubroot resistance from ECD 02 (male parent) into two susceptible female *B. rapa* genotypes CR 2599 and CR 1505, to evaluate the genetic basis of resistance to three isolates of *P. brassicae* representing different pathotypes and identify molecular markers associated with resistance to the new pathotypes.

The *Brassica rapa* subsp. *rapifera* (ECD 02) which exhibits broad-spectrum resistance to many Canadian *P. brassicae* isolates was crossed with two clubroot-susceptible *B. rapa* accessions to produce two F2 populations. TheF2 plants were screened against *P. brassicae* pathotypes 3H, 5X, and 5G. The Chi-square goodness of fit test showed that the vast majority ($\approx$75%) of the crosses that produced the F2 populations showed segregation ratios of 9R:7S, 7R:9S, 13R:3S, 3R:13S, 5R:11S,

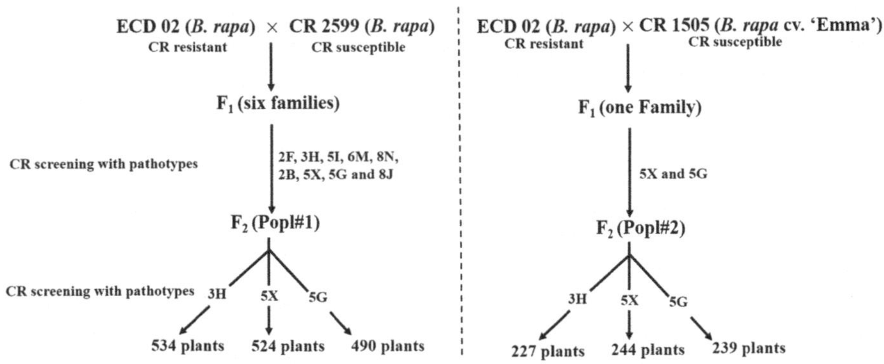

Fig. 3.7 Development of F2 mapping population for the study of the inheritance pattern of clubroot resistance introgressed from ECD 02(Fedua–Agyeman et al. 2020)

11R:5S, and 1R:15S. These were modifications of the 15R:1S ratio expected for the inheritance of two dominant major clubroot resistance (CR) genes from ECD 02. The distorted segregation ratios suggest that the two resistance genes are on different chromosomes and those two genes interact in an epistatic manner to confer resistance. Genotyping was conducted with 144 PCR-based markers in the two F2 populations. Linkage and QTL analysis with the polymorphic markers identified two QTLs on chromosome A03 to be associated with resistance to *P. brassicae* pathotypes 5X and 5G in Popl #1 while only the second QTL on chromosome A03 was associated with resistance to pathotypes 5X and 5G in Popl#2. The QTLs clustered in genomic regions on the A03 chromosome of *B. rapa* where the *CRa*/*CRb*^{Kato} gene(s) are mapped. In addition, the *Crr1* gene on the A08 chromosome of *B. rapa* was detected in the two F2 populations. Therefore, the phenotypic and molecular data confirm the existence of two CR genes in ECD 02 (Figs. 3.7, 3.8, and 3.9). The study shows that major dominant genes in *Brassica* interact in a non-additive manner to confer resistance to different *P. brassicae* pathotypes (Fedua–Agyeman et al. 2020).

3.7.5 Inheritance and Mapping of Clubroot Resistance Genes in *Brassica rapa* Using Bulk Segregant RNA Sequencing

The pathogen recognition systems that provide strong resistance in plants generally function in one of two ways. In one system, a pattern recognition receptor (PRR) in the plant interacts with a microbial/pathogen-associated molecular pattern (MAMP, PAMP) from the pathogen inside the apoplast, resulting in PAMP-triggered immunity (PTI). In the other system, an R-gene encoded NBS-LRR (nucleotide-binding site leucine-rich repeat) protein in the plant interacts with an effector from the pathogen, usually inside the cytoplast, resulting in effector-triggered immunity (ETI). The NBS-LRR protein family can be subdivided according to their functional

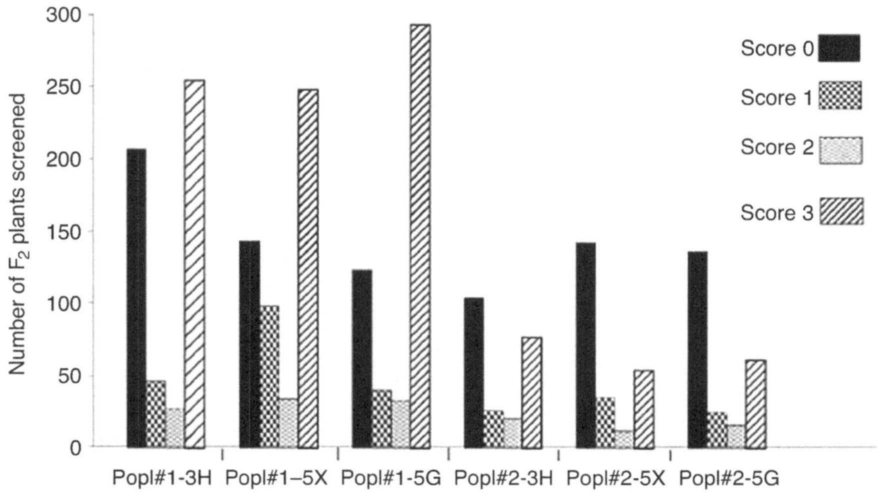

Fig. 3.8 Frequency distribution of disease scores in two F2 populations (Popl#1 and Popl#2) derived from ECD 02 × CR 2599 and ECD 02 × CR 1505 ("Emma") to three *Plasmodiophora brassicae* pathotypes 3H, 5X, and 5G (Fedua–Agyeman et al. 2020)

domain as toll/interleukin-1receptor (TIR)-domain-containing (TIR-NBS-LRR protein) and coiled-coil (CC)-domain-containing (CC-NBS-LRR protein) subfamilies. NBS-LRR related disease resistance is effective against obligate and hemibiotrophic pathogens (Glazebrook 2005).

Brassica rapa ssp. *rapifera* has been the sole source of resistance to clubroot in commercial cultivars of *B. rapa, B. napus*, and *B. oleracea*. Across all *Brassica* spp., at least 15 CR loci from *B. rapa*, 22 quantitative trait loci (QTLs) from *B. oleracea*, and 16 QTLs from *B. napus* have been identified. In *B. rapa*, chromosome A03 carried the highest number of loci, including *CRa*, *CRb*, *Crr3*, *CRk*, *PbBa3*.1 and *PbBa3*.3, *Rcr1*, *Rcr2*, *Rcr4*, and *Rcr5*. Another four chromosomes were identified as carrying CR loci: *Crr1*, *Rcr9*, and *CRs* on A08; *Crr2*on A01; *Crr4* on A06; and *CRc* and *Rcr8* on A02. *Crr1* was demonstrated to consist of two gene loci; the major locus, *Crr1a*, and the minor locus, *Crr1b*. *Crr1a* is expressed as a TIR-NB-LRR disease resistance protein in the stele and roots, and was shown to have originated from the European fodder turnip "Siloga." The CR genes *CRa* and CRb^{kato} also encode TIR-NB-LRR class proteins. Several new sources of CR have been identified. Additionally, resistance locus CRs, resistant to a Korean isolate of *P. brassicae* was recently reported from turnip line SCNU-T2016 on chromosome A08 (10.7–11.5 Mb). Two genes, *Bra020918* and *Bra020876*, were reported as candidate genes for *CRs* (Laila et al. 2019; Suwabe et al. 2006, 2012). Pathotype 5X was the first pathotype identified that overcame resistance in the first generation of CR canola cultivars in Canada, although many more resistance-breaking pathotypes have subsequently been identified. Studies to identify sources of resistance to pathotype 5X identified two independent QTLs, *Rcr8* and *Rc9*, in the A genome

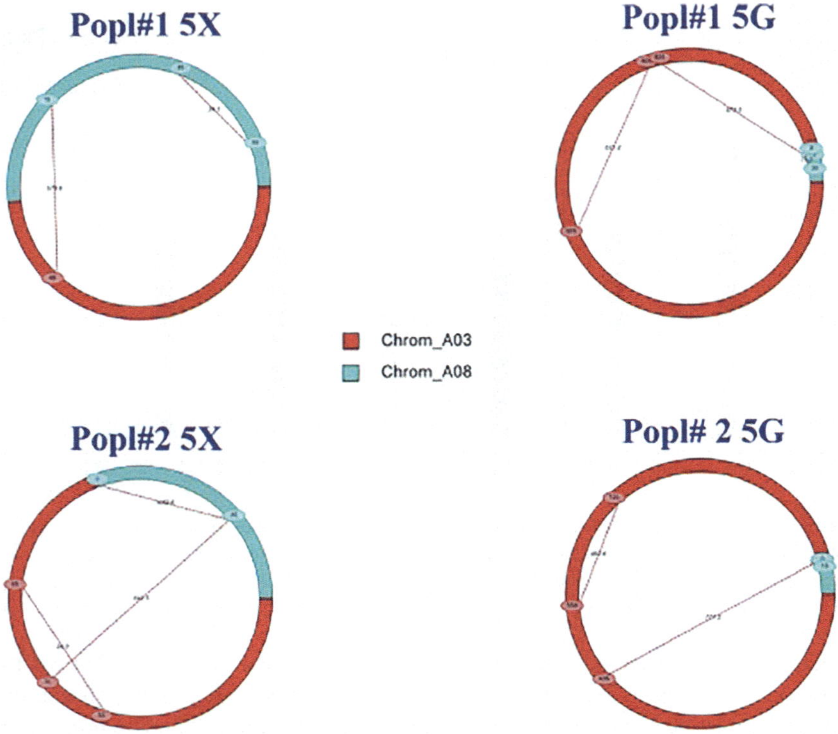

Fig. 3.9 Epistatic QTLs conferring clubroot resistance in two F2 mapping populations (Popl#1 and Popl#2) as revealed by QTL Ici Mapping software V4.1. The dashed lines represent epistatic interaction pairs of CR QTLs. The red and cyan colors indicate A03 and A08 QTLs, respectively (Fedua–Agyeman et al. 2020)

on chromosomes A02 and A08. In addition, a single dominant gene, *Rcr7*, was identified that provided resistance to pathotypes 3H and 5X in the C genome on chromosome C07, which was homologous to a region on the A03 chromosome (Strelkov et al. 2018). In several cases, a modifier gene has been reported to increase the efficacy of a dominant CR gene against a virulent pathotype. The *Crr1* alone provided resistance to a moderately aggressive field collection, but resistance against a more aggressive field collection required the presence of the modifier locus *Crr2*. Enhancers and locus control regions are occasionally located away from candidate genes, with the result that communications may occur via histone modification or intra- and interchromosomal loops. Such unusual relationships can complicate use of marker-assisted selection for CR genes (Dean 2006).

Next-generation sequencing (NGS) technology is being used to accelerate conventional breeding. For example, high-throughput, sequence-based, single-nucleotide polymorphism (SNP) markers have largely replaced PCR and hybridization-based markers in the last decade. SNPs are the most frequent and the most abundant DNA sequence variant type in plant genomes, human genomes, and

Table 3.23 Clubroot reaction (resistant, R; susceptible, S) to pathotypes 3H and 5X of *Plasmodiophora brassicae* in parental lines, F_1, and backcross (BC_1) progeny in crosses between *Brassica rapa* lines 96-6990-2 and ACDC, compared to an expected ratio of 1:1 (Karim et al. 2020)

		Clubroot rating (0-3)					No. of plants and analysis			
Parents and crosses	Type	0	1	2	3	Total	*R*	S	2	*P*
Pathotype 3H										
96-6990-2	R parent	14	0	0	0	14	14	0		
ACDC	S parent	0	0	0	14	14	0	14		
ACDC 96-6990-2	F_1	7	0	0	0	7	7	0		
96-6990-2 ACDC	F_1	7	0	0	0	7	7	0		
ACDC F_1	BC_1	118	0	0	122	240	118	122	0.06	0.81
Pathotype 5X										
96-6990-2	R parent	7	0	0	0	7	7	0		
ACDC	S parent	0	0	0	2	7	0	7		
ACDC 96-6990-2	F_1	14	0	0	0	14	14	0		
96-6990-2 ACDC	F_1	14	0	0	0	14	14	0		
ACDC F_1	BC_1	77	0	0	60	137	77	60	2.10	0.15

other organisms (Van Nimwegen et al. 2016). The advantage of RNA sequencing (RNA-Seq) over other NGS technologies is that RNA-Seq can be applied to both genetic mapping and expression analysis. RNA-Seq also identifies higher numbers of functional variants from coding regions. Coding sequences represent only 1–2% of the genome, making downstream data handling and analysis of RNA much easier than with DNA. RNA-Seq has been used to identify functional variants related to disease resistance. Bulk-segregant analysis (BSA) identifies differences between two distinct characters. Using BSA together with RNA Seq (BSR-Seq) provides an efficient approach for the identification of differential gene expression and a high-density of SNP markers for gene mapping and marker-assisted selection (MAS). A *B. rapa* line that was resistant to several pathotypes of *P. brassicae* but genetically distant from genotypes known to carry the major CR gene *Rcr1* was selected to (1) map the CR gene(s) in the *B. rapa* line using BSR-Seq through analysis of percentage of polymorphic variants (PPV) described by Dakouri et al. (2018), (2) use genome-wide DNA variant analysis to identify SNP markers for MAS, and (3) identify the most probable candidate gene(s) responsible for the resistance in the *B. rapa* line(Karim et al. 2020)

BC1 plants of *Brassica rapa* from a cross of line 96-6990-2 (clubroot resistance originating from turnip cultivar "Waaslander") and a susceptible doubled-haploid line, ACDC, exhibited a 1:1 segregation for resistance against pathotypes 3H and 5X (Table 3.23). A resistance gene designated as *Rcr3* was mapped initially based on the percentage of polymorphic variants using bulked segregant RNA sequencing (BSR-Seq) and further mapped using Kompetitive Allele Specific PCR. DNA variants were identified by assembling short reads against a reference genome of *B. rapa*. *Rcr3* was mapped into chromosome A08. It was flanked by single nucleotide polymorphisms (SNP) markers (A90_A08_SNP_M12 and M16) between 10.00

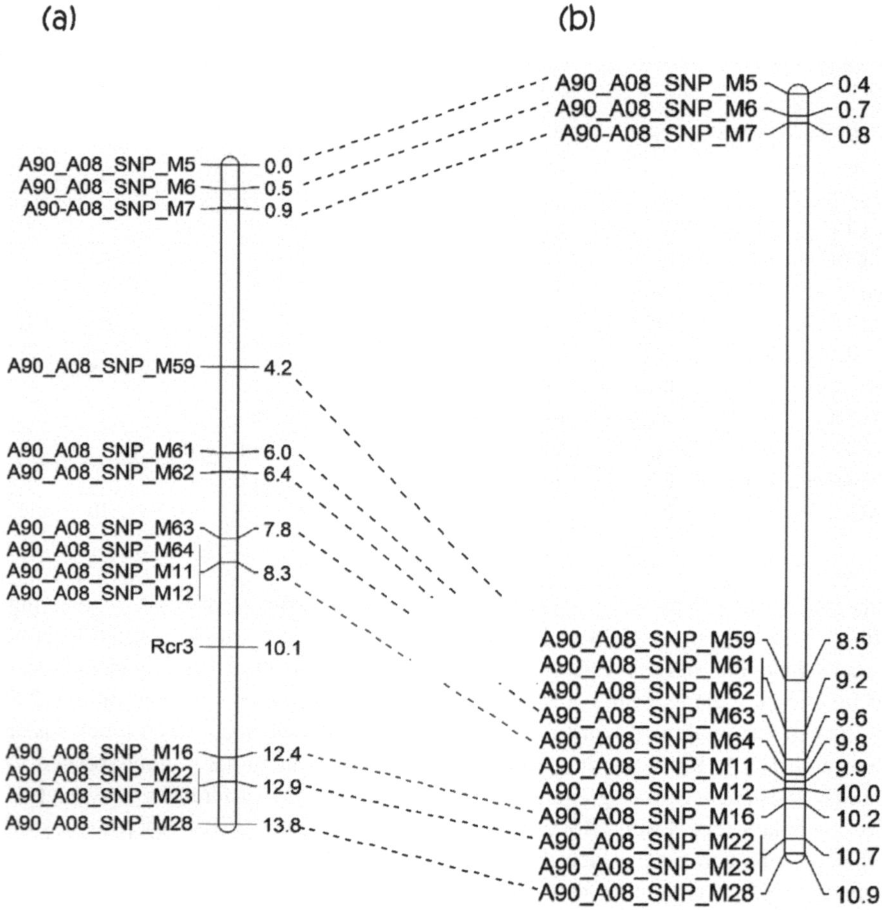

Fig. 3.10 Genetic linkage map based on BC1 populations of *Brassica rapa* line 96-6990-2; (**a**) genetic map of the region in which the Rcr3 gene is located (genetic distance on right), and (**b**) physical location of the Rcr3 region (in bases, on right), with SNP markers connected with a broken line between the maps (Karim et al. 2020)

and 10.23 Mb, in an interval of 231.6 Kb. There were 32 genes in the *Rcr3* interval. Three genes (*Bra020951*, *Bra020974*, and *Bra020979*) were annotated with disease resistance mechanisms, which are potential candidates for *Rcr3*. Another resistance gene, designated as $Rcr9^{wa}$, for resistance to pathotype 5X was mapped, with the flanking markers (A90_A08_SNP_M28 and M79) between 10.85 and 11.17 Mb using the SNP sites identified through BSR-Seq for *Rcr3*. There were 44 genes in the $Rcr9^{wa}$ interval, three of which (*Bra020827*, *Bra020828*, *Bra020814*) were annotated as immune-system-process related genes (Figs. 3.10 and 3.11), which are potential candidates for $Rcr9^{wa}$ (Karim et al. 2020).

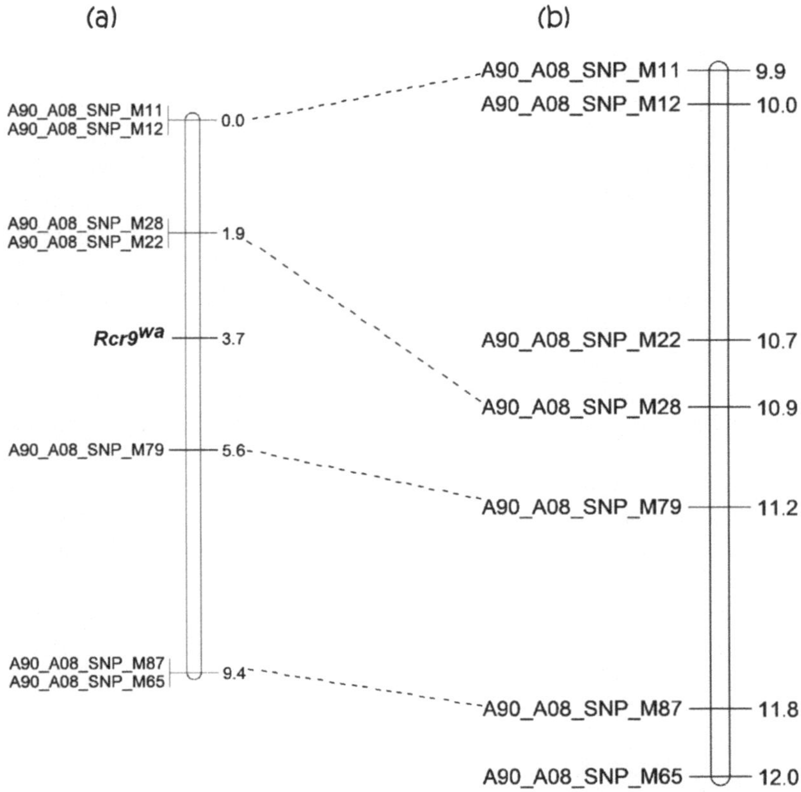

Fig. 3.11 Genetic linkage map based on BC1 populations of *Brassica rapa* line 96-6990-2; (**a**) genetic map of the region in which the Rcr9wa gene is located (genetic distance on right), and (**b**) physical location of the Rcr9wa region (in bases, on right), with SNP markers connected with a broken line between the maps (Karim et al. 2020)

3.8 *Brassica-Sclerotinia* Inheritance

3.8.1 Inheritance of Leaf and Stem Rot Resistance to *Sclerotinia* in *Brassica*

The inheritance of *Sclerotinia* resistance is complex, and no single consistent inheritance mode exists which is resistant to the disease. Several studies using rapeseed and other *Brassicas* have identified partial, dominant or polygenic control, including additive by additive epistatic interactions, which confer *Sclerotinia* resistance (Li et al. 2004; Wang et al. 2004; Zhao et al. 2004, 2006, 2009). One to ten quantitative trait loci (QTLs) responsible for *Sclerotinia* resistance has been reported (Yin et al. 2010). Some studies have shown differences in the number of QTLs in different populations and variations depending on the inoculation techniques applied. Working with F2:3 populations, Zhao and Meng (2003) reported a total of nine QTLs (six for leaf and three for stem) suggesting that *Sclerotinia* resistance in *B. napus* is tissue specific. A similar study using two populations of Double Haploid (DH) lines also found that different QTLs control the same trait in different populations (Zhao et al. 2006). These general results reflect the nature of disease resistance inheritance, indicating resistance is under strong environmental influence. Although *B. oleracea* was used to develop and improve *B. napus*, understanding of the genetics of *S. sclerotiorum* resistance in *B. oleracea* still remains poor. To understand the inheritance of *S. sclerotiorum* resistance in *B. oleracea*, a generation means analysis (Mather and Jinks 1982) was conducted using six generations including two parental lines (*Brassica incana* and *B. oleracea* var. *alboglabra*) (Mei et al. 2011) and four of their progeny.

The inheritance of leaf and stem rot resistance involving a single interspecific cross between *Brassica incana* and *Brassica oleracea* var. *alboglabra* was determined. Detached leaves and stems of parental lines, F1, F2, and the backcrosses were inoculated with *S. sclerotiorum* to determine resistance. Significant differences were detected across the two growing seasons between parents and some of the progeny for measures of both leaf and stem resistance. Continuous variation patterns among the segregating generations suggest the quantitative nature of resistance in both leaf and stem. Dominant and additive × additive epistatic interactions controlled the genetic effects for both traits. Broad- and narrow-sense heritability estimates were moderately high for leaf resistance, but were intermediate to high for stem resistance in the two field seasons. Low estimates of the minimum number of genes for leaf and stem resistance were recorded in the two field seasons. The results indicate that selection gains and the identification of quantitative trait loci can be maximized in marker-assisted-selection through differential selection (tissue-based selection) on a replicated plot basis (Disi et al. 2014).

3.8.2 Inheritance of Leaf Resistance to *Sclerotinia* in *Brassica* and Its Correlation with Cotyledon Resistance

Three *B. napus* breeding populations, C2 (NC-8 9 RQ-001-NCA-8 NC2-7), C5 (cv. Charlton 9 RQ-001- NCA-8 NC2-7), and C6 (cv. Charlton 9 NC4-5), were screened for leaf resistance (based on mean lesion diameter) under controlled environment conditions. Each population consisted of parents (P1 and P2), F1, F2, BC1P1 and BC2P2, except for C5 that lacked BC1P1. Moderate broad sense heritability for leaf resistance (0.45) to *S. sclerotiorum* was only found in population C6, where genetic variance was mostly non-additive. Analyses of generation means and variances indicated that both dominance and complex epistatic interactions were present in C6. Bivariate analysis revealed a positive genetic covariance between the non-additive effects for mean leaf lesion and cotyledon lesion diameters, and significant negative covariance of residuals, which supports a common genetic control of cotyledon and leaf resistance to *S. sclerotiorum*. Brassica breeders in selection and development of genotypes with both cotyledon and leaf resistance against Sclerotinia can plan accordingly (Khan et al. 2020).

3.9 *Brassica-Turnip Mosaic Virus* Inheritance

3.9.1 Inheritance of Disease Resistance to *TuMV* in *Brassica*

Extensive studies have revealed various types of inheritance of resistance to TuMV in *Brassicas*, including *B. rapa* (AA genome), *B. oleracea* (CC), *B. napus* (AACC), and *B. juncea* (AABB), which indicated that the genetic mechanism depends on the host, as well as the pathogen strain. However, inheritance of TuMV resistance has not been reported in *B. nigra* (BB genome) or *B. carinata* (BBCC genome) (Li et al. 2019).

3.9.1.1 *Brassica rapa*

There are dominant and recessive types of inheritance in *B. rapa* (Table 2.20, Chap. 2). Niu et al. (1983) used F2 and backcross (BC) segregating populations to demonstrate that resistance to the TuMV C1 strain was controlled by two independent dominant genes. In contrast, Yoon et al. (1993) found that the resistance of Chinese cabbage 022 to TuMV C1–C5 was conferred by two recessive genes. Suh et al. (1995) used resistant accessions BP058 and 022 as well as susceptible accessions ssd31 and B18 to generate segregating populations and demonstrated that resistance to individual strains was conferred by one or two dominant genes, while resistance to mixed strains was controlled by more than two major-effect genes and several minor-effect alleles.

In subsequent years, further genetic analysis conducted using DH, F2, or BC segregating populations led to the identification of new individual dominant genes, TuMVR (Chung et al. 2014), TuRB07 (Jin et al. 2014), and TuRBCS01 (Li et al. 2015; Gao et al. 2016). In addition to the QTLs and dominant genes for resistance, a few recessive genes have been identified. Rusholme et al. (2007) screened a recessive gene (*retr01*) on the upper portion of chromosome R4 along with the dominant gene ConTR01 and found that retr01 was epistatic to ConTR01. Fujiwara et al. (2011) found that resistance symptoms such as systemic lethal necrosis, leaf malformation, and mosaic were differentially expressed in *B. rapa* infected by the *TuMV* isolate UK1, according to the genotype at the host locus rnt1-2/rnt1-3 (Li et al. 2019).

3.9.1.2 *Brassica oleracea*

Some studies have revealed a dominant genetic architecture in *B. oleracea*. Pound and Walker (1945) were the first to perform inheritance analysis that revealed the resistance to TuMV is an incomplete dominant trait controlled by multiple loci; this is similar to the inheritance in Brussels sprouts, which contains at least four loci closely related to this resistance (Pink and Walkey 1986). Walkey and Pink (1988) estimated the TuMV resistance of four white cabbage cultivars using TuMV isolates from the UK, Germany, Denmark, and Greece and found that three cultivars (Decema Extra, Vitala, Winter White III) had high levels of resistance, and that the Polinius cultivar was highly susceptible. Pink and Walkey (1990) inoculated three white cabbage lines with the UK-NVRS strain of TuMV to analyze the inheritance effect and found that heritability of resistance ranged from 41% to 48% (Li et al. 2019).

3.10 *Brassica-Xanthomonas* Inheritance

3.10.1 Inheritance of Disease Resistance to *Xanthomonas* in *Brassica*

The inheritance of resistance to *X. campestris* pv. *campestris* has been studied mainly in *B. oleracea* and without knowledge of existing races. Bain (1955) attributed resistance in cabbage cv. Huguenot to one or more dominant genes. Williams et al. (1972) proposed a trigenic model to explain the segregation of progenies derived from the Japanese cabbage cv. Early Fuji. According to this model, resistance was controlled by one major recessive gene and two modifiers. Resistance in seedlings of the cabbage accession PI 436606 (cv. HehYeh da Ping Tou) was attributed to one recessive gene, while the resistance of several accessions of cabbage and cauliflower appeared to be controlled by multiple genes. Quantitative trait loci (QTL), assumed to control resistance in Badger Inbred-16. A line derived from cv. Early Fuji were mapped by Camargo et al. (1995). Ignatov et al. (1998) have attributed resistance to two *X. campestris* pv. *campestris* races (probably races 1 and 3) in *B. oleracea* to a recessive gene and to a dominant gene. Guo et al. (1991) attributed the strong resistance found in accessions PI 199947 and PI 199949 to a single dominant gene and the moderate level of resistance of PI 273640 to a recessive gene. These accessions originally were thought to be *B. napus*, but have been re-identified as *B. carinata*. Ignatov et al. (2000a, 2000b) suggested that a single dominant gene controlled resistance to race 4 in *B. rapa* (cvs. Just Right Turnip and Seven Top Green Turnip) and in *B. napus* (cv. Cobra). Vicente et al. (2002) elucidated the inheritance of race-specific resistance to *X. campestris* pv. *campestris* in a number of *Brassica* spp. accessions. These accessions were selected after the broad screening exercise reported previously.

The inheritance of resistance to three *Xanthomonas campestris* pv. *campestris* races was studied in crosses between resistant and susceptible lines of *Brassica oleracea* (C genome), *B. carinata* (BC genome), and *B. napus* (AC genome). Resistance to race 3 in the *B. oleracea* doubled haploid line BOH 85c and in PI 436606 was controlled by a single dominant locus (*Xca3*). Resistance to races 1 and 3 in the *B. oleracea* line Badger Inbred-16 was quantitative and recessive. Strong resistance to races 1 and 4 was controlled by a single dominant locus (*Xca1*) in the *B. carinata* line PI 199947. This resistance probably originates from the B genome. Resistance to race 4 in three *B. napus* lines, cv. Cobra, the rapid cycling line CrGC5, and the doubled haploid line N-o-1, was controlled by a single dominant locus (*Xca4*). A set of doubled haploid lines, selected from a population used previously to develop a restriction fragment length polymorphism map, was used to map this locus. *Xca4* was positioned on linkage group N5 of the *B. napus* A genome, indicating that this resistance originated from *B. rapa*. *Xca4* is the first major locus to be mapped that controls race-specific resistance to *X. campestris* pv. *campestris* in *Brassica* spp. (Vicente et al. 2002).

The genetics of major gene resistance has been studied by different groups in the diploid and allopolyploid *Brassica* species (Saha et al. 2014, 2016; Sharma et al. 2016; Table 2.22, Chap. 2). Contrary results on the genetics of Xcc have been observed using different genetic sources. Such as, both polygenic dominant (Sharma et al. 1972) and monogenic dominant resistance (Jamwal and Sharma 1986; Thakur et al. 2003) to black rot has been reported in Indian cauliflower. Similarly, monogenic dominant resistance to an Xcc isolate in the accession Columbia reported by Tsuji et al. (1991). The strong resistance to Xcc in accessions PI 199947 and PI 199949 contributed by a single dominant gene and the moderate level of resistance in PI 273640 attributed to a recessive gene has been reported (Guo et al. 1991). These accessions originally were thought to be of *B. napus* by Guo et al. 1991, but later have been re-identified as of *B. carinata* (Tonguc and Griffiths 2004). A study by Ignatov et al. (2000a, b) also suggested single dominant resistance to race 4 in *B. rapa* (e.g., cvs. Just Right Turnip and Seven Top Green Turnip) and in *B. napus* (cv. Cobra). Further, a single dominant race specific black rot resistance gene was mapped to the A genome in *B. napus* (Vicente et al. 2002), and QTLs controlling resistance to at least two of the most predominant races of Xcc were mapped in a Chinese cabbage accession of *B. rapa* (Soengas et al. 2007). Many reports of oligogenic and additive nature of resistance to black rot have also been published in *B. oleracea*. Vicente et al. (2002) reported that resistance to race 3 in the *B. oleracea* doubled haploid line BOH 85c and in the accession PI 436606 is controlled by a single dominant locus (*Xca3*). They also reported polygenic recessive control of resistance to races 1 and 3 in the *B. oleracea* line Badger Inbred-16 and high level of resistance to races 1 and 4 in the *B. carinata* line PI 199947 controlled by a single dominant locus (*Xca1*). Resistance to race 4 in three *B. napus* lines, cv. Cobra, the rapid cycling line CrGC5, and the DH line N-o-1, is controlled by a single dominant locus (*Xca4*). Kaur et al. (2009) found monogenic resistance to black rot in the cabbage line January King. Hence, by these results it is clear that the genetics of black rot resistance is very complex and genotype and race dependent (Singh et al. 2018).

References

Adam L, Somerville SC (1996) Genetic characterization of five powdery mildew disease resistance loci in *Arabidopsis thaliana*. Plant J 9(3):341–356

Adam L, Ellwood S, Wilson I, Saenz G, Xiao S, Oliver RP, Turner JG, Somerville S (1999) Comparison of *Erysiphe cichoracearum* and *E. cruciferarum* and a survey of 360 *Arabidopsis thaliana* accessions for resistance to these two powdery mildew pathogens. Mol Plant-Microbe Interact 12:1031–1043

Allen RL, Bittner-Eddy PD, Grenville-Briggs LJ, Meitz JC, Rehmany AP, Rose LE, Jl B (2004) Host-parasite co-evolutionary conflict between *Arabidopsis* and downy mildew. Science 306: 1957–1960. https://doi.org/10.1126/science.1104022

Ayers GW (1972) Clubroot resistance in tutabagas, turnips and cole crops. Agric Can Cnadex 163: 630. Loose leaf, 1 p

Ayers GW, Lelacheur KE (1972) Genetics of resistance in rutabaga to races of *Plasmodiophora brassicae*. Can J Plant Sci 52:897–900

Bailey K, Cevik V, Holton N, Byrne-Richardson J, Sohn KH, Coates M, Woods-Tor A, Aksoy HM, Hughes L, Baxter L, Jones JDG, Beynon J, Holub EB, Tor M (2011) Molecular cloning of ATR5 Emoy2 from *Hyaloperonospora arabidopsidis*, an avirulence determinant that triggers RPP5-mediated defense in *Arabidopsis*. Mol Plant-Microbe Interact 24:827–838. https://doi.org/10.1094/MPMI-12-10-0278

Bain D (1955) Resistance of cabbage to black rot. Phytopathology 45:35–37

Bains SS (1993) Differential reaction of leaves and young flowers of different cruciferous crops to *A. candida*. Plant Dis Res 8:70–72

Bansal VK, Thiagarajah MR, Stringam GR, Tewari JP (1999) Inheritance of partial resistance to race 2 of *Albugo candida* in canola-quality mustard (*Brassica juncea*) and its role in resistance breeding. Plant Pathol 48:817–822

Bansal VK, Tewari JP, Stringam GR, Thiagarajah MR (2005) Histological and inheritance studies of partial resistance in the *Brassica napus–Albugo candida* host–pathogen interaction. Plant Breed 124:27–32

Baxter L, Tripathy S, Ishaque N, Boot N, Cabral A, Kemen E, Thines M, Ah-Fong A, Anderson R, Badejoko W, Bittner-Eddy P, Boore JL, Chibucos MC, Coates M, Dehal P, Dele-haunty K, Dong S, Downton P, Dumas B, Fabro G, Fronick C, Fuerstenberg SI, Fulton L, Gaulin E, Govers F, Hughes L, Humphray S, Jiang RHY, Judelson H, Kamoun S, Kyung K, Meijer H, Minx P, Morris P, Nelson J, Phuntumart V, Qutob D, Rehmany A, Rougon-Cardoso A, Ryden P, Torto-Alalibo T, Studholme D, Wang Y, Win J, Wood J, Clifton SW, Rogers J, Van den Ackerveken G, Jones JDG, McDowell JM, Beynon J, Tyler BM (2010) Signatures of adaptation to obligate biotrophy in the *Hyaloperonospora arabidopsidis* genome. Science 330: 1549–1551. https://doi.org/10.1126/science.1195203.

Bell CJ, Ecker JR (1994) Assignment of 30 microsatellite loci to the linkage map of *Arabidopsis*. Genomics 19:137–144

Bisgrove SR, Simonich MT, Smith NM, Sattler A, Innes RW (1994) A disease resistance gene in *Arabidopsis* with specificity for two different pathogen avirulence genes. Plant Cell 6:927–933

Bonnet A (1981) Resistance to white rust in radish (*Raphanus sativus* L.). Cruciferae NewsLett 6: 60

Bonnet A, Blancard D (1987) Resistance of radish (*Raphanus sativus* L.) to downy mildew, *Peronospora parasitica*. Cruciferae NewsLett 12:98–99

Borhan MH, Gunn N, Cooper A, Gulden S, Tor M, Rimmer SR, Holub EB (2008) WRR4 encodes a TIR-NB-LRR protein that confers broad-spectrum white rust resistance in *Arabidopsis thaliana* to four physiological races of *Albugo candida*. MPMI 21:757–768

Botella MA, Parker JE, Frost LN, Bittner-Eddy PD, Beynon JL, Daniels MJ, Holub EB, Jones JDG (1998) Three genes of the *Arabidopsis* RPP1 complex resistance locus recognize distinct *Peronospora parasitica* avirulence determinants. Plant Cell 10:1847–1860

Buczacki ST, Toxopeus H, Mattusch P, Johnston TD, Dixon GR, Hobolth LA (1975) Study of physiologic specialization in *Plasmodiophora brassicae*: proposals for attempted rationalization through an international approach. Trans Br Mycol Soc 65:295–303. https://doi.org/10.1016/s0007-1536(75)80013-1

Camargo LEA, Williams PH, Osborn TC (1995) Mapping of quantitative loci controlling resistance to *Brassica oleracea*to *Xanthomonas campestris* pv. *campestris* in the field and green house. Phytopathology 85:1296–1300

Caravalho T, Monteiro A (1996) Preliminary study on the inheritance of resistance to downy mildew (*Peronospora parasitica* (Pers. ex. Fr.) at cotyledon stage in Tranchuda cabbage "Algarvia". Cruciferae NewsLett 18: 104.

Carlsson M, Bothmer RV, Merker A (2004) Screening and evaluation of resistance to downy mildew (*Peronospora parasitica*) and clubroot (*Plasmodiophora brassicae*) in genetic resources of *Brassica oleracea*. Hereditas 141:293–300. https://doi.org/10.1111/j.1601-5223.2004.01818.x

Carver TLW (1986) Histology of infection by *E. graminis* f. sp. *hordei* in spring barley lines with various levels of partial resistance. Plant Pathol 35:232–240

Chiang MS, Crête R (1970) Inheritance of clubroot resistance in cabbage (*Brassica oleracea* L. var. *capitata*). Can J Genet Cytol 12:253–256

Chiang MS, Chiang BY, Grant WF (1977) Transfer of resistance to race 2 of *Plasmodiophora brassicae* from *Brassica napus* to cabbage (*B. oleracea* var. *capitata*). I. Interspecific hybridization between *B. napus* and *B. oleracea* var. *capitata*. Euphytica 26:319–336

Cho KS, Han YH, Lee JT, Hur EJ, Yang TJ, Woo JG (2002) Pathogenic differentiation of *Plasmodiophora brassicae* and selection of Chinese cabbage cultivars resistant to clubroot disease in highland. Korean J Breed 34:168–173

Chu M, Song T, Falk KC, Zhang X, Liu X, Chang A, Lahlali R, McGregor L, Gossen BD, Yu F, Peng G (2014) Fine mapping of *Rcr1* and analyses of its effect on transcriptome patterns during infection by *Plasmodiophora brassicae*. BMC Genomics 15:1166. https://doi.org/10.1186/1471-2164-15-1166

Chung H, Jeong YM, Mun JH, Lee SS, Chung WH, Yu HJ (2014) Construction of a genetic map based on high-throughput SNP genotyping and genetic mapping of a TuMV resistance locus in *Brassica rapa*. Mol Gen Genomics 289:149–160

Comstock RE, Robinson HF (1952) Estimation of average dominance of genes. In: Gowen JW (ed) Heterosis. Iowa State College Press, Ames, pp 494–516

Crisp P, Crute IR, Sutherland RA, Angell SM, Bloor K, Burgess H, Gordon PL (1989) The exploitation of genetic resources of *Brassica oleracea* in breeding for resistance to clubroot (*Plasmodiophora brassicae*). Euphytica 42:215–226

Crute IR, Pink DAC (1989) The characteristics and inheritance of *resistance* to *clubroot* in *Brassica oleracea*. Aspects Appl Biol 23:57–60

Crute IR, Gray AR, Crisp P, Buczacki ST (1980) Variation in *Plasmodiophora brassicae* and resistance to clubroot disease in *Brassicas* and allied crops—a critical review. Plant Breed Abstr 50:91–104

Crute IR, Phelps K, Barnes A, Buczacki ST, Crisp P (1983) The relationship between genotypes of three *Brassica* species and collections of *Plasmodiophora brassicae*. Plant Pathol 32:405–420

Crute IR, Beynon J, Dangl J, Holub E, Mauch-Mani B, Slusarenko A, Staskawicz B, Ausubel F (1994) Microbial pathogenesis of *Arabidopsis*. Cold Spring Harbor Laboratory Press:705–747

Dakouri A, Zhang X, Peng G, Falk KC, Gossen BD, Strelkov SE, Yu F (2018) Analysis of genome-wide variants through bulked segregant RNA sequencing reveals a major gene for resistance to *Plasmodiophora brassicae* in *Brassica oleracea*. Sci Reptr 8:17657. https://doi.org/10.1038/s41598-018-36187-5

Dean A (2006) On a chromosome far, far away: LCRs and gene expression. Trends Genet 22:38–45

Delwiche PA, Williams PH (1974) Resistance to *Albugo candida* race 2 in *Brassica* sp. Proc Am Phytopathology Soc 1:66. (Abstr)

Delwiche PA, Williams PH (1976) Identification of marker genes in *Brassica nigra*. Proc Am Phytopathol Soc 3: 234. (Abstr.)

Delwiche PA, Williams PH (1977) Genetic studies in *Brassica nigra* (L.) Koch. Cruciferae NewsLett 2:39

Delwiche PA, Williams PH (1981) Thirteen marker genes in *Brassica nigra*. J Hered 72:289–290

Dias JS, Ferrcira ME, Williams PH (1993) Screening of Portuguese cole landraces (*Brassica oleracea* L.) with *Peronospora parasitica* and *Plasmodiophora brassicae*. Euphytica 67:135–141

Dickson MH, Petzoldt R (1996) Seedling of downy mildew resistance in *B. oleracea*. Cruciferae Newslett 18:108

Disi JO, Mei J, Wei D, Ding Y, Qian W (2014) Inheritance of leaf and stem resistance to *Sclerotinia sclerotiorum* in a cross between *Brassica incana* and *Brassica oleracea* var. *alboglabra*. J Agril Sci 152:146–152

Dixon GR, Doodson JK, Beeney BW, Davies H, Jemoiett JL, Moxon RH (1972) Studies of resistance in swede stocks to clubroot (*Plasmodiophora brassicae* Woronin). J Natl Inst Agril Bot 12:456–463

Ebrahimi AG, Delwiche PA, Williams PH (1976) Resistance in *Brassica juncea* to *Peronospora parasitica* and *Albugo candida* Race 2. Proc Am Phytopathol Soc 3:273. (Abstr)

Edwards M, Williams PH (1982) Selection for quantitatively inherited resistance to *Albugo candida* race 2 in *B. campestris*, CGS-1. Cruciferae NewsLett 7:66–67

Edwards MD, Williams PH (1987) Selection for minor gene resistance to *Albugo candida* in a rapid-cycling population of *Brassica campestris*. Phytopathology 77:527–532

Fan Z, Rimmer SR, Stefansson BR (1983) Inheritance of resistance to *Albugo candida* in rape (*Brassica napus* L). Can J Genet Cytol 25:420–424

Ferreira ME, Williams PH, Osborn TC (1995) Mapping of locus controlling resistance to *Albugo candida* in *Brassica napus* using molecular markers. Phytopathology 85:218–220

Frauen M (1999) A new clubroot resistant variety in winter oilseed rape. In: Proc 10th Intl rapeseed Cong, Canberra, ACT.

Fredua-Agyeman R, Rahman H (2016) Mapping of the clubroot disease resistance in spring *Brassica napus* canola introgressed from European winter canola cv. 'Mendel.' Euphytica 211:201–213. https://doi.org/10.1007/s10681-016-1730-2.

Fredua-Agyeman R, Jiang J, Hwang SF, Strelkov SE (2020) QTL mapping and inheritance of clubroot resistance genes derived from *Brassica rapa* subsp. *rapifera* (ECD 02) reveals resistance loci and distorted segregation ratios in two F2 populations of different crosses. Front Plant Sci 11:899. https://doi.org/10.3389/fpls.2020.00899

Frye CA, Innes RW (1998) An Arabidopsis mutant with enhanced resistance to powdery mildew. Plant Cell 10:947–956

Fuchs H, Sacristan MD (1996) Identification of a gene in *Arabidopsis thaliana* controlling resistance to clubroot (*Plasmodiophora brassicae*) and characterization of the resistance response. Mol Plant-Microbe Interact 9:91–97

Fujiwara A, Inukai T, Kim BM, Masuta C (2011) Combinations of a host resistance gene and the CI gene of Turnip mosaic virus differentially regulate symptom expression in *Brassica rapa* cultivars. Archiv Virol 156:1575–1581

Gadewadikar PN, Bhadouria SS, Bartaria AM (1993) Inheritance of resistance to white rust (*Albugo candida*) disease in Indian mustard (*Brassica juncea*). In: Natl Sem, oilseeds Research and Development in India: status and strategies. Aug 2–4, 1993.

Gao Y, Zhang Y, Yi J, Zhou J, Huang X, Shi X, Xiao S, Lin D (2016) A longitudinal study on urinary cadmium and renal tubular protein excretion of nickel-cadmium battery workers after cessation of cadmium exposure. Intl Arch Occup Environ Health 89:1137–1145

Gates FE, Coastock RE, Robinson HF (1957) Generalized genetic variation and covariance formulae for self-fertilized crop assuming linkage. Genet 42:749–763

Geiger HH, Heun M (1989) Genetics of quantitative resistance to fungal diseases. Annu Rev Phytopathol 27:317–341

Glazebrook J (2005) Contrasting mechanisms of defense against biotrophic and necrotrophic pathogens. Annu Rev Phytopathol 43:205–227

Gollner K, Schweizer P, Bai Y, Panstruga R (2008) Natural genetic resources of *Arabidopsis thaliana* reveals a high prevalence and unexpected phenotypic plasticity of RPW8-mediated powdery mildew resistance. New Phytol 1776(1):725–742

Goritschnig S, Krasileva KV, Dahlbeck D, Staskawicz BJ (2012) Computational prediction and molecular characterization of an oomycete effector and the cognate *Arabidopsis* resistance gene. PLoS Genet 8:e1002502

Grant MR, Godiard L, Straube E, Ashfield T, Lewald J, Sattler A, Innes RW, Dangl JL (1995) Structure of the *Arabidopsis* RPM1 gene enabling dual-specificity disease resistance. Sci 269: 843–846

Guo H, Dickson MH, Hunter JE (1991) *Brassica napus* sources of resistance to black rot in crucifers and inheritance of resistance. Hort Sci 26:1545–1547

Gupta K, Saharan GS, Singh D (2002) Effective selection criteria to assess disease tolerance in Indian mustard [*Brassica juncea* (L.) Czern & Coss.]. J Mycol Plant Pathol 32:72–80

Gustafsson M, Falt AS (1986) Genetic studies on resistance to clubroot in *Brassica napus*. Ann Appl Biol 108:409–415

Hall D (1994) Interactions of Arabidopsis with fungal pathogens. Ph.D. thesis. University of East Anglia, Norwich

Hasan MJ, Rahman H (2013) Genetics of resistance in rutabaga (*Brassica napus* var. *napobrassica*) to Canadian *Plasmodiophora brassicae* pathotypes. In: Proc 2013 Intl Clubroot workshop, Edmonton, MB.

Hasan MJ, Rahman H (2011) Genetics of resistance in rutabaga (*Brassica napus* var. *napobrassica*) to Canadian *Plasmodiophora brassicae* pathotype 3. In: Proc 13th Intl rapeseed Cong, Prague, 691–694.

Hasan MJ, Rahman H (2016) Genetics and molecular mapping of resistance to *Plasmodiophora brassicae* pathotypes 2, 3, 5, 6, and 8 in rutabaga (*Brassica napus* var. *napobrassica)*. Genome 59:805–815. https://doi.org/10.1139/gen-2016-0034

Hatakeyama K, Suwabe K, Tomita RN, Kato T, Nunome T, Fukuoka H, Matsumoto S (2013) Identification and characterization of Crr1a, a gene for resistance to clubroot disease (*Plasmodiophora brassicae* Woronin) in *Brassica rapa* L. PLoS One 8:e54745. https://doi.org/10.1371/journal.pone.0054745.

Hill C, Crute I, Sherriff C, Williams PH (1988) Specificity of *Albugo candida* and *Peronospora parasitica* pathotypes toward rapid-cycling crucifers. Cruciferae Newslett 13:112–113

Hirai M (2006) Genetic analysis of clubroot resistance in *Brassica* crops. Breed Sci 56:223–229. https://doi.org/10.1270/jsbbs.56.223

Hirai M, Harada T, Kubo N, Tsukada M, Suwabe K, Matsumoto S (2004) A novel locus for clubroot resistance in *Brassica rapa* and its linkage markers. Theor Appl Genet 108:639–643. https://doi.org/10.1007/s00122-003-1475-x

Hirani AH, Gao F, Liu J, Fu G, Wu C, McVetty PBE, Duncan RW, Li G (2018) Combinations of independent dominant loci conferring clubroot resistance in all four turnip accessions (*Brassica rapa*) from the European clubroot differential set. Front Plant Sci 9:1628

Holub EB (2006) Evolution of parasitic symbiosis between plants and filamentous microorganisms. Curr Opin Plant Biol 9:397–405. https://doi.org/10.1016/j.pbi.2006.05.011

Holub EB, Beynon JL, Crute IR (1994) Phenotypic and genotypic characterization of interactions between isolates of *Peronospora parasitica* and accessions of *Arabidopsis thaliana*. Mol Plant-Microbe Interact 7:223–239

Holub EB, Brose E, Tor M, Clay C, Crute IR, Beynon JL (1995) Phenotypic and genotypic variation in the interaction between *Arabidopsis thaliana* and *Albugo candida*. Mol Plant-Microbe Interact 8:916–928

Hoser-Krauze J, Lakowska-Ryk E, Antosik J (1991) Resistance to some *Brassica oleracea* L. plant introductions to downy mildew (*Peronospora parasitica*). Cruciferae Newslett 14:144–145

Hoser-Krauze J, Lakowska-Ryk E, Antosik J (1995) The inheritance of broccoli (*Brassica oleracea botrytis*) leaf resistance to downy mildew– *Peronospora parasitica* (Pers. ex. Fr.). J Appl Genet 36:27–33

Hougas RW, Rieman GH, Stokes GW (1952) Resistance to white rust in horseradish seedlings. Phytopathology 42:109–110

Humaydan HS, Williams PH (1976) Inheritance of seven characters in *Raphanus sativus*. Hort Sci 11:146–147

Ignatov A, Kuginuki Y, Hida K (1998) Race-specific reaction of resistance to black rot in *Brassica oleracea*. Eur J Plant Pathol 104:821–827

Ignatov A, Kuginuki Y, Hidam K (2000a) Distribution and inheritance of race-specific resistance to *Xanthomonas campestris* pv. *campestris* in *Brassica rapa* and *B. napus*. J Russ Phytopathol Soc 1:89–94

Ignatov AN, Kuginuki NY, Suprunova TP, Pozmogova GE, Seitova A, Dorokhov DB (2000b) RAPD-markers linked to the locus for resistance to the race 4 pathogen for black rot, *Xanthomonas campestris* pv. *campestris* (Pamm.) Dowson in *Brassica rapa* L. Genetika 36(3):357–360

James RV, Williams PH, Maxwell DP (1978) Inheritance and linkage studies related to resistance in *Brassica campestris* L. to *Plasmodiophora brassicae* race 6. Eucarpia Cruciferae Newslett 3:27

Jamwal RS, Sharma PP (1986) Inheritance of resistance to black rot (*Xanthomonas campestris* pv. *campestris*) in cauliflower (*Brassica oleracea* var. *botrytis*). Euphytica 35:941–943

Jat RR (1999) Pathogenic variability and inheritance of resistance to *Albugo candida* in oilseed *Brassica*. PhD Thesis, CCSHAU, Hisar, p 129

Jensen BD, Hockenhull J, Munk L (1999a) Seedling and adult plant resistance to downy mildew (*Peronospora parasitica*) in cauliflower (*Brassica oleracea* convar. *Botrytis* var. *botrytis*). Plant Pathol 48:604–612. https://doi.org/10.1046/j.1365-3059.1999.00388.x

Jensen BD, Værbak S, Munk L, Andersen SB (1999b) Characterization and inheritance of partial resistance to downy mildew, *Peronospora parasitica*, in breeding material of broccoli, *Brassica oleracea* con var. *botrytis* var. *italica*. Plant Breed 118:549–554

Jin F, Hu L, Yuan D, Xu J, Gao W, He L, Yang X, Zhang X (2014) Comparative transcriptome analysis between somatic embryos (SEs) and zygotic embryos in cotton: evidence for stress response functions in SE development. Plant Biotechnol J 12:161–173. https://doi.org/10.1111/pbi.12123

Johnston TD (1970) A new factor for resistance to club root in *Brassica napus* L. Plant Pathol 12: 156–158

Jones JDG, Dangl JL (2006) The plant immune system. Nat 444:323–329

Joos HJ, Mauch-Mani B, Slusarenko AJ (1996) Molecular mapping of the *Arabidopsis* locus RPP11 which conditions isolate-specific hypersensitive resistance against downy mildew in ecotype RLD. Theor Appl Genet 92:281–284

Jorgensen JH (1994) Genetics of powdery mildew resistance in barley. Crit Rev Plant Sci 13:97–119

Kamoun S (2006) A catalogue of the effector secretome of plant pathogenic oomycetes. Annu Rev Phytopathol 44:41–60

Karim MM, Dakouri A, Zhang Y, Chen Q, Peng G, Strelkov SE, Gossen BD, Yu F (2020) Two clubroot-resistance genes, Rcr3 and Rcr9wa, mapped in *Brassica rapa* using bulk segregant RNA sequencing. Intl J Mol Sci 21:5033. https://doi.org/10.3390/ijms21145033

Karling JS (1968) The plasmodiophorales. Hafner Publishing, New York

Kato T, Hatakeyama K, Fukino N, Matsumoto S (2012) Identification of a clubroot resistance locus conferring resistance to a *Plasmodiophora brassicae* classified into pathotype group 3 in Chinese cabbage (*Brassica rapa* L.). Breed Sci 62:282–287. https://doi.org/10.1270/jsbbs.62.282

Kato T, Hatakeyama K, Fukino N, Matsumoto S (2013) Fine mapping of the clubroot resistance gene CRb and development of a useful selectable marker in *Brassica rapa*. Breed Sci 63:116–124. https://doi.org/10.1270/jsbbs.63.116

Kaur R, Saxena Shivani B, Kanwar HS, Dohroo NP, Majeed S, Sharma DR (2009) Detecting RAPD markers associated with black rot resistance in cabbage (*Brassica oleracea* var. *capitata*). Fruit Veg Cereal Sci Biotech 13:12–15

Khan MA, Cowling W, Banga SS, You MP, Tyagi V, Bharti B, Barbetti MJ (2020) Patterns of inheritance for cotyledon resistance against *Sclerotinia sclerotiorum* in *Brassica napus*. Euphytica 216(5):79. https://doi.org/10.1007/s10681-020-02612-y

Kole C, Teutonico R, Mengistu A, Williams PH, Osborn TC (1996) Molecular mapping of a locus controlling resistance to *Albugo candida* in *Brassica rapa*. Phytopathology 86:367–369

Kole C, Williams PH, Rimmer SR, Osborn TC (2002) Linkage mapping of genes controlling resistance to white rust (*Albugo candida*) in *Brassica rapa* (syn. *Campestris*) and comparative mapping to *Brassica napus* and *Arabidopsis thaliana*. Genome 45:22–27

Kolte SJ (1987) Important diseases of rapeseed and mustard in India—present research progress and future research needs. Proc 3rd oil crops network workshop, Ethiopia, pp 91–106.

Konieczny A, Ausubel FM (1993) A procedure for mapping *Arabidopsis* mutations using co-dominant ecotype-specific PCR-based markers. Plant J 4:403–410

Krishnia SK, Saharan GS, Singh D (2000) Genetic variation for multiple disease resistance in the families of interspecific cross of *Brassica juncea* × *B. carinata*. Cruciferae Newslett 22:51–53

Kuginuki Y, Yoshikawa H, Hirai M (1999) Variation in virulence of *Plasmodiophora brassicae* in Japan tested with clubroot resistant cultivars of Chinese cabbage (*Brassica rapa* L. spp. *pekinensis*). Eur J Plant Pathol 105:327–332

Kumar B, Kolte SJ (2001) Progression of Alternaria blight of mustard in relation to components of resistance. Indian Phytopath 54:329–331

Kumar B, Kolte SJ (2006) Development of Alternaria blight in genotypes of Indian mustard [*B. juncea* (L.) Czern & Cross.] under field. Indian Phytopath 39:314–317

Kumar S, Saharan GS, Singh D (2002) Inheritance of resistance in inter and intraspecific crosses of *Brassica juncea* and *Brassica carinata* to *Albugo candida* and *Erysiphe cruciferarum*. J Mycol Pl Pathol 32(1):59–63

Kunkel BN (1996) A useful weed put to work – genetic analysis of disease resistance in *Arabidopsis thaliana*. Trends Genet 12:63–69

Laila R, Park JI, Robin AHK, Natarajan S, Vijayakumar H, Shirasawa K, Isobe S, Kim HT, Nou IS (2019) Mapping of a novel clubroot resistance QTL using ddRAD-seq in Chinese cabbage (*Brassica rapa* L.). BMC Plant Biol 19:13. https://doi.org/10.1186/s12870-018-1615-8

Lammerink J (1967) The inheritance of clubroot resistance in *Brassica napus* L. N Z J Agric Res 10:109–115

Laurens F, Thomas G (1993) Inheritance of resistance to clubroot (*Plasmodiophora brassicae* Worn.) in kale (*Brassica oleracea* ssp. *acephala*). Hereditas 119:253–262

Leboldus JM, Manolii VP, Turkington TK, Strelkov SE (2012) Adaptation to *brassica* host genotypes by a single-spore isolate and population of *Plasmodiophora brassicae* (clubroot). Plant Dis 96:833–838. https://doi.org/10.1094/pdis-09-11-0807

Leonelli L, Pelton J, Schoeffler A, Dahlbeck D, Berger J, Wemmer DE, Staskawicz B (2011) Structural elucidation and functional characterization of the *Hyaloperonospora arabidopsidis* effector protein ATR13. PLoS Pathog 7:e1002428

Leung H, Williams PH (1983) Cytoplasmic male-sterile *Brassica campestris* breeding lines with resistance to clubroot, turnip mosaic and downy mildew. Hort Sci 18:774–775

Li R, Rimmer R, Buchwaldt L, Sharpe AG, Seguinswartz G, Coutu C, Hegedus DD (2004) Interaction of *Sclerotinia sclerotiorum* with a resistant *Brassica napus* cultivar: expressed sequence tag analysis identifies genes associated with fungal pathogenesis. Fungal Genet Biol 41:735–753

Li J, Zhao Z, Hayward A, Cheng H, Fu D (2015) Integration analysis of quantitative trait loci for resistance to *Sclerotinia sclerotiorum* in *Brassica napus*. Euphytica 205:483–489. https://doi.org/10.1007/s10681-015-1417-0

Li L, Luo Y, Chen B, Xu K, Zhang F, Li H, Huang Q, Xiao X, Zhang T, Hu J, Li F, Wu X (2016) A genome-wide association study reveals new loci for resistance to clubroot disease in *Brassica napus*. Front Plant Sci 7:1483. https://doi.org/10.3389/fpls.2016.01483

Li LS, Luo C, Huang F, Liu ZL, An ZY, Dong L, He XH (2019) Identification and characterization of the mango eIF gene family reveals MieIF1A-a, which confers tolerance to salt stress in transgenic *Arabidopsis*. Sci Horticult 248:274–281

Liu Q, Rimmer SR (1992) Inheritance of resistance in *Brassica napus* to an Ethiopian isolate of *Albugo candida* from *Brassica carinata*. Can J Plant Pathol 14:116–120

Liu O, Rimmer SR, Scarth R, McVetty PBE (1987) Confirmation of a digenic model of inheritance of resistance to *Albugo candida* race 7 in *Brassica napus*. Proc 7th Intl rapeseed Cong, Poznam, Polland, pp 1204–1209

Liu JQ, Parks P, Rimmer SR (1996) Development of monogenic lines for resistance to *Albugo candida* from a Canadian *Brassica napus* cultivar. Phytopathology 86:1000–1004

Liu X, Xing M, Kong C, Fang Z, Yang L, Zhang Y, Wang Y, Ling J, Yang Y, Lv H (2019) Genetic diversity, virulence, race profiling, and comparative genomic analysis of the *Fusarium oxysporum* f. sp. *conglutinans* strains infecting cabbages in China. Front Microbiol 10:1373. https://doi.org/10.3389/fmicb.2019.01373

Lucas JA, Crute IR, Sheriff C, Gordon PL (1988) The identification of a gene for race specific resistance to *Peronospora parasitica* (downy mildew) in *Brassica napus* var. *oleifera* (oilseed rape). Plant Pathol 37:538–545

Lv H, Wang Q, Yang L, Fang Z, Liu Y, Zhuang M, Zhang Y, Yang Y, Xie B, Wang X (2014) Breeding of cabbage (*Brassica oleracea* L. var. *capitata*) with Fusarium wilt resistance based on microspore culture and marker–assisted selection. Euphytica 200:465–473. https://doi.org/10.1007/s10681-014-1197-y

Lv H, Fang Z, Yang L, Zhang Y, Wang Y (2020) An update on the arsenal: mining resistance genes for disease management of *Brassica* crops in the genomic era. Hortic Res 7:34. https://doi.org/10.1038/s41438-020-0257-9

Mahajan V, Gill HS, More TA (1995) Inheritance of downy mildew resistance in Indian cauliflower (group III). Euphytica 86:1–3

Mahajan V, Gill HS, Sharma SR, Singh R (1996) Combining ability studies in Indian cauliflower (*Brassica oleracea* var. *botrytis* L.) group III. Veg Sci 23:166–170

Mani N, Gulati SC, Raman R, Raman R (1996) Breeding for genetic resistance to white rust in Indian mustard. Crop Improv 23:75–79

Manzanares-Dauleux MJ, Divaret I, Baron F, Thomas G (2000a) Evaluation of French *Brassica oleracea* landraces for resistance to *Plasmodiophora brassicae*. Euphytica 113:211–218. https://doi.org/10.1023/A:1003997421340.

Manzanares-Dauleux MJ, Delourme R, Baron F, Thomas G (2000b) Mapping of one major gene and of QTLs involved in resistance to clubroot in *Brassica napus*. Theor Appl Genet 101:885–891. https://doi.org/10.1007/s001220051557

Mather K, Jinks JL (1982) Biometrical genetics, 3rd edn. Chapman and Hall, London

Matsumoto E, Yasui C, Ohi M, Tsukada M (1998) Linkage analysis of RFLP markers for clubroot resistance and pigmentation in Chinese cabbage (*Brassica rapa* ssp. *pekinensis)*. Euphytica 104: 79–86. https://doi.org/10.1023/A:1018370418201

Matsumoto E, Ueno H, Aruga D, Sakamoto K, Hayashida N (2012) Accumulation of three clubroot resistance genes through marker-assisted selection in Chinese cabbage (*Brassica rapa* ssp. *pekinensis)*. J Jpn Soc Hortic Sci 81:184–190. https://doi.org/10.2503/jjshs1.81.184

Matzinger DF, Cockerham CC (1963) Simultaneous selfing and partial diallel set crossing I. estimation of genetic and environmental parameters. Crop Sci 3:309–314

Mei J, Qian L, Disi JO, Yang X, Li Q, Li J, Frauen M, Cai D, Qian W (2011) Identification of resistant sources against *Sclerotinia sclerotiorum* in *Brassica* species with emphasis on *B. oleracea*. Euphytica 177:393–399

Mitchell-Olds T, James RV, Palmer MJ, Williams PH (1995) Genetics of *Brassica rapa* (syn. *campestris*). 2. Multiple disease resistance to three fungal pathogens: *Peronospora parasitica*, *Albugo candida* and *Leptosphaeria maculans*. Heredity 75:362–369

Monteiro AA, Williams PH (1989) The exploration of genetic resources of Portuguese cabbage and kale for resistance of several *Brassica* diseases. Euphytica 41:15–225. https://doi.org/10.1007/BF00021588

Moss NA, Crute IR, Lucas JA, Gordon PL (1988) Requirements for analysis of host-species specificity in *Peronospora parasitica* (downy mildew). Cruciferae Newslett 13:114–116

Nashaat NI, Awasthi RP (1995) Evidence for differential resistance to *Peronospora brassicae* (downy mildew) in accessions of *Brassica juncea* (mustard) at the cotyledon stage. J Phytopathol 143:157–159

Nashaat NI, Mitchell S, Awasthi RP (1995a) Genetic analysis of a noval factor for resistance to *Peronospora parasitica* in *Brassica napus* spp. *oleifera*. Indian J Mycol Plant Pathol 25:55

Nashaat NI, Mitchell SE, Awasthi RP (1995b) Inheritance of two new factors for resistance to downy mildew (*Peronospora parasitica*) in oilseed rape (*Brassica napus* spp. *oleifera*). Proc 9th Intl Cong, Cambridge, July 4-7, 1995, 4: pp 1283–1285

Nashaat NI, Heran A, Mitchell S (1996) Resistance to downy mildew and its interaction with white rust in rapeseed-mustard. In: 2nd Intl Crop Sci Cong, New Delhi, India, November 17–24, 1996, p 199

Nashaat NI, Heran A, Mitchell SE, Awasthi RP (1997) New genes for resistance to downy mildew (*Peronospora parasitica*) in oilseed rape (*Brassica napus* ssp. *oleifera*). Plant Pathol 46:964–968

Nashaat NI, Heran A, Awasthi RP, Kolte SJ (2004) Differential response and genes for resistance to *Peronospora parasitica* (downy mildew) in *Brassica juncea* (mustard). Plant Breed 123:512–515

Natti JJ, Dickson MH, Atkin JD (1967) Resistance of *Brassica oleracea* varieties to downy mildew. Phytopathology 57:144–147

Newton AC (1993) The effect of humidity on the expression of partial resistance to powdery mildew in barley. Plant Pathol 42:364–367

Nguyen ML, Monakhos GF, Komakhin RA, Monakhos SG (2018) The new Clubroot resistance locus is located on chromosome A05 in Chinese cabbage (*Brassica rapa* L.). Russ J Genet 54: 296–304. https://doi.org/10.1134/S1022795418030080

Niu XK, Leung H, Williams PH (1983) Sources and nature of resistance to downy mildew and turnip mosaic in Chinese cabbage. J Am Soc Hortic Sci 108:775–778

Orgil U, Araki H, Tangchaiburana S, Berkey R, Xiao S (2007) Intraspecific genetic variations, fitness cost and benefit of RPW8, a disease resistance locus in *Arabidopsis thaliana*. Genet 1766(1):2317–2333

Pal Y, Singh H, Singh D (1991) Genetic components of variation for white rust resistance in Indian × Exotic crosses of Indian mustard. Crop Res 4:280–283

Paladhi MM, Prasad RC, Dass B, Dass B (1993) Inheritance of field reaction to white rust in Indian mustard. Indian J Genet Plant Breed 53:327–328

Pang W, Fu P, Li X, Zhan Z, Yu S, Piao Z (2018) Identification and mapping of the clubroot resistance gene *CRd* in Chinese cabbage (*Brassica rapa* ssp. *pekinensis*). Front. Plant Sci 9:653. https://doi.org/10.3389/fpls.2018.00653

Parker JE, Szabo V, Staskawicz BJ, Lister C, Dean C, Daniels MJ, Jones JDG (1993) Phenotypic characterization and molecular mapping of the *Arabidopsis thaliana* locus *RPP5*, determining disease resistance to *Peronospora parasitica*. Plant J 4:821–831

Parlevliet JE (1992) Selecting components of partial resistance. In: Stalker HT, Murphy JP (eds) Plant breeding in the 1990s, proc symposium on plant breeding in the 1990s. CAB International, Wallingford

Piao ZY, Deng YQ, Choi SR, Park YJ, Lim YP (2004) SCAR and CAPS mapping of *CRb*, a gene conferring resistance to *Plasmodiophora brassicae* in Chinese cabbage (*Brassica rapa* ssp. *pekinensis)*. Theor Appl Genet 108:1458–1465. https://doi.org/10.1007/s00122-003-1577-5

Piao Z, Ramchiary N, Lim YP (2009) Genetics of clubroot resistance in *brassica* species. J Plant Growth Regul 28:252–264. https://doi.org/10.1007/s00344-009-9093-8

Pink DAC, Walkey DGA (1986) Resistance to turnip mosaic virus and cauliflower mosaic virus in horticultural Brassicas. Aspects Appl Biol 12:69–73

Pink DAC, Walkey DGA (1990) Resistance to turnip mosaic virus in white cabbage. Euphytica 51: 101–107

Pound GS, Walker JC (1945) Differentiation of certain crucifer viruses by the use of temperature and host immunity reactions. Aust J Agric Res 71:255–278

Pound GS, Williams PH (1963) Biologal races of *Albugo candida*. Phytopathology 53:1146–1149

Prabhu KV, Somers DJ, Rakow G, Gugel RK (1998) Molecular markers linked to white rust resistance in mustard *Brassica juncea*. Theor Appl Genet 97:865–870

Rao MVB, Raut RN (1994) Inheritance of resistance to white rust (*Albugo candida*) in an interspecific cross between Indian mustard (*Brassica juncea*) and rapeseed (*Brassica napus*). Indian J Agril Sci 64:249–251

Reddy KS, Pawar SE, Bhatia CR (1994) Inheritance of powdery mildew resistance (*Erysiphe polygoni* DC) resistance in mungbean. Theor Appl Genet 88:945–948

Rehmany AP, Gordon A, Rose LE, Allen RL, Armstrong MR, Whisson SC, Kamoun S, Tyler BM, Birch PR, Beynon JL (2005) Differential recognition of highly divergent downy mildew avirulence gene alleles by *RPP1* resistance genes from two *Arabidopsis* lines. Plant Cell 17: 1839–1850

Reignault P, Frost LN, Richardson H, Daniels MJ, Jones JDG, Parker JE (1996) Four *Arabidopsis* RPP loci controlling resistance to the Noco2 isolate of *Peronospora parasitica* map to regions known to contain other RPP recognition specificities. Mol Plant-Microbe Interact 9:464–473

Rusholme RE, Higgins E, Walsh JA, Lydiate DJ (2007) Genetic control of broad-spectrum resistance to turnip mosaic virus in *Brassica rapa* (Chinese cabbage). J Gen Virol 88:3177–3186

Sachan JN, Kolte SJ, Singh B (1995) Genetics of resistance to white rust (*Albugo candida* race-2) in mustard (*Brassica juncea*). In: GCIRC 9th Intl rapeseed Cong. UK, Cambridge, pp 1295–1297

Saha P, Kalia P, Sharma P, Sharma TR (2014) Race-specific genetics of resistance to black rot disease [*Xanthomonas campestris* pv. *campestris* (*Xcc*) (Pammel) Dowson] and the development of three random amplified polymorphic DNA markers in cauliflower. J Hort Sci Biotechnol 89:480–486

Saha P, Kalia P, Sharma M, Singh D (2016) New source of black rot disease resistance in *Brassica oleracea* and genetic analysis of resistance. Euphytica 207:35–48

Saharan GS, Kadian AK (1983) Analysis of components of horizontal resistance in rapeseed and mustard cultivars against *Alternaria brassicae*. Indian Phytopathol 36:503–507

Saharan GS, Krishnia SK (2001) Multiple disease resistance in rapeseed and mustard. In: Nagarajan S, Singh DP (eds) Role of resistance in intensive agriculture. Kalyani Pub, New Delhi, pp 98-108

Saharan GS, Verma PR, Meena PD, Kumar A (2014) White rust of crucifers: biology, ecology and disease management. Springer, p 244

Saito M, Kubo N, Matsumoto S, Suwabe K, Tsukada M, Hirai M (2006) Fine mapping of the clubroot resistance gene, Crr3, in *Brassica rapa*. Theor Appl Genet 114:81–91. https://doi.org/10.1007/s00122-006-0412-1

Sakamoto K, Saito A, Hayashida N, Taguchi G, Matsumoto E (2008) Mapping of isolate-specific QTLs for clubroot resistance in Chinese cabbage (*Brassica rapa* L. ssp. *pekinensis*). Theor Appl Genet 117:759–767. https://doi.org/10.1007/s00122-008-0817-0

Schiff CL, Wilson IW, Somerville SC (2001) Polygenic powdery mildew disease resistance in *Arabidopsis thaliana*: quantitative trait analysis of the accession Warschau-1. Plant Pathol 506(1):690–701

Sharma BR, Swarup V, Chatterjee SS (1972) Inheritance of resistance to black rot in cauliflower. Can J Gen Cytol 14:363–370

Sharma BB, Kalia P, Yadava DK, Singh D, Sharma TR (2016) Genetics and molecular mapping of black rot resistance locus Xca1bc on chromosome B-7 in Ethiopian mustard (*Brassica carinata* A. Braun). PLoS One 11:e0152290

Shimizu M, Pu ZJ, Kawanabe T, Kitashiba H, Matsumoto S, Ebe Y, Sano M, Funaki T, Fukai E, Fujimoto R, Okazaki K (2015) Map–based cloning of a candidate gene conferring *fusarium* yellows resistance in *Brassica oleracea*. Theor Appl Genet 128:119–130. https://doi.org/10.1007/s00122-014-2416-6

Singh D, Singh H (1987) Genetic analysis of resistance to white rust in Indian mustard. Proc 7th Intl Rapeseed Cong, Poznan, Poland, 11–14 May, 1987, Poznan, Poland, p 126

Singh H, Singh D (1989) Studies on genetic control of resistance to *Alternaria brassicae* in Indian mustard. In: Proc XII Eucarpia congress, 11-13, Abstr.

Singh D, Lodhi GP, Singh H (1986) Studies on genetic variability for various attributes in different mating system F2 and F3 populations of rapeseed. Oil Crops Newslett 58:60

Singh D, Singh H, Yadava TP (1988) Performance of white rust (*A. candida*) resistance genotypes developed from interspecific crosses of *B. juncea* L. × *B. carinata* L. Cruciferae NewsLett 13: 110

Singh US, Doughty KJ, Nashaat NI, Bennett RN, Kolte SJ (1999) Induction of systemic resistance to *Albugo candida* in *Brassica juncea* by pre- or co-inoculation with an incompatible isolate. Phytopathology 89:1226–1232

Singh S, Dey SS, Bhatia R, Batley J, Kumar R (2018) Molecular breeding for resistance to black rot [*Xanthomonas campestris* pv. *campestris* (Pammel) Dowson] in *Brassicas*: recent advances. Euphytica 214:196

Smith EF (1899) The fungus infection of agricultural soils in the United States. Sci Am Sup 48: 19981–19982

Soengas P, Hand P, Vicente JG, Pole JM, Pink DAC (2007) Identification of quantitative trait loci for resistance to *Xanthomonas campestris* pv. *campestris* in *Brassica rapa*. Theor Appl Genet 114:637–645

Song J, Win J, Tian M, Schornack S, Kaschani F, Ilyas M, van der Hoorn RA, Kamoun S (2009) Apoplastic effectors secreted by two unrelated eukaryotic plant pathogens target the tomato defense protease Rcr3. Proc Natl Acad Sci U S A 106:1654–1659

Sridhar K, Raut RN (1998) Differential expression of white rust resistance in Indian mustard (*Brassica juncea*). Indian J Genet Plant Breed 58:319–322

Steinbrenner AD, Goritschnig S, Staskawicz BJ (2015) Recognition and activation domains contribute to allele-specific responses of an *Arabidopsis* NLR receptor to an oomycete effector protein. PLoS Pathog 11:e1004665

Strandberg JO, Williams PH (1967) Inheritance of clubroot resistance in Chinese cabbage. Phytopathology 57:330

Strelkov SE, Tewari JP, Smith-Degenhardt E (2006) Characterization of *Plasmodiophora brassicae* populations from Alberta, Canada. Can J Plant Pathol 28:467–474. https://doi.org/10.1080/07060660609507321

Strelkov SE, Manolii VP, Cao T, Xue S, Hwang SF (2007) Pathotype classification of *Plasmodiophora brassicae* and its occurrence in *Brassica napus* in Alberta, Canada. J Phytopathol 155:706–712. https://doi.org/10.1111/j.1439-0434.2007.01303.x

Strelkov SE, Hwang SF, Manolii VP, Cao T, Feindel D (2016) Emergence of new virulence phenotypes of *Plasmodiophora brassicae* on canola (*Brassica napus*) in Alberta, Canada. Eur J Plant Pathol 145:517–529. https://doi.org/10.1007/s10658-016-0888-8

Strelkov SE, Hwang SF, Manolii VP, Cao T, Fredua-Agyeman R, Harding MW, Peng G, Gossen BD, McDonald MR, Feindel D (2018) Virulence and pathotype classification of *Plasmodiophora brassicae* populations collected from clubroot resistant canola (*Brassica napus*) in Canada. Can J Plant Pathol 40:284–298

Subudhi PK, Raut RN (1994) White rust resistance and its association with parental species type and leaf waxiness in *Brassica juncea* × *Brassica napus* crosses under the action of EDTA and gamma-ray. Euphytica 74:1–7

Suh SK, Green SK, Park HG (1995) Genetics of resistance to five strains of turnip mosaic virus in Chinese cabbage. Euphytica 81:71–77

Suwabe K, Tsukazaki H, Iketani H, Hatakeyama K, Fujimura M, Nunome T, Fukuoka H, Matsumoto S, Hirai M (2003) Identification of two loci for resistance to clubroot (*Plasmodiophora brassicae* Woronin) in *Brassica rapa* L. Theor Appl Genet 107:997–1002. https://doi.org/10.1007/s00122-003-1309-x

Suwabe K, Tsukazaki H, Iketani H, Hatakeyama K, Kondo M, Fujimura M, Nunome T, Fukuoka H, Hirai M, Matsumoto S (2006) Simple sequence repeat-based comparative genomics between

Brassica rapa and *Arabidopsis thaliana*: the genetic origin of clubroot resistance. Genetics 173: 309–319. https://doi.org/10.1534/genetics.104.038968

Suwabe K, Suzuki G, Kondo M, Tomita RN, Mukai Y, Fukuoka H, Matsumoto S (2011) Microstructure of the *Brassica rapa* genome segment that are homeologous to resistance gene cluster in *Arabidopsis* chromosome 4. Breed Sci 62:170–177

Suwabe K, Suzuki G, Nunome T, Hatakeyama K, Mukai Y, Fukuoka H, Matsumoto S (2012) Microstructure of a *Brassica rapa* genome segment homoeologous to the resistance gene cluster on *Arabidopsis* chromosome 4. Breed Sci 62:170–177

Thakur BS, Korla BN, Khosla K (2003) Inheritance of black rot resistance in late cauliflower. Ann Agric Res 24:244–248

Thomas CE, Jourdain L (1990) Evaluation of broccoli and cauliflower germplasm for resistance to race 2 of *Peronospora parasitica*. Hort Sci 25:1429–1431

Thomas CE, Jourdain EL (1992) Resistance to race 2 of *Peronospora parasitica* in US. Plant introductions of *Brassica oleracea* var. *capitata*. Hort Sci 27:1120–1122

Thukral SK, Singh H (1986) Inheritance of white rust resistance in *Brassica juncea*. Plant Breed 97: 75–77

Tiwari AS, Petrie GA, Downey RK (1988) Inheritance of resistance to *Albugo candida* race 2 in mustard [*Brassica juncea* (L.) Czem.]. Can J Plant Sci 68:297–300

Tonguc M, Griffiths PD (2004) Evaluation of *Brassica carinata* accessions for resistance to black rot (*Xanthomonas campestris* pv. *campestris*). Hort Sci 39:952–954

Tor M, Holub EB, Brose E, Muskar R, Gunn N, Can C, Crute IR, Beynon JL (1994) Map positions of three loci in *Arabidopsis thaliana* associated with isolate-specific recognition of *Peronospora parasitica* (downy mildew). Mol Plant-Microbe Interact 7:214–222

Toxopeus H, Janssen AMP (1975) Clubroot resistance in turnip. II. The "slurry" screening method and clubroot races in the Netherlands. Euphytica 24:751–755

Toxopeus H, Dixon GR, Mattusch P (1986) Physiological specialization in *Plasmodiophora brassicae*, an analysis by international experimentation. Trans Br Mycol Soc 87:279–287

Tsuji J, Somerville SC, Hammerschmidt R (1991) Identification of a gene in *Arabidopsis thaliana* that controls resistance to *Xanthomonas campestris* pv. *campestris*. Physiol Mol Plant Pathol 381(1):57–65

Tyler BM, Kale SD, Wang Q, Tao K, Clark HR, Drews K, Antignani V, Rumore A, Hayes T, Plett JM, Fudal I, Gu B, Chen Q, Affeldt KJ, Berthier E, Fischer GJ, Dou D, Shan W, Keller NP, Martin F, Rouxel T, Lawrence CB (2013) Microbe-independent entry of oomycete R × LR effectors and fungal R × LR-like effectors into plant and animal cells is specific and reproducible. Mol Plant-Microbe Interact 26:611–616. https://doi.org/10.1094/MPMI-02-13-0051-IA

Van Nimwegen K, Van Soest RA, Veltman JA, Nelen MR, Van Der Wilt GJ, Vissers LE, Grutters JP (2016) Is the $1000 genome as near as we think? A cost analysis of next-generation sequencing. Clin Chem 62:1458–1464

Verma U, Bhowmik TP (1989) Inheritance of resistance to a *Brassica juncea* pathotype of *Albugo candida* in *Brassica napus*. Can J Plant Pathol 11:443–444

Vicente JG, Taylor JD, Sharpe AG, Parkin IAP, Lydiate DJ, King GJ (2002) Inheritance of race-specific resistance to *Xanthomonas campestris* pv. *campestris* in *Brassica* genomes. Phytopathology 92:1134–1141

Vicente JG, Gunn ND, Bailey L, Pink DAC, Holub EB (2012) Genetics of resistance to downy mildew in *Brassica oleracea* and breeding towards durable disease control for UK vegetable production. Plant Pathol 61:600–609

Vogel J, Somerville S (2000) Isolation and characterization of powdery mildew resistant *Arabidopsis* mutants. Proc Natl Acad USA 976(1):1897–1902

Voorrips RE, Kanne HJ (1997a) Genetic analysis of resistance to clubroot (*Plasmodiophora brassicae*) in *Brassica oleracea*. I Analysis of symptom grades. Euphytica 93:31–39

Voorrips RE, Kanne HJ (1997b) Genetic analysis of resistance to clubroot (*Plasmodiophora brassicae*) in *Brassica oleracea*. II Quantitative analysis of root system measurements. Euphytica 93(1):41–48

Voorrips RE, Visser DL (1993) Examination of resistance to clubroot in accessions of *Brassica oleracea* using a glasshouse seedling test. Netherlands J Plant Pathol 99:269–276

Voorrips RE, Jongerius MC, Kanne HJ (1997) Mapping of two genes for resistance to clubroot (*Plasmodiophora brassicae*) in a population of doubled haploid lines of *Brassica oleracea* by means of RFLP and AFLP markers. Theor Appl Genet 94:75–82. https://doi.org/10.1007/s001220050384

Walker JC, Larson RH (1951) Progress in the development of clubroot-resistant cabbage. Phytopathology 41:37

Walker JC, Williams PH (1965) The inheritance of powdery mildew resistance in cabbage. Plant Dis Rep 49:198–201

Walkey DGA, Pink DAC (1988) Reactions of white cabbage (*Brassica oleracea* var. *capitata*) to four different strains of turnip mosaic virus. Ann Appl Biol 112(2):273–284. https://doi.org/10.1111/j.1744-7348.1988.tb02063.x

Wang H, Liu G, Zheng Y, Wang X, Yang Q (2004) Breeding of the *Brassica napus* cultivar Zhongshuang 9 with high-resistance to *Sclerotinia sclerotiorum* and dynamics of its important defense enzyme activity. Sci Agric Sin 37:23–28

Werner S, Diederichsen E, Frauen M, Schondelmaier J, Jung C (2008) Genetic mapping of clubroot resistance genes in oilseed rape. Theor Appl Genet 116:363–372. https://doi.org/10.1007/s00122-007-0674-2

Whisson SC, Boevink PC, Moleleki L, Avrova AO, Morales JG, Gilroy EM, Armstrong MR, Grouffaud S, van West P, Chapman S, Hein I, Toth IK, Pritchard L, Birch PRJ (2007) A translocation signal for delivery of oomycete effector proteins into host plant cells. Nature 450:115–118

Williams PH, Hill CB (1986) Rapid cycling populations of *brassica*. Sci 232:1385–1389

Williams PH, Pound GS (1963) Nature and inheritance of resistance to *Albugo candida* in radish. Phytopathology 53:1150–1154

Williams PH, Walker JC, Pound GS (1968) Hybelle and Sanibel, multiple disease-resistant F1 hybrid cabbages. Phytopathology 58:791–796

Williams PH, Staub T, Sutton JC (1972) Inheritance of resistance in cabbage to black rot. Phytopathology 62:247–252

Wilson IW, Schiff CL, Hughes DE, Somerville SC (2001) Quantitative trait loci analysis of powdery mildew disease resistance in the *Arabidopsis thaliana* accession Kashmir-1. Genet 1586(1):1301–1309

Woods-Tor A, Studholme DJ, Cevik V, Telli O, Holub EB, Tor M (2018) A suppressor/avirulence gene combination in *Hyaloperonospora arabidopsidis* determines race specificity in *Arabidopsis thaliana*. Front Plant Sci 9:265

Xiao SY, Ellwood S, Findlay K, Oliver RP, Turner JG (1997) Characterization of three loci controlling resistance of *Arabidopsis thaliana* accession Ms-0 to two powdery mildew diseases. Plant J 126(1):757–768

Yin X, Yi B, Chen W, Zhang W, Tu J, Fernando D, Fu TD (2010) Mapping of QTLs detected in a *Brassica napus* DH population for resistance to *Sclerotinia sclerotiorum* in multiple environments. Euphytica 173:25–35

Yoon JY, Green SK, Opena RT (1993) Inheritance of resistance to turnip mosaic virus in Chinese cabbage. Euphytica 69:103–108

Yoshikawa H (1981) Breeding for clubroot resistance in Chinese cabbage. In: Talekar NS, Griggs TD (eds) Chinese cabbage proc 1st Intl Symp. Tsukuba, Japan, pp 405–413

Yoshikawa H (1993) Studies on breeding of clubroot resistance in cole crops. Bull Natl Res Inst Veg Ornam Plants Tea Jpn Ser A 7:1–165

Yu F, Zhang X, Huang Z, Chu M, Song T, Falk KC, Deora A, Chen Q, Zhang Y, McGregor L, Gossen BD, McDonald MR, Peng G (2016) Identification of genome-wide variants and discovery of variants associated with *Brassica rapa* clubroot resistance gene *Rcr1* through bulked segregant RNA sequencing. PLoS One 11:e0153218. https://doi.org/10.1371/journal.pone.0153218

Yuen JE (1991) Resistance to *Peronospora parasitica* in Chinese cabbage. Plant Dis 75:10–13

Zeiler FJ, Lutz J, Stephan U (1993) Chromosomal location of genes for powdery mildew resistance in common wheat (*Triticum aestivum* L.) 1. M/k and other alleles at the Pro3 locus. Euphytica 68:223–229

Zhang H, Feng J, Hwang SF, Strelkov SE, Falak I, Huang X, Sun R (2016) Mapping of clubroot (*Plasmodiophora brassicae*) resistance in canola (*Brassica napus*). Plant Pathol 65:435–440. https://doi.org/10.1111/ppa.12422

Zhao J, Meng J (2003) Genetic analysis of loci associated with partial resistance to *Sclerotinia sclerotiorum* in oilseed *Brassica napus*. Theor Appl Genet 106:759–764

Zhao J, Peltier AJ, Meng J, Osborn TC, Grau CR (2004) Evaluation of Sclerotinia stem rot resistance in oilseed *Brassica napus* using a petiole inoculation technique under greenhouse conditions. Plant Dis 88:1033–1039

Zhao J, Udall JA, Quijada PA, Grau CR, Meng J, Osborn TC (2006) Quantitative trait loci for resistance to *Sclerotinia sclerotiorum* and its association with a homeologous non-reciprocal transposition in *Brassica napus* L. Theor Appl Genet 112:509–516. https://doi.org/10.1007/s00122-005-0154-5

Zhao J, Buchwaldt L, Rimmer R, Sharpe A, McGregor L, Bekkaoui D, Hegedus D (2009) Patterns of differential gene expression in *Brassica napus* cultivars infected with *Sclerotinia sclerotiorum*. Mol Plant Pathol 10:635–649. https://doi.org/10.1111/j.1364-3703.2009.00558.x

Transfer of Disease Resistance

4

Abstract

Substantial progress has been made in transfer of R-genes from different sources and developing of disease-resistant cultivars of major cruciferous crops against major pathogens using conventional and molecular biotechnological approaches. However, problems in transfer of R-genes arise due to diverse nature of pathogens which are physiologically and taxonomically highly variable; therefore no single gene product and approach can be expected to have a direct effect on all types of pathogens. The advent of molecular genetics has provided new opportunities for mapping and tracking of genes of disease resistance leading to more efficient marker-assisted selection. The improvement in plant transformation protocols and high-throughput gene attenuation methods like RNA interference and virus induced gene silencing (VIGS) will accelerate the identification and transfer of novel genes for durable resistance in crucifers. The introgression of R-genes to major pathogens has been widely used to breed disease-resistant cultivars in *Brassica* crops through inter- or intraspecific crosses and embryo rescue. The conventional and modern approaches have been used to exploit R-genes sources from wild relatives to breed disease-resistant cultivars of *Brassica* crops. For developing durable Alternaria blight-resistant *Brassica* crops, strategies like inter- or intraspecific introgression of R-genes from non-host plant to *Brassica* crops, microarray-based gene identification and transfer gene silencing, and transfer of an endo-chitinase gene from *Trichoderma* approaches have been used. R-genes sources of multiple disease resistance have been utilized to breed resistant cultivars of *Brassica* to *Albugo, Alternaria, Erysiphe, Hyaloperonospora, Leptosphaeria,* and *Xanthomonas* . To combine the traits from distantly related *Brassica* species for development of superior, durable, and tolerant cultivars of *Brassica* to major diseases, approaches like in vitro embryo rescue, somatic hybridization, interspecific hybridization, distant hybridization, somaclonal variations, genetic transformations, molecular markers, mutagenic

G. S. Saharan et al., *Genomics of Crucifer's Host-Resistance*,
https://doi.org/10.1007/978-981-16-0862-9_4

resistance, induction of systemic resistance, and genetic engineering have been exploited. Substantial progress has been made in breeding for clubroot resistance genotypes/cultivars of *B. rapa, B. oleracea, B, napus, Raphanus sativus*, and *Arabidopsis thaliana*. Several Clubroot loci or QTL's have been mapped on the chromosomes of these species. There is substantial genetic and pathotype variation in *Plasmodiophora brassicae* populations all over the world and the deployment of a cultivar with single gene resistance imposes a strong selection pressure to evolve new virulences to overcome this resistance. The successful development of *Sclerotinia* resistant *Brassica* cultivars has been obtained by introducing multiple resistant loci (QTL's) from wild relatives using marker-assisted introgression approach. Black rot (*Xanthomonas*) resistance has been introgressed from *B. carinata* to *B. oleracea* through embryo rescue. Somatic hybrids by protoplast fusion between rapid cycling *B. oleracea* and *Xanthomonas* resistant lines of *B. napus* have been developed. Understanding the evolutionary potential between the pathogen and the host is essential in developing and transfer of R-genes for durable *Brassica* resistant cultivars. In the present era, the breeding for disease-resistant *Brassica* cultivars, application of molecular approaches like marker-assisted selection, genome-wide marker-assisted selection, genetic diversity analysis and comparative genomics have opened new avenues for superior, durable, and multiple disease-resistant cultivars in different *Brassica* crops.

Keywords

Disease resistance · Mode of transfer of resistance · Introgression breeding · Embryo rescue · Somatic hybridization · Gene silencing · Transfer of an endochitinase gene from *Trichoderma* · Multiple disease resistance · Strategies and methods of screening · Genetic transformation · Somaclonal variations · Systemic resistance · Mechanisms of resistance · Gene pyramiding · Molecular markers · Genetic engineering · Distant hybridization · Factors affecting transfer of disease resistance · Novel sources of resistance · Genome-wide marker-assisted selection · Durable resistance · Bottle necks in resistance breeding · Biotechnological approaches

4.1 Introduction

The transfer of disease resistance is achieved after identification of R-genes sources with well-characterized mode of inheritance of genes in the progenies to get a better output as disease resistance cultivars with all desired traits of yield and quality. Various methods including conventional and molecular approaches have been used either alone or in combination to incorporate desired traits to develop improved cultivars of crucifers. Hybridization combined with backcrossing either singly or in combination with pedigree selection is the most widely used and accepted conventional method to transfer disease resistance in crucifers. The level of resistance in developed cultivars has been achieved through multiple crosses, recurrent selection, diallele crossing, and selective mating system. The resistance has been introgressed with intraspecific pedigree, backcross, modified recurrent mass selection, interspecific genome substitution, chromosome substitution, and gene introgression. The molecular and genetic engineering approaches like genome manipulation, cytoplasmic genome manipulation, transformation, foreign gene expression, and embryo rescue for wide hybridization or distant hybridization have been exploited. The introgression and pyramiding of major R-QTL's have been attempted including transgenics development. In crucifers, substantial problems have been faced in the introgression and pyramiding of major QTL's including transgenic development because of diverse nature of pathogens infecting these crops with different modes of infection, pathogenesis and survival as biotrophs, hemi-biotrophs and necrotrophs.

4.2 Transfer of Disease Resistance in *Brassica* Through Introgression Breeding

The introgression of genes for disease resistance between species has been widely studied in *Brassica*. The utilization of the B genome as a source of resistance against *Leptosphaeria maculans* (blackleg) from diploid and tetraploid species has been studied. For instance, chromosome B4 from *B. nigra* was introgressed into rapeseed variety "Darmor," which showed high resistance with the addition of this chromosome. Similarly, high resistance from *B. juncea* was obtained in selected recombinant lines of *B. napus* carrying a resistance gene located on chromosome B8. Similarly, successfully introgressed a B-genome chromosome from *B. carinata* to *B. napus*, with plants carrying this chromosome showing variation in traits such as blackleg resistance, days to flowering, days of maturity, and fatty acid composition. Another example is the improvement of resistance against *Erysiphe polygoni*. Resistance in 100% of BC1 progeny was successfully demonstrated in hybrids obtained by hand crossing and embryo rescue between *B. carinata* (donor) × *B. oleracea*. Other cases of resistance transfer include transfer of black rot resistance from *B. carinata* to *B. oleracea*, resistance to *Brassica* leaf blight caused by *Alternaria brassicae* from *B. hirta* to *B. juncea* and transfer of powdery mildew resistance from *B. carinata* to *B. oleracea* through embryo rescue followed by backcrossing to *B. oleracea* (Katche et al. 2019; Saharan et al. 2019).

4.3 Transfer of Disease Resistance from Germplasm Sources

Several attempts have been made to exploit resistance sources from the wild relatives of *Brassica* and transfer them into cultivated Brassica (Aneja and Agnihotri 2013). Efforts have been made in screening blight-resistant cultivars; however, high levels of resistance to A. brassicae have not been observed among cultivated Brassica species. The species show varying degrees of resistance responses against Alternaria blight, and accordingly, *B. rapa* and *B. juncea* are classified as being more susceptible to the disease than *B. napus* and *B. carinata* (Skoropad and Tewari 1977). These related wild species have been found to harbor genes conferring resistance to Alternaria blight that can be used for resistance breeding (Ghose et al. 2008; Chatterjee et al. 2013).

In the past few decades, various conventional and modern approaches have been used to develop blight-resistant cultivars by exploiting resistance sources from host germplasm. Conventional approaches involve the screening of germplasm for blight-resistant plant materials, followed by the transfer of resistance traits from tolerant species into susceptible cultivars by hybridization combined with multiple crosses. Inter- and intraspecific crosses of *B. juncea* and *B. carinata* were established to understand the inheritance pattern of resistance to Alternaria blight (Krishnia et al. 2000), which showed that dominant and additive genes govern the inheritance of resistance to Alternaria blight. Thus, a successful breeding method should utilize both the gene effects. However, there is a limitation in such breeding approaches owing to the low availability of suitable resistance sources/genes within the available germplasm of cultivated *Brassica* and insurmountable self-incompatibility barriers in combining the resistance traits from distantly related wild *Brassica* species. Modern biotechnological interventions involving tissue culture and genetic transformation approaches have also been established for the development of improved varieties. In vitro ovary culture and ovule culture have been attempted for transferring resistance traits from S. alba cv. Carine to *B. napus* cv. Brutor (Chevre et al. 1994) and *Erucastrum cardaminoides* to *B. oleracea* var. *alboglabra* (Mohanty et al. 2009). Intergeneric hybrids of *B. campestris* and *B. spinescens* have also been developed (Agnihotri et al. 1991). Moreover, the somatic hybridization technique has been applied for transferring the resistance traits from blight-tolerant species to susceptible ones, for example, *S. alba* to *B. napus* (Primard et al. 1988), *B. carinata* to *B. juncea* (Sharma and Singh 1992), and *B. nigra* and *B. oleracea* (Jouradan and Salazar 1993). Attempts have also been made for the introduction of somaclonal variations for the incorporation of disease resistance/tolerance to Alternaria blight through mutagenesis using gamma-rays, ethyl methane sulfonate, and ethyl nitrosourea (Verma and Rai 1980; Agnihotri et al. 2009). In addition, the defense response might be activated by treatment with biotic and abiotic agents for combating *A. brassicae* infection in oilseed Brassicas. Vishwanath et al. (1999) showed that resistance in a susceptible cultivar of *B. juncea* against the extremely virulent *A. brassicae* isolate A and reasonably virulent isolate C, both from *B. carinata*, was induced after inoculating the plants with the avirulent *A. brassicae* isolate D. As another example, the application of β-aminobutyric acid on the leaves of *B. carinata* was found to induce resistance to *A. brassicae* (Chavan et al. 2013). In future, the activation of the defense response without gene manipulation may act as a better substitute for the conventional use of fungicides. Unfortunately, till date, none of the conventional or modern biotechnological methods have been found to be feasible or stable in developing blight-resistant *Brassica* species. Under such circumstances, it is pragmatic to use novel sources of resistance from plant species that are not hosts to the pathogen. The transfer of genes from non-host plants conferring resistance to the pathogen is a promising strategy in parallel to the current efforts towards the development of oilseed *Brassica* varieties resistant to Alternaria blight (Fatima et al. 2019).

4.4 Strategies for Developing Durable Alternaria Blight-Resistant *Brassica* Crops

Based on a literature survey on intercropping systems, Fatima et al. (2019) have shortlisted the non-host plants for *A. brassicae* (Table 4.1) that can be chosen for dissecting the NHR mechanisms. The NHR to *A. brassicae* may involve four different phases based on fungal invasion and penetration events (Fig. 4.1). Prior to any specific study, it is important to classify the non-host plants based on these four phases by using pathological assays. This will provide a clear idea about

Table 4.1 List of the potential non-host plants towards *Alternaria brassicae* (Fatima et al. 2019)

Sl. no.	Name of the non-host plant	Name of the host plant intercropped with non-host plant[a]	Criteria for categorization of non-host plant	Information source
1.	Chickpea (*Cicer arietinum* and *C. kabulium*)	*B. juncea, B. napus, B. rapa*	Intercropping system	Gangasaran and Giri (1985), Tewari et al. (1992), Ahlawat and Sharma (2002), Ahlawat et al. (2005), Singh et al. (2010), Shekhawat et al. (2012), Jeromela et al. (2017)
2.	Lentils (*Lens culinaris*)	*B. juncea, B. rapa*	Intercropping system	Gangasaran and Giri (1985), Tewari et al. (1992), Shyam et al. (2007), Shekhawat et al. (2012), Ao et al. (2016), Jeromela et al. (2017)
3.	Barley (*Hordeum vulgare*)	*B. juncea, B. rapa*	Intercropping system	Gangasaran and Giri (1985), Singh et al. (2010), Shekhawat et al. (2012)
4.	Wheat (*Triticum aestivum*)	*B. juncea, B. napus, B. rapa*	Intercropping system; Literature	McRoberts and Lennard (1996), Ahlawat and Sharma (2002), Singh et al. (2010), Shekhawat et al. (2012), Ebrahimi et al. (2016)
5.	Tomato (*Solanum lycopersicum*)		Literature	McRoberts and Lennard (1996)
6.	Potato (*Solanum tuberosum*)	*B. juncea, B. napus*	Intercropping system	Ahlawat and Sharma (2002), Singh et al. (2010), Shekhawat et al. (2012)
7.	Sugarcane (*Saccharum officinarum*)	*B. juncea, B. rapa, B. napus, B. campestris*	Intercropping system	Singh et al. (1986), Toor et al. (2000), Suman et al. (2006), Ahlawat and Sharma (2002), Singh et al. (2010)

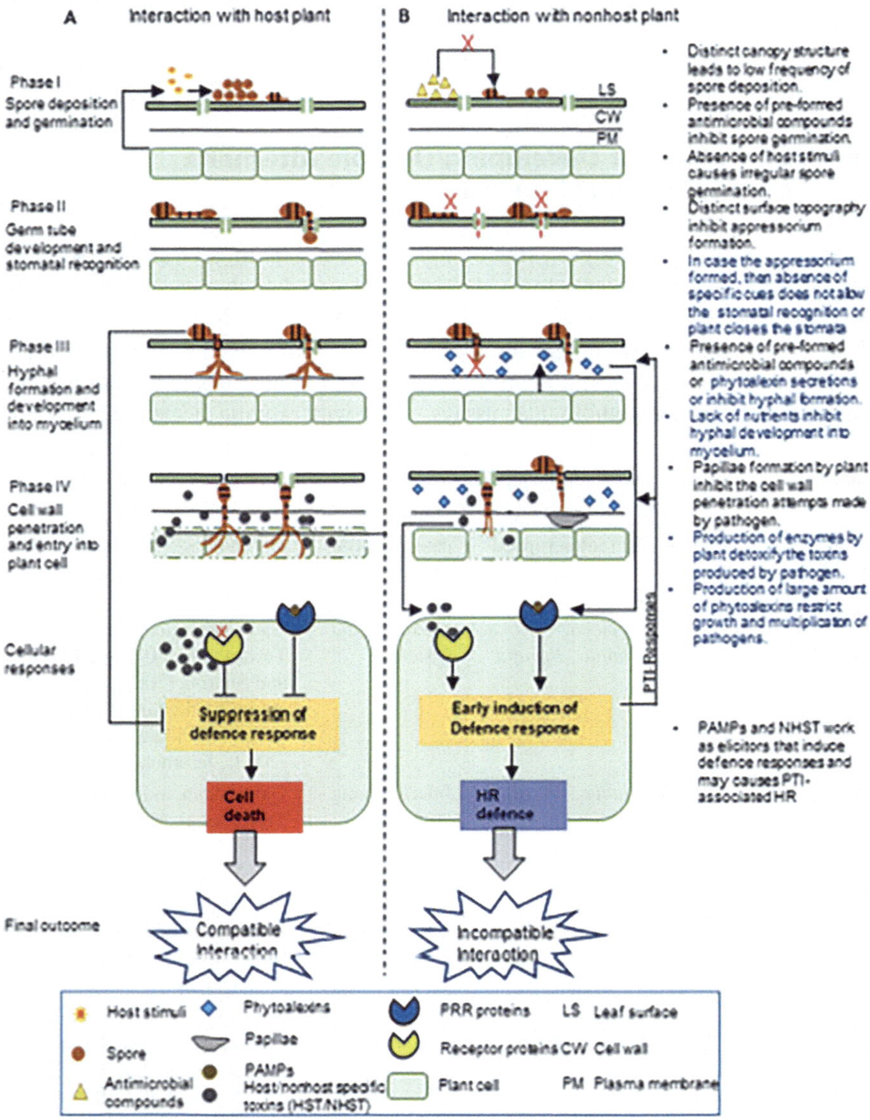

Fig. 4.1 Proposed stages of non-host resistance against *Alternaria brassicae*. This schematic representation depicts the possible events during the interaction of Alternaria brassicae with the host (**a**) and non-host plants (**b**), which is divided into four different phases. The NHR towards *A. brassicae* may involve the pre-formed and inducible defense barrier at different phases. (*I*) The first phase involves the deposition and germination of *A. brassicae* spores on the leaf surface. Non-host plants have a low frequency of spore deposition on the leaf surface due to the differences in their canopy structure pre-formed antimicrobial compounds on the leaf surface and irregular germination of spores due to the absence of host stimuli. (*II*) The second phase involves germ tube development from the spores and the recognition of the stomata. The difference in the leaf surface topography of the non-host plant compared to the host plant may not allow the proper development of the germ tube into the appressorium. At the next level, even if the appressorium is formed, either stomata are closed by plant or the pathogen may fail to recognize the stomata due to the absence of

whether the NHR operates at the pre-invasion or post-invasion level. The following strategies and approaches can be used for developing durable Alternaria blight-resistant *Brassica* crops.

4.4.1 Inter- or Intraspecific Introgression of Candidate Genes from Non-host Plants to *Brassica* Crops

The first strategy involves the identification of genes conferring NHR by utilizing conventional or molecular breeding approaches, followed by the introgression of candidate genes from non-host plants to *Brassica* crops. This can be achieved in two ways. First, where possible, interspecific crosses

of the host with a non-host may be performed to transfer resistance traits from the non-host plant to a host plant. The unwanted background in the progeny can be removed by crossing it with wild-type host plants and selecting for the targeted traits only. Then, the progeny can be screened for response to *A. brassicae*. Plants showing resistance to the pathogen can be subsequently subjected to QTL mapping to identify the genes involved in NHR. Once the genes are identified, they can be successfully introgressed into the desired varieties to provide resistance to *A. brassicae*. The major limitation with this approach, however, is that the progeny generally suffers from sterility and abnormal segregation, which can hinder the identification of genes involved in resistance. The only available example of an interspecific cross made between a host and non-host plant is that of *Lactuca* and *Bremia* (Jeuken et al. 2008).

The second approach is to cross two different varieties of non-host plants (intraspecific cross: completely resistant genotype x moderately resistant genotype) that differ in the degree of resistance towards *A. brassicae*. This will allow the inheritance of NHR-specific QTL of the resistant donor into the moderately susceptible genotypes of the same species of the non-host plant (Jafary et al. 2008; Niks and Marcel 2009). The progeny obtained can be similarly studied as described above for

Fig. 4.1 (continued) specific cues, generally derived from the distinct composition and amount of cuticular waxes in the non-host leaf, which might be entirely different in the case of the host leaf. (*III*) The third phase involves fungal hyphal formation and its development into mycelium. The non-host plant may inhibit hyphal formation and its development into mycelium by secreting inducible chemicals by pre-formed antimicrobial compounds or by limiting nutrient availability in the intercellular spaces. (*IV*) The fourth phase involves the penetration of the fungal pathogen into the plant cell wall and its subsequent entry inside the plant cell. The non-host plant may inhibit the penetration of the fungal hyphae into the plant cell wall by inducing the formation of papillae. The defense responses are induced by non-host plants after recognizing PAMPs or elicitor molecules. This may lead to the production of enzymes and a large amount of phytoalexins by the non-host plants, which detoxify the toxins produced by the fungal pathogen and restrict fungal growth and multiplication. Note: The text on the right in the figure indicates the NHR mechanisms, and the highlighted blue text indicates the most exploitable NHR mechanisms to develop durable disease-resistant *Brassica* crops (Fatima et al. 2019) identifying the genes involved in NHR and transferring them into target crops. However, if the non-host plant does not have significant variation in the degree of resistance towards the pathogen, the inheritance of NHR-specific genes might be difficult (Fig. 4.2) (Fatima et al. 2019).

4.4.2 Microarray-Based Gene Identification and Transfer

The second strategy is to perform microarray analysis after infecting the non-host plant (e.g., chickpea or barley) with *A. brassicae* and then identify the candidate genes from the list of up-regulated or down-regulated genes. If a homologous gene is identified, mutant or overexpression lines can be generated for the identified genes in the susceptible model plant. Alternatively, the candidate genes can be cloned from the non-host plant and expressed in the susceptible model plant. These transgenic plants can then be tested for their response to *A. brassicae* infection. Finally, the genes imparting NHR to *A. brassicae* can be successfully transferred to *Brassica* crops.

4.4.3 Gene Silencing

The third strategy to identify genes underlying NHR involves gene silencing in the non-host plant. Virus-induced gene silencing (VIGS)-based forward genetic screening can be performed in non-host plants that have established VIGS protocols and cDNA libraries to identify the genes involved in NHR to *A. brassicae*. For example, for wheat and barley, the Barley stripe mosaic virus (BSMV)-based vector and a well-developed VIGS protocol for gene silencing are available (Ma et al. 2012). Such a strategy will be useful in identifying important genes involved in NHR by screening a VIGS population for its response to *A. brassicae*. Furthermore, high-throughput transient expression and gene silencing methods, e.g., transient-induced gene silencing (TIGS), have been used in a non-host plant such as barley. This technique is based on the transformation of leaf epidermal cells by the biolistic method (Douchkov et al. 2014a, b). TIGS constructs may be made for the genes that are up-regulated in the non-host plant after *A. brassicae* infection. These TIGS constructs can be bombarded into leaf epidermal cells of the non-host plant to induce gene silencing. After bombardment, leaf discs can be infected with *A. brassicae* and screened for the response. TIGS will allow the identification of potentially important candidate genes involved in NHR to *A. brassicae*. The genes identified using any of these functional genomic methods can be successfully transferred to Brassica crops to ensure resistance to *A. brassicae*.

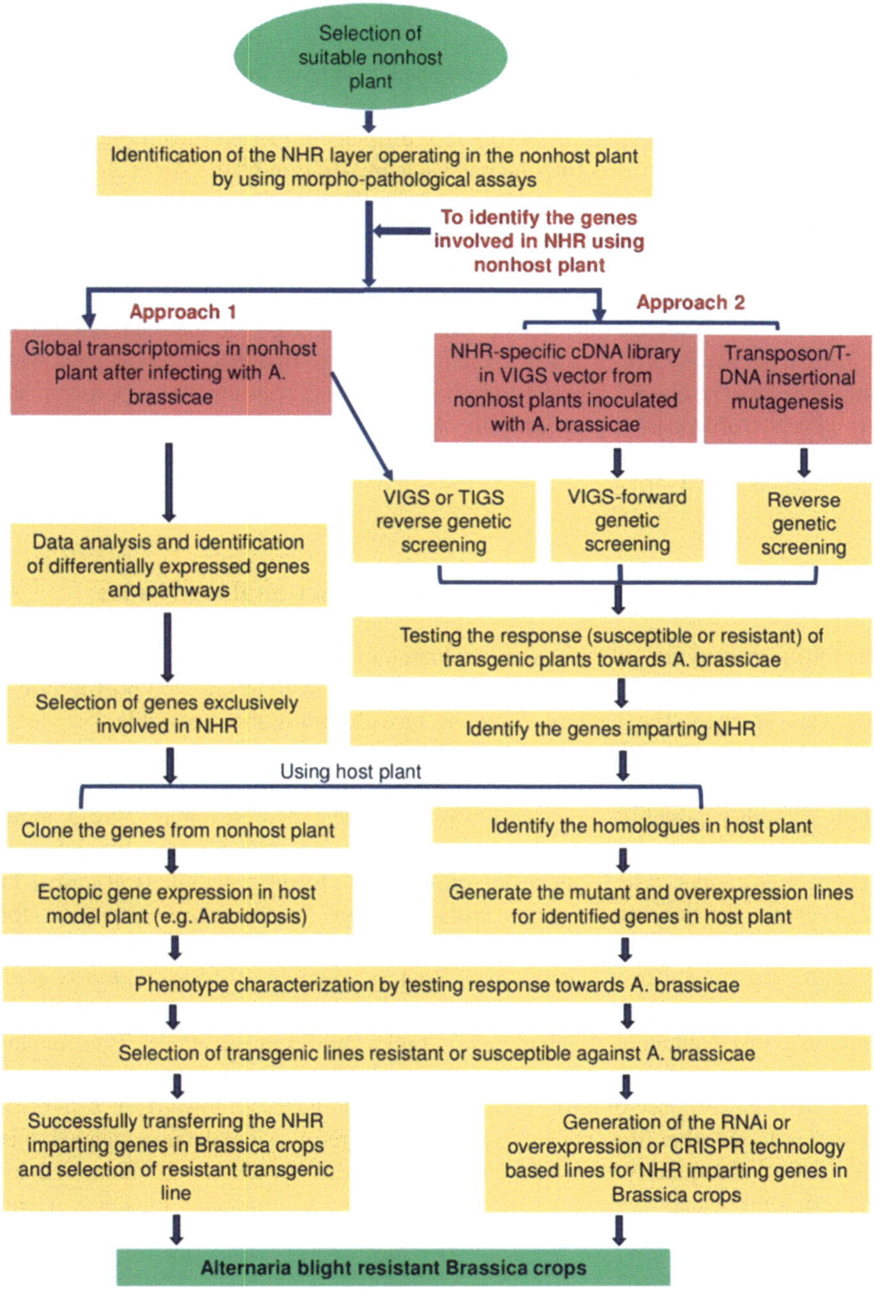

Fig. 4.2 Schematic representation of the possible strategies for exploiting the NHR mechanisms to develop durable blight-resistant *Brassica* crops. The selection of a suitable non-host plant is essential for exploiting NHR mechanisms. NHR may operate at two different layers, i.e., pre-invasive or post-invasive layers. Once the exploitable NHR layer is identified in the selected non-host plant, several strategies can be suitably utilized for identifying the genes and pathways involved in the NHR mechanisms. The first approach involves global transcriptome profiling

4.4.4 Identification and Use of Model Plants as Genetic Resources

The fourth strategy is to work with model plants that have well-developed molecular and genetic resources and a large population of mutants as non-hosts for *A. brassicae*. For example, *Brachypodium distachyon* can be directly utilized for identifying genes and dissecting the pathways involved in NHR (An et al. 2016). If the mutants are not readily available, then with the help of well-developed genetic tools, transposon or T-DNA insertional mutagenesis can be attempted in the model non-host plant. The mutant population can be screened for the response to *A. brassicae* to identify the candidate genes involved in NHR. These candidate genes can then be successfully transferred to *Brassica* crops for durable resistance to *A. brassicae*. *Arabidopsis* is a well-established, easy-to-manipulate model plant with ample genetic resources. Another approach can be screening the response of all available *Arabidopsis* accessions to *A. brassicae*. This may help identify *Arabidopsis* accessions imparting resistance to *A. brassicae* (Rajarammohan et al. 2017; Mandal et al. 2018).

4.4.5 Literature Mining and Targeted Functional Genomics

The fifth strategy is solely based on literature mining and targeted functional genomics to dissect the different layers of NHR involved in pre- and post-invasive defense responses. The available literature provides hints about possible

candidates specifically involved in pre- and post-invasive defenses. This will provide the opportunity to directly start studying these candidates for their role in NHR. Another possibility is that the post-invasive defenses may play a more important role in restricting the fungal pathogen rather than pre-invasive defenses against *A. brassicae*, which can be determined by performing pathological assays. In such cases, we can use mutants with mutations in the genes important for pre-invasion defense to identify the genes involved in post-invasion defenses. For example, the mutant of the barley gene Ror1, a homolog of the *Arabidopsis* gene Pen1, is important for pre-invasive defense against many non-host pathogens, including rust pathogens (Collins et al. 2003; Trujillo et al. 2004). This mutant line can be used for identifying genes involved in the post-invasion defense response. The response of a ror1 mutant line should be tested towards *A. brassicae* to ascertain whether this gene is involved in pre-invasive defense against this pathogen. Subsequently, the NHR-specific candidate genes can be identified by transcriptomic analysis in wild-type barley after *A. brassicae* infection. Gene

Fig. 4.2 (continued) specific to *Alternaria brassicae*-induced responses from the non-host plant to identify genes playing a role in two different layers of NHR. The data can be enriched by using proper controls, e.g., non-adapted pathogens, and the genes exclusively involved in the NHR can be selected. The second approach involves the use of functional genomic tools, e.g., VIGS, TIGS, and T- DNA/transposon insertional mutagenesis, for identifying the genes (Fatima et al. 2019).

silencing approaches can be applied to silence the candidate genes in the ror1 mutant line for identifying their role in post-invasion immunity by screening them against *A. brassicae*. The validation experiments can be carried out by constructing double mutants of candidate genes and Ror1 in barley. The double mutants compromising NHR and showing a higher degree of susceptibility to *A. brassicae* compared to the single mutant ror1 can be likely candidates that contribute to NHR at the post-invasion level. The important genes identified can be further transferred, or their homologs can be overexpressed in the susceptible *Brassica* crops to test their resistance to *A. brassicae* (Fatima et al. 2019).

4.5 Transfer of an Endo-Chitinase Gene from *Trichoderma virens* Confers Enhanced Tolerance to *Alternaria* in Transgenic *Brassica*

Genetic engineering is an alternative for developing disease-resistant plants where the resistance source is not available or there are sexual compatibilities between the source and the cultivar. (Grover and Gowthaman 2003; Chhikara et al. 2012). Genes from the myco-parasitic *Trichoderma* spp. have been used to impart tolerance to biotic and abiotic stresses in many crops. Among the Trichoderma genes transferred to plants, an endo-chitinase of 42 kDa has been very popular due to its broad spectrum of activity against many fungal pathogens. Endochitinases are known to degrade fungal chitin by hydrolyzing the glycosidic bond between chitin monomers. In addition, the endo-chitinases also release chitin oligomers which act as elicitor molecules triggering broad-spectrum plant defense. There are very few reports on development of transgenic *Brassica* species for resistance to fungal pathogens. For example, *B. napus* transformed with two genes *sporamin* and *chitinase PjChi*-1 (derived from sweet potato and *Paecilomyces javanicus*, respectively) showed enhanced resistance against both *Plutella xylostella* and *Sclerotinia sclerotiorum* (Liu et al. 2011). Wu et al. (2009) introduced a plant defensin gene Ovd, cloned from *Orychophragmus violaceus* (L.) into *B. napus*, and the transgenic plants showed resistance to *S. sclerotiorum*. Transgenic *B. napus* expressing *B. napus* mitogen-activated protein kinase (*BnMPK4*) showed resistance to *S. sclerotiorum* (Wang et al. 2009). Chitinase and glucanase gene from tomato have been introduced into *B. juncea* and the transgenic plants showed resistance to *A. brassicae* (Mondal et al. 2003, 2007). Kamble et al. (2016) observed expression of an endo-chitinase "*ech42*" gene from *Trichoderma virens* in transgenic *B. juncea* lines and also assessed tolerance of transgenic lines against two fungal pathogens, *A. brassicicola* and *A. brassicae*, causing Alternaria blight.

An endo-chitinase gene "*ech42*" from the biocontrol fungus "*Trichoderma virens*" was introduced to *Brassica juncea* (L). Czern and Coss via *Agrobacterium tumefaciens* mediated genetic transformation method. Integration and expression of the "*ech42*" gene in transgenic lines were confirmed by PCR, RT-PCR, and Southern hybridization. Transgenic lines (T1) showed expected 3:1 Mendelian segregation ratio when segregation analysis for inheritance of transgene "hpt" was carried out. Fluorimetric analysis of transgenic lines (T0 and T1) showed sevenfold higher endo-chitinase activity than the non-transformed plant. Fluorimetric zymogram showed presence of endo-chitinase (42 kDa) in crude protein extract of transgenic lines. In detached leaf bioassay with fungi *Alternaria brassicae* and *Alternaria brassicicola*, transgenic lines (T0 and T1) showed delayed onset of lesions as well as 30–73% reduction in infected leaf area compared to non-transformed plant (Kamble et al. 2016).

4.6 Sources of Resistance

Di-genomic *Brassica* species such as *B. napus* and *B. juncea* have better sources of resistance than the mono-genomic (Tables 2.1, 2.2, and 2.4, Chap. 2) species like *B. rapa.* Some lines of *B. alba, B. carinata, B. spinescenes, B. maurorum, Eruca sativa, Camelina sativa, Capsella bursa-pastoris, Diplotaxis* sp., and *N. paniculata* have also shown resistance to *A. brassicae* in different areas of the world (Brun et al. 1987a, b; Conn et al. 1991; Grontoft 1986; Jejelowo et al. 1991; Saharan 1992a, 1992b, 1997; Tewari and Conn 1993; Hansen and Earle 1997; Chrungu et al. 1999; Sharma et al. 2002). Sources of resistance to *A. brassicae* and *A. brassicicola* in different host species from different areas of the world are given in Table 2.4, Chap. 2. In India, *Brassica* genotypes, viz. CSR 43, CSR I42, CSR- I42-2, CSR 343 , CSR 448, CSR 622, CSR 741, Gulivar, KRV-Tall, Midas, PHR 1, RC-781, TMV2, Tower, and YRT3, were consistently found to have field resistance proven after testing for several years at different locations under uniform disease nursery trials (Kolte 1985b, 1987; Saharan 1984, 1992b, 1997; Saharan and Chand 1988; Saharan et al. 2003, 2005, Saharan et al. 2016, 2017, 2019; Verma and Saharan 1994). *Brassica coenospecies* are rich reservoir for genetic resistance to *A. brassicae* (Sharma et al. 2002; Table 2.5, Chap. 2). *Alternaria brassicae* pathotype specific resistance has been recorded in genotype GS-05-1 of *B. napus* to pathotypes Abr 1 and Abr 5, whereas genotypes RH-8544, Pusa Swarnim, and HC-9605 of *B. juncea* showed moderately resistant reaction to three pathotypes (Kumar et al. 2014).

4.7 Sources of Disease Resistance from Cruciferous Relatives

Sources with high level of resistance against *A. brassicicola* and *A. brassicae* have not been identified among the cultivated *Brassica* species, although individual cabbage varieties differ considerably in their level of susceptibility to black spot (Otani et al. 2001). The highest level of *Alternaria* resistance among the oilseed *Brassica* crops is displayed by the Ethiopian mustard (*B. carinata*). Among the wild cruciferous plants closely related to the *Brassica* genus, the highest resistance against *A. brassicae* is found in white mustard (*Sinapus alba*), (Kolte 1985a; Brun et al. 1987a, b; Ripley et al. 1992; Sharma and Singh 1992; Verma and Saharan 1994; Hansen and Earle 1995, 1997). The highest overall *Alternaria* spp. resistance, however, has been identified in the crucifers more distant from the *Brassica*, such as

Camelina (*Camelina sativa*; false flax), shepherd's purse (*Capsella bursa-pestoris*), taramira (*Eruca sativa*), and ball mustard (*Neslia paniculata*) (Conn and Tewari 1986; Conn et al. 1988, Tewari 1991). Resistance against *A. brassicae* and *A. brassicicola* has been reported among other wild members of the Brassicaceae family (Sharma et al. 2002; Tewari and Conn 1993; Warwick 2011), viz. *Alliaria petiolata*, *Barbarea vulgaris*, *B. elongate*, *B. desnottessi*, *B. fruticulosa*, *B. maurorum*, *B. nigra*, *B. souliei*, *B. spinescens*, *C. sativa*, *C. bursa-pestoris*; *Coincya* spp., *Diplotaxis catholica*, *D. berthautii*, *D. creacea*, *D. erucoides*, *D. tenuifolia*; *Erucastrum gallicum*, *E. vesicaria* subsp. *sativa*; *Hemicrambe fruticulosa*, *H. matronalis*, *N. paniculata*; *R. sativus*; *S. alba*, and *S. arvensis*. The completely immune plants remain symptom free both, under natural field infection as well as under controlled artificial inoculation (Sharma et al. 2002). Comparatively, broccoli and cauliflower varieties exhibit moderate *Alternaria* resistance, while the cabbage as susceptible.

4.8 Sources of Multiple Disease Resistance

Brassica alba and genotypes, viz. HC-1, PCC-2 (*B. carinata*), GSL-1501, GS-7027, Midas, Tower (*B. napus*), and DIR-1507, DIR-1522 (*B. juncea*), have been identified as sources of multiple disease resistance against white rust, Alternaria blight, downy mildew, and powdery mildew diseases (Dang et al. 2000; Kumar and Saharan 2002). Earlier *B. napus* genotype GSL-1501 was reported resistant to white rust and powdery mildew (Gupta and Singh 1994). The variations in the incidence of powdery mildew scores among different *Brassica* genotypes differ considerably due to variations in date of sowing, stage of the crop at the time of observations and prevailing environmental conditions. Hence, early maturing genotypes or the genotypes planted early in the season would help in escaping the powdery mildew infection.

The genotypes EC-322090, EC-322092, EC-322093, and RC-781 of *B. juncea*, HC-1 and PCC-2 of *B. carinata*, and GS-7027, GSL-1, GSL-1501, Gulivar 1, HNS-4, and Tower sel.1 of *B. napus* are resistant to *A. candida* and *A. brassicae*. Whereas genotypes EC-129126-1, EC-129126-2, PR-8805 of *B. juncea*, Candle, Tobin, Torch of *B. rapa*, DHC-1, HDC-4, DHC-1960, HC-1, PCC-2 of *B. carinata*, and GSL-1501, N 20-12-2, and Wester of *B. napus* are resistant to *A. candida* and *Erysiphe cruciferarum*. However, HC-1, PCC-2 of *B. carinata* and GSL-1501 (*B. napus*) are resistant to both *A. brassicae* and *E. cruciferarum*. GSL-1501 and Gulivar-1 of *B. napus* were found resistant to *A. candida* and *E. cruciferarum* by Gupta and Singh (1994). The genotypes Wester (*B. napus*), Tobin and Candle (*B. rapa*) are reported completely free from *A. candida* and *E. cruciferarum* (Shivpuri et al. 1997). The variations in the observations may be attributed to the occurrence of new races of the pathogen and congenial weather factors (Table 2.6, Chap. 2; Kumar and Saharan 2002).

4.8.1 Relationship Between Major Foliar Diseases

The increase in level of resistance to different pathogens is positively correlated, the improvement in the accumulation of resistance to multiple diseases may be facilitated easily in agricultural or natural populations; negative association among the level of resistance will hinder plant defense against pathogens attack and leads to difficulties in development of multiple disease resistance. The non-availability of single resistance source line with different mode of inheritance further complicates the work of genetic amelioration for multiple disease resistance. The significant and positive association between disease score of white rust, Alternaria blight, and powdery mildew has been recorded except, white rust stag head and powdery mildew infection in genotypes of different *Brassica* species. A high degree of correlation is observed between leaf phases of white rust and Alternaria blight (Table 4.2). Similarly, the positive genetic correlation is observed between *Hyaloperonospora parasitica* and *Leptosphaeria maculans* and *Hyaloperonospora*

Table 4.2 Relationship among major foliar diseases of *Brassica* (Kumar and Saharan 2002)

Disease	WR (L)	WR (S)	AB (L)	AB(S)	PM	ADL	ADI
			B. juncea				
WR (L)	1.00						
WR (S)	0.32**	1.00					
AB (L)	0.39**	0.12	1.00				
AB(S)	0.14	0.09	0.75**	1.00			
PM	0.10	0.09	0.35**	0.30**	1.00		
ADL	0.81**	0.28*	0.69**	0.44**	0.50**	1.00	
ADI	0.13	0.11	0.61**	0.71**	0.82**	0.55**	1.00
B. rapa, B. carinata and *B. juncea*							
WR (L)	1.00						
WR (S)	0.19	1.00					
AB (L)	0.01	0.30	1.00				
AB (S)	0.21	0.30	0.96**	1.00			
PM	0.46**	0.05	0.07	0.15	1.00		
ADL	0.69**	0.25	0.53**	0.64**	0.79**	1.00	
ADI	0.43*	0.31	0.73**	0.79**	0.71**	0.94**	1.00
Brassica species							
WR (L)	1.00						
WR (S)	0.32**	1.00					
AB (L)	0.39**	0.26**	1.00				
AB(S)	0.34**	0.27**	0.93**	1.00			
PM	0.33**	0.14	0.36**	0.33**	1.00		
ADL	0.82**	0.31**	0.68**	0.62**	0.68**	1.00	
ADI	0.42**	0.29**	0.75**	0.77**	0.85**	0.81**	1.00

WR (L) = White rust at leaf; WR (S) = White rust at staghead; AB (L) = Alternaria blight at leaf; AB(S) = Alternaria blight at siliquae; PM = powdery mildew; ADL = Average disease on leaf; ADI = Average disease on inflorescence *=Significant at ($P = 0.05$); **= Significant at ($P = 0.01$)

parasitica and *Albugo candida* in *B. rapa* (Mitchell-Olds et al. 1995). Sharma et al. (1991) reported significant positive correlation between Alternaria blight and black rot in cauliflower. The strong evidence for a positive genetic correlation in the level of resistance against *A. candida, A. brassicae,* and *Erysiphe cruciferarum* may indicate the presence of plant resistance genes that provide defense against some fundamental characteristics common to different taxonomic orders of fungal pathogens as observed by Bruns et al. (1991). Genetic correlations may arise from several factors like I). linkage disequilibrium, a nonrandom association among genotypes at several loci, which is rarely found in cross-pollinated species (Crow and Kimura 1970). However, the linkage disequilibrium may arise from previous generations of selection or small population size; II). Pleiotropy, and III). tight linkage between resistance genes (Martin et al. 1993).

4.9 Development of Resistant Cultivars

Various methods including conventional as well as biotechnological approaches have been utilized either alone or in combination to incorporate the desired traits, and to steer the genetic variability into new improved cultivars of crucifers. The conventional approach involves the identification and selection of disease-resistant genotypes and transfer of the desired traits into agronomically superior genotypes by hybridization. Hybridization combined with backcrossing, either singly or in combination, with pedigree selection is the most widely used and accepted conventional method to transfer disease resistance in crucifers.

Attempts to obtain resistance in agronomically superior genotypes of *B. juncea* through mutagenic agents such as gamma-rays have met with little success. Verma and Rai (1980) exposed the seeds of *B. juncea* cultivars Varuna, PR-5, and RS-3 to gamma-rays and obtained a mutant with 60 KR doses which is promising in yield and resistant to *A. brassicae.* Rahman et al. (1987) selected mutant BINA 1 resistant to *A. brassicae* from *B. rapa* cultivar YS52 treated with EMS in Bangladesh. Das and Rahman (1989) obtained the highest frequency of variation for Alternaria blight in M2 by irradiating seeds of Yellow Sarson cultivar YS52 with 70 KR gamma-rays. Sharma (1990) identified mutants in M2 population of cultivars Varuna and Kranti as resistant to Alternaria blight on the basis of small lesion size under natural epiphytotic conditions. MacDonald and Ingram (1985) selected and regenerated plants from calli resistant to partially purified culture filtrates of *A. brassicicola.* According to Katiyar and Chopra (1990), a somaclonal variant obtained from an exotic yellow-seeded accession of *B. juncea* retained resistance to disease similar to the parental material. The transfer of resistance from different sources in different *Brassica* crops is possible and is being done through the following conventional and modern techniques (Verma and Saharan 1994, Saharan 1992a, 1997; Saharan et al. 2003, 2005).

1. Germplasm evaluation for sources of resistance at National and International levels.
2. Selection for disease resistance through (a) pure line selection, (b) mass selection, (c) modified recurrent mass selection, and (d) recurrent selection.
3. Breeding for disease resistance by increasing the level of resistance through (a) multiple crosses, (b) recurrent selection, (c) diallele crossing, and (d) selective mating system.
4. Transfer of resistance by (a) intraspecific pedigree, backcross, and modified recurrent mass selection methods, and (b) interspecific genome substitution, chromosome substitution, and gene introgression.
5. Transfer of resistance through mutation breeding.
6. Use of biotechnological and genetic engineering techniques such as (a) genome manipulation, (b) manipulation of cytoplasmic genomes, (c) use of transformation and foreign gene expression techniques, and (d) embryo rescue techniques for wide hybridization.

4.9.1 Strategies and Methods of Screening for Resistance

The pre-requisites to obtain resistant (R) cultivars (cvs.) are:

(a) Knowledge of the pathogenic variability.
(b) Development of a screening method able to mimic the conditions required by the plants when exposed to natural sources of inoculum in the diverse field environment, and
(c) Availability of usable sources of resistance.

Screening methods for disease resistance (R) should be developed within the framework of a general strategy for resistance. The changes in the frequency of virulence genes among the populations of pathogens inciting disease of the above-ground parts of plants are very frequent. Populations of such pathogens vary in time and space because of the air borne, and seed borne nature of inoculum which facilitates long distance dispersal of their variants. As a result of these situations, breeding for resistance to foliar pathogens is, in general, more difficult than in the case of less mobile pathogens, e.g., soil borne fungi which are, therefore, more stable. Screening as part of a strategy for developing R cultivars require good planning and understanding of the processes involved in resistance. A screening program should be initiated with a clear statement of the type of resistance, which is sought, i.e., complete resistance (No sporulation of the pathogen) or partial resistance (reduced sporulation of the pathogen) or both, and with at least some knowledge of pathogenicity and virulence patterns in pathogen. The durability of resistance can be practically tested only when the R cultivars are widely used in space and time. Multilocation cultivar testing or the challenge of cultivars to a large amount of pathogen population can help verify resistance and give timely warning of the possibility of resistance breakdown, but cannot actually be considered as a test for durability of resistance. Care must be taken in interpreting results of glasshouse or laboratory tests, as the expression of resistance in the field may be considerably different because of interaction between microorganism, pathogens, and environmental conditions.

For foliar pathogens, the plant material must be adequately challenged with a single race or a pathotype at a realistic inoculum dose to allow disease development, but at the same time, not obscure minor differences in host response required to identify partial resistance. Use of inoculum composed of a mixture of known races or unknown races/pathotypes in the naturally infested crop debris will not be adequate to achieve this objective. Illustrations of three cultivars each having a single gene for complete resistance to a given race would be identified only when inoculated singly with each isolates but not if the isolates were used in a mixture. The use of a single race provides the best condition for the selection of partial resistance in the presence of complete resistance, and that the selected races should have the broadest possible virulence spectrum to suppress the expression of as many complete R genes as possible. Genotypes with resistance to one virulent race should then be systematically tested with a collection of other isolates. The identification of cultivars with complete resistance is a first step in the development of effective and durable resistance, and genetic analysis of resistance reaction is essential to reveal similarities and differences in the gene(s) that confer resistance in each genotype. Several genes known to be effective against a given race or genes effective against all prevalent races in a productive region are then recombined in single cultivars. Large amount of genetic variability sources are available in cruciferous crops in the world, which can be exploited by researchers to develop a resistant cultivars against pathogens in each *Brassica* crop (Table 4.3). Figure 4.3 illustrates a scheme to develop resistant cultivars of *Brassicas* (Verma and Saharan, 1994; Saharan et al. 2003, 2005).

4.10 Bottlenecks in Resistance Breeding

Since resistance against Alternaria black spot is generally governed by polygenes, breeding for resistance should involve pyramiding minor genes to provide additive/polygene resistance. Rapid advances in techniques of tissue culture, protoplast fusion, embryo rescue, and genetic engineering have made possible the transfer of disease resistance traits across the otherwise impassable self-incompatibility barriers. Disease-resistant transgenic plants which overexpress different antifungal compounds like pathogenesis related (PR) proteins (chitinase, glucanase, osmotin, etc.) and ribosome inhibiting proteins (RIPs) (thionins, defensins, and phytoalexins) to inhibit growth of the pathogen seem less efficacious (Zhou et al. 1999).

For introduction of *Camelina*-derived *A. brassicicola* resistance into commercial varieties, although somatic hybrids between *C. sativa* and *B. carinata* have been procured, the researchers failed to multiply the resulting hybrids (Narasimhulu et al. 1994). Similarly, strategy of protoplast fusion between *C. sativa* and *B. oleracea* with subsequent hybrid regeneration has also proved unsuccessful (Hansen 1998). Several research groups, although attempted but not succeeded in introducing the *E. sativa*- derived black spot resistance into various species of cultivated crucifers

Table 4.3 Sources of cruciferous genetic variability in the world (Rapeseed-mustard germplasm at ICAR-DRMR, Bharatpur)

Species	Exotic lines	Indigenous collections	Adv. Breeding lines	Registered lines	Released varieties	Others	Total
Brassica juncea	189	636	656	19	73	263	1836
B. rapa var Toria	12	67	59	-	14	74	226
B. rapa var Yellow Sarson	07	78	50	01	12	24	172
B. rapa var Brown Sarson	01	01		-	02	08	12
B. napus	61		30	09	06	14	120
Eruca sativa	27		19	-	03		49
B. carinata	02		10	03	05	01	21
Others				01	–	15	16
Total	299	782	824	33	115	399	2452

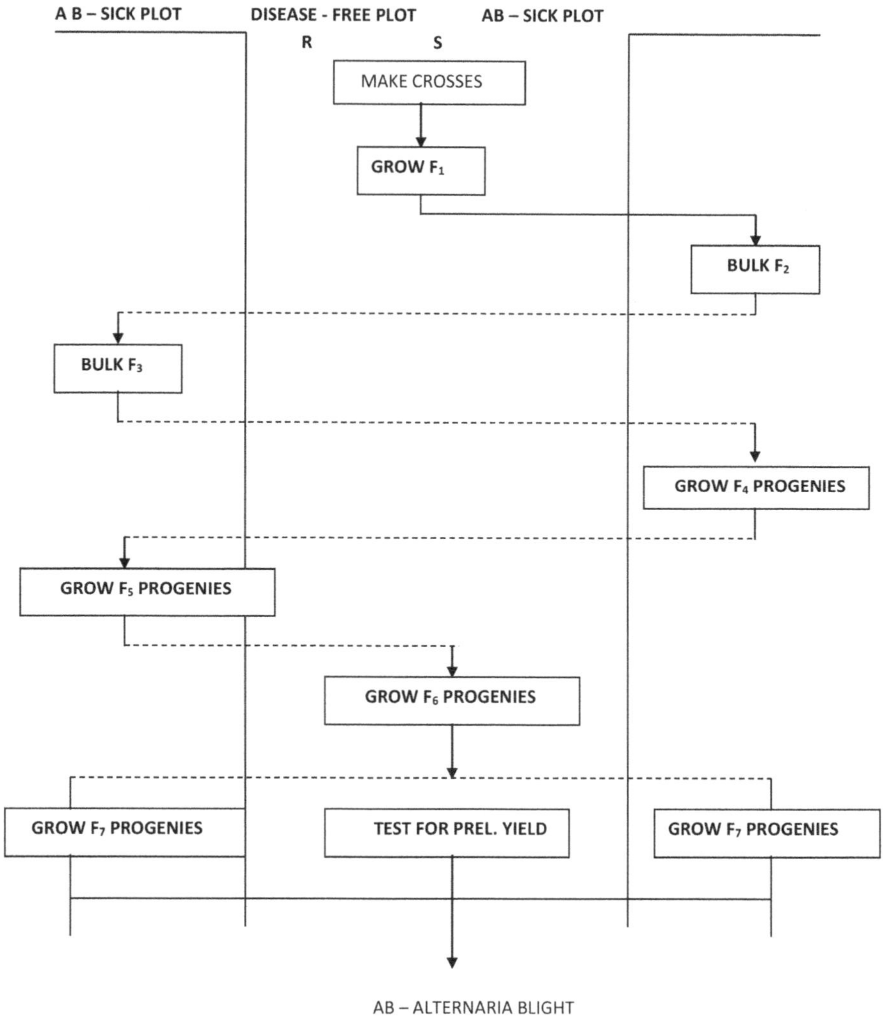

Fig. 4.3 Scheme to develop high-yielding disease-resistant cultivars (Saharan et al. 2003)

(Fahleson et al. 1988; Sikdar et al. 1990; Sigareva and Earle 1997). First somatic hybrids to be obtained as a result of protoplast fusion were those of *B. napus* (rapeseed) and *S. alba* (Primard et al. 1988). None of the hybrids procured that way showed *A. brassicae* resistance comparable to that exhibited by *S. alba*. Chevre et al. (1991) used these species for interspecies crosses through somatic hybridization and bio-directional crosses using the embryo rescue technique; these researchers have succeeded in regeneration of *B. napus* plants carrying 38 chromosomes and displayed *A. brassicae* resistance at levels close to those of *S. alba*, *B. oleracea* var. *Botrytis*, or *B. carinata* (Ryschka et al. 1996). Seeds of developed intertribal somatic hybrids between *B. napus* and *C. sativa* (by means of protoplast electrofusion) exhibited phenotype intermediate to that of the parental species. Although the hybrid plants exhibited higher levels of linolenic and eicosanoic acids, their resistance against *Alternaria* still awaits confirmation (Jiang et al. 2009).

Researchers, in general, now realize that introduction of *Alternaria* resistance genes into commercial crucifer cultivars is dependent on cumulation of horizontal resistance genes (Sharma et al. 2002). It is therefore imperative to identify sources of horizontal resistance among the various *Brassica* species, and subsequently combine them to increase durable *Alternaria* resistance. Strong cross-incompatibility, polygenic background of the resistance (additive and dominant gene interactions), as well as the differences in ploidy (differing number of chromosomes) between different *Brassicaceae* species render the *Alternaria* resistance transfer from the wild to the cultivated species difficult. Additionally, researchers often need to employ advanced *in vitro* hybridization techniques, including somatic hybridization, embryo and ovary rescue or protoplast fusion (Nowicki et al. 2012).

4.11 Use of Biotechnological Approaches

Besides utilizing conventional methods of selection and sexual reproduction, biotechnological techniques involving tissue culture and genetic transformation-based methods have revolutionized the search and development of economically promising disease-resistant crops. These tools have enabled the researchers to combine the traits from distantly related *Brassica* species for development of superior *Alternaria* blight-tolerant varieties of *Brassica* (Aneja et al. 2012; Aneja and Agnihotri 2013; Saharan et al. 2003, 2016; Meena et al. 2019)

4.11.1 In Vitro Embryo Rescue

Embryo rescue, documented for the first time in the eighteenth century, is one of the most successful techniques for the production of interspecific and intergeneric hybrids from naturally incompatible crosses that do not result in the production of viable offspring. The technique involves excision of the ovary/ovule/embryo resulting from an interspecific cross, its culturing and maturing on suitable media for the production of a hybrid plant with desired traits. The technique has proven very efficient in transferring *Alternaria* blight tolerance in oilseed *Brassicas*. In vitro ovary culture was used to transfer the resistance against *Alternaria* blight from *B. tourneforti* to *B. juncea* cv. RH 30 (Yadav et al. 1991). Agnihotri et al. (1991) developed intergeneric hybrids of *B. campestris* and *B. spinenscens* through sequential ovary, ovule, and embryo culture. The hybrid plants were then multiplied using somatic embryogenesis. The resistance trait from *Sinapis alba* cv. Carine was transferred to *B. napus* cv. Brutor using in vitro fertilized ovary culture (Chevre et al. 1994). Intergeneric hybrids with *Alternaria* blight resistance were developed between *Erucastrum cardaminoides* and *B. oleracea* var. *Alboglabra* using sequential ovary and ovule culture (Mohanty et al. 2009). Gupta et al. (2010) also used interspecific hybridization in combination with *in vitro* ovule culture to incorporate high tolerance to *Alternaria* blight and white rust from *B. carinata* cv. Kiran to low erucic acid TERl (OE) M21 lines; hybrids with high tolerance to *Alternaria* blight and white rust contained negligible amount of erucic acid (1.3%) and high oleic acid (42.5%).

4.11.2 Somatic Hybridization

Somatic hybridization refers to the enzymatic removal of the cell wall and isolation of protoplasts which are then fused together to produce hybrids. Due to the absence of the cell wall, there occurs a nonspecific fusion between the two protoplasts without any barrier to intergeneric/interspecific crosses. It also bypasses both the pre- and post-fertilization barriers. This technique has been utilized effectively in *Brassicas*, but the studies are limited to *B. napus* and *B. oleracea.* Efforts have been made to transfer *Alternaria* resistant trait from *Moricandida arvensis* to *B. oleracea* (Toriyama et al. 1987), and from *S. alba* to *B. napus* (Primard et al. 1988). Sharma and Singh (1992) attempted the transfer of *Alternaria* resistance trait from *B. carinata* to *B. juncea.* Leaf mesophyll protoplasts from *M. arvensis* and *B. napus* were hybridized to produce disease-resistant hybrid plants (O'Neill et al. 1996). Jouradan and Salazar (1993) resynthesized *B. carinata* by protoplast fusion between *B. nigra* and *B. oleracea,* and the hybrids thus obtained were fertile and grew into robust plants. Hybridization between *S. alba* and *B. oleracea* and between *Camelina sativa* and *B. oleracea* has also shown to produce resistant hybrids (Hansen and Earle 1997; Hansen 1998). Sigareva and Earle (1999) developed somatic hybrids between *S. alba* x *B. oleracea* by protoplast fusion followed by embryo rescue, and recovered four highly resistant hybrid progenies after repeated backcrosses.

4.11.3 Somaclonal Variations

These are regarded as the impulsive, inheritable changes in gene expression which often appear under *in vitro* conditions especially in tissue culture experiments. Sharma and Singh (1995) used somaclonal variations for incorporation of disease resistance/ tolerance against *Alternaria* blight. Resistance in Alternaria blight susceptible *B. juncea* varieties was achieved by inducing variations through mutagenesis (exposing seeds to gamma radiations) (Verma and Rai 1980), and by treating the microspores derived embryos with chemical mutagens (Agnihotri et al. 2009) including ethyl methane-sulfonate (EMS) and ethyl nitrosourea (ENU).

4.11.4 Genetic Transformation

Production of transgenic plants through genetic transformation forms another biotechnological approach to integrate the disease resistance genes from resistant/tolerant genotypes to the economically important susceptible ones. The genetic transformations in *B. juncea* have been reported to achieve delayed onset of the *Alternaria* blight. Kanrar et al. (2002) isolated cDNA sequence for "hevein-rubber tree lectin" from *Hevea brasiliensis* and transferred it into *B. juncea* cv. RLM-198 through binary vector *pBinAR*. The transgenic plants thus produced supported delayed onset of *Alternaria* blight with considerably low disease intensity. Mondal et al. (2003) identified the role of "chitinase" in plant defense responses and transformed *B. juncea* (cv. RLMI98) with chitinase gene tagged with over expressing promoter 35S *CaMV*. The transformed plants resulted in delayed disease onset and about 12–56% reduction in fungal colony size. Osmotin protein, known for its role in signal transduction, when transferred to *B. juncea* also showed tolerance to *A. brassicae* in transgenic plants (Taj et al. 2004). In another study, transgenic Indian mustard expressing "*class I glucanase*" gene under the control of *CaMV* 35S promoter was developed (Mondal et al. 2007). For undertaking the genetic transformation, cDNA encoding "*class I glucanase*" gene was isolated from tomato and cloned into *Bam* HI-*Sal I* restriction site of binary vector, *pBinAR* flanking *CaMV* 35S constitutive promoter and an *ocs* terminator region. The resulting recombinant $pBinG_B$ clone was transferred into *Agrobacterium tumefaciens* strain GV 2260, which was further used to transform *B. juncea* (cv. RLM 198). A study by Verma et al. 2012 reported the introgression of cysteine-rich antimicrobial peptide, *PmAMP1* from *Pinus monticola* into *B. napus,* thus conferring enhanced protection against *A. brassicae, Leptosphaeria maculans,* and *Sclerotinia sclerotiorum.* Combined expression of type I ribosome inactivating protein (RIP; *AAA32951*) and Barley "*class II chitinase*" gene in *B. juncea* not only reduced the growth of fungal hyphae by 44%, but also affected the onset of disease (Chhikara et al. 2012). Efforts have also been made to study the signal transduction pathways in *Brassica* species. Taj et al. (2011) reported interaction of MAPK3 and Lox genes during pathogenesis of *A. brassicae,* which is suggested to play a role in biosynthesis of jasmonic acid and jasmonic acid-induced expression of defense genes in *B. juncea.*

4.11.5 Molecular Markers

As the efficient management of plant diseases relies on the accurate identification of the plant pathogen, knowledge of the variability in the pathogen population, both at the pathological as well as genetic level, is very important. Characterization of variability in the pathogens including *Alternaria* spp. has been done earlier using traditional methods of morphological markers (Goyal et al. 2011), biochemical tests (Vishwanath and Kolte 1997), and cultural and pathogenic assays (Goyal et al. 2011; Singh et al. 2011). These markers differentiate the pathogens on the basis of their growth behavior, physiological characters and microscopic appearance. However, these criteria are greatly influenced by different factors including age and quality of the inoculum, age and type of host, culture media used, and incubation conditions. Also, these techniques are more laborious, time consuming, and often give non-reproducible results. To overcome these limitations of traditional markers, detection of genetic variability through DNA fingerprinting technology by the use of molecular markers has come to the aid of plant breeders and pathologists. The main advantage of using these fingerprinting techniques is that, they do not require any prior knowledge or availability of variable sequences in the database, as they deal with the genome-wide biodiversity.

Various molecular markers being used for assessing variability include, internal transcribed spacer regions (ITS), restriction fragment length polymorphism (RFLP), randomly amplified polymorphic DNA (RAPD), amplified fragment length polymorphism (AFLP), inter-simple sequence repeats (ISSRs), microsatellite, sequence tagged sites (STS), and single nucleotide polymorphism (SNPs). Due to the presumed dependency on traditional characterization only a few studies are reported on molecular characterization of *Alternaria* species pathogenic to oilseed *Brassicas*. The ITS regions (also called the conserved regions) in the fungal genome are considered the most popular locus for DNA based mycological studies at species level. Chou and Wu (2002) and Berbee et al. (2003) studied the ITS region of rDNA of *A. brassicae* and *A. brassicicola* to evaluate the phylogeny of the pathogen. Molecular techniques like RAPD, that do not require any prior knowledge of the DNA sequences have been efficiently used to analyze the genetic variations in *Alternaria* species (Cooke et al. 1998; Sharma and Tewari 1995, 1998; Gherbawy 2005). However, due to the limitation of reproducibility in RAPD-generated data, the assaying of genetic variability in *Alternaria* species has shifted to more sensitive techniques like AFLP (Bock et al. 2002) and microsatellite markers. Twelve polymorphic microsatellite markers were isolated from *A. brassicicola*. The number of alleles detected in 12 loci range from 2 to 10 (mean 3.5). Cross-species amplification shows that the designated primers are specific to the pathogen (Avenot et al. 2005). Though these molecular tools have revealed huge amount of genetic variations in *Alternaria* at the sub-generic as well as sub-species level, this field still remains largely unexplored, and demands an extensive study into the variability at different geographic levels among different hosts.

4.11.6 Induction of Systemic Resistance

Induced host resistance refers to activation of resistance in otherwise susceptible plants, through biotic as well as abiotic agents (Stitcher et al. 1997; Kessman et al. 1994), without changing their basic genetic makeup. Foliar application of β-aminobutyric acid induces resistance against *A. brassicae* in *B. carinata* (Chavan et al. 2013). *GLIP1* protein in association with ethylene signaling is a critical component in plant resistance to *A. brassicicola* (Oh et al. 2005). The resistance/defense-related genes in the vulnerable plants can be activated by inoculating the plant either by an avirulent form of the pathogen, or by limited inoculation with the pathogen (Deverall 1995). The infecting avirulent pathogen triggers natural defense responses in the plant through the release of the elicitors, which then result in the expression of novel anti-pathogenic proteins. In order to harp the benefits of induced host resistance and build up a stable, long-term resistance mechanism in the host plant against the pathogen, there is a need to identify the pathogen and understand its behavior under diversified conditions. In case of Alternaria blight, considerable morphological and pathological variations have been observed in *A. brassicae* under different environmental conditions (Goyal et al. 2011; Kaur et al. 2007; Singh et al. 2007; Patni et al. 2005; Meena et al. 2005, 2012; Ansari et al. 1989; Awasthi and Kolte 1989; Saharan et al. 2015, 2016; Meena et al. 2019). Vishwanath et al. (1999) reported the induction of resistance in susceptible *B. juncea* cv. PR-15 against extremely virulent *A. brassicae* isolate A (AbA) and moderately virulent isolate C (AbC) from *B. carinata* cv. PPCS-1 by inoculating with an avirulent isolate D (AbD) isolated from same *B. carinata* cultivar. About 60 % reduction in the disease severity was observed against AbA and AbC by the induction of resistance; However, due to the limited studies in molecular characterization of the pathogen, the clear outcome is still awaited.

Variations in *A. brassicae* isolates have been recorded in traits including colony morphology, colony diameter ranging from 32–68 mm, surface texture from velvety to wooly, color from cream to dark olive green, and sporulation from very less to intense. The host-pathogen interaction studies using four varieties of *B. juncea* and two of *B. rapa* have shown pathological variation among the isolates that differ in their aggressiveness from avirulent (few) to highly virulent (Aneja et al. 2012; Aneja and Agnihotri 2013). A thorough knowledge of diversity in the pathogen will not only help in characterizing and analyzing it phylogenetically, but also in inducing durable resistance against *Alternaria* blight, which can act as an environment friendly substitute to the traditional use of fungicides.

4.11.7 Genetic Engineering

The development of *Alternaria* resistant genetically modified plants (GM) is an excellent approach to manage Alternaria diseases of crucifers. Resistant genes from other plants or bacteria encoding enzymes like chitinase or glucanase can be introduced in crucifers. Both chitinase and glucanase enzymes breakdown chitin and glucan, which are essential components of fungal cell wall. Introduction of plant genes to enhance innate plant defense mechanisms like activating phytoalexins, proteinase inhibitors, and detoxification of phytotoxins is another useful way of developing resistant varieties. Invoking hypersensitive reaction when a fungal pathogen is attacking can also be an innovative approach.

4.12 Transfer of R-Genes Approaches for Clubroot

4.12.1 Transferring R-Genes by Molecular Marker

To develop resistant *Brassica* cultivars, several resistant sources such as European turnips, Chinese cabbage, and rutabaga cultivars are available and the resistance in these sources is dominant, which makes it easier to transfer clubroot resistance through interspecific and intraspecific hybridization. Rutabaga cultivars have been identified as clubroot-resistant sources. However, the genetics of clubroot resistance in rutabaga is complicated, so it is difficult to develop molecular markers that are closely linked to the dominant clubroot resistance genes (Ayers and Lelacheur 1972).

In Chinese cabbage, the dominant clubroot resistance from European turnips has been successfully used to develop clubroot-resistant Chinese cabbage. Since the gene mapping has been performed extensively in Chinese cabbage, molecular markers closely linked to clubroot resistance loci that are used in gene mapping can be easily selected to transfer clubroot resistance genes in the development of Chinese cabbage cultivars through molecular marker-assisted selection. Since the amphidiploid *B. napus*, has a very complex genome, most of the molecular markers developed in *B. rapa* may not be polymorphic and cannot be directly used in *B. napus*. Additional efforts are required to develop molecular markers in *B. napus* when the mapped clubroot resistance loci in Chinese cabbage are transferred into *B. napus*. Currently, most clubroot resistance genes in European turnips have not been intensively investigated, and mapping and cloning of these clubroot resistance genes in European turnips will allow using these genes effectively and efficiently in *Brassica* breeding (Hirani and Li 2015).

4.12.2 Transfer of R-Genes by Interspecific Hybridization

A dominant clubroot resistance gene from *B. rapa* (Chinese cabbage) was fine mapped, and molecular markers were developed in Chinese cabbage that could be used for marker-assisted selection (MAS) in other *Brassica* crops. To transfer this clubroot resistance gene to *B. napus* (canola), an interspecific hybridization was made between *B. napus* (canola), and *B. rapa* (Chinese cabbage) (Fig. 4.4). Subsequently, the F_1 was backcrossed to the canola recurrent parent for three generations to produce BC_1, BC_2, and BC_3 progenies (Table 4.4). Using these populations, simple sequence repeat (SSR) markers flanking the clubroot resistance gene were used to perform MAS in canola. These molecular markers were then evaluated in 13 different canola and rapeseed quality genotypes in *B. napus* and *B. rapa*. These markers exhibited high reliability in identifying clubroot resistance in this diverse set of *Brassica* genotypes. Clubroot resistance also co-segregated with the SSR markers flanking the clubroot resistance gene in the BC_3 and BC_3S1. The segregation ratio of resistant and susceptible individuals in the BC_3 supported the expected 1:1 ratio for the segregation of a single Mendelian gene. BC_3S1 families with homozygous

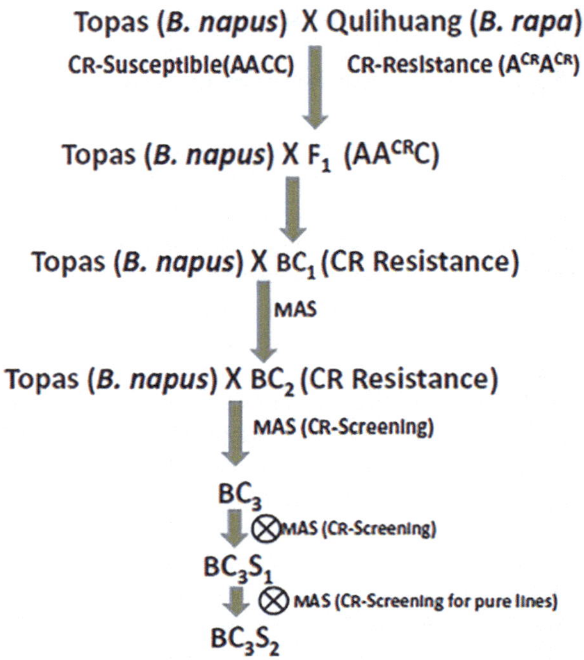

Fig. 4.4 Schematic diagram of interspecific backcross breeding strategy to introgress clubroot resistance gene from *Brassica rapa* (Chinese cabbage) to *Brassica napus* (canola) (Hirani et al. 2016).

clubroot resistance were developed during this process, and will be valuable sources of clubroot resistance in *B. napus* breeding activities (Hirani et al. 2016).

4.12.3 Transfer of R-Genes by Distant Hybridization

The three clubroot-resistant Chinese cabbages (1003, 1007, and 1008) were used to transfer the clubroot resistance genes to *B. napus* by distant hybridization combined with embryo rescue. Three methods including morphological identification, cytology identification, and molecular marker-assisted selection were used to determine hybrid authenticity, and 0, 2, and 4 false hybrids were identified by these three methods. In total, 297 true hybrids were identified. Clubroot resistance markers and artificial inoculation were utilized to determine the source of clubroot resistance in the true hybrids. Two simple sequence repeat (SSR), and two intron polymorphic (IP) markers linked to clubroot resistance genes were identified. The clubroot resistance genes 1007 and 1008 were mapped to A03. At last, 159 clubroot-resistant hybrids were obtained by clubroot resistance markers and artificial inoculation. Four molecular markers associated with clubroot resistance were identified, which were located in a region from 23.77 to 25.56 Mb on A03 of *B. rapa*. In the adjacent area of this region, two clubroot-resistant genes *CRa* (Ueno et al. 2012) and *CRb* (Hatakeyama et al. 2017) have been cloned, in which both genes encode the TIR-NBS-LRR (TNL) protein. Through analyzing the genes structure of the region surrounding the *CRa* and *CRb*, it was found that six genes *Bra012540*,

Table 4.4 Analysis of molecular markers for clubroot resistance gene introgression in canola using marker-assisted selection in a cross between the clubroot-resistant *Brassica rapa* cv. Qulihuang and the susceptible *Brassica napus* cv. Topas. The BC1, BC2, BC3, and BC3S1 populations developed using "Topas" as the recurrent parent and screened using a set of SSR markers linked with a clubroot resistance allele (Hirani et al. 2016)

Populations	Name of markers	Total[a]	CR resistance allele	Susceptible allele	χ^2	P^b
BC_1	SScr062	69	41	28	2.45	0.12
	SScr045b	69	40	29	1.74	0.19
	SScr079b	69	31	38	0.71	0.4
BC_2	SScr062	27	15	12	0.33	0.56
	SScr045b	30	18	12	1.2	0.27
	SScr079b	29	18	11	1.69	0.19
BC_3	SScr071	87	38	49	1.39	0.24
	SScr072	87	44	43	0.01	0.91
	SScr082b	86	45	41	0.19	0.67
	SScr062	82	42	40	0.05	0.83
	SScr045b	82	43	39	0.2	0.66
	SScr079b	83	41	42	0.01	0.91
		AA	**AB**	**BB**	$\boldsymbol{\chi^2}$	$\mathbf{P^c}$
BC_3S_1	SScr071	7	16	9	0.25	0.88
	SScr072	7	15	7	0.31	0.86
	SScr045b	7	13	7	0.81	0.67
	SScr079b	7	16	8	0.13	0.94

[a]Indicates total number of individual plants excluding missing values
[b]Expected segregation ratio was 1:1
[c]Expected segregation ratio was 1:2:1 in BC_3S_1

Bra012541, *Bra019409*, *Bra019410*, *Bra019412*, and *Bra019413* have the structure of TNL (Liu et al. 2018).

4.12.4 Transfer of R-Genes from European Clubroot Differential (ECD) Set

European turnips are the most resistant sources and their resistance genes have been introduced into other crops such as oilseed rape (*Brassica napus* L.), Chinese cabbage, and other *Brassica* vegetables. The European clubroot differential (ECD) set contains four turnip accessions (ECD1–4). These ECD turnips exhibit high levels of resistance to clubroot when they are tested under controlled environmental conditions with Canadian field isolates. Gene mapping of the clubroot resistance genes in ECD1–4 was performed and three independent dominant resistance loci were identified. Two resistance loci were mapped on chromosome A03, and the third on chromosome A08. Each ECD turnip accession contained two of these three resistance loci. Some resistance loci were homozygous in ECD accessions while others showed heterozygosity based on the segregation of clubroot resistance in

20 BC_1 families derived from ECD1 to 4. Molecular markers were developed and linked to each clubroot resistance loci for the resistance gene introgression in different germplasm (Hirani et al. 2018).

4.12.5 Transferring of Powdery Mildew R-Genes Through Embryo Rescue

Interspecific hybrid plants, and backcross 1 (BC1) progeny were produced through sexual crosses, and embryo rescue between *Brassica carinata* accession PI 360883, and *B. oleracea* cvs. Titleist and Cecile to transfer resistance to powdery mildew to *B. oleracea*. Four interspecific hybrids were obtained through application of embryo rescue from crosses with *B. carinata* as the maternal parent, and their interspecific nature confirmed through plant morphology, and random amplified polymorphic DNA (RAPD) analysis (Plate 4.1). Twenty-one BC1 plants were obtained through sexual crosses and embryo rescue although embryo rescue was not necessary to produce first backcross generation plants between interspecific hybrids and *B. oleracea*. All interspecific hybrids and eight of the BC1 plants were resistant to powdery mildew (Tonguc and Griffiths 2004a).

4.12.6 Mutagenic Resistance to Powdery Mildew

The method of induced mutagenesis is a powerful tool for creating new genotypes with an increased yield, quality, and resistance to diseases, and pests (Ahloowalia et al. 2004). Upon irradiation with gamma-rays, the variation of the resulting genetic alterations has high frequency and wider range of morphological and biochemical changes (Skoric et al. 2008). There is evidence that the mutations may also change the expression of certain enzymes in the biosynthetic pathway of fatty acids (Auld

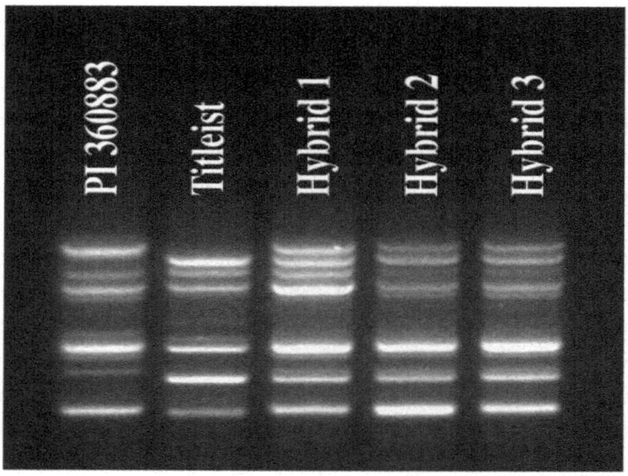

Plate 4.1 Confirmation of interspecific hybrids using random amplified polymorphic DNA (RAPD) polymorphisms from *Brassica carinata* parent (PI 360883) and broccoli parent (Titleist) with 1.5% agarose gel electrophoresis (Tonguc and Griffiths 2004a)

Table 4.5 Total fat, saturated, and unsaturated fatty acids in the seeds of Abacus, and M0, M1, and M2 segregating generations of Abacus, treated with absorbed dose of 100 Gy and 150 Gy gamma-rays (Petkova et al. 2014)

Fats and fatty acids (%)	Abacus control	M0 A-100	M0 A-150	M1 A-100	M1 A-150	M2 A-100	M2 A-150
Content of crude fat	35.9	40.7	39.5	43.2	45.1	42.5	47.1
Saturated fatty acids	10.1	9.3	7.5	6.3	6.8	6.3	6.9
Unsaturated fatty acids	89.7	82.3	83.5	85.2	86.2	92.7	93.7

et al. 1992). The irradiation of rapeseed with low doses of gamma-rays of 100 Gy and 150Gy has a stimulating effect on the synthesis of linoleic and linolenic acids. Conversely, at higher doses of the mutagen factor, the concentration of these fatty acids decreases while the amount of oleic acid increases (Rahimi and Bahrani 2011). The fatty acids synthesized in plants play an important role by participating in the composition of tissue hormones, and in cell, and nuclear membrane structures (Kolattukudy 1974).

It is assumed that unsaturated fatty acids in plant cells are involved in cutin and suberin synthesis (Pinot et al. 1999). Cutin, which acts as first barrier and protects plant organs (leaves and fruits) from dehydration providing protection against pathogens and pests, is composed of C16 and C18 fatty acids. Suberin, which is predominantly composed of hydroxyl and epoxy unsaturated fatty acids, is a structural component of the plant underground part roots and tubers (Kolattukudy 1980). C18 fatty acids in plants play a role in building the defense mechanism against diseases and pests (Kachroo and Kachroo 2009). It has been found that hydroxyl fatty acids also inhibit the development of *Erysiphe graminis, Puccinia recondita,* and *Phytophthora infestans* (Hou and Forman 2000). The powdery mildew development is favored by meteorological factors, such as temperature and high humidity (Penaud 1999; Enright and Cipollini 2007). Fungicide application is often not efficient enough, due to the pathogen's rapid spread in favorable conditions. The effects of gamma radiation on the fatty acid composition of rapeseed oil and the increased resistance to *E. cruciferarum* have been determined by Petkova et al. (2014). Biochemical changes in the oil after gamma-ray irradiation are also associated with changes in the susceptibility to powdery mildew.

Through the gas chromatography, the differences in the content of saturated and unsaturated fatty acids in the seeds of the non-irradiated variants and the ones irradiated with two doses of gamma radiation were assessed (Petkova et al. 2014). A summary of the data for all three years of the fatty acid composition of the Abacus hybrid oil and its M0, M1, and M2 was decomposing the populations. The chemical analysis showed generally that the irradiated plants showed a lower content of oleic acid and a higher content of linoleic acid and linolenic acid compared with the non-irradiated variants. The data in Table 4.5 showed a tendency towards increase in the total content of raw fats in mutant populations of M1 and M2 of Abacus as

Table 4.6 Fatty acid composition of rapeseed oil from the seeds of Abacus, and M0, M1, and M2 generation of irradiated plants with 100 and 150 Gy gamma-rays (Petkova et al. 2014)

Fats and fatty acids (%)	Abacus control	M0 A-100	M0 A-150	M1 A-100	M1 A-150	M2 A-100	M2 A-150
C12:0 Lauric acid	-	0.1	0.1	0.1	0.1	0.1	0.2
C14:0 Myristic acid	0.1	0.1	0.2	0.2	0.2	0.1	0.1
C16:0 Palmic acid	7.5	6.7	6.3	6.5	6.7	5.9	5.7
C18:0 stearic acid	2.3	1.8	2.0	1.9	1.2	0.3	0.3
C18:1 oleic acid	66.3	69.3	67.0	67.8	68.5	67.5	73.0
C18:2 linoleic acid	17.2	15.5	17.4	17.2	18.3	19.1	15.9
C18:3 linoleic acid	4.4	3.1	3.5	4.2	5.9	6.9	4.6

compared with non-irradiated plants. This, in turn, is associated with a reduction in the total content of saturated fatty acids, and an increase in the content of unsaturated fatty acids. There is an inverse relationship between the values of the total fat content and the content of saturated and unsaturated fatty acids (Table 4.5). The rapeseed oil of the Abacus hybrid showed higher content of saturated fatty acids as compared with the M2 and M3 generations of the irradiated variants. In contrast, the percentage content of unsaturated fatty acids is influenced positively by irradiation with gamma-rays, and is comparatively high in M2 A1-100 and M2 A-150, 92.7% and 93.7%, compared with the baseline 89.7 %.

The data in Table 4.6 shows the extent of damage by the powdery mildew agent on plants of the Abacus variety, and its gamma-ray irradiation-treated variants. It also shows the arithmetic mean values, and their errors, the accuracy indicators, and the variation coefficients. The data show that the highest extent of damage on the stems and branches in the range of 98.9%– 99% ($p = 0.1\%$) is observed with the M0 Abacus generation, irradiated with absorbed dose 100 Gy and 150 Gy gamma-rays, compared with the non-irradiated variant (Table 4.7).

The irradiation did not have a positive impact on the sustainability of pods to *E. cruciferarum.* The differences found are insignificant compared with the corresponding baseline group. In contrast, the M1 and M2 generations were shown to have a proven lower extent of damage by the phytopathogen as compared with the baseline. It is assumed that the increased resistance demonstrated is due to the higher content of unsaturated fatty acids in the seeds of the irradiated plants. Unsaturated fatty acids can be used as a means of biological control of plant diseases on global scales. Increasing the dose of the mutagen leads to an increase in the concentration of unsaturated fatty acids with 18 carbon atoms, and to a decrease in the extent of damage by *Erysiphe cruciferarum* (Petkova et al. 2014).

Table 4.7 Assessing the extent of powdery mildew damage on stems, branches, and pods, as a percentage of the infected plant area on rapeseed plants from Abacus, and M0, M1, and M2 generation of gamma irradiated plants (Petkova et al. 2014)

Variants	Extent of powdery mildew damage %											
	Stems				Branches				Pods			
	Mean ± SD	*D*	*t*	Sign.	Mean ± SD	*D*	*t*	Sign.	Mean ± SD	*D*	*t*	Sign.
Abacus control	79.3 ± 1.31				90.5 ± 1.07				97.6 ± 0.49			
M0-Abascus-100 Gy	99.3 ± 0.18	20.09	17.62	+++	98.9 ± 0.22	8.4	16.43	+++	99.12 ± 0.2	1.52	1.36	ns
M0-Abascus-150 Gy	99 ± 0.21	19.69	17.27	+++	99 ± 0.24	8.5	15.71	+++	98.9 ± 0.31	1.3	1.04	ns
M1-Abascus-100 Gy	28 ± 1.52	−51.31	19.23	− − −	40.5 ± 1.34	−50	32.53	− − −	49.9 ± 1.17	−47.7	26.2	− − −
M1-Abascus-150 Gy	26.7 ± 1.26	−52.61	34.1	− − −	34.8 ± 1.23	−55.7	29.46	− − −	47.4 ± 1.02	−50.2	29.01	− − −
M2-Abascus-100 Gy	10.3 ± 0.58	−79.31	74.6	− − −	21.1 ± 0.77	−69	42.37	− − −	27.1 ± 1.04	−70.5	40.5	− − −
M2-Abascus-150 Gy	11.5 ± 0.56	−67.81	39.8	− − −	22.7 ± 1.31	67.8	35.12	− − −	27.5 ± 1.11	−70.1	39.8	− − −

Mean average value in % from 30 repetion; D–difference against control; *t*-Student's t distribution

ns - non-significant

$P > 5\%$ +, $P > 1\%$ ++, $P > 0.1\%$ +++, $P < 5\%$ −, $P < 1\%$ − −, $P < 0.1\%$ − − −

4.13 Clubroot Resistance Breeding in Cruciferous Crops

Substantial progress has been made in breeding for clubroot resistance genotypes/ cultivars in cruciferous crops like *B. rapa, B. oleracea, B. napus, Raphanus sativus,* and *Arabidopsis thaliana*. However, work on race-specific resistance cvs. needs more attention of researchers.

4.13.1 *Brassica rapa*

The perusal of data in Table 4.8 summarizes the available information about eight CR loci or QTL that have been mapped in *B. rapa*. Four of these reside on chromosome A03, probably representing two complex regions that control CR. One region, originally identified by mapping *CRb* and its marker TCR05 (Piao et al. 2004), is assumed to be located in the top part of A03 because a QTL derived from cv. Gelria could be allocated around TCR05 and was mapped between 12.5 cM (BRMS124) and 30.2 cM (BRMS206)(Suwabe et al. 2006). Locus *CRa* was identified as being linked with the SCAR marker HC352b (Matsumoto et al. 1998, 2005; Hayashida et al. 2008), which has been mapped by Sakamoto et al. (2008) on A03 in the vicinity of markers that were mapped in this region. Therefore, *CRa* and *CRb* might be allelic or closely linked CR loci. So far, it is not known whether these differ in their race specificity, but they differ in their mode of inheritance. Another region on A03 is highlighted by the *Crr3* locus, which is

Table 4.8 Mapped CR loci in *Brassica rapa* and their linked markers (Diederichsen et al. 2009)

CR locus[a]	Identified as	Linkage group	Linked markers	Inheritance	References
Cra[b]	Major gene	A03	HC352b-SCAR	Dominant	Matsumoto et al. (1998), Hayashida et al. (2008)
Crr1	Major gene	A08	BRMS-088	Codominant	Suwabe et al. (2003, 2006)
Crr2	Modifier gene	A01	BRMS-096	Codominant	Suwabe et al. (2003, 2006)
Crr3	Major gene	A03	OPC11-2S	Codominant	Hirai et al. (2004), Saito et al. (2006)
CRb	Major gene	A03	TCR05	Codominant	Piao et al. (2004)
Crr4	QTL	A06	BRMS-125	Codominant	Suwabe et al. (2003, 2006)
CRk[c]	QTL, possibly major gene	A03	OPC11-2S	Codominant	Sakamoto et al. (2008)
CRc	QTL, possibly major gene	A02	m6R	Dominant	Sakamoto et al. (2008)

[a]Suffixes, a, b, and c in the locus name do not correspond to A, B, and C genes of ECD set, but indicate chronological order of discovery
[b]Very near or identical to *CRb*
[c]Very near or identical to *Crr3*

located between 34.6 cM (BRMS158) and 73.5 cM (BRMS058) (Suwabe et al. 2006; Saito et al. 2006). Sakamoto et al. (2008) reported a large QTL (*CRk*) around OPC11-2S, a marker of *Crr3*. The locus *CRk* was derived from Chinese cabbage cv. CR Kanko and possibly it is identical, allelic, or closely linked to *Crr3*. Other QTL or major loci that have a strong impact on CR were found on chromosomes A02 [*CRc* (Sakamoto et al. 2008)] and A08 [*Crr1* (Suwabe et al. 2003)]. Because only a few resistant turnip cultivars have been used as CR sources, it is likely that additional CR loci will be identified in this crop type. The QTLs for CR residing in the *Brassica* A genome and originating from turnip (ECD-04) were also mapped in *B. napus* (Werner et al. 2008). A QTL on A03 had a strong effect on CR for three *P. brassicae* isolates. This QTL was located at 65.3 cM in the map (Werner et al. 2008); it might correspond to *Crr3* or eventually to *Crb*. Map did not have enough markers in common with the previously mentioned maps; however, the identity of these loci/QTLs remains to be determined. Differential race-specific responses of the *CRk*, and *CRc* loci were reported by Sakamoto et al. (2008). According to the available information, about race-specific effects of CR genes from *B. rapa*, it is most likely that the mapped CR loci confer race-specific resistance, as in Japanese CR cultivars. Kuginuki et al. (1999) and Hatakeyama et al. (2004) suggested the use of two commercial CR Chinese cabbage cultivars showing opposite interactions with most Japanese field isolates of *P. brassicae* for race designation.

Most mapping studies have used a CR cultivar or CR breeding line as a resistant parent. These breeding lines have already been selected and may have only a proportion of those CR genes that might have been present in the original CR source. Suwabe et al. (2003) in their study used a doubled haploid (DH) line, G004, as a resistant parent. This line was selected from DH lines derived from an F_1 between the CR turnip cv. Siloga and a susceptible Chinese cabbage line. Kuginuki et al. (1997) selected DH lines with CR similar to that of cv. Siloga. This selection was essential for finding *Crr2* in addition to *Crr1*. The effect of *Crr2* was visible only in the presence of *Crr1*. The *Crr2* locus was lost during the breeding of commercial CR Chinese cabbage, but it was shown to be essential for resistance towards highly virulent pathotypes (Suwabe et al. 2003). These CR loci will become important in breeding cultivars with more durable resistance (Diederichsen et al. 2009).

4.13.2 *Brassica oleracea*

The cabbage cv. Badger Shipper was one of the first CR cabbage cultivars introduced in North America in the 1960s (Crute et al. 1980). The resistance was thought to be based on a single gene group with an unknown number of genes (Chiang and Crete 1976). This resistance, however, was readily broken after introduction in the field. However, in the absence of the resistant cultivar for 3 years, the pathogen population seemed to lose pathogenicity to cv. Badger Shipper, suggesting that there are adverse fitness costs associated with pathogenicity (Crute et al. 1980). Other attempts to introduce resistant *B. oleracea* cultivars include broccoli and cabbage lines developed in North America (Baggett 1983; Baggett and Kean

1985) that are resistant to various pathotypes present in North America. Resistant cabbage cultivars released in Norway were based upon resistance from the primitive cabbage cultivar Bohmerwaldkohl (Weisaeth 1977), and a cabbage landrace from the Shetland Islands (Dixon 1988; Crute et al. 1980; Voorrips 1995).

In Canada, the *B. napus* swede cv. Wilhelmsburger was chosen as a means to introduce resistance into *B. oleracea* (Chiang et al. 1977a; Chiang and Crete 1983), as the prevalent races from the USA and Canada seem to lack virulence to most of the *B. napus* differential hosts (Chiang and Crete 1989). A single resistant BC1 plant with n = 26 was selected for introgression of CR into the cabbage cultivar Badger Shipper and the cabbage line 8-41 (Chiang et al. 1977b, 1980; Landry et al. 1992). They both possess resistance against race 6, similar to cv. Wilhelms burger (Chiang and Crete 1976; Williams 1966). It resulted in the release of two CR cultivars, Acadie and Richelain (Chiang and Crete 1985, 1989). The first disease resistance QTLs that were mapped in *Brassica* was reported in 1992 by Landry et al. who found two QTLs for CR.

Toxopeus et al. (1986) concluded that the turnip *B. rapa* ECD hosts had the strongest and broadest resistance. Within *B. oleracea,* many resistance sources have been described (Crute et al. 1980; Crisp et al. 1989; Hansen 1989; Voorrips and Visser 1993; Manzanares-Dauleux et al. 2000a, b). Only a few lines or cultivars have reached the commercial state (Crute et al. 1980; Voorrips 1995). Linkage drag between CR and undesired traits such as reduced shelf-life has been reported for some *B. oleracea* CR sources (Crute et al. 1980). Although CR from *B. oleracea* is regarded as mainly non differential, it is less used in breeding than race-specific CR from *B. napus* or *B. rapa*. The CR *B. oleracea* cultivars carry race-specific CR that originates from *B. rapa*. CR from *B. oleracea* is genetically more complex, mainly recessive, and therefore difficult to use in breeding particularly in hybrid breeding. In addition, *P. brassicae* isolates that can completely overcome the resistance of all *B. oleracea* ECD hosts are very common (Crute et al. 1983; Toxopeus et al. 1986).

In different QTL analyses of CR in *B. oleracea*, major and minor QTLs were found (Voorrips et al. 1997; Moriguchi et al. 1999; Rocherieux et al. 2004; Werner et al. 2008). It is unknown whether some of these QTLs are allelic or identical because most of them have not been mapped onto a reference map. Rocherieux et al. (2004) identified several isolate-specific QTLs and one major QTL that conferred resistance to all five *P. brassicae* isolates used in their study. Werner et al. (2008) mapped QTLs from *B. rapa* and *B. oleracea* after introduction into *B. napus* and identified eight QTLs in the *B. oleracea* C genome, all showing race specificity. Despite its non differential appearance, it has to be assumed that CR from *B. oleracea* is likely to be based on a mixture of race-specific and race-independent CR loci.

The first interspecific crosses were made in 1987 between a broccoli inbred line and Chinese cabbage cv. Parkin, which possesses a single dominant resistance gene (Fig. 4.5). Embryo rescue (Harbert 1969) was used to obtain F_1 and BC_1 plants. Nine resistant BC_2 plants were derived from a single BC_1 plant. The expected 1:1 segregations were found only in the BC_4, suggesting that the CR gene had been introgressed in the C genome. A backcross program in cauliflower, cabbage, and

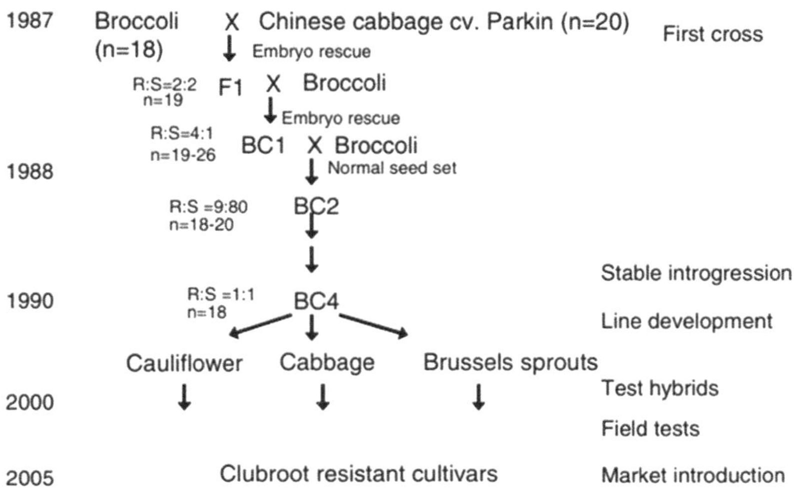

Fig. 4.5 Introgression of clubroot resistance from *Brassica rapa* cv. Parkin into *Brassica oleracea* crops at Syngenta Seeds B.V.; chromosome numbers and segregation ratios for resistance are given for initial generations (Diederichsen et al. 2009)

Brussels sprouts was then initiated, it included after several years of backcrosses, selections, line development, and test crosses. Field tests evaluating the resistance in *B. oleracea* background under field conditions were carried out for several years at naturally infested locations in Europe. In all tests the CR entries remained resistant, whereas the susceptible controls were heavily infected. Trials performed in India, Australia (Donald et al. 2006a, b), and other countries yielded similar results. These results indicated that the resistance gene was as effective in the *B. oleracea* background as in the original *B. rapa* cultivar, and encouraged introduction of cultivars into the market. The CR cultivars were introduced to the market in 2005 (cauliflower cvs. Clapton and Clarify and cabbage cvs. Kilaton and Tekila), and well received by growers. The resistance from *B. rapa* does not cover all isolates of *P. brassicae* (Toxopeus et al. 1986); however, a small number of isolated infections were observed. Medium to severe attacks occurred at some locations in Germany, Poland, and Brittany, France. This situation had already been encountered with *B. rapa* cv. Parkin before introduction of the new cultivars.

Often infections of the CR cultivars are quantitative with small clubs as reported for CR Chinese cabbage cultivars (Osaki et al. 2008). Compatible isolates sometimes have a reduced fitness on susceptible *B. oleracea* hosts as compared with a standard isolate not capable of overcoming *B. rapa* genes. An isolate that was capable of infecting CR cv. Parkin from Indonesia did show a strongly reduced reproduction rate on three susceptible cabbage cultivars (Table 4.9). Donald et al. (2007) studied the histology of the Clapton-like resistance and showed that the pathogen was inhibited in the cortex; for some isolates this interaction may be more quantitative than qualitative. Fitness costs associated with certain pathogenicity factors, as has

Table 4.9 Reproduction rate[a] of a standard isolate compared to an Indonesian isolate capable of infecting CR Chinese cabbage cv. Parkin on three cabbage entries with different levels of susceptibility (Diederichsen et al. 2009)

	Spores/g infected root		Spores/plant		Mean score (0–5)	
Entry	Stand.	Ind.	Stand.[b]	Ind.[c]	Stand	Ind.
Marathon	595	199b*	1809	347b	5.0	4.3b
Cab.1	563	248b	802	158a	4.9	3.4ab
Cab 2	468	68a	792	64a	4.6	2.5a
Mean	542b	172a	1134b	189a	4.8b	3.4a

[a]Reproduction rate = No. of spores × 10^6 per gram of infected root or per plant
[b]Stand. = race 4, Nord-Ditmar Germany
[c]Ind. = Indonesian
*Different letters indicate significant differences between means according to two-way ANOVA and Tukey test after square root transformation; $n = 14, p < 0.05$. Scoring scale: 0 = no symptoms, 5 = fully diseased

been observed with cv. Badger Shipper (Crute et al. 1980), or interaction between pathotypes may be present.

4.13.3 *Brassica napus*

Clubroot disease was originally most important to the storage-root-forming *B. napus* type spp. *napobrassica* (swede or rutabaga), and to fodder rape (*B. napus* ssp. *oleifera* var. annuum) compared with winter oilseed rape because their growing period better matches the pathogen's temperature requirements. As a result of the spread of oilseed rape into areas where soil conditions are more conducive, clubroot incidences are more frequent in this crop. In a Swedish survey, 72% of 190 soil samples from oilseed rape fields tested positive for the presence of *P. brassicae* (Wallenhammar 1996). In naturally infested fields, Wallenhammar et al. (1999) found up to 50% yield reduction in spring oilseed rape. In Canada, the first incidences of clubroot in *B. napus* were recorded in 2003 in Alberta and the disease is still spreading in this region (Strelkov et al. 2005; Xue et al. 2008).

Breeding for CR in *B. napus* focused initially on swedes, where resistant types had been identified by Colhoun (1958) and Ayers and Lelacheur (1972). Swedes have been used to introduce CR into *B. oleracea* (Chiang et al. 1977a) and received CR genes from *B. rapa* by interspecific hybridization (Lammerink 1969; Williamson and Bradshaw 1989; Bradshaw et al. 1997). Fodder rape cultivars with CR were bred after selection of resistant individuals (Lammerink 1964; Johnston 1970); some of these are included in the ECD set. The genetics of CR in the ECD *B. napus* differentials (ECD-06 to ECD-10) has been studied by Gustafsson and Falt (1986), who postulated that four race-specific genes are present in these hosts (Table 4.10). Based on differential pathogenicity patterns with many isolates, Crute et al. (1983) postulated a slightly different allocation of CR loci. Genes with the same numbers in both models may not be identical. The difference between both models is most

Table 4.10 CR genotypes postulated for *Brassica napus* ECD hosts (Diederichsen et al. 2009)

B. napus ECD host	Gustafsson and Falt (1986)	Crute et al. (1983)
ECD-06	CR 2	R2+R3
ECD-08	CR 1+ CR 3	R1
ECD-09	CR 1+ CR 2	R1+R2
ECD-10	CR 1+ CR 4	R1+R2+R4

probably due to the choice of *P. brassicae* isolates that were used in the Swedish study. By using an isolate with the capability to overcome the gene "R3" sensu Crute et al. (1983), the effect of this gene cannot be observed, this would explain the identification of only one gene in ECD-06 and two in ECD-10 (Gustafsson and Falt 1986). Several attempts were made to introduce CR from both ancestral species into *B. napus*. Diederichsen and Sacristan (1994) combined different sources of CR from *B. rapa* and *B. oleracea*. The expression of CR in resynthesized *B. napus* forms was compared (Diederichsen and Sacristan 1996). Both CR sources were less efficient in the *B. napus* background compared with the original parent, in particular, CR from *B. oleracea* appeared to be strongly diluted, indicating the presence of epistatic factors in the *B. rapa* genome. Combining the most resistant ancestral parents led to resynthesized *B. napus* forms that were resistant against all isolates of *P. brassicae* that were available in their study, representing a level and broadness of CR that had not been present previously in *B. napus*. Bradshaw et al. (1997) crossed an auto tetraploid of the *B. rapa* host ECD-04 with a tetraploid kale and selected a CR swede after backcrossing, in 1996 the CR cultivar "Invitation" was released in the UK seed market. Two CR oilseed rape cultivars, cv. Mendel and cv. Tosca, were introduced to the European seed market in 2000. Both cultivars possess race-specific CR (Diederichsen et al. 2003) originating from a resynthesized *B. napus* form (Diederichsen et al. 2003; Happstadius et al. 2003). The hybrid cultivar Mendel has yields comparable with other cultivars from the same breeding period. In Germany, its market share was increased from 0.7% in 2001 to 2% in 2006. Its breeding process is illustrated in Fig. 4.6. Resistance was initially combined in a resynthesized *B. napus* by crossing two ECD hosts, *B. rapa* ECD-04, and *B. oleracea* ECD-15, with the aid of ovule culture (Diederichsen and Sacristan 1994).

After backcrossing, di-haploidization, and selection with a highly virulent isolate of *P. brassicae*, a resistant DH line was identified, used for breeding and genetic analysis. The latter confirmed the presence of three dominant, race-specific CR genes from ECD-04 in a backcross population (Diederichsen et al. 1996), as expected. Further backcrosses and selections with undefined *P. brassicae* isolates originating from infested crops led to the resistant parent of the hybrid cultivar Mendel. Subsequent mapping of CR in cv. Mendel revealed one major locus plus two recessive genes (Diederichsen et al. 2006), indicating that two dominant CR genes were lost during breeding. The resistance of cv. Mendel is efficient in most cropping areas. Compatible populations of *P. brassicae* were known to exist before its release and occasionally infected crops of cv. Mendel have been observed. In

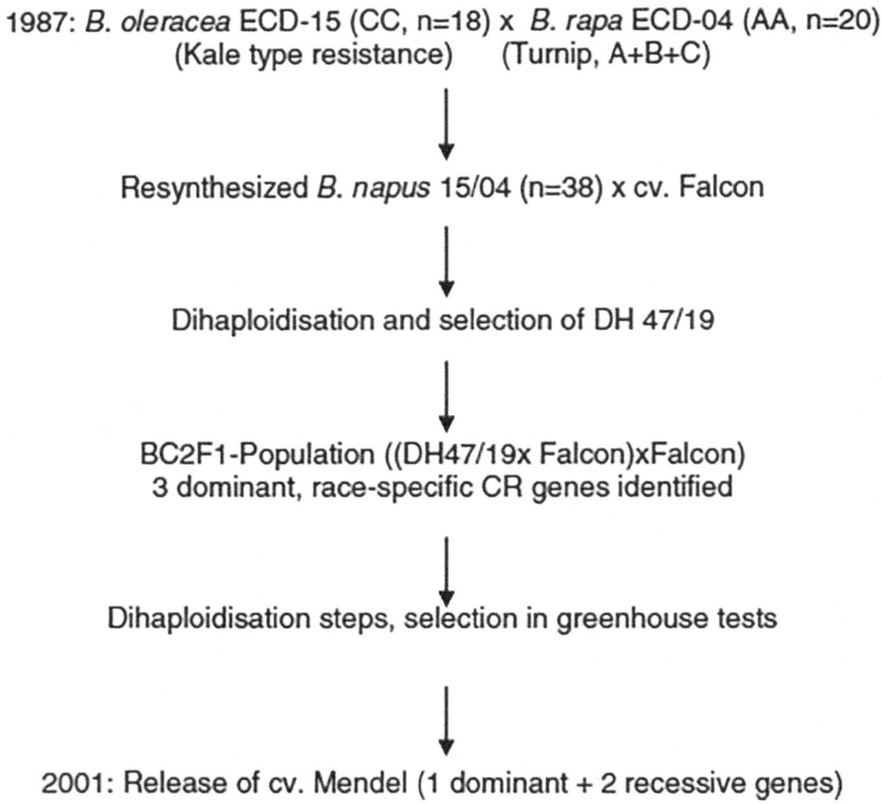

Fig. 4.6 Breeding of *Brassica napus* cv. Mendel by Norddeutsche Pflanzenzucht H.G. Lembke KG (Diederichsen et al. 2009)

many cases, the infected individuals were identified to be volunteers of previous crops. However, because cv. Mendel's resistance is based on only one major gene (recessive genes are unlikely to contribute in a hybrid cultivar). It is recommended only in those cases where clubroot infections have already been encountered and should not be grown as insurance against potential disease incidence. Control strategies involving the integrated use of pyramids of control methods will help to prolong the usefulness of this cultivar. Identification of race-independent CR genes from *B. oleracea* in *B. napus* was attempted by using as the CR source a resynthesized *B. napus* form that originated from a cross between ECD-04 and the cabbage landrace Bohmerwaldkohl. Nineteen QTLs were detected in a DH population (Werner et al. 2008), all showing race specificity. There was no indication of race-independent QTLs, which were assumed to be present in the *B. oleracea* C genome. The only strong QTL in the C genome that could be detected in this material originated from the susceptible parent cv. Express and conferred CR to only one isolate of *P. brassicae*. Some QTLs had overlapping confidence intervals on chromosomes A03 and A08 and conferred CR against more than one isolate. Based on this study, the QTLs with the broadest effects against *P. brassicae* pathotypes in *B. napus* could be expected to be located in the *B. rapa* A genome. Manzanares-Dauleux et al. (2000b) identified resistance against a SSI in the *B. napus* dwarf line Darmor *bzh*. They mapped one major gene and two QTLs in a DH population. As the linkage groups of their map had not been allocated to either the A or the C genome, it cannot be concluded whether CR from Darmor-*bzh* originated from *B. rapa* or *B. oleracea*. As CR from Darmor-*bzh* has not been observed in the field so far, it seems likely that the major gene has a race-specific effect that has already been overcome in the field.

4.13.4 Radish (*Raphanus sativus* L.)

Radish, a cruciferous crop closely related to *Brassica*, originated in the Mediterranean region (Kitamura 1958; Tsunoda 1980), but today is more important in East Asia where a wide variety of genotypes are found. Yoshikawa (1993) surveyed CR in radishes and found that most Japanese cultivars are highly resistant, except for some cultivars with bolting resistance, such as cvs. Tokinashi or Ninengo. Some northern Chinese cultivars were highly susceptible to this disease. European radish cultivars such as Comet and White Icicle showed mild symptoms or were resistant (Yoshikawa 1993). Rat-tail radish, which is grown in tropical Asia for the consumption of its pods, is resistant to the pathogen. Resistance to clubroot seems to be widespread in this species, but there are reports of clubroot outbreaks in radish in the USA (Thuma et al. 1983) and more recently in Japan (Horikoshi and Tairako 2002). In Japan, radish was believed for a long time to be totally immune to clubroot. Therefore, radish is often used as a cleaning crop for fields contaminated with *P. brassicae* (Murakami et al. 2000). The susceptible cultivars were newly bred, so it is possible that some CR genes were lost during breeding. Kamei et al. (2007) reported on a QTL analysis of radish CR using a Japanese radish as the resistant parent and a Chinese cultivar as the susceptible parent. They found that the trait is controlled by one locus or closely linked loci in a single linkage group. Using nine types of monosomic addition lines of radish, Akaba et al. (2008) showed that only one addition type had clear resistance, suggesting that CR is controlled predominantly by a single radish chromosome. Therefore, it seems likely that only a single locus controls a large part of radish CR. Considering the broadness of CR from radish, simple inheritance was unexpected. Resistant radishes can be bred by a simple backcross method starting with a Japanese radish as a donor plant. Mass selection is risky, however, when the trait is dominant. If CR from radish could be transferred into *Brassica* species, it might be valuable for breeding highly resistant *Brassica* cultivars. Some efforts to transfer CR to *Brassica* have been reported using sexual or somatic hybrids between radish and *Brassica* species (Hagimori et al. 1992; Hagimori 1995; Xing et al. 1989). Recombination between *Brassica* and *Raphanus* chromosomes would be a precondition for a successful introgression. Attempts to introgression other traits such as restorer genes for cytoplasmic male sterility into *Brassica* did show, however, this is a rare event (Sakai et al. 1996; Delourme et al. 1998).

4.14 Development of Clubroot-Resistant Cultivars

Resistant cultivars are the only solution for protection against *P. brassicae* in combination with suitable crop rotation and management strategies (Peng et al. 2014).

4.14.1 Genetics and Breeding

The clubroot host-plant resistance can be found in main cruciferous crops (except *Brassica juncea* and *B. carinata*) (Diederichsen et al. 2009). The resistance against *P. brassicae* is polygenic (quantitative) and also monogenic (qualitative) (Crute et al. 1980; Voorrips 1996). Many of the resistance types are pathotype specific and do not cover all pathotypes of *P. brassicae* (Diederichsen et al. 2009). The studies proved the existence of 8 resistance loci, some of them are pathotype specific major genes, and some are quantitative loci (QTL) with specific or nonspecific resistance responses (Voorrips 1996; Hwang et al. 2012). Some studies also proved that resistance is qualitative and is controlled by dominant and recessive genes (Chiang and Crete 1983; Yoshikawa 1993). The oilseed rape resistant cultivar Mendel was bred from ECD hosts-resistant turnip (*B. rapa,* ECD-04) and kale (*B. oleracea* ECD-15) through backcross with oilseed rape (Diederichsen et al. 2009). The F_1 backcrosses were inoculated with *P. brassicae* and the most resistant ones were selected for further breeding and genetic analysis (Diederichsen et al. 2009). Later the presence of resistant genes was confirmed. There is one major locus and two recessive genes in the cultivar Mendel whereas other genes were lost in the breeding process and negative selection (Diederichsen et al. 2009). This is why cv. Mendel is recommended to use wisely together with other management strategies (Diederichsen et al. 2009; Dixon 2009). The same situation exists in other resistant cultivars.

4.14.2 Mechanism of Resistance

The resistance mechanisms are based on suppressing the pathogen development in a host rather than on infection avoidance. The term resistance is used because the infection inhibition is based on resistance genes, not on tolerance genes. The zoospore penetration occurs in the root hairs of a resistant host (the primary phase of the pathogen life) (Hwang et al. 2012). The secondary phase of infection in the root cortex also follows, but formed plasmodia do not mature and resting spores are not created (Hwang et al. 2011; Deora et al. 2013). Slight suppression of primary infection and significant suppression of secondary infection occur during the resistance response, the disruption of the cell walls is lesser and the secondary thickening of the stem is not stopped (Donald et al. 2008). The immune response includes the activation of metabolic pathways for synthesis of phytoalexins, phenolic compounds, lignin, and other substances that block the pathogen invasion (Siemens et al. 2006).

4.14.3 R-Genes Cultivars in the Field

The first resistant cultivars were Mendel and Tosca, marketed at the beginning of the new millennium. These cultivars are heterozygous hybrids with one gene of pathotype specific resistance. Unfortunately, the Mendel host plant resistance was broken in some parts of Germany and France, where it was grown since 2000 (Diederichsen et al. 2014). In Great Britain, resistance was broken quite rapidly in the field, where cultivars were grown under strong selection pressure (Diederichsen et al. 2003). These cultivars were also tested on local pathotypes in Canada. Cultivar Mendel showed resistance against pathotypes 3, 5, 6, and 8 (Williams set) (Rahman et al. 2014). Cultivar Tosca was susceptible to pathotype 3 and resistant to pathotype 2 (Williams set) (Peng et al. 2013). These results showed that the host-plant resistance is quite variable and some genes may be ineffective in some regions with specific pathotypes (Peng et al. 2014). New resistant cultivars of spring rape with genes from European cultivars were introduced into Canada in 2009 (Peng et al. 2014). In the European market, the available resistant cultivars which are derived from the first resistant cultivar Mendel (Diederichsen et al. 2014). It is expected that a wider range of resistant varieties will come to the market in the next years (Diederichsen et al. 2014). However, these varieties cannot be grown without suitable agronomic measures and pathogen-resistance management strategies (Diederichsen et al. 2014). Field experiments with resistant cultivars in infested fields were also conducted in the Czech Republic. Eight resistant cultivars were tested, some of them are normally available on the market and some were test cultivars in the breeding process (new breeds). Normally traded cultivars had stable performance in resistance results and good yield. Cultivars from new breeding showed some variations in resistance performance and yield (Ricarova et al. 2016).

4.14.4 Deployment of R-Genes Cultivars

Clubroot-resistant canola cultivars have been the dominant component of the club-root management strategy in western Canada in recent years. The first resistant cultivar for western Canada, Pioneer "45H29," was registered in 2009 and its release was quickly followed by several other resistant cultivars (Strelkov et al. 2011; Gossen et al. 2013). However, the durability of this resistance is not known. The genetic basis for the resistance in these cultivars also is proprietary knowledge that is not available to the public. This has complicated efforts aimed at resistance steward-ship, since it is not possible to develop rational strategies for the rotation of resistance sources in infested fields. The number of resistance genes also that are currently available to breeders is limited (Hirai 2006). Recent studies indicated that there were no detectable differences in pathotype reactions among clubroot-resistant canola cultivars available in western Canada (Deora et al. 2012, 2013). This represents one line of evidence that resistance genes in all these lines originated from a similar source. In contrast, another study by LeBoldus et al. (2012) revealed that pathogen populations that had been repeatedly cycled on and become adapted to one *Brassica* host did not exhibit a similar adaptation to other hosts, when the cycled populations were inoculated on the latter, indicating that distinct sources of resistance may occur in at least some genotypes. Regardless of the relationships between resistance sources, there is substantial genetic and pathotype variation present in the *P. brassicae* populations in western Canada (Strelkov et al. 2006; Xue et al. 2008; Cao et al. 2009), and the deployment of a cultivar with single-gene resistance against a genetically diverse pathogen on a large scale imposes a strong selection pressure for pathogen genotypes that are able to overcome this resistance (Hwang et al. 2012). Similar mechanisms may operate in other countries also.

4.15 Factors Affecting Transfer of Plant Disease Resistance

The different cultivars in biology of the several types of plant pathogens (fungi, bacteria, viruses, and nematodes) infecting crucifers presents a substantial problems in transfer of R-genes and developing GM-resistant plants. Firstly, the kind of organisms causing disease is taxonomically highly diverse; the major groups include cellular pathogens (bacteria, fungi, and the algal Oomycetes) and molecular pathogens (viruses). These are physiologically very different from each other, and therefore no single gene product can be expected to have a direct toxic effect on all types of pathogens. Secondly, pathogens use two major life strategies, namely biotrophic and necrotrophic. Biotrophic pathogens essentially act as a sink for the host's anabolic assimilates, and therefore keep it alive. Mean while, necrotrophic pathogens consume the host tissues on invasion. Hemi-biotrophs combine both strategies in their life cycle. Consequently, plants have developed quite different ways for dealing with these two strategies.

4.16 Accessing and Exploiting Genetic Diversity

The advent of molecular genetics provided new opportunities for mapping and tracking genes of agronomic interest, leading to more efficient marker-assisted selection. Whole-genome sequencing, starting with *Arabidopsis* and rice as models for dicotyledons and monocotyledons, respectively, and followed by a rapidly increasing number of crop plant genomes, have led to a quantum leap not only in understanding plant genetic diversity, but also for accessing this enormous resource. For many crop plant species, for example, tomato, barley, maize, wheat, and various *Brassica* species, either the entire genome or the generic parts of the genome are now emerging. As bioinformatic tools for analyzing the exponential increase in genome data improve, the practical utility of such data will also be enhanced. This will extend the options for breeding disease resistance.

The presence of conserved motifs in plant resistance (R genes), such as the nucleotide-binding site leucine-rich repeat (NBS-LRR) domains, has facilitated the identification of gene families, and resistance gene analogs in other plants. The *Arabidopsis* genome has around 150 NBS-LRR encoding genes and rice c. 400 (McHale *et al.* 2006). Studies of genome structure have shown that many putative R genes are clustered and have undergone duplication and evolution due to diversifying selection. Functional analysis of these entire candidate genes is a demanding task, but improvement in plant transformation protocols, and high-throughput gene attenuation methods, such as RNA interference (RNAi) and virus induced gene silencing (VIGS) should accelerate the identification of novel genes of practical utility.

4.17 Introgression and Pyramiding of Major R-QTL's into *Brassica napus* Against *Sclerotinia*

To maintain stable production in the face of damage from diseases, the development of resistant lines or cultivars is a cost-effective strategy. However, due to the lack of highly resistant resource in *B. napus* gene pool, the breeding of *Sclerotinia*-resistant rapeseed has been difficult and rather unsuccessful. Alternatively, diploid progenitors which always provide precious gene resources could be considered in rapeseed improvement. *Sclerotinia*-resistant sources were reported in *B. oleracea* germplasm (Mei et al. 2011). In particular, a wild *B. oleracea* genotype (*B. incana*) identified as strong resistance to *S. sclerotiorum* (Mei et al. 2011), provided a promising basis to improve *Sclerotinia* resistance in present rapeseed cultivars. However, this raised problems in utilization of the *B. oleracea* resource by re-synthesizing rapeseed technique, i.e., resynthesized allotetraploid lines are genetically unstable (Hurgobin et al. 2018; Wu et al. 2018; Zou et al. 2018), and the synthetic rapeseed lines contain A genome from *B. rapa* which is highly susceptible to *S. sclerotiorum* (Ding et al. 2013; Mei et al. 2011). Therefore, novel breeding strategies are needed to avoid these problems (Mei et al. 2020). There is another bottleneck in *Sclerotinia*-resistant breeding of rapeseed, as it is a typical quantitative trait controlled by multiple genetic loci (Mei et al. 2013; Wei et al. 2016; Wu et al. 2013, 2016). Till date, no resistant gene has been cloned from any of those reported QTLs, and most importantly a single locus always provides very limited contribution to resistance. These factors have largely limited the molecular marker-assisted breeding for *Sclerotinia*-resistant rapeseed.

Resistance QTLs were identified from *Sclerotinia*-resistant wild *B. oleracea* (Mei et al. 2013), and several molecular markers were proved effective in selecting resistant individuals (Ding et al. 2015, 2019; Mei et al. 2015). A hexaploidy strategy was preliminarily proposed to introduce *Sclerotinia* resistance from the wild *B. oleracea* into rapeseed (Mei et al. 2015). During the genotypic (marker-assisted selection) and phenotypic selection in high generations generated from the hexaploidy strategy, successfully yielded resistant rapeseed lines with multiple resistant QTLs, improved *Sclerotinia* resistance, high seed yield and quality. Resistant rapeseed lines pyramided with multiple resistant QTLs derived from *Brassica oleracea* were developed via a hexaploidy strategy. Interspecific hexaploids were generated between rapeseed variety "Zhongshuang 9" and a wild *B. oleracea* which was highly resistant to *S. sclerotiorum*, followed by backcrossing with Zhongshuang 9 and successive selfing. By molecular marker-assisted selection, three major resistant QTLs were transferred and pyramided from *B. oleracea* into two BC1F8 lines which exhibited ~35% higher resistance level than Zhongshuang 9 and produced good seed yield and seed quality. The successful development of *Sclerotinia*-resistant rapeseed lines was obtained by introducing multiple resistant loci from wild *B. oleracea*. This revealed the effectiveness of pyramiding multiple QTLs in improving *Sclerotinia* resistance in rapeseed and provided a novel breeding strategy on utilization of *B. oleracea* in rapeseed improvement (Fig. 4.7; Mei et al. 2020).

4.18 Introgression of Resistance from Wild *Brassica* Species into *Brassica juncea* to *Sclerotinia*

A wild *Brassicaceae* species with excellent resistance against stem rot to develop a set of *B. juncea-B. fruticulosa* introgression lines (ILs) was used by Atri et al. (2019). These were assessed for resistance using a highly reproducible stem inoculation technique against a virulent pathogen isolate. Over 40% of ILs showed higher levels of resistance. IL-43, IL-175, IL-215, IL-223, and IL-277 were most resistant ILs over three crop seasons. Sequence reads (21x) from the three most diverse ILs were then used to create *B. juncea* pseudo molecules, by replacing SNPs of reference *B. juncea* with those of re-sequenced ILs. Genotyping by sequencing (GBS) was also carried out for 88 ILs. Resultant sequence tags were then mapped on to the *B. juncea* pseudo molecules, and SNP genotypes prepared for each IL. Genome-wide association studies helped to map resistance responses to stem rot. A total of 13 significant loci were identified on seven *B. juncea* chromosomes (A01, A03, A04, A05, A08, A09, and B05). Annotation of the genomic region around identified SNPs allowed identification of 20 candidate genes belonging to major disease resistance protein families, including *TIR-NBS-LRR* class, Chitinase, Malectin/receptor-like protein kinase, defensin-like (*DEFL*), desulfoglucosinolate sulfotransferase protein, and lipoxygenase. A majority of the significant SNPs could be validated using whole-genome sequences (21x) from five advanced generation lines being bred for *Sclerotinia* resistance as compared to three susceptible *B. juncea* germplasm lines. The understanding of the defensive pathway of *B. fruticulosa* resistance will also enable development of marker candidates for assisted transfer of introgressed resistant loci into agronomically superior cultivars of crop *Brassica* (Atri et al. 2019).

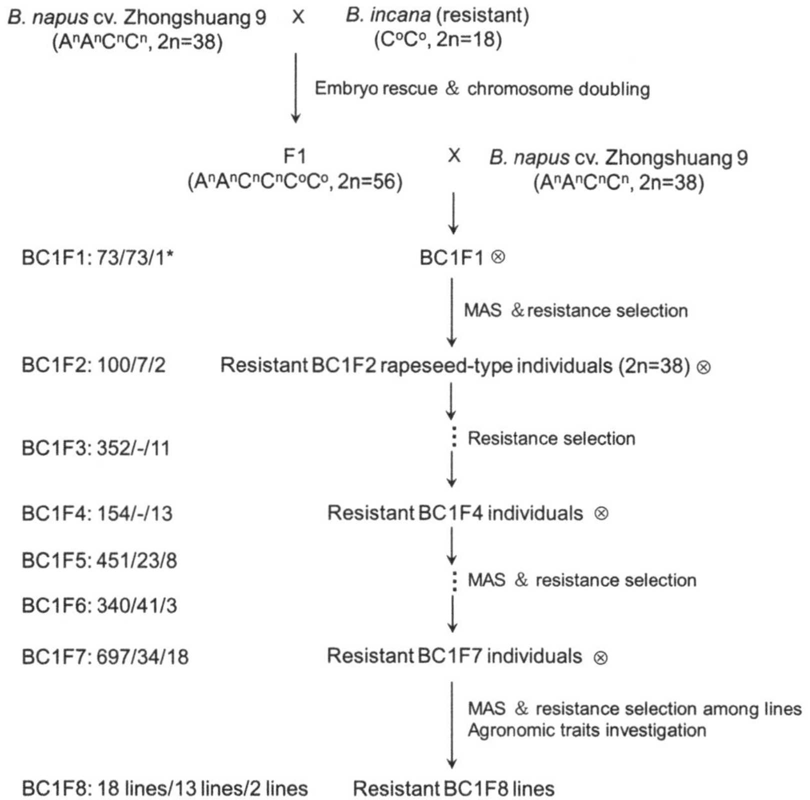

Fig. 4.7 A diagram of breeding process of *Sclerotinia*-resistant *Brassica napus* lines by using *Brassica incana* as the donor of resistance. Asterisk represents the number of plants for selection/the number of plants elected by MAS/the number of plants planted for the next generation (Mei et al. 2020)

4.18.1 Marker-Assisted Introgression of *Sclerotinia* Resistance into *Brassica juncea*

A set of 96 *Brassica juncea–Erucastrum cardaminoides* introgression lines (ILs) were developed with genomic regions associated with Sclerotinia stem rot (*Sclerotinia sclerotiorum*) resistance from a wild Brassicaceous species *E. cardaminoides*. ILs were assessed for their resistance responses to stem inoculation with *S. sclerotiorum*, over three crop seasons. Initially, ILs was genotyped with transferable SSR markers and subsequently through genotyping by sequencing. SSR-based association mapping identified six marker loci associated to resistance in both A and B genomes. Subsequent genome-wide association analysis (GWAS) of 84 ILs recognized a large number of SNPs associated to resistance, in chromosomes A03, A06, and B03. Chromosomes A03 and A06 harbored the maximum number of resistance related SNPs. Annotation of linked genomic regions highlighted an array of resistance mechanisms in terms of signal transduction pathways, hypersensitive responses, and production of antifungal proteins and metabolites. Of major importance was the clustering of SNPs, encoding multiple resistance genes on small regions spanning approximately 885 kb region on chromosome A03 and 74 kb on B03. Five SNPs on chromosome A03 (6,390,210-381) were associated with LRR-RLK (receptor like kinases) genes that encode LRR-protein kinase family proteins. Genetic factors associated with pathogen-associated molecular patterns (PAMPs) and effector triggered immunity (ETI) were predicted on chromosome A03, exhibiting 11 SNPs (6,274,763-994). These belonged to three R-Genes encoding TIR-NBS-LRR proteins. Marker trait associations (MTAs) identified will facilitate marker-assisted introgression of these critical resistances, into new cultivars of B. juncea initially and, subsequently, into other crop *Brassica* species (Rana et al. 2019). Association mapping allowed detection of ten markers trait associations (MTAs) after Benferroni correction. These were CNU-M-157-2, RA2-G05, CNU-M-353-3, CNU-M-442-5, ACMP00454-2, ACMP00454-3, EIN2-3-1, M 641-1, Na10D09-1, and Na 10D11-1 (Rana et al. 2017)

4.19 Introgression of Black Rot Resistance from *Brassica carinata* to *Brassica oleracea* Through Embryo Rescue

The black rot resistance sources have been reported in A and B genomes of *Brassica* species (Westman et al. 1999; Taylor et al. 2002) and genes present in these genomes can be pyramided in Cole crops to provide a durable solution to this malady. The monogenomic and digenomic species in Uthant triangle carry resistance gene(s) for Xcc, which can be utilized for guarding *B. oleracea* genotypes against this disease using embryo rescue and somatic hybridization techniques. Genetic transformation option can be explored when resistance is not available within and related gene pools. Besides, a number of disease resistance genes in *Brassicas* and *A. thaliana* have been postulated and mapped, but none has been cloned as yet (Vicente and Holub 2013). Therefore, there is no information on genetic transformation for resistance to black rot disease in *B. oleracea* group.

Brassica carinata (Ethiopian mustard, 2n = 4x = 34, BBCC genome) is an important oilseed crop thought to have originated in the Ethiopian plateau (Warwick 2011). Its leaves are also consumed as vegetable in North East Africa. The *Brassica* B genome is known to carry several important traits, yet there has been limited analyses of its underlying genome structure (Navabi et al. 2013). *B. carinata* harbors genes controlling valuable resistance/tolerance traits for biotic and abiotic stresses, viz. resistance to black leg, black rot, and tolerance to aluminum, salinity, heat, and drought (Pan et al. 2012; Enjalbert et al. 2013). Moreover, strong resistance to Xcc race 1 and 4 has been reported in high proportion of *B. nigra* (B genome) and *B. carinata* (BC genome) accessions (Taylor et al. 2002; Vicente et al. 2002). A single dominant gene conferring high resistance to *Xcc* race 1 was reported by Tonguc et al. (2003). A single dominant gene (Xca1bc) controlling black rot resistance was mapped with intron length polymorphic markers at B-7 genome of *B. carinata* (Sharma et al. 2016). Therefore, it is important and feasible to transfer this black rot resistance gene into cauliflower cultivars. Alien gene transfer in crop plants has emerged as a potent approach to broaden gene pools and increase genetic variability in existing germplasm. Hybridization between a diploid and a tetraploid species is difficult and failure occurs at many stages starting from pollination incompatibility to pre/post-fertilization barriers.

Search for new gene(s) for disease resistance in closer genome related species, viz. *B. napus* (AACC) and *B. carinata* (BBCC), for incorporation into *B. oleracea* group via interspecific (Tonguc et al. 2003) or somatic hybridization using embryo rescue (Hansen and Earle 1995) has to be a continuous process. Of the three *Brassica* genomes, A and C are very closely related while the B-genome is phylogenetically distant. Zhou et al. (1997) transferred resistance to Xcc from *B. napus* to broccoli (*B. oleracea*) through protoplast fusion. Tonguc and Griffiths (2004b), however, explored the half B genome *B. juncea* (AABB) for the introgression of black rot resistance into *B. oleracea* using *in vitro* embryo rescue. Somatic hybridization was attempted to transfer black rot resistance from B. nigra into *B. oleracea* (Wang et al. 2011, 2016).

Utilization of embryo rescue or somatic hybridization techniques, or combination of both, could help to overcome natural reproductive barriers in the development of interspecific hybrids in *Brassica* (Ayotte et al. 1987; Hansen and Earle 1995; Momotaz et al. 1998; Weerakoon et al. 2009; Niemann et al. 2013). Wide crosses between crop plants and their wild relatives have now become routinely possible through the use of the embryo rescue technique. Different techniques of plant cell and tissue culture, such as ovary, ovule, and embryo culture as well as protoplast fusion, have proved useful for production of interspecific hybrids. Rescue of hybrid embryos and their culture in vitro helps to overcome post-fertilization barriers in interspecific crosses. Since its first use in *Brassica* by Nishi et al. (1959), extensive investigations have been carried out to improve the techniques for obtaining higher seed set (Inomata 1993, 2002; Zhang et al. 2003, 2004). The successful application of this technique depends on the stage of the embryo being rescued and cultured in vitro. Several attempts were made to transfer desirable gene(s) from alien *Brassica* spp. to *B. oleracea* such as powdery mildew resistance (Tonguc and Griffiths 2004a), downy mildew (Chiang et al. 1977b), male sterility (Chiang and Crete 1987), and atrazine resistance (Jourdan et al. 1989). The progress for marker-assisted Xcc resistance gene transfer from *B. carinata* to cauliflower has been very slow (Tonguc et al. 2003; Tonguc and Griffiths 2004b). The transfer of monogenic resistance to Xcc race 1 from an accession "NPC-9" of *B. carinata* into a commercial susceptible cauliflower variety "Pusa Sharad" using embryo rescue technique has been attempted by Sharma et al. (2017).

Black rot caused by *X. campestris* pv. *campestris* (*Xcc*) is a very important disease of cauliflower (*Brassica oleracea botrytis* group) resulting into 10–50% yield losses every year. Since there is a dearth of availability of resistance to black rot disease in *B. oleracea* (C genome), therefore exploration of A and B genomes was inevitable as they have been reported to be potential reservoirs of gene(s) for

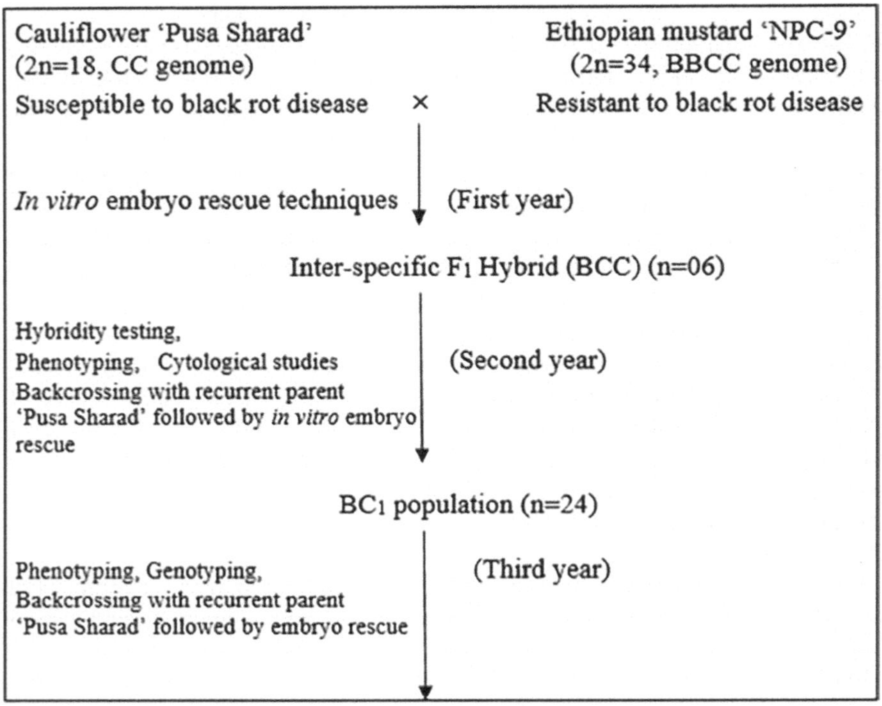

Fig. 4.8 Flow chart of development of plant materials using *in vitro* embryo rescue technique (Sharma et al. 2017)

resistance to black rot. To utilize these sources, interspecific hybrid and backcross progeny (B1) were generated between cauliflower "Pusa Sharad" and Ethiopian mustard "NPC-9" employing *in vitro* embryo rescue technique. Direct ovule culture method was better than siliqua culture under different temperature regime periods. Hybridity testing of F1 interspecific plants was carried out using codominant SSR marker and *Brassica* B and C genome-specific (DB and DC) primers. Meiosis in the di-genomic (BCC) interspecific hybrid of *B. oleracea botrytis* group ($2n = 18$, CC) × *B. carinata* ($2n = 4x = 34$, BBCC) was highly disorganized and cytological analysis of pollen mother cells revealed chromosomes $2n = 26$ at metaphase-I. Fertile giant pollen grain formation was observed frequently in interspecific F1 hybrid and BC1 plants. The F1 interspecific plants were found to be resistant to *Xcc* race 1. Segregation distortion was observed in BC1 generation for black rot resistance and for different morphological traits. The At1g70610 marker analysis confirmed successful introgression of black rot resistance in interspecific BC1 population (Fig. 4.8). This effort will go a long way in pyramiding gene(s) for resistance against black rot in Cole crops, especially cauliflower and cabbage for developing durable resistance, thus minimize dependency on bactericides (Sharma et al. 2017).

4.20 Distant Hybridization for *Xcc* Resistance Through Somatic Hybridization and Embryo Rescue

Wide hybridization or distant hybridization is a powerful approach to combine the diverged genomes into a single nucleus. Wide hybridization is an important driving force in broadening the gene pools, the evolution of novel plant forms and speciation, transferring the genomes of one species to another, enhancing the genetic diversity and improving the quality of crops (Soltis and Soltis 2000; Barton 2001; Jiao et al. 2011). The *Brassica* vegetables and oilseed crops in the cruciferae family have undergone multiple genomic duplications (Schranz et al. 2006). The classic genome polyploidization event is the interspecific hybridization of *B. rapa* (A genome), *B. nigra* (B genome), and *B. oleracea* (C genome) to form the allotetraploids *B. juncea* (AB genome), *B. napus* (AC genome), and *B. carinata* (BC genome) (Nagaharu 1935). Since the first intergeneric hybrid between *R. sativus* and *B. oleracea*, "*Raphanobrassica*," reported by Karpechenko (1928), numerous intergeneric and interspecific hybrids have been produced between *Brassica* species of the U triangle (Nagaharu 1935) and *R. sativus*. The related or wild species are a rich reservoir of several useful genes having resistance to various biotic or abiotic stresses. In the family Brassicaceae, interspecific hybridization is a potential approach for transfer of desirable genes from one species to another, enhancing genetic diversity and creation of hexaploids for the improvement of *Brassica* crops (Prakash 1973; Prakash and Chopra 1988; Choudhary et al. 1999; Siemens et al. 2002; Niemann et al. 2012; Kaneko and Bang 2014).

The sexual incompatibility or reproductive barriers can be overcome by utilization of embryo rescue or somatic hybridization techniques, or a combination of both, in the development of interspecific hybrids in crucifer crops (Hansen and Earle 1995; Weerakoon et al. 2009; Niemann et al. 2012; Dey et al. 2015). The development of black rot resistant hybrids is one of the major objectives of *Brassica* breeders. Since the resistance sources in *B. oleracea* are insufficient, interspecific hybridization with related or distant species followed by embryo rescue or somatic hybridization for the development of black rot resistant hybrids has been practiced and accomplished by many workers (Table 4.11).

Some lines of *B. napus* have very high resistance to black rot (*X. campestris* pv. *campestris*) and resistance is governed by a dominant gene, but the successful crossing between two species (*B. oleracea* and *B. napus*) is very difficult (Hansen and Earle 1995). Therefore, somatic hybrids by protoplast fusion between rapid cycling *B. oleracea* and *X. campestris* pv. *campestris* resistant line of *B. napus* was developed. Hybrid identity was confirmed by morphological studies, flow cytometric estimation of nuclear DNA content, and analysis of RAPDs. In another study, Wang et al. (2011) generated new alien addition lines having high resistance to black rot (*X. campestris* pv. *campestris*) by employing a somatic hybridization technique between cauliflower and black mustard. It has been evident that success of interspecific hybridization is also influenced by the genotype of the parents (Meng and Lu 1993; Tonguc and Griffiths 2004a; Mason et al. 2011). Tonguc and Griffiths (2004b) utilized two *B. juncea* accessions (A 19182 and A 19183) having resistance

Table 4.11 Wide hybridization for introgression of black rot resistance in different *Brassica* crops (Singh et al. 2018)

Type of hybridization	Cross combination	Technique	Reference(s)
Interspecific hybridization	*Brassica oleracea x Brassica napus*	Protoplast fusion	Hansen and Earle (1995)
Interspecific hybridization	*Brassica oleracea* var. *botrytis x Brassica carinata*	Embryo rescue	Sharma et al. (2017)
Interspecific hybridization	*Brassica oleracea x Brassica juncea* (A1982)	Embryo rescue	Tonguc and Griffiths (2004a)
Interspecific hybridization	*Brassica oleracea* var. *botrytis x Brassica nigra*	Somatic hybridization	Wang et al. (2011)
Interspecific hybridization	*Brassica oleracea x Brassica carinata*	Protoplast fusion	Tonguc et al. (2003)
Interspecific hybridization	*Brassica oleracea* var. *botrytis x Brassica carinata*	Embryo rescue	Dey et al. (2015)

to Xcc race 1 and 4, as maternal and paternal parents to develop interspecific hybrids with *B. oleracea* cultivars. They employed embryo rescue to recover interspecific hybrids and further confirmed by inheritance of paternal molecular markers. They obtained twenty-six interspecific hybrid plants between A 19182 and *B. oleracea* cultivars, but no interspecific hybrids were obtained using A 19183, suggesting influence of genotype in the success of interspecific hybridization. The interspecific hybrid plants were male-sterile and they were used successfully as maternal parents to develop backcross plants employing embryo rescue. All hybrid and BC1 plants were found resistant to races 1 and 4 of Xcc. Dey et al. (2015) reported introgression of black rot resistance genes into Indian cauliflower (*B. oleracea* var. *botrytis* L.) through interspecific hybridization (*B. carinata* 9 *B. oleracea*) followed by embryo rescue. They utilized two *B. carinata* genotypes, NPC-9, and HCA-6, having very high resistance to Xcc race 1 and 4, for interspecific hybridization to introgress resistance genes from *B. carinata* into Indian cauliflower employing three different embryo rescue approaches. They reported that all the F1 hybrids and backcross were resistant to *Xcc* races 1 and 4 (Singh et al. 2018).

4.21 Novel Sources and Transfer of R-Genes

Novel sources of R genes for Blackleg can be sourced from other A- and C-genome *Brassica* species (Chen et al. 2010) or from the B genomes in *B. nigra* (Chevre et al. 1996), *B. juncea* (Christianson et al. 2006), and *B. carinata* (Plieske et al. 1998). Wild Brassicaceae species, such as *Sinapis arvensis*, also serve as a good source of R genes for Blackleg (Snowdon et al. 2000). However R genes from other wild species remain to be tapped. Besides identifying candidate R genes through comparative genomic analysis, more screening work on wild and land races, including species from the Brassicaceae family, should be performed against the main pathogens to identify resistant plants. This would broaden the gene pool for introgression of novel R genes or alleles into *Brassica* cultivars. Combining this information with genome sequencing will enable characterization of these novel resistance genes. Although effector genes are usually unique and do not often share structural similarity but encode very diverse proteins, it is still possible to identify an R gene that could recognize common or near identical effector gene(s) in different pathogen pathotypes/ races/ isolates, which already exists naturally in some plants/ crops (Dangl et al. 1992). This type of R gene is useful and can be transferred between hosts that are closely related or within the same genus or family. Some R genes can also be transferred from host to non-host using GM technologies, providing broad-spectrum immunity against one or more pathogens. One such successful example is the pepper *Bs2* gene against bacterial spot disease transferred from its host pepper into non-host tomato (both members of Solanaceae family; Tai et al. 1999). The Bs2 gene encodes the R gene NBS-LRR class (Tai et al. 1999) while the gene avrBs2 in *Xanthomonas campestris* pv. *vesicatoria*, causal pathogen of bacterial spot disease, is essential for virulence activity and is highly conserved between and within the pathovars of *X. campestris* (Kearney and Staskawicz 1990). This has opened up new possibilities to protect economically valuable crops, which lack R gene resources, from disastrous pathogens (Neik et al. 2017).

4.22 Breeding Resistant Cultivars

Continuous efforts are ongoing to breed resistant *B. napus* cultivars through introgression. *B. rapa* remains a good source of race-specific R genes that often result in high resistance against Clubroot (Strelkov et al. 2006) and Blackleg (Yu et al. 2008). For Clubroot, a single dominant gene from resistant *B. rapa* turnip morphotypes was successfully introduced into Chinese cabbage (Yoshikawa 1980). Transfer of a Clubroot resistance gene was also performed from *B. napus* into *B. oleracea*, with the resistance gene believed to also be carried in the A genome of *B. rapa* (Chiang et al. 1980). In Canada, efforts to transfer a Clubroot resistance gene from the *B. napus* hybrid "Mendel" into a local susceptible spring *B. napus* cultivar was promising (Rahman et al. 2011). Transfer of a Clubroot R gene was also made from *B. rapa* Chinese cabbage "Qulihuang" into *B. napus* "Topas" using marker-assisted selection (Hirani et al. 2016). For *L. maculans*, R genes have been genetically mapped and identified in the A genome of *B. napus* (Alamery et al. 2017). For *S. sclerotiorum*, introgression lines of *B. juncea* carrying genomic segments from the wild Brassicaceae *B. fruticulosa* demonstrated marker trait association for Sclerotinia Stem Rot resistance (Rana et al. 2017). Within breeding programs, it is important to make informed decisions about the source of novel R genes and which parental genomes to cross, because the R gene from the parents may be masked in the progenies of the resynthesized genomes due to one or more epistatic interactions, such as was found for a Clubroot R gene (Werner et al. 2007). These challenges can be overcome by the latest genome editing techniques, such as CRISPR-Cas9

(Scheben et al. 2017), as was recently reported in maize with precise mutation on the targeted allele (Svitashev et al. 2016; Liang et al. 2017).

The B genome offers good sources of resistance against two other pathogens besides *P. brassicae* and *S. sclerotiorum*; White Rust and White Leaf Spot (*A. candida* and *P. capsellae*, respectively). For White Rust, potential sources of resistance genes come from *B. nigra* and *B. juncea* (Westman et al. 1999), while resistance against White Leaf Spot potentially comes from *B. carinata* (Gunasinghe et al. 2016). For White Leaf Spot, in contrast with Clubroot and Blackleg, *B. rapa* is more susceptible than *B. napus* (Gunasinghe et al. 2014), but as with Downy Mildew, there is good resistance to white leaf spot in *B. oleracea*. In fact, complete resistance was recorded in *B. oleracea* var. *capitata* (Gunasinghe et al. 2016). However, *B. juncea* originating in India or Australia is highly susceptible to White Leaf Spot (Gunasinghe et al. 2014).

To breed resistant cultivars, it is important to consider the fitness cost to pathogen isolates when they do overcome host resistances. The pathogen avirulence/virulence genes in the field that are adapted to the host cultivar and determine the genetic changes in the pathotype that cause heavy fitness consequences (Vera Cruz et al. 2000; Delmas et al. 2016). It is also important to consider the host R gene as selection pressure on the pathogen can affect the stability of the R gene. In terms of the linked genes in *L. maculans*, *AvrLm1*, and *AvrLm6*, high selection pressure on one *Avr* gene can cause a change in allele frequency of the other (Brun et al. 2010; Van De Wouw et al. 2010). Understanding the evolutionary potential between the pathogen and the host is essential in developing a durable resistant *Brassica* crop (McDonald and Linde 2002; Neik et al. 2017).

4.23 Application of Molecular Markers in Breeding for Disease Resistance

Genetic markers can be used to study patterns of heredity, genomic variation, evolutionary and selection phenomena, allele–allele linkages, and allele–phenotype associations. The application to which a genetic marker is best suited depends on its physical properties and genomic location, the cost involved, ease of use, and degree of throughput required. Molecular markers have been successfully applied in plant science towards the genetic and physical mapping of genomes, the identification of genes controlling various processes and phenotypes (trait association), genetic diversity and evolutionary analyses, and in marker-assisted breeding for crop improvement. With advances in genome sequencing technologies, genetic markers with a known genomic location and environment are becoming more popular and applicable to an increasingly diverse and high-throughput range of objectives. DNA sequence-based markers are inherently able to capture vast amounts of variation at single-base resolution, making them particularly useful for the detection of perfect markers (DNA polymorphisms causally linked to traits of interest) and discovery and analysis of the alleles involved.

4.23.1 Marker-Assisted Selection

Markers provide the potential to fine map important genetic loci with high resolution through the use of mapping populations. Where these populations are phenotyped for traits, such as disease resistance, the inheritance of particular marker loci or haplotypes in the population can be linked to such phenotypes. Genotyping of markers tightly linked to traits can then rapidly predict the phenotypes of a large selection of segregating individuals at an early stage of development, often well before phenotypic screening would be possible, and at reduced cost. The application of single marker–trait associations to crop breeding is known as marker-assisted selection (MAS). MAS enables efficient selection of breeding lines for the introgression of desirable traits into commercial crop accessions as well as the high-throughput screening of the resulting progeny.

An effective marker for MAS must generally be located within 1 cM of a desired trait and able to be genotyped at high throughput and reproducibility. Low polymorphism, poor genomic distribution, and/or poor reproducibility of marker types including RFLPs (Restriction Fragment Length Polymorphisms) and RAPDs (Randomly Amplified Polymorphic DNA) limit their application to MAS. Microsatellites (also known as Simple Sequence Repeats, SSRs) are highly polymorphic, reproducible alternatives, but are often poorly linked to genes. Nonetheless, SSRs and RAPDs have been applied in *B. napus* (canola) MAS programs for selection for major gene disease resistance, yellow seed coat color, male fertility restorer lines, and improvement of oil quality. SNPs are currently the best markers for MAS due to their high prevalence and polymorphism in the genome and their potential for strong, or even perfect, linkage to traits of interest. Perfect linkage is possible where the polymorphism is directly responsible for variation in the desired trait. The development of high-throughput sequencing technologies has greatly assisted association studies that utilize SNP markers (Snowdon and Friedt 2004; Varshney et al. 2009; Syvanen 2005; Hayward et al. 2015).

4.23.2 Genome-Wide Marker-Assisted Selection

The availability of phenotypic data along with genotypic data permits the association of loci, or haplotypes, at a genome-wide scale, which may be used to mine an entire genome for genotype–phenotype correlations. When there are enough markers, spanning the entire genome in a dense manner, it is expected that the gene, or genes, of interest will be in linkage disequilibrium with at least one or some of the markers, leading to marker-assisted selection on a genomic scale. Genome-wide marker-assisted selection studies will be an important way of safeguarding global food supplies into the future. Genome-wide association studies (GWAS) based on the markers identified was then carried out to identify novel loci underlying variations in agroclimatic traits. Another study, utilizing restriction-site associated DNA (RAD) sequencing identified 8,207 SNP markers across the Lupin genome, which once filtered, led to the discovery of 38 molecular markers linked to the *Lanr1* disease resistance gene. Sequences involved in the analysis were derived from 20 informative plants resultant of a cross between a disease-resistant and a disease-susceptible line.

Marker-assisted breeding programs implement the introgression of genomic fragments to deliver a desired trait. In some instances, gene pyramiding is utilized, by which one, two, three, or even more genes carrying a particular trait (i.e., pathogen resistance) are introgressed into a hybrid line. Analysis of the resulting degrees of resistance to the pathogen can then be performed. Jiang et al. (2012) carried out marker-assisted gene pyramiding on rice cultivars to introgress rice blast resistance genes. Results from this study indicated that the greater the number of resistance genes contained in the improved lines, the higher the resistance to the pathogen and a subsequent growth benefit. In canola, MAS has enabled selection of intervarietal substitution lines and enrichment of genomic introgression lines (Howell et al. 1996; Zou et al. 2010).

4.23.3 Genetic Diversity Analyses

The development and implementation of molecular marker technology has paved the way for large-scale analyses of genetic diversity in and between species. This is valuable for clarifying evolutionary relationships and taxonomies as well as providing an understanding of genome change rates within and between different species. Importantly, the ability to assess genetic diversity in crops also has implications for crop breeding and sustainability. SSRs and SNPs have been widely applied to crop genetic diversity analyses. SNPs, as the most common form of highly heritable genetic variation across the genome, are superior indicators of genetic diversity and phylogeny, particularly in crop species with ancient genome duplications. Moreover, genomic SNPs are most often free of selective pressures, allowing a more complete estimate of diversity levels based on random genetic drift. This makes them highly useful in identifying regions of LD and then in tracking chromosome segments to identify recombination events that break up such regions (Cowling 2007)

4.23.4 Comparative Genomics

The ability to compare genomic properties of various evolutionarily related individuals can provide a wealth of information regarding the mechanisms underlying genome evolution, hybridization, polyploidization, and speciation. Molecular markers facilitate rapid and high-throughput comparative genomics analyses and enable analysis of presence/absence variation (PAV), copy number variation (CNV), and, for physically mapped markers such as SNPs, genomic rearrangements between individuals or species. In addition to elucidating the mechanisms and patterns of genome evolution, this information can then be linked to phenotype to better understand the influence of various selective pressures on genome stability and phenotype expression. Many genomic regions associated with disease resistance in plants are rapidly evolving due to constant selective pressure from rapidly evolving pathogens. This information can be highly useful for elucidating the genetic basis for disease resistance and co-evolution of the pathogen and host plant. By comparing genomic structural change and any associated effects on agricultural vigor in hybrids expressing heterosis for any trait of interest can be very useful in *Brassica* breeding programs (Hayward et al. 2012).

4.24 Development of Transgenic *Brassica* Crops Against Biotic Stresses

The *Brassica* genus includes one of the 10 most agronomically and economically important crop plant groups in the world. Within this group, crops such as broccoli, cabbage, cauliflower, kale, Brussels sprouts, turnip or rapeseed and mustard are widely cultivated at global level. Their cultivation and postharvest are continually threatened by significant stresses of biotic origin, such as pathogens (fungal, bacterial, viral, and nematodes) and pests. In recent years, numerous research groups around the world have developed transgenic lines within the *Brassica* genus that are capable of defending themselves effectively against these enemies. Some of the main transgenic lines developed include coding genes for chitinases, glucanases or cry proteins, which show effective results against pathogens such as *Alternaria brassicae, Leptosphaeria maculans, Sclerotinia sclerotiorum*, pests such as *Lipaphis erysimi* and *Plutella xylostella*.

The transformation of *Brassica* crops with chitinases has been achieved that are resistant to fungal pathogens in a very effective way. Against bacteria and fungi, the transformation with antimicrobial peptides, such as defensins, represents a very effective strategy in the control of these pathogens. On the other hand, modification of plant defense responses through transformation with genes involved in intermediate steps between the recognition of the pathogen and the defensive response has very significant role in reductions and development of diseases in *Brassica* transgenic crops. In the development of resistant lines against insect pests, cry genes from Bt are the most widely used, although there are other genes used in *Brassica* crops. Based on their ability to block intestinal activity, carbohydrate-binding proteins or lectins have been used with significant reductions in the attack of Hemipteran and Lepidopteran pests. The interactions of transgenic lines of *Brassica* crops with beneficial microorganisms from the agro-system, effects of an increasing mutualistic interaction have been found, without any negative effects.

4.24.1 Improvement of *Brassica* Species by Transgenesis

For the improvement of *Brassica* species, main strategies are use of molecular breeding and genetic transformation technology. The development of different genetic engineering tools has opened up vast opportunities for the introduction of novel and useful genes of agronomical importance. Many agronomical important genes have been identified and transferred into Brassica species. Transgenic lines developed so far in the Brassica genus are focused on tolerance to salt stress, phytoremediation of heavy metals, oil quality improvement, insect resistance, herbicide resistance, and development of male-sterile lines and the production of pharmacological and industrial products (Poveda et al. 2020).

The genetic transformation of *Brassica* has been carried out in almost all the economically important species, but far more work has been conducted with *B. napus* (rapeseed or canola) than with any of the others. In *B. rapa*, both oilseed and Chinese cabbage forms have been transformed. *B. oleracea* transgenic plants include all the major vegetables like cabbage, broccoli, Brussels sprouts, and cauliflower. A few reports dealing with *B. nigra, B. juncea,* and *B. carinata* have also been conducted. A fundamental step in the development of transgenic lines is regeneration, which has been achieved in *Brassica* lines by somatic embryogenesis and organogenesis using different explants. As far as transformation methods are concerned, the Agrobacterium-mediated gene uptake has been most commonly used for *Brassica*. With regard to Agrobacterium tumefaciens, effective transformations in cotyledonary or hypocotyl explants of Chinese cabbage, or by floral dip in rapeseed have been reported. In addition, there are perfectly developed protocols for this from seedling explants, even having developed an effective transient expression system, and also with *Agrobacterium rhizogenes*. However, nowadays, methods are beginning to be developed with other vector bacteria, such as the rhizospheric bacterium of rapeseed *Ensifer adhaerens*, which is capable of genetically transforming its original host (Poveda et al. 2020).

An aspect that is always linked to the development of transgenic crops is their safety for the environment, which is based on the capacity of new crops to survive outside the agro-system and the potential exchange of genetic information from these crops to related plants. This is particularly relevant in the case of *B. napus* because of its known potential for genetic exchange with wild or weedy relatives. In this sense, it has been reported how a glyphosate-resistant rapeseed line, which has been cultivated in Australian fields since 2009, is able to grow naturally outside the agro-system.

However, it was also shown how it ends up becoming extinct from the area within 3 years, without being able to become an invasive plant. On the other hand, there are numerous examples of hybridizations between various transgenic crops of the genus *Brassica* and nearby wild species, such as *B. rapa, B. juncea, B. oleracea, B. nigra, Hirschfeldia incana,* and *Raphanus raphanistrum*. The recombination of sets of different chromosomes may make not all hybrids viable, but there are reports of hybrids showing good field performance, although they produce fewer seeds than their wild parent (Poveda et al. 2020).

4.24.2 Effective Transgenic Lines of *Brassica* Species

At global level, 16 biotic stresses have been used extensively for molecular studies but only few of them have been exploited for development of effective transgenic *Brassica* crops. The bacterial disease, *Xanthomonas campestris* pv. *campestris* (Xcc), the causal agent of black rot, is considered to be one of the most important pathogens affecting *Brassica* vegetables worldwide. There are nine races of Xcc, but races 1 and 4 are considered the most virulent and widespread. *Xcc* penetrates leaves through hydathodes or wounds. Subsequently, the pathogen travels through the vascular system, thus invading the xylem and colonizing the mesophyll. This causes the appearance of typical symptoms, which include V-shaped chlorosis from the edges of the leaves, necrosis and darkening of leaf veins and the stem vascular tissue. As the disease progresses, wilting and necrosis throughout the plant appear. When the pathogen spreads along the veins and petioles to the plant stems and roots, the vascular tissue within the roots and stems turns black, and the whole plant dies, sometimes accompanied by a secondary soft rot. The other most common disease of Brassicaceae is soft rot, mainly caused by highly pectinolytic bacteria *Pectobacterium carotovorum* (Pc) (formerly *Erwinia carotovora*). Pc may infect the plant through natural openings or wounds caused by insects, other diseases, or by abiotic factors, such as low temperatures or mechanical damages. The bacterial soft rot on cabbage includes water-soaked lesions, which later become a rotted mass of macerated tissue.

The most widely distributed fungal pathogen in all *Brassica* crops are the four species described of genus *Alternaria*: *Alternaria alternata, Alternaria brassicae, Alternaria brassicicola,* and *Alternaria raphani. A. brassicae* is a major pathogen of oil-yielding *Brassica*, while the other three are more common on vegetable crops. All the four Alternaria species cause symptoms on cotyledons at the seedling stage and on leaves, leaf petiole, stem, inflorescence, siliquae, and seeds at the adult stage. Quantitative and qualitative losses in the yield of oil seeds and vegetables range from 11 to 100%, depending on the time of infection, prevailing environmental conditions after infection and the strategies used for its control.

Sclerotinia stem rot, caused by *Sclerotinia sclerotiorum*, is one of the most devastating diseases for *B. napus* and *B. juncea* worldwide, including Australia, China, Europe, and North America. In Australia, this disease causes up to 24% yield loss. The pathogen attacks cotyledons and leaves in seedlings and stems and leaves in adult plants, causing some key symptoms like water-soaked lesions, necrotic tissues with fluffy white mycelium and sclerotia inside of stems. In rapeseed, a major constraint in production is blackleg disease, caused by ascomycete fungus *Leptosphaeria maculans* (Lm). Blackleg has been reported in all canola-growing regions except for China, and causes annual yield losses of 10–20%. In seedlings, the fungus colonizes intercellular spaces, thus causing necrotic cotyledon and leaf lesions before entering a biotrophic phase while growing down the petiole and into the stem. At plant maturity, the fungus causes blackening of the stem and cankering, thus, restricting the nutrient flow up the plant, and in severe situations, completely killing the plant.

The *Albugo candida* (Ac) is a biotrophic plant pathogen that causes white rust disease in Brassicaceae. Zoospores enter their host through stomata, where they germinate and begin colonization of mesophyll cells. Finally, the pathogen forms zoosporangia that appear as white pustules rupturing the epidermis, hence constituting the visible white rust symptoms of the disease. The downy mildew, which is caused by *Hyaloperonospora brassicae* (Hb) (*Peronospora parasitica*), is commonly associated with *Albugo* and mustard mosaic virus. This pathogen causes destructive damages on *Brassica, Raphanus,* and *Sinapis* species, including on many economically relevant crops, such as broccoli, cabbage, radish, rapeseed, tatsoi, and wasabi. In *Brassica* seedlings, it appears on cotyledons and leaves as a pale green, yellowish growth on leaf undersides, while it causes irregular angular yellow blotches in older plants, which may have dark speckling.

Turnip mosaic virus (TuMV), family Potyviridae, causes disease in many kinds of *Brassica* and other crops worldwide, which results in losses in yield and quality of produce. TuMV is transmitted non-persistently by more than 40 different aphid species, having a wide natural host range. The disease symptoms often include vein clearing, mosaic, necrosis, plant stunting, and plant death, although they depend on the TuMV strain, host plant, and environmental conditions.

Among the insect-pests, aphids represent a major constraint to the production of many crops worldwide, especially *Brassica* crops. Mustard or turnip aphid *Lipaphis erysimi pseudobrassicae* (*Lep*) is one of the most destructive pests for *Brassica*, causing over 50% yield loss. These aphids feed by sucking sap from their host plants, which leads to stunted growth, and they excrete honeydew, which leads to fungal growth (sooty mold). Furthermore, aphids of the *L. erysimi* group transmit over 13 different viruses, including important viruses of the Brassicaceae, such as TuMV.

The Lepidopteran pest, diamondback moth, *Plutella xylostella* (Px), is the most destructive insect pest for *Brassica* crops. Larvae feed on host plants' leaves, causing substantial crop losses. Similarly, the large white butterfly, *Pieris brassicae* (Pb), also known as cabbage white butterfly, can be very destructive, and up to 90% loss in yield of certain *Brassica* spp. has been reported, especially in developing countries where chemical insecticides are too expensive to be used regularly against this pest.

4.24.3 Transgenics Developed in *Brassica* Species Against Different Pathogens

The transgenic *Brassica* crops developed against various pathogenic diseases are of very great importance owing to the diverse nature of pathogens from biotrophs to necrotrophs and of wide nature of pathogens from nucleic acid (virus), unicellular (bacteria) to multicellular (fungal) pathogens and nematodes. Table 4.12 contains the information on the transgenic lines developed, and indicates information such as the target pathogen or the mechanism involved in resistance.

The greatest number of transgenic lines of *Brassica* crops against fungal pathogens was made by transformation with genes that code for chitinases. Chitinase

Table 4.12 Development of transgenic *Brassica* crops against pathogens (Poveda et al. 2020)

Pathogen		Brassica	Gene			
Group	Species	Species	Name	Protein	Origin	Mechanism
Viruses	*Turnip mosaic virus*	*B. napus*	CP	Coat protein	*Turnip mosaic virus*	RNA silencing mechanism
Bacteria	*Pectobacterium carotovorum* ssp. *carotovorum*	*B. rapa* ssp. *pekinensis*	–	Anti-bacterial peptide	Not indicated	Bacteriostasis action
		B. rapa ssp. *pekinensis*	BAA1	Bromelain 1	Pineapple	Plant programed cell death
		B. rapa ssp. *pekinensis*	PinII	Proteinase inhibitor II	Potato	Inhibition of bacterial cell communication.
		B. rapa ssp. *pekinensis*	BrPGIP2	Polygalacturonase-inhibiting protein 2	*B. rapa* ssp. *pekinensis*	Inhibition of bacterial polygalacturones
	Xanthomonas campestris pv. *campestris*	*B. oleracea* var. *botrytis*	–	Cecropin B	*Antheraea polyphemus*	Lyse bacterial cell membranes
			–	Magainin II	*Xenopus laevis*	Disruption of microbe membranes
	Ralstonia solanacearum	*B. rapa*	RRS1	*R. solanacearum* resistance protein 1	*Arabidopsis thaliana*	Activation of defensive response by JA/ET route
			RPS4	*Pseudomonas syringae* resistance protein 4		
	Alternaria brassicae	*B. juncea*	–	Lectin	*Hevea brasiliensis*	Fungal cell wall carbohydrate binding (immobilization)
		B. juncea	–	Class I chitinase	Tobacco	Fungal cell wall degradation
		B. juncea	–	Class I basic glucanase	Tomato	Fungal cell wall degradation
		B. juncea	–	Class II chitinase	Barley	Fungal cell wall degradation
			RIP	Ribosome inactivating protein		Inactivation of foreign ribosomes in distantly related species and in other eukaryotes, including fungi
		B. napus	PmAMP1	Cysteine-rich antimicrobial peptide 1	*Pinus monticola*	Unidentified

(continued)

Table 4.12 (continued)

Pathogen		Brassica	Gene			
Group	Species	Species	Name	Protein	Origin	Mechanism
		B. juncea	MsrA1	Cecropin–melittin cationic peptide	–	Powerful membrane antagonism
		B. juncea	–	Lectin	Chickpea	Fungal cell wall carbohydrate binding (immobilization)
		B. juncea	ech42	Endochitinase 42	*Trichoderma virens*	Fungal cell wall degradation
		B. juncea	NPR1	Non-expressor of pathogenesis-related	*B. juncea*	Activation of the SA-mediated plant-defense
		B. juncea	MPK3	Mitogen-activated protein kinase 3	*B. juncea*	Activation of the SA-mediated plant-defense
Fungi	*A. brassicicola*	*B. oleracea* var. *italica*	–	Endochitinase	*Trichoderma harzianum*	Fungal cell wall degradation
		B. juncea	ech42	Endochitinase 42	*Trichoderma virens*	Fungal cell wall degradation
		B. juncea	NIC	Synthetic chitinase	–	Fungal cell wall degradation
	A. solani	*B. napus*	Chit42	Endochitinase 42	*Trichoderma atroviride*	Fungal cell wall degradation
	Botrytis cinerea	*B. napus*	MAM1	Methylthioalkylmalate synthase 1	*B. napus*	Glucosinolate biosynthesis
			CYP83A1	Cytochrome P450 83A1		
			UGT74B1	Glucosyltransferase 74B1		
	Erysiphe polygoni	*B. napus*	katE	Catalase E	*Escherichia coli*	Control of the activation of plant defense responses through the H2O2 dismutation

		B. juncea	NPR1	Non-expressor of pathogenesis-related	*B. juncea*	Activation of the SA-mediated plant-defense
	Fusarium oxysporum	*B. napus*	Chit42	Endochitinase 42	*Trichoderma atroviride*	Fungal cell wall degradation
	Leptosphaeria maculans	*B. napus*	Cf9	*Cladosporium fulvum* resistance protein 9	Tomato	Hypersensitive response activation
			Avr9	Dominant pathogen avirulence protein 9	Tomato	Hypersensitive response activation
		B. napus	DRR206	Dirigent protein	*Pisum sativum*	Unidentified
		B. napus	MiAMP1	Antimicrobial peptide 1	*Macadamia integrifolia*	Powerful membrane antagonism
		B. napus	Lm1	*L. maculans* 1	*B. nigra*	Unidentified
		B. napus	PmAMP1	Cysteine-rich antimicrobial peptide 1	*Pinus monticola*	Unidentified
		B. napus	DWF4	C-22 hydroxylase	*A. thaliana*	Brassinosteroids biosynthesis
		B. napus	Thkel1	Kelch domain protein	*Trichoderma harzianum*	β-Glucosidase activity
	Peronospora parasitica	*B. napus*	katE	Catalase E	*E. coli*	Control of the activation of plant defense responses through the H2O2 dismutation
	Rhizoctonia solani	*B. napus*	DRR206	Dirigent protein	*P. sativum*	Unidentified
		B. napus	pgip2	Polygalacturonase-inhibiting protein 2	*Proteus vulgaris*	Inhibition of fungal endo-polygalacturones
		B. napus	Chit42	Endochitinase 42	*Trichoderma atroviride*	Fungal cell wall degradation

(continued)

Table 4.12 (continued)

Pathogen		Brassica	Gene			
Group	Species	Species	Name	Protein	Origin	Mechanism
	Sclerotinia sclerotiorum	*B. napus*	–	Chitinase	*B. napus*	Fungal cell wall degradation
			–	Beta-1,3-glucanase	*B. napus*	Fungal cell wall degradation
		B. napus	DRR206	Dirigent protein	*P. sativum*	Unidentified
		B. napus	OXO	Oxalate oxidase	Wheat	Breakdown of the oxalic acid produced by the fungus.
		B. napus	MPK4	Mitogen-activated protein kinase 4	*B. napus*	Activation of the JA-mediated plant-defense
		B. napus	Ovd	Defensin	*Orychophragmus violaceus*	Permeabilization of fungal membranes
		B. napus	scFv	*S. sclerotiorum* antibody	*S. sclerotiorum*	Binding to the cell wall (immobilization, activation of plant defenses, etc.)
		B. napus	PjChi-1	Chitinase 1	*Paecilomyces javanicus*	Fungal cell wall degradation
			–	Sporamin	Sweet potato	Proteases inhibition
		B. napus	PmAMP1	Cysteine-rich antimicrobial peptide 1	*Pinus monticola*	Unidentified
		B. napus	LTP	Lipid transfer protein	*Oryza sativa*	Powerful membrane antagonism
		B. napus	LJAMP2	nsLTPs-like antimicrobial protein	*Leonurus japonicus*	Powerful membrane antagonism
		B. napus	MSI-99m	Magainin II analog	*Xenopus laevis*	Powerful membrane antagonism
		B. napus	bgn13.1	β-1,3-glucanase	*Trichoderma virens*	Fungal cell wall degradation
		B. juncea	MsrA1	Cecropin–melittin cationic peptide	–	Powerful membrane antagonism
		B. napus	WRKY33	Protein containing WRKY zinc-finger motifs 33	*B. napus*	Activation of the SA- and JA-mediated plant-defense

		B. napus	Chit42	Endochitinase 42	*Trichoderma atroviride*	Fungal cell wall degradation
		B. napus	Chit42	Endochitinase 42	*Trichoderma atroviride*	Fungal cell wall degradation
			–	Defensin	*Raphanus sativus*	Permeabilization of fungal membranes
		B. napus	Chit42	Endochitinase 42	*Trichoderma atroviride*	Fungal cell wall degradation
		B. napus	Chit42	Endochitinase 42	*Trichoderma atroviride*	Fungal cell wall degradation
			PG1P2	Polygalacturonase-inhibiting protein 2	*P. vulgaris*	Inhibition of fungal endo-polygalacturones
		B. napus	OsPGIP2	Polygalacturonase-inhibiting protein 2	*O. sativa*	Inhibition of fungal endo-polygalacturones
		B. napus	GDSL1	GDSL lipase	*A. thaliana*	Release of phosphatidic acid from fungal cell membrane and activation of plant defenses
		B. napus	NPR1	Non-expressor of pathogenesis-related	*B. napus*	Activation of the SA-mediated plant-defense
	Verticillium dahlia	*B. napus*	Chit42	Endochitinase 42	*Trichoderma atroviride*	Fungal cell wall degradation
	-	*B. juncea*	ChiC	Chitinase C	*Streptomyces griseus*	Fungal cell wall degradation
Oomycetes	*Albugo candida*	*B. napus*	WRR4	TIR-NB-LRR protein	*A. thaliana*	Activation of defensive response by JA/ET route
Nematodes	*Rotylenchulus* sp.	*B. oleracea* var. *capitata*	Bt	δ-Endotoxin	*Bacillus thuringiensis*	Intestinal toxicity for consumption

(continued)

Table 4.12 (continued)

Pathogen		Brassica	Gene			
Group	Species	Species	Name	Protein	Origin	Mechanism
	Heterodera schachtii	*B. napus*	Hs1pro-1	*Heterodera schachtii* resistance protein	*Beta procumbens*	Activation of a nematode-responsive and feeding site-specific gene expression
			cZR3	CC-NBS-LRR resistance protein		Necrotic hypersensitive response
Necrotrophic pathogens	–	*B. rapa* ssp. *oleifera*	entC	Isochorismate synthase	*E. coli*	SA biosynthesis

is an enzyme (a pathogenesis-related protein) that catalyzes the degradation of chitin, which is the main component of the fungal cell wall. In this sense, the most used transformation methodology is from hypocotyls using *A. tumefaciens*, as it was done in *B. juncea*, which was transformed with the gene that encodes chitinase C (*ChiC*) from *Streptomyces griseus* (Ahmad et al. 2015).

Different species belonging to the *Trichoderma* are widely studied and used as biocontrol agents in agriculture due to their ability to parasitize different fungal pathogens, by degrading their cell wall through the release of powerful chitinases. This host-parasite system has been used as a resource in the transformation of *Brassica* crops for obtaining plants that are resistant to *Alternaria brassicicola* and *A. brassicae* using endochitinase 42 from *Trichoderma harzianum* and *T. virens*. Transgenic lines showed a delayed onset of lesions, as well as a 30–73% reduction in the infected leaf area compared to non-transformed plants. In addition, this chitinolytic effect is increased when a chimeric chitinase produced by the fusion of the *Chit42* gene from *Trichoderma atroviride* and a chitin-binding domain from *Serratia marcescens* is used, hence causing a great inhibition of fungal growth (Poveda et al. 2020).

Other organisms have been used to obtain genes that encode chitinases, such as class I chitinase from tobacco in *B. juncea* against *A. brassicae*, which inhibited the fungal colony size by 12–56% over the non-transgenic control. The *NIC* gene has even been used, coding for a synthetic chitinase, in *B. juncea* transformation, for which leaf tissue extracts showed considerable resistance and antifungal activity against *A. brassicae*. The double transformation of *B. juncea* with two genes from barley, class II chitinase and the gene coding for a ribosome inactivating protein (RIP), which inactivates foreign ribosomes in distantly related species and in other eukaryotes, including fungi are of great importance (Chhikara et al. 2012).

Against *S. sclerotiorum*, transgenic lines of *B. napus* have been developed that co express the *Chit42* gene from *T. atroviride* and a defensin from *Raphanus sativus*, involved in increasing the fungal membranes permeability, or a polygalacturonase inhibiting protein 2 from *Phaseolus vulgaris* (a protein inhibiting fungal endo-polygalacturonases), showing greater resistance to the disease by delayed onset of restricted size and expansion of lesions when compared to wild-type plants. The chitinase I gene from *Paecilomyces javanicus* together with a sporamin gene from sweet potato, which is a protein involved in trypsin inhibition have been used (Liu et al. 2011). However, in isolation, these genes are also capable of improving the defensive capacity of *Brassica* crops against different pathogens. For example, the defensive capacity of *B. napus* against *S. sclerotiorum* is improved by the transformation with a defensin gene from *Orychophragmus violaceus* or with a polygalacturonase inhibiting protein 2 gene (*PGIP2*) from *Oryza sativa*. *PGIP2* genes are not only effective against fungal polygalacturonases, as it has also been observed in the transformation of *B. napus* with a *PGIP2* gene from *Proteus vulgaris* against *Rhizoctonia solani*, but they inhibit the disease up to 37%. They are also effective against bacterial polygalacturonases, as in the overexpression of the *PGIP2* gene from Chinese cabbage in the same plant, hence exhibiting improved resistance to bacterial soft rot caused by Pc (up to 54%) (Hwang et al. 2010).

The transformations of *Brassica* crops with genes that code for defense-related enzymes are found in glucanases, lipases, or oxalate oxidases. Since β-1, 3-glucan is a structural polymer present in the fungal cell wall, among fungal-resistant genes, glucanases that are potential antifungal agents to their glucan degradation activity are excellent candidates for controlling fungal pathogens development. In this sense, transformation of *B. napus* with a β-1, 3-glucanase from *Trichoderma virens* showed a stronger inhibition against S. sclerotiorum hyphal growth, and the transformation of *B. juncea* with the class I basic glucanase from tomato showed a restricted number, size and spread of lesions caused by *A. brassicae*. Similarly, lipases act against the cellular membrane of pathogenic fungi. In *B. napus*, lipase GDSL1 from *Arabidopsis thaliana* was characterized as an extracellular GDSL lipase functioning in *S. sclerotiorum* resistance. By releasing fragments of the fungal membrane, such as phosphatidic acid, GDSL1 activates the defensive responses of the plant through an increase in reactive oxygen species (ROS) and salicylic acid (SA) levels. However, this plant defense response could also be a consequence of the release of phosphatidic acid from the plant plasma membrane, instead of the fungal plasma membrane. During plant infection, *S. sclerotiorum* secretes oxalic acid, thus acidifying the plant tissue surrounding the site of infection and causing tissue damage, chelating divalent cations and sequestration of calcium that may weaken cell walls, and inhibiting the plant defense route by phenolic compounds. By transforming *B. napus* plants with a coding gene for enzyme oxalate oxidase from wheat, the breakdown of oxalic acid to CO_2 and H_2O_2 was achieved, and also a significant reduction in infection caused by the pathogenic fungus (Ding et al. 2013, 2020).

There has been an increased interest in producing transgenic plants that are strongly resistant to a broader spectrum of microbial phytopathogens through the expression of cationic antimicrobial peptides (CAPs). Although CAPs are structurally diverse, most of them fall into two general structure types: α-helical peptides, such as cecropins and magainins, and β-sheet peptides, such as defensins, protegrins, and tachyplesins. In *Brassica* crops, there are several transformations with genes coding for this type of protein, which have resulted in important inhibitions of pathogen growth and the development of the disease, by powerful membrane antagonism. Regarding α-helical peptides, found effective rapeseed transgenic lines against *L. maculans*, the transformation of *B. juncea* with the gene coding for a cecropin-melittin cationic peptide has been effective against *A. brassicae* and *S. sclerotiorum*, and also *B. napus* with a magainin II analog against *S. sclerotiorum*. The simultaneous transformation of cauliflower with a cecropin from *Antheraea polyphemus* and a magainin from *Xenopus laevis* against *Xcc* is effective as well. Another group of antimicrobial proteins is found in those known as lipid transfer proteins, which also cause a powerful membrane antagonism against different pathogens. Transgenic lines of *B. napus* that overexpress these type of genes from *Leonurus japonicus* and *Oryza sativa* show greater resistance against *S. sclerotiorum*, as shown in *in vitro* experiments with crude leaf extracts from transgenic lines, which significantly inhibit mycelial growth (Fan et al. 2013; Sang et al. 2013).

Although *Brassica* crops have been transformed with genes directly involved in defensive responses against pathogens, they have also been transformed with intermediaries in the signaling of plant responses. Once the plant recognizes the pathogen attack, it activates the plant defense responses through a signaling cascade, where mitogen-activated protein kinases (MAPKs) play a key intermediate role. In *Brassica* crops, overexpression of MAPKs genes, such as MPK3 and MPK4, increases the resistance of *B. juncea* and *B. napus* against *A. brassicae* and *S. sclerotiorum* through the activation of the SA- and JA-mediated plant-defenses, respectively. WRKY transcription factors are a type of DNA-binding proteins, whose domain is defined by the conserved amino acid sequence WRKYGQK. They are activated by MAPKs and are responsible for activating the gene expression related to the response to SA and JA pathogens. Overexpression of WRKY33 in *B. napus* has shown significant increases in plant resistance against *S. sclerotiorum*. On the other hand, the non-expressor of pathogenesis-related genes 1 (*NPR1*) is a master regulator of SA-mediated systemic acquired resistance (SAR), which is a broad-spectrum disease resistance mechanism in plants. *NPR1* overexpression in *B. juncea* and *B. napus* led to an increase in the SA-mediated defensive response and a greater resistance against *A. brassicae*, *E. polygoni* and *S. sclerotiorum* (Poveda 2020; Poveda et al. 2020; Wang et al. 2020a).

Against viruses, one of the most widely used mechanisms in plant transgenesis is the use of RNA interference that blocks the pathogen molecular machinery. By analyzing a fragment of a 130 bp coat protein, it was possible to develop an efficient hairpin construct against TuMV. The transformation of *B. napus* with this construction assumes the total absence of viral infection (Jafari and Shams-Bakhsh 2018).

4.24.4 Development of Transgenics in *Brassica* Species Against Pests

Genes used in the transformation of *Brassica* crops for resistance against pests are less diverse than in the case of defense against pathogens. Although almost all of the transgenic lines developed to date against insects are *Bacillus thuringiensis* (Bt) crops, there are other examples that share the strategies used against pathogens, which are represented in Table 4.13. Information such as the target arthropod pest or the mechanism involved in resistance is reflected. As with viruses, the use of interfering RNA has shown good results in controlling herbivorous arthropods. The transformation of Chinese cabbage with complementary fragments of the gene encoding coatomer protein complex subunit 2 (*COPB2*) derived from mite *Tetranichus urticae* was achieved. A high mite resistance with nearly a 100% mortality rate was reported. This study demonstrated the effectiveness of the plant-mediated RNAi technique in developing mite-resistant plants (Shin et al. 2020).

One of the most important direct defense responses in plants against the attack of phytophagous insects is the production of insecticidal peptides or proteins. Among the entomotoxic proteins, the group of carbohydrate-binding proteins or lectins is most effective. These proteins, after being ingested by phytophagous insects, are released from disrupted cellular structures and come into contact with carbohydrate

Table 4.13 Development of transgenic *Brassica* crops against insects and mites pests (Poveda et al. 2020)

Pest		*Brassica*	Gene			
Group	Species	species	Name	Protein	Origin	Mechanism
Insects-Hemiptera	*Lipaphis erysimi*	*B. juncea*	WGA	Wheat germ agglutinin	Wheat	Not indicated
		B. juncea	ASAL	Leaf agglutinin	*Allium sativum*	Blockage of the insect gut epithelial membrane
		B. juncea	ACA	Agglutinin	*Allium cepa*	Blockage of the insect gut epithelial membrane
		B. juncea	Ebf	(E)- β-farnesene	*Myzus arvensis*	Volatile sesquiterpene compound acting as the main component of aphid alarm pheromones
		B. juncea	LL	Lentil lectin	Lentil	Blockage of the insect gut epithelial membrane
			CPPI	Chickpea protease inhibitor	Chickpea	Disruption in assimilation of dietary protein
		B. juncea	RiD	Defensin	*Rorippa indica*	Inhibition of nutrient uptake
		B. juncea	HSPRO2	Nematode resistance protein-like homolog	*R. indica*	Activation of basal plant resistance
Insects-Lepidoptera	*Helicoverpa armigera*	*B. napus*	cry1Ac	Cry1Ac protein	*Bacillus thuringiensis*	Lysis of the gut epithelial cells
		B. juncea	cry1Ac	Cry1Ac protein	*B. thuringiensis*	Lysis of the gut epithelial cells
	Helicoverpa zea	*B. napus*	cry1Ac	Cry1Ac protein	*B. thuringiensis*	Lysis of the gut epithelial cells
	Pieris rapae	*B. rapa* ssp. *pekinensis*	cry1C	Cry1C protein	*B. thuringiensis*	Lysis of the gut epithelial cells
		B. rapa subsp. *pekinensis*	CpTI	Cowpea trypsin inhibitor	*Vigna unguiculata*	Inhibition of insect digestive activity
		B. napus	cry1Ac	Cry1Ac protein	*B. thuringiensis*	Lysis of the gut epithelial cells
		B. oleracea var. *capitata*	cry1Ac1	Cry1Ac1 protein	*B. thuringiensis*	Lysis of the gut epithelial cells
		B. oleracea var. *capitata*	cry1Ia8	Cry1Ia8 protein	*B. thuringiensis*	Lysis of the gut epithelial cells

	Plutella xylostella	*B. oleracea* var. *capitata*	cry1Ab3	Cry1Ab3 protein	*B. thuringiensis*	Lysis of the gut epithelial cells
		B. rapa ssp. *pekinensis*	cry1Ab and cry1Ac	Cry1Ab and Cry1Ac proteins	*B. thuringiensis*	Lysis of the gut epithelial cells
		Brassica oleracea var. *italica*	cry1Ab	Cry1Ab protein	*B. thuringiensis*	Lysis of the gut epithelial cells
		B. rapa ssp. *pekinensis*	cry1C	Cry1C protein	*B. thuringiensis*	Lysis of the gut epithelial cells
		B. oleracea var. *botrytis*	cryIA(b)	CryIA(b) protein	*B. thuringiensis*	Lysis of the gut epithelial cells
		B. napus	cry1Ac	Cry1Ac protein	*B. thuringiensis*	Lysis of the gut epithelial cells
		B. oleracea var. *italica*	cryIA(b)	CryIA(b) protein	*B. thuringiensis*	Lysis of the gut epithelial cells
		B. oleracea var. *acephala*	cry1Ac and cry1C	Cry1Ac and Cry1C proteins	*B. thuringiensis*	Lysis of the gut epithelial cells
		B. oleracea var. *capitata*	cry1B and cry1Ab	Cry1B and Cry1Ab proteins	*B. thuringiensis*	Lysis of the gut epithelial cells
		B. napus	Chi	Chitinase	*Manduca sexta*	Cuticle degradation
			BmkIT	Insect-specific neurotoxin	*Buthus martensii*	Neurotoxin of contractive paralysis type
		B. juncea	cry1Ac	Cry1Ac protein	*B. thuringiensis*	Lysis of the gut epithelial cells
		B. juncea	cry1Ac and cry1C	Cry1Ac and Cry1C proteins	*B. thuringiensis*	Lysis of the gut epithelial cells
		B. napus	cry1Ac	Cry1Ac protein	*B. thuringiensis*	Lysis of the gut epithelial cells

(continued)

Table 4.13 (continued)

Pest		*Brassica* species	Gene			Mechanism
Group	Species		Name	Protein	Origin	
		B. oleracea var. capitata	cry1Ab	Cry1Ab protein	*B. thuringiensis*	Lysis of the gut epithelial cells
		B. juncea	cry1C	Cry1C protein	*B. thuringiensis*	Lysis of the gut epithelial cells
		B. napus	cry1Ac	Cry1Ac protein	*B. thuringiensis*	Lysis of the gut epithelial cells
		B. napus	PjChi-1	Chitinase 1	*Paecilomyces javanicus*	Cuticle degradation
			–	Sporamin	Sweet potato	Proteases inhibition
		B. oleracea var. *capitata*	cry1Ba3	Cry1Ba3 protein	*B. thuringiensis*	Lysis of the gut epithelial cells
		B. juncea	cry1Ac	Cry1Ac protein	*B. thuringiensis*	Lysis of the gut epithelial cells
		B. rapa subsp. *pekinensis*	–	Sporamin	Sweet potato	Disruption in assimilation of dietary protein
		B. oleracea var. *capitata*	cry1Ia8 and cry1Ba3	Cry1Ia8 and Cry1Ba3 proteins	*B. thuringiensis*	Lysis of the gut epithelial cells
		B. napus	cry1C*	Cry1C* protein	*B. thuringiensis*	Lysis of the gut epithelial cells
		B. oleracea var. *capitata*	cry1Ac1	Cry1Ac1 protein	*B. thuringiensis*	Lysis of the gut epithelial cells
		B. napus	cry1Ac	Cry1Ac protein	*B. thuringiensis*	Lysis of the gut epithelial cells
		B. oleracea var. *capitata*	cry1Ia8	Cry1Ia8 protein	*B. thuringiensis*	Lysis of the gut epithelial cells
		B. oleracea var. *capitata*	cry1Ia8	Cry1Ia8 protein	*B. thuringiensis*	Lysis of the gut epithelial cells
		Brassica oleracea var. *italica*	cryIAa	CryIAa protein	*B. thuringiensis*	Lysis of the gut epithelial cells

	Trichoplusia ni	*B. rapa* ssp. *pekinensis*	cry1C	Cry1C protein	*B. thuringiensis*	Lysis of the gut epithelial cells
		B. rapa	cry1Ac	Cry1Ac protein	*B. thuringiensis*	Lysis of the gut epithelial cells
	Not indicated	*B. napus*	cry1Ac	Cry1Ac protein	*B. thuringiensis*	Lysis of the gut epithelial cells
		B. napus	cry1Ac	Cry1Ac protein	*B. thuringiensis*	Lysis of the gut epithelial cells
		B. oleracea var. *italica*	cry1Aa	Cry1Aa protein	*B. thuringiensis*	Lysis of the gut epithelial cells
Arachnids-mites	*Tetranychus urticae*	*B. rapa* subsp. *pekinensis*	COPB2	Coatomer protein complex subunit 2	*T. urticae*	RNA silencing mechanism

structures present in the midgut of insects. In the development of transgenic lines with lectins, there are several examples of the control of aphid *Lipaphis erysimi* by means of the blockage of the insect gut epithelial membrane and the inhibition of nutrient uptake. In this sense, genes coding for agglutinins (from wheat, garlic, or onion) have been the most used in the transformation of *B. juncea*, reporting a high mortality and significantly reduced fecundity of aphids. In conjunction with a lentil lectin, the transformation of *B. juncea* with a chickpea protease inhibitor was achieved, thus reducing aphid survival by 40%. This is because the protease inhibitors of plant origin are basically competitive inhibitors, acting as pseudo-substrates that bind to the active site of respective proteases, thus causing a disruption in assimilation of dietary protein in herbivorous insect pests, which primarily delays their significant growth and development (Rani et al. 2017).

Sporamins is a widely used group in inhibitor protease transgenesis, and in the transformation of Chinese cabbage and rapeseed against the diamondback moth (*Plutella xylostella*), in conjunction with chitinase 1 from *Paecilomyces javanicus*. The cuticle of insect species consists largely of chitin. Therefore, chitinase production has been used as a criterion for selecting potential biocontrol agents against insects. Therefore, because of their ability to interfere with chitin deposition, microbial chitinolytic enzymes have been considered to be important in the biological control of many insects. However, in the transformation of B. napus against *P. xylostella*, chitinases from Lepidoptera have also been used, such as Manduca sexta, together with the gene encoding for a neurotoxin of contractive paralysis type insect-specific from scorpion *Buthus martensii* (Wang et al. 2020b; Qiu et al. 2013).

Bacillus thuringiensis (Bt) is the main microorganism used in the biological control of insect pests. Bt produces a variety of entomotoxic crystalline proteins (Cry) known as δ-endotoxins. When ingested by susceptible insects, crystals dissolve in the insect gut and protoxins are liberated and activated proteolytically to a toxic fragment. This fragment binds to a specific cadherin on the brush border membrane of gut epithelial cells, and inserts into the membrane, thus generating pores. The change in membrane permeability leads to a colloid osmotic lysis of the gut epithelial cells and, ultimately, to insect death. In the last 30 years, numerous transgenic crops expressing Bt insecticidal crystals have been developed. The potential benefits of Bt crops include increased crop yields, a reduction in the use of broad-spectrum insecticide and the associated application costs and energy input, a reduced need for scouting, an improvement of the health conditions of farm workers, and savings in time.

In *Brassica* crops, transgenic-Bt lines of the most cultivated species have been developed against Lepidoptera-pests, such as *P. xylostella*, *Trichoplusia ni, Pieris rapae,* and *Helicoverpa armigera* (Table 4.13), which showed significant resistance. Moreover, it has been proven that the presence of cry genes in the genome of *Brassica* crops does not modify the synthesis or mode of action of constitutive and herbivore-inducible glucosinolates, hence maintaining an important strategy for these crops against herbivores (Himanen et al. 2008; Van Rie and Jansens 2019).

4.24.5 Use of Transgenics in Beneficial Biotic Interactions of *Brassica* Species

The potential negative effects of this new technology on the soil and air environments effecting different beneficial organisms (soil microorganisms, pollinating insects, have been a matter of great concern. As far as rhizospheric microorganisms are concerned, studies have only been carried out with the most widely cultivated transgenic *Brassica* in the world, which is *B. napus*. Transgenic *B. napus* harboring the synthetic chitinase (NiC) gene exhibits broad-spectrum antifungal resistance, for example, against Alternaria brassicicola. Since rhizosphere microorganisms play an important role in element cycling and nutrient transformation, biosafety assessment of NiC containing transgenic plants on the soil ecosystem is a regulatory requirement. By analyzing the rhizosphere enzyme activities and the microbial community structure, it was demonstrated how NiC *B. napus* lines may not affect rhizospheric microorganisms. The effects produced by herbicide-tolerant transgenic lines must be addressed in a different way, since the application of an herbicide must be analyzed as an indirect effect of the use of a transgenic line. In this sense, by Illumina MiSeq sequencing method, the effect of transgenic glufosinate-tolerant rapeseed and the associated herbicide application on rhizospheric bacterial communities were studied. The results showed that growing glufosinate-resistant transgenic rapeseed line Z7B10 and the application of glufosinate herbicide had no adverse effects on the rhizospheric bacterial community composition (Tang et al. 2019).

Moreover, several studies have been carried out in soils contaminated with heavy metals in order to determine potential modifications in the beneficial interaction between B. napus and plant growth promoting bacteria (PGPRs). A transgenic rapeseed line that overexpresses the gene for enzyme 1-aminocyclopropane-1-carboxylate (ACC) deaminase from *Pseudomonas putida* inoculated with the *P. putida* bacterium, developed better than its non-transgenic control in nickel-contaminated soils. This shows that the interaction between *P. putida* and transgenic *B. napus* leads to greater plant growth and higher heavy metal tissue accumulation (Farwell et al. 2006, 2007).

In *B. napus*, the ability of different species of the fungal genus *Trichoderma* to increase their productivity and tolerance against abiotic stresses such as drought or salinity has been described. In rapeseed plants transformed with the Thkel1 gene from *T. harzianum*, which codes for a Kelch domain protein, it is pointed out how this protein can modify glucosinolate hydrolysis by binding myrosinase enzymes, hence preventing the formation of toxic compounds such as isothiocyanates. As a consequence, the transgenic rapeseed plants in interaction with *T. harzianum* significantly increased the levels of fungal root colonization and their productivity. In addition, the presence of this Kelch domain protein reported increases in the β-glucosidase activity, which implies a greater resistance against foliar infection by *L. maculans* in rapeseed plants (Poveda 2020; Poveda et al. 2019, 2020).

The present work compiled the existing studies to date on the interaction between transgenic plants and biotic factors, focusing on those of greater relevance in recent years, the choice of the gene of interest and the mechanisms involved in increasing plant defenses.

References

Agnihotri A, Lakshmikumaran MS, Jagannathan V, Shivanna KR (1991) Wide hybridization for improvement in cultivated *Brassicas*. Acta Horti (ISHS) 289:213–214

Agnihotri A, Gupta K, Prem D, Sarkar, Mehra VS, Zargar SM (2009) Genetic enhancement in rapeseed-mustard for quality and disease resistance through in vitro techniques. In: Proceedings of 16th Australian Research Assembly on Brassicas, Ballarat, Australia, Sept 14–16, p 28

Ahlawat IPS, Sharma RP (2002) Agronomic terminology, vol 50. Indian Society of Agronomy, Division of Agronomy, Indian Agriculture Research Institute, New Delhi, p 132

Ahlawat IPS, Gangaiah B, Ompal S (2005) Production potential of chickpea (*Cicer arietinum*) based intercropping systems under irrigated conditions. Indian J Agron 50:27–30

Ahloowalia BS, Maluszynski M, Nichterlein K (2004) Global impact of mutation-derived varieties. Euphytica 135:187–201

Ahmad B, Ambreen MSKAH, Khan I (2015) *Agrobacterium* mediated transformation of *Brassica juncea* (L.) Czern. with chitinase gene conferring resistance against fungal infections. Pak J Bot 47:211–216

Akaba M, Kaneko Y, Hatakeyama K, Ishida M, Bang SW, Matsuzawa Y (2008) Evaluation of clubroot resistance in *Raphanus sativus* monosomic addition lines. In: Proceedings 113th meeting of Japanese Society of Breeding. Breed res, vol 10, p 205.

Alamery S, Tirnaz S, Bayer P, Tollenaere R, Chaloub B, Edwards D, Batley J (2017) Genome-wide identification and comparative analysis of NBS-LRR resistance genes in *Brassica napus*. Crop Pasture Sci 69:79–94. https://doi.org/10.1071/CP17214

An T, Cai Y, Zhao S, Zhou J, Song B, Bux H, Qi X (2016) *Brachypodium distachyon* T-DNA insertion lines: a model pathosystem to study nonhost resistance to wheat stripe rust. Sci Reptr s 6:25510. https://doi.org/10.1038/srep25510

Aneja JK, Agnihotri A (2013) Alternaria blight of oilseed *Brassicas*: epidemiology and disease control strategies with special reference to use of biotechnological approaches for attaining host resistance. J Oilseed Brassica 4:1–10

Aneja JK, Goel P, Agnihotri A (2012) Diversity assessment in *Alternaria brassicae*. In: Proc. of the 1st National Brassica Conference on production barriers and technological options in oilseed *Brassicas*, organized by Society for Rapeseed-Mustard Research, Bharatpur and Indian Council of Agricultural Research, New Delhi, at CCS Haryana agriculture university Hisar, mar 2-3, 2012, pp. 149 (Abstr).

Ansari NA, Khan MW, Muheet A (1989) Effect of some factors on growth and sporulation of *Alternaria brassicae* causing Alternaria blight of rapeseed and mustard. Acta Bot Indica 17:49–53

Ao E, Thakuria K, Saud RK (2016) Production potential of rapeseed (*Brassica rapa* var. *dichotoma*)—based intercropping systems under rain fed conditions. J Oilseed Brassica 7 (1):91–97

Atri C, Akhatar J, Gupta M, Gupta N, Goyal A, Rana K, Kaur R, Mittal M, Sharma A, Singh MP, Sandhu PS, Barbetti MJ, Banga SS (2019) Molecular and genetic analysis of defensive responses of *Brassica juncea – B. fruticulosa* introgression lines to Sclerotinia infection. Sci Reptr 9:17089 https://doi.org/10.1038/s41598-019-53444-3

Auld D, Heikkinen M, Erickson D, Sernyk J, Romero J (1992) Rapeseed mutants with reduced levels of polyunsaturated fatty acids and increased levels of oleic acid. Crop Sci 32:657–662

Avenot H, Dongo A, Simoneau NB, Iacomivasilescu B, Hamon B, Peltiers D, Simoneau P (2005) Isolation of 12 polymorphic microsatellite loci in the phytopathogenic fungus *Alternaria brassicicola*. Mol Ecol Notes 5:948–950

Awasthi RP, Kolte SJ (1989) Variability in *Alternaria brassicae* affecting rapeseed and mustard. Indian Phytopath 42:275

Ayers GW, Lelacheur KE (1972) Genetics of resistance in rutabaga to races of *Plasmodiophora brassicae*. Can J Plant Sci 52:897–900

Ayotte R, Harney PM, Souza MV (1987) The transfer of atriazine resistance from *Brassica napus* L. to *B. oleracea* L. production of F1 hybrids through embryo rescues. Euphytica 36:615–624. https://doi.org/10.1007/BF00041511

Baggett JR (1983) Clubroot-resistant cabbage breeding lines Oregon 100, 123, 140, and 142. Hortic Sci 18:112–114

Baggett JR, Kean D (1985) Clubroot-resistant broccoli breeding lines OSU CR-2 to OSU CR-8. Hortic Sci 20:784–785

Barton NH (2001) The role of hybridization in evolution. Mol Ecol 10:551–568

Berbee ML, Payne BP, Zhang G, Roberts RG, Turgeon BG (2003) Shared ITS DNA substitutions in isolates of opposite mating type reveal a recombining history for three presumed species in the filamentous ascomycete genus *Alternaria*. Mycol Res 107(2):169–182

Bock CH, Thrall PH, Brubaker CL, Burdon JJ (2002) Detection of genetic variation in *Alternaria brassicicola* using AFLP fingerprinting. Mycol Res 106(4):428–434

Bradshaw JE, Gemmell DJ, Wilson RN (1997) Transfer of resistance to clubroot (*Plasmodiophora brassicae*) to swedes (*Brassica napus* L. var. *napobrassica* Peterm) from *B. rapa*. Ann Appl Biol 130:337–348

Brun H, Plessis J, Renard M (1987a) Resistance of some crucifers to Alternaria brassicae (Berk.) Sacc. Proc. 7th Int. Rapeseed Congr. Poznan, Poland. pp 1222–1227

Brun H, Renard M, Plessis J (1987b) Resistance of some crucifers to Alternaria brassicae (Berk.) Sacc. Proc. 7th Int. Rapeseed Congr. Poznan, Poland, p. 247.

Brun H, Chevre AM, Fitt BD, Powers S, Besnard AL, Ermel M, Huteau V, Marquer B, Eber F, Renard M, Andrivon D (2010) Quantitative resistance increases the durability of qualitative resistance to *Leptosphaeria maculans* in *Brassica napus*. New Phytol 185:285–299. https://doi.org/10.1111/j.1469-8137.2009.03049.x

Bruns TD, White TJ, Taylor JW (1991) Fungal molecular systematic. Ann Rev Ecol Syst 22:525–564

Cao T, Manolii VP, Hwang SF, Howard RJ, Strelkov SE (2009) Virulence and spread of *Plasmodiophora brassicae* [clubroot] in Alberta. Canada Can J Plant Pathol 31:321–329

Chatterjee M, Mazumder M, Basu D (2013) Functional analysis of the promoter of a glycosyl hydrolase gene induced in resistant *Sinapis alba* by *Alternaria brassicicola*. Phytopathology 103:841–850. https://doi.org/10.1094/PHYTO-11-12-0303-R

Chavan V, Bhargava S, Kamble A (2013) Temporal modulation of oxidant and anti-oxidative responses in *Brassica carinata* during β-amino-butyric acid-induced resistance against *Alternaria brassicae*. Physiol Mol Plant Pathol 83:35–39. https://doi.org/10.1016/j.pmpp.2013.03.002

Chen S, Zou J, Cowling WA, Meng J (2010) Allelic diversity in a novel gene pool of canola-quality *Brassica napus* enriched with alleles from *B. rapa* and *B. carinata*. Crop Pasture Sci 61:483–492. https://doi.org/10.1071/CP09327

Chevre AM, This P, Eber F, Deschamps M, Renard M, Delseny M, Quiros CF (1991) Characterization of disomic addition lines *Brassica napus-Brassica nigra* by isozyme, fatty acid and RFLP markers. Theor Appl Genet 81:43–49

Chevre AM, Eber F, Margale E, Kerlan MC (1994) Comparison of somatic and sexual *Brassica napus-Sinapis alba* hybrids and their progeny by cytogenetic studies and molecular characterization. Genome 37:367–371

Chevre AM, Eber F, This P, Barret P, Tanguy X, Brun H, Dlseny M, Renard M (1996) Characterization of *Brassica nigra* chromosomes and of blackleg resistance in *B. napus–B. nigra* addition lines. Plant Breed 115:113–118. https://doi.org/10.1111/j.1439-05231996.tb00884.x

Chhikara S, Chaudhury D, Dhankher OP, Jaiwal PK (2012) Combined expression of a barley class II chitinase and type I ribosome inactivating protein in transgenic *Brassica juncea* provides protection against *Alternaria brassicae*. Plant Cell Tissue Organ Cult 108:83–89

Chiang MS, Crete R (1976) Diallel analysis of the inheritance of resistance to 6 of *Plasmodiophora brassicae* in cabbage. Can J Plant Sci 56:865–868

Chiang MS, Crete R (1983) Transfer of resistance to race 2 of *Plasmodiophora brassicae* from *Brassica napus* to cabbage (*B. oleracea* spp. *capitata*). V. The inheritance of resistance. Euphytica 32:479–483

Chiang MS, Crete R (1985) Acadie: a clubroot resistant cabbage cultivar. Can J Plant Sci 65:223–235

Chiang MS, Crete R (1987) Cytoplasmic male sterility in *Brassica oleracea* induced by *B*. napus cytoplasm-female fertility and restoration of male fertility. Can J Plant Sci 672:891–897. https://doi.org/10.4141/cjps87-126

Chiang MS, Crete R (1989) Richelain: a clubroot resistant cabbage cultivar. Can J Plant Sci 69:337–340

Chiang MS, Chiang BY, Grant WF (1977a) Transfer of resistance to race 2 of *Plasmodiophora brassicae* from *Brassica napus* to cabbage (*B. oleracea* var. *capitata*). I. Interspecific hybridization between *B. napus* and *B. oleracea* var. *capitata*. Euphytica 26:319–336

Chiang MS, Chiang BY, Grant WF (1977b) Transfer of resistance to race 2 of *Plasmodiophora brassicae* from *Brassica napus* to cabbage (*B. oleracea* var. *capitata*). I. Interspecific hybridization between *B. napus* and *B. oleracea* var. *capitata*. Euphytica 26:319–336

Chiang BY, Chiang MS, Grant WF, Crete R (1980) Transfer of resistance to race 2 of *Plasmodiophora brassicae* from *Brassica napus* to cabbage (*B. oleracea* spp. capitata). IV. A resistant 18-chromosome B1 plant and its B2 progenies. Euphytica 29:47–55. https://doi.org/10.1007/BF00037248

Chou HH, Wu WS (2002) Phylogenetic analysis of internal transcribed spacer regions of the genus *Alternaria*, and the significance of filament-beaked conidia. Mycol Res 106:164–169

Choudhary BR, Joshi P, Ramarao S (1999) Interspecific hybridization between *Brassica carinata* and *Brassica rapa*. Plant Breed 119:417–420

Christianson JA, Rimmer SR, Good AG, Lydiate DJ (2006) Mapping genes for resistance to *Leptosphaeria maculans* in *Brassica juncea*. Genome 49:30–41. https://doi.org/10.1139/g05-085

Chrungu B, Verma N, Mohanty A, Pradhan A, Shivanna KR (1999) Production and characterization of interspecific hybrids between *Brassica maurorum* and crop *Brassicas*. Theor Appl Genet 98:608–613

Colhoun J (1958) Clubroot disease of crucifers caused by *Plasmodiophora brassicae* Woron. A monograph. Phytopathol Paper 3:1–109

Collins NC, Thordal-Christensen H, Lipka V, Bau S, Kombrink E, Qiu JL, Huckelhoven R, Stein M, Freialdenhoven A, Somerville SC, Schulze-Lefert P (2003) SNARE-protein-mediated disease resistance at the plant cell wall. Nat 4256(1):973–977

Conn KL, Tewari JP (1986) Hypersensitive reaction induced by *Alternaria brassicae* in Eruca sativa, an oil-yielding crucifer. Can J Plant Pathol 8:348. (Abstr)

Conn KL, Tewari JP, Dahiya JS (1988) Resistance to *Alternaria brassicae* and phytoalexin-elicitation in rapeseed and other crucifers. Plant Sci 55:21–25. https://doi.org/10.1016/0168-9452(88)90037-4

Conn KL, Jejelowa OA, Tewari JP, Bains PS (1991) *Camelina sativa* phytoalexins (Calmexin and methoxy camalexin) provide resistance against *Alternaria brassicae*. Abstract 274, Can J Plant Pathol 13

Cooke DEL, Forster JW, Jenkins PD, Gareth DJ, Lewis DM (1998) Analysis of intraspecific and interspecific variation in the genus Alternaria by the use of RAPD-PCR. Ann App Biol 132:197–209

Cowling WA (2007) Genetic diversity in Australian canola and implications for crop breeding for changing future environments. Field Crop Res 104:103–111

Crisp P, Crute IR, Sutherland RA, Angell SM, Bloor K, Burgess H, Gordon PL (1989) The exploitation of genetic resources of *Brassica oleracea* in breeding for resistance to clubroot (*Plasmodiophora brassicae*). Euphytica 42:215–226

Crow JF, Kimura M (1970) An introduction to population genetics theory. Harper and Row, New York

Crute IR, Gray AR, Crisp P, Buczacki ST (1980) Variation in *Plasmodiophora brassicae* and resistance to clubroot disease in *Brassicas* and allied crops-a critical review. Plant Breed Abstr 50:91–104

Crute IR, Phelps K, Barnes A, Buczacki ST, Crisp P (1983) The relationship between genotypes of three *Brassica* species and collections of *Plasmodiophora brassicae*. Plant Pathol 32:405–420

Dang JK, Sangwan MS, Naresh M, Kaushik CD (2000) Multiple disease resistance against four fungal foliar diseases of rapeseed-mustard. Indian Phytopath 53:455–458

Dangl JL, Ritter C, Gibbon MJ, Mur LA, Wood JR, Goss S, Mansfield J, Taylor JD, Vivian A (1992) Functional homologs of the Arabidopsis RPM1 disease resistance gene in bean and pea. Plant Cell 4:1359–1369. https://doi.org/10.1105/tpc.4.11.1359

Das ML, Rahman A (1989) Alternaria resistance of Brassica campestris L. improved by induced mutation. Mutation Breed Newslett 33:9–10

Delmas CEL, Fabre F, Jolivet J, Mazet ID, Richart Cervera S, Delière L, Delmotte F (2016) Adaptation of a plant pathogen to partial host resistance: selection for greater aggressiveness in grapevine downy mildew. Evol Appl 9:709–725. https://doi.org/10.1111/eva.12368

Delourme R, Foisset N, Horvais R, Barret P, Champagne G, Cheung WY, Landry BS, Renard M (1998) Characterization of the radish introgression carrying the Rfo restorer gene for the Ogu-INRA cytoplasmic male sterility in rapeseed (*Brassica napus* L.). Theor Appl Genet 97:129–134

Deora A, Gossen BD, Mcdonald MR (2012) Infection and development of Plasmodiophora brassicae in resistant and susceptible canola cultivars. Can J Plant Pathol 34(2):239–247. https://doi.org/10.1080/07060661.2012.681071

Deora A, Gossen BD, McDonald MR (2013) Expression of resistance to *Plasmodiophora brassicae* in four resistant canola cultivars. Can J Plant Pathol35(1):106–106

Deverall BJ (1995) Plant protection using natural defense systems of plants. Adv Plant Pathol 11:211–228

Dey SS, Sharma K, Dey RB, Kumar GMS, Singh D, Kumar R, Parkash C (2015) Inter specific hybridization (*Brassica carinata* × *Brassica oleracea*) for introgression of black rot resistance genes into Indian cauliflower (*B. oleracea* var. botrytis L.). Euphytica 204:149–162. https://doi.org/10.1007/s10681-015-1352-0

Diederichsen E, Sacristan MD (1994) The use of ovule culture in reciprocal hybridization between *B. campestris* L. and *B. oleracea* L. Plant Breed 113:79–82

Diederichsen E, Sacristan MD (1996) Disease response of resynthesized *Brassica napus* L. lines carrying different combinations of resistance to *Plasmodiophora brassicae* Worn. Plant Breed 115:5–10. https://doi.org/10.1111/j.1439-0523.1996.tb00862.x

Diederichsen E, Wagenblatt B, Schallehn V, Deppe U, Sacristan MD (1996) Transfer of clubroot resistance from re-synthesized *Brassica napus* into oilseed rape-identification of race-specific interactions with *Plasmodiophora brassicae*. In: Proceedings of the international symposium on *Brassicas*. Acta Hortic 407:423–429

Diederichsen E, Deppe U, Sacristan MD (2003) Characterization of clubroot resistance in recent winter oilseed rape material. In: Proceedings of the 11th international rapeseed congress, vol 1. Copenhagen, Denmark, pp 68–70

Diederichsen E, Beckmann J, Schondelmeier J, Dreyer F (2006) Genetics of clubroot resistance in *Brassica napus* 'Mendel'. Acta Hortic 706: 307–312. https://doi.org/10.17660/ActaHortic.2006.706.35

Diederichsen E, Frauen M, Linders EGA, Hatakeyama K, Hirai M (2009) Status and perspectives of clubroot resistance breeding in crucifer crops. J Plant Growth Regul 28:265–281. https://doi.org/10.1007/s00344-009-9100-0

Diederichsen E, Frauen M, Ludwig-Muller J (2014) Clubroot disease management challenges from a German perspective. Can J Plant Pathol 36:85–98. https://doi.org/10.1080/07060661.2013.861871

Ding Y, Mei J, Li Q, Liu Y, Wan H, Wang L, Becker HC, Qian W (2013) Improvement of *Sclerotinia sclerotiorum* resistance in *Brassica napus* by using *B. oleracea*. Genet Resour Crop Evol 60:1615–1619. https://doi.org/10.1007/s10722-013-9978-z

Ding Y, Mei J, Liu Y, Wang L, Li Y, Wan H, Li J, Qian W (2015) Transfer of Sclerotinia stem rot resistance from wild *Brassica oleracea* into *B. rapa*. Mol Breed 35:225

Ding Y, Mei J, Wu Q, Xiong Z, Li Y, Shao C, Wang L, Qian W (2019) Synchronous improvement of sub-genomes in allopolyploid: a case of *Sclerotinia* resistance improvement in *Brassica napus*. Mol Breed 39:10

Ding L, Li M, Guo X, Tang M, Cao J, Wang Z, Liu R, Zhu K, Guo L, Liu S, Tan X (2020) *Arabidopsis* GDSL1 overexpression enhances rapeseed *Sclerotinia sclerotiorum* resistance and the functional identification of its homolog in *Brassica napus*. Plant Biotechnol J 18:1255–1270

Dixon GR (1988) Shetland cabbage as a source of resistance to *Plasmodiophora brassicae*. In: Proceedings of the 5th international congress of plant pathology, 20–27. Int Soc Plant Pathol, Kyoto, Japan, p 178

Dixon GR (2009) Husbandry – the sustainable means of controlling soil borne pathogens:- a synoptic review. In: Dixon GR, editor. Proceedings of the symposium of horticulture in Europe, Vienna, Austria; 17–21. February 2008. Acta Hortic. 817, pp 233–242

Donald EC, Porter IJ, Faggian R, Lancaster RA (2006a) An integrated approach to the control of clubroot in vegetable *Brassica* crops. Acta Hortic 706:283–300

Donald EC, Cross SJ, Lawrence JM, Porter IJ (2006b) Pathotypes of *Plasmodiophora brassicae*, the cause of clubroot in Australia. Ann Appl Biol 148:239–244

Donald EC, Jaudzems G, Porter IJ (2007) Pathology of cortical invasion by *Plasmodiophora brassicae* in clubroot resistant and susceptible*Brassica oleracea* hosts. Plant Pathol 57:201–209

Donald EC, Jaudzems G, Porter IJ (2008) Pathology of cortical invasion by *Plasmodiophora brassicae* in clubroot resistant and susceptible *Brassica oleracea* hosts. Plant Pathol 57:201–209. https://doi.org/10.1111/j.1365-3059.2007.01765.x

Douchkov D, Luck S, Johrde A, Nowara D, Himmelbach A, Rajaraman J, Schweizer P (2014a) Discovery of genes affecting resistance of barley to adapted and non-adapted powdery mildew fungi. Genome Biol 15(12). https://doi.org/10.1186/gb-2014-15-1-r1

Douchkov D, Baum T, Ihlow A, Schweizer P, Seiffert U (2014b) Microphenomics for interactions of barley with fungal pathogens. In: Tuberosa R, Graner A, Frison E (eds) Genomics of plant genetic resources. 2nd edn. Springer, Heidelberg, pp 123–148

Ebrahimi E, Kaul HP, Neugschwandtner RW, Dabbagh Mohammadi Nassab A (2016) Productivity of wheat (*Triticum aestivum L.*) intercropped with rapeseed (*Brassica napus L*). Can J Plant Sci 97:557–568

Enjalbert JN, Zheng S, Johnson JJ, Mullen JL, Byrne PF, McKay JK (2013) Brassicaceae germplasm diversity for agronomic and seed quality traits under drought stress. Indian Crops Prod 47:176–185

Enright S, Cipollini D (2007) Infection by powdery mildew Erysiphe cruciferarum (Erysiphaceae) strongly affects growth and fitness of *Alliaria petiolata* (Brassicaceae). Am J Bot 94:1813–1820

Fahleson J, Rahlen L, Glimelius K (1988) Analysis of plants regenerated from protoplast fusions between *Brassica napus* and *Eruca sativa*. Theor Appl Genet 76(4):507–512. https://doi.org/10.1007/BF00260900

Fan Y, Du K, Gao Y, Kong Y, Chu C, Sokolov V, Wang Y (2013) Transformation of LTP gene into *Brassica napus* to enhance its resistance to *Sclerotinia sclerotiorum*. Russ J Gen 49:380–387

Farwell AJ, Vesely S, Nero V, Rodriguez H, Shah S, Dixon DG, Glick BR (2006) The use of transgenic canola (*Brassica napus*) and plant growth-promoting bacteria to enhance plant biomass at a nickel-contaminated field site. Plant Soil 288:309–318

Farwell AJ, Vesely S, Nero V, Rodriguez H, McCormack K, Shah S, Dixon DG, Glick BR (2007) Tolerance of transgenic canola plants (*Brassica napus*) amended with plant growth-promoting bacteria to flooding stress at a metal-contaminated field site. Environ Pollut 147:540–545

Fatima U, Bhorali P, Borah S, Senthil-Kumar M (2019) Perspectives on the utilization of resistance mechanisms from host and nonhost plants for durable protection of *Brassica* crops against Alternaria blight. Peer J 7:e7486. https://doi.org/10.7717/peerj.7486

Gangasaran G, Giri G (1985) Intercropping of mustard with chickpea, lentil and barley in drylands. Indian J Agron 30:244–250

Gherbawy YAMH (2005) Genetic variation among isolates of *Alternaria* spp. from selected Egyptian crops. Arch Phytopath Plant Protec 38(2):77–89

Ghose K, Dey S, Barton H, Loake GJ, Basu D (2008) Differential profiling of selected defense-related genes induced on challenge with *Alternaria brassicicola* in resistant white mustard and their comparative expression pattern in susceptible India mustard. Mol Plant Pathol 9 (6):763–775. https://doi.org/10.1111/j.1364-3703.2008.00497.x

Gossen BD, Mcdonald MR, Hwang SF, Strelkov SE, Peng G (2013) A comparison of clubroot development and management on canola and *Brassica* vegetables. Can J Plant Pathol 35 (2):175–191. https://doi.org/10.1080/07060661.2013.763293

Goyal P, Chahar M, Mathur AP, Kumar A, Chattopadhyay C (2011) Morphological and cultural variation in different oilseed *Brassica* isolates of *Alternaria brassicae* from different geographical regions of India. Indian J Agril Sci 81(11):1052–1058

Grontoft M (1986) Resistance to *Alternaria* spp. in oil crops. Sveriges Utsadesforenings Tidskrift 96(3):263

Grover A, Gawthaman R (2003) Strategies for development of fungus-resistant transgenic plants. Curr Sci 84:330–340

Gunasinghe N, You MP, Banga SS, Barbetti MJ (2014) High level resistance to *Pseudocercosporella capsellae* offers new opportunities to deploy host resistance to effectively manage white leaf spot disease across major cruciferous crops. Eur J Plant Pathol 138:873–890. https://doi.org/10.1007/s10658-013-0360-y

Gunasinghe N, You MP, Li XX, Banga SS, Banga SK, Barbetti MJ (2016) New host resistances to *Pseudocercosporella capsellae* and implications for white leaf spot management in Brassicaceae crops. Crop Prot 86:69–76. https://doi.org/10.1016/j.cropro.2016.04.014

Gupta RBL, Singh M (1994) Source of resistance to white rust and powdery mildew of mustard. Intl J Trop Plant Dis 12:225–227

Gupta K, Prem D, Agnihotri A (2010) Pyramiding white rust and Alternaria resistance in low erucic *Brassica juncea* using *B. carinata*. J Oilseed Brassica 1(2):55–65

Gustafsson M, Falt AS (1986) Genetic studies on resistance to clubroot in *Brassica napus*. Ann Appl Biol 108:409–415

Hagimori M (1995) Clubroot disease resistance and other characteristics of the progeny of somatic hybrid between Japanese radish and cauliflower. Acta Hortic 392:77–80

Hagimori M, Nagaoka M, Kato N, Yoshikawa H (1992) Production and characterization of somatic hybrids between Japanese radish and cauliflower. Theor Appl Genet 84:819–824

Hansen M (1989) Genetic variation and inheritance of tolerance to clubroot (*Plasmodiophora brassicae* Wor.) and other quantitative characters in cabbage (*Brassica oleracea* L). Hereditas 110:13–22

Hansen LN (1998) Intertribal somatic hybridization between rapid cycling *Brassica oleracea* and *Camelina sativa*. Euphytica 104:173–179

Hansen LN, Earle ED (1995) Transfer of resistance to *Xanthomonas campestris* pv. *campestris* into *Brassica oleracea* L. by protoplast fusion. Theor Appl Genet 91:1293–1300. https://doi.org/10.1007/BF00220944

Hansen LH, Earle ED (1997) Somatic hybrids between *Brassica oleracea* (L.) and *Sinapis alba* (L.) with resistance to *Alternaria brassicae* (Berk) Sacc. Theor Appl Genet 94:1078–1085

Happstadius I, Ljungberg A, Kristiansson B, Dixelius C (2003) Identification of *Brassica oleracea* germplasm with improved resistance to *Verticillium* wilt. Plant Breed 122:30–34

Harbert DJ (1969) A simple effective embryo culture technique for *Brassica*. Euphytica 18:425–429

Hatakeyama K, Fujimura M, Ishida M, Suzuki T (2004) New classification methods for *Plasmodiophora brassicae* field isolates in Japan based on resistance of F1 cultivars of Chinese cabbage (*Brassica rapa* L.) to clubroot. Breed Sci 54:197–201

Hatakeyama K, Niwa T, Kato T, Ohara T, Kakizaki T, Matsumoto S (2017) The tandem repeated organization of NB-LRR genes in the clubroot-resistant *CRb* locus in *Brassica rapa* L. Mol Genet Genome 292:397–405. https://doi.org/10.1007/s00438-016-1281-1

Hayashida N, Takabatake Y, Nakazawa N, Aruga D, Nakanishi H, Taguchi G, Sakamoto K, Matsumoto E (2008) Construction of a practical SCAR marker linked to clubroot resistance in Chinese cabbage, with intensive analysis of HC352b genes. J Jpn Soc Hortic Sci 77:150–154

Hayward A, McLanders J, Campbell E, Edwards D, Batley J (2012) Genomic advances will herald new insights into the *Brassica: Leptosphaeria maculans* pathosystem. Plant Biol 14:1–10

Hayward AC, Tollenaere R, Dalton-Morgan J, Batley J (2015) Molecular marker application in plants. . In: Batley J (ed) Plant genotyping methods and protocols, methods in molecular biology, vol 1245. Springer Science+Business Media, New York. https://doi.org/10.1007/978-1-4939-1966-6_2

Himanen SJ, Nissinen A, Auriola S, Poppy GM, Stewart CN Jr, Holopainen JK, Nerg A-M (2008) Constitutive and herbivore-inducible glucosinolate concentrations in oilseed rape (*Brassica napus*) leaves are not affected by *Bt Cry1Ac* insertion but change under elevated atmospheric CO_2 and O_3. Planta 227:427–437

Hirai M (2006) Genetic analysis of clubroot resistance in *brassica* crops. Breed Sci 56:223–229

Hirai M, Harada T, Kubo N, Tsukada M, Suwabe K, Matsumoto S (2004) A novel locus for clubroot resistance in *Brassica rapa* and its linkage markers. Theor Appl Genet 108:639–643

Hirani AH, Li G (2015) Understanding the genetics of clubroot resistance for effectively controlling this disease in *Brassica* species. Plants for the Future Intech Open Doi. https://doi.org/10.5772/60936

Hirani AH, Gao F, Liu J, Fu G, Li G (2016) Transferring clubroot resistance from Chinese cabbage (*Brassica rapa*) to canola (*B. napus*). Can J Plant Pathol 38(1):82–90. https://doi.org/10.1080/07060661.2016.1141799

Hirani AH, Gao F, Liu J, Fu G, Wu C, McVetty PBE, Duncan RW, Li G (2018) Combinations of independent dominant loci conferring Clubroot resistance in all four turnip accessions (Brassica rapa) from the European Clubroot differential set. Front Plant Sci 9:1628. https://doi.org/10.3389/fpls.2018.01628

Horikoshi N, Tairako K (2002) Occurrence of clubroot in Japanese radish caused by *Plasmodiophora brassicae* Woron in Fukushima. Annu Rep Soc Plant Prot North Jpn 53:58–60

Hou CT, Forman RJ (2000) Growth inhibition of plant pathogenic fungi by hydroxy fatty acids. J Indust Microbiol Biochem 24:275–276

Howell PM, Marshall DF, Lydiate DJ (1996) Towards developing inter-varietal substitution lines in *Brassica napus using* marker-assisted selection. Genome 39:348–358

Hurgobin B, Golicz AA, Bayer PE, Chan CKK, Tirnaz S, Dolatabadian A, Schiessl SV, Samans B, Montenegro JD, Parkin IAP, Pires JC, Chalhoub B, King GJ, Snowdon R, Batley J, Edwards D (2018) Homoeologous exchange is a major cause of gene presence/absence variation in the amphidiploid *Brassica napus*. Plant Biotechnol J 16:1265–1274

Hwang BH, Bae H, Lim HS, Kim KB, Kim SJ, Im MH, Park BS, Kim J (2010) Overexpression of polygalacturonase-inhibiting protein 2 (PGIP2) of Chinese cabbage (*Brassica rapa* ssp. *pekinensis*) increased resistance to the bacterial pathogen *Pectobacterium carotovorum* ssp. *carotovorum*. PCTOC 103:293–305

Hwang SF, Ahmed HU, Zhou Q, Strelkov SE, Gossen BD, Peng G, Turnbull D (2011) Influenceof cultivar resistance and inoculum density on root hair infection of canola (*Brassica napus*) by *Plasmodiophora brassicae*. Plant Pathol 60:820–829

Hwang SF, Strelkov SE, Feng J, Gossen BD, Howard RJ (2012) *Plasmodiophora brassicae*: a review of an emerging pathogen of the Canadian canola (*Brassica napus*) crop. Mol Plant Pathol 13:105–113. https://doi.org/10.1111/j.1364-3703.2011.00729.x

Inomata N (1993) Crossability and cytology of hybrid progenies in the cross between *Brassica campestris* and three wild relatives of *B. oleracea, B. bourgeaui, B. cretica and B. montana*. Euphytica 69:7–17. https://doi.org/10.1007/BF00021721

Inomata N (2002) A cytogenetic study of the progenies of hybrids between *Brassica napus* and *B. oleracea, B. bourgeaui, B. cretica* or *B. montana*. Plant Breed 121:174–176. https://doi.org/10.1046/j.1439-0523.2002.00695.x

Jafari M, Shams-Bakhsh M (2018) Preliminary results of an attempt to produce resistance to turnip mosaic virus in transgenic canola (*Brassica napus*). Iran J Virol 12:25–33

Jafary H, Albertazzi G, Marcel TC, Niks RE (2008) High diversity of genes for nonhost resistance of barley to heterologous rust fungi. Genet 178:2327–2339. https://doi.org/10.1534/genetics.107.077552

Jejelowo OA, Conn KL, Tewari JP (1991) Relationship between conidial concentration, germling growth, and phytoalexin production by *Camelina sativa* leaves inoculated with *Alternaria brassicae*. Mycol Res 95:928–934. https://doi.org/10.1016/S0953-7562(09)80089-0

Jeromela AM, Mikic AM, Vujic S, Cupina B, Krstic Đ, Dimitrijevic A, Vasiljevic S, Mihailovic V, Cvejic S, Miladinovic D (2017) Potential of legume–Brassica intercrops for forage production and green manure: encouragements from a temperate southeast European environment. Front Plant Sci 8:312

Jeuken MJW, Pelgrom K, Stam P, Lindhout P (2008) Efficient QTL detection for nonhost resistance in wild lettuce: backcross inbred lines versus F2 population. Theor Appl Genet 116:845–857. https://doi.org/10.1007/s00122-008-0718-2

Jiang JJ, Zhao XX, Tian W, Li TB, Wang YP (2009) Intertribal somatic hybrids between *Brassicanapus* and *Camelina sativa* with high linolenic acid content. Plant Cell Tissue Organ Culture (PCTOC) 99:91–95

Jiang HC, Feng YT, Bao L, Li X, Gao GJ, Zhang QL, Xiao JH, Xu CG, He YQ (2012) Improving blast resistance of Jin 23B and its hybrid rice by marker-assisted gene pyramiding. Mol Breed 30:1679–1688

Jiao Y, Wickett NJ, Ayyampalayam S, Chanderbali AS, Landherr L, Ralph PE, Tomsho LP, Hu Y, Liang H, Soltis PS, Soltis DE, Clifton SW, Schlarbaum SE, Schuster SC, Ma H, Leebens-Mack J, de Pamphili CW (2011) Ancestral polyploidy in seed plants and angiosperms. Nature 473:97–100

Johnston TD (1970) A new factor for resistance to club root in *Brassica napus* L. Plant Pathol 12:156–158

Jouradan PS, Salazar E (1993) *B. carinata* resynthesized by protoplast fusion. Theor Appl Genet 86:567–572

Jourdan PS, Earle ED, Mutscher MA (1989) Atrazine-resistant cauliflower obtained by somatic hybridization between *Brassica oleracea* and ATR-*B. napus*. Theor Appl Genet 78:271–279. https://doi.org/10.1007/BF00288810

Kachroo A, Kachroo P (2009) Fatty acid-derived signals in plant defense. Ann Rev Phytopathology 47:153–176

Kamble S, Mukherjee PK, Eapen S (2016) Expression of an endochitinase gene from *Trichoderma virens* confers enhanced tolerance to Alternaria blight in transgenic *Brassica juncea* (L.) Czern & Coss lines. Physiol Mol Biol Plants 22(1):69–76. https://doi.org/10.1007/s12298-016-0340-8

Kamei A, Tsuro M, Kubo N, Hirai M (2007) QTL analysis of clubroot resistance in radish (*Raphanus sativus* L.). In: Proceedings of the 111th meeting of the Japanese Society of Breeding. Breed Res 9, p 83

Kaneko Y, Bang SW (2014) Interspecific and intergeneric hybridization and chromosomal engineering of Brassicaceae crops. Breed Sci 64:14–22

Kanrar S, Venkateswari JC, Kirti PB, Chopra VL (2002) Transgenic expression of hevein, the rubber tree lectin, in Indian mustard confers protection against *Alternaria brassicae*. Plant Sci 162:441–448

Karpechenko GD (1928) Polyploid hybrids of Raphanus sativus L. × Brassica oleracea L. Z Indukt Abstamm Vererbungsl 48:1–85

Katche E, Quezada-Martinez D, Katche EI, Vasquez-Teuber P, Mason AS (2019) Interspecific hybridization for *Brassica* crop improvement. Crop Breed Genet Genom 1(1–32):e190007. https://doi.org/10.20900/cbgg20190007

Katiyar RK, Chopra VL (1990) Somaclonally induced earliness in a *Brassica juncea* germplasm accession with field resistance to important diseases. Plant Breed 104:262–264

Kaur S, Singh G, Banga SS (2007) Documenting variation in *Alternaria brassicae* isolates based on conidial morphology, fungicidal sensitivity and molecular profile. In Proceeding of the 12th international rapeseed congress, 26–30 march. Wuhan, China 4, pp 87–89

Kearney B, Staskawicz BJ (1990) Widespread distribution and fitness contribution of *Xanthomonas campestris* avirulence gene *avrBs2*. Nature 346:385–386. https://doi.org/10.1038/346385a0

Kessman H, Staub T, Ligon J, Oostendorp M, Ryals J (1994) Activation of systemic acquired disease resistance in plants. Eur J Plant Pathol 100:359–369

Kitamura S (1958) Cultivars of radish and their change. In: Nishiyama I (ed) Japanese radish. Jpn Soc Prom Sci, Tokyo, pp 1–19

Kolattukudy PE (1974) Biosynthesis of a hydroxy fatty acid polymer, cutin. Identification and biosynthesis of 16-oxo-9- or 10- hydroxypalmitic acid, a novel compound in *Vicia faba*. Biochemistry 13:1354–1363

Kolattukudy PE (1980) Biopolyester membranes of plants: cutin and suberin. Science 208:990–1000

Kolte SJ (1985a) Disease management strategies for rapeseed-mustard crops in India—a review. Agril Rev 6:81–88

Kolte SJ (1985b) Diseases of annual edible oilseed crops.Rapeseed-mustard and sesame diseases Vol. II, CRC Press Inc; Boca Raton, p 135.

Kolte SJ (1987) Important diseases of rapeseed and mustard in oilseed crops: Niger and Rapeseed Mustard. Proceedings of the Third Oil Crops Network, held in Addis Ababa, Ethiopia, 6-10 October, 1986, 99–106 pp

Krishnia SK, Saharan GS, Singh D (2000) Genetics of Alternaria blight resistance in inter and intraspecific crosses of *Brassica juncea* and *B. carinata*. Ann Bio 16(2):212–216

Kuginuki Y, Ajisaka H, Yui M, Yoshikawa H, Hida K, Hirai M (1997) RAPD markers linked to a clubroot-resistance locus in *Brassica rapa* L. Euphytica 98:149–154

Kuginuki Y, Yoshikawa H, Hirai M (1999) Variation in virulence of *Plasmodiophora brassicae* in Japan tested with clubroot-resistant cultivars of Chinese cabbage (*Brassica rapa* L. ssp. *pekinensis*). Eur J Plant Pathol 105:327–332

Kumar S, Saharan GS (2002) Sources of multiple disease resistance in *Brassica* spp. J Mycol Plant Pathol 32:184–188

Kumar D, Maurya N, Bharati YK, Kumar A, Kumar K, Srivastava K, Chand G, Singh SK, Mishra RK, Kumar A (2014) Alternaria blight of oilseed *Brassicas*: a comprehensive review. African J Microbiol Res 8:2816–2829. https://doi.org/10.5897/AJMR2013.6434

Lammerink J (1964) Pathologic specialisation of *Plasmodiophora brassicae* Worn. NZ J Agric Res 7:37–41

Lammerink J (1969) Inter-specific transfer of clubroot resistance from *Brassica campestris* L to *B. napus* L. NZ J Agric Res 13:105–110

Landry BS, Hubert N, Crete R, Chiang MS, Lincoln SE, Etoh T (1992) A genetic map for *Brassicaoleracea* based on RFLP markers detected with expressed DNA sequences and mapping of resistance genes to race 2 of *Plasmodiophora brassicae* (Worn). Genome 35:409–420

LeBoldus JM, Manolii VP, Turkington TK, Strelkov SE (2012) Adaptation to *Brassica* host genotypes by a single-spore isolate and population of *Plasmodiophora brassicae* (clubroot). Plant Dis 96:833–838

Liang Z, Chen K, Li T, Zhang Y, Wang Y, Zhao Q, Liu J, Zhang H, Liu C, Ran Y, Gao C (2017) Efficient DNA-free genome editing of bread wheat using CRISPR/Cas9 ribonucleoprotein complexes. Nat Commun 8:14261. https://doi.org/10.1038/ncomms14261

Liu H, Guo X, Naeem MS, Liu D, Xu L, Zhang W, Tang G, Zhou W (2011) Transgenic *Brassica napus* L. lines carrying a two gene construct demonstrate enhanced resistance against *Plutella xylostella* and *Sclerotinia sclerotiorum*. PCTOC 106:143–151

Liu C, Yang Z, He P, Munir S, Wu Y, Ho H, He Y (2018) Deciphering the bacterial and fungal communities in clubroot-affected cabbage rhizosphere treated with *Bacillus subtilis* XF-1. Agric Ecos Environ 256:12–22

Ma M, Yan Y, Huang L, Chen M, Zhao H (2012) Virus-induced gene-silencing in wheat spikes and grains and its application in functional analysis of HMW-GS-encoding genes. BMC Plant Biol 12:1. https://doi.org/10.1186/1471-2229-12-1

MacDonald MV, Ingram DS (1985) *In vitro* selection for resistance to *Alternaria brassicicola* in *Brassica napus* spp. oleifera (winter oilseed rape) using partially purified culture filtrates. Cruciferae Newsl 10:97–100

Mandal S, Rajarammohan S, Kaur J (2018) *Alternaria brassicae* interactions with the model *Brassicaceae* member *Arabidopsis thaliana* closely resembles those with mustard (*Brassica juncea*). Physiol Mol Biol Plants 24:51–59. https://doi.org/10.1007/s12298-017-0486-z

Manzanares-Dauleux MJ, Delourme R, Baron F, Thomas G (2000a) Mapping of one major gene and of QTLs involved in resistance to clubroot in *Brassica napus*. Theor Appl Genet 101:885–891. https://doi.org/10.1007/s001220051557

Manzanares-Dauleux MJ, Delourme R, Baron F, Thomas G (2000b) QTLs involved in mapping of one major gene and of resistance to clubroot in *Brassica napus*. Theor Appl Genet 101:885–891. https://doi.org/10.1007/s001220051557

Martin GB, Brommonschenkel SH, Chunwongse J, Frary A, Ganal MW, Spivy R, Wu T, Earle ED, Tanksley SD (1993) Map-based cloning of a protein kinase gene conferring disease resistance in tomato. Science 262:1432–1436

Mason AS, Nelson MN, Yan G, Cowling WA (2011) Production of viable male unreduced gametes in Brassica interspecific hybrids is genotype specific and stimulated by cold temperatures. BMC Plant Biol 11:103. https://doi.org/10.1186/1471-2229-11-103

Matsumoto E, Yasui C, Ohi M, Tsukada M (1998) Linkage analysis of RFLP markers for clubroot resistance and pigmentation in Chinese cabbage (*Brassica rapa* ssp. *pekinensis*). Euphytica 104:79–86

Matsumoto E, Hayashida N, Sakamoto K, Ohi M (2005) Behavior of DNA markers linked to a clubroot resistance gene in segregating populations of Chinese cabbage (*Brassica rapa* ssp *pekinensis*). J Jpn Soc Hortic Sci 74:367–373. https://doi.org/10.2503/jjshs.74.367

McDonald BA, Linde C (2002) Pathogen population genetics, evolutionary potential and durable resistance. Ann Rev Phytopathology 40:349–379. https://doi.org/10.1146/annurev.phyto.40.120501.101443

McHale L, Tan X, Koehl P, Michelmore RW (2006) Plant NBS-LRR proteins: adaptable guards. Genome Biol 7:212. https://doi.org/10.1186/gb-2006-7-4-212

McRoberts N, Lennard JH (1996) Pathogen behaviour and plant cell reactions in interactions between Alternaria species and leaves of host and nonhost plants. Plant Pathol 45(4):742–752

Meena PD, Chattopadhyay C, Kumar VR, Meena RL, Rana US (2005) Spore behaviour in atmosphere and trends in variability of *Alternaria brassicae* population in India. J Mycol Plant Pathol 35:511

Meena PD, Rani A, Meena R, Sharma P, Gupta R, Chowdappa P (2012) Aggressiveness, diversity and distribution of *Alternaria brassicae* isolates infecting oilseed *brassica* in India. African J Microbiol Res 6:5249–5258

Meena PD, Mehta Naresh, Saharan GS (2019) *Alternaria-Brassica* associations: research progress and prospects. Agric Res J 56(4): 600-612.

Mei J, Qian L, Disi JO, Yang X, Li Q, Li J, Frauen M, Cai D, Qian W (2011) Identification of resistant sources against *Sclerotinia sclerotiorum* in *Brassica* species with emphasis on *B. oleracea*. Euphytica 177:393–399

Mei J, Ding Y, Lu K, Wei D, Liu Y, Disi JO, Li J, Liu L, Liu S, McKay J, Qian W (2013) Identification of genomic regions involved in resistance against *Sclerotinia sclerotiorum* from wild *Brassica oleracea*. Theor Appl Genet 126:549–556. https://doi.org/10.1007/s00122-012-2000-x

Mei J, Liu Y, Wei D, Wittkop B, Ding Y, Li Q, Li J, Wan H, Li Z, Ge X, Frauen M, Snowdon RJ, Qian W, Friedt W (2015) Transfer of Sclerotinia resistance from wild relative of *Brassica oleracea* into *Brassica napus* using a hexaploidy step. Theor Appl Genet 128:639–644

Mei J, Shao C, Yang R, Feng Y, Gao Y, Ding Y, Li J (2020) Qian W (2020) Introgression and pyramiding of genetic loci from wild *Brassica oleracea* into *B. napus* for improving *Sclerotinia* resistance of rapeseed. Theor Appl Genet 133:1313–1319. https://doi.org/10.1007/s00122-020-03552-w

Meng J, Lu M (1993) Genotype effects of *Brassica napus* on its reproductive behaviour after pollination with *B. juncea*. Theor Appl Genet 87:238–242. https://doi.org/10.1007/BF00223771

Mitchell-Olds T, James RV, Palmer MJ, Williams PH (1995) Genetics of *Brassica rapa* (syn. *Campestris*). 2. Multiple disease resistance to three fungal pathogens: *Peronospora parasitica*, *Albugo candida* and *Leptosphaeria maculans*. Heredity 75:362–369

Mohanty A, Chrungu B, Verma N, Shivanna KR (2009) Broadening the genetic base of crop *Brassicas* by production of new intergeneric hybrid. Czech J Genet Pl Breed 45:117–122. https://doi.org/10.17221/35/2009-CJGPB

Momotaz A, Kato M, Kakihara F (1998) Production of inter-generic hybrids between *Brassica* and *Sinapis* species by means of embryo rescue techniques. Euphytica 103:123–130. https://doi.org/10.1023/A:1018331528368

Mondal KK, Chatterjee SC, Viswakarma N, Bhatacharya RC, Grover A (2003) Chitinase mediated inhibitory activity of *Brassicas* transgenic on growth of *Alternaria brassicae*. Curr Microbiol 47:171–173

Mondal KK, Bhattacharya RC, Koundal KR, Chatterjee SC (2007) Transgenic Indian mustard (*Brassica juncea*) expressing tomato glucanase leads to arrested growth of *Alternaria brassicae*. Plant Cell Reptr 26:247–252. https://doi.org/10.1007/s00299-006-0241-3

Moriguchi K, Kimizuka-Takagi C, Ishii K, Nomura K (1999) A genetic map based on RAPD, RFLP, isozyme, morphological markers and QTL analysis for clubroot resistance in *Brassica oleracea*. Breed Sci 49: 257–265. https://doi.org/10.1270/jsbbs.49.257

Murakami H, Tsushima S, Akimoto T, Murakami K, Goto I, Shishido Y (2000) Effects of growing leafy daikon (*Raphanus sativus*) on populations of *Plasmodiophora brassicae* (clubroot). Plant Pathol 49:584–589

Nagaharu U (1935) Genome analysis in *Brassica* with special reference to the experimental formation of *B. napus* and peculiar mode of fertilization. Japan J Bot 7:389–452

Narasimhulu SB, Kirti PB, Bhatt SR (1994) Intergeneric protoplast fusion between *Brassica carinata* and *Camelina sativa*. Plant Cell Rep 13:657–660

Navabi ZK, Huebert T, Sharpe AG, Carmel M, Bancroft I, Isobel Parkin AP (2013) Conserved microstructure of the *Brassica* B genome of *Brassica nigra* in relation to homologous regions of *Arabidopsis thaliana, B. rapa* and *B. oleracea*. BMC Genom 14:250. https://doi.org/10.1186/1471-2164-14-250

Neik TX, Barbetti MJ, Batley J (2017) Current status and challenges in identifying disease resistance genes in *Brassica napus*. Front Plant Sci 8:1788. https://doi.org/10.3389/fpls.2017.01788

Niemann J, Wojciechowski A, Janowicz J (2012) Broadening the variability of quality traits in rapeseed through inter specific hybridization with an application of immature embryo culture. J Biotech Comput Biol Bionanotechnol 93:109–115

Niemann J, Wojciechowski A, Jedryczka M, Kaczmarek J (2013) Interspecific hybridization as a tool for broadening the variability of useful traits in rapeseed (*Brassica napus* L.). In: Branca F, Tribulato A (eds) Proceedings of the VIth. IS on *Brassicas* and XVIIIth crucifer genetics workshop, Catania: Acta Hort, p 1005

Niks RE, Marcel TC (2009) Nonhost and basal resistance: how to explain specificity? New Phytol 182:817–828. https://doi.org/10.1111/j.1469-8137.2009.02849.x

Nishi S, Wata JK, Toda M (1959) On the breeding of interspecific hybrids between two genomes. 'C' and 'B' of *Brassica* through the application of embryo culture techniques. Japan J Breed 8:215–222

Nowicki M, Nowakowska M, Niezgoda A, Kozik EU (2012) Alternaria black spot of crucifers: symptoms, importance of disease, and perspectives of resistance breeding. Veg Crops Res Bull 76:5–19

O'Neill CM, Murata T, Morgan CL, Mathia RJ (1996) Expression of the C3-C4 intermediate character in somatic hybrids between *B. napus* and the C3-C4 species *Moricandida arvensis*. Theor Appl Genet 93:1234–1241

Oh IS, Park AR, Bae MS, Kwon SJ, Kim YS, Lee JE, Kang NY, Lee S, Cheong H, Park OK (2005) Secretome analysis reveals an *Arabidopsis* lipase involved in defense against *Alternaria brassicicola*. Plant Cell 17:2832–2847

Osaki K, Tanaka S, Ito S (2008) Pathogenicity of *Plasmodiophora brassicae* populations from small, spheroid, resistant-type clubroot galls on roots of clubroot-resistant cultivars of Chinese cabbage (*Brassica rapa* L. subsp. *pekinensis*). J Gen Plant Pathol 74:242–245

Otani H, Kohnobe A, Narita M, Shiomi H, Kodama M, Kohmoto K (2001) A new type of host-selective toxin, a protein from *Alternaria brassicicola*. In: Mayama S, Leach JE, Tsuyumu S, Keen NT (eds) Delivery and perception of pathogen signals in plants. APS Press, St. Paul, MN, pp 68–76

Pan X, Caldwell CD, Falk KC, Lada R (2012) The effect of cultivar, seeding rate and applied nitrogen on *Brassica carinata* seed yield and quality in contrasting environments. Can J Plant Sci 92:961–971. https://doi.org/10.4141/cjps2011-169

Patni CS, Kolte SJ, Awasthi RP (2005) Cultural variability of *Alternaria brassicae*, causing Alternaria blight of mustard. Ann Plant Physiol 19:231–242

Penaud A (1999) Chemical control and yield losses caused by *Erysiphe cruciferarum* on oilseed rape in France. In: 10th International rapeseed congress. Canberra, Australia: GCIRC. http://www.regional.org.au/au/gcirc/3/327.htm

Peng G, Pageau D, Strelkov SE, Lahlali R, Hwang SF, Hynes RK (2013) Assessment of crop rotation, cultivar resistance and *Bacillus subtilis* biofungicide for control of clubroot on canola. Acta Hortic 1005:591–598

Peng G, Falk KC, Gugel RK, Franke C, Yu F, James B, Strelkov SE, Hwang SF, McGregor L (2014) Sources of resistance to *Plasmodiophora brassicae* (clubroot) pathotypes virulent on canola. Can J Plant Pathol 36:89–99. https://doi.org/10.1080/07060661.2013.863805

Petkova M, Dimova M, Dimova D, Bistrichanov S (2014) Effect of gamma-irradiation on the fatty acid composition and susceptibility to powdery mildew (*Erysiphe cruciferarum*) of oilseed rape plants. Agril Sci Technol 6(4):413–416

Piao ZY, Deng YQ, Choi SR, Park YJ, Lim YP (2004) SCAR and CAPS mapping of *CRb*, a gene conferring resistance to *Plasmodiophora brassicae* in Chinese cabbage (*Brassica rapa* ssp. *pekinensis)*. Theor Appl Genet 108:1458–1465. https://doi.org/10.1007/s00122-003-1577-5

Pinot F, Benveniste IJ, Salau P, Loreau O, Noe J, Schreiber L, Durst F (1999) Production in vitro by the cytochrome P450 CYP94A1 of major C18 cutin monomers and potential messengers in plant–pathogen interactions: enantio selectivity studies. Biochem J 342:27–32

Plieske J, Struss D, Robbelen G (1998) Inheritance of resistance derived from the B-genome of *Brassica* against *Phoma lingam* in rapeseed and the development of molecular markers. Theor Appl Genet 97:929–936. https://doi.org/10.1007/s001220050973

Poveda J (2020) *Trichoderma parareesei* favors the tolerance of rapeseed (*Brassica napus* L.) to salinity and drought due to a *chorismate mutase*. Agron 10:118

Poveda J, Hermosa R, Monte E, Nicolás C (2019) *Trichoderma harzianum* favours the access of arbuscular mycorrhizal fungi to non-host Brassicaceae roots and increases plant productivity. Sci Rep 9:1–11

Poveda J, Francisco M, Cartea ME, Velasco P (2020) Development of transgenic *Brassica* crops against biotic stresses caused by pathogens and arthropod pests. Plan Theory 9:1664–1689

Prakash S (1973) Non-homologous meiotic pairing in the A and B genomes of *Brassica*: its breeding significance in the production of variable amphidiploids. Genet Res 21:133–137

Prakash S, Chopra VL (1988) Introgression of resistance to shattering in *Brassica napus* from *Brassica juncea* through non-homologous recombination. Plant Breed 101:167–168

Primard C, Vedel F, Mathieu C, Pelletier G, Chevre AM (1988) Interspecific somatic hybridization between *Brassica napus* and *Brassica hirta* (*Sinapis alba* L.). Theor Appl Genet 75:546–552

Qiu L, Wu T, Dong H, Wu L, Cao J, Huang L (2013) High-level expression of sporamin in transgenic Chinese cabbage enhances resistance against diamondback moth. Plant Mol Biol Rep 31:657–664

Rahimi M, Bahrani A (2011) Effect of gamma irradiation on qualitative and quantitative characteristics of canola. Middle East J Sci Res 8:519–525

Rahman A, Das ML, Howlider MAR, Mansur MA (1987) Promising mutants in *Brassica campestris*. Mutat Breed Newsl 29:14–15

Rahman H, Shakir A, Jakir Hasan M (2011) Breeding for clubroot resistant spring canola (*Brassica napus* L.) for the Canadian prairies: can the European winter canola cv. Mendel be used as a source of resistance? Can J Plant Sci 91:447–458. https://doi.org/10.4141/cjps10073

Rahman H, Peng G, Yu F, Falk KC, Kulkarni M, Selvaraj G (2014) Genetics and breeding for clubroot resistance in Canadian spring canola (*Brassica napus* L.). Can J Plant Pathol 36:122–134. https://doi.org/10.1080/07060661.2013.862571

Rajarammohan S, Kumar A, Gupta V, Pental D, Pradhan AK, Kaur J (2017) Genetic architecture of resistance to *Alternaria brassicae* in *Arabidopsis thaliana*: QTL mapping reveals two major resistance-conferring loci. Front Plant Sci 8:260

Rana K, Atri C, Gupta M, Akhatar J, Sandhu PS, Kumar N, Jaswal R, Barbetti MJ, Banga SS (2017) Mapping resistance responses to Sclerotinia infestation in introgression lines of *Brassica juncea* carrying genomic segments from wild Brassicaceae *B. fruticulosa*. Sci Reptr 7:5904. https://doi.org/10.1038/s41598-017-05992-9

Rana K, Atri C, Akhatar J, Kaur R, Goyal A, Singh MP, Kumar N, Sharma A, Sandhu PS, Kaur G, Barbetti MJ, Banga SS (2019) Detection of first marker trait associations for resistance against *Sclerotinia sclerotiorum* in *Brassica juncea–Erucastrum cardaminoides* introgression lines. Front Plant Sci 10:1015. https://doi.org/10.3389/fpls.2019.01015

Rani S, Sharma V, Hada A, Bhattacharya RC, Koundal KR (2017) Fusion gene construct preparation with lectin and protease inhibitor genes against aphids and efficient genetic transformation of *Brassica juncea* using cotyledons explants. Acta Physiol Plant 39:115

Ricarova V, Kazda J, Singh K, Rysanek P (2016) Clubroot caused by *Plasmodiophora brassicae* Worn.: a review of emerging serious disease of oilseed rape in the Czech Republic. Plant Prot Sci 52:71–86. https://doi.org/10.17221/87/2015-PPS

Ripley V, Thorpe M, Iler S, Mizier K, Beversdorf WD (1992) Isozyme analysis as a tool for introgression of *Sinapis alba* germplasm into *Brassica napus*. Theor Appl Genet 84:403–410. https://doi.org/10.1007/BF00229500

Rocherieux J, Glory P, Giboulot A, Boury S, Barbeyron G, Thomas G, Manzanares-Dauleux MJ (2004) Isolate-specific and broad-spectrum QTLs are involved in the control of clubroot in *Brassica oleracea*. Theor Appl Genet 108:1555–1563. https://doi.org/10.1007/s00122-003-1580-x

Ryschka U, Schumann G, Klocke E, Scholze P, Neumann M (1996) Somatic hybridization in Brassicaceae. Acta Hortic 407:201–208. http://www.actahort.org/books/407/407_24.htm

Saharan GS (1984) A review of research on rapeseed-mustard pathology in India. Ann Workshop AICORPO, ICAR, Jaipur, 6–10 (August 1984)

Saharan GS (1992a) Disease resistance. In: Labana KS, Banga SS, Banga SK (eds) Breeding oilseed *Brassicas*. Narosa Publishing House, New Delhi. pp 181–200

Saharan GS (1992b) Management of rapeseed and mustard diseases. Advancements of oilseed research, Vol. 1. Scientific Publication, India: pp 152–533

Saharan GS (1997) Disease resistance. In: Kalia HR, Gupta SC (eds) Recent advances in oilseed *Brassica*, vol 12. Kalyani Pub, Ludhiana, pp 233–259

Saharan GS, Chand JN (1988) Diseases of rapeseed and mustard. In: Directorate of oilseed crops (in Hindi). Directorate of Publication, Haryana Agric Univ Press, Hisar, India. Chapter 3, pp 84–91

Saharan GS, Naresh M, Sangwan MS (2003) Nature and mechanism of resistance to Alternaria blight in rapeseed-mustard system. Ann Rev Plant Pathol 2:85–128

Saharan GS, Naresh M, Sangwan MS (2005) Development of disease resistance in rapeseed-mustard. In: Saharan GS, Naresh M, Sangwan MS (eds) Diseases of oilseed crops. Indus Publishing Co., New Delhi, pp 561–617

Saharan GS, Naresh M, Verma PR, Meena PD (2015) Analysis of variability determinants in *Alternaria-Brassica*-interactions. J Oilseed Brassica 6:126–141

Saharan GS, Mehta Naresh, Meena PD (2016) Alternaria blight of crucifers: biology, ecology and management. Springer Verlag, Singapore: p 326. ISBN:978-981-10-0019-5.

Saharan GS, Mehta Naresh, Meena PD (2017) Downy mildew disease of crucifers: biology, ecology and disease management. Springer Verlag, Singapore, LVI, p 357, ISBN 978-981-10-7499-8.

Saharan GS, Naresh M, Meena PD (2019) Powdery mildew disease of crucifers: biology. Ecology and disease management. Springer Verlag, Singapore, p 362. isbn:978-981-13-9852-0

Saito M, Kubo N, Matsumoto S, Suwabe K, Tsukada M, Hirai M (2006) Fine mapping of the clubroot resistance gene, Crr3, in *Brassica rapa*. Theor Appl Genet 114:81–91. https://doi.org/10.1007/s00122-006-0412-1

Sakai T, Liu HJ, Iwabuchi M, Kohno-Murase J (1996) Introduction of a gene from fertility restored radish (*Raphanus sativus*) into *Brassica napus* by fusion of X-irradiated protoplasts from a radish restorer line and iodacetoamide-treated protoplasts from a cytoplasmic male-sterile cybrid of *B. napus*. Theor Appl Genet 93:373–379

Sakamoto K, Saito A, Hayashida N, Taguchi G, Matsumoto E (2008) Mapping of isolate-specific QTLs for clubroot resistance in Chinese cabbage (*Brassica rapa* L. ssp. *pekinensis*). Theor Appl Genet 117:759–767. https://doi.org/10.1007/s00122-008-0817-0

Sang X, Jue D, Yang L, Bai X, Chen M, Yang Q (2013) Genetic transformation of *Brassica napus* with MSI-99m gene increases resistance in transgenic plants to *Sclerotinia sclerotiorum*. Mol Plant Breed 4:247–253

Scheben A, Wolter F, Batley J, Puchta H, Edwards DC (2017) Towards CRISPR/Cas crops – bringing together genomics and genome editing. New Phytol 216:682–698. https://doi.org/10.1111/nph.14702

Schranz ME, Lysak MA, Mitchell-Olds T (2006) The ABC's of comparative genomics in the Brassicaceae: building blocks of crucifer genomes. Trends Plant Sci 11:535–542. https://doi.org/10.1016/j.tplants.2006.09.002

Sharma TR (1990) Induction and transfer of resistance to Alternaria blight of mustard caused by *Alternaria brassicae* and *A. brassicicola*. Ph D thesis, Himachal Pradesh Krishi Vishva Vidyalaya, Palampur.

Sharma TR, Singh BM (1992) Transfer of resistance to *Alternaria brassicae* in *Brassica juncea* through interspecific hybridization among *Brassicas*. J Genet Breed 46(4):373–378

Sharma TR, Singh BM (1995) Generation and evaluation of somaclones of Brassica juncea for resistance to Albugo candida and Alternaria brassicae. Proc Indian Natl Sci Acad B 61 (2):155–162

Sharma TR, Tewari JP (1995) Detection of genetic variation in *Alternaria brassicae* by RAPD fingerprints. J Plant Biochem Biotech 4:105–107

Sharma TR, Tewari JP (1998) RAPD analysis of three Alternaria species pathogenic to crucifers. Mycol Res 102:807–814

Sharma BR, Dhiman JS, Thakur JC, Singh A, Bajaj KL (1991) Multiple disease resistance in cauliflower. Adv Hort Sci 5:30–34

Sharma G, Kumar DV, Haque A, Bhat SR, Shyam P, Chopra VL (2002) *Brassica* coenospecies: a rich reservoir for genetic resistance to leaf spot caused by *Alternaria brassicae*. Euphytica 125:411–419. https://doi.org/10.1023/A:1016050631673

Sharma BB, Kalia P, Yadava DK, Dingh D, Sharma TR (2016) Genetics and molecular mapping of black rot resistance locus Xca1bc on chromosome B-7 in Ethiopian mustard (*Brassica carinata* A. Braun). PLoS One 11:e0152290. https://doi.org/10.1371/journal.pone.0152290

Sharma BB, Kalia P, Singh D (2017) Sharma TR (2017) Introgression of black rot resistance from *Brassica carinata* to cauliflower (*Brassica oleracea botrytis* group) through embryo rescue. Front Plant Sci 8:1255. https://doi.org/10.3389/fpls.2017.01255

Shekhawat K, Rathore SS, Premi OP, Kandpal BK, Chauhan JS (2012) Advances in agronomic management of Indian mustard (*Brassica juncea* (L.) Czernj. Cosson): an overview. Intl J Agron 2012:1–14

Shin YH, Lee SH, Park YD (2020) Development of mite (*Tetranychus urticae*)-resistant transgenic Chinese cabbage using plant-mediated RNA interference. Hortic Environ Biotechnol 61:305–315

Shivpuri A, Chipa HP, Gupta RBL, Sharma KN (1997) Field evaluation of mustard genotypes against white rust, powdery mildew and stem rot. Ann Arid Zone Res 36:387–390

Shyam SY, David L, Philip C (2007) Lentil an ancient crop for modern times, vol 472. Springer, Dordrecht, p 259

Siemens J, Nagel W, Ludwig-Muller J, Sarcristán MD (2002) The interaction of *Plasmodiophora brassicae* and *Arabidopsis thaliana*: parameters for disease quantification and screening mutant lines. J Phytopathol 150:592–605

Siemens J, Keller I, Sarx J, Kunz S, Schuller A, Nagel W, Schmulling T, Parniske M, Ludwig-Muller J (2006) Transcriptome analysis of *Arabidopsis* clubroots indicate a key role for cytokinins in disease development. Mol Plant-Microbe Interact 19:480–494. https://doi.org/10.1094/MPMI-19-0480

Sigareva MA, Earle ED (1997) Direct transfer of a cold-tolerant Ogura male sterile cytoplasm into cabbage (*Brassica oleracea* ssp. *capitata*) via protoplast fusion. Theor Appl Genet 94:213–220. https://doi.org/10.1007/s001220050402

Sigareva MA, Earle ED (1999) Production and characterization of somatic hybrids between *Camelina sativa* and rapid-cycling *B. oleracea*. Theor Appl Genet 98:164–170

Sikdar SR, Chatterjee G, Das S, Sen SK (1990) "Erussuca" the intergeneric fertile somatic hybrid developed through protoplast fusion between *Eruca sativa* lam. And *Brassica juncea* (L) Czern. Theor Appl Genet 79:561–567. https://doi.org/10.1007/BF00226168

Singh D, Lodhi GP, Singh H (1986) Studies on genetic variability for various attributes in different mating system F2 and F3 populations of rapeseed. Oil Crops Newsl 58:60

Singh D, Singh R, Singh H, Yadav RC, Yadav N, Barbetti MJ, Salisbury P, Nimbal S, Chattopadhyay C, Kumar A (2007) Cultural and morphological variability in *Alternaria brassicae* isolates of Indian mustard (*Brassica juncea* L. Czern & Coss.). In Proc 12th Intl rapeseed congress, 26–30 march. Wuhan, China, pp 158–160

Singh R, Singh D, Salisbury P, Barbetti MJ (2010) Field evaluation of Indian and exotic oilseed *Brassica napus* and *B. juncea* germplasm against Sclerotinia stem rot. Indian J Agric Sci 80:1067–1071

Singh R, Singh D, Singh H, Caxia L, Sivasithampram K, Salisbury P, Barbetti MJ (2011) Pathogenic variability in *Alternaria brassicae* incitant of Alternaria blight of *Brassica* species. In: Proc 13th Intl rapeseed congress, 5–9 June. Czech Republic pp, Prague, pp 1265–1270

Singh S, Dey SS, Bhatia R, Batley J, Kumar R (2018) Molecular breeding for resistance to black rot [*Xanthomonas campestris* pv. *Campestris* (Pammel) Dowson] in *Brassicas*: recent advances. Euphytica 214:196

Skoric D, Jocic S, Sakac Z, Lecic N (2008) Genetic possibilities for altering sunflower oil quality to obtain novel oils. Can J Physio Pharmacol 86:215–221

Skoropod WP, Tewari JP (1977) Field evaluation of the role of epicuticular wax in rapeseed and mustard in resistance to Alternaria blackspot. Can J Plant Sci 57:1001–1003

Snowdon RJ, Friedt W (2004) Molecular markers in *Brassica* oilseed breeding: current status and future possibilities. Plant Breed 123:1–8

Snowdon RJ, Winter H, Diestel A, Sacristan MD (2000) Development and characterization of *Brassica napus-Sinapis arvensis* addition lines exhibiting resistance to *Leptosphaeria maculans*. Theor Appl Genet 101:1008–1014. https://doi.org/10.1007/s001220051574

Soltis PS, Soltis DE (2000) The role of genetic and genomic attributes in the success of polyploids. Proc Natl Acad Sci U S A 97:7051–7057

Stitcher L, Mauch-Mani B, Metraux JP (1997) Systemic acquired resistance. Annu Rev Phytopathol 35:235–270

Strelkov SE, Tewari JP, Hartmann M, Orchard D (2005) Clubroot on canola in Alberta in 2003 and 2004. Can Plant Dis Surv 85:72–73

Strelkov SE, Tewari JP, Smith-Degenhardt E (2006) Characterization of *Plasmodiophora brassicae* populations from Alberta, Canada. Can J Plant Pathol 28:467–474. https://doi.org/10.1080/07060660609507321

Strelkov SE, Hwang SF, Howard RJ, Hartman M, Turkington TK (2011) Progress towards the sustainable management of clubroot [*Plasmodiophora brassicae*] of canola on the Canadian prairies. Prairie Soils Crops J 4:114–121

Suman A, Lal M, Singh AK, Gaur A (2006) Microbial biomass turnover in Indian subtropical soils under different sugarcane intercropping systems. Agron J 98:698–704. https://doi.org/10.2134/agronj2005.0173

Suwabe K, Tsukazaki H, Iketani H, Hatakeyama K, Fujimura M, Nunome T, Fukuoka H, Matsumoto S, Hirai M (2003) Identification of two loci for resistance to clubroot (*Plasmodiophora brassicae* Woronin) in *Brassica rapa* L. Theor Appl Genet 107:997–1002

Suwabe K, Tsukazaki H, Iketani H, Hatakeyama K, Kondo M, Fujimura M, Nunome T, Fukuoka H, Hirai M, Matsumoto S (2006) Simple sequence repeat-based comparative genomics between *Brassica rapa* and *Arabidopsis thaliana*: the genetic origin of clubroot resistance. Genetics 173:309–319

Svitashev S, Schwartz C, Lenderts B, Young JK, Mark Cigan A (2016) Genome editing in maize directed by CRISPR–Cas9 ribonucleoprotein complexes. Nat Commun 7:13274. https://doi.org/10.1038/ncomms13274

Syvanen AC (2005) Toward genome-wide SNP genotyping. Nat Genet 37:S5–S10

Tai TH, Dahlbeck D, Clark ET, Gajiwala P, Pasion R, Whalen MC, Stall RE, Staskawicz BJ (1999) Expression of the Bs2 pepper gene confers resistance to bacterial spot disease in tomato. Proc Natl Acad Sci U S A 96:14153–14158. https://doi.org/10.1073/pnas.96.24.14153

Taj G, Kumar A, Bansal KC, Garg GK (2004) Introgression of osmotin gene for creation of resistance against Alternaria blight by perturbation of cell cycle machinery. Indian J Biotechnol 3:291–298

Taj G, Agarwal P, Grant M, Kumar A (2011) Co-expression and in-silico interaction studies for inter-linking the activation of MAPK3 and*LOX* genes during pathogenesis of *Alternaria brassicae* in *Brassica juncea*. J Oilseed Brassica 2:13–20

Tang T, Chen G, Liu F, Bu C, Liu L, Zhao X (2019) Effects of transgenic glufosinate-tolerant rapeseed (*Brassica napus* L.) and the associated herbicide application on rhizospheric bacterial communities. Physiol Mol Plant Pathol 106:246–252

Taylor JD, Conway J, Roberts SJ, Astley D, Vicente JG (2002) Source and origin of resistance to *Xanthomonas campestris* pv. *Campestris* in *Brassica* genomes. Phytopathology 92:105–111

Tewari JP (1991) Structural and biochemical bases of black spot disease of crucifers. Adv Strl Biol 1:325–349

Tewari JP, Conn KL (1993) Reaction of some wild crucifers to *Alternaria brassicae*. Bull OILB SROP 16:53–58

Tewari KP, Tomar RKS, Mishra GL, Raghu JS (1992) Intercropping of mustard with gram and lentil. J Oilseeds Res 9:248–252

Thuma BA, Rowe RC, Madden LV (1983) Relationships of soil temperature and moisture to clubroot (*Plasmodiophora brassicae*) severity on radish in organic soil. Plant Dis 67:758–762

Tonguc M, Griffiths PD (2004a) Transfer of powdery mildew resistance from *Brassica carinata* to *Brassica oleracea* through embryo rescue. Plant Breed 123:587–589. https://doi.org/10.1111/j.1439-0523.2004.00987.x.

Tonguc M, Griffiths PD (2004b) Development of black rot resistant interspecific hybrids between *Brassica oleracea* L. cultivars and *Brassica* accession A 19182, using embryo rescue. Euphytica 136:313–318. https://doi.org/10.1023/B:EUPH.0000032733.47031.5f

Tonguc M, Earle ED, Griffiths PD (2003) Segregation distortion of *Brassica carinata* derived black rot resistance in *Brassica oleracea*. Euphytica 134:269–276

Toor SS, Singh S, Garch AIS, Kapur ML, Singh A (2000) Gobhi sarson as an intercrop in autumn sugarcane for higher returns. Intens Agric 38:29–30

Toriyama K, Hinata K, Kameya T (1987) Production of somatic hybrid plants "*Brassica moricandida*", through protoplast fusion between *Moricandida arvensis* and *Brassica oleracea*. Plant Sci 48(2):123–128

Toxopeus H, Dixon GR, Mattusch P (1986) Physiological specialization in *Plasmodiophora brassicae*: an analysis by international experimentation. Trans Br Mycol Soc 87:279–287

Trujillo M, Troeger M, Niks RE, Kogel KH, Huckelhoven R (2004) Mechanistic and genetic overlap of barley host and non-host resistance to *Blumeria graminis*. Mol Plant Pathol 5:389–396. https://doi.org/10.1111/j.1364-3703.2004.00238.x

Tsunoda S (1980) Eco-physiology of wild and cultivated forms in *Brassica* and allied genera. In: Tsunoda S, Hinata K, Gomez Campo C (eds) *Brassica* crops and wild allies. Japan Scientific Societies Press, Tokyo, pp 109–132

Ueno H, Matsumoto E, Aruga D, Kitagawa S, Matsumura H, Hayashida N (2012) Molecular characterization of the CRa gene conferring clubroot resistance in *Brassica rapa*. Plant Mol Biol 80:621–629. https://doi.org/10.1007/s11103-012-9971-5

Van de Wouw AP, Cozijnsen AJ, Hane JK, Brunner PC, McDonald BA, Oliver RP, Howlett BJ (2010) Evolution of linked avirulence effectors in *Leptosphaeria maculans* is affected by genomic environment and exposure to resistance genes in host plants. PLoS Pathog 6:e1001180

Van Rie J, Jansens S (2019) Midgut-transgenic crops expressing *bacillus thuringiensis* cry proteins. Modern Crop Prot Compd 3:1103–1136

Varshney RK, Nayak SN, May GD, Jackson SA (2009) Next-generation sequencing technologies and their implications for crop genetics and breeding. Trends Biotechnol 27:522–530

Vera Cruz CM, Bai J, Ona I, Leung H, Nelson RJ, Mew TW, Leach JE (2000) Predicting durability of a disease resistance gene based on an assessment of the fitness loss and epidemiological consequences of avirulence gene mutation. Proc Natl Acad Sci U S A 97:13500–13505. https://doi.org/10.1073/pnas.250271997

Verma VD, Rai B (1980) Note on induced mutagenesis for spotting out the sources of resistance to Alternaria leaf spot in Indian mustard. Indian J Agril Sci 50(3):278

Verma PR, Saharan GS (1994) Monograph on Alternaria diseases of crucifers. Agri-Food Canada pp:1–146

Verma S, Yajima WR, Rahman MH, Shah S, Liu JJ, Ekramoddoullah AKM, Kav NNV (2012) A cysteine-rich antimicrobial peptide from *Pinus monticola* (PmAMP1) confers resistance to multiple fungal pathogens in canola (*Brassica napus*). Plant Mol Biol 79(1-2):61–74

Vicente JG, Holub EB (2013) *Xanthomonas campestris* pv. *campestris* (cause of black rot of crucifers) in the genomic era is still a worldwide threat to *Brassica* crops. Mol Plant Pathol 14:2–18. https://doi.org/10.1111/j.1364-3703.2012.00833.x

Vicente JG, Taylor JD, Sharpe AG, Parkin IA, Lydiate PDJ, King GJ (2002) Inheritance of race-specific resistance to *Xanthomonas campestris* pv. *campestris* in *Brassica* genomes. Phytopathology 92:1134–1141. https://doi.org/10.1094/PHYTO.2002.92.10.1134

Vishwanath, Kolte SJ (1997) Biochemical variation among three isolates of *Alternaria brassicae*. Indian Phytopath 50(3):437–438

Vishwanath KSJ, Singh MP, Awasthi RP (1999) Induction of resistance in mustard (*Brassica juncea*) against Alternaria black spot with an avirulent *Alternaria brassicae* isolate-D. Eur J Plant Pathol 105:217–220. https://doi.org/10.1023/A:1008717323002

Voorrips RE (1995) *Plasmodiophora brassicae*: aspects of pathogenesis and resistance in *Brassica oleracea*. Euphytica 83:139–146

Voorrips RE (1996) Production, characterization and interaction of single-spore isolates of *Plasmodiophora brassicae*. Eur J Plant Pathol 102:377–383

Voorrips RE, Visser DL (1993) Examination of resistance to clubroot in accessions of *Brassica oleracea* using a glasshouse seedling test. Neth J Plant Pathol 99:269–276

Voorrips RE, Jongerius MC, Kanne HJ (1997) Mapping of two genes for resistance to clubroot (*Plasmodiophora brassicae*) in a population of doubled haploid lines of *Brassica oleracea* by means of RFLP and AFLP markers. Theor Appl Genet 94:75–82. https://doi.org/10.1007/s001220050384

Wallenhammar AC (1996) Prevalence of *Plasmodiophora brassicae* in a spring oilseed rape growing area in central Sweden and factors influencing soil infestation levels. Plant Pathol 45:710–719

Wallenhammar AC, Johnsson L, Gerhardson B (1999) Clubroot resistance and yield loss in spring oilseed turnip rape and spring oilseed rape. In: Wratten N, Salisbury PA (eds) New horizons of an old crop: proceedings of the 10th international rapeseed congress. Canberra, Australia, 1999. http://www.regional.org.au/au/gcirc/3/438.htm#P1_88.

Wang Z, Mao H, Dong C, Ji R, Cai L, Fu H, Liu S (2009) Over expression of *Brassica napus MPK4* enhances resistance to *Sclerotinia sclerotiorum* in oilseed rape. Mol Pl Microbiol Inter 22:235–244. https://doi.org/10.1094/MPMI-22-3-0235

Wang X, Wang H, Wang J, Sun R, Wu J, Liu S, Bai Y, Mun JH, Bancroft I, Cheng F, Huang S, Li X, Hua W, Wang J, Wang X, Freeling M, Pires JC, Paterson AH, Chalhoub B, Wang B, Hayward A, Sharpe AG, Park BS, Weisshaar B, Liu B, Li B, Liu B, Tong C, Song C, Duran C, Peng C, Geng C, Koh C, Lin C, Edwards D, Mu D, Shen D, Soumpourou E, Li F, Fraser F, Conant G, Lassalle G, King GJ, Bonnema G, Tang H, Wang H, Belcram H, Zhou H, Hirakawa H, Abe H, Guo H, Wang H, Jin H, Parkin IA, Batley J, Kim JS, Just J, Li J, Xu J, Deng J, Kim JA, Li J, Yu J, Meng J, Wang J, Min J, Poulain J, Wang J, Hatakeyama K, Wu K, Wang L, Fang L, Trick M, Links MG, Zhao M, Jin M, Ramchiary N, Drou N, Berkman PJ, Cai Q, Huang Q, Li R, Tabata S, Cheng S, Zhang S, Zhang S, Huang S, Sato S, Sun S, Kwon SJ, Choi SR, Lee TH, Fan W, Zhao X, Tan X, Xu X, Wang Y, Qiu Y, Yin Y, Li Y, Du Y, Liao Y, Lim Y, Narusaka Y, Wang Y, Wang Z, Li Z, Wang Z, Xiong Z, Zhang Z (2011) *Brassica rapa* genome sequencing project consortium. The genome of the mesopolyploid crop species *Brassica rapa*. Nat Genet 43:1035–1039. https://doi.org/10.1038/ng.919

Wang GX, Lv J, Zhang J, Han S, Zong M, Guo N, Zeng XY, Zhang YY, Wang YP, Liu F (2016) Genetic and epigenetic alterations of *Brassica nigra* introgression lines from somatic hybridization: a resource for cauliflower improvement. Front Plant Sci 7:1258. https://doi.org/10.3389/fpls.2016.01258

Wang Z, Zhang WH, Ma LY, Li X, Zhao FY, Tan XL (2020a) Overexpression of *Brassica napus* NPR1 enhances resistance to *Sclerotinia sclerotiorum* in oilseed rape. Physiol Mol Plant Pathol 110:101460

Wang Z, Zhao FY, Tang MQ, Chen T, Bao LL, Cao J, Li YL, Yang YH, Zhu KM, Liu S, Tan XL (2020b) *BnaMPK6* is a determinant of quantitative disease resistance against *Sclerotinia sclerotiorum* in oilseed rape. Plant Sci 291:110362

Warwick SI (2011) Brassicaceae in agriculture. In: Bancroft I, Schmidt R (eds) Genetics and genomics of the Brassicaceae. Springer, New York, pp 33–67. https://doi.org/10.1007/978-1-4419-7118-0_2

Weerakoon SR, Si P, Zili W, Meng J, Yan G (2009) Production and confirmation of hybrids through interspecific crossing between tetraploid *B. juncea* and diploid *B. oleracea* towards a hexaploid *Brassica* population. In: Proc 16th Australian res assembly on Brassicas. Ballarat, VIC

Wei L, Jian H, Lu K, Filardo F, Yin N, Liu L, Qu C, Li W, Du H, Li J (2016) Genome-wide association analysis and differential expression analysis of resistance to Sclerotinia stem rot in *Brassica napus*. Plant Biotechnol J 14:1368–1380. https://doi.org/10.1111/pbi.12501

Weisaeth G (1977) Hodelka lsortene 'Respla' og 'Resista' foredlinger med klumprotresistens. Gartneryrket 67:937–939

Werner S, Diederichsen E, Frauen M, Schondelmaier J, Jung C (2007) Genetic mapping of clubroot resistance genes in oilseed rape. Theor Appl Genet 116:363–372. https://doi.org/10.1007/s00122-007-0674-2

Werner S, Diederichsen E, Frauen M, Schondelmaier J, Jung C (2008) Genetic mapping of clubroot resistance genes in oilseed rape. Theor Appl Genet 116:363–372. https://doi.org/10.1007/s00122-007-0674-2.

Westman AL, Kresovich S, Dickson MH (1999) Regional variation in *Brassica nigra* and other weedy crucifers for disease reaction to *Alternaria brassicicola* and *Xanthomonas campestris* pv. *campestris*. Euphytica 106:253–259. https://doi.org/10.1023/A:1003544025146

Williams PH (1966) A system for the determination of races of *Plasmodiophora brassicae* that infect cabbage and rutabaga. Phytopathology 56:624–626

Williamson CJ, Bradshaw JE (1989) Sources of resistance to *Plasmodiophora brassicae* used in breeding programmes at SCRI. XIIth Eucarpia Congress. Vort Pflanzenz 15:17/12

Wu XM, Chen BY, Lu G, Wang HZ, Xu K, Guizhan G, Song Y (2009) Genetic diversity in oil and vegetable mustard (*Brassica juncea*) landraces revealed by SRAP markers. Genet Resour Crop Evol 56:1011–1022. https://doi.org/10.1007/s10722-009-9420-8

Wu J, Cai G, Tu J, Li L, Liu S, Luo X, Zhou L, Fan C, Zhou Y (2013) Identification of QTLs for resistance to Sclerotinia stem rot and BnaC.IGMT5. A as a candidate gene of the major resistant QTL SRC6 in *Brassica napus*. PLoS One 8:e67740. https://doi.org/10.1371/journal.pone.0067740

Wu J, Zhao Q, Yang Q, Liu H, Li Q, Yi X, Cheng Y, Guo L, Fan C, Zhou Y (2016) Comparative transcriptomic analysis uncovers the complex genetic network for resistance to *Sclerotinia sclerotiorum* in *Brassica napus*. Sci Rep 6:19007. https://doi.org/10.1038/srep19007

Wu J, Lin L, Xu M, Chen P, Liu D, Sun Q, Ran L, Wang Y (2018) Homoeolog expression bias and expression level dominance in resynthesized allopolyploid *Brassica napus*. BMC Genomics 19 (1):586

Xing GM, Long MH, Tanaka S, Fujieda K (1989) Clubroot resistance in *Brassica raphanus* (hybrid of cabbage 9 radish). Plant Dis 33:189–194

Xue S, Cao T, Howard RJ, Hwang SF, Strelkov SE (2008) Isolation and variation in virulence of single-spore isolates of *Plasmodiophora brassicae* from Canada. Plant Dis 92:456–462

Yadav RC, Sareen PK, Chowdhury JB (1991) Inter-specific hybridization in *Brassica juncea* X *Brassica tournefortii* using ovary culture. Cruciferae Newslett 14/15:84

Yoshikawa H (1980) Breeding for clubroot resistance in Chinese cabbage. In: Talekar NS, Selleck GW (eds) First international symposium on Chinese cabbage, Tsukuba Science City, Japan, Hong Wen Printing Works, Tainan

Yoshikawa H (1993) Studies on breeding of clubroot resistance in cole crops. Bull Natl Res Inst Veg Ornam Plants Tea Jpn Ser A 7:1–165

Yu F, Lydiate DJ, Rimmer SR (2008) Identification and mapping of a third blackleg resistance locus in *Brassica napus* derived from *B. rapa* subsp. *sylvestris*. Genome 51:64–72. https://doi.org/10.1139/G07-103

Zhang Y, Tessaro MJ, Lassner M, Li X (2003) Knockout analysis of *Arabidopsis* transcription factors TGA2, TGA5, and TGA6 reveals their redundant and essential roles in systemic acquired resistance. Plant Cell 15:2647–2653. https://doi.org/10.1105/tpc.014894

Zhang GQ, Tang GX, Song WJ, Zhou WJ (2004) Re-synthesizing *Brassica napus* from interspecific hybridization between *Brassica rapa* and *B. oleracea* through ovary culture. Euphytica 140:181–187. https://doi.org/10.1007/s10681-004-3034-1

Zhou JM, Tang XY, Martin GB (1997) The Pto kinase conferring resistance to tomato bacterial speck disease interacts with proteins that bind a cis-element of pathogenesis-related genes. EMBO J 16:3207–3218. https://doi.org/10.1093/emboj/16.11.3207

Zhou N, Tootle TL, Glazebrook J (1999) Arabidopsis PAD3, a gene required for camalexin biosynthesis, encodes a putative cytochrome P450 monooxygenase. Plant Cell 11:2419–2428. https://doi.org/10.1105/tpc.11.12.2419

Zou J, Zhu JL, Huang SM, Tian ET, Xiao Y, Fu DH, Tu JX, Fu TD, Meng JL (2010) Broadening the avenue of intersubgenomic heterosis in oilseed *Brassica*. Theor Appl Genet 120:283–290

Zou P, Lu X, Jing C, Yuan Y, Lu Y, Zhang C, Meng L, Zhao H, Li Y (2018) Low-molecular-Weightt polysaccharides from *Pyropia yezoensis* enhance tolerance of wheat seedlings (*Triticum aestivum L.)* to salt stress. Front Plant Sci 9:427. https://doi.org/10.3389/fpls.2018.00427

Host Resistance Signaling Network System to Multiple Stresses

5

Abstract

Cruciferous plants have developed numerous molecular mechanisms to respond to various environmental and multiple biotic and abiotic stresses. Whenever plant encounters any kind of stress, first of all there is perception of the signals by the potential sensors. This further generates secondary messengers (e.g., ABA, ROS, Ca^{2+}, inositol phosphates). The secondary messengers can alter the levels of intracellular Ca^{2+}. Any change in cytoplasmic Ca^{2+} is sensed by Ca^{2+} sensors, i.e., calcium binding proteins, and after interacting with corresponding interacting partners they initiate a protein phosphorylation cascade. This signaling cascade ultimately regulates the transcription factors that manage specific suite of stress-regulated genes and the functional proteins involved in protecting the cell against stress. The expression of stress-regulated genes leads to the production of various kinds of regulatory molecules like the phytohormones ABA, ET, and SA, transcription factors, etc. A second round of signaling can get started by these regulatory molecules which may further enhance the initial signaling pathway. These signaling networks are very complex and interconnected at many levels for transferring information to generate robust host resistance. Moreover, during signaling other components are also involved. A successful signal transduction pathway demands proper coordination of all signaling molecules. This is brought about by specific molecules whose functions are to transport, bring together, or modify the signaling molecules. They themselves are not involved in relaying the signal. These proteins consist of scaffolding proteins, protein modifiers (e.g., enzymes needed for protein glycosylation, methylation, ubiquitination, and lipidation), and adaptors. Cruciferous interactions with multiple stresses trigger a wide range of defense reactions at molecular and cellular levels to ensure that host cells are insulated from these challenges they encounter. The production of ROS, MAP Kinase cascades, and hormone signaling are triggered by multiple stresses and play crucial roles as common components of the host immune

G. S. Saharan et al., *Genomics of Crucifer's Host-Resistance*,
https://doi.org/10.1007/978-981-16-0862-9_5

system. GbRLK, a receptor like kinase (RLK) gene, is involved in response to multiple stresses as positive regulator Transcription factors (TFs) are the central core of the genes as defense signaling pathways to multiple stress. MYB TFs play essential roles in multiple stress responses. ATAF1 is a multi-faceted TF with versatile capabilities and plays an important role in mediating crosstalk between multiple stress responses. HSPs gene regulate the function of NBS-LRR genes in response to multiple stresses. miRNAs may target only one locus or several loci in response to multiple stresses. Cell signaling in response to multiple stresses largely depends on the SnRK family of protein kinases in plants. Calcium is a key signaling pathway in host responses to multiple stresses through calcium sensor proteins like calcineurin-B-like protein (CBls), the calmodulin (CaM), and calmodulin-like protein (CMLs).

Transcriptome analyses based on microarrays and proteomic studies have given rise to rapid progress in the field of plant signal transduction and gene regulation. Gene Chip and cDNA microarrays together with massive whole genome sequencing proved to be useful in identifying novel signaling determinants on a whole-genome scale in response to various stresses. This has increased our understanding towards the complex mechanisms involved in various aspects of plant responses to environmental cues. Proteomics approach is useful in studying post-translational modifications of the proteins. It can also be used to clone unique genes by means of differential analysis which will expedite our understanding of stress signaling mechanisms in plants. Therefore, stress-tolerant plants can be genetically engineered by the combined efforts of plant molecular biologists, physiologists, and breeders.

Keywords

Signaling network system · Biotic stress · Abiotic stress · Weed infestation · Insects infestation · Nutrient stress · Multiple stress · Genome-wide marker assisted selection · Receptor proteins as frontline defense signals · MicroRNAs analysis in response to defense signaling · Systemic signaling process · Calcium signaling · CMLS in multiple stresses · CPK's signaling molecules

5.1 Introduction

Multiple stresses in the form of biotic and abiotic stresses, as a part of the natural ecosystem, seriously impact crucifer's crop productivity and threaten global oilseed and vegetable food security. Plants in their natural habitats are persistently and simultaneously confronted with a range of biotic and other internal adverse changes imposed by multiple interactions of soil environment (Tables 5.1, 5.2, 5.3, and 5.4). Environmental pollution, global warming, and climate change exacerbate the impact of biotic and abiotic stresses on plant growth and yield. Plants have evolved sophisticated defense network, also called innate immune system, in response to ever-changing environmental conditions. Recent evidence highlights the complex

Table 5.1 Biotic stresses of crucifers (Saharan 1984, 1992; Kolte 1985; Williams 1985; Saharan et al. 2014, 2016, 2017, 2019; up dated 2020)

Sr. no.	Common name	Symbol[a]	Causal organism	Geographical distribution	Yield loss (World average %)
1.	Alternaria disease, black spot, leaf stem or pod spots and blight	AB	*Alternaria* spp. [*A. brassicae* (Berk.) Sacc.; *A. brassicicola* (Schw.) Wiltsh.; *A. raphani* Groves and Skolko; *A. alternata, Alternaria citri* Ellis and Pears]	Africa, Argentina, Australia, Austria, Bangladesh, Brazil, Britain, Bhutan, Bulgaria, Burma, Canada, Chile, China, Cyprus, Czechoslovakia, Denmark, Egypt, England, Ethiopia, Finland, France, Germany, Ghana, Guinea, Holland, Hong Kong, Hungary, India, Iran, Ireland, Italy, Jamaica, Japan, Kenya, Libya, Malawi, Malaya, Mauritius, Morocco, Mozambique, Nepal, Netherland, New Guinea, New Zealand, Nicaragua, Nigeria, Norway, Pakistan, Philippines, Poland, Rhodesia, Romania, Russia, Sabah, Saudi Arabia, Scotland, Sierra Leone, Singapore, Spain, South Africa, Sri Lanka, Sudan, Sweden, Switzerland, Taiwan, Tanganyika, Tanzania, Trinidad, Turkey, Uganda,	Rapeseed (60), *Brassica juncea* (30), *B. napus* (30), *B. carinata* (11), *B. rapa* var. Yellow Sarson (36), *Eruca sativa* (56), Cabbage- cauliflower (73), Radish (18)

(continued)

Table 5.1 (continued)

Sr. no.	Common name	Symbol[a]	Causal organism	Geographical distribution	Yield loss (World average %)
				USA, USSR, and Zambia. (Anonymous 1980, 1981, 1983, 1985a, 1985b, 1985c, 1985d; Ellis 1971; Kolte 1985; Saharan 1992; Verma and Saharan 1993; 1994; Nourani et al. 2008)	
2.	Anthracnose	CH	*Colletotrichum higginsianum* Sacc./ *Colletotrichum brassicicola* (*Brassica oleracea*)	USA, West Indies, Japan, and Southeast Asia, Germany, China, Italy, Korea, Trinidad and Tobago, New Zealand (Higgins 1917; Scheffer 1950; Sutton 1980; Moriwaki et al. 2002; Huser et al. 2009; He et al. 2016; Damm et al. 2012, 2014; Jayawardena et al. 2016)	DNA[a]
3.	Aster yellows	AY	Mycoplasma Like Organism or Phytoplasma Like Organism (MLO)	South America, (Argentina, Brazil, Chile, Costa Rica, and Bolivia, Brazil, China), Greece, Hungary, Iran, India, Italy, Poland, Serbia, UK, USA (Rappussi et al. 2012; Mou et al. 2012, Gkavaleka et al. 2012; Fodor et al. 1999; Salehi et al. 2007a, b; Bindra and	DNA[a]

				Bakhetia 1967; Marcone and Ragozzino 1995; Kaminska et al. 2012a, b; Duduk et al. 2007; Keane et al. 1996; Lee et al. 2004)	
4.	Autogenic necrosis	AN(G)	Genetic disorder	DNA[a]	DNA[a]
5.	Bacterial leaf spot	PSM	*Xanthomonas campestris* pv. Armoraciae/ *Pseudomonas syringae* pv. *maculicola* (Mc Culloch) Young	United States of America (Zhao et al. 2000), Japan (Tamura et al. 1994), Turkey, and India (Bradbury 1986), Pakistan (Muhammad Ehetisham-ul-Haq et al. 2017), United States, Japan, (Tamura et al. 1994), Brazil, China (Zeng and Huang 1994), Turkey, and India (Sherf and MacNab 1986)	DNA[a]
6.	Bacterial soft rot, Erwinia stalk rot	EC	*Erwinia carotovora* pv. *carotovora* (Jones) Bergy	India (Burkholder 1957)	Upto 80 (Meena et al. 2010; Singh et al. 2016)
7.	Bacterial soft rot, Pseudomonas rot	PMM	*Pseudomonas marginalis* pv. *marginalia* (Brown) Stevens	China (Mew et al. 1976), USA (Canaday et al. 1991)	DNA[a]
8.	Black leg and phoma root rot	LM	*Leptosphaeria maculans* (Desm.) Ces. and de not. (anamorph: *Phoma lingam* (Tode: Fr.) Desm. *L. biglobosa*	**Africa:** Egypt, Ethiopia, Kenya, Mozambique, Nigeria, South Africa, Zambia, Zimbabwe; (EPPO 2020; CAB International 1978; Michail et al. 1979; Gugel and Petrie 1992)	18–99 (Wang et al. 2019a, b)

(continued)

Table 5.1 (continued)

Sr. no.	Common name	Symbol[a]	Causal organism	Geographical distribution	Yield loss (World average %)
				Asia: Armenia, China – Henan, -Sichuan, -Yunnan, Georgia, India -Himachal Pradesh, -Rajasthan, Iran, Israel, Japan, Kazakhstan, Kyrgyzstan, Malaysia-Sabah, North Korea, Pakistan, Philippines, South Korea, Taiwan, Thailand, Turkey; (EPPO 2020; CAB International 1978) **Europe**: Austria, Belarus, Belgium, Bulgaria, Czechia, Denmark, Estonia, Finland, France, Germany, Hungary, Ireland, Italy, Latvia, Lithuania, Malta, Netherlands, Norway, Poland, Portugal, Romania, Russia, Slovakia, Spain, Sweden, Switzerland, Ukraine, United Kingdom – Scotland; (EPPO 2020; CAB International 1978; Buonassisi and MacDonald 1990; Gladders 1978; EPPO 2020; CAB International 1978; Lacoste et al. 1969; Kachlicki and	

				Jetaildryczka 1994) **North America**: Canada – Alberta, -British Columbia, -Manitoba, -New Brunswick, -Newfoundland and Labrador, -Nova Scotia, -Ontario, -Prince Edward, Island –Quebec, -Saskatchewan, Costa Rica, El Salvador, Guadeloupe, Mexico, Panama, Puerto Rico, United States – California, –Georgia, -Hawaii, –Idaho, -Illinois, -Kentucky, -New York, -North Dakota, -Oregon, -Washington, -Wisconsin; (Henderson 1918; Jesperson 1989; Gugel and Petrie 1992; Hill 1992; Hall et al. 1993; EPPO 2020; CAB International 1978; Farr et al. 1989) **Oceania**: Australia - New South Wales, -Northern Territory, -Queensland, -South Australia, -Tasmania, -Victoria, -Western Australia, New Caledonia, New Zealand, Papua New Guinea; (Cunningham 1927, Bokor	

(continued)

Table 5.1 (continued)

Sr. no.	Common name	Symbol[a]	Causal organism	Geographical distribution	Yield loss (World average %)
				1972; Bokor et al. 1975; Chambers 1959; McGee 1973; EPPO 2020; CAB International 1978) **South America**: Argentina, Brazil -Rio Grande do Sul (EPPO 2020; CAB International 1978)	
9.	Black mold rot	RST	*Rhizopus stolonifer* (Ehr.: Fr.) Vuill.	India (Bhajbhuje 2014)	DNA[a]
10.	Black root	AR	*Aphanomyces raphani* Kendrick	India (Singh and Pavgi 1977), USA (Ghafoor 1964), Japan (Suzuki et al. 1980)	DNA[a]
11.	Black rot	XCC	*Xanthomonas campestris* pv. *campestris* (Pammel) Dowson/*X. arboricola*	**Africa:** Angola, Ethiopia, Ghana, Kenya, Libya, Malawi, Mauritius, Morocco, Mozambique, Seychelles, Somalia, Tanzania, Uganda (CAB International 1987) Togo (Bradbury 1986), Zimbabwe (Mguni 1987; CAB International 1987; Mguni 1995; Mguni 1996; Mguni et al. 1996). **Asia:** Brunei, Cambodia, China (CAB International	10–50 (Williams 1980)

				1987), Shaanxi (Yan et al. 2008), -Zhejiang (Shen and Chen 1990), India (Sharma et al. 1992), Haryana, Himachal Pradesh (Shyam et al. 1994), Jammu and Kashmir (Bhat and Masoodi 2000), Karnataka (Kishun 1984), Manipur (Gupta 1991), Odisha (Beura et al. 2006), Sikkim (Gupta and Choudhary 1995), Tamil Nadu (Raju and Sivaprakasam 1989), Assam, Punjab, Rajasthan (CAB International 1987), Indonesia, Irian Jaya, Java, Sumatra, Iran (CAB International 1987), Israel (Kritzman and Ben-Yephet 1990), Japan (Goto 1992), Jordan (CAB International 1987), Malaysia (Ho 1985), Nepal (Shrestha 1983, Shakya and Malla 1988), Philippines (CAB International 1987), South Korea (Kim 1986), Sri Lanka Taiwan, Thailand, Turkey, Uzbekistan, Vietnam, (CAB International 1987).	

(continued)

Table 5.1 (continued)

Sr. no.	Common name	Symbol[a]	Causal organism	Geographical distribution	Yield loss (World average %)
				Europe: Bosnia and Herzegovina (Trkulja 2006), Bulgaria, Czechoslovakia, Yugoslavia, Denmark, Finland, France, Germany, Greece (CAB International 1987), Hungary (Nemeth and Laszlo 1983), Iceland, Ireland (CAB International 1987), Italy (Caponero and Iacobellis 1994; Scortichini et al. 1994), Netherlands (Kocks and Zadoks 1996), Norway, Portugal, Romania, Russia (CAB International 1987), Serbia (Popovic et al. 2013), Sweden, Switzerland, United Kingdom, (CAB International 1987). **North America:** Barbados, Bermuda, Canada, Costa Rica, Cuba, Dominica, El Salvador, Guadeloupe, Guatemala, Haiti, Honduras, Jamaica, Martinique, Mexico, Nicaragua, Panama, Puerto	

				Rico, Saint Kitts and Nevis, Saint Lucia, Trinidad and Tobago, United States (Hunter et al. 1975, Schaad and Dianese 1981, CAB International 1987). **Oceania:** American Samoa, Australia (Fahy and Persley 1983; UK, CAB International 1987), Cook Islands, Fiji, New Caledonia, New Zealand, Norfolk Island, Papua New Guinea (CAB International 1987), Samoa (Gerlach 1988), Tonga, Vanuatu (CAB International 1987). **South America:** Argentina (Gaetan et al. 1995), Bolivia (CAB International 1987), Brazil (Mariano et al. 1985), Suriname (CAB International 1987).	
12.	Black speck	BS	Physiological disorder	DNA[a]	DNA[a]
13.	Bottom rot, damping off, head rot, seedling root rot, wire stem, basal stem rot	RS	*Rhizoctonia solani* Kuhn (teleomorph: *Thanatephorus cucumeris* (Frank) Donk)	Canada (Ellis 1983), USA (Cappelli et al. 1999, Paulitz et al. 2006), Finland (Tahvonen et al. 1984), India (Chauhan and Saksena 1974), Australia	30 (Sippell et al. 1985)

(continued)

Table 5.1 (continued)

Sr. no.	Common name	Symbol[a]	Causal organism	Geographical distribution	Yield loss (World average %)
				(Khangura et al. 1999), UK (Sturrock et al. 2015)	
14.	Broomrape	OA	*Orobanche aegyptiaca* Pers.	**Africa:** Egypt (Chater and Webb 1972, Holm et al. 1979), Mauritania, Morocco (Parker and Wilson 1986), **Asia:** Afghanistan, Bahrain, India, Iran, Iraq, Israel, Jordan, Kuwait, Lebanon, Pakistan, Saudi Arabia, UAE (Holm et al. 1979, Parker and Wilson 1986), Nepal (Acharya et al. 2002), Syria, Turkmenistan (Beck-Mennagetta 1930), Turkey (Chater and Webb 1972) **Europe:** Bulgaria (Chater and Webb 1972), Greece (Vagelas and Gravanis 2014), Hungary, Italy, Russia (Holm et al. 1979) **North America:** Cuba (Oviedo Prieto et al. 2012)	5–100 (Habimana et al. 2014)
15.	Cercospora leaf spot	CB	*Cercospora brassicicola* Henn.	DNA[a]	DNA[a]
16.	Clubroot	PB	*Plasmodiophora brassicae* Worn.	**Africa:** Angola, South Africa, Sao Tome	All Cruciferous crops (29), Saharan et al. (2021)

				and Principe (CABI/EPPO 2011; EPPO 2014), Malawi (Mazuma et al. 2008), Uganda, Zambia (Dixon 2009) **ASIA:** Brunei, China, Hong Kong, Iraq, Israel, Kyrgyzstan, Malaysia, Nepal, Philippines, Sri Lanka, Taiwan, Turkey, Russia, Omsk, Irkutsk, Kemerovo, Yakutsk Republic, Siberia (CABI/EPPO 2011; EPPO 2014), India (Sen 2005, Dixon 2009; Bhattacharya et al. 2014), Indonesia (Yamada et al. 2004), Iran, Thailand, Vietnam (Dixon 2009), Japan (Yoshikawa and Buczacki 1978), Korea (Kim et al. 2009), Pakistan (Muhammad and Khan 2008), Singapore (AVA 2001) **Australasia and Oceania:** Australia (Khangura and Wright 2012), Hawaii, New Zealand, Papua New Guinea (CABI/EPPO 2011; EPPO 2014), **Europe:** Austria, Belarus,	

(continued)

Table 5.1 (continued)

Sr. no.	Common name	Symbol[a]	Causal organism	Geographical distribution	Yield loss (World average %)
				Belgium, Bosnia-Hercegovina, Bulgaria, Croatia, Czech Republic, Denmark (Faeroe Islands) Finland, France, Germany, Greece, Hungary, Iceland, Ireland, Italy, Latvia, Lithuania, Luxembourg, Montenegro, Netherlands, Norway, Poland, Portugal, Romania, Slovakia, Spain, Sweden, Switzerland, Russia, Ukraine, Yugoslavia (CABI/EPPO 2011; EPPO 2014), Serbia (Todorovic 2007), UK (Buczack and Ockendon 1979), **North America:** Canada (Tewari et al. 2004), Mexico, USA (CABI/EPPO 2011), Costa Rico, Ecuador (Dixon 2009) **Central America and West Indies:** Guatemala, Puerto Rico, Trinidad (CABI/EPPO 2011; EPPO 2014) **South America**, Argentina,	

				Brazil, Chile, Colombia, Guyana, Venezuela (CABI/ EPPO 2011; EPPO 2014), Bolivia (Dixon 2009)	
17.	Collar rot	FE	*Fusarium equiseti* (Corda) Sacc.	India (Prasad et al. 2017; Meena et al. 2019)	DNA[a]
18.	Crown gall	AT	*Agrobacterium tumefaciens* (Smith and Townsend) Conn.	USA (Kado et al. 1972; Cleene and Ley 1976)	DNA[a]
19.	Damping off	F	*Fusarium* spp.	India (Ashraf and Basu-Choudhary 2008)	Ashraf and Basu-Choudhary (2008)
20.	Damping off, *Pythium*	PT	*Pythium* spp.	India (Singh 1974, Singh and Pavgi 1978), South Africa (Wager 1940), USA (Drechsler 1925, Tompkins et al. 1939), Germany (Kryopoulus 1916)	Middleton (1943)
21.	Downy mildew, stag head	HP	*Hyaloperonospora parasitica* (Gaum) Gokar	Argentina (Lindquist 1946), Australia (Samuel 1925), Austria (Glaeser 1970), Bavaria (Boning 1936), Bermuda (Waterston 1940), Brazil (Viegas and Teixeira 1943), Britain (UK) (Foister 1948; Moore 1959), Brunei (Channon 1981), Bulgaria (Khristov 1979), Canada (Jones 1944; Downey and Bolton 1961), Chile (Mujica and Vergara 1960),	35, (*Brassica juncea*)

(continued)

Table 5.1 (continued)

Sr. no.	Common name	Symbol[a]	Causal organism	Geographical distribution	Yield loss (World average %)
				China (Porter 1926), Costa Rica (McGuire and Crandall 1967), Cuba (Fernandez 1973), Cypress (Channon 1981), Czechoslovakia (Rydl 1968), Denmark (Gram and Rostrup 1924), Egypt (Baka 1996), Ethiopia (Channon 1981), Finland (Channon 1981), France (Darpoux 1945), Germany (Klemm 1938; Raabe 1939), Greece (Channon 1981), Guatemala (Muller 1950), Holland (Thung 1926a), Hong Kong (Johnston 1963), Hungary (Lehoczky 1957), Iberia (Spain) (Gonzalez 1924), India (Butler 1918; Thind 1942); Iran (Etebarian 1989), Iraq (Channon 1981), Ireland (McKee 1961), Israel (Peleg 1953), Italy (Ciferri 1961), Jamaica (Leather 1967), Japan (Hiura and Kanegae 1934), Kampuchea (Soonthronpoct 1969),	

				Korea (So et al. 1981), Libya (Channon 1981), Ludlow (Preston 1929), Malawi (Peregrine and Siddiqi 1972), Malaysia (Mcintosh 1951), Malta (Channon 1981), Mauritius (Orian 1951), Mexico (Ponce and Mendoza 1983), Montpellier (Kuhnholtz and Gastuad 1943), Moravia (Baudys 1928), Morocco (Channon 1981), Mozambique (De Carvalho 1948), Nepal (Bhatt 1966), Netherlands (Thung 1926b), New South Wales (Australia) (Bertus 1968), New Zealand (Channon 1981), Norway (Semb 1969), Palestine (Rayss 1938; Chorin 1946), Panama (McGuire and Crandall 1967), Pakistan (Perwaiz et al. 1969), Papua New Guinea (Channon 1981), Peru (Icochea et al. 1994), Philippines (Ocfemia 1925), Poland (Zarzycka 1970), Portugal (Da Costa and Da-Camara 1953), Puerto Rico	

(continued)

Table 5.1 (continued)

Sr. no.	Common name	Symbol[a]	Causal organism	Geographical distribution	Yield loss (World average %)
				(Channon 1981), Queensland (Langdon 1948), Romania (Savulescu and Rayss 1930), Russia (Timina et al. 1989), Samoa (Firman 1975), Saxony (Wiese 1927), South Africa (Dippenaar 1934), Spain (Gonzalez 1924), Sri Lanka (Park 1932), Sweden (Hammarlund 1931; Bjorling 1944), Switzerland (Gaumann 1923), Taiwan (Lo 1961), Tanzania (Channon 1981), Thailand (Chandrasrikul 1962), Trinidad and Tobago (Stell 1922), Turkey (Channon 1981), Uganda (Channon 1981), USA (Farlow 1883), Uruguay (Koch de Brotus and Boasse 1955), USSR (Pimenova and Maslennikov 1955), Venezuela (Channon 1981), Victoria (Smith and Price 1996), Vietnam (My 1966), Yugoslavia (Sutic and Klijajic 1954)	

22.	Gray mold	BC	*Botrytis cinerea* Pers.: Fr. (Telemorph: *Botryotinia fuckeliana* (de Bary) Whetz.	China (Wu et al. 2007)	DNA[a]
23.	Gummosis	G	Physiological disorder	India	DNA[a]
24.	Leaf curl virus Cabbage leaf curl virus	LCV CLCV	Begomo virus-Beta satellite complex Gemini virus	India (Roy et al. 2013, Kumar et al. 2018) USA (Walsh and Tomlinson 1985)	DNA[a]
25.	Light leaf spot	PYB	*Pyrenopeziza brassicae* Sut. and Rawl	Australia, Denmark, France, Germany, Ireland, Japan, Latvia, Netherlands, New Zealand, Norway, Philippines, 17 Portugal, Romania, Tasmania, and the UK (Hickman et al. 1955; Karolewski 2010; Rawlinson et al. 1978; Sutton 1977)	DNA[a]
26.	Nematode				
	a. Awl b. Cyst c. Pin d. Root-knot e. Root lesion f. Sting	D(N) H(N) PA(N) M(N) P(N) B(N)	*Dolichodorus* spp. *Heterodera schactii* Schmidt *Heterodera cruciferae* Franklin *Paratylenchus* spp. *Meloidogyne* spp. *Paratylenchus pratensis* (de Man) Filipjev *Paratylenchus* spp. *Belonolaimus* spp.	DNA[a]	DNA[a]

(continued)

Table 5.1 (continued)

Sr. no.	Common name	Symbol[a]	Causal organism	Geographical distribution	Yield loss (World average %)
27.	Phyllody	P	*Phytoplasma*	India, Iran, Greece, Poland, China	Salehi et al. (2007a, b), Gkavaleka et al. (2012), Kaminska et al. (2012a, 2012b), Mou et al. (2012)
28.	Pod malformation	GP	Genetical and Physiological disorder	DNA[a]	DNA[a]
29.	Powdery mildew	EC	*Erysiphe cruciferarum* Opiz. ex. Junell	Argentina (Gaetan and Madia 2004); Australia (Shivas 1989), Bulgaria (Negrean and Denchev 2000); Canada (Shattuck and Parry 1990); China (Alkooranee et al. 2015); Czech and Slovak (Pastircakova and Pastircak 2013), Egypt (Melchers 1931), Finland (Farr et al. 1989), France (Ale-Agha et al. 2008), Greece (Vellios et al. 2017); Germany (Koch and Slusarenko 1990); India (Sankhla et al. 1967); Iran (Mohammadi-Doustdar 1967); Japan (Amano 1986); Korea (Shin and La 1992); Mexico (Morales et al. 2009); New Zealand (Boesewinkel 1977); Poland (Sadowski et al.	*Brassica juncea* (28.5), *Eruca sativa* (17)

				2002); Russia (Bunkina 1991); Sweden (Junell 1967); Switzerland (Bolay et al. 2005); Turkey (Mert-Turk et al. 2008); UK (Ellis and Ellis 1985); USA (Ellett 1966) and Vietnam (Tam et al. 2016)	
30.	Ring spot	MB	*Mycosphaerella brassicicola* (Duby) Lindau	USA (Weimer 1926), Canada (Vanterpool 1960)	DNA[a]
31.	Root rot *Phymatotrichum*	PO	*Phymatotrichum omnivorum* (Shear) Dug.	USA (Worf et al. 1985)	DNA[a]
32.	Root rot *Phytophthora*	PM	*Phytophthora megasperma* Drechs.	USA (Worf et al. 1985)	DNA[a]
33.	Scab	STS	*Streptomyces scabies* (Thaxter) Waksman and Henrici	DNA[a]	DNA[a]
34.	Sclerotinia stem rot Watery soft rot	SS	*Sclerotinia sclerotiorum* (Lib.) de Bary	**Africa:** Algeria (Maire 1927), Republic of the Congo (Foucart 1954), Egypt (Tarabeih and Abou-el-Fadl 1979), Kenya (Robinson 1960); Martens et al. 1970), Mauritius (Felix 1962), Morocco (Achbani and Tourvieille 1993), Nigeria (Akem and Dashiell 1992), South Africa (Denner et al. 1992), Tanzania (Fivawo 1987), Zimbabwe	Rapeseed (11.3), *Brassica juncea* (60)

(continued)

Table 5.1 (continued)

Sr. no.	Common name	Symbol[a]	Causal organism	Geographical distribution	Yield loss (World average %)
				(Rothwell 1972), Ethiopia, Libya, Malawi, Saint Helena, (CABI and EPPO 2005) **Asia:** Azerbaijan, China, Georgia, Hong Kong, Japan, Lebanon, Sri Lanka, Tajikistan, Thailand, Uzbekistan (CABI and EPPO 2005), Bangladesh (Prova et al. 2014, Ahmed and Akhond 2015), India (Mukerji 1973), Iran (Moghaddam and Izadyar 1994), Israel (Ben-Yephet et al. 1993), Jordan (Abu-Blan and Abu-Gharbieh 1994), Nepal (Duwadi et al. 1993), Pakistan (Mirza and Ahmad 1988), Palestine (Elazari-Volcani 1944), Singapore (Adair 1971), South Korea (Bai 1977), Syria (Haware 1990), Taiwan (Tsai 1991), Turkey (Yucer and Karaca 1978), Vietnam (Trinh et al. 2012) **Europe:** Albania (Kola	

				1980), Austria (Krexner 1969), Belgium (Couvreur and Herman 1994), Bosnia and Herzegovina (Pagliarini et al. 2007), Bulgaria (Ivancheva-Gabrovska et al. 1978), Croatia (Vrataric et al. 1991), Czechoslovakia (Ondrej 1983), Federal Republic of Yugoslavia (Acimovic 1977), Estonia (Ilumae et al. 2007), Finland (Mukula 1957), France (Grill 1979), Germany (Kruger 1983), Greece (Thanassoulopoulos 1987), Hungary (Folk and Tusnádi 1985), Ireland (Oethybridge 1911, Cassells and Walsh 1995), Italy (Zazzerini and Tosi 1985), Netherlands (Ester and Gerlagh 1984), North Macedonia (Tanas 2004), Norway (Tronsmo 1989), Poland (Tylkowska 1984), Portugal (de Dias and Lucas 1978), Romania (Şesan and Csep 1992), Slovakia (Vavrac and Kubova 1992), Spain (Tores and Moreno 1991), Switzerland (Winter	

(continued)

Table 5.1 (continued)

Sr. no.	Common name	Symbol[a]	Causal organism	Geographical distribution	Yield loss (World average %)
				et al. 1993), Ukraine (Petrenkova 1994), United Kingdom (Jellis et al. 1984), Belarus, Cyprus, Czechia, Denmark, Iceland, Latvia, Lithuania, Malta, Moldova, Russia, Serbia and Montenegro, Sweden, (CABI and EPPO 2005) **North America:** Bermuda (Waterston 1947), Canada (Baribeau 1923, Bisby 1924, Creelman 1962), Costa Rica (Laguna and Salazar 1984), Cuba (Martinez-de la Parte et al. 2013), Mexico (Crispin and Campos 1976), United States (Brown and Butler 1936, Greathead and Greathead 1992), El Salvador, Guatemala, Nicaragua, Panama (CABI and EPPO 2005) **Oceania:** American Samoa (Brooks 2002), Australia (Blackford 1944, Henderson 1962), Fiji (Morwood 1956),	

				New Zealand (Brien 1932), Samoa (Gerlach 1988, CABI and EPPO 2005) **South America:** Argentina (Pontis and Feldman 1959), Brazil (Gasparotto and Chaves 1982), Chile (Acuna et al. 1975), Ecuador (Velastegui and Ball 1991), Peru (Bazan De Segura and Onsuelo 1951), Uruguay (Sackston 1957), Venezuela (Pons et al. 1979), Bolivia, Colombia (CABI and EPPO 2005)	
35.	Southern blight, root rot	SR	*Sclerotinia rolfsii* Sacc. (teleomorph: *Athelia rolfsii* (Curzi) Tu and Kimbrough)	India	DNA[a]
36.	Stem blight	NO	*Nigrospora oryzae* (Berk. and Broome) Petch	Canada (Vaartnou et al. 1974), India (Sharma et al. 2013)	Minor
37.	Verticillium wilt	Vl (VD,VA)	*Verticillium longisporum* (*Verticillium dahliae* Kleb. *V. albo-atrum* Reinke and Berth).	Denmark, France, Germany, Poland, Sweden, Ukraine, USA (Karapapa and Typas 2001)	*B. napus* (10–50) (Daebeler et al. 1988)
38.	Virus diseases a. a. Cauliflower b. mosaic virus c. b. Mustard mosaic d. virus e. c. Radish mosaic	CAM (V) MMV RM(V) (Comovirus) TUM(V) BMY(V)	Cauliflower mosaic virus (Caulimovirus) Mustard mosaic virus Radish mosaic virus Turnip mosaic virus Beet western yellows virus	Asia, Canada, Europe, Kenya, North America, UK (Li et al. 2019)	*B. napus* (70) (Hardwick et al. 1994) *B. oleracea* (50) (Spence et al. 2007)

(continued)

Table 5.1 (continued)

Sr. no.	Common name	Symbol[a]	Causal organism	Geographical distribution	Yield loss (World average %)
	f. virus g. d. Turnip mosaic h. virus i. e. Beat yellows j. virus k. f. Cucumber mosaic l. virus m. g. Turnip yellow n. mosaic virus- o. h. Turnip crinkle p. virus q. i. Radish ilar virus	Cucumovirus Tymovirus Carmo virus Ilar virus			
39.	White rust, stag head	AC	*Albugo candida* (Pers. Ex. Lev.) Kuntze (*Hyaloperonospora* sp. commonly present in staghead phase)	**Africa:** Egypt, Kenya, Libya, Malawi, Mauritius, Sierra Leone, Sudan, Tanzania (CABI and EPPO 2001), Ethiopia (Liu and Rimmer 1991) , South Africa (Holtzhausen 1978, Morris and Knox-Davies 1980), Tunisia (Djebali et al. 2009) **Asia:** Bhutan, Hong Kong (CABI and EPPO 2001), China (Zhang et al. 1984), India (Rai et al. 1974, Singh et al. 1993), Iran (Etebarian 1993), Iraq, Israel, Japan, Malaysia, Nepal,	Rapeseed (29–47.5), *B. rapa* var. *toria* (34), *B. juncea* (23–54.5) Saharan et al. (2014)

				Philippines, Taiwan, Turkey, Yemen (CABI and EPPO 2001), Pakistan (Jalaluddin et al. 1993), Saudi Arabia (Baka 2008), South Korea (So et al. 1981) **Europe:** Austria (Glaeser 1973; Szith and Furlan 1979), Belgium (Vanparys et al. 1993, Vanparys and Calus 1994), Bulgaria (Stefcheva et al. 2005), Croatia (Cvjetkovic 2008), Cyprus, Denmark, Finland, France, Hungary, Latvia, Malta, Romania, Russia, Serbia and Montenegro, Slovakia, Spain, Sweden, Switzerland, Union of Soviet Socialist Republics (CABI and EPPO 2001), Germany (Bedlan 1987, Paul and Rawlinson 1992), Greece (Vakalounakis 1991), Ireland (Ryan 1986), Italy (Minuto et al. 1997), Netherlands (Biesheuvel 1991, Meier 1996), Poland (Kurzawinska and Poniedziałek 1987, Macias 1996), Portugal (de Dias and Lucas 1978), Ukraine	

(continued)

Table 5.1 (continued)

Sr. no.	Common name	Symbol[a]	Causal organism	Geographical distribution	Yield loss (World average %)
				(Burdyukova and Dudka 1982, Bogarada et al. 1984), United Kingdom (Brooks 1975, Humpherson-Jones 1991) **North America:** Barbados, Bermuda, Canada, Cuba, Dominican Republic, El Salvador, Jamaica, Puerto Rico, Trinidad and Tobago, (CABI and EPPO 2001), Mexico (Ponce and Mendoza 1983), United States (Farr et al. 1995) **Oceania:** Australia, Cook Islands, Fiji, French Polynesia, New Caledonia, New Zealand, Papua New Guinea, Samoa, Vanuatu (CABI and EPPO 2001) **South America:** Argentina, Brazil, Falkland Islands, Guyana, Suriname (CABI and EPPO 2001)	
40.	White leaf spot, gray stem	PC	*Pseudocercosporella capsellae* (Ell. and Ev.) Deighton	Australia (Barbetti and Sivasithampuram 1981, Eshraghi et al. 2005), Algeria, Antigua, Australia, Bhutan, Belgium, Canada,	DNA[a]

				Chile, Denmark. Ethiopia, France, Germany, India, Irish Republic, Israel, Japan, Kenya, Malaysia, Nepal, New Zealand, Norway, Romania, South Africa, Sri Lanka, Sweden, Taiwan, Turkey, UK, USA, Russia (Commonwealth Mycological Institute 1986)	
41.	Wilt	FO	*Fusarium oxysporum* f. sp. *conglutinans/ Fusarium oxysporum* Schlecht. spp.	Argentina (Gaetan 2005); AFRICA, Cameroon, Morocco, South Africa, Zaire, Zimbabwe, ASIA, China (Canton), India (West Bengal), (Assam), Iraq, Japan, Philippines, Thailand, Vietnam, AUSTRALASIA & OCEANIA, Australia (Queensland), (NT), New Caledonia, New Zealand, Samoa (Am.), EUROPE, France, Hungary, Italy (Sardinia), Netherlands, USSR (Lithuania), (Ukraine), NORTH AMERICA, Canada (Quebec), (Manitoba, Ontario), USA, CENTRAL AMERICA and WEST	DNA[a]

(continued)

Table 5.1 (continued)

Sr. no.	Common name	Symbol[a]	Causal organism	Geographical distribution	Yield loss (World average %)
				INDIES, Costa Rica, Cuba, Panama, Puerto Rico, Salvador, Trinidad, SOUTH AMERICA, Brazil (CAB International 1986)	
42.	*Xanthomonas* black rot	XCC	*Xanthomonas campestris* pv. *Xanthomonas arboricola* strain 3004 on *Brassica*	Russia (Kyrova et al. 2020)	DNA[a]
43.	Yellows	FOC	*Fusarium oxysporum* Schlecht. spp.	DNA[a]	DNA[a]

[a]*DNA* data not available

Table 5.2 Abiotic stresses of crucifers (Saharan 1984, 1992; Kolte 1985; Williams 1985; Saharan et al. 2014, 2017, 2019; up dated 2020)

Sr. no.	Common cause	Stress/ Problem	Symbol	Cause	Symptoms	Distribution	Yield losses (World average %)	References
1.	Abiotic stress	Frost	F	Mechanical strain on the cell wall and membrane leading to cell rupture and death of the tissue	Reduced leaf expansion, wilting, chlorosis and necrosis and membrane damage due to acute dehydration associated with freezing	Canada	70	Cerkauskas et al. (1998), Olien and Smith (1997)
2.		Draught	D	Induced adverse effects by secondary or even tertiary pathways. Affects growth and physiological processes like leaf area, root growth, shoot growth, excised leaf water loss (ELWL), relative water content (RWC), photosynthesis and water use efficiency (WUE) by disturbing stomatal adjustment, producing reactive oxygen species (ROS),	Depressed seedling emergence, early seedling mortality, sub-optimal plant stand, decline in photosynthesis, delayed growth and removal of sinks in extreme situation	India	31	Sharma et al. (2015), Chinnusamy et al. (2004), Elferjani and Soolanayakanahally (2018)

(continued)

Table 5.2 (continued)

Sr. no.	Common cause	Stress/ Problem	Symbol	Cause	Symptoms	Distribution	Yield losses (World average %)	References
				lowering stomatal conductance and destroying the chlorophyll and other pigments				
3.		High temperature	HT	Increases in temperature, Photosynthetic assimilation altered, stomatal conductance reduced	Reduced oil content, and fatty acid composition	Canada	85.3	Canvin (1965), Elferjani and Soolanayakanahally (2018)
4.		Salt	S	Saline contain chloride and sulfates of Na, K (Electrical conductivity >4 dS/m, pH 7.5–8.0); Alkali contain Carbonate and bicarbonate of Na (Electrical conductivity <2 dS/m, pH 7.5–14.0)	Change the response of plants, limits growth through enhanced ion toxicity, photosynthetic activity, transpiration rate, stomatal conductance, assimilation rate and internal CO_2 affected	DNA[a]	Upto 84	Sharma et al. (2015), Chinnusamy et al. (2004), Elferjani and Soolanayakanahally (2018)
5.	Phytohormones stress	Jasmonate-	JA	C-repeat binding factor (CBF) pathway during	regulating plant defense against	DNA[a]	DNA[a]	Kazan (2015)

				cold stress, have complex regulatory roles, mediated signaling	abiotic stresses, pests and pathogens			
6.		ethylene-	ET	have complex regulatory roles	regulating plant defense against abiotic stresses, pests and pathogens	DNA[a]	DNA[a]	Kazan (2015)
7.		Proline	PR	accumulation of proline	contents in young seedlings greatly increased under drought and salinity stress conditions	DNA[a]	DNA[a]	DNA[a]
8.	Macronutrient stress	Nitrogen	N	Initially show anthocyanin pigmentation on leaves	Gradually fully mature older leaves turned to a light purple color and finally to uniform purple yellow color followed by dried up and withered	DNA[a]	30	Meena et al. (2016), Sorin et al. (2016), Grant and Bailey (1993)
9.		Phosphorus	P	Plants light green with leaves light green or yellow, no necrotic spotting	Initially bluish dark patches appeared on matured leaves of phosphorus deficient plant and gradually turned into pale color and finally necrosis of upper half leaves.	DNA[a]	DNA[a]	Meena et al. (2016), Grant and Bailey (1993)

(continued)

Table 5.2 (continued)

Sr. no.	Common cause	Stress/ Problem	Symbol	Cause	Symptoms	Distribution	Yield losses (World average %)	References
10.		Potassium	K	Leaf edge scorch in sever deficiency. The chlorosis followed by necrosis starts from margins of lower and/or middle and gradually spread to the inner surface of leave.	K-deficient plants showed stunted growth, leaves become small, fleshy, deformed and curled downward.	DNA[a]	DNA[a]	Meena et al. (2016), Grant and Bailey (1993)
11.		Sulfur	S	Young leave light green, typically no chlorotic spotting or striping	General, chlorosis, necrosis, yellowing, shrinking, cupping or purpling of leaves, reddish-purple discoloration of stems or pods, delay of flowering and maturation, and small and poorly filled pods.			Meena et al. (2016), Schnug and Haneklaus (2005), Grant and Bailey (1993)
12.		Calcium	Ca	important in structural and metabolic functions	Deficiency causes membrane leakage of low-molecular weight compounds,	DNA[a]	DNA[a]	Meena et al. (2016), Grant and Bailey (1993)

13.		Magnesium	Mg	important in enzymatic reactions and as a component of chlorophyll		Canada	DNA[a]	Cerkauskas et al. (1998), Grant and Bailey (1993)
14.	Micronutrient stress	Iron	Fe	Sharp distinction between veins and chlorotic areas	Chlorosis starts from the base and spread towards lamina. Bleached leaves, silique number and size decrease and appear chlorotic	DNA[a]	DNA[a]	Meena et al. (2016)
15.		Manganese	Mn	No sharp distinction between veins and chlorotic areas, spotty appearance	Chlorosis in leaves, followed by developing small grayish-brown spots which coalesce to form large necrotic lesions.	DNA[a]	DNA[a]	Meena et al. (2016)
16.		Copper	Cu	Chlorosis of young leaves, tips appear withered and will eventually die	Young leaves develop interveinal chlorosis. Reduced growth at the time of flowering. Inflorescence is very poorly developed. Pod formation and seed setting are also reduced.	DNA[a]	DNA[a]	Meena et al. (2016), Billard et al. (2014)
17.		Zinc	Zn	Middle leaves with interveinal chlorosis stunted growth	There is retardation of growth, after 20 days of sowing.	DNA[a]	DNA[a]	Meena et al. (2016)

(continued)

Table 5.2 (continued)

Sr. no.	Common cause	Stress/ Problem	Symbol	Cause	Symptoms	Distribution	Yield losses (World average %)	References
					Leaves are small in size with pinkish margin. Interveinal tissues turn yellowish white to papery white, with veins remaining green. Severely affected leaves die. Flowering and fruiting is delayed.			
18.		Molybdenum	Mo	Plants light green, necrotic spotting on leaves, pale leaves sometimes scorched, cupped or rolled	Growth is markedly reduced and plants develop foliar symptoms like cupping, marginal scorching and loss of lamina.	DNA[a]	DNA[a]	Meena et al. (2016), Billard et al. (2014)
19.		Boron	B	Young leaves of terminal bud become light green at bases, leaves become twisted and brittle and die back at growing point, chlorosis of young leaves	Cell walls are dramatically affected by B deficiency. This shows up as cracked, hollow or corky stems. The cell wall diameter and proportion of plant	DNA[a]	India, Canada	Meena et al. (2016), Grant and Bailey (1993), Cerkauskas et al. (1998), Wei et al. (2002)

					dry weight increases under B-deficiency. Most plant B is complexes with organic compounds in the cell walls, apparently serving a nonspecific structural role.			

[a]*DNA* data not available

Table 5.3 Weed infestation in the field affecting crucifer's yields

Sr. no.	Common name	Scientific name	Yield loss (World average %)	Geographical distribution	References
1.	Perennial sow thistle	*Sonchus arvensis* L.	23 to 64	Finland	Salonen et al. (2011), Bagherani and Shimi (2001)
2.	Bathua, fat hen, lamb's quarters	*Chenopodium album* L.	45	Iran, India	Valizadeh and Mirshekari (2011)
3.	Hemp-nettle, Brittle stem hemp nettle	*Galeopsis* L. spp.	DNA[a]	DNA[a]	
4.	Scentless or false mayweed	*Tripleurospermum inodorum* (L.) Sch. Bip.	DNA[a]	DNA[a]	
5.	Creeping thistle or Kandai	*Cirsium arvense* (L.) Scop.	DNA[a]	DNA[a]	
6.	Chickweed	*Stellaria media* (L.) Vill.	DNA[a]	DNA[a]	
7.	Quackgrass, Couch grass, Dog grass	*Elymus repens* (L.) Gould	DNA[a]	DNA[a]	
8.	Black bindweed	*Fallopia convolvulus* (L.) À. Löve	DNA[a]	DNA[a]	
9.	Curlytop knotweed	*Persicaria lapathifolia* (L.) Gray	DNA[a]	DNA[a]	
10.	Dandelion	*Taraxacum officinale* Weber in Wigg.	DNA[a]	DNA[a]	
11.	Field penny cress	*Thlaspi arvense* L.	DNA[a]	DNA[a]	
12.	Mugwort	*Artemisia vulgaris* L.	DNA[a]	DNA[a]	
13.	Disc mayweed	*Matricaria matricarioides* (Less.) Porter	DNA[a]	DNA[a]	
14.	Knotweed	*Polygonum aviculare* L.	DNA[a]	DNA[a]	
15.	Marsh Woundwort, Marsh hedge nettle	*Stachys palustris* L.	DNA[a]	DNA[a]	
16.	grass-leaf starwort	*Stellaria graminea* L.	DNA[a]	DNA[a]	

(continued)

Table 5.3 (continued)

Sr. no.	Common name	Scientific name	Yield loss (World average %)	Geographical distribution	References
17.	field violet	*Viola arvensis* Murray	DNA[a]	DNA[a]	
18.	Garden cornflower, Blue Bottle,	*Centaurea cyanus* L.	DNA[a]	DNA[a]	
19.	Bottle brush, bull pipes	*Equisetum arvense* L.	DNA[a]	DNA[a]	
20.	Drug fumitory	*Fumaria officinalis* L.	DNA[a]	DNA[a]	
21.	Red Dead Nettle, Purple dead nettle	*Lamium purpureum* L.	DNA[a]	DNA[a]	
22.	Nipplewort	*Lapsana communis* L.	DNA[a]	DNA[a]	
23.	Mouseear cress	*Arabidopsis thaliana* (L.) Heynh.	DNA[a]	DNA[a]	
24.	Shepherd's Purse	*Capsella bursa-pastoris* (L.) Medik.	DNA[a]	DNA[a]	
25.	Alpine Cress, Alpine Bitter	*Cardamine* spp.	DNA[a]	DNA[a]	
26.	Madwoman's Milk	*Euphorbia helioscopia* L.	DNA[a]	DNA[a]	
27.	Field forget-me-not	*Myosotis arvensis* (L.) Hill.	DNA[a]	DNA[a]	
28.	Broadleaf plantain	*Plantago major* L.	DNA[a]	DNA[a]	
29.	Broad Leaf Plantain	*Polygonum* spp.	DNA[a]	DNA[a]	
30.	Wild Radish	*Raphanus raphanistrum* L.	DNA[a]	DNA[a]	
31.	Door-yard dock	*Rumex longifolius* DC.	DNA[a]	DNA[a]	
32.	Potato	*Solanum tuberosum* L.	DNA[a]	DNA[a]	
33.	Corn spurry	*Spergula arvensis* L.	DNA[a]	DNA[a]	
34.	Alsike Clover	*Trifolium hybridum* L.	DNA[a]	DNA[a]	
35.	White clover	*Trifolium* spp.	DNA[a]	DNA[a]	
36.	Coltsfoot	*Tussilago farfara* L.	DNA[a]	DNA[a]	

(continued)

Table 5.3 (continued)

Sr. no.	Common name	Scientific name	Yield loss (World average %)	Geographical distribution	References
37.	California nettle	*Urtica dioica* L.	DNA[a]	DNA[a]	

[a]*DNA* data not available

nature of interactions between biotic and abiotic stress responses, significant aberrant signaling crosstalk in response to combined stresses and a degree of overlap, but unique response to each environmental stimulus. Further, the results of simultaneous combined biotic and abiotic stress studies indicate that abiotic stresses particularly heat and drought enhance plant susceptibility to plant pathogens. Plant-biotic and/or abiotic stress interactions trigger a wide range of defense reactions at molecular and cellular levels to ensure that plant cells are insulated from the challenges they encounter (Kim et al. 2014) (Fig. 5.1).

A sophisticated defense network is activated to orchestrate transcriptional reprogramming using a two-tiered defense approach (Dodds and Rathjen 2010; Tsuda and Katagiri 2010). Pathogen/microbe associated molecular patterns (PAMP/MAPM)-triggered immunity (PTI) is the first line of defense. It is a general and nonspecific defense response and provides basal resistance not only against entire classes of microbial pathogens but also abiotic stresses. PTI constitutes the inducible defense through an array of early cellular responses comprising of ion flux across the membrane, production of reactive oxygen species (ROS) and mitogen-activated protein kinase (MAPK) cascades phosphorylation and long-term later responses that induce callose deposition (Nakagami et al. 2005; Dodds and Rathjen 2010; Tsuda and Katagiri 2010; Schwessinger et al. 2015).

Defense-related phytohormones comprise salicylic acid (SA), jasmonic acid (JA), ethylene (ET) and abscisic acid (ABA) as the core immune signaling components. They are produced in response to both, biotic and abiotic stresses (Dodds and Rathjen 2010; Tsuda and Katagiri 2010). Among those, ROS, MAPkinase cascades and hormone signaling pathways are of paramount importance as they lead the defense responses to biotic and abiotic stresses (Nakagami et al. 2005; Fujita et al. 2006; Atkinson and Erwin 2012; Kissoudis et al. 2014; Perez and Brown 2014; Schwessinger et al. 2015) (Fig. 5.2). Nejat and Mantri (2017) proposed that PTI which is activated either by PAMP or MAPM is the "stress triggered immunity" as it has evolutionary conserved molecular signatures across similar types of microbes or abiotic stresses. The effector-triggered immunity (ETI) is the second, prolonged and robust layer of plant immune response that is specific to the pathogen infection. It is activated by resistance (*R*) genes when pathogen virulence factors, called effectors, are released into plant cells (Cui et al. 2015). If the first line of inducible nonspecific defense, so-called PTI, is successfully overcome by pathogens through suppression of PTI response, then the second line of plant immunity is activated to produce defense proteins that cause programmed cell death. This is also referred to as the

Table 5.4 Insect-pests' stresses on the yield of crucifers

Sr. no.	Common name of the insect	Scientific name	Geographical distribution	Yield losses (World average %)	References
	Coleoptera				
1.	Bronzed field beetle	*Adelium brevicorne*	Australia	10–19	Ahuja et al. (2009), Kumar and Singh (2015)
2.	Black headed pasture cockchafer	*Aphodius tasmaniae*	Australia	DNA[a]	Kumar and Singh (2015)
3.	Cabbage seed weevil	*Ceutorhynchus assimilis*	Europe, N. America (Harmon and McCaffrey 1997), Canada (Dolinski 1979)	DNA[a]	Free and Williams (1978)
4.	Rape stem weevil	*Ceutorhynchus napi*	Europe, N. America	DNA[a]	
5.	Cabbage curculio	*Ceutorhynchus rapae*	Europe, N. America, Russia	DNA[a]	
6.	Cabbage seed pod weevil	*Ceutorhynchus obstrictus*	Europe	DNA[a]	Evans and Bergeron (1994)
7.	Cabbage gall weevil	*Ceutorhynchus pleurostigma*	Europe	DNA[a]	
8.	Cabbage stem weevil	*(Ceutorhynchus quadridens)*	Europe	DNA[a]	Graham and Gould (1980)
9.	Desiantha weevil	*Desiantha diversipes*	Australia	DNA[a]	
10.	Red turnip beetle	*Entomoscelis americana*	N. America, Canada	DNA[a]	Gerber (1976)

(continued)

Table 5.4 (continued)

Sr. no.	Common name of the insect	Scientific name	Geographical distribution	Yield losses (World average %)	References
11.	Grey false wireworm	*Isopteron punctatissimus*	Australia	DNA[a]	
12.	Vegetable weevil	*Listroderes difficilis*	Australia	DNA[a]	
13.	Pollen beetle	*Meligethes aeneus*	Asia, Europe, N. America, N. Africa, Russia, Australia	80 (Hensen 2004)	Nilsson (1987), Buchi (2002), Cook et al. (1999)
14.	Leaf beetle	*Phyllotreta aerea*	Europe, N. America	DNA[a]	
15.	Crucifer flea beetle	*Phyllotreta cruciferae*	Europe, India, N. America, N. Africa, Canada	DNA[a]	Cerkauskas et al. (1998), Kumar and Singh (2015)
16.	Yellow striped flea beetle	*Pyllotreta nemorum*	Europe	DNA[a]	
17.	Striped flea beetle	*Phyllotreta striolata*	Europe, Canada	DNA[a]	Burgess (1977), Bakhetia and Labana (1978), Augustin (1986)
18.	Turnip flea beetle	*Phyllotreta undulata*	Australia, Europe, N. America, Canada	DNA[a]	Augustin (1986)
19.	Cabbage stem flea beetle	*Psylliodes chrysocephala*	Asia, Europe, N. America	DNA[a]	Alford (1979)
20.	Sitona weevil	*Sitona discoideus*	Australia	DNA[a]	
	Diptera				
21.	Cabbage maggot	*Contarinia nasturtii*	Europe, Canada	DNA[a]	
22.	Brassica pod midge	*Dasineura brassicae*	Europe	DNA[a]	Axelsen (1992)

23.	Turnip maggot	*Delia floralis*	Europe, Canada, China, Japan, Russia	DNA[a]	
24.	Root maggot	*Delia platura/ planipalpis/ florilega*	Canada	DNA[a]	
25.	Cabbage maggot	*Delia radicum*	Canada	DNA[a]	Griffiths (1986)
26.	Cruciferous leaf miner	*Phytomyza horticola*	Asia, Europe, India	DNA[a]	Cerkauskas et al. (1998), Kumar and Singh (2015)
27.	Agromyzid leaf miner	*Chromatomyia horticola Gourear*	Kashmir, India	DNA[a]	Bhat and Bhagat (2009)
	Heteroptera				
28.	Painted bug	*Bagrada hilaris, B. cruciferarum*	Africa, Arabia, India, Sri Lanka	DNA[a]	Cerkauskas et al. (1998); Kumar and Singh (2015)
29.	Cabbage bug	*Eurydema oleracea*	Europe, Russia	DNA[a]	
30.	Orange stink and shield bugs	*Eurydema* spp.	Asia, Australia, Europe, India, N. Africa, Russia, Philippines	DNA[a]	
31.	Lygus bug	*Lygus borealis*	Canada	3–5	Turnock et al. (1995)
32.	Pale legume bug	*Lygus elisus*	N. America	DNA[a]	
33.	Western tarnished plant bug	*Lygus hesperous*	N. America	DNA[a]	
34.	Tarnished plant bug	*Lygus lineolaris*	N. America	DNA[a]	

(continued)

Table 5.4 (continued)

Sr. no.	Common name of the insect	Scientific name	Geographical distribution	Yield losses (World average %)	References
35.	European tarnished plant bug	*Lygus rugulipennis*	Europe, Russia	DNA[a]	
36.	Harlequin bug	*Murgantia histrionica*	USA	DNA[a]	
37.	Rutherglen bug	*Nysius vinitor*	Australia	DNA[a]	
	Homoptera				
38.	White flies	*Aleyrodes* spp.	Europe, N. America, Russia, South Pacific Australia	DNA[a]	
39.	Cowpea aphid	*Aphis craccivora*	Australia	DNA[a]	
40.	Mealy Cabbage aphid	*Brevicoryne brassicae L.*	Worldwide	DNA[a]	Cerkauskas et al. (1998); Kumar and Singh (2015)
41.	Turnip or Mustard Aphid	*Lipaphis erysimi)*	Worldwide	50/10–90 (Ahuja et al. 2009), 75 (Sekhon 1999), 26.9 Indian mustard	Cerkauskas et al. (1998); Kumar and Singh (2015), Bakhetia (1983), Kumar (2017)
42.	Green peach aphid	*Myzus persicae)*	Worldwide (warmer climates)	DNA[a]	Cerkauskas et al. (1998), Kumar and Singh (2015)
43.	Turnip root aphid	*Pemphigus populitransversus*	USA	DNA[a]	
	Hemiptera				
44.	Silver leaf whitefly	*Bemisia tabaci*	India	DNA[a]	Singh et al. (2005)

45.	*Lygus elisus*		Canada	DNA[a]	Lamb (1989)
46.	*Lygus lineolaris*		Canada	DNA[a]	Lamb (1989)
47.	Painted bug	*Bagrada cruciferarum*	India	DNA[a]	Bakhetia and Labana (1978)
	Hymenoptera				
48.	Mustard Sawfly	*Athalia lugens proxima*	India	DNA[a]	Bakhetia and Labana (1978)
49.	Turnip sawfly	*Athalia rosae*	Africa, Asia, Europe, India, Japan, N. America	DNA[a]	
	Lepidoptera				
50.	Common cutworm	*Agrotis infusa*	Australia	DNA[a]	
51.	Brown cutworm	*Agrotis munda*	Australia	DNA[a]	
52.	Pale western cutworm	*Agrotis orthogonia*)	Canada	DNA[a]	
53.	Turnip moth	*Agrotis segetum*	Worldwide	DNA[a]	Singh et al. (2005)
54.	Alfalfa looper	*Autographa californica*	N. America	DNA[a]	
55.	Tobacco loop caterpillar	*Chrysodeixis argentifera*	Australia	DNA[a]	
56.	Brown pasture looper	*Ciampa arietaria*)	Australia	DNA[a]	
57.	Cabbage head caterpillar	*Crocidolomia binotalis*	Asia, India	DNA[a]	

(continued)

Table 5.4 (continued)

Sr. no.	Common name of the insect	Scientific name	Geographical distribution	Yield losses (World average %)	References
58.	Cabbage cluster caterpillar	*Crocidolomia pavonana*	Africa, India, South Pacific	DNA[a]	
59.	Jute hairy caterpillar	*Spilosoma obliqua*	Asia, India	DNA[a]	
60.	Clover cutworm	*Discestra trifolii*	Canada	DNA[a]	
61.	Bihar hairy Caterpillar	*Diacrisia obliqua*	India	DNA[a]	Bakhetia and Labana (1978)
62.	Red backed cutworm	*Euxoa ochrogaster*	Canada	DNA[a]	
63.	Garden pebble moth	*Evergestis forficalis*	Europe	DNA[a]	
64.	Purple-backed cabbage worm	*Evergestis pallidata*	Europe, N. America	DNA[a]	
65.	Cross-striped cabbage worm	*Avergestis rimosalis*	N. America	DNA[a]	
66.	Native budworm	*Helicoverpa punctigera*	Australia	DNA[a]	
67.	Cotton bollworm	*Helicoverpa armigera*	India	DNA[a]	
68.	Cabbage centre-grub	*Hellula hydralis*	Australia	DNA[a]	
69.	Cabbage head borer	*Hellula undalis*	India	DNA[a]	

70.	Cabbage webworm	*Hellula rogatalis*	USA	DNA[a]	
71.	Beet webworm	*Loxostege sticticalis*	Canada	DNA[a]	
72.	Cabbage moth	*Mamestra brassicae*	Asia, Europe, Japan, Russia	80	Kumar and Singh (2015)
73.	Bertha armyworm	*Mamestra configurata*	Canada, N. America	DNA[a]	Bracken (1987)
74.	Large white butterfly	*Pieris brassicae*	China, Europe, India, Russia, N. Africa	18.3	Bakhetia and Labana (1978), Cerkauskas et al. (1998), Kumar and Singh (2015), Kumar (2017)
75.	Imported cabbage worm or small white butterfly	*Pieris rapae*	Asia, Australia, Europe, Japan, New Zealand, N. America	DNA[a]	
76.	Mustard white butterfly	*Pieris napi*	Europe, Japan, N. America	DNA[a]	
77.	Indian cabbage white	*Pieris canidia*	India and adjoining countries	DNA[a]	
78.	Diamond back moth	*Plutella xylostella* L.	India, Canada, Australia, Worldwide	90 (Charleston and Kfir 2000)	Palaniswamy et al. (1986), Bakhetia and Labana (1978), Cerkauskas et al. (1998), Kumar and Singh (2015)
79.	Beet armyworm	*Spodoptera exigua*	Worldwide	DNA[a]	
80.	Tobacco cutworm	*Spodotera litura*	India, Australia	DNA[a]	
81.	Cabbage looper	*Trichoplusia ni*	Worldwide	DNA[a]	
82.	Painted lady butterfly	*Vanessa cardui*	Canada	DNA[a]	

(continued)

Table 5.4 (continued)

Sr. no.	Common name of the insect	Scientific name	Geographical distribution	Yield losses (World average %)	References
83.	Jute hairy caterpillar or Bihar hairy caterpillar	*Spilosoma obliqua*	India	DNA[a]	Singh et al. (2005)
84.	Pea semilooper/ golden-wing moth	*Thysanoplusia orichalcea*	Asia, Africa, Europe, America, Oceania, Kenya	DNA[a]	
	Thysanoptera				
85.	Plague thrips	*Thrips imaginis*	Australia	DNA[a]	
86.	Onion thrips	*Thrips tabaci*	Worldwide	DNA[a]	
	Isoptera				
87.	Termite	*Odontotermus obesus*	Worldwide	DNA[a]	Singh et al. (2005)

[a]*DNA* data not available

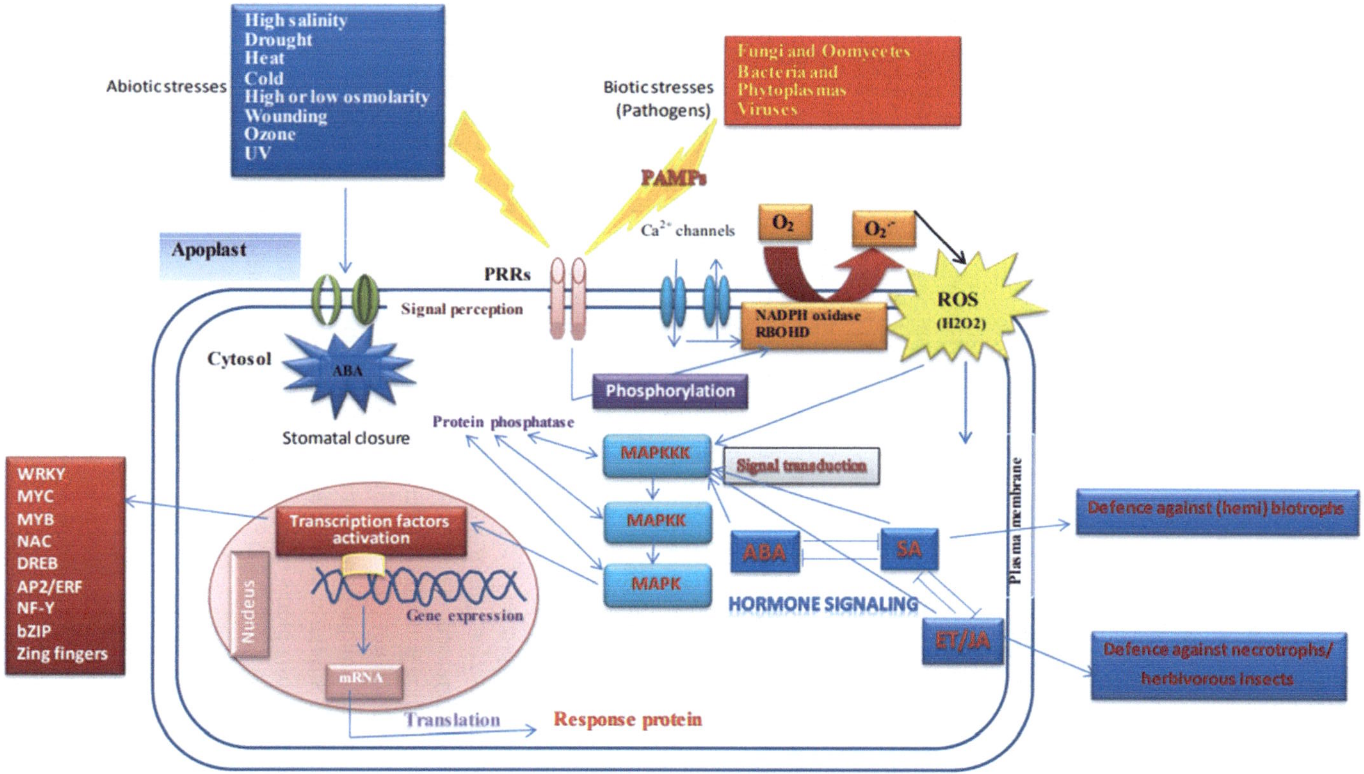

Fig. 5.1 A model of host resistance response to multiple stresses. Perception of pathogen-associated molecular patterns (PAMPs), sensing abiotic stress and/or any extracellular signals by surface-localized pattern recognition receptors (PRRs) trigger phosphorylation of the RBOHD (respiratory burst oxidase-D) and activate the NADPH oxidase RBOHD for rapid production of reactive oxygen species (ROS) in calcium-independent or -dependent manner, which

hypersensitive response (HR) (Thomma et al. 2011). Hence, after stress recognition, the immune system triggers a wide range of defense mechanisms to orchestrate transcriptional reprogramming through receptor proteins, signal transduction cascades, kinase cascades, reactive oxygen species, hormone signaling pathways, heat shock proteins, and transcription factors to protect plants against multiple stresses (Fig. 5.1). However, recent advances suggest significant aberrant signaling crosstalk in plant resistance response to multiple stresses compared with an individual stress (Atkinson et al. 2013; Prasch and Sonnewald 2013; Suzuki et al. 2014).

5.2 Complex Network of Signaling Pathways in Host Resistance to Multiple Stresses

Host responses to multiple stresses involve complex network of signaling pathways. Increasing evidence suggest that there is significant overlap and several defense genes are commonly involved in response to multiple biotic and abiotic stresses as shared or general stress-responsive genes (Table 5.5) (Sham et al. 2014,2015; Zhang et al. 2016). Nevertheless, in response to combination of stresses, plants orchestrate extensive transcriptional reprogramming and exhibit a unique program of transcript response that is not similar to that of either stresses individually, although there is significant overlap in response pathways to combined environmental stresses (Mantri et al. 2010a, 2010b; Atkinson 2011; Prasch and Sonnewald 2013; Rasmussen et al. 2013; Sham et al. 2015). There is crosstalk between responses to combined stress treatment synergistically or antagonistically. Of note, abiotic stresses have a negative impact on biotic stress resistance and can lead to enhanced plant susceptibility to biotic stresses (Wang et al. 2009; Atkinson 2011; Prasch and Sonnewald 2013; Sewelam et al. 2014; Kissoudis et al. 2015).

The molecular responses of *A. thaliana* to the concurrent drought stress and plant-parasitic nematode, *Heterodera schachtii,* infection using microarrays revealed that drought stress increased the susceptibility of *A. thaliana* to nematode infection. Moreover, induction patterns of differentially expressed genes following these stress treatments displayed not only a specific response to each stress but also a particular response that was uniquely activated by the combination of biotic/abiotic stress (Atkinson 2011). Similarly, combined heat and drought stress increased the susceptibility of *Arabidopsis* plants to *Turnip mosaic virus* infections through suppression of defense responses to the biotic stress (Prasch and Sonnewald 2013).

Fig. 5.1 (continued) subsequently trigger mitogen-activated protein kinase (MAPK) phosphorylation as general defense response. MAPK transducer extracellular signals to nucleus leading to activation of transcription factors that regulate immunity gene expression. There is antagonistic and synergistic crosstalk between hormone signal transduction pathways in response to various attackers. *ABA* abscisic acid, *SA* salicylic acid, *JA* jasmonic acid, *ET* ethylene (Nejat and Mantri 2017)

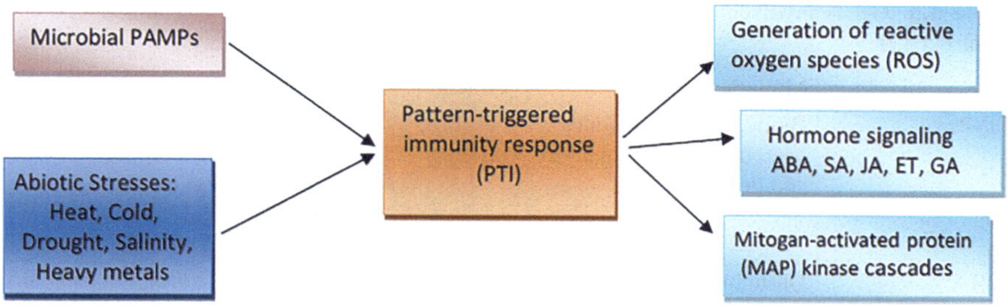

Fig. 5.2 Activation of the first line of defense in a nonspecific approach. *ABA* Abscisic acid, *SA* Salicylic acid, *JA* Jasmonic acid, *ET* Ethylene, *GA* Gibberellins (Nejat and Mantri 2017)

In *A. thaliana*, using microarray mRNA expression profiling, it was found that the genes regulated in response to *Botrytis cinerea* were more similar to those induced by osmotic stress than genes differentially expressed in response to heat and salt stresses. Conversely, 13 stress-inducible and 29 stress-repressible genes were responsive to all four stresses (Sham et al. 2015). The transcriptome analysis of potato using RNAseq in response to multiple stresses revealed several genes. The results exhibited that pathogen-related genes are more specifically induced than abiotic- and hormone-related genes and hormones and abiotic stresses share proportionately more differentially expressed genes than each does with the biotic stress. Further, a weighted gene co-expression network analysis revealed that 50% of the differentially expressed genes were not specific and expressed in response to two or more stress conditions, while the 50% remaining were stress specific genes that were only regulated in response to single environmental factor (Massa et al. 2013). The results exhibited a degree of overlap in the defense response to pathogen infection and abiotic stresses, but a unique response to each environmental stress. Weston et al. (2011) using a weighted gene co-expression network approach found that heat shock proteins respond to heat stress in a species-specific manner. A gene co-expression network analysis of 1094 microarrays of *Arabidopsis* using a non-target approach revealed that genes involved in photosynthesis, protein synthesis, and response to oxidative stress are also largely involved in response to environmental stresses (Mao et al. 2009).

The course of day affected transcriptome response of the plant to drought and the response can vary with the time of day (Wilkins et al. 2010). In addition, responses of different plant species to drought are not just affected by the time of day but also clone history and location (Raj et al. 2011). The impact of individual environmental history on response to drought stress and their results showed that differences in transcript abundance patterns in response to drought were based on differences in geographic origin of clones, suggesting the interplay between genotype and environment.

The level of crosstalk between different defense response pathways and plant immune responses to single stresses and the combination of different environmental stimuli depend upon several factors such as plant species, tissues, developmental

Table 5.5 Genes involved in response to multiple stresses to *Arabidopsis thaliana* (Nejat and Mantri 2017)

Gene	Multiple stresses	Function	References
HSPs	Biotic, drought, salinity, heat, osmotic[a]; Nematode and drought[b]	Stabilize proteins	Sham et al. (2014, 2015), Ma and Bohnert (2007), Rasmussen et al. (2013), Sewelam et al. (2014), Atkinson (2011)
ABA	Drought, cold, salinity, cold, osmotic[a]	Defense Against abiotic stresses	Seki et al. (2002), Kreps et al. (2002)
Ethylene biosynthesis genes	Drought, cold, salinity[a]	Defense signals	Seki et al. (2002)
Senescence-related genes	Nematode, virus, heat and drought[a]	Defense signals	Atkinson (2011), Prasch and Sonnewald (2013), Kreps et al. (2002)
PR proteins	Virus and heat[b]	Resistance	Prasch and Sonnewald (2013)
Photosynthesis-related genes	Biotic (Fungi, bacteria, virus, nematodes), drought, cold, salinity, heat[b]; drought[a]	Decreased metabolic rate	Mantri et al. (2010b), Prasch and Sonnewald (2013), Shaik and Ramakrishna (2013), Atkinson (2011), Zhang et al. (2016), Narsai et al. (2013)
Cytochrome p450	Nematode and drought[b]; Bacteria, fungi, drought, salt and heat[c]	Cell wall reinforcement	Atkinson (2011), Narsai et al. (2013), Shaik and Ramakrishna (2013)
Peroxidase	Biotic, cold, drought[b]	Generating H_2O_2	Narsai et al. (2013), Shaik and Ramakrishna (2013)

[a]Up-regulated
[b]Down-regulated
[c]Some members were up-regulated and some were down-regulated

stage, stressors, stress intensity, time series, time of day, and geographic origin (Wu et al. 2009; Mantri et al. 2010b; Skibbe et al. 2010; Cramer et al. 2011; Weston et al. 2011; Massa et al. 2013; Shaik and Ramakrishna 2013; Asai and Shirasu 2015; Li et al. 2015; Pandey et al. 2015). The complex nature of interactions and exhibited the complex mechanisms comprising overlap and specific signaling pathways fine-tune plant responses. The multi-parallel single stress along with the combination of single stresses is required to be investigated at spatial and temporal resolution, as comparative spatiotemporal gene expression analysis, to identify a unique response modulated by multiple stresses.

5.3 Molecular Mechanisms of Signaling System

Plants have evolved complex molecular mechanisms enabling them to survive not only against single stresses but also against a combination of stresses and multiple stresses. Understanding complexity of plant defense along with molecular mechanisms and multiple stress signaling is indispensable in order to minimize the impact of stresses on plant growth, development, and productivity and maximize their life span. Studies on plant immune responses to individual and combination of biotic and abiotic stress along with multiple stresses have revealed the complex nature and high degree of signaling crosstalk in responses to various attackers. Nevertheless, understanding signaling crosstalk between components of the plant defense pathways under combined stress conditions is still limited, although it is continually evolving due to remarkable advances in high-throughput sequencing technologies and powerful bioinformatics tools for genome-wide study, transcriptome, and proteome analysis. Studies demonstrate that production of ROS, MAPkinase cascades, and hormone signaling are triggered by both biotic and abiotic environmental stimuli in a nonspecific manner and play crucial roles in the response to biotic and abiotic stresses as common components of the immune system. Receptor proteins as a frontline of the immune system are key regulators of crosstalk between biotic and abiotic signaling. Transcription factors as positive and negative regulators of plant defense also play a significant complex role in crosstalk and fine-tuning responses to combined biotic and abiotic stresses. Though, interactions between TFs involved in the regulation of stress- and hormone responsive genes are still needed to be elucidated. Further, identification of both, common and specific components of the signaling crosstalk in plants, will allow development of novel tools to combat the combination of biotic and abiotic stresses usually encountered in the field conditions. Hence, identification of multifunctional genes and genes responsible for multiple stress resistance would help accelerate development of resistant crop plants through precise genome editing using emerging technologies for targeted genome editing such as Transcription Activator-Like Effector Nucleases (TALENs) and CRISPR (Clustered Regulatory Interspaced Short Palin-dromic Repeats)/Cas9 systems (Nejat and Mantri 2017).

5.3.1 Receptor Proteins as Frontline Defense Signals

Plant pattern-recognition receptors (PRRs) as a frontline of the immune system are an extremely diverse family of cell surface-localized receptors. They are synthesized in the endoplasmic reticulum and transported to the plasma membrane via the secretary pathway (Frescatada-Rosa et al. 2015). PRRs have been classified into receptor-like kinases (RLKs) and receptor-like proteins (RLPs) containing extracellular ligand-binding domains that perceive a wide range of extracellular signals such as PAMPs and abiotic stresses. They trigger the downstream intracellular stress signaling cascades by phosphorylation of intracellular serine/threonine kinase domains (Osakabe et al. 2013; Zipfel 2014; Trdá et al. 2015). RLKs form one of the largest gene super family in the genomes of different plant species with 610 and 1131 members in *Arabidopsis* and rice, respectively (Gish and Clark 2011). They are classified into 44 subfamilies including lysine motifs, lectin, epidermal growth factor-like repeats, self-compatibility domain (S-domain), wall associated kinase (WAK), and leucine-rich repeats (LRRs) (Frescatada-Rosa et al. 2015; Trdá et al. 2015). The latter constitute the largest group of RLKs in plants which play essential roles in a wide range of processes governing growth and development as key regulators, grain yield components improvement, hormone perception, initiating innate immune defense at front-line against microbial pathogens and adaptation to abiotic stresses (Osakabe et al. 2013; Macho and Zipfel 2014; Zou et al. 2015).

There is growing evidence indicating that RLKs can have positive and negative regulatory role in both biotic and abiotic stress response (Tanaka et al. 2012; Zhao et al. 2013; Jun et al. 2015). It has been revealed that *GbRLK*, a receptor like kinase (RLK) gene, involved in response to both biotic and abiotic stresses is a positive regulator. *GbRLK* increased salinity and drought stress tolerance through reduction of water loss rate in leaves and increased sensitivity to ABA and enhanced tolerance to Verticillium wilt in *GbRLK* transgenic cotton and *Arabidopsis* lines (Zhao et al. 2013; Jun et al. 2015). Jun et al. (2015) used transgenic *GbRLK* cotton and *Arabidopsis* lines to reveal that overexpression of *GbRLK* modulated expression of several genes involved in biotic and abiotic stresses and suggested that the increased resistance to *Verticillium dahliae* infection in transgenic plants could be a result of reduction in the water damage losses and regulation of defense-related gene expression (Jun et al. 2015). These findings suggest that *GbRLK* could possibly be a key regulator of both biotic and abiotic signaling.

The ABA and osmotic stress signal transduction in *A. thaliana* is negatively regulated by a receptor-like cytosolic kinase (RLCK) gene, *ARCK1*, that encodes a cytosolic protein kinase belonging to the RLCK family and cysteine-rich repeat (CRR) RLK subfamily (Tanaka et al. 2012). The *arck1* knockout mutant showed higher sensitivity to ABA and osmotic stress than the wild type. Microarray analysis using *CRK36* RNAi transgenic plants exhibited significantly increased ABA-responsive gene expression. *CRK36* is another member of the CRR RLK subfamily that interacts with *ARCK1* to form a complex to fine-tune defense responses against abiotic stresses and adjust growth via hormone signaling according to the growth phase. Hence, *ARCK1* is not involved in the suppression of ABA-responsive genes *per se* (Tanaka et al. 2012). The RLKs are multifunctional receptors which are involved in regulation of plant responses to both biotic and abiotic stresses. It suggests a conserved mechanism for signal perception and activating downstream signaling networks, and hence PRRs could be signal recognition receptors (SRRs). However, a mechanism for crosstalk of signal perception in response to environmental stimuli remains to be addressed.

5.3.2 Transcription Factors as Defense Signaling Pathways

Transcription factors (TFs) include members of the APETALA2/ETHYLENE-RESPONSE ELEMENT BINDING FACTOR (AP2/ERF), basic helix-loop-helix (bHLH), basic domain leucine zipper (bZIP), MYB, NAC, and WRKY families. TFs are the central core of the gene regulatory networks and precisely modulate the transcription rate through either repression or activation of gene expression. They are multifunctional and mediate diverse aspects of developmental processes and responses to various biotic and abiotic stimuli in plants (Tsuda and Somssich 2015). Carrera et al. (2009) performed a gene co-expression network analysis of 1436 microarray experiments of *Arabidopsis* using Infer Gene software. Their results showed that several TFs were the most central regulatory hubs including KAN3 involved in auxin, MYB29 in gibberellin, MYB121 in abscisic acid, and ERF1 in ethylene responses. The *Arabidopsis* BOTRYTIS SUSCEPTIBLE 1 (*BOS1*) gene encoding an R2R3MYB transcription factor conferred resistance to pathogens and tolerance to salt and drought abiotic stress factors through a JA-mediated defense signaling pathway. *BOS1*-deficient mutant *bos1* displayed increased susceptibility to necrotrophic fungi *B. cinerea* and *Alternaria brassicicola* and biotrophic bacterial pathogen *Pseudomonas syringae* and impaired tolerance to abiotic stresses salinity, drought and oxidative stress (Mengiste et al. 2003). The results of this study indicate that MYB TFs play essential roles in multiple stress responses. Functional studies of the *Arabidopsis Botrytis Susceptible1 Interactor* (*BOI*) and three *BOIRELATED GENES* (*BRGs*) encoding a subclass of RING E3 ligases have revealed that *BOI* interacts with *BOS1* and ubiquitinates it to confine the extent of cell death induced by pathogens and abiotic stresses (Luo et al. 2010). *B. cinerea* and salt induced the expression of *BOI* but its expression was intriguingly suppressed by gibberellins. *BOI* RNA interference (RNAi) plants showed more susceptibility to fungal pathogen and salt stress and reduced growth responsiveness to gibberellin. Hence, *BOI* and *BRGs* are associated with resistance to both biotic and abiotic stresses *via* suppression of cell death which is different from HR cell death mediated by the disease resistance genes *RPM1* and *RPS2* (Luo et al. 2010). This finding suggests that the expression of *BOI* gene is complex and regulated by multiple factors and there is hormonal crosstalk between JA and GA to regulate the expression of *BOS1*. Further, GA may act as a negative regulator of *BOI*. However, more comprehensive analysis is required to dissect the complex interacting pathways with crosstalk at different levels.

A transcription factor of the NAC family, one of the largest TFs family that is only found in plants, *Arabidopsis* thaliana activating factor1 (*ATAF1*) is induced in response to various biotic and abiotic stresses. *ATAF1* is drought-inducible gene and it was significantly expressed in an ABA-deficient mutant *aba2* in response to drought and high salinity, demonstrating that *ATAF1* expression is ABA-independent. *ATAF1* is also involved in stomatal regulation and extensively expressed in stomatal guard cells (Lu et al. 2007; Wu et al. 2009). Therefore, *ATAF1* may augment plant drought tolerance by stomatal closure via both ABA-dependent and ABA-independent signaling pathway. On the other hand, ABA biosynthesis

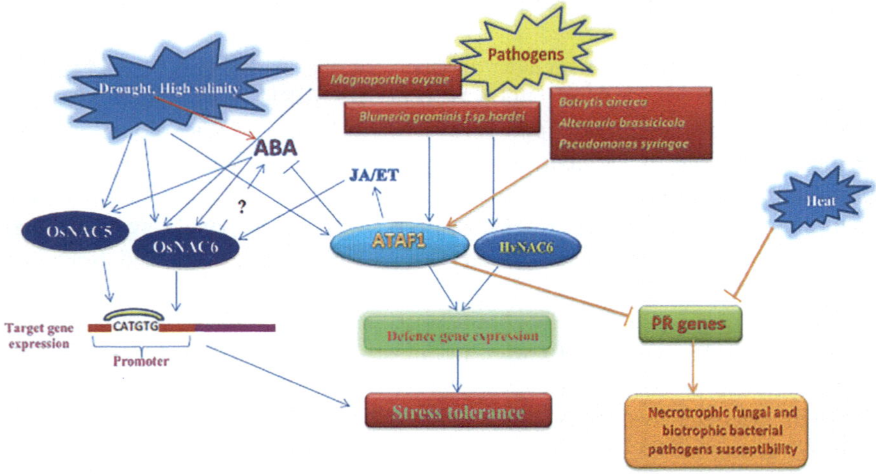

Fig. 5.3 A schematic representation of transcriptional regulatory networks of NAC transcription factors in biotic and abiotic stresses responses (Nejat and Mantri 2017)

aldehyde oxidase gene (*AAO3*) was induced in *ataf1* mutant plants upon biotrophic powdery mildew fungal pathogen *Blumeria graminis* f. sp. *hordei* (*Bgh*) infection and compromised penetration resistance towards *Bgh* (Jansen et al. 2008). By contrast, ABA biosynthesis-deficient mutant was highly resistant to attempted *Bgh* penetration. These data suggest that *ATAF1* is a stimulus-dependent transcriptional regulator of ABA biosynthesis and ABA is a negative regulator of penetration resistance. Moreover, the expression of JA/ET-activated transcription of plant defensin genes was suppressed in *ataf1* mutants whereas induced in the wild-type plants (Jansen et al. 2008). Further, barley *ATAF1* homologue HvNAC6 is also a positive regulator of basal defense and offers penetration resistance towards virulent *Bgh* attack (Jensen et al. 2007) (Fig. 5.3).

Collectively, *ATAF1* as a point of crosstalk plays a critical role in the regulation of antagonistic interplay between ABA- and JA/ET-mediated signaling and regulates the inhibitory effect of ABA on JA/ET activated transcription of plant defensin genes to enhance plant tolerance in response to biotrophic fungus. Moreover, *ATAF1* knockout mutant positively regulates the expression of several stress responsive genes such as *COR47*, *ERD10*, *KIN1*, *RD22,* and *RD29A* under drought stress (Lu et al. 2007; Wu et al. 2009). By contrast, overexpression of *ATAF1* in transgenic plants ATAF1-OE increased susceptibility to *B. cinerea*, *Alternaria brassicicola,* and *P. syringae* pv. Tomato, while *ATAF1* chimeric repressor construct enhanced disease resistance to these pathogens *via* up-regulation of pathogenesis related (PR) genes, PR-1 and PR-5, and JA/ET mediated defense responsive gene plant defensin1.2 (PDF1.2) (Lu et al. 2007). In line with this, Prasch and Sonnewald (2013) found that the expression of *PR* genes and *R* genes-mediated disease resistance was suppressed and virus resistance was compromised in virus infected *A. thaliana* under simultaneous double drought and heat stress, whereas *PR1*, *PR2,* and *PR5* were induced in response to single virus infection (Prasch and Sonnewald 2013).

The *ATAF1* is a repressor of basal defense and negative regulator of disease resistance, which seems to be ABA mediated, in response to both necrotrophic fungi and biotrophic bacterial pathogen through repression of PRs (Wang et al. 2009). A novel role has been identified for *ATAF1* as a positive regulator of senescence *via* activating senescence-promoting TF ORESARA1 (*AtNAC092*) and repressing MYB transcription factor GOLDEN2-LIKE1 (GLK1) by directly binding to their promoters (Garapati et al. 2015). These findings demonstrate that *ATAF1* is a multifaceted TF with versatile capabilities and plays an important role in mediating crosstalk between biotic and abiotic stress responses, though further research is required to unravel the role of senescence-associated genes in defense response and the correlation between *ATAF1*, hormone signaling, and disease resistance.

The *OsNAC6* gene, another member of the NAC family which has high similarity to genes in the ATAF subfamily, is a positive regulator of defense in response to both biotic and abiotic stresses (Nakashima et al. 2007). *OsNAC6* was induced by cold, drought, high salinity, wounding, ABA, JA, and blast disease in rice (Ohnishi et al. 2005; Nakashima et al. 2007; Takasaki et al. 2010). Transgenic rice plants overexpressing *OsNAC6* exhibited enhanced tolerance to drought and high salinity and some tolerance to hemibiotrophic fungal pathogen *Magnaporthe oryzae*. Further, microarray analysis revealed that many stress-responsive genes were up-regulated in OsNAC6-overexpressing rice plants, though *OsNAC6* has a negative effect on growth and productivity in rice (Nakashima et al. 2007) (Fig. 5.3). Interestingly, *OsNAC5* homolog of *OsNAC6* was induced by drought, cold, high-salinity, abscisic acid and methyl jasmonic acid and augmented stress tolerance by overexpression of stress-responsive genes such as OsLEA3 with no negative effect on rice growth (Takasaki et al. 2010). Notwithstanding, the crosstalk between *OsNAC6/OsNAC5* and hormone signaling underlying tolerance to multiple biotic and abiotic stresses requires further investigation.

5.3.3 Heat Shock Proteins and Heat Shock Transcription Factors as Signaling Pathways

A family of highly conserved genes encodes heat shock proteins (HSPs) across all cells of both prokaryotes and eukaryotes. These stress response proteins are constitutively expressed in cells under normal conditions, in the absence of environmental stresses, and function as molecular chaperones regulating protein synthesis, folding, assembly, translocation, and degradation (Wang et al. 2004). Their expression is increased upon exposure to a wide variety of environmental stimuli to protect cells against stress by preventing protein aggregation and stabilizing misfolded proteins and cellular homeostasis (Wang et al. 2004). The overexpression of *OsHSP18.6* augments tolerance to diverse types of abiotic stresses including heat, drought, salt, and cold in rice (Wang et al. 2015a). In addition, genes encoding HSPs were induced in *Arabidopsis* by necrotrophic fungus *Botrytis cinerea* infection, cold, drought, and oxidative stress (Sham et al. 2014). Therefore, HSPs may regulate pathogen defense as well as abiotic stress tolerance.

The transcription of HSPs is primarily regulated by heat shock transcription factors (HSFs) which bind to highly conserved motifs of the promoter regions of *HSP* genes known as heat stress-elements (HSEs; 5′-AGAAnnTTCT-3′) resulting in increased tolerance to biotic and abiotic stresses (Hu et al. 2015; Virdi et al. 2015). The genome-wide transcriptional analyses in a number of plants have shown that HSFs have an integral role in response to biotic and abiotic stresses such as heat, cold, salt, drought, and oxidative stress (Chung et al. 2013; Xue et al. 2015). Several *HSF* genes were significantly induced in strawberry *Fragaria vesca* by heat, drought, cold, and salt stresses and by biotic stress powdery mildew infection, and biotrophic fungus *Podosphaera aphanis* (Hu et al. 2015). Further, the level of *AtHsfA6a* transcript highly increased under high salinity and dehydration conditions (Hwang et al. 2014). In addition, *HSF4* was induced by salinity and/or osmotic stress and fungal pathogen *Botrytis cinerea* in *A. thaliana* (Sham et al. 2015) while *HsfA3* was reported to be associated with response to drought and salt stress (Li et al. 2013). Moreover, several *Hsf* genes were highly overexpressed by exogenous ABA (Hwang et al. 2014; Hu et al. 2015). These studies thus provide a link between ABA and *Hsfs* activation and indicate that HSFs, as a point of cross-tolerance, act as the transcriptional activators of several stress-responsive genes which could be *via* ABA-dependent signaling pathway. They also suggest the involvement of HSFs-activated stress responsive genes in HSFs-mediated stress tolerance.

Studies have identified melatonin (N-acetyl-5-methoxytryptamine) as an important secondary messenger that confers resistance to heat, salt, drought, cold and pathogen *Pseudomonas syringe* pv. tomato in *Arabidopsis* (Shi et al. 2015a, b). It positively regulates the expression of *class A1 heat*-shock factors (HSFA1s) as the master regulators of heat stress response and several stress-responsive genes such as COR15A, RD22, and KIN1. Hence, melatonin may remarkably impact crosstalk among stress responses, and control biotic and abiotic stress response as a regulator of HSFs (Shi et al. 2015a, b). There is also evidence to suggest a crosstalk between HSFs and oxidative stress signaling, and HSPs and TIR-NBS-LRR genes (Scarpeci et al. 2008; Prasch and Sonnewald 2013). Scarpeci et al. (2008) identified heat shock transcription factors HsfA2 and HsfA4A that can sense oxidative stress caused during heat and pathogen attack via direct sensing of H2O2 in *Arabidopsis* (Scarpeci et al. 2008). Prasch and Sonnewald investigated the response of *Arabidopsis* plants to heat, drought, and Turnip mosaic virus (TuMV) by transcriptome and metabolomic analysis. Their results revealed that the highest number of TIR-NBS-LRR resistance genes is induced under heat and combined heat and drought stress, while only a few TIR-NBS-LRR genes were regulated under combined triple stress. These findings suggest that HSPs regulate the function of NBS-LRR genes in response to biotic and abiotic stresses (Prasch and Sonnewald 2013).

5.3.4 MicroRNAs Analysis in Response to Defense Signaling

MicroRNAs (miRNAs) are a class of endogenous small non-coding, single-stranded RNAs of approximately 20–24 nucleotides in length. They regulate expression of their target genes in plants as key post-transcriptional gene regulators through translational repression or transcript cleavage (Mantri et al. 2013; Kamthan et al. 2015). miRNAs have been implicated in plant growth and development, cell proliferation and cell death, organogenesis, auxin signaling and also established to perform key roles in response to a wide array of biotic and abiotic stresses. (Khraiwesh et al. 2012; Mantri et al. 2013). miRNAs target transcription factors comprising AP2, bHLH, MYB, TCP, NAC, NFY, and HD-Zip and diverse range of defense-related genes such as receptor-like kinase, MAPK genes, peroxidase, NB-LRRs, and ABC transporters (Kulcheski et al. 2011; Khraiwesh et al. 2012; Kumar 2014; Raghuram et al. 2014; Omidvar et al. 2015). The studies revealed that miRNAs respond to abiotic stresses in a genotype-, tissue-, stress-, and miRNA-dependent manner (Omidvar et al. 2015; Zhang 2015). Negative correlation, however, has prevailed between the expression levels of miRNAs implicated in the response to biotic and abiotic stresses and their target genes (Xin et al. 2010; Raghuram et al. 2014; Omidvar et al. 2015).

A few studies have been conducted on the role of miRNA in plant responses to both biotic and abiotic stresses. Using Solexa high-throughput sequencing, Xin et al. (2010) defined the role of miRNAs in regulating the response of wheat to heat stress and powdery mildew infection caused by *Erysiphe graminis* f. sp. *tritici* (Egt). They identified 24 and 12 miRNAs that were expressed with different expression patterns in response to powdery mildew infection and heat stress, respectively. Further, 9 miRNAs were co-regulated by both powdery mildew and heat, of which miR827 and miR2005 were overexpressed in both Egt and heat stress. This exhibits their important role in wheat response to biotic and abiotic stresses. Further, miR156 was down-regulated in Egt susceptible (Jindong8) and resistant (Jindong8-Pm30) wheat lines after Egt inoculation while it was up-regulated in both heat susceptible (Chinese Spring) and tolerant (TAM107) genotypes, after heat treatment. In addition, negative correlations were identified between miR156 and its target genes Ta3711 and Ta7012 (Xin et al. 2010). These studies reveal crosstalk and functional interactions among stress-induced wheat miRNAs and corroborate the stress-specific role of miRNAs.

With the use of Solexa sequencing technology (Kulcheski et al. 2011) analyzed the miRNA expression pattern of soybean cultivars susceptible and resistant to Asian soybean rust, *Phakopsora pachyrhizi*, to pathogen infection and drought stress. Their results revealed that many miRNAs are involved in response to both biotic and abiotic stresses, though they exhibited completely different and contrasting expression profiles in the response to rust infection and drought stress. In addition, some miRNAs respond to the same stress differently depending on the soybean genotypes. Of these, MIR-Seq11 predicted to target peroxidase precursor mRNAs was significantly induced in the roots of drought susceptible genotype (BR 16) compared to the controls, although it responds similarly in the treated and untreated drought-tolerant cultivar (Embrapa 48). Further, expression of MIR-Seq11 was remarkably different between the mock and rust inoculated susceptible genotype (Embrapa 48), while it responds in a similar manner in the soybean rust resistant genotype (PI561356) (Kulcheski et al. 2011). These findings suggest that miRNAs may respond to stresses in a not only stress- but also genotype-specific manner. The results also indicate that miRNAs may target only one locus or several loci in response to biotic and abiotic stresses. MIR-Seq10, MIR-Seq15, and MIR-Seq18 targeted only one locus, although MIRSeq05, MIR-Seq11, MIR-Seq16, and MIR-Seq19 presented several loci as targets (Kulcheski et al. 2011). Omidvar et al. (2015) used Illumina Hiseq2000 platform to investigate the simultaneous impact of ABA and mannitol treatments on the wild type and abiotic stress-tolerant *7B-1* tomato mutant. They also analyzed the expression of 6 selected miRNAs (including miR159, miR166, miR472, miR482, miRNA, and miRND) in response to ABA, mannitol, NaCl, and cold treatments. Their results revealed that miRNAs were differentially expressed in response to abiotic stresses. miRNA159 and miRNA166 were inversely expressed in response to cold, while they showed the same pattern in response to ABA, mannitol, and NaCl. Likewise, different levels of expression were revealed between hypocotyl and root tissues. miR159, miR472, and miR482 were remarkably down-regulated in roots compared with hypocotyl. These results support previous studies that found miRNAs react to abiotic stresses in a stress-, miRNA-specific and plant tissue-dependent manner.

5.3.5 Epigenetics and Stress Response Signaling

The term epi-genetics describes heritable changes in gene expression patterns that transpire without altering the underlying DNA sequence (Iwasaki and Paszkowski 2014). Numerous studies have corroborated the crucial role of epigenetic mechanisms in plant immunity and stress adaptation (Ding and Wang 2015; Iwasaki and Paszkowski 2014; Kim et al. 2015; Kinoshita and Seki 2014) and ample progress has been made in elaborating epigenetic regulation of plant responses to stressors (Álvarez-Venegas and De-la-Pena 2015; Ding and Wang 2015; Dowen et al. 2012; Su et al. 2015). It has been demonstrated that plants have evolved intricate epigenetic mechanisms in regulating plant responses to environmental factors comprising DNA methylation, histone posttranslational modification, chromatin remodeling complexes, small non-coding RNAs (miRNAs and small interfering RNAs, siRNAs), and long non-coding RNAs. DNA methylomes profiling of *Arabidopsis thaliana* plants challenged with a bacterial pathogen, *Pseudomonas syringae* pv. *tomato*, nonpathogenic bacteria or salicylic acid (SA) revealed that stress induced differentially methylated regions (DMRs) can regulate the expression of proximal stress responsive genes as general regulatory hubs for defense-responsive genes. Moreover, their results exhibited that SA-induced demethylation, along with up-regulation of transposable elements and 21-nt small RNAs produced from these elements increased the expression of neighboring protein coding genes in SA-treated plants (Dowen et al. 2012).

There is involvement of AhHDA1, an RPD3/HDA1-like super family histone deacetylase (HDAC), in the regulation of stress resistance genes in response to osmotic stress and ABA in peanut (*Arachis hypogaea*) (Su et al. 2015). Liu et al. (2015) described the major epigenetic mechanisms contributing to the control of plant heat responses, highlighting epigenetic modifications comprising DNA methylation, histone modifications, histone chaperones, histone variants, ATP-dependent chromatin remodeling, small non-coding RNAs, and long non-coding RNAs in modulating the expression of heat-responsive genes under high temperature. Kim et al. (2015) highlight the role of histone modification in stress memory and modulating stress-responsive gene networks in response to drought, salinity, heat, and cold. The finding has unraveled that when plants confront stressful environmental conditions, the first experience of stress can be memorized and inherited to the next generation as a transgenerational epigenetic inheritance (Molinier et al. 2006). The studies have revealed epigenetic mechanisms underlying this stress-driven phenomenon. The evidence suggests full-scale reprogramming of the epigenome does not occur during gametogenesis due to escaping reprogramming of epigenetic marks (Bilichak and Kovalchuk 2016). Nevertheless, stress-responsive epigenomes have thus far been studied in response to single biotic/or abiotic stresses. Studies of the epigenome, however, are scarce in simultaneous biotic and abiotic stresses. Extending our knowledge of epigenetic mechanisms to simultaneous combined stresses is not only essential for understanding the crosstalk between epigenetic components, and epigenome and transcriptome but to also providing vital information for epigenetic engineering and crop improvement. Though, we are still far from being able to incorporate it in plant breeding for crop improvement (Nejat and Mantri 2017).

5.3.6 Systemic Signaling Process

Pathogen infection and wounding trigger systemic responses in plants. Similarly, abiotic stresses such as drought, salt, cold, heat, and high light also elicit systemic responses such that locally applied stress causes responses not only locally but also in distal tissues, resulting in systemic acquired acclimation (SAA). SAA involves long-distance electrical and hydraulic signals as well as calcium and ROS waves. Stress-triggered calcium and ROS waves can move at speeds that exceed 1000 mm/s as visualized in transgenic plants expressing calcium-sensitive fluorescent protein and luciferase reporter driven by a ROS responsive promoter, respectively. Calcium and ROS waves have been shown to cause transcriptional responses in distal tissues (Choi et al. 2014; Miller et al. 2009). ROS waves require the plasma-membrane NADPH oxidase RbohD (Miller et al. 2009), whereas calcium waves depend on the vacuolar ion channel TPC1, which may be involved in calcium-induced calcium release (Choi et al. 2014). RbohD is phosphorylated by the calcium-dependent protein kinase CPK5, which can be activated by ROS and is required for systemic defense responses (Dubiella et al. 2013). A model that emerges is that ROS generated by NADPH oxidases triggers a cytosolic calcium signal to activate more

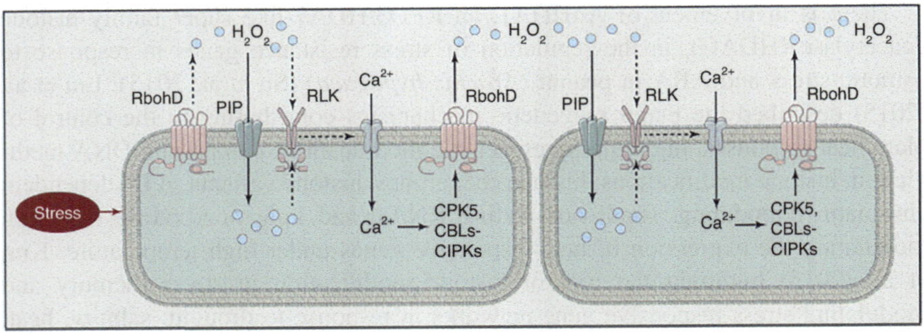

Fig. 5.4 Model of systemic stress signaling

NADPH oxidases via calcium responsive kinases, thus generating a self-propagating mutual activation circuit between ROS and calcium signals (Fig. 5.4). H_2O_2 generated by RbohD may enter cells through a plasma membrane-intrinsic protein/water channel (PIP) (Grondin et al. 2015), and its activation of the plasma-membrane calcium channel likely requires GHR1 (Hua et al. 2012) or a similar RLK. The plasma-membrane calcium channel that is activated by ROS has not been identified. In addition, it is unclear how the calcium and ROS waves may be linked to long-distance electrical signals.

Local exposure to stress generates H_2O_2 and Ca^{2+} signals. The Ca^{2+} signal can activate CPKs and CBLs-CIPKs, which phosphorylate and activate RbohD. Activated RbohD generates H_2O_2 that diffuses through the cell wall to neighboring cells, where it induces a Ca^{2+} signal through RLKs like GHR1. H_2O_2 may activate Ca^{2+} signaling at the cell surface and may also enter the cell through PIP water channels and then activate Ca^{2+} signaling intracellularly. The mutual activation between the Ca^{2+} and H_2O_2 signals generates self-propagating calcium and ROS waves that can travel to distant tissues to cause systemic acquired acclimation responses. Dashed lines indicate postulated regulation (Zhu 2016).

5.3.7 Signaling System in Response to Abiotic and Hormonal Stresses

Cell signaling in response to salt, drought, and the stress hormone ABA largely depends on the SnRK family of protein kinases in plants. SnRKs are related to the yeast SNF1 and mammalian AMPK, which are key sensors of cellular energy status (Hardie et al. 2016). In plants, abiotic stresses diminish the energy supply by inhibiting photosynthesis and energy-releasing catabolic reactions. So, SNF1/AMPK-related kinases proliferate and diversified during evolution to mediate the signaling of various abiotic stresses. SnRK1s are SNF1/AMPK orthologs that function in regulating metabolism in plants. All SnRK2s participate in osmotic stress and ABA signaling, whereas SnRK3s are key regulators of ion homeostasis required to cope with salt and nutrient stress in soil. Many of these stress-signaling pathways also involve the calcium-dependent protein kinase CPKs, which share homology to SnRKs in their kinase domains (Hrabak et al. 2003). In addition, virtually all of the stress pathways also involve MAPKs, which is a conserved feature of stress signaling in organisms from fungi to plants and metazoans. Other conserved features include the widespread use of calcium, ROS, NO, and lipid molecules as second messengers, although the generation and signal transduction of the second messengers are different in plants.

Identifying stress sensors remains an important but challenging goal for abiotic stress research in plants. Efficient gene-editing technologies and chemical genetic approaches will help overcome the gene redundancy problems that prevent the genetic identification of stress sensors. The growing appreciation of the importance of various cell organelles in stress sensing and responses and the dispersed stress-sensing model (Fig. 5.5) will also help us understand stress sensing and stress resistance, although the integration of signals from perturbed organelles is still poorly understood. Because plant stress responses must be coordinated with growth and development, it is important to understand the crosstalk between stress signaling pathways and hormonal as well as growth and developmental signaling pathways. Similarly, more attention should be directed towards plant responses to simultaneous, multiple abiotic stresses and to the crosstalk between abiotic and biotic stress signaling because much of the abiotic stress research thus far has been carried out on sterile plants grown in culture media in the laboratory; in nature, however, plants co-exist with insects and microorganisms. The root and shoot microbiomes presumably include many beneficial bacteria and fungi that help plants resist stress. Understanding how bacteria and fungi enhance plant stress resistance should increase ability to use these beneficial organisms and should also increase understanding of stress resistance in plants (Zhu 2016).

5.4 Calcium as Key Signaling Pathways in Host Responses to Multiple Stresses

The Ca^{2+} ion is recognized as a crucial second messenger in signaling pathways coupling the perception of environmental stimuli to plant adaptive responses. Indeed, one of the earliest events following the perception of environmental changes (temperature, salt stress, drought, pathogen, or herbivore attack) is intracellular variation of free calcium concentrations. These calcium variations differ in their spatio-temporal characteristics (subcellular location, amplitude, kinetics) with the nature and strength of the stimulus and, for this reason, they are considered as signatures encrypting information from the initial stimulus. This information is believed to drive a specific response by decoding via calcium-binding proteins. Plants have evolved sophisticated signaling strategies allowing them, in most cases, to withstand these stresses and survive deleterious conditions. Among the strategies employed, plants use an intricate signaling network involving calcium as a second messenger. It is now well acknowledged that the Ca^{2+} ion plays a crucial role, as a mediator, in regulating and specifying the cellular responses to

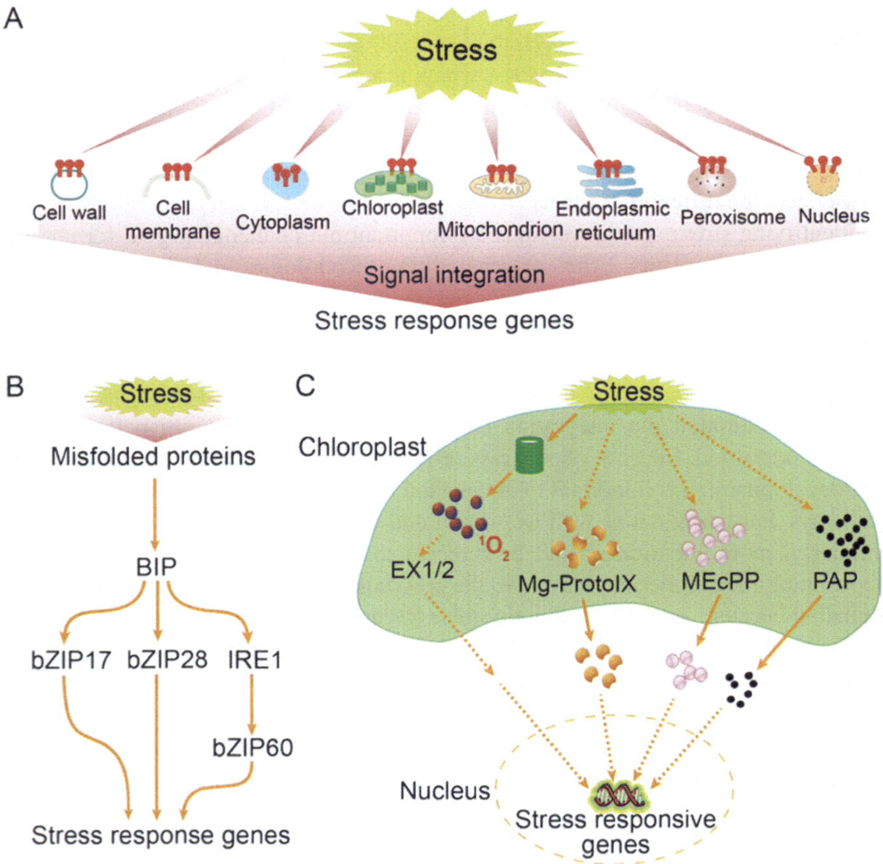

Fig. 5.5 Stress Sensing and Signaling in Different Cell Organelles. (**a**) Model of dispersed stress sensing by organelles. Stress causes perturbations in various organelles, generating signals that are integrated to regulate nuclear gene expression and other cellular activities, which helps to restore cellular homeostasis; (**b**) ER stress sensing and signaling; (**c**) Chloroplast stress sensing and signaling. Dashed lines indicate postulated regulation (Zhu 2016)

environmental stresses (Sanders et al. 2002; White and Broadley 2003; Dodd et al. 2010). Each stimulus perceived by the plant is generally followed by an immediate increase in intracellular Ca^{2+} concentration occurring either simultaneously or after a lag time in a single or several intracellular compartments of the cell (cytosol, nucleus, mitochondria, etc.). Such Ca^{2+} transients, each linked to a particular stimulus, differ in their spatio-temporal patterns, being therefore considered as signatures (McAinsh and Hetherington 1998). Each Ca^{2+} signature contributes to a first layer of specificity through its tissue and subcellular location, amplitude and frequency and the calcium pool involved in producing it. However, to be fully significant in terms of signaling, decoding Ca^{2+} signals by Ca^{2+}modulated proteins (usually termed calcium sensors) is mandatory. Calcium sensor proteins estimated to number over 250 in *Arabidopsis* (Day et al. 2002) are represented by three main families, i.e., the calcineurin-B-like proteins (CBLs) (Luan 2009), the calmodulin (CaM), and calmodulin-like proteins (CMLs) (Yang and Poovaiah 2003; Bender and Snedden 2013), the calcium-dependent protein kinases (CPKs) and the calcium and calmodulin-dependent protein kinase (CCaMK) (Cheng et al. 2002; Wang et al. 2015b).

These proteins display various affinities for calcium ions and this property, combined with their subcellular location within the cell, will control their behavior. Calcium binding to Ca^{2C} sensors will induce a conformational change that triggers either their association to downstream target proteins or a direct stimulation of the kinase activity when CPKs are considered (Harmon et al. 2000). The diversity of Ca^{2C} sensors and their downstream targets contributes to a second layer of specificity, allowing the transduction of various primary stimuli into distinct biological responses (Hashimoto and Kudla 2011). In addition, the hypothesis of signaling microdomains gathering calcium signaling components as described in animal cells (Good et al. 2011; Gueguinou et al. 2015) can also be proposed to contribute to response specificity in plants (Ranty et al. 2016).

5.4.1 Role of CMLS in Multiple Stresses

In addition to CaM, which has been well conserved through evolution, plant genomes are predicted to encode a broad range of CML proteins (Zhu et al. 2015). Like CaM, most CMLs contain 4 EF-hand calcium-binding motifs and no other functional domains; they share at least 16% overall amino acid identity with CaM in *Arabidopsis* (McCormack and Braam 2003). Several CMLs were shown to display biochemical properties typical of proteins that function as Ca^{2+} sensors, including, upon calcium binding, a shift in their electrophoretic mobility, changes in secondary structure and in exposed surface hydrophobicity (Bender et al. 2014; Scholz et al. 2015). Thus, CMLs are believed to act as calcium sensors, and the Ca^{2+}-induced conformational changes likely increase their interaction affinity to downstream effectors, as described for CaM. Search for CML targets has allowed the identification of diverse CML-binding proteins including protein kinases, transcription factors, metabolic enzymes, and transporters (Yamaguchi et al. 2005; Popescu et al. 2007; Perochon et al. 2010). Although regulation of these targets by CMLs is most often presumptive, CMLs likely function as calcium sensors/relays by tuning the activity of downstream effectors. Diverse roles for CMLs in plant development and stress responses have been reported (Cheval et al. 2013; Zeng et al. 2015).

The Ca^{2+}-mediated stress signaling in the model plant *Arabidopsis* has been identified AtCML9 as both a positive and negative regulator of plant immune response and drought stress tolerance, respectively (Magnan et al. 2008; Leba et al. 2012). *AtCML9* was found to be induced early in plants exposed to salt, cold, or dehydration treatment, to a bacterial pathogen and to application of stress-associated phytohormones including abscisic acid (ABA) and salicylic acid (SA). Salt-responsive expression of *AtCML9* is dependent on ABA production, while its expression in response to a virulent strain of *Pseudomonas syringae* is dependent on SA synthesis, suggesting that this CML is involved in stress hormone-mediated responses. In this respect, *Atcml9* null mutants exhibit a hypersensitive response to ABA during early growth stages that could be correlated with the enhanced tolerance to drought and salt stress in adult plants. *Atcml9* mutants also show alterations in the expression of several stress and ABA-responsive genes as well as defense-related genes. The AtCML9 contributes to both abiotic and biotic stress responses in conjunction with hormone signaling, and suggests a role for AtCML9 in the regulation of stress/defense-related genes. AtCML9 was reported to interact with diverse transcription factors including WRKY53 and TGA3 two factors that participate in plant defense responses (Popescu et al. 2007). However, the biological relevance of these physical interactions requires further analysis.

Among other CMLs shown to be linked to stress signaling, AtCML37 and 42, appear to play dual roles in abiotic stress responses and defense against herbivorous insects. Loss of function of AtCML42 in *Arabidopsis* mutants results in enhanced resistance to *Spodoptera littoralis*, up-regulation of jasmonic acid (JA)-responsive genes and an increased accumulation of aliphatic glucosinolates (Vadassery et al. 2012). Thus, AtCML42 acts negatively on herbivore resistance by decreasing the expression of JA-responsive genes and the accumulation of defense secondary metabolites. *Atcml42* null mutants also show a reduced content in flavonol glucosides that play a role in UV-B protection. As a consequence, *Atcml42* mutant seedlings are less resistant to UV-B exposure. In contrast to AtCML42, AtCML37 acts positively on defense against *S. littoralis* (Scholz et al. 2014). AtCML37 loss-of-function mutants exhibit an enhanced susceptibility to herbivory correlated with a lower level of the bioactive form of jasmonate (JA-Ile), an important hormone in plant defense against herbivores, and lower expression of JA-responsive genes encoding proteins involved in the synthesis of molecules toxic for insects. In addition, *Atcml37* null mutants show a decrease in drought stress tolerance correlated with a low content in ABA, whereas *Atcml42* null mutants are not impaired in drought stress response (Scholz et al. 2015). Collectively, these data reveal opposite roles for AtCML37 and 42 in insect herbivory resistance and distinct functions of these two CMLs in abiotic stress responses. It can also be noted that these particular CMLs, at the crossroads of biotic and abiotic signaling pathways, play antagonistic roles for the plant, being generally protective in one pathway and deleterious in the second. Due to their dual role in distinct signaling pathways, these Ca^{2+} sensors are interesting tools that can help in understanding how specificity is achieved in plant responses to environmental cues. Identification of their interacting partners will be helpful to clarify how these CMLs exert their action at a molecular level and will shed light on the mechanisms controlling their antagonistic effects in defense against herbivores (Ranty et al. 2016).

5.4.2 CPKs as Signaling Molecules Mediating Plant Responses to Multiple Stresses

Calcium-dependent protein kinase and CMLs share a similar broad distribution in the plant kingdom (Valmonte et al. 2014; Zhu et al. 2015). They possess a CaM-like and a kinase domain that make them direct effectors upon activation by Ca^{2+} binding. Their conserved molecular structure consists of a variable N-terminal domain joined to a Ser/Thr kinase domain associated to a CPK activation domain (CAD). CAD comprises an inhibitory junction domain (also termed pseudo-substrate region) fused to a CaM-like domain (Harper et al. 2004). The current *in vivo* activation model of CPKs states that, upon Ca^{2+} binding, the conformational change induced will release the pseudo-substrate region from the active site of the kinase domain (Liese and Romeis 2013). Like CMLs, and according to their tissue and subcellular location or Ca^{2+} affinity, CPKs participate as Ca^{2+} sensors in the regulation of various plant functions ranging from plant growth and development to plant defense and adaptation against pathogens, pests, and environmental cues (Asano et al. 2012a; Boudsocq and Sheen 2013; Schulz et al. 2013; Romeis and Herde 2014). The NADPH oxidase RBOHD, activated by AtCPK5 in *Arabidopsis* (Dubiella et al. 2013), is involved in ROS-mediated long distance signaling in response to several stimuli (Miller et al. 2009; Gilroy et al. 2014; Romeis and Herde 2014). Other targets are located in guard cells where they display ion transport activity (Brandt et al. 2012; Scherzer et al. 2012; Zhang et al. 2014). Interestingly, as for CMLs, some members of the CPK family appear to play a dual role in both biotic and abiotic stress responses.

In understanding how calcium can regulate sphingolipid-induced programmed cell death (PCD), Ranty et al. (2016) identified AtCPK3 as a crucial effector in this process (Lachaud et al. 2013). Long chain bases (LCBs) which are precursors of more complex sphingolipids, have been proposed to mediate immune responses in plants (Peer et al. 2010). Some natural analogs mimicking LCBs, such as mycotoxins produced by necrotrophic *Fusarium* fungi, likely interfere with the sphingolipid signaling pathway and plant immunity processes. Indeed, fumonisin B1 (FB1) produced by *Fusarium monoliforme* is able to quickly increase the level of free LCBs in the cell, thus mimicking a LCB treatment (Saucedo-Garcia et al. 2011). In addition, FB1 induces a Ca^{2+} burst that activates AtCPK3 which forms a complex with dimeric 14-3-3 proteins in resting conditions. Upon activation, AtCPK3 phosphorylates a conserved serine residue located at the N-terminal part of each 14-3-3 monomer. Phosphorylation of 14-3-3 proteins at the dimer interface results in disruption of their dimeric structure, promoting AtCPK3 release and cell death induction. Using *Atcpk3* null lines, Ranty et al. (2016) demonstrated that AtCPK3 is required for full development of PCD symptoms in leaves infiltrated with this mycotoxin (Lachaud et al. 2013). In parallel studies, using co-expression assays in *Nicotiana benthamiana*, AtCPK3 was also shown to be involved in biotic stress responses during plant-insect interactions. In wild-type plants exposed to *Spodoptera littoralis*, AtCPK3 is able to phosphorylate the heat-shock transcription factor B2a (HsfB2a) that likely activates the transcription of the defensin gene

PDF1.2, known to be induced as a resistance mechanism (Kanchiswamy et al. 2010). Besides participation of AtCPK3 in signaling processes associated with biotic stresses, various studies also describe its involvement in the transduction of abiotic signals. Indeed, genetic approaches have shown that this kinase, together with AtCPK6, mediates ABA responses in guard cells. They participate in Ca^{2+} activation of plasma membrane slow-type anion channels (Mori et al. 2006). In the *Atcpk3/ Atcpk6* double mutant, ABA-induced stomatal closure is impaired while long-term Ca^{2+}-programmed stomatal closure is not. In addition, AtCPK3 binds and can phosphorylate the *Arabidopsis* vacuolar two-pore K^C channel TPK1, thereby contributing to its regulation during salt stress response (Latz et al. 2013) whereas during the same stress, AtCPK3 was shown to modulate the membrane phosphoproteome likely in a MAPK-independent pathway (Mehlmer et al. 2010). Another example illustrating the dual role of CPKs in plant responses to biotic and abiotic stimuli concerns *AtCPK1*, which when overexpressed confers a SA-mediated resistance phenotype in *Arabidopsis* towards both necrotrophic fungi (*Botrytis cinerea, Fusarium oxysporum*) and biotrophic bacteria (*Pseudomonas syringae*), while its suppression (null mutants) results in an enhanced susceptibility to these pathogens (Coca and San Segundo, 2010). Like AtCPK3, AtCPK1 is thought to mediate responses to other environmental physical cues such as light.

In an exhaustive work, Boudsocq et al. (2010) identified several CPKs activated by flg22, a 22-amino acid peptide derived from flagellin known to induce plant immune responses (Boudsocq et al. 2010). By transiently expressing constitutively active *Arabidopsis* CPKs, five AtCPKs (AtCPK4,5, 6,11, 26) were found able to activate the expression of the Ca^{2+} dependent flg22-responsive reporter gene NHL10-LUC (NDR1/Hin1-*Like10*-LUCiferase), in the same manner as flg22 (Boudsocq et al. 2010). In the light of the presumed functions of AtCPKs (Boudsocq and Sheen 2013), it is interesting to note that the AtCPKs mentioned above play a dual role being also involved in abiotic signaling pathways. AtCPK4 and AtCPK11 were shown to promote the expression of the reporter luciferase (LUC) under the control of an ABA-responsive promoter (*RD29A)*. The activity of LUC was synergistically increased upon co-expression of AtCPK4 with the ABA-responsive bZIP transcription factor ABF2, whereas AtCPK11 which shares more than 95% identity with AtCPK4 was unable to enhance the LUC response through ABF2 (Lu et al. 2013).

5.4.3 A Single Calcium Sensor in Specific Responses to Multiple Stresses

It appears that CPKs as CMLs, share the property of being hubs, able to integrate signals triggered by both biotic and abiotic stimuli and to drive the signaling pathway towards the appropriate response. This property that seems common to related Ca^{2+} sensors in other plants species, including monocots (Asano et al. 2012b), raises the question of the mechanisms involved in the switching functions that these CPKs or CMLs are expected to play. While it is quite difficult to answer

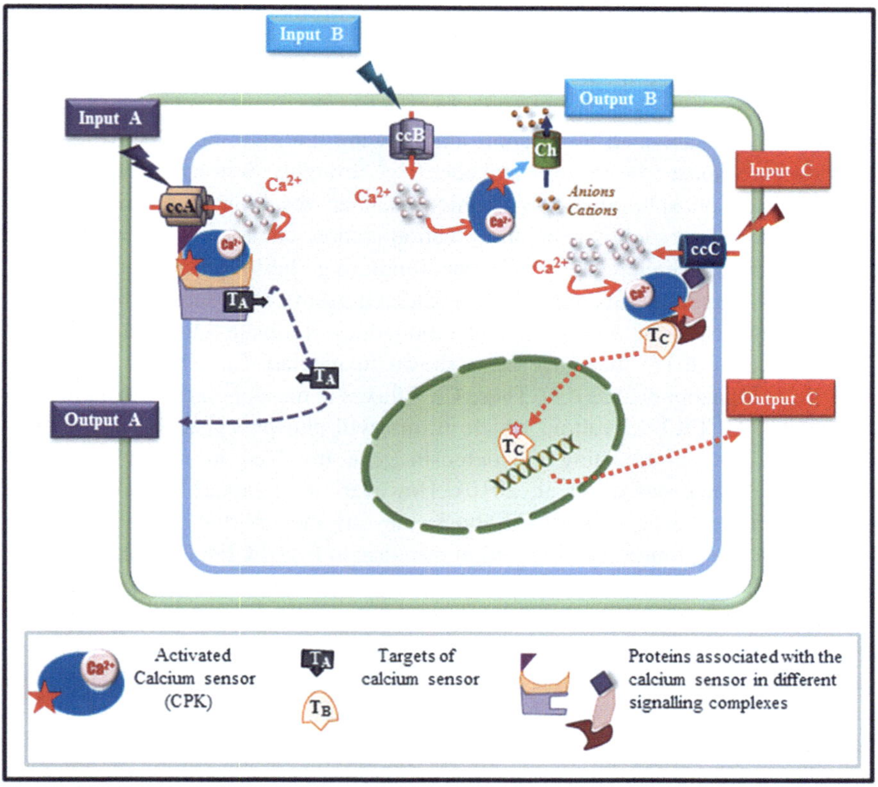

Fig. 5.6 Figure depicts various possible scenarios explaining how a single calcium sensor such as a calcium-dependent protein kinase (CPK) can drive signaling pathways towards specific responses in a single cell exposed to different stimuli. In response to a stimulus A (input A), a specific calcium channel (ccA) is activated and generates a calcium flux that will activate the CPK that will phosphorylate target A (TA) involved in output A. In response to a second stimulus (input B), a specific calcium channel is activated (ccB) and calcium released into the cytosol will activate the CPK that can directly regulate an effector (Ch) such as an ion transporter (channel, pump, exchanger) leading, according to the charge of the transported ions, to membrane depolarization or hyperpolarization contributing to the final response (output B). The same calcium sensor can be associated, in the same cell, with a different scaffold tethered to a different calcium channel (ccC) that is activated only by stimulus C (input C). Upon ccC activation, the calcium sensor is activated and can lead, through this specific signaling module, to the activation of a different target (T_C) migrating to the nucleus in order to control the adaptive response C through gene reprogramming for instance (output C) (Ranty et al. 2016)

this question from studies performed on CMLs, mainly due to our lack of knowledge concerning their downstream targets, different hypotheses can be drawn from existing knowledge on CPKs (Fig. 5.6). Taking AtCPK3, as a model, Ranty et al. (2016) speculated on the way in which a single Ca^{2+} sensor can wire the flow of signaling information towards the right adaptive response upon challenging the plant with different stimuli. Beyond the importance of the specific Ca^{2C} signature generated in response to a given stimulus, it is clear that specificity of the routing will be achieved only if the calcium sensor, its downstream targets and putative partner proteins are co-localized and expressed at the same time. In guard cells, AtCPK3 can be activated by calcium ions flowing through non-selective calcium channels in response to environmental cues (e.g., drought) (Schroeder and Hagiwara 1990) and can phosphorylate S-type anion channels resulting in stomatal closure. When located in mesophyll cells, the situation is more complex if we consider that these cells can be challenged by different stimuli (e.g., herbivores and mycotoxins) leading to different outputs (e.g., defense and cell death). Exposure of these cells to the chewing action of *S. littoralis* is reported to stimulate glutamate receptors (Mousavi et al. 2013) that have been shown to mediate Ca^{2+} influx (Kwaaitaal et al. 2011; Manzoor et al. 2013). These Ca^{2+} fluxes in mesophyll cells will probably account for AtCPK3 activation, which in turn will phosphorylate HsfB2a, a transcription factor up-regulating the defensin gene involved in responses against herbivores (Kanchiswamy et al. 2010). However, if glutamate channels can be suspected in the case of *S. littoralis* attack, they are most probably not involved in the cytosolic Ca^{2+} transients observed in response to FB1/LCBs. Indeed, ionotropic glutamate channel inhibitors did not inhibit the cytosolic Ca^{2+} responses when cells were challenged with LCBs (Lachaud et al. 2010). A different type of Ca^{2+} channel could be involved in FB1-induced calcium increase leading to PCD. Therefore, in this particular case, the switching mode of AtCPK3, leading either to PCD or basal defense mechanisms, can rather be related to the origin of Ca^{2+} and/or the nature of the channels involved in shaping the Ca^{2+} signatures triggered by each stimulus (insect or sphingolipid). It can then be hypothesized that these different classes of activated channels can engage AtCPK3 in different signaling modules that are specifically associated to each type of channel. Such an association will favor the interaction of AtCPK3 with its downstream substrates specifically dedicated to one output (defense or cell death). Similar assemblies of signaling proteins defined as signaling complexes or scaffolds have been shown for a few signaling proteins in plants. Indeed AtCPK21 was shown to interact in an ABA-dependent manner with the slow anion channel 1 (SLAC1) homolog 3 (SLAH3) in microdomains of the guard cell plasma membrane. The activity of AtCPK21, which is crucial for SLAH3 activation, seems to be regulated by a protein scaffold involving protein phosphatase 2C (PP2C) ABI1 and an ABA receptor of the RCAR1/PYR/PYL family (Demir et al. 2013). This scaffold hypothesis has got a much larger scientific audience in animals (Good et al. 2011) where clustering of Ca^{2+} channels in membrane microdomains and association with their downstream signaling complexes have been reported (Dart 2010; Gueguinou et al. 2015).

5.5 Crosstalk and Specificity Between Signaling Pathways

Through the information generated on signaling pathways in plants against multiple stresses, it is quite apparent that there is existence of complex signaling network which is interconnected at many levels (Fig. 5.7). During crosstalk, with the help of common components there is transfer of important information between signaling pathways. The purpose of this information transfer is to modify the output of the different signaling pathways. But at the same time, there is clear demarcation

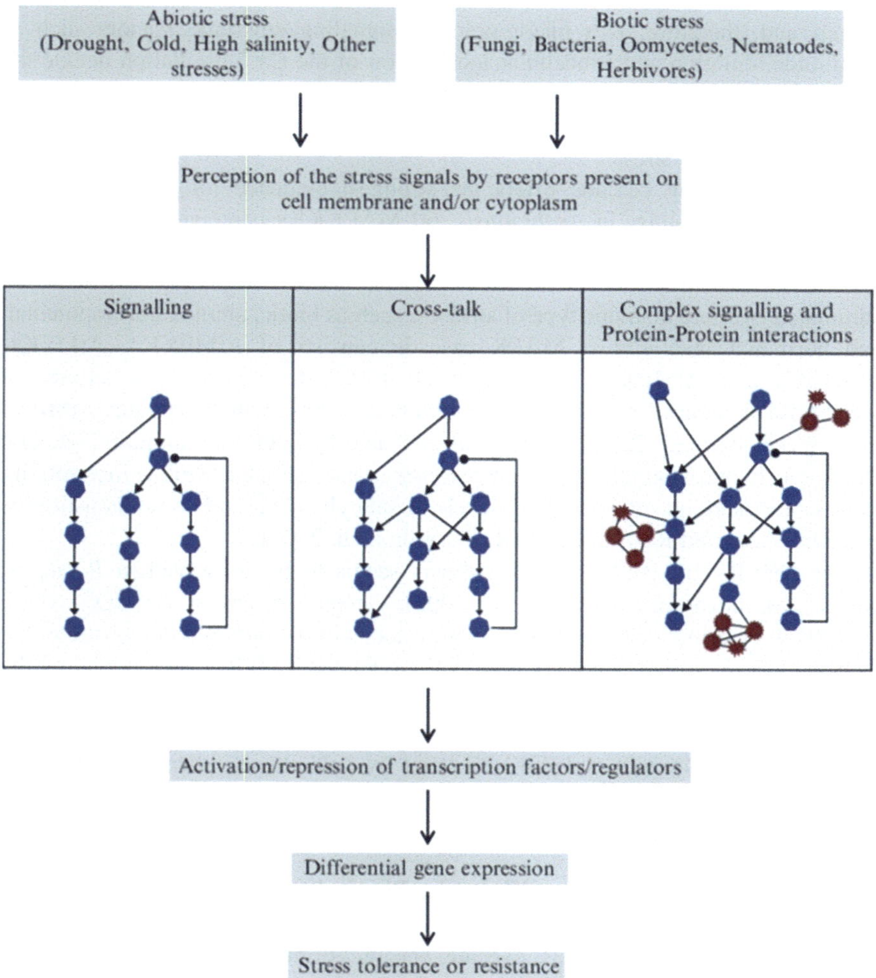

Fig. 5.7 Schematic representation of crosstalk among various signaling pathways. Large-omics experiments have revealed more complex pathway maps with a large degree of crosstalk and interaction with proteins other than signaling components (Verma et al. 2013)

between two or more potential consequences of a signaling pathway. Therefore, a particular stimulus ends up at a particular end response and not at some other responses. This is termed as "specificity" (Knight and Knight 2001). In any signaling pathway, there are chances of existence of crosstalk and specificity.

5.6 Crosstalk and Specificity in Abiotic Stress Signaling

During abiotic stress signaling, Ca^{2+} serves as a second messenger in plant cells. Ca^{2+} can be stimulated by various abiotic, biotic, developmental and hormonal signals and, therefore, is a major point of signaling crosstalk. Factors such as magnitude, duration and subcellular localization of the Ca^{2+} oscillation decide the specificity and/or crosstalk in Ca^{2+} signaling. Ca^{2+} dependent protein kinases (CDPKs) are found to be involved in signaling pathways in response to drought, wounding and cold stress.

MAPK signaling cascade offers the strongest evidence for crosstalk during abiotic stress signaling. In *Arabidopsis*, 60 MAPKKKs perceive the signals and transduce them to 10 MAPKKs and then to 20 MAPKs. This provides great scope for crosstalk between different stress signals. MAPK signaling cascades are stimulated by more than one type of stresses, such as biotic, abiotic, developmental, and hormonal. *Arabidopsis* MAPK cascade consists of AtMEKK1, AtMEK1/AtMKK2, and AtMPK4 (Ichimura et al. 1998). Expression and activity of AtMEKK1 is induced by salinity (Ichimura et al. 2000), which activates AtMPK4 in vitro (Huang et al. 2000). AtMPK4 is activated by cold, low humidity, osmotic stress, touch, and wounding. The *Arabidopsis* mutant atmpk4 exhibits insensitivity to jasmonic acid, accumulates high levels of salicylic acid, and shows constitutive Systemic Acquired Resistance (SAR, Peterson et al. 2000).

The DREB/CBF cold-response pathway seems to be conserved in *B. napus*, wheat, rye, and tomato (Hsieh et al. 2002). Genetic studies of the HOS1 (high expression of osmotically responsive genes) locus in *Arabidopsis* demonstrate that COR (cold responsive) genes are negatively regulated by HOS1 by modulating the expression level of the DREBs/CBFs (Lee et al. 2001). The DREB/CBF pathway is thought to be regulated in an ABA-independent manner, but DREB4/CBF4 gene gets induced by drought and ABA, but not by cold stress indicating a potential crosstalk.

5.7 Crosstalk and Specificity in Biotic Stress Signaling

During pathogen attack plants execute specific primary defense responses by generating the alarm signals SA, JA, and ET in appropriate balance (De Vos et al. 2005). Endogenous accumulation of these signals initiates an array of signaling pathways which generate different defense responses against partially distinct classes of attackers (Koornneef and Pieterse 2008). Generally, biotrophic pathogens are more sensitive to SA-mediated induced defenses, whereas necrotrophic pathogens and herbivorous insects are sensitive to JA/ET-mediated defenses (Glazebrook 2005), although few exceptions are there (Thaler et al. 2004). To cope up with simultaneous or subsequent invasion by multiple challengers, a powerful regulatory potential is provided by crosstalk between induced defense-signaling pathways.

SA- and JA-dependent signaling mostly exhibit a mutually antagonistic interaction, but there are reports demonstrating their synergistic interactions also (Mur et al. 2006). Along with SA/JA crosstalk there are interactions between SA and ET, JA and ABA, and JA and ET that function in the adaptive response of plants against various pathogens. There are several regulatory components which act as key molecular players of SA/JA crosstalk, such as NONEXPRESSOR OF PR GENES1 (NPR1), WRKY transcription factors, MAP kinases, and glutaredoxin GRX480. The regulatory protein NPR1 is essential for SA signal transduction which activates the genes encoding pathogenesis-related (PR) proteins (Van Loon et al. 2006). In *Arabidopsis*, when NPR1 was mutated, the suppression of JA-inducible gene expression by SA was obstructed. This indicates a crucial role of NPR1 in the crosstalk between SA and JA signaling (Spoel et al. 2007). Rayapuram and Baldwin (2007) proposed that in wild-type plants of tobacco, during herbivore attack NPR1 negatively regulates SA production and thus results in suppression of SA/JA crosstalk to allow induction of JA-mediated defenses against herbivores.

WRKY transcription factors are one of the significant regulatory components. They regulate SA-dependent defense responses and some of them are involved in SA/JA crosstalk (Wang et al. 2006). WRKY70 positively regulates SA-mediated defenses and inhibits the JA response (Li et al. 2004). Interestingly, WRKY11 and WRKY17 acts as negative regulators of WRKY70 and, thus, of SA-mediated defense responses, thereby positively regulating the JA response (Journot-Catalino et al. 2006). In another study, Mao et al. (2007) found that WRKY62 functions as a repressor of JA response.

MAP kinases relay information from sensors to the cellular responses and are involved in plant defense signaling (Nakagami et al. 2005). In *Arabidopsis*, Peterson et al. (2000) demonstrated MAP KINASE4 (MPK4) as a negative regulator of SA signaling which positively regulated JA signaling. In contrast to it, downstream effectors of MPK4, i.e., ENHANCED DISEASE SUSCEPTIBILITY1 (EDS1) and PHYTOALEXIN-DEFICIENT4 (PAD4), show reverse action to that of MPK4 by acting as activators of SA signaling and repressors of JA signaling (Brodersen et al. 2006).

5.8 Crosstalk Among Multiple Stresses

Transcription factors and kinases are key players involved in crosstalk between signaling pathways. ABA, SA, JA, ethylene (ET) and ROS regulated signaling pathways play significant roles in the crosstalk between multiple stresses signaling. ABA is mainly involved in responses against abiotic stresses such as drought, low temperature, and osmotic stress, whereas SA, JA, and ET have important roles in biotic stress signaling. In certain cases, ABA is found to negatively regulate disease resistance (Mauch-Mani and Mauch 2005). In *Arabidopsis*, when ABA was exogenously applied the expression of JA- or ET-responsive defense gene get repressed (Anderson et al. 2004). This indicates that ABA functions antagonistically to JA and ET. A basic helix-loop-helix (bHLH) transcription factor, AtMYC2, functions as a transcriptional activator in the ABA-mediated drought stress signaling pathway (Abe et al. 2003). It acts as a negative regulator of JA/ET-mediated defense gene expression (Lorenzo et al. 2004).

Therefore, for crosstalk between biotic and abiotic stress responses, AtMYC2 might act as a key regulator via hormone signaling (Fujita et al. 2006). Another feasible candidate molecule that can regulate both biotic and abiotic signaling aspects is dehydration-responsive NAC transcription factor, RD26. The factors inducing the expression of RD26 include JA, hydrogen peroxide (H_2O_2), and pathogen infections, as well as by drought, high salinity, and ABA treatment (Fujita et al. 2004). RD26-regulated genes function in the detoxification of ROS, defense, or senescence and, hence, prove to be a convergence point for the pathogen defense, senescence, and ABA-mediated signaling pathways. In tobacco, CDPKs also regulate biotic and abiotic stress responses with the help of SA, ET, JA, and ABA-mediated signaling pathways (Ludwig et al. 2005).

In *Arabidopsis*, MEKK1-MAPK kinase 2 (MKK2)-MPK4/MPK6 cascade has been shown to activate in response to cold and salt stress signaling. In contrast to it, MEKK1-MKK4/MKK5-MPK3/MPK6 cascades were shown to be involved in the regulation of the pathogen defense response pathway via the expression of WRKY22 and WRKY29 (Asai et al. 2002). MPK3 and MPK6 are stimulated by abiotic stresses also and are involved in hormone signaling pathways (Droillard et al. 2002). In *Arabidopsis*, MPK3 and MPK6 are activated by oxidative stress (Yuasa et al. 2001). The components which affect their activities in ROS signaling are OXI1 (serine/threonine kinase), ANP1 (MAPKKK), and NUCLEOSIDE- DIPHOSPHATE KINASE 2 (AtNDPK2) (Rentel et al. 2004; Moon et al. 2003). Therefore, MAPK cascades seem to intermediate ROS signaling and make the plant tolerant to environmental stresses by improving ROS scavenging capacity (Fujita et al. 2006). ROS is an important signaling molecule (Apel and Hirt 2004) and is a AU3 key player in both ABA signaling and disease resistance responses (Laloi et al. 2004). This suggests the role of ROS in mediating crosstalk between biotic and abiotic stress-responsive gene-expression networks.

5.9 Abiotic Stress Diagnosing as Nutrient Deficiencies

A first step in diagnosing nutrient deficiencies is to describe the symptoms. Each deficiency symptom is related to some function of the nutrient in the plant. Symptoms caused by nutrient deficiencies are generally grouped into five categories: (1) stunted growth, (2) chlorosis, (3) interveinal chlorosis, (4) purplish-red coloring, and (5) necrosis. Stunting is a common symptom for many deficient nutrients due to their varied roles in the plant. For example, when nutrients involved in plant functions such as stem elongation, photosynthesis, and protein production are deficient, plant growth is typically slow and plants are small in stature. Chlorosis and interveinal chlorosis are found in plants deficient of nutrients necessary for photosynthesis and/or chlorophyll (green leaf pigment involved in photosynthesis) production. Chlorosis can result in either the entire plant or leaf turning light green to yellow, or appear more localized as white or yellow spotting. Interveinal chlorosis occurs when certain nutrients [B, Fe, magnesium (Mg), Mn, nickel (Ni), and Zn] are deficient. Purplish-red discolorations in plant stems and leaves are due to above normal levels of anthocyanin (a purple-colored pigment) that can accumulate when plant functions are disrupted or stressed. This symptom can be particularly difficult to diagnose because cool temperatures, disease, drought, and even maturation of some plants can also cause anthocyanin to accumulate. Certain plant cultivars may also exhibit this purple coloring. Necrosis generally happens in later stages of a deficiency and causes the parts of the plant first affected by the deficiency to brown and die. Since a number of nutrient deficiencies can produce similar symptoms, further evaluation of symptoms related to particular leaf patterns or locations on the plant are needed to diagnose nutrient specific deficiencies.

5.9.1 Mobile and Immobile Nutrients

Another step in identifying deficiency symptoms is to determine whether the deficiency is the result of a mobile or immobile nutrient based on where the symptom appears in the plant. Mobile nutrients are nutrients that are able to move out of older leaves to younger plant parts when supplies are inadequate. Mobile nutrients include N, P, K, Cl, Mg, and molybdenum (Mo). Because these nutrients are mobile, visual deficiencies will first occur in the older or lower leaves and effects can be either localized or generalized. In contrast, immobile nutrients [B, calcium (Ca), Cu, Fe, Mn, Ni, S, and Zn] cannot move from one plant part to another and deficiency symptoms will initially occur in the younger or upper leaves and be localized. Zn is a partial exception to this as it is only somewhat immobile in the plant, causing Zn deficiency symptoms to initially appear on middle leaves and then affect both older and younger leaves as the deficiency develops.

5.9.2 Identification Key

The "key" given in flow chart (Fig. 5.8) can be used for identifying nutrient deficiencies based on common symptoms. It consists of alternative statements about the appearance of different plant structures. If possible, it may be helpful to have a healthy plant on hand for comparison purposes. Beginning at the top of the key, read the first statement and determine whether the statement applies to the plant (s) being evaluated. If the statement describes the plant's symptoms, proceed along the "YES" arrow to the next statement. If not, follow the "NO" arrow to an alternative statement. Continue this process until the probable nutrient responsible for the deficiency is identified. Although most descriptions in this key are

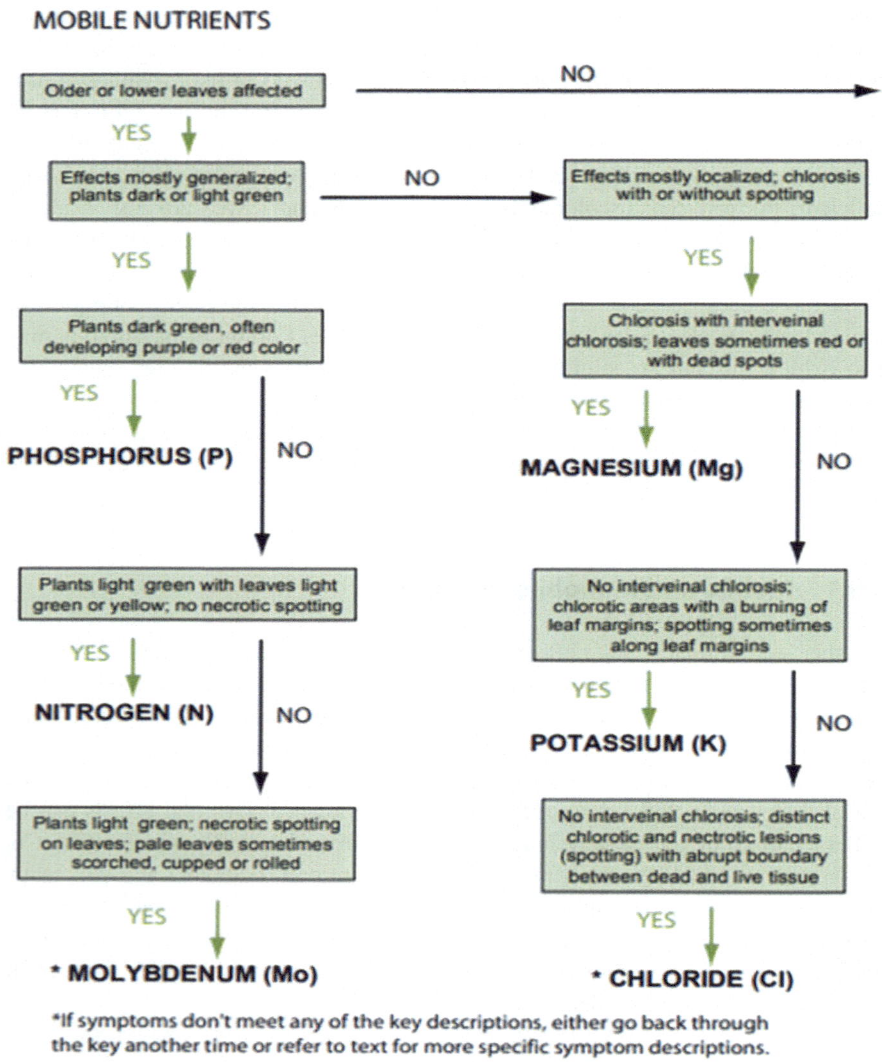

Fig. 5.8 Identifying key for abiotic stresses as nutrients deficiency in crucifers (McCauley et al. 2011)

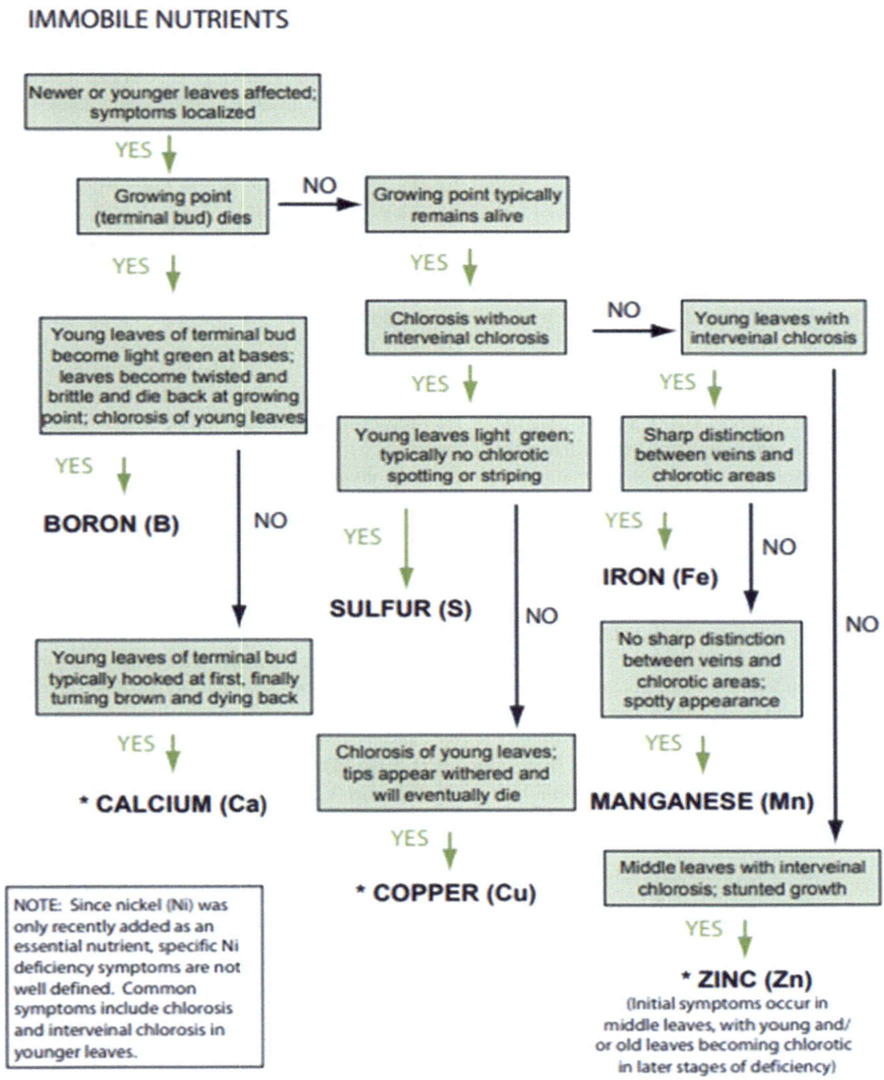

Fig. 5.8 (continued)

generalized to accommodate common symptoms seen in various crops, bear in mind that deficiency symptoms differ among crop types and plant specific.

References

Abe H, Urao T, Ito T, SekiM SK, Yamaguchi-Shinozaki K (2003) *Arabidopsis* AtMYC2 (bHLH) and AtMYB2 (MYB) function as transcriptional activators in abscisic acid signaling. Plant Cell 15(1):63–78

Abu-Blan HA, Abu-Gharbieh WI (1994) Effect of soil solarization on winter planting of potato, cauliflower and cucumber in the central Jordan Valley. Dirasat Ser B Pure Appl Sci 21 (3):203–213

Acharya BD, Khattri GB, Chettri MK, Srivastava SC (2002) Effect of *Brassica campestris* var. Toria as a catch crop on *Orobanche aegyptiaca* seed bank. Crop Prot 21(7):533–537. https://doi.org/10.1016/S0261-2194(01)00137-5

Achbani EH, Tourvieille D (1993) Phytosanitary situation of sunflower in Morocco. (La situation phytosanitaire du tournesol au Maroc.). Al Awamia, pp 117–138

Acimovic M (1977) Distribution of important sunflower diseases in Yugoslavia. In: Distribution of important sunflower diseases in Yugoslavia. Novi Sad, Yugoslavia: Inst. Field Veg Crops, 17 pp

Acuna PR, Latorre GB, Cancino EL (1975) Translocation of systemic fungicides in sunflower (*Helianthus annuus* L.) and its action on *Sclerotinia sclerotiorum* (Lib.) de Bary. (Translocacion de fungicidas sistemicos en maravilla (*Helianthus annuus* L.) y su accion sobre *Sclerotinia sclerotiorum* (Lib.) de Bary.). Agricultura Tecnica 35(1):35–40

Adair CN (1971) Influence of controlled-atmosphere storage conditions on Cabbage postharvest decay fungi. Plant Dis Reptr 55(10):864–868. https://doi.org/10.5197/j.2044-0588.2015.031.023

Ahmed AU, Akhond MAY (2015) First report of Sclerotinia rot caused by *Sclerotinia sclerotiorum* on *Lens culinaris* in BangladeshNew Dis Reptr:23

Ahuja I, Rohloff J, Bones AM (2009) Defense mechanisms of Brassicaceae: implications for plant – insect interactions and potential for integrated pest management- a review. Agron Sust Develop. https://doi.org/10.1051/agro/2009025

Akem CN, Dashiell KE (1992) Sclerotinia stem rot caused by *Sclerotinia sclerotiorum* on soybeans in Nigeria. Plant Dis 76(1):101. https://doi.org/10.1094/PD-76-0101C

Ale-Agha N, Boyle H, Braun U, Butin H, Jage H, Kummer V, Shin H-D (2008) Taxonomy, host range and distribution of some powdery mildew fungi (*Erysiphales*). Schlechtendalia 17:39–54

Alford DV (1979) Observations on the cabbage stem flea beetle, *Psylliodes chrysocephala*, on winter oil-seed rape in Cambridgeshire. Ann Appl Biol 93(2):117–123. https://doi.org/10.1111/j.1744-7348.1979.tb06521.x

Alkooranee JT, Liu S, Aledan TR, Yin Y, Li M (2015) First report of powdery mildew caused by *Erysiphe cruciferarum* on *Brassica napus* in China. Plant Dis 99(11):1651

Álvarez-Venegas R, De-la-Pena C (2015) Editorial: recent advances of epigenetics in crop biotechnology. Front Plant Sci 7:413

Amano K (1986) Host range and geographical distribution of the powdery mildew fungi. Tokyo

Anderson JP, Badruzsaufari E, Schenk PM, Manners JM, Desmond OJ, Ehlert C, Maclean DJ, Ebert PR, Kazan K (2004) Antagonistic interaction between abscisic acid and jasmonate-ethylene signaling pathways modulates defense gene expression and disease resistance in *Arabidopsis*. Plant Cell 16(12):3460–3479

Anonymous (1980) Diseases of rape. Commonwealth Agricultural Bureaux-Annotated Bibliography No. M3:38

Anonymous (1981) *Brassica* diseases. Annual Report Edinburgh School Agriculture, 1981. Edinburgh, UK 131-132:137–138

Anonymous (1983) Edinburgh. UK:39–40

Anonymous (1985a) Diseases of vegetables. Annual Report Edinburgh School Agriculture, 1984. Edinburgh, UK 54:57–58

Anonymous (1985b) Foliar diseases of Brassicas. Research Report 1984, Horticulture. Dublin, Ireland; An Foras Taluntais, pp 45–46

Anonymous (1985c) Oilseed rape and other Brassicas. Annual Report Edinburgh School Agriculture, 1985. Edinbourgh, UK 57-59:102–105

Anonymous (1985d) Oilseed rape. Annual Report Edinburgh School Agriculture (1984) Edinbourgh. UK 53:108–109

Apel K, Hirt H (2004) Reactive oxygen species: metabolism, oxidative stress, and signal transduction. Annu Rev Plant Biol 55:373–399

Asai S, Shirasu K (2015) Plant cells under siege: plant immune system versus pathogen effectors. Curr Opin Plant Biol 28:1–8

Asai T, Tena G, Plotnikova J, Willmann MR, Chiu WL, Gomez-Gomez L, Boller T, Ausubel FM, Sheen J (2002) MAP kinase signaling cascade in *Arabidopsis* innate immunity. Nature 415 (6875):977–983

Asano T, Hayashi N, Kikuchi S, Ohsugi R (2012a) CDPK-mediated abiotic stress signaling. Plant SignalBehav 7:817–821. https://doi.org/10.4161/psb.20351

Asano T, Hayashi N, Kobayashi M, Aoki N, Miyao A, Mitsuhara I et al (2012b) A rice calcium-dependent protein kinase OsCPK12 oppositely modulates salt-stress tolerance and blast disease resistance. Plant J 69:26–36. https://doi.org/10.1111/j.1365-313X.2011.04766

Ashraf SS, Basu-Chaudhary KC (2008) Effect of seed-borne Fusarium species on the physico-chemical properties of rapeseed oil. Phytopathology 117(2):107–112

Atkinson NJ (2011) Plant molecular response to combined drought and nematode stress. Ph.D. thesis, University of Leeds. http://etheses.whiterose.ac.uk/2131/

Atkinson NJ, Erwin PE (2012) The interaction of plant biotic and abiotic stresses: from genes to the field. J Exp Bot 63:3523–3544

Atkinson NJ, Lilley CJ, Urwin PE (2013) Identification of genes involved in the response of *Arabidopsis* to simultaneous biotic and abiotic stresses. Plant Physiol 162:2028–2041

Augustin A, Tulisalo U, Korpela S (1986) Flea beetles (Coleoptera: Chrysomelidae) (Halticinae) on rapeseed and sugar beet in Finland. Jf Agril Sci Finland 58:69–82

AVA (2001) Diagnostic records of the Plant Health Diagnostic Services, Plant Health Centre. Agri-food & Veterinary Authority, Singapore

Axelsen J (1992) The population dynamics and mortalities of the pod gall midge (*Dasyneura brassicae* Winn.) (Dipt., Cecidomyiidae) in winter rape and spring rape (*Brassica napus* L.) in Denmark. J Appl Entomol 114(5):463–471

Bagherani N, Shimi P (2001) Evaluation of the efficiency of different herbicides on weed control in rapeseed fields. J Agric Nat Res 8:157–167

Bai DH (1977) Survey on the fruit rot occurrence and damages of shipping mandarin. Korean J Plant Prot 16(4):245–247

Baka ZAM (1996) A fine-structural study of the downy mildew fungus *Peronospora parasitica* in the new host *Cakile maritima*. Archiv Phytopathol Plant Protect 30(1):41–51

Baka ZAM (2008) Occurrence and ultrastructure of *Albugo candida* on a new host, *Arabis alpina* in Saudi Arabia. Micron 39(8):1138–1144

Bakhetia DRC (1983) Losses in rapeseed and mustard due to Lipaphis erysimi (Kalt.) in India-a literature study. Proc. 6th Intl Rapeseed Conf, 17-19 May 1983. Paris, France 6:1142–1147

Bakhetia DRC, Labana KS (1978) Insect resistance in *Brassica* crops. Crop Improv 5:95–103

Barbetti MJ, Sivasithampuram K (1981) *Pseudocercosporella capsellae* and *Myrothecium verrucaria* on rapeseed in Western Australia. Aus Plant Pathol Soc Proc 10:43–44

Baribeau (1923) Stem wilt and rot of Sunflower in the province of Quebec. (Flétrissure pourriture de la tige du urnesal dans la province de Quebec.). Sci Agric 3(11):397–400

Baudys E (1928) Phytopathologyogical notes W. Ochrana Rostlin viii 6:151–162

Bazan De Segura C, Onsuelo L (1951) Plant diseases new to Peru. Plant Dis Reptr 35(10):465–466

Beck-Mennagetta G (1930) Das Pflanzenreich.96 (IV-261) [ed. by Engler H G A]. 275 pp.

Bedlan G (1987) The most important diseases of radishes. (Wichtigste Krankheiten an Rettich und Radieschen.). Pflanzenschutz 3(3-4):9–10

Bender KW, Snedden WA (2013) Calmodulin-related proteins step out from the shadow of their namesake. Plant Physiol 163:486–495. https://doi.org/10.1104/pp.113.221069

Bender KW, Dobney S, Ogunrinde A, Chiasson D, Mullen RT, Teresinski HJ et al (2014) The calmodulin-like protein CML43 functions as a salicylic-acid-inducible root-specific Ca($^{2+}$) sensor in *Arabidopsis*. Biochem J 457:127–136. https://doi.org/10.1042/bj20131080

Ben-Yephet Y, Genizi A, Siti E (1993) Sclerotial survival and apothecial production by *Sclerotinia sclerotiorum* following outbreaks of lettuce drop. Phytopathology 83(5):509–513. https://doi.org/10.1094/Phyto-83-509

Bertus AL (1968) Downy mildew of stocks. The Agril Gazette New South Wales 79:178–179.

Beura SK, Mohapatra KB, Paul PK, Nandi A (2006) Integrated management of Black-rot disease of cauliflower in Orissa. J Mycopathol Res 44(2):293–296

Bhajbhuje MN (2014) Biodiversity of seed mycoflora in storage of *Brassica campestris* L. Intl J Life Sci 2(4):289–303

Bhat DM, Bhagat RC (2009) Natural parasitism of leaf miner, Chromatomyia horticola (Goureau) (Diptera: Agromyzidae) on Vegetable Crops in Kashmir (India). World J Agric Sci 5(S):888–891

Bhat NA, Masoodi SD (2000) Efficacy of various antibiotics against *Xanthomonas campestris*pv. *campestris* the casual pathogen of black rot of cabbage. Appl Biolog Res 2(1/2):161–163

Bhatt DD (1966) Preliminary list of plant diseases recorded in the Katmandu Valley. J Sci Tri-Chandra College Sci Assoc 2 (1):13–20

Bhattacharya I, Dutta S, MondalS MB (2014) Clubroot disease on *Brassica* crops in India. Can J Plant Pathol 36:154–160

Biesheuvel AR (1991) Information on white cabbages. Compromise between diseases and quality. (Voorlichting - witte kool. Schipperen tussen ziekten en kwaliteit.). Groenten + Fruit, Vollegrondsgroenten 1(9):14–15

Bilichak A, Kovalchuk I (2016) Transgenerational response to stress in plants and its application for breeding. J Exp Bot 67:2081–2092

Billard V, Ourry A, Maillard A, Garnica M (2014) Copper-deficiency in *Brassica napus* induces copper remobilization, molybdenum accumulation and modification of the expression of chloroplastic proteins. PLoS One 9(10):e109889. https://doi.org/10.1371/journal.pone.0109889

Bindra OS, Bakhetia DRC (1967) A note on the natural incidence of sesamum phyllody virus disease in *Brassica* spp. at Ludhiana. J Res Punjab Agril Univ 4:406–408

Bisby GR (1924) The Sclerotinia disease of sunflowers and other plants. Sci Agric 4(12):381–384

Bjorling K (1944) Some fungus diseases of colza and white mustard. Vaxt sky dds noti servaxt skyddsanst Stockh 7:73–77

Blackford FW (1944) Sclerotinia or cottony rot. Queensland Agril J 53(3):161–163

Boesewinkel HJ (1977) Identification of Erysiphaceae by conidial characteristics. Rev Mycol 41:493–507

Bogarada AP, Spiridonova VP, Martynovskaya NM (1984) Pests and diseases of *Erysimum diffusum* in the Ukraine. Byulleten' Glavnogo Botanicheskogo Sada:94–97

Bokor A (1972) Diseases of rape. J Agric, WA 13:45–48

Bokor A, Barbetti MJ, Brown AGP, Mac Nish GC, PMCR W (1975) Blackleg of rapeseed. J Agric WA 16:7–10

Bolay A, Braun U, Delhey R, Kummer V, Piatek M, Wolczanska A (2005) *Erysiphe deutziae*—a new epidemic spread in Europe. Cryptogam Mycol 26: 293–298.

Boning K (1936) Investigations on horseradish disease and their control. J Appl Bot 18:482–494

Boudsocq M, Sheen J (2013) CDPKs in immune and stress signaling. Trends Plant Sci 18:30–40. https://doi.org/10.1016/j.tplants.2012.08.008

Boudsocq M, Willmann MR, Mccormack M, Lee H, Shan L, He P et al (2010) Deferential innate immune signaling via Ca(2C) sensor protein kinases. Nature 464:418–422. https://doi.org/10.1038/nature08794

Bracken GK (1987) Relation between pod damage caused by larvae of bertha armyworm, *Mamestra configurata* Walker (Lepidoptera: Noctuidae), and yield loss, shelling and seed quality in canola. Can Entomol 119(4):365–369

Bradbury JF (1986) Guide to Plant Pathogenic Bacteria. CAB International, Wallingford, UK

Brandt B, Brodsky DE, Xue S, Negi J, Iba K, Kangasjarvi J et al (2012) Reconstitution of abscisic acid activation of SLAC1 anion channel by CPK6 and OST1 kinases and branched ABI1 PP2C phosphatase action. Proc Natl Acad Sci USA 109:10593–10598. https://doi.org/10.1073/pnas.1116590109

Brien RM (1932) Host range of *Sclerotinia sclerotiorum* in New Zealand. New Zealand J Agric 44 (2):127–129

Brodersen P, Petersen M, Bjorn Nielsen H, Zhu S, Newman MA, Shokat KM, Rietz S, Parker J, Mundy J (2006) Arabidopsis MAP kinase 4 regulates salicylic acid- and jasmonic acid/ ethylene dependent responses via EDS1 and PAD4. Plant J 47(4):532–546

Brooks AV (1975) Notes from Wisley. Some disease problems of 1974. J Royal Hortl Soc 100 (5):188–191

Brooks F (2002) List of Plant Diseases in American Samoa 2002. In: Land Grant Technical Report No. 44, Pago Pago, American Samoa, American Samoa Community College Land Grant Program.

Brown JG, Butler KD (1936) Sclerotinioise of lettuce in Arizona. Technical Bulletin of the Arizona Agriculture Experiment Station:474–506

Buchi R (2002) Mortality of pollen beetle (*Meligethes* spp.) larvae due to predators and parasitoids in rape fields and the effect of conservation strips. Agric Ecosys Environ 90(3):255–263. https://doi.org/10.1016/S0167-8809(01)00213-4

Buczack ST, Ockendon JG (1979) Preliminary observations on variation in susceptibility to clubroot among collections of some wild crucifers. Ann Appl Biol 92(1):113–118

Bunkina IA (1991) *Erysiphales*. In: Nizshie rastenija, griby i mohoobraznye Sovetskogo Dal'nego Vostoka, Griby, Azbukina ZM (ed), Tom 2: Askomicety, Erizifal'nye, Klavicipital'nye, Gelocial'nye: 11–142, Leningrad. [In Russian]

Buonassisi A, MacDonald LS (1990) 1988 and 1989 canola disease survey in British Columbia. Can Plant Dis Surv 70:62

Burdyukova LI, Dudka IA (1982) Species of the genus Albugo on new and rare host plants in the Ukrainian SSR. Mikol Fitopatol 16(4):289–294

Burgess L (1977) Flea beetles (Coleoptera: Chrysomelidae) attacking rape crops in the Canadian Prairie Provinces. Can Entomologist 109(1):21–32

Burkholder WH (1957) Genus *Erwinia*. In: Berge's Manual of Determinate Bacteriology, 7th edn. The Williams and Wilkins Co, Baltimore, pp 349–359

Butler EJ (1918) Fungi and diseases in plant. Thaker Spink and Co., Calcutta, India, pp 297–300

CAB International (1978) *Leptosphaeria maculans*. [Distribution map]. In: Distribution maps of plant diseases, CAB Inter., Wallingford, UK, p 73

CAB International (1986) Distribution maps of plant diseases (Edition 5) UK, p 54

CAB International (1987) *Xanthomonas campestris* pv. *campestris*. [Distribution map]. Distribution Maps of Plant Diseases, May (Edition 5). CAB International, Wallingford, UK, p 136

CABI, EPPO (2001) *Albugo candida*. [Distribution map]. In: Distribution maps of plant diseases. CAB International, Wallingford, UK, p 821

CABI, EPPO (2005) *Sclerotinia sclerotiorum*. [Distribution map]. In: Distribution maps of plant diseases, Wallingford, UK: CAB International. Map 971.

CABI/EPPO (2011) *Plasmodiophora brassicae*. [Distribution map]. Distribution maps of plant diseases, No. April. CABI, Wallingford, UK, (Edition 4), p 101

Canaday CH, Wyatt JE, Mullins JA (1991) Resistance in Broccoli to bacterial soft rot caused by *Pseudomonas marginalis* and fluorescent *Pseudomonas* species. Plant Dis 75:715–720. https://doi.org/10.1094/PD-75-0715

Canvin T (1965) The effect of temperature on the oil content and fatty acid composition of the oils from several oil seed crops. Can J Bot 43:63–69

Caponero A, Iacobellis NS (1994) Foci of 'black rot' on cauliflower in Basilicata. Informatore Agrario 50(25):67–68

Cappelli C, Corazza L, Luongo L, Stravato VM (1999) Interactions between crucifers and *Rhizoctonia solani* AG 2-1, AG 2-2IIIB, AG 2-2IV, AG 4. Phytopathol Medit 38:37–39

Carrera J, Rodrigo G, Jaramillo A, Elena SF (2009) Reverse-engineering the *Arabidopsis thaliana* transcriptional network under changing environmental conditions. Genome Biol 10:R96, Cell Physiol 55:1859–1863

Cassells AC, Walsh M (1995) Screening for Sclerotinia resistance in *Helianthus tuberosus* L. (Jerusalem artichoke) varieties, lines and somaclones, in the field and in vitro. Plant Pathol 44(3):428–437. https://doi.org/10.1111/j.1365-3059.1995.tb01665.x

Cerkauskas RF, Stobbs LW, Lowery DT, Van Driel L, Liu W, Vanschagen J (1998) Diseases, pests, and abiotic problems associated with oriental cruciferous vegetables in southern Ontario in 1993–1994. Can J Plant Pathol 20:87–94

Chambers SC (1959) A revised list of vegetable diseases recorded in Western Australia. J Agric WA 8:427–431

Chandrasrikul A (1962) Tech Bull Dept Agric. Bangkok 6:23

Channon AG (1981) Downy mildew of *Brassicas*. In: Spencer DM (ed) The downy mildews. Academic Press, London, pp 321–339

Charleston DS, Kfir R (2000) The possibility of using Indian mustard, *Brassica juncea*, as a trap crop or the diamondback moth *Plutella xylostella* in South Africa. Crop Protect 19:455–460

Chater AO, Webb DA (1972) Orobanche. In: Tutin TG, Heywood VH, Burgess NA, Morre DM, Valentine DH, Walters SM, Webb DM (eds.) Flora Europaea 3. Diapensiaceae to Myoporaceae. University Press, Cambridge, UK, pp 286–293

Chauhan LS, Saksena HK (1974) A new Rhizoctonia leaf blight of rapes and mustard. Indian J Farm Sci 2:98–99

Cheng SH, Willmann MR, Chen HC, Sheen J (2002) Calcium signaling through protein kinases. The *Arabidopsis* calcium-dependent protein kinase gene family. Plant Physiol 129:469–485. https://doi.org/10.1104/pp.005645

Cheval C, Aldon D, Galaud JP, Ranty B (2013) Calcium/calmodulin-mediated regulation of plant immunity. Biochem Biophys Acta 1833:1766–1771. https://doi.org/10.1016/j.bbamcr.2013.01.031

Chinnusamy V, Schumaker K, Zhu JK (2004) Molecular genetic perspectives on cross-talk and specificity in abiotic stress signaling in plants. J Exp Bot 55:225–236. https://doi.org/10.1093/jxb/erh005

Choi WG, Toyota M, Kim SH, Hilleary R, Gilroy S (2014) Salt stress-induced Ca2+ waves are associated with rapid, long-distance root to-shoot signaling in plants. Proc Natl Acad Sci U S A 111:6497–6502

Chorin M (1946) Downy mildew on cauliflower curds. Palestine J Bot Jerusalem Series-I 2:258–259

Chung E, Kim KM, Lee JH (2013) Genome-wide analysis and molecular characterization of the heat shock transcription factor family in *Glycine max*. J Genet Genomics 40:127–135

Ciferri R (1961) Lista delle Peronosporaceae italiane. Riv Patol Veg (Pavia) Ser (J Plant Pathol) 1:333–348

Cleene MD, Ley JD (1976) The host range of crown gall. Bot Rev 42(4):389–466

Coca M, San Segundo B (2010) AtCPK1 calcium-dependent protein kinase mediates pathogen resistance in *Arabidopsis*. Plant J 63:526–540. https://doi.org/10.1111/j.1365-313X.2010.04255.x

Commonwealth Mycological Institute (1986) Distribution map of plant diseases, pathogen: *Pseudocercosporella capsellae (EII.and Ev.)* Deighton. Commonwealth Mycological Institute, Wallingford

Cook SM, Murray DA, Williams IH (1999) Pollen beetle, *Meligethes aeneus* Fabricius, incidence in the composite hybrid winter oilseed rape, Synergy. Proceedings 10th Intl Rapeseed Congr: New Horizons for an Old Crop, Canberra, 26-29 September 1999. CD ROM and Abstract Book 3:311–436

Couvreur L, Herman JL (1994) Protection of winter rape against diseases in Belgium. (La protection du colza d'hiver contre les maladies en Belgique.). In: Mededelingen - Faculteit

Landbouwkundige en Toegepaste Biologische Wetenschappen, Universiteit Gent [46th International symposium on crop protection, Gent, Belgium, 3 May, 1994.], 59 (3a): 1133-1139.
Cramer GR, Urano K, Delrot S, Pezzotti M, Shinozaki K (2011) Effects of abiotic stress on plants: a systems biology perspective. BMC Plant Biol 11:163
Creelman DW (1962) Diseases of vegetables and field crops-potato. Can Plant Dis Sur 42:57–65
Crispin A, Campos J (1976) Bean diseases of importance in Mexico in 1975. Plant Dis Reptr 60 (6):534–535
Cui H, Tsuda K, Parker JE (2015) Effector triggered immunity: from pathogen perception to a robust defense. Annu Rev Plant Biol 66:487–511
Cunningham GH (1927) New Zealand. Bulletin, Department of Agriculture
Cvjetkovic B (2008) Less frequent diseases of oilseed rape. (Rjeđe ucestale bolesti uljane repice). Glasilo Biljne Zastite 8(5):318–320
Da Costa MEAP, Da-Camara ES (1953) Some Portuguese mycological species II. Protugaliae Acta Biol 4:162–176
Daebeler F, Amelung D, Zeise K (1988) *Verticillium*-welke an winterraps—auftreten und bedeutung. Nachrichtenblatt Pflanzenschutzdienst DDR 42:71–73
Damm U, Cannon PF, Woudenberg JHC, Johnston PR, Weir BS, Tan YP, Shivas RG, Crous PW (2012) The *Colletotrichum boninense* species complex. Studies Mycol 73:1–36
Damm U, O'Connell RJ, Groenewald JZ, Crous PW (2014) The *Colletotrichum destructivum* species complex – hemibiotrophic pathogens of forage and field crops. Studies Mycol 79:49–84
Darpoux H (1945) A contribution to the study of the disease of Oleaginous plants in France. Ann Epiphyt NS 11:71–103
Dart C (2010) Lipid microdomains and the regulation of ion channel function. J Physiol 588:3169–3178. https://doi.org/10.1113/jphysiol.2010.191585
Day I S, Reddy V S, Shad Ali G, Reddy A S (2002) Analysis of EF-hand-containing proteins in *Arabidopsis*. Genome Biol 3. doi: https://doi.org/10.1186/gb-2002-3-10-research0056
De Carvalho T (1948) Preliminary report on the diseases observed in plants and insects with Phytopathological annotations. Colonia de Mocambique, Reparticao de Agricultura, Seccao de Micologia: 84.
de Dias MRS, Lucas MT (1978) Portuguese fungi. XXVI Agronomia Lusitana 38(4):285–295
De Vos M, Van Oosten VR, Van Poecke RM, Van Pelt JA, Pozo MJ, Mueller MJ, Buchala AJ, Metraux JP, Van Loon LC, Dicke M, Pieterse CM (2005) Signal signature and transcriptomechanges of *Arabidopsis* during pathogen and insect attack. Mol Plant-Microbe Interact 18(9):923–937
Demir F, Horntrich C, Blachutzik J O, Scherzer S, Reinders Y, Kierszniowska S, et al. (2013) *Arabidopsis* nanodomain-delimited ABA signaling pathway regulates the anion channel SLAH3. Proc Natl Acad Sci USA 110:8296–8301. doi: https://doi.org/10.1073/pnas.1211667110
Denner FDN, Theron DJ, Serfontein S (1992) *Sclerotinia sclerotiorum* on potatoes in South Africa. Phytophylactica 24(2):221–223
Ding B, Wang GL (2015) Chromatin versus pathogens: the function of epigenetics in plant immunity. Front Plant Sci 6:675
Dippenaar BJ (1934) Downy mildew of cabbage and related plants. Farming South Africa 9:279–280
Dixon GR (2009) The occurrence and economic impact of *Plasmodiophora brassicae* and clubroot disease. In: Dixon GR (ed) *Plasmodiophora brassicae* (clubroot) – a plant pathogen that alters host growth and productivity, J Plant Growth Regul, vol 28(Spl), pp 194–202
Djebali N, Scott JK, Jourdan M, Souissi T (2009) Fungi pathogenic on wild radish (*Raphanus raphanistrum* L.) in northern Tunisia as potential biocontrol agents. Phytopathol Mediterr 48 (2):205–213
Dodd AN, Kudla J, Sanders D (2010) The language of calcium signaling. Annu Rev Plant Biol 61:593–620. https://doi.org/10.1146/annurev-arplant-070109-104628

Dodds PN, Rathjen JP (2010) Plant immunity: towards an integrated view of plant pathogen interactions. Nat Rev Genet 11:539–548

Dolinski MG (1979) The cabbage seedpod weevil, Ceutorhynchus assimilis (Payk) (Coleoptera: Curuliondae), as a potential pest of rape production in Canada, MPM Thesis. Simon Fraser University, Burnaby, Canada

Dowen RH, Pelizzola M, Schmitz RJ, Lister R, Dowen JM, Nery JR, Dixon JE, Ecker JR (2012) Widespread dynamic DNA methylation in response to biotic stress. Proc Natl Acad Sci U S A 109:E2183–E2191

Downey RK, Bolton JL (1961) Production of rape in western Canada. Publ Dept Agric Canada No 1021:19

Drechsler C (1925) Pythium infection of cabbage heads. Phytopathology 15:482–485

Droillard M, Boudsocq M, Barbier-Brygoo H, Lauriere C (2002) Different protein kinase familiesare activated by osmotic stresses in *Arabidopsis* thaliana cell suspensions. Involvement of the MAP kinases AtMPK3 and AtMPK6. FEBS Lett 527(1–3):43–50

Dubiella U, Seybold H, Durian G, Komander E, Lassig R, Witte CP et al (2013) Calcium-dependent protein kinase/NADPH oxidase activation circuit is required for rapid defense signal propagation. Proc Natl Acad Sci U S A 110:8744–8749. https://doi.org/10.1073/pnas.1221294110

Duduk B, Bulajić A, Duduk N, Calari A, Paltrinier S, Krstić B, Bertaccinin (2007) Identification of phytoplasmas belonging to aster yellows ribosomal group (16SrI) in vegetables in Serbia. Bull Insectol 60:341–342

Duwadi VR, Paneru RB, Bhattarai MR (1993) An estimate of yield loss of cauliflower (cv. Kibo Giant) seed caused by *Sclerotinia sclerotiorum* (Lib.) de Bary. In: PAC Working Paper - Pakhribas Agricultural Centre. 4 pp

Elazari-Volcani Z (1944) Observations on *Sclerotinia sclerotiorum* in Palestine. Palestine J Bot (Rehovot Series). 206–207

Elferjani R, Soolanayakanahally R (2018) Canola responses to drought, heat, and combined stress: shared and specific effects on carbon assimilation, seed yield, and oil composition. Front Plant Sci 9:1224. https://doi.org/10.3389/fpls.2018.01224

Ellett CW (1966) Host Range of the *Erysiphaceae* of Ohio. The Ohio J Sci 66(6):570–581

Ellis MB (1971) Dematiaceous hyphomycetes. Commonwealth Mycological Institute, Kew, Surrey. England No 608:464–497

Ellis PJ (1983) Diseases of rapeseed in the Peace River region of Alberta. M.P.M. Thesis. Simon Fraser University, Burnaby, BC, Canada. 89 pp.

Ellis MB, Ellis JP (1985) Microfungi on land plants: an identification handbook. Macmillan Publishing Co., New York, 818 pages

EPPO (2014) PQR database. European and Mediterranean Plant Protection Organization, Paris, France. http://www.eppo.int/DATABASES/pqr/pqr.htm

EPPO (2020) EPPO global database. In: EPPO global database. EPPO, Paris, France

Eshraghi L, You MP, Barbetti MJ (2005) First report of white leaf spot caused by *Pseudocercosporella capsellae* on *Brassica juncea* in Australia. Plant Dis 89(10):1131. https://doi.org/10.1094/PD-89-1131B

Ester A, Gerlagh M (1984) Resistance breeding. Zaadbelangen 38(4):73–74

Etebarian HR (1989) Studies on downy mildew of radish and its chemical control. Iranian J Agril Sci 20:23–31

Etebarian HR (1993) Studies on white rust of radish and cress in Varamin and Shahr-e-Ray areas. Iranian J Plant Pathol 29(3-4):51–52

Evans KA, Bergeron J (1994) Behavioral and electrophysiological response of cabbage seed weevils (*Ceutorhynchus assimilis*) to conspecific odor. J Chem Ecol 20:979–989

Fahy PC, Persley GJ (1983) Plant bacterial diseases. A diagnostic guide. Plant bacterial diseases .A diagnostic guide. vii + 393 pp.

Farlow WG (1883) Enumeration of the Peronosporeae of the United States. Bot Gaz 8 (305-315):327–337

Farr DF, Bills GF, Chamuris GP, Rossman AY (1989) Fungi on plants and plant products in the United States. APS Press, St. Paul, MN, p 1252

Farr DF, Bills GF, Chamuris GP, Rossman AY (1995) Fungi on Plants and Plant Products in the United States. APS Press, St. Paul, MN

Felix S (1962) Investigations on *Sclerotinia sclerotiorum* (Lib.) de Bary, the fungus causing sclerotial disease of vegetables and ornamentals in Mauritius. Rev Agric Sucriere Lile Maurice 41:205–207

Fernandez RM (1973) Serie Agricola Academia de ciencias de Cuba, Instituto de Investigaciones Tropicoles. 27:78

Firman ID (1975) Information Document South Pacific Commission, No. 38: 7+3.

Fivawo NC (1987) Diseases and pests of sunflower in Tanzania. TARO Newsletter 2(3):6–8

Fodor M, Viczian O, Mergenthaler E, Sule S (1999) Cabbage infected with phytoplasma from the aster yellows group in Hungary. Acta Phytopathol Entomol Hung 34(1-2):1–6

Foister CE (1948) Report of the Annual Conference of the Crytogamic section, 1947. Trans Bot Soc Edinburgh 35:103–107

Folk G, Tusnádi CSK (1985) A new disease of Gerbera in Hungary, Sclerotinia wilt. (A Gerbera új betegsége Magyarországon, a szklerotíniás hervadás.). Novenyvedelem 21(12):557–561

Foucart G (1954) Observations on some fungus diseases of pyrethrum (*Chrysanthemum cinerariifolium* (Trev.) Bocc). Bulletin agricole du Congo belge 45(3):599–614

Free JB, Williams IH (1978) The responses of the pollen beetle, *Meligethes aeneus*, and the seed weevil, *Ceutorhynchus assimilis* to oilseed rape, and other plants. J Appl Ecol 15:761–764

Frescatada-Rosa M, Robatzek S, Kuhn H (2015) Should I stay or should I go? Traffic control for plant pattern recognition receptors. Curr Opin Plant Biol 28:23–29

Fujita M, Fujita Y, Maruyama K, Seki M, Hiratsu K, Ohme-Takagi M, Tran LS, Yamaguchi-Shinozaki K, Shinozaki K (2004) A dehydration induced NAC protein, RD26, is involved in a novel ABA-dependent stress-signaling pathway. Plant J 39(6):863–876

Fujita M, Fujita Y, Noutoshi Y, Takahashi F, Narusaka Y, Yamaguchi-Shinozaki K, Shinozaki K (2006) Crosstalk between abiotic and biotic stress responses: a current view from the points of convergence in the stress signaling networks. Curr Opin Plant Biol 9:436–442

Gaetan S, Madia M (2004) First report of canola powdery mildew caused by *Erysiphe polygoni* in Argentina. Plant Dis 88(10):1163–1163. https://doi.org/10.1094/PDIS.2004.88.10.1163C

Gaetan SA, Garbagnoli C, Irigoyen ED (1995) Microorganisms isolated from rape seeds (*Brassica napus* L. subsp. *oleifera* (Metzg.) Sinsk.)in Argentina. VIII Latin American congress of plant pathology. XIV Venezuelan congress of plant pathology. Faculty of forest sciences, University of Los Andes, Merida, Venezuela, 22-26 October 1995. Fitopatologfa, 30:10.

Gaetan SA (2005) Occurrence of Fusarium wilt on canola caused by *Fusarium oxysporum* f. sp. *conglutinans* in Argentina. Plant Dis 89(4):432–432. https://doi.org/10.1094/PD-89-0432C

Garapati P, Xue GP, Munné-Bosch S, Balazadeh S (2015) Transcription factor ATAF1 in *Arabidopsis* promotes senescence by direct regulation of key chloroplast maintenance and senescence transcriptional cascades. Plant Physiol 168:1122–1139

Gasparotto L, Chaves GM (1982) Chemical control of lettuce rot. (Controle quimico da podridao da alface). Fitopatol Bras 7(2):261–267

Gaumann E (1923) Contributions toward a monograph of the genus *Peronospora corda*. Beitrage zur kryptogamenflora de schweiz 5:360

Gerber GH (1976) Effects of feeding by adults of the red turnip beetle, *Entomoscelis americana* Brown (Coleoptera: Chrysomelidae), during late July and August on the yield of rapeseed (Cruciferae). Manitoba Entomol 10:31–35

Gerlach WWP (1988) Plant diseases of Western Samoa. Samoan German Crop Protection Project, Apia, Western Samoa, 215 pp

Ghafoor A (1964) Radish black-root fungus: Host range, nutrition and oospore production and germination. Phytopathology 54:1167–1171

Gilroy S, Suzuki N, Miller G, Choi WG, Toyota M, Devireddy AR et al (2014) A tidal wave of signals: calcium and ROS at the forefront of rapid systemic signaling. Trends Plant Sci 19:623–630. https://doi.org/10.1016/j.tplants.2014.06.013

Gish LA, Clark SE (2011) The RLK/Pelle family of kinases. Plant J 66:117–127

Gkavaleka S, Efthimiou K, Lotos L, Katis NI (2012) First report of a 'Candidatus Phytoplasma asteris'-related strain associated with cabbage stunting in Greece. J Plant Pathol 94:84–85

Gladders P (1978) The disease threat to oilseed rape. Arable Farming 5:47–49

Glaeser G (1970) The occurrence of important causes of injury to cultivated plants in Austria in 1969. Pflanzenschutzberichte 41:49–62

Glaeser G (1973) White rust also on radishes. (Weisser Rost auch an Radieschen.). Pflanzenarzt 26 (3):20.

Glazebrook J (2005) Contrasting mechanisms of defense against biotrophic and necrotrophic pathogens. Annu Rev Phytopathol 43:205–227

Gonzalez FR (1924) Peronosporaceae at present known to occur in the Iberian flora. Boletín de la Real Sociedad Española de Historia Natural 24:305–312

Good MC, Zalatan JG, Lim WA (2011) Scaffold proteins: hubs for controlling the flow of cellular information. Science 332:680–686. https://doi.org/10.1126/science.1198701

Goto M (1992) Fundamentals of bacterial plant pathology. Academic Press Inc., San Diego, USA

Graham CW, Gould HJ (1980) Cabbage stem weevil (*Ceutorhynchus quadridens*) on spring oilseed rape in Southern England and its control. Ann Appl Biol 95(1):1–10. https://doi.org/10.1111/j.1744-7348.1980.tb03964.x

Gram E, Rostrup S (1924) Survey of diseases of agricultural and horticultural cultivated plants in 1923. Tidsskr for Planteavl 30:361–414

Grant CA, Bailey LD (1993) Fertility management in canola production. Can J Plant Sci 73:651–670

Greathead DJ, Greathead AH (1992) Biological control of insect pests by insect parasitoids and predators: the BIOCAT database. Biocontr News Inform 13(4):61N–68N

Griffiths GCD (1986) Relative abundance of the root maggots *Delia radicum* (L.) and *D. floralis* (Fallén) (Diptera: Anthomyiidae) as pests of canola in Alberta. Quaest Entomol 22:253–260

Grill D (1979) *Botrytis* and *Sclerotinia* on lettuce. (*Botrytis* et *Sclerotinia* de la laitue.). Revue Horticole 47–48.

Grondin A, Rodrigues O, Verdoucq L, Merlot S, Leonhardt N, Maurel C (2015) Aquaporins contribute to ABA-triggered stomatal closure through OST1-mediated phosphorylation. Plant Cell 27:1945–1954

Gueguinou M, Gambade A, Felix R, Chantome A, Fourbon Y, Bougnoux P, et al. (2015) Lipid rafts, KCa/ClCa/Ca^{2+}channel complexes and EGFR signaling: novel targets to reduce tumor development by lipids? Biochem Biophys Acta 1848:2603–2620. doi:https://doi.org/10.1016/j.bbamem.2014.10.036

Gugel RK, Petrie GA (1992) History, occurrence, impact, and control of blackleg of rapeseed. Can J Plant Pathol 14:36–45

Gupta DK (1991) Studies on black rot of cabbage in Manipur. Indian J Mycol Plant Pathol 21 (2):203–204

Gupta DK, Choudhary KCB (1995) Infection of radish and raya sag seeds by *Xanthomonas campestris* pv. *campestris*. Indian J Mycol Plant Pathol 25(3):332

Habimana S, Nduwumuremyi A, Chinama RJD (2014) Management of orobanche in field crops-a review. J Soil Sci Plant Nutr 14 Epub 19-Ene-2014. https://doi.org/10.4067/S0718-95162014005-000004.

Hall R, Peters R, Assabgui R (1993) Occurrence and impact of blackleg on oilseed rape in Ontario. Can J Plant Pathol 15:305–313

Hammarlund C (1931) Shorter mycological notices II. A giant form of *Peronospora brassicae* Gaumann (=*P. parasitica* (Fries) Tulasne) on *Raphanus sativus* f. *radicula*. Bot Notiser 5:392–393

Hardie DG, Schaffer BE, Brunet A (2016) AMPK: an energy-sensing pathway with multiple inputs and outputs. Trends Cell Biol 26:190–201

Hardwick NV, Davies JML, Wright DM (1994) The incidence of three virus diseases of winter oilseed rape in England and Wales in the 1991/92 and 1992/93 growing seasons. Plant Pathol 43:1045–1049

Harmon BL, McCaffrey JP (1997) Laboratory bioassay to assess *Brassica* spp. Germplasm for resistance to the cabbage seedpod weevil (Coleoptera: Curculionidae). J Econ Entomol 90:1392–1399

Harmon AC, Gribskov M, Harper JF (2000) CDPKs – a kinase for every Ca^{2+} signal? Trends Plant Sci 5:154–159. https://doi.org/10.1016/S1360-1385(00)01577-6

Harper JF, Breton G, Harmon A (2004) Decoding Ca(2C) signals through plant protein kinases. AnnuRev. Plant Biol 55:263–288. https://doi.org/10.1146/annurev.arplant.55.031903.141627

Hashimoto K, Kudla J (2011) Calcium decoding mechanisms in plants. Biochimie 93:2054–2059. https://doi.org/10.1016/j.biochi.2011.05.019

Haware MP (1990) Fusarium wilt and other important diseases of chickpea in the Mediterranean area. In: Options Mediterraneennes Serie A: Seminaires Medit 61–64.

He YY, Chen Q, Shu C, Yang M, Zhou E (2016) *Colletotrichum truncatum*, a new cause of anthracnose on Chinese flowering cabbage (*Brassica parachinensis*) in China. Trop Plant Pathol 41:183–192. https://doi.org/10.1007/s40858-016-0086-4

Henderson MP (1918) The black-leg disease of cabbage caused by *Phoma lingam* (Tode) Desmaz. Phytopathology 8:379–431

Henderson RM (1962) Some aspects of the life cycle of the plant pathogen *Sclerotinia sclerotiorum* in western Australia. Royal Society of Western Australia J 45:133–135

Hensen LM (2004) Economic damage threshold model for pollen beetles (*Meligethes aeneus* F) in spring oilseed rape (*Brassica napus* L) crops. Crop Prot 23:43–46

Hickman CJ, Schofield ER, Taylor ER (1955) Light leaf spot of *Brassica*. Plant Pathol 4:129–131

Higgins BB (1917) Notes on some diseases of collards. Georgia Agric Expt Stn Annu Reprt 29:21–27

Hill CB (1992) Blackleg of crucifers. In: Chaube HS, Kumar J, Mukhopadhyay AN, Singh US (eds) Plant diseases of international importance, Vol. 2: Diseases of vegetables and oilseed crops. Prentice Hall, Inc., Englewood Cliffs, pp 253–271

Hiura M, Kanegae H (1934) Studies on the downy mildews of cruciferous vegetables in Japan. Trans Sapporo Nat Hist Soc13: 125-133.

Ho BL (1985) The etiology of soft rot, black rot and bacterial blight diseases of cabbage (*Brassica oleracea* var. *capitata* L.). MARDI Res Bull 13(1):28–37

Holm L, Pancho JV, Herberger JP, Plucknett DL (1979) A geographical atlas of world weeds. John Wiley and Sons, New York, Chichester, Brisbane, Toronto, UK, p 391

Holtzhausen MA (1978) Seed-borne fungal pathogens and diseases of Japanese radish and their control in South Africa. Phytophylactica 10(4):107–113

Hrabak EM, Chan CW, Gribskov M, Harper JF, Choi JH, Halford N, Kudla J, Luan S, Nimmo HG, Sussman MR et al (2003) The *Arabidopsis* CDPK-SnRK super family of protein kinases. Plant Physiol 132:666–680

Hsieh TH, Lee JT, Yang PT, Chiu LH, Charng YY, Wang YC, Chan MT (2002) Heterology expression of the *Arabidopsis* C-repeat/dehydration response element binding factor 1 gene confers elevated tolerance to chilling and oxidative stresses in transgenic tomato. Plant Physiol 129(3):1086–1094.

Hu Y, Han YT, Wei W, Li YJ, Zhang K, Gao YR, Zhao FL, Feng JY (2015) Identification, isolation, and expression analysis of heat shock transcription factors in the diploid woodland strawberry *Fragaria vesca*. Front Plant Sci 6:736

Hua D, Wang C, He J, Liao H, Duan Y, Zhu Z, Guo Y, Chen Z, Gong Z (2012) A plasma membrane receptor kinase, GHR1, mediates abscisic acid- and hydrogen peroxide-regulated tomato movement in *Arabidopsis*. Plant Cell 24:2546–2561

Huang Y, Li H, Gupta R, Morris P, Luan S, Kieber JJ (2000) ATMPK4, an Arabidopsis homolog of mitogen-activated protein kinase, is activated in vitro by AtMEK1 through threonine phosphorylation. Plant Physiol 122(4):1301–1310

Humpherson-Jones FM (1991) The development of weather-related disease forecasts for vegetable crops in the UK. Problems and prospects. Bulletin OEPP 21(3):425–429. https://doi.org/10.1111/j.1365-2338.1991.tb01272.x

Hunter JE, Abawi GS, Becker RF (1975) Observations on the source and spread of *Xanthomonas campestris* in an epidemic of black rot in New York. Plant Dis Reptr 59(5):384–387

Huser A, Takahara H, Schmalenbach W, O'Connell R (2009) Discovery of pathogenicity genes in the crucifer anthracnose fungus, *Colletotrichum higginsianum*, using random insertional mutagenesis. Mol Plant-Microbe Interact 22:143–156. https://doi.org/10.1094/MPMI-22-2-0143

Hwang SM, Kim DW, Woo MS, Jeong HS, Son YS, Akhter S, Choi GJ, Bahk JD (2014) Functional characterization of *Arabidopsis* HsfA6a as a heat-shock transcription factor under high salinity and dehydration conditions. Plant Cell Environ 37:1202–1222

Ichimura K, Mizoguchi T, Irie K, Morris P, Giraudat J, Matsumoto K, Shinozaki K (1998) Isolation of ATMEKK1 (a MAP Kinase kinase kinase)-interacting proteins and analysis of a MAP kinase cascade in *Arabidopsis*. Biochem Biophys Res Commun 253(2):532–543

Ichimura K, Mizoguchi T, Yoshida R, Yuasa T, Shinozaki K (2000) Various abiotic stresses rapidly activate *Arabidopsis* MAP kinases ATMPK4 and ATMPK6. Plant J 24(5):655–665

Icochea T, Torres H, Perez W (1994) Downy mildew of Moca (*Lepidium meyenii*): symptoms and identification of causal agent. Fitapatologia 29:156–159

Ilumae E, Kastanje V, Hansson A, Akk E (2007) Distribution and control of Sclerotinia stem rot. (Valgemädaniku (*Sclerotinia sclerotiorum* (Lib.) De Bary) levik suvirapsil (*Brassica napus*) ja selle tõrje.). In: Kadaja J (ed) Agronomy 2007. Estonian Research Institute of Agriculture, Sacu, Estonia, pp 117–120

Ivancheva-Gabrovska T, Ilieva E, Kadir SR (1978) Pathogens of damping-off of tobacco seedlings in Bulgaria. (Prichiniteli na secheneto po tyutyuneviya razsad v B"lgariya.) Rasteniev"dni Nauki 15(8):113–120

Iwasaki M, Paszkowski J (2014) Epigenetic memory in plants. EMBO J 33:1987–1998

Jalaluddin M, Ali A, Anwar QMK (1993) Foliar and floral diseases of rapeseed and mustard: their vectors and control. Pakistan J Phytopathol 5(1–2):29–39

Jansen MK, Hangedorn PH, De Torres-Zabala M, Grant MR, Rung JH, Colling D, Lyngkjaer MF (2008) Transcriptional regulation by an NAC (NAM-ATAF1, 2-CUC2) transcription factor attenuates ABA signalling for efficient basal defense towards *Blumeria graminis* f. sp. *hordei* in *Arabidopsis*. Plant J 56:867–880

Jayawardena RS, Hyde KD, Damm U, Cai L, Liu M, Li XH, Zhang W, Zhao WS, Yan JY (2016) Notes on currently accepted species of *Colletotrichum*. Mycosphere 7(8):1192–1260. https://doi.org/10.5943/mycosphere/si/2c/9

Jellis GJ, Scott ES, Smith DB (1984) The causal organism of stem rot disease of faba beans. FABIS Newsletter, Faba Bean Information Service, ICARDA:24–27

Jensen MK, Rung JH, Gregersen PL, Gjetting T, Fuglsang AT, Hansen M, Joehnk N, Lyngkjaer MF, Collinge DB (2007) The HvNAC6 transcription factor: a positive regulator of penetration resistance in barley and *Arabidopsis*. Plant Mol Biol 65:137–150

Jesperson GD (1989) Survey of blackleg, Sclerotinia and foot rot in Saskatchewan canola crops, 1986. Can Plant Dis Surv 69:60–61

Johnston A (1963) A preliminary plant disease survey in Hong Kong. Plant Prod Prot Div FAO, Rome 32

Jones W (1944) Downy mildew disease of cauliflower seed plants. Scientia Agricola 24:282–284.

Journot-Catalino N, Somssich IE, Roby D, Kroj T (2006) The transcription factors WRKY11 and WRKY17 act as negative regulators of basal resistance in *Arabidopsis thaliana*. Plant Cell 18 (11):3289–3302

Jun Z, Zhang Z, Gao Y, Zhou L, Fang L, Chen X, Ning Z, Chen T, Guo W, Zhang T (2015) Overexpression of GbRLK, a putative receptor-like kinase gene, improved cotton tolerance to *Verticillium* wilt. Sci Rep 5:15048

Junell L (1967) Erysiphaceae of Sweden. Symbolae Botanicae Upsalienses 19:1–117

Kachlicki P, Jetaildryczka M (1994) Properties of *Phoma lingam* (Tode ex Fr.) Desm. isolates from Poland II. Secondary metabolites production. Phytopathol Polonica. 1994:81–86

Kado CI, Heskett MG, Langley RA (1972) Studies on *Agrobacterium tumefaciens*.Characterization of strains 1D135 and B6, and analysis of the bacterial chromosome, transfer RNA and ribosomes for tumor-inducing ability. Physiol Plant Pathol 2:47–57

Kaminska M, Berniak H, Kaminski P (2012a) Detection and identification of '*Candidatus* Phytoplasma asteris' in *Brassica rapa* subsp. *pekinensis* in Poland. J Phytopathol. https://doi.org/10.1111/jph.12011

Kaminska M, Berniak H, Kaminski P (2012b) Failure of lower bud formation in Brassica plants associated with phytoplasma infection. J Agric Sci 4:219–226

Kamthan A, Chaudhuri A, Kamthan M, Datta A (2015) Small RNAs in plants: recent development and application for crop improvement. Front Plant Sci 6:208

Kanchiswamy CN, Takahashi H, Quadro S, Maffei ME, Bossi S, Bertea C et al (2010) Regulation of *Arabidopsis* defense responses against *Spodopteralittoralis* by CPK-mediated calcium signaling. BMC Plant Biol10:97. https://doi.org/10.1186/1471-2229-10-97

Karapapa VK, Typas MA (2001) Molecular characterization of the host-adapted pathogen *Verticillium longisporum* on the basis of a group-i intron found in the nuclear SSU-rRNA Gene. Current Microbiol 42:217–224. https://doi.org/10.1007/s002840010207

Karolewski Z (2010) Development of light leaf spot (*Pyrenopeziza brassicae*) on Brassicas. Phytopathology 55:13–20

Kazan K (2015) Diverse roles of jasmonates and ethylene in abiotic stress tolerance. Trends Plant Sci 20:219–229. https://doi.org/10.1016/j.tplants.2015.02.001

Keane G, Edwards A, Clark MF (1996) Differentiation of group 16Sr-IB aster yellows phytoplasmas with monoclonal antibodies. In: Proceedings of the Diagnostics in Crop Production Symposium. No. 65. April 1-3, Coventry, UK: Br Crop Prot Council, pp 263–268

Khangura RK, Wright DG (2012) First report of club root caused by Plasmodiophora brassicae on canola in Australia. Plant Dis 96(7):1075–1076

Khangura RK, Barbetti MJ, Sweetingham MW (1999) Characterization and pathogenicity of *Rhizoctonia* species on canola. Plant Dis 83:714–721

Khraiwesh B, Zhu JK, Zhu J (2012) Role of miRNAs and siRNAs in biotic and abiotic stress responses of plants. Biochem Biophys Acta 1819:137–148

Khristov A (1979) Diseases of turnip in Bulgaria. Rastitelna Zaschita 27:27–28

Kim BS (1986) Testing for detection of *Xanthomonas campestris*pv. *campestris*in crucifer seeds and seed disinfection. Korean J Plant Pathol 2(2):96–101

Kim WG, Moon MH, Kim JH, Choi HW, Hong SK (2009) Occurrence of Clubroot on Pak-Choi Caused by *Plasmodiophora brassicae*. Mycobiol 37(1):69–71

Kim Y, Tsuda K, Igarashi D, Hillmer RA, Sakakibara H, Myers CL, Katagiri F (2014) Signaling mechanisms underlying the robustness and tenability of the plant immune network. Cell Host Microbe 15:84–94

Kim JM, Sasaki T, Ueda M, Sako K, Seki M (2015) Chromatin changes in response to drought, salinity, heat, and cold stresses in plants. Front Plant Sci 6:114

Kinoshita T, Seki M (2014) Epigenetic memory for stress response and adaptation in plants. Plant Cell Physiol 55:1859–1863

Kishun R (1984) Seed treatment for control of cabbage black rot. J Turkish Phytopathol 13 (2/3):81–86

Kissoudis C, van de Wiel C, Visser RGF, van der Linden G (2014) Enhancing crop resilience to combined abiotic and biotic stress through the dissection of physiological and molecular crosstalk. Front Plant Sci 5:207

Kissoudis C, Chowdhury R, van Heusden S, van de Wiel C, Finkers R, Visser RGF, Bai Y, van der Linden G (2015) Combined biotic and abiotic stress resistance in tomato. Euphytica 202:317–332

Klemm M (1938) The most important diseases and pests of colza and rape. Dtsch Landw Pr IXV 19: 239; 20: 251–252

Knight H, Knight MR (2001) Abiotic stress signaling pathways: specificity and cross-talk. Trends Plant Sci 6(6):262–267

Koch de Brotus L, Boasse C (1955) Publ Minist Gonad Agric Montevideo 106:65

Koch E, Slusarenko AJ (1990) *Arabidopsis* is susceptible to infection by a downy mildew fungus. Plant Cell 2:437–445

Kocks CG, Zadoks JC (1996) Cabbage refuse piles as sources of inoculum for black rot epidemics. Plant Dis 80(7):789–792

Kola K (1980) Comparative study of sunflower varieties to assess their resistance to fungal diseases in the Zadrime area. (Studim krahasues i llojeve te lulediellit per te pare qendrueshmerine ndaj semundjeve kerpudhore ne zonen e Zadrimes.). Buletini i Shkencave Bujqesore 19(1):87–94

Kolte SJ (1985) Diseases of annual edible oilseed crops, vol II. CRC Press, Inc

Koornneef A, Pieterse CMJ (2008) Cross talk in defense signaling. Plant Physiol 146(3):839–844

Kreps JA, Wu Y, Chang HS, Zhu T, Wang X, Harper JF (2002) Transcriptome changes for *Arabidopsis* in response to salt, osmotic, and cold stress. Plant Physiol 130:2129–2141

Krexner R (1969) Sclerotial disease of Sunflower. (Sklerotienkrankheit der Sonnenblume.). Pflanzenarzt 22(3):20–22

Kritzman G, Ben-Yephet Y (1990) Control by metham-sodium of *Xanthomonas campestris*pv. *campestris* and the pathogen's survival in soil. Phytoparasitica 18(3):217–227

Kruger W (1983) White stalk (rape canker). Information or integrated plant protection. (Weissstengeligkeit (Rapskrebs). Information zum Integrierten Pflanzenschutz.). Nachrichtenblatt des Deutschen Pflanzenschutzdienstes 35(10):159–160

Kryopoulus P (1916) Failure of seedlings, particularly of crucifers. Centbl Bakt 2 Abt 45:6–12, 244–257, pls 3

Kuhnholtz LM, Gastuad JM (1943) Plant Pathological Notes. Ann Epiphyt N.S. 9: 207–219

Kulcheski FR, de Oliveira LF, Molina LG, Almerao MP, Rodrigues FA, Marcolino J, Barbosa JF, Stolf-Moreira R, Nepomuceno AL, Marcelino-Guimaraes FC, Abdelnoor RV, Nascimento LC, Carazzolle MF, Pereira GAG, Margis R (2011) Identification of novel soybean microRNAs involved in abiotic and biotic stresses. BMC Genomics 12:307

Kumar R (2014) Role of microRNAs in biotic and abiotic stress responses in crop plants. Appl Biochem Biotechnol 174:93–115

Kumar S (2017) Assessment of avoidable yield losses in crop brassicas by insect-pests. J Entomol Zool Stud 5(3):1814–1818

Kumar S, Singh YP (2015) Insect-Pests. In: Arvind Kumar S, Banga SS, Meena PD, Kumar PR (eds) *Brassica* oilseeds: breeding and management. CAB International, London, UK, p 261

Kumar A, Bag MK, Singh R, Jailani AAK, Mandal B, Roy A (2018) Natural infection of croton yellow vein mosaic virus and its cognate betasatellite in germplasm of different *Crambe* spp. in India. Virus Res 243:60–64

Kurzawinska H, Poniedziałek M (1987) Diseases of horse radish with a particular consideration for brittle root. Bull Polish Acad Sci, Biol Sci 35(7-9):237–244

Kwaaitaal M, Huisman R, Maintz J, Reinstadler A, Panstruga R (2011) Ionotropic glutamate receptor (iGluR)-like channels mediate MAMP-induced calcium influx in *Arabidopsis thaliana*. Biochem J 440:355–365. https://doi.org/10.1042/BJ20111112

Kyrova E, Egorova M, Ignatov A (2020) Species of the genus *Xanthomonas* infecting cereals and oilseeds in the Russian Federation and its diagnostics. BIO Web of Conferences 18, 00017, IV All-Russian Plant Prot Cong. https://doi.org/10.1051/bioconf/20201800017

Lachaud C, Da Silva D, Cotelle V, Thuleau P, Xiong T C, Jauneau A et al. (2010) Nuclear calcium controls the apoptotic-like cell death induced by derythro-sphinganine in tobacco cells. CellCalcium 47: 92–100. doi: https://doi.org/10.1016/j.ceca.2009.11.011

Lachaud C, Prigent E, Thuleau P, Grat S, Da Silva D, Briere C et al (2013) 14-3-3-regulated Ca (2C)-dependent protein kinase CPK3 is required for sphingolipid-induced cell death in *Arabidopsis*. Cell Death Differ 20:209–217. https://doi.org/10.1038/cdd.2012.114

Lacoste L, Louvet J, Anselme C, Alabouvette C, Brunin B, Pierre JG (1969) Role of P. Ungam and of its perfect form L. maculans in epidemics of black leg of Colza (*B. napus* var. *oleifera*). (Role de Phoma Ungam (Tode) Desm. et de sa forme parfaite *Leptospbaeria maculans* (Desm.) Ces.et de Not. dans les épidémies de nécrose du collet de Colza (*Brassica napus* L. var. *oleifera* Metzer).). Compte Rendu Hebdomadaire des Seances de l'Academie d'Agriculture de France 55 (14):981–989

Laguna IG, Salazar LG (1984) Banana fruit rot caused by *Sclerotinia sclerotiorum* (Lib.) de Bary (*Whetzelinia sclerotiorum*) in Costa Rica. (Pudrición del fruto del banano causada por *Sclerotinia sclerotiorum* (Lib.) de Bary (*Whetzelinia sclerotiorum*) en Costa Rica.). Turrialba 34(1):105–106

Laloi C, Apel K, Danon A (2004) Reactive oxygen signaling: the latest news. Curr Opin Plant Biol7 (3):323–328.

Lamb RJ (1989) Entomology of oilseed *Brassica* crops. Annu Rev Entomol 34:211–229

Langdon RFN (1948) Records of Queensland fungi v. Pap. Dept. Biol., Univ. Qd, 2: 9

Latz A, Mehlmer N, Zapf S, Mueller TD, Wurzinger B, Pfister B et al (2013) Salt stress triggers phosphorylation of the *Arabidopsis* vacuolar KC channel TPK1 by calcium-dependent protein kinases (CDPKs). Mol Plant 6:1274–1289. https://doi.org/10.1093/mp/sss158

Leather RI (1967) A catalogue of some plant diseases and fungi in Jamaica. Bull Minist Agric Lands Jamaica 61:92

Leba LJ, Cheval C, Ortiz-Martin I, Ranty B, Beuzon CR, Galaud JP et al (2012) CML9, an *Arabidopsis* calmodulin-like protein, contributes to plant innate immunity through a flagellin-dependent signaling pathway. Plant J 71:976–989. https://doi.org/10.1111/j.1365-313X.2012.05045.x

Lee H, Xiong L, Gong Z, Ishitani M, Stevenson B, Zhu JK (2001) The *Arabidopsis* HOS1 gene negatively regulates cold signal transduction and encodes a RING finger protein that displayscold-regulated nucleo-cytoplasmic partitioning. Genes Dev 15:912–924

Lee IM, Gundersen-Rindal DE, Davis RE, Bottner KD, Marcone C, Seemüller E (2004) *'Candidatus Phytoplasma asteris'*, a novel phytoplasma taxon associated with aster yellows and related diseases. Intl J Syst Evol Microbiol 54(4):1037–1048

Lehoczky J (1957) The occurrence of some economically significant parasitic micro-fungi. Ann Acad Hort Vitic Budapest 21:3–14

Li J, Brader G, Palva ET (2004) The WRKY70 transcription factor: a node of convergence forjasmonate-mediated and salicylate-mediated signals in plant defense. Plant Cell 16:319–331

Li Z, Zhang L, Wang A, Xu X, Li J (2013) Ectopic overexpression of SlHsfA3, a heat stress transcription factor from tomato, confers increased thermo tolerance and salt hypersensitivity in germination in transgenic *Arabidopsis*. PLoS One 8:e54880

Li Y, Varala K, Coruzzi M (2015) From milliseconds to lifetimes; tracking the dynamic behavior of transcription factors in gene networks. Trends Genet 31:509–515

Li G, Lv H, Zhang S, Zhang S, Li F, Zhang H, Qian W, Fang Z, Sun R (2019) TuMV management for Brassica crops through host resistance: retrospect and prospects. Plant Pathol 68:1035–1044. https://doi.org/10.1111/ppa.13016

Liese A, Romeis T (2013) Biochemical regulation of in vivo function of plant calcium-dependent protein kinases (CDPK). Biochem Biophys Acta 1833:1582–1589. https://doi.org/10.1016/j.bbamcr.2012.10.024

Lindquist JC (1946) Argentine species of the genus *Peronospora*. Physics at University of Buenos Aires 15:13–20

Liu Q, Rimmer SR (1991) Inheritance of resistance in *Brassica napus* to an Ethiopian isolate of *Albugo candida* from *Brassica carinata*. Can J Plant Pathol 13(3):197–201

Liu J, Feng L, Li J, He Z (2015) Genetic and epigenetic control of plant heat responses. Front Plant Sci 6:267

Lo TT (1961) Plant Ind. Ser. Chin-AM Jt. Comm Rur Reconstr 23:52
Lorenzo O, Chico JM, Sanchez-Serrano JJ, Solano R (2004) JASMONATE-INSENSITIVE1 encodes a MYC transcription factor essential to discriminate between different jasmonate-regulated defense responses in *Arabidopsis*. Plant Cell 16(7):1938–1950
Lu PL, Chen NZ, An R, Su Z, Qi BS, Ren F, Chen J, Wang XC (2007) A novel drought-inducible gene, ATAF1, encodes a NAC family protein that negatively regulates the expression of stress-responsive genes in *Arabidopsis*. Plant Mol Biol 63:289–305
Lu Y, Chen X, Wu Y, Wang Y, He Y, Wu Y (2013) Directly transforming PCR-amplified DNA fragments into plant cells is a versatile system that facilitates the transient expression assay. PLoS One 8:e57171. https://doi.org/10.1371/journal.pone.0057171
Luan S (2009) The CBL-CIPK network in plant calcium signaling. Trends Plant Sci 14:37–42. https://doi.org/10.1016/j.tplants.2008.10.005
Ludwig AA, Saitoh H, Felix G, Freymark G, Miersch O, Wasternack C, Boller T, Jones JD, Romeis T (2005) Ethylene-mediated cross-talk between calcium-dependent protein kinaseand MAPK signaling controls stress responses in plants. Proc Natl Acad Sci U S A 102(30):10736–10741
Luo H, Laluk K, Lai Z, Veronesa P, Song F, Mengiste T (2010) The *Arabidopsis botrytis* susceptible interactor defines a subclass of RING E3 ligases that regulate pathogen and stress responses. Plant Physiol 154:1766–1782
Ma S, Bohnert HJ (2007) Integration of *Arabidopsis thaliana* stress-related transcript profiles, promoter structures, and cell-specific expression. Genome Biol 8:R49
Macho AP, Zipfel C (2014) Plant PRRs and the activation of innate immune signaling. Mol Cell 54:263–272
Macias W (1996) *Albugo candida* – a serious disease of horseradish. (Bielik krzyzowych – grozna choroba chrzanu). Ochrona Roslin. 40 (11):12
Magnan F, Ranty B, Charpenteau M, Sotta B, Galaud JP, Aldon D (2008) Mutations in AtCML9, a calmodulin-like protein from *Arabidopsis thaliana*, alter plant responses to abiotic stress and abscisic acid.Plant J56:575–589. https://doi.org/10.1111/j.1365-313X.2008.03622.x
Maire R (1927) Algeria: plant disease observed in March, 1927. Intl Bull Plant Prot 51-52.
Mantri N, Pang ECK, Ford R (2010a) Molecular biology for stress management. In: Yadav SS, McNeil DN, Weeden N, Patil SS (eds) Climate change and management of cool season grain legume crops. Springer, Heidelberg, pp 377–408. isbn:978-90-481-3708-4
Mantri NL, Ford R, Coram TE, Pang ECK (2010b) Evidence of unique and shared responses to major biotic and abiotic stresses in chickpea. Environ Exp Bot 69:286–292
Mantri N, Basker N, Ford R, Pang E, Pardeshi V (2013) The role of micro-ribonucleic acids in legumes with a focus on abiotic stress response. Plant Genome 6:1–14
Manzoor H, Kelloniemi J, Chiltz A, Wendehenne D, Pugin A, Poinssot B et al (2013) Involvement of the glutamate receptor AtGLR3.3 in plant defense signaling and resistance to *Hyaloperonospora arabidopsidis*. Plant J 76:466–480. https://doi.org/10.1111/tpj.12311
Mao P, Duan M, Wei C, Li Y (2007) WRKY62 transcription factor acts downstream of cytosolicNPR1 and negatively regulates jasmonate responsive gene expression. Plant Cell Physiol 48(6):833–842
Mao L, Van HJL, Dash S, Dickerson JA (2009) *Arabidopsis* gene co-expression network and its functional modules. BMC Bioinformatics 10:346
Marcone C, Ragozzino A (1995) Detection of phytoplasmas in *Brassica* spp. In: southern Italy and their characterization by RFLP analysis. J Plant Dis Protect 102:449–460
Mariano RLB, Barros ST, Menezes M, Pio-Ribeiro G (1985) Detection of *Xanthomonas campestris* pv. *campestris* in batches of commercial *Brassica* seeds. Fitopatol Bras 10(3):423–426
Martens JW, Ravagan G, Mcuonald WC (1970) Diseases of sunflowers in kenya. East African Agril Forestry J 35(4):389–395
Martínez-de la Parte E, Trujillo M, Cantillo-Pérez T, Garcia D (2013) First report of white mould of beans caused by *Sclerotinia sclerotiorum* in Cuba. New Dis Reptr 5. https://doi.org/10.5197/j.2044-0588.2013.027.005

Massa AN, Childs KL, Buell R (2013) Abiotic and biotic stress responses in *Solanum tuberosum* group Phureja DM1-3 516 R44 as measured through whole transcriptome sequencing. Plant Genome 6:1–10

Mauch-Mani B, Mauch F (2005) The role of abscisic acid in plant–pathogen interactions. Curr Opin Plant Biol 8(4):409–414

Mazuma EDL, Theu MPKJ, Likongwe MC (2008) Participatory on-farm evaluation of Ridomil Gold MZ 68 WP for its efficacy for clubroot disease control in cabbage. In: Theu MPKJ (ed) Plant protection progress report for the 2007/2008 season, presented at the Department of Agricultural Research Services Planning and Review Meeting, Andrews Hotel, Mangochi, 14-20 September, 2008. Ministry of Agriculture and Food Security, Lilongwe, Malawi, pp 127–134

McAinsh MR, Hetherington AM (1998) Encoding specificity in Ca^{2+} signaling systems. Trends PlantSci 3:32–36. https://doi.org/10.1016/S1360-1385(97)01150-3

McCauley A, Jones C, Jacobsen J (2011) Plant nutrient function and deficiency and toxicity symptoms. Montana State University. Nutr Manag Module 9:1–16

McCormack E, Braam J (2003) Calmodulins and related potential calcium sensors of *Arabidopsis*. New Phytol 159:585–598. https://doi.org/10.1046/j.1469-8137.2003.00845.x

McGee DC (1973) Losses in rapeseed caused by black leg in Victoria, Australia, 2nd Intl Congr Plant Pathol Minneapolis. MN Abstr 0825

McGuire JV, Crandall BS (1967) USDA Intern. Agric Develop Service 157

Mcintosh AES (1951) Vegetable diseases. Annual Report for the year 1949 of the Department of Agriculture, Malaya: p 55

Mckee RK (1961) Plant pathology research division-report research and technology. Work Minist Agric N Ireland 1970:105–113

Meena PD, Mondal K, Jha SK, Chattopadhyay C, Kumar A (2010) Bacterial rot: a new threat for rapeseed-mustard production system in India. J Oilseed Brassica 1:39–41

Meena MC, Meena PD, Shukla AK, Barman M, Meena MK, Singh P, Kumar A (2016) Role of plant nutrients in enhancing productivity and managing diseases of oilseed *Brassica*. J Oilseed Brassica 7(1):1–20

Meena PD, Gupta R, Sharma P, Rai PK, Meena HS, Singh SK (2019) Isolation and characterization of a new fungal species, *Fusarium equiseti* (Corda) Sacc., from *Brassica juncea* in Bharatpur. J Oilseed Brassica 10(2):130–133

Mehlmer N, Wurzinger B, Stael S, Hofmann-Rodrigues D, Csaszar E, Pfister B et al (2010) The Ca (2+) -dependent protein kinase CPK3 is required for MAPK-independent salt-stress acclimation in *Arabidopsis*. Plant J 63:484–498. https://doi.org/10.1111/j.1365-313X.2010.04257.x

Meier R (1996) Research into controlling white blister (*Albugo candida*) in Brussels sprouts. (Onderzoek naar de bestrijding van witte roest (*Albugo candida*) in spruitkool.). Publicatie - Proefstation voor de Akkerbouw en de Groenteteelt in de Vollegrond, Lelystad. pp 137–139

Melchers LE (1931) A check list of plant diseases and fungi occurring in Egypt. Trans Kansas Acad Sci 34:41–106

Mengiste T, Chen X, Salmeron JM, Dietrich RA (2003) The *BOS1* gene encodes an R2R3MYB transcription factor protein that is required for biotic and abiotic stress responses in *Arabidopsis*. Plant Cell 15:2551–2565

Mert-Turk F, Kemal Gül M, Egesel C (2008) Nitrogen and fungicide applications against *Erysiphe cruciferarum* affect quality components of oilseed rape. Mycopathologia 165(1):27–35

Mew TW, Ch WC, Chu L (1976) Infectivity and survival of soft-rot bacteria in Chinese cabbage. Phytopathology 66:1325–1327

Mguni CM (1987) Diseases of crops in the semiarid areas of Zimbabwe: can they be controlled economically? In: Cropping in the Semiarid Areas of Zimbabwe. Workshop Proceedings, Harare. Agritex 2: pp 417–433

Mguni CM (1995) Cabbage research in Zimbabwe. *Brassica* planning workshop for east and south Africa region. Lilongwe, Malawi, 15-18 May. GTZ/IPM Hortic, Nairobi, Kenya, p 31

Mguni CM (1996) Bacterial Black Rot (*Xanthomonas campestris* pv. *campestris*) of Vegetable Brassicas in Zimbabwe. Ph.D thesis, Department of Plant Biology.The Royal Veterinary and Agricultural University, Copenhagen and Danish Government Institute for Developing Countries, Hellerup, Denmark.

Mguni CM, Mortensen CN, Hockenhull J, Keswani CL (1996) Pathogenicity of strains of *Xanthomonas campestris*pv. *campestris* from Zimbabwe belonging to Biolog types Xcc, XccA and XccB on *Brassica* spp. Abstracts, IX international conference on plant pathogenic bacteria, August 26–29, 1996, Centre for Advanced Study in Botany, University of Madras, Guindy Campus, India.

Michail SH, Al-Menoufi OA, Abo-Taleb EA (1979) Seed health testing, leaf spot and damping-off of certain crucifers in Egypt. Acta Phytopathol Acad Sci Hung 14:41–48

Middleton JT (1943) The taxonomy, host range and geographic distribution of the Genus *Pythium*. Memoirs Torrey Bot Club 20(1):1–171

Miller G, Schlauch K, Tam R, Cortes D, Torres MA, Shulaev V et al (2009) The plant NADPH oxidase RBOHD mediates rapid systemic signaling in response to diverse stimuli. Sci Signal 2: ra45. https://doi.org/10.1126/scisignal.2000448

Minuto G, Minuto A, Garibaldi A (1997) *Eruca sativa*: chemical control of downy mildew and white rust. (Rucola: lotta chimica contro Peronospora e ruggine bianca.). Colture Protette 26 (6):61–66

Mirza MS, Ahmad Y (1988) Cabbage: a new host of *Sclerotinia sclerotiorum* for Pakistan. Agronomie 8(1):85–87. https://doi.org/10.1051/agro:19880111

Moghaddam SAMH, Izadyar M (1994) Bud and twig dieback of mulberry in Gilan. Iranian J Plant Pathol 30 (1-4): 5-6 En; 18-24.

Mohammadi-Doustdar E (1967) Mycology, powdery mildew of Iran, Tehran University no. 1262.

Molinier J, Ries G, Zipfel C, Hohn B (2006) Trans-generation memory of stress in plants. Nature 442:1046–1049

Moon H, Lee B, Choi G, Shin D, Prasad DT, Lee O, Kwak SS, Kim DH, Nam J, Bahk J, Hong JC, Lee SY, Cho MJ, Lim CO, Yun DJ (2003) NDP kinase 2 interacts with two oxidative stress-activated MAPKs to regulate cellular redox state and enhances multiple stress tolerance in transgenic plants. Proc Natl Acad Sci U S A 100(1):358–363

Moore WC (1959) British parasitic fungi. Cambridge University Press 9:236

Morales Y, de Jesús M, Braun U, Minnis AM, Tovar-Pedraza JM (2009) Some new records and new species of powdery mildew fungi from Mexico. Schlechtendalia 19:47–61

Mori IC, Murata Y, Yang Y, Munemasa S, Wang YF, Andreoli S et al (2006) CDPKs CPK6 and CPK3 function in ABA regulation of guard cell S-type anion- and Ca(2C)-permeable channels and stomatal closure. PLoS Biol 4:e327. https://doi.org/10.1371/journal.pbio.0040327

Moriwaki J, Tsukiboshi T, Sato T (2002) Grouping of *Colletotrichum* species in Japan based on rDNA sequences. J Gen Plant Pathol 68:307–320

Morris MJ, Knox-Davies PS (1980) *Raphanus raphanistrum* as a weed host of pathogens of cultivated cruciferae in the Western Cape Province of South Africa. Phytophylactica 12 (2):53–55

Morwood RB (1956) Notes on plant diseases listed for Fiji. Agric J Fiji 27(3-4):83–86

Mou HQ, Zhang YJ, Li HX, Zhu SF, Li ZH, Zhano WJ (2012) Molecular Identification of a *Candidatus Phytoplasma asteris* associated with cabbage witches'-broom in China. J Phytopathol 160:304–307

Mousavi SA, Chauvin A, Pascaud F, Kellenberger S, Farmer EE (2013) GLUTAMATE RECEPTOR-LIKE genes mediate leaf-to-leaf wound signaling. Nature 500:422–426. https://doi.org/10.1038/nature12478

Muhammad Ehetisham-ul-Haq M, Kamran M, Idrees M, Ali S, Iqbal M, Abbas H, Khan QAT, Rashid A, Yasin SI (2017) Management of bacterial leaf spot disease of mustard through resistant germplasm. Plant Prot 01(01):07–10

Muhammad S, Khan SH (2008) Effect of abiotic factors on incidence and severity of diseases on cabbage (*Brassica oleracea* L.) in Dera Ghazi Khan. Pakistan J Phytopathol 20 (1):108–112. Papers presented at the 3rd Intl Conf Plant Pathol, Lahore, Pakistan, 19-21 November 2007.
Mujica F, Vergara C (1960) Balm tec. No. 6. Dept Invest Agric Chili 6: 60.
Mukerji KG (1973) Additions to plant diseases of Delhi. Angew Bot 47(5/6):205–213
Mukula J (1957) On the decay of stored Carrots in Finland. Acta Agric Scand 2(Suppl):132 pp
Muller AS (1950) A preliminary survey of plant diseases in Guatemala. Plant Dis Reptr 34:161
Mur LA, Kenton P, Atzorn R, Miersch O, Wasternack C (2006) The outcomes of concentration-specific interactions between salicylate and jasmonate signaling include synergy, antagonism, and oxidative stress leading to cell death. Plant Physiol 140(1):249–262
My HT (1966) A preliminary list of plant diseases in south Vietnam. Directorate of Research, Saigon 142
Nakagami H, Pitzchke A, Hirt H (2005) Emerging MAP kinase pathways in plant stress signaling. Trends Plant Sci 10:339–346
Nakashima K, Tran LS, Van Nguyen D, Fujita M, Maruyama K, Todaka D, Hayashi N, Shinozaki K, Yamaguchi-Shinozaki K (2007) Functional analysis of a NAC-type transcription factor OsNAC6 involved in abiotic and biotic stress-responsive gene expression in rice. Plant J 51:617–630
Narsai R, Wang C, Chen J, Wu J, Shou H, Whelan J (2013) Antagonistic, overlapping and distinct responses to biotic stress in rice (*Oryza sativa*) and interactions with abiotic stress. BMC Genomics 14:93
Negrean G, Denchev CM (2000) New records of Bulgarian parasitic fungi. Fl Medit 10:101–108
Nejat N, Mantri N (2017) Plant immune system: Crosstalk between responses to biotic and abiotic stresses the missing link in understanding plant defenses. Curr Issues Mol Biol 23:1–16
Nemeth J, Laszlo EM (1983) Bacterial black rot (*Xanthomonas campestris* (Pammel) Dowson 1939) of *Brassica* species. Novenyvedelem 19(9):391–397
Nilsson C (1987) Yield losses in summer rape caused by pollen beetles (Meligethes spp.). Swedish J Agril Res 17:105–111
Nourani SL, Minassian V, Safaie N (2008) Identification, pathogenicity and distribution of *Alternaria* spp. of canola in Iran. Iran J Plant Path 44:33–36
Ocfemia GO (1925) The occurrence of the white rust of crucifers and its associated downy mildew in the Philippines. Philippines Agric Sci J 14:289–296
Oethybridge GH (1911) Potato diseases in Ireland. J Dept Agric 11(3):431
Ohnishi T, Sugahara S, Yamada T, Kikuchi K, Yoshiba HY, Hirano HY, Tsutsumi N (2005) OsNAC6, a member of the NAC gene family, is induced by various stresses in rice. Genes Genet Syst 80:135–139
Olien CR, Smith MN (1997) Ice adhesions in relation to freeze stress. Plant Physiol 60:499–503
Omidvar V, Mohorianu I, Dalmay T, Fellner M (2015) MicroRNA regulation of abiotic stress response in *7B-1* male-sterile tomato mutant. Plant Genome 8:1–13
Ondrej M (1983) The occurrence of fungi on caraway (Carum carvi L.) in Czechoslovakia. (Vyskyt hub no kmínu (Carum carvi L.) v CSSR.). Sborník ÚVTIZ, Ochrana Rostlin 19(3):235–237
Orian G (1951) Division of plant pathology. Report Dept Agric Mauritius, 1949: 66-72; 1950: 80-85.
Osakabe Y, Yamaguchi-Shinozaki K, Shinozaki K, Phan Tran L (2013) Sensing the environment: key roles of membrane-localized kinases in plant perception and response to abiotic stress. J Exp Bot 64:445–458
Oviedo Prieto R, Herrera Oliver P, Caluff MG et al (2012) National list of invasive and potentially invasive plants in the Republic of Cuba - 2011. Bissea: Boletín sobre Conservación de Plantas del Jardín Botánico Nacional de Cuba. 6 (Special Issue No. 1), 22-96.
Pagliarini NC, Berberovic H, Okanovic M (2007) Pests and diseases of tomatoes under hydroponic cultivation. (Stetnici i bolesti rajcice u hidroponskom uzgoju). Glasilo Biljne Zastite 7 (4):220–226

Palaniswamy PC, Gillot C, Slater GP (1986) Attraction of diamondback moths, *Plutella xylostella* (L.) (Lepidoptera: Plutellidae), by volatile compounds of canola, white mustard, and faba bean. Can Entomol 118:1279–1285

Pandey P, Ramegowda V, Senthil-Kumar M (2015) Shared and unique responses of plants to multiple individual stresses and stress combinations: physiological and molecular mechanisms. Front Plant Sci 6:723

Park M (1932) Ceylon Adm. Rep Agric 1931:D103–D111

Parker C, Wilson AK (1986) Parasitic weeds and their control in the Near East. FAO Plant Prot Bull 34(2):83–98

Pastircakova K, Pastircak M (2013) A powdery mildew (*Pseudoidium* sp.) found on *Chelidonium majus* in the Czech Republic and Slovakia. Czech Mycol 65(1):125–132

Paul VH, Rawlinson CJ (1992) Diseases and pests of rape. Verlag Th. Mann, Gelsenkirchen-Buer, Germany

Paulitz TC, Okubara PA, Schillinger WF (2006) First report of damping-off of canola caused by *Rhizoctonia solani* AG2-1 in Washington State. Plant Dis 90:829. https://doi.org/10.1094/pd-90-0829b

Peer M, Stegmann M, Mueller MJ, Waller F (2010) *Pseudomonas syringae* infection triggers de novo synthesis of phytosphingosine from sphinganine in *Arabidopsis thaliana*. FEBS Lett 584:4053–4056. https://doi.org/10.1016/j.febslet.2010.08.027

Peleg J (1953) Outbreaks and new records. Israel FAO Pl Protect Bull 1-4:60–61

Peregrine WTH, Siddiqi (1972) Phytopathological Papers No. 16. Commonwealth Mycological Institute, Kew. U.K. p 51

Perez IB, Brown P (2014) The role of ROS signaling in cross-tolerance: from model to crop. Front Plant Sci 5:754

Perochon A, Dieterle S, Pouzet C, Aldon D, Galaud JP, Ranty B (2010) Interaction of a plant pseudo-response regulator with a calmodulin-like protein. Biochem Biophys Res Commun398: 747–751. doi:https://doi.org/10.1016/j.bbrc.2010.07.016

Perwaiz MS, Moghal SM, Kamal M (1969) Studies on the chemical control of white rust and downy mildew of rape (Sarson). W Pak J Agric Res 7:71–75

Peterson M, Brodersen P, Naested H, Andreasson E, Lindhart U, Johansen B, Nielsen HB, Lacy M, Austin MJ, Parker JE, Sharma SB, Klessig DF, Martienssen R, Mattsson O, Jensen AB, Mundy J (2000) *Arabidopsis* MAP kinase 4 negatively regulates systemic acquired resistance. Cell 103 (7):1111–1120

Petrenkova VP (1994) Survival of the sclerotia of the causal agent of white rot. Zashchita Rasteniĭ (Moskva) 20

Pimenova A, Maslennikov I (1955) Diseases and pests of cabbage seedlings and their control. Orchard Garden 6:30–33

Ponce GF, Mendoza ZC (1983) Fungal diseases and insect pests of rapeseed in the high valleys of Mexico. (Enfermedades fungosas y plagas insectiles de la colza, *Brassica napus* L. y *B. campestris* L., en los Valles Altos de México.). Revista Chapingo 8(39):53–61

Pons N, de Alvarado H, Castro JL (1979) *Whetzelinia sclerotiorum*, a new pathogen in Venezuela, on lettuce, cabbage and bean. (*Whetzelinia sclerotiorum* en lechuga, repollo yaraoata, nuevo patogeno en Venezuela.). Fitopatologia 14(1):10–14

Pontis VRE, Feldman JM (1959) A new host of *Sclerotinia sclerotiorum* in Argentina. Plant Dis Reptr 43(3):421 p

Popescu SC, Popescu GV, Bachan S, Zhang Z, Seay M, Gerstein M et al (2007) Differential binding of calmodulin-related proteins to their targets revealed through high-density *Arabidopsis* protein microarrays. Proc Natl Acad Sci USA 104:4730–4735. https://doi.org/10.1073/pnas.0611615104

Popovic T, Balaz J, Starovic M, Trkulja N, Ivanovic Z, Ignjatov M, Josic D (2013) First report of *Xanthomonas campestris* pv. *campestris* as the causal agent of black rot on oilseed rape (*Brassica napus*) in Serbia. Plant Dis 97(3):418.

Porter RH (1926) Preliminary report of surveys for plant diseases in East China. Plant Dis Reptr 46 (S):153–166

Prasad L, Kamil D, Singh N, Singh OW, Yadava DK, Prameela Devi T (2017) First report of *Fusarium equiseti* causing stem and root rot on *Brassica juncea* in India. J Plant Pathol 99 (3):799–818

Prasch CM, Sonnewald U (2013) Simultaneous application of heat, drought and virus to *Arabidopsis* plants reveals significant shifts in signaling networks. Plant Physiol 162:1849–1866

Preston NC (1929) Part I. Report of the Advisory Mycologist (April 1928 March) 1929-Rept Advisory Dept Harper Adams Agric Coll New Port. Salop 4:4–8

Prova A, Akanda MAM, Islam S, Sultana F, Islam MT, Hossain MM (2014) First report of stem and pod blight of hyacinth bean caused by *Sclerotinia sclerotiorum* in Bangladesh. J Plant Pathol 96 (3):607

Raabe A (1939) Investigations on parasitic fungal diseases of colza and rape. Zbl Bakt Abt 2Cl-3:35–52

Raghuram B, Sheikh AH, Sinha AK (2014) Regulation of MAP kinase signaling cascade by micro RNAs in *Oryza sativa*. Plant Signal Behav 9:e972130

Rai JN, Tewari JP, Singh RP, Saxena VC (1974) Fungal diseases of Indian crucifers. Beihefte zur Nova Hedwigia:477–486

Raj S, Brautigam K, Hamanishi ET, Wilkins O, Thomas BR, Schroeder W, Mansfield SD, Plant AL, Campbell MM (2011) Clone history shapes Populus drought responses. Proc Natl Acad Sci U S A 108:12521–12526

Raju UJ, Sivaprakasam K (1989) Survey on pests and diseases of cabbage. Madras Agril J 76 (4):192–196

Ranty B, Aldon D, Gotelle V, Galaud JP, Thuleau P, Mazars C (2016) Calcium sensors as key hubs in plant responses to biotic and abiotic stresses. Front Plant Sci 7:327. https://doi.org/10.3389/fpls.2016.00327

Rappussi MCC, Eckstein B, Flôres D, Haas ICR, Amorim L, Bedendo IP (2012) Cauliflower stunt associated with a phytoplasma of subgroup 16SrIII-J and the spatial pattern of disease. Eur J Plant Pathol 133:829–840. https://doi.org/10.1007/s10658-012-0004-7

Rasmussen S, Barah P, Suarez-Rodriguez MC, Bressendorff S, Friis P, Costantino P, Bones AM, Nielsen HB, Mundy J (2013) Transcriptome responses to combinations of stresses in *Arabidopsis*. Plant Physiol:1611783–1611794

Rawlinson CJ, Sutton BC, Muthyalu G (1978) Taxonomy and biology of *Pyrenopeziza brassicae* sp. nov. (*Cylindrosporium concentricum*), a pathogen of winter oilseed rape (*Brassica napus* spp. *oleifera*). Trans Br Mycol Soc 71:425–439

Rayapuram C, Baldwin IT (2007) Increased SA in NPR1-silenced plants antagonizes JA and JA-dependent direct and indirect defenses in herbivore-attacked Nicotiana *attenuata* in nature. Plant J 52(4):700–715

Rayss T (1938) A new contribution to the study of the Palestinian mycoflora. Palestine J Bot Jerusalem Series-I 2:143–160

Rentel MC, Lecourieux D, Ouaked F, Usher SL, Petersen L, Okamoto H, Knight H, Peck SC, Grierson CS, Hirt H, Knight MR (2004) OXI1 kinase is necessary for oxidative burst-mediated signaling in *Arabidopsis*. Nature 427:858–861

Robinson RA (1960) Notes on Kenya agriculture VIII: important plant diseases. East African Agril Forestry J 25(3):131–146

Romeis T, Herde M (2014) From local to global: CDPKs in systemic defense signaling upon microbial and herbivore attack. Curr Opin Plant Biol 20:1–10. https://doi.org/10.1016/j.pbi.2014.03.002

Rothwell A (1972) Groundnut diseases in Rhodesia. Rhodesia Agril J 69(2):35–38

Roy A, Spoorthi P, Bag MK, Prasad TV, Singh R, Dutta M, Mandal B (2013) A leaf curl disease in germplasm of rapeseed-mustard in India: Molecular evidence of a weed-infecting Begomovirus–Betasatellite complex emerging in a new crop. J Phytopathol 161(7-8):522–535

Ryan EW (1986) Research Report 1985, Horticulture. Dublin, Irish Republic, An Foras Taluntais, pp 35–36
Rydl R (1968) Cabbage mould on kale. Uroda 16:186–187
Sackston WE (1957) Diseases of Sunflowers in Uruguay. Plant Dis Reptr 41(10):885–889
Sadowski C, Dakowska S, Lukanowski A, Jedryczka M (2002) Occurrence of fungal diseases on spring rape in Poland. In: Paul VH, Foller I (eds) Working Group on Integrated Control in Oilseed Crops. Soest, Germany, IOBC/WPRS Bulletin, pp 1–12
Saharan GS (1984) A review of research on rapeseed mustard pathology in India. Annual Workshop AICORPO ICAR, Jaipur, 6–10 August 1984.
Saharan GS (1992) Management of rapeseed and mustard diseases. In: Kumar D, Rai M (eds) Advances in oil seed research. Sci. Pub, Jodhpur
Saharan GS, Verma PR, Meena PD, Kumar A (2014) White rust of crucifers: Biology, ecology and disease management. Springer Verlag, Frankfurt, Germany, p 244. isbn:978-81-322-1791-6
Saharan GS, Mehta Naresh, Meena PD (2016) Alternaria blight of crucifers: Biology, ecology and disease management. Springer Verlag, Singapore: 326, ISBN 978-981-10-0019-5.
Saharan GS, Mehta N, Meena PD (2017) Downy mildew disease of crucifers: biology, ecology and disease management. Springer Verlag, Singapore, LVI, 357, ISBN 978-981-10-7499-8.
Saharan GS, Mehta N, Meena PD (2019) Powdery mildew disease of crucifers: Biology, ecology and disease management. Springer Verlag, Singapore, p 362. isbn:978-981-13-9852-0
Saharan GS, Mehta N, Meena PD (2021) Clubroot disease of crucifers: biology, ecology and disease management. Springer Verlag, Singapore. (In Press)
Salehi M, Izadpanah K, Nejat N, Siampour M (2007a) Partial characterization of phytoplasmas associated with lettuce and wild lettuce phyllodies in Iran. Plant Pathol 56:669–676. https://doi.org/10.1111/j.1365-3059.2007.01616.x
Salehi M, Izadpanah K, Siampour M (2007b) Characterization of a phytoplasma associated with cabbage yellows in Iran. Plant Dis 91:625–630
Salonen J, Hyvonen T, Jalli H (2011) Composition of weed flora in spring cereals in Finland - a fourth survey. Agri Food Sci 20:245–261
Samuel G (1925) Rep. Dep. Agric. S. Australia for year ending 30th June, 1924: pp 76–78.
Sanders D, Pelloux J, Brownlee C, Harper JF (2002) Calcium at the crossroads of signaling. Plant Cell 14:S401–S417. https://doi.org/10.1105/tpc.002899
Sankhla HS, Dalela GG, Mathur RL (1967) Occurrence of perithecial stage of *E. poligoni* on *B. campestris* var. Sarson and *B. juncea.* Plant Dis Rep 51:800.
Saucedo-Garcia M, Guevara-Garcia A, Gonzalez-Solis A, Cruz-Garcia F, Vazquez-Santana S, Markham JE et al (2011) MPK6, sphinganine and the LCB2a gene from serine palmitoyl transferase are required in the signaling pathway that mediates cell death induced by long chain bases in *Arabidopsis*. New Phytol 191:943–957. https://doi.org/10.1111/j.1469-8137.2011.03727.x
Savulescu T, Rayss T (1930) Contribution to the knowledge of the Peronosporaceae of Rumania. Ann Mycologici 18:297–320
Scarpeci TE, Zanor MI, Valle EM (2008) Investigating the role of plant heat shock proteins during oxidative stress. Plant Signal Behav 3:856–857
Schaad NW, Dianese JC (1981) Cruciferous weeds as sources of inoculum of *Xanthomonas campestris* in black rot of crucifers. Phytopathology 71(11):1215–1220
Scheffer RP (1950) Anthracnose leaf spot of crucifers. Tech Bull NC Agric Expt Sta 92:26
Scherzer S, Maierhofer T, Al-Rasheid KA, Geiger D, Hedrich R (2012) Multiple calcium-dependent kinases modulate ABA-activated guard cell anion channels. Mol Plant 5:1409–1412. https://doi.org/10.1093/mp/sss084
Schnug E, Haneklaus S (2005) Sulphur deficiency symptoms in oilseed rape (*Brassica napus* L.) – the aesthetics of starvation. Phyton 45:79–95
Scholz SS, Vadassery J, Heyer M, Reichelt M, Bender K W, Snedden W A et al. (2014) Mutation of the *Arabidopsis* calmodulin-like protein CML37 deregulates the jasmonate pathway and

enhances susceptibility to herbivory. Mol Plant7: 1712–1726. doi: https://doi.org/10.1093/mp/ssu102

Scholz SS, Reichelt M, Vadassery J, Mithofer A (2015) Calmodulin-like protein CML37 is a positive regulator of ABA during drought stress in *Arabidopsis*. Plant Signal Behav 10: e1011951. https://doi.org/10.1080/15592324.2015.1011951

Schroeder JI, Hagiwara S (1990) Repetitive increases in cytosolic Ca^{2+} of guard cells by abscisic acid activation of nonselective Ca^{2+} permeable channels. Proc Natl Acad Sci USA87:9305–9309. https://doi.org/10.1073/pnas.87.23.9305

Schulz P, Herde M, Romeis T (2013) Calcium-dependent protein kinases: hubs in plant stress signaling and development. Plant Physiol 163:523–530. https://doi.org/10.1104/pp.113.222539

Schwessinger B, Bart R, Krasileva K, Coaker G (2015) Focus issue on plant immunity: from model systems to crop species. Front Plant Sci 6:195

Scortichini M, Rossi MP, Ruggini L, Cinti S (1994) Recurrent infections of *Xanthomonas campestris* pv. *campestris* on Cruciferae in some areas of central-southern Italy. Informatore Fitopatologico 44(12):48–50

Sekhon BS (1999) Population dynamics of *Lipaphis erysimi* and *Myzus persicae* on different species of *Brassica*. In: Proceedings of the 10th Intl rapeseed Congr, 26–29 Sep 1999, Canberra, Australia

Seki M, Narusaka M, Ishida J, Nanjo T, Fujita M, Oono Y, Kamiya A, Nakajima M, Enju A, Sakuri T, Satou M, Akiyama K, Taji T, Yamaguchi-Shinozaki K, Carninci P, Kawai J, Hayashizaki Y, Shinozaki K (2002) Monitoring the expression profiles of 7000 *Arabidopsis* genes under drought, cold, and high salinity stresses using a full-length cDNA microarray. Plant J 31:279–292

Semb L (1969) Cabbage downy mildew. Jord Avling 12:32–35

Sen P (2005) Antagonistic effect of Ca, B and Mo on club-root disease of rape-mustard. Indian Agricul 49(1/2):13–16

Şesan T, Csep N (1992) Prevention of white rot (*Sclerotinia sclerotiorum* (Lib.) de Bary) of sunflower and soybean by the biological control agent *Coniothyrium minitans* Campbell. In: Bulletin OILB/SROP 15(1):60–63

Sewelam N, Oshima Y, Mitsuda N, Ohme- Takagi M (2014) A step towards understanding plant responses to multiple environmental stresses: a genome-wide study. Plant Cell Environ 37:2024-2035.

Shaik R, Ramakrishna W (2013) Genes and co-expression modules common to drought and bacterial stress responses in *Arabidopsis* and rice. PLoS One 8:e77261

Shakya DD, Malla G (1988) Studies on black rot of crucifers. In: Proceedings of National Conference on Science and Technology. Royal Nepal Acad Sci. Tech, April 24-29, 1988, Kathmandu, Nepal

Sham A, Al-Azzawi A, Al-Ameri S, Al-Mahmoud B, Awwad F, Al-Rawashdeh A, Iratni R, AbuQamar S (2014) Transcriptome analysis reveals genes commonly induced by *Botrytis cinerea* infection, cold, drought and oxidative stresses in *Arabidopsis*. PLoS One 25:e113718

Sham A, Moustafa K, Al-Ameri S, Al-Azzawi A, Iratni R, AbuQamar S (2015) Identification of *Arabidopsis* candidate genes in response to biotic and abiotic stresses using comparative microarrays. PLoS One 10:e0125666

Sharma J, Agrawal K, Singh D (1992) Detection of *Xanthomonas campestris*pv. *campestris* (Pammel) dowson infection in rape and mustard seeds. Seed Res 20(2):128–133

Sharma P, Meena PD, Chauhan JS (2013) First report of *Nigrospora oryzae* (Berk. and Broome) Petch causing stem blight on *Brassica juncea* in India. J Phytopathol 161(6):439–441. https://doi.org/10.1111/jph.12081

Sharma DK, Kumar D, Sharma PC (2015) Abiotic stresses with emphasis on *Brassica juncea*. In: Kumar A, Banga SS, Meena PD, Kumar PR (eds) *Brassica* oilseeds breeding and management. CBAI International, Oxfordshire, UK, pp 233–250

Shattuck VI, Parry R (1990) The occurrence of powdery mildew on rutabagas in Southern Ontario. Can Pl Dis Sur 70:75–16

Shen FJ, Chen WL (1990) Preliminary study on cruciferous black rot in Zhejiang. Acta Agriculturae Universitatis Zhejiangensis 16(Suppl. 2):252–257

Sherf AF, MacNab AA (1986) Vegetable diseases and their control. John Wiley and Sons

Shi H, Qian Y, Tan DX, Reiter RJ, He C (2015a) Melatonin induces the transcripts of CBF/DREB1s and their involvement in both abiotic and biotic stresses in *Arabidopsis*. J Pineal Res 59:334–342

Shi H, Tan DX, Reiter RJ, Ye T, Yang F, Chan Z (2015b) Melatonin induces class A1 heat-shock factors (HSFA1s) and their possible involvement of thermo tolerance in *Arabidopsis*. J Pineal Res 58:335–342

Shin HD, La YJ (1992) Addition to the new records of host plants of powdery mildews in Korea. Korean J Plant Pathol 8:57–60

Shivas RG (1989) Fungal and bacterial diseases of plants in Western Australia. J Royal Soc W Aus 72:1–62

Shrestha K (1983) Major diseases and control of vegetable crops. In: Second Vegetable and Seed Production Workshop, February 1, 1983, Kathmandu.

Shyam KR, Gupta SK, Mandraia RK (1994) Prevalence of different types of curd rots and extent of yield loss due to plant mortlity in cauliflower seed crop. Indian J Mycol Plant Pathol 24:172–175

Singh SL (1974) Studies on some diseases of cruciferous crops in Varanasi. PhD Thesis, Banaras Hindu University. 177 p

Singh SL, Pavgi MS (1977) Aphanomyces root rot of cauliflower. Mycopathologia 61:167–172

Singh SL, Pavgi MS (1978) Pythium root rot of crucifers. Mycopathologia 63:155–159. https://doi.org/10.1007/BF00490930

Singh RM, Agarwal DK, Sarbhoy AK, Chatterjee SC, Sharma J (1993) HCIO descriptions of plant pathogenic fungi. Set 11, Nos. 61-66. In: HCIO descriptions of plant pathogenic fungi. Set 11, Nos. 61-66. set 11 New Delhi, India: Malhotra Publishing House. 17 pp

Singh YP, Kumar A, Singh NB (2005) Rapeseed-Mustard Integrated Insect-Pests management. Technical Bulletin No. 13, National Research Centre on Rapeseed-Mustard (ICAR), Bharatpur (Rajasthan) India, pp 34

Singh R, Bhagawati R, Baruah S, Sharma PK (2016) Bacterial Stalk Rot: an emerging disease of mustard in Arunachal Pradesh. Env Ecol 34(2):451–453

Sippell DW, Davidson JGN, Cox M (1985) Factors affecting severity of root rot of canola in the Peace River region. Proc 6th Annu. Meet. Plant Pathol Soc Alberta, Lethbridge, Alberta. 7–8 November 1985 p. 34. (Abstract)

Skibbe DS, Doehlemann G, Fernandes J, Walbot V (2010) Maize tumors caused by *Ustilago maydis* require organ-specific genes in host and pathogen. Science 328:89–92

Smith PA, Price TV (1996) Victorian vegetable growers' estimates of the incidence of downy mildew on *Brassica* crops, crop losses and fungicidal control. Aus Plant Pathol 25(2):99–105

So IY, Lee SH, Kim HM, Lee WH (1981) Disease incidence in radish and Chinese cabbage. I Major diseases in radish and Chinese cabbage grown in alpine areas in Jeonbug Province. Korean J Plant Prot 20(3):135–145

Soonthronpoct P (1969) Tech Docum. FAO Pl Prot Comm SE Asia 70:23

Sorin C, Leport L, Cambert M, Bouchereau A, Mariette F, Musse M (2016) Nitrogen deficiency impacts on leaf cell and tissue structure with consequences for senescence associated processes in *Brassica napus*. Bot Stud 57(1):11. https://doi.org/10.1186/s40529-016-0125-y

Spence NJ, Phiri NA, Hughes SL, Mwaniki A, Simons S, Oduor G, Chacha D, Kuria A, Ndirangu S, Kibata GN, Marris GC (2007) Economic impact of Turnip mosaic virus, Cauliflower mosaic virus and Beet mosaic virus in three Kenyan vegetables. Plant Pathol 56:317–323

Spoel SH, Johnson JS, Dong X (2007) Regulation of tradeoffs between plant defenses against pathogens with different lifestyles. Proc Natl Acad Sci U S A 104(47):18842–18847

Stefcheva M, Denchev CM, Karadjova J, Dimitrova T, Milanova SD, Bakalova GG (2005) Indigenous fungal pathogens – a possibility for biological control of economically important weeds in Bulgaria. Herbologia 6(3):43–51

Stell F (1922) Some common diseases of kitchen garden crops. Proc Agric Socences 12:779–785

Sturrock CJ, Woodhall J, Brown M, Walker C, Mooney SJ, Ray RV (2015) Effects of damping-of caused by *Rhizoctonia solani* anastomosis group 2-1 on roots of wheat and oil seed rape quantified using X-ray computed tomography and real-time PCR. Front Plant Sci 6:461

Su LC, Deng B, Liu S, Li LM, Hu B, Zhong YT, Li L (2015) Isolation and characterization of an osmotic stress and ABA induced histone deacetylase in Arachishypogaea. Front Plant Sci 6:512

Sutic C, Klijajic R (1954) A contribution to the knowledge of the parasitic flora of Deliblatska pescara. Plant Prot Beograd 24:104–108

Sutton (1977) *Cylindrosporium concentricum*. CMI Descr. Fungi Bact. No. 536. Commonwealth Mycological Institute, Wallingford

Sutton BC (1980) The coelomycetes. Fungi imperfecti with Pycnidia, Acervuli and Stromata. CABI, Kew, UK: 523–537

Suzuki K, Takahashi K, Kataoka M, Kyoto-fu (1980) Occurrence of Japanese turnip (*Brassica campestris*) root injury by *Aphanomyces raphani* and its control. Bulletin-of-the-Kyoto-Prefectural-Institute-of-Agriculture (Japan) 9:1–7. (Japanese)

Suzuki N, Rivero RM, Shulaev V, Blumwald E, Mittler R (2014) Abiotic and biotic stress combinations. New Phytol 203:32–43

Szith R, Furlan H (1979) New experiences with the chemical treatment of the white rust on horseradish. (Neue Erfahrungen bei der chemischen Behandlung des Weissen Rostes (*Albugo candida*) [Pers.] Ktze.bei Kren (Meerrettich)). Pflanzenarzt 32(7/8):70–72

Tahvonen R, Hollo J, Hannukkala A, Kurppa A (1984) *Rhizoctonia solani* damping-off on spring turnip rape and spring rape (*Brassica* spp.) in Finland. J Agric Sci Finland 56:143–154

Takasaki H, Maruyama K, Kidokoro S, Ito Y, Fujita Y, Shinozaki K, Yamaguchi-Shinozaki K, Nakashima K (2010) The abiotic stress responsive NAC-type transcription factor OsNAC5 regulates stress-inducible genes and stress tolerance in rice. Mol Gen Genomics 284:173–183

Tam LTT, Dung PN, Liem NV (2016) First report of powdery mildew caused by *Erysiphe cruciferarum* on *Brassica juncea* in Vietnam. Plant Dis 100(4):856

Tamura K, Takikawa Y, Tsuyumu S, Goto M (1994) Bacterial spot of crucifers caused by *Xanthomonas campestris* pv. *raphani*. Ann Phytopathol Soc Jpn 60:281–287

Tanaka H, Osakabe Y, Katsura S, Mizuno S, Maruyama K, Kusakabe K, Mizoi J, Shinozaki K, Yamaguchi-Shinozaki K (2012) Abiotic stress-inducible receptor-like kinases negatively control ABA signaling in *Arabidopsis*. Plant J 70:599–613

Tanas T (2004) The results of examination safeguard cucumbers in plastic houses from attack by *Sclerotinia sclerotiorum* (Lib.) de Bary. (Rezultati istraživanja mjera zaštite krastavca u plastenicima od napada: *Sclerotinia sclerotiorum* (Libert) de Bary.). Glasnik Zaštite Bilja 27 (4):63–69

Tarabeih AM, Abou-el-Fadl IA (1979) Effect of *Sclerotinia sclerotiorum* on the volatile oil content of some medicinal plants. Acta Phytopath Acad Sci Hung 14(1/2):31–35

Tewari JP, Orchard D, Hartman M, Lange R, Turkington TK, Strelkov S (2004) First report of clubroot of canola caused by *Plasmodiophora brassicae* in the Canadian Prairies. Can J Plant Pathol 26:228–229 (Abstr.)

Thaler JS, Owen B, Higgins VJ (2004) The role of the jasmonate response in plant susceptibility to diverse pathogens with a range of lifestyles. Plant Physiol 135(1):530–538

Thanassoulopoulos CC (1987) Outbreaks and new records. Greece. New diseases on various crops in Greece. FAO Plant Prot Bull 35 (4):164.

Thind KS (1942) The genus *Peronospora* in the Punjab. J Indian Bot Soc 21:197–215

Thomma BP, Nurnberger T, Joosten MH (2011) Of PAMPs and effectors: the blurred PTIETI dichotomy. Plant Cell 23:4–15

Thung TH (1926a) Observations on *Peronospora parasitica* on cabbage. Tijdschr Over Plantenziekten 32:161–179

Thung TH (1926b) *Peronospora parasitica* (Pers.) de Bary attacking cabbage heads. Phytopathology 16:365–366

Timina LT, Korganova NN, Dobrorutskaya EG (1989) Downy mildew on radish roots. Zashchita Rasenii (Moskua) No. 10:39.

Todorovic D (2007) Quarter-century of presence and spread of *Plasmodiophora brassicae*, cabbage hernia in Leskovac Region. (Cetvrt veka prisustva i irenja kile kupusa (*Plasmodiophora brassicae*) u okolini Leskovca.). Biljni Lekar (Plant Doctor) 35(5):505–511

Tompkins CM, Ark PA, Tucker CM, Middleton JT (1939) Soft rot of pumpkin and watermelon fruits caused by *Pythium ultimum*. J Agric Res 58:461–475

Tores JA, Moreno R (1991) *Sclerotinia sclerotiorum*: epidemiological factors affecting infection of greenhouse aubergine crops. J Phytopathol 132(1):65–74. https://doi.org/10.1111/j.1439-0434.1991.tb00094.x

Trdá L, Boutrot F, Claverie J, Brulé D, Dorey S, Poinssot B (2015) Perception of pathogenic or beneficial bacteria and their evasion of host immunity: pattern recognition receptors in the frontline. Front Plant Sci 6:219

Trinh HX, Quan MV, Groenewald JZ, Burgess LW (2012) First report of stub dieback of poinsettia (*Euphorbia pulcherrima*) caused by *Sclerotinia sclerotiorum* in Vietnam. Aust Plant Dis Notes 7(1):55–57. https://doi.org/10.1007/s13314-012-0047-9

Trkulja V (2006) The most important bacterial diseases of cole crops in Bosnia-Hercegovina and possibilities for their control. (Najznacajnije bakterioze kupusnjaca u BiH i mogucnosti njihovog suzbijanja.). Glasnik Zatite Bilja 29(4):41–53

Tronsmo A (1989) Trichoderma harzianum used for biological control of storage rot on carrots. Norwegian J Agric Sci 3(2):157–161

Tsai YP (1991) List of plant diseases in Taiwan. Taichung, Taiwan; China: The Plant Protection Society of the Republic of China, 604 pp

Tsuda K, Katagiri F (2010) Comparing signaling mechanisms engaged in patterntriggered and effector-triggered immunity. Curr Opin Plant Biol 13:459–465

Tsuda K, Somssich I (2015) Transcriptional networks in plant immunity. New Phytol 206:932–947

Turnock WJ, Gerber GH, Timlick BH, Lamb RJ (1995) Losses of canola seeds from feeding by *Lygus* species (Heteroptera: Miridae) in Manitoba. Can J Plant Sci 75:731–736

Tylkowska K (1984) Occurrence of fungi on bean seeds reproduced in different regions of Poland. (Występowanie grzybów na nasionach fasoli reprodukowanych w różnych rejonach Polski.). Biuletyn Instytutu Hodowli i Aklimatyzacji Roślin, pp 185–202

Vaartnou H, Tewari I, Horricks J (1974) Fungi associated with diseases on Polish-type rape in Alberta. Mycopathol Mycol Appl 52:255–260

Vadassery J, Reichelt M, Hause B, Gershenzon J, Boland W, Mithofer A (2012) CML42-mediated calcium signaling coordinates responses to *Spodoptera* herbivory and abiotic stresses in *Arabidopsis*. Plant Physiol 159:1159–1175. https://doi.org/10.1104/pp.112.198150

Vagelas I, Gravanis F (2014) *Phelipanche nana* (Reut.) Sojak parasitism on lentil (*Lens culinaris*) and parasitism of *P. aegyptiaca* on *Carduus marianus* in Thessalia region, Greece. Archiv Phytopathol Plant Prot 47(16):1956–1962

Vakalounakis DJ (1991) *Albugo candida*, the cause of a serious white rust of *Lunaria biennis* in Crete. Plant Pathol 40(4):637–638. https://doi.org/10.1111/j.1365-3059.1991.tb02430.x

Valizadeh N, Mirshekari B (2011) Determination of economical yield loss threshold of *Chenopodium album* at interference with rapeseed (*Brassica napus*). J Food Agric Env 9 (2):409–412

Valmonte GR, Arthur K, Higgins CM, Macdiarmid RM (2014) Calcium-dependent protein kinases in plants: evolution, expression and function. Plant Cell Physiol 55:551–569. https://doi.org/10.1093/pcp/pct200

Van Loon LC, Rep M, Pieterse CM (2006) Significance of inducible defense-related proteins in infected plants. Annu Rev Phytopathol 44:135–162

Vanparys L, Calus A (1994) Brussels sprouts. A review of the research in (1992) Beitem-Roeselare. Provinciaal Onderzoek-en Voorlichtingscentrum voor Land-en Tuinbouw, Belgium, 48 pp

Vanparys L, Seynnaeve M, Calus A (1993) Brussels sprouts. A review of research in 1990. (Spruitkool. Overzicht van het onderzoek in 1990). Beitem Roeselare, Belgium: Belgium, Provinciaal Onderzoek en Voorlichtingscentrum voor Land en Tuinbouw

Vanterpool TC (1960) The ring spot disease of rape in an inland parkland region. Plant Dis Reptr 44:362–363

Vavrac J, Kubova A (1992) The successes and the problems of the growing of soybeans in Czechoslovakia. Eur Secur:58–61

Velastegui JR, Ball SFL (1991) First record of *Sclerotinia sclerotiorum* on tree tomato in Ecuador. Plant Pathol 40(3):476–477. https://doi.org/10.1111/j.1365-3059.1991.tb02406.x

Vellios E, Karkanis A, Bilalis D (2017) Powdery mildew (*Erysiphe cruciferarum*) infection on Camelina (*Camelina sativa*) under Mediterranean conditions and the role of wild mustard (*Sinapis arvensis*) as alternative host of this pathogen. Emirates J Food Agric 29(8):639–642

Verma PR, Saharan GS (1993) *Alternaria brassicae* (Berk.) Sacc., *A. brassicicola* (Schwein.) Wiltsh. and *A. raphani* Groves and Skolko: Introduction, bibliography and subject index. Tech. Bull. No. 93–002, Agriculture and Agri-food Canada, Res. Centre Saskatoon, p 81

Verma PR, Saharan GS (1994) Monograph on *Alternaria* Diseases of Crucifers. Saskatoon Research Centre Technical Bulletin 1994–6E, Agriculture and Agri-Food Canada, Saskatoon, SK, Canada.

Verma S, Nizam S, Verma PK (2013) Biotic and abiotic stress signaling in plants. Chapter 2. In: Sarwat M, Ahmad A, Abdin MZ (eds) Stress signaling in plants: genomics and proteomics perspective, vol 1, pp 25–50. https://doi.org/10.1007/978-1-4614-6372-6_2

Viegas AP, Teixeira AR (1943) Some fungi of Brazil Phycomycetos. Bragantia S Paulo 3:223–245

Virdi A, Singh S, Singh P (2015) Abiotic stress responses in plants: roles of calmodulin regulated proteins. Front Plant Sci 6:809

Vrataric M, Bilandzic M, Jurkovic D, Prodanovic G, Krizmanic M, Sudaric A (1991) A study of principal diseases affecting several soyabean varieties and genotypes and their control with fungicide combinations. (Proucavanje vaznijih bolesti na nekoliko novih sorata i genotipova soje te mogućnosti suzbijanja s fungicidima u Osijeku.). Znanost i Praksa u Poljoprivredi i Prehrambenoj Tehnologiji 21 (Suppl):85–94

Wager VA (1940) Descriptions of the South African Pythiaceae with records of their occurrence. Bothalia 4:3–35

Walsh JA, Tomlinson JA (1985) Viruses infecting winter oilseed rape (*Brassica napus* ssp. *oleifera*). Ann Appl Biol 107(3):485–495

Wang W, Vinocur B, Shoseyov O, Altman A (2004) Role of plant heat-shock proteins and molecular chaperones in the abiotic stress response. Trends Plant Sci 9:244–252

Wang D, Amornsiripanitch N, Dong X (2006) A genomic approach to identify regulatory nodes in the transcriptional network of systemic acquired resistance in plants. PLoS Pathol 2 (11):1042–1050

Wang X, Basnayake B, Zhang H, Li G, Li W, Virk N, Mengiste T, Song F (2009) The *Arabidopsis* ATAF1, a NAC transcription factor, is a negative regulator of defense responses against necrotrophic fungal and bacterial pathogens. Mol Plant-Microbe Interact 22:1227–1238

Wang A, Yu X, Mao Y, Liu G, Liu Y, Niu X (2015a) Overexpression of a small heat-shock-protein gene enhances tolerance to abiotic stresses in rice. Plant Breed 134:384–393

Wang JP, Munyampundu JP, Xu YP, Cai XZ (2015b) Phylogeny of plant calcium and calmodulin-dependent protein kinases (CCaMKs) and functional analyses of tomato CCaMK in disease resistance. Front Plant Sci 6:1075. https://doi.org/10.3389/fpls.2015.01075

Wang Y, Hwang SF, Akhavan A, Ahmed H, Turnbull G, Strelkov S (2019a) Yield losses in blackleg of canola and pyraclostrobin sensitivity in populations of *Leptosphaeria maculans*. Presentation http://hdl.handle.net/10388/12048, University of Saskatchewan, Canada.

Wang Y, Strelkov SE, Hwang SF (2019b) Yield losses in canola in response to blackleg disease. Can J Plant Sci

Waterston JM (1940) Report of the plant pathologist, 1939. Report Dept. Agric, Bermuda, pp 19–31

Waterston JM (1947) The fungi of Bermuda. Bull Dep Agric Bermuda 23:305 pp

Wei F, Wing RA, Wise RP (2002) Genome dynamics and evolution of the *Mla* (powdery mildew) resistance locus in barley. Plant Cell 14:1903–1917

Weimer JL (1926) Ringspot of crucifers caused by *Mycosphaerella brassicicola*. J Agric Res 32:97–132
Weston DJ, Karve AA, Gunter LE, Jawdy SS, Yang X, Allen SM, Wullschleger SD (2011) Comparative physiology and transcriptional networks underlying the heat shock response in *Populus trichocarpa*, *Arabidopsis thaliana*, and *Glycine max*. Plant Cell Environ 34:1488–1506
White PJ, Broadley MR (2003) Calcium in plants. Ann Bot 92:487–511. https://doi.org/10.1093/aob/mcg164
Wiese E (1927) A fungus disease of wallflowers and stocks. Gartenwelt 31:486
Wilkins O, Brautigam K, Campbell MM (2010) Time of day shapes *Arabidopsis* drought transcriptomes. Plant J 63:715–727
Williams PH (1980) Black rot: a continuing threat to world crucifers. Plant Dis 64:736–742
Williams PH (1985) Crucifer genetics co-operative resource book Downy mildew. University of Wisconsin, Madison, pp D5–D7
Winter W, Burkhard L, Bänziger I, Krebs H, Gindrat D, Frei P, Brändle G, Hirner M, Forrer HR, Hogger C, Schwarz A (1993) Rape diseases: occurrence on rape varieties, effect of fungicides and preventive control measures. (Krankheitsanfälligkeit von Rapssorten, Fungizidwirkung und Massnahmen zur Krankheitseindämmung.). Landwirtschaft Schweiz 6(10):589–596
Worf GL, McCain AH, Covey RP, Castanho BJ (1985) Common names for plant diseases. Plant Dis 69(8):649–676
Wu MD, Zhang L, Li GQ, Jiang DH, Hou MS, Huang HC (2007) Hypovirulence and double-stranded RNA in *Botrytis cinerea*. Phytopathology 97:1590–1599
Wu Y, Deng Z, Lai J, Zhang Y, Yang C, Yin B, Zhao Q, Li Y, Yang C, Xie Q (2009) Dual function of *Arabidopsis* ATAF1 in abiotic and biotic stress responses. Cell Res 19:1279–1290
Xin M, Wang Y, Yao Y, Xie C, Peng H, Ni Z, Sun Q (2010) Diverse set of microRNAs are responsive to powdery mildew infection and heat stress in wheat. BMC Plant Biol 10:123
Xue GP, Drenth J, Mcintyre CL (2015) TaHsfA6f is a transcriptional activator that regulates a suite of heat stress protection genes in wheat (*Triticum aestivum* L.) including previously unknown Hsf targets. J Exp Bot 66:1025–1039
Yamada M, Asandhi AA, Purwati E, Dianawati M (2004) Control of cabbage clubroot disease by employing one-year crop rotation with three vegetable combinations in the West Java highlands of Indonesia. JIRCAS Working Report, No. 43:175-184.
Yamaguchi T, Aharon GS, Sottosanto JB, Blumwald E (2005) Vacuolar NaC/HC antiporter cation selectivity is regulated by calmodulin from within the vacuole in a Ca^{2C}- and pH-dependent manner. Proc Natl Acad Sci U S A 102:16107–16112. https://doi.org/10.1073/pnas.0504437102
Yan L, Zhang LG, Hui MX, Zhang MK (2008) Pathogenesis division of black rot bacteria of Chinese cabbage in Shaanxi central productive area. J Northwest A & F University Natl Sci Edn 36(10):132–138
Yang T, Poovaiah BW (2003) Calcium/calmodulin-mediated signal network in plants. Trends Plant Sci 8:505–512. https://doi.org/10.1016/j.tplants.2003.09.004
Yoshikawa H, Buczacki ST (1978) Club root in Japan: research and problems. Rev Plant Pathol 57 (7):253–257
Yuasa T, Ichimura K, Mizoguchi T, Shinozaki K (2001) Oxidative stress activates ATMPK6, an *Arabidopsis* homologue of MAP kinase. Plant Cell Physiol 42(9):1012–1016
Yucer MM, Karaca I (1978) Investigations on sunflower diseases in Thrace, their rate of existence, their fungal pathogens and their pathogenicities. J Turkish Phytopathol 7(1):39–50
Zarzycka H (1970) Investigations on sources of infection of *Camelina* by *Peronospora camelinae*. Acta Mycologica Warszawa 6:7–19
Zazzerini A, Tosi L (1985) Tests of antagonism of some bacterial isolates towards *Sclerotinia sclerotiorum* (Lib.) de Bary. (Prove di antagonismo di alcuni isolati batterici nei confronti della *Sclerotinia sclerotiorum* (Lib.) de Bary.). Informatore Fitopatologico 35(1):27–30
Zeng X, Huang Q (1994) Identification on the pathogenic bacteria of a new tabacco disease-Black rot. J South China Agril Univ 15:46–49

Zeng H, Xu L, Singh A, Wang H, Du L, Poovaiah BW (2015) Involvement of calmodulin and calmodulin-like proteins in plant responses to abiotic stresses. Front Plant Sci 6:600. https://doi.org/10.3389/fpls.2015.00600

Zhang B (2015) MicroRNA: a new target for improving plant tolerance to abiotic stress. J Exp Bot 66:1749–1761

Zhang ZY, Wang YX, Liu YL (1984) Taxonomic studies on the family Albuginaceae of China II. A new species of *Albugo* on Acanthaceae and known species of *Albugo* on Cruciferae. Acta Mycol Sin 3(2):65–71

Zhang T, Chen S, Harmon AC (2014) Protein phosphorylation in stomatal movement. Plant Signal Behav 9:e972845. https://doi.org/10.4161/15592316.2014.972845

Zhang F, Zhang F, Huang L, Vera Cruz C, Ali J, Xu J, Zhou Y, Li Z (2016) Overlap between signaling pathways responsive to *Xanthomonas oryzae* pv. *oryzae* infection and drought stress in rice introgression line revealed by RNA-Seq. J Plant Growth Regul 35:345–356

Zhao Y, Damicone JP, Demezas DH, Bender CL (2000) Bacterial leaf spot diseases of leafy crucifers in Oklahoma caused by pathovars of *Xanthomonas campestris*. Plant Dis 84:1008–1014

Zhao J, Gao Y, Zhang Z, Chen T, Guo W, Zhang T (2013) A receptor-like kinase gene (GbRLK) from *Gossypium barbadense* enhances salinity and drought-stress tolerance in *Arabidopsis*. BMC Plant Biol 13:110

Zhu JK (2016) Abiotic Stress signaling and responses in plants. Cell 167:313–324. https://doi.org/10.1016/j.cell.2016.08.029

Zhu X, Dunand C, Snedden W, Galaud JP (2015) CaM and CML emergence in the green lineage. Trends Plant Sci 20, 483:–489. https://doi.org/10.1016/j.tplants.2015.05.010

Zipfel C (2014) Plant pattern-recognition receptors. Trends Immuno 35:345–351

Zou X, Qin Z, Zhang C, Liu B, Liu J, Zhang C, Lin C, Li H, Zhao T (2015) Overexpression of an S-domain receptor-like kinase extracellular domain improves panicle architecture and grain yield in rice. J Exp Bot 66:7197–7209

Glimpses of Molecular Mechanisms of Host Resistance

6

Abstract

Cruciferous crops are attacked by several pathogens which deteriorate yield and quality of major *Brassica* crops affecting global food security. The pathogens infecting these crops have diverse lifestyle, genomic plasticity with rapid evolutionary mechanisms, prolific reproduction, and longer survival nature which is impedimental to manage efficiently under field conditions. Moreover pathogens secrete small protein known as effector genes, which counter plant defense mechanisms and modulate host physiology to promote pathogen growth and reproduction. However, plants evolve disease resistance (R) genes encoding receptor proteins that detect and respond to pathogen effector gene, leading to the activation of effector-triggered immunity. This discovery with the use of multi-omics approaches has revealed molecular mechanisms of crucifers' host resistance to manage major pathogens of *Brassica* crops. Host resistance in crucifers' to biotic and abiotic stresses is multilayered and multicomponents. Resistance is activated either through SA signaling and simultaneous perception of ethylene and jasmonic acid depending on biotrophic, hemibiotrophic, and necrotrophic *Brassica* host–pathogen interactions. Molecular mechanisms of host resistance against biotrophs have been revealed through identification, functionally characterization, and fine mapping of R-genes and QTLs on different chromosomes of *Brassica* species. The molecular mechanisms of R-genes resistance imparted by NBS-LRR are effective against biotrophs or hemibiotrophs but are not effective against necrotrophs. JA- and ET-mediated signaling is predominantly involved in general defense to necrotrophs. Similarly, R-genes and QTLs have been identified from *B. rapa, B. juncea, B. nigra*, and *B. napus* to provide resistance to hemibiotrophs and necrotrophs. Cysteine-rich protein kinase genes and multi-environment stable six QTLs suitable for quantitative blackleg resistance in *B. napus* have been identified. Overexpression of *BnNPR1*, *OsPGIP2*, *BnWRKY33*, and *BnMPK4* genes enhances resistance in *B. napus* to *Sclerotinia*.

G. S. Saharan et al., *Genomics of Crucifer's Host-Resistance*,
https://doi.org/10.1007/978-981-16-0862-9_6

Seven NBS-LRR encoding genes linked to black rot resistance in *B. oleracea* have been identified on four QTL regions and chromosomes Co1, Co3, CO6, CO7, and CO8, respectively. Biometabolomics mechanisms of *Brassica* host resistance is initiated soon after perception of pathogens by signaling molecules and results in execution of broad range of defense responses to stop the infection and pathogenesis of invading pathogens. Biometabolomics and biomolecular mechanisms of host resistance is expressed in the form of different metabolites (primary and secondary) and biomolecules involved in the induction of biochemical changes and pathways in the host to confer resistance at pre- and post-invasion stages to biotrophs, hemibiotrophs, and necrotrophs. Transcriptomics and proteomics have discovered novel biochemical pathways and proteins/genes involved in conferring resistance to virulent pathotypes. With the use of multi-omics supplemented with functional genomics studies, key resistance pathways involved in Brassica host-pathosystem and induction of differential signaling molecules like SA, JA, and ET in response to diverse pathogens have been revealed.

Novel molecular approaches have been exploited to breed disease-resistant cultivars of *Brassica*. Through introgression breeding high level of resistance has been obtained from *B. nigra*, and *B. carinata* to *B. napus* against blackleg (*Leptosphaeria*). Similarly transfer of resistance to black rot from *B. carinata* to *B. oleracea*, from *B. hirta* to *B. juncea* against *Alternaria*, and from B. carinata to B. oleracea against powdery mildew has been obtained. The approaches like in vitro embryo rescue, somatic hybridization, somaclonal variations, genetic transformation, molecular markers, use of systemic resistance induction, and genetic engineering have great promise but have been used against few pathogens. Clubroot resistance in *B. napus* and *B. rapa* has been transferred through interspecific and distant hybridization using molecular markers. R-genes from European Clubroot-Differentials (ECD) set have been introduced into *B. napus*, and other species to provide resistance to clubroot. Powdery mildew resistance in *B. oleracea* has been obtained from *B. carinata* through embryo rescue. The mutagenic agent gamma radiation increases resistance to *E. cruciferarum* in *Brassica*. Durable resistance in *Brassica* species to Alternaria can be obtained through introgression of genes from non-host plants by inter- or intraspecific crosses, identification and transfer of NHR based on microarray, identification and transfer of NHR genetic resources from model host plants, identification of functional genes based on bioinformatics, and use of endochitinase gene from *Trichoderma virens*. Marker-assisted introgression of resistance from wild relatives has been achieved in *B. juncea* to *Sclerotinia*. Resistant *B. napus* lines pyramided with multiple R-QTL derived from wild *B. oleracea* have been developed via hexaploidy strategy. Introgression of black rot resistance from *B. carinata* to *B. oleracea* through embryo rescue has been attempted successfully. The *B. carinata* genotypes, NPC-9 and HCA-6 having very high resistance to *XCC* race 1 and 4, have been used to introgress resistance in Indian cauliflower through embryo rescue approaches.

The advances made in the molecular genetics through whole-genome sequencing of *Brassica* crops have led to quantum jump not only in understanding plant genetic diversity but also for accessing these enormous resources. Novel sources of R-genes for blackleg has been tapped for other A- and C-genome of *Brassica* species from the B-genomes in *B. nigra, B. juncea, B. carinata*, and a wild species *Sinapis arvensis*. Several sources of resistance to 15 major pathogens of crucifers except nematodes have been identified from different *Brassica* species and wild relatives. Some of these sources have been utilized to breed disease-resistant cultivars of *Brassica* through conventional and molecular approaches. From *Brassica* species, multiple pathogens resistance sources have been identified which provide resistance to *Albugo, Alternaria, Erysiphe*, and *Hyaloperonospora*. There is strong evidence for positive genetic correlation among *Albugo, Alternaria, Erysiphe, Hyaloperonospora, Leptosphaeria,* and *Xanthomonas* to develop multiple disease-resistant cultivars of *Brassica* crops. Clubroot-resistant cultivars with R-genes and QTLs have been developed in *B. rapa* and *B. napus* with pathotype specificity at different counties but most of them have scummed through evolution of new virulences of *Plasmodiophora*. In *Brassica* crops, transfer of resistance through different approaches is facing problems because of diverse nature of infecting pathogens for which there is no single gene product to have a direct toxic effect on all kinds of pathogens. Consequently, host has to develop quite different strategies to deal with these pathogens.

Keywords

Molecular mechanism of host resistance · Biotrophs · Necrotrophs · Hemibiotrophs · Crucifers' pathogens · Biomolecular mechanisms of resistance · Molecular approaches for breeding disease resistance · Durable resistance

6.1 Introduction

Cruciferous crop plants (canola/oilseed rape/rapeseed/mustard/cabbage/cauliflower) are economically important crops, widely used as edible oil vegetables. These crops are mostly cultivated in temperate and sub-tropical regions. Brassica crops are always challenged by major diseases such as sclerotinia stem rot, white rust, downy mildew, Alternaria blight, blackleg, and clubroot. These diseases cause significant yield and economic losses in rapeseed-mustard and vegetable crops worldwide, which has been aggravated by global climate change. If it is not resolved meritoriously, will impend global food security. To advance further comprehensions into the *Brassica*–pathogen interactions, it is advisable to analyze critically the current knowledge in this area and how advanced skills and technologies can offer encouraging consequences to understand the resistance mechanisms. Resistant varieties development is the utmost important, economical, environmentally sustainable solution and effective to cope with diseases management in *Brassica* species.

Transfer of disease resistance remains attained once sources of R-genes have recognized with sound characters, mode of inheritance of genes in the progenies along with all desired traits of yield and quality. Conventional and molecular approaches have been used either alone or in combination to incorporate desired and popular traits to develop improved cultivars of crucifers. Hybridization combined with backcrossing either singly or in combination with pedigree selection is the most widely used and accepted conventional method to transfer disease resistance in crucifers. The level of resistance in developed cultivars has been accomplished through multiple crosses, recurrent selection, diallele crossing, and selective mating system. The resistance has been introgressed with intraspecific pedigree, backcross, modified recurrent mass selection, interspecific genome substitution, chromosome substitution, and gene introgression against different pathogens. The molecular and genetic engineering techniques like genome manipulation, cytoplasmic genome manipulation, transformation, foreign gene expression, and embryo rescue for wide hybridization have been exploited in *Brassica*. The introgression and pyramiding major R-QTLs have been attempted including transgenics development. To develop stress tolerance and/or resistant *Brassica* cultivars, there is need to apprehend both of the plant and pathogenic disease development mechanisms.

The genomics of crucifers' host resistance has been strengthened significantly after the sequencing of all the six *Brassica* species genome and major crucifers' pathogens genome. The development and application of omics technology and approaches have facilitated to discover molecular mechanisms of host–pathogen interaction and host resistance in major crucifers' host-pathosystem. During the course of literature search, it was revealed that voluminous research data have been generated on genomics of crucifers' host-pathosystem which needs separate treatment on the genomics of host resistance and molecular mechanisms of host resistance. In this chapter to make the story of host resistance complete, briefly molecular mechanisms of host resistance has been covered. The readers are advised to consult the other book entitled *Molecular Mechanisms of Crucifers' Host Resistance* for detailed information.

6.2 Molecular Mechanisms of Disease Resistance

Biotic and abiotic stress of crucifers' crop plants cause significant yield losses and affect global food security, particularly in the face of increased demand due to a growing world population (Fisher et al. 2012). Diverse lifestyle, genome plasticity and evolution, prolific reproduction, and longevity of spores under harsh environment impede the efficient control of pathogen causing crucifers' crop diseases. Bacterial and fungal pathogens secrete small proteins, known as effectors, which counter plant defense and modulate plant physiology to promote pathogen growth and reproduction (Presti et al. 2015; Stergiopoulos and de Wit 2009). Accumulating evidences show that the secreted effectors are key players in suppressing pathogen-associated molecular patterns (PAMPs)-triggered immunity (PTI) induced upon fungal recognition by plant (Presti et al. 2015). The lysine motif (LysM) effector

ECP6 secreted by the tomato pathogen *Cladosporium fulvum* suppresses chitin-triggered immunity. ECP6 binds chitin via the intramolecular chitin-binding groove formed by the LysM domains, resulting in the sequestering of chitin released from the cell walls of invading hyphae (de Jonge et al. 2010; Sánchez-Vallet et al. 2013). Plants, on the other hand, have evolved disease resistance (R) genes encoding receptor proteins that detect and respond to pathogen effector proteins, leading to the activation of effector-triggered immunity (ETI) (Jones and Dangl 2006). Effectors that trigger ETI are referred to as avirulence (*Avr*) genes. Discovery of the host targets of effectors is essential for our understanding of effector activity and molecular mechanisms of plant defense (Win et al. 2012). There are many examples of plant proteins that are targeted by bacterial effectors secreted by the type III secretion system (Block and Alfano 2011; Deslandes and Rivas 2012). However, despite the economic importance of diseases caused by the fungal plant pathogens, for most fungal effectors, the host targets and underlying molecular mechanisms remain unknown. Mitogen-activated protein (MAP) kinase cascades are highly conserved signaling modules in eukaryotes and have a central role in plant immunity against pathogen attack (Meng and Zhang 2013; Pitzschke et al. 2009). An MAP kinase cascade is commonly composed of three types of kinases including an MAP kinase kinase kinase (MAPKKK), an MAP kinase kinase (MKK), and an MAP kinase (MPK). In *Arabidopsis*, pathogen-responsive MPK signaling has been reported to be involved in both PTI and ETI, in which the best characterized MPKs are MPK3, MPK4, and MPK6 (Asai et al. 2002; Rasmussen et al. 2012). PAMPs, such as flg22 (a conserved 22-amino acid flagellin peptide) and elf18 (elongation factor-Tu peptide), activate the kinase signaling cascades involving MPK3, MPK4, and MPK6 (Felix et al. 1999; Gao et al. 2008; Zipfel et al. 2006). In contrast, plant pathogens have evolved mechanisms to target the MPK signaling pathways to enhance plant susceptibility. *Pseudomonas syringae* HopAI1 targets MPK3 and MPK6 and inactivates their kinase function to suppress plant defense responses (Zhang et al. 2007). *P. syringae* HopF2 targets MKK5 and can inactivate MKK5 via ADP-ribosylation of the C terminus of MKK5 in vitro (Wang et al. 2010). Avr B is a *P. syringae* effector that interacts with MPK4 to perturb hormone signaling and promote infection (Cui et al. 2010). *Phytophthora infestans* RXLR effector PexRD2 as a virulence factor interacts with the kinase domain of the host MAPKKKε to suppress MAPKKKε-dependent phosphorylation of MPKs to modulate plant immunity (King et al. 2014). These findings highlight the importance of MPK pathways in plant immunity and as targets of bacterial and biotrophs pathogen effectors. However, to date no MAPKs have been identified as targets of effectors from fungal plant pathogens (Ma et al. 2018).

Plant MPKs can be classified into four groups (A, B, C, D) based on the conserved amino acid sequences of the T × Y motif present in the activation loop of MPKs (Ichimura et al. 2002). Among the best studied MPKs, MPK3 and MPK6 are in group A and MPK4 belongs to group B. However, MPK9 belongs to the D group with the plant-specific TDY phosphorylation motif and is distinct from the A, B, and C groups. In *Arabidopsis*, studies have reported that AtMPK9, together with AtMPK12, is involved in stomatal closure (Lee et al. 2016). Both abscisic acid

(ABA)- and methyl jasmonate-induced stomatal closure are impaired in *Arabidopsis* double mpk9/mpk12 mutants, but not in mpk9 or mpk12 single mutants, indicating a functional redundancy (Jammes et al. 2009; Khokon et al. 2015). The mpk9/mpk12 double mutant is highly susceptible to *P. syringae* pv. tomato (Pst) DC3000 infection (Jammes et al. 2011). However, Montillet et al. (2013) reported that ABA signaling, including open stomata 1 (OST1) protein kinase, AtMPK9, or AtMPK12, plays a limited role in response to Pst early infection. This is based on the observations that there is no difference between double mpk9/mpk12 mutants and wild-type *Arabidopsis* Col in response to flg22-induced stomatal closure and expression of ABA-specific marker genes is not affected by Pst treatment (Montillet et al. 2013). Thus, it is unclear whether MPK9 acts in ABA-mediated guard cell immune signaling in response to biotic stresses. A study by Nagy et al. (2015) revealed that AtMPK9 is activated through intramolecular autophosphorylation independent of any upstream MAPKKs, which is similar to MAPKK-independent activation mechanisms reported for the mammalian atypical MAPKs, such as extracellular signal-regulated kinase (ERK)7/8 (Klevernic et al. 2006; Nagy et al. 2015).

Host resistance in crucifers to biotic and abiotic stresses is multilayered and multicomponents. The perception of defense response and molecular mechanisms in a host against the pathogens are of a great significance in the breeding of resistant cultivars. The sequencing of host and pathogen genome during the last decade has facilitated the researchers using modern omics approaches to reveal molecular mechanisms of host–pathogen interaction and induction of signal molecules for expression of R-genes in the host at different levels of host growth. The intricate immune responses are evolved through accumulation of ROS, H_2O_2, deposition of callose, pectin, cellulose, waxes, silicon, ion-influxes, formation of papilla, cell wall apposition, phenolic compounds, overexpression of R-genes, PR proteins, protein phosphorylation, biosynthesis of phytoalexins, fungal enzymes inhibitors, triggering of HR, induction of SAR, non-host resistance mechanisms, the production of secondary metabolites, calcium and hormones signaling and homeostasis, the induction of transcriptional factors and mitogen-activated protein kinases. Resistance is activated either through SA signaling and simultaneous perception of ethylene and jasmonic acid depending on biotrophic and necrotrophic host–pathogen interactions. The majority of plant R-genes encode nucleotide-binding site leucine-rich repeat (NB-LRR)-type protein, which can be further, grouped into two sub-classes based on their N-terminal sequence. First, those containing coiled coil (CC) domain (CC-NB-LLR) and second those containing a domain with interleukin-1 receptor (TIR) (TIR-NB-LRR). The NBS-LRR proteins are thought to recognize effectors or effector-specific signals from the pathogen and activate the host immune system to provide hypersensitive responses.

Cruciferous plants evolve R-genes in response to pathogens attack which encodes receptors proteins that detect and respond to pathogen effector protein leading to the activation of effector-triggered host immunity. Effectors that trigger ETI are referred to as avirulence gene (*Avr*). Mitogen-activated protein (MAP) kinase cascades are highly conserved signaling molecules in eukaryotes and have a central role in plant immunity against pathogen attack. Plants are very often exposed to a variety of biotic stresses and thus have evolved multidimensional defense approaches to survive or retain their fitness. The plant used both pre-formed and inducible defense mechanisms to overcome pathogen challenges. Much stronger and long lasting is inducible defense response such as systemic acquired resistance (SAR), which confers enhanced disease resistance to broad range of pathogens. NPR1 gene in *Brassica* plants mediates the SA-induced expression of pathogenesis-related (PR) genes and SAR. Overexpression of NPR1 results in the increased transcript levels of antifungal genes like PR1, PR2 (glucanase), and PR5 (thaumatin). Many of the underlying resistance genes have been cloned and their downstream signaling pathways have been characterized using *Arabidopsis* host-pathosystem in molecular mechanisms of host resistance. Plants have evolved intracellular receptors with nucleotide-binding, leucine-rich repeat domains (NLRs) to detect fungal effectors, or their activity to trigger immunity. The high polymorphic information content (PIC) and number of phylogenetically informative bands established RAPD as a useful tool for phylogenetic reconstruction, quantification of genetic diversity for conservation, cultivar classification, and molecular disease resistance cultivars breeding in *Brassica*. The clustering and duplication of R-genes can cause problems in R-genes identification as these genes tend to collapse in genome sequence assemblies. The outcome of host–pathogen interactions is determined by R-genes that enable host to recognize invading pathogens and activate inducible defense mechanisms. A typical R-genes allele encodes "race specific" resistance to only one or a few strains of a single pathogen species. R-genes loci are functionally polymorphic within a plant species and encode alternate alleles that either recognize different strains of the same pathogen or do not recognize any tested pathogen. The molecular studies have revealed that R-genes often reside in complex loci consisting of the R-genes and lightly linked homologs. The cloning of R-genes against diverse pathogen provides the tools for comparative analysis of R-genes alleles within and between species, which will provide insight into the evolutionary history of R-genes. The functional polymorphism at R-genes loci can arise from gene deletions. The majority of plant PRRs identified belongs to the family of leucine-rich repeat receptor-like kinases (LRR-RLKs) or receptor-like proteins (LRR-RLPs) that lack a c-terminal kinase domain and interact with RLKs for transducing signals. Several numbers of genes have well documented that they are differentially expressed against different biotic stresses. Several techniques and approaches have been used to reveal molecular mechanisms of crucifers' host resistance. These techniques include: genomics, bioinformatics, epigenetics, transcriptomics, proteomics, metabolomics, metagenomics, pathogenomics, phylogenomics, and pangenomics.

6.3 Molecular Mechanisms of Host Resistance to Biotrophs

The genome sequencing of a small cruciferous weed, *Arabidopsis thaliana* and its establishment as a model for host-pathosystem to reveal molecular mechanisms of host–pathogen interactions have transformed traditional sciences like Genetics, Plant Breeding, and Plant Pathology into Molecular Genetics, Molecular Plant Breeding, and Molecular Plant Pathology under the umbrella of Molecular Biology. During the last three decades, significant investigations have been carried out on molecular mechanisms of crucifers' host resistance. Out of 44 pathogens known to infect crucifers, 16 pathogens are of significant important at global level on the basis of geographical distribution, host range, yield losses, and investigations on the molecular studies. On the basis of mode of acquisition of nutrients from the host (lifestyle) and mode of infection, differences in molecular mechanisms of host–pathogen interactions can be divided as biotrophs, hemibiotrophs, and necrotrophs. Biotrophs essentially act as a sink for the host, anabolic assimilates, and keep it alive. Necrotrophs consume the host tissues on invasion. Hemibiotrophs combine both strategies in their life cycle. Therefore, host plants have quite different molecular mechanisms and strategies to deal with these groups of pathogens. Pathogen infection produces a variety of phytohormones depending mainly on the pathogen lifestyle and their mode of infection. Salicylic acid pathway generally provides resistance to biotrophic pathogens whereas JA/ET pathways are commonly associated with resistance to necrotrophic pathogens and herbivorous pests. Generally, SA and JA signaling pathways operate antagonistically and thus, elevated resistance against biotrophs is often related with increased susceptibility to necrotrophs and vice versa.

Molecular mechanisms of crucifers' host resistance against biotrophs have been revealed in different *Brassica* species during host–pathogen interactions at different stages of host growth and disease development. The R-genes and QTLs have been identified, functionally characterized, and molecularly mapped on the chromosomes of *Brassica* species expressed during pathogens interactions. Timing of the expression of defense-related genes plays a crucial role during pathogenesis and host resistance for mounting a successful defense response. A single gene (*Acr*) responsible for conferring resistance to *Albugo* was mapped on a densely populated *B. juncea* RFLP map. A single dominant *RAC1* gene in accession Ksk 1 of *Arabidopsis thaliana* confers resistance to *A. candida* isolate ACeM1. Two accessions of *A. thaliana* (Ksk1 and Ksk2) were used to identify and map three loci (RAC1, RAC2, and RAC3) of genes that confer resistance to *A. candida*. In *B. rapa*, the ACA1 locus was mapped to linkage group 4 and was flanked by RFLP marker loci. *WRR* 4 genes encode a cytoplasmic TIR-NBL-LRR receptor-like protein in Columbia *A. thaliana* and confer broad spectrum white rust resistance to four races of *A. candida*. Three WRR (*WRR4BCol*-0, *WRR8Sf*-2, and *WRR9Hi*-0) genes against *Ac2V*, and a gene *WRR12* (*SOC3*), conferring NHR to AiBoT have been identified. Analysis of allelic variants of BjuA046215 in Indian, east Europe, and Chinese gene pool suggested the presence of three types of alleles. Allele 1 is present in the Chinese Tumida, Allele 2 in Indian, and Allele 3 in east European gene pools. Host resistance in *Brassica* species to powdery mildew is multilayered and multicomponent at pre- and post-penetration both stages. Host resistance is activated either through SA signaling and simultaneous perception of ethylene and jasmonic acid. MLO genes encoding seven transmembrane calmodulin-binding proteins confer broad-spectrum resistance to *Arabidopsis* powdery mildew. The Pmr mutants confer resistance to powdery mildew through altered cell wall composition of host.

Salicylic acid enhances the expression of RPW 8.1, and RPW 8.2 leading to HR or SHL and resistance. *BjNPR1* gene activates SAR to confer broad-spectrum resistance to powdery mildew of *B. juncea*. There is role of WRKY transcription factors and overexpression of R-genes like PMR, MLO, PEN, EDR, MAPK, MAPK 65-3, NPR1, PAD3, PAD4, ED5, SNARE, RLKs, and KDL (AtCEP1) to confer resistance to powdery mildew of crucifers. Powdery mildew resistance genes of *Arabidopsis* have been mapped on chromosomes II (RPW 1), III (RPW 2, 3, 7, 8), IV (RPW 4), and V (RPW 5, 6). The papain-type KDEL Cys EP CEP1 fulfills function in plant defense during late development of *Erysiphe cruciferarum* in close spatial association with the fungal haustorium and haustorial callose encasement.

Resistance to *Hyaloperonospora* in *Brassica* is controlled by dominant gene at seedling and adult plant stages. White rust R-loci on three *Arabidopsis* chromosomes are closely linked to downy mildew R-loci. Downy mildew R-genes are differentially expressed. The identified 129 genes of which 121 TDFs are up-regulated and eight are down-regulated. R-genes of *Arabidopsis* are mapped on five chromosomes in MRC-A, MRC-B, MRC-F, MRC-H, and MRC-J as RPP1 to RPP 27. Single transduction pathways used by the different RPP genes are complex and diverse. Stable and transient gene silencing approaches have been applied in downy mildew of crucifers. Strong dependence on genes EDS1 and NDR1 is governed by R-genes structural type rather than pathogens of *Arabidopsis*. Several clubroot-resistant (CR) genes have been identified and mapped on *B. rapa, B. oleracea, B. nigra, B. juncea, B. napus,* and *A. thaliana* chromosomes. *Brassica* species genome sequencing has allowed rapid identification of genome-wide QTLs and GWAS regions for R-genes of clubroot. With the use of transcriptomic and proteomic approaches, more than 1000 *Arabidopsis* genes have been identified which are differentially expressed in *Plasmodiophora brassicae*-infected roots. More than ten CR-genes in *B. rapa* have been identified with pathotypes specificity. The genetic analysis, genetic mapping, and fine mapping on chromosomes of *Brassica* species with molecular markers and pathotypes specificity have been studied in *B. rapa, B. oleracea, B. napus, B. napus* var. *napobrassicae*, and *A. thaliana*. The genes encoding TIR-NBS-LRR class proteins impart resistance against *P. brassicae*. The R-genes are present in clusters on the chromosomes of *Brassica* species. R-genes QTLs using dd RAD-seq has been mapped in *B. rapa* against clubroot pathotypes. The CR QTLs have been identified through GWAS analysis in the natural populations of *B. napus*. Bioinformatic analysis has been used to predict CR QTLs of *Brassica*. For the determination of loss of R-genes, linkage markers of CR in *Brassica* have been identified. Marker-assisted selection of CR-genes has reinforced the accumulation of varied R-genes in *Brassica*. The genetic origin of CR genes has been revealed in *Brassica*. Quantitative resistance to clubroot is mediated by trans-generational epigenetic variation in *Arabidopsis*. The use of SNP microarray, proteomic, and transcriptome approaches in *Brassica* has revealed multicomponent molecular mechanisms conferred by CR QTLs to *P. brassicae*. In *Brassica,* R-genes and QTLs have been identified with molecular markers and mapped on chromosomes to TuMV isolates. The retr01/retr02/retr03 genes have broad-spectrum resistance to TuMV. Alternative oxidase gene (*BjAOX1a*) in

Brassica enhances resistance to turnip mosaic virus. The techniques such as yeast two hybrid (Y2H), biomolecular fluorescence complement (BiFC), and co-immune precipitation (COIP) methods to analyze the resistance interaction between host and TuMV isolates have been employed.

6.4 Molecular Mechanisms of Host Resistance to Hemibiotrophs and Necrotrophs

The development in the science and technology through modern approaches like omics and genomic have facilitated to investigate molecular mechanisms of crucifers' host resistance against different biotic and abiotic stresses. The molecular mechanisms of host resistance have been investigated in *Brassica* crops using three different categories of host-pathosystem, viz., biotrophs, hemibiotrophs, and necrotrophs. The molecular mechanisms of R-genes resistance imparted by NBS-LRR are effective against biotrophs or hemibiotrophs but are not effective against necrotrophs. Salicylic acid (SA), jasmonic acid (JA), and abscisic acid (ABA) are the key modulators involved in signaling pathways during host defense. Jasmonic acid and ethylene-mediated signaling is predominantly involved in general defense against necrotrophs. NBS-LRR genes are intracellular receptors that can recognize the pathogen invasion by binding to pathogen effector proteins in host plant and trigger various defense signal transduction to yield hypersensitive response and inhibit pathogen development. The NBS domain binds and hydrolyzes ATP and GTP while LRR domain is liable for protein–protein interaction. NBS domains are of various motifs like P-loop and kinase 2 motif. NBS-LRR genes are grouped into TIR-BSS-LRR (TNL) and CC-NBS-LRR (ONL). Many NBS-LRR genes have been identified in *Brassica* species for the defense against different pathogens infecting *Brassica* crops. The first line of plant defense (pre-penetration) against fungal pathogen relies on cell surface-mediated defense signaling initiated by recognition of PAMPs, secondary (post-penetration) defense responses are often induced by the salicylic acid (SA) or jasmonic acid/ethylene (JA/ET) phytohormone signaling pathways.

The molecular mechanisms of non-host resistance to *Alternaria* in *Brassica* indicated activation of defense-related genes earlier in non-host plants than the host plants. Six QTLs governing resistance to *Alternaria brassicae* have been identified, five of them are population specific while one QTL is common between all the three mapping populations. In *Brassica* defense-related genes (PR proteins) including defensins (β-glucanases and chitinases) have been identified against Alternaria. Different defensin genes in *B. juncea* have been characterized for their structures, evolution, cellular location, and regulation in response to SA, JA, and Alternaria infection. Chitinase genes in *Brassica* have been identified to breed *Alternaria*-resistant cultivars. A single R-locus RCH-1 has been identified at the tip of chromosomes 4 in *Arabidopsis thaliana* ecotype. Eil-O against *Colletotrichum higginsianum.* The locus RCHz maps to an extensive cluster of R-loci known as MRC-J in the *Arabidopsis* ecotype Ws-O. In *B. oleracea*, two types of R-genes have been identified. A type with single dominant genes is effective against *Fusarium* race 1 and stable under high or low temperatures. B type is polygenic resistance and it is unstable under high temperature above 24 °C. Type A resistance to *FOC* 1 race 1 conferred by a dominant single gene, *FOC1* has been mapped and molecular markers have developed. R-genes and QTLs have been identified from *B. rapa, B. juncea, B. nigra,* and *B. napus* to provide resistance to *Leptosphaeria.* The LepR1 and lepR2 genes provide cotyledon and adult plant resistance in *B. napus* to *Leptosphaeria.* The background is useful markers to select the most suitable genotype as a recipient parent while developing *Brassica* cvs. with *Lm* resistance. The combined effect of R-genes and host genetic background generates stronger defense responses through induction of genes related to callose deposition, up-regulation of ROS, and SA and JA pathways.

Expression of *Cf9* and *Avr* 9 genes in *Brassica* induces resistance to *Leptosphaeria maculans* (*Lm*) R-genes and QTLs from *B. rapa, B. juncea*, and *B. napus* have been molecularly mapped with markers to provide resistance to *Lm*. Two R-genes, *LepR3* and *Rlm2* have been cloned along with seven corresponding avirulence (*Avr*) genes (*Avr lm1, AvrLm6, AvrLm3, AvrLm5/AvrLmJ1, AvrLm4-7, Avr Lm6*, and *AvrLm11*). Molecular basis of assessment of R-genes breakdown in *Brassica* cvs. can be predicted by frequencies of avirulence/virulence alleles in populations of *Lm*. In resistant and susceptible lines of *B. napus,* gene expression profile differs to confer resistance to *Lm*. Cystein-rich protein kinase genes and multi-environment stable six QTLs suitable for quantitative blackleg resistance in *B. napus* have been identified. Marker-associated mapping of QTLs in *Brassica* will help identification of desirable alleles and facilitate QTL introgression. For stable resistance to *Lm*, eight QTL regions for resistance on AO2, AO9, AO10, CO1, and CO9 has been identified in *B. napus* under diverse growing environments. Nine of the 18 blackleg R-genes from *Brassica* species correspond to genetically defined *Avr* genes from *Lm*. Mechanisms of resistance to light leaf spot (*Pyrenopeziza*) in *Brassica* is controlled by two R-genes. QTLs controlling partial resistance in *Brassica* at leaf and stem infection stages by *Sclerotinia* have been identified. Resistance to *Sclerotinia* operates at host tissue breakdown, pre-formed antifungal compounds, and production of phytoalexins. Resistance in *Brassica* to *Sclerotinia* is governed by a complex interplay of minor effect genes designated as quantitative disease resistance (QDR). The candidates defense-related genes have been mapped by genome-wide association combined with other omics approaches. Overexpression of genes like *BnNPR1, OsPGIP2, BnWRKY33*, and *BnMPK4* enhances resistance in *B. napus* to *Sclerotinia* whereas overexpression of *AtWRKY28* and AtWRKY75 genes in *Arabidopsis* enhanced resistance to oxalic acid and Sclerotinia. In *B. napus* with the microsatellite marker for genome-wide associations has identified three loci DSRC4, DSRC6, and DSRC8 for Sclerotinia stem rot resistance. SSR-based association mapping identified six markers loci associated to *Sclerotinia* resistance in A- and B-genomes of *Brassica*. Genome-wide association analysis of 84ILS recognized a large number of SNPs associated to *Sclerotinia* resistance in chromosomes AO3, AO6, and BO3.

In *B. napus*, one major R-QTL contributed by R53 on chromosome C5 and minor QTL contributed by Express 617 on chromosome C1 confer resistance to *Verticillium longisporum*. Black rot (*XCC*) resistance genes and QTLs in different *Brassica* species have been identified on different chromosomes to *Xanthomonas* races 1 and 4. Seven NBS-LRR encoding genes linked to black rot resistance in *B. oleracea* have been identified on four QTL regions and chromosomes CO1, CO3, CO6, CO7, and CO8, respectively. MicroRNAs (miR156, miR 167, miR169, and miR 90) play a role in *B. oleracea* resistance enhancement to *XCC*. R-gene 1 conferring resistance to race 1 has been identified in *B. carinata, B. juncea,* and *B. nigra*.

6.5 Biometabolomics of Disease Resistance to Biotrophs

Biometabolomics of disease resistance plays a significant role in induction of immunity signaling pathways and defense responses. With the perception of pathogens, signaling is initiated that result in the execution of a broad range of defense responses to stop the infection and pathogenesis of invading pathogens. Polar metabolites of *Brassica* plants including glucosinolates, polar indole metabolites, soluble compounds, and wall-bound phenolics are induced to confer resistance to *Albugo*. The phytoalexins cyclobrassinin, rapalexin, and brassilexin are inhibitory to *Albugo* zoospores release.

Strong and accelerated activation of defense-related transcriptional programs in *mlo2 mlo6 mlo12* triple mutants of Arabidopsis has been observed against powdery mildew with jasmonic acid/ethylene (JA/ET) signaling and biosynthesis of indole-derived antimicrobial compounds. Defense compounds like camalexin, indole glucosinolates, and indole cyanogenic metabolites provide full immunity in *mlo2 mlo6 mlo12* mutant to powdery mildew. Salicylic acid (SA) plays an important role in resistance to powdery mildew of *Arabidopsis* required for signaling of mutant's edr1, edr2, and edr4. Genes encoding camalexin biosynthesis for powdery mildew resistance have been determined. Pre-penetration resistance to powdery mildew of *Arabidopsis* operates through cuticle, and epicuticular wax when pathogen exerts pressure to penetrate cell wall, host activates, mechanosensors to activate immune signaling gene *CERK1* to contribute basal resistance. Resistance is expressed through papilla formation which contains callose, silicon, ROS, and phenolic compounds. Induction of salicylate, NPR1, PAD4, and EDS5 in *Arabidopsis* provides resistance to powdery mildew. Chito-oligomers in *Brassica* induce defense responses to powdery mildew. Salicylic acid-dependent transcription of *RPW* 8.1 and *RPW* 8.2 genes leads to accumulation of SA, and formation of SHL or HR in *Brassica* to confer powdery mildew resistance. *BjNPR1* transgenic *B. juncea* plants exhibit partial resistance to powdery mildew through activation of SAR. The expression of gene encoding AtMAP 65-3 in *A. thaliana* down-regulates SA signaling for powdery mildew infection. Loss of AtMAP 65-3 function in the map65-3 mutant reduces infection at the stage of leaf penetration. There is role of AtROP-regulated AtRLCK VI A3 in basal resistance to *Erysiphe cruciferarum*. The PAD 3 overexpressing plants accumulates higher camalexin and MiMick-resistant eyp

8301-3 mutant to powdery mildew. Powdery mildew resistance of atg 2 plants depends on SA signaling. Autophagy contributes to suppression of cell death and defense response to powdery mildew. JA/Et-induced defense responses are effective against virulent powdery mildew fungi if stimulated constitutively, artificially or systemically despite biotrophic nature of the interaction. Expression of WRKY TFs in *Brassica* enhances resistance to powdery mildew in edrl plants. SR 1 plays a critical role in powdery mildew resistance by regulating EIN 3 and NDR1 expression. Identification of defense-related TFs in the regulation indicates recognition of pathogen, response initiation, and defense execute an efficient immune output. In *Arabidopsis,* WRKY 70 is transcriptionally induced by SA signaling and mediates basal resistance to powdery mildew. Transcriptional regulation and expression of several genes in response to powdery mildew interaction with host and non-hosts is regulated by signaling molecules like SA, JA, and ET pathways differentially. The application of *Trichoderma harzianum* and cultural filtrate provides systemic resistance to powdery mildew in *Brassica.* Phytohormones abscisic acid (ABA) is involved in NHR against powdery mildew of *Brassica.* Two distinct pre-haustorial non-host defense mechanisms mediated by genes *PEN1, SNAP 33, VAMP 721, VAMP 722, PEN 2,* and *PEN* 3 have in common that they are subject to pathogen-induced cell polarization and mediate the induction of antimicrobial compounds. *Arabidopsis* NHR to powdery mildew is based on multicomponent and independently effective defense layers. *PEN* gene-mediated pre-invasion resistance and post-invasion immunity is controlled by *EDS1, PAD4,* and *SAG* 101 genes. Mutagenic resistance to powdery mildew is expressed by biochemical changes in the form of fatty acids, enzymes, and biosynthetic pathways.

Induction of several biomolecules in *Brassica* confers resistance to downy mildew. The synthesis of phytoalexins in *A. thaliana* is mediated by genes *PAD1* to *PAD4* to provide resistance against downy mildew. In resistant genotypes of *Brassica,* cell wall lignification provides resistance to downy mildew. Salicylic acid is important for defense against most *Hyaloperonospora parasitica* races as signal amplifier to induce anti-pathogenic molecules to provide downy mildew resistance. Downy mildew resistance signal pathways for the expression of R-genes are very complex.

R-genes pyramiding in *Brassica* for clubroot resistance strongly trigger multiple resistance pathways by hormonal signal transduction, secondary metabolites, transcription factors, and other process via multi-gene network controlled by SA, ROS, PCD, and R line *ZHE-226* gene. The pyramided line 618 R showed higher number of DEGs than the parental lines. PbBa 8.1 shared higher number of DEGs than CRb. SA and ROS signaling pathways are significant in pyramided and parental lines. There is significant indication of PR proteins, thaumatin, and lipid transfer proteins in the pyramided line as compared to parental lines. Lignin biosynthesis enhances resistance to *Plasmodiophora brassicae* at early stage of infection. Flavonoids act as defense compounds or antioxidants for clubroot. MET serve as the major energy source required for defense responses mediated by Rcr1 against clubroot. To activate defense response in *Brassica* to clubroot, there is induction of signal molecules like SA, BA, SABATH enzymes, JA, IAA, GA, and MeSA. At the later stages of

P. brassicae infection, the gene expression is different in *B. napus* cvs. differing in resistance for induction of hormones like CK, IAA, SA, JA, ABA, and ET. In general, gene expression was low in resistant cultivar and higher in susceptible ones. There is differential expression of hormones in leaves versus roots and formation of galls in resistant and susceptible cultivars indicating organ and host resistance specificity in production of hormones. However, SA signal transduction is up-regulated mainly in the R-genotypes as comparison to susceptible ones. The biometabolomic resistance to clubroot in *Brassica* is expressed by NBS-LRR proteins of PbBa 1 and CRb which recognize the elicitors released by *P. brassicae* and trigger defense signal transduction. The key pathways are SA, JA, MAPK, ET, Ca^{++}, and ROS. RBOH and phenylpropanoid biosynthesis activate ROS, SA signaling, PCD, HR, and antioxidant enzymes to enhance defense responses in *Brassica* to clubroot. Numerous genes for immune-related functional categories are associated with SA-mediated responses. JA-mediated responses are mostly inhibited especially in clubroot-resistant genotypes.

The biometabolomics of disease resistance plays a significant role in induction of immunity signaling pathways and defense responses. Following the perception of pathogens, signaling is initiated that ultimately results in the execution of a broad range of defense responses that stop the invading pathogens. In *Arabidopsis*, many genetic screens have been performed to identify genes that are important for defense signaling. Some of these screens were focused on signaling downstream of specific R proteins, while others were aimed to identify more general signal transduction components. The important plant proteins, EDS1 and NPR1, play crucial roles in the interaction with pathogens and are well studied in *Arabidopsis*. EDS1 and NPR1 are closely linked to the immunity-related hormone salicylic acid (SA) that is also key for resistance to several pathogens. Multi-omics approaches combining genomics, transcriptomics, proteomics, and biometabolomics using computational strategies will allow us to identify suitable mimicking molecules in the pathogen and/or host species that trigger stronger plant defense system during plant–pathogen interaction.

EDS1 (ENHANCED DISEASE SUSCEPTIBILITY 1) is a lipase-like protein that is required for the function of a subgroup of NLRs, those with a Toll–interleukin 1 receptor domain called *TIR-NLR*. In addition, EDS1 has a function in oxidative stress responses and as a positive regulator in basal resistance to virulent pathogens making the *eds1* mutant enhanced disease-susceptible (Wiermer et al. 2005). Two sequence-related signaling partners, PAD4 (PHYTOALEXIN-DEFICIENT 4) and SAG101 (SENESCENCE-ASSOCIATED GENE 101), form complexes with EDS1 and have signaling functions in the cytoplasm and nucleus. The NLR helper protein NRG1 also resides in this complex and plays a role in cell death signaling. Interestingly, EDS1 was found to associate with AvrRps4, a *Pseudomonas* effector, and to interact with TNLs, like RPS4 that mediate recognition of AvrRps4. It was hypothesized that EDS1, being important for basal resistance, could be a target for pathogen effectors and thus serve as a molecular bridge to mediate recognition by R proteins that guard the EDS1 protein in addition to its signaling functions. This hypothesis is currently debated and under further investigation. The *eds1* mutant was originally identified in a screen for loss of resistance to downy mildew (Parker et al.

1996). EDS1 was shown to be required for the function of *TIR-NLR* genes such as *RPP1*, *RPP4*, *RPP5*, and *WRR4*. Virulent strains of *H. parasitica* and *Albugo* are also able to cause more severe infections on *eds1* mutant plants because of the reduced level of basal resistance. *Arabidopsis eds1* mutants also enhance susceptibility to a number of pathogens including *H. parasitica*. The lowered basal immunity is associated with reduced production of the defense hormone SA that is essential for defense against biotrophic pathogens. EDS1 positively regulates SA accumulation, acting upstream of SA, but the *EDS1* gene is also activated at the transcriptional level by SA. EDS1 thus seems to be a part of an SA-associated positive feedback loop of plant defense (Lapin et al. 2019).

SA is not only crucial for resistance to downy mildew, but also affects basal and R-gene-mediated resistance to many other pathogens of crucifers. Exogenous application of the hormone SA is sufficient to trigger efficient defense responses to a broad range of pathogens, and in particular to downy mildew and hemibiotrophic species. SA (2-hydroxybenzoic acid) is synthesized via two main routes, the isochorismate (IC) and phenylalanine ammonia lyase (PAL) route. The *Arabidopsis* IC route goes via two IC synthases (ICS) and downstream steps that differ from the canonical pathway defined in bacteria (Torrens-Spence et al. 2019). The *ics1* (*sid2*) mutant has received the majority of experimental attention to date. *Ics1* mutants exhibit 90% lower levels of SA, and levels are ~95% lower in the *ics1 ics2* double mutant following powdery mildew infection. The *ICS1* gene is induced by MAMPs and by NLR-mediated pathogen recognition leading to high local levels of SA that activate defense. Besides local defenses, SA plays an important role, together with pipecolic acid, in systemic acquired resistance (SAR) that enhances immunity of distant plant tissues to protect against future infections. The export protein EDS5 has been implicated in synthesis of pipecolic acid, in addition to its well-known role in SA biosynthesis, revealing a surprising instance of immune convergence (Rekhter et al. 2019).

Downstream of SA, the central NPR1 (NONEXPRESSOR of PR GENES 1) protein plays a crucial role in the transcriptional activation of SA-induced defense in *Arabidopsis*. Mutation of *NPR1* leads to loss of basal resistance to (hemi-) biotrophic pathogens and of SAR. NPR1 proteins form a multimeric complex in the plant cell cytoplasm. Upon triggering of the plant immune system SA accumulates and subsequently elicits the thioredoxin-mediated reduction of a cysteine residue in NPR1. This results in the release of monomeric NPR1 that can travel into the nucleus where it interacts with TGA transcription factors to activate transcription of defense genes, like pathogenesis-related (*PR*) genes. NPR1 and the two paralogous *Arabidopsis* proteins NPR3 and NPR4 have been shown to bind SA and could thus constitute the SA immune receptors (Ding et al. 2018; Innes 2018).

The RPP proteins vary significantly in their requirements for functionality of these signaling components. For example, resistance mediated by the TIR-NLR protein RPP5 is strongly compromised by mutations that nullify EDS1, NPR1, or SA biosynthesis. On the other hand, the CC-NLR proteins RPP7 and RPP13 retain almost full functionality in these backgrounds. These differences underscore the complexity of ETI regulation. It is well known that many proteins are specifically produced during plant defense. In particular, many secreted and vacuolar proteins have been classified as pathogenesis-related proteins. Also in *Arabidopsis* many PR proteins have been identified, e.g., PR1, PR2, and PR5 that were found in apoplastic fluid of *Arabidopsis* treated with 2,6-dichloroisonicotinic acid (INA), which is a mimic of SA. Several of these PR proteins have anti-pathogenic activities, while others contribute to resistance. One of the most used marker genes of SA-induced plant defense is *PR1*. Plants have multiple *PR1* genes encoding different protein variants with different activities. Overexpression of several *PR1* genes in different plant species also increased their resistance to pathogens. PR1 has been shown to bind sterols, indicative of defensive mode of action that is based on restriction of this important nutrient for pathogens (Gamir et al. 2017). The PR2 proteins are ß-1, 3-beta-glucanases that are thought to attack the cell wall of invading fungal pathogens. Several cases of enhanced resistance of plants overexpressing *PR2* to pathogens are well known.

Other non-proteinaceous compounds that are important for resistance to pathogens in *Arabidopsis* are small secondary metabolites. Important anti-pathogenic compounds produced by *Arabidopsis* are, for instance, indole glucosinolates (iGS) and the phytoalexin camalexin. Mutant plants that no longer produce either iGS or camalexin are only slightly more susceptible to *P. brassicae* and other pathogens. A very strong gain in susceptibility was observed in the double mutant *cyp79b2 cyp79b3* that is blocked in the production of indole-3-aldoxime, which is a common precursor for iGS and camalexin. Camalexin appears to contribute to resistance to *Arabidopsis* downy mildew, as several *phytoalexin-deficient* (*pad*) mutants are slightly more susceptible to *H. parasitica*.

A combination of metabolomics with quantitative genetics was used to discover the potential role of gluconasturtiin in the *B. napus* resistance response against clubroot and the underlying QTL controlling the trait on chromosome C03 and C09. Gluconasturtiin is a type of glucosinolate compound associated with the biotic resistance responses of *Brassica* species. With the use of multi-omics supplemented with functional studies key resistance pathways involved in the *Brassica* host-pathosystem have been revealed (Singh 2017; Wagner et al. 2019).

6.6 Biometabolomics of Disease Resistance to Hemibiotrophs and Necrotrophs

The mechanisms of non-host resistance (NHR) in plants to *Alternaria* operate at pre-invasion stage mainly by both pre-formed and inducible defense responses. Pre-formed defenses involved structural features like trichomes, chemical compounds (antimicrobial enzymes, secondary metabolites), cuticular waxes, formation of phytoalexins, and detoxification of toxins. Defense responses mainly involve ROS accumulation and defense-induced cell death where fungal proliferation is restricted by rapid hypersensitive reaction (HR). Resistance in *Brassica* to *Alternaria* depends on phytoalexins, detoxification, and simultaneous phytoalexins (camalexin) elicitation. MAPK4 gene activates defense in *Brassica* to *Alternaria*.

The biomolecules like phenols, polyphenol oxidase, peroxidase, PAL, and catalase are higher in Alternaria-tolerant genotypes of *Brassica*. *GLIP1* gene in association with ethylene signaling is critical factor in host resistance to *Alternaria*. β-amino butyric acid (BABA) and Zeatin up-regulated host immunity via elevation of MAPK 5 which antagonizes the effect of *A. brassicae*. Resistant *Brassica* genotypes accumulate phytoalexins like camalexin, brassinin, allyl (Al TIC), and benzyl (Bz ITC) isothiocynates to inhibit *Alternaria* infection. Foliar sprays of calcium compounds sequester the organic acids at the site of *Alternaria* infection to boost calcium content of plants to provide resistance. The MAPK signaling cascade leading to biosynthesis of camalexin triggers defense responses in *B. rapa* to *Alternaria*. The up-regulation of genes such as WARKY, peroxidase, p450 oxidase, and chitinase in *Brassica* mediated with JA and SA pathways enhances defense response to *Alternaria*. Differential defense signaling pathways of JA and SA operates in different species of *Brassica* to *Alternaria*.

Mutation of lysine histidine transporter 1 (lhr1) in *Arabidopsis* reduces contents of glutamine, alanine, and proline to enhance resistance against *Colletotrichum* mediated with SA signaling and increased callose deposition and PR1. Defense responses in *Arabidopsis* to *Colletotrichum* are activated through SA signaling at early stage of interaction and by JA signaling subsequently.

The TF homologs associated with JA (MYC3, JAM1), ABA (AKS2, ABF3), and calcium (CAMTA5, CAMTA 3/SR1) signaling regulates defense response processes in resistant host of *Brassica* to *Leptosphaeria*. At the site of host–pathogen interface, DEGs are down-regulated in resistant plants in association with response to cytokinin, photosynthesis, amino acid biosynthesis, metabolisms, and unsaturated fatty acid biosynthesis. Defense response in resistant cvs. is faster activated by complex network of WRKY, BZIP, bHLH, CAMTA, and host shock TFs. Calcium signaling is regulated for resistance and basal defense against *L. maculans*. There is differential response of resistant and susceptible cvs. of *Brassica* for expression, biosynthesis, and signaling pathways of different hormones in response to *L. maculans*. Genes associated with surface receptors and MAPK activities are up-regulated in *Leptosphaeria*-resistant hosts but not in susceptible ones.

NADPH oxidases are critical components of resistance to *R. solani* AG8 in *Arabidopsis*. There is a role of glucosinolates in the defense of *Brassica* against pathogens. Defense response against *Sclerotinia* are regulated by a complex network of hormonal and signaling pathways via JA, ET, abscisic acid in promoting the production of GSLs and other metabolites working together to enhance host resistance. Several pathways and biological processes are associated with the defense against *Sclerotinia*, such as oxidative burst, cell wall enforcement or modification, secondary metabolisms and calcium binding, signaling, antioxidation, glutathione metabolism, glucosinolate metabolisms, and biosynthesis of lignin. In *Brassica*, resistance is expressed by quick accumulation of ROS via activating Ca^{2+} signaling and suppression of oxalic acid generation in *Sclerotinia*. Overexpression of AtGDSL1 in *B. napus* enhances resistance to *Sclerotinia* with increased ROS and SA levels and reduced JA level. In *B. napus*, mitogen SA-activated protein kinase gene *BnaMPK3* plays an important role in the defense against *Sclerotinia*. Host resistance to *Sclerotinia* involves various biological process like redox homeostasis; signaling by lipids, calcium, SA, JA, ET, and ABA; histone, DNA methylation-mediated transcription regulation, defensing-like proteins, cyanate lyase, miRNA, PTGs, and glycolate oxidase genes encoding crucial enzymes in photorespiration. The crosstalk of signaling molecules, viz., JA, SA, ET, ABA, Auxin, NO, ROSs, and OA is very complex to induce defense against *Sclerotinia* on *Brassica* and it varies with interaction of pathogen with different genotypes. Differential alternative slicing genes and isoform regulates resistance in *Brassica* to *Sclerotinia*.

The phenyl propanoids are involved in defense responses of *Arabidopsis* against *Verticillium*. Tryptophan-derived secondary metabolites like camalexin provides defense against *Verticillium* to *Brassica*. The application of BABA induces biomolecules to confer resistance in *Brassica* to *Verticillium*. The increased level of proteins like annexin, AIG2, ferredoxin, ferredoxin–NADP, and epithiospecifier like in resistant cvs. of *Brassica* provides resistance to *Xanthomonas*. The role of several proteins, in resistant and susceptible cvs. of *Brassica* has been characterized with several signal molecules against *Xanthomonas* and responses indicated up- and down-regulated levels. *CHI-B4* like gene encoding an endochitinase has been identified in *Arabidopsis* to confer resistance to *Xanthomonas*. The overexpression of *BoCHB4* in *Brassica* confers resistance to *Xanthomonas*.

Biometabolomics and biomolecular mechanisms of host resistance are expressed in the form of different metabolites and biomolecules involved in the induction of biochemical changes and pathways in the host to confer resistance against biotic and abiotic stresses. Biometabolomics and systems biology determines a variety of molecular targets for breeding disease resistance cvs. in *Brassica* crops. Transcriptomics and proteomics help to discover novel biochemical pathways involved in conferring resistance to virulent pathogens and reveal mechanisms of biometabolomics host resistance. With the use of multi-omics supplemented with functional studies, key resistance pathways involved in *Brassica*-hemibiotrophs pathosystem have been discovered, where the induction of JA signaling, elevated ROS with reprogramming of the phenyl propanoid pathways to impart resistance. The hormone SA plays a major role in plant resistance to hemibiotrophs. In *A. thaliana*, SA synthesis occurs in plastids via isochorismate synthesis (ICS1 or SID2) and is triggered by pathogens. SA accumulated in the cytoplasm can be converted to SA glycoside (SAG) which is stored in the vacuole and hydrolyzed back to SA when required. Elevated levels of total SA (free SA plus SAG) are correlated with induction of defense gene expression and enhanced host resistance. Different from SA, JA, and ET accumulate mainly in response to necrotrophs. Loss of susceptibility is frequently caused by deregulation of SA- or JA-dependent host defense signaling by the impairment of cellular rearrangement, or by the limitation of nutrient supply for the pathogens. There is role of MAPK 4 gene in host defense against *Alternaria*. MAP kinase signaling cascade is known to be involved in synthesis of camalexin which contributes to disease resistance against necrotrophs. The lifestyle transition of the pathogens from biotrophics to necrotrophics is regulated by signaling molecules from SA to JA and ET.

6.7 Techniques to Study Molecular Mechanisms of Host Resistance

To understand the molecular mechanisms of crucifers' host resistance, useful protocols have been developed to analyze the host–pathogen interactions through omics approaches like genomics, proteomics, transcriptomics, and bioinformatics. Novel R-genes have been identified, molecularly characterized, and have been fine mapped on the chromosomes of different *Brassica* species. Techniques have been developed to introgress high yielding and disease resistance genes through traditional as well as transgenic approaches. Protocols for molecular characterization of R-genes in *Brassica* species, DNA extraction, purification, and quantification, construction of a linkage map and mapping of R-loci and proteome analysis of *Brassica-Albugo* pathosystem have been developed. The procedures for analysis of molecular and biochemical mechanisms of resistance in *Brassica* to *Alternaria* and molecular characterization of *Alternaria* genes showing fungicidal resistance have been described. Genome-wide identification of defensin genes in *B. juncea* and *Camelina sativa*; Expression of defensin genes analysis; and distribution of chitinase genes identification against *Alternaria*, techniques are available. *Brassica-Erysiphe* molecular techniques include identification of molecular markers linked powdery mildew R-genes and DNA sequence analysis of *Erysiphe* isolates. In *Brassica–Hyaloperonospora* interactions, assessment of small RNA role; R-genes overexpression; and c-DNA-AFLP analysis to reveal gene expression have been developed. *Brassica-Leptosphaeria* pathosystem has revealed, phylogenetic relationship of R-loci; transcriptome analysis; cloning and transformation of *Leptosphaeria* avirulence genes; identification of QTLs; molecular mapping of R-genes, and identification of NBS encoding genes in *Brassica* species. Molecular techniques developed with the *Brassica–Plasmodiophora* system are mapping of R-genes and QTLs; SNP array, mapping, population structure, and linkage disequilibrium analysis; genome-wide association study; and the methodology for transcriptome analysis of *Brassica–Sclerotinia* pathosystem has been developed. The protocols for TuMV detection, preservation, inoculation, and identification by ELISA have been developed. In *Brassica* species, RNA sequencing and NBS domain, and NBS-associated conserved domains protocols have been developed.

In science and technology to get reliable and widely acceptable results, one of the basic impacts is use of standard, repeatable, and high-throughput methodology, procedure, protocols, and techniques. The techniques included in this chapter have been developed by the experts of molecular biology who have long experience of working with crucifers' host-pathosystem. The development of molecular techniques was initiated during 1990 after the sequencing of *Arabidopsis thaliana* genome using *Arabidopsis* host-pathosystem. With the use of this system, rapid progress has been made on molecular mechanisms of host resistance through analysis of crucifers' host–pathogen genomics, proteomics, and transcriptomics. These techniques can be used for demonstrating biomolecular events by the teachers for the graduate students and by research scientists to pursue their research programs to get widely acceptable results on molecular mechanisms of host resistance. Only selected techniques required to study molecular mechanisms of host resistance have been described. For more detailed procedures, readers are advised to consult cross references given in each technique.

6.8 Exploitation of Molecular Approaches for Breeding Disease Resistance

6.8.1 Through Introgression Breeding Approaches

The introgression of genes for disease resistance between species has been widely studied in *Brassica*. The utilization of the B-genome as a source of resistance against *Leptosphaeria maculans* (blackleg) from diploid and tetraploid species has been studied. For instance, chromosome B4 from *B. nigra* was introgressed into rapeseed variety "Darmor," which exhibited high resistance with the addition of this chromosome. Similarly, the high resistance from *B. juncea* was obtained in selected recombinant lines of *B. napus* carrying a resistance gene located on chromosome B8. Alike, successfully introgressed a B-genome chromosome from *B. carinata* to *B. napus*, with plants carrying this chromosome showing variation in traits such as blackleg resistance, days to flowering, days of maturity, and fatty acid composition. Transfer of black rot resistance from *B. carinata* to *B. oleracea*, resistance to Alternaria leaf blight caused by *Alternaria brassicae* from *B. hirta* to *B. juncea* and transfer of powdery mildew resistance from *B. carinata* to *B. oleracea* through embryo rescue followed by backcrossing to *B. oleracea has been* successfully demonstrated (Saharan et al. 2019).

Besides utilizing conventional methods of selection and sexual reproduction, biotechnological techniques involving tissue culture and genetic transformation-based methods have revolutionized the search and development of economically promising disease-resistant crops. These tools have enabled the researchers to combine the traits from distantly related *Brassica* species for the development of superior *Alternaria* blight tolerant varieties of *Brassica* (Aneja et al. 2012; Aneja and Agnihotri 2013; Saharan et al. 2003, 2016).

6.8.2 In Vitro Embryo Rescue

Embryo rescue, documented for the first time in the eighteenth century, is one of the most successful techniques for the production of interspecific and intergeneric hybrids from naturally incompatible crosses that do not result in the production of viable offspring. The technique involves excision of the ovary/ovule/embryo resulting from an interspecific cross, its culturing and maturing on suitable media for the production of a hybrid plant with desired traits. The technique has proven very efficient in transferring *Alternaria* blight tolerance in oilseed *Brassicas*. In vitro ovary culture was used to transfer the resistance against *Alternaria* blight from *B. tourneforti* to *B. juncea* cv. RH 30 (Yadav et al. 1991). Agnihotri et al. (1991)

developed intergeneric hybrids of *B. campestris* and *B. spinenscens* through sequential ovary, ovule, and embryo culture. The hybrid plants were then multiplied using somatic embryogenesis.

The resistance trait from *Sinapis alba* cv. Carine was transferred to *B. napus* cv. Brutor using in vitro fertilized ovary culture (Chevre et al. 1994). Intergeneric hybrids with *Alternaria* blight resistance were developed between *Erucastrum cardaminoides* and *B. oleracea* var. *Alboglabra* using sequential ovary and ovule culture (Mohanty et al. 2009). Gupta et al. (2010) also used interspecific hybridization in combination with in vitro ovule culture to incorporate high tolerance to *Alternaria* blight and white rust from *B. carinata* cv. Kiran to low erucic acid TERl (OE) M21 lines; hybrids with high tolerance to *Alternaria* blight and white rust contained negligible amount of erucic acid (1.3%) and high oleic acid (42.5%).

6.8.3 Somatic Hybridization

Somatic hybridization refers to the enzymatic removal of the cell wall and isolation of protoplasts which are then fused together to produce hybrids. Due to the absence of the cell wall, there occurs a nonspecific fusion between the two protoplasts without any barrier to intergeneric/interspecific crosses. It also bypasses both the pre- and post-fertilization barriers. This technique has been utilized effectively in *Brassicas*, but the studies are limited to *B. napus* and *B. oleracea.* Efforts have been made to transfer *Alternaria*-resistant trait from *Moricandia arvensis* to *B. oleracea* (Mohanty et al. 2009) *B. carinata* to *B. juncea* (Gupta et al. 2010), and from *S. alba* to *B. napus* (Primard et al. 1988). Sharma and Singh (1992) attempted the transfer of *Alternaria* resistance trait from *B. carinata* to *B. juncea.* Leaf mesophyll protoplasts from *M. arvensis* and *B. napus* were hybridized to produce disease-resistant hybrid plants (O'Neill et al. 1996). Jourdan and Salazar (1993) resynthesized *B. carinata* by protoplast fusion between *B. nigra* and *B. oleracea,* and the hybrids thus obtained were fertile and grew into robust plants. Hybridization between *S. alba* and *B. oleracea* and between *Camelina sativa* and *B. oleracea* has also shown to produce resistant hybrids (Hansen and Earle 1997; Hansen 1998). Sigareva and Earle (1999) developed somatic hybrids between *S. alba* × *B. oleracea* by protoplast fusion followed by embryo rescue and recovered four highly resistant hybrid progenies after repeated back crosses.

6.8.4 Somaclonal Variations

These are regarded as the impulsive, inheritable changes in gene expression which often appear under in vitro conditions especially in tissue culture experiments. Sharma and Singh (1995) used somaclonal variations for incorporation of disease resistance/tolerance against *Alternaria* blight. Resistance in Alternaria blight susceptible *B. juncea* varieties was achieved by inducing variations through mutagenesis (exposing seeds to gamma radiations) (Verma and Rai 1980), and by treating the microspores derived embryos with chemical mutagens (Agnihotri et al. 2009) including ethyl methane-sulfonate (EMS) and ethyl nitrosourea (ENU).

6.8.5 Genetic Transformation

Production of transgenic plants through genetic transformation forms another biotechnological approach to integrate the disease resistance genes from resistant/tolerant genotypes to the economically important susceptible ones. The genetic transformations in *B. juncea* have been reported to achieve delayed onset of the *Alternaria* blight. Kanrar et al. (2002) isolated cDNA sequence for "hevein-rubber tree lectin" from *Hevea brasiliensis* and transferred it into *B. juncea* cv. RLM-198 through binary vector *pBinAR*. The transgenic plants thus produced supported delayed onset of *Alternaria* blight with considerably low disease intensity. Mondal et al. (2003) identified the role of "chitinase" in plant defense responses and transformed *B. juncea* (cv. RLMI98) with chitinase gene tagged with overexpressing promoter 35S *CaMV*. The transformed plants resulted in delayed disease onset and about 12–56% reduction in fungal colony size. Osmotin protein, known for its role in signal transduction, when transferred to *B. juncea* also showed tolerance to *A. brassicae* in transgenic plants (Taj et al. 2004). In another study, transgenic Indian mustard expressing "*class I glucanase*" gene under the control of *CaMV* 35S promoter was developed (Mondal et al. 2007). For undertaking the genetic transformation, cDNA encoding "*class I glucanase*" gene was isolated from tomato and cloned into *Bam*HI-*Sal I* restriction site of binary vector, *pBinAR* flanking *CaMV* 35S constitutive promoter and an *ocs* terminator region. The resulting recombinant $pBinG_B$ clone was transferred into *Agrobacterium tumefaciens* strain GV 2260, which was further used to transform *B. juncea* (cv. RLM 198). A study by Verma et al. (2012) reported the introgression of cysteine-rich antimicrobial peptide, *PmAMP1* from *Pinus monticola* into *B. napus,* thus conferring enhanced protection against *A. brassicae, Leptosphaeria maculans,* and *Sclerotinia sclerotiorum.* Combined expression of type I ribosome inactivating protein (RIP; *AAA32951*) and Barley "*class II chitinase*" gene in *B. juncea* not only reduced the growth of fungal hyphae by 44%, but also affected the onset of disease (Chhikara et al. 2012). Efforts have also been made to study the signal transduction pathways in *Brassica* species. Taj et al. (2011) reported interaction of MAPK3 and Lox genes during pathogenesis of *A. brassicae,* which is suggested to play a role in biosynthesis of jasmonic acid and jasmonic acid-induced expression of defense genes in *B. juncea.*

6.8.6 Molecular Marker-Assisted Breeding

As the efficient management of plant diseases relies on the accurate identification of the plant pathogen, knowledge of the variability in the pathogen population, both at the pathological and genetic level, is very important. Characterization of variability in the pathogens including *Alternaria* spp. has been done earlier using traditional methods of morphological markers (Meena et al. 2012), biochemical tests (Vishwanath and Kolte 1997), and cultural and pathogenic assays (Meena et al. 2012; Singh et al. 2011). These markers differentiate the pathogens on the basis of their growth behavior, physiological characters, and microscopic appearance. However, these criteria are greatly influenced by different factors including age and quality of the inoculum, age and type of host, culture media used, and incubation conditions. Also, these techniques are more laborious, time consuming, and often give non-reproducible results. To overcome these limitations of traditional markers, detection of genetic variability through DNA fingerprinting technology by the use of molecular markers has come to the aid of plant breeders and pathologists. The main advantage of using these fingerprinting techniques is that, they do not require any prior knowledge or availability of variable sequences in the database, as they deal with the genome-wide biodiversity.

Various molecular markers being used for assessing variability include, internal transcribed spacer regions (ITS), restriction fragment length polymorphism (RFLP), randomly amplified polymorphic DNA (RAPD), amplified fragment length polymorphism (AFLP), intersimple sequence repeats (ISSRs), microsatellite, sequence-tagged sites (STS), and single nucleotide polymorphism (SNPs). Due to the presumed dependency on traditional characterization, only a few studies are reported on molecular characterization of *Alternaria* species pathogenic to oilseed Brassicas. The ITS regions (also called the conserved regions) in the fungal genome are considered the most popular locus for DNA-based mycological studies at species level. Chou and Wu (2002) and Berbee et al. (2003) studied the ITS region of rDNA of *A. brassicae* and *A. brassicicola* to evaluate the phylogeny of the pathogen. Molecular techniques like RAPD, that do not require any prior knowledge of the DNA sequences, have been efficiently used to analyze the genetic variations in *Alternaria* species (Cooke et al. 1998; Sharma and Tewari 1995, 1998; Gherbawy 2005). However, due to the limitation of reproducibility in RAPD-generated data, the assaying of genetic variability in *Alternaria* species has shifted to more sensitive techniques like AFLP (Bock et al. 2002) and microsatellite markers. Twelve polymorphic microsatellite markers were isolated from *A. brassicicola*. The number of alleles detected in 12 loci range from 2 to 10 (mean 3.5). Cross-species amplification shows that the designated primers are specific to the pathogen (Avenot et al. 2005). Though these molecular tools have revealed huge amount of genetic variations in *Alternaria* at the sub-generic as well as subspecies level, this field still remains largely unexplored, and demands an extensive study into the variability at different geographic levels among different hosts.

6.8.7 Use of Systemic Resistance Induction

Induced host resistance refers to activation of resistance in otherwise susceptible plants, through biotic as well as abiotic agents (Stitcher et al. 1997; Kessman et al. 1994), without changing their basic genetic make-up. Foliar application of β-amino-butyric acid induces resistance against *A. brassicae* in *B. carinata* (Chavan et al.

2013). *GLIP1* protein in association with ethylene signaling is a critical component in plant resistance to *A. brassicicola* (Oh et al. 2005). The resistance/defense-related genes in the vulnerable plants can be activated by inoculating the plant either by an avirulent form of the pathogen, or by limited inoculation with the pathogen (Deverall 1995). The infecting avirulent pathogen triggers natural defense responses in the plant through the release of the elicitors, which then result in the expression of novel anti-pathogenic proteins. In order to harp the benefits of induced host resistance and build up a stable, long-term resistance mechanism in the host plant against the pathogen, there is a need to identify the pathogen and understand its behavior under diversified conditions. In case of Alternaria blight, considerable morphological and pathological variations have been observed in *A. brassicae* under different environmental conditions (Kaur et al. 2007; Singh et al. 2007; Patni et al. 2005; Meena et al. 2005, 2012; Ansari et al. 1989; Awasthi and Kolte 1989; Saharan et al. 2015, 2016). Vishwanath et al. (1999) reported the induction of resistance in susceptible *B. juncea* cv.PR-15 against extremely virulent *A. brassicae* isolate A (AbA) and moderately virulent isolate C (AbC) from *B. carinata* cv. PPCS-1 by inoculating with an avirulent isolate D (AbD) isolated from same *B. carinata* cultivar. About 60% reduction in the disease severity was observed against AbA and AbC by the induction of resistance; However, due to the limited studies in molecular characterization of the pathogen, the clear outcome is still awaited.

Variations in *A. brassicae* isolates have been recorded in traits including colony morphology, colony diameter ranging from 32 to 68 mm, surface texture from velvety to wooly, color from cream to dark olive green, and sporulation from very less to intense. The host–pathogen interaction studies using four varieties of *B. juncea* and two of *B. rapa* have shown pathological variation among the isolates that differ in their aggressiveness from avirulent (few) to highly virulent (Aneja et al. 2012; Aneja and Agnihotri 2013). A thorough knowledge of diversity in the pathogen will not only help in characterizing and analyzing it phylogenetically, but also in inducing durable resistance against *Alternaria* blight, which can act as an environment friendly substitute to the traditional use of fungicides.

6.8.8 Use of Genetic Engineering

The development of *Alternaria*-resistant genetically modified plants (GM) is an excellent approach to manage Alternaria diseases of crucifers. Resistant genes from other plants or bacteria encoding enzymes like chitinase or glucanase can be introduced in crucifers. Both chitinase and glucanase enzymes breakdown chitin and glucan, which are essential components of fungal cell wall. Introduction of plant genes to enhance innate plant defense mechanisms like activating phytoalexins, proteinase inhibitors, and detoxification of phytotoxins is another useful way of developing resistant varieties. Invoking hypersensitive reaction when a fungal pathogen is attacking can also be an innovative approach.

6.9 Transferring R-Genes by Molecular Marker

To develop resistant *Brassica* cultivars, several resistant sources such as European turnips, Chinese cabbage, and rutabaga cultivars are available and the resistance in these sources is dominant, which makes it easier to transfer clubroot resistance through interspecific and intraspecific hybridization. Rutabaga cultivars have been identified as clubroot-resistant sources. However, the genetics of clubroot resistance in rutabaga is complicated, so it is difficult to develop molecular markers that are closely linked to the dominant clubroot resistance genes (Ayers and Lelacheur 1972).

In Chinese cabbage, the dominant clubroot resistance from European turnips has been successfully used to develop clubroot-resistant Chinese cabbage. Since the gene mapping has been performed extensively in Chinese cabbage, molecular markers closely linked to clubroot resistance loci that are used in gene mapping can be easily selected to transfer clubroot resistance genes in the development of Chinese cabbage cultivars through molecular marker-assisted selection. Since the amphidiploid *B. napus* has a very complex genome, most of the molecular markers developed in *B. rapa* may not be polymorphic and cannot be directly used in *B. napus*. Additional efforts are required to develop molecular markers in *B. napus* when the mapped clubroot resistance loci in Chinese cabbage are transferred into *B. napus*. Currently, most clubroot resistance genes in European turnips have not been intensively investigated, and mapping and cloning of these clubroot resistance genes in European turnips will allow using these genes effectively and efficiently in *Brassica* breeding (Hirani and Li 2015).

6.10 Transfer of R-Genes by Interspecific Hybridization

A dominant clubroot resistance gene from *B. rapa* (Chinese cabbage) was fine mapped, and molecular markers were developed in Chinese cabbage that could be used for marker-assisted selection (MAS) in other *Brassica* crops. To transfer this clubroot resistance gene to *B. napus* (canola), an interspecific hybridization was made between *B. napus* (canola) and *B. rapa* (Chinese cabbage). Subsequently, the F_1 was backcrossed to the canola recurrent parent for three generations to produce BC_1, BC_2, and BC_3 progenies. Using these populations, simple sequence repeat (SSR) markers flanking the clubroot resistance gene were used to perform MAS in canola. These molecular markers were then evaluated in 13 different canola and rapeseed quality genotypes in *B. napus* and *B. rapa*. These markers exhibited high reliability in identifying clubroot resistance in this diverse set of *Brassica* genotypes. Clubroot resistance also co-segregated with the SSR markers flanking the clubroot resistance gene in the BC_3 and BC_3S1. The segregation ratio of resistant and susceptible individuals in the BC_3 supported the expected 1:1 ratio for the segregation of a single Mendelian gene. BC_3S1 families with homozygous clubroot resistance were developed during this process and will be valuable sources of clubroot resistance in *B. napus* breeding activities (Hirani et al. 2016).

6.11 Transfer of R-Genes by Distant Hybridization

The three clubroot-resistant Chinese cabbages (1003, 1007, and 1008) were used to transfer the clubroot resistance genes to *B. napus* by distant hybridization combined with embryo rescue. Three methods including morphological identification, cytology identification, and molecular marker-assisted selection were used to determine hybrid authenticity, and 0, 2, and 4 false hybrids were identified by these three methods. In total, 297 true hybrids were identified. Clubroot resistance markers and artificial inoculation were utilized to determine the source of clubroot resistance in the true hybrids. Two simple sequence repeat (SSR) and two intron polymorphic (IP) markers linked to clubroot resistance genes were identified. The clubroot resistance genes 1007 and 1008 were mapped to A03. At last, 159 clubroot-resistant hybrids were obtained by clubroot resistance markers and artificial inoculation. Four molecular markers associated with clubroot resistance were identified, which were located in a region from 23.77 to 25.56 Mb on A03 of *B. rapa*. In the adjacent area of this region, two clubroot-resistant genes *CRa* (Ueno et al. 2012) and *CRb* (Hatakeyama et al. 2017) have been cloned, in which both genes encode the TIR-NBS-LRR (TNL) protein. Through analyzing the genes structure of the region surrounding the *CRa* and *CRb*, it was found that six genes *Bra012540*, *Bra012541*, *Bra019409*, *Bra019410*, *Bra019412,* and *Bra019413* have the structure of TNL (Liu et al. 2018).

6.12 Transfer of R-Genes from European Clubroot Differential (ECD) Set

European turnips are the most resistant sources and their resistance genes have been introduced into other crops such as oilseed rape (*Brassica napus* L.), Chinese cabbage, and other *Brassica* vegetables. The European clubroot differential (ECD) set contains four turnip accessions (ECD1–4). These ECD turnips exhibit high levels of resistance to clubroot when they are tested under controlled environmental conditions with Canadian field isolates. Gene mapping of the clubroot resistance genes in ECD1–4 were performed and three independent dominant resistance loci were identified. Two resistance loci were mapped on chromosome A03, and the third on chromosome A08. Each ECD turnip accession contained two of these three resistance loci. Some resistance loci were homozygous in ECD accessions while others showed heterozygosity based on the segregation of clubroot resistance in 20 BC_1 families derived from ECD1 to 4. Molecular markers were developed linked to each clubroot resistance loci for the resistance gene introgression in different germplasm (Hirani et al. 2018).

6.13 Transferring of Powdery Mildew R-Genes Through Embryo Rescue

Interspecific hybrid plants and backcross 1 (BC1) progeny were produced through sexual crosses, and embryo rescue between *Brassica carinata* accession PI 360883, and *B. oleracea* cvs. Titleist and Cecile to transfer resistance against powdery mildew of *B. oleracea*. Four interspecific hybrids were obtained through application of embryo rescue from crosses with *B. carinata* as the maternal parent, and their interspecific nature confirmed through plant morphology, and random amplified polymorphic DNA (RAPD) analysis. Twenty-one BC1 plants were obtained through sexual crosses, and embryo rescue although embryo rescue was not necessary to produce first backcross generation plants between interspecific hybrids and *B. oleracea*. All interspecific hybrids, and eight of the BC1 plants were resistant to powdery mildew (Tonguc and Griffiths 2004).

6.14 Use of Mutagenic Resistance to Powdery Mildew

The method of induced mutagenesis is a powerful tool for creating new genotypes with an increased yield, quality, and resistance to diseases, and pests (Ahloowalia et al. 2004). Upon irradiation with gamma rays, the variation of the resulting genetic alterations has high frequency, and wider range of morphological and biochemical changes (Skoric et al. 2008). There is evidence that the mutations may also change the expression of certain enzymes in the biosynthetic pathway of fatty acids (Auld et al. 1992; Pleines and Friendt 1989). The irradiation of rapeseed with low doses of gamma rays of 100 Gy, and 150 Gy has a stimulating effect on the synthesis of linoleic and linolenic acids. Conversely, at higher doses of the mutagen factor, the concentration of these fatty acids decreases while the amount of oleic acid increases (Rahimi and Bahrani 2011). The fatty acids synthesized in plants play an important role by participating in the composition of tissue hormones, and in cell, and nuclear membrane structures (Kolattukudy 1974). It is assumed that unsaturated fatty acids in plant cells are involved in cutin and suberin synthesis (Pinot et al. 1999). Cutin, which acts as first barrier, protects plant organs (leaves and fruits) from dehydration providing protection against pathogens, and pests are composed of C16 and C18 fatty acids. Suberin, which is predominantly composed of hydroxyl and epoxy unsaturated fatty acids is a structural component of the plant underground part roots and tubers (Kolattukudy 1980). C18 fatty acids in plants play a role in building the defense mechanism against diseases and pests (Kacharoo and Kacharoo 2009). It has been found that hydroxyl fatty acids also inhibit the development of *Erysiphe graminis, Puccinia recondita,* and *Phytophthora infestans* (Hou and Forman 2000). The powdery mildew development is favored by meteorological factors, such as temperature and high humidity (Penaud 1999; Enright and Cipollini 2007). Fungicide application is often not efficient enough, due to the pathogen's rapid spread in favorable conditions. The effects of gamma radiation on the fatty acid composition of rapeseed oil, and the increased resistance to *E. cruciferarum* have been determined by Petkova et al. (2014). Biochemical changes in the oil after gamma-ray irradiation are also associated with changes in the susceptibility to powdery mildew.

Petkova et al. (2014) assessed through the gas chromatography, the difference in the content of saturated and unsaturated fatty acids in the seeds of the non-irradiated variants, and the ones irradiated with two doses of gamma radiation. A summary of the data for all 3 years of the fatty acid composition of the Abacus hybrid oil, and its M0, M1, and M2 was decomposing the populations. The chemical analysis showed generally that the irradiated plants showed a lower content of oleic acid, and a higher content of linoleic acid, and linolenic acid compared with the non-irradiated variants. There is tendency towards increase in the total content of raw fats in mutant populations of M1, and M2 of Abacus as compared with non-irradiated plants. This, in turn, is associated with a reduction in the total content of saturated fatty acids and an increase in the content of unsaturated fatty acids. There is an inverse relationship between the values of the total fat content and the content of saturated and unsaturated fatty acids. The rapeseed oil of the Abacus hybrid showed higher content of saturated fatty acids as compared with the M2 and M3 generations of the irradiated variants. In contrast, the percentage content of unsaturated fatty acids is influenced positively by irradiation with gamma rays and is comparatively high in M2 A1-100 and M2 A-150 with 92.7%, and 93.7%, respectively, compared with the baseline 89.7%.

The extent of damage by the powdery mildew on plants of the Abacus variety, and its gamma-ray irradiation-treated variants shows the arithmetic mean values, and their errors, the accuracy indicators, and the variation coefficients. The highest extent of damage on the stems, and branches in the range of 98.9–99% ($p = 0.1\%$) was observed with the M0 Abacus generation, irradiated with absorbed dose 100 Gy, and 150 Gy gamma rays, compared with the non-irradiated variant. The irradiation did not have a positive impact on the sustainability of pods to *E. cruciferarum.* The differences found are insignificant compared with the corresponding baseline group. In contrast, the M1 and M2 generations were shown to have a proven lower extent of damage by the phytopathogen as compared with the baseline. It is assumed that the increased resistance demonstrated is due to the higher content of unsaturated fatty acids in the seeds of the irradiated plants. Unsaturated fatty acids can be used as a means of biological control of plant diseases on global scales. Increasing the dose of the mutagen leads to an increase in the concentration of unsaturated fatty acids with 18 carbon atoms and to a decrease in the extent of damage by *Erysiphe* cruciferarum (Petkova et al. 2014).

6.15 Strategies for Developing Durable Alternaria Blight-Resistant *Brassica* Crops

Based on a literature survey on intercropping systems, Fatima et al. (2019) have shortlisted the non-host plants for *A. brassicae* that can be chosen for dissecting the NHR mechanisms. The NHR to *A. brassicae* is based on fungal invasion and penetration events. Prior to any specific study, it is important to classify the non-host plants based on these four phases by using pathological assays. This will provide a clear idea about whether the NHR operates at the pre-invasion or post-invasion level. The following strategies and approaches can be used for developing durable Alternaria blight-resistant *Brassica* crops.

6.15.1 Inter- or Intraspecific Introgression of Candidate Genes from Non-Host Plants to *Brassica* Crops

The identification of genes conferring NHR by utilizing conventional or molecular breeding approaches, followed by the introgression of candidate genes from non-host plants to *Brassica* crops. This can be achieved by interspecific crosses of the host with a non-host may be performed to transfer resistance traits from the non-host plant to a host plant. The unwanted background in the progeny can be removed by crossing it with wild-type host plants and selecting for the targeted traits only. Then, the progeny can be screened for response to *A. brassicae*. Plants showing resistance to the pathogen can be subsequently subjected to QTL mapping to identify the genes involved in NHR. Once the genes are identified, they can be successfully introgressed into the desired varieties to provide resistance to *A. brassicae*. The major limitation with this approach, however, is that the progeny generally suffers from sterility and abnormal segregation, which can hinder the identification of genes involved in resistance. The only available example of an interspecific cross made between a host and non-host plant is that of *Lactuca* and *Bremia* (Jeuken et al. 2008).

Secondly by crossing two different varieties of non-host plants (intraspecific cross: completely resistant genotype × moderately resistant genotype) that differ in the degree of resistance towards *A. brassicae*. This will allow the inheritance of NHR-specific QTL of the resistant donor into the moderately susceptible genotypes of the same species of the non-host plant (Jafary et al. 2008; Niks and Marcel 2009). The progeny obtained can be similarly studied for identifying the genes involved in NHR and transferring them into target crops. However, if the non-host plant does not have significant variation in the degree of resistance towards the pathogen, the inheritance of NHR-specific genes might be difficult (Table 6.1) (Fatima et al. 2019).

6.15.2 Microarray-Based Gene Identification and Transfer

The second strategy is to perform microarray analysis after infecting the non-host plant (e.g., chickpea or barley) with *A. brassicae* and then identify the candidate genes from the list of up-regulated or down-regulated genes. If a homologous gene is identified, mutant or overexpression lines can be generated for the identified genes in the susceptible model plant. Alternatively, the candidate genes can be cloned from the non-host plant and expressed in the susceptible model plant. These transgenic plants can then be tested for their response to *A. brassicae* infection. Finally, the

Table 6.1 List of the potential non-host plants towards *Alternaria brassicae* (Fatima et al. 2019)

Name of the non-host plant	Name of the host plant intercropped with non-host plant	Criteria for categorization of non-host plant	Reference
Chickpea (*Cicer arietinum* and *C. kabulium*)	*B. juncea, B. napus, B. rapa*	Intercropping system	Ahlawat and Sharma (2002); Ahlawat et al. (2005); Singh et al. (2010); Shekhawat et al. (2012); Jeromela et al. (2017)
Lentils (*Lens culinaris*)	*B. juncea, B. rapa*	Intercropping system	Shyam et al. (2007); Shekhawat et al. (2012); Ao and Saud (2016); Jeromela et al. (2017)
Barley (*Hordeum vulgare*)	*B. juncea, B. rapa*	Intercropping system	Singh et al. (2010); Shekhawat et al. (2012)
Wheat (*Triticum aestivum*)	*B. juncea, B. napus, B. rapa*	Intercropping system; Literature	Singh et al. (2010); Shekhawat et al. (2012); Ebrahimi et al. (2016)
Tomato (*Solanum lycopersicum*)		Literature	McRoberts and Lennard (1996)
Potato (*Solanum tuberosum*)	*B. juncea B. napus*	Intercropping system	Singh et al. (2010); Shekhawat et al. (2012)
Sugarcane (*Saccharum officinarum*)	*B. juncea, B. rapa, B. napus, B. campestris*	Intercropping system	Singh et al. (1986); Toor et al. (2000); Suman et al. (2006); Ahlawat and Sharma (2002); Singh et al. (2010)

genes imparting NHR to *A. brassicae* can be successfully transferred to *Brassica* crops.

6.15.3 Gene Silencing

The third strategy to identify genes underlying NHR involves gene silencing in the non-host plant. Virus-induced gene silencing (VIGS)-based forward genetic screening can be performed in non-host plants that have established VIGS protocols and cDNA libraries to identify the genes involved in NHR to *A. brassicae*. For example, for wheat and barley, the Barley stripe mosaic virus (BSMV)-based vector and a well-developed VIGS protocol for gene silencing are available (Ma et al. 2012). Such a strategy will be useful in identifying important genes involved in NHR by screening a VIGS population for its response to *A. brassicae*. Furthermore, high-throughput transient expression and gene silencing methods, e.g., transient-induced gene silencing (TIGS), have been used in a non-host plant such as barley. This technique is based on the transformation of leaf epidermal cells by the biolistic method (Douchkov et al. 2014a, b). TIGS constructs may be made for the genes that are up-regulated in the non-host plant after *A. brassicae* infection. These TIGS constructs can be bombarded into leaf epidermal cells of the non-host plant to induce gene silencing. After bombardment, leaf discs can be infected with *A. brassicae* and screened for the response. TIGS will allow the identification of potentially important candidate genes involved in NHR to *A. brassicae*. The genes identified using any of these functional genomic methods can be successfully transferred to Brassica crops to ensure resistance to *A. brassicae*.

6.15.4 Identification and Use of Model Plants as Genetic Resources

The fourth strategy is to work with model plants that have well-developed molecular and genetic resources and a large population of mutants as non-hosts for *A. brassicae*. For example, *Brachypodium distachyon* can be directly utilized for identifying genes and dissecting the pathways involved in NHR (An et al. 2016). If the mutants are not readily available, then with the help of well-developed genetic tools, transposon or T-DNA insertional mutagenesis can be attempted in the model non-host plant. The mutant population can be screened for the response to A. brassicae to identify the candidate genes involved in NHR. These candidate genes can then be successfully transferred to *Brassica* crops for durable resistance to *A. brassicae*. *Arabidopsis* is a well-established, easy-to-manipulate model plant with ample genetic resources. Another approach can be screening the response of all available *Arabidopsis* accessions to *A. brassicae*. This may help identify *Arabidopsis* accessions imparting resistance to A. brassicae (Rajarammohan et al. 2017; Mandal et al. 2018).

6.15.5 Targeted Functional Genomics Based on Literature Mining

The fifth strategy is solely based on literature mining and targeted functional genomics to dissect the different layers of NHR involved in pre- and post-invasive defense responses. The available literature provides hints about possible candidates specifically involved in pre- and post-invasive defenses. This will provide the opportunity to directly start studying these candidates for their role in NHR. Another possibility is that the post-invasive defenses may play a more important role in restricting the fungal pathogen rather than pre-invasive defenses against *A. brassicae*, which can be determined by performing pathological assays. In such cases, we can use mutants with mutations in the genes important for pre-invasion defense to identify the genes involved in post-invasion defenses. For example, the mutant of the barley gene Ror1, a homolog of the *Arabidopsis* gene Pen1, is important for pre-invasive defense against many non-host pathogens, including rust pathogens (Collins et al. 2003; Trujillo et al. 2004). This mutant line can be used for identifying genes involved in the post-invasion defense response. The response of a ror1 mutant line should be tested towards *A. brassicae* to ascertain whether this gene is involved in pre-invasive defense against this pathogen.

Subsequently, the NHR-specific candidate genes can be identified by transcriptomic analysis in wild-type barley after *A. brassicae* infection. Gene silencing approaches can be applied to silence the candidate genes in the ror1 mutant line for identifying their role in post-invasion immunity by screening them against *A. brassicae*. The validation experiments can be carried out by constructing double mutants of candidate genes and Ror1 in barley. The double mutants compromising NHR and showing a higher degree of susceptibility to *A. brassicae* compared to the single mutant ror1 can be likely candidates that contribute to NHR at the post-invasion level. The important genes identified can be further transferred, or their homologs can be overexpressed in the susceptible *Brassica* crops to test their resistance to *A. brassicae* (Fatima et al. 2019).

6.15.6 Transfer of an Endochitinase Gene from *Trichoderma virens* Confers Enhanced Tolerance to *Alternaria* in Transgenic *Brassica*

Genetic engineering is an alternative for developing disease-resistant plants where the resistance source is not available or there are sexual compatibilities between the source and the cultivar (Grover and Gawthaman 2003; Chhikara et al. 2012). Genes from the mycoparasitic Trichoderma spp. have been used to impart tolerance to biotic and abiotic stresses in many crops (Nicolas et al. 2014). Among the *Trichoderma* genes transferred to plants, an endochitinase of 42 kDa has been very popular due to its broad spectrum of activity against many fungal pathogens. Endochitinases are known to degrade fungal chitin by hydrolyzing the glycosidic bond between chitin monomers. In addition, the endochitinases also release chitin oligomers which act as elicitor molecules triggering broad-spectrum plant defense. There are very few reports on the development of transgenic *Brassica* species for resistance to fungal pathogens. For example, *B. napus* transformed with two genes *sporamin* and *chitinase PjChi*-1 (derived from sweet potato and *Paecilomyces javanicus*, respectively) showed enhanced resistance against both *Plutella xylostella* and *Sclerotinia sclerotiorum* (Liu et al. 2011). Wu et al. (2009) introduced a plant defensin gene Ovd, cloned from *Orychophragmus violaceus* (L.) into *B. napus*, and the transgenic plants showed resistance to *S. sclerotiorum*. Transgenic *B. napus* expressing *B. napus* mitogen-activated protein kinase (*BnMPK4*) showed resistance to *S. sclerotiorum* (Wang et al. 2009). Chitinase and glucanase gene from tomato have been introduced into *B. juncea* and the transgenic plants showed resistance to *A. brassicae* (Mondal et al. 2003, 2007). Kamble et al. (2016) observed expression of an endochitinase "*ech42*" gene from *Trichoderma virens* in transgenic *B. juncea* lines and also assessed tolerance of transgenic lines against two fungal pathogens, *A. brassicicola* and *A. brassicae*, causing Alternaria blight.

An endochitinase gene "*ech42*" from the biocontrol fungus "*Trichoderma virens*" was introduced to *Brassica juncea* via *Agrobacterium tumefaciens*-mediated genetic transformation method. Integration and expression of the "*ech42*" gene in transgenic lines were confirmed by PCR, RT-PCR, and Southern hybridization. Transgenic lines (T1) showed expected 3:1 Mendelian segregation ratio when segregation analysis for inheritance of transgene "hpt" was carried out. Fluorimetric analysis of transgenic lines (T0 and T1) showed sevenfold higher endochitinase activity than the non-transformed plant. Fluorimetric zymogram showed the presence of endochitinase (42 kDa) in crude protein extract of transgenic lines. In detached leaf bioassay with fungi *Alternaria brassicae* and *Alternaria brassicicola*, transgenic lines (T0 and T1) showed delayed onset of lesions as well as 30–73% reduction in infected leaf area compared to non-transformed plant (Kamble et al. 2016).

References

Agnihotri A, Lakshmikumaran MS, Jagannathan V, Shivanna KR (1991) Wide hybridization for improvement in cultivated Brassicas. Acta Hortic (ISHS) 289:213–214

Agnihotri A, Gupta K, Prem D, Sarkar, Mehra VS, Zargar SM (2009) Genetic enhancement in rapeseed-mustard for quality and disease resistance through *in vitro* techniques. In: Proceedings of 16th Australian research assembly on Brassicas, Ballarat, Australia, 14–16 Sept, p 28

Ahlawat IPS, Sharma RP (2002) Agronomic terminology. 50. New Delhi: Indian Society of Agronomy, Division of Agronomy, Indian Agriculture Research Institute, p 132

Ahlawat IPS, Gangaiah B, Ompal S (2005) Production potential of chickpea (*Cicer arietinum*) based intercropping systems under irrigated conditions. Indian J Agron 50:27–30

Ahloowalia BS, Maluszynski M, Nichterlein K (2004) Global impact of mutation-derived varieties. Euphytica 135:187–201

An T, Cai Y, Zhao S, Zhou J, Song B, Bux H, Qi X (2016) *Brachypodium distachyon* TDNA insertion lines: a model pathosystem to study nonhost resistance to wheat stripe rust. Sci Rep 6:25510. https://doi.org/10.1038/srep25510

Aneja JK, Agnihotri A (2013) Alternaria blight of oilseed *Brassicas*: epidemiology and disease control strategies with special reference to use of biotechnological approaches for attaining host resistance. J Oilseed Brassica 4:1–10

Aneja JK, Goel P, Agnihotri A (2012) Diversity assessment in *Alternaria brassicae*. In: Proc. of the 1st National Brassica Conference on production barriers and technological options in Oilseed *Brassica*, organized by Society for Rapeseed-Mustard Research, Bharatpur and Indian Council of Agricultural Research, New Delhi, at CCS Haryana Agriculture University Hisar, 2–3 Mar 2012, Abstr, p 149

Ansari NA, Khan MW, Muheet A (1989) Effect of some factors on growth and sporulation of *Alternaria brassicae* causing Alternaria blight of rapeseed and mustard. Acta Bot Ind 17:49–53

Ao E, Saud KTR (2016) Production potential of rapeseed (*Brassica rapa* var. *dichotoma*)—based intercropping systems under rain fed conditions. J Oilseed Brassica 1:91–97

Asai T, Tena G, Plotnikova J, Willmann MR, Chiu WL, Gomez-Gomez L, Boller T, Ausubel FM, Sheen J (2002) MAP kinase signaling cascade in *Arabidopsis* innate immunity. Nature 415:977–983

Auld D, Heikkinen M, Erickson D, Sernyk J, Romero J (1992) Rapeseed mutants with reduced levels of polyunsaturated fatty acids and increased levels of oleic acid. Crop Sci 32:657–662

Avenot H, Dongo A, Simoneau NB, Iacomivasilescu B, Hamon B, Peltiers D, Simoneau P (2005) Isolation of 12 polymorphic microsatellite loci in the phytopathogenic fungus *Alternaria brassicicola*. Mol Ecol Notes 5:948–950

Awasthi RP, Kolte SJ (1989) Variability in *Alternaria brassicae* affecting rapeseed and mustard. Indian Phytopathol 42:275

Ayers GW, Lelacheur KE (1972) Genetics of resistance in rutabaga to races of *Plasmodiophora brassicae*. Can J Plant Sci 52:897–900

Berbee ML, Payne BP, Zhang G, Roberts RG, Turgeon BG (2003) Shared ITS DNA substitutions in isolates of opposite mating type reveal a recombining history for three presumed species in the filamentous ascomycete genus Alternaria. Mycol Res 107(2):169–182

Block A, Alfano JR (2011) Plant targets for *Pseudomonas syringae* type III effectors: virulence targets or guarded decoys? Curr Opin Microbiol 14:39–46

Bock CH, Thrall PH, Brubaker CL, Burdon JJ (2002) Detection of genetic variation in *Alternaria brassicicola* using AFLP fingerprinting. Mycol Res 106(4):428–434

Chavan V, Bhargava S, Kamble A (2013) Temporal modulation of oxidant and anti-oxidative responses in *Brassica carinata* during β-aminobutyric acid-induced resistance against *Alternaria brassicae*. Physiol Mol Plant Pathol 83:35–39. https://doi.org/10.1016/j.pmpp.2013.03.002

Chevre AM, Eber F, Margale E, Kerlan MC (1994) Comparison of somatic and sexual *Brassica napus-Sinapis alba* hybrids and their progeny by cytogenetic studies and molecular characterization. Genome 37:367–371

Chhikara S, Chaudhury D, Dhankher OP, Jaiwal PK (2012) Combined expression of a barley class II chitinase and type I ribosome inactivating protein in transgenic *Brassica juncea* provides protection against *Alternaria brassicae*. Plant Cell Tissue Organ Cult 108:83–89

Chou HH, Wu WS (2002) Phylogenetic analysis of internal transcribed spacer regions of the genus *Alternaria*, and the significance of filament-beaked conidia. Mycol Res 106:164–169

Collins NC, Thordal-Christensen H, Lipka V, Bau S, Kombrink E, Qiu JL (2003) SNARE-protein-mediated disease resistance at the plant cell wall. Nature 425:973–977. https://doi.org/10.1038/nature02076

Cooke DEL, Forster JW, Jenkins PD, Gareth DJ, Lewis DM (1998) Analysis of intraspecific and interspecific variation in the genus *Alternaria* by the use of RAPD-PCR. Ann Appl Biol 132:197–209

Cui H, Wang Y, Xue L, Chu J, Yan C, Fu J, Chen M, Innes RW, Zhou JM (2010) *Pseudomonas syringae* effector protein AvrB perturbs *Arabidopsis* hormone signaling by activating MAP kinase 4. Cell Host Microbe 7:164–175

de Jonge R, Peter van Esse H, Kombrink A, Shinya T, Desaki Y, Bours R, van der Krol S, Shibuya N, Joosten MHAJ, Thomma BPHJ (2010) Conserved fungal LysM effector Ecp6 prevents chitin-triggered immunity in plants. Science 329:953–955

Deslandes L, Rivas S (2012) Catch me if you can: bacterial effectors and plant targets. Trends Plant Sci 17:644–655

Deverall BJ (1995) Plant protection using natural defense systems of plants. Adv Plant Pathol 11:211–228

Ding Y, Sun T, Ao K, Peng Y, Zhang Y, Li X, Zhang Y (2018) Opposite roles of salicylic acid receptors NPR1 and NPR3/NPR4 in transcriptional regulation of plant immunity. Cell 173:1454–1467.e1415

Douchkov D, Luck S, Johrde A, Nowara D, Himmelbach A, Rajaraman J, Schweizer P (2014a) Discovery of genes affecting resistance of barley to adapted and non-adapted powdery mildew fungi. Genome Biol 15(12):518. https://doi.org/10.1186/gb-2014-15-1-r1

Douchkov D, Baum T, Ihlow A, Schweizer P, Seiffert U (2014b) Microphenomics for interactions of barley with fungal pathogens. In: Tuberosa R, Graner A, Frison E (eds) Genomics of plant genetic resources, 2nd edn. Springer, Heidelberg, pp 123–148

Ebrahimi E, Kaul HP, Neugschwandtner RW, Dabbagh Mohammadi Nassab A (2016) Productivity of wheat (*Triticum aestivum* L.) intercropped with rapeseed (*Brassica napus* L). Can J Plant Sci 97:557–568

Enright S, Cipollini D (2007) Infection by powdery mildew *Erysiphe cruciferarum* (Erysiphaceae) strongly affects growth and fitness of *Alliaria petiolata* (Brassicaceae). Am J Bot 94:1813–1820

Fatima U, Bhorali P, Borah S, Senthil-Kumar M (2019) Perspectives on the utilization of resistance mechanisms from host and non-host plants for durable protection of *Brassica* crops against Alternaria blight. PeerJ 7:e7486. https://doi.org/10.7717/peerj.7486

Felix G, Duran JD, Volko S, Boller T (1999) Plants have a sensitive perception system for the most conserved domain of bacterial flagellin. Plant J 18:265–276

Fisher MC, Henk DA, Briggs CJ, Brownstein JS, Madoff LC, McCraw SL, Gurr SJ (2012) Emerging fungal threats to animal, plant and ecosystem health. Nature 484:186–194

Gamir J, Darwiche R, Van't Hof P, Choudhary V, Stumpe M, Schneiter R, Mauch F (2017) The sterol-binding activity of PATHOGENESIS-RELATED PROTEIN 1 reveals the mode of action of an antimicrobial protein. Plant J 89:502–509

Gao M, Liu J, Bi D, Zhang Z, Cheng F, Chen S, Zhang Y (2008) MEKK1, MKK1/MKK2 and MPK4 function together in a mitogen-activated protein kinase cascade to regulate innate immunity in plants. Cell Res 18:1190–1198

Gherbawy YAMH (2005) Genetic variation among isolates of *Alternaria* spp. from selected Egyptian crops. Arch Phytopathol Plant Prot 38(2):77–89

Grover A, Gawthaman R (2003) Strategies for development of fungus-resistant transgenic plants. Curr Sci 84:330–340

Gupta K, Prem D, Agnihotri A (2010) Pyramiding white rust and Alternaria resistance in low erucic *Brassica juncea* using *B. carinata*. J Oilseed Brassica 1(2):55–65

Hansen LN (1998) Intertribal somatic hybridization between rapid cycling *Brassica oleracea* and *Camelina sativa*. Euphytica 104:173–179

Hansen LH, Earle ED (1997) Somatic hybrids between *Brassica oleracea* (L.) and *Sinapis alba* (L.) with resistance to *Alternaria brassicae* (Berk) Sacc. Theor Appl Genet 94:1078–1085

Hatakeyama K, Niwa T, Kato T, Ohara T, Kakizaki T, Matsumoto S (2017) The tandem repeated organization of *NB-LRR* genes in the clubroot-resistant *CRb* locus in *Brassica rapa* L. Mol Genet Genomics 292:397–405. https://doi.org/10.1007/s00438-016-1281-1

Hirani AH, Li G (2015) Understanding the genetics of clubroot resistance for effectively controlling this disease in *Brassica* species. In: Plants for the future. IntechOpen. https://doi.org/10.5772/60936

Hirani AH, Gao F, Liu J, Fu G, Li G (2016) Transferring clubroot resistance from Chinese cabbage (*Brassica rapa*) to canola (*B. napus*). Can J Plant Pathol 38(1):82–90. https://doi.org/10.1080/07060661.2016.1141799

Hirani AH, Gao F, Liu J, Fu G, Wu C, McVetty PBE, Duncan RW, Li G (2018) Combinations of independent dominant loci conferring clubroot resistance in all four turnip accessions (*Brassica rapa*) from the European clubroot differential set. Front Plant Sci 9:1628. https://doi.org/10.3389/fpls.2018.01628

Hou CT, Forman RJ (2000) Growth inhibition of plant pathogenic fungi by hydroxy fatty acids. J Ind Microbiol Biotechnol 24(4):275–276

Ichimura K, Shinozaki K, Tena G, Sheen J, Henry Y, Champion A, Kreis M, Zhang S, Hirt H, Wilson C, Heberle-Bors E, Ellis BE, Morris PC, Innes RW, Ecker JR, Scheel D, Klessig DF, Machida Y, Mundy J, Ohashi Y, Walker JC (2002) Mitogen-activated protein kinase cascades in plants: a new nomenclature. Trends Plant Sci 7:301–308

Innes R (2018) The positives and negatives of NPR: a unifying model for salicylic acid signaling in plants. Cell 173:1314–1315

Jafary H, Albertazzi G, Marcel TC, Niks RE (2008) High diversity of genes for nonhost resistance of barley to heterologous rust fungi. Genet 178:2327–2339. https://doi.org/10.1534/genetics.107.077552

Jammes F, Song C, Shin D, Munemasa S, Takeda K, Gu D, Cho D, Lee S, Giordo R, Sritubtim S, Leonhardt N, Ellis BE, Murata Y, Kwak JM (2009) MAP kinases MPK9 and MPK12 are preferentially expressed in guard cells and positively regulate ROS-mediated ABA signaling. Proc Natl Acad Sci U S A 106:20520–20525

Jammes F, Yang X, Xiao S, Kwak JM (2011) Two Arabidopsis guard cell-preferential MAPK genes, MPK9 and MPK12, function in biotic stress response. Plant Signal Behav 6:1875–1877

Jeromela AM, Mikic AM, Vujic S, Cupina B, Krstic D, Dimitrijevic A, Vasiljevic S, Mihailovic V, Cvejic S, Miladinovic D (2017) Potential of legume–*Brassica* intercrops for forage production and green Manure: encouragements from a temperate southeast European environment. Front Plant Sci 8:312

Jeuken MJW, Pelgrom K, Stam P, Lindhout P (2008) Efficient QTL detection for nonhost resistance in wild lettuce: backcross inbred lines versus F2 population. Theor Appl Genet 116:845–857. https://doi.org/10.1007/s00122-008-0718-2

Jones JD, Dangl JL (2006) The plant immune system. Nature 444:323–329

Jourdan PS, Salazar E (1993) *B. carinata* resynthesized by protoplast fusion. Theor Appl Genet 86:567–572

Kacharoo A, Kacharoo P (2009) Fatty acid–derived signals in plant defense. Annu Rev Phytopathol 47:153–176

Kamble S, Mukherjee PK, Eapen S (2016) Expression of an endochitinase gene from *Trichoderma virens* confers enhanced tolerance to Alternaria blight in transgenic *Brassica juncea* (L.) Czern & Coss lines. Physiol Mol Biol Plants 22(1):69–76. https://doi.org/10.1007/s12298-016-0340-8

Kanrar S, Venkateswari JC, Kirti PB, Chopra VL (2002) Transgenic expression of hevein, the rubber tree lectin, in Indian mustard confers protection against *Alternaria brassicae*. Plant Sci 162:441–448

Kaur S, Singh G, Banga SS (2007) Documenting variation in *Alternaria brassicae* isolates based on conidial morphology, fungicidal sensitivity and molecular profile. In: Proc 12th international rapeseed congress, Wuhan, China, 26–30 Mar, vol 4, pp 87–89

Kessman H, Staub T, Ligon J, Oostendorp M, Ryals J (1994) Activation of systemic acquired disease resistance in plants. Eur J Plant Pathol 100:359–369

Khokon MAR, Salam MA, Jammes F, Ye W, Hossain MA, Uraji M, Nakamura Y, Mori IC, Kwak JM, Murata Y (2015) Two guard cell mitogen-activated protein kinases, MPK9 and MPK12, function in methyl jasmonate-induced stomatal closure in *Arabidopsis thaliana*. Plant Biol (Stuttg) 17:946–952

King SRF, McLellan H, Boevink PC, Armstrong MR, Bukharova T, Sukarta O, Win J, Kamoun S, Birch PRJ, Banfield MJ (2014) *Phytophthora infestans* RXLR effector PexRD2 interacts with host MAPKKKε to suppress plant immune signaling. Plant Cell 26:1345–1359

Klevernic IV, Stafford MJ, Morrice N, Peggie M, Morton S, Cohen P (2006) Characterization of the reversible phosphorylation and activation of ERK8. Biochem J 394:365–373

Kolattukudy PE (1974) Biosynthesis of a hydroxy fatty acid polymer, cutin. Identification and biosynthesis of 16-oxo-9- or 10-hydroxypalmitic acid, a novel compound in *Vicia faba*. Biochemistry 13:1354–1363

Kolattukudy PE (1980) Biopolyester membranes of plants: cutin and suberin. Science 208:990–1000

Lapin D, Kovacova V, Sun X, Dongus JA, Bhandari DD, vonBorn P, Bautor J, Guarneri N, Stuttmann J, Beyer A, Parker JE (2019) A coevolved EDS1-SAG101-NRG1 module mediates cell death signaling by TIR-domain immune receptors. BioRxiv. https://doi.org/10.1101/572826

Lee Y, Kim YJ, Kim MH, Kwak JM (2016) MAPK cascades in guard cell signal transduction. Front Plant Sci 7:80

Liu H, Guo X, Naeem MS, Liu D, Xu L, Zhang W, Tang G, Zhou W (2011) Transgenic *Brassica napus* L. lines carrying a two gene construct demonstrate enhanced resistance against *Plutella xylostella* and *Sclerotinia sclerotiorum*. Plant Cell Tissue Organ Cult 106(1):143–151. https://doi.org/10.1007/s11240-010-9902-6

Liu C, Yang Z, He P, Munir S, Wu Y, Ho H, He Y (2018) Deciphering the bacterial and fungal communities in clubroot-affected cabbage rhizosphere treated with *Bacillus subtilis* XF-1. Agric Ecosyst Environ 256:12–22

Ma M, Yan Y, Huang L, Chen M, Zhao H (2012) Virus-induced gene-silencing in wheat spikes and grains and its application in functional analysis of HMW-GS-encoding genes. BMC Plant Biol 12:1. https://doi.org/10.1186/1471-2229-12-1

Ma L, Djavaheri M, Wang H, Larkan NJ, Haddadi P, Beynon E, Gropp G, Borhan MH (2018) *Leptosphaeria maculans* effector protein *AvrLm1* modulates plant immunity by enhancing MAP Kinase 9 phosphorylation. iScience 3:177–191

Mandal S, Rajarammohan S, Kaur J (2018) *Alternaria brassicae* interactions with the model Brassicaceae member *Arabidopsis thaliana* closely resembles those with Mustard (*Brassica juncea*). Physiol Mol Biol Plants 24:51–59. https://doi.org/10.1007/s12298-017-0486-z

McRoberts N, Lennard JH (1996) Pathogen behaviour and plant cell reactions in interactions between *Alternaria* species and leaves of host and nonhost plants. Plant Pathol 45(4):742–752

Meena PD, Chattopadhyay C, Kumar VR, Meena RL, Rana US (2005) Spore behaviour in atmosphere and trends in variability of *Alternaria brassicae* population in India. J Mycol Plant Pathol 35:511

Meena PD, Rani A, Meena R, Sharma P, Gupta R, Chowdappa P (2012) Aggressiveness, diversity and distribution of *Alternaria brassicae* isolates infecting oilseed *Brassica* in India. Afr J Microbiol Res 6:5249–5258

Meng X, Zhang S (2013) MAPK cascades in plant disease resistance signaling. Annu Rev Phytopathol 51:245–266

Mohanty A, Chrungu B, Verma N, Shivanna KR (2009) Broadening the genetic base of crop Brassicas by production of new intergeneric hybrid. Czech J Genet Plant Breed 45(3):117–122

Mondal KK, Chatterjee SC, Viswakarma N, Bhatacharya RC, Grover A (2003) Chitinase mediated inhibitory activity of Brassicas transgenic on growth of *Alternaria brassicae*. Curr Microbiol 47:171–173

Mondal KK, Bhattacharya RC, Koundal KR, Chatterjee SC (2007) Transgenic Indian mustard (*Brassica juncea*) expressing tomato glucanase leads to arrested growth of *Alternaria brassicae*. Plant Cell Rep 26:247–252. https://doi.org/10.1007/s00299-006-0241-3

Montillet JL, Leonhardt N, Mondy S, Tranchimand S, Rumeau D, Boudsocq M, Garcia AV, Douki T, Bigeard J, Laurière C, Chevalier A, Castresana C, Hirt H (2013) An abscisic acid-independent oxylipin pathway controls stomatal closure and immune defense in Arabidopsis. PLoS Biol 11:e1001513

Nagy SK, Darula Z, Kállai BM, Bögre L, Bánhegyi G, Medzihradszky KF, Horváth GV, Mészáros T (2015) Activation of AtMPK9 through autophosphorylation that makes it independent of the canonical MAPK cascades. Biochem J 467:167–175

Nicolas C, Hermosa R, Rubio B, Mukherjee P, Monte E (2014) Trichoderma genes in plants for stress tolerance—status and prospects. Plant Sci 228:71–78

Niks RE, Marcel TC (2009) Nonhost and basal resistance: how to explain specificity? New Phytol 182:817–828. https://doi.org/10.1111/j.1469-8137.2009.02849.x

O'Neill CM, Murata T, Morgan CL, Mathia RJ (1996) Expression of the C3-C4 intermediate character in somatic hybrids between *B. napus* and the C3-C4 species *Moricandia arvensis*. Theor Appl Genet 93:1234–1241

Oh IS, Park AR, Bae MS, Kwon SJ, Kim YS, Lee JE, Kang NY, Lee S, Cheong H, Park OK (2005) Secretome analysis reveals an Arabidopsis lipase involved in defense against *Alternaria brassicicola*. Plant Cell 17:2832–2847

Parker JE, Holub EB, Frost LN, Falk A, Gunn ND, Daniels MJ (1996) Characterization of eds1, a mutation in *Arabidopsis* suppressing resistance to *Peronospora parasitica* specified by several different RPP genes. Plant Cell 8:2033–2046

Patni CS, Kolte SJ, Awasthi RP (2005) Cultural variability of *Alternaria brassicae*, causing Alternaria blight of mustard. Ann Plant Physiol 19:231–242

Penaud A (1999) Chemical control and yield losses caused by *Erysiphe cruciferarum* on oilseed rape in France. In: 10th international rapeseed congress, Canberra, Australia, GCIRC

Petkova M, Dimova M, Dimova D, Bistrichanov S (2014) Effect of gamma-irradiation on the fatty acid composition and susceptibility to powdery mildew (*Erysiphe cruciferarum*) of oilseed rape plants. Agric Sci Technol 6(4):413–416

Pinot F, Benveniste IJ, Salau P, Loreau O, Noe J, Schreiber L, Durst F (1999) Production in vitro by the cytochrome P450 CYP94A1 of major C18 cutin monomers and potential messengers in plant–pathogen interactions: enantioselectivity studies. Biochem J 342:27–32

Pitzschke A, Schikora A, Hirt H (2009) MAPK cascade signaling networks in plant defense. Curr Opin Plant Biol 12:421–426

Pleines S, Friendt W (1989) Genetic control of linolenic acid concentrations in seed oil of rapeseed (*Brassica napus* L.). Theor Appl Genet 78:793–797

Presti LL, Lanver D, Schweizer G, Tanaka S, Liang L, Tollot M, Zuccaro A, Reissmann S, Kahmann R (2015) Fungal effectors and plant susceptibility. Ann Rev Plant Biol 66:513–545

Primard C, Vedel F, Mathieu C, Pelletier G, Chevre AM (1988) Interspecific somatic hybridization between *Brassica napus* and *Brassica hirta* (*Sinapis alba* L.). Theor Appl Genet 75:546–552

Rahimi M, Bahrani A (2011) Effect of gamma irradiation on qualitative and quantitative characteristics of canola. Middle East J Sci Res 8:519–525

Rajarammohan S, Kumar A, Gupta V, Pental D, Pradhan AK, Kaur J (2017) Genetic architecture of resistance to *Alternaria brassicae* in *Arabidopsis thaliana*: QTL mapping reveals two major resistance-conferring loci. Front Plant Sci 8:260

Rasmussen MW, Roux M, Petersen M, Mundy J (2012) MAP kinase cascades in *Arabidopsis* innate immunity. Front Plant Sci 3:169

Rekhter D, Mohnike L, Feussner K, Zienkiewicz K, Zhang Y, Feussner I (2019) Enhanced disease susceptibility 5 (EDS5) is required for N-hydroxy pipecolic acid formation. BioRxiv a:630723

Saharan GS, Naresh M, Sangwan MS (2003) Nature and mechanism of resistance to Alternaria blight in rapeseed-mustard system. Annu Rev Plant Pathol 2:85–128

Saharan GS, Naresh M, Verma PR, Meena PD (2015) Analysis of variability determinants in *Alternaria-Brassica* Interactions. J Oilseed Brassica 6:126–141

Saharan GS, Naresh M, Meena PD (2016) Alternaria blight of crucifers: biology, ecology and disease management. Springer, Singapore, p 326. ISBN: 978-981-10-0019-5

Saharan GS, Naresh M, Meena PD (2019) Powdery mildew disease of crucifers: biology, ecology and disease management. Springer, Singapore, p 362. ISBN 978-981-13-9852-0

Sánchez-Vallet A, Saleem-Batcha R, Kombrink A, Hansen G, Valkenburg DJ, Thomma BPHJ, Mesters JR (2013) Fungal effector Ecp6 outcompetes host immune receptor for chitin binding through intrachain LysM dimerization. Elife 2:e00790

Sharma TR, Singh BM (1992) Transfer of resistance to *Alternaria brassicae* in *Brassica juncea* through interspecific hybridization among Brassicas. J Genet Breed 46(4):373–378

Sharma TR, Singh BM (1995) Generations and evaluations of somaclones of Brassica juncea for resistance to *Albugo candida* and *Alternaria brassicae*. Proc Indian Natl Sci Acad Part B Biol Sci 61(2):155–161

Sharma TR, Tewari JP (1995) Detection of genetic variation in *Alternaria brassicae* by RAPD fingerprints. J Plant Biochem Biotechnol 4:105–107

Sharma TR, Tewari JP (1998) RAPD analysis of three *Alternaria* species pathogenic to crucifers. Mycol Res 102:807–814

Shekhawat K, Rathore SS, Premi OP, Kandpal BK, Chauhan JS (2012) Advances in agronomic management of Indian mustard (*Brassica juncea* (L.) Czernj. Cossọn): an overview. Int J Agron 2012:1–14

Shyam SY, David L, Philip C (2007) Lentil an ancient crop for modern times, vol 472. Springer, Dordrecht, p 259

Sigareva MA, Earle ED (1999) Production and characterization of somatic hybrids between *Camelina sativa* and rapid-cycling *B. oleracea*. Theor Appl Genet 98:164–170

Singh A (2017) Glucosinolates and plant defense. In: Mérillon JM, Ramawat KG (eds) Glucosinolates. Springer International Publishing, Cham, pp 237–246

Singh VS, Kothari K, Tripathi HN (1986) Studies on intercropping in sugarcane in central Uttar Pardesh. Indian Sugarcane J 35:559–562

Singh D, Singh R, Singh H, Yadav RC, Yadav N, Barbetti MJ, Salisbury P, Nimbal S, Chattopadhyay C, Kumar A (2007) Cultural and morphological variability in *Alternaria brassicae* isolates of Indian mustard (*Brassica juncea* L. Czern & Coss.). In: Proceeding of the 12th international rapeseed congress, 26–30 Mar, Wuhan, China, vol 4, pp 158–160

Singh RK, Kumar H, Singh AK (2010) *Brassica* based intercropping systems—a review. Agric Rev 3:253–266

Singh R, Singh D, Singh H, Caxia L, Sivasithampram K, Salisbury P, Barbetti MJ (2011) Pathogenic variability in *Alternaria brassicae* incitant of Alternaria blight of *Brassica* species. In: Proc 13th international rapeseed congress, 5–9 June, Prague, Czech Republic, pp 1265–1270

Skoric D, Jocic S, Sakac Z, Lecic N (2008) Genetic possibilities for altering sunflower oil quality to obtain novel oils. Can J Physiol Pharmacol 86:215–221

Stergiopoulos I, de Wit P (2009) Fungal effector proteins. Annu Rev Phytopathol 47:233–263

Stitcher L, Mauch-Mani B, Metraux JP (1997) Systemic acquired resistance. Annu Rev Phytopathol 35:235–270

Suman A, Lal M, Singh AK, Gaur A (2006) Microbial biomass turnover in Indian subtropical soils under different sugarcane intercropping systems. Agron J 98:698–704. https://doi.org/10.2134/agronj2005.0173

Taj G, Kumar A, Bansal KC, Garg GK (2004) Introgression of osmotin gene for creation of resistance against Alternaria blight by perturbation of cell cycle machinery. Indian J Biotechnol 3:291–198

Taj G, Agarwal P, Grant M, Kumar A (2011) Co-expression and in-silico interaction studies for inter-linking the activation of MAPK3 and LOX genes during pathogenesis of *Alternaria brassicae* in *Brassica juncea*. J Oilseed Brassica 2:13–20

Tonguc M, Griffiths PD (2004) Transfer of powdery mildew resistance from *Brassica carinata* to *Brassica oleracea* through embryo rescue. Plant Breed 123:587–589. https://doi.org/10.1111/j.1439-0523.2004.00987.x

Toor SS, Singh S, Garch AIS, Kapur ML, Singh A (2000) Gobhi sarson as an intercrop in Autumn Sugarcane for higher returns. Inten Agric 38:29–30

Torrens-Spence MP, Bobokalonova A, Carballo V, Glinkerman CM, Pluskal T, Shen A, Weng JK (2019) PBS3 and EPS1 complete salicylic acid biosynthesis from isochorismate in Arabidopsis. Mol Plant 12(12):1577–1586. https://doi.org/10.1016/j.molp.2019.11.005

Trujillo M, Troeger M, Niks RE, Kogel KH, Huckelhoven R (2004) Mechanistic and genetic overlap of barley host and non-host resistance to *Blumeria graminis*. Mol Plant Pathol 5:389–396. https://doi.org/10.1111/j.1364-3703.2004.00238.x

Ueno H, Matsumoto E, Aruga D, Kitagawa S, Matsumura H, Hayashida N (2012) Molecular characterization of the CRa gene conferring clubroot resistance in *Brassica rapa*. Plant Mol Biol 80:621–629. https://doi.org/10.1007/s11103-012-9971-5

Verma VD, Rai B (1980) Note on induced mutagenesis for spotting out the sources of resistance to Alternaria leaf spot in Indian mustard. Indian J Agric Sci 50(3):278

Verma S, Yajima WR, Rahman MH, Shah S, Liu JJ, Ekramoddoullah AKM, Kav NNV (2012) A cysteine-rich antimicrobial peptide from *Pinus monticola* (PmAMP1) confers resistance to multiple fungal pathogens in canola (*Brassica napus*). Plant Mol Biol 79(1–2):61–74

Vishwanath, Kolte SJ (1997) Biochemical variation among three isolates of *Alternaria brassicae*. Indian Phytopathol 50(3):437–438

Vishwanath, Kolte SJ, Singh MP, Awasthi RP (1999) Induction of resistance in mustard (*Brassica juncea*) against Alternaria black spot with an avirulent *Alternaria brassicae* isolate-D. Eur J Plant Pathol 105:217–220. https://doi.org/10.1023/A:1008717323002

Wagner G, Laperche A, Lariagon C, Marnet N, Renault D, Guitton Y, Bouchereau A, Delourme R, Manzanares-Dauleux MJ, Gravot A (2019) Resolution of quantitative resistance to clubroot into QTL-specific metabolic modules. J Exp Bot 70:5375–5390

Wang Z, Mao H, Dong C, Ji R, Cai L, Fu H, Liu S (2009) Over expression of *Brassica napus MPK4* enhances resistance to *Sclerotinia sclerotiorum* in oilseed rape. Mol Plant Microbe Interact 22:235–244. https://doi.org/10.1094/MPMI-22-3-0235

Wang Y, Li J, Hou S, Wang X, Li Y, Ren D, Chen S, Tang X, Zhou JM (2010) A *Pseudomonas syringae* ADP-ribosyl transferase inhibits *Arabidopsis* mitogen-activated protein kinase kinases. Plant Cell 22:2033–2044

Wiermer M, Feys BJ, Parker JE (2005) Plant immunity: the EDS1 regulatory node. Curr Opin Plant Biol 8:383–389

Win J, Chaparro-Garcia A, Belhaj K, Saunders DGO, Yoshida K, Dong S, Schornack S, Zipfel C, Robatzek S, Hogenhout SA, Kamoun S (2012) Effector biology of plant-associated organisms: concepts and perspectives. Cold Spring Harb Symp Quant Biol 77:235–247

Wu XM, Chen BY, Lu G, Wang HZ, Xu K, Guizhan G, Song Y (2009) Genetic diversity in oil and vegetable mustard (*Brassica juncea*) landraces revealed by SRAP markers. Genet Resour Crop Evol 56:1011–1022. https://doi.org/10.1007/s10722-009-9420-8

Yadav RC, Sareen PK, Chowdhury JB (1991) Interspecific hybridization in *Brassica juncea* x *Brassica tournefortii* using ovary culture. Cruciferae Newslett 14/15:84

Zhang J, Shao F, Li Y, Cui H, Chen L, Li H, Zou Y, Long C, Lan L, Chai J, Chen S, Tang X, Zhou JM (2007) A *Pseudomonas syringae* effector inactivates MAPKs to suppress PAMP-induced immunity in plants. Cell Host Microbe 1:175–185

Zipfel C, Kunze G, Chinchilla D, Caniard A, Jones JDG, Boller T, Felix G (2006) Perception of the bacterial PAMP EF-Tu by the receptor EFR restricts Agrobacterium-mediated transformation. Cell 125:749–760

Management of Disease Resistance 7

Abstract

In the integrated disease management strategies of cruciferous crops, use of R-genes through development of disease resistance cultivars is one of the most important ecofriendly and cost-effective components. Major R-genes impose strong selection pressure on pathogen populations and in combination with a number of factors can act to cause "boom-and-bust" cycle of pathotype diversification within a population leading to breakdown or erosion of R-genes. Strong selection imposed by widespread deployment of one R-gene also may lead to breakdown of resistance conferred by another gene, if the two complementary avirulence genes are closely linked. Therefore, the effective and sustainable deployment of resistant cultivars to manage crucifers' diseases requires the integrated approaches to manage disease resistance with other control measures. The deployment of R-genes cultivars based on qualitative and quantitative pyramiding is considered as a key lever to increase the durability of R-genes against rapidly evolving pathogen populations. The disease resistance can be effectively managed in crucifers' crops against different pathogens under field conditions by fortification with different approaches. To manage white rust under field conditions, in addition to major gene resistant cvs. with slow white rusting and systemic resistance induction approaches may be exploited. Non-host resistance (NHR) is the most durable form of plant immunity effective against all genetic variants of all pathogens and it is multilayered. NHR-related genes can be transferred to host plants to confer durable resistance to *Alternaria brassicae* and other pathogens of crucifers. Multilayered and multicomponent mechanisms of resistance to crucifers' powdery mildew pathogen may be more durable along with major gene resistance under field conditions. The competitive ability of the disease resistance genotype increases significantly in the presence of *Hyaloperonospora parasitica* isolate when competing with susceptible genotype. Small RNA inhibits infection of *Hyaloperonospora*. The deployment of different

G. S. Saharan et al., *Genomics of Crucifer's Host-Resistance*,
https://doi.org/10.1007/978-981-16-0862-9_7

R-genes cvs. of *Brassica* against *Leptosphaeria* through space and time derives the genetic composition of the pathogen populations. The deployment of R-genes cvs. of *Brassica* to manage clubroot has been the dominant component to manage *Plasmodiophora* in several countries. The evolution of new virulence in the pathogen has eroded resistance in these countries. R-genes pyramiding in *Brassica* cvs. may provide protection against different *P. brassicae* pathotypes. The erosion of R-genes in *Brassica* may be managed through integrated approaches such as use of non-host and basal resistance, use of bacterial and fungal biocontrol agents, use of ProCa, nitrogen and arginase gene to modulate resistance positively. Introgression of R-genes from *B. carinata* against *Pseudocercosporella* (white leaf spot) in *Brassica* crops has enormous potential. Hypovirulent isolates of Sclerotinia with the potential in reducing virulence in pathogen populations can help to manage stem rot disease. The breeding of TuMV *Brassica* cvs. can be achieved through transformation of R-genes into crops, molecular marker-assisted selection (MAS), pyramiding of R-genes, and host-induced gene silencing (HIGS). Interspecific hybridization with related or distant species followed by embryo rescue or somatic hybridization for the development of black rot (*Xanthomonas*)-resistant hybrids has been accomplished. Beet cyst nematode has been managed by pyramiding of R-genes in *Brassica*. Multiple pathogen disease resistance of *Brassica* has been managed by development of transgenics, use of fungal and bacterial biocontrol agents, and use of other defense-related genes like WRKY and CAMTA. The breakdown of resistance in *Brassica* can be assessed with molecular approaches to prevent another breakdown of disease resistance. The phenomena of net cost of R-genes resistance in the diseased crucifers operates under natural cultivation system.

Keywords

Management · Disease resistance · Deployment of R-genes · R-genes pyramiding · Integrated approaches · Molecular markers · Marker-assisted breeding · Omics approaches · Systemic resistance induction approach · Multilayered and multicomponent mechanisms · Management of R-genes resistance · Multiple pathogen disease resistance management

7.1 Introduction

Management of crucifers' disease resistance for effective utilization under field conditions is very important and challenging. The *Brassica* genomes (A-, B-, and C-genome) are crucially important to provide novel genetic inheritance for economically important traits that can be used for the overall improvement of crop production and quality. For example, single genome of diploid *B. rapa* (Agenome) holds more than 230 R-gene sequences in 16 gene families (Cheng et al. 2011), among which, over 8 genetic loci have been identified in different research studies, which have functional properties for crucifers' disease resistance. There are possibilities of the existence of more R-genes specific to the disease resistance and their allelic variations in different genetic pools or wild relative species. Effective utilization of resistance loci and their allelic variations may enhance the durability of resistance against different diseases in different *Brassica* species. As a long history of host and pathogen association, it has been revealed relatively high evolutionary patterns of the pathogens. In various cultivating geographical regions of *Brassica* crops, persistence of different pathotypes with high levels of pathogenicity poses challenges to breed durable disease-resistant cultivars. The breakdown of disease-resistant cultivars has become a serious problem in various countries of the world. Effective management of resistance genetic resources in breeding novel cultivars could enhance the performance of resistance loci in different *Brassica* species for sustainable, more durable, and cost-effective control of the crucifers' diseases. Understanding the evolutionary potential between the pathogen and the host is essential in developing a durable resistant *Brassica* crops.

7.2 *Brassica-Albugo*-White Rust Management

7.2.1 Use of Slow White Rusting Genotypes/Cultivars

Rate of infection or disease spread is influenced by both incubation and latent periods of *A. candida* in its compatible host. In WR, the sporangia become visible after the host epidermis is ruptured as a white powdery mass which can readily be dispersed by wind or water-splash to cause secondary infection. According to Liu et al. (1989), WR pustules become visible 5–6 days after inoculation. However, Coffey (1975) observed symptoms after 8 days on the under surface of the cabbage leaves. Longer incubation and latent periods have been observed in slow-rusting *B. juncea* cvs. Rajat (11/14 days) and cv. RC 781 (11/15 days). Similarly, *B. rapa* cvs. Candle (11/15 days), Tobin (15/18 days), and Span (11/18 days) also have longer incubation and latent periods (Lakra and Saharan 1988; Jat 1999; Gupta and Saharan 2002; Saharan et al. 2014). There is a need to identify genotypes with slow-rusting attributes to curb the epidemic development of WR in field. Slow-rusting resistance in crucifer genotypes can be accessed through low infection frequency, low spore production, and a long latent and a short infection period of *A. candida* (Saharan 2010).

7.2.2 Use of Systemic Resistance Induction Approach

Induction of systemic resistance in *B. juncea* to WR by pre- or co-inoculation with an incompatible isolate of *A. candida* has been observed by Singh et al. (1999). In an investigation of the interaction between two isolates of *A. candida* that were compatible and incompatible on a *B. juncea* accession, the incompatible isolate induced both local and systemic protection of cotyledons and true leaves against the compatible isolate. The extent of the protection was proportional to the zoosporangial concentration used in inducing infection by incompatible isolate. Protection was greatest locally on cotyledons, and least on true leaves (the most remote tissue from the point of inoculation). Protection occurred only when the two isolates were inoculated together, and it was maximum when the interval between the incompatible and compatible isolate inoculations was longer. The incompatible isolate induced only slight protection when it was inoculated after the compatible isolate. No induced-susceptibility to the incompatible isolate occurred with any treatment. There was some evidence of competition between compatible and incompatible zoospores for infection sites (stomata). The occurrence of systemic protection, changes in phenylalanine ammonia lyase, and total soluble peroxidase activities in inoculated cotyledons, particularly after the inoculation of incompatible isolate, suggested that host-mediated factors may also be involved in the interaction between the two isolates (Plate 7.1; Figs. 7.1 and 7.2).

However, altered phenotypic response to *Hyaloperonospora parasitica* in *B. juncea* seedlings following prior inoculation with an avirulent or virulent isolate of *A. candida* was observed by Singh et al. (2002). Prior inoculation with the incompatible isolate of *A. candida*-induced resistance to subsequently inoculated *H. parasitica.* The degree of resistance was proportional to the zoosporangial concentration of the incompatible isolate, and induced resistance was more marked in the cotyledons receiving the inducing inoculum compared to the opposite cotyledons, and subsequently emerging true leaves which had not been pre-inoculated. Induction of resistance was also observed if the incompatible isolate of *A. candida* and *H. parasitica* were co-inoculated simultaneously. The effect, however, was greater if the interval period between inoculations was up to 4 days. When the incompatible isolate of *A. candida* was inoculated 4 h after *H. parasitica*, there was no marked effect on resistance to the later. In contrast, prior inoculation with the compatible *A. candida* isolate increased susceptibility to *H. parasitica.* Pre- or co-inoculation with *H. parasitica* suppressed the development of the compatible *A. candida* isolate. A spectrum of responses was observed when one cotyledon was inoculated simultaneously with both the incompatible and compatible isolates of *A. candida,* and followed subsequently with *H. parasitica* after different time intervals. This range of altered responses appeared to reflect the outcome of the differing kinetics and counter-effects of resistance and susceptibility induction.

7.3 *Brassica-Alternaria* Blight Management

7.3.1 Exploring Non-host Disease Resistance Mechanisms

Non-host resistance (NHR) is the most durable form of plant immunity effective against all genetic variants of a pathogen. NHR is multilayered and can be divided into two major types: the pre-invasion layer and the post-invasion layer. The pre-invasion layer has both pre-formed and inducible defense mechanisms. Pre-formed defense involves the physical and chemical barriers that prevent fungal

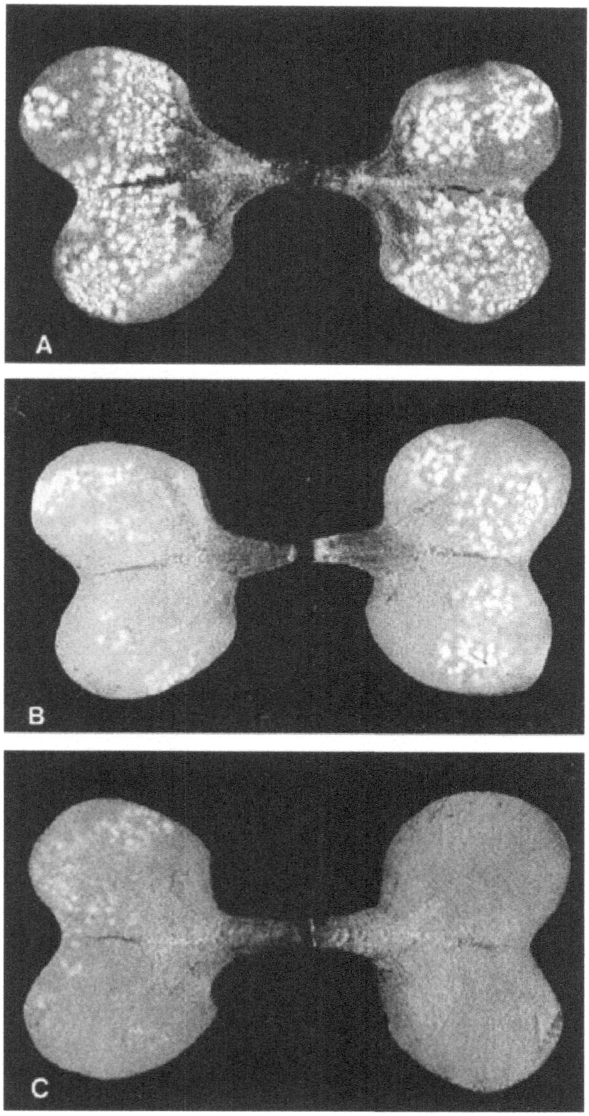

Plate 7.1 Symptom development on cotyledons of *Brassica juncea* accession PPBJ-1 at 7 days after inoculation with incompatible (IN) and compatible (CO) isolates of *Albugo candida*. (**a**) CO isolate inoculated on both cotyledons; (**b**) CO and IN isolates inoculated together on left cotyledon; CO isolate applied alone to right cotyledon; (**c**) CO and IN isolates inoculated together on left cotyledon; IN isolate applied alone to right cotyledon (Singh et al. 1999)

entry inside the plant. In addition, the plant recognizes pathogen-associated molecular patterns (PAMPs) and elicitor molecules from fungal pathogens and induces defense responses against them. Meanwhile, the post-invasion layer mainly comprises inducible defense responses from the plant. In the post-invasion layer, the defense response culminates in a hypersensitive response (HR) and reactive oxygen species (ROS)-mediated cell death. These defense responses in the post-invasion layer can further restrict fungal growth inside the plant (Lee et al. 2017; Fonseca and Mysore 2018). These NHR layers together offer durable resistance and

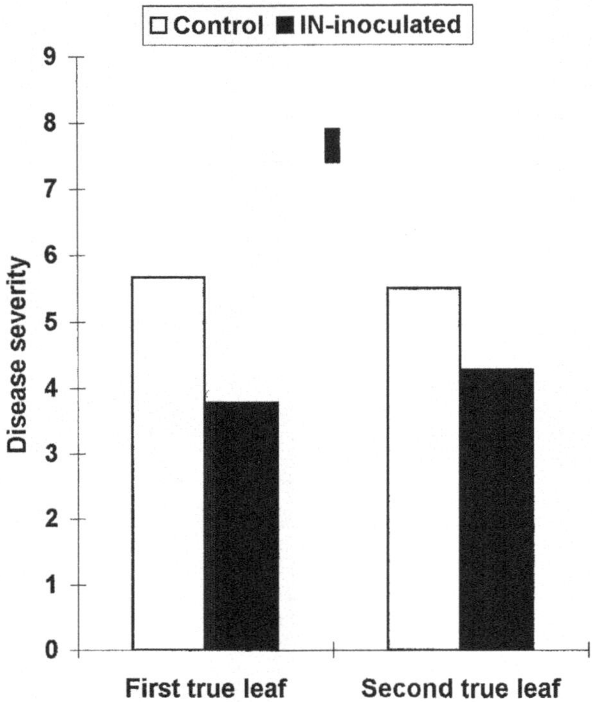

Fig. 7.1 White rust severity on first and second true leaves of *Brassica juncea* when the compatible *Albugo candida* isolate was applied 5 days after both cotyledons had been sprayed either with sterile distilled water (control), or inoculated with the incompatible (IN) isolate. Bar represents LSD ($P = 0.05$, df = 8) (Singh et al. 1999)

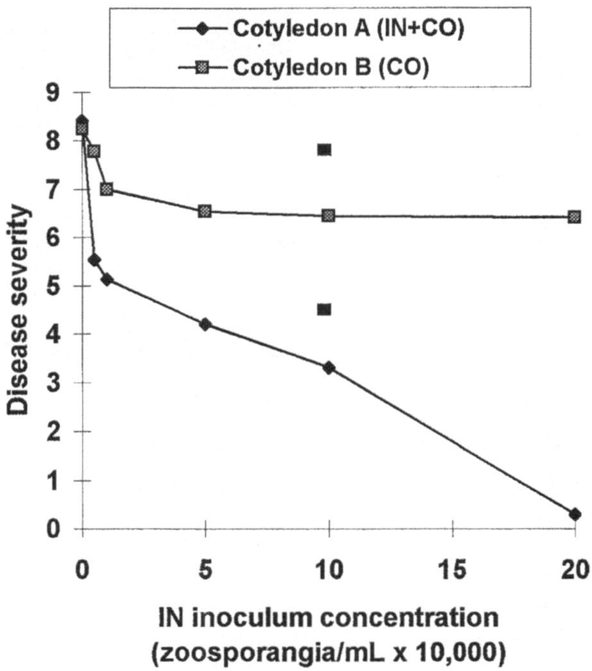

Fig. 7.2 Effect of incompatible (IN) isolate (*A. candida*) inoculum concentration on disease severity on *Brassica juncea* after inoculation with the compatible (CO) isolate. On each seedling, cotyledon A was inoculated with both the IN and CO isolates; at the same time, cotyledon B was inoculated with the CO isolate alone. The top bar represents LSD ($P = 0.05$, df = 24) for comparisons at different inoculum concentrations. The bar between the lines represents LSD ($P = 0.05$, df = 12) for comparisons of disease reaction on the two cotyledons at a particular IN inoculum concentration (Singh et al. 1999)

can be utilized for developing disease-resistant crops. The identification of the genes involved in NHR and the elucidation of the NHR mechanisms against *A. brassicae* should be exploited. Eventually, NHR-related genes can be transferred to host plants to confer durable resistance to *A. brassicae.*

NHR is expressed by all the members of a plant species to a particular pathogen. Exploiting NHR has been suggested as an option for developing broad spectrum disease resistance in plants (Ellis 2006; Lee et al. 2017). Considerable progress has been made in understanding the interaction of several host species and *A. brassicae*, but there are few and indirect reports on studies examining the NHR mechanisms involved during the interaction of *A. brassicae* with non-host plants. Understanding the mechanisms and identifying the genes involved in NHR to *A. brassicae* may provide an opportunity to engineer transgenic *Brassica* plants that provide durable resistance to *A. brassicae.* Some success in genetic transformation involving the transfer of resistance genes from non-host plants into oilseed *Brassicas* has been reported. Mondal et al. (2003) found that the overexpression of tobacco chitinase conferred resistance to Alternaria leaf spot in transgenic *Brassica*, as demonstrated by the inhibitory effect of the transgenics on the hyphal growth of *A. brassicae.* They further reported the development of transgenic *B. juncea* (cv. RLM 198) constitutively expressing a class I basic glucanase of tomato, which arrested the hyphal growth of *A. brassicae* by 15–54% (Mondal et al. 2007). Moreover, the introduction of the chitin-binding lectin hevein from rubber plants into *B. juncea* led to resistance to *A. brassicae* (Kanrar et al. 2002). Such studies highlight the necessity to dissect the mechanisms and reveal the genes involved in NHR for the development of resistant cultivars.

For this, the first step is to identify potential non-host plants that can reveal robust and exploitable mechanisms for suitable transfer to a host plant for imparting broad spectrum resistance. Besides, the non-host plant should be distinct from the non-adapted plant, where, for example, the fungal spores fail to adhere, deposit, and remain on the leaf surface for germination (Niks and Rubiales 2002). Some of the non-host plants for *A. brassicae* have been identified by intercropping systems followed in different parts of world. In many intercropping systems, species of the Brassicaceae family are cultivated with certain other plant species in the same field for many decades (Singh et al. 2010; Shekhawat et al. 2012; Ao et al. 2016; Ebrahimi et al. 2016; Jeromela et al. 2017). In the same field, *A. brassicae* infects *Brassica* plants but cannot infect the other plant species (http://www.plantwise.org/KnowledgeBank). The Plant wise Knowledge Bank database (http://www.plantwise.org/KnowledgeBank) lists the plant species infected by *A. brassicae* in different parts of the world. Through this database, ensured that the identified non-host plants have not been reported to be infected by *A. brassicae.* Hence, those species can act as potential non-host plants against this pathogen. Curiously, in the course of evolution, the pathogen gets enough opportunities to infect these non-host plants, but they are unable to overcome the NHR. A few examples of such non-host plants, which are intercropped with host plants such as mustard (*B. juncea*), are chickpea (*Cicer arietinum* and *C. kabulium*), lentil (*Lens culinaris*), cowpea (*Vigna unguiculata*), barley (*Hordeum vulgare*), sugarcane (*Saccharum *officinarum*), and potato (*Solanum tuberosum*) (Table 4.1, Chap. 4). It would be interesting to determine the factors that dictate the failure of *A. brassicae* to infect these non-host plants. The NHR of these plants to *A. brassicae* may involve multiple layers of defense that could be distinct for each plant species (Fatima et al. 2019).

7.3.2 Use of Host Resistance

Four species of *Alternaria*, i.e., *A. brassicae, A. brassicicola, A. raphani,* and *A. alternata* are pathogenic on Brassicas. Out of these, *A. brassicae,* and *A. brassicicola* are widespread, and economically most important since they cause losses of more than 70%. All commercial Brassicas are essentially susceptible to these pathogens. However, there are some differences in their degree of susceptibility. *Brassica napus* and *B. carinata* are less susceptible than *B. rapa* and *B. juncea*. In addition to these, some wild crucifers have high degree of resistance to *A. brassicae,* and attempts have been made to transfer this resistance to cultivated Brassicas. Alternaria blight management strategies are to incorporate the host resistance as one of the components into the cultivated Brassicas. One should have the knowledge of pathogenic variability development, and standardization of effective germplasm screening techniques from diverse sources to broaden the genetic base; availability of usable source of resistance, and nature, and mechanism of resistance. The various techniques are available to screen germplasm/varieties/lines, etc. against *A. brassicae* in the field, screen house/glass house, and in the laboratory for resistance. Various studies have also been conducted which indicate the existence of *A. brassicae* pathotypes (Saharan 1992; Saharan and Kadian 1983b; Saharan et al. 2003, 2016). Inheritance of resistance in inter- and intraspecific crosses of *B. juncea* × *B. carinata* genotypes reveals polygenic, additive gene effects, dominance gene effects, additive × dominance epistasis, and non-allelic interactions. Intermating between resistance plants helps in increasing the level of resistance through resistance gene pyramiding. Disease tolerance attributes in *B. juncea* genotypes have been identified. Identification of different components of horizontal resistance in *B. napus* cv. Tower, and *B. juncea* cv. RC-781 can be exploited to breed for horizontal/slow blighting resistance. Rapid progress has been made in identification, cloning, and sequencing of resistant genes in *B. juncea*, *B. napus*, *R. sativus*, and *A. thaliana* (Krishnia et al. 2000; Saharan and Krishnia 2001; Gupta et al. 2002; Saharan and Kadian 1983a; Kumar and Kolte 2001, 2006; Kanrar et al. 2002; Kamble and Bhargava 2007; Mondal et al. 2007; Minami et al. 2008; Mishra et al. 2010; Verma et al. 2012; Chhikara et al. 2012; Sellam et al. 2007).

The effect of host-specific phytotoxins produced by *Alternaria* species on cruciferous plants at physiological, biochemical, and molecular levels have been determined. The role of phytotoxins, and other metabolites produced during *Alternaria–Brassica* interactions on pathogenesis, and host defense along with activity of MAPK 4 cascade gene in plant defense against biotic stress has been identified (Bains and Tewari 1989; Verma and Saharan 1994; Pedras et al. 2009; Thomma 2003; Marmath et al. 2013). The nature and mechanisms of resistance in cruciferous crops is multilayered. It is in the form of morphological barriers/features like fewer numbers of stomata with narrow aperture in some genotypes of *B. napus* and *B. juncea*. The presence of epicuticular wax layers in *Brassica* genotypes provides a hydrophobic leaf surface, which hinders *Alternaria* spore germination, and infection. A large number of biochemical constituents govern biochemical resistance. Proteome, and enzymes activity in resistant genotypes activates resistant genes. Hypersensitive response genes for induced resistance have been identified. Expression of MAPK4 (Mitogen-activated protein kinase) cascade gene antagonizes the effects of pathogens in resistant genotypes. Elicitation of phytoalexins by *Brassica* species inhibits fungal growth on host leaf surface. Calcium sequestration character of Brassicas gives ample possibilities of enhancing resistance to *Alternaria* spp. infection. Sources of resistance in cruciferous species (closely as well as distantly related) against *Alternaria* spp. have been identified. But, very few have been explored for commercial use to develop resistant cultivars. The information generated so far on parameters of host resistance is being utilized through conventional, and biotechnological approaches (Verma and Saharan 1994; Saharan et al. 2003; Saharan and Kadian 1983a, b; Kumar and Kolte 2006; Gupta et al. 1990; Dhingra et al. 2004; Oh et al. 2005; Mishra et al. 2010; Marmath et al. 2013; Conn et al. 1991; Sellam et al. 2007; Aneja and Agnihotri 2013).

7.4 *Brassica-Erysiphe* Powdery Mildew Management

7.4.1 Multilayered and Multicomponent Mechanisms to Manage Powdery Mildew Resistance

Host resistance in crucifers to powdery mildews is multilayered, and multicomponents both at pre- and post-penetration stages. The intricate immune responses are evolved through accumulation of ROS, H_2O_2, deposition of callose, pectin, cellulose, waxes, silicon, ion fluxes, formation of papilla, cell wall apposition, phenolic compounds, overexpression of R-genes, PR proteins, protein phosphorylation, biosynthesis of phytoalexins, fungal enzymes inhibitors, chito-octamers, triggering of HR, induction of SAR, and non-host resistance mechanisms. These cytoskeleton components have very important and crucial functional and structural roles in host resistance to powdery mildew pathogens of crucifers. Defenses are activated either through SA signaling, and simultaneous perception of ethylene, and jasmonic acid (JA). The overexpression of several R-genes in crucifers–powdery mildew host-pathosystem induces host resistance. MLO genes encoding seven-transmembrane, calmodulin-binding protein confers broad spectrum resistance to adapted powdery mildews of *Arabidopsis*. *edr* mutants of *Arabidopsis* have a general link between SA-mediated resistance, mitochondrial function, and programmed cell death. The pmr mutants confer resistance to powdery mildew through altered cell wall composition of host. Increased SA enhances the expression of RPW 8.1, and RPW 8.2 leading to HR, or SHL, and resistance. Bj NPR1 gene activates SAR to confer broad spectrum resistance to powdery mildew of *B. juncea*.

At ROP regulated At RLCK V1 A3 has a role in basal resistance to powdery mildews. The At MLO2, At MLO6, and At MLO12 triple mutants are resistance to G. *orontii*. CPR5 controls resistance to powdery mildews, and PCD in response to infection by *E. cruciferarum*. There is a role of WRKY transcription factors, and overexpression of R-genes like PMR, MLO, PEN, EDR, MAPK, MAPK 65-3, NPR1, PAD3, PAD4, ED5, SNARE, RLCKs, and KDL (At CEP1) to confer R to powdery mildews of crucifers. Higher levels of camalexin contribute to the enhanced R to powdery mildew in Cyp83 a1-3 mutants of *Arabidopsis*. SR1 plays a critical role in powdery mildew resistance by regulating EIN3 and NDR1 expression. There is harmonious coordination between transcriptional regulation and resistance to powdery mildews. The application of *Trichoderma harzianum* and its CF induces (ISR) resistance in crucifers. Mechanisms of non-host R in crucifers to powdery mildews have been unrevealed which is strong and durable. Non-host resistance is PEN-gene-mediated at pre-invasion, and controlled by genes EDS1, PAD4, and SAG (101) at post-invasion of powdery mildew pathogens. In cabbage, a single dominant gene with modifiers controls R to powdery mildew. A single R-gene controls R to powdery mildew in HC-1, and PCC-2 with complete dominance. In *Arabidopsis*, R to PM is polygenic, and based on R-gene *RPW8* or on combination of *RPW* 8 gene complex loci. Powdery mildew resistance genes of *Arabidopsis* have been mapped on chromosomes II (RPW1), III (RPW2, RPW3, RPW7, RPW8), IV (RPW4), and V (RPW 5, RPW 6). In rapeseed gamma rays mutagenic plants exhibit R to powdery mildew due to an increase in concentration of unsaturated fatty acids with 18 carbon atoms. Induction of glucosinolates and camalexin plays important roles for resistance to powdery mildew of crucifers. Camalexin biosynthesis and accumulation are affected by WRKY 18, WRKY 40 transcription factor of *Arabidopsis*, and enhance upon *G. orontii* infection to confer resistance. Transfer of powdery mildew resistance to *B. oleracea* from *B. carinata* through embryo rescue has been confirmed. Major gene sources of resistance against powdery mildew of crucifers have been identified (Saharan et al. 2019).

7.5 *Brassica-Hyaloperonospora* Downy Mildew Management

7.5.1 Disease Resistance Increases Competitive Ability of Host Plants

Disease resistance is believed to play a role in the dynamics of plant communities and plant populations. The competitive ability of two *Arabidopsis thaliana* genotypes, one exhibiting resistance, and the other exhibiting susceptibility to an isolate of *H. parasitica* was determined by Damgaard and Jensen (2002). The *A. thaliana* genotypes were grown in competition experiments under controlled conditions at three densities in pure and mixed stands both in the presence and absence of *H. parasitica*. After seed set, the dry weight of the plants was determined, and the amount of seeds produced was calculated from an established linear relationship. The two genotypes were found to be ecologically different; the susceptible

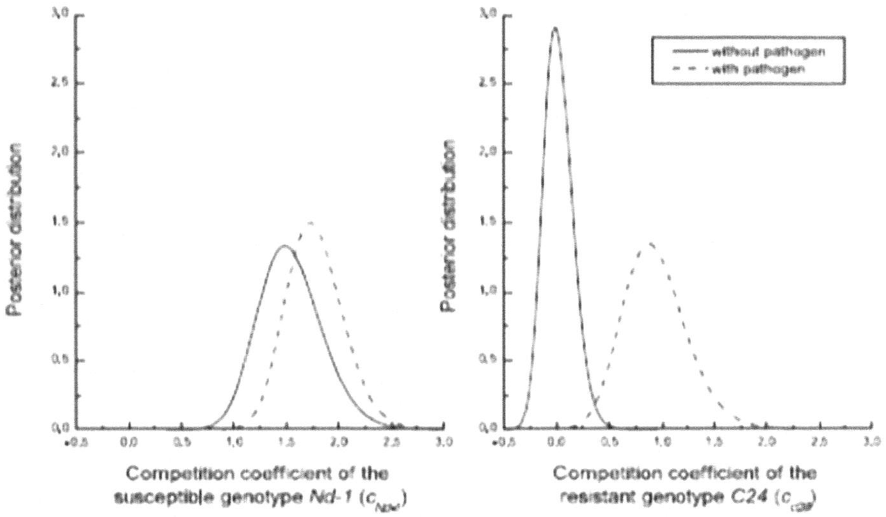

Fig. 7.3 The Bayesian posterior distribution of the competition coefficient of two *Arabidopsis thaliana* genotypes, *Nd*-1 (susceptible) and *C*24 (resistant), competing against each other. Solid line: without pathogens. Dashed line: with pathogens. Percentiles (2.5%, 50%, and 97.5%) in the posterior distributions: *cC*24: without pathogen: (−0.23, 0.01, 0.31), with pathogen: (0.40, 0.92, 1.60), *cNd*-1: without pathogen: (0.99, 1.52, 2.19), with pathogen: (1.28, 1.76, 2.34). The Bayesian posterior distribution of a parameter provides information on our degree of beliefs in the different possible values of the parameter. The mode of the posterior distribution corresponds to the maximum likelihood value, and increasing variance of the distribution corresponds to an increasing degree of uncertainty about the true value of the parameter (increasing experimental variation) (Damgaard and Jensen 2002)

genotype was the strongest competitor (significantly so in the absence of the disease), whereas the resistant genotype produced most seeds. The competitive ability of the disease-resistant *A. thaliana* genotype increased significantly in the presence of a *H. parasitica* isolate when competing with the susceptible *A. thaliana* genotype, whereas the disease did not affect the competitive ability of the susceptible genotype significantly. Assuming that the population only consisted of the two genotypes, and mainly was affected by the disease it was possible to predict the probabilities of four long-term ecological scenarios. Without the disease the most likely long-term ecological scenario was that the two *A. thaliana* genotypes would co-exist, whereas, in the presence of the disease, the resistant genotype most likely would outcompete the susceptible genotype (Fig. 7.3).

7.5.2 Small RNA Inhibits Infection of *Hyaloperonospora arabidopsidis*

Non-coding 20–30 nucleotide (nt)-long small RNAs (sRNAs) have been known to be involved in the regulation of gene expression and defense in eukaryotes (Chen et al. 2018; Qin et al. 2017; Zhang et al. 2019). Different types of RNAs, such as double-stranded RNA (dsRNA) and small interfering RNA (siRNA), can trigger homologous RNA degradation or inhibit mRNA translation (Huang et al. 2016; Nejat and Mantri 2018). This process is known as RNA silencing and plays a significant role in various biological processes, including innate immunity (Brant and Budak 2018; Deng et al. 2018) and development (Li et al. 2017; Qin et al. 2017).

Movement of sRNAs from plant to pathogens has been explored using the host-induced gene silencing (HIGS) technique where the sRNAs are generally made by producing dsRNA in transgenic plants using *Agrobacterium* or in viruses that replicate through dsRNA. HIGS has been successfully used to suppress essential pathogen genes in various plant–pathogen interaction systems including barley–*Fusarium* (Koch et al. 2013), *Arabidopsis*–and tomato–*Verticillium* (Song and Thomma 2018), barley–and wheat–*Blumeria* (Nowara et al. 2010), and lettuce–*Bremia* (Govindarajulu et al. 2015). Similarly, HIGS has also been used against nematodes in *Arabidopsis* (Huang et al. 2006). Several recent studies have shown that exogenously applied sRNAs can be taken up by fungal or plant cells and trigger RNA silencing. The exogenous application of RNA by spraying it directly onto plants has been referred to as spray-induced gene silencing (SIGS) (Koch et al. 2016). The induction of gene silencing by spraying, or otherwise applying RNA, avoids the need to develop transgenic plants (Wang and Jin 2017). Several different methods have been used to deliver sRNAs onto plants. For example, high-pressure spraying siRNAs was found to efficiently silence transgenic GFP gene expression in *Nicotiana tabacum* (Dalakouras et al. 2016).

Hyaloperonospora arabidopsidis (*Hpa*) is an obligate biotrophic oomycete pathogen that causes downy mildew disease on *Arabidopsis thaliana*. The *Hpa*–*Arabidopsis* system has been used as a model to investigate pathogen effectors (Woods-Tor et al. 2018) and plant immunity (Holub 2007). This pathogen has both sexual and asexual reproduction. For infection, an asexual conidiospore germinates on the surface of plant leaves, forms an appressorium, then a penetration hypha grows between the walls of neighboring epidermal cells (Koch and Slusarenko 1990). In susceptible plants, the hypha branches out into the intercellular space and forms haustoria in the epidermal and mesophyll cells. Within 1–2 weeks, conidiophores develop through stomata, carrying conidiospores that begin new rounds of infection. Sexual spores, called oospores, are produced within the cotyledons or on the leaves of the infected plant (Koch and Slusarenko 1990). Bilir et al. (2019) used *in vitro* synthesized sRNAs targeting the *Hpa*-CesA3 gene and report that antisense or double-stranded capped sRNAs of 25 nt or longer inhibit spore germination and hence infection.

Gene silencing exists in eukaryotic organisms as a conserved regulation of the gene expression mechanism. In general, small RNAs (sRNAs) are produced within the eukaryotic cells and incorporated into an RNA-induced silencing complex (RISC) within cells. However, exogenous sRNAs, once delivered into cells, can also silence target genes via the same RISC. Bilir et al. (2019) explored this concept by targeting the cellulose synthase A3 (CesA3) gene of *Hyaloperonospora arabidopsidis* (*Hpa*), the downy mildew pathogen of *Arabidopsis thaliana*. *Hpa* spore suspensions were mixed with sense or antisense sRNAs and inoculated onto susceptible *Arabidopsis* seedlings. While sense sRNAs had no obvious effect on *Hpa* pathogenicity, antisense sRNAs inhibited spore germination and hence infection. Such inhibition of infection was not race-specific, but dependent on the length and capping of sRNAs. Inhibition of infection by double-stranded sRNA was more efficient than that observed with antisense sRNA. Thus, exogenous sRNA targeting conserved CesA3 could suppress *Hpa* infection in *Arabidopsis*, indicating the potential of this simple and efficient sRNA-based approach for deciphering gene functions in obligate biotrophic pathogens as well as for R-gene-independent control of diseases in plants (Bilir et al. 2019).

7.6 *Brassica-Leptosphaeria* Blackleg Management

7.6.1 Deployment of *Brassica R*-Genes Cultivars to *Leptosphaeria*

The deployment of different resistance genes through space and time drives the genetic composition of the pathogen population. Because epidemics of successive cropping seasons are not independent, the adaptation of pathogen populations to host resistances proceeds at the scale of a network of fields on which the selection pressures are not homogeneous, during a succession of cropping seasons. The resistance genes in varieties exert directional selection for isolate-host compatibility. Predictable changes in pathotype frequencies were observed following commercial use of genes, or in field experiments (Hovmoller et al. 1993; Brun et al. 2010; Daverdin et al. 2012). In *B. napus,* there are more than 15 qualitative resistance genes conferring resistance to *L. maculans*. Their presence in host varieties can be detected using sets of *L. maculans* isolates to inoculate seedlings under controlled conditions. Phoma stem canker populations are diversified worldwide, with contrasted histories regarding the use of plant resistance genes (Zhang et al. 2016) and subsequent adaptation of pathogen populations (Van de Wouw et al. 2017). Populations of *L. maculans* collected from hosts with different resistance gene combinations harbor different infectivity allele frequencies (Van de Wouw et al. 2014, 2017). When resistance genes have been deployed in oilseed rape varieties, phoma stem canker populations have repeatedly become adapted over a few years to the *B. napus* genes *Rlm1* in Europe and Australia (Rouxel et al. 2003); *LepR3* (*Rlm1*) and *LepR1* in Australia (Van de Wouw et al. 2014, 2017); *Rlm7* in Europe (Leflon 2013; Winter and Koopmann 2016); and *Rlm3* in Canada (Zhang et al. 2016). Thus, *L. maculans* has repeatedly evolved virulence against nearly all major resistance genes released in oilseed rape so far, sometimes with almost complete yield loss.

On *B. napus*, phoma stem canker is one of the many examples of loss of efficacy of genes following their use in varieties of this crop. However, a slightly different concern appears when the same fungus can cause epidemics on several related hosts, as are *B. juncea* and *B. napus*. In Australia, "*juncea* canola" varieties were commercialized in 2007 (Burton et al. 2008). Their resistance genes were unknown, but assumed to be different from those of the *B. napus* varieties (Marcroft et al. 2012b; Elliott et al. 2015). *Rlm5* and *Rlm6* are reported to be present in *B. juncea* (Chèvre et al. 1997; Barret et al. 1998; Balesdent et al. 2002, 2005), as well as a recessive resistance gene (*LMJR2*) (Elliott et al. 2015). Prior to the commercialization of "juncea canola" there were only small areas of *B. juncea* for condiment use; however, phoma stem canker infection was already occurring. Isolates virulent on *B. juncea* are widespread throughout the Australian canola growing regions (Elliott et al. 2015; Van de Wouw et al. 2017).

In France, *B. juncea* is not grown over wide acreages; however, in the Dijon area it has both a patrimonial importance and contributes significantly to the local economy, for the production of Dijon mustard. Formerly imported from Canada, mustard seed production has been significantly increased since 2010, up to 6000 ha in Bourgogne in 2016–2017 (http://www.cote-dor.chambagri.fr) and stable since. Phoma stem canker is not currently a disease problem on *B. juncea*. Intercrop *Brassica* species used to prevent nitrogen leaching are mainly *Sinapis alba*, or *Raphanus sativus* and only few are *B. juncea* selected for the biofumigating action of sinigrin. As opposed to Canada and Australia where virulent isolates have been detected in *L. maculans* populations (Liban et al. 2016; Van de Wouw et al. 2017), the frequency of isolates virulent on *Rlm6* is very low in France, and the one on *Rlm5* intermediate, locally as low as 43% (Rouxel et al. 2003; Balesdent et al. 2005, 2006). Due to this, there is potentially only a very low frequency of isolates that can attack *B. juncea,* compared to *B. napus*. *Rlm6* gene has been introduced from *B. juncea* into winter *B. napus* (Chèvre et al. 2008). Bousset et al. (2018) assessed the presence of most common genes against phoma stem canker, including *Rlm5* and *Rlm6,* in a set of the 12 current French *B. juncea* condiment mustard varieties and breeding lines.

Breeding varieties for increased disease resistance is a major means to control epidemics. However, the deployment of resistance genes through space and time drives the genetic composition of the pathogen population, with predictable changes in pathotype frequencies. In France, *Leptosphaeria maculans* causes disease on *Brassica napus* oilseed rape crops but not on *B. juncea* condiment mustard. Prior to the deployment of winter *B. napus* varieties with *Rlm6* resistance gene introduced from *B. juncea*, Bousset et al. (2018) investigated if this deployment can impact disease control in condiment mustard. They assessed the presence of resistance genes against phoma stem canker in a set of current French *B. juncea* varieties and breeding lines. *Rlm6* was detected in all the 12 condiment mustard varieties. *Rlm5* was also detected in 8 varieties. No additional resistance genes were detected with the set of isolates used (Table 7.1). Because frequency of isolates virulent on *Rlm6* is very low, these results indicate that *Rlm6* gene is a major component of disease control in the French *B. juncea* mustards tested. Using *Rlm6* in oilseed rape varieties will very likely induce an increase in frequency of *Rlm6* virulent isolates. This raises

Table 7.1 Presence/absence of the avirulence genes in the fungal isolate of *Leptosphaeria* (Bousset et al. 2018)

Isolate	*AvrLm1*	*AvrLm2*	*AvrLm3*	*AvrLm4*	*AvrLm5*	*AvrLm6*	*AvrLm7*	*AvrLm8*	*AvrLm9*	*AvrLm10*	*AvrLmS*	Avirulence genotype	Ref.
P27d	+	–	–	–	+	+	+	nd	–	+	nd	*AvrLm1,5, 6,7,(8),10,(S)*	[a]
05-30	–	+	–	+	–	+	+	nd	–	+	–	*AvrLm2,4,6,7, (8),10, S*	[b]
IBCN16	–	–	–	–	+	–	–	–	–	nd	–	*AvrLm5, (10)*	[c,d]
v34.4.4	–	–	–	–	+	–	+	–	–	nd	nd	*AvrLm5,7,(10),(S)*	[e]
IBCN18	+	+	–	+	–	–	+	–	–	nd	+	*AvrLm1,2,4,7, (10),S*	[c,d]

Avirulence genotypes of *Leptosphaeria maculans* isolates based on disease scores on seedlings of *Brassica* lines. "+" refers to the presence of the avirulence allele; "–" refers to the presence of the virulence allele; "nd" refers to alleles that were not determined. The avirulence genotype indicates the avirulence loci for which the isolate has been characterized as avirulent; this nomenclature is as used by Balesdent et al. (2005)

[a]Delourme et al. (2008)

[b]Kutcher, personnal communication

[c]Balesdent et al. (2005)

[d]Marcroft et al. (2012b)

[e]Balesdent et al. (2002)

the acute concern of a wise deployment of oilseed rape around the condiment mustard growing area. Scientific knowledge on adaptation dynamics, spatial segregation of crops and cooperation between actors is currently available in order to mitigate the risk and advert negative consequences of the introduction of *Rlm6* resistance gene in oilseed rape varieties (Bousset et al. 2018).

7.6.2 Marker-Assisted Selection of Avirulence Gene in *Leptosphaeria* for Effective Deployment of *Brassica*-Resistant Cultivars

Blackleg, which is caused by the fungus *Leptosphaeria maculans* (*L. maculans*), is a major disease of canola in western Canada and worldwide. Long-term use of one source of resistance could cause the breakdown of its effectiveness. Therefore, appropriate use of R-genes is very important, and knowledge about the distribution of avirulence genes is a prerequisite for effectively deploying resistance. Of the 14 avirulence genes identified in *L. maculans*, *AvrLm5,* and *AvrLm9* were recognized as the two alleles of the same gene based on two single nucleotide polymorphisms, C85T and G164A/C. A specific marker was developed to identify *AvrLm5* and *AvrLm9* based on two single nucleotide polymorphisms, C85T and G164A/C, which are responsible for the function of *AvrLm9*. The specific marker can be used to discriminate the *AvrLm9* from *avrLm9* accurately in *L. maculans* isolates, which is consistent with inoculation tests in isolates without *AvrLm4-7*. This specific marker was used to screen 1229 isolates collected from fields in the years 2014 through 2016 in Manitoba. From 68 to 84% of the isolates were found to contain the *AvrLm9* allele; while 4–7% of them were avirulent on the variety Goéland with *Rlm9* loci. Furthermore, no isolates having both *AvrLm9* and *AvrLm7* were detected using a cotyledon test, while 67–84% of isolates contained both avirulence genes via PCR detection, implying suppression of *AvrLm9* by *AvrLm7*. In addition, avirulence gene profiles of the other 10 avirulence alleles were examined with the 1229 isolates using cotyledon tests or PCR amplifications. Taken together, this research enables the fast identification of *AvrLm5/9*, provides the *Avr* genes' landscape of western Canada and elaborates the relationship between *AvrLm9* and *AvrLm7* using isolates from grower fields (Liu et al. 2020).

7.7 *Brassica-Plasmodiophora* Clubroot Management

7.7.1 R-Genes Cultivars of *Brassica* to Control Clubroot in the Field

The first resistant cultivars were Mendel and Tosca, marketed at the beginning of the new millennium. These cultivars are heterozygous hybrids with one gene of pathotype-specific resistance. Unfortunately, the Mendel host plant resistance was broken in some parts of Germany and France, where it was grown since 2000 (Diederichsen et al. 2014). In Great Britain, resistance was broken quite rapidly in the field, where cultivars were grown under strong selection pressure (Diederichsen et al. 2003). These cultivars were also tested with local pathotypes in Canada. Cultivar Mendel showed resistance against pathotypes 3, 5, 6, and 8 (Williams set) (Rahman et al. 2014). Cultivar Tosca was susceptible to pathotype 3 and resistant to pathotype 2 (Williams set) (Peng et al. 2013). These results showed that the host–plant resistance is quite variable and some genes may be ineffective in some regions with specific pathotypes (Peng et al. 2014). New resistant cultivars of spring rape with genes from European cultivars were introduced into Canada in 2009 (Peng et al. 2014). In the European market, the available resistant cultivars are derived from the first resistant cultivar Mendel (Diederichsen et al. 2014). It is expected that a wider range of resistant varieties will come to the market in the next years (Diederichsen et al. 2014). However, these varieties cannot be grown without suitable agronomic measures and pathogen resistance management strategies (Diederichsen et al. 2014). Field experiments with resistant cultivars in infested fields were also conducted in the Czech Republic. Eight resistant cultivars were tested, some of them are normally available on the market and some were test cultivars in the breeding process (new breeds). Normally traded cultivars had stable performance in resistance results and good yield. Cultivars from new breeding showed some variations in resistance performance and yield (Ricarova et al. 2016).

7.7.2 Deployment of R-Genes Cultivars of *Brassica* to Manage Pathogen Selection

Clubroot-resistant canola cultivars have been the dominant component of the clubroot management strategy in western Canada in recent years. The first resistant cultivar for western Canada, Pioneer "45H29," was registered in 2009 and its release was quickly followed by several other resistant cultivars (Strelkov et al. 2011; Gossen et al. 2013). However, the durability of this resistance is not known. The genetic basis for the resistance in these cultivars also, is proprietary knowledge that is not available to the public. This has complicated efforts aimed at resistance stewardship since it is not possible to develop rational strategies for the rotation of resistance sources in infested fields. The number of resistance genes also, that are currently available to breeders is limited (Hirai 2006). Studies indicated that there were no detectable differences in pathotype reactions among clubroot-resistant canola cultivars available in western Canada (Deora et al. 2012, 2013). This represents one line of evidence that resistance genes in all these lines originated from a similar source. In contrast, another study by LeBoldus et al. (2012) revealed that pathogen populations that had been repeatedly cycled on and become adapted to one *Brassica* host did not exhibit a similar adaptation to other hosts, when the cycled populations were inoculated on the latter, indicating that distinct sources of resistance may occur in at least some genotypes. Regardless of the relationships between resistance sources, there is substantial genetic and pathotype variation present in the *P. brassicae* populations in western Canada (Strelkov et al. 2006; Xue et al. 2008; Cao et al. 2009), and the deployment of a cultivar with single-gene resistance against a genetically diverse pathogen on a large scale imposes a strong selection pressure for pathogen genotypes that are able to overcome this resistance (Hwang et al. 2012). Similar mechanisms may operate in other countries also.

7.7.3 R-Genes Pyramiding in *Brassica* Cultivars

The success of gene pyramiding in a cultivar depends on several factors, such as the number of genes to be transferred and the ease of genotyping and phenotyping the plants. Validation of the number of resistance genes pyramided in a single plant or a cultivar is a challenging task, as a plant carrying one or multiple resistance genes can show similar resistance phenotype unless the different resistance genes confer resistance to different pathotypes. Also, testing a plant for resistance to multiple pathotypes is not feasible in a backcross breeding program. In this regard, development of gene-specific markers for use in marker-assisted selection would facilitate the gene pyramiding task greatly. Among the cultivated Brassicas, resistance to *P. brassicae* in *B. rapa* often shows race-specific reaction controlled by major genes, while resistance in *B. oleracea* is often controlled by multiple QTL. In the case of *B. napus*, some of the swede or rutabaga (*B. napus* var. *napobrassica*) genotypes have been reported to possess resistance to different *P. brassicae* pathotypes.

RNA-Seq is a technique that allows fast and extensive study of transcriptional levels of gene expression. A number of published transcriptome analysis results of the defense responses of Brassica crops against *P. brassicae* at early infection stages at different time points, and resistant and susceptible lines were compared (Chen et al. 2016; Wang et al. 2019). The results have shown that despite the presence of more robust effector-triggered immunity, the salicylic acid (SA) signaling pathway plays a pivotal role in clubroot resistance in *Brassica rapa*, which contains the *CRb* resistance gene (Chen et al. 2016), and strongly and rapidly accelerates receptor kinase and G proteins in *B. napus* (ZHE-226) resistance lines that contain the *PbBa8*.1 resistance gene, which may trigger reactive oxygen species (ROS), programmed cell death (PCD), and the hormonal signal transduction signaling pathway (Mei et al. 2019). Previous studies were based on lines that were resistant or susceptible to clubroot isolates, and the resistant lines consisted of single resistance genes, in either a homozygous or heterozygous state at the locus. Shah et al. (2020) pyramided two resistant genes (*CRb* and *PbBa8*.1) and developed a homozygous line. The lines show strong resistance against a number of *P. brassicae* field isolates compared with heterozygous pyramided *B. napus* and single resistant homozygous and heterozygous lines (Shah et al. 2019). Historically, the molecular mechanism of the strong resistance of pyramided lines has been poorly understood. Therefore, a comparative transcriptomic analysis between 618R and its parental lines (305R and 409R) was performed in order to understand the superior resistance mechanisms of the pyramided line and to elucidate the genetic compatibility of the two resistant genes in a single line.

The pyramiding of *PbBa8*.1 and *CRb*-resistant genes strongly induces resistance against a number of *P. brassicae* field isolates (Shah et al. 2019). To understand the molecular mechanism of how these two resistance genes elevate resistance in a single line, Shah et al. (2020) selected only one time point to comparatively investigate the genetic coordination of the pyramided line (618R) with its parental lines (305R and 409R), rather than focus on the change in the transcriptome at different time points after clubroot pathogen inoculation. It was shown that *PbBa8*.1 shares a greater number of DEGs than *CRb* and plays a dominant role, with a higher number of non-additive genes than the pyramided line. This insight revealed that the pyramided line strongly triggers multiple resistance pathways, including the plant–pathogen interaction, plant hormone signal transduction, secondary metabolites, transcription factors, and other processes, indicating that the plant has responded via a multi-gene network to pathogen attack. Most interestingly, it was shown that the SA signaling pathway was higher in 409R than 305R, and the ROS response was greater than signaling in 305R. However, higher expressions were evaluated in 618R controlled by SA and ROS in the pyramided line. It will provide comprehensive transcriptomic insight into uncovering the interaction of pyramiding resistance genes in a single line, and this architecture further facilitates breeding programs with improved resistance efficiency in *B. napus*.

7.7.4 Non-host and Basal Resistance to Clubroot

Non-host resistance is the resistance shown by an entire plant species to all genetic variants of a pathogen (Mysore and Ryu 2004). In this context, the pathogen is referred to as non-adapted to the plant. Basal resistance is the first line of defense of a host to the adapted pathogen, which is achieved through a set of defined receptors, also referred to as pattern recognition receptors (PRRs), that recognize and conserved microbe-associated molecular patterns (MAMPs) (Jones and Dangl 2006; de Wit 2007). Upon MAMP recognition, primary defense responses are induced such as cell wall alterations, deposition of cellulose, and the accumulation of defense-related proteins (van Loon et al. 2006). A new conceptual framework has been recently proposed that non-host resistance and basal resistance to adapted pathogens may rest on similar principles (Niks and Marcel 2009; Fan and Doerner 2012). Based on studies of other pathosystems, plant primary metabolism has been demonstrated to play a significant role in non-host and basal resistance (Stuttmann et al. 2011).

The involvement of plant primary metabolism in clubroot pathogenesis has been studied, with the induction of sink metabolism regarded as a nonspecific response of host roots to *P. brassicae* infection (Ludwig-Muller et al. 2009). In *P. brassicae*-infected *Arabidopsis* roots, expression of sucrose synthase and starch synthase is induced (Siemens et al. 2006). The lower rate of photosynthesis is associated with an increase in invertase activity in the roots, and an accumulation of hexoses with a down-regulation of photosynthetic gene expression is observed (Siemens et al. 2011). It is postulated that plant genotypes with down-regulated sucrose/starch synthase or invertase activity may carry a basal resistance to the pathogen (Gravot et al. 2011). The induction of plant trehalase during clubroot disease in *Arabidopsis* is a defense mechanism, which can help the plant to cope with the accumulation of pathogen-synthesized trehalose. This mechanism of resistance is also found in the susceptible *Arabidopsis* accession Col-0, and the inhibition of trehalase activity during clubroot infection results in enhanced disease symptoms, suggesting that trehalase induction is a basal defense mechanism and act as a disease resistance mechanism activated by a virulent pathogen on a susceptible host (Jones and Dangl 2006).

7.7.5 Bacterial Biocontrol Agents to Manage Clubroot

Bacillus subtilis XF-1 is an effective strain with great biocontrol capabilities. It has a broad antimicrobial spectrum and proved to be capable of controlling the clubroot disease. *B. subtilis* XF-1 has been patented as a biological control agent and commercialized in China. Previously, it was very hard to control the clubroot disease as the resting spores have the ability to survive in soil for more than 10 years, which upon getting favorable conditions germinate as zoospores to infect the roots of cruciferous crops. The PBT1 protein produced by XF-1 can rupture the wall of resting spore, thus preventing them from initiating the disease. Based on the results of molecular chromatography and regenerated affinity chromatography, the protein purified from *B. subtilis* XF-1 was most likely a form of chitinase, or protein similar to chitinase. However, the chitinase activity test showed PBT1 protein is not able to hydrolyze the chitin. From the whole genome sequence of XF-1 (Guo et al. 2015), no chitinase gene was discovered although Schwelm et al. (2015) stated that *P. brassicae* contained 16 GH18 domain-containing chitinases. The precise composition of PBT1 protein remains to be elucidated yet. Conceivably, the protein can then be synthesized to provide a more efficient means of clubroot disease control, either as a seed dressing or as a soil fungicide. In addition to PBT1, XF-1 also produced antimicrobial cyclic lipopeptides that can greatly inhibit the mycelial growth of *Fusarium solani* and also causes the deformation by rupturing the resting spores of *P. brassicae* (Li et al. 2015). In general, Bacillus *subtilis* could produce a variety of proteins possessing antagonistic activity (Stein 2005; Vlamakis et al. 2013). Gel electrophoresis analysis of the antimicrobial proteins showed different strains with different characteristics. The protein with a molecular weight of 25 kDa was weakly acidic, heat resistant, and very stable but became unstable under alkaline condition. Based on published reports (Aishaha et al. 2014; Howie and Suslow 1991), the inhibitory mechanism of antimicrobial proteins involved: (a) rupture of the spores due to swelling, and (b) distortion and lysis of the hyphae resulting in the breaking of the mycelium into sections (Zhao et al. 2016).

The antimicrobial action on agar plates seeded with various plant-pathogenic fungi, inhibition zones of different sizes were found around the Oxford cups containing the antimicrobial protein, but not on the control with only phosphate buffer in the cup. The antagonistic effect could be due to the lysis of hyphae and spores by the microbial protein within the inhibition zone. The PBT1 not only inhibited P. brassicae but also other plant-pathogenic fungi, including *Fusarium solani* and *Alternaria alternata*. Compared with *B. subtilis* 168, PBT1 revealed changes at five amino acid sites. *B. subtilis* 168 has no antagonistic effect against *P. brassicae*. It may be revealed that the mutation in the five amino acids sites in *B. subtilis* 168 results in non-inhibitory effects against the pathogen and new strain of *B. subtilis* showed inhibition against the pathogen, but further investigation is needed to elucidate the biocontrol mechanism of the protein PBT1.

There have been many reports about the utilization of *Pseudomonas* as the biological control of plant diseases (Zhang et al. 2009), due to its secondary metabolites that played an important role in protecting plants from infection by pathogens. *Pseudomonas fluorescens* strain 2P24 was a high-yield producer of phenolic antibiotics and had significant antimicrobial effects on many soil-borne pathogens, except *P. brassicae*. Zhao et al. (2016) observed that the gene Pbt1 was transformed into strain 2P24 and the fusion protein showed detrimental effects on the resting spores of *P. brassicae*. In addition, the fusion protein of PBT1 expressed in *E. coli* BL21 also had similar inhibitory effects. Thus, PBT1 is most likely one of the main biocontrol factors against *P. brassicae*. The successful transformation of the gene Pbr1 into foreign host strain 2P24 proved to be feasible and effective method to expand the control of potential range of soil-borne diseases with this improved strain. But the expression level of PBT1 in *Pseudomonas fluorescens* and *E. coli* is not yet clear. More work is required to use this biocontrol strain is filed to control the plant diseases and to maximize the expression of protein PBT1 in the foreign host strains that might lead to large-scale commercial production (Zhao et al. 2016).

7.7.6 Molecular Mechanism of Biocontrol Agent Against Clubroot

Biological control has received more attention recently because it is environmentally friendly, cost-effective, and sustainable (Jaschke et al. 2010). The antagonism mechanisms of biocontrol strains include the competition for nutrients and space, mycoparasitism, the production of antifungal metabolites, plant growth promotion, and the induction of defense responses in plants (Howell 2003).

A transcriptome analysis of *Brassica rapa* using two near-isogenic lines carrying clubroot-resistant and clubroot-susceptible alleles showed that the differently expressed genes were mainly associated with signal transduction, defense, transport, and metabolism (Chen et al. 2016). Luo et al. (2018) screened an antagonistic bacteria *Zhihengliuella aestuarii* B18 against *P. brassicae*, which was obtained from the rhizospheric soil of *B. juncea* var. *tumida* grown in Fuling, Chongqing. The biocontrol effects of the antagonistic strain against clubroot on *B. juncea* var. *tumida* reached 63.4 and 49.7% in the greenhouse experiment and field trial, respectively (Luo et al. 2018). To better understand the biocontrol mechanism, a transcriptome analysis of *B. juncea* var. *tumida* Tsen infected by *P. brassicae* was conducted after the application of biocontrol agent *Z. aestuarii*. Luo et al. (2018) constructed three cDNA libraries using Illumina HiSeq 4000, one from *B. juncea* only inoculated with *P. brassicae* (P), one inoculated with *P. brassicae* and the biocontrol agent *Z. aestuarii* (deposit number CICC11045s in the China Center of Industrial Culture Collection) at the same time (P + B), and the other was the control (H), in which *P. brassicae* was replaced by sterile water. A total of 107,617 unigenes were obtained and these unigenes' functional annotations were determined. A differential expression analysis was performed among the biocontrol-treated (P + B), the infected (P), and the control (H) samples, and resistance-related genes were identified among the differentially expressed genes. The transcriptome results were validated by reverse transcription quantitative PCR (qRT-PCR) (Luo et al. 2018).

To better understand the biocontrol mechanism, three transcriptome analyses of *B. juncea* var. *tumida* Tsen were conducted using Illumina HiSeq 4000, one from *B. juncea* only inoculated with *P. brassicae* (P), one inoculated with *P. brassicae* and the biocontrol agent *Z. aestuarii* at the same time (P + B), and the other was the control (H), in which *P. brassicae* was replaced by sterile water. A total of 19.94 Gb was generated by Illumina HiSeq sequencing. The sequence data were de novo assembled, and 107,617 unigenes were obtained. In total, 5629 differentially expressed genes between biocontrol-treated (P + B) and infected (P) samples were assigned to 126 KEGG pathways. Using multiple testing corrections, 20 pathways were significantly enriched with Q value ≤ 0.05. The resistance-related genes, involved in the production of pathogenesis-related proteins, pathogen-associated molecular pattern-triggered immunity, and effector-triggered immunity signaling pathways, calcium influx, salicylic acid pathway, reactive oxygen intermediates, and mitogen-activated protein kinase cascades, and cell wall modification, were obtained. The various defense responses induced by the biocontrol strain combated the *P. brassicae* infection. The genes and pathways involved in plant resistance were induced by a biocontrol strain. The transcriptome data explained the molecular mechanism of the potential biocontrol strain against *P. brassicae*. The data will also serve as an important public information platform to study *B. juncea* var. *tumida* Tsen and will be useful for breeding mustard plants resistant to *P. brassicae* (Luo et al. 2018).

7.7.7 Role of ProCa in Modulation of Clubroot

Plant growth hormones, especially cytokinins and auxins, which are both up-regulated during the secondary infection phase (Devos et al. 2005; Siemens et al. 2006), are involved in symptom development. A role for auxin during later stages might be specifically associated with the expansion of host cells. Consequently, modulation of auxin levels resulted in reduced clubroot symptoms. Auxin homeostasis can be controlled by biosynthesis, degradation or hydrolysis/synthesis of inactive conjugates, and transport. Also, secondary metabolites accumulate during clubroot development such as glucosinolates (Ludwig-Muller 2009) and flavonoids (Pasold et al. 2010). Flavonoids are thought to modulate auxin transport (Geisler and Murphy 2006) and genetic and molecular evidence point to a role for flavonoids in gall formation by changing auxin transport properties, even though they are not essential compounds for gall formation (Pasold et al. 2010; Pasold and Ludwig-Muller 2013).

Flavonoids are a group of secondary products with an array of biological functions, including apparent roles in stress protection such as UV radiation. In addition, they are best known as the characteristic red, blue, and purple anthocyanin pigments of plant tissues, which serve essential functions in plant reproduction by recruiting pollinators and seed dispersers (Winkel-Shirley 2001). The phenylpropanoid pathway from which the flavonoids are derived links to plant defense (Ryder et al. 1987; Dixon 2001). Some plant species synthesize 3-deoxyanthocyanins, which are involved both in defense (Snyder and Nicholson 1990) and in pigmentation (Grotewold et al. 1994). Flavonoids with phytoalexin-like properties accumulate (Halbwirth et al. 2003; Spinelli et al. 2005; Rademacher et al. 2006) if plants are treated with prohexadione-calcium (ProCa), a plant growth regulator, which inhibits ascorbic acid and 2-oxoglutaric acid-dependent dioxygenases such as flavanone 3-hydroxylase (Fig. 7.4).

Being an inhibitor of several dioxygenases involved in gibberellin, ethylene, and flavonoid synthesis, ProCa has an inhibitory effect on shoot growth via gibberellins, stimulatory effect on fruit set via ethylene inhibition and increased effect on biotic stress resistance via altered flavonoid patterns (Rademacher et al. 2006). The effect of ProCa application on the flavonoid patterns of several plant species has been shown. While an effect of ProCa as growth regulator, mainly in inhibiting gibberellin biosynthesis, is already well established (Rademacher 2000), the influence on the patterns of flavonoids was only recently discovered (Puhl et al. 2008). If an unspecific flavanone reductase was present, treatment with ProCa resulted in the accumulation of the phytoalexin flavonoids eriodictyol and luteoliflavane (Puhl et al. 2008). Various disorders, caused both by microbes and insects, were shown to be reduced after application of the compound, probably due to the accumulation of the antimicrobial flavonoids (Bazzi et al. 2003; Halbwirth et al. 2003; Spinelli et al. 2005; Rademacher et al. 2006). The reduction of diseases after ProCa treatment has been shown for several plant species including *Arabidopsis*. As flavonoids are linked to altered club development in this species (Pasold et al. 2010).

ProCa has an effect on: (1) clubroot development, (2) flavonoid accumulation, and (3) auxin responsiveness because flavonoids are thought to modulate auxin transport by inhibiting auxin efflux (Geisler and Murphy 2006). In addition, oxoglutaric acid-dependent dioxygenases, which are inhibited by ProCa, are also involved in gibberellin (GA) biosynthesis, but there are other inhibitors of GA biosynthesis with different targets (Fig. 7.4). Chlormequat chloride (CCC) targets the steps from geranylgeranyl diphosphate via copalyl diphosphate to ent-kaurene catalyzed by two enzymes, copalyldiphosphate synthase and ent-kaurene synthase (Rademacher 2000). Clubroot disease in the Brassicaceae alters many biosynthetic pathways. It was shown that flavonoids accumulate in root galls and play a role in gall formation. The most probable explanation given was the modulation of auxin efflux, thereby increasing local auxin levels in roots, which would in consequence lead to larger galls. This hypothesis was corroborated by investigating flavonoid biosynthesis mutants, of which some showed altered clubroot formation (Pasold

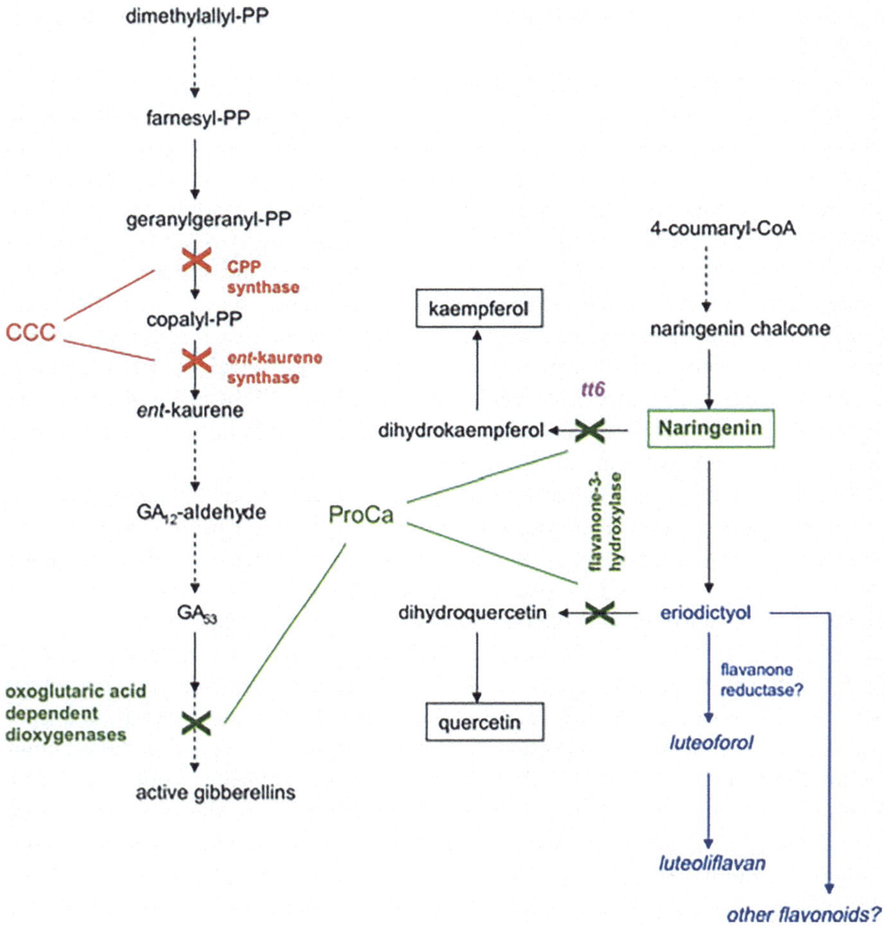

Fig. 7.4 Main steps in the biosynthesis of gibberellins and flavonoids indicating the enzymes which are inhibited by prohexadione-calcium (ProCa) and chlormequat chloride (CCC). Inhibition of oxoglutaric acid-dependent dioxygenases by ProCa is indicated by a cross and the target enzymes and the compound(s) accumulating are also shown in bold (colored green online); inhibition by CCC is indicated with a cross and targets are also shown in bold (colored red online). Metabolites shown in italics (colored blue online) are those which could putatively be formed when eriodictyol is not converted to dihydroquercetin by flavanone-3-hydroxylase and a flavanone reductase is present. The position of the transparent testa6 (*tt6*) mutation is indicated. Dashed arrows indicate more than one reaction step, solid arrows indicate one enzymatic reaction. Compounds in boxes have been determined (Pasold and Ludwig-Muller 2013)

et al. 2010). Flavonoids generally play a significant role as defense compounds. The flavonoid type phytoalexins inhibit fungal growth (Dixon 2001). The growth regulator ProCa was shown to influence the flavonoid biosynthesis pathway in different plant species by inhibition of the enzyme flavanone-3-hydroxylase (Puhl et al. 2008), so that phytoalexin-like substances accumulated, which led to more pathogen and pest-tolerant plants (Rademacher et al. 2006). After ProCa application, clubroot symptoms were indeed reduced, indicated by a lower disease index, larger total root system, and an increase in the shoot index (Fig. 7.5; Table 7.2).

ProCa has not previously been used on root diseases in practical applications, mainly due to its instability in soil. In the mutant tt6, the F3H is genetically altered, so that these plants do not accumulate wild-type amounts of quercetin and kaempferol and therefore should be a genetic mimic of ProCa treatment. Even though the disease index is reduced in the tt6 mutants (Table 7.2), the observed effects of ProCa on *Arabidopsis* roots cannot be explained by this mutant phenotype alone. After ProCa treatment even the infected roots were larger (Plate 7.2), whereas this was not the case for *tt6* plants.

This might be due to the observation that the mutation in *F3H* is leaky (Peer et al. 2001). However, in *Arabidopsis*, naringenin accumulated strongly (Fig. 7.6), suggesting that the pathway is not altered to produce other compounds, or that these flavonoids are present only in trace amounts. These data suggest that *Arabidopsis* does not possess an unspecific enzyme for the conversion of flavonones to phytoalexin-type flavonols as was found in other plant species. Secondly, because flavonoids are involved in the modulation of polar auxin transport (Buer and Muday 2004), it was investigated whether ProCa treatment altered the auxin response using the promoter-reporter line DR5:: GUS. If polar auxin transport were influenced by flavonoids and ProCa, changes in auxin distribution would be expected, which could be measured indirectly using the auxin-responsive promoter-reporter line. It was shown that auxin responsiveness was confined to the root parts harboring the galls in all treatments, so that an effect of ProCa directly or via altered flavonoid patterns on auxin can be ruled out (Plate 7.2). Thirdly, the specificity of ProCa on flavonoid synthesis was tested by using another growth regulator, chlormequatchloride (CCC), inhibiting specifically gibberellin biosynthesis (Rademacher et al. 2006). In general, plant growth promoting hormones are suggested to be important for clubroot disease (Ludwig-Muller et al. 2009). The plants treated with CCC did not show any gall size reduction, showing for the first time experimentally that intact GA biosynthesis is not necessary for clubroot development and that low GA doses, which were shown to promote root growth (Tanimoto 2012), do not account for the effect on clubroot reduction. ProCa also targets the last step in ethylene synthesis, amino cyclopropanecarboxylic acid oxidase (Rademacher et al. 2006). For the ethylene pathway, it was shown that elevated ethylene levels did not alter disease development in *Arabidopsis*, while ethylene perception was necessary to restrict gall growth (Knaust and Ludwig-Muller 2013), making it unlikely that ProCa has an effect on the ethylene biosynthetic pathway in *Arabidopsis*.

However, in the complex management of clubroot in the field, calcium treatment together with acidic pH values plays a major role (Donald and Porter 2009). The short period between germination of resting spores and penetration of zoospores is the phase where survival could be affected by soil factors, among them calcium. These factors influence the inoculum potential and its viability and invasive capacity (Dixon 2009a, b). Dixon and Webster (1988) demonstrated that treatment with calcium resulted in the inhibition of the total numbers of root-hair infections and

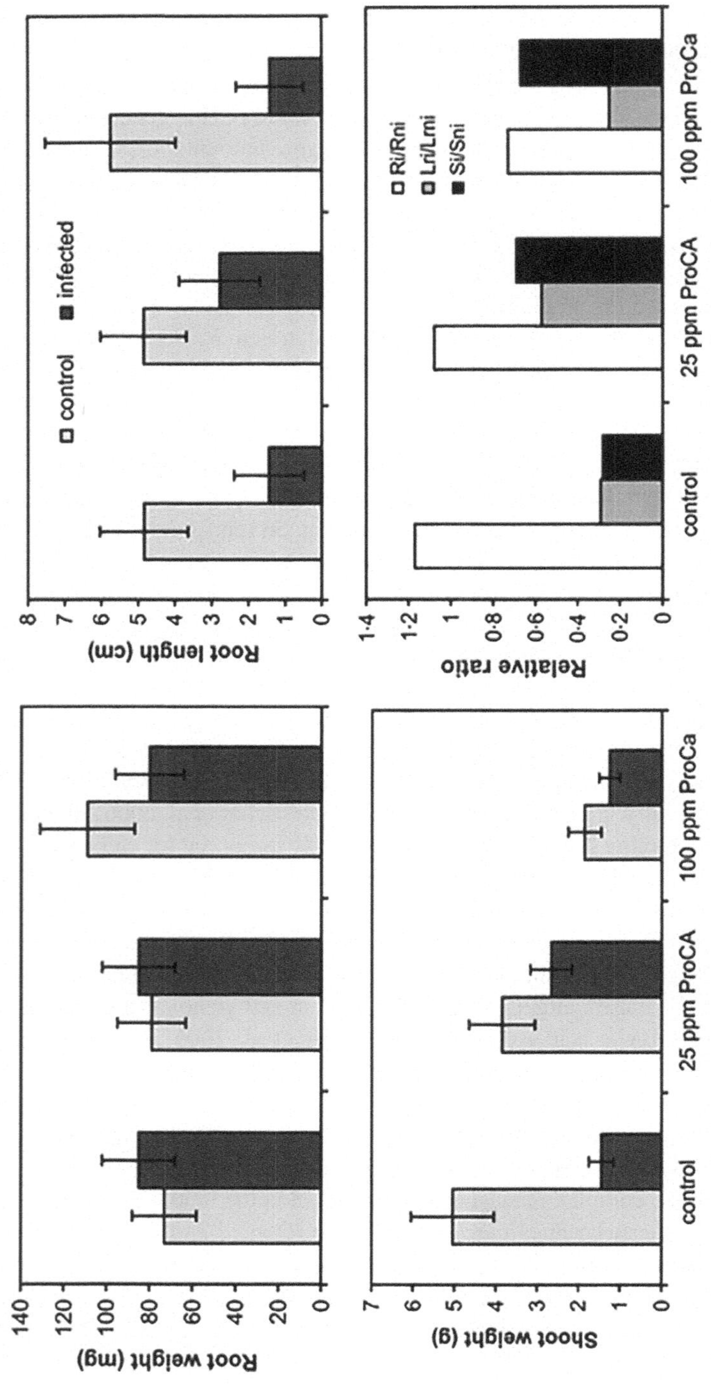

Fig. 7.5 Detailed phytopathological analysis of infected roots with or without prohexadione-calcium (ProCa) treatment. A low root index (Ri/Rni) means small galls, a high shoot index (Si/Sni) means vigorous plants. The total root length (Lri/Lmi) was affected by low concentrations of ProCa (Pasold and Ludwig-Muller 2013)

Table 7.2 Disease indices (DI) after prohexadione-calcium and chlormequatchloride treatments in *Arabidopsis thaliana* infected with *Plasmodiophora brassicae*, in comparison to control infection and infection of the tt6 mutant, which results in naringenin accumulation. Inoculation was done with 10^7 spores mL^1 (Pasold and Ludwig-Muller 2013)

Plant/treatment	Concentration	DI control plant	DI treated/mutant
Prohexadione-calcium	25 ppm	90	48*
Prohexadione-calcium	100 ppm	90	68*
Chlormequatchloride	100 ppm	90	89
tt6[a]	–	82	63*
Naringenin[b]	10 lM	65	61

*Significant differences at the *P* 0.05 level are indicated by an asterisk

[a]Data for tt6 and Ler (corresponding wild type) taken from Pasold et al. (2010)

[b]Naringenin treatment itself did not induce tolerance; the DI was calculated from the same experiment that is depicted for root and shoots indices in Pasold et al. (2010)

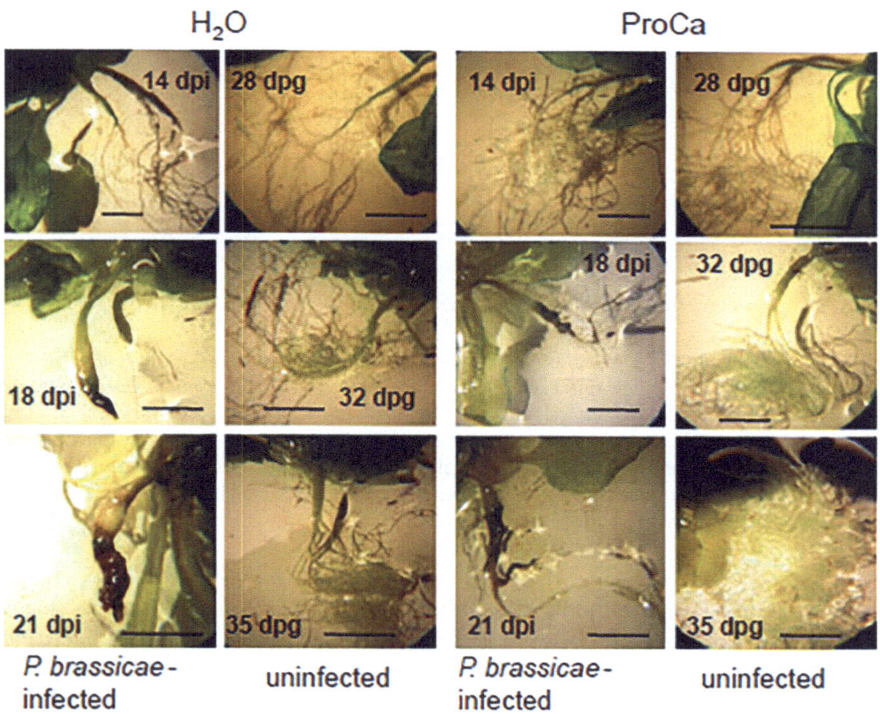

Plate 7.2 Pictures of typical root galls and control roots with and without prohexadione-calcium (ProCa) treatment at three time points after inoculation (14, 18, and 21 days post-inoculation, dpi, for infected plants and 28, 32, and 35 days post-germination, dpg, for uninfected plants). The distribution of DR5::GUS staining is shown which indicates responsiveness to IAA. The bar represents 5 mm in all panels (Pasold and Ludwig-Muller 2013)

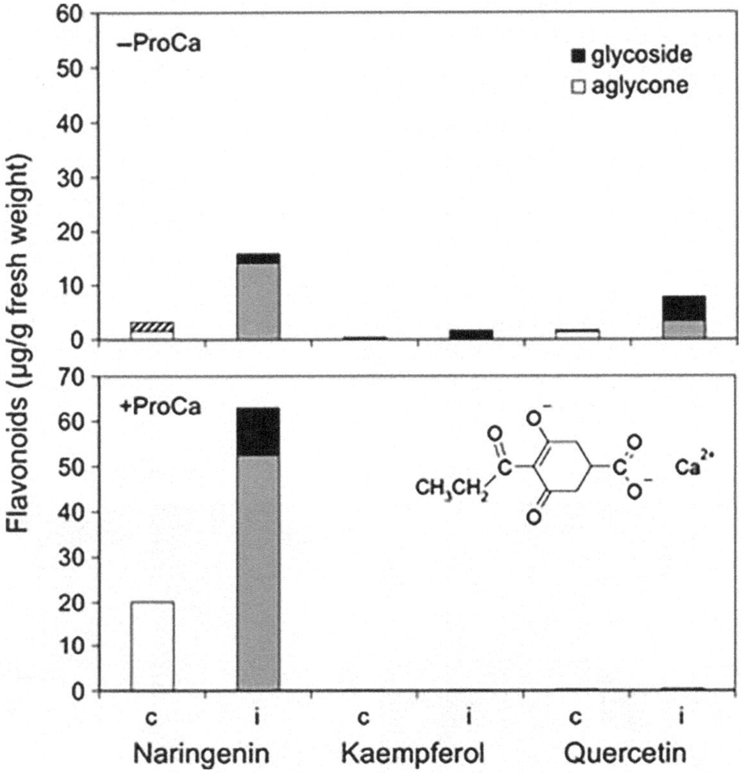

Fig. 7.6 Accumulation of flavonoids in *Arabidopsis thaliana* control roots (Pasold et al. 2010, the data were recorded at the same time) and root galls after infection with *Plasmodiophora brassicae* and treatment with prohexadione-calcium (ProCa) determined by HPLC analysis. The control samples were taken 37 days post-germination, and the infected samples at 23 days post-inoculation (equivalent age to control plants). Upper panel: control roots and *Plasmodiophora brassicae* inoculated roots; lower panel: same tissues but plants treated simultaneously with 100 ppm ProCa (Pasold and Ludwig-Muller 2013)

the rate of maturation through plasmodial, sporangial, and zoosporangial stages compared to controls. Elevated concentrations of calcium completely inhibit the later stages of *P. brassicae* development in the root hair (Dixon and Webster 1988; Webster and Dixon 1991; Donald and Porter 2004). Niwa et al. (2008) provided direct evidence that spore germination and subsequent root-hair colonization was retarded by the presence of calcium and alkaline pH values. Because calcium is a signal affecting many plant responses and is also involved directly in disease resistance (Romeis et al. 2001; Navazio et al. 2007), stimulation of the plant's defense responses by calcium (or the complex with prohexadione) cannot be ruled out. It was shown that ProCa is an inhibitor of clubroot formation. However, a direct inhibitory effect of naringenin can be ruled out because treatment of clubroots with naringenin did not result in the reduction of disease (Pasold et al. 2010). Either the two observations are not linked, or endogenous changes in naringenin have a different effect than application of the compound. Perhaps the ProCa-treated root systems are larger and therefore might simply be more robust against pathogens. A more general role of ProCa on roots can be further tested by using other root pathogens of *Arabidopsis* (Pasold and Ludwig-Muller 2013).

7.7.8 Role of Arginase Gene in Modulation of Clubroot

Conversion of arginine to ornithine and urea is catalyzed by arginase (EC 3.5.3.1), an Mn-dependent enzyme that functions as an arginine aminohydrolase. In mammals, this reaction has been mostly studied as a key final step in the urea cycle, which allows potentially toxic ammonium to be detoxified into urea. In addition to this major role in nitrogen waste management, arginine catabolism can positively modulate several subsequent ornithine-dependent pathways, including proline, glutamate, and ornithine-dependent polyamine biosynthesis (Li et al. 2001). Plants have not developed any significant mechanism for nitrogen excretion, and although all of the enzymes of the urea cycle are present in plants there is no report that it actually functions as a cycle. Instead, arginase activity has been studied in plants mostly for its involvement in nitrogen remobilization processes from protein degradation, especially during seed germination (de Ruiter and Kolloffel 1983; Cao et al. 2010). Arginase induction was reported to be a jasmonate-driven defense mechanism against phytophageous insects in tomato. In this system, plant arginase activity acts as an anti-nutritional factor that reduces the assimilation of arginine in the insect digestive tract (Chen and Wang 2005). *Arabidopsis thaliana* synthesizes two functional arginases, each predicted to be targeted to the mitochondrion and encoded by two duplicated tandem genes, ARGAH1 (At4g08900), and ARGAH2 (At4g08870). In *Arabidopsis*, the mitochondrial localization of arginase was demonstrated by immunolocalization (Flores et al. 2008), which confirmed findings from subcellular fractionation experiments previously conducted in other plant species (Goldraij and Polacco 2000). *ARGAH1* is predominantly expressed in pollen grains, whereas ARGAH2 expression is induced by methyl-jasmonate treatment. The induction of arginase activity in the single mutant line argah2 is nearly abolished in germinating seedlings and in jasmonate-treated tissues, whereas argah1 displays strongly reduced arginase activity in pollen, supporting a distinct functional specialization for the two paralogs (Brownfield et al. 2008). ARGAH2 was also reported to exert a negative control over the auxin response, as argah2 displays both enhanced auxin-induced rhizogenesis and enhanced auxin-triggered induction of DR5, an auxin-responsive gene marker (Flores et al. 2008).

In a study on the *Arabidopsis–Plasmodiophora brassicae* pathosystem, the induction of ARGAH2 and high arginase activity in clubroot-infected tissues of the susceptible accession Col-0 has been observed (Jubault et al. 2008). This biochemical phenomenon reaches a maximum 3 weeks after inoculation, i.e., during the last stages of the infection process. At this time point, pathogen-induced deregulation of auxin and cytokinin has already led to an almost complete conversion of the plant root system into galls filled with P. brassicae spores (Siemens et al. 2006). This pathological root structure is no longer capable of providing water and nutrients to the plant, ultimately leading to the late, sudden, and severe depression of plant shoots. Arginase induction in clubroot is accompanied by modest regulation of the arginine-derived polyamine content (Jubault et al. 2008). In contrast, clubroot is associated with pronounced changes to a few amino acids, including the accumulation of proline. High levels of arginase activity, in combination with ornithine aminotransferase activity, might potentially feed a biosynthetic pathway towards proline, which could play a significant osmoprotecting role in root galls. Alternatively, it was proposed that plant arginase could play a role in the regulation of other primary amino acids, thus controlling nitrogen availability for the pathogen (Jubault et al. 2008). Gravot et al. (2012) investigated the potential role of the jasmonate signaling pathway in the triggering of arginase induction and analyzed the potential impact of argah2 mutation on amino acid contents in infected roots. Because arginase had also been proposed to exert a negative control on auxin-triggered rhizogenesis, it was also tested whether arginase could also be involved in cellular proliferation stimulated by exogenous auxin/cytokinin treatments.

Induction of arginase activity during clubroot infection essentially depends on the induction of the plant gene ARGAH2. The arginase activity is induced in clubroot galls during the last steps of the infection (Jubault et al. 2008). The induction of this enzyme paralleled clubroot-triggered transcriptional induction of the plant arginase-encoding gene ARGAH2. However, in such a heterogeneous biological system where the *P. brassicae* and *A. thaliana* gene products intimately co-exist, it was necessary to verify that induction of the plant ARGAH2 gene is entirely responsible for the enhancement of arginase activity. Based on the low levels of arginase activity in argah2 clubroot galls, it was concluded that the clubroot-triggered increase in arginase activity is essentially due to up-regulation of the plant ARGAH2 gene. The slight enhancement of arginase activity which remains in clubroot galls on the argah2 mutant could be explained by either the pathogen's enzymatic machinery or the protein encoded by ARGAH1, the second arginase-encoding gene copy in the *Arabidopsis* genome. This latter possibility supposes that arginase activity enhancement is related to post-transcriptional regulation since the ARGAH1 gene is not induced during clubroot development (Jubault et al. 2008). Arginase induction exerts a negative control on the development of gall growth. (1) Plant arginase activity is one of the plant's metabolic functions targeted by pathogen effectors, among the pathogen's global strategy for controlling plant metabolism; (2) arginase is a plant defense response; or (3) induction of plant arginase is a side effect of the plant–pathogen warfare, for example, one among many stereotypical responses to jasmonate, without biological significance in the context of clubroot infection. The fact that arginase is induced in infected Col-0 roots at higher levels than in the partially resistant accession Bur-0 was initially considered to support the hypothesis that arginase serves the infectious process (Jubault et al. 2008). However, the present study gave support to the opposite second hypothesis; arginase is hardly induced in the argah2 mutant, resulting in more symptoms compared with the control Col-0, and arginase induction is then a defense response limiting clubroot development.

Table 7.3 Clubroot-induced variations in the content of jasmonate-related compounds in roots of the accessions Col-0, argah2, and jar1 (Gravot et al. 2012)

Genotype	Treatment	OPDA	JA	OH-JA	JA-Ile	COOH-JA-Ile
Col-0	NI 21 dpi	7112	126	D	2914	279
	eH 21 dpi	44,170*	349*	733*	1540	1358*
argah2	NI 21 dpi	8868	159	D	1926	329
	eH 21 dpi	36,417*	398	748*	2485	1704*
jar1	NI 21 dpi	9011	175	D	ND	ND
	eH 21 dpi	37,528	427	977*	D	D

Hormone contents were analyzed in infected and non-infected roots at 14 or 21 dpi. Results are expressed in arbitrary units calculated from mass signals relative to sample biomass. Only peaks areas >10 times the chromatographic background noise were considered for quantitative measurements. Data are means of three replicates, each one consisting of at least 15 plants. *ND* not detected, *D* detected (peak areas between three and 10 times the background noise). Asterisks indicate significant differences between eH and NI treatments for one given genotype, inferred from Student's t-tests ($^*P < 0.05$). *OPDA* 12-oxo-phytodienoic acid, *JA* jasmonic acid, *OH-JA* hydroxyjasmonic acid, *JA-Ile* jasmonoyl-L-isoleucine, *COOH-JA-Ile* carboxylated jasmonoyl-L-isoleucine.

Arginase induction by clubroot infection depends on jasmonate signaling. ARGAH2 induction was reported to be under jasmonate signaling control in non-pathological conditions (Flores et al. 2008). JA accumulation during clubroot infection has been reported (Grsic et al. 1999). Jasmonate profiling confirms this idea, and additionally shows a significant accumulation of both a jasmonate precursor (OPDA) and a jasmonate-derived metabolite. Non-detection of any significant regulation of JA-Ile, but the accumulation of its catabolite COO-JA-Ile suggests an induction of the activity of the JA-Ile conjugating enzyme JAR1 in infected roots before the sampling time point. In parallel, it showed that the jar1 mutation strongly reduces JA-Ile and JA-Ile derivative (Table 7.3), and almost abolishes arginase induction by clubroot infection; supporting the idea that arginase is regulated by JA-Ile. According to the proposed model, clubroot symptoms were enhanced in jar1 when compared with Col-0. However, symptoms were greater in the jar1 mutant compared with argah2, suggesting that—independently of ARGAH2 induction—other jasmonate-driven mechanisms also contribute to reduce gall growth. Phenotypic analysis of jar1 agrees with the results obtained by others which previously described the increased susceptibility of jar1 to the e3 monospore isolate of *P. brassicae* (Siemens et al. 2002). In contrast, Agarwal et al. (2011) reported an absence of effect of jar1 and coi1 mutations on clubroot symptoms when using an Australian field isolate. This discrepancy could be related either to differences between isolates or to differences in the methods used for the evaluation of clubroot symptoms (Gravot et al. 2012).

7.7.9 Host Resistance Mechanisms of *Brassica* Cultivars

A commercial clubroot-resistant F_1 cultivar of Chinese cabbage (*Brassica rapa* L. subsp. *pekinensis*), Kukai 70, is resistant to an isolate of populations of *Plasmodiophora brassicae* from Hagi (HG) city and is susceptible to another from Yamaguchi (YMG) city, Japan. The degree and frequency of primary and secondary cycle colonization by the isolates in the root hairs and root tissues of cv. Kukai 70 were compared. Seedlings of cv. Kukai 70 were grown in soils amended with inoculum of either HG or YMG and harvested 10 days after inoculation to observe the primary cycle (number of root-hair infections) and 20, 30, and 40 days after inoculation to observe of the secondary cycle (frequency of infected cells and degree of plasmodial development based on the number of nuclei in infected cells). Although more root hairs were infected in HG than in YMG, fewer cells in root tissues including the cortex, and medullary rays were infected in HG than in YMG. In addition, YMG developed plasmodia with many nuclei and formed resting spores, whereas plasmodia remained immature with a small number of nuclei in HG and did not form resting spores even by 40 days after inoculation (Plate 7.3). These results suggest that suppression of plasmodial development during secondary colonization is associated with resistance mechanisms to HG in cv. Kukai 70. Starch did not accumulate (i.e., development of amyloplasts) in HG-infected cells. This may be involved in the suppression of secondary colonization of *P. brassicae* in the cultivar (Tanaka et al. 2006).

7.7.10 Erosion of R-Genes Management Through Integrated Approaches

Resistance to *P. brassicae* has broken down in various cruciferous crops, including oilseed rape and vegetable Brassicas (Seaman et al. 1963; Kuginuki et al. 1999; Oxley 2007; Diederichsen et al. 2009). While *P. brassicae*-resistant cultivars of *B. napus*, *B. oleracea,* and *B. rapa* are available, their use is often limited by the short durability and the pathotype specificity of the resistance (Voorrips 1995; Diederichsen et al. 2009). Each of the resistant canola cultivars available for production in western Canada will develop small galls under high inoculum pressure, and repeated cycles of inoculation can result in a rapid decline in the level of resistance (LeBoldus et al. 2012). The effective and sustainable deployment of genetically resistant cultivars to manage clubroot disease requires that the use of these cultivars be integrated with various other approaches for control (Diederichsen et al. 2009; Donald and Porter 2009). As such, regardless of the nature of the resistance, it is prudent to ensure that resistance is carefully managed to maximize its durability (Hwang et al. 2012). Erosion and breakdown of clubroot-resistant (CR) cultivars to the old and evolution of new virulent pathogen has been assessed from very few countries of the world. The identification in 2013 of a few *P. brassicae* populations able to overcome the resistance in CR canola, only 3–4 years after the introduction of this trait on the Canadian prairies, underscored

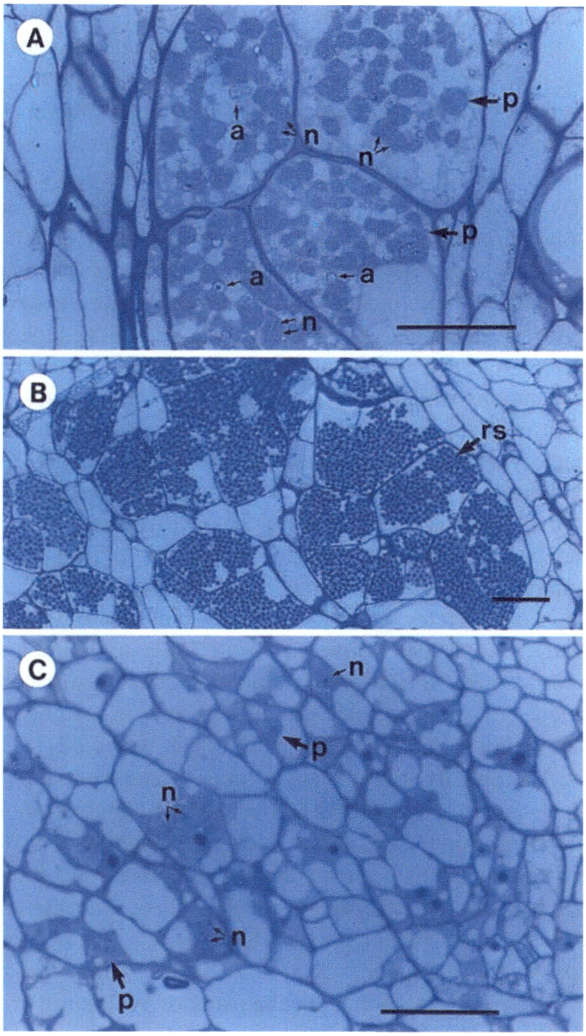

Plate 7.3 Light micrographs of plasmodia and resting spores within infected cells of root tissues of clubroot-resistant cultivar Kukai 70 grown in pots of soil amended with either isolate Yamaguchi (YGM) or Hagi (HG) of *Plasmodiophora brassicae*. (**a**) Plasmodia (*p*) of YGM with many nuclei (*n*) and starch-accumulating amyloplasts (*a*) in infected cells at 30 days after inoculation. (**b**) Resting spores (*rs*) of YGM at 40 days after inoculation. (**c**) Small plasmodia (*p*) of HG with a small number of nuclei (*n*) at 40 days after inoculation. Bars = 50 μm (Tanaka et al. 2006)

the continued vulnerability of *Brassica* to clubroot (Strelkov et al. 2016). This threat was further highlighted in the study conducted by Strelkov et al. (2018) with the identification of an additional 61 *P. brassicae* populations in Alberta between 2014 and 2016 capable of overcoming the clubroot resistance in the existing suite of CR cultivars. In a greenhouse study, rapid increases in clubroot severity were observed on some *Brassica* host genotypes after repeated cultivation in a *P. brassicae*-inoculated potting mix, which indicated that adaptation by the pathogen to selection pressure had occurred quickly (LeBoldus et al. 2012). Likewise, the loss or erosion of clubroot resistance was reported under field conditions in cabbage in the USA (Seaman et al. 1963), Dutch stubble turnip (*B. rapa* ssp. *rapifera*) in the UK, Chinese cabbage in Japan (Osaki et al. 2004; Hatakayema et al. 2006), and "Mendel" oilseed rape in the UK and Germany (Oxley 2007; Diederichsen et al. 2014). In Canada, all of the CR canola (*Brassica*) cultivars evaluated were susceptible to at least some of the new, virulent *P. brassicae* field populations tested. Also, if a population of the pathogen could cause clubroot on one of these cultivars, it most often was virulent on all of them. Indeed, there was little differentiating capacity between individual CR canola cultivars, and all were susceptible to pathogen populations that could induce IDs >75% on at least one of the CR cultivars. Moreover, while a few pathogen populations were virulent on only one or a few hosts, such populations produced IDs <75% on even the most susceptible CR canola cultivars. This suggests that those populations were either less aggressive, or represented a mix of virulent and avirulent resting spores (Xue et al. 2008). All of the pathogen populations caused IDs >90% on the universally susceptible ECD 05, which confirmed that these populations were highly virulent on a susceptible host.

It is interesting to note the burst of pathotype diversity observed in 2014 (nine pathotypes) and its apparent reduction in 2015 (five pathotypes) and 2016 (three pathotypes), with pathotype A becoming increasingly predominant. The phenomenon of diversifying selection in response to host resistance is not novel in agricultural pathosystems. Major resistance genes impose strong selection pressures on pathogen populations, and in combination with a number of factors, can act to cause "boom-and-bust" cycles of pathotype diversification within a population (Wolfe and McDermott 1994; Kolmer 1996; de Vallavieille-Pope et al. 2012; Lof and van der Werf 2017). Protection from infection due to a single major gene is short-lived and often brings on a measurable increase in pathotype diversity followed by a stabilizing shift towards a few dominant genotypes.

Additionally, genome environment/context and the "two speed" genome concept also can accommodate an effective, sustainable amount of allele diversity in a pathogen population under selection. Since, *P. brassicae* is an obligate parasite, elucidating the mechanisms of virulence diversification is more challenging and thus mechanistic details are not well-known or characterized in the *Brassica* clubroot pathosystem. Similarly, the amount of *P. brassicae* pathotype diversity prior to applying the R-gene selection is virtually impossible to measure. Nonetheless, *P. brassicae* clearly has mechanisms of generating diversity and maintaining low levels of rare pathotypes within field populations over extended periods, which can then respond to the varying selection pressures imposed by deployment of CR genes.

Given its prevalence and virulence on the CR cultivars, pathotype A would make a logical target for the resistance screening activities associated with canola development. Another novel pathotype, B (a variant of the original pathotype 2), also may be of interest in clubroot resistance breeding, since, it shares almost the same host range as pathotype A, but additionally is virulent on the *B. oleracea* host ECD 11, which was resistant to all of the other *P. brassicae* pathotypes characterized (Strelkov et al. 2018). Nonetheless, pathotype B was identified only once during 2014–2016. The second most common pathotype in this study, after A, was pathotype D. This pathotype, which was another variant of Williams' pathotype 3, shared an identical virulence pattern with pathotype A except that it was avirulent on var. "Mendel." Pathotype X (a variant of pathotype 5), which corresponded to the first populations found to be able to overcome resistance in 2013 (Strelkov et al. 2016), was not identified again in 2014–2016. Since, it was first identified, pathotype X has been one of the most commonly used for clubroot resistance screening, given that it represented the first pathotype found to overcome resistance in CR canola. In light of its relative rarity, however, and its inability to cause clubroot on all of the CCD hosts susceptible to pathotypes A, B, and D, its current prevalent use should be reconsidered.

Fields in which clubroot resistance appeared to be overcome often occurred as distinct clusters. This may indicate that new populations of *P. brassicae* with novel virulence patterns have emerged independently in multiple fields, via the selection pressure imposed by the planting of CR *Brassica*, and then spread locally by the movement of infested soil on farm and other machinery (Cao et al. 2009), and/or by the movement of wind-blown dust infested with the pathogen (Gossen et al. 2015; Rennie et al. 2015). Nonetheless, the occurrence of clusters of virulent clubroot pathotypes highlights the need for growers accessing *Brassica* fields to take precautions, such as cleaning farm machinery, to avoid spread of these new pathotypes.

The emergences of multiple new virulence phenotypes of *P. brassicae*, which are able to overcome genetic resistance in *Brassica,* have a serious impact on the long-term production of this crop. Collectively, CR *Brassica* cultivars represent the most important clubroot management tool available to farmers, but genetic resistance should be deployed carefully to maximize its longevity. As such, resistance must be used as part of an integrated clubroot management approach that includes longer rotations out of *Brassica* in infested fields, cleaning/sanitization of machinery, and potentially the use of liming or other strategies to reduce disease pressure.

7.7.11 Clubroot QTL's Modulation by Nitrogenes

The positive or negative modulation of plant disease development resulting from variations in nitrogen supply has been reported for many pathosystems (Dordas 2008; Fagard et al. 2014). In case of clubroot disease, a few studies have suggested that a high-nitrogen supply tends to reduce the damage caused by *P. brassicae* infection (Dixon 2009b; Gossen et al. 2014), while another drew the opposite conclusion. None of these reports, however, took into account the possible influence of plant genotype (mainly the presence of genetic resistance factors) on the observed interactions between disease development and nitrogen supply. Genetic sources of both qualitative and quantitative resistance to *P. brassicae* have been identified in *B. rapa*, *B. oleracea,* and *B. napus*. Nonetheless, only a few commercial varieties with clubroot resistance are available, and most of these ("Mendel," "Mendelsson," "Cracker," "Andromeda," "SY Alister") possess qualitative resistance under oligogenic or major gene control. The deployment of clubroot-resistant varieties based on qualitative resistance can impose strong selection pressure on pathogen populations, favoring strains capable of overcoming resistance. In oilseed rape or canola, resistance defeating strains of *P. brassicae* were detected within 3–4 years of the introduction of clubroot-resistant host cultivars (Strelkov et al. 2016). Indeed, many examples exist in which the effectiveness of major resistance genes has been broken, including various cases involving major genes for clubroot resistance. The pyramiding of qualitative and quantitative resistance is considered today as a key lever to increase the durability of plant resistance against rapidly evolving pathogen populations (Brun et al. 2010; Tomita et al. 2013). As with most quantitative traits, the effect of quantitative resistance is dependent on other biotic and abiotic environmental factors. The stability of quantitative resistance to clubroot under different fertilization regimes, including low-nitrogen input was determined. A preliminary assay was conducted with non-inoculated (healthy) plants grown under standard clubroot resistance screening conditions, in order to assess the effect of two different NO_3^- levels on plant growth and leaf nitrogen content. A segregating host population and a collection of oilseed rape genotypes then were used to identify and compare clubroot resistance QTL under high- and low-nitrogen supply following inoculation with two different isolates of *P. brassicae.* To assess the variability of clubroot resistance under different nitrogen levels and to characterize the impact of nitrogen supply on genetic resistance factors a linkage analysis and a genome-wide association study were conducted to detect QTL for clubroot resistance and evaluate their sensitivity to nitrogen. The clubroot response of a set of 92 diverse oilseed rape accessions and 108 lines derived from a cross between "Darmor-*bzh*" (resistant) and "Yudal" (susceptible) was studied in the greenhouse under high- and low-nitrogen conditions, following inoculation with the *P. brassicae* isolates eH and K92-16. Resistance to each isolate was controlled by a major QTL and a few small-effects QTL. While the same QTL were detected under both high and low nitrogen, their effects were altered. Clubroot resistance to isolate eH, but not K92-16, was greater under a low-N supply versus a high-N supply. New sources of resistance were found among the oilseed rape accessions under both low- and high-N conditions (Laperche et al. 2017).

7.7.12 Management of R-Genes Resistance to Clubroot

To develop clubroot-resistant germplasm with a pool of CR genes for long-term management of this disease, diverse genetic sources must be found and characterized. This is important to enable rotation of cultivars with different CR genes, or for pyramiding CR genes within a single cultivar. Pyramiding the major resistance genes in Chinese cabbage is known to confer high resistance against multiple pathotypes (Matsumoto et al. 2012). With advances in molecular technologies in the last two decades, at least nine CR loci in the A-genome of *B. rapa* (Piao et al. 2004; Suwabe et al. 2003, 2006, 2012; Saito et al. 2006; Sakamoto et al. 2008; Ueno et al. 2012; Hatakeyama et al. 2013), and several quantitative trait loci (QTL) in *B. oleracea* (C-genome) (Figdore et al. 1993; Voorrips et al. 1997; Moriguchi et al. 1999; Rocherieux et al. 2004; Nagaoka et al. 2010), and *B. napus* (AC genome) (Manzanares-Dauleux et al. 2000; Werner et al. 2008) have been identified. Chu et al. (2013) mapped a CR gene of *B. rapa* ssp.

chinensis (Pak Choy) "Flower Nabana" using 318 SSR markers in a mapping population of 1299 backcross plants segregating for resistance to pathotype 3. A 1: 1 segregation ($\chi 2 = 2.98$, $P = 0.084$) in the mapping population suggested that the resistance in "Flower Nabana" is controlled by a dominant gene. This gene was designated as *Rpb1* and genetically mapped to *B. rapa* chromosome A3. Six molecular markers flanking this resistance gene, with three markers on each side, were identified. The markers sR6340I and sN8591 are the closest ones from each side, at a distance of 0.77 and 0.54 cM to *Rpb1*, respectively. Based on the location of these markers, and a sequence BLAST search of the four CR loci *CRa*, *CRb*, *CRk*, and *Crr3* reported to be in the linkage group A3 (Piao et al. 2004, 2009; Saito et al. 2006; Sakamoto et al. 2008), the *Rpb1* gene is seemingly located very close to *CRa* (Ueno et al. 2012) or *CRb* (Kato et al. 2012).

Often R-genes afford resistance against diverse pathogens through interactions with pathogen-encoded proteins, but individual R-genes can be pathogen/pathotype/ race-specific. The Toll/interleukin-1 receptor (TIR)-nucleotide-binding site (NBS)–leucine-rich repeat (NBS-LRR) class is the largest among the five well-characterized classes of R-genes (Pto; TIR-NBS-LRR; Coiled-Coil-NBS-NRR; Cf; Xa21) (Martin et al. 2003; Bernoux et al. 2011). Two *TIR-NBS-LRR* genes have been identified as the genes underlying two distinct CR loci in *B. rapa*. Ueno et al. (2012) cloned *CRa* present on chromosome A3. An intact *CRa* confers resistance to *P. brassicae* (isolate M85) in *B. rapa* while insertions/deletions that abrogate the gene also abolish the resistance. Even a single amino acid change in critical regions abolished the resistance trait. In a loss-of-function analysis, randomly mutagenized derivatives of the resistant line that were found to be sensitive to *P. brassicae* were also found to have undergone mutations in the *CRa* located *NBS-LRR* gene. The other resistant *B. rapa* accessions, such as "Debra," "Gelria R," and two commercially grown Chinese cabbage cultivars, "Ryutoku" and "CR Shinki," are likely to carry the functional *TIR-NBS-LRR* gene. The second *TIR-NBSLRR* associated with *Crr1a* in chromosome A8 was isolated by Hatakeyama et al. (2013). The transgenic expression of this gene confers clubroot resistance in an otherwise susceptible *B. rapa* line. There are some common features in the two functionally characterized CR genes of *B. rapa,* both are expressed at higher levels in the resistant lines than in the susceptible types, and interestingly, both genes showed alternative splicing. Alternative splicing of *TIR-NBS-LRR* genes, in addition to the production of normal transcripts, has been found to be necessary for complete resistance against tobacco mosaic virus in tobacco (Dinesh-Kumar and Baker 2000) and against *Pseudomonas syringae* pv. tomato strain DC3000 in *Arabidopsis* (Zhang and Gassmann 2003). As R-genes are race or pathotype-specific, a pathotype can evolve to "defeat" an R-gene until a new R-gene(s) arises by mutation(s) to overcome the defeated gene. Some *NBS-LRR* genes mediate resistance in a more complex manner than the classical gene–gene interaction (one host R-gene–one pathogen gene). An example is the requirement of two *NBS-LRR* genes for resistance against wheat leaf rust (*Puccinia triticina*) (Loutre et al. 2009).

Given that R-genes are central players in adaptive resistance in plants, there are hundreds of *NBS-LRR* genes in a typical plant genome, and these appear to have

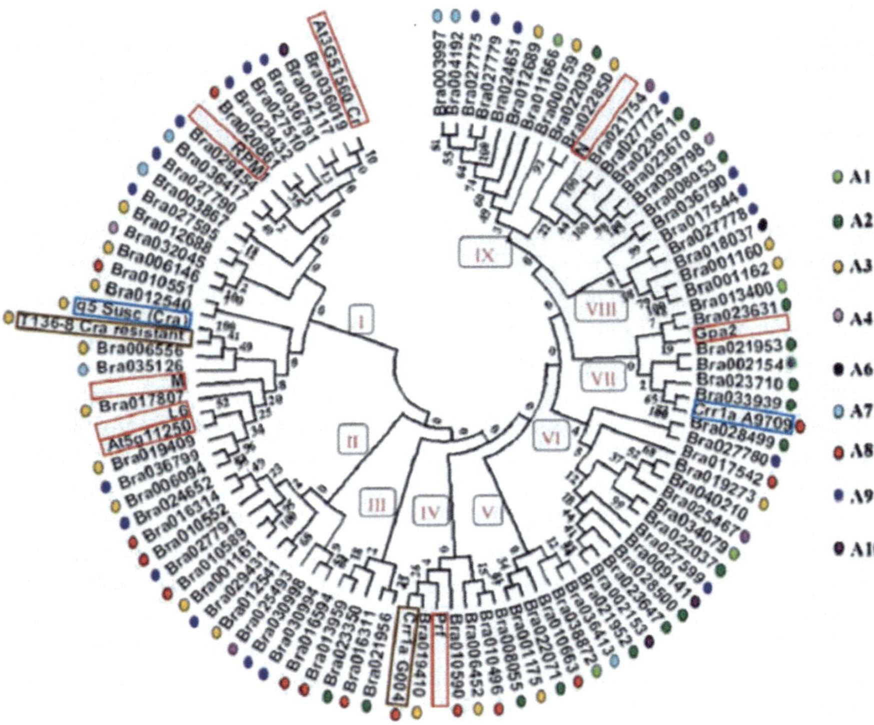

Fig. 7.7 (Color online) Phylogenetic relationship of *TIR-NBS-LRR* gene products in the *Brassica rapa* genome (www.brassicadb.org) along with some R-genes from other species: *AT3G51560, AT5G11250, N (U15605), L6 (U27081), M (U73916), Prf (U65391), Gpa2 (AF195939),* and *RPM1 (X87851).* Known R-genes are marked by red rectangles. *CRa* and *Crr1a* are indicated by brown rectangles; their counterparts in susceptible lines are indicated by blue rectangles. Neighbor joining method in MEGA5 (Tamura et al. 2011) was used in the analysis. The chromosomal associations (A1 through A10) of the NBS-LRR of *Brassica rapa* are as indicated on the right (Rahman et al. 2014)

arisen by gene duplications and dispersions in the genome. It is very likely that additional CR loci will also be identified to carry *NBS-LRR*-encoding genes. This has implications for breeding resistance in *Brassica*, and especially for gene pyramiding. To manage clubroot in *Brassica* crops, there is requirement of specific R-gene alleles in various combinations. In this regard, it is useful to have a genome-wide view of the *TIR-NBS-LRR* genes as the landscape of potential sources of CR genes. Mun et al. (2009) identified 62 *TIR-NBS-LRR* genes in the *B. rapa* genome. Since then, additional *TIR-NBS-LRR* genes have been found in the genome (Rahman et al. 2014). Out of nine, two functionally known CR genes *CRa* and *Crr1a* are in Group I and Group IV, respectively. Group I contains nine of the 18 *TIR-NBS-LRR* genes found in chromosome A3. It is likely that additional NBS-LRR will be positionally matched to the other CR loci that have been mapped (*Crr2* at A1, *CRc* at

A2, *Crr4* at A6) (Piao et al. 2009) but not yet characterized on a molecular level (Fig. 7.7).

Arabidopsis is a well-established model plant for *Brassica* crops. The chromosomal regions are syntenic and often the deduced gene products appear very similar. Although most of the *Arabidopsis* accessions are susceptible to *P. brassicae*, some resistant ecotypes have been found (Fuchs and Sacristan 1996; Siemens et al. 2002; Alix et al. 2007; Agarwal et al. 2009; Sharma et al. 2013). Pathotype 3 is an aggressive race on *Arabidopsis* (Sharma et al. 2013). With reference to pathotype 1 resistance, Fuchs and Sacristan (1996) found that *Arabidopsis* ecotypes Tsu-0 and Ze-0 show pathotype-specific dominant resistance to *P. brassicae* isolate "e" and mapped a dominant gene (named *RPB1*) to chromosome 1. A major QTL for CR has been mapped on chromosome 5 along with four other QTL with minor effects. This region has many R-genes which are already associated with resistance to other pathogens (*TTR1, RPS4,* and *RRS1*) (Jubault et al. 2008).

Given the absence of information on *P. brassicae* effector and host R interactions, *Arabidopsis* is likely to be useful for discerning the molecular aspects of the clubroot pathosystem. This knowledge may in turn provide a direct source of diverse CR genes for transgene deployment or for allele identification in primary and secondary gene pools. In genetic management of diseases, qualitative loci (R genes) are most commonly deployed. Such resistance to blackleg or stem canker in *B. napus* was found to be more durable when combined with quantitative resistance afforded by small-effects loci that are not race-specific for resistance (Brun et al. 2010). Alix et al. (2007) found partial resistance to pathotype 1 in Bur-0 and Kn-0 ecotypes and in some mutants which were affected in auxin metabolism. Pasold et al. (2010) reported an association between flavonoid content and disease, but it is unknown if auxin transport, a general stress response (antioxidant effects), or some other aspects of flavonoid functions are involved. Gravot et al. (2012) proposed that the enhanced arginase levels found in clubs is a form of negative control over hormone-induced hyperplasia. Transcriptome analysis has been conducted in pathotype 3-susceptible and resistant *B. napus* that showed expression modulation in clusters of biochemically related genes. The emerging patterns from *Arabidopsis* and *B. napus* are complex, underscoring the challenges in deciphering the genes contributing to resistance. Nevertheless, there are opportunities to learn about CR in *Arabidopsis*, and from the use of modern genomic approaches in *Arabidopsis* and *Brassica* crop plants, to find genetic networks that are pertinent to resistance at qualitative and quantitative levels and to build durable resistance in the long term (Rahman et al. 2014).

7.7.13 Systemic Acquired Resistance in *Brassica* to Clubroot

Salicylic acid (SA) biosynthesis is down-regulated in the compatible (susceptible) interaction, so exogenous application of SA will reduce the susceptibility of *A. thaliana* Col-0 to *P. brassicae* (Agarwal et al. 2011). The dipping of 14-day-old *A. thaliana* seedlings in 500 μM unbuffered SA for 60 s before planting and inoculation with *P. brassicae* reduces the disease index from 81.5 (inoculated/untreated) to 20 (inoculated/SA treated) out of a possible 100 (Agarwal et al. 2011). Lovelock et al. (2013) reported systemic transportation of exogenously applied SA from roots to leaves in *B. oleracea*. This is associated with up-regulation of pathogenesis-related genes PR-1, and PR-2 in a "systemic acquired resistance like" response as early as 24 h after treatment. Application of concentrations greater than 5 mM significantly reduces root galling in between 25 and 65% measured 6 weeks after inoculation, but concentrations in excess of 20 mM are phytotoxic, significantly reducing root and shoot weights. The use of *Arabidopsis* mutants confirmed the role of SA in the host defense against *P. brassicae* (Lovelock et al. 2013). The mutants cpr1, and dnd1 overexpress SA. Both shows reduced root galling after inoculation with *P. brassicae* compared with the SA-deficient NahG mutant and the wild-type Col-0. Elevated levels of PR-1 were measured in the overexpresser mutants and in the wild type treated with 0.5 or 1 mM SA.

7.7.14 Impact of Clubroot-Resistant Cultivars on Disease Management

The management of clubroot is complicated by the longevity of *P. brassicae* resting spores. Infection of *Brassica* root tissue results in the formation of plasmodia within the host root cortical cells. As the plasmodia mature, *P. brassicae* resting spores form and when host roots decay, these resting spores are released into the soil. In many cases, farmers find it difficult to abstain from planting *Brassica* crops long enough to significantly reduce soil inoculum levels, particularly in heavily infested fields (Wallenhammar et al. 2012). When soil infestation levels exceed c. 105 *P. brassicae* resting spores per g soil under field conditions, even the planting of resistant *Brassica* cultivars is not recommended (Wallenhammar et al. 2012). Nonetheless, Peng et al. (2015) demonstrated the potential effectiveness of an extended rotation away from canola in heavily infested fields. After a ≥2-year break from canola, resting spore loads dropped by up to 90% from initial levels, resulting in yield gains of 32–76% (Peng et al. 2015). Even at very high concentrations of *P. brassicae* resting spores, resistant host cultivars contribute significantly less inoculum to the soil in subsequent years than do susceptible cultivars (Hwang et al. 2011). Resistant *Brassica* spp. may still undergo primary infection (Donald et al. 2008; Hwang et al. 2011), but secondary infection and the production of resting spores may be prevented or reduced significantly (Hwang et al. 2011; Rennie et al. 2013). It is reasonable to hypothesize that, if initial spore loads are low enough, resistant cultivars may induce resting spore germination while supporting limited production of new inoculum, thereby serving to deplete spore loads in the soil.

The impact of growing CR canola cultivars on *P. brassicae* resting spore dynamics under field conditions in commercial cropping systems was evaluated. This included examining the effect of CR canola on *P. brassicae* inoculum within a single season, as well as its effect over various cropping sequences and multiple years. The study attempted to determine what effect initial *P. brassicae* resting spore concentrations in the soil had on clubroot development in CR canola, in order to identify a threshold resting spore concentration at which it may be acceptable to plant CR canola (Ernst et al. 2019).

The soil-borne pathogen, *Plasmodiophora brassicae*, is difficult to manage due to the longevity of its resting spores, ability to produce large amounts of inoculum, and the lack of effective fungicides. The cropping of clubroot-resistant (CR) canola cultivars is one of the few effective strategies for clubroot management. Ernst et al. (2019) evaluated the impact of the cultivation of CR canola on *P. brassicae* resting spore concentrations in commercial cropping systems in Alberta, Canada. Soil was sampled pre-seeding and post-harvest at multiple geo-referenced locations within 17 *P. brassicae*-infested fields over periods of up to 4 years in length. Resting spore concentrations were measured by quantitative PCR analysis, with a subset of samples also evaluated in greenhouse bioassays with a susceptible host. The cultivation of CR canola in soil with quantifiable levels of *P. brassicae* DNA resulted in increased inoculum loads. There was a notable lag in the release of inoculum after harvest, and quantifiable *P. brassicae* inoculum peaked in the year following cultivation of CR canola. Rotations that included a $\geq$2-year break from *P. brassicae* hosts resulted in significant declines in soil resting spore concentrations. A strong positive relationship was found between the bioassays and qPCR-based estimates of soil infestation. Results suggest that CR canola should not be used to reduce soil inoculum loads, and crop rotations in *P. brassicae*-infested fields should include breaks of at least 2 years away from *B. napus*; otherwise, the risk of selecting for virulent pathotypes may increase (Ernst et al. 2019).

7.8 *Brassica-Pseudocercosporella* White Leaf Spot Management

7.8.1 Deployment of Resistant Cultivars of *Brassica* Against *Pseudocercosporella*

Brassica napus genotypes carrying B-genome introgression from *B. carinata*, highlight the enormous potential for incorporating high levels of resistance against white leaf spot disease into commercial germplasm of cruciferous oilseed, forage, and vegetable crops, particularly for *Brassica* species. Wide variation was observed in response to *P. capsellae* infection among different genotypes under field conditions, ranging from highly susceptible to completely immune. Twenty eight genotypes from *B. carinata* (viz. 22 genotypes with AUDPC 0 and six genotypes with AUDPC <5) demonstrated high levels of resistance, highlighting the tremendous potential to use such species in breeding programs for improvement of resistance in commercial *Brassica* crops against this important pathogen. Although there was wide variation across different hosts in response to *P. capsellae*, and while some forage or vegetable types had sources of excellent resistance, such as *B. oleracea* var. *acephala* (totally resistant), and *B. oleracea* var. *capitata,* and *B. oleracea* var. *italic* (both highly resistant), this was not the case within other vegetable *Brassicas*. For example, *B. rapa* var. *rosularis*, *B. campestris* var. *chinensis,* and *B. rapa* contained genotypes that were the most susceptible. While as a group, in general, Asian leafy *Brassicas*, particularly *B. rapa*, are unlikely to provide excellent sources of resistance against this pathogen, wider exploration among even this species is warranted. Similarly, within B. *napus, B. juncea,* and *B. carinata*, while some highly resistant genotypes were identified (e.g., Scoop, MPIR, and ATC 94129, respectively), there were also some highly susceptible genotypes (e.g., Trilogy, Rohini, and INIA 0572-69, respectively).

It was confirmed that *B. napus* Trilogy and *B. napus* Surpass 400 are susceptible genotypes (Eshraghi et al. 2007). While genotypes showing high-level resistance is highly desirable for breeding programs targeting more disease-resistant cultivars. Major variations in resistance responses for *B. juncea* and *B. napus* genotypes based on country of origin were also observed in these studies. *B. juncea* and *B. napus* originating from China were less susceptible to white leaf spot, while genotypes of Indian origin were highly susceptible. Australian genotypes were intermediate for each species group. While the reason for such differences is not immediately apparent, Chen et al. (2008, 2011) working with a wider range of *B. juncea* genotypes from these same three countries, showed that there was a primary center of origin of *B. juncea* in central Asia, and that China and India are secondary centers of diversity. They also presented evidence to support the proposition that there were two independent migrations of *B. juncea* into China and India. Despite *B. napus* in Australia and China having similar origins, there is both a greater genetic diversity and genetic distinctiveness of Chinese germplasm, compared with Australian germplasm, after 50 years of breeding from similar origins (Chen et al. 2008). A small group of genotypes from India was distinctly separated from all other *B. napus* genotypes in that same study (Chen et al. 2008). Hence, it is not surprising that the responses to *P. capsellae* were different across genotypes from these three countries, despite the small number of genotypes from India were tested (Gunasinghe et al. 2014).

Due to variable environmental conditions, concern has been that field screening may fail to provide consistent disease screening results. However, when comparing performances of the four genotypes of *E. vesicaria*, *B. napus* Ag-Spectrum, *B. juncea* Xingyou 9, and *B. juncea* JM06006 that were in common to the Crawley and Shenton Park field sites; and when comparing the two genotypes *B. napus* Charlton and *B. juncea* JO006 in common to all three field experiments, relatively similar performances were noted between experiments and across field sites. For example, *E. sativa* Eruc-01 and *E. vesicaria* Yellow rocket were highly resistant genotypes in both field sites while *B. juncea* JM06006 was highly susceptible to *P. capsellae* in all three field experiments over two different locations during the 2012 growing season. While *B. napus* Charlton, was located in what could be loosely called an "intermediate resistance group" across all three field experiments, it is not always the case that genotypes expressing intermediate levels of resistance expression maintain consistent relative rankings across different field experiments.

Many canola growing countries worldwide, including Australia, have now recognized the potential benefits of introducing more drought and high temperature-tolerant *Brassica* species, particularly the mustards *B. juncea* and *B. carinata*, over *B. napus* (Cardone et al. 2003; Burton et al. 2003; Potts et al. 1999; Malik 1990). In particular, a high priority has been given to developing *B. juncea* as a commercial oilseed crop well adapted to low rainfall areas across southern Australian grain-belt (Burton et al. 2003). Further, some genotypes within these alternative *Brassica* species show enhanced resistance to other *Brassica* diseases, such as blackleg disease (Rimmer and Van den Berg 1992) and *Sclerotinia* Stem rot disease (Uloth et al. 2013), encouraging wider deployment of such commercial varieties. Indian- and Australian-derived genotypes of *B. juncea* were highly susceptible to white leaf spot disease and will likely suffer significant yield losses from this disease under Australian conditions unless improved resistance is incorporated into future varieties. The evident development of pod lesions in almost all of Australian and Indian *B. juncea* genotypes tested at the Crawley field site demonstrates the potential for significant and widespread distribution of pathotypes of *P. capsellae* across Australia, particularly into new areas where oilseed or other *Brassicas* are currently not widely cultivated.

B. carinata genotypes (with purple colouration) from Ethiopia and some other *B. carinata* genotypes from Pakistan and Sweden as carrying high level of resistance to white leaf spot disease have been identified. There is a strong indication that the high-level resistance present in some genotypes of *B. carinata* is likely major gene resistance conferring complete resistance (in contrast to the intermediate levels of resistance in some other genotypes that is more likely polygenic in nature). The identification of major gene resistance in *B. carinata*, highlights the value of evaluating additional genotypes to further explore the value of this species in its own right and the durability of such major gene resistance for utilization by breeding programs in Australia and elsewhere. Genotypes with high level of resistance will be of great value for developing new cultivars of oilseed, forage, and vegetable crucifers with much improved levels of resistance to white leaf spot. Even genotypes identified with moderate levels of resistance, if deployed commercially, offer prospects of improving current management strategies compared with current reliance upon cultural and/or chemical controls, often in conjunction with cultivars lacking effective levels of resistance (Gunasinghe et al. 2014).

7.9 *Brassica-Sclerotinia* Stem Rot Management

7.9.1 Reduction of Pathogen Virulence

Hypovirulence refers to the reduced virulence of selected isolates within a population of a plant pathogen. In many cases, hypovirulence has been associated with the presence of double-stranded ribonucleic acids (dsRNA) characteristics of fungal viruses (Nuss and Koltin 1990) and at least three of these agents from *Cryphonectria parasitica*, the causal agent of chestnut blight, have been classified within the new virus family, Hypoviridae (Hillman et al. 1995); however, other factors such as mitochondrial mutations (Mahanti et al. 1993), nuclear mutations, and encapsidated fungal viruses have been associated with hypovirulent isolates. Therefore, hypovirulence is best considered as a phenotypic response characterized by reduced virulence but also may be associated with other characters such as reduced growth rate, sporulation, and survival. The potential in utilizing hypovirulent isolates of fungal pathogens in a biological control strategy resides in the ability to transfer hypovirulence from hypovirulent isolates to virulent isolates and thereby, reduce the mean disease severity of the population through overall reductions in virulence, growth, sporulation, and/or survival.

Hypovirulent isolates of *S. sclerotiorum* and *S. minor* with their role in reducing virulence in populations of these pathogens and their potential application in disease management have been identified. One isolate of *S. sclerotiorum* has been reported that is hypovirulent and contains dsRNA (Boland 1992). This isolate grows slowly in culture, develops a typical colony morphology, produces significantly smaller lesions on celery than virulent isolates, and contains dsRNA. The hypovirulent phenotype and dsRNA are transferred to vegetatively compatible recipient isolates through hyphal anastomosis and recipient isolates develop the hypovirulent phenotype and become hypovirulent. Ultra-structural examination of hyphal and sclerotial cells from the virulent isolate of *S. sclerotiorum* reveal pleomorphic vesicles surrounded by a double membrane associates with the nuclear envelope (Boland et al. 1993). No such vesicles are found in a near-isogenic, virulent, dsRNA-free isolate. The evidence indicates that dsRNA in this hypovirulent isolate is not associated with virus particles, but instead is surrounded by a lipid membrane. This suggests that this agent may be a member of the newly established family of un-encapsidated dsRNA viruses termed Hypoviridae (Hillman et al. 1995).

Despite the discovery of a hypovirulent isolate of *S. sclerotiorum* that can transmit hypovirulence to selected, virulent isolates, further attempts to utilize hypovirulent isolates of this pathogen for biological control have not been actively pursued because of the known diversity of vegetative (Ford et al. 1995) and mycelial compatibility groups (Kohli et al. 1992; Kohn et al. 1991) within this pathogen; however, the number of MCGs in *S. minor* is relatively small (Patterson 1986) and therefore, the use of transmissible hypovirulence in this pathogen represents a more promising approach as a management strategy for disease in controlled environments. Mycelial suspensions of a hypovirulent isolate of *S. minor* applied to leaf lesions initiate by compatible isolates on lettuce suppresses lesion expansion upto 100% and reduces the development of sclerotia on diseased tissues upto 100% (Melzer and Boland 1996). The earlier in lesion development the hypovirulent isolate is applied, the more it arrests the growth of the lesion of the virulent isolate and reduces the numbers of sclerotia produced. Sclerotia that are produced on leaves treated with the hypovirulent isolate tends to be nonviable, or to grow atypically. Application of the hypovirulent isolates to lesions initiated by incompatible, virulent isolates does not suppress disease or reduce the development of survival structures. The results suggest that hypovirulent isolates of *S. minor* applied to actively growing lesions on lettuce plants in the field may reduce inoculum production and increase the level of hypovirulence in the population (Saharan and Naresh 2008).

7.10 *Brassica-Turnip Mosaic Virus* (*TuMV*) Management

7.10.1 *TuMV* Resistance Breeding in *Brassica*

Understanding the molecular interactions between hosts and pathogens could provide information to assist plant breeding programs. Much work still needs to be done to develop and apply the knowledge of resistance genes and virus–*Brassica* interactions to breeding. This can be achieved in many ways, including transformation of resistance genes into crops, molecular marker-assisted selection (MAS), pyramiding of resistance genes and host-induced gene silencing (HIGS).

7.10.2 Transformation of R-Genes into *Brassicas*

As more and more resistance genes are cloned, transforming these genes into *Brassica* crops will become an important approach for creation of new germplasm material. Kim et al. (2014) demonstrated that W95L, K150L, and W95L/K150L amino acid mutations of *B. rapa eIF (iso) 4E* interrupted the interaction with TuMV VPg. They overexpressed these mutants of *eIF (iso) 4E* and the wild type in a susceptible Chinese cabbage cultivar and found that these mutations could confer resistance to multiple strains of TuMV. Nellist et al. (2014) clarified that the resistance genes *retr01/retr02* could encode eIF (iso) 4E in *B. rapa*, so the different copies of eIF(iso) 4E from a resistant *B. rapa* line were transformed into an *eIF(iso) 4E* knockout line of *Arabidopsis thaliana* to analyze the function of the *retr01/retr02* genes. Shopan et al. (2017) cloned the recessive resistance gene *retr03* from *B. juncea*; when they transformed the *retr03* gene from a TuMV-susceptible line into a TuMV-resistant mustard plant line, they confirmed the new resistance mechanism. Obviously, genetic engineering methods can improve *Brassica* crops rapidly; however, generation of transgenic *Brassica* crop plants has proven difficult, especially *B. rapa* (Li et al. 2018). The improvement of transformation technology would enhance the use of TuMV-resistant gene transformation for applied breeding (Li et al. 2019).

7.10.3 Molecular Marker-Assisted Breeding in *Brassica*

Molecular markers are specific, inheritable, and detectable DNA segments that can be used for linkage map construction, gene mapping, and MAS. In the genomic era, marker types and mapping methods have improved greatly. In the 1990s, low-efficiency RAPD, AFLP, and RFLP markers were mainly used. For example, Walsh et al. (1999) mapped the first dominant resistance gene to TuMV (*TuRB01*) in

B. napus. Since the 2000s, convenient and easily detectable SSR, microsatellite, and indel markers have gained popularity and have been used for the identification of resistance genes to TuMV (*ConTR01, retr01*, TuMV-QTLs; Rusholme et al. 2007; Zhang et al. 2008, 2009). During the 2010s, high-throughput methods of mapping have become popular, including the use of SNP-based markers such as BSA-developed markers, competitive allele-specific PCR (KASP) markers and markers from genome-wide association studies (Qian et al. 2013; Li et al. 2016; Cheng et al. 2016). For example, Li et al. (2016) developed a KASP marker based on the single recessive gene, *retr02*, to detect the *retr02* allele in Chinese cabbage accessions and used in breeding programs. This marker will probably be beneficial to establishing high-throughput assays for marker-assisted selection (Li et al. 2019).

7.10.4 Pyramiding R-Genes for Durable Resistance to *Brassica*

Resistance from individual genes, such as isolate-specific R-genes, is highly effective but easily broken by pathogens, as well as by global climate changes. Additionally, polygenic quantitative resistance is considered to be more durable than qualitative resistance, but its effectiveness varies between cropping seasons because of environmental conditions (Lindhout 2002). Undoubtedly, more durable resistance is urgently needed to safeguard *Brassica* crops. Thus, the pyramiding of qualitative resistance with a high level of quantitative resistance in cultivars may be an ideal way to maximize the durability of resistance. Now, 20 genes/loci have been mapped in *Brassica* crops, including 16 dominant genes/loci and four recessive genes, as well as some QTLs (Table 2.1, Chap. 2). After mapping these resistance genes, breeders began to incorporate them into *Brassica* lines via hybridization, backcrossing, and inbreeding to create new germplasm materials (Rusholme et al. 2007; Qian et al. 2013; Shopan et al. 2017). In addition, breeders developed several resistant line hybrids for gene pyramiding to impart resistance to different TuMV isolates. A good example comes from clubroot resistance breeding. Specifically, Matsumoto et al. (2012) attempted to accumulate various resistance genes (three CR genes: *CRa*, *CRk*, and *CRc*) through marker-assisted selection in *B. rapa* and found that clubroot resistance could be reinforced (Li et al. 2019).

7.10.5 Host-Induced Gene Silencing in *Brassica*

In contrast to the use of host resistance genes against TuMV, genes from the pathogen can be used to develop resistance via HIGS (Nowara et al. 2010). In particular, the CP gene of TuMV has been used via HIGS to inhibit virus replication, thereby conferring resistance. Jan et al. (2000) transformed the CP gene of TuMV linked to the 218 or 110 bp N gene (N gene expressed alone did not confer resistance) to tobacco and the generated plants were resistant to multiple viruses. These findings demonstrate that the fusion of viral segments to DNAs can confer resistance to multiple viruses. Successful enhancement of resistance by the CP gene strategy has also been reported for oilseed rape and Chinese cabbage. Lehmann et al. (2003) transformed the CP gene of *TuMV* into oilseed rape to clarify the effectiveness of resistance to TuMV mediated by CP and found that the CP could change the resistance, providing broad spectrum resistance to *TuMV*. The P3 protein of *TuMV* was also reported to be associated with avirulence when interacting with *TuRB03* (a dominant resistance gene to TuMV isolate CDN1) and *TuRB04* in *B. napus* (Jenner et al. 2002; Hughes et al. 2003). In addition, Jenner et al. (2002) found that the *TuRB05* gene could interact with the wild-type CI protein of *TuMV*, causing an HR-like phenotype with limited viral replication and spread, and that mutation at position +5447 in the CI gene would break the TuRB05 resistance. Expression of *TuMV* genes in plants has also been used to investigate plant responses. For example, similar to the effects of *TuMV* infection, the expression of a single TuMV protein, nuclear inclusion a-protease domain (NIa-Pro), in plants increased aphid arrestment (attraction), suppressed callose deposition and increased the abundance of free amino acids, the major source of nitrogen for aphids (Casteel et al. 2014). Overall, individual genes from *TuMV* expressed in plants can repress or change the virulence of *TuMV* to plants, opening new avenues of research for breeders to control *TuMV* (Li et al. 2019).

7.11 *Brassica-Xanthomonas* Black Rot Management

7.11.1 Distant Hybridization for *Xcc* Resistance Through Somatic Hybridization and Embryo Rescue

Wide hybridization or distant hybridization is a powerful approach to combine the diverged genomes into a single nucleus. Wide hybridization is an important driving force in broadening the gene pools, the evolution of novel plant forms and speciation, transferring the genomes of one species to another, enhancing the genetic diversity and improving the quality of crops (Soltis and Soltis 2000; Barton 2001; Jiao et al. 2011). The *Brassica* vegetables and oilseed crops in the cruciferae family have undergone multiple genomic duplications (Schranz et al. 2006). The classic genome polyploidization event is the interspecific hybridization of *B. rapa* (A-genome), *B. nigra* (B-genome), and *B. oleracea* (C-genome) to form the allotetraploids *B. juncea* (AB genome), *B. napus* (AC genome), and *B. carinata* (BC genome) (Nagaharu 1935). Since the first intergeneric hybrid between *R. sativus* and *B. oleracea*, "*Raphanobrassica*," reported by Karpechenko (1928), numerous intergeneric and interspecific hybrids have been produced between *Brassica* species of the U triangle (Nagaharu 1935) and *R. sativus*. The related or wild species are a rich reservoir of several useful genes having resistance to various biotic or abiotic stresses. In the family Brassicaceae, interspecific hybridization is a potential approach for transfer of desirable genes from one species to another, enhancing genetic diversity and creation of hexaploids for the improvement of *Brassica* crops (Prakash 1973; Prakash and Chopra 1988; Choudhary et al. 2000; Siemens et al. 2002; Niemann et al. 2012; Kaneko and Bang 2014). The sexual incompatibility or reproductive barriers can be overcome by utilization of embryo rescue or somatic hybridization techniques, or a combination of both, in the development of interspecific hybrids in crucifer crops (Weerakoon et al. 2009; Niemann et al. 2012; Dey et al. 2015). The development of black rot-resistant hybrids is one of the major objectives of *Brassica* breeders. Since the resistance sources in *B. oleracea* are insufficient, interspecific hybridization with related or distant species followed by embryo rescue or somatic hybridization for the development of black rot-resistant hybrids has been practiced and accomplished by many workers (Table 4.11, Chap. 4).

The lines of *B. napus* having very high resistance to black rot (*X. campestris* pv. *campestris*) and resistance is governed by a dominant gene, but the successful crossing between two species (*B. oleracea* and *B. napus*) is very difficult (Hansen and Earle 1995). Therefore, the development of somatic hybrids by protoplast fusion between rapid cycling *B. oleracea* and *X. campestris* pv. *Campestris*-resistant line of *B. napus* was attempted. Hybrid identity was confirmed by morphological studies, flow cytometric estimation of nuclear DNA content, and analysis of RAPDs. In another study, Wang et al. (2011) generated new alien addition lines having high resistance to black rot (*X. campestris* pv. *campestris*) by employing a somatic hybridization technique between cauliflower and black mustard. It has been evident that success of interspecific hybridization is also influenced by the genotype of the parents (Tonguc and Griffiths 2004; Mason et al. 2011). Tonguc and Griffiths (2004) utilized two *B. juncea* accessions (A 19182 and A 19183) having resistance to Xcc race 1 and 4, as maternal and paternal parents to develop interspecific hybrids with *B. oleracea* cultivars. They employed embryo rescue to recover interspecific hybrids and further confirmed by inheritance of paternal molecular markers. They obtained 26 interspecific hybrid plants between A 19182 and *B. oleracea* cultivars, but no interspecific hybrids were obtained using A 19183, suggesting influence of genotype in the success of interspecific hybridization. The interspecific hybrid plants were male-sterile, and they were used successfully as maternal parents to develop backcross plants employing embryo rescue. All hybrid and BC1 plants were found resistant to races 1 and 4 of Xcc. Dey et al. (2015) reported introgression of black rot resistance genes into Indian cauliflower (*B. oleracea* var. *botrytis* L.) through interspecific hybridization (*B. carinata* x *B. oleracea*) followed by embryo rescue.

They utilized two *B. carinata* genotypes, NPC-9 and HCA-6, having very high resistance to Xcc race 1 and 4, for interspecific hybridization to introgress resistance genes from *B. carinata* into Indian cauliflower employing three different embryo rescue approaches. They reported that all the F1 hybrids and backcross were resistant to *Xcc* races 1 and 4 (Singh et al. 2018).

7.11.2 Use of Omics Approaches for Black Rot Resistance Management

In present century, rapid advances have been made in "omic" technologies to furnish new information through the genomic, transcriptomic, proteomic, or metabolomic levels. The development of genomic resources like whole genome sequences, genomic survey sequences (GSS), large number of expressed sequence tags (ESTs), large-insert genomic libraries, high density genetic maps, millions of molecular markers and high-throughput genome sequences (HTGs) are essential to study the complex defense mechanisms, metabolic pathways, and proteins involved in plant pathogen resistance in different crops. The genomic resources have been used for sequencing and annotation, genomic selection, mapping and cloning of genes or QTL, association mapping, breeding by design, and MAS in important horticultural crops (Lorenz et al. 2011; Sonah et al. 2011; Witzel et al. 2015; Roohie and Umesha 2015; Huang et al. 2017; Rahman et al. 2018). The advent of next-generation sequencing (NGS) technologies like FLX-454, Illumina, SOLiD/ion torrent, pacbio, HiC, bionano, oxford nanopore, and helicose have enhanced the possibilities to generate genomic resources and their utilization in *Brassica* improvement programs (Sharma et al. 2012, 2014; Hu et al. 2012; Izzah et al. 2014; Witzel et al. 2015; Cheng et al. 2016; Perumal et al. 2017; Yang et al. 2018).

With the advent of transcriptome sequencing approach large amount of transcribed sequence data, such as expressed sequence tags (ESTs) have been generated and that can be efficiently utilized for study of gene expression, genome annotation, comparative genomics, and physical mapping of genes (Rudd 2003). Izzah et al. (2014) performed the transcriptome sequencing of two parental lines of cabbage, C1184 (susceptible to black rot, Xcc) and C1234 (resistant to black rot, Xcc), using the 454 platform, Roche GS FLX Titanium series and generated large-scale EST data and developed EST-based functional markers; EST-based SSR and EST-based dCAPS markers. They constructed a genetic linkage map of cabbage by exploiting these EST-based markers which will facilitate the QTL mapping of desirable traits including resistance to Xcc. Proteomics analysis is carried out to characterize the full protein complement present in a particular organism, organ, tissue, or cell, and for the identification of functions of proteins expressed under different conditions (Graves and Haystead 2002; Mehta et al. 2008). In the past few years, the interest in proteome analysis has rapidly increased. The proteome analysis can be carried out by exploring the high resolution of two-dimensional electrophoresis (2-DE) (Jungblut and Wittmann-Liebold 1995) coupled to MS. Besides 2-DE, current proteomics platforms exploit more versatile LC (liquid chromatography)-based techniques (Rabilloud et al. 2010; Matros et al. 2011), coupled with mass spectrometry (MS). In this regard, Andrade et al. (2008) performed in vivo proteome analysis for the characterization of the protein expression of *Xcc* in close interaction with the host plant B. oleracea and suggested that this approach can contribute to a better understanding of the mechanisms associated with the infection process of Xanthomonas species. In continuation to this work, the group of researchers employed this approach of proteome analysis and compared the proteome of Xcc during the interaction with the resistant and the susceptible cultivars of *B. oleracea* (Villeth Jr et al. 2009). They identified the differentially expressed proteins, of which some were specifically expressed during the interaction with the resistant cultivar of B. oleracea. De Bernonville et al. (2014) studied and defined the proteomic impact in the xylem sap induced by infection of Xcc in *B. oleracea*. Functional genomics, proteomics, and metabolomics studies in both resistant and susceptible host plants of *Brassica* species infected with black rot will generate key information on the host–pathogen interaction. Furthermore, in the near future, many advancements and refinements in all the 'omics approaches can be expected to come out, and these will provide exciting new avenues of research in the cruciferous crops (Singh et al. 2018).

7.12 *Brassica*-Cyst Nematode Management

7.12.1 Pyramiding of R-Genes in *Brassica* Against Beet Cyst Nematode

Beet cyst nematode (*Heterodera schachtii* Schm.) is an important pest of sugar beet that can cause significant reductions in yield. Unlike most cyst nematode, the beet cyst nematode (BCN) has a wide host range and can infect more than 218 plant species, including family Brassicaceae and Chenopodiaceae such as sugar beet (*Beta vulgaris* L.), oilseed rape (*Brassica napus* L.), and spinach (*Spinacia oleracea*). In general, nematodes can be controlled by treating with nematicides, growing resistant cultivars, and crop rotation with non-host or trap crops. However, control via nematicides application is difficult and expensive because nematode larvae and eggs are well protected. In addition, the use of nematicides increases the threat of environmental pollution and some are prohibited worldwide. Growing resistant cultivars with a monogenic resistance can induce the emergence of more virulent nematode pathotypes. Oilseed rape (*Brassica napus* L.) is a good host for BCN and always rotates with sugar beet in the agro-farming system. Therefore, breeding for BCN-resistant oilseed rape is of great value and importance (Hemayati et al. 2017).

Investigation on 111 *Brassica* germplasm lines revealed that all lines are susceptible to *H. schachtii*. BCN resistance is found only in a few spices of Brassiceae including oil radish (*Raphanus sativus* L. ssp. *oleiferus* DC.) and white mustard (*Sinapis alba* L.). Oil radish shows complete resistance and is often used as a trap crop to mitigate the degree of damage in infested fields. Resistant/trap crop could stimulate the hatching of larvae, which invaded the roots, but prevents larvae from fulfilling their life cycle and thus lowers BCN populations. A dominant nematode gene $Hs1^{Rph}$, which is located in the radish chromosome d, confers beet cyst nematode resistance. Efforts have been made to transfer this resistance gene into oilseed rape genome by intergeneric hybridization. However, expression of the resistance gene in a hybrid species remains a great challenge due to the extremely low and unstable inheritance. Cloning of the cyst nematode resistance genes and their subsequent transfer into the rapeseed could be one of the most feasible strategies to induce nematode resistance in oilseed rape (Zhong et al. 2019; Zhang et al. 2008).

$Hs1^{pro\text{-}1}$ gene is the first beet cyst nematode resistance gene cloned from sugar beet translocation line (A906001) by a map-based cloning strategy. The transloca-tion line carried the locus from *B. procumbens* chromosome 1 that confers resistance to *H. schachtii* in sugar beet (*Beta vulgaris* L.). The $Hs1^{pro-1}$ gene is different from other nematode resistance genes amid the presence of the NBS-LRR structure, such as *Mi*, *Gpa2*, and *Hero*, due to the presence of the NBS-LRR structure. The resistance mechanism of $Hs1^{pro-1}$ gene is based on the gene-for-gene relationship. The transcript of $Hs1^{pro-1}$ gene is upregulated about fourfold after 1 day of nematode infection. However, no considerable change in the transcript accumulation of $Hs1^{pro-1}$ is recorded in uninfected roots of resistant beet plants. McLean et al. (2007) published a complete sequence of the $Hs1^{pro-1}$ protein, which included an additional 176 amino acid at N-terminal extension conferred resistance to soybean (*Glycine max* L.) cyst nematode (*Heterodera glycines*). Hunger et al. (2003) cloned 47 resistance gene analogs (RGAs) from genomic DNA of sugar beet. Most cloned resistance genes belong to nucleotide-binding site leucine-rich repeat (NBS-LRR) gene family. Members of this family have been isolated in both dicot and monocot plants, exhibiting resistance to a variety of plant pathogens, including bacteria, fungi, viruses, and nematodes. The cloning of conserved sequence of NBS domain has been successfully applied to isolate resistance gene candidates or resistance gene analogs (RGAs) from plant genomes, using a degenerate primer-based PCR strategy. For example, from sugar beet translocation line (A906001), cZR3 was cloned using a degenerated primer-based PCR strategy. The gene is similar to a subset of CC-NBS-LRR resistance proteins, including nematode resistance genes *Mi* from tomato and *Gpa2* from potato. The phylogenetic analysis shows that this gene originates from the ancestral gene from which I2C1 (vascular wilt disease resistance), *Xa1* (bacterial blight resistance) and *Cre3* (nematode resistance) were also originated. For CC-NBS domains, RGAs are required for initiation of a necrotic hypersensitive response, as well as for cell death on the initial infection sites. RGAs are ubiquitous in plant genomes, and often cluster with close linkage to active resistance genes. Because no complete resistance could be observed so far by transgenic sugar beet plants, it is proposed that additional genes are required to confer full resistance. Zhong et al. (2019) hypothesized that the *cZR3* gene may interact with the $Hs1^{pro-1}$ gene to confer resistance against sugar beet nematode in oilseed rape. Therefore, Zhong et al. (2019) transferred the $Hs1^{pro-1}$ and *cZR3* genes into oilseed rape using hypocotyl explants by *A. tumefaciens*-mediated transformation method and pyramided the $cZR3Hs1^{pro-1}$ genes by crossing homozygous transgenic *cZR3* and *Hs1* plants. The beet cyst nematode resistance tests *in vitro* and *in vivo* were detected in *cZR3*, $Hs1^{pro-1}$, and *cZR3* $Hs1^{pro-1}$ pyramiding plants. Possible resistance mechanisms of $Hs1^{pro-1}$, *cZR3*, and $cZR3Hs1^{pro-1}$ gene pyramiding were revealed.

Beet cyst nematode (*Heterodera schachtii* Schm.) is one of the most damaging pests in sugar beet growing areas around the world. The $Hs1^{pro-1}$ and *cZR3* genes confer resistance to the beet cyst nematode, and both were cloned from sugar beet translocation line (A906001). The translocation line carried the locus from *B. procumbens* chromosome 1 including $Hs1^{pro-1}$ gene and resistance gene analogs (RGA), which confer resistance to *Heterodera schachtii*. The $BvHs1^{pro-1}$ and *BvcZR3* genes were transferred into oilseed rape to obtain different transgenic lines by *A. tumefaciens*-mediated transformation method. The $cZR3Hs1^{pro-1}$ gene was pyramided into the same plants by crossing homozygous *cZR3* and $Hs1^{pro-1}$

plants to identify the function and interaction of *cZR3* and *Hs1* $^{pro-1}$ genes. *In vitro* and *in vivo* cyst nematode resistance tests showed that *cZR3* and *Hs1* $^{pro-1}$ plants could be infested by beet cyst nematode (BCN) juveniles; however, a large fraction of penetrated nematode juveniles was not able to develop normally and stagnated in roots of transgenic plants, consequently resulting in a significant reduction in the number of developed nematode females. A higher efficiency in inhibition of nematode females was observed in plants expressing pyramiding genes than in those only expressing a single gene. Molecular analysis demonstrated that *BvHs1* $^{pro-1}$ and *BvcZR3* gene expressions in oilseed rape constitutively activated transcription of plant defense-related genes such as *NPR1* (non-expresser of *PR1*), *SGT1b* (enhanced disease resistance 1), and *RAR1* (suppressor of the G2 allele of *skp1*). Transcript of *NPR1* gene in transgenic *cZR3* and *Hs1* $^{pro-1}$ plants were slightly up-regulated, while its expression was considerably enhanced in *cZR3Hs1*$^{pro-1}$ gene pyramiding plants. The expression of *EDS1* gene did not change significantly among transgenic *cZR3*, *Hs1* $^{pro-1}$, and *cZR3Hs1* $^{pro-1}$ gene pyramiding plants and wild type. The expression of *SGT1b* gene was slightly up-regulated in transgenic *cZR3* and *Hs1* $^{pro-1}$ plants compared with the wild type; however, its expression was not changed in *cZR3Hs1* $^{pro-1}$ gene pyramiding plant and had no interaction effect. *RAR1* gene expression was significantly up-regulated in transgenic *cZR3* and *cZR3Hs1*$^{pro-1}$ genes pyramiding plants, but almost no expression was found in *Hs1* $^{pro-1}$ transgenic plants. These results show that nematode resistance genes from sugar beet were functional in oilseed rape and conferred BCN resistance by activation of a CC-NBS-LRR R-gene-mediated resistance response. The gene pyramiding had enhanced resistance, thus offering a novel approach for the BCN control by preventing the propagation of BCN in oilseed rape. The transgenic oilseed rape could be used as a trap crop to offer an alternative method for beet cyst nematode control (Zhong et al. 2019).

7.13 Multiple Pathogen Disease Resistance Management

7.13.1 Transgenic Approach for Biotic Stress Management

The environmental stresses, both biotic and abiotic, drastically affect the capacity of crop plants to produce their maximum yields. The various biotic factors like insects, fungi, virus, bacteria, and nematodes are a major threat to the productivity of a large number of crop plants. The biggest limitation of traditional breeding is its notoriously slow nature of transferring a desired trait which is often associated with linkage drag (undesirable traits) into an otherwise superior crop cultivar. The time needed to transfer a desired gene into a crop plant depends on the source of the gene and the evolutionary distance of that source to the recipient crop plant. If the gene source is a landrace or a related species, forming a primary gene pool with the crop species in question, the gene transfer may take 5–8 years. Less related wild species belonging to the secondary or even tertiary gene pool may be rich reservoirs of genes for agronomic traits like disease or pest resistance, but to transfer such genes into crop cultivars may take 10–15 years. Pre- and post-fertilization barriers may impede sexual hybridization between the donor and the crop species with the problem of alien gene transfer (Jauhar 2006). Great challenge of food security has directed plant scientists towards gene revolution which involves direct modification of qualitative and quantitative traits in an organism by transferring desired genes using "transgenic approach" or "genetic transformation." In contrast to classical breeding, genetic engineering offers an excellent tool for asexually inserting gene(s) of unrelated organisms into plant cells. This process may take less time thus accelerating the process of genetic improvement of crop plants. In addition, this exciting technology allows access to an unlimited gene pool without the constraint of sexual compatibility. Crop plants engineered to suit the environment better through incorporation of genes for tolerance to biotic and abiotic stresses have been suggested to represent an improvement in crop production. Over the past few decades, breeding possibilities have been broadened by genetic engineering and gene transfer technologies, as well as by gene mapping and identification of the genome sequences of model plants and crops which resulted in efficient transformation and generation of transgenic lines in a number of crop species (Gosal et al. 2009). Further, pyramiding of desirable genes with similar effects can also be achieved by using these approaches. Genetic transformation by *Agrobacterium*-mediated approaches (Sanghera et al. 2009) has been successfully demonstrated for economically important traits like biotic resistance (Wani 2010; Wani and Sanghera 2010), abiotic stress tolerance (Wani et al. 2008; Gosal et al. 2009; Wani and Gosal 2010), in different crop plants. Use of agrochemicals for control of crop pests and diseases is however inefficient and environmentally and ethically unsound (Pimentel 1995; Robinson 1999). Genetic engineering could offer a remedy through more precise targeting of pest and viral diseases (Wani and Sanghera 2010).

Transgenic plants have been produced with genes involved in different pathways to enhance disease resistance against fungal pathogens. The first report on developing fungus-resistant transgenic plant (Broglie et al. 1991) expressed bean chitinase in tobacco and *Brassica napus* to enhance resistance towards *Rhizoctonia solani*. Similarly, antifungal genes have been engineered in various crop plants to render resistance against fungal pathogens (Jauhar and Khush 2002; Sahrawat et al. 2003).

Mondal et al. (2003) developed transgenic Indian mustard plants with *chitinase* gene to incorporate resistance against *A. brassicae*. The appearance of symptoms of the disease was delayed by a period of 10–15 days in transgenic plants than the control plants. Taj et al. (2004) transformed *B. juncea* with *osmotin* gene and reported that the transformed calli exhibited resistance to the purified *A. brassicae* toxin. It was found that introgression of *osmotin* gene in transformed calli expressed resistance by perturbation of cell cycle machinery. The transformed calli suppressed the expression of P53, thus inhibiting programmed cell death, while caspase 1 was not at all affected. Mondal et al. (2007) incorporated a PR protein-encoding *glucanase* from tomato into Indian mustard plants to have enhanced resistance towards *A. brassicae*. Glucanase kills the invading fungal pathogens by hydrolyzing glucan, a major component of their cell wall. Transgenic plants arrested growth of *Alternaria* hyphae by 15–54% during in vitro bioassay. Under greenhouse challenged conditions, reduced number, size, and spread of lesion/black spots were observed in transgenic plants along with the delay in symptom appearance time in comparison to the control plants. Borhan et al. (2010) transferred *A. thaliana*-derived *WRR4* gene into Indian mustard and the transgenic plants conferred significant level of white rust resistance. Chhikara et al. (2012) co-expressed a barley antifungal class II *chitinase* gene and type I ribosome inactivating protein (*RIP*) gene in Indian mustard to combat *Alternaria* blight disease. During in vitro bioassays, transgenic mustard plants exhibited 44% reduction in *A. brassicae* hyphal growth than the control plants. During greenhouse screening, when transgenic events were sprayed with fungal spores, delayed onset of the disease along with lesser number of lesions with reduced size spread was reported. Bangash et al. (2013) also optimized an efficient genetic transformation protocol in Indian mustard cv. NIFA RAYA with *WD* (Wasabi defensin) gene using *Agrobacterium*-mediated gene transfer technique. Antimicrobial peptides (AMP) have now become the target of genetic engineering of plants for incorporating disease resistance traits. Cationic antimicrobial peptides (CAPs) and their synthetic versions are found to express resistance against a broad spectrum of phytopathogens. Rustagi et al. (2014) transformed *B. juncea* (cv. Varuna) with *MSRA1* gene to incorporate resistance against *Alternaria* blight and stem rot diseases. MSRA1 is a synthetic CAP having a fusion of cecropin A and mellitin CAPs with antimicrobial properties. In *in vitro* bioassays with *A. brassicae*, transgenic lines inhibited *Alternaria* hyphae growth by 44–62% and delayed and restricted infection protection varying from 69 to 85%. For stem rot disease, a protection rate of 56–71.5% was exhibited by the transgenic lines with less number and smaller size of sclerotia formation than the control plants. Ahmad et al. (2015) also transferred *chitinase* gene isolated from *Streptomyces griseus* HUT6037 into Indian mustard cv. Raya Anmol for incorporation of fungal tolerance. Lectins are ubiquitous carbohydrate-binding proteins, which are found as secondary metabolites in plants and are found in conferring resistance to various biotic and abiotic stresses besides protecting plants against aphids. Kumar et al. (2015) transferred chickpea *lectin* gene into Indian mustard cv. Varuna to impart resistance against *A. brassicae* in transgenic lines as observed by the number of lesions and time taken for appearance of leaf necrosis, with the protection varying in the range of 36–60% in different transgenic plants as compared to the non-transformed plants. The transgenic plants also exhibited tolerance to salt stress (NaCl) and mannitol-induced water stress. Ali et al. (2017) reported that the constitutive overexpression of *NPR1* isolated from *B. juncea* in transgenic *B. juncea* cv. Varuna exhibited an enhanced resistance to both *Alternaria* blight pathogen, *A. brassicae* and powdery mildew pathogen, *Erysiphe cruciferarum* as inferred delay in symptom appearance and lesser disease severity comparatively. Tasleem et al. (2017) transformed *B. juncea* with *MPK3* gene and analyzed its role in imparting tolerance against *A. brassicae*. The estimation of activity of antioxidative enzyme was determined in transgenic plants. It was found that both ascorbate peroxidase (APX) and guaiacol peroxidase activity as well as proline content were higher in transgenic plants, which led to scavenging of ROS in transgenic plants produced as a result of *Alternaria* infection, which is responsible for inducing resistance trait in transgenic mustard plants. Arora et al. (2019) identified and isolated a CC-NB-LRR protein-coding

Table 7.4 Genetic transformation in *Brasssica juncea* for disease resistance (Thakur et al. 2020)

Sr. no.	Genotype	Gene transferred	Target trait	References
1.	*B. juncea* cv. Pusa Jaikisan	Chitinase	Resistance against *A. brassicae*	Mondal et al. (2003)
2.	*B. juncea*	*osmotin*	Resistance against *A. brassicae*	Taj et al. (2004)
3.	*B. juncea*	*nptII* and *BAR*	Resistance against *A. brassicicola*	Munir et al. (2016)
4.	*B. juncea* cv. RLM	Tomato *glucanase*	198 Resistance against *A. brassicae*	Mondal et al. (2007)
5.	*B. juncea* cv. Pusa Jaikisan	Osmotin-ferritin	Resistance against *A. brassicae*	Nirupa et al. (2007)
6.	*B. juncea*	*WRR4*	Resistance against *A. candida*, white rust fungus	Borhan et al. (2010)
7.	*B. juncea*	A barley antifungal class II *chitinase* gene and type I ribosome inactivating protein (*RIP*)	Resistance against *A. brassicae*	Chhikara et al. (2012)
8.	*B. juncea* cv. Varuna	*MSRA1*	Resistance against Alternaria blight and stem rot fungi	Rustagi et al. (2014)
9.	*B. juncea* cv. Raya Anmol	*Chitinase*	Resistance against *A. brassicae*	Ahmad et al. (2015)
10.	*B. juncea* cv. Varuna	Chickpea *lectin*	Resistance against *A. brassicae*	Kumar et al. (2015)
11.	*B. juncea* cv. Varuna	*NPR1*	Resistance against *A. brassicae* and *Erysiphe cruciferarum*	Ali et al. (2017)
12.	*B. juncea* cv. Varuna	*BjuWRR1*	Resistance against *A. candida*, white rust fungus	Arora et al. (2019)

BjuWRR1 gene from Donskaja-IV, an exotic line of Indian mustard and transferred it into Indian mustard by *Agrobacterium*-mediated gene transfer technique to validate its role in providing resistance against white rust pathogen, *A. candida*. Transgenic mustard plants exhibited resistance against six isolates of white rust fungus collected from different Indian mustard growing zones of India. The perusal of data in Table 7.4 contains information on genetic transformation of *B. juncea* genotypes for disease resistance.

Expression of genes for suitable pathogenesis-related proteins and defensins offers a suitable approach for controlling diseases. Some proteins, called defensins, are small cysteine-rich peptides with antimicrobial activity. Transgenic expression of plant defensins has been reported to enhance protection in vegetative tissues against pathogen attack. Canola expressing a pea defensin showed enhanced resistance against blackleg disease caused by *Leptosphaeria maculans* (Wang et al. 1999). The constitutive expression of an alfalfa defensin in potato provided robust resistance against the agronomically important fungus *Verticillium dahlia* under field conditions (Gao et al. 2000).

The transgenic technology will be considered as the "silver bullet" that will solve the problems of twenty-first century arising from biotic stress. To realize the full potential of this technology, concerted efforts in research and development are further needed to enhance the efficiency of various procedures/techniques used in different crops against multiple stresses, as witnessed with the success stories Bt cotton, brinjal, maize, canola, soybean, tobacco, etc. that have been transformed with one or more economic traits so far. The advances in the field of molecular biology and transgenic technologies have enabled us to know the expression of transgenes, emerging trends involving fusion proteins, deployment of antimicrobial peptides, stacked genes, and various genes regulating metabolic pathways. Although the introduced genes are well-defined, the field trials also provide the opportunity to ascertain whether any unexpected or undesirable consequences have resulted from the transformation procedure. Genetically engineered crops are just beginning to make their way into the hands of breeders. Their potential for changing the characteristics of plants has already been demonstrated to a great extent. It remains for the market to put a value on those traits and the technology that makes these changes possible (Sanghera et al. 2011).

7.13.2 Development of Transgenics for Multiple Pathogenic Resistances

Generally, resistance (R) genes function by activating numerous defense genes, which carry out various defense responses, such as the production of antifungal, antiviral, or antibacterial compounds, strengthening the cell wall, papilla formation, or cell death to contain the pathogen. Use of defense genes for genetic engineering seeks to bypass R-gene regulation, either by constitutive expression of one or more defense genes, or expression under the control of a pathogen-inducible promoter. The disadvantage of this strategy is that one cannot always predict the range of pathogens against which a particular defense gene may be effective.

It was shown that constitutive expression of pea defense gene DRR206 confers resistance to *Leptosphaeria maculans* pathogenicity group PG2 (Wang et al. 1999). DRR206 is strongly induced in pea (*Pisum sativum*) in response to both fungal and bacterial pathogens and elicitors (Fristensky et al. 1985; Daniels et al. 1987; Cody et al. 1988). In non-transformed *Brassica*, compatible isolates of *L. maculans* invade tissue through stomatas or wounds and grow intercellularly in an asymptomatic biotrophic phase (Hammond and Lewis 1987). Subsequently, necrotic cell death occurs, which is required for sporulation of the fungus. In DRR206 transgenic

Brassica, necrosis is limited to the site of inoculation and no sporulation is seen. Microscopic examination shows that little biotrophic hyphal growth is seen even at the inoculation site (Wang et al. 1999). Wang et al. (1999) have shown that water-soluble extracts from DRR206 transgenic plants have inhibitory activity towards germination of *L. maculans* pycnidiospores. Wang and Fristensky (2001) have tested DRR206 transgenic plants for resistance to *L. maculans* pathogenicity groups PG3 and PG4, and to *Rhizoctonia solani* and *Sclerotinia sclerotiorum*. In contrast to the hemibiotrophic ascomycetes *L. maculans*, *R. solani* is a basidiomycete which grows saprophytically in the soil and infects the roots and hypocotyls, rather than the leaves and stems. *Rhizoctonia* causes wilting of seedlings at the cotyledon stage. *Sclerotinia sclerotiorum* is distinct both in its mode of action and its broad host range. *Sclerotinia* is a necrotrophic ascomycetes that kills tissue as it spreads and proliferates in necrotic tissue. Its mode of pathogenicity includes the production of oxalic acid, which is toxic to most plants, and polygalacturonase, which breaks down plant cell walls. Whereas most plant pathogens are highly specialized for their hosts, *Sclerotinia* can parasitize a very wide range of species. Despite its ability to infect most parts of a plant, *Sclerotinia* preferentially invades petals and leaves in *B. napus*.

DRR206-mediated resistance is effective in stems and cotyledons against *L. maculans,* in roots and hypocotyls against *R. solani*, and, to some extent, in leaves against *S. sclerotiorum*. There is therefore no evidence that DRR206 activity is limited to any specific organ. In this context, it is interesting to note that in pea roots, DRR206 is expressed in response to inoculation by mycorrhizal fungi such as *Glomus mosseae* (Ruiz-Lozano et al. 1999). Roots of Myc–mutants which resist mycorrhizal symbiosis show elevated levels of DRR206 and other defense transcripts, compared to wild-type plants. Taken together with previous work (Wang et al. 1999), the results establish that constitutive expression of pea DRR206 can provide resistance to fungi with a range of modes of pathogenicity, in most parts of the plant, and that this resistance can be maintained stably over several generations. These data point to DRR206 as a promising tool for genetic engineering of disease resistance in a range of plant hosts against a range of plant pathogens. While sequence identity strongly suggests that DRR206 may be involved in lignan synthesis, its precise function or functions in plant defense could fall in several areas. DRR206 may (1) produce (+) pinoresinol or other lignans, (2) produce some other phenolic product. In response to pathogenic attack, the product may be a precursor for lignin synthesis to strengthen cell walls. Alternatively, the product may drive the lignification of pathogen structures, inhibiting hyphal growth. Third, the product may have other fungitoxic or fungistatic properties. Finally, the product might have epistatic or pleiotropic effects which activate or enhance other defenses (Wang and Fristensky 2001).

7.13.3 Fungal Biocontrol Agents to Manage Multiple Pathogens

The application of beneficial rhizospheric or endophytic microbes (most suitable and environmentally friendlier alternative strategy) can be used to induce systemic resistance. Beneficial endophytic fungi *Piriformospora indica* has been extensively reported to mitigate a number of abiotic and biotic stresses in many host plants (de Souza et al. 2016; Unnikumar et al. 2013). The symbiotic relationship of *P. indica* with *A. thaliana* is a unique system. It has been reported that fungal exudates affect the root transcriptome after the establishment of physical contact (Vahabi et al. 2015). *P. indica* has the ability to deliver effector proteins that are not only involved in the manipulation of host plant defense response but also in the establishment of symbiotic relationship (Akum et al. 2015; Plett et al. 2014). This *P. indica* interaction has revealed the upregulation of defense-related genes including pathogenesis-related PR genes, jasmonate JA (PDF1.2, VSP, LOX2), and ethylene ET (*ERF1*) signaling genes in response to pathogen attack (Camehl et al. 2010; Molitor et al. 2011). In barley, *P. indica* root colonization also improved the resistance to root and leaf pathogens (*Fusarium culmorum*, *Blumeria graminis*) (Waller et al. 2005). *P. indica* was also found to be effective against take-all disease of *Triticum aestivum*, root-pathogenic *Cochliobolus sativus* in *Hordeum vulgare*, powdery mildew in *A. thaliana* and Fusarium wilt disease of lentil (Ghahfarokhi and Goltapeh 2010; Kari Dolatabadi et al. 2011; Oelmuller et al. 2009). Several beneficial fungi (including *Arbuscular mycorrhiza* and *Acremonium alternatum*) have been reported to confer tolerance against different plant diseases (Saravanakumar et al. 2016). Several experiments on *A. thaliana* and *Brassica rapa* revealed that symptoms of clubroot can be reduced by co-inoculation with beneficial fungi or bacteria (Auer and Ludwig-Muller 2014; Yu et al. 2015). *B. campestris* sp. *chinensis* was used to elucidate the interaction between the beneficial endosymbiont *P. indica* and clubroot causing pathogen *P. brassicae*. *P. indica* has been reported to colonize a variety of plants from pteridophytes to higher vascular plants and deliver multifaceted amenities such as disease resistance, nutrient uptake, and stress tolerance (Gill et al. 2016). Colonization assay showed successful colonization of *B. campestris* sp. *chinensis* roots by *P. indica*. It was expected that there might be direct interaction between the beneficial endosymbiont *P. indica* and *P. brassicae* which is an obligate pathogen of the host plant root system. Moreover, a significant increase in plant biomass was observed such as fresh and dry weight, leaf number, and leaf area when the plants were inoculated with *P. indica* either alone or with the pathogen *P. brassicae* comparatively. Such biomass enhancement features of *P. indica* have been reported previously in *A. thaliana* which might be due to higher expression of genes involved in plant developmental processes (Sherameti et al. 2005; Waller et al. 2008). Additionally, the positive effect on growth may also be correlated with the enhancement of nutrient uptake especially phosphorous and nitrogen from the soil (Kumar et al. 2011; Varma et al. 2001). Furthermore, a significant difference was observed in flavonoid biosynthesis when experimental plants were inoculated with *P. indica* and *P. brassicae* either alone or in combination. Flavonoids are polyphenolic secondary metabolites produced by the plants with diverse array of biological functions including protection against unfavorable environmental conditions, defense against pathogens, and recruitment of beneficial symbionts. The elevation of major flavonoids was higher in plants inoculated with *P. indica* either alone or in combination with *P. brassicae*. The enhancement of flavonoids has also been reported in *A. thaliana* root during infection with *P. brassicae*. Some flavonoids may be involved directly or indirectly to inhibit pathogen's enzymes, particularly the plant cell wall digesting enzymes by chelating metals required for their activities (Treutter 2005). Flavonols, such as quercetin have been reported to provide resistance to pathogen attack by suppressing ATPase activity of DNA gyrase (Babii et al. 2016; Naoumkina et al. 2010). Similarly, kaempferide (flavonol triglycoside) produced by *Dianthus caryophyllus* is a fungitoxic compound against *Fusarium oxysporum* (Curir et al. 2005; Monazzah et al. 2016). Flavonoids may be involved in the auxin transportation to the infected root by *P. brassicae* for gall formation (Pasold et al. 2010). In addition, disease severity was analyzed by measuring disease index and it was observed that co-inoculation of *P. indica* with *P. brassicae* reduced the disease index as compared to the *P. brassicae* treatment alone. Moreover, the low disease index was correlated to the lower expression level of genes in *P. indica* and *P. brassicae* co-cultivated plants comparatively to the plants inoculated with *P. brassicae* only (Fig. 7.8a, b) Data, from both the molecular analysis and disease index, showed that *P. indica* suppresses the pathogenesis of *P. brassicae* by delaying its life cycle. However, this suppression may be the result for nutrient/space competition. Additionally, it has been reported that *P. indica* can induce the antioxidative system of the host plant which further enhance plant defense system. It has been claimed that the regulatory proteins (NPR1) are responsible for precise resistance induced by *P. indica* to powdery mildew in *A. thaliana* (Stein et al. 2008). The important signaling component to induce resistance pathways was identified as *NPR1* (Pieterse and Van Loon 2004) while mutants of *NPR1* were reported with no altered responses to clubroot formation (Siemens et al. 2002). Endosymbiont *P. indica* colonizes the root system of host plant and in terms of overall performance; it may have a direct suppressive effect on *P. brassicae* which is also a biotrophic, obligate root pathogen (Khalid et al. 2020).

7.13.4 Bacterial Biocontrol Agents to Manage Multiple Pathogens

The bacterium secretes antifungal compounds. The major bioactive molecules produced by the *Bacillus* species are lipopeptides (LPs,) which display a wide spectrum of activity. LPs are genetically and biochemically well characterized and important in plant colonization, direct antagonism to phytopathogens and stimulation of plant immunity (Farace et al. 2014; Rahman et al. 2015). An LP-enriched extract from *Bacillus* amyloliquefaciens UCMB5113 culture broth retained antifungal activity as observed also when intact bacteria were used. Bioassays using Arabidopsis plants showed that treatment of leaves with a *Bacillus* UCMB5113 LP extract provided disease protection. Furthermore, treatment of plant roots with the LP extract resulted in disease suppression when challenged with Alternaria, indicating primed systemic defense as ISR. Signaling has mostly been studied in *Pseudomonas* strains (van Loon et al. 1998; Verhagen et al. 2010) for which the ISR concept was coined although *Bacillus*-derived ISR has also been demonstrated (Falardeau et al. 2013; Farace et al. 2014; Rahman et al. 2015).

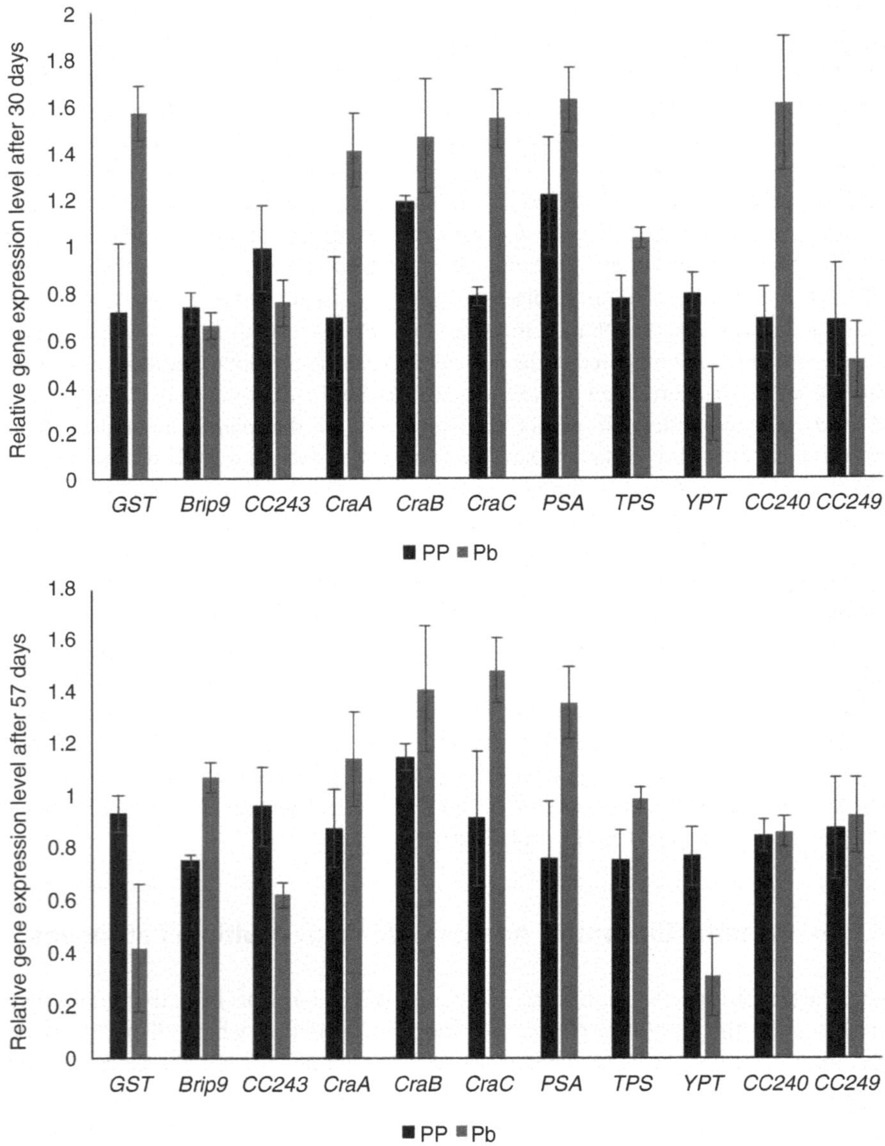

Fig. 7.8 Q-PCR analysis of *Plasmodiophora brassicae* genes in samples without and with *Piriformospora indica* at different times after inoculation with *Plasmodiophora brassicae*. P = Inoculated with *P. indica* only; PP = Inoculated with *Plasmodiophora indica* and *Plasmodiophora brassicae*. All samples were from plants co-inoculated at the same time with *Plasmodiophora indica* and *Plasmodiophora brassicae* (Khalid et al. 2020)

Analysis with signaling-compromised plants as well as JA-marker GUS reporter lines enabled evaluation of how plant hormones involved in different defense responses responded to *Bacillus* UCMB5113 LPs. Growth of *Alternaria* was suppressed by *Bacillus* UCMB5113 LPs in wild type, jar1 and myc2 lines, while coi1-16, myb72 and sid2 plants did not respond and the npr1 accession even displayed increased Alternaria levels after LP treatment. Priming of ISR by *Pseudomonas* acts locally in roots through MYB72 and systemically through MYC2 (Pieterse et al. 2014). AR1 produces the JaIle conjugate that stimulates COI-dependent degradation of JAZ proteins allowing MYC2 to stimulate G-box containing promoters of JA-responsive genes (Kazan and Manners 2013).

The growth of *A. brassicicola* in plants is known to be affected by JA (Thomma et al. 1998). Asari et al. (2017) showed that the LP-enriched extract suppressed *Alternaria* without the need for JA–Ile formation but could not rescue plants with compromised JAZ degradation and did not work very well in the absence of MYC2 (where redundancy may assuage the effect). Root treatment of VSP2: GUS plants with the LP extract followed by fungal inoculation resulted in distinct staining systemically in leaves and roots while fungal inoculation alone only raised local VSP2 expression. In contrast, plants expressing PDF1.2: GUS showed low local expression around the fungal inoculation site on leaves both in the control and in the LP extract-treated samples. These results imply JA involvement in the effects observed by the LP extract but with no obvious dependence on ethylene while ABA may contribute to a reaction involving MYC to activate VSP2 transcription. Involvement of ABA in plant defense has been demonstrated for other JA-dependent pathogens with a necrotrophic life style (Adie et al. 2007). These results point to a complex signaling caused by the *Bacillus* UCMB5113 LPs involving JA.

MS analysis of the most active fractions from the *Bacillus* UCMB5113 LP extract allowed identification of novel linear forms of antifungal fengycins. Linear forms are unusual, as the standard form is a cyclic structure formed by the multicomplex fengycin biosynthetic apparatus. Many studies have shown the ability of *Bacillus* species to produce different cyclic fengycins (Vanittanakom et al. 1986; Sun et al. 2006; Deleu et al. 2008; Eeman et al. 2009; Malfanova et al. 2012; Pathak et al. 2012; Farace et al. 2014; O'Connor et al. 2014). However, Pathak et al. (2012) provided data in support of the presence of linear forms of fengycin in LP extracts of the banyan tree endophyte Bacillus subtilis K1, but no biological activity was reported. That report also found linear fengycins upon direct analysis of cell suspensions although at low levels. The time-course analysis suggested strongly that the linear C15 fengycin B1 is co-produced by *Bacillus* UCMB5113 along with several other structural variants. pH increases slowly during growth of *Bacillus* UCMB5113 in LB, which could cause non-enzymatic hydrolysis of the bond involved in cyclization. However, in vitro tests where cyclic fengycins were incubated at higher pH did not result in ring opening. In addition to NRPS templated pathways, non-templated reactions can occur where the arrangement of the peptide is more difficult to predict and various post-synthesis modifications may further change the structure (Giessen and Marahiel 2012). The linear forms could thus be synthesized as such or result from subsequent modifications (Aleti et al. 2015).

The mechanisms forming these linear but highly active forms of fengycins warrant further investigation. Use of synthetic peptide mimics of the identified *Bacillus* UCMB5113 linear fengycin showed that the fatty acid side chain is necessary for direct antagonistic activity to phytopathogens and priming of ISR in *Arabidopsis* plants since replacement with an acetyl group gave no significant inhibition of *A. brassicicola* and *V. longisporum* in plate assays or with plants. LPs can serve as membrane-disturbing bio-surfactants providing a direct protective effect against phytopathogens (Ongena and Jacques 2008). The membrane-disturbing properties of LPs imply that many kinds of micro-organisms can be targeted especially when a mixture of LPs is produced. Other surfactin and fengycin LPs have been shown to act as elicitors of ISR in plants (Ongena et al. 2007; Cawoy et al. 2014). Estimates of up to 10% of the *B.* amyloliquefaciens genomes may be designated for production of secondary metabolites, providing a plethora of potentially antagonistic molecules (Chen et al. 2009; Niazi et al. 2014). In addition, *Bacillus* strains can interact with a wide range of plant species, providing potential protection to many different kinds of pathogens (Alvarez et al. 2012; Garcia-Gutierrez et al. 2013; Adam et al. 2014; Kalai-Grami et al. 2014). These results indicate that the LP extract from *Bacillus* UCMB5113 provides systemic resistance in plants and is dependent on JA, indicative of a primed ISR response. *Bacillus* UCMB5113 has the ability to produce many secondary metabolites, resulting in a complex cocktail of various antibiotics that can act directly and indirectly support efficient plant protection to phytopathogens. Novel forms of such potentially antifungal linear fengycins may serve as a basis for the development of more powerful plant protective agents. The results obtained suggest that *Bacillus* UCMB5113 shows potential as a BCA to counteract Brassica phytopathogens. The LP compounds produced by this bacterium are stable under high temperatures, indicating long-term efficacy. Use of ecosystem services such as *Bacillus* UCMB5113 that could support plant protection through priming of ISR is an important step towards more sustainable crop production, reducing the need for chemical pesticides (Asari et al. 2017).

7.13.5 Abiotic Conditions Modulate Diseases

Climatic conditions can influence the interaction between plants and pathogens and the pathogen distribution, contributing directly to the disease outcome on the host (Barbetti et al. 2012). Abiotic conditions including temperature and soil nutrient content could influence the resistance spectrum on the host *A. thaliana* against powdery mildew pathogens (Jorgensen 2012). In the *L. maculans-B. napus* pathosystem, temperature effect influences the phenotypic outcome at different plant growth stages. In general, during infection at nearly all growth stages from cotyledon to pod at higher temperature regimes (e.g., 18/24 °C), resistance expressed at the seedling stage was positively correlated with the adult stage, as compared to lower temperature regimes (e.g., 11/18 °C) (Li et al. 2006). Warmer temperature and high humidity levels favored *L. maculans* colonization of the host during an *Rlm6*-mediated resistance study on *B. napus* (Huang et al. 2006), confirming a related study on *Rlm1*-mediated resistance by Badawy et al. (1992). Cold temperature and low soil acidity slow down or inhibit *P. brassicae* infection on *B. napus,* while the influence from soil micronutrient stops at high inoculum concentration (Gossen et al. 2014).

7.13.6 Use of Defense-Related Genes Other Than *R*-Genes Confers Resistance to Multiple Pathogens

In addition to the up-regulation of HR-related genes characterized by signaling hormones (Chu et al. 2014; Nováková et al. 2016; Uloth et al. 2016), regulatory genes such as the transcription factors WRKY (Zhou et al. 1997) and calmodulin-binding transcription activator (CAMTA) (Rahman et al. 2016) have also been implicated to regulate the disease response in *Brassica* pathosystems (Zhou et al. 1997). The WRKY gene was studied in *B. napus-S. sclerotiorum* and *Alternaria brassicae* pathosystems, where gene expression was increased upon pathogen infection alongside defense-related hormones (Yang et al. 2009; Zhao et al. 2009). This gene was shown to enhance resistance against Downy Mildew in the Chinese broccoli *B. oleracea* var. *italica* (Jiang et al. 2016). Besides the WRKY gene, for the same host species, a pathogenicity-related defensin gene, which encodes defense-related cysteine-rich protein, has also been shown to improve resistance against Downy Mildew (Jiang et al. 2012) and *S. sclerotiorum* (Zarinpanjeh et al. 2016). Identification of candidate genes for clubroot resistance revealed Rho proteins may be involved (Kato et al. 2013). In Head Smut in maize, *ZmWAK* was successfully identified as a regulator gene with quantitative effects on host resistance expression (Balint-Kurti and Holland 2015; Zuo et al. 2015). Further characterization of these genes should be performed in *Brassica* species, through genome-wide identification and gene expression studies to determine their role in disease resistance.

7.13.7 Fitness Cost to Multiple Disease Resistance in *Brassica*

Multiple disease resistance (MDR) in *B. rapa* to three pathogens showed heritable genetic variation or resistance to each pathogen, and a positive genetic correlation between resistance to *H. parasitica,* and *Leptosphaeria maculans* (Mitchell-Olds et al. 1995). However, in a later study, Mitehell-Olds and Bradley (1996) found that *Hyaloperonospora*-resistant genotypes grew 6% slower than *Hyaloperonospora* susceptible genotypes in pathogen-free environments, indicating a significant genetic fitness cost to *Hyaloperonospora* resistance. Genetic costs of resistance to pathogens may be an important factor maintaining heritable variation for resistance in natural populations. Pleiotropic fitness trade-offs occur when genetic resistance causes reduction in other components of fitness. Although costs of resistance have an important influence on plant–pathogen interactions, few previous studies have detected pleiotropic costs of resistance in the absence of confounding effects of linkage disequilibrium. To avoid this potential problem, Mitehell-Olds and Bradley (1996) performed artificial selection experiments on resistance to two fungal pathogens, *L. maculans*, and *H. parasitica*, and compared growth rates of resistant and susceptible genotypes of *Brassica rapa* in the absence of pathogens. *Leptosphaeria* resistance had no effect on growth rate, indicating cost-free defense. In contrast, *Hyaloperonospora*-resistant genotypes grow 6% slower than *Hyaloperonospora*-susceptible genotypes in pathogen-free environments, indicating a significant genetic fitness cost to *Hyaloperonospora* resistance. Such genetic trade-offs could maintain genetic variation in the wild. Another factor that might explain heritable variation for resistance is ecological trade-offs, in which genetic resistance to one species causes susceptibility to another. Such ecological trade-offs do not exist for the pathogens studied in this system.

Mitehell-Olds and Bradley (1996) observed that genes conferring resistance can have pleiotropic effects on other aspects of plant performance (Simms 1992). If resistant plants have reduced growth or fecundity in the absence of pathogen attack, then selection may favor intermediate levels of disease resistance. When allocation to constitutive resistance mechanisms varies among genotypes, then resistant genotypes may experience a fitness benefit in the presence of a pathogen, relative to susceptible plants. However, resistant plants would have reduced relative fitness when pathogen attack is reduced or absent because resources would be expended on unneeded defensive mechanisms. For constitutive defense pathways, a genetic cost to resistance is indicated by a significant negative genetic correlation between growth and fitness in a pathogen-free environment, and resistance measured in the presence of pathogens. Few published studies provide clear evidence for pleiotropy causing genetic fitness costs. Berenbaum et al. (1986) showed a negative genetic correlation between insect resistance, and fecundity in *Pastinaca sativa*. Han and Lincoln (1994) found a significant negative genetic correlation between growth rate, and levels of secondary metabolites in *Diplacus aurantiacus*. In *Escherichia coli*, mutations causing resistance to virus T4 carry pleiotropic costs of resistance, which can be ameliorated by epistatic modifiers (Lenski 1988a, b). Several studies from habitually inbreeding species or near-isogenic lines are consistent with the existence of pleiotropic fitness costs (Bergelson 1994; Brinkmamn and Frey 1977; Burdon and Muller 1987; Chaplin 1970; Frey and Browning 1971; Simons 1979), but linkage disequilibrium and pleiotropy are confounded in these experiments. Additional evidence from ecological and physiological studies is also suggestive of genetic costs of resistance (Simms 1992).

7.13.8 Molecular Bases for Assessment of Breakdown of R-Genes in *Brassica*

The *AvrLm6* gene, which was adjacent to a single-copy non-coding sequence at the 3′ end, had six different RIP alleles conferring virulence. The degree (intensity) of RIP mutation in these single-copy sequences was proportional to the proximity of flanking repetitive DNA that had undergone RIP mutation. The potential leakage of

RIP mutations into closely linked genes highlights the power of RIP to lead to major evolutionary changes to genes such as effectors that play an important role in fungal living strategies. The breakdown of "sylvestris" resistance was associated with an eightfold increase in frequencies of isolates lacking *AvrLm1*. Although this gene was embedded in repetitive DNA, no RIP alleles were identified. The frequencies of virulence alleles (both deletion and RIP mutation) of *AvrLm6* increased sixfold, even though cultivars with the complementary resistance gene, *Rlm6*, had not been sown on Eyre Peninsula (van de Wouw et al. 2010a). The close linkage and genomic location of *AvrLm1* and *AvrLm6* might have led to a selective sweep, whereby selection at *AvrLm1* affected the frequency of alleles of *AvrLm6* through hitchhiking (Barton et al. 2013). Thus strong selection imposed by widespread deployment of one resistance gene also may lead to breakdown of resistance conferred by another gene, if the two complementary avirulence genes are closely linked. Clearly the genomic environment, as well as extent of exposure to resistance genes in *B. napus* cultivars, affects evolution of avirulence effectors in this fungal pathogen.

Monitoring changes in virulence frequencies in Australian populations of the blackleg fungus is armed with knowledge about the field biology and genomic analysis of *L. maculans*. The monitoring of changes in frequency of virulence of fungal populations across Australia is with the aim of preventing another breakdown of disease resistance. Since 2009, cultivars with different complements of resistance genes sown in 32 trial sites covering *Brassica* growing regions of Australia have been assessed for disease severity. Resistance genes in canola cultivars are identified by the use of 12 differential *L. maculans* isolates. On this basis of responses (susceptible or resistant) to these isolates, seven resistance groups (A–G) that include a range of resistance genes have been defined (Marcroft et al. 2012a). Stubble from each site represents local fungal populations, which can be analyzed in high-throughput laboratory assays to quantify regional frequencies of alleles of avirulence genes. In these assays, blackleg-infested stubble is wetted and placed in a wind tunnel. This triggers release of ascospores, which are deposited onto tape from which ascospore DNA is extracted and analyzed. The quantitative polymerase chain reaction (qPCR) detects the frequency of alleles of particular avirulence genes (*AvrLm1*, *AvrLm6*), where virulence is due to deletion, while PCR and pyrosequencing detects single nucleotide base-pair changes where virulence is due to a particular point mutation (*AvrLm4*) (van de Wouw and Howlett 2012; van de Wouw et al. 2010b). The total number of ascospores is estimated by qPCR of the internal transcribed spacer region of the ribosomal DNA, a genomic region present in all isolates. This is the first example of a high-throughput molecular assay that can distinguish genotypes of airborne spores. This type of assay can be applied to other diseases that involve airborne inocula and where the genetic basis of virulence in the pathogen has been identified. Analysis of data on changes over time in disease severity of cultivars, and in frequencies of avirulence/virulence alleles in populations released from canola stubble, allows predictions of risk of disease outbreaks in different geographic regions. If an epidemic is predicted, farmers are advised to plant a different canola cultivar with a different complement of resistance genes (Howlett et al. 2015).

7.13.9 Net Cost of R-Gene Resistance in the Diseased Crucifers

Despite obvious benefits of resistance to pests, diseases, and many resistance traits are highly variable within plant species. The most common hypothesis for the persistence of susceptible individuals is a trade-off between benefits of resistance in the presence of pests and costs of resistance in their absence. Conventionally, a cost of resistance has been defined as a reduction in fitness in the absence of the targeted pest (Simms and Rausher 1987). There is now substantial evidence for costs of resistance in the absence of pests for several types of resistance although costs are not always. In addition to conventional costs in the absence of attack, costs manifest in the presence of attack may contribute to the maintenance of variation for some resistance traits. The most likely candidates are induced resistance traits that are produced on attack by natural enemies. Because induced defenses are expressed only when triggered by the presence of pests, costs of producing these defenses should only occur when pests are present. In some environments, resistant plants may have lower fitness than susceptible plants when pests are present. This potential is referred as outcome of a net cost with attack.

There are several reasons why net costs with attack may exist for induced resistance traits. First, recent evidence suggests that the artificial induction of defenses in the absence of attack can entail substantial costs. The expression of resistance can reduce resources available for growth and reproduction (allocation costs) and/or reduce a plant's ability to resist or tolerate other enemies. When enemies are not particularly voracious or virulent, resistance responses may be excessive. Susceptibility gains a further advantage whenever resistance is not completely effective at preventing damage. Finally, tolerance mechanisms, such as increased photosynthetic rates, growth, and lateral branching, can enable substantial compensation for attack by herbivores. Indeed, if tolerance mechanisms were more effective at improving fitness than resistance responses, a net cost with attack would result even in the absence of allocation or ecological costs.

R-gene resistance is a likely candidate for the presence of net costs with attack. R-genes are disease resistance genes that enable plants to recognize specific molecules secreted by pathogens and subsequently trigger a defense response. R-genes conform to a gene-for-gene interaction in which plants lacking a particular allele are susceptible to the corresponding pathogen. The induced R-gene resistance response is extensive, including local cell death, cell wall reinforcement, the production of many proteins and secondary compounds, and systemic acquired resistance. Studies indicate that constitutive and artificially induced systemic acquired resistance incurs fitness costs. Because the resistance response can be costly, susceptibility may be favored over resistance if a pathogen is weak or if environmental conditions do not favor the development of severe disease symptoms. In addition, *Arabidopsis thaliana* plants are known to vary in their capacity to tolerate pathogen infection (Kover and Schaal 2002). This tolerance may be associated with a suite of developmental responses, including accelerated flowering and changes in branch architecture that occur when *A. thaliana* is attacked by a variety of pathogens (Korves and Bergelson 2003).

For the presence of net costs of R-gene resistance with infection in *A. thaliana,* Korves and Bergelson (2004) defined net costs with infection as a fitness advantage of infected, susceptible plants over infected, resistant plants. They examined two R-genes, *RPS2* and *RPP5* that confer resistance to two different pathogens, *Pseudomonas syringae*, a bacterial pathogen, and *Hyaloperonospora parasitica.* Both R-genes are polymorphic for resistance/susceptibility in wild *A. thaliana*. For each R-gene, the control for genetic background was done by using mutant or near-isogenic lines. First, compared the fitness of R-gene resistant and susceptible lines when infected with a pathogen in two different growth environments, and later on evaluated the contribution of induction costs of the resistance response, ineffective resistance, and tolerance to the fitness differences. Finally, a model to explore the role of net costs with infection in maintaining resistance/susceptibility polymorphism was prepared.

The expression of induced resistance can reduce fitness in the absence of pests. Korves and Bergelson (2004) suggested that not only may costs of induced resistance be important in the absence of attack, but also net costs of induced resistance can occur in the presence of attack. They found that susceptible plants out-performed resistant plants when infected for two different R-genes and pathogens in two different growth environments. In the case of *RPS2*, a net cost of attack occurred in the absence of competition because susceptible plants over compensated for infection. In the case of *RPP5*, a net cost occurred in the presence of competition because susceptible plants compensated for infection, while resistant plants tended to have reduced fitness with infection. In partitioning the net effect of resistance with infection into components, it was found that the compensatory ability of susceptible plants was an important contributor to net costs with infection. In the *RPP5–Hyaloperonospora parasitica* experiment, there was complete compensation for pathogen infection both with and without competition, suggesting that *Arabidopsis thaliana* can tolerate *H. parasitica* infection at the 2-weeks-old growth stage. In the *RPS2–Pseudomonas syringae*, infected susceptible plants produced more seed than mock-treated susceptible plants in the absence of competition but undercompensated in the presence of competition. Similarly, other studies have reported that the compensatory ability of plants depends on environmental quality (Maschinski and Whitham 1989; Juenger and Bergelson 1997).

There are several published examples of systemic fungal pathogens increasing plant fitness; however, in most cases the mechanism is that infection protects hosts from herbivore attack (Clay 1996). The evidence of pathogen-induced overcompensation due to changes in branch architecture has been reported in response to herbivores. Although these responses are superficially similar (Tiffin 2000; Weinig et al. 2003), the physiological mechanism appears to be different. In particular, regrowth in response to herbivore damage is associated with damage to the apical meristem (loss of apical dominance), whereas basal branching in response to pathogens is not (Korves and Bergelson 2003).

Another factor that affected the net fitness effects of resistance was reduced fitness in resistant plants with infection. In the competition treatment, infection tended to reduce the fitness of resistant plants compared with the mock treatment.

These fitness reductions may have occurred because of an allocation (or auto toxicity) cost of the R-gene resistance response, as has been shown for systemic acquired resistance. Heil (2002) found that the costs of artificially induced SAR in wheat were greater with low resources, which is consistent with result that infection reduced the fitness of resistant lines in the competition treatment only. In contrast, Cipollini (2002) found that costs of artificially induced SAR did not vary with intraspecific competition in *A. thaliana*. The allocation costs may have been further accentuated by the additional stress of disease symptoms (despite resistance).

Infection may also have reduced the fitness of resistant plants because of incompletely effective resistance. R-gene-resistant plants supported some pathogen growth that may have directly harmed plant fitness in the competition treatment. It is interesting to note that *rpp5* plants endured even greater pathogen growth than *RPP5* plants and did not suffer reduced fitness with infection in competition. While this may reflect induction costs of the *RPP5* resistance response, it is also possible that disease symptoms reduced fitness more in *RPP5* lines than in *rpp5* lines because of genetic differences other than at *RPP5*. The introgressed segment contained other genes that might affect fitness in the presence of disease symptoms. In this case, the net cost of *RPP5* resistance would be caused by tight physical linkage of *RPP5* to an allele that decreases fitness when plants are infected. In natural populations, such associations could persist for long periods of time in a highly selfing organism with extensive linkage disequilibrium such as *A. thaliana*. Though the segment, introgressed into Columbia was no bigger than 295 kb, *RPP5*, and *rpp5* lines in the mock treatment differed in flowering time and had significant differences in branch architecture. These results suggest that the introgressed segment includes genetic variation that affects plant development and leaves open the possibility that one or more of these traits interact with disease to affect fitness. Indeed, the introgressed segment falls within a Columbia-Landsberg quantitative trait locus on chromosome 4 for bolting time under short days (Weinig et al. 2002).

There was no evidence of a cost of R-gene resistance in the absence of infection. This is in contrast to a study of *RPM1* resistance, where, in the absence of disease, a fitness cost of 9% was detected (Tian et al. 2003). For *RPP5*, such costs may not be expected because lines did not differ in the presence of a single gene but had differences in a whole gene family. For *RPS2*, however, susceptible lines had an *RPS2* gene with a premature stop codon. The difference in results between *RPM1* and *RPS2* could reflect differences in the costs of these two R-genes in the absence of disease or differences in the way in which the experiments were performed.

Results suggest that net costs are not restricted to a single environmental condition. Net costs were observed both in the presence of competition and in the absence of competition, and these conditions produced plants with very different statures and amounts of seed. Often, costs are detected in relatively stressful conditions (Bergelson 1994), as was the net cost in *RPP5–H. parasitica*. In contrast, in the *RPS2–P. syringae* experiment, the net cost was driven by overcompensation in the high resource, no competition condition. Belsky et al. (1993) noted that many of the examples of overcompensation for herbivory come from greenhouse or agricultural studies that utilize unusually high resource levels. It is worth noting that *A. thaliana* in the field can produce amounts of seed comparable to those associated with overcompensation. In addition, overcompensation does not have to be frequent to substantially impact resistance/susceptibility polymorphism because plants in high resource environments are likely to produce many more seeds per plant than in low resource conditions and thereby disproportionately impact selection on R-genes.

Molecular evolutionary studies indicate that some R-gene resistance/susceptibility polymorphisms have persisted for millions of years and have been maintained by natural selection (Caicedo et al. 1999; Stahl et al. 1999; Tian et al. 2002; Mauricio et al. 2003). Korves and Bergelson (2004) model suggests that net costs with attack could contribute to this persistence of resistance/susceptibility polymorphism. Previous models have explained the maintenance of polymorphism at R-gene loci as a trade-off between benefits of resistance in the presence of disease and costs of resistance in the absence of disease (Leonard 1977; Frank 1992, 1993; Stahl et al. 1999; Roy and Kirchner 2000). The distinction between costs in the absence of disease and net costs with infection is important because they might produce different co-evolutionary and ecological dynamics. At sufficiently high levels, costs of R-genes in the absence of disease are expected to produce epidemic cycles (Leonard and Czochor 1980; Stahl et al. 1999). However, selection for susceptibility in the presence of disease in some environments, like selection for tolerance, could promote the persistence of substantial pathogen populations (Roy and Kirchner 2000). Of course, costs in the absence of disease and net costs with infection could both accrue to a single resistance trait and affect its evolution. These provides evidence that, where fitness costs of resistance in the absence of disease are not detected, net costs of resistance with attack can occur. These net costs can occur as a result of compensatory mechanisms on the part of susceptible plants and reduced fitness in resistant plants with infection. Our model shows that variation in resistance/susceptibility can, in theory, be maintained without a cost of resistance in the absence of disease, if there are net costs with attack in some environments. Further work is necessary to determine whether net costs of R-gene resistance with infection occur in natural plant populations and whether similar net costs occur in other plant-enemy systems (Korves and Bergelson 2004).

References

Adam M, Heuer H, Hallmann J (2014) Bacterial antagonists of fungal pathogens also controls root-knot nematodes by induced systemic resistance of tomato plants. PLoS One 9:e90402

Adie BA, Perez-Perez J, Perez-Perez MM, Godoy M, Sanchez-Serrano JJ, Schmelz EA, Solano R (2007) ABA is an essential signal for plant resistance to pathogens affecting JA biosynthesis and the activation of defenses in *Arabidopsis*. Plant Cell 19(5):1665–1681. https://doi.org/10.1105/tpc.106.048041

Agarwal A, Kaul V, Faggian R, Cahill DM (2009) Development and use of a model system to monitor clubroot disease progression with an Australian field population of *Plasmodiophora brassicae*. Australas Plant Pathol 38:120–127. https://doi.org/10.1071/AP08093

Agarwal A, Kaul V, Faggian R, Rookes JE, Ludwig-Muller J, Cahill DM (2011) Analysis of global host gene expression during the primary phase of the *Arabidopsis thaliana–Plasmodiophora brassicae* interaction. Funct Plant Biol 38:462–478

Ahmad B, Ambreen, Khan MS, Haiser A, Khan I (2015) *Agrobacterium* mediated transformation of *Brassica juncea* (L.) Czern. with chitinase gene conferring resistance against fungal infections. Pak J Bot 47(1):211–216

Aishaha AWN, Hanina MN, Asyikin IN, Shahril MH, Salina MR, Maryama MR, Jalil AKA, Rosfarizan M (2014) Antimicrobial protein produced by *Bacillus subtilis* ATCC 21332 in the presence of *Allium sativum*. Asian Pac J Trop Dis 4(3):245. https://doi.org/10.1016/S2222-1808(14)60555-7

Akum FN, Steinbrenner J, Biedenkopf D, Imani J, Kogel KH (2015) The *Piriformospora indica* effector PIIN_08944 promotes the mutualistic *Sebacinalean symbiosis*. Front Plant Sci 6:906. https://doi.org/10.3389/fpls.2015.00906

Aleti G, Sessitsch A, Brader G (2015) Genome mining: prediction of lipopeptides and polyketides from *Bacillus* and related *Firmicutes*. Comput Struct Biotechnol J 13:192–203

Ali S, Mir ZA, Tyagi A, Mehari H, Meena RP, Bhat JA, Yadav P, Papalou P, Rawat S, Grover A (2017) Overexpression of *NPR1* in *Brassica juncea* confers broad spectrum resistance to fungal pathogens. Front Plant Sci 8:1693. https://doi.org/10.3389/fpls.2017.01693

Alix K, Lariagon C, Delourme R, Manzanares-Dauleux MJ (2007) Exploiting natural genetic diversity and mutant resources of *Arabidopsis thaliana* to study the *Arabidopsis thaliana-Plasmodiophora brassicae* interaction. Plant Breed 126:218–221

Alvarez F, Castro M, Principe A, Borioli G, Fischer S, Mori G, Jofre E (2012) The plant-associated *Bacillus amyloliquefaciens* strains MEP218 and ARP23 capable of producing the cyclic lipopeptides iturin or surfactin and fengycin are effective in biocontrol of *Sclerotinia* stem rot disease. J Appl Microbiol 112:159–174

Andrade AE, Silva LP, Pavin J, Noronha EF (2008) In vivo proteome analysis of *Xanthomonas campestris* pv. *campestris* in the interaction with the host plant *Brassica oleracea*. FEMS Microbiol Lett 281(2):167–174

Aneja JK, Agnihotri A (2013) Alternaria blight of oilseed *Brassicas*: epidemiology and disease control strategies with special reference to use of biotechnological approaches for attaining host resistance. J Oilseed Brassica 4:1–10

Ao E, Thakuria K, Saud RK (2016) Production potential of rapeseed (*Brassica rapa* var. *dichotoma*)—based intercropping systems under rain fed conditions. J Oilseed Brassica 7(1):91–97

Arora H, Lakshmi Padmaja K, Paritosh K, Mukhi N, Tewari AK, Mukhopadhyay A, Gupta V, Pradhan AK, Pental D (2019) *BjuWRR1*, a *CC-NB-LRR* gene identified in *Brassica juncea*, confers resistance to white rust caused by *Albugo candida*. Theor Appl Genet 132:2223–2236

Asari S, Ongena M, Debois D, De Pauw E, Chen K, Bejai S, Meijer J (2017) Insights into the molecular basis of biocontrol of *Brassica* pathogens by *Bacillus amyloliquefaciens* UCMB5113 lipopeptides. Ann Bot 120(4):551–562. https://doi.org/10.1093/aob/mcx089

Auer S, Ludwig-Muller J (2014) Effects of the endophyte *Acremonium alternatum* on oilseed rape (*Brassica napus*) development and clubroot progression. Albanian J Agric Sci 15

Babii C, Bahrin L, Neagu AN, Gostin I, Mihasan M, Birsa L, Stefan M (2016) Antibacterial activity and proposed action mechanism of a new class of synthetic tricyclic flavonoids. J Appl Microbiol 120:630–637

Badawy HMA, Kakau J, Hoppe HH (1992) Temperature and ageing of host tissue affect the interactions between different oilseed rape cultivars and pathotype groups of *Leptosphaeria maculans*. J Phytopathol 134:255–263. https://doi.org/10.1111/j.1439-0434.1992.tb01234.x

Bains PS, Tewari JP (1989) Bioassay and mode of action of an Alternaria brassicae toxin. Can J Plant Pathol 11:184–189

Balesdent MH, Attard A, Kuhn ML, Rouxel T (2002) New avirulence genes in the phytopathogenic fungus *Leptosphaeria maculans*. Phytopathology 92:1122–1133. https://doi.org/10.1094/PHYTO.2002.92.10.1122

Balesdent MH, Barbetti M, Hua L, Sivasithamparam K, Gout L, Rouxel T (2005) Analysis of *Leptosphaeria maculans* race structure in a worldwide collection of isolates. Phytopathology 95:1061–1071

Balesdent MH, Louvard K, Pinochet X, Rouxel T (2006) A large-scale survey of races of *Leptosphaeria maculans* occurring on oilseed rape in France. Eur J Plant Pathol 114:53–65

Balint-Kurti P, Holland JB (2015) New insight into a complex plant–fungal pathogen interaction. Nat Genet 47(2):101–103. https://doi.org/10.1038/ng.3203

Bangash SAK, Khan MS, Ambreen KSH, Siddique AN (2013) Genetic transformation of *Brassica juncea* with an antimicrobial Wasabi defensin gene. Pak J Bot 45(3):993–998

Barbetti MJ, Banga SS, Salisbury PA (2012) Challenges for crop production and management from pathogen biodiversity and diseases under current and future climate scenarios—case study with oilseed *Brassicas*. Field Crop Res 127:225–240. https://doi.org/10.1016/j.fcr.2011.11.021

Barret P, Delourme R, Foisset N, Renard M (1998) Development of a SCAR (sequence characterized amplified region) marker for molecular tagging of the dwarf BREIZH (Bzh) gene in *Brassica napus* L. Theor Appl Genet 97:828–833. https://doi.org/10.1007/s001220050962

Barton NH (2001) The role of hybridization in evolution. Mol Ecol 10:551–568

Barton NH, Etheridge AM, Kelleher J, Veber A (2013) Genetic hitchhiking in spatially extended populations. Theor Pop Biol 87:757–789. https://doi.org/10.1016/j.tpb.2012.12.001

Bazzi C, Messina C, Tortoreto L, Stefani E, Bini F, Brunelli A, Andreotti C, Sabatini E, Spinelli F, Costa G, Hauptmann S, Stammler G, Doerr S, Marr J, Rademacher W (2003) Control of pathogen incidence in pome fruits and other horticultural crop plants with prohexadione-Ca. Eur J Hortic Sci 68:8–14

Belsky AJ, Carson WP, Jensen CL, Fox GA (1993) Overcompensation by plants: herbivore optimization or red herring? Evol Ecol 7:109–121

Berenbaum MR, Zangerl AR, Nitao JK (1986) Constraints on chemical evolution: wild parsnip and the parsnip webworm. Evolution 40:1215–1228

Bergelson J (1994) The effects of genotype and the environment on costs of resistance in lettuce. Am Nat 143:349–359

Bernoux M, Ve T, Williams S, Warren C, Hatters D, Valkov E, Zhang X, Ellis JG, Kobe B, Dodds PN (2011) Structural and functional analysis of a plant resistance protein TIR domain reveals interfaces for self-association, signaling, and auto regulation. Cell Host Microbe 9:200–211. https://doi.org/10.1016/j.chom.2011.02.009

Bilir O, Telli O, Norman C, Budak H, Hong Y, Tor M (2019) Small RNA inhibits infection by downy mildew pathogen *Hyaloperonospora arabidopsidis*. Mol Plant Pathol 20 (11):1523–1534. https://doi.org/10.1111/mpp.12863

Boland GJ (1992) Hypovirulence and double-stranded RNA in *Sclerotinia sclerotiorum*. Can J Plant Pathol 14:10–17

Boland GJ, Mould MJR, Robb J (1993) Ultrastructure of a hypovirulent isolate of *Sclerotinia sclerotiorum* containing double-stranded RNA. Physiol Mol Plant Pathol 43:21–32

Borhan MH, Holub EB, Kindrachuk C et al (2010) *WRR4*, a broad spectrum *TIR-NB-LRR* gene from *Arabidopsis thaliana* that confers white rust resistance in transgenic oilseed *Brassica* crops. Mol Plant Pathol 11:283–291. https://doi.org/10.1111/j.1364-3703.2009.00599.x

Bousset L, Sprague SJ, Thrall PH, Barrett LG (2018) Spatio-temporal connectivity and host resistance influence evolutionary and epidemiological dynamics of the canola pathogen *Leptosphaeria maculans*. Evol Appl 11:1354–1370. https://doi.org/10.1111/eva.12630

Brant EJ, Budak H (2018) Plant small non-coding RNAs and their roles in biotic stresses. Front Plant Sci 9:1038

Brinkmamn A, Frey KJ (1977) Growth analysis of isoline-recurrent parent grain yield differences in oats. Crop Sci 17:426–430

Broglie K, Chet I, Holliday M, Cressman R, Biddle P, Knowlton S, Mauvais CJ, Broglie R (1991) Transgenic plants with enhanced resistance to the fungal pathogen *Rhizoctonia solani*. Science 254:1194–1197

Brownfield DL, Todd CD, Deyholos MK (2008) Analysis of *Arabidopsis* arginase gene transcription patterns indicates specific biological functions for recently diverged paralogs. Plant Mol Biol 67:429–440

Brun H, Chèvre AM, Fitt BD, Powers S, Besnard AL, Ermel M, Huteau V, Marquer B, Eber F, Renard M, Andrivon D (2010) Quantitative resistance increases the durability of qualitative resistance to *Leptosphaeria maculans* in *Brassica napus*. New Phytol 185:285–299. https://doi.org/10.1111/j.1469-8137.2009.03049.x

Buer CS, Muday GK (2004) The transparent testa4 mutation prevents flavonoid synthesis and alters auxin transport and the response of Arabidopsis roots to gravity and light. Plant Cell 16:1191–1205

Burdon JJ, Muller WJ (1987) Measuring the cost of resistance to *Puccinia coronata* in *Avena fatua*. J Appl Ecol 24:191–200

Burton W, Salisbury P, Potts D (2003) The potential of canola quality *Brassica juncea* as an oilseed crop for Australia. In: Proc. 13th Australian research assembly on *Brassicas* Tamworth, NSW, Australia, pp 62–64

Burton WA, Flood RF, Norton RM, Field B, Potts DA, Robertson MJ, Salisbury PA (2008) Identification of variability in phenological responses in canola-quality *Brassica juncea* for utilization in Australian breeding programs. Aust J Agric Res 59:874–881

Caicedo AL, Schaal BA, Kunkel BN (1999) Diversity and molecular evolution of the RPS2 resistance gene in *Arabidopsis thaliana*. PNAS 96:302–306. https://doi.org/10.1073/pnas.96.1.302. PMID: 9874813

Camehl I, Sherameti I, Venus Y, Bethke G, Varma A, Lee J, Oelmuller R (2010) Ethylene signaling and ethylene-targeted transcription factors are required to balance beneficial and non-beneficial traits in the symbiosis between the endophytic fungus *Piriformospora indica* and *Arabidopsis thaliana*. New Phytol 185:1062–1073

Cao T, Manolii VP, Hwang SF, Howard RJ, Strelkov SE (2009) Virulence and spread of *Plasmodiophora brassicae* [clubroot] in Alberta, Canada. Can J Plant Pathol 31:321–329

Cao FQ, Werner AK, Dahncke K, Romeis T, Liu LH, Witte CP (2010) Identification and characterization of proteins involved in rice urea and arginine catabolism. Plant Physiol 154:98–108

Cardone M, Mazzoncini M, Menini S, Rocco V, Senatore A, Seggiani M, Vitolo S (2003) *Brassica carinata* as an alternative oil crop for the production of biodiesel in Italy: agronomic evaluation, fuel production by trans esterification and characterization. Biomass Bioenergy 25(6):623–636. https://doi.org/10.1016/S0961-9534(03)00058-8

Casteel CL, Yang C, Nanduri A, Jong HD, Whitham SA, Jander G (2014) The NIa-pro protein of turnip mosaic virus improves growth and reproduction of the aphid vector, *Myzus persicae* (Green Peach Aphid). Plant J 77(4):653–663. https://doi.org/10.1111/tpj.12417

Cawoy H, Mariutto M, Henry G, Fisher C, Vasilyeva N, Thonart P, Mommes J, Ongena M (2014) Plant defense stimulation by natural isolates of *Bacillus* depends on efficient surfactin production. Mol Plant-Microbe Interact 27:87–100

Chaplin JF (1970) Associations among disease resistance, agronomic characteristics, and chemical constituents in flue-cured tobacco. Agron J 62:87–91

Chen Y, Wang D (2005) Two convenient methods to evaluate soybean for resistance to *Sclerotinia sclerotiorum*. Plant Dis 89:1268–1272

Chen XF, Hou XL, Zhang JY, Zheng JQ (2008) Molecular characterization of two important antifungal proteins isolated by downy mildew infection in non-heading Chinese cabbage. Mol Biol Rep 35(4):621–629

Chen XH, Koumoutsi A, Scholz R, Borriss R (2009) More than anticipated—production of antibiotics and other secondary metabolites by *Bacillus amyloliquefaciens* FZB42. J Mol Microbiol Biotechnol 16:14–24

Chen S, Wan Z, Nelson MN, Chauhan JS, Lin P, Banga SS, Salisbury PA, Redden RJ, Fu T, Cowling W (2011) Two distinct genetic diversity groups of oilseed *Brassica juncea* in both China and India. In: Proc 13th international rapeseed congress

Chen L, Zhang X, Xu H, Song B, Fan X (2016) Introgression of clubroot resistance into an elite pak choi inbred line through marker-assisted introgression breeding. Plant Breed 135:471–475. https://doi.org/10.1111/pbr.12379

Chen W, Zhang X, Fan Y, Li B, Ryabov E, Shi N, Zhao M, Yu Z, Qin C, Zheng Q, Zhang P (2018) A genetic network for systemic RNA silencing in plant. Plant Physiol 176:2700–2719

Cheng F, Liu S, Wu J, Fang L, Sun S, Liu B, Li P, Hua W, Wang X (2011) BRAD, the genetics and genomics database for *Brassica* plants. BMC Plant Biol 11:136. https://doi.org/10.1186/1471-2229-11-136

Cheng F, Wu J, Cai C, Fu F, Liang J, Borm T, Zhuang M, Zhang Y, Zhang F, Bonnema G, Wang X (2016) Genome resequencing and comparative variance analysis in a *Brassica rapa* and *Brassica oleracea* collection. Sci Data 3:160119

Chèvre AM, Barret P, Eber F, Dupuy P, Brun H, Tanguy X, Renard M (1997) Selection of stable *Brassica napus-B. juncea* recombinant lines resistant to blackleg (*Leptosphaeria maculans*). 1. Identification of molecular markers, chromosomal and genomic origin of the introgression. Theor Appl Genet 95:1104–1111

Chèvre AM, Brun H, Eber F, Letanneur JC, Vallee P, Ermel M, Glais I, Li H, Sivasithamparam K, Barbetti MJ (2008) Stabilization of resistance to *Leptosphaeria maculans* in *Brassica napus–B. juncea* recombinant lines and its introgression into spring-type *Brassica napus*. Plant Dis 92:1208–1214

Chhikara S, Chaudhury D, Dhankher OP, Jaiwal PK (2012) Combined expression of a barley class II chitinase and type I ribosome inactivating protein in transgenic *Brassica juncea* provides protection against *Alternaria brassicae*. Plant Cell Tissue Organ Cult 108:83–89. https://doi.org/10.1007/s11240-011-0015-7

Choudhary BR, Joshi P, Ramarao S (2000) Interspecific hybridization between *Brassica carinata* and *Brassica rapa*. Plant Breed 119(5):417–420. https://doi.org/10.1046/j.1439-0523.2000.00503.x

Chu M, Yu F, Falk KC, Liu X, Zhang X, Chang A, Peng G (2013) Identification of the clubroot resistance gene Rpb1 and introgression of the resistance gene into canola breeding lines using a marker-assisted selection approach. Acta Hortic 1005:599–605. https://doi.org/10.17660/ActaHortic.2013.1005.74

Chu M, Song T, Falk KC, Zhang X, Liu X, Chang A, Lahlali R, McGregor L, Gossen BD, Peng G, Yu F (2014) Fine mapping of Rcr1 and analyses of its effect on transcriptome patterns during infection by *Plasmodiophora brassicae*. BMC Genomics 15:1166. https://doi.org/10.1186/1471-2164-15-1166

Cipollini DF (2002) Does competition magnify the fitness costs of induced responses in *Arabidopsis thaliana*? A manipulative approach. Oecologia 131:514–520

Clay K (1996) Fungal endophytes, herbivores, and the structure of grassland communities. In: Gange AC (ed) Multitrophic interactions in terrestrial systems. British ecological symposium. Blackwell Scientific Pub., pp 151–169

Cody YS, Daniels CH, Hadwiger LA (1988) Host responses in peas challenged by wall components of *Pseudomonas syringae* pv. *pisi* races 1, 2 and 3. Phytopathology 78:1451–1453

Coffey MD (1975) Ultra-structural features of the haustorial apparatus of the white blister fungus *Albugo candida*. Can J Bot 53:1285–1299

Conn KL, Jejelowa OA, Tewari JP, Bains PS (1991) *Camelina sativa* phytoalexins (Camalexin and methoxy camalexin) provide resistance against *Alternaria brassicae*. Abstract 274. Can J Plant Pathol 13

Curir P, Dolci M, Galeotti F (2005) A phytoalexin-like flavonol involved in the carnation (*Dianthus caryophyllus*)-*Fusarium oxysporum* f. sp. *dianthi* pathosystem. J Phytopathol 153:65–67

Dalakouras A, Wassenegger M, McMillan JN, Cardoza V, Maegele I, Dadami E, Runne M, Krczal G, Wassenegger M (2016) Induction of silencing in plants by high-pressure spraying of *in vitro*-synthesized small RNAs. Front Plant Sci 7:356

Damgaard C, Jensen BD (2002) Disease resistance in *Arabidopsis thaliana* increases the competitive ability and the predicted probability of long-term ecological success under disease pressure. Oikos 98:459–466

Daniels C, Fristensky B, Wagoner W, Hadwiger LA (1987) Pea genes associated with non-host resistance to fungi are also active in race-specific resistance to bacteria. Plant Mol Biol 8:309–316

Daverdin G, Rouxel T, Gout L, Aubertot JN, Fudal I, Meyer M, Parlange F, Carpezat J, Balesdent MH (2012) Genome structure and reproductive behaviour influence the evolutionary potential of a fungal phytopathogen. PLoS Pathog 8:e1003020. https://doi.org/10.1371/journal.ppat.1003020

De Bernonville T, Albenne C, Arlat M, Hoffmann L, Lauber E, Jamet E (2014) Xylem sap proteomics. Plant Proteomics: methods and protocols, Jomin-Novo JV et al. (eds). Methods Mol Biol 1072:391–405. https://doi.org/10.1007/978-1-62703-631-3_28

de Ruiter H, Kolloffel C (1983) Arginine catabolism in the cotyledons of developing and germinating pea seeds. Plant Physiol 73:525–528

de Souza ALSR, De Souza SA, De Oliveira MVV, Ferraz TM, Figueiredo F, Da Silva ND, Rangel PL, Panisset CRS, Olivares FL, Campostrini E, De Souza Filho GA (2016) Endophytic colonization of *Arabidopsis thaliana* by Gluconacetobacter diazotrophicus and its effect on plant growth promotion, plant physiology, and activation of plant defense. Plant Soil 399:257–270

de Vallavieille-Pope C, Ali S, Leconte M, Enjalbert J, Delos M, Rouzet J (2012) Virulence dynamics and regional structuring of *Puccinia striiformis* f. sp. *tritici* in France between 1984 and 2009. Plant Dis 96:131–140

De Wit CT (2007) On competition. Versl Landbouwkd Onderz 66:1e82

Deleu M, Paquot M, Nylander T (2008) Effect of fengycin, a lipopeptide produced by *Bacillus subtilis*, on model biomembranes. Biophys J 94:2667–2679

Delourme R, Brun H, Ermel M, Lucas MO, Vallee P, Domin C, Walton G, Li H, Sivasithamparam K, Barbetti MJ (2008) Expression of resistance to *Leptosphaeria maculans* in *Brassica napus* double haploid lines in France and Australia is influenced by location. Ann Appl Biol 153:259–269

Deng Y, Wang J, Tung J, Liu D, Zhou Y, He S, Du Y, Baker B, Li F (2018) A role for small RNA in regulating innate immunity during plant growth. PLoS Pathog 14:e1006756

Deora A, Gossen BD, McDonald MR (2012) Infection and development of *Plasmodiophora brassicae* in resistant and susceptible canola cultivars. Can J Plant Pathol 34(2):239–247. https://doi.org/10.1080/07060661.2012.681071

Deora A, Gossen BD, McDonald MR (2013) Expression of resistance to *Plasmodiophora brassicae* in four resistant canola cultivars. Can J Plant Pathol 35(1):106–106

Devos M, Van-Oosten VR, van Poecke RMP, Van Pelt JA, Pozo MJ, Mueller MJ, Buchala AJ, Métraux JP, Van Loon LC, Dicke M, Pieterse CMJ (2005) Signal signature and transcriptome changes of *Arabidopsis* during pathogen and insect attack. Mol Plant-Microbe Interact 18:923–937. https://doi.org/10.1094/MPMI-18-0923

Dey SS, Sharma K, Dey RB, Kumar GMS, Singh D, Kumar R, Parkash C (2015) Inter specific hybridization (*Brassica carinata* × *Brassica oleracea*) for introgression of black rot resistance genes into Indian cauliflower (*B. oleracea* var *botrytis* L.). Euphytica. https://doi.org/10.1007/s10681-015-1352-0

Dhingra A, Portis AR Jr, Daniell H (2004) Enhanced translation of a chloroplast-express *RbcS* gene restores small subunit levels and photosynthesis in nuclear *RbcS* antisense plants. Proc Natl Acad Sci U S A 101:6315–6320

Diederichsen E, Deppe U, Sacristan MD (2003) Characterization of clubroot resistance in recent winter oilseed rape material. In: Proc 11th international rapeseed congress, vol 1, Copenhagen, Denmark, pp 68–70

Diederichsen E, Frauen M, Linders EGA, Hatakayema K, Hirai M (2009) Status and perspectives of clubroot resistance breeding in crucifer crops. J Plant Growth Regul 28:265–281

Diederichsen E, Frauen M, Ludwig-Muller J (2014) Clubroot disease management challenges from a German perspective. Can J Plant Pathol 36(S1):85–98

Dinesh-Kumar SP, Baker BJ (2000) Alternatively spliced N resistance gene transcripts: their possible role in tobacco mosaic virus resistance. Proc Natl Acad Sci U S A 97:1908–1913. https://doi.org/10.1073/pnas.020367497

Dixon RA (2001) Natural products and plant disease resistance. Nature 411:843–847

Dixon GR (2009a) The occurrence and economic impact of *Plasmodiophora brassicae* and clubroot disease. J Plant Growth Regul 28:194–202. https://doi.org/10.1007/s00344-009-9090-y

Dixon GR (2009b) Husbandry–the sustainable means of controlling soil borne pathogens: a synoptic review. In: Proc symposium of horticulture in Europe, Dixon GR, (ed.) Vienna, Austria; 17–21 February 2008. Acta Hortic 817:233–242

Dixon GR, Webster MA (1988) Antagonistic effects of boron, calcium and pH on pathogenesis caused by *Plasmodiophora brassicae* (clubroot)—a review of recent work. Crop Res (Hortic Res) 28:83–95

Donald C, Porter IJ (2004) A sand solution culture technique used to observe the effect of calcium and pH on root hair and cortical stages of infection by *Plasmodiophora brassicae*. Australas Plant Pathol 33:585–589

Donald C, Porter I (2009) Integrated control of clubroot. J Plant Growth Regul 28:194–202

Donald EC, Jaudzems G, Porter IJ (2008) Pathology of cortical invasion by *Plasmodiophora brassicae* in clubroot resistant and susceptible *Brassica oleracea* hosts. Plant Pathol 57:201–209. https://doi.org/10.1111/j.1365-3059.2007.01765.x

Dordas C (2008) Role of nutrients in controlling plant diseases in sustainable agriculture—a review. Agron Sustain Dev 28:33–46. https://doi.org/10.1051/agro:2007051

Ebrahimi E, Kaul HP, Neugschwandtner RW, Dabbagh Mohammadi Nassab A (2016) Productivity of wheat (*Triticum aestivum* L.) intercropped with rapeseed (*Brassica napus* L). Can J Plant Sci 97:557–568

Eeman M, Pegado L, Dufrene YF, Paquot M, Deleu M (2009) Influence of environmental conditions on the interfacial organization of fengycin, a bioactive lipopeptide produced by *Bacillus subtilis*. J Colloid Interface Sci 329:253–264

Elliott VL, Norton RM, Khangura RK, Salisbury PA, Marcroft SJ (2015) Incidence and severity of blackleg caused by *Leptosphaeria* spp. in *juncea* canola (*Brassica juncea* L.) in Australia. Australas Plant Pathol 44:149–159

Ellis J (2006) Insights into nonhost disease resistance: can they assist disease control in agriculture? Plant Cell 18:523–528. https://doi.org/10.1105/tpc.105.040584

Ernst TW, Kher S, Stanton D, Rennie DC, Hwang SF, Strelkov SE (2019) *Plasmodiophora brassicae* resting spore dynamics in clubroot resistant canola (*Brassica napus*) cropping systems. Plant Pathol 68(2):399–408. https://doi.org/10.1111/ppa.12949

Eshraghi L, Barbetti MJ, Li H, Danehloueipour N, Sivasithamparam K (2007) Resistance in oilseed rape (*Brassica napus*) and Indian mustard (*Brassica juncea*) to a mixture of *Pseudocercosporella capsellae* isolates from Western Australia. Field Crop Res 101:37–43

Fagard M, Launay A, Clement G, Courtial J, Dellagi A, Farjad M, Krapp A, Soulie MC, Masclaux-Daubresse C (2014) Nitrogen metabolism meets phytopathology. J Exp Bot 65:5643–5656. https://doi.org/10.1093/jxb/eru323

Falardeau J, Wise C, Novitsky L, Avis TJ (2013) Ecological and mechanistic insights into the direct and indirect antimicrobial properties of *Bacillus subtilis* lipopeptides on plant pathogens. J Chem Ecol 39:869–878

Fan J, Doerner P (2012) Genetic and molecular basis of non-host disease resistance: complex, yes; silver bullet, no. Curr Opin Plant Biol 15(4):400–406. https://doi.org/10.1016/j.pbi.2012.03.001

Farace G, Fernandez O, Jacquens L, Coutte F, Krier F, Jacques P, Clement C, Barka EA, Jacquard C, Dorey S (2014) Cyclic lipopeptides from *Bacillus subtilis* activate distinct patterns of defense responses in grapevine. Mol Plant Pathol 16:177–187

Fatima U, Bhorali P, Borah S, Senthil-Kumar M (2019) Perspectives on the utilization of resistance mechanisms from host and nonhost plants for durable protection of *Brassica* crops against Alternaria blight. PeerJ 7:e7486. https://doi.org/10.7717/peerj.7486

Figdore SS, Ferreira ME, Slocum MK, Williams PH (1993) Association of RFLP markers with trait loci affecting clubroot resistance and morphological characters in *Brassica oleracea* L. Euphytica 69(1/2):33–44

Flores T, Todd CD, Tovar-Mendez A, Dhanoa PK, Correa-Aragunde N, Hoyos ME, Brownfield DM, Mullen RT, Lamattina L, Polacco JC (2008) Arginase-negative mutants of *Arabidopsis* exhibit increased nitric oxide signaling in root development. Plant Physiol 147:1936–1946. https://doi.org/10.1104/pp.108.121459

Fonseca JP, Mysore KS (2018) Genes involved in nonhost disease resistance as a key to engineer durable resistance in crops. Plant Sci 279:108–116

Ford E, Miller R, Gray H, Sherwood J (1995) Heterokaryon formation and vegetative compatibility in *Sclerotinia sclerotiorum*. Mycol Res 99(2):241–247. https://doi.org/10.1016/s0953-7562(09)80893-9

Frank SA (1992) Models of plant pathogen co-evolution. Trends Genet 8:213–219

Frank SA (1993) Coevolutionary genetics of plants and pathogens. Evol Ecol 7:45–75

Frey KJ, Browning JA (1971) Association between genetic factors for crown rust resistance and yield in oats. Crop Sci 11:757–760

Fristensky B, Riggleman CR, Wagoner W, Hadwiger LA (1985) Gene expression in susceptible and disease resistant interactions of peas induced with *Fusarium solani* pathogens and chitosan. Physiol Plant Pathol 27:15–28

Fuchs H, Sacristan MD (1996) Identification of a gene in *Arabidopsis thaliana* controlling resistance to clubroot and characterization of the resistance response. Mol Plant-Microbe Interact 9:91–97

Gao AG, Hakimi SM, Mittanck CA, Wu Y, Woerner BM, Stark DM, Shah DM, Liang J, Rommens CM (2000) Fungal pathogen protection in potato by expression of a plant defensin peptide. Nat Biotechnol 18(12):1307–1310. https://doi.org/10.1038/82436

Garcia-Gutierrez L, Zeriouh H, Romero D, Cubero J, de Vicente A, Perez-Garcia A (2013) The antagonistic strain *Bacillus subtilis* UMAF6639 also confers protection to melon plants against cucurbit powdery mildew by activation of jasmonate and salicylic acid-dependent defense responses. Microb Biotechnol 6:264–274

Geisler M, Murphy AS (2006) The ABC of auxin transport: the role of p-glycoproteins in plant development. FEBS Lett 580:1094–1102

Ghahfarokhi R, Goltapeh M (2010) Potential of the root endophytic fungus *Piriformospora indica*; *Sebacina vermifera* and *Trichoderma* species in biocontrol of take-all disease of wheat *Gaeumannomyces graminis* var. *tritici in vitro*. J Agric Technol 6:11–18

Giessen TW, Marahiel MA (2012) Ribosome-independent biosynthesis of biologically active peptides: application of synthetic biology to generate structural diversity. FEBS Lett 586:2065–2075

Gill SS, Gill R, Trivedi DK, Anjum NA, Sharma KK, Ansari MW, Ansari AA, Johri AK, Prasad R, Pereira E, Verma A, Tuteja N (2016) *Piriformospora indica*: potential and significance in plant stress tolerance. Front Microbiol 7:332

Goldraij A, Polacco JC (2000) Arginine degradation by arginase in mitochondria of soybean seedling cotyledons. Planta 210:652–658

Gosal SS, Wani SH, Kang MS (2009) Biotechnology and drought tolerance. J Crop Improv 23:19–54

Gossen BD, Mcdonald MR, Hwang SF, Strelkov SE, Peng G (2013) A comparison of clubroot development and management on canola and *Brassica* vegetables. Can J Plant Pathol 35 (2):175–191. https://doi.org/10.1080/07060661.2013.763293

Gossen BD, Deora A, Peng G, Hwang SF, Mcdonald MR (2014) Effect of environmental parameters on clubroot development and the risk of pathogen spread. Can J Plant Pathol 36:37–48. https://doi.org/10.1080/07060661.2013.859635

Gossen BD, Strelkov SE, Manolii VP, Rennie DC, Cao T, Hwang SF, Peng G, McDonald MR (2015) Spread of *Plasmodiophora brassicae* on canola in Canada, 2003–2014: old pathogen, new home. Can J Plant Pathol 37:403–413

Govindarajulu M, Epstein L, Wroblewski T, Michelmore RW (2015) Host-induced gene silencing inhibits the biotrophic pathogen causing downy mildew of lettuce. Plant Biotechnol J 13:875–883

Graves PR, Haystead TAJ (2002) Molecular biologist's guide to proteomics. Microbiol Mol Biol Rev 66:39–63

Gravot A, Grillet L, Wagner G, Jubault M, Lariagon C, Baron C, Deleu C, Delourme R, Bouchereau A, Maanzanares-Dauleux MJ (2011) Genetic and physiological analysis of the relationship between partial resistance to clubroot and tolerance to trehalose in *Arabidopsis thaliana*. New Phytol 191:1083–1094

Gravot A, Deleu C, Wagner G, Lariagon C, Lugan R, Todd C, Wendehenne D, Delourme R, Bouchereau A, Manzanares-Dauleux MJ (2012) Arginase induction represses gall development during clubroot infection in *Arabidopsis*. Plant Cell Physiol 53(5):901–911. https://doi.org/10.1093/pcp/pcs037

Grotewold E, Drummond BJ, Bowen B, Peterson T (1994) The myb-homologous P gene controls phlobaphene pigmentation in maize floral organs by directly activating a flavonoid biosynthetic gene subset. Cell 76:543–553

Grsic S, Kirchheim B, Pieper K, Fritsch M, Hilgenberg W, Ludwig-Muller J (1999) Induction of auxin biosynthetic enzymes by jasmonic acid and in clubroot diseased Chinese cabbage plants. Physiol Plant 105:521–531. https://doi.org/10.1034/j.1399-3054.1999.105318.x

Gunasinghe N, You MP, Banga SS, Barbetti MJ (2014) High level resistance to *Pseudocercosporella capsellae* offers new opportunities to deploy host resistance to effectively manage white leaf spot disease across major cruciferous crops. Eur J Plant Pathol 138:873–890. https://doi.org/10.1007/s10658-013-0360-y

Guo SY, Li XY, He PF, Ho HH, Wu Y, He Y (2015) Whole-genome sequencing of *Bacillus subtilis* XF-1 reveals mechanisms for biological control and multiple beneficial properties in plants. J Ind Microbiol Biotechnol 42:925–937

Gupta K, Saharan GS (2002) Identification of pathotypes of *Albugo candida* with stable characteristic symptoms on Indian mustard. J Mycol Plant Pathol 32:46–51

Gupta SK, Gupta PP, Yadav TP, Kaushik CD (1990) Metabolic changes in mustard due to *Alternaria* leaf blight. Indian Phytopathol 43:64–69

Gupta K, Saharan GS, Singh D (2002) Effective selection criteria to assess disease tolerance in Indian mustard [*Brassica juncea* (L.) Czern & Coss.]. J Mycol Plant Pathol 32:72–80

Halbwirth H, Fischer TC, Roemmelt S, Spinelli F, Schlangen K, Peterek S, Sabatini E, Messina C, Speakman JB, Andreotti C, Rademacher W, Bazzi C, Costa G, Treutter D, Forkmann G, Stich K (2003) Induction of antimicrobial 3-deoxyflavonoids in pome fruit trees control fire blight. Z Naturforsch C J Biosci 58(11–12):765–770. https://doi.org/10.1515/znc-2003-11-1201

Hammond KE, Lewis BG (1987) The establishment of systemic infection in leaves of oilseed rape by Leptosphaeria maculans. Plant Pathol 36:135–147

Han K, Lincoln DE (1994) The evolution of carbon allocation to plant secondary metabolites: a genetic analysis of cost in *Diplacus aurantiacus*. Evolution 48:1550–1563

Hansen LH, Earle ED (1995) Transfer of resistance to *Xanthomonas campestris* pv. *campestris* (L.) by protoplast fusion. Theor Appl Genet 91:1293–1300. https://doi.org/10.1007/BF00220944

Hatakayema K, Fujimura M, Ishida M, Suzuki T, Sato T (2006) Classification of pathogenicity of *Plasmodiophora brassicae* field isolates in Japan based on resistance of F1 cultivars of Chinese cabbage (*Brassica rapa* L.) to clubroot. Acta Hortic 706:323–328. https://doi.org/10.17660/ActaHortic.2006.706.38

Hatakeyama K, Suwabe K, Tomita RN, Kato T, Nunome T, Fukuoka H, Matsumoto S (2013) Identification and characterization of Crr1a, a gene for resistance to clubroot disease (*Plasmodiophora brassicae* Woronin) in *Brassica rapa* L. PLoS One 8:e54745. https://doi.org/10.1371/journal.pone.0054745

Heil M (2002) Ecological costs of induced resistance. Curr Opin Plant Biol 5:345–350

Hemayati SS, Akbar JE, Ghaemi AR, Fasahat P (2017) Efficiency of white mustard and oilseed radish trap plants against sugar beet cyst nematode. Appl Soil Ecol 119:192–196. https://doi.org/10.1016/j.apsoil.2017.06.017

Hillman BI, Fulbright DW, Nuss DL, Van Alfen NK (1995) Hypoviridae. In: Murphy FA (ed) Sixth report of the international committee for the taxonomy of viruses. Springer, New York

Hirai M (2006) Genetic analysis of clubroot resistance in *Brassica* crops. Breed Sci 56:223–229. https://doi.org/10.1270/jsbbs.56.223

Holub EB (2007) Natural variation in innate immunity of a pioneer species. Curr Opin Plant Biol 10:415–424

Hovmoller MS, Munk L, Ostergard H (1993) Observed and predicted changes in virulence gene frequencies at 11 loci in a local barley powdery mildew population. Phytopathology 83:253–260. https://doi.org/10.1094/Phyto-83-253

Howell CR (2003) Mechanisms employed by *Trichoderma* species in the biological control of plant diseases: the history and evolution of cur-rent concepts. Plant Dis 87:4–10

Howie WJ, Suslow TV (1991) Role of antibiotic synthesis in the inhibition of *Pythium ultimum* in the cotton spermosphere and rhizosphere by *Pseudomonas fluorescence*. Mol Plant-Microbe Interact 4:393–399

Howlett BJ, Rohan GT, Lowe Stephen J, Marcroft Angela P, van de Wouw AP (2015) Evolution of virulence in fungal plant pathogens: exploiting fungal genomics to control plant disease. Mycologia 107(3):441–451. https://doi.org/10.3852/14-317

Hu Z, Huang S, Sun M, Wang H, Hua W (2012) Development and application of single nucleotide polymorphism markers in the polyploid *Brassica napus* by 454 sequencing of expressed sequence tags. Plant Breed 131:293–299

Huang YJ, Evans N, Li ZQ, Eckert M, Chèvre AM, Renard M, Fitt BDL (2006) Temperature and leaf wetness duration affect phenotypic expression of *Rlm6*-mediated resistance to *Leptosphaeria maculans* in *Brassica napus*. New Phytol 170:129–141

Huang YJ, Jestin C, Welham SJ, King GJ, Manzanares-Dauleux MJ, Fitt BD, Delourme R (2016) Identification of environmentally stable QTL for resistance against *Leptosphaeria maculans* in oilseed rape (*Brassica napus*). Theor Appl Genet 129:169–180. https://doi.org/10.1007/s00122-015-2620-z

Huang Z, Peng G, Liu X, Deora A, Falk KC, Gossen BD, McDonald MR, Yu F (2017) Fine mapping of a clubroot resistance gene in Chinese cabbage using SNP markers identified from bulked segregant RNA sequencing. Front Plant Sci 8:1448. https://doi.org/10.3389/fpls.2017.01448

Hughes SL, Hunter PJ, Sharpe AG, Kearsey MJ, Lydiate DJ, Walsh JA (2003) Genetic mapping of the novel Turnip mosaic virus resistance gene TuRB03 in *Brassica napus*. Theor Appl Genet 107:1169–1173

Hunger S, Di Gaspero G, Mohring S, Bellin D, Schafer-Pregl R, Borchardt DC, Durel CE, Werber M, Weisshaar B, Salamini F, Schneider K (2003) Isolation and linkage analysis of expressed disease-resistance gene analogues of sugar beet (*Beta vulgaris* L.). Genome 46:70–82. https://doi.org/10.1139/g02-10612669798

Hwang SF, Ahmed HU, Zhou Q, Strelkov SE, Gossen BD, Peng G, Turnbull D (2011) Influence of cultivar resistance and inoculum density on root hair infection of canola (*Brassica napus*) by *Plasmodiophora brassicae*. Plant Pathol 60:820–829

Hwang SF, Strelkov SE, Feng J, Gossen BD, Howard RJ (2012) *Plasmodiophora brassicae*: a review of an emerging pathogen of the Canadian canola (*Brassica napus*) crop. Mol Plant Pathol 13:105–113. https://doi.org/10.1111/j.1364-3703.2011.00729.x

Izzah NK, Lee J, Jayakodi M, Perumal S, Jin M, Park B-S, Ahn K, Yang T-J (2014) Transcriptome sequencing of two parental lines of cabbage (*Brassica oleracea* var. *capitata* L.) and construction of an EST-based genetic map. BMC Genomics 15:149. http://www.biomedcentral.com/1471-2164/15/149

Jan FJ, Fagoaga C, Pang SZ, Gonsalves D (2000) A single chimeric transgene derived from two distinct viruses confers multi-virus resistance in transgenic plants through homology-dependent gene silencing. J Gen Virol 81:2103–2109

Jaschke D, Dugassa-Gobena D, Karlovsky P, Vidal S, Ludwig-Muller J (2010) Suppression of clubroot development in *Arabidopsis thaliana* by the endophytic fungus *Acremonium alternatum*. Plant Pathol 59:100–111

Jat RR (1999) Pathogenic variability and inheritance of resistance to *A. candida* in oilseed *Brassica*. Ph.D. Thesis, CCSHAU, Hisar, p 129

Jauhar PP (2006) Modern biotechnology as an integral supplement to conventional plant breeding: the prospects and challenges. Crop Sci 46:1841–1859

Jauhar PP, Khush GS (2002) Importance of biotechnology in global food security. In: Lal R et al (eds) Food security and environmental quality in the developing world. CRC Press, Boca Raton, pp 107–128

Jenner CE, Tomimura K, Ohshima K, Hughes SL, Walsh JA (2002) Mutations in Turnip mosaic virus P3 and cylindrical inclusion proteins are separately required to overcome two *Brassica napus* resistance genes. Virology 300:50–59

Jeromela AM, Mikic AM, Vujic S, Cupina B, Krstic Đ, Dimitrijevic A, Vasiljevic S, Mihailovic V, Cvejic S, Miladinovic D (2017) Potential of legume–*Brassica* intercrops for forage production and green Manure: encouragements from a temperate southeast European environment. Front Plant Sci 8:312

Jiang M, He CM, Miao LX, Zhang YC (2012) Overexpression of a broccoli defensin gene BoDFN enhances downy mildew resistance. J Integr Agric 11:1137–1144. https://doi.org/10.1016/S2095-3119(12)60107-5

Jiang Z, Dong X, Zhang Z (2016) Network-based comparative analysis of *Arabidopsis* immune responses to *Golovinomyces orontii* and *Botrytis cinerea* infections. Sci Rep 6:19149. https://doi.org/10.1038/srep19149

Jiao Y, Wickett NJ, Ayyampalayam S, Chanderbali AS, Landherr L, Ralph PE, Tomsho LP, Hu Y, Liang H, Soltis PS, Soltis DE, Clifton SW, Schlarbaum SE, Schuster SC, Ma H, Leebens-Mack J, de Pamphili CW (2011) Ancestral polyploidy in seed plants and angiosperms. Nature 473:97–100

Jones JD, Dangl JL (2006) The plant immune system. Nature 444:323–329

Jorgensen TH (2012) The effect of environmental heterogeneity on RPW8-mediated resistance to powdery mildews in *Arabidopsis thaliana*. Ann Bot 109:833–842. https://doi.org/10.1093/aob/mcr320

Jubault M, Hamon C, Gravot A, Lariagon C, Delourme R, Bouchereau A, Manzanares-Dauleux MJ (2008) Differential regulation of root arginine catabolism and polyamine metabolism in clubroot-susceptible and partially resistant *Arabidopsis* genotypes. Plant Physiol 146:2008–2019

Juenger T, Bergelson J (1997) Pollen and resource limitation of compensation to herbivory in scarlet gilia, *Ipomopsis aggregata*. Ecology 78:1684–1695

Jungblut P, Wittmann-Liebold B (1995) Protein analysis on a genomic scale. J Biotechnol 41:111–120

Kalai-Grami L, Ben Slimane I, Mnari-Hattab M, Rezgui S, Aouani MA, Hajlaoui MR, Limam F (2014) Protective effect of *Bacillus amyloliquefaciens* against infections of *Citrus aurantium* seedlings by *Phoma tracheiphila*. World J Microbiol Biotechnol 30:529–538

Kamble A, Bhargava S (2007) β-aminobutyric acid-induced resistance in *Brassica juncea* against the necrotrophic pathogen *Alternaria brassicae*. J Phytopathol 155(3):152–158. https://doi.org/10.1111/j.1439-0434.2007.01209.x

Kaneko Y, Bang SW (2014) Interspecific and intergeneric hybridization and chromosomal engineering of brassicaceae crops. Breed Sci 64:14–22

Kanrar S, Venkateswari JC, Kirti PB, Chopra VL (2002) Transgenic expression of hevein, the rubber tree lectin, in Indian mustard confers protection against *Alternaria brassicae*. Plant Sci 162:441–448

Kari Dolatabadi H, Mohammadi Goltapeh E, Mohammadi N, Rabiey M, Rohani N, Varma A (2011) Biocontrol potential of root endophytic fungi and *Trichoderma* species against Fusarium wilt of lentil under *in vitro* and greenhouse conditions. J Agric Sci Technol 14:407–420

Karpechenko GD (1928) Polyploid hybrids of *Raphanus sativus* L. × *Brassica oleracea* L. Z Indukt Abstamm Vererbungsl 48:1–85

Kato T, Hatakayema K, Fukino N, Matsumoto S (2012) Identification of a clubroot resistance locus conferring resistance to a *Plasmodiophora brassicae* classified into pathotype group 3 in Chinese cabbage (*Brassica rapa* L.). Breed Sci 62:282–287. https://doi.org/10.1270/jsbbs.62.282

Kato T, Hatakeyama K, Fukino N, Matsumoto S (2013) Fine mapping of the clubroot resistance gene CRb and development of a useful selectable marker in *Brassica rapa*. Breed Sci 63:116–124. https://doi.org/10.1270/jsbbs.63.116

Kazan K, Manners JM (2013) MYC2: the master in action. Mol Plant 6:686–703

Khalid M, Hui N, Saeed-ur-Rahman, Hayat K, Huang D (2020) Suppression of clubroot (*Plasmodiophora brassicae*) development in *Brassica campestris* sp. *chinensis* L. via exogenous inoculation of *Piriformospora indica*. J Radiat Res Appl Sci 13:180–190

Kim H, O'Connell R, Maekawa-Yoshikawa M, Uemura T, Neumann U, Schulze-Lefert P (2014) The powdery mildew resistance protein RPW8.2 is carried on VAMP721/722 vesicles to the extrahaustorial membrane of haustorial complexes. Plant J 79:835–847

Knaust A, Ludwig-Muller J (2013) The ethylene signaling pathway is needed to restrict root gall growth in *Arabidopsis* after infection with the obligate biotrophic protist *Plasmodiophora brassicae*. J Plant Growth Regul 32:9–21

Koch E, Slusarenko AJ (1990) Fungal pathogens of *Arabidopsis thaliana* (L.) Heyhn. Bot Helv 100:257–268

Koch A, Kumar N, Weber L, Keller H, Imani J, Kogel KH (2013) Host-induced gene silencing of cytochrome P450 lanosterol C14 alpha-demethylase-encoding genes confers strong resistance to *Fusarium* species. Proc Natl Acad Sci U S A 110:19324–19329. https://doi.org/10.1073/pnas.1306373110

Koch A, Biedenkopf D, Furch A, Weber L, Rossbach O, Abdellatef E, Linicus L, Johannsmeier J, Jelonek L, Goesmann A, Cardoza V, McMillan J, Mentzel T, Kogel KH (2016) An RNAi-based control of *Fusarium graminearum* infections through spraying of long dsRNAs involves a plant passage and is controlled by the fungal silencing machinery. PLoS Pathog 12:e1005901

Kohli Y, Morrall RAA, Anderson JB, Kohn LM (1992) Local and trans-Canadian clonal distribution of *Sclerotinia sclerotiorum* on canola. Phytopathology 82:875–880

Kohn LM, Stasovski E, Carbone I, Royer J, Anderson JB (1991) Mycelial incompatibility and molecular markers identify genetic variability in field populations of *Sclerotinia sclerotiorum*. Phytopathology 81:480–485. https://doi.org/10.1094/Phyto-81-480

Kolmer JA (1996) Genetics of resistance to wheat leaf rust. Annu Rev Phytopathol 34:435–455

Korves T, Bergelson J (2003) A developmental response to pathogen infection in *Arabidopsis*. Plant Physiol 133:339–347

Korves T, Bergelson J (2004) A novel cost of R gene resistance in the presence of disease. Am Nat 163(4):489–504. https://doi.org/10.1086/382552

Kover PX, Schaal BA (2002) Genetic variation for disease resistance and tolerance among *Arabidopsis thaliana* accessions. Proc Natl Acad Sci U S A 99:11270–11274

Krishnia SK, Saharan GS, Singh D (2000) Genetic variation for multiple disease resistance in the families of interspecific cross of *Brassica juncea* x *B. carinata*. Cruciferae Newslett 22:51–53

Kuginuki Y, Yoshikawa H, Hirai M (1999) Variation in virulence of *Plasmodiophora brassicae* in Japan tested with clubroot resistant cultivars of Chinese cabbage (*Brassica rapa* L. spp. *pekinensis*). Eur J Plant Pathol 105:327–332

Kumar B, Kolte SJ (2001) Progression of Alternaria blight of mustard in relation to components of resistance. Indian Phytopathol 54:329–331

Kumar B, Kolte SJ (2006) Development of Alternaria blight in genotypes of Indian mustard [*B. juncea* (L.) Czern & Cross.] under field. Indian Phytopathol 39:314–317

Kumar M, Yadav V, Kumar H, Sharma R, Singh A, Tuteja N, Johri AK (2011) *Piriformospora indica* enhances plant growth by transferring phosphate. Plant Signal Behav 6:723–725

Kumar D, Shekhar S, Bisht S, Kumar V, Varma A (2015) Ectopic overexpression of lectin in transgenic *Brassica juncea* plants exhibit resistance to fungal phytopathogen and showed alleviation to salt and drought stress. J Bioeng Biomed Sci 5:147. https://doi.org/10.4172/2155-9538.1000147

Lakra BS, Saharan GS (1988) Influence of host resistance on colonization and incubation period of *Albugo candida* in mustard. Cruciferae NewsLett 13:108–109

Laperche A, Aigu Y, Jubault M, Ollier M, Guichard S, Glory P, Strelkov S, Gravot A, Manzananares-Dauleux M (2017) Clubroot resistance QTL are modulated by nitrogen input in *Brassica napus*. TAG Theor Appl Genet 130(4):669–684. https://doi.org/10.1007/s00122-016-2842-8

LeBoldus JM, Manolii VP, Turkington TK, Strelkov SE (2012) Adaptation to *Brassica* host genotypes by a single-spore isolate and population of *Plasmodiophora brassicae* (clubroot). Plant Dis 96:833–838. https://doi.org/10.1094/pdis-09-11-0807

Lee HA, Lee HY, Seo E, Lee J, Kim SB, Oh S, Choi E, Choi E, Lee SE, Choi D (2017) Current understandings of plant nonhost resistance. Mol Plant-Microbe Interact 30:5–15. https://doi.org/10.1094/MPMI-10-16-0213-CR

Leflon M (2013) Phoma du colza: la résistance specifique Rlm7 perd de son efficacite. Perspect Agricoles 399:58–61

Lehmann P, Jenner CE, Kozubek E, Greenland AJ, Walsh JA (2003) Coat protein-mediated resistance to Turnip mosaic virus in oilseed rape (*Brassica napus*). Mol Breed 11:83–94

Lenski RE (1988a) Experimental studies of pleiotropy and epistasis in *Escherichia coli*. I. Variation in competitive fitness among mutants resistant to virus T4. Evolution 42:425–432

Lenski RE (1988b) Experimental studies of pleiotropy and epistasis in *Escherichia coli*. II. Maladaptive effects associated with resistance to virus T4. Evolution 42:433–440

Leonard KJ (1977) Selection pressures and plant pathogens. Ann N Y Acad Sci 287:207–222

Leonard KJ, Czochor RJ (1980) Theory of genetic interactions among populations of plants and their pathogens. Annu Rev Phytopathol 18:237–258

Li Y, Mei D, Huang Y, Xu Y, Roger W, Guy K, Richard B (2001) Biochemical changes of Zhongyou 821 after inoculation with mycelia of *Sclerotinia sclerotiorum*. Chin J Oil Crop Sci 23:63–65

Li CX, Li H, Sivasithamparam K, Fu TD, Li YC, Liu SY, Barbetti MJ (2006) Expression of field resistance under Western Australian conditions to *Sclerotinia sclerotiorum* in Chinese and Australian *Brassica napus* and *Brassica juncea* germplasm and its relation with stem diameter. Aust J Agric Res 57:1131–1135

Li J, Zhao Z, Hayward A, Cheng H, Fu D (2015) Integration analysis of quantitative trait loci for resistance to *Sclerotinia sclerotiorum* in *Brassica napus*. Euphytica 205:483–489. https://doi.org/10.1007/s10681-015-1417-0

Li GL, Qian W, Zhang SJ, Zhang SF, Li F, Zhang H, Wu J, Wang ZW, Sun RF (2016) Development of gene-based markers for the Turnip mosaic virus resistance gene retr02 in *Brassica rapa*. Plant Breed 135:466–470

Li S, Castillo-Gonzalez C, Yu B, Zhang X (2017) The functions of plant small RNAs in development and in stress responses. Plant J 90:654–670

Li G, Yue L, Li F, Zhang S, Zhang H, Qian W, Fang Z, Wu J, Wang X, Zhang S, Sun R (2018) Research progress on *Agrobacterium tumefaciens*-based transgenic technology in *Brassica rapa*. Hortic Plant J 4(3):126–132. https://doi.org/10.1016/j.hpj.2018.03.006

Li G, Lv H, Zhang S, Zhang S, Li F, Zhang H, Qian W, Fang Z, Sun R (2019) TuMV management for *Brassica* crops through host resistance: retrospect and prospects. Plant Pathol 68:1035–1044. https://doi.org/10.1111/ppa.13016

Liban SH, Cross DJ, Kutcher HR, Peng G, Fernando WGD (2016) Race structure and frequency of avirulence genes in the Western Canadian *Leptosphaeria maculans* pathogen population, the causal agent of blackleg in *Brassica* species. Plant Pathol 65:1161–1169

Lindhout P (2002) The perspectives of polygenic resistance in breeding for durable disease resistance. Euphytica 124:217–226

Liu Q, Rimmer SR, Scarth R (1989) Histopathology of compatibility and incompatibility between oilseed rape and *Albugo candida*. Plant Pathol 38:176–182

Liu F, Zou Z, Huang S, Parks P, Fernando WGD (2020) Development of a specific marker for detection of a functional *AvrLm9* allele and validating the interaction between *AvrLm7* and *AvrLm9* in *Leptosphaeria maculans*. Mol Biol Rep 47(9):7115–7123. https://doi.org/10.1007/s11033-020-05779-8

Lof ME, van der Werf W (2017) Modeling the effect of gene deployment strategies on durability of plant resistance under selection. Crop Prot 97:10–17

Lorenz AJ, Chao S, Asoro FG, Heffner EL, Hayashi T, Iwata H, Smith KP, Sorrells MK, Jannink JL (2011) Genomic selection in plant breeding: knowledge and prospects. Adv Agron 110:77–123

Loutre C, Wicker T, Travella S, Galli P, Scofield S, Fahima T, Feuillet C, Keller B (2009) Two different CC-NBS-LRR genes are required for Lr10-mediated leaf rust resistance in tetraploid and hexaploid wheat. Plant J 60(6):1043–1054

Lovelock D, Donald C, Conlan X, Cahill D (2013) Salicylic acid suppression of clubroot in broccoli (*Brassica oleracea* var. *italica*) caused by the obligate biotroph *Plasmodiophora brassicae*. Australas Plant Pathol 42:141–153

Ludwig-Muller J (2009) Glucosinolates and the clubroot disease: defense compounds or auxin precursors? Phytochem Rev 8:135–148. https://doi.org/10.1007/s11101-008-9096-2

Ludwig-Muller J, Prinsen E, Rolfe S, Scholes J (2009) Metabolism and plant hormone action during clubroot disease. J Plant Growth Regul 28(3):229–244

Luo Y, Dong D, Su Y, Wang X, Peng Y, Peng J, Zhou C (2018) Transcriptome analysis of *Brassica juncea* var. *tumida* Tsen responses to *Plasmodiophora brassicae* primed by the biocontrol strain *Zhihengliuella aestuarii*. Funct Integr Genomics 18:301–314. https://doi.org/10.1007/s10142-018-0593-0

Mahanti N, Bertrand H, Monteiro-Vitorella C, Fulbright DW (1993) Elevated mitochondrial alternative oxidase activity in ds RNA free, hypovirulent isolates of *Cryphonectria parasitica*. Physiol Mol Plant Pathol 42:455–463

Malfanova N, Franzil L, Lugtenberg B, Chebotar V, Ongena M (2012) Cyclic lipopeptide profile of the plant-beneficial endophytic bacterium *Bacillus subtilis* HC8. Arch Microbiol 194:893–899

Malik RS (1990) Prospects for *Brassica carinata* as an oilseed crop in India. Exp Agric 26:125–129

Manzanares-Dauleux MJ, Delourme R, Baron F, Thomas G (2000) Mapping of one major gene and of QTLs involved in resistance to clubroot in *Brassica napus*. Theor Appl Genet 101:885–891. https://doi.org/10.1007/s001220051557

Marcroft SJ, Elliott VL, Cozijnsen AJ, Salisbury PA, Howlett BJ, Van De Wouw AP (2012a) Identifying resistance genes to *Leptosphaeria maculans* in Australian *Brassica napus* cultivars based on reactions to isolates with known avirulence genotypes. Crop Pasture Sci 63:338–350. https://doi.org/10.1071/CP11341

Marcroft SJ, Van De Wouw AP, Salisbury PA, Potter TD, Howlett BJ (2012b) Effect of rotation of canola (*Brassica napus*) cultivars with different complements of blackleg resistance genes on disease severity. Plant Pathol 61:934–944. https://doi.org/10.1111/j.1365-3059.2011.02580.x

Marmath KK, Giri P, Taj G, Pandey D, Kumar A (2013) Effect of zeatin on the infection process and expression of MAPK-4 during pathogenesis of *Alternaria brassicae* in non-host and host *Brassica* plants. Afr J Biotechnol 12(17):2164–2174. https://doi.org/10.5897/AJB12.2725

Martin GB, Bogdanove AJ, Sessa G (2003) Understanding the functions of plant disease resistance proteins. Annu Rev Plant Biol 54:23–61

Maschinski J, Whitham TG (1989) The continuum of plant responses to herbivory: the influence of plant association, nutrient availability, and timing. Am Nat 134:1–19

Mason AS, Nelson MN, Yan G, Cowling WA (2011) Production of viable male unreduced gametes in *Brassica* interspecific hybrids is genotype specific and stimulated by cold temperatures. BMC Plant Biol 11:103. https://doi.org/10.1186/1471-2229-11-103

Matros A, Kaspar S, Witzel K, Mock HP (2011) Recent progress in liquid chromatography-based separation and label-free quantitative plant proteomics. Phytochemistry 72(10):963–974

Matsumoto E, Ueno H, Aruga D, Sakamoto K, Hayashida N (2012) Accumulation of three clubroot resistance genes through marker-assisted selection in Chinese cabbage (*Brassica rapa* ssp. *pekinensis*). Engei Gakkai Zasshi 81:184–190. https://doi.org/10.2503/jjshs1.81.184

Mauricio R, Stahl EA, Korves T, Tian D, Kreitman M, Bergelson J (2003) Natural selection for polymorphism in the disease resistance gene RPS2 of *Arabidopsis thaliana*. Genetics 163:735–746

McLean MD, Hoover GJ, Bancroft B, Makhmoudova A, Clark SM, Welacky T, Simmonds DH, Shelp BJ (2007) Identification of the full-length $Hs1^{pro\text{-}1}$ coding sequence and preliminary evaluation of soybean cyst nematode resistance in soybean transformed with $Hs1^{pro\text{-}1}$ cDNA. Can J Bot 85:437–441. https://doi.org/10.1139/B07-038

Mehta N, Singh K, Sangwan MS (2008) Assessment of yield losses and evaluation of different varieties/genotypes of mustard against powdery mildew in Haryana. Plant Dis Res 23:55–59

Mei J, Guo Z, Wang J, Feng Y, Ma G, Zhang C, Qian W, Chen G (2019) Understanding the resistance mechanism in *Brassica napus* to clubroot caused by *Plasmodiophora brassicae*. Phytopathology 109:810–818. https://doi.org/10.1094/PHYTO-06-18-0213-R

Melzer MS, Boland GJ (1996) Transmissible hypo-virulence in *Sclerotinia minor*. Can J Plant Pathol 18:19–28

Minami H, Kim JS, Ikezawa N, Takemura T, Katayama T, Kumagai H, Sato F (2008) Microbial production of plant benzylisoquinoline alkaloids. Proc Natl Acad Sci U S A 105:7393–7398

Mishra A, Pandey D, Goel A, Kumar A (2010) Molecular cloning and in-silico analysis of functional homologous of hypersensitive response genes induced during pathogenesis of Alternaria Blight in two genotypes of *Brassica*. J Proteomics Bioinformatics 3:244–248

Mitchell-Olds T, James RV, Palmer MJ, Williams PH (1995) Genetics of *Brassica rapa* (syn. *campestris*). 2. Multiple disease resistance to three fungal pathogens: *Peronospora parasitica*, *Albugo candida* and *Leptosphaeria maculans*. Heredity 75:362–369

Mitehell-Olds T, Bradley RD (1996) Genetics of *Brassica rapa*. 3. Cost of disease resistance to three fungal pathogens. Evolution 50(5):1859–1865. https://doi.org/10.1111/j.1558-5646.tb03572.x

Molitor A, Zajic D, Voll LM, Pons-Kuhnemann J, Samans B, Kogel KH, Waller F (2011) Barley leaf transcriptome and metabolite analysis reveals new aspects of compatibility and *Piriformospora indica*-mediated systemic induced resistance to powdery mildew. Mol Plant-Microbe Interact 24:1427–1439

Monazzah M, Enferadi ST, Soleimani MJ, Rabiei Z (2016) An unspecific phytotoxin oxalic acid and its effect on sunflower proteome. Aust J Bot 64:219–226

Mondal KK, Chatterjee SC, Viswakarma N, Bhatacharya RC, Grover A (2003) Chitinase mediated inhibitory activity of *Brassicas* transgenic on growth of *Alternaria brassicae*. Curr Microbiol 47:171–173

Mondal KK, Bhattacharya RC, Koundal KR, Chatterjee SC (2007) Transgenic Indian mustard (*Brassica juncea*) expressing tomato glucanase leads to arrested growth of *Alternaria brassicae*. Plant Cell Rep 26:247–252. https://doi.org/10.1007/s00299-006-0241-3

Moriguchi K, Kimizuka-Takagi C, Ishii K, Nomura K (1999) A genetic map based on RAPD, RFLP, isozyme, morphological markers and QTL analysis for clubroot resistance in *Brassica oleracea*. Breed Sci 49:257–265. https://doi.org/10.1270/jsbbs.49.257

Munir I, Hussan W, Kazi MSK, Farhatullah MAA, Iqbal A, Munir R (2016) Production of transgenic *B rassica juncea* with the synthetic chitinase gene (nic) conferring resistance to *Alternaria brassicicola*. Pak J Bot 48(5):2063–2070

Mun JH, Yu HJ, Park S, Park B-S (2009) Genome-wide identification of NBS-encoding resistance genes in *Brassica rapa*. Mol Gen Genomics 282:617–631

Mysore KS, Ryu CM (2004) Non-host resistance how much do we know? Trends Plant Sci 9:97–104

Nagaharu U (1935) Genome analysis in *Brassica* with special reference to the experimental formation of *B. napus* and peculiar mode of fertilization. Jpn J Bot 7:389–452

Nagaoka T, Doullah MA, Matsumoto S, Kawasaki S, Ishikawa T, Hori H, Okazaki K (2010) Identification of QTLs that control clubroot resistance in *Brassica oleracea* and comparative analysis of clubroot resistance genes between *B. rapa* and *B. oleracea*. Theor Appl Genet 120:1335–1346. https://doi.org/10.1007/s00122-010-1259-z

Naoumkina MA, Zhao Q, Gallego-Giraldo L, Dai X, Zhao PX, Dixon RA (2010) Genome-wide analysis of phenylpropanoid defense pathways. Mol Plant Pathol 11:829–846

Navazio L, Baldan B, Moscatiello R, Zuppini A, Woo SL, Mariani P, Lorito M (2007) Calcium-mediated perception and defense responses activated in plant cells by metabolite mixtures secreted by the biocontrol fungus *Trichoderma atroviride*. BMC Plant Biol 7:41

Nejat N, Mantri N (2018) Emerging roles of long non-coding RNAs in plant response to biotic and abiotic stresses. Crit Rev Biotechnol 38:93–105

Nellist CF, Qian W, Jenner CE, Moore JD, Zhang S, Wang X, Briggs WH, Barker GC, Sun R, Walsh JA (2014) Multiple copies of eukaryotic translation initiation factors in *Brassica rapa* facilitate redundancy, enabling diversification through variation in splicing and broad-spectrum virus resistance. Plant J 77:261–268

Niazi A, Manzoor S, Asari S, Bejai S, Meijer J, Bongcam-Rudloff E (2014) Genome analysis of *Bacillus amyloliquefaciens* subsp. *plantarum* UCMB5113: a rhizobacterium that improves plant growth and stress management. PLoS One 9:e104651

Niemann J, Wojciechowski A, Janowicz J (2012) Broadening the variability of quality traits in rapeseed through interspecific hybridization with an application of immature embryo culture. J Biotechnol Comput Biol Bionanotechnol 93:109–115

Niks RE, Marcel TC (2009) Nonhost and basal resistance: how to explain specificity? New Phytol 182:817–828. https://doi.org/10.1111/j.1469-8137.2009.02849.x

Niks RE, Rubiales D (2002) Potentially durable resistance mechanisms in plants to specialized fungal pathogens. Euphytica 124:201–216. https://doi.org/10.1023/A:1015634617334

Nirupa N, Prasad MNV, Jami SK, Kirti PB (2007) Optimization of *Agrobacterium*-mediated overexpression of osmotin-ferritin genes in *Brassica juncea*. Transgenic Plant J 1(2):384–392

Niwa R, Nomura Y, Osaki M, Ezawa T (2008) Suppression of clubroot disease under neutral pH caused by inhibition of spore germination of *Plasmodiophora brassicae* in the rhizosphere. Plant Pathol 57:445–452

Nováková M, Sasek V, Trda L, Krutinova H, Mongin T, Valentova O, Balesdent MH, Rouxel T, Burketova L (2016) *Leptosphaeria maculans* effector AvrLm4-7 affects salicylic acid (SA) and ethylene (ET) signalling and hydrogen peroxide (H2O2) accumulation in *Brassica napus*. Mol Plant Pathol 17:818–831. https://doi.org/10.1111/mpp.12332

Nowara D, Gay A, Lacomme C, Shaw J, Ridout C, Douchkov D, Hensel G, Kumlehn J, Schweizer P (2010) HIGS: host-induced gene silencing in the obligate biotrophic fungal pathogen *Blumeria graminis*. Plant Cell 22:3130–3141

Nuss DL, Koltin Y (1990) Significance of dsRNA genetic elements in plant pathogenic fungi. Annu Rev Phytopathol 28:37–58

O'Connor NK, Hudson AS, Cobb SL, O'Neil D, Robertson J, Duncan V, Murphy CD (2014) Noval flourinated lipopeptides from *Bacillus* sp. CS93 via precursor-directed biosynthesis. Amino Acids 46:2745–2752

Oelmuller R, Sherameti I, Tripathi S, Varma A (2009) *Piriformospora indica*, a cultivable root endophyte with multiple biotechnological applications. Symbiosis 49:1–17

Oh SJ, Song SI, Kim YS et al (2005) Arabidopsis CBF3/DREB1A and ABF3in transgenic rice increased tolerance to abiotic stress without stunting growth. Plant Physiol 138:341–351

Ongena M, Jacques P (2008) *Bacillus lipopeptides*: versatile weapons for plant disease biocontrol. Trends Microbiol 16:115–125

Ongena M, Jourdan E, Adam A, Paquot M, Brans A, Joris B, Arpigny JL, Thonart P (2007) Surfactin and fengycin lipopeptides of *Bacillus subtilis* as elicitors of induced systemic resistance in plants. Environ Microbiol 9:1084–1090

Osaki K, Tanaka S, Ito S (2004) A case of breakdown in clubroot-resistant cultivars of Chinese cabbage. Jpn J Phytopathol 70:66

Oxley S (2007) Clubroot disease of oilseed rape and other *Brassica* crops. Technical Note 602, Scottish Agricultural College, Edinburgh, UK

Pasold S, Ludwig-Muller J (2013) Reduction of clubroot (Plasmodiophora brassicae) formation in *Arabidopsis thaliana* after treatment with prohexadionecalcium, an inhibitor of oxoglutaric acid-dependent dioxygenases. Plant Pathol 62:1357–1365. https://doi.org/10.1111/ppa.12049

Pasold S, Siegel I, Seidel C, Ludwig-Muller J (2010) Flavonoid accumulation in *Arabidopsis thaliana* root galls caused by the obligate biotrophic pathogen *Plasmodiophora brassicae*. Mol Plant Pathol 11:545–562. https://doi.org/10.1111/j.1364-3703.2010.00628.x

Pathak KV, Keharia H, Gupta K, Thakur SS, Balaram P (2012) Lipopeptides from the banyan endophyte, *Bacillus subtilis* K1: mass spectrometric characterization of a library of fengycins. J Am Soc Mass Spectrom 23:1716–1728

Patterson CL (1986) The comparative biology, epidemiology and control of lettuce drop caused by *Sclerotinia minor* and *S. sclerotiorum* and the genetic analysis of vegetative and sexual compatibility in *S. minor*. Ph.D. thesis, University of California, Davis, 76 p

Pedras MSC, Chumala PB, Jin W, Islam MS, Hauck DW (2009) The phytopathogenic fungus *Alternaria brassicicola*: phytotoxin production and phytoalexin elicitation. Phytochemistry 70:394–402

Peer WA, Brown DE, Tague BW, Muday GK, Taiz L, Murphy AS (2001) Flavonoid accumulation patterns of transparent testa mutants of *Arabidopsis*. Plant Physiol 126:536–548

Peng G, Pageau D, Strelkov SE, Lahlali R, Hwang SF, Hynes RK (2013) Assessment of crop rotation, cultivar resistance and *Bacillus subtilis* biofungicide for control of clubroot on canola. Acta Hortic 1005:591–598

Peng G, Falk KC, Gugel RK, Franke C, Yu F, James B, Strelkov SE, Hwang SF, McGregor L (2014) Sources of resistance to *Plasmodiophora brassicae* (clubroot) pathotypes virulent on canola. Can J Plant Pathol 36:89–99. https://doi.org/10.1080/07060661.2013.863805

Peng G, Pageau D, Strelkov SE, Gossen BD, Hwang SF, Lahlali R (2015) A >2-year crop rotation reduces resting spores of *Plasmodiophora brassicae* in soil and the impact of clubroot on canola. Eur J Agron 70:78–84

Perumal S, Waminal NE, Lee J, Lee J, Choi BS, Kim HH, Grandbastien MA, Yang TJ (2017) Elucidating the major hidden genomic components of the A, C and AC genomes and their influence on *Brassica* evolution. Sci Rep 7:17986

Piao ZY, Deng YQ, Choi SR, Park YJ, Lim YP (2004) SCAR and CAPS mapping of *CRb*, a gene conferring resistance to *Plasmodiophora brassicae* in Chinese cabbage (*Brassica rapa* ssp. *pekinensis).* Theor Appl Genet 108:1458–1465. https://doi.org/10.1007/s00122-003-1577-5

Piao Z, Ramchiary N, Lim YP (2009) Genetics of clubroot resistance in *Brassica* species. J Plant Growth Regul 28:252–264. https://doi.org/10.1007/s00344-009-9093-8

Pieterse CM, Van Loon L (2004) NPR1: the spider in the web of induced resistance signaling pathways. Curr Opin Plant Biol 7:456–464

Pieterse CM, Zamioudis C, Berendsen RL, Weller DM, Van Wees SC, Bakker PA (2014) Induced systemic resistance by beneficial microbes. Annu Rev Phytopathol 52:347–375

Pimentel D (1995) Amounts of pesticides reaching target hosts: environmental impacts and ethics. J Agric Environ Ethics 8:17–29

Plett JM, Daguerre Y, Wittulsky S, Vayssieres A, Deveau A, Melton SJ, Kohler A, Morrel-Falvey JL, Brun A, Veneault-Fourrey C, Martin F (2014) Effector MiSSP7 of the mutualistic fungus *Laccaria bicolor* stabilizes the populus JAZ6 protein and represses jasmonic acid (JA) responsive genes. Proc Natl Acad Sci U S A 111:8299–8304. https://doi.org/10.1073/pnas.1322671111

Potts DJ, Rakow GW, Males DR (1999) Canola-quality *Brassica juncea*, a new oilseed crop for the Canadian prairies. In: Proc 10th international rapeseed congress, Canberra, Australia. CD Rom

Prakash S (1973) Non-homologous meiotic pairing in the A and B genomes of *Brassica*: its breeding significance in the production of variable amphidiploids. Genet Res 21:133–137

Prakash S, Chopra VL (1988) Introgression of resistance to shattering in *Brassica napus* from *Brassica juncea* through non-homologous recombination. Plant Breed 101:167–168

Puhl I, Stadler F, Treutter D (2008) Alterations of flavonoid biosynthesis in young grapevine (*Vitis vinifera* L.) leaves, flowers, and berries induced by the dioxygenase inhibitor Prohexadione-Ca. J Agric Food Chem 56:2498–2504

Qian W, Zhang SJ, Zhang SF, Li F, Zhang H, Wu J, Wang X, Walsh JA, Sun R (2013) Mapping and candidate-gene screening of the novel Turnip mosaic virus resistance gene retr02 in Chinese cabbage (*Brassica rapa* L.). Theor Appl Genet 126:179–188

Qin C, Li B, Fan Y, Zhang X, Yu Z, Ryabov E, Zhao M, Wang H, Shi N, Zhang P, Jackson S, Tor M, Cheng Q, Liu Y, Gallusci P, Hong Y (2017) Roles of dicer-like proteins 2 and 4 in intra- and intercellular antiviral silencing. Plant Physiol 174:1067–1081

Rabilloud T, Chevallet M, Luche S, Lelong C (2010) Two-dimensional gel electrophoresis in proteomics: past, present and future. J Proteome 73:2064–2077

Rademacher W (2000) Growth retardants: effects on gibberellin biosynthesis and other metabolic pathways. Annu Rev Plant Physiol Plant Mol Biol 51:501–531

Rademacher W, Spinelli F, Costa G (2006) Prohexadione-Ca: modes of action of a multifunctional plant bioregulator for fruit trees. Acta Hortic 727:97–106

Rahman H, Peng G, Yu F, Falk KC, Kulkarni M, Selvaraj G (2014) Genetics and breeding for clubroot resistance in Canadian spring canola (*Brassica napus* L.). Can J Plant Pathol 36:122–134. https://doi.org/10.1080/07060661.2013.862571

Rahman A, Uddin W, Wenner NG (2015) Induced systemic resistance responses in perennial ryegrass against *Magnaporthe oryzae* elicited by semi-purified surfactin lipopeptides and live cells of *Bacillus amyloliquefaciens*. Mol Plant Pathol 16:546–558

Rahman H, Xu YP, Zhang XR, Cai XZ (2016) *Brassica napus* genome possesses extraordinary high number of CAMTA Genes and CAMTA3 contributes to PAMP triggered immunity and resistance to *Sclerotinia sclerotiorum*. Front Plant Sci 7:581. https://doi.org/10.3389/fpls.2016.00581

Rahman H, Bennett RA, Kebede B (2018) Molecular mapping of QTL alleles of *Brassica oleracea* affecting days to flowering and photosensitivity in spring *Brassica napus*. PLoS One 13: e0189723

Rennie DC, Manolii VP, Plishka M, Strelkov SE (2013) Histological analysis of spindle and spheroid root galls caused by *Plasmodiophora brassicae*. Eur J Plant Pathol 135:771–781

Rennie DC, Holtz MD, Turkington TK, LeBoldus JM, Hwang SF, Howard RJ, Strelkov SE (2015) Movement of *Plasmodiophora brassicae* resting spores in windblown dust. Can J Plant Pathol 37:188–196

Ricarova V, Kazda J, Singh K, Rysanek P (2016) Clubroot caused by *Plasmodiophora brassicae* Worn: a review of emerging serious disease of oilseed rape in the Czech Republic. Plant Prot Sci 52:71–86. https://doi.org/10.17221/87/2015-PPS

Rimmer SR, van den Berg CGJ (1992) Resistance of oilseed *Brassica* spp. to blackleg caused by *Leptosphaeria maculans*. Can J Plant Pathol 14:56–66. https://doi.org/10.1080/07060669209500906

Robinson J (1999) Ethics and transgenic crops: a review. Electron J Biotechnol 2:71–81

Rocherieux J, Glory P, Giboulot A, Boury S, Barbeyron G, Thomas G, Manzanares-Dauleux MJ (2004) Isolate-specific and broad-spectrum QTLs are involved in the control of clubroot in *Brassica oleracea*. Theor Appl Genet 108:1555–1563. https://doi.org/10.1007/s00122-003-1580-x

Romeis T, Ludwig AA, Martin R, Jones JDG (2001) Calcium-dependent protein kinases play an essential role in a plant defense response. EMBO J 20:5556–5567

Roohie RK, Umesha S (2015) Identification of genes associated with black rot resistance in cabbage through suppression subtractive hybridization. 3 Biotech 5:1089–1100

Rouxel T, Willner E, Coudard L, Balesdent MH (2003) Screening and identification of resistance to *Leptosphaeria maculans* (stem canker) in *Brassica napus* accessions. Euphytica 133:219–231. https://doi.org/10.1023/A:1025597622490

Roy BA, Kirchner JW (2000) Evolutionary dynamics of pathogen resistance and tolerance. Evolution 54:51–63

Rudd S (2003) Expressed sequence tags: alternative or complement to whole genome sequences? Trends Plant Sci 8:321–329

Ruiz-Lozano JM, Roussel H, Gianinazzi S, Gianinazzi-Pearson V (1999) Defense genes are differentially induced by a mycorrhizal fungus and *Rhizobium* sp. in wild-type and symbiosis-defective pea genotypes. Mol Plant-Microbe Interact 12:976–984

Rusholme RE, Higgins E, Walsh JA, Lydiate DJ (2007) Genetic control of broad-spectrum resistance to Turnip mosaic virus in *Brassica rapa* (Chinese cabbage). J Gen Virol 88:3177–3186

Rustagi A, Kumar D, Shekhar S, Yusuf MA, Misra S, Sarin NB (2014) Transgenic *Brassica juncea* plants expressing *MsrA1*, a synthetic cationic antimicrobial peptide, exhibit resistance to fungal phytopathogens. Mol Biotechnol 56:535–545. https://doi.org/10.1007/s12033-013-9727-8

Ryder TB, Hedrick SA, Bell JN, Liang X, Clouse SD, Lamb CJ (1987) Organization and differential activation of a gene family encoding the plant defense enzyme chalcone synthase in *Phaseolus vulgaris*. Mol Gen Genomics 210:219–233

Saharan GS (1992) Disease resistance. In: Labana KS, Banga SS, Banga SK (eds) Breeding oilseed *Brassicas*, vol 12. Narosa Pub House, New Delhi, pp 181–205

Saharan GS (2010) Analysis of genetic diversity in *Albugo*-crucifer system. J Mycol Plant Pathol 40:1–13

Saharan GS, Kadian AK (1983a) Analysis of components of horizontal resistance in rapeseed and mustard cultivars against *Alternaria brassicae*. Indian Phytopathol 36:503–507

Saharan GS, Kadian AK (1983b) Physiologic specialization in *Alternaria brassicae*. Cruciferae Newslett 8:32–33

Saharan GS, Krishnia SK (2001) Multiple disease resistance in rapeseed and mustard. In: Nagarajan S, Singh DP (eds) Role of resistance in intensive agriculture. Kalyani Pub, New Delhi, pp 98–108

Saharan GS, Naresh M (2008) Sclerotinia diseases of crop plants: biology, ecology and disease management. Springer Science + Busines Media B.V., Dordrecht, 485 p

Saharan GS, Naresh M, Sangwan MS (2003) Nature and mechanism of resistance to Alternaria blight in rapeseed-mustard system. Annu Rev Plant Pathol 2:85–128

Saharan GS, Verma PR, Meena PD, Kumar A (2014) White rust of crucifers: biology, ecology and disease management. Springer, Frankfurt, p 244. ISBN 978-81-322-1791-6

Saharan GS, Naresh M, Meena PD (2016) Disease forewarning and management of blight of crucifers-a review. J Mycol Plant Pathol 46(3):193–223

Saharan GS, Mehta N, Meena PD (2019) Powdery mildew disease of crucifers: biology, ecology and disease management. Springer, Singapore, p 362. ISBN 978-981-13-9852-0

Sahrawat AK, Becker D, Lütticke S, Lörz H (2003) Genetic improvement of wheat via alien gene transfer, an assessment. Plant Sci 165:1147–1168

Saito M, Kubo N, Matsumoto S, Suwabe K, Tsukada M, Hirai M (2006) Fine mapping of the clubroot resistance gene, Crr3, in *Brassica rapa*. Theor Appl Genet 114:81–91. https://doi.org/10.1007/s00122-006-0412-1

Sakamoto K, Saito A, Hayashida N, Taguchi G, Matsumoto E (2008) Mapping of isolate-specific QTLs for clubroot resistance in Chinese cabbage (*Brassica rapa* L. ssp. *pekinensis*). Theor Appl Genet 117:759–767. https://doi.org/10.1007/s00122-008-0817-0

Sanghera GS, Wani SH, Gill MS, Gosal SS (2009) Recent approaches for optimization of transgene expression in plants. In: Malik CP, Wadhwani C, Kaur B (eds) Crop breeding and biotechnology. Pointer Publishers, Jaipur, pp 1–27

Sanghera GS, Wani SH, Singh G, Kashyap PL, Singh NB (2011) Designing crop plants for biotic stresses using transgenic approach. Vegetos 24(1):1–25

Saravanakumar K, Yu C, Dou K, Wang M, Li Y, Chen J (2016) Synergistic effect of Trichoderma-derived antifungal metabolites and cell wall degrading enzymes on enhanced biocontrol of *Fusarium oxysporum* f. sp. *cucumerinum*. Biol Control 94:37–46

Schranz ME, Lysak MA, Mitchell-Olds T (2006) The ABC's of comparative genomics in the Brassicaceae: building blocks of crucifer genomes. Trends Plant Sci 11:535–542. https://doi.org/10.1016/j.tplants.2006.09.002

Schwelm A, Fogelqvist J, Knaust A, Jülke S, Lilja T, Bonilla-Rosso G, Karlsson M, Shevchenko A, Dhandapani V, Choi SR, Kim HG, Park JY, Lim YP, Ludwig-Muller J, Dixelius C (2015) The *Plasmodiophora brassicae* genome reveals insights in its life cycle and ancestry of chitin synthases. Sci Rep 5:11153

Seaman WL, Walker JC, Larson R (1963) A new race of *Plasmodiophora brassicae* affecting Badger Shipper cabbage. Phytopathology 94:33–43

Sellam A, Dongo A, Guillemette T, Hudhomme P, Simoneau P (2007) Transcriptional responses to exposure to the brassicaceous defense metabolites camalexin and allyl-isothiocyanate in the necrotrophic fungus *Alternaria brassicicola*. Mol Plant Pathol 8:195–208. https://doi.org/10.1111/j.1364-3703.2007.00387.x

Shah N, Sun J, Yu S, Yang Z, Wang Z, Huang F, Dun B, Gong J, Liu Y, Li Y, Li Q, Yuan L, Baloch A, Li G, Li S, Zhang C (2019) Genetic variation analysis of field isolates of clubroot and their responses to *Brassica napus* lines containing resistant gene CRb, PbBa8.1 and their combination in homozygous and heterozygous state. Mol Breed 39:153. https://doi.org/10.1007/s11032-019-1075-3

Shah N, Li Q, Xu Q, Liu J, Huang F, Zhan Z, Qin P, Zhou X, Yu W, Zhu L, Zhang C (2020) *CRb* and *PbBa8.1* synergically increases resistant genes expression upon infection of *Plasmodiophora brassicae* in *Brassica napus*. Genes (Basel) 11(2):202. https://doi.org/10.3390/genes11020202

Sharma P, Kumar A, Chauhan JS (2012) Evaluation of Indian and exotic *Brassica* germplasm for tolerance to stem rot caused by *Sclerotinia sclerotiorum*. J Mycol Plant Pathol 42:297–302

Sharma K, Gossen BD, Greenshields D, Selvaraj G, Strelkov SE, McDonald MR (2013) Reaction of lines of the rapid cycling *Brassica* collection and *Arabidopsis thaliana* to four pathotypes of *Plasmodiophora brassicae*. Plant Dis 97:720–727

Sharma S, Padmaja KL, Gupta V, Paritosh K, Pradhan AK, Pental D (2014) Two plastid DNA lineages—Rapa/Oleracea and Nigra—within the tribe Brassiceae can be best explained by reciprocal crosses at hexaploidy: evidence from divergence times of the plastid genomes and R-block Genes of the A and B genomes of *Brassica juncea*. PLoS One 9:e93260

Shekhawat K, Rathore SS, Premi OP, Kandpal BK, Chauhan JS (2012) Advances in agronomic management of Indian mustard (*Brassica juncea* (L.) Czernj. Cosson): an overview. Int J Agron 2012:1–14

Sherameti I, Shahollari B, Venus Y, Altschmied L, Varma A, Oelmuller R (2005) The endophytic fungus *Piriformospora indica* stimulates the expression of nitrate reductase and the starch-degrading enzyme glucan-water dikinase in tobacco and *Arabidopsis* roots through a homeodomain transcription factor that binds to a conserved motif in their promoters. J Biol Chem 280:26241–26247

Shopan J, Mou H, Zhang L, Zhang C, Ma W, Walsh JA, Hu Z, Yang J, Zhang M (2017) Eukaryotic translation initiation factor 2B-beta (eIF2Bβ), a new class of plant virus resistance gene. Plant J 90:929–940

Siemens J, Nagel W, Ludwig-Muller J, Sarcristán MD (2002) The interaction of *Plasmodiophora brassicae* and *Arabidopsis thaliana*: parameters for disease quantification and screening mutant lines. J Phytopathol 150:592–605

Siemens J, Keller I, Sarx J, Kunz S, Schuller A, Nagel W, Schmulling T, Parniske M, Ludwig-Muller J (2006) Transcriptome analysis of *Arabidopsis* clubroots indicate a key role for cytokinins in disease development. Mol Plant-Microbe Interact 19:480–494. https://doi.org/10.1094/MPMI-19-0480

Siemens J, Gonzalez MC, Wolf S, Hofmann C, Greiner S, Du Y, Rausch T, Roitsch T, Ludwig-Muller J (2011) Extracellular invertase is involved in the regulation of clubroot disease in *Arabidopsis thaliana*. Mol Plant Pathol 12:247–262. https://doi.org/10.1111/j.1364-3703.2010.00667.x

Simms EL (1992) Costs of plant resistance to herbivory. In: Fritz RS, Simms EL (eds) Plant resistance to herbivores and pathogens: ecology, evolution, and genetics. University of Chicago Press, Chicago, pp 392–425

Simms EL, Rausher MD (1987) Costs and benefits of plant resistance to herbivory. Am Nat 130:570–581

Simons MD (1979) Influence of genes for resistance to *Puccinia coronata* from *Arena sterilis* on yield and rust reaction of cultivated oats. Phytopathology 69:450–452

Singh DN, Singh NK, Srivastava S (1999) Biochemical and morphological characters in relation to Alternaria blight resistance in rape-seed mustard. Ann Agric Res 20:472–477

Singh KB, Foley RC, Onate-Sanchez L (2002) Transcription factors in plant defense and stress responses. Curr Opin Plant Biol 5:430–436

Singh R, Singh D, Salisbury P, Barbetti MJ (2010) Field evaluation of indigenous and exotic *Brassica juncea* genotypes against blight, white rust, downy mildew and powdery mildew diseases in India. Indian J Agric Sci 80:155–159

Singh S, Dey SS, Bhatia R, Batley J, Kumar R (2018) Molecular breeding for resistance to black rot [*Xanthomonas campestris* pv. *campestris* (Pammel) Dowson] in *Brassicas*: recent advances. Euphytica 214:196

Snyder BA, Nicholson RL (1990) Synthesis of phytoalexins in sorghum as a site-specific response to fungal ingress. Science 248:1637–1639

Soltis PS, Soltis DE (2000) The role of genetic and genomic attributes in the success of polyploids. Proc Natl Acad Sci U S A 97:7051–7057

Sonah H, Deshmukh RK, Singh VP, Gupta DK, Singh NK, Sharma TR (2011) Genomic resources in horticultural crops: status, utility and challenges. Biotechnol Adv 29:199–209

Song Y, Thomma BPHJ (2018) Host-induced gene silencing compromises Verticillium wilt in tomato and *Arabidopsis*. Mol Plant Pathol 19:77–89

Spinelli F, Speakman JB, Rademacher W, Halbwirth H, Stich K, Costa G (2005) Luteoforol, a flavan-4-ol, is induced in pome fruits by Prohexadione-Calcium and shows phytoalexin-like properties against *Erwinia amylovora* and other plant pathogens. Eur J Plant Pathol 112:133–142

Stahl EA, Dwyer G, Mauricio R, Kreitman M, Bergelson J (1999) Dynamics of disease resistance polymorphism at the Rpm1 locus of *Arabidopsis*. Nature 400:667–671

Stein T (2005) *Bacillus subtilis* antibiotics: structure, syntheses and specific functions. Mol Microbiol 56:845–857

Stein E, Molitor A, Kogel KH, Waller F (2008) Systemic resistance in *Arabidopsis* conferred by the mycorrhizal fungus *Piriformospora indica* requires jasmonic acid signaling and the cytoplasmic function of NPR1. Plant Cell Physiol 49:1747–1751. https://doi.org/10.1093/pcp/pcn147

Strelkov SE, Tewari JP, Smith-Degenhardt E (2006) Characterization of *Plasmodiophora brassicae* populations from Alberta, Canada. Can J Plant Pathol 28:467–474. https://doi.org/10.1080/07060660609507321

Strelkov SE, Hwang SF, Howard RJ, Hartman M, Turkington TK (2011) Progress towards the sustainable management of clubroot (*Plasmodiophora brassicae*) of canola on the Canadian Prairies. Prairie Soils Crops J 4:114–121

Strelkov SE, Hwang SF, Manolii VP, Cao T, Feindel D (2016) Emergence of new virulence phenotypes of *Plasmodiophora brassicae* on canola (*Brassica napus*) in Alberta, Canada. Eur J Plant Pathol 145:517–529. https://doi.org/10.1007/s10658-016-0888-8

Strelkov SE, Hwang SF, Manolii VP, Cao T, Fredua-Agyeman R, Harding MW, Peng G, Gossen BD, McDonald MR, Feindel D (2018) Virulence and pathotype classification of *Plasmodiophora brassicae* populations collected from clubroot resistant canola (*Brassica napus*) in Canada. Can J Plant Pathol 40:284–298

Stuttmann J, Hubberten HM, Rietz S, Kaur J, Muskett P, Guerois R, Bednarek P, Hoefgen R, Parker JE (2011) Perturbation of *Arabidopsis* amino acid metabolism causes incompatibility with the adapted biotrophic pathogen *Hyaloperonospora arabidopsidis*. Plant Cell Online 23:2788–2803

Sun L, Zhaoxin L, Bie X, Fengxia L, Yang S (2006) Isolation and characterization of a co-producer of fengycins and surfactins, endophytic *Bacillus amyloliquefacines* ES-2, from *Scutellaria baicalensis Gerorgi*. World J Microbiol Biotechnol 22:1259–1266

Suwabe K, Tsukazaki H, Iketani H, Hatakeyama K, Fujimura M, Nunome T, Fukuoka H, Matsumoto S, Hirai M (2003) Identification of two loci for resistance to clubroot (*Plasmodiophora brassicae* Woronin) in *Brassica rapa* L. Theor Appl Genet 107:997–1002

Suwabe K, Tsukazaki H, Iketani H, Hatakeyama K, Kondo M, Fujimura M, Nunome T, Fukuoka H, Hirai M, Matsumoto S (2006) Simple sequence repeat-based comparative genomics between *Brassica rapa* and *Arabidopsis thaliana*: the genetic origin of clubroot resistance. Genetics 173:309–319

Suwabe K, Suzuki G, Nunome T, Hatakeyama K, Mukai Y, Fukuoka H, Matsumoto S (2012) Microstructure of a *Brassica rapa* genome segment homoeologous to the resistance gene cluster on *Arabidopsis* chromosome 4. Breed Sci 62:170–177

Taj G, Kumar A, Bansal KC, Garg GK (2004) Introgression of *osmotin* gene for creation of resistance against *Alternaria* blight by perturbation of cell cycle machinery. Indian J Biotechnol 3:291–298

Tamura K, Peterson D, Peterson N, Stecher G, Nei M, Kumar S (2011) MEGA5: molecular evolutionary genetics analysis using maximum likelihood, evolutionary distance, and maximum parsimony methods. Mol Biol Evol 28(10):2731–2739. https://doi.org/10.1093/molbev/msr121

Tanaka S, Mido H, Ito SI (2006) Colonization by two isolates of *Plasmodiophora brassicae* with differing pathogenicity on a clubroot-resistant cultivar of Chinese cabbage (*Brassica rapa* L. subsp. *pekinensis*). J Gen Plant Pathol 72:205–209. https://doi.org/10.1007/s10327-006-0276-x

Tanimoto E (2012) Tall or short? Slender or thick? A plant strategy for regulating elongation growth of roots by low concentrations of gibberellins. Ann Bot 110:373–381

Tasleem M, Baunthiyal M, Taj G (2017) Induction of MPK3, MPK6 and MPK4 mediated defense signaling in response to Alternaria blight in transgenic *Brassica juncea*. Biosci Biotech Res Asia 14(4):1469–1474. https://doi.org/10.13005/bbra/2593

Thakur AK, Parmar N, Singh KH, Nanjundan J (2020) Current achievements and future prospects of genetic engineering in Indian mustard (*Brassica juncea* L. Czern & Coss). Plants 252:56. https://doi.org/10.1007/s00425-020-03461-8

Thomma BPHJ (2003) *Alternaria* spp.: from general saprophyte to specific parasite. Mol Plant Pathol 4(4):225–236. https://doi.org/10.1046/j.1364-3703.2003.00173.x

Thomma BP, Eggermont K, Penninckx IA, Mauch-Mani B, Vogelsang R, Cammue BP, Broekaert WF (1998) Separate jasmonate-dependent and salicylate-dependent defense-response pathways in *Arabidopsis* are essential for resistance to distinct microbial pathogens. Proc Natl Acad Sci U S A 95(25):15107–15111. https://doi.org/10.1073/pnas.95.25.15107

Tian D, Araki H, Stahl E, Bergelson J, Kreitman M (2002) Signature of balancing selection in Arabidopsis. Proc Natl Acad Sci U S A 99:11525–11530. https://doi.org/10.1073/pnas.172203599

Tian D, Traw MB, Chen JQ, Kreitman M, Bergelson J (2003) Fitness costs of R-gene-mediated resistance in *Arabidopsis thaliana*. Nature 423:74–77. https://doi.org/10.1038/nature01588

Tiffin P (2000) Mechanisms of tolerance to herbivore damage: what do we know? Evol Ecol 14:523–536

Tomita H, Shimizu M, Doullah MA, Fujimoto R, Okazaki K (2013) Accumulation of quantitative trait loci conferring broad-spectrum clubroot resistance in *Brassica oleracea*. Mol Breed 32 (4):889–900. https://doi.org/10.1007/s11032-013-9918-9

Tonguc M, Griffiths PD (2004) Evaluation of *Brassica carinata* accessions for resistance to black rot (*Xanthomonas campestris* pv. *campestris*). Hortic Sci 39:952–954

Treutter D (2005) Significance of flavonoids in plant resistance and enhancement of their biosynthesis. Plant Biol 7:581–591

Ueno H, Matsumoto E, Aruga D, Kitagawa S, Matsumura H, Hayashida N (2012) Molecular characterization of the CRa gene conferring clubroot resistance in *Brassica rapa*. Plant Mol Biol 80:621–629. https://doi.org/10.1007/s11103-012-9971-5

Uloth MB, Ming PY, Finnegan PM, Banga SS, Banga SK, Sandhu PS, Phillip HY, Salisbury A, Barbetti MJ (2013) New sources of resistance to *Sclerotinia sclerotiorum* for crucifer crops. Field Crop Res 154:40–52. https://doi.org/10.1016/j.fcr.2013.07.013

Uloth MB, Ming PY, Barbetti MJ (2016) Cultivar resistance offers the first opportunity for effective management of the emerging powdery mildew (*Erysiphe cruciferarum*) threat to oilseed Brassicas in Australia. Crop Pasture Sci 67:1179–1187

Unnikumar K, Sree KS, Varma A (2013) *Piriformospora indica*: a versatile root endophytic symbiont. Symbiosis 60:107–113

Vahabi K, Sherameti I, Bakshi M, Mrozinska A, Ludwig A, Oelmuller R (2015) Microarray analyses during early and later stages of the *Arabidopsis/Piriformospora indica* interaction. Genomics Data 6:16–18

van de Wouw AP, Howlett BJ (2012) Estimating frequencies of virulent isolates in field populations of a plant pathogenic fungus, *Leptosphaeria maculans*, using high-throughput pyrosequencing. J Appl Microbiol 113:1145–1153. https://doi.org/10.1111/jam.2012.113.issue-5

Van de Wouw AP, Cozijnsen AJ, Hane JK, Brunner PC, McDonald BA, Oliver RP, Howlett BJ (2010a) Evolution of linked avirulence effectors in *Leptosphaeria maculans* is affected by genomic environment and exposure to resistance genes in host plants. PLoS Pathog 6: e1001180. https://doi.org/10.1071/CP16411

Van de Wouw AP, Stonard JF, Howlett BJ, West JS, Fitt BDL, Atkins SD (2010b) Determining frequencies of avirulent alleles in airborne *Leptosphaeria maculans* inoculum using quantitative PCR. Plant Pathol 59:809–818. https://doi.org/10.1111/j.1365-3059.2010.02311.x

Van de Wouw AP, Lowe RGT, Elliott CE, Dubois DJ, Howlett BJ (2014) An avirulence gene, *AvrLmJ1*, from the blackleg fungus, *Leptosphaeria maculans*, confers avirulence to *Brassica juncea* cultivars. Mol Plant Pathol 15:523–530

Van de Wouw AP, Elliott VL, Chang S, López-Ruiz FJ, Marcroft SJ, Idnurm A (2017) Identification of isolates of the plant pathogen *Leptosphaeria maculans* with resistance to the triazole fungicide fluquinconazole using a novel in planta assay. PLoS One 12:e0188106. https://doi.org/10.1371/journal.pone.0188106

van Loon LC, Bakker PA, Pieterse CM (1998) Systemic resistance induced by rhizosphere bacteria. Annu Rev Phytopathol 36:453–483

van Loon LC, Rep M, Pieterse CMJ (2006) Significance of inducible defense-related proteins in infected plants. Annu Rev Phytopathol 448(1):135–162

Vanittanakom N, Loeffler W, Koch U, Jung G (1986) Fengycin—a novel antifungal lipopeptide antibiotic produced by *Bacillus subtilis* F-29–3. J Antibiot (Tokyo) 39:888–901

Varma A, Singh A, Sahay NS, Sharma J, Roy A, Kumari M, Rana D, Thakran S, Deka D, Bharti K, Hurek T, Blechert O, Rexer KH, Kost G, Hahn A, Maier W, Walter M, Strack D, Kranner I (2001) *Piriformospora indica*: an axenically culturable mycorrhiza-like endosymbiotic fungus. In: Fungal associations. Springer, Berlin, Heidelberg, pp 125–150

Verhagen BW, Trotel-Aziz P, Couderchet M, Hofte M, Aziz A (2010) *Pseudomonas* spp.-induced systemic resistance to *Botrytis cinerea* is associated with induction and priming of defense responses in grapevine. J Exp Bot 61:249–260

Verma PR, Saharan GS (1994) Monograph on *Alternaria* diseases of crucifers. Saskatoon Research Centre Technical Bulletin 1994–6E, Agriculture and Agri-Food Canada, Saskatoon, SK, Canada

Verma S, Yajima WR, Rahman MH, Shah S, Liu JJ, Ekramoddoullah AKM, Kav NNV (2012) A cysteine-rich antimicrobial peptide from *Pinus monticola* (PmAMP1) confers resistance to multiple fungal pathogens in canola (*Brassica napus*). Plant Mol Biol 79(1–2):61–74

Villeth GR Jr, Reis FB, Tonietto A, Huergo L, de Souza EM, Pedrosa FO, Franco OL, Mehta A (2009) Comparative proteome analysis of *Xanthomonas campestris* pv. *campestris* in the interaction with the susceptible and the resistant cultivars of *Brassica oleracea*. FEMS Microbiol Lett 298:260–266

Vlamakis H, Chai Y, Beauregard P, Losick R, Kolter R (2013) Sticking together: building a biofilm the *Bacillus subtilis* way. Nat Rev Microbiol 11:157–168

Voorrips RE (1995) *Plasmodiophora brassicae*: aspects of pathogenesis and resistance in *Brassica oleracea*. Euphytica 83:139–146

Voorrips RE, Jongerius MC, Kanne HJ (1997) Mapping of two genes for resistance to clubroot (*Plasmodiophora brassicae*) in a population of doubled haploid lines of *Brassica oleracea* by means of RFLP and AFLP markers. Theor Appl Genet 94:75–82. https://doi.org/10.1007/s001220050384

Wallenhammar AC, Almquist C, Soderstrom M, Jonsson A (2012) In-field distribution of *Plasmodiophora brassicae* measured using quantitative real-time PCR. Plant Pathol 61:16–28

Waller F, Achatz B, Baltruschat H, Fodor J, Becker K, Fischer M, Heier T, Huckelhoven R, Neumann C, von Wettstein D, Franken P, Kogel KH (2005) The endophytic fungus *Piriformospora indica* reprograms barley to salt-stress tolerance, disease resistance, and higher yield. Proc Natl Acad Sci U S A 102:13386–13391. https://doi.org/10.1073/pnas.0504423102

Waller F, Mukherjee K, Deshmukh SD, Achatz B, Sharma M, Schafer P, Kogel KH (2008) Systemic and local modulation of plant responses by *Piriformospora indica* and related Sebacinales species. J Plant Physiol 165:60–70

Walsh JA, Sharpe AG, Jenner EC, Lydiate DJ (1999) Characterization of resistance of Turnip mosaic virus in oilseed rape (*Brassica napus*) and genetic mapping of TuRB01. Theor Appl Genet 99:1149–1154

Wang Y, Fristensky B (2001) Transgenic canola lines expressing pea defense gene DRR206 have resistance to aggressive blackleg isolates and to *Rhizoctonia solani*. Mol Breed 8:263–271

Wang M, Jin H (2017) Spray-induced gene silencing: a powerful innovative strategy for crop protection. Trends Microbiol 25:4–6

Wang Y, Nowak G, Culley D, Hadwiger LA, Fristensky B (1999) Constitutive expression of pea defense gene DRR206 confers resistance to blackleg (*Leptosphaeria maculans*) disease in transgenic canola (*Brassica napus*). Mol Plant-Microbe Interact 12:410–418

Wang X, Wang H, Wang J, Sun R, Wu J, Liu S, Bai Y, Mun JH, Bancroft I, Cheng F, Huang S, Li X, Hua W, Wang J, Wang X, Freeling M, Pires JC, Paterson AH, Chalhoub B, Wang B, Hayward A, Sharpe AG, Park BS, Weisshaar B, Liu B, Li B, Liu B, Tong C, Song C, Duran C, Peng C, Geng C, Koh C, Lin C, Edwards D, Mu D, Shen D, Soumpourou E, Li F, Fraser F, Conant G, Lassalle G, King GJ, Bonnema G, Tang H, Wang H, Belcram H, Zhou H, Hirakawa H, Abe H, Guo H, Wang H, Jin H, Parkin IA, Batley J, Kim JS, Just J, Li J, Xu J, Deng J, Kim JA, Li J, Yu J, Meng J, Wang J, Min J, Poulain J, Wang J, Hatakeyama K, Wu K, Wang L, Fang L, Trick M, Links MG, Zhao M, Jin M, Ramchiary N, Drou N, Berkman PJ, Cai Q, Huang Q, Li R, Tabata S, Cheng S, Zhang S, Zhang S, Huang S, Sato S, Sun S, Kwon SJ, Choi SR, Lee TH, Fan W, Zhao X, Tan X, Xu X, Wang Y, Qiu Y, Yin Y, Li Y, Du Y, Liao Y, Lim Y, Narusaka Y, Wang Y, Wang Z, Li Z, Wang Z, Xiong Z, Zhang Z (2011) *Brassica rapa* genome sequencing project consortium. The genome of the mesopolyploid crop species *Brassica rapa*. Nat Genet 43:1035–1039. https://doi.org/10.1038/ng.919

Wang Z, Bao L-L, Zhao F-Y, Tang M-Q, Chen T, Li Y, Wang B-X, Fu B, Fang H, Li G-Y, Cao J, Ding L-N, Zhu K-M, Liu S-Y, Tan X-L (2019) BnaMPK3 Is a key regulator of defense responses to the devastating plant pathogen *Sclerotinia sclerotiorum* in oilseed rape. Front Plant Sci 10:91. https://doi.org/10.3389/fpls.2019.00091

Wani SH (2010) Inducing fungus resistance in plants through Biotechnology. Notula Sci Biol 2 (2):14–21

Wani SH, Gosal SS (2010) Genetic engineering for osmotic stress tolerance in plants-role of proline. IUP J Genet Evol 3:14–25

Wani SH, Sanghera GS (2010) Genetic engineering for viral disease management in plants. Notulae Sci Biol 2(1):20–28

Wani SH, Sandhu JS, Gosal SS (2008) Genetic engineering of crop plants for abiotic stress tolerance. In: Malik CP, Kaur B, Wadhwani C (eds) Advanced topics in plant biotechnology and plant biology. MD Publications, New Delhi, pp 149–183

Webster MA, Dixon GR (1991) Calcium, pH and inoculum concentration as factors limiting root hair colonization by *Plasmodiophora brassicae* Worn. Mycol Res 95:64–73

Weerakoon SR, Si P, Zili W, Meng J, Yan G (2009) Production and confirmation of hybrids through interspecific crossing between tetraploid *B. juncea* and diploid *B. oleracea* towards a hexaploid *Brassica* population. In: 16th Australian research assembly on Brassicas, Ballarat, Victoria

Weinig C, Ungerer MC, Dorn LA, Kane NC, Toyonaga Y, Halldorsdottir SS, Mackay TFC, Purugganan DD, Schmitt J (2002) Novel loci control variation in reproductive timing in *Arabidopsis thaliana* in natural environments. Genetics 162:1875–1884

Weinig C, Stinchcombe JR, Schmitt J (2003) QTL architecture of resistance and tolerance traits in *Arabidopsis thaliana* in natural environments. Mol Ecol 12:1153–1163

Werner S, Diederichsen E, Frauen M, Schondelmaier J, Jung C (2008) Genetic mapping of clubroot resistance genes in oilseed rape. Theor Appl Genet 116:363–372. https://doi.org/10.1007/s00122-007-0674-2

Winkel-Shirley B (2001) Flavonoid biosynthesis: a colorful model for genetics, biochemistry, cell biology and biotechnology. Plant Physiol 126:485–493

Winter M, Koopmann B (2016) Race spectra of *Leptosphaeria maculans*, the causal agent of blackleg disease of oilseed rape, in different geographic regions in northern Germany. Eur J Plant Pathol 145:629–641

Witzel K, Neugart S, Ruppel S, Schreiner M, Wiesner M, Baldermann S (2015) Recent progress in the use of 'omics technologies in Brassicaceous vegetables. Front Plant Sci 6:244. https://doi.org/10.3389/fpls.2015.00244

Wolfe MS, McDermott JM (1994) Population genetics of plant pathogen interactions: the example of the *Erysiphe graminis-Hordeum vulgare* pathosystem. Annu Rev Phytopathol 32:89–113. https://doi.org/10.1146/annurev.py.32.090194.000513

Woods-Tor A, Studholme DJ, Cevik V, Telli O, Holub EB, Tor M (2018) A suppressor/avirulence gene combination in *Hyaloperonospora arabidopsidis* determines race specificity in *Arabidopsis thaliana*. Front Plant Sci 9:1957

Xue S, Cao T, Howard RJ, Hwang SF, Strelkov SE (2008) Isolation and variation in virulence of single-spore isolates of *Plasmodiophora brassicae* from Canada. Plant Dis 92:456–462

Yang Y, Zhang H, Li G, Li W, Wang X, Song F (2009) Ectopic expression of MgSM1, a Ceratoplatan in family protein from *Magnaporthe grisea*, confers broad-spectrum disease resistance in *Arabidopsis*. Plant Biotechnol J 7(8):763–777. https://doi.org/10.1111/j.1467-7652.2009.00442.x

Yang K, Nath UK, Biswas MK, Kayum MA et al (2018) Whole-genome sequencing of *Brassica oleracea* var. *capitata* reveals new diversity of the mitogenome. PLoS One 13:e01194356

Yu XX, Zhao YT, Cheng J, Wang W (2015) Biocontrol effect of *Trichoderma harzianum* T4 on Brassica clubroot and analysis of rhizosphere microbial communities based on T-RFLP. Biocontrol Sci Tech 25:1493–1505

Zarinpanjeh N, Motallebi M, Zamani MR, Ziaei M (2016) Enhanced resistance to *Sclerotinia sclerotiorum* in *Brassica napus* by co-expression of defensin and chimeric chitinase genes. J Appl Genet 57:417–425. https://doi.org/10.1007/s13353-016-0340-y

Zhang XC, Gassmann W (2003) *RPS4*-mediated disease resistance requires the combined presence of *RPS4* transcripts with full-length and truncated open reading frames. Plant Cell 15:2333–2342

Zhang FL, Wang M, Liu XC, Zhao XY, Yang JP (2008) Quantitative trait loci analysis for resistance against Turnip mosaic virus based on a doubled-haploid population in Chinese cabbage. Plant Breed 127:82–86

Zhang XW, Yuan YX, Wang XW (2009) QTL mapping for TuMV resistance in Chinese cabbage [*Brassica campestris* L. ssp. *pekinensis* (Lour.) Olssom]. Acta Hortic Sin 36:731–736

Zhang H, Feng J, Hwang SF, Strelkov SE, Falak I, Huang X, Sun R (2016) Mapping of clubroot (*Plasmodiophora brassicae*) resistance in canola (*Brassica napus*). Plant Pathol 65:435–440. https://doi.org/10.1111/ppa.12422

Zhang X, Lai T, Zhang P, Zhang X, Yuan C, Jin Z, Li H, Yu Z, Qin C, Tor M, Ma P, Cheng Q, Hong Y (2019) Mini review: revisiting mobile RNA silencing in plants. Plant Sci 278:113–117

Zhao J, Buchwaldt L, Rimmer R, Sharpe A, McGregor L, Bekkaoui D, Hegedus D (2009) Patterns of differential gene expression in *Brassica napus* cultivars infected with *Sclerotinia sclerotiorum*. Mol Plant Pathol 10:635–649. https://doi.org/10.1111/j.1364-3703.2009.00558.x

Zhao J, Wu YX, Ho HH, Chen ZJ, Li XY, He YQ (2016) PBT1, a novel antimicrobial protein from the biocontrol agent *Bacillus subtilis* XF-1 against *Plasmodiophora brassicae*. Eur J Plant Pathol 145:583–590. https://doi.org/10.1007/s10658-016-0905-y

Zhong X, Zhou Q, Cui N, Cai D, Tang G (2019) *BvcZR3* and *BvHs1pro-1* genes pyramiding enhanced beet cyst nematode (*Heterodera schachtii* Schm.) resistance in oilseed rape (*Brassica napus* L.). Int J Mol Sci 20(7):1740. https://doi.org/10.3390/ijms20071740

Zhou JM, Tang XY, Martin GB (1997) The Pto kinase conferring resistance to tomato bacterial speck disease interacts with proteins that bind a cis-element of pathogenesis-related genes. EMBO J 16:3207–3218. https://doi.org/10.1093/emboj/16.11.3207

Zuo W, Chao Q, Zhang N, Ye J, Tan G, Li B, Xing Y, Zhang B, Liu H, Fengler KA, Zhao J, Zhao X, Chen Y, Lai J, Yan J, Xu M (2015) A maize wall-associated kinase confers quantitative resistance to head smut. Nat Genet 47:151–157. https://doi.org/10.1038/ng.3170

Genomics of Host Resistance at a Glance

8

Abstract

Cruciferous crops are challenged by numerous biotic and abiotic stresses under field conditions. Some of the biotic stresses are of major consequences causing heavy losses to the economically very important *Brassica* crops at global level. The management practices employed to control these stresses are delivered through agronomic practices, chemical application, use of biological agents, and integration of all with growing of resistant cultivars. The use of resistant cultivars is most economical, ecofriendly practices to get higher yield with better quality of *Brassica* produce under the threat of various biotic stresses of fungal, bacterial, viral, and nematodes origin. The breeding of resistant cultivars of *Brassica* crops have received significant attention of scientists all over the world after the advances made in genomics of host resistance using multi-omics technology and analysis of transcriptomics profiles through next-generation sequencing approaches to reveal molecular mechanisms of host–parasitic interaction and expression of host resistance. The resistance conferred by R-genes products is of five major kinds, i.e., NTL (TIR-NBS-LRR), CNL (CC-NBS-LRR), RLK (Receptor-like kinase), RLP (Receptor-like proteins), and Pto (a Ser/Thr kinase proteins). Most of the R-genes in plants are members of NBS-LRR proteins. Identification of R-genes sources is the most essential and crucial step to breed Brassica crops with strong, durable, and effective resistance against all pathogens and their pathotypes. Several R-genes effective against major pathogens of *Brassica* with quantitative and qualitative resistance have been identified, cloned, and molecularly mapped on different chromosomes of *Brassica* species. These R-genes have been functionally characterized to use them for breeding resistant cultivars of *Brassica* through different breeding approaches including molecular technology. R-genes governing resistance to multiple pathogens and seedling as well as adult plant stages has been characterized. Pathotype/isolate/strain specific R-genes effective against fungal,

G. S. Saharan et al., *Genomics of Crucifer's Host-Resistance*,
https://doi.org/10.1007/978-981-16-0862-9_8

bacterial, and viral pathogens of *Brassica* have been identified and molecularly characterized to use through appropriate introgression approaches. The genetics of resistance in the form of inheritance of R-genes in *Brassica* against different pathogens have been determined. Inheritance of resistance in crosses of *B. juncea* × *B. carinata* to *Alternaria brassicae* is governed by adaptive genes, dominant genes, additive × additive type epistatic genes, additive × dominance, and dominance × dominance type of non-allelic interaction genes. Partial resistance has been identified against *Alternaria* and *Sclerotinia* pathogens. The resistance to *Albugo* in *Brassica* crops is governed by one or two dominant genes with epistatic effects, additive dominant with epistatic effects and partial resistant single recessive genes *wpr*. In interspecific crosses of *B. juncea* × *B. carinata*, single resistance gene with complete dominance confers resistance to *Erysiphe*. In *Brassica*, resistance to *Fusarium* is conferred by a dominant gene. Inheritance to *Hyaloperonospora* is governed by one to four dominant complementary genes, additive genes, one to two partial dominant genes, partial dominant epistatic and gene dosage effects along with quantitative genes depending on host species and pathotype interaction. In *Brassica*, most resistant cultivars contain major genes/QTLs conferring resistance to *Leptosphaeria* pathotypes. Resistant to *Plasmodiophora* in *Brassica* is governed by two to four dominant genes, additive and recessive genes and QTLs. *Plasmodiophora brassicae* pathotype 5 × identified two independent QTLs *Rer* 8 and *Lc* 9 in the A-genome on chromosomes AO2 and AO8 whereas a single dominant gene, *Rcr* 7 has been identified to provide resistance to pathotypes 3h and 5x in the C-genome on chromosome Co7. In *Brassica*, partial dominant, polygenic, additive epistatic interactions confer resistance to *Sclerotinia*. In leaf and stem, resistance is conferred by QTLs dominant and additive × additive epistatic interactions of genetic effects. In *B. rapa*, resistance to *TuMV* is controlled by dominant and recessive genes. In *B. oleracea*, resistance to *TuMV* is controlled by both dominant genes and multiple loci. In *B. oleracea*, *B. carinata*, and *B. napus*, resistance to *Xanthomonas* is controlled by single dominant genes. Substantial progress has been achieved in transfer of R-genes to develop disease-resistant cultivars of *Brassica* crops against major pathogens using conventional and molecular techniques. To develop durable *Alternaria*-resistant *Brassica* cultivars, inter- or intraspecific introgression of R-genes from non-host, microarray-based gene identification, and transfer by gene silencing, and transfer of an endochitinase gene from *Trichoderma* approaches have been used. R-genes sources of multiple disease resistance have been used to breed resistant cultivars of *Brassica* to *Albugo, Alternaria, Erysiphe*, *Hyaloperonospora*, *Leptosphaeria*, and *Xanthomonas*. Clubroot-resistant cultivars of *B. rapa, B. oleracea, B. napus, Raphanus sativus*, and *Arabidopsis thaliana* have been developed. *Sclerotinia*-resistant *Brassica* cultivar has been obtained by introducing QTLs from wild relatives using marker-assisted introgression approach. Resistance to *Xanthomonas* has been introgressed from *B. carinata* to *B. oleracea* through embryo rescue. Crucifers' interactions with multiple stresses trigger a wide range of defense reactions at molecular and cellular levels to ensure that host cells are insulated from these challenges they encounter. The production of ROS, MAP kinase cascades, and hormonal signaling are triggered by multiple stresses and play crucial roles as common components of the immune system. GbRLK, a receptor-like kinase (RLK) gene, TFs' MYBTFs, ATAF1, HSPs gene, miRNA, SnRK, and calcium are involved in defense signaling pathways to multiple stresses.

Management of crucifers' disease resistance for its effective utilization under field conditions is very important and challenging. Effective utilization of R-loci and their allelic variations enhances the durability of resistance cultivars against different pathogens and their pathotypes. The deployment of R-genes cultivars based on quantitative and qualitative pyramiding alongwith integrated disease management is the key lever to increase the durability of R-genes against rapidly evolving pathogen populations with new virulences. In addition to major gene-resistant cultivars, use of horizontal resistance, non-host resistance, multilayered and multicomponent resistance, and use of bacterial and fungal biocontrol agents; use of Proca, nitrogen, and arginase gene to modulate resistance positively, use of genes like WRKY and CAMA should be exploited to fortify the resistance of cvs. under field conditions. Standardized techniques on germplasm screening, identification of R-genes, introgression of R-genes, and methodology for breeding disease-resistant cultivars against different pathogen included here will be helpful for students, teachers, and researchers to plan their research projects on crucifers' host–pathogen interactions.

Keywords

Functional characterization of R-genes · Use of R-genes · Exploitation of R-genes · Genetics of disease resistance · Regulation of genomics and biometabolomics to confer molecular mechanisms of host resistance

8.1 Introduction

Crucifers are significant components of cultivated crop biodiversity of the world. Out of total cultivated area in the world, crucifers occupy 2.69% area with 1384.8 mha total cultivated area producing 147.69 mmt vegetables and oil yielding food and fodder crop for feeding human and animal population at global level. Large number of biotic and abiotic stresses are known to deteriorate quantity and quality of crops in the largest group of crucifers encompassing Brassicaceae family. The biostresses have been managed with several approaches including disease-resistant cvs. as a major component to save resources and environment. The genomics studies accompanied with molecular omics approaches have strengthened our understanding on molecular mechanisms of host–parasite interactions and functional characterization of valuable genes to breed better cvs. of *Brassica* crops with high yield, quality and biotic and abiotic stress management attributes. During this century, multi-omics approaches like genomics, proteomics, transcriptomics, biometabolomics, pangenomics, and bioinformatics have been extensively used to comprehend the genomics of host resistance through identification, isolation, cloning, and molecular mapping of R-genes with functional characterization to confer high level of resistance to *Brassica* pathogens. It has facilitated breeding of disease resistance *Brassica* crop with strong and durable resistance against major pathogens with high level of precision and efficacy.

8.2 Functional Characterization of Genes for *Brassica* Crops Improvement

Cruciferous crop plants are always challenged by various biotic and abiotic stresses. In agriculture cropping system, plant protection is being delivered using different approaches such as chemical control, various agronomic practices, biological control, integrated pest management (IPM), and cultivation of resistance cultivars. Among these approaches, resistance cultivars are the most economical, environmentally sustainable solution to control different diseases in *Brassica* species. Previous studies suggested that the inheritance of disease resistance is either qualitative or quantitative in *Brassica* species. Recently mapped disease resistance genetic loci and closely linked molecular marker to these loci can be used for marker-assisted selection in cruciferous resistance breeding of *Brassica* species. Extensive use of recently available resistance sources can be combined with molecular tools and new technologies such as gene/QTL mapping, fine mapping, gene cloning, comparative genomics, and analysis of transcriptomic profiles through next-generation sequencing approaches to enhance our understanding of resistance mechanism. Novel information can help controlling diseases in an effective way, so yield losses would be reduced and the quality of Brassica crop product would be improved. The principles involved in the effective utilization of genomics technology largely depend on the following approaches including molecular genetics.

In natural environ, cruciferous plants are challenged by 167 biotic (44 pathogen, 87 pests, and 36 weeds) and 19 abiotic stresses (environmental, soil salinity and alkality, nutrients deficiency, and hormonal changes) (Tables 5.1–5.4, Chap. 5), and some of them have successfully invaded crop plants to cause diseases which resulted in deterioration of crop quality and yield. The major pathogens of crucifers can be ranked at global level on the basis of host range, geographical distribution, yield losses, and emphasis given by the *Brassica* scientists on genomic investigations for improving the yield and quality of *Brassica* crops. On the basis of average rank at global level eight pathogen (1-*Plasmodiophora brassicae*; 2-*Alternaria brassicae*; 3-*Sclerotinia sclerotiorum*; 4-*Leptosphaeria maculans*; 5-*Erysiphe cruciferarum*; 6-*Albugo candida*; 7-*Turnip mosaic virus*; 8-*Xanthomonas campestris* pv. *campestris*) can be considered as of major consequences with 18.25–68.25 rank (Table 1.1, Chap. 1). In order to cope with disease attacks, the plants have developed multiple layers of defense mechanisms. Plant disease resistance (R) genes which specifically interact/recognize with corresponding pathogen avirulence (*avr*) genes are considered as plant genetic factors of a major layer. The interactions of this gene-for-gene (or genes-for-genes) manner activate the signal transduction cascades that turn on complex defense responses against pathogen attack, and this is called incompatible interaction. The interaction between a host species and a pathogenic species is dynamic where a host variety often lost the R-gene-dependent resistance due to its pathogen race evolution for a virulent gene, and thus a new R-gene was selected against this new race. R-genes provide innate immunity whereas outcomes of defense responses lacking R-genes are partial resistance. Therefore, identification of R-genes is crucial for resistant variety development and relevant mechanism investigation.

More than one hundred R-genes, which are reported in PRGdb (http://prgdb.crg.eu/wiki), were functionally identified and comprise a super family in plants. Sequence composition analysis of R-genes indicates that they share high similarity and contain seven different conserved domains like NBS (nucleotide-binding site), LRR (leucine-rich repeat), TIR (Toll/Interleukin-1 receptor), CC (coiled coil), LZ (leucine zipper), TM (transmembrane), and STK (serine-threonine kinase). Based on domain organization, R-gene products can be categorized into five major types: TNL (TIR-NBS-LRR), CNL (CC-NBS-LRR), RLK (Receptor-like kinases), RLP (Receptor-like proteins), and Pto (a Ser/Thr kinase protein). Most of the R-genes in plant kingdom are members of NBS-LRR (nucleotide-binding site-leucine-rich repeat) proteins. "NBS" and "LRR" domains play different roles in plant–microbe interaction, where the former have the ability to bind and hydrolyze ATP or GTP and the latter is involved in protein–protein interactions. NBS-LRR proteins in plants share sequence similarity with the mammalian NOD-LRR containing proteins which play a role in inflammatory and immune responses. On the basis of presence or absence of N-terminal domains (TOLL/interleukin-1 receptor (TIR) and the coiled coil (CC) motif), NBS-LRR class can be further divided into two major types, TNL (TIR-NBS-LRR) and CNL (CC-NBS-LRR). TNL type share homology with the Drosophila toll and human interleukin-1 receptor (TIR). The two types show divergence in their sequence and signaling pathways. Several partial NBS-LRR variants like TIR, TIR-NBS (TN), CC, CC-NBS (CN), and NBS (N) have also been identified in plant species (Wan et al. 2012; Inohara et al. 2005; Meyers et al. 2002, 2003).

The whole-genome sequence data enabled the genome-wide identification, mapping and characterization of candidate NBS-containing R-genes in economically important plants. For example, the approximate arrays of 159 NBS-encoding R-genes in *A. thaliana*, have been identified. Earlier genome-wide studies have demonstrated that TNL subfamily is abundant in dicots while absent in cereals (monocots). The presence of the full length of TNL and CNL types in the common ancestor (mosses) of both angiosperms and gymnosperms and exceptional presence of truncated domains of TN or TX type proteins in cereals indicate that the TNL class might have been lost in monocot plants. On the chromosomes, the NBS-LRR R-genes are arranged in clusters. The genes in the clusters could be homogenous (often tandem duplicated from single ancestor gene) or heterogenous (with different protein domains). However, the variation of the number and sequences of the

R-genes presented in the *Brassica* lineage since split from the *Arabidopsis* lineage and their distributions in chromosomes are unknown.

The genera *Brassica* and *Arabidopsis*, both belong to the mustard family Brassicaceae (Cruciferae), are a model plant and a model crop, respectively. The two genera shared a latest and obviously detectable alpha genome duplication event before their divergence ~20 million years ago (MYA) and subsequently *Brassica* ancestor underwent a whole-genome triplication event (common to the tribe Brassicaceae) ~16 MYA. In *Brassica*, interspecific cytogenetic relationship between important crops (oilseed and vegetables) is well described by a "U" triangle where each two diploid species [*B. rapa* (AA, $2n = 20$), *B. oleracea* (CC, $2n = 18$), and *B. nigra* (BB, $2n = 16$)] formed a tetraploidy species [*B. napus* (AACC, $2n = 38$), *B. juncea* (AABB, $2n = 36$), or *B. carinata* (BBCC, $2n = 34$)]. This well-established phylogenetic relationship provides a chance to trace evolution of the R-genes between wild plants and their relative crops. Yu et al. (2014) identified R-genes on genome-wide scale in *B. oleracea* and *B. rapa* and provided insights into their evolutionary history and disease resistance in crucifers.

Plant disease resistance (R) genes with the nucleotide-binding site (NBS) play an important role in offering resistance to pathogens. The availability of complete genome sequences of *Brassica oleracea* and *Brassica rapa* provides an important opportunity for researchers to identify and characterize NBS-encoding R-genes in *Brassica* species and to compare with analogs in *Arabidopsis thaliana* based on a comparative genomics approach. However, little is known about the evolutionary fate of NBS-encoding genes in the *Brassica* lineage after split from *A. thaliana*. Yu et al. (2014) presented genome-wide analysis of NBS-encoding genes in *B. oleracea, B. rapa,* and *A. thaliana*. Through the employment of HMM search and manual curation, they have identified 157, 206, and 167 NBS-encoding genes in *B. oleracea, B. rapa,* and *A. thaliana* genomes, respectively. Phylogenetic analysis among three species classified NBS-encoding genes into six subgroups. Tandem duplication and whole-genome triplication (WGT) analyses revealed that after WGT of the *Brassica* ancestor, NBS-encoding homologous gene pairs on triplicated regions in *Brassica* ancestor were deleted or lost quickly, but NBS-encoding genes in *Brassica* species experienced species-specific gene amplification by tandem duplication after divergence of *B. rapa* and *B. oleracea*. Expression profiling of NBS-encoding orthologous gene pairs indicated the differential expression pattern of retained orthologous gene copies in *B. oleracea* and *B. rapa*. Furthermore, evolutionary analysis of CNL type NBS-encoding orthologous gene pairs among three species suggested that orthologous genes in *B. rapa* species have undergone stronger negative selection than those in *B. oleracea* species. But for TNL type, there are no significant differences in the orthologous gene pairs between the two species. The identification and characterization of NBS-encoding genes in *B. rapa,* and *B. oleracea* based on whole-genome sequences will be landmark in genomics research. Through tandem duplication and whole-genome triplication analysis in *B. oleracea, B. rapa,* and *A. thaliana* genomes, the study provides insight into the evolutionary history of NBS-encoding genes after divergence of *A. thaliana* and the *Brassica* lineage. These results together with expression pattern analysis of NBS-encoding orthologous genes provide useful resource for functional characterization of these genes and genetic improvement of relevant crops (Yu et al. 2014).

Molecular genetics investigations of disease resistance have been at the vanguard in *Arabidopsis* research of utilizing natural genetic variation. Numerous disease resistance genes (the so-called R-genes) have been revealed that confer genotype or race-specific resistance to a wide assortment of pathogens. A majority of these genes are of the same structural class, now widely known by the abbreviated name of NB-LRR (**n**ucleotide-**b**inding site and **l**eucine-**r**ich **r**epeat domains). Genes of this class have also been described in a wide diversity of crops such as maize, lettuce, sugar beet, tomato and tobacco. An obvious spin-off, therefore would be to use DNA sequence from NB-LRR genes and other known R-gene classes to develop more effective genetic markers for traditional breeding of resistance in crops. The prospect of using *Arabidopsis* R-genes themselves to confer resistance in crops has also been demonstrated. Most *Arabidopsis* research has been advanced using genetic variation that was derived artificially from mutagenesis induced chemically, by irradiation, or by transposon or T-DNA insertion. For investigation of disease resistance, numerous mutants classified broadly as ones that enhance susceptibility to avirulent pathogens or else enhance resistance to virulent pathogens, have been molecularly characterized. A complex impression of defense signaling has emerged from these mutational analyses, and consequently redefines the frontline for *Arabidopsis* pathology as an effort to identify the intermediate components of defense regulation and dynamic relationships among distinct responses.

The underlying defense responses appear to be widely conserved among plant species. In some cases, genes have been conserved across Kingdoms, such as *SGT1* which has recently been identified as a major regulator of powdery mildew resistance in barley and downy mildew resistance in *Arabidopsis*, having first been described as a critical gene for cell cycling and proteolysis in yeast. Disease resistance can potentially be improved in crops by altering the expression of defense regulation, as suggested by broad spectrum resistance that was achieved by overexpression in *Arabidopsis* of *NPR1,* a regulator of acquired disease resistance.

Wild plant species represent a gene pool that can potentially provide a rich source of durable, broad spectrum disease resistance for use in crops. Species that are inter-fertile with a particular crop have often been used for transferring monogenic resistance that is often genotype or race specific. Unfortunately, such resistance derived from either crop or wild germplasm often lacks durability when deployed in field production because of selection it imposes that force the pathogen to adapt and overcome the resistance. Non-durable resistance is potentially still useful; however, plant breeders are usually challenged to continue the search for new sources of resistance. Broad spectrum resistance from wild germplasm, which may be polygenic and inherently more difficult to transfer and reconstruct in a crop, will be increasingly more accessible as a result of recent advances in our knowledge of candidate resistance genes. The underlying basis for this resistance is that the pathogen as a taxonomic group (species or subspecies) is highly specialized to cause disease in the crop, but not in the wild crop relative. Broad spectrum resistance that occurs universally in the wild relative is referred to below as *species level ("non-host") resistance* to the pathogen.

Some of the advances demonstrate clearly that signaling mechanisms governing genotype-specific resistance can also govern species level resistance. For example, mutations in one of the defense regulators discovered in *Arabidopsis*, designated *EDS1* (**e**nhanced **d**isease **s**usceptibility), exhibit broad spectrum susceptibility to several pathogens including isolates of *Albugo candida* (white rust) and *Hyaloperonospora parasitica* (downy mildew) from *Brassica oleracea*. Isolates of both pathogens collected from either *Brassica* or *Arabidopsis* represent distinct subspecies, based on their restrictive host specialization and molecular divergence. Therefore, *EDS1* is an essential regulator of species level resistance to at least two *Brassica* pathogens. The highly conserved protein, *SGT1* has also recently been reported as a key regulator for species level resistance.

Two bacterial R-genes, *RPS4* and *RPS5,* have been described in *Arabidopsis* which demonstrate that NB-LRR genes can provide necessary components of species level resistance. Both of these genes were characterized using corresponding avirulence determinants that were derived from legume isolates of *Pseudomonas syringae*. Arabidopsis exhibits species level resistance, at least in symptomology, to these pathogens. The avirulence determinants, however, were introduced into an *Arabidopsis*-virulent isolate of *Ps. syringae* to facilitate molecular characterization of the resistance genes. Transient suppression of species level resistance has also been observed in *Arabidopsis*. For example, *Bremia lactuca* (lettuce downy mildew) can reproduce asexually in cotyledons or true leaves following pre-infection with *A. candida* (Holub et al. 2002).

Crucifers include genus *Brassica* in the Brassicaceae family with 325 genera with 3740 species with 75 accepted species. Most of the species in the genus are cultivated for their edible roots, leaves, stems, buds, flower, and seeds for human consumption as raw, cooked, edible oil, and as fodder for animals. Genetic diversity within *Brassica* species has been broadly studied with a special focus on six crop species widely cultivated for vegetables and oilseed production. Out of 44 pathogens known to infect crucifers, 16 pathogens have received attention for genomics investigations. The principles involved in the effective utilization of genomics technology largely depends on the several omics approaches for identification, inheritance, transfer, and management of R-genes for developing durable resistant cultivars of crucifers. Wild plant species represent a gene pool that can potentially provide a rich source of durable, broad spectrum disease resistance for use in crops. Hybridization between *Brassica* species and wild relatives' coenospecies and cytodemes can be possible under controlled conditions with molecular technology. Interspecific hybridization for developing disease resistance in *Brassica* crops has potential since they share genomics in common. Genes containing a nucleotide-binding site (NBS) constitute one of the largest R-genes families which are co-located with resistance QTLs and facilitate multiple disease resistance breeding in *Brassica* crops. Elevated resistance against biotrophs is often related with increased susceptibility to necrotrophs and vice versa. Plants have evolved intracellular receptors with nucleotide-binding leucine-rich repeat domain (NLRs) to detect pathogens effectors or their activity to trigger immunity. Understanding the inter- and intraspecific relationship among cultivated *Brassica* species will be useful for better parental selection to widen the genetic base for multiple disease resistance breeding programs of *Brassica* crops. Despite the genome complexicity of *Brassica* species, homoeologous rearrangements/recombination occurs more frequently in resynthesized species contributing towards genetic diversity to provide good sources of R-genes for breeding resistant cultivars. The complexity of R-genes resistance levels vary in *Brassica* species for major pathogens due to evolution of pathogenic variability. The genomic era is characterized by high efficiency integrated breeding (HIB) in which multiple methods are combined including traditional ways such as microspore culture, backcrossing, and distant introgression with modern ways such as MAS, gene editing, and genome design. During HIB genome back ground analysis is helpful in eliminating undesirable linkage drags and rapidly identifying desirable individuals. Several numbers of genes have well documented that they are differentially expressed against different pathogens. Several omics technologies and approaches have provided tool to look at the differences in DNA, RNA, proteins, and other cellular molecules between species and among individuals of species to get better insight into the genomics of crucifers' disease resistance breeding programs. Using the various omics and multi-omics tools, supplemented with further fine tuning of the bioinformatics methods will speed up the screening of favorable alleles in *Brassica* germplasm promoting resistance against major pathogens along with identification and cloning of favorable genes with increased precision.

8.3 Exploitation of R-Genes Sources

Identification of R-genes sources is the first most essential and crucial step towards breeding of disease resistance cultivars through conventional and molecular approaches. Crucifer's pathogens are of diverse nature from biotrophs and hemibiotrophs to necrotrophs with lot of pathogenic variability in the form of races, pathotypes, and strains. For introgression of durable resistance both qualitative and quantitative, R-genes sources are required to plan a breeding of cultivars with desired traits of yield and quality in *Brassica* crops. Considerable efforts have been made to take the R-genes from cultivated diploid and amphidiploid *Brassica* species in addition to *Arabidopsis thaliana*, *Brassica* wild relatives, and land races. With the use of omics approaches, several R-genes effective against major pathogens of crucifers have been identified, cloned, and mapped on chromosomes of *Brassica* with functional characterization using the various omics or multi-omics tools, supplemented with further fine tuning of the bioinformatics method, will not only speed up the screening of favorable alleles in *Brassica* germplasm promoting resistance against the major pathogens, but also expedite the identification and cloning of favorable genes with increased precision. Efforts have been made to incorporate and introgress desired R-genes in *Brassica* crops through different breeding approaches including molecular technology. Now, after genome sequencing of both host and pathogen, it has become easy to use the R-genes pyramiding in a cultivar with desirable gene editing and gene silencing approach.

Identification of effective and durable sources of R-genes with amicable introgression in the crucifers' genotypes using conventional as well as molecular/biotechnological approaches is essential to breed disease-resistant cultivars of *Brassica* to ameliorate production and productivity at global level. R-genes identification is the first and foremost step towards the disease-resistant hybrid/varieties development through appropriate breeding programs. R-genes conferring qualitative and quantitative (QTLs) loci have been identified from *Brassica* species, cruciferous wild relatives, land races, and *Brassica* coenospecies against major pathogens of crucifers. Several sources in *Brassica* species with R-genes and QTLs for inter- and intraspecific hybridization disease resistance cultivars development with multiple pathogens protection like *Albugo*, *Alternaria*, *Erysiphe,* and *Hyaloperonospora* have been identified R-genes governing seedling and adult plant resistance have been determined. Some R-genes show differential interaction on *Brassica* genotypes at different locations because of environmental and pathogenic variations. Multi-R-genes genotypes have been identified from *Brassica* species against *Albugo candida*. R-genes in *Brassica* species to *Colletotrichum higginsianum* have been characterized using molecular and biochemical approaches. Several genotypes of *Brassica* species has slow mildewing attributes to provide protection against *Erysiphe cruciferarum*. In *B. oleracea*, a single R-genes and polygenic resistance to Fusarium wilt has been characterized. Several R-genes sources in *Arabidopsis thaliana* have been characterized as RPP genes against *H. arabidopsidis*.

In *Brassica* species, several *Leptosphaeria* qualitative race-specific R-genes and quantitative (QTLs) have been identified. However, it is uncertain if some of these genes are same genes with different nomenclature or allelic variants of the same gene since researchers have used different crosses, different isolates, and different markers system. R-genes effective against clubroot (*Plasmodiophora brassicae*) have been identified from cruciferous species as pathotypes specific as well as QTLs. GBS identified QTLs provide resistance to multiple pathotypes. Multiple wild *Brassica* species have shown high levels of resistance against stem rot pathogen (*Sclerotinia*). Isolate specific resistant sources to TuMV have been identified from *Brassica* species. Race-specific sources of resistance to *Xanthomonas* have been identified from cultivated *Brassica* species and *A. thaliana*. black rot (*Xanthomonas*) R-genes have been identified in cauliflower through molecular approaches. In general, the *Brassica* A-genome is rich source for resistance to turnip mosaic virus, black rot, downy mildew, and clubroot, whereas B-genome possesses resistance against black rot and blackleg, and C-genome harbors stem rot, Fusarium wilt, and downy mildew resistance. Well-characterized R-genes/QTLs in *Brassica* crops have been identified and genetically mapped against clubroot, sclerotinia stem rot, downy mildew, and *Leptosphaeria*. However, only limited R-genes/alleles have been cloned from *Brassica* species. Eight different bonafide R-genes-like (RGL) PCR fragments have been obtained from *A. thaliana* accessions *Col* and L*er*. The application of omics technology in R-genes identification has widen the scope of *Brassica* species genome sequencing, use of pangenomics, use of NGS-based SNP methods, *In Silico* method, NGS-based bulked segregant analysis (BSA), resistance gene enrichment and sequencing (Ren Seq), and effectromics approaches to detect R-genes.

Genetic resistance is one of the most important attribute of the host defense against pathogens. Host resistance provides an economical, environmentally benign, and widely accepted method of managing disease especially downy mildew in crucifers. Present day's emphasis on organic agriculture relies mainly on large-scale deployment of disease-resistant crop cultivars in addition to the use of techniques such as cultural practices (sowing time, soil solarization, crop rotation, use of organic manures, plant population, clean cultivation, soil and plant moisture regulation), and biological disease control. *Hyaloperonospora–Brassica* interaction can be classified into two types, compatible and incompatible interaction, leading to the phenotypes of susceptibility, and resistance to certain pathogens. Obviously, the incompatible interaction has been extensively exploited by the *Brassica* breeders to develop resistant cultivars for increasing crop production. Large-scale deployment of elite cultivars carrying an R-gene (monoculture) imposes higher selection pressure on the pathogen carrying the cognate *Avr* gene to survive, resulting in the modification or deletion of the recognized *Avr* gene or generation of novel effector genes that can escape the recognition of the old "*R*" gene. This *R-Avr* interactive co-evolution explains why many resistant cultivars in the field lose their resistance in a relatively short period of time. Hence, knowledge on the molecular mechanisms of *Hyaloperonospora–Brassica,* i.e., host–pathogen interaction, and use of horizontal, durable, and broad spectrum resistance has become very important in disease resistance breeding programs. The interplay of host resistance and pathogen virulence in latter system is likely to be quite different to major gene-based system. Major gene resistance in crucifers to downy mildew has been exploited for breeding resistant cultivars.

To assess the nature of powdery mildew resistance in crops, particularly in context to slow mildewing components. The studied revealed that the incubation period was ranged from 3 to 4 days. The maximum incubation period of 4 days was recorded on the varieties RH-9304 and RH-9801 whereas in other cases, it was 3 days. The latent period of test cvs. ranged between 1 and 3 days. The latent period was 1–2 days in varieties belonging to the *Brassica juncea*. However, it was 3 days in case of variety GSL-1. The results also revealed that there was no significant difference in the latent period in the varieties belonging to *Brassica juncea*. However, it differed in case of *Brassica napus* where it was slightly higher. The number of powdery mildew specks/leaf was also recorded and revealed that the variety GSL-1 showed minimum number of specks/leaf (5.52) whereas the variety RH-9801 contracted maximum number of the specks/leaf (39.40). The variety GSL-1 contracted less number of specks per leaf which significantly differed from the other varieties (less than ten specks/leaf) which may be considered as resistant whereas other varieties such as RH-9801, RH-9304, and RH-30 had higher number of specks/leaf (more than 30) and may be termed as susceptible. The other varieties such as RH-9901, RH-8812, RC-781, and Purple Mutant contracted the powdery mildew specks ranging between 10 and 30 specks/leaf and considered as moderately susceptible.

The minimum number of conidia/speck was recorded on the variety GSL-1 (10 conidia/speck). It was followed by the variety RC-781 where it was 41.58 conidia/speck. In other varieties such as RH-30, RH-9801, RH-9304, RH-8812, RH-9901, and Purple Mutant, the conidial production ranged between 50.08 and 83.75 conidia per speck being maximum on the variety RH-30 (83.8) and minimum on RH-8812 (50.1). It was revealed that the considerable amount of conidia per speck was produced in all the susceptible varieties belonging to except GSL-1 and HC-9603. The variety GSL-1 and HC-9603 appeared as resistant to powdery mildew with the expression of slow mildewing components whereas other varieties belonging to group appeared as susceptible to the disease showing faster powdery mildew development under field conditions (Singh 1994; Meena et al. 2018).

8.4 Genetics of Disease Resistance

In the absence of host resistance, intensive cultivation has resulted in perpetuation, build-up, and dissemination of *Alternaria* species on crucifers in areas where these crops are grown. The main factors responsible for the increase of inoculum of pathogenic *Alternaria* to the present level include the lack of genetic sources of resistance, cultivation of susceptible crops with high plant density, use of irrigation with high rates of nitrogen fertilizers, contiguous areas under monoculture, staggered sowing dates, poor weed management, and poor plant protection strategies. The information gathered on *Alternaria* diseases of crucifers indicates that some gaps and bottle necks still exist in our comprehensive understanding of various dimensions of Alternaria-host-pathosystem. Although there is some information on genetic sources of resistance, the information on the nature, mechanism, and inheritance of resistance is still not satisfactory and/or lacking. The results of this area of investigation are an important prerequisite to incorporate stable and effective sources of resistance in breeding programs. In diseases caused by *Alternaria* species where major gene resistance is not known, the potential of minor genes/horizontal resistance, tolerance, and slow blighting should be exploited. Feasibility of induced resistance needs attention. Biotechnology or molecular biology and genetic engineering techniques can be used to transfer resistant genes from distantly related sources. Identification and sequencing of resistant genes along with identification of suitable gene combinations may be taken up. Hybrids and GM cultivars with multiple disease resistance is need of the hour.

Characterization of secondary metabolite biosynthetic genes along with signal transduction and their role in pathogenicity and fungal development are important areas of investigation. Disruption of Aso-1, a gene required for hyphal fusion (anastomosis) which is important for *Alternaria* spores to become pathogenic. Studies on the nature and mechanism of inheritance of virulence genes in different species of *Alternaria* are required. Determination of virulence spectrum of pathotypes is another aspect to be look into. Exploitation of morphological, structural, and biochemical basis of resistance needs in-depth study in crucifer since lot of genetic variability is available in related and distantly related crucifers all over the world. Comparative studies on all aspects of host–parasite interaction in crucifers with all the four species of *Alternaria* may reveal some fruitful results for comprehensive and management of disease.

Host resistance in crucifers against *Alternaria* species has various components. Inheritance of resistance in inter- and intraspecific crosses of *B. juncea* and *B. carinata* to *A. brassicae* is governed by adaptive genes, dominant genes, additive × additive type epistatic genes, additive × dominance, and dominance × dominance type of non-allelic interaction genes (Singh and Singh 1989; Krishnia et al. 2000b). Inter-mating between resistance plants helps in increasing the level of resistance against *A. brassicae* through pyramiding of resistant genes. Elevated horizontal resistance in oilseed *Brassicas* genotypes has been recorded (Saharan and Kadian 1983; Saharan and Krishnia 2001). *Brassica* genotypes, Rajat, Kranti, RH-781, and RL-1359 have been identified with disease tolerance attributes (Gupta et al. 2002). Genotypes PR-8988 and PR-9024 show higher degree of partial resistance or slow blighting (Kumar and Kolte 2001). Epicuticular wax (Candle, Tobin, Altex, Midas, Tower), low number and narrow stomatal aperture (Tower, RC-781) provide resistance to Alternaria infection in *Brassica* species (Saharan and Kadian 1983; Conn et al. 1984; Tewari 1991b). The concentration of phenolic compounds, activation of polyphenol oxidase and catalase is higher in tolerance genotype of mustard (Gupta et al. 1990).

Chitinase modifying proteins, *Cmps* are secreted by fungal pathogens of crucifers, which hamper with fungalysin *cmp* activity to develop plant resistance to multiple fungal diseases (Naumann and Wicklow 2013). *GLIP1* in association with ethylene signaling may be a critical component in plant resistance (*A. thaliana*) to *A. brassicicola* (Oh et al. 2005). *B. juncea* plants transformed with chitinase gene tagged with an overexpressing promoters, 35S CaMV gave defense response by degrading the cell walls of invading fungi (Mondal et al. 2003). Increased level of PAL, PPO, and peroxidase play an important role in the defense mechanisms of *B. juncea* genotypes against *Alternaria* pathogenesis (Parihar et al. 2012). B-aminobutyric acid treatment leads to proper balance of oxidant and antioxidants suitable for expression of resistance in *B. carinata* against *A. brassicae* by curtailing pathogen entrance during early stages of colonization (Chavan et al. 2013). Zeatin a cytokinin up-regulates plant immunity via an elevation of *MAPK-4* and antagonizes the effects of *A. brassicae* (Marmath et al. 2013). Transgenic expression of hevein, the rubber tree lectin, in *B. juncea* cv. RLM-198 confers defense against *A. brassicae* (Kanrar et al. 2002), β-aminobutyric acid pretreatment of *B. juncea* plants induces *A. brassicae* resistance mediated through an enhanced expression of pathogenesis-related protein genes, independent of SA and JA accumulation (Kamble and Bhargava 2007). Hypersensitive response gene (hsr 203J) like homologs of *Brassica* play important role in differential defense response against *A. brassicae* (Mishra et al. 2010). A highly contiguous genome assembly of *A. brassicae* has been generated using Nanopore MinION sequencing with an N50 of 2.98 Mb, yielding nine full chromosome-level sequences. This study adds to the current genomic resources available for the genus Alternaria and will provide opportunities for further analyses to unravel the mechanisms underlying pathogenicity of this important necrotrophic pathogen (Rajarammohan et al. 2019).

The cDNA encoding Pm AMP 1 has been successfully incorporated into the genome of *B. napus*, and it is in planta expression confers greater protection against *A. brassicae*. Pm AMP1 is a cysteine-rich antimicrobial peptide isolated from *Pinus monticola* (Verma et al. 2012). Combined expression of a barley class II chitinase and Type I ribosome inactivating protein in transgenic *B. juncea* provides resistance against *A. brassicae* (Chhikara et al. 2012). Transcriptional responses to exposure to the brassicaceous defense metabolites camalexin and allyl-isothiocyanate in *A. brassicicola* have been recorded (Sellam et al. 2007).

Elicitation and accumulation of phytoalexins in crucifers after exposure to Alternaria and their role in disease resistance have been demonstrated (Verma and Saharan 1994). Calcium sequestration property of *A. brassicae* can be used in enhancing resistance to this pathogen in rapeseed by soil or foliar application of calcium compounds (Tewari 1991a, b). Many sources of resistance to *Alternaria* species have been identified from different *Brassica* species, but very few have been utilized to develop resistant cultivars (Verma and Saharan 1994; Sharma et al. 2002). Strong and positive correlation in the increased level of resistance to different pathogens reported in *Brassica* species will facilitate improvement in accumulation of resistance to multiple disease resistance (Kumar and Saharan 2002). In spite of several bottlenecks in the development of resistance cultivars, various methods/ techniques together with conventional as well as biotechnological approaches have been utilized to incorporate desired traits in cruciferous crops against black spot disease (Nowicki et al. 2012; Aneja and Agnihotri 2013). The resistance sources available in host germplasm, the exploitation of NHR mechanisms, and their applications in protecting Brassica crops from Alternaria blight needs attention promptly. The insights might also be applicable to a wider repertoire of plant pathogens (Fatima et al. 2019).

Sclerotinia sclerotiorum (Lib.) de Bary, the causal organism of stem rot of *Brassica* alongwith over 500 host plants distributed worldwide. *Sclerotinia* rot is menace to cultivation of oilseed *Brassica* crops all over the world. Infection occurs on leaves, stems, and pods at different developmental stages, causing seed yield losses of up to 80%, as well as significant reductions in oil content and quality. The initial mycelial infection at the base of the stem is an appearance of elongated water-soaked lesions that expand rapidly. Ascosporic (carpogenic) infection is quite general and occurs on the leaves or leaf axil. Effective pathogenesis by *S. sclerotiorum* requires the secretion of pathogenicity factors including oxalic acid and extracellular lytic enzymes. Germination of overwintered sclerotia, and release, survival, and germination of ascospores are important factors for the development of disease and in the life cycle of this pathogen. Isolates of *S. sclerotiorum* show a high level of morphological variability and molecular diversity. Management of *S. sclerotiorum* is a major challenge, and the best being the integration of various IPM measures. Partial resistance has been identified in some *Brassica napus* and,

B. juncea, genotypes, though; wild *Brassicas* show better resistant reactions (Sharma et al. 2015).

Identification of effective R-genes sources along with the nature of inheritance in *Brassica* genotypes is the basic impact in breeding of disease resistance cultivars of crucifers against different range of pathogens. There are two different genetic mechanisms for disease resistance traits being inherited in generations. One is monogenic resistance which is based on single gene and second is quantitative which is governed by two or more genes. In most cases, single R-genes confer complete resistance but are active against few races of the pathogen. R-genes are clustered at some loci in the genome or exist as different alleles conferring resistance towards specific pathogen races. Quantitative resistance shows efficacy against more number of pathogenic races. Resistance breeding in crucifers depends on both types of resistance. Monogenic resistance is easy to incorporate but it is not durable. Mostly quantitative resistance is preferred. The use of molecular markers has allowed the isolation, cloning, and gene pyramiding to develop more durable resistance cvs. of *Brassica* crops. Qualitative resistance mostly provides defense against biotrophs, whereas quantitative resistance is involved in defense response to all plant pathogens from biotrophs, hemibiotrophs to necrotrophs pathogenic to crucifers.

The inheritance of disease resistance in *Brassica* crops is by qualitative one gene and polygenes as well as quantitative conferred by QTLs. The resistance to white rust (*Albugo*) in different *Brassica* species is governed by single dominant gene, one or two dominant genes with epistatic effects, additive dominant with epistatic effects, and partial resistant single recessive genes *wpr*. In *Arabidopsis thaliana*, *WRR* 4 genes confer broad spectrum resistance to races AC-2, 4, 7, and 9. In *Brassica* genotypes, significant additive gene effects and dominant gene effects govern resistance to Alternaria blight disease. In some crosses, additive × dominance epistatic effects are significant. In *Brassica juncea* genotypes, *Alternaria* tolerance with components of horizontal resistance attributes has been identified. In the interspecific crosses of *B. juncea* × *B. carinata*, single resistance gene with complete dominance confers resistance to *Erysiphe*. In *A. thaliana*, accessions resistance to powdery mildew is either polygenic based on atypical R-gene RPW8 or combination thereof with RPW8 representing major QTL. The inheritance of enhanced R-genes in *edr* 1 mutant is conferred by recessive genes to powdery mildew. In *Brassica,* resistance to *Fusarium* is conferred by a dominant gene. Inheritance of resistance in crucifers to downy mildew has been observed to be governed by one to four dominant complementary genes, additive genes, one to two partial dominant genes, partial dominant epistatic and gene dosage affects along with quantitative genes depending on host species and pathotype interactions. In *Brassica*, most resistant cultivars contain major genes (QTLs) conferring resistance to specific blackleg pathotypes (*Leptosphaeria maculans*). Resistant loci to *L. maculans* are distributed in A- and B-genome of *B. napus*. Clubroot (CR) resistance in *B. oleracea* is governed by 2–4 dominant genes, additive and recessive genes and more than 22 QTLs. In *B. napus*, clubroot resistance is governed by one to two dominant genes and more than 19 QTLs, whereas in *B. rapa*, two dominant genes interact in a non-additive manner to confer resistance to different *Plasmodiophora brassicae* pathotypes. Across all *Brassica* spp. at least 15 CR loci from *B. rapa*, 22 QTLs from *B. oleracea*, and 16 QTLs from *B. napus* have been identified. In *B. rapa*, chromosomes AO3 carry the highest number of loci. *P. brassicae* pathotype 5 × identified two independent QTLs, *Rcr* 8 and *Lc9*, in the A-genome on chromosomes AO2 and AO8. A single dominant gene, *Rcr* 7 has been identified to provide resistance to pathotypes 3H and 5X in the C-genome on chromosome CO7. A modifier gene has been identified to increase efficiency of a dominant CR gene against virulent pathotypes. The inheritance of resistance to *Sclerotinia* is complex since no single consistent mode exists to provide resistance to *Brassica* crops. In *Brassica*, partial dominant, polygenic, additive epistatic interactions confer resistance to *Sclerotinia*. Quantitative resistance in both leaf and stem by dominant and additive × additive epistatic interactions controls the genetic effects. In *B. rapa*, resistance to turnip mosaic virus (TuMV) is controlled by dominant and recessive genes. In *B. oleracea*, resistance to TuMV has been observed to be controlled by both dominant genes and multiple loci. The inheritance of resistance in *Brassica* species to black rot (*Xanthomonas*) is very complex and depends on genotype–race interactions. In *B. oleracea*, resistance is controlled by single dominant gene to race 1 and 4. It is polygenic recessive to races 1 and 3 in *B. oleracea* line Badger Inbred-16 and high level of resistance to races 1 and 4 in *B. carinata* controlled by a single dominant locus (Xca1). Resistance to race 4 in three *B. napus* lines is controlled by a single dominant locus (Xca4).

Latent period and sporulation capacity are two major components of partial resistance and governed by the same mechanism in the interaction between broccoli (*B. oleracea*) and *H. parasitica*. Twenty doubled haploid broccoli lines from breeding material were evaluated for resistance to *H. parasitica* at the seedling stage. All lines supported sporulation of the pathogen, but to varying extents, and intensities. Partial resistance of the more resistant lines "br8" and "br9" reduced conidia production on cotyledons by 50–70% compared with the most susceptible lines. Inoculation of the two most resistant lines with 13 isolates of different geographical origins revealed that "br9" showed a rather uniform level of resistance to all isolates while "br8" showed some isolate specificity. Partial resistance was evaluated in six of the broccoli lines in a half diallel set of crosses. Disease assessment of seedlings showed that additive genetic effects explained 45.8% and 31.8% of the total variation of sporulation score and conidia production, respectively. This suggests that recurrent selection for partial resistance to *H. parasitica* in early generation inbreds or in populations of broccoli will be efficient to obtain cotyledon resistance (Jensen et al. 1999).

Seedling resistance conditioned by a single major gene is isolate specific and may lose the effect through selection of virulent types of the pathogen, if resistant cultivars are produced on large areas (Lucas et al. 1988). Partial quantitative disease resistance by definition provides reduced host invasion without complete protection and is generally anticipated to be based on several genes. It may be a source of durable resistance with less isolate specificity than resistance based on single genes (Parlevliet 1992). In *Brassica oleracea*, partial resistance to *H. parasitica* at the seedling stage has been reported in Portuguese coles (Monteiro and Williams 1989; Dias et al. 1993) and in gene bank accessions of cabbage and broccoli (Thomas and Jourdain 1990, 1992). Jensen et al. (1999) characterizes partial resistance to *H. parasitica* at the cotyledon stage within 20 doubled haploid broccoli lines originating from a wide range of breeding material. Inheritance of the resistance was analyzed in progeny from a half diallel set of crosses.

8.5 Breeding of Disease Resistance *Brassica* Crops

Cruciferous crop plants (canola/oilseed rape/rapeseed/mustard/cabbage/cauliflower) are economically important oilseed crop, widely used as edible oil. These crops are mostly cultivated in temperate and sub-tropical regions. To advance further comprehensions into the *Brassica*–pathogen interactions, it is critical to have current knowledge about the advanced skills which can offer encouraging consequences and help to impetus limitations in understanding of the resistance mechanisms. Resistance development is the utmost economical, environmentally sustainable resolution to cope diseases in *Brassica* species. Transfer of disease resistance remains attained once sources of R-genes recognized with sound characterized mode of inheritance of genes in the progenies with all desired traits of yield and quality. To develop stress tolerance and/or resistant *Brassica* cultivars, there is need to apprehend both of the plant and pathogenic disease development mechanisms. Hence, there is need to focuses on (1) major and devastating stresses on *Brassica* crops, (2) role of genes to overcome the stresses, and (3) differential genes expressed under biotic stresses in *Brassica* species for getting insight into the mechanisms of development of resistance cultivars.

The transfer of disease resistance is achieved after identification of R-genes sources with well-characterized mode of inheritance of genes in the progenies to get a better output as disease resistance cultivars with all desired traits of yield and quality. Various methods including conventional and molecular approaches have been used either alone or in combination to incorporate desired traits to develop improved cultivars of crucifers. Hybridization combined with backcrossing either singly or in combination with pedigree selection is the most widely used and accepted conventional method to transfer disease resistance in crucifers. The level of resistance in developed cultivars has been achieved through multiple crosses, recurrent selection, diallele crossing, and selective mating system. The resistance has been introgressed with intraspecific pedigree, backcross, modified recurrent mass selection, interspecific genome substitution, chromosome substitution, and gene introgression. The molecular and genetic engineering approaches like genome manipulation, cytoplasmic genome manipulation, transformation, foreign gene expression, and embryo rescue for wide hybridization or distant hybridization have been exploited. The introgression and pyramiding of major R-QTLs have been attempted including transgenics development. In crucifers, substantial problems have been faced in the introgression and pyramiding of major QTLs including transgenic development because of diverse nature of pathogens infecting these crops with different modes of infection, pathogenesis, and survival as biotrophs, hemibiotrophs, and necrotrophs.

Substantial progress has been made in transfer of R-genes from different sources and developing of disease-resistant cultivars of major cruciferous crops against major pathogens using conventional and molecular biotechnological approaches. However, problems in transfer of R-genes arise due to diverse nature of pathogens which are physiologically and taxonomically highly variable; therefore, no single-gene product and approach can be expected to have a direct effect on all types of pathogens. The advent of molecular genetics has provided new opportunities for mapping and tracking of genes of disease resistance leading to more efficient marker-assisted selection. The improvement in plant transformation protocols and high-throughput gene attenuation methods like RNA interference and virus-induced gene silencing (VIGS) will accelerate the identification and transfer of novel genes for durable resistance in crucifers. The introgression of R-genes to major pathogens have been widely used to breed disease-resistant cultivars in *Brassica* crops through inter- or intraspecific crosses and embryo rescue. The conventional and modern approaches have been used to exploit R-genes sources from wild relatives to breed disease-resistant cultivars of *Brassica* crops. For developing durable Alternaria blight-resistant *Brassica* crops, strategies like inter- or intraspecific introgression of R-genes from non-host plant to *Brassica* crops, microarray-based gene identification, and transfer by gene silencing, and transfer of an endochitinase gene from *Trichoderma* approaches have been used. R-genes sources of multiple disease resistance have been utilized to breed resistant cultivars of *Brassica* to *Albugo, Alternaria, Erysiphe, Hyaloperonospora, Leptosphaeria,* and *Xanthomonas.* To combine the traits from distantly related *Brassica* species for the development of superior, durable, and tolerant cultivars of *Brassica* to major diseases, approaches like in vitro embryo rescue, somatic hybridization, interspecific hybridization, distant hybridization, somaclonal variations, genetic transformations, molecular markers, mutagenic resistance, induction of systemic resistance, and genetic engineering have been exploited. Substantial progress has been made in breeding for clubroot resistance genotypes/cultivars of *B. rapa, B. oleracea, B, napus, Raphanus sativus*, and *Arabidopsis thaliana.* Several clubroot loci or QTLs have been mapped on the chromosomes of these species. There is substantial genetic and pathotype variation in *Plasmodiophora brassicae* populations all over the world and the deployment of a cultivar with single-gene resistance imposes a strong selection pressure to evolve new virulences to overcome this resistance. The successful development of *Sclerotinia*-resistant *Brassica* cultivars have been obtained by introducing multiple resistant loci (QTLs) from wild relatives using marker-assisted introgression approach. Black rot (*Xanthomonas*) resistance has been introgressed from *B. carinata* to *B. oleracea* through embryo rescue. Somatic hybrids by protoplast fusion between rapid cycling *B. oleracea* and *Xanthomonas*-resistant lines of *B. napus* have been developed. Understanding the evolutionary potential between the pathogen and the host is essential in developing and transfer of R-genes for durable *Brassica*-resistant cultivars. In the present era, the breeding for disease-resistant *Brassica* cultivars, application of molecular approaches like marker-assisted selection, genome-wide marker-assisted selection, genetic diversity analysis, and comparative genomics have opened new avenues for superior, durable, and multiple disease-resistant cultivars in different *Brassica* crops (Saharan et al. 2021).

8.6 Signaling Network System of Host Resistance to Multiple Stress Responses

Cruciferous crops can suffer from stress too due to various biotic and abiotic. Biotic stress is caused by living organisms like fungi, bacteria, oomycetes, viruses, insects, weeds, and parasitic animals and plants. These organisms can infect plants to draw plant produced nutrients thus causing stress in plants. These stresses can significantly reduce crop yield and sometimes they kill the plants. Abiotic stresses are caused by nonliving factors such as high temperature, low temperature (frost), excess water, drought, salinity, mineral toxicity, and lack of nutrients and minerals. These stresses can affect plant growth and development, causing significant reductions in crop yield. Some of these stresses can co-exist increasing the overall magnitude of stress on plants. In fact, stress can seriously affect the plant growth, which can have a severe implication which further reduces the growth and ultimately the loss in food production. With the world's growing population, there is need to increase food production. One significant way to do this is to find ways to reduce stress in plants. There is need to develop or engineering the plant for multiple stress tolerance. The molecular plant breeding can help in the development of stress-tolerant plants with the help of microbe interactions, particularly with the help of non-host resistance and for developing genetic resources for plant functional genomics.

Multiple stresses in the form of biotic and abiotic stresses, as a part of the natural ecosystem, seriously impact crucifer's crop productivity and threaten global oilseed and vegetable food security. Plants in their natural habitats are persistently and simultaneously confronted with a range of biotic and other internal adverse changes imposed by multiple interactions of soil environment (Tables 5.1–5.4, Chap. 5). Environmental pollution, global warming, and climate change exacerbate the impact of biotic and abiotic stresses on plant growth and yield. Plants have evolved sophisticated defense network, also called innate immune system, in response to ever changing environmental conditions. Recent evidence highlights the complex nature of interactions between biotic and abiotic stress responses, significant aberrant signaling crosstalk in response to combined stresses and a degree of overlap, but unique response to each environmental stimulus. Further, the results of simultaneous combined biotic and abiotic stress studies indicate that abiotic stresses particularly heat and drought enhance plant susceptibility to plant pathogens. Plant biotic and/or abiotic stress interactions trigger a wide range of defense reactions at molecular and cellular levels to ensure that plant cells are insulated from the challenges they encounter (Kim et al. 2014) (Fig. 5.1, Chap. 5).

A sophisticated defense network is activated to orchestrate transcriptional reprogramming using a two-tiered defense approach (Dodds and Rathjen 2010;

Tsuda and Katagiri 2010). Pathogen/microbe-associated molecular patterns (PAMP/MAPM)-triggered immunity (PTI) is the first line of defense. It is a general and nonspecific defense response and provides basal resistance not only against entire classes of microbial pathogens but also abiotic stresses. PTI constitutes the inducible defense through an array of early cellular responses comprising of ion flux across the membrane, production of reactive oxygen species (ROS), and mitogen-activated protein kinase (MAPK) cascades phosphorylation and long-term later responses that induce callose deposition (Nakagami et al. 2005; Dodds and Rathjen 2010; Tsuda and Katagiri 2010; Schwessinger et al. 2015).

Defense-related phytohormones comprise salicylic acid (SA), jasmonic acid (JA), ethylene (ET), and abscisic acid (ABA) as the core immune signaling components. They are produced in response to both, biotic and abiotic stresses (Dodds and Rathjen 2010; Tsuda and Katagiri 2010). Among those, ROS, MAPkinase cascades and hormone signaling pathways are of paramount importance as they lead the defense responses to biotic and abiotic stresses (Nakagami et al. 2005; Fujita et al. 2006; Atkinson and Urwin 2012; Kissoudis et al. 2014; Perez and Brown 2014; Schwessinger et al. 2015) (Fig. 5.2, Chap. 5). Nejat and Mantri (2017) proposed that PTI which is activated either by PAMP or MAPM is the "stress-triggered immunity" as it has evolutionary conserved molecular signatures across similar types of microbes or abiotic stresses. The effector-triggered immunity (ETI) is the second, prolonged and robust layer of plant immune response that is specific to the pathogen infection. It is activated by resistance (*R*) genes when pathogen virulence factors, called effectors, are released into plant cells (Cui et al. 2015). If the first line of inducible nonspecific defense, the so-called PTI, is successfully overcome by pathogens through suppression of PTI response, then the second line of plant immunity is activated to produce defense proteins that cause programmed cell death. This is also referred to as the hypersensitive response (HR) (Thomma et al. 2011). Hence, after stress recognition, the immune system triggers a wide range of defense mechanisms to orchestrate transcriptional reprogramming through receptor proteins, signal transduction cascades, kinase cascades, reactive oxygen species, hormone signaling pathways, heat shock proteins, and transcription factors to protect plants against multiple stresses (Fig. 5.1, Chap. 5). However, recent advances suggest significant aberrant signaling crosstalk in plant resistance response to multiple stresses compared with an individual stress (Atkinson et al. 2013; Prasch and Sonnewald 2013; Suzuki et al. 2014).

Cruciferous plants have developed numerous molecular mechanisms to respond to various environmental and multiple biotic and abiotic stresses. Whenever plant encounters any kind of stress, first of all there is perception of the signals by the potential sensors. This further generates secondary messengers (e.g., ABA, ROS, Ca^{2+}, inositol phosphates). The secondary messengers can alter the levels of intracellular Ca^{2+}. Any change in cytoplasmic Ca^{2+} is sensed by Ca^{2+} sensors, i.e., calcium-binding proteins and after interacting with corresponding interacting partners they initiate a protein phosphorylation cascade. This signaling cascade ultimately regulates the transcription factors that manage specific suite of stress-regulated genes and the functional proteins involved in protecting the cell against stress. The expression of stress-regulated genes leads to the production of various kinds of regulatory molecules like the phytohormones ABA, ET, and SA and transcription factors. A second round of signaling can get started by these regulatory molecules which may further enhance the initial signaling pathway. These signaling networks are very complex and interconnected at many levels for transferring information to generate robust host resistance. Moreover, during signaling other components are also involved. A successful signal transduction pathway demands proper coordination of all signaling molecules. This is brought about by specific molecules whose functions are to transport, bring together, or modify the signaling molecules. They themselves are not involved in relaying the signal. These proteins consist of scaffolding proteins, protein modifiers (e.g., enzymes needed for protein glycosylation, methylation, ubiquitination, and lipidation), and adaptors. Cruciferous interactions with multiple stresses trigger a wide range of defense reactions at molecular and cellular levels to ensure that host cells are insulated from these challenges they encounter. The production of ROS, MAP kinase cascades and hormone signaling are triggered by multiple stresses and play crucial roles as common components of the host immune system. GbRLK, a receptor-like kinase (RLK) gene is involved in response to multiple stresses as positive regulator transcription factors (TFs) are the central core of the genes as defense signaling pathways to multiple stress. MYB TFs play essential roles in multiple stress responses. ATAF1 is a multifaceted TF with versatile capabilities and plays an important role in mediating crosstalk between multiple stress responses. HSPs gene regulate the function of NBS-LRR genes in response to multiple stresses. miRNAs may target only one locus or several loci in response to multiple stresses. Cell signaling in response to multiple stresses largely depends on the SnRK family of protein kinases in plants. Calcium is a key signaling pathway in host responses to multiple stresses through calcium sensor proteins like calcineurin-B-like protein (CBls), the calmodulin (CaM), and calmodulin-like protein (CMLs).

Transcriptome analyses based on microarrays and proteomic studies have given rise to rapid progress in the field of plant signal transduction and gene regulation. GeneChip and cDNA microarrays together with massive whole-genome sequencing proved to be useful in identifying novel signaling determinants on a whole-genome scale in response to various stresses. This has increased our understanding towards the complex mechanisms involved in various aspects of plant responses to environmental cues. Proteomics approach is useful in studying post-translational modifications of the proteins. It can also be used to clone unique genes by means of differential analysis which will expedite our understanding of stress signaling mechanisms in plants. Therefore, stress-tolerant plants can be genetically engineered by the combined efforts of plant molecular biologists, physiologists, and breeders.

8.7 Regulation of Genomics and Biometabolomics to Confer Molecular Mechanisms of Host Resistance

Cruciferous crops are attacked by several pathogens which deteriorate yield and quality of major *Brassica* crops affecting global food security. The pathogens infecting these crops have diverse lifestyle, genomic plasticity with rapid evolutionary mechanisms, prolific reproduction, and longer survival nature which is impedimental to manage efficiently under field conditions. Moreover, pathogens secrete small protein known as effector genes, which counter plant defense mechanisms and modulate host physiology to promote pathogen growth and reproduction. However, plants evolve disease resistance (R) genes encoding receptor proteins that detect and respond to pathogen effector gene, leading to the activation of effector-triggered immunity. This discovery with the use of multi-omics of approaches has revealed molecular mechanisms of crucifers' host resistance to manage major pathogens of *Brassica* crops. Host resistance in crucifers to biotic and abiotic stresses is multilayered and multicomponents. Resistance is activated either through SA signaling and simultaneous perception of ethylene and jasmonic acid depending on biotrophic, hemibiotrophic, and necrotrophic *Brassica* host–pathogen interactions. Molecular mechanisms of host resistance against biotrophs have been revealed through identification, functionally characterization, and fine mapping of R-genes and QTLs on different chromosomes of *Brassica* species. The molecular mechanisms of R-genes resistance imparted by NBS-LRR are effective against biotrophs or hemibiotrophs but are not effective against necrotrophs. JA- and ET-mediated signaling is predominantly involved in general defense to necrotrophs. Similarly, R-genes and QTLs have been identified from *B. rapa, B. juncea, B. nigra*, and *B. napus* to provide resistance to hemibiotrophs and necrotrophs. Cysteine-rich protein kinase genes and multi-environment stable six QTLs suitable for quantitative blackleg resistance in *B. napus* have been identified. Overexpression of *BnNPR1*, *OsPGIP2*, *BnWRKY33*, and *BnMPK4* genes enhances resistance in *B. napus* to *Sclerotinia*. Seven NBS-LRR-encoding genes linked to black rot resistance in *B. oleracea* have been identified on four QTL regions and chromosomes Co1, Co3, CO6, CO7, and CO8, respectively. Biometabolomics mechanisms of *Brassica* host resistance is initiated soon after perception of pathogens by signaling molecules and results in execution of broad range of defense responses to stop the infection and pathogenesis of invading pathogens. Biometabolomics and biomolecular mechanisms of host resistance is expressed in the form of different metabolites (primary and secondary) and biomolecules involved in the induction of biochemical changes and pathways in the host to confer resistance at pre- and post-invasion stages to biotrophs, hemibiotrophs, and necrotrophs. Transcriptomics and proteomics have discovered novel biochemical pathways and proteins/genes involved in conferring resistance to virulent pathotypes. With the use of multi-omics supplemented with functional genomics studies, key resistance pathways involved in Brassica host-pathosystem and induction of differential signaling molecules like SA, JA, and ET in response to diverse pathogens have been revealed.

Novel molecular approaches have been exploited to breed disease-resistant cultivars of *Brassica*. Through introgression breeding high level of resistance has been obtained from *B. nigra*, and *B. carinata* to *B. napus* against blackleg (*Leptosphaeria*). Similarly transfer of resistance to black rot from *B. carinata* to *B. oleracea*, from *B. hirta* to *B. juncea* against *Alternaria*, and from B. carinata to B. oleracea against powdery mildew has been obtained. The approaches like in vitro embryo rescue, somatic hybridization, somaclonal variations, genetic transformation, molecular markers, use of systemic resistance induction, and genetic engineering have great promise but have been used against few pathogens. Clubroot resistance in *B. napus* and *B. rapa* has been transferred through interspecific and distant hybridization using molecular markers. R-genes from European clubroot-differentials (ECD) set have been introduced into *B. napus*, and other species to provide resistance to clubroot. Powdery mildew resistance in *B. oleracea* has been obtained from *B. carinata* through embryo rescue. The mutagenic agent gamma radiation increases resistance to *E. cruciferarum* in *Brassica*. Durable resistance in *Brassica* species to Alternaria can be obtained through introgression of genes from non-host plants by inter- or intraspecific crosses, identification, and transfer of NHR based on microarray, identification, and transfer of NHR genetic resources from model host plants, identification of functional genes based on bioinformatics, and use of endochitinase gene from *Trichoderma virens*. Marker-assisted introgression of resistance from wild relatives has been achieved in *B. juncea* to *Sclerotinia*. Resistant *B. napus* lines pyramided with multiple R-QTL derived from wild *B. oleracea* have been developed via hexaploidy strategy. Introgression of black rot resistance from *B. carinata* to *B. oleracea* through embryo rescue has been attempted successfully. The *B. carinata* genotypes, NPC-9 and HCA-6 having very high resistance to *XCC* race 1 and 4 have been used to introgress resistance in Indian cauliflower through embryo rescue approaches.

The advances made in the molecular genetics through whole-genome sequencing of *Brassica* crops have led to quantum jump not only in understanding plant genetic diversity but also for accessing these enormous resources. Novel sources of R-genes for blackleg have been tapped for other A- and C-genome of *Brassica* species from the B-genomes in *B. nigra, B. juncea, B. carinata*, and a wild species *Sinapis arvensis*. Several sources of resistance to 15 major pathogens of crucifers except nematodes have been identified from different *Brassica* species and wild relatives. Some of these sources have been utilized to breed disease-resistant cultivars of *Brassica* through conventional and molecular approaches. From *Brassica* species, multiple pathogens resistance sources have been identified which provide resistance to *Albugo, Alternaria, Erysiphe*, and *Hyaloperonospora*. There is strong evidence for positive genetic correlation among *Albugo, Alternaria, Erysiphe, Hyaloperonospora, Leptosphaeria,* and *Xanthomonas* to develop multiple disease-resistant cultivars of *Brassica* crops. Clubroot-resistant cultivars with R-genes and QTLs have been developed in *B. rapa* and *B. napus* with pathotype specificity at different counties but most of them have scummed through evaluation of new virulences of *Plasmodiophora*. In *Brassica* crops, transfer of resistance through different approaches is facing problems because of diverse nature of infecting pathogens for which there is no single-gene product to have a direct toxic effect on all kinds of pathogens. Consequently, host has to develop quite different strategies to deal with these pathogens.

8.8 Management of Disease Resistance Under Field Conditions

Management of crucifers' disease resistance for its effective utilization under field conditions is very important and challenging. The *Brassica* genomes (A-, B-, and C-genome) are crucially important to provide novel genetic inheritance for economically important traits that can be used for the overall improvement of crop production and quality. For example, single genome of diploid *B. rapa* (A-genome) holds more than 230 R-gene sequences in 16 gene families (Cheng et al. 2011), among which, over 8 genetic loci have been identified in different research studies, which have functional properties for crucifers' disease resistance. There are possibilities of the existence of more R-genes specific to the disease resistance and their allelic variations in different genetic pools or wild relative species. Effective utilization of resistance loci and their allelic variations may enhance the durability of resistance against different diseases in different *Brassica* species. As a long history of host and pathogen association, it has been revealed relatively high evolutionary patterns of the pathogens. In various cultivating geographical regions of *Brassica* crops, persistence of different pathotypes with high levels of pathogenicity poses challenges to breed durable disease-resistant cultivars. The breakdown of disease-resistant cultivars has become a serious problem in various countries of the world. Effective management of resistance genetic resources in breeding novel cultivars could enhance the performance of resistance loci in different *Brassica* species for sustainable, more durable, and cost-effective control of the crucifers' diseases. Understanding the evolutionary potential between the pathogen and the host is essential in developing a durable resistant *Brassica* crops.

In the integrated disease management strategies of cruciferous crops, use of R-genes through the development of disease resistance cultivars is one of the most important ecofriendly and cost-effective components. Major R-genes impose strong selection pressure on pathogen populations and in combination with a number of factors can act to cause "boom-and-bust" cycle of pathotype diversification within a population leading to break down or erosion of R-genes. Strong selection imposed by widespread deployment of one R-genes also may lead to breakdown of resistance conferred by another gene, if the two complementary avirulence genes are closely linked. Therefore, the effective and sustainable deployment of resistant cultivars to manage crucifers' disease requires the integrated approaches to manage disease resistance with other control measures. The deployment of R-genes cultivars based on qualitative and quantitative pyramiding is considered as a key lever to increase the durability of R-genes against rapidly evolving pathogen populations. The disease resistance can be effectively managed in crucifers' crops against different pathogens under field conditions by fortification with different approaches. To manage white rust under field conditions, in addition to major gene resistant cvs. with slow white rusting and systemic resistance induction approaches may be exploited. Non-host resistance (NHR) is the most durable form of plant immunity effective against all genetic variants of all pathogens and it is multilayered. NHR-related genes can be transferred to host plants to confer durable resistance to *Alternaria brassicae* and other pathogens of crucifers. Multilayered and multicomponent mechanisms of resistance to crucifers' powdery mildew pathogen may be more durable along with major gene resistance under field conditions. The competitive ability of the disease resistance genotype increases significantly in the presence of *Hyaloperonospora parasitica* isolate when competing with susceptible genotype. Small RNA inhibits infection of *Hyaloperonospora*. The deployment of different R-genes cvs. of *Brassica* against *Leptosphaeria* through space and time derives the genetic composition of the pathogen populations. The deployment of R-genes cvs. of *Brassica* to manage clubroot has been the dominant component to manage *Plasmodiophora* in several countries. The evolution of new virulence in the pathogen has eroded resistance in these countries. R-genes pyramiding in *Brassica* cvs. may provide protection against different *P. brassicae* pathotypes. The erosion of R-genes in *Brassica* may be managed through integrated approaches such as use of non-host and basal resistance, use of bacterial and fungal biocontrol agents; use of ProCa; nitrogen and arginase gene to modulate resistance positively. Introgression of R-genes from *B. carinata* against *Pseudocercosporella* (white leaf spot) in *Brassica* crops has enormous potential. Hypovirulent isolates of sclerotinia with the potential in reducing virulence in pathogen populations can help to manage stem rot disease. The breeding of TuMV *Brassica* cvs. can be achieved through transformation of R-genes into crops, molecular marker-assisted selection (MAS), pyramiding of R-genes, and host-induced gene silencing (HIGS). Interspecific hybridization with related or distant species followed by embryo rescue or somatic hybridization for the development of black rot (*Xanthomonas*)-resistant hybrids has been accomplished. Beet cyst nematode has been managed by pyramiding of R-genes in *Brassica*. Multiple pathogen disease resistance of *Brassica* has been managed by the development of transgenics, use of fungal and bacterial biocontrol agents, and use of other defense-related genes like WRKY and CAMTA. The breakdown of resistance in *Brassica* can be assessed with molecular approaches to prevent another breakdown of disease resistance. The phenomena of net cost of R-genes resistance in the diseased crucifers operates under natural cultivation system.

8.9 Techniques to Develop Resistant Cultivars of *Brassica*

Techniques, methodology, procedures, and protocols developed are essential and prerequisites for research and teaching in science and technology as well as for the development of resistant varieties of crucifers. To initiate any research project, the first and foremost requirement is a standardized, validated, and reproducible technique. A scientific technique is a systematic way of obtaining information about a scientific nature in the areas of explored and unexplored material or product or biotic and abiotic interactions. Scientific techniques can be divided preparative (cultures purification, growth, and analysis) and measurement techniques like molecular characterization, isolation, cloning, and other genomics protocols. Some selected techniques have been included in this book to facilitate research workers to use, refine, and obtain desired results in the areas of genomics of crucifers and host resistance. For the benefit of graduate students and research scientists of *Brassica*, these selected techniques with methodology, procedures, and protocols will be helpful in planning disease resistance breeding programs. These techniques are standardized and have been developed by experts working on genomics of crucifers' host-pathosystem. They can be used for identification of sources of resistance with R-genes by screening germplasm from different sources. The procedures and methodology for transfer of resistance through various approaches have also been included for developing and breeding disease resistance cultivars of *Brassica* against range of pathogens. To get reliable and widely acceptable results in research pursuits is the primary object of any research programs which can be met only through standardized techniques included in the chapter.

8.10 Exploitation of Novel Protocols for Breeding Disease-Resistant Cultivars of Crucifers

Cruciferous crop plants (canola/oilseed rape/rapeseed/mustard/cabbage/cauliflower) are economically important crops, widely used as edible oil vegetables. These crops are mostly cultivated in temperate and sub-tropical regions. Brassica crops are always challenged by major diseases such as sclerotinia stem rot, white rust, downy mildew, Alternaria blight, blackleg, black rot, and clubroot. These diseases cause significant yield and economic losses in rapeseed-mustard and vegetable crops worldwide, which has aggravated by global climate change. If it is not resolved meritoriously, will impend global food security. To advance further comprehensions into the *Brassica*–pathogen interactions, it is advisable to analyze critically the current knowledge in this area and how advanced skills and technologies can offer encouraging consequences to understand the resistance mechanisms. Resistant varieties development is the utmost important, economical, environmentally sustainable solution and effective to cope with diseases management in *Brassica* species. Transfer of disease resistance remains attained once sources of R-genes have recognized with sound characters, mode of inheritance of genes in the progenies alongwith all desired traits of yield and quality. Conventional and molecular approaches have been used either alone or in combination to incorporate desired and popular traits to develop improved cultivars of crucifers. Hybridization combined with backcrossing either singly or in combination with pedigree selection is the most widely used and accepted conventional method to transfer disease resistance in crucifers. The level of resistance in developed cultivars has been accomplished through multiple crosses, recurrent selection, diallele crossing, and selective mating system. The resistance has been introgressed with intraspecific pedigree, backcross, modified recurrent mass selection, interspecific genome substitution, chromosome substitution, and gene introgression against different pathogens. The molecular and genetic engineering techniques like genome manipulation, cytoplasmic genome manipulation, transformation, foreign gene expression, and embryo rescue for wide hybridization have been exploited in *Brassica*. The introgression and pyramiding major R-QTLs have been attempted including transgenics development. To develop stress tolerance and/or resistant *Brassica* cultivars, there is need to apprehend both of the plant and pathogenic disease development mechanisms.

8.11 Use of Germplasm Sources to Transfer Disease Resistance

Several attempts have been made to exploit resistance sources from the wild relatives of *Brassica* and transfer them into cultivated *Brassica* (Aneja and Agnihotri 2013). Efforts have been made in screening blight-resistant cultivars; however, high levels of resistance to *A. brassicae,* have not been observed among cultivated *Brassica* species. The species show varying degrees of resistance responses against Alternaria blight, and accordingly, *B. rapa* and *B. juncea* are classified as being more susceptible to the disease than *B. napus* and *B. carinata* (Skoropad and Tewari 1977). These related wild species have been found to harbor genes conferring resistance to Alternaria blight that can be used for resistance breeding (Ghose et al. 2008; Chatterjee et al. 2013).

In the past few decades, various conventional and modern approaches have been used to develop blight-resistant cultivars by exploiting resistance sources from host germplasm. Conventional approaches involve the screening of germplasm for blight-resistant plant materials, followed by the transfer of resistance traits from tolerant species into susceptible cultivars by hybridization combined with multiple crosses. Inter- and intraspecific crosses of *B. juncea* and *B. carinata* were established to understand the inheritance pattern of resistance to Alternaria blight (Krishnia et al. 2000a), indicated that dominant and additive genes govern the inheritance of resistance to Alternaria blight. Thus, a successful breeding method should utilize both the gene effects. However, there is a limitation in such breeding approaches owing to the low availability of suitable resistance sources/genes within the available germplasm of cultivated *Brassica* and insurmountable self-incompatibility barriers in combining the resistance traits from distantly related wild *Brassica* species. Modern biotechnological interventions comprising tissue culture and genetic transformation approaches have also been recognized for the development of improved varieties. In vitro ovary culture and ovule culture have been attempted for transferring resistance traits from *S. alba* cv. Carine to *B. napus* cv. Brutor (Chevre et al. 1994) and *Erucastrum cardaminoides* to *B. oleracea* var. *alboglabra* (Mohanty et al. 2009). Intergeneric hybrids of *B. campestris* and *B. spinescens* have also been developed (Agnihotri et al. 1991). Moreover, the somatic hybridization technique has been applied for transferring the resistance traits from blight-tolerant species to susceptible ones, for example, *S. alba* to *B. napus* (Primard et al. 1988), *B. carinata* to *B. juncea* (Sharma and Singh 1992), and *B. nigra* and *B. oleracea* (Jourdan and Salazar 1993). Attempts have also been made for the introduction of somaclonal variations for incorporation of disease resistance/tolerance to Alternaria blight through mutagenesis using gamma rays, ethyl methanesulfonate, and ethyl nitrosourea (Verma and Rai 1980; Agnihotri et al. 2009; Meena et al. 2020). In addition, the defense response might be activated by treatment with biotic and abiotic agents for combating *A. brassicae* infection in oilseed Brassicas. Vishwanath et al. (1999) showed that resistance in a susceptible cultivar of *B. juncea* against the extremely virulent *A. brassicae* isolate A and reasonably virulent isolate C, both from *B. carinata*, was induced after inoculating the plants with the avirulent *A. brassicae* isolate D. Similarly, the application of β-aminobutyric acid on the leaves of *B. carinata* was found to induce resistance to *A. brassicae* (Chavan et al. 2013). In future, the activation of the defense response without gene manipulation may act as a better substitute for the conventional use of fungicides. Unfortunately, till date, none of the conventional or modern biotechnological methods have been found to be feasible or stable in developing blight-resistant *Brassica* species. Under such circumstances, it is pragmatic to use the novel sources of resistance from plant species that are not hosts to the pathogen. The transfer of genes from non-host plants conferring resistance to the pathogen is a promising strategy in parallel to the current efforts towards the development of oilseed *Brassica* varieties resistant to Alternaria blight (Fatima et al. 2019).

8.12 Sources of Resistance Identified at Different Locations

Digenomic *Brassica* species such as *B. napus* and *B. juncea* have better sources of resistance than the monogenomic species like *B. rapa*. Some lines of *B. alba, B. carinata, B. spinescens, B. maurorum, Eruca sativa, Camelina sativa, Capsella bursa-pastoris, Diplotaxis* sp., and *N. paniculata* have also shown resistance to *A. brassicae* in different areas of the world (Brun et al. 1987a, b; Conn et al. 1991; Grontoft 1986; Jejelowa et al. 1991; Saharan 1992, 1997; Tewari and Conn 1993; Hansen and Earle 1997; Chrungu et al. 1999; Sharma et al. 2002). Sources of resistance to *A. brassicae* and *A. brassicicola* in different host species from different areas of the world have been reported. In India, *Brassica* genotypes viz., CSR 43, CSR I42, CSR-I42-2, CSR 343, CSR 448, CSR 622, CSR 741, Gulivar, KRV-Tall, Midas, PHR 1, RC-781, TMV2, Tower, and YRT3 were consistently found to have field resistance proven after testing for several years at different locations under uniform disease nursery trials (Kolte 1985b, 1987; Saharan 1984, 1992, 1997; Saharan and Chand 1988; Saharan et al. 2003, 2005, 2014, 2016, 2018, 2019, 2021; Kumar et al. 2015; Verma and Saharan 1994; Meena et al. 2019b). *Brassica coenospecies* are rich reservoir for genetic resistance to *A. brassicae* (Sharma et al. 2002). *Alternaria brassicae* pathotype-specific resistance has been recorded in genotype, GS-05-1 of *B. napus* to pathotypes, Abr 1 and Abr 5 whereas genotypes, RH-8544, Pusa Swarnim, and HC-9605 of *B. juncea* showed moderately resistant reaction to three pathotypes (Kumar et al. 2014). Pravel Pratap et al. (2015) reported tolerant to Alternaria blight genotypes of *B. juncea* DRMR 2805, and DRMR 2806. Tolerance to *Sclerotinia* rot was reported in *B. juncea* genotypes, EC 597317, EC 597328, RH 1222-28, DRMR 261, DRMR 360, DRMR 1493,

DRMR 1034, and DRMR 2035 (Sharma et al. 2018). Resistant to white rust was reported by Meena et al. (2019a, b) in DRMR 2019 and DRMR 2035 genotypes of B. juncea while DRMR 2035 also showed tolerance to sclerotinia rot.

8.12.1 Use of Cruciferous Relatives as R-Genes Sources

Sources with high level of resistance against *A. brassicicola* and *A. brassicae* have not been identified among the cultivated *Brassica* species although individual cabbage varieties differ considerably in their level of susceptibility to black spot (Otani et al. 2001). The highest level of *Alternaria* resistance among the oilseed *Brassica* crops is displayed by the Ethiopian mustard (*B. carinata*). Among the wild cruciferous plants closely related to the *Brassica* genus, the highest resistance against *A. brassicae* is found in white mustard (*Sinapis alba*), (Kolte 1985a; Brun et al. 1987a, b; Ripley et al. 1992; Sharma and Singh 1992; Verma and Saharan 1994; Hansen and Earle 1995, 1997). The highest overall *Alternaria* spp. resistance, however, has been identified in the crucifers more distant from the *Brassica*, such as *Camelina* (*Camelina sativa*; false flax), shepherd's purse (*Capsella bursa-pastoris*), taramira (*Eruca sativa*), and ball mustard (*Neslia paniculata*) (Conn and Tewari 1986; Conn et al. 1988; Tewari 1991b). Resistance against *A. brassicae* and *A. brassicicola* has been reported among other wild members of the Brassicaceae family (Sharma et al. 2002; Tewari and Conn 1993; Warwick 2011) viz., *Alliaria petiolata*, *Barbarea vulgaris*, *B. elongate*, *B. desnottessi*, *B. fruticulosa*, *B. maurorum*, *B. nigra*, *B. souliei*, *B. spinescens*, *C. sativa*, *C. bursa-pestoris*; *Coincya* spp., *Diplotaxis catholica*, *D. berthautii*, *D. creacea*, *D. erucoides*, *D. tenuifolia*; *Erucastrum gallicum*, *E. vesicaria* subsp. *sativa*; *Hemicrambe fruticulosa*, *H. matronalis*, *N. paniculata*; *R. sativus*; *S. alba,* and *S. arvensis*. The completely immune plants remain symptomfree both under natural field infection and under controlled artificial inoculation (Sharma et al. 2002). Comparatively, broccoli and cauliflower varieties exhibit moderate *Alternaria* resistance, while the cabbage as susceptible.

8.12.2 Multiple Disease Resistance Sources Identified from *Brassica* Species

Brassica alba and genotypes viz., HC-1, PCC-2 (*B. carinata*), GSL-1501, GS-7027, Midas, Tower (*B. napus*), and DIR-1507, DIR-1522 (*B. juncea*) have been identified as sources of multiple disease resistance against white rust, Alternaria blight, downy mildew, and powdery mildew diseases (Dang et al. 2000; Kumar and Saharan 2002). Earlier *B. napus* genotype, GSL-1501 was reported resistant to white rust and powdery mildew (Gupta and Singh 1994). The variations in the incidence of powdery mildew scores among different Brassica genotypes differ considerably due to variations in date of sowing, stage of the crop at the time of observations and prevailing environmental conditions. Hence, early maturing genotypes or the genotypes planted early in the season would help in escaping the powdery mildew infection.

The genotypes, EC-322090, EC-322092, EC-322093, EC-399301, EC-399296, EC 399299, PHR-2, RC-781 and of *B. juncea*, HC-1 and PCC-2 of *B. carinata,* and GS-7027, GSL-1, GSL-1501, Gulivar 1, HNS-4, and Tower sel. 1 of *B. napus* are resistant to *A. candida* and *A. brassicae* (Awasthi et al. 2012; Meena et al. 2014; Sandhu et al. 2015). Whereas genotypes, EC-129126-1, EC-129126-2, PR-8805 of *B. juncea*, Candle, Tobin, Torch of *B. rapa*, DHC-1, HDC-4, DHC-1960, HC-1, PCC-2 of *B. carinata*, and GSL-1501, N 20-12-2 and Wester of *B. napus* are resistant to *A. candida* and *Erysiphe cruciferarum.* However, HC-1, PCC-2 of *B. carinata,* and GSL-1501 (*B. napus*) are resistant to both *A. brassicae* and *E. cruciferarum*. GSL-1501 and Gulivar-1 of *B. napus* were found resistant to *A. candida* and *E. cruciferarum* by Gupta and Singh (1994). The genotypes Wester (*B. napus*), Tobin and Candle (*B. rapa*) are reported completely free from *A. candida* and *E. cruciferarum* (Shivpuri et al. 1997). The variations in the observations may be attributed to the occurrence of new races of the pathogen and congenial weather factors (Kumar and Saharan 2002).

8.12.3 Assessing Relationship Between Major Foliar Diseases

The increase in level of resistance to different pathogens is positively correlated, the improvement in the accumulation of resistance to multiple diseases may be facilitated easily in agricultural or natural populations; negative association among the level of resistance will hinder plant defense against pathogens attack and leads to difficulties in the development of multiple disease resistance. The non-availability of single resistance source line with different mode of inheritance further complicates the work of genetic amelioration for multiple disease resistance. The significant and positive association between disease score of white rust, Alternaria blight and powdery mildew has been recorded except, white rust stag head and powdery mildew infection in genotypes of different *Brassica* species. A high degree of correlation is observed between leaf phases of white rust and Alternaria blight. Similarly, the positive genetic correlation is observed between *Hyaloperonospora parasitica* and *Leptosphaeria maculans* and *Hyaloperonospora parasitica* and *Albugo candida* in *B. rapa* (Mitchell-Olds et al. 1995; Mehta et al. 2018). Sharma et al. (1991) reported significant positive correlation between Alternaria blight and black rot in cauliflower. The strong evidence for a positive genetic correlation in the level of resistance against *A. candida, A. brassicae,* and *Erysiphe cruciferarum* may indicate the presence of plant resistance genes that provide defense against some fundamental characteristics common to different taxonomic orders of fungal pathogens as observed by Bruns et al. (1991). Genetic correlations may arise from several factors like (1) linkage disequilibrium, a non-random association among genotypes at several loci, which is rarely found in cross-pollinated species (Crow and Kimura 1970). However, the linkage disequilibrium may arise from previous generations of selection or small population size; (2) Pleiotropy, and (3) tight linkage between resistance genes (Martin et al. 1993).

8.13 Use of R-Genes Sources for the Development of Clubroot-Resistant Cultivars

Resistant cultivars are the only solution for protection against *P. brassicae* in combination with suitable crop rotation and management strategies (Peng et al. 2014). The clubroot host–plant resistance can be found in main cruciferous crops (except *Brassica juncea* and *B. carinata*) (Diederichsen et al. 2009). The resistance against *P. brassicae* is polygenic (quantitative) and also monogenic (qualitative) (Crute et al. 1980; Voorrips 1996). Many of the resistance types are pathotype specific and do not cover all pathotypes of *P. brassicae* (Diederichsen et al. 2009). The studies proved the existence of eight resistance loci, some of them are pathotype-specific major genes, and some are quantitative loci (QTL) with specific or nonspecific resistance responses (Voorrips 1996; Hwang et al. 2012). Some studies also proved that resistance is qualitative and is controlled by dominant and recessive genes (Chiang and Crete 1983; Yoshikawa 1993). The oilseed rape-resistant cultivar Mendel was breed from ECD hosts-resistant turnip (*B. rapa,* ECD-04) and kale (*B. oleracea* ECD-15) through backcross with oilseed rape (Diederichsen et al. 2009). The F_1 backcrosses were inoculated with *P. brassicae* and the most resistant ones were selected for further breeding and genetic analysis (Diederichsen et al. 2009). Later the presence of resistant genes was confirmed. There is one major locus and two recessive genes in the cultivar Mendel whereas other genes were lost in the breeding process and negative selection (Diederichsen et al. 2009). This is why cv. Mendel is recommended to use wisely together with other management strategies (Diederichsen et al. 2009; Dixon 2009). The same situation exists in other resistant cultivars.

8.13.1 Defense Mechanisms of Resistant Cultivars

The resistance mechanisms are based on suppressing the pathogen development in a host rather than on infection avoidance. The term resistance is used because the infection inhibition is based on resistance genes, not on tolerance genes. The zoospore penetration occurs in the root hairs of a resistant host (the primary phase of the pathogen life) (Hwang et al. 2012). The secondary phase of infection in the root cortex also follows, but formed plasmodia do not mature and resting spores are not created (Hwang et al. 2011; Deora et al. 2013). Slight suppression of primary infection and significant suppression of secondary infection occur during the resistance response, the disruption of the cell walls is lesser, and the secondary thickening of the stem is not stopped (Donald et al. 2008). The immune response includes the activation of metabolic pathways for synthesis of phytoalexins, phenolic compounds, lignins, and other substances that block the pathogen invasion (Siemens et al. 2006).

8.13.2 Performance of R-Genes Cultivars in the Field

The first resistant cultivars were Mendel and Tosca, marketed at the beginning of the new millennium. These cultivars are heterozygous hybrids with one gene of pathotype-specific resistance. Unfortunately, the Mendel host plant resistance was broken in some parts of Germany and France, where it was grown since 2000 (Diederichsen et al. 2014). In Great Britain, resistance was broken quite rapidly in the field, where cultivars were grown under strong selection pressure (Diederichsen et al. 2003). These cultivars were also tested on local pathotypes in Canada. Cultivar Mendel showed resistance against pathotypes 3, 5, 6, and 8 (Williams set) (Rahman et al. 2014). Cultivar Tosca was susceptible to pathotype 3 and resistant to pathotype 2 (Williams set) (Peng et al. 2013). These results showed that the host–plant resistance is quite variable and some genes may be ineffective in some regions with specific pathotypes (Peng et al. 2014). New resistant cultivars of spring rape with genes from European cultivars were introduced into Canada in 2009 (Peng et al. 2014). In the European market, the available resistant cultivars are derived from the first resistant cultivar Mendel (Diederichsen et al. 2014). It is expected that a wider range of resistant varieties will come to the market in the next years (Diederichsen et al. 2014). However, these varieties cannot be grown without suitable agronomic measures and pathogen-resistance management strategies (Diederichsen et al. 2014). Field experiments with resistant cultivars in infested fields were also conducted in the Czech Republic. Eight resistant cultivars were tested, some of them are normally available on the market and some were test cultivars in the breeding process (new breeds). Normally, traded cultivars had stable performance in resistance results and good yield. Cultivars from new breeding showed some variations in resistance performance and yield (Ricarova et al. 2016).

8.13.3 Field Deployment of R-Genes Cultivars

Clubroot-resistant canola cultivars have been the dominant component of the clubroot management strategy in western Canada in recent years. The first resistant cultivar for western Canada, Pioneer "45H29," was registered in 2009 and its release was quickly followed by several other resistant cultivars (Strelkov et al. 2011; Gossen et al. 2013). However, the durability of this resistance is not known. The genetic basis for the resistance in these cultivars also is proprietary knowledge that is not available to the public. This has complicated efforts aimed at resistance stewardship since it is not possible to develop rational strategies for the rotation of resistance sources in infested fields. The number of resistance genes also that are currently available to breeders is limited (Hirai 2006). Recent studies indicated that there were no detectable differences in pathotype reactions among clubroot-resistant canola cultivars available in western Canada (Deora et al. 2012, 2013). This represents one line of evidence that resistance genes in all these lines originated from a similar source. In contrast, another study by LeBoldus et al. (2012) revealed that pathogen populations that had been repeatedly cycled on and become adapted to one *Brassica* host did not exhibit a similar adaptation to other hosts, when the cycled populations were inoculated on the latter, indicating that distinct sources of resistance may occur in at least some genotypes. Regardless of the relationships between resistance sources, there is substantial genetic and pathotype variation present in the *P. brassicae* populations in western Canada (Strelkov et al. 2006; Xue et al. 2008; Cao et al. 2009), and the deployment of a cultivar with single-gene resistance against a genetically diverse pathogen on a large scale imposes a strong selection pressure for pathogen genotypes that are able to overcome this resistance (Hwang et al. 2012). Similar mechanisms may operate in other countries also.

8.14 Introgression of *Sclerotinia* Resistance into *Brassica juncea* Through Marker-Assisted Breeding

A set of 96 *Brassica juncea–Erucastrum cardaminoides* introgression lines (ILs) were developed with genomic regions associated with sclerotinia stem rot (*Sclerotinia sclerotiorum*) resistance from a wild Brassicaceous species *E. cardaminoides*. ILs was assessed for their resistance responses to stem inoculation with *S. sclerotiorum*, over three crop seasons. Initially, ILs was genotyped with transferable SSR markers and subsequently through genotyping by sequencing. SSR-based association mapping identified six marker loci associated to resistance in both A- and B-genomes. Subsequent genome-wide association analysis (GWAS) of 84 ILs recognized a large number of SNPs associated to resistance, in chromosomes A03, A06, and B03. Chromosomes A03 and A06 harbored the maximum number of resistance-related SNPs. Annotation of linked genomic regions highlighted an array of resistance mechanisms in terms of signal transduction pathways, hypersensitive responses, and production of antifungal proteins and metabolites. Of major importance was the clustering of SNPs, encoding multiple resistance genes on small regions spanning approximately 885 kb region on chromosome A03 and 74 kb on B03. Five SNPs on chromosome A03 (6,390,210-381) were associated with LRR-RLK (receptor-like kinases) genes that encode LRR-protein kinase family proteins. Genetic factors associated with pathogen-associated molecular patterns (PAMPs) and effector-triggered immunity (ETI) were predicted on chromosome A03, exhibiting 11 SNPs (6,274,763-994). These belonged to three R-genes encoding TIR-NBS-LRR proteins. Marker trait associations (MTAs) identified will facilitate marker-assisted introgression of these critical resistances, into new cultivars of B. juncea initially and, subsequently, into other crop *Brassica* species (Rana et al. 2019). Association mapping allowed detection of ten markers trait associations (MTAs) after Benferroni correction. These were CNU-M-157-2, RA2-G05, CNU-M-353-3, CNU-M-442-5,

ACMP00454-2, ACMP00454-3, EIN2-3-1, M 641-1, Na10D09-1, and Na 10D11-1 (Rana et al. 2017).

8.14.1 *Sclerotinia* Resistance Introgression from Wild *Brassica* Species into *Brassica juncea*

A wild *Brassicaceae* species with excellent resistance against stem rot to develop a set of *B. juncea—B. fruticulosa* introgression lines (ILs) was used by Atri et al. (2019). These were assessed for resistance using a highly reproducible stem inoculation technique against a virulent pathogen isolate. Over 40% of ILs showed higher levels of resistance. IL-43, IL-175, IL-215, IL-223, and IL-277 were most resistant ILs over three crop seasons. Sequence reads (21x) from the three most diverse ILs were then used to create *B. juncea* pseudo molecules, by replacing SNPs of reference *B. juncea* with those of resequenced ILs. Genotyping by sequencing (GBS) was also carried out for 88 ILs. Resultant sequence tags were then mapped on to the *B. juncea* pseudo molecules, and SNP genotypes prepared for each IL. Genome-wide association studies helped to map resistance responses to stem rot. A total of 13 significant loci were identified on seven *B. juncea* chromosomes (A01, A03, A04, A05, A08, A09, and B05). Annotation of the genomic region around identified SNPs allowed identification of 20 candidate genes belonging to major disease resistance protein families, including *TIR-NBS-LRR* class, Chitinase, Malectin/receptor-like protein kinase, defensin-like (*DEFL*), desulfoglucosinolate sulfotransferase protein, and lipoxygenase. A majority of the significant SNPs could be validated using whole-genome sequences (21x) from five advanced generation lines being bred for *Sclerotinia* resistance as compared to three susceptible *B. juncea* germplasm lines. The understanding of the defensive pathway of *B. fruticulosa* resistance will also enable the development of marker candidates for assisted transfer of introgressed resistant loci in to agronomically superior cultivars of crop *Brassica* (Atri et al. 2019).

8.15 Use of Hexaploidy Approach for Pyramiding of Major R-QTLs into *Brassica napus* Against *Sclerotinia*

To maintain stable production in the face of damage from diseases, the development of resistant lines or cultivars is a cost-effective strategy. However, due to the lack of highly resistant resource in *B. napus* gene pool, the breeding of *Sclerotinia*-resistant rapeseed has been difficult and rather unsuccessful. Alternatively, diploid progenitors which always provide precious gene resources could be considered in rapeseed improvement. *Sclerotinia*-resistant sources were reported in *B. oleracea* germplasm (Mei et al. 2011). In particular, a wild *B. oleracea* genotype (*B. incana*) identified as strong resistance to *S. sclerotiorum* (Mei et al. 2011), provided a promising basis to improve *Sclerotinia* resistance in present rapeseed cultivars. However, this raised problems in utilization of the *B. oleracea* resource by resynthesizing rapeseed technique, i.e., resynthesized allotetraploid lines are genetically unstable (Hurgobin et al. 2018; Wu et al. 2018; Zou et al. 2018), and the synthetic rapeseed lines contain A-genome from *B. rapa* which is highly susceptible to *S. sclerotiorum* (Ding et al. 2013; Mei et al. 2011). Therefore, novel breeding strategies are needed to avoid these problems (Mei et al. 2020). There is another bottleneck in *Sclerotinia*-resistant breeding of rapeseed, as it is a typical quantitative trait controlled by multiple genetic loci (Mei et al. 2013; Wei et al. 2016; Wu et al. 2013, 2016). Till date, no resistant gene has been cloned from any of those reported QTLs, and most importantly, a single locus always provides very limited contribution to resistance. These factors have largely limited the molecular marker-assisted breeding for *Sclerotinia*-resistant rapeseed.

Resistance QTLs were identified from *Sclerotinia*-resistant wild *B. oleracea* (Mei et al. 2013), and several molecular markers were proved effective in selecting resistant individuals (Ding et al. 2015, 2019; Mei et al. 2015). A hexaploidy strategy was preliminarily proposed to introduce *Sclerotinia* resistance from the wild *B. oleracea* into rapeseed (Mei et al. 2015). During the genotypic (marker-assisted selection) and phenotypic selection in high generations generated from the hexaploidy strategy, successfully yielded resistant rapeseed lines with multiple resistant QTLs, improved *Sclerotinia* resistance, high seed yield and quality. Resistant rapeseed lines pyramided with multiple resistant QTLs derived from *Brassica oleracea* were developed via a hexaploidy strategy. Interspecific hexaploids were generated between rapeseed variety "Zhongshuang 9" and a wild *B. oleracea* which was highly resistant to *S. sclerotiorum*, followed by backcrossing with Zhongshuang 9 and successive selfing. By molecular marker-assisted selection, three major resistant QTLs were transferred and pyramided from *B. oleracea* into two BC1F8 lines which exhibited ~35% higher resistance level than Zhongshuang 9 and produced good seed yield and seed quality. The successful development of *Sclerotinia*-resistant rapeseed lines by introducing multiple resistant loci from wild *B. oleracea* has been obtained. This revealed the effectiveness of pyramiding multiple QTLs in improving *Sclerotinia* resistance in rapeseed and provided a novel breeding strategy on utilization of *B. oleracea* in rapeseed improvement (Wang et al. 2020).

8.16 Black Rot Resistance Introgression from *Brassica carinata* to *Brassica oleracea* Through Embryo Rescue

The black rot resistance sources have been reported in A- and B-genomes of *Brassica* species (Westman et al. 1999; Taylor et al. 2002) and genes present in these genomes can be pyramided in Cole crops to provide a durable solution to this malady. The monogenomic and digenomic species in Uthant triangle carry resistance gene(s) for *Xcc*, which can be utilized for guarding *B. oleracea* genotypes against this disease using embryo rescue and somatic hybridization techniques. Genetic transformation option can be explored when resistance is not available within and related gene pools. Besides, a number of disease resistance genes in *Brassicas* and *A. thaliana* have been postulated and mapped, but none has been cloned as yet (Vicente and Holub 2013). Therefore, there is no information on genetic transformation for resistance to black rot disease in *B. oleracea* group.

Brassica carinata (Ethiopian mustard, $2n = 4\times = 34$, BBCC genome) is an important oilseed crop thought to have originated in the Ethiopian plateau (Warwick 2011). Its leaves are also consumed as vegetable in North East Africa. The *Brassica* B-genome is known to carry several important traits, yet there has been limited analyses of its underlying genome structure (Navabi et al. 2013). *B. carinata* harbors genes controlling valuable resistance/tolerance traits for biotic and abiotic stresses viz., resistance to blackleg, black rot, and tolerance to aluminum, salinity, heat, and drought (Pan et al. 2012; Enjalbert et al. 2013). Moreover, strong resistance to Xcc race 1 and 4 has been reported in high proportion of *B. nigra* (B-genome) and *B. carinata* (BC genome) accessions (Taylor et al. 2002; Vicente et al. 2002). A single dominant gene conferring high resistance to *Xcc* race 1 was reported by Tonguc et al. (2003). A single dominant gene (Xca1bc) controlling black rot resistance was mapped with intron length polymorphic markers at B-7 genome of *B. carinata* (Sharma et al. 2016). Therefore, it is important and feasible to transfer this black rot resistance gene into cauliflower cultivars. Alien gene transfer in crop plants has emerged as a potent approach to broaden gene pools and increase genetic variability in existing germplasm. Hybridization between a diploid and a tetraploid species is difficult and failure occurs at many stages starting from pollination incompatibility to pre/post-fertilization barriers.

Search for new gene(s) for disease resistance in closer genome-related species viz., *B. napus* (AACC) and *B. carinata* (BBCC) for incorporation into *B. oleracea* group via interspecific (Tonguc et al. 2003) or somatic hybridization using embryo rescue (Hansen and Earle 1995) has to be a continuous process. Of the three *Brassica* genomes, A and C are very closely related while the B-genome is phylogenetically distant. Zhou et al. (1997) transferred resistance to Xcc from *B. napus* to broccoli (*B. oleracea*) through protoplast fusion. Tonguc and Griffiths (2004b), however, explored the half B-genome *B. juncea* (AABB) for the introgression of black rot resistance into *B. oleracea* using *in vitro* embryo rescue. Somatic hybridization was attempted to transfer black rot resistance from *B. nigra* into *B. oleracea* (Wang et al. 2011, 2016).

Utilization of embryo rescue or somatic hybridization techniques, or combination of both, could help to overcome natural reproductive barriers in the development of interspecific hybrids in *Brassica* (Ayotte et al. 1987; Hansen and Earle 1995; Momotaz et al. 1998; Weerakoon et al. 2009; Niemann et al. 2013). Wide crosses between crop plants and their wild relatives have now become routinely possible through the use of the embryo rescue technique. Different techniques of plant cell and tissue culture, such as ovary, ovule, and embryo culture as well as protoplast fusion, have proved useful for production of interspecific hybrids. Rescue of hybrid embryos and their culture in vitro helps to overcome post-fertilization barriers in interspecific crosses. Since its first use in *Brassica* by Nishi et al. (1959), extensive investigations have been carried out to improve the techniques for obtaining higher seed set (Inomata 1993, 2002; Zhang et al. 2003, 2004). The successful application of this technique depends on the stage of the embryo being rescued and cultured in vitro. Several attempts were made to transfer desirable gene(s) from alien *Brassica* spp. to *B. oleracea* such as powdery mildew resistance (Tonguc and Griffiths 2004a), downy mildew (Chiang et al. 1977), male sterility (Chiang and Crete 1987), and atrazine resistance (Jourdan et al. 1989). The progress for marker-assisted Xcc resistance gene transfer from *B. carinata* to cauliflower has been very slow (Tonguc et al. 2003; Tonguc and Griffiths 2004b). The transfer of monogenic resistance to Xcc race 1 from an accession "NPC-9" of *B. carinata* into a commercial susceptible cauliflower variety "Pusa Sharad" using embryo rescue technique has been attempted by Sharma et al. (2017).

Black rot caused by *X. campestris* pv. *campestris* (*Xcc*) is a very important disease of cauliflower (*Brassica oleracea botrytis* group) resulting into 10–50% yield losses every year. Since there is a dearth of availability of resistance to black rot disease in *B. oleracea* (C-genome), therefore exploration of A- and B-genomes was inevitable as they have been reported to be potential reservoirs of gene(s) for resistance to black rot. To utilize these sources, interspecific hybrid and backcross progeny (B1) were generated between cauliflower "Pusa Sharad" and Ethiopian mustard "NPC-9" employing *in vitro* embryo rescue technique. Direct ovule culture method was better than siliqua culture under different temperature regime periods. Hybridity testing of F1 interspecific plants was carried out using co-dominant SSR marker and *Brassica* B- and C-genome-specific (DB and DC) primers. Meiosis in the digenomic (BCC) interspecific hybrid of *B. oleracea botrytis* group ($2n = 18$, CC) × *B. carinata* ($2n = 4x = 34$, BBCC) was higly disorganized and cytological analysis of pollen mother cells revealed chromosomes $2n = 26$ at metaphase-I. Fertile giant pollen grain formation was observed frequently in interspecific F1 hybrid and BC1 plants. The F1 interspecific plants were found to be resistant to *Xcc* race 1. Segregation distortion was observed in BC1 generation for black rot resistance and for different morphological traits. The At1g70610 marker analysis confirmed successful introgression of black rot resistance in interspecific BC1 population. This effort will go a long way in pyramiding gene(s) for resistance against black rot in Cole crops, especially cauliflower and cabbage for developing durable resistance, thus minimize dependency on bactericides (Sharma et al. 2017).

8.17 Black Rot Resistance Breeding Through Somatic Hybridization and Embryo Rescue

Wide hybridization or distant hybridization is a powerful approach to combine the diverged genomes into a single nucleus. Wide hybridization is an important driving force in broadening the gene pools, the evolution of novel plant forms and speciation, transferring the genomes of one species to another, enhancing the genetic diversity and improving the quality of crops (Soltis and Soltis 2000; Barton 2001; Jiao et al. 2011). The *Brassica* vegetables and oilseed crops in the cruciferae family have undergone multiple genomic duplications (Schranz et al. 2006). The classic genome polyploidization event is the interspecific hybridization of *B. rapa* (A-genome), *B. nigra* (B-genome), and *B. -oleracea* (C-genome) to form the allotetraploids *B. juncea* (AB genome), *B. napus* (AC genome), and *B. carinata* (BC genome) (Nagaharu 1935). Since the first intergeneric hybrid between *R. sativus* and *B. oleracea*, "*Raphanobrassica*," reported by Karpechenko (1928), numerous intergeneric and interspecific hybrids have been produced between *Brassica* species of the U triangle (Nagaharu 1935) and *R. sativus*. The related or wild species are a rich reservoir of several useful genes having resistance to various biotic or abiotic stresses. In the family Brassicaceae, interspecific hybridization is a potential approach for transfer of desirable genes from one species to another, enhancing genetic diversity and creation of hexaploids for the improvement of *Brassica* crops (Prakash 1973; Prakash and Chopra 1988; Choudhary et al. 2000; Siemens et al. 2002; Niemann et al. 2012; Kaneko and Bang 2014). The sexual incompatibility or reproductive barriers can be overcome by utilization of embryo rescue or somatic hybridization techniques, or a combination of both, in the development of interspecific hybrids in crucifer crops (Hansen and Earle 1995; Weerakoon et al. 2009; Niemann et al. 2012; Dey et al. 2015). The development of black rot-resistant hybrids is one of the major objectives of *Brassica* breeders. Since the resistance sources in *B. oleracea* are insufficient, interspecific hybridization with related or distant species followed by embryo rescue or somatic hybridization for the development of black rot-resistant hybrids has been practiced and accomplished by many workers.

The lines of *B. napus* have very high resistance to black rot (*X. campestris* pv. *campestris*) and resistance is governed by a dominant gene, but the successful crossing between two species (*B. oleracea* and *B. napus*) is very difficult (Hansen and Earle 1995). Therefore, the development of somatic hybrids by protoplast fusion between rapid cycling *B. oleracea* and *X. campestris* pv. *Campestris*-resistant line of *B. napus* has been attempted. Hybrid identity was confirmed by morphological studies, flow cytometric estimation of nuclear DNA content, and analysis of RAPDs. Wang et al. (2011) generated new alien addition lines having high resistance to black rot (*X. campestris* pv. *campestris*) by employing a somatic hybridization technique between cauliflower and black mustard. It has been evident that success of interspecific hybridization is also influenced by the genotype of the parents (Meng and Lu 1993; Tonguc and Griffiths 2004a, b; Mason et al. 2011). Tonguc and Griffiths (2004b) utilized two *B. juncea* accessions (A 19182 and A 19183) having resistance to Xcc race 1 and 4, as maternal and paternal parents to develop interspecific hybrids with *B. oleracea* cultivars. They employed embryo rescue to recover interspecific hybrids and further confirmed by inheritance of paternal molecular markers. They obtained 26 interspecific hybrid plants between A 19182 and *B. oleracea* cultivars, but no interspecific hybrids were obtained using A 19183, suggesting influence of genotype in the success of interspecific hybridization. The interspecific hybrid plants were male-sterile, and they were used successfully as maternal parents to develop backcross plants employing embryo rescue. All hybrid and BC1 plants were found resistant to races 1 and 4 of *Xcc*. Dey et al. (2015) reported introgression of black rot resistance genes into Indian cauliflower (*B. oleracea* var. *botrytis* L.) through interspecific hybridization (*B. carinata* 9 *B. oleracea*) followed by embryo rescue. They utilized two

B. carinata genotypes, NPC-9 and HCA-6, having very high resistance to Xcc race 1 and 4, for interspecific hybridization to introgress resistance genes from *B. carinata* into Indian cauliflower employing three different embryo rescue approaches. They reported that all the F1 hybrids and back cross were resistant to *Xcc* races 1 and 4 (Singh et al. 2018).

8.18 Accessing and Exploiting Genetic Diversity

The advent of molecular genetics provided new opportunities for mapping and tracking genes of agronomic interest, leading to more efficient marker-assisted selection. Whole-genome sequencing, starting with *Arabidiopsis* and rice as models for dicotyledons and monocotyledons, respectively, and followed by a rapidly increasing number of crop plant genomes, have led to a quantum leap not only in understanding plant genetic diversity, but also for accessing this enormous resource. For many crop plant species, for example, tomato, barley, maize, wheat, and various *Brassica* species, either the entire genome or the generic parts of the genome are now emerging. As bioinformatic tools for analyzing the exponential increase in genome data improve, the practical utility of such data will also be enhanced. This will extend the options for breeding disease resistance.

The presence of conserved motifs in plant resistance (R genes), such as the nucleotide-binding site-leucine-rich repeat (NBS-LRR) domains, has facilitated the identification of gene families, and resistance gene analogs in other plants. The *Arabidopsis* genome has around 150 NBS-LRR encoding genes and rice c. 400 (McHale et al. 2006). Studies of genome structure have shown that many putative R-genes are clustered and have undergone duplication, and evolution due to diversifying selection. Functional analysis of these entire candidate genes is a demanding task, but improvement in plant transformation protocols, and high-throughput gene attenuation methods, such as RNA interference (RNAi) and virus-induced gene silencing (VIGS) should accelerate the identification of novel genes of practical utility.

8.19 Identification of Novel Sources and Transfer of R-Genes

Novel sources of R-genes for blackleg can be sourced from other A- and C-genome of *Brassica* species (Chen et al. 2010) or from the B-genomes in *B. nigra* (Chevre et al. 1996), *B. juncea* (Christianson et al. 2006), and *B. carinata* (Plieske et al. 1998). Wild Brassicaceae species, such as *Sinapis arvensis*, also serve as a good source of R-genes for blackleg (Snowdon et al. 2000). However, R-genes from other wild species remain to be tapped. Besides identifying candidate R-genes through comparative genomic analysis, more screening work on wild and land races, including species from the Brassicaceae family, should be performed against the main pathogens to identify resistant plants. This would broaden the gene pool for introgression of novel R-genes or alleles into *Brassica* cultivars. Combining this information with genome sequencing will enable characterization of these novel resistance genes. Although effector genes are usually unique and do not often share structural similarity but encode very diverse proteins, it is still possible to identify R-gene that could recognize common or near identical effector gene(s) in different pathogen pathotypes/races/isolates, which already exists naturally in some plants/crops (Dangl et al. 1992). This type of R-gene is useful and can be transferred between hosts that are closely related or within the same genus or family. Some R-genes can also be transferred from host to non-host using GM technologies, providing broad spectrum immunity against one or more pathogens. One such successful example is the pepper *Bs2* gene against bacterial spot disease transferred from its host pepper into non-host tomato (both members of Solanaceae family; Tai et al. 1999). The Bs2 gene encodes the R-gene NBS-LRR class (Tai et al. 1999) while the gene avrBs2 in *Xanthomonas campestris* pv. *vesicatoria*, causal pathogen of bacterial spot disease, is essential for virulence activity and is highly conserved between and within the pathovars of *X. campestris* (Kearney and Staskawicz 1990). This has opened up new possibilities to protect economically valuable crops, which lack R-gene resources, from disastrous pathogens (Neik et al. 2017).

8.20 Difficulties in Transfer of Plant Disease Resistance Trait

The difference in biology of the several types of plant pathogens (fungi, bacteria, viruses, and nematodes) infecting crucifers presents a substantial problems in transfer of R-genes and developing GM-resistant plants. Firstly, the kinds of organisms causing disease are taxonomically highly diverse; the major groups include cellular pathogens (bacteria, fungi, and the algal Oomycetes) and molecular pathogens (viruses). These are physiologically very different from each other, and therefore, no single-gene product can be expected to have a direct toxic effect on all types of pathogens. Secondly, pathogens use three major life strategies, namely biotrophic, hemibiotrophic, and necrotrophic. Biotrophic pathogens essentially act as a sink for the host's anabolic assimilates, and therefore keep it alive. Meanwhile, necrotrophic pathogens consume the host tissues on invasion. Hemibiotrophs combine both strategies in their life cycle. Consequently, plants have developed quite different ways for dealing with these strategies.

References

Agnihotri A, Lakshmikumaran MS, Jagannathan V, Shivanna KR (1991) Wide hybridization for improvement in cultivated Brassicas. Acta Hortic (ISHS) 289:213–214

Agnihotri A, Gupta K, Prem D, Sarkar, Mehra VS, Zargar SM (2009) Genetic enhancement in rapeseed-mustard for quality and disease resistance through *in vitro* techniques. In: Proceedings of 16th Australian research assembly on Brassicas, Ballarat, Australia, 14–16 Sept, p 28

Aneja JK, Agnihotri A (2013) Alternaria blight of oilseed *Brassicas*: epidemiology and disease control strategies with special reference to use of biotechnological approaches for attaining host resistance. J Oilseed Brassica 4:1–10

Atkinson NJ, Urwin PE (2012) The interaction of plant biotic and abiotic stresses: from genes to the field. J Exp Bot 63(10):3523–3543. https://doi.org/10.1093/jxb/ers100

Atkinson NJ, Lilley CJ, Urwin PE (2013) Identification of genes involved in the response of Arabidopsis to simultaneous biotic and abiotic stresses. Plant Physiol 162:2028–2041. https://doi.org/10.1104/pp.113.222372

Atri C, Akhatar J, Gupta M, Gupta N, Goyal A, Rana K, Kaur R, Mittal M, Sharma A, Singh MP, Sandhu PS, Barbetti MJ, Banga SS (2019) Molecular and genetic analysis of defensive responses of *Brassica juncea—B. fruticulosa* introgression lines to *Sclerotinia* infection. Sci Rep 9:17089. https://doi.org/10.1038/s41598-019-53444-3

Awasthi RP, Nashaat NI, Kolte SJ, Tewari AK, Meena PD, Bhatt R (2012) Screening of putative resistant sources against Indian and exotic isolates of *Albugo candida* inciting white rust in rapeseed-mustard. J Oilseed Brassica 3:27–37

Ayotte R, Harney PM, Souza MV (1987) The transfer of atrazine resistance from *Brassica napus* L. to *B. oleracea* L. Production of F1 hybrids through embryo rescues. Euphytica 36:615–624. https://doi.org/10.1007/BF00041511

Barton NH (2001) The role of hybridization in evolution. Mol Ecol 10:551–568

Brun H, Plessis J, Renard M (1987a) Resistance of some crucifers to *Alternaria brassicae* (Berk.) Sacc. In: Proc. 7th international rapeseed congress, Poznan, Poland, pp 1222–1227

Brun H, Renard M, Plessis J (1987b) Resistance of some crucifers to Alternaria brassicae (Berk.) Sacc. In: Proc. 7th international rapeseed congress, Poznan, Poland, p 247

Bruns TD, White TJ, Taylor JW (1991) Fungal molecular systematic. Annu Rev Ecol Syst 22:525–564

Cao T, Manolii VP, Hwang SF, Howard RJ, Strelkov SE (2009) Virulence and spread of *Plasmodiophora brassicae* [clubroot] in Alberta, Canada. Can J Plant Pathol 31:321–329

Chatterjee M, Mazumder M, Basu D (2013) Functional analysis of the promoter of a glycosyl hydrolase gene induced in resistant *Sinapis alba* by *Alternaria brassicicola*. Phytopathology 103:841–850. https://doi.org/10.1094/PHYTO-11-12-0303-R

Chavan V, Bhargava S, Kamble A (2013) Temporal modulation of oxidant and anti-oxidative responses in *Brassica carinata* during β-aminobutyric acid-induced resistance against *Alternaria brassicae*. Physiol Mol Plant Pathol 83:35–39. https://doi.org/10.1016/j.pmpp.2013.03.002

Chen S, Zou J, Cowling WA, Meng J (2010) Allelic diversity in a novel gene pool of canola-quality *Brassica napus* enriched with alleles from *B. rapa* and *B. carinata*. Crop Pasture Sci 61:483–492. https://doi.org/10.1071/CP09327

Cheng F, Liu S, Wu J, Fang L, Sun S, Liu B, Li P, Hua W, Wang X (2011) BRAD, the genetics and genomics database for *Brassica* plants. BMC Plant Biol 11:136

Chevre AM, Eber F, Margale E, Kerlan MC (1994) Comparison of somatic and sexual *Brassica napus-Sinapis alba* hybrids and their progeny by cytogenetic studies and molecular characterization. Genome 37:367–371

Chevre AM, Eber F, This P, Barret P, Tanguy X, Brun H, Dlseny M, Renard M (1996) Characterization of *Brassica nigra* chromosomes and of blackleg resistance in *B. napus–B. nigra* addition lines. Plant Breed 115:113–118. https://doi.org/10.1111/j.1439-05231996.tb00884.x

Chhikara S, Chaudhury D, Dhankher OP, Jaiwal PK (2012) Combined expression of a barley class II chitinase and type I ribosome inactivating protein in transgenic *Brassica juncea* provides protection against *Alternaria brassicae*. Plant Cell Tissue Organ Cult 108(1):83–89

Chiang MS, Crete R (1983) Transfer of resistance to race 2 of *Plasmodiophora brassicae* from *Brassica napus* to cabbage (*B. oleracea* spp. *capitata*). V. The inheritance of resistance. Euphytica 32:479–483

Chiang MS, Crete R (1987) Cytoplasmic male sterility in *Brassica oleracea* induced by *B.* napus cytoplasm-female fertility and restoration of male fertility. Can J Plant Sci 672:891–897. https://doi.org/10.4141/cjps87-126

Chiang MS, Chiang BY, Grant WF (1977) Transfer of resistance to race 2 of *Plasmodiophora brassicae* from *Brassica napus* to cabbage (*B. oleracea* var. *capitata*). I. Interspecific hybridization between *B. napus* and *B. oleracea* var. *capitata*. Euphytica 26:319–336

Choudhary BR, Joshi P, Ramarao S (2000) Interspecific hybridization between *Brassica carinata* and *Brassica rapa*. Plant Breed 119:417–420

Christianson JA, Rimmer SR, Good AG, Lydiate DJ (2006) Mapping genes for resistance to *Leptosphaeria maculans* in *Brassica juncea*. Genome 49:30–41. https://doi.org/10.1139/g05-085

Chrungu B, Verma N, Mohanty A, Pradhan A, Shivanna KR (1999) Production and characterization of interspecific hybrids between *Brassica maurorum* and crop Brassicas. Theor Appl Genet 98:608–613

Conn KL, Tewari JP (1986) Hypersensitive reaction induced by *Alternaria brassicae* in *Eruca sativa*, an oil-yielding crucifer. Can J Plant Pathol 8:348. (Abstr)

Conn KL, Tewari JP, Hadziyev D (1984) The role of epicuticular wax in canola in resistance to *Alternaria brassicae*. Phytopathology 74:851

Conn KL, Tewari JP, Dahiya JS (1988) Resistance to *Alternaria brassicae* and phytoalexin-elicitation in rapeseed and other crucifers. Plant Sci 55:21–25. https://doi.org/10.1016/0168-9452(88)90037-4

Conn KL, Jejelowa OA, Tewari JP, Bains PS (1991) *Camelina sativa* phytoalexins (Calmexin and methoxy camalexin) provide resistance against *Alternaria brassicae*. Abstract 274. Can J Plant Pathol 13

Crow JF, Kimura M (1970) An introduction to population genetics theory. Harper and Row, New York

Crute IR, Gray AR, Crisp P, Buczacki ST (1980) Variation in *Plasmodiophora brassicae* and resistance to clubroot disease in Brassicas and allied crops—a critical review. Plant Breed Abstr 50:91–104

Cui H, Tsuda K, Parker JE (2015) Effector-triggered immunity: from pathogen perception to robust defense. Annu Rev Plant Biol 66:487–511. https://doi.org/10.1146/annurev-arplant-050213-040012

Dang JK, Sangwan MS, Mehta N, Kaushik CD (2000) Multiple disease resistance against four fungal foliar diseases of rapeseed-mustard. Indian Phytopathol 53:455–458

Dangl JL, Ritter C, Gibbon MJ, Mur LA, Wood JR, Goss S, Mansfield J, Taylor JD, Vivian A (1992) Functional homologs of the *Arabidopsis* RPM1 disease resistance gene in bean and pea. Plant Cell 4:1359–1369. https://doi.org/10.1105/tpc.4.11.1359

Deora A, Gossen BD, Mcdonald MR (2012) Infection and development of *Plasmodiophora brassicae* in resistant and susceptible canola cultivars. Can J Plant Pathol 34(2):239–247. https://doi.org/10.1080/07060661.2012.681071

Deora A, Gossen BD, McDonald MR (2013) Cytology of infection, development and expression of resistance to *Plasmodiophora brassicae* in canola. Ann Appl Biol 163:56–71

Dey SS, Sharma K, Dey RB, Kumar GMS, Singh D, Kumar R, Parkash C (2015) Inter specific hybridization (*Brassica carinata* × *Brassica oleracea*) for introgression of black rot resistance genes into Indian cauliflower (*B. oleracea* var. botrytis L.). Euphytica 204:149–162. https://doi.org/10.1007/s10681-015-1352-0

Dias JS, Ferreira ME, Williams PW (1993) Screening of Portuguese cole landraces (*Brassica oleracea* L.) with *Peronospora parasitica* and *Plasmodiophora brassicae*. Euphytica 67:135–141

Diederichsen E, Deppe U, Sacristan MD (2003) Characterization of clubroot resistance in recent winter oilseed rape material. In: Proceedings of the 11th international rapeseed congress, vol 1, Copenhagen, Denmark, pp 68–70

Diederichsen E, Frauen M, Linders EGA, Hatakeyama K, Hirai M (2009) Status and perspectives of clubroot resistance breeding in crucifer crops. J Plant Growth Regul 28:265–281. https://doi.org/10.1007/s00344-009-9100-0

Diederichsen E, Frauen M, Ludwig-Muller J (2014) Clubroot disease management challenges from a German perspective. Can J Plant Pathol 36:85–98. https://doi.org/10.1080/07060661.2013.861871

Ding Y, Mei J, Li Q, Liu Y, Wan H, Wang L, Becker HC, Qian W (2013) Improvement of *Sclerotinia sclerotiorum* resistance in *Brassica napus* by using *B. oleracea*. Genet Resour Crop Evol 60:1615–1619. https://doi.org/10.1007/s10722-013-9978-z

Ding Y, Mei J, Liu Y, Wang L, Li Y, Wan H, Li J, Qian W (2015) Transfer of sclerotinia stem rot resistance from wild *Brassica oleracea* into *B. rapa*. Mol Breed 35:225

Ding Y, Mei J, Wu Q, Xiong Z, Li Y, Shao C, Wang L, Qian W (2019) Synchronous improvement of subgenomes in allopolyploid: a case of *Sclerotinia* resistance improvement in *Brassica napus*. Mol Breed 39:10

Dixon GR (2009) Husbandry—the sustainable means of controlling soil borne pathogens: a synoptic review. In: Dixon GR (ed). Proceedings of the symposium of horticulture in Europe, Vienna, Austria; 17–21. February 2008. Acta Hortic 817:233–242

Dodds PN, Rathjen JP (2010) Plant immunity: towards an integrated view of plant-pathogen interactions. Nat Rev 11:539–548. https://doi.org/10.1038/nrg2812

Donald EC, Jaudzems G, Porter IJ (2008) Pathology of cortical invasion by *Plasmodiophora brassicae* in clubroot resistant and susceptible *Brassica oleracea* hosts. Plant Pathol 57:201–209. https://doi.org/10.1111/j.1365-3059.2007.01765.x

Enjalbert JN, Zheng S, Johnson JJ, Mullen JL, Byrne PF, McKay JK (2013) Brassicaceae germplasm diversity for agronomic and seed quality traits under drought stress. Indian Crops Prod 47:176–185

Fatima U, Bhorali P, Borah S, Senthil-Kumar M (2019) Perspectives on the utilization of resistance mechanisms from host and nonhost plants for durable protection of *Brassica* crops against Alternaria blight. PeerJ 7:e7486. https://doi.org/10.7717/peerj.7486

Fujita M, Fujita Y, Noutoshi Y, Takahashi F, Narusaka Y, Yamaguchi-Shinozaki K, Shinozaki K (2006) Crosstalk between abiotic and biotic stress responses, a current view from the points of convergence in the stress signaling networks. Curr Opin Plant Biol 9:436–442. https://doi.org/10.1016/j.pbi.2006.05.014

Ghose K, Dey S, Barton H, Loake GJ, Basu D (2008) Differential profiling of selected defense-related genes induced on challenge with *Alternaria brassicicola* in resistant white mustard and their comparative expression pattern in susceptible India mustard. Mol Plant Pathol 9 (6):763–775. https://doi.org/10.1111/j.1364-3703.2008.00497.x

Gossen BD, Mcdonald MR, Hwang SF, Strelkov SE, Peng G (2013) A comparison of clubroot development and management on canola and *Brassica* vegetables. Can J Plant Pathol 35 (2):175–191. https://doi.org/10.1080/07060661.2013.763293

Grontoft M (1986) Resistance to *Alternaria* spp. in oil crops. Sver Utsadesforenings Tidskr 96 (3):263

Gupta RBL, Singh M (1994) Source of resistance to white rust and powdery mildew of mustard. Int J Trop Plant Dis 12:225–227

Gupta SK, Gupta PP, Yadav TP, Kaushik CD (1990) Metabolic changes in mustard due to Alternaria leaf blight. Indian Phytopathol 43:64–69

Gupta K, Saharan GS, Singh D (2002) Effective selection criteria to assess disease tolerance in Indian mustard [*Brassica juncea* (L.) Czern & Coss.]. J Mycol Plant Pathol 32:72–80

Hansen LN, Earle ED (1995) Transfer of resistance to *Xanthomonas campestris* pv. *campestris* into *Brassica oleracea* L. by protoplast fusion. Theor Appl Genet 91:1293–1300. https://doi.org/10.1007/BF00220944

Hansen LH, Earle ED (1997) Somatic hybrids between *Brassica oleracea* (L.) and *Sinapis alba* (L.) with resistance to *Alternaria brassicae* (Berk) Sacc. Theor Appl Genet 94:1078–1085

Hirai M (2006) Genetic analysis of clubroot resistance in *Brassica* crops. Breed Sci 56:223–229

Holub EB, Tor M, Cooper A, Gordon P, Gunn N (2002) Genetics of disease resistance in *Arabidopsis* to crop pathogens. Plant Prot Sci 38:151–154. Proc 6th conf EFPP 2002, Prague

Hurgobin B, Golicz AA, Bayer PE, Chan CKK, Tirnaz S, Dolatabadian A, Schiessl SV, Samans B, Montenegro JD, Parkin IAP, Pires JC, Chalhoub B, King GJ, Snowdon R, Batley J, Edwards D (2018) Homoeologous exchange is a major cause of gene presence/absence variation in the amphidiploid *Brassica napus*. Plant Biotechnol J 16:1265–1274

Hwang SF, Ahmed HU, Zhou Q, Strelkov SE, Gossen BD, Peng G, Turnbull D (2011) Influence of cultivar resistance and inoculum density on root hair infection of canola (*Brassica napus*) by *Plasmodiophora brassicae*. Plant Pathol 60:820–829

Hwang SF, Strelkov SE, Feng J, Gossen BD, Howard RJ (2012) *Plasmodiophora brassicae*: a review of an emerging pathogen of the Canadian canola (*Brassica napus*) crop. Mol Plant Pathol 13:105–113. https://doi.org/10.1111/j.1364-3703.2011.00729.x

Inohara N, Chamaillard M, McDonald C, Nunez G (2005) NOD-LRR proteins: role in host-microbial interactions and inflammatory disease. Annu Rev Biochem 74:355–383. https://doi.org/10.1146/annurev.biochem.74.082803.133347

Inomata N (1993) Cross ability and cytology of hybrid progenies in the cross between *Brassica campestris* and three wild relatives of *B. oleracea, B. bourgeaui, B. cretica and B. montana.* Euphytica 69:7–17. https://doi.org/10.1007/BF00021721

Inomata N (2002) A cytogenetic study of the progenies of hybrids between *Brassica napus* and *B. oleracea, B. bourgeaui, B. cretica* or *B. montana*. Plant Breed 121:174–176. https://doi.org/10.1046/j.1439-0523.2002.00695.x

Jejelowa OA, Conn KL, Tewari JP (1991) Relationship between conidial concentration, germling growth, and phytoalexin production by Camelina sativa leaves inoculated with *Alternaria brassicae*. Mycol Res 95:928–934

Jensen BD, Værbak S, Munk L, Andersen SB (1999) Characterization and inheritance of partial resistance to downy mildew, *Peronospora parasitica*, in breeding material of broccoli, *Brassica oleracea* convar. *botrytis* var. *italica*. Plant Breed 118:549–554. https://doi.org/10.1046/j.1439-0523.1999.00409.x

Jiao Y, Wickett NJ, Ayyampalayam S, Chanderbali AS, Landherr L, Ralph PE, Tomsho LP, Hu Y, Liang H, Soltis PS, Soltis DE, Clifton SW, Schlarbaum SE, Schuster SC, Ma H, Leebens-Mack J, de Pamphili CW (2011) Ancestral polyploidy in seed plants and angiosperms. Nature 473:97–100

Jourdan PS, Salazar E (1993) *B. carinata* resynthesized by protoplast fusion. Theor Appl Genet 86:567–572

Jourdan PS, Earle ED, Mutscher MA (1989) Atrazine-resistant cauliflower obtained by somatic hybridization between *Brassica oleracea* and ATR-*B. napus*. Theor Appl Genet 78:271–279. https://doi.org/10.1007/BF00288810

Kamble A, Bhargava S (2007) β–aminobutyric acid-induced resistance in *Brassica juncea* against the necrotrophic pathogen *Alternaria brassicae*. J Phytopathol 155:152–158

Kaneko Y, Bang SW (2014) Interspecific and intergeneric hybridization and chromosomal engineering of brassicaceae crops. Breed Sci 64:14–22

Kanrar S, Venkateswari JC, Kirti PB, Chopra VL (2002) Transgenic expression of hevein, the rubber tree lectin, in Indian mustard confers protection against *Alternaria brassicae*. Plant Sci 162:441–448

Karpechenko GD (1928) Polyploid hybrids of *Raphanus sativus* L. × *Brassica oleracea* L. Z Indukt Abstamm Vererbungsl 48:1–85

Kearney B, Staskawicz BJ (1990) Widespread distribution and fitness contribution of *Xanthomonas campestris* avirulence gene avrBs2. Nature 346:385–386. https://doi.org/10.1038/346385a0

Kim Y, Tsuda K, Igarashi D, Hillmer RA, Sakakibara H, Myers CL, Katagiri F (2014) Signaling mechanisms underlying the robustness and tunability of the plant immune network. Cell Host Microbe 15:84–94

Kissoudis C, van de Wiel C, Visser RGF, van der Linden G (2014) Enhancing crop resilience to combined abiotic and biotic stress through the dissection of physiological and molecular crosstalk. Front Plant Sci 5:207

Kolte SJ (1985a) Disease management strategies for rapeseed-mustard crops in India—a review. Agric Rev 6:81–88

Kolte SJ (1985b) Diseases of annual edible oilseed crops, vol II. Rapeseed-mustard and sesame diseases. CRC Press Inc., Boca Raton, p 135

Kolte SJ (1987) Important diseases of rapeseed and mustard in oilseed crops: niger and rapeseed mustard. In: Proceedings of the third oil crops network, held in Addis Ababa, Ethiopia, 6–10 Oct 1986, pp 99–106

Krishnia SK, Saharan GS, Singh D (2000a) Genetics of Alternaria blight resistance in inter and intraspecific crosses of *Brassica juncea* and *B. carinata*. Ann Biol 16(2):212–216

Krishnia SK, Saharan GS, Singh D (2000b) Genetic variation for multiple disease resistance in the families of interspecific cross of *Brassica juncea* x *B. carinata*. Cruciferae Newslett 22:51–53

Kumar B, Kolte SJ (2001) Progression of Alternaria blight of mustard in relation to components of resistance. Indian Phytopathol 54:329–331

Kumar S, Saharan GS (2002) Sources of multiple disease resistance in *Brassica* spp. J Mycol Plant Pathol 32:184–188

Kumar D, Maurya N, Bharati YK, Kumar A, Kumar K, Srivastava K, Chand G, Singh SK, Mishra RK, Kumar A (2014) Alternaria blight of oilseed *Brassicas*: a comprehensive review. Afr J Microbiol Res 8:2816–2829. https://doi.org/10.5897/AJMR2013.6434

Kumar A, Banga SS, Meena PD, Kumar PR (2015) *Brassica* oilseeds breeding and management. CABI International, UK, 280 p. ISBN-13: 9781780644837

LeBoldus JM, Manolii VP, Turkington TK, Strelkov SE (2012) Adaptation to *Brassica* host genotypes by a single-spore isolate and population of *Plasmodiophora brassicae* (clubroot). Plant Dis 96:833–838

Lucas JA, Crute IR, Sherriff C, Gordon PL (1988) The identification of a gene for race-specific resistance to *Peronospora parasitica* (downy mildew) in *Brassica napus* var. oleifera (oilseed rape). Plant Pathol 37:538–545. https://doi.org/10.1111/j.1365-3059.1988.tb02112.x

Marmath KK, Giri P, Taj G, Pandey D, Kumar A (2013) Effect of zeatin on the infection process and expression of MAPK-4 during pathogenesis of *Alternaria brassicae* in non-host and host *Brassica* plants. Afr J Biotechnol 12:2164–2174

Martin GB, Brommonschenkel SH, Chunwongse J, Frary A, Ganal MW, Spivy R, Wu T, Earle ED, Tanksley SD (1993) Map-based cloning of a protein kinase gene conferring disease resistance in tomato. Science 262:1432–1436

Mason AS, Nelson MN, Yan G, Cowling WA (2011) Production of viable male unreduced gametes in Brassica interspecific hybrids is genotype specific and stimulated by cold temperatures. BMC Plant Biol 11:103. https://doi.org/10.1186/1471-2229-11-103

McHale L, Tan X, Koehl P, Michelmore RW (2006) Plant NBS-LRR proteins: adaptable guards. Genome Biol 7:212

Meena PD, Bala M, Sharma P, Singh D (2014) Interaction and tolerance to *Alternaria brassicae* in Indian mustard (*Brassica juncea*) genotypes. J Oilseeds Res 31:136–139

Meena PD, Mehta N, Rai PK, Saharan GS (2018) Geographical distribution of rapeseed-mustard powdery mildew disease in India. J Mycol Plant Pathol 48(3):284–302

Meena SS, Meena PD, Singh VV, Meena HS, Singh D, Yadav R, Singh KH, Sharma P, Nanjundan J, Singh BK, Singh YP, Ram B (2019a) DRMR-2019 (IC0598622; INGR17077), An Indian mustard (*Brassica juncea*) germplasm with white rust resistance. Indian J Plant Genet Resour 32(2):281

Meena SS, Meena PD, Singh VV, Meena HS, Dhiraj S, Yadav R, Singh KH, Sharma P, Nanjundan J, Singh BK, Singh YP, Ram B (2019b) DRMR-2035 (IC0598623; INGR17078), An Indian mustard (*Brassica juncea*) germplasm with white rust resistance. Indian J Plant Genet Resour 32(2):281

Meena PD, Sujith Kumar MS, Meena HS, Jambhulkar S, Gohartaj, Pathak D, Srivastava S, Gupta R, Singh D, Gurung B, Rai PK (2020) Confirmation of induced tolerance to Alternaria blight disease in Indian mustard (*Brassica juncea* L.). Appl Biochem Biotechnol 192 (3):965–978. https://doi.org/10.1007/s12010-020-03362-2

Mei J, Qian L, Disi JO, Yang X, Li Q, Li J, Frauen M, Cai D, Qian W (2011) Identification of resistant sources against *Sclerotinia sclerotiorum* in *Brassica* species with emphasis on *B. oleracea*. Euphytica 177:393–399

Mei J, Ding Y, Lu K, Wei D, Liu Y, Disi JO, Li J, Liu L, Liu S, McKay J, Qian W (2013) Identification of genomic regions involved in resistance against *Sclerotinia sclerotiorum* from wild *Brassica oleracea*. Theor Appl Genet 126:549–556. https://doi.org/10.1007/s00122-012-2000-x

Mei J, Liu Y, Wei D, Wittkop B, Ding Y, Li Q, Li J, Wan H, Li Z, Ge X, Frauen M, Snowdon RJ, Qian W, Friedt W (2015) Transfer of *Sclerotinia* resistance from wild relative of *Brassica oleracea* into *Brassica napus* using a hexaploidy step. Theor Appl Genet 128:639–644

Mei J, Shao C, Yang R, Feng Y, Gao Y, Ding Y, Li J (2020) Qian W (2020) Introgression and pyramiding of genetic loci from wild *Brassica oleracea* into *B. napus* for improving *Sclerotinia* resistance of rapeseed. Theor Appl Genet 133:1313–1319. https://doi.org/10.1007/s00122-020-03552-w

Meng J, Lu M (1993) Genotype effects of *Brassica napus* on its reproductive behaviour after pollination with *B. juncea*. Theor Appl Genet 87:238–242. https://doi.org/10.1007/BF00223771

Meyers BC, Morgante M, Michelmore RW (2002) TIR-X and TIR-NBS proteins: two new families related to disease resistance TIR-NBS-LRR proteins encoded in *Arabidopsis* and other plant genomes. Plant J 32(1):77–92. https://doi.org/10.1046/j.1365-313X.2002.01404.x

Meyers BC, Kozik A, Griego A, Kuang H, Michelmore RW (2003) Genome-wide analysis of NBS-LRR-encoding genes in *Arabidopsis*. Plant Cell 15(4):809–834. https://doi.org/10.1105/tpc.009308

Mishra A, Pandey D, Goel A, Kumar A (2010) Molecular cloning and in silico analysis of functional homologues of hypersensitive response gene(s) induced during pathogenesis of Alternaria blight in two genotypes of *Brassica*. J Proteomic Bioinformat 3:244–248

Mitchell-Olds T, James RV, Palmer MJ, Williams PH (1995) Genetics of *Brassica rapa* (syn. *campestris*). 2. multiple disease resistance to three fungal pathogens: *Peronospora parasitica*, *Albugo candida* and *Leptosphaeria maculans*. Heredity 75:362–369

Mohanty AL, Chrungu B, Verma N, Shivanna KR (2009) Broadening the genetic base of crop Brassicas by production of new intergeneric hybrid. Czech J Genet Plant Breed 45(3):117–122

Momotaz A, Kato M, Kakihara F (1998) Production of inter-generic hybrids between *Brassica* and *Sinapis* species by means of embryo rescue techniques. Euphytica 103:123–130. https://doi.org/10.1023/A:1018331528368

Mondal KK, Chattejee SC, Viswakarma N, Bhattacharya RC, Grover A (2003) Chitinase-mediated inhibitory activity of *Brassica* transgenic on growth of *Alternaria brassicae*. Curr Microbiol 47:171–173

Monteiro AA, Williams PW (1989) The exploration of genetic sources of Purtuguese cabbage and kale for resistance to several Brassica diseases. Euphytica 41:215–225

Nagaharu U (1935) Genome analysis in *Brassica* with special reference to the experimental formation of *B. napus* and peculiar mode of fertilization. Jpn J Bot 7:389–452

Nakagami H, Pitzchke A, Hirt H (2005) Emerging MAP kinase pathways in plant stress signaling. Trends Plant Sci 10:339–346

Mehta N, Saharan GS, Meena PD (2018) Expression of disease resistance in *Brassica-Hyaloperonospora* host-patho system—a review. Plant Dis Res 33(2):112–141

Naumann TA, Wicklow DT (2013) Chitinase modifying proteins from phylogenetically distinct lineages of *Brassica* pathogens. Physiol Mol Plant Pathol 82:1–9

Navabi ZK, Huebert T, Sharpe AG, Carmel M, Bancroft I, Isobel Parkin AP (2013) Conserved microstructure of the *Brassica* B Genome of *Brassica nigra* in relation to homologous regions of *Arabidopsis thaliana, B. rapa* and *B. oleracea*. BMC Genomics 14:250. https://doi.org/10.1186/1471-2164-14-250

Neik TX, Barbetti MJ, Batley J (2017) Current status and challenges in identifying disease resistance genes in *Brassica napus*. Front Plant Sci 8:1788. https://doi.org/10.3389/fpls.2017.01788

Nejat N, Mantri N (2017) Plant immune system: crosstalk between responses to biotic and abiotic stresses the missing link in understanding plant defense. Curr Issues Mol Biol 23:1–16. https://doi.org/10.21775/cimb.023.001

Niemann J, Wojciechowski A, Janowicz J (2012) Broadening the variability of quality traits in rapeseed through interspecific hybridization with an application of immature embryo culture. J Biotechnol Comput Biol Bionanotechnol 93:109–115

Niemann J, Wojciechowski A, Jedryczka M, Kaczmarek J (2013) Interspecific hybridization as a tool for broadening the variability of useful traits in rapeseed (*Brassica napus* L.). In: Branca F, Tribulato A (eds) Proceedings of the VIth IS on Brassicas and XVIIIth crucifer genetics workshop, (Catania: Acta Hort), p 1005

Nishi S, Wata JK, Toda M (1959) On the breeding of interspecific hybrids between two genomes. 'C' and 'B' of *Brassica* through the application of embryo culture techniques. Jpn J Breed 8:215–222

Nowicki M, Nowakowska M, Niezgoda A, Kozik EU (2012) Alternaria black spot of crucifers: symptoms, importance of disease, and perspectives of resistance breeding. Veg Crops Res Bull 76:5–19

Oh IS, Park AR, Bae MS, Kwon SJ, Kim YS, Lee JE, Kang NY, Lee S, Cheong H, Park OK (2005) Secretome analysis reveals an *Arabidopsis* lipase involved in defense against *Alternaria brassicicola*. Plant Cell 17:2832–2847

Otani H, Kohnobe A, Narita M, Shiomi H, Kodama M, Kohmoto K (2001) A new type of host-selective toxin, a protein from *Alternaria brassicicola*. In: Keen NT, Mayama S, Leach JE, Tsuyumu S (eds) Delivery and perception of pathogen signals in plants. American Phytopathological Soc, Press, St Paul, pp 68–76

Pan X, Caldwell CD, Falk KC, Lada R (2012) The effect of cultivar, seeding rate and applied nitrogen on *Brassica carinata* seed yield and quality in contrasting environments. Can J Plant Sci 92:961–971. https://doi.org/10.4141/cjps2011-169

Parihar PS, Prakash O, Punetha H (2012) Investigation on defensive enzymes activity of *Brassica juncea* genotypes during pathogenesis of Alternaria blight. Nature Sci 10(2):63–68

Parlevliet JE (1992) Selecting components of partial resistance. In: Stalker HT, Murphy JP (eds) Plant breeding in the 1990s. pp 281–302

Peng G, Pageau D, Strelkov SE, Lahlali R, Hwang SF, Hynes RK (2013) Assessment of crop rotation, cultivar resistance and *Bacillus subtilis* biofungicide for control of clubroot on canola. Acta Hortic 1005:591–598

Peng G, Falk KC, Gugel RK, Franke C, Yu F, James B, Strelkov SE, Hwang SF, McGregor L (2014) Sources of resistance to *Plasmodiophora brassicae* (clubroot) pathotypes virulent on canola. Can J Plant Pathol 36:89–99. https://doi.org/10.1080/07060661.2013.863805

Perez IB, Brown P (2014) The role of ROS signaling in cross-tolerance: from model to crop. Front Plant Sci 5:754

Plieske J, Struss D, Robbelen G (1998) Inheritance of resistance derived from the B-genome of Brassica against *Phoma lingam* in rapeseed and the development of molecular markers. Theor Appl Genet 97:929–936. https://doi.org/10.1007/s001220050973

Prakash S (1973) Non-homologous meiotic pairing in the A and B genomes of *Brassica*: its breeding significance in the production of variable amphidiploids. Genet Res 21:133–137

Prakash S, Chopra VL (1988) Introgression of resistance to shattering in *Brassica napus* from *Brassica juncea* through non-homologous recombination. Plant Breed 101:167–168

Prasch CM, Sonnewald U (2013) Simultaneous application of heat, drought and virus to *Arabidopsis* plants reveals significant shifts in signaling networks. Plant Physiol 162:1849–1866

Pratap P, Thakur AK, Meena PD, Meena HS, Sharma P, Singh D, Majumdar R (2015) Genetic diversity assessment in Indian mustard (*Brassica juncea* L.) for Alternaria blight tolerance using SSR markers. J Oilseed Brassica 6(1):175–182

Primard C, Vedel F, Mathieu C, Pelletier G, Chevre AM (1988) Interspecific somatic hybridization between *Brassica napus* and *Brassica hirta* (*Sinapis alba* L.). Theor Appl Genet 75:546–552

Rahman H, Peng G, Yu F, Falk KC, Kulkarni M, Selvaraj G (2014) Genetics and breeding for clubroot resistance in Canadian spring canola (*Brassica napus* L.). Can J Plant Pathol 36:122–134. https://doi.org/10.1080/07060661.2013.862571

Rajarammohan S, Pental D, Kaur J (2019) Near-complete genome assembly of *Alternaria brassicae*—a necrotrophic pathogen of *Brassica* crops. MPMI 32:928–930

Rana K, Atri C, Gupta M, Akhatar J, Sandhu PS, Kumar N, Jaswal R, Barbetti MJ, Banga SS (2017) Mapping resistance responses to *Sclerotinia* infestation in introgression lines of *Brassica juncea* carrying genomic segments from wild Brassicaceae *B. fruticulosa*. Sci Rep 7:5904. https://doi.org/10.1038/s41598-017-05992-9

Rana K, Atri C, Akhatar J, Kaur R, Goyal A, Singh MP, Kumar N, Sharma A, Sandhu PS, Kaur G, Barbetti MJ, Banga SS (2019) Detection of first marker trait associations for resistance against *Sclerotinia sclerotiorum* in *Brassica juncea–Erucastrum cardaminoides* introgression lines. Front Plant Sci 10:1015. https://doi.org/10.3389/fpls.2019.01015

Ricarova V, Kazda J, Singh K, Ryšánek P (2016) Clubroot caused by *Plasmodiophora brassicae* Worn.: a review of emerging serious disease of oilseed rape in the Czech Republic. Plant Prot Sci 52:71–86. https://doi.org/10.17221/87/2015-PPS

Ripley V, Thorpe M, Iler S, Mizier K, Beversdorf WD (1992) Isozyme analysis as a tool for introgression of *Sinapis alba* germplasm into *Brassica napus*. Theor Appl Genet 84:403–410. https://doi.org/10.1007/BF00229500

Saharan GS (1984) A review of research on rapeseed-mustard pathology in India. In: Annual workshop, AICORPO, ICAR, Jaipur, 6–10 Aug 1984

Saharan GS (1992) Disease resistance. In: Labana KS, Banga SS, Banga SK (eds) Breeding oilseed Brassicas. Narosa Publishing House, New Delhi, pp 181–200

Saharan GS (1997) Disease resistance. In: Kalia HR, Gupta SC (eds) Recent advances in oilseed Brassicas, vol 12. Kalyani Publications, Ludhiana, pp 233–259

Saharan GS, Chand JN (1988) Chapter 3: Diseases of rapeseed and mustard. In: Directorate of oilseed crops (In Hindi). Directorate of Publication, Haryana Agric Univ Press, Hisar, India, pp 84–91

Saharan GS, Kadian AK (1983) Analysis of components of horizontal resistance in rapeseed and mustard cultivars against *Alternaria brassicae*. Indian Phytopathol 36:503–507

Saharan GS, Krishnia SK (2001) Multiple disease resistance in rapeseed and mustard. In: Nagarajan S, Singh DP (eds) Role of resistance in intensive agriculture. Kalyani Publications, New Delhi, pp 98–108

Saharan GS, Naresh M, Sangwan MS (2003) Nature and mechanism of resistance to Alternaria blight in rapeseed-mustard system. Annu Rev Plant Pathol 2:85–128

Saharan GS, Naresh M, Sangwan MS (2005) Development of disease resistance in rapeseed-mustard. In: Saharan GS, Mehta N, Sangwan MS (eds) Diseases of oilseed crops. Indus Publishing Co., New Delhi, pp 561–617

Saharan GS, Verma PR, Meena PD, Kumar A (2014) White rust of crucifers: biology, ecology and disease management. Springer, Frankfurt, p 244. ISBN 978-81-322-1791-6

Saharan GS, Naresh M, Meena PD (2016) Alternaria blight of crucifers: biology, ecology and disease management. Springer, Singapore, p 326. ISBN: 978-981-10-0019-5

Saharan GS, Mehta N, Meena PD (2018) Downy mildew disease of crucifers: biology, ecology and disease management. Springer, Singapore, LVI, p 357. ISBN 978-981-10-7499-8

Saharan GS, Naresh M, Meena PD (2019) Powdery mildew disease of crucifers: biology, ecology and disease management. Springer, Singapore, p 362. ISBN 978-981-13-9852-0

Saharan GS, Mehta N, Meena PD (2021) Clubroot disease of crucifers: biology, ecology and disease management. Springer, Singapore (in press)

Sandhu PS, Brar KS, Chauhan JS, Meena PD, Awasthi RP, Rathi AS, Kumar A, Gupta JC, Kolte SJ, Manhas SS (2015) Host–pathogen interactions of *Brassica* genotypes for white rust (*Albugo candida*) disease severity under aided epiphytotic conditions in India. Phytoparasitica 43 (2):197–207. https://doi.org/10.1007/s12600-014-0432-3

Schranz ME, Lysak MA, Mitchell-Olds T (2006) The ABC's of comparative genomics in the Brassicaceae: building blocks of crucifer genomes. Trends Plant Sci 11:535–542. https://doi.org/10.1016/j.tplants.2006.09.002

Schwessinger B, Bart R, Krasileva K, Coaker G (2015) Focus issue on plant immunity: from model systems to crop species. Front Plant Sci 6:195

Sellam A, Dongo A, Guillemette T, Hudhomme P, Simoneau P (2007) Transcriptional responses to exposure to the brassicaceous defense metabolites camalexin and allyl-isothiocyanate in the necrotrophic fungus *Alternaria brassicicola*. Mol Plant Pathol 8:195–208

Sharma TR, Singh BM (1992) Transfer of resistance to *Alternaria brassicae* in *Brassica juncea* through interspecific hybridization among Brassicas. J Genet Breed 46(4):373–378

Sharma BR, Dhiman JS, Thakur JC, Singh A, Bajaj KL (1991) Multiple disease resistance in cauliflower. Adv Hortic Sci 5:30–34

Sharma G, Kumar DV, Haque A, Bhat SR, Shyam P, Chopra VL (2002) *Brassica* coenospecies: a rich reservoir for genetic resistance to leaf spot caused by *Alternaria brassicae*. Euphytica 125:411–419. https://doi.org/10.1023/A:1016050631673

Sharma P, Meena PD, Verma PR, Saharan GS, Mehta N, Singh D, Kumar A (2015) *Sclerotinia sclerotiorum* (Lib.) de Bary causing Sclerotinia rot in oilseed Brassicas: a review. J Oilseed Brassica 6(Spl):1–44

Sharma BB, Kalia P, Yadava DK, Dingh D, Sharma TR (2016) Genetics and molecular mapping of black rot resistance locus Xca1bc on chromosome B-7 in Ethiopian mustard (*Brassica carinata* A. Braun). PLoS One 11:e0152290. https://doi.org/10.1371/journal.pone.0152290

Sharma BB, Kalia P, Singh D, Sharma TR (2017) Introgression of black rot resistance from *Brassica carinata* to cauliflower (*Brassica oleracea botrytis* Group) through embryo rescue. Front Plant Sci 8:1255. https://doi.org/10.3389/fpls.2017.01255

Sharma P, Singh S, Singh VV, Gupta NC, Prasad L, Meena PD, Ambawat S, Rai PK (2018) Identification and characterization of new resistance sources against Sclerotinia stem rot (*Sclerotinia sclerotiorum*) in oilseed Brassica. J Mycol Plant Pathol 48(1):35–44

Shivpuri A, Chippa HP, Gupta RLB, Sharma KN (1997) Field evaluation of mustard genotypes against white rust, powdery mildew, and stem rot. Ann Arid Zone Res 36:387–390

Siemens J, Nagel W, Ludwig-Muller J, Sarcristán MD (2002) The interaction of *Plasmodiophora brassicae* and *Arabidopsis thaliana*: parameters for disease quantification and screening mutant lines. J Phytopathol 150:592–605

Siemens J, Keller I, Sarx J, Kunz S, Schuller A, Nagel W, Schmulling T, Parniske M, Ludwig-Muller J (2006) Transcriptome analysis of *Arabidopsis* clubroots indicate a key role for cytokinins in disease development. Mol Plant-Microbe Interact 19:480–494. https://doi.org/10.1094/MPMI-19-0480

Singh K (1994) Studies on the ecofriendly management of powdery mildew (*Erysiphe cruciferarum* Opiz ex. Junell) of mustard [*Brassica juncea* (L.) Czern & Coss]. M.Sc Thesis, Department of Plant Pathology, CCS HAU, Hisar, p 106+ xvii

Singh H, Singh D (1989) Studies on genetic control of resistance to *Alternaria brassicae* in Indian mustard. In: XII Eucarpia congress, pp 11–13

Singh S, Dey SS, Bhatia R, Batley J, Kumar R (2018) Molecular breeding for resistance to black rot [*Xanthomonas campestris* pv. *campestris* (Pammel) Dowson] in *Brassicas*: recent advances. Euphytica 214:196

Skoropad WP, Tewari JP (1977) Field evaluation of the role of epicuticular wax in rapeseed and mustard in resistance to Alternaria black spot. Can J Plant Sci 57:1001–1003

Snowdon RJ, Winter H, Diestel A, Sacristan MD (2000) Development and characterization of *Brassica napus-Sinapis arvensis* addition lines exhibiting resistance to *Leptosphaeria maculans*. Theor Appl Genet 101:1008–1014. https://doi.org/10.1007/s001220051574

Soltis PS, Soltis DE (2000) The role of genetic and genomic attributes in the success of polyploids. Proc Natl Acad Sci U S A 97:7051–7057

Strelkov SE, Tewari JP, Smith-Degenhardt E (2006) Characterization of *Plasmodiophora brassicae* populations from Alberta, Canada. Can J Plant Pathol 28:467–474. https://doi.org/10.1080/07060660609507321

Strelkov SE, Hwang SF, Howard RJ, Hartman M, Turkington TK (2011) Progress towards the sustainable management of clubroot [*Plasmodiophora brassicae*] of canola on the Canadian prairies. Prairie Soils Crops J 4:114–121

Suzuki N, Rivero RM, Shulaev V, Blumwald E, Mittler R (2014) Abiotic and biotic stress combinations. New Phytol 203:32–43

Tai TH, Dahlbeck D, Clark ET, Gajiwala P, Pasion R, Whalen MC, Stall RE, Staskawicz BJ (1999) Expression of the Bs2 pepper gene confers resistance to bacterial spot disease in tomato. Proc Natl Acad Sci U S A 96:14153–14158. https://doi.org/10.1073/pnas.96.24.14153

Taylor JD, Conway J, Roberts SJ, Astley D, Vicente JG (2002) Source and origin of resistance to *Xanthomonas campestris* pv. *campestris* in *Brassica* genomes. Phytopathology 92:105–111

Tewari JP (1991a) Structural and biochemical bases of black spot disease of crucifers. Adv Struct Biol 1:325–349

Tewari JP (1991b) Current understanding of resistance to *Alternaria brassicae* in crucifers. In: GCIRC 8th international rapeseed congress, 9–11 July, Saskatoon, Canada, p 84 (Abst.); Proc. 2, pp 471–476

Tewari JP, Conn KL (1993) Reaction of some wild crucifers to *Alternaria brassicae*. Bull OILB SROP 16:53–58

Thomas CE, Jourdain EL (1990) Evaluation of broccoli and cauliflower germplasm for resistance to Race 2 of *Peronospora parasitica*. Hortic Sci 25:1429–1431

Thomas CE, Jourdain EL (1992) Resistance to race 2 of *Peronospora parasitica* in U.S. plant introductions of *Brassica oleracea* var. *capitata*. Hortic Sci 27:1120–1122

Thomma BPHJ, Nürnberger T, Joosten MHAJ (2011) Of PAMPs and effectors: the blurred PTI-ETI dichotomy. Plant Cell 23:4–15. https://doi.org/10.1105/tpc.110.082602

Tonguc M, Griffiths PD (2004a) Transfer of powdery mildew resistance from *Brassica carinata* to *Brassica oleracea* through embryo rescue. Plant Breed 123:587–589. https://doi.org/10.1111/j.1439-0523.2004.00987.x

Tonguc M, Griffiths PD (2004b) Development of black rot resistant interspecific hybrids between *Brassica oleracea* L. cultivars and *Brassica* accession A 19182, using embryo rescue. Euphytica 136:313–318. https://doi.org/10.1023/B:EUPH.0000032733.47031.5f

Tonguc M, Earle ED, Griffiths PD (2003) Segregation distortion of *Brassica carinata* derived black rot resistance in *Brassica oleracea*. Euphytica 134:269–276

Tsuda K, Katagiri F (2010) Comparing signaling mechanisms engaged in pattern triggered and effector-triggered immunity. Curr Opin Plant Biol 13:459–465

Verma VD, Rai B (1980) Note on induced mutagenesis for spotting out the sources of resistance to Alternaria leaf spot in Indian mustard. Indian J Agric Sci 50(3):278

Verma PR, Saharan GS (1994) Monograph on Alternaria diseases of crucifers. In: Agriculture and agri-food Canada, pp 1–146

Verma SS, Yajima WR, Rahman MH, Shah S, Liu JJ, Ekramoddoullah AK, Kav NN (2012) A cysteine-rich antimicrobial peptide from *Pinus monticola* (PmAMP1) confers resistance to multiple fungal pathogens in canola (*Brassica napus*). Plant Mol Biol 79:61–74

Vicente JG, Holub EB (2013) *Xanthomonas campestris* pv. *campestris* (cause of black rot of crucifers) in the genomic era is still a worldwide threat to *Brassica* crops. Mol Plant Pathol 14:2–18. https://doi.org/10.1111/j.1364-3703.2012.00833.x

Vicente JG, Taylor JD, Sharpe AG, Parkin IA, Lydiate PDJ, King GJ (2002) Inheritance of race-specific resistance to *Xanthomonas campestris* pv. *campestris* in *Brassica* genomes. Phytopathology 92:1134–1141. https://doi.org/10.1094/PHYTO.2002.92.10.1134

Vishwanath, Kolte SJ, Singh MP, Awasthi RP (1999) Induction of resistance in mustard (*Brassica juncea*) against Alternaria black spot with an avirulent *Alternaria brassicae* isolate-D. Eur J Plant Pathol 105:217–220. https://doi.org/10.1023/A:1008717323002

Voorrips RE (1996) Production, characterization and interaction of single-spore isolates of *Plasmodiophora brassicae*. Eur J Plant Pathol 102:377–383

Wan H, Yuan W, Ye Q, Wang R, Ruan M, Li Z, Zhou G, Yao Z, Zhao J, Liu S, Yang Y (2012) Analysis of TIR- and non-TIR-NBS-LRR disease resistance gene analogous in pepper: characterization, genetic variation, functional divergence and expression patterns. BMC Genomics 13:502. https://doi.org/10.1186/1471-2164-13-502

Wang X, Wang H, Wang J, Sun R, Wu J, Liu S, Bai Y, Mun JH, Bancroft I, Cheng F, Huang S, Li X, Hua W, Wang J, Wang X, Freeling M, Pires JC, Paterson AH, Chalhoub B, Wang B, Hayward A, Sharpe AG, Park BS, Weisshaar B, Liu B, Li B, Liu B, Tong C, Song C, Duran C, Peng C, Geng C, Koh C, Lin C, Edwards D, Mu D, Shen D, Soumpourou E, Li F, Fraser F, Conant G, Lassalle G, King GJ, Bonnema G, Tang H, Wang H, Belcram H, Zhou H, Hirakawa H, Abe H, Guo H, Wang H, Jin H, Parkin IA, Batley J, Kim JS, Just J, Li J, Xu J, Deng J, Kim JA, Li J, Yu J, Meng J, Wang J, Min J, Poulain J, Wang J, Hatakeyama K, Wu K, Wang L, Fang L, Trick M, Links MG, Zhao M, Jin M, Ramchiary N, Drou N, Berkman PJ, Cai Q, Huang Q, Li R, Tabata S, Cheng S, Zhang S, Zhang S, Huang S, Sato S, Sun S, Kwon SJ, Choi SR, Lee TH, Fan W, Zhao X, Tan X, Xu X, Wang Y, Qiu Y, Yin Y, Li Y, Du Y, Liao Y, Lim Y, Narusaka Y, Wang Y, Wang Z, Li Z, Wang Z, Xiong Z, Zhang Z (2011) *Brassica rapa* genome sequencing project consortium. The genome of the mesopolyploid crop species *Brassica rapa*. Nat Genet 43:1035–1039. https://doi.org/10.1038/ng.919

Wang GX, Lv J, Zhang J, Han S, Zong M, Guo N, Zeng XY, Zhang YY, Wang YP, Liu F (2016) Genetic and epigenetic alterations of *Brassica nigra* introgression lines from somatic hybridization: a resource for cauliflower improvement. Front Plant Sci 7:1258. https://doi.org/10.3389/fpls.2016.01258

Wang Z, Zhao FY, Tang MQ, Chen T, Bao LL, Cao J, Li YL, Yang YH, Zhu KM, Liu S, Tan XL (2020) *BnaMPK6* is a determinant of quantitative disease resistance against *Sclerotinia sclerotiorum* in oilseed rape. Plant Sci 291:110362

Warwick SI (2011) Brassicaceae in agriculture. In: Bancroft I, Schmidt R (eds) Genetics and genomics of the Brassicaceae. Springer, New York, pp 33–67. https://doi.org/10.1007/978-1-4419-7118-0_2

Weerakoon SR, Si P, Zili W, Meng J, Yan G (2009) Production and confirmation of hybrids through interspecific crossing between tetraploid *B. juncea* and diploid *B. oleracea* towards a hexaploid *Brassica* population. In: 16th Australian research assembly on Brassicas, Ballarat, VIC

Wei L, Jian H, Lu K, Filardo F, Yin N, Liu L, Qu C, Li W, Du H, Li J (2016) Genome-wide association analysis and differential expression analysis of resistance to Sclerotinia stem rot in *Brassica napus*. Plant Biotechnol J 14:1368–1380. https://doi.org/10.1111/pbi.12501

Westman AL, Kresovich S, Dickson MH (1999) Regional variation in *Brassica nigra* and other weedy crucifers for disease reaction to *Alternaria brassicicola* and *Xanthomonas campestris* pv. *campestris*. Euphytica 106:253–259. https://doi.org/10.1023/A:1003544025146

Wu J, Cai G, Tu J, Li L, Liu S, Luo X, Zhou L, Fan C, Zhou Y (2013) Identification of QTLs for resistance to Sclerotinia stem rot and BnaC.IGMT5.a as a candidate gene of the major resistant QTL SRC6 in *Brassica napus*. PLoS One 8:e67740. https://doi.org/10.1371/journal.pone.0067740

Wu J, Zhao Q, Yang Q, Liu H, Li Q, Yi X, Cheng Y, Guo L, Fan C, Zhou Y (2016) Comparative transcriptomic analysis uncovers the complex genetic network for resistance to *Sclerotinia sclerotiorum* in *Brassica napus*. Sci Rep 6:19007. https://doi.org/10.1038/srep19007

Wu J, Lin L, Xu M, Chen P, Liu D, Sun Q, Ran L, Wang Y (2018) Homoeolog expression bias and expression level dominance in resynthesized allopolyploid *Brassica napus*. BMC Genomics 19 (1):586

Xue S, Cao T, Howard RJ, Hwang SF, Strelkov SE (2008) Isolation and variation in virulence of single-spore isolates of *Plasmodiophora brassicae* from Canada. Plant Dis 92:456–462

Yoshikawa H (1993) Studies on breeding of clubroot resistance in cole crops. Bull Natl Res Inst Veg Ornam Plants Tea Jpn Ser A 7:1–165

Yu J, Tehrim S, Zhang F, Tong C, Huang J, Cheng X, Dong C, Zhou Y, Qin R, Hua W, Liu S (2014) Genome-wide comparative analysis of NBS-encoding genes between *Brassica* species and *Arabidopsis thaliana*. BMC Genomics 15:3. https://doi.org/10.1186/1471-2164-15-3

Zhang Y, Tessaro MJ, Lassner M, Li X (2003) Knockout analysis of *Arabidopsis* transcription factors TGA2, TGA5, and TGA6 reveals their redundant and essential roles in systemic acquired resistance. Plant Cell 15:2647–2653. https://doi.org/10.1105/tpc.014894

Zhang GQ, Tang GX, Song WJ, Zhou WJ (2004) Resynthesizing *Brassica napus* from interspecific hybridization between *Brassica rapa* and *B. oleracea* through ovary culture. Euphytica 140:181–187. https://doi.org/10.1007/s10681-004-3034-1

Zhou JM, Tang XY, Martin GB (1997) The Pto kinase conferring resistance to tomato bacterial speck disease interacts with proteins that bind a cis-element of pathogenesis-related genes. EMBO J 16:3207–3218. https://doi.org/10.1093/emboj/16.11.3207

Zou P, Lu X, Jing C, Yuan Y, Lu Y, Zhang C, Meng L, Zhao H, Li Y (2018) Low-molecular-weightt polysaccharides from *Pyropia yezoensis* enhance tolerance of wheat seedlings (*Triticum aestivum L.*) to salt stress. Front Plant Sci 9:427. https://doi.org/10.3389/fpls.2018.00427

Development of Resistance Cultivars' Techniques

9

Abstract

For the benefit of graduate students and research scientists of *Brassica*, some selected techniques with methodology, procedures, and protocols have been included in this chapter. These techniques are standardized and have been developed by experts working on genomics of crucifers' host-pathosystem. They can be used for identification of sources of resistance with R-genes by screening germplasm from different sources. The procedures and methodology for transfer of resistance through various approaches have also been included for developing and breeding disease resistance cultivars of *Brassica* against range of pathogens. To get reliable and widely acceptable results in research pursuits is the primary object of any research programs which can be met only through standardized techniques included in this chapter. These techniques are of basic as well as advance nature on molecular aspects of crucifers' host–pathogen interactions to reveal mechanisms of host resistance and breed high yielding disease resistance cultivars of crucifers against major pathogens of fungal, bacterial, and viral origins. The methodology and protocols of rapid molecular detection, identification, *in vitro* and *in vivo* culturing; maintenance, multiplication, single spore isolation, identification of R-genes, and QTLs, screening for sources of resistance are included for demonstration and use in research projects of students and researchers. Other pathogen specific and genomic techniques like inoculum preparation, pathotypes/races/strains maintenance, identification of partial/induced resistance/QTL's development of near-isogenic lines, DNA extraction, use of PCR, qPCR; transfer of R-genes; breeding for multiple disease resistance; mapping of R-genes; cloning and transformation of genes; pyramiding of R-QTLs; marker-assisted selection and breeding; genome-wide association study; introgression of R-genes from wild relatives; virus detection, preservation, and identification, and *Brassica* species genomics techniques have been included to get reliable and high-throughput research results.

G. S. Saharan et al., *Genomics of Crucifer's Host-Resistance*,
https://doi.org/10.1007/978-981-16-0862-9_9

Keywords

Technology for disease resistance · Pathogens · Isolation · Inoculation · Collection · Preservation · Maintenance · Screening · Characterization of genes · Molecular identification · Biochemical and molecular analysis · RNA isolation · RNA sequencing · PCR assay

9.1 Introduction

Techniques, methodology, procedures, and protocols developed are essential and prerequisites for research and teaching in science and technology. To initiate any research project, the first and foremost requirement is a standardized, validated, widely acceptable, and reproducible technique. A scientific technique is a systematic way of obtaining information about a scientific nature in the areas of explored and unexplored material or product or biotic and abiotic interactions. Scientific techniques can be divided preparative (cultures purification, growth, and analysis) and measurement techniques like molecular characterization, isolation, cloning, and other genomics protocols. In the present chapter, some selected techniques have been included to facilitate research workers to use, refine, and obtain desired and largely acceptable results in the areas of genomics of crucifers and host resistance.

9.2 *Brassica-Albugo* Techniques

9.2.1 Maintenance of *Albugo candida* Isolates and Inoculum Preparation

The *A. candida* isolates used in the present investigation were maintained separately on 1-week-old seedlings of the susceptible host raised from untreated seeds sown in pots filled with moist-sterilized soil-less compost placed in propagators (16 × 20 × 5 cm). The propagators were placed in an air-flow-bench under spore-free conditions in the glasshouse at 18 °C ± 2 °C (Jenkyn et al. 1973) with supplementary light to maintain a 16 h light/8 h dark; day/night cycle. The seedlings were sprayed with sterilized distilled water (SDW) to clean the surface of cotyledons 24 h prior to inoculation.

Sporangial suspension was prepared by adding 1–2 mL SDW to glass vial containing the excised, sporulating cotyledons from the previous stock (kept in the freezer or excised from freshly sporulating material). The vial was then closed and shaken vigorously on a vortex to rupture the pustules and to facilitate the release of the sporangia from the sporangiophores. The concentration of the suspension was then adjusted to 2.5×10^4 sporangia/mL with the aid of a hemocytometer. Each cotyledon was inoculated with two 5 μL droplets of sporangial suspension using a micropipette. Alternatively, the propagator was sprayed to run-off with the spore suspension using an atomizer. After inoculation, the propagators were covered with clear plastic lids and sealed with insulation tape to allow relative humidity (RH) to increase to approximately 100%. They were then placed in a growth chamber for 12 days at 16 °C with 8 h darkness initially, followed by 16 h photoperiod with 70–120 μmol/m^2/s irradiance. After this period, peak sporulation was expected to occur. Infected cotyledons were excised and placed into 5 mL glass vials, used immediately for the preparation of spore suspensions, or stored in the freezer for future use. All the work pertaining to preparation of inoculum and inoculation was carried out in a sterile environment inside a Class II Laminar Flow Cabinet.

9.2.2 Pathogen Culture and Inoculation Method

Sporangial suspension of *A. candida* is prepared by adding 1–2 mL SDW to glass vial containing the excised, sporulating cotyledons from the previous stock (kept in the freezer or excised from freshly sporulating material). The vial was then closed and shaken vigorously on a vortex to rupture the pustules and to facilitate the release of the sporangia from the sporangiophores. The concentration of the suspension was adjusted to 2.5×10^4 sporangia/mL with the aid of a hemocytometer. Prior to inoculation, sporangial concentration is adjusted to 1×10^4 sporangia/mL as opposed to 2.5×10^4 sporangia/mL used for maintenance because fresh sporangia are used rather than frozen sporangia. The sporangia burst to release eight zoospores so the concentration of zoospores is therefore 8×10^4 per mL.

Seedlings are inoculated 7 days after sowing by spraying to run-off with sporangial suspension using an atomizer (spray inoculation) or by placing 10 μL drops of zoosporangial suspension onto the adaxial surface of each of the two lobes of each cotyledon (Plate 9.1). When preparing a sporangial suspension containing more than one isolate, the concentration is adjusted using a formula: (number of isolates/2) × original sporangial concentration. For example, if six isolates are combined in one suspension: $(6/2) \times 1 \times 10^4 = 3 \times 10^4$ sporangia/mL giving a zoospore concentration of 2.4×10^5 zoospores/mL. After inoculation, the pots are covered with clear plastic lids and the ventilators and margins are sealed with insulation tape to allow the RH to increase to almost 100%. The seedlings are then placed in a growth chamber at 16 °C, 8 h dark period, followed by 16 h light/8 h dark; day/night cycle, at 70–120 μmol/m^2/s irradiations for 12 days to reach peak sporulation. The WR pustules are collected, 10 days after inoculation, and immediately frozen in liquid nitrogen, and stored at −80 °C for further analysis (Meena 2007). Based on the percentage of cotyledon area covered with pustules, disease is assessed on 0 (resistant) to 9 (highly susceptible) scale (Williams 1985a). Leckie et al. (1996) developed a disease scoring system to assess the host plant response as well as the growth of the pathogen, 14 days after inoculation. Accordingly, seedlings were categorized into eight interaction phenotype classes. In a screen house tests using oospore-infested-seeds, Kolte et al. (1998) observed, both in *B. juncea cv.* Varuna and *B. rapa* var. Toria cv. PT 303, very distinctive WR pustules in 20 days after seeding. The pustules on cv. Varuna were more conspicuous than cv. PT 303. The cotyledonary-infected plants were 79.5% in cv. Varuna compared to only 4–12% in cv. PT 303. The 2–15%

Plate 9.1 Seedlings are inoculated by placing 10 μL drops of zoosporangial suspension onto the adaxial surface of each of the two lobes of each cotyledon (Garg et al. 2010)

of infected plants of cv. Varuna produced stagheads compared to none in cv. PT 303. Results of these preliminary experiments suggest that in India two different biological races of *A. candida* exist on *B. juncea* and *B. rapa;* more comprehensive studies are needed to confirm these findings.

9.2.3 Sporangial Viability Test

A fluorescent microscopy technique to test sporangial viability was suggested by Vyalykh and Lanetskii (1974). This method provides information on the number of viable sporangia of *Albugo* in a population within minutes. Sporangia are stained with a 0.01% aqueous solution of acridine orange for 5–10 min, and viewed under a fluorescence microscope. The dead sporangia are fluorescence green.

9.2.4 Sporangial Preservation

Freezing (Zalewski 1883) has been used to maintain the WR fungus on living plants. O'Brien and Webb (1958) suggested that *A. occdentalis* can be preserved for 5–6 months by storing infected spinach leaves at −10 °C. According to Lakra and Saharan (1989b), sporangia of *A. candida* from *B. juncea* survive for 4.5 days at 15 °C on detached infected leaves, but lose their viability in 18 h if stored as a dry powdered mass. However, sporangia can be stored as a dry powdered mass for 105 days at −40 °C.

9.2.5 Inoculation Applicator

A micro-ULVA-controlled droplet applicator was developed as a tool for inoculating *A. candida* (race 2) in the field and greenhouse by Perry and Williams (1984). The severity of the disease is greater in the field or greenhouse when large spray droplets (60–70 μL) are used instead of small droplets (30–40 μL). Since relatively small volumes of inoculum are required, this inoculator is potentially a valuable tool for field use.

9.2.6 Components of Partial Resistance

Partial resistance may be the result of low infection frequency, low spore production, and a long incubation period, and/or a short infectious period. Fox and Williams (1984) stated that spore production by *A. candida* on *B. rapa* is highly correlated ($r = 0.93$) with a visual WR interaction phenotype (IP) rating scale. Plants are rated from 0 to 9 according to the amount of leaf necrosis or area covered by white rust pustules. Means of spore production on plants rated as 1, 3, 5, and 7 are significantly different ($P = 0.05$) from each other. These results support the use of a visual scale for selection of components of partial resistance to the WR disease. Black and Dainello (1986) made comparisons of percent leaf area with WR lesions, and used two other methods for evaluating partial resistance to *A. occidentalis* in spinach. Lakra and Saharan (1988), using length of incubation period for determining level of resistance in several *B. juncea* cultivars, found that the decision between compatible (susceptible) and incompatible (resistant) reaction of *Albugo-B. juncea* system is made within 80 h of inoculation. Verma et al. (1975) also reported that this decision is made within 72 h of inoculation when the single haustorium formed in the resistant host is encapsulated by a marked hallo. In a histopathological study, Liu et al. (1989) and Verma et al. (1975) showed that the earliest event distinguishing a compatible from an incompatible interaction occurs after formation of the first haustorium, and that resistance is not manifested until the host mesophyll cell had come into contact with the first haustorium.

9.2.7 Germplasm Screening

Many procedures have been described for evaluating host germplasm for resistance to WR (Fox and Williams 1984; Pound and Williams 1963; Verma et al. 1975; Verma and Petrie 1980).

9.2.7.1 Preparation for Sowing of Seeds

Seeds are sown into steam-treated potting mix (composted pine bark: cocoa peat: river sand, 2:1:1 by vol.) in 8-cell plastic trays, and plants grown at 13/18 °C night/day temperatures with a 16 hphotoperiod for screening against pathogen under artificially controlled conditions. Untreated seeds are sown 5 mm deep in 5 ×

5 cm jiffy-pots containing soil-less peat-based compost mix and placed inside a propagator (57 × 29 × 21 mm). Approximately 5 mm diam. hole was made in the base of the Jiffy-pots to allow water drainage. Five to six pots are used for each line. Each propagator contained 11–12 lines including susceptible controls. Five or more seeds are sown in each jiffy-pot, but later thinned to five prior to inoculation to maintain a uniform number of seedlings per pot at the same growth stage. Some lines only had a few seeds, so a lesser number of seeds are sown. After sowing, the propagators are immersed in water-filled trays for 1 h to allow absorption of water through the base. After water absorption, holes in the base of each propagator are cleared of debris and perforations made between the junctions of jiffy-pots, so that air would pass through them when they were placed on the airflow bench to provide the plants with a spore-free environment in the glasshouse. The air is exhausted through the two ventilators on the top of the propagator lids and through the junction between the lids and the propagator base. After 4–5 days, each propagator is watered again, if needed in the same manner for 30 min to maintain proper soil moisture prior to inoculation. The propagators are then returned to the airflow bench. The seedlings are grown under the same conditions and inoculated using the same method as used for isolate maintenance. Approximately 20 mL and 40 mL of spore suspension is used to spray inoculate the seedlings and five to seven-leaf stage plants, respectively, in each propagator (Meena 2007).

9.2.8 Growth Chamber and Greenhouse Screening

Verma and Petrie (1978, 1979, 1980), Verma et al. (1975, 1983, 1999), Singh and Singh (1983) Singh et al. (1999), and Goyal et al. (1995, 1996a, b) in their several comprehensive studies described growth chamber or growth room inoculation technique, with some variation between studies, as follows: Seeds of host plants are sown 2 cm deep in a modified cornell soil-less mix (Stringam 1971) in 16 cm diam. plastic pots and placed in 7 cm deep trays in a growth chamber with an 18 h photoperiod (312 μEM^2S^{-1}) and day-night temperatures of 18 °C and 14 °C, respectively. The plants are watered by flooding and draining the trays. Two weeks after seeding, plants are inoculated by spraying zoospore (100,000–150,000 per mL) suspension to run-off. Control plants are sprayed with distilled/tap water. Following inoculation, plants are exposed to continuous misting for 72 h with day-night temperature of 18 °C and 14 °C, respectively. Except for an initial 24 h dark period, an 18 h day was maintained for the duration of the experiment. Two weeks after inoculation the number of diseased plants per pot and disease severity ratings on the most heavily infected leaf on each diseased plant are recorded using 0–4 disease severity scale (Verma et al. 1999). Sporangial spray inoculation technique has also been reported by several other researchers (Fan et al. 1983; Fox and Williams 1984; Pound and Williams 1963; Singh and Singh 1983; Stone et al. 1987; Verma and Bhowmik 1986). This technique has also been used successfully for producing hypertrophies of inflorescence (stagheads) in artificially inoculated plants by Verma and Petrie (1980) and Goyal et al. (1996a).

9.2.8.1 Field Screening

Verma and Petrie (1980) demonstrated the potential importance of using oospore-infested seeds for initiating WR infection. In their 2 years of field experiments, seeding oospores of *A. candida* race 7 with seeds of turnip rape (*B. rapa*) cv. Torch resulted in a significant increase over the control in both foliar and systemically infected (stagheads) plants. Saharan et al. (1988) and Lakra and Saharan (1989c) also established a WR disease nursery by infesting soil with a staghead powder continuously for 3–5 years by broadcasting, and seeding oospores-infested plant material, and also by sowing oospores-infested seeds. Frequent irrigations are given to maintain high humidity. Sporangial spray inoculations were also made at the seedling, branching, and flowering stages of the crop. Genotypes were usually grown in paired rows between highly susceptible "infector" rows of plants. Flowers were inoculated by plucking the main shoot about 6″ from the tip of the test row genotype. This allowed the sporangial inoculum of the taller infector plants to fall on to the test plants. Rapid identification of resistant genotypes can be reliably identified under suitable controlled environment conditions at the cotyledonary, seedling, or flowering stages (Li et al. 2007a).

9.2.9 Induction of Stagheads

A number of plant pathologists believe that the hypertrophies or stagheads are produced as a result of early infection of young seedlings and systemic development of the fungus in the plant. This theory was rejected when Verma and Petrie (1980) and Goyal et al. (1996b) routinely obtained stagheads by artificially inoculating flower buds of plants grown under growth chamber and greenhouse conditions. The results of several field experiments (Verma and Petrie 1979, 1980) conclusively proved that a large percentage of stagheads in the field are produced as a result of secondary infection of flower buds rather than a systemic development of the fungus in the plant. The flower bud inoculation technique at growth stage 3.1 is routinely used for screening advanced breeding lines (Goyal et al. 1996b). Results of these studies have also proved useful in determining actual time of application of both protectant and systemic fungicides. Verma and Petrie (1980) suggested two possible mechanisms for staghead development.

Early infection of young cotyledons, and leaves with oospore-infested seeds, and infection progressing systemically throughout the development of the plant, and infection of young flower buds by zoospores arising from wind- or water-borne sporangia needs to be recorded. According to Verma and Petrie (1980), oospores of *A. candida* race 7 mixed with seeds of turnip rape prior to sowing results in high levels of leaf and staghead infections. Over 55% staghead infections were induced by flower bud inoculation. Lakra and Saharan (1989a) although failed to induce stagheads in artificially inoculated *B. juncea* plants, they were able to induce more than 66% stagheads in plants grown in soil infested with 5 g/pot of oosporic inoculum at sowing time, and with subsequent sporangial spray inoculation at the seedling, branching, and flowering stages. Systemic infection was found to be directly correlated with the host age.

9.2.10 Identification of R-Genes in *Brassica* to *Albugo*

1. **Development of AcB1-A5.1 near-isogenic lines (NILs):** NILs for white rust resistance-conferring locus AcB1-A5.1 were developed in susceptible Indian lines Varuna, Pusa bold, Pusa Jai Kisan, and Rohini by marker-assisted backcross breeding (MAB). Locus AcB1-A5.1 was introgressed from the donor parent, Donskaja-IV into susceptible lines through positive selection by a PCR-based intron polymorphism (IP) marker At2g36360 (Panjabi-Massand et al. 2010) tightly linked to the resistance phenotype and negative selection by IP marker At2g44970 on one side and SSR marker At2g34700 on the other side. Detailed information on the markers is available in previous studies (Panjabi et al. 2008; Dhaka et al. 2017; Rout et al. 2018). The entire scheme took five generations of backcrossing followed by selfing to develop homozygous NILs.
2. **BAC isolation, sequencing, and annotation:** A 5 × pooled and amplified Hind III BAC library was developed using genomic DNA extracted from young leaves of Donskaja-IV. Partially digested, size-selected DNA fragments (150–350 kb) were legated into Hind III digested vector pIndigoBAC-5 and transformed into *E. coli* DH10B cells. Colonies were distributed into 1 × 96-well plate (Pooled BAC library). Each well contained about 573 primary clones (total clones >55,000). The BAC library was screened using PCR-based IP marker At2g36360. Two additional IP markers (At2g36480 and At2g36310) flanking At2g36360 were developed using sequence information of *B. rapa* line Chiifu-401 obtained from the BRAD database (Cheng et al. 2011). BAC clones giving amplifications with all the three markers were sequenced on a PacBio RSII instrument using single-molecule real-time (SMRT) technology—followed by assembly of the reads using a hierarchical genome assembly process. BAC sequences were annotated for ORFs using Augustus (Stanke et al. 2004), a web server for gene prediction in eukaryotes. ORFs were screened for conserved domains using NCBI Batch CD-Search (Marchler-Bauer and Bryant 2004; Marchler-Bauer et al. 2011). The domains of BjuWRR1 protein were also predicted using an online tool—InterPro (Finn et al. 2017) and COILS (Ex PASy Bioinformatics Resource portal). The conserved motifs of the NB domain were identified following Meyers et al. (2003). Molecular weight and the isoelectric point of the protein were computed by Compute pI/MW online (ExPASy Bioinformatics Resource portal https://web.expasy.org/compute_pi/).
3. **Vector construction and genetic transformation of *B. juncea*:** PCR amplification of the candidate gene using BAC clone as a template was carried out with Q5 High-Fidelity DNA Polymerase. The amplified product was cloned into a modified binary vector pCGMCP22 (Arumugam et al. 2007) containing the *bar* gene, conferring resistance to phosphinothricin (PPT), as a selectable marker between the *loxP* sites. The *bar* gene was driven by the CaMV35S promoter with

an AMV leader sequence (Jobling and Gehrke 1987) and ocspA as the terminator. The modified vector was digested with restriction enzymes Hind III and PstI and amplified product was cloned using In-Fusion HD Cloning Kit followed by sequencing on a 3730 DNA Analyzer. The binary vector containing the cloned gene was transferred into disarmed *Agrobacterium tumefaciens* strain GV3101 by electroporation. Genetic transformation of *B. juncea* line Varuna was carried out using hypocotyl explants following a protocol described by Mehra et al. (2000). Positive transformants were selected in vitro on a medium containing PPT at a concentration of 10 mg L^{-1}. For segregation analysis of the progeny of the T0 transgenics for resistance and susceptibility to PPT, a concentration of 200 mg L^{-1} was used in the field. For Southern blot analysis, genomic DNA was isolated from well-expanded leaves collected from the field grown transgenic lines (T0) following the protocol of Rogers and Bendich (1994). Six micrograms of DNA was digested with restriction enzyme EcoRI. Digested DNA was separated on 0.8% agarose gel and blotted onto Hybond-N+ nylon membrane. The blots were hybridized with probes that were labeled using PCR DIG Probe Synthesis Kit, and hybridization was detected with DIG Luminescent Detection Kit following the manufacturer's instructions.

4. **Gene expression and cDNA analysis:** For the gene expression studies, RNA was extracted from the cotyledons of 1-week-old seedlings of Donskaja-IV, Varuna, Heera, and Tumida using Spectrum Plant Total RNA Kit. RNA was treated with RNase-free DNaseI to remove any contaminating genomic DNA. Two micrograms of total RNA was used to synthesize the first-strand cDNA using High-Capacity cDNA Reverse Transcription Kit according to the manufacturer's instructions. For studying changes in transcript accumulation in response to pathogen infection, cotyledons of 1-week-old seedlings of Donskaja-IV were inoculated with 10 μL of sporangial suspension (2 × 104 sporangia mL^{-1} in water) of *A. candida* isolate AcB1 using the drop inoculation method. The cotyledon samples inoculated with the pathogen and distilled sterile water (mock) were collected 4 h, 48 h, 72 h, and 14 days after the inoculations—with four seedlings for each sample. Total RNA was extracted from the collected samples, and first-strand cDNA was synthesized as mentioned above. Real-time quantitative reverse transcription-PCR (qRT-PCR) analyses were carried out using a Power Up SYBR Green Master Mix on Quant Studio 6 Flex Real-Time PCR System. Comparisons of relative gene expression were made between the pathogen and mock-inoculated samples collected at the same time points. The ubiquitin gene, UBQ9 (Chandna et al. 2012) was used as an internal control. The $2^{-\Delta\Delta C}T$ method (Livak and Schmittgen 2001) was followed in evaluating the relative gene expression levels. To obtain a cDNA sequence, RT-PCR was carried out using total RNA isolated from 1-week-old cotyledons of Donskaja-IV. The first-strand cDNA was synthesized as described above and was used as a template for PCR using Phusion High-Fidelity DNA Polymerase with a gene-specific primer pair. The PCR product was cloned in the pGEM-T Easy vector and sequenced on a 3730 DNA Analyzer.

5. **Sequence information of syntenic regions in *B. juncea* lines and sequenced Brassicaceae genomes:** The sequence of the syntenic region in line Heera was obtained by screening a genomic DNA BAC library of the line reported earlier (Sharma et al. 2014). Identified BAC clone was sequenced by the procedure reported for the Donskaja-IV BAC. The sequence of line Varuna was obtained from an SMRT-based genome assembly of the line. Sequences present in Cutlass, Skorospieka, Kranti, *B. nigra* (Sangam), and the B-genome of *B. juncea* (Donskaja-IV) were obtained from Illumina short-read sequencing of the genomes of these lines, the sequence of Tumida was downloaded from the BRAD database (Wang et al. 2015). Sequence information of other members of the Brassicaceae family was obtained from the BRAD database. Multiple sequence alignment of the alleles was done with Clustal W using MegAlign software (Arora et al. 2019).

9.3 *Brassica-Alternaria* Techniques

9.3.1 Greenhouse Method for Testing Resistance

Small differences in infection of *A. brassicae* on *Brassica* species are difficult to detect in the field. Therefore, a more detailed method under greenhouse conditions was developed by Grontoft and O'Connor (1990). Plants of test genotypes are grown in pots. When the first true leaf emerges, the plants are thinned to one healthy plant per pot. From the time of sowing to the time of evaluations of genotypes, the temperature is kept constant at approximately 20 °C. The plants are inoculated at the three-leaf stage using *Alternaria* inoculum grown on PDA for 3 weeks at 18 °C. Inoculation is performed using small agar discs (10 mm) cut from just behind the active mycelial growth front with a cork borer. The discs are placed with the fungus growth sides against the upper side of the second leaf over a 4 mm hole. A second hole of similar size is cut on the other side of the mid vein. This hole is used to control if the size of necrotic lesions are in any way influenced by leaf surface area growth. Immediately after inoculation, each plant is covered with plastic bag in order to increase relative humidity and enhance fungal growth. The genotypes are evaluated 7 days later by measuring the diameter of the necrotic lesions on the leaves.

9.3.2 *Brassica* Germplasm Screening for Resistance Through AB-Toxin

Sensitivity of test genotypes to the Alternaria blight toxin is correlated with the disease reaction in this technique assuming that the level of sensitivity of different *Brassica* species to the toxin is similar to the *in vivo* susceptibility of these to *A. brassicae* (Bains and Tewari 1987). It can cause symptoms of varying severity, depending upon the host genotype, which range from severe chlorosis and necrosis to almost no visible chlorosis (Tewari and Bains 1988). Test genotypes are grown under controlled conditions, 3 mm^2 area is gently scratched with a pasture pipette on third or fourth detached leaf (3.3–4.2 growth stage), and 20 μL of the purified AB toxin from the single spore culture of *A. brassicae* is applied (Bains and Tewari 1987). The leaves are incubated on a moist filter paper in a Petri dish on laboratory benches under continuous fluorescent light, and the reaction of genotypes recorded 48 h of incubation.

In vitro screening of protoplasts of different *Brassica* genotypes, or screening of secondary embryoids, originally derived from a single anther embryoid through AB-toxin, have been employed since plant protoplasts are known to be more sensitive to host-specific toxins (MacDonald and Ingram 1985, 1986; MacDonald et al. 1986). Secondary embryogenic culture lines are maintained on MS medium (Murashige and Skoog 1962) containing 0.8% Difco agar and 2% sucrose, but without plant growth regulators. When regenerates' are required, secondary embryoids are transferred for 3–7 days to MS medium supplemented with kinetin and then transferred to fresh medium without growth substances for root development. Plantlets are potted in peat balls and placed in high humidity in the greenhouse until they are established. Leaves from such plants are surface-sterilized and placed in plastic boxes on moist filter paper. Two drops (10 μL each of suspension of approximately 5×10^5 spores mL^{-1} of *A. brassicicola*) are placed on each detached leaf, and incubated in low continuous light (approx. 1.7 m W cm^{-2}) at 25 °C. Secondary embryoids have exhibited a wide range of reactions, including a number of resistant plants (MacDonald and Ingram 1985, 1986; MacDonald et al. 1986).

In vitro pollen selection (microspore culture) in *Brassicas* for resistance to toxins from *A. brassicae* and *A. brassicicola* has been applied (Hodgkin 1990; Tewari and Bains 1988). The AB-toxin from *A. brassicae* at 12.5 μg/mL completely inhibits the germination of pollens of highly susceptible *B. rapa* var. Yellow Sarson, whereas pollens of highly resistant *C. sativa* germinates even at 100 μg/mL of the toxin (Tewari and Bains 1988). Hodgkin (1990) described *in vitro* pollen selection in *B. napus* for resistance to phytotoxic compounds from *A. brassicicola.* Pollen samples from *B. napus* cultivars "Arran" and "Hercules" are incubated for 1 h in a germination medium or in a medium to which 20 mg mL^{-1} of a *A. brassicicola* toxic extract has been added. The pollen samples are then used to pollinate cultivar Primar. A number of plants, obtained from such pollinations, produce pollen with a significantly increased ability to germinate in medium containing 10 mg mL^{-1} of the extract. Evidence that some selection for resistance to the toxic compounds produced by *A. brassicicola* has been obtained (Hodgkin and MacDonald 1986).

9.3.3 Characterization of a Gene from *Alternaria* for Fungicidal Resistance

1. **Method of estimating *Alternaria brassicicola* in seed:** Naturally infected cabbage seeds are incubated for 4 days at 18–21 °C with alternating light and dark periods on filter papers moistened with 2000 ppm aqueous solution of Dow

sodium salt of 2,4-dichlorophenoxyacetate (2,4-D). When seeds are placed closer than 6 mm apart, or incubated longer than 4 days, higher pathogen counts occur. Two thousand ppm 2, 4-D and efficiently inhibits seed germination and does not reduce pathogen count. Incubation temperature above the optimum temperature range of 20–25 °C increases problem with the contaminant fungi. Continuous light or alternate light and dark is preferable to continuous dark. Excessive moisture is detrimental. Pre-treating seeds with ethanol (EtOH) or sodium hypochlorite (NaOCl) effectively reduces rapid growing contaminant fungi, but significantly reduces pathogen count (Bassey and Gabrielson 1983a, b). However, other common methods in use for detection of seed-borne inoculum are (a) deep-freezing blotter method, (b) germination test, (c) blotter method, and (d) agar plate method (Vannacci 1981).

2. **Identification of fungicide antagonists in leaf exudates:** Attempts have been made to identify the fungicide antagonists in exudate extracts of leaves of wallflower (*Cheiranthus cheiri)* (Beynon and Brown 1969). In extracts, prepared by immersing wallflower leaves in water, the antagonistic activity of the fungicide ethylene thiuram disulfide (ETD) against the spores of *A. brassicicola* has been shown to be due to the presence of glucose and fructose. These compounds are present in the extracts at concentrations corresponding to 0.015 μg of glucose and 0.002–0.005 μg of fructose for every square centimeter of leaf extracted. Dunn et al. (1969) showed that water-soluble leaf exudates extracted from apple, vine, tomato, potato, and wallflower reduced the fungi toxicity of ETD to *A. brassicicola.* The spectrum of the leaf exudate effect is wide, and the magnitude of the effect can vary in each host–parasite–fungicide relationship. It seems that a large proportion of fungicide efficacy may be reduced by antagonistic materials, which either occur on the plant surface or are readily, and quickly available from intact plant tissue. It is likely that the active principle affects the fungal spore, either by reducing its permeability to fungicides or by influencing some part of the metabolism involved in spore germination. Stimulation of the spore metabolism could activate systems capable of detoxifying alien chemicals. The major part of the exudate effect observed is due to glucose.
3. **Strains and media:** All strains used were isolated from radish or cabbage during quality controls of commercial seed lots (French isolates) or seeds from small local markets (Romanian isolates). All strains were purified by mono spore isolation and maintained in tubes of malt agar medium (malt 20 g, agar 20 g in 1 L) at 4 °C. The laboratory-resistant Abra20MLR and Abra43MLR strains were obtained after transferring, onto selective medium, mycelial tips from fast-growing sectors observed with some sensitive isolates on iprodione-amended media. Fresh subcultures were made by transferring hyphal plugs to dishes of potato dextrose agar (PDA) medium amended or not with iprodione (dicarboximide) or fludioxonil (phenylyprrole) at a final concentration of 100 mg L^{-1}. The effects of fungicides on spore germination and germ tube elongation were tested as described by Iacomi-Vasilescu et al. (2004). Mycelial plugs were transferred onto PDA containing 2–4% NaCl (w/v) in order to test sensitivity to osmotic stress.

4. **DNA procedures:** Genomic DNA was extracted from mycelium according to Adachi et al. (1993). For Southern analysis, DNA fragments resulting from genomic DNA digestion with EcoRI and Hind III were separated on 1% agarose gels and vacuum-transferred to Hybond-N membranes. Blots were then probed with PCR fragments purified with Gene-Clean II and labeled with a random primer labeling kit.

Fungal RNA was extracted, DNAse-treated and reverse-transcribed as described by Guillemette et al. (2004). PCR reactions using degenerated primers similar to those designed by Alex et al. (1996) and based on the conserved N and H boxes (N1A, 5¢-GARAAY TTI AII GCR TT-3¢; H1A, 5¢-CAYGAI HTI MGI ACI CC-3¢) were used to amplify a putative histidine kinase gene fragment from Abra43FS genomic DNA. The amplification was performed in 50 mM KCl, 1.5 mM $MgCl_2$, 10 mM Tris-HCl (pH 9.0), 0.1% Triton X-100, containing 0.2 mmol of dATP, dCTP, dTTP, and dGTP, 0.4 Lmol of each primer, 50 ng of DNA, and 1 unit of Taq polymerase. in a final volume of 50 ll. PCR was carried out in a Thermo jet with 35 cycles at 95 °C (1 min), 40 °C (1 min), and 72 °C (1 min), followed by a 30-min extension at 72 °C. Amplification products were separated on 1.2% agarose gel, purified from gel by Gene-Clean II, cloned into a pGEM-T vector (Promega), and subsequently sequenced. Specific primers were designed from this sequence to perform PCR-walking according to the method described by Siebert et al. (1995). PCR amplifications (initial denaturation for 2 min at 94 °C, then 35 cycles of 1 min at 94 °C, 1 min at 55 °C, 1–4 min at 72 °C, with a final extension for 10 min at 72 °C) with specific primers were also performed on genomic DNA and first-strand cDNA in order to verify the presence of introns in coding sequences and to confirm the physical linkage between the 5¢ and 3¢ amplified parts of the histidine kinase gene. A set of 11 primer pairs (Table 9.1) that span the whole coding sequence of the AbNIK1 gene was designed for PCR amplification of corresponding portions of this gene using genomic DNA from three fungicide-sensitive strains (Abra43FS, AbraCMFS, Abra20FS) and from the seven resistant mutants. Nucleotide sequences were obtained directly from these PCR products by Genome Express (France) and deposited in GenBank (accessions AY700092–AY00098). Sequences were compared using the Multalin program (Corpet 1988).

5. **PCR-RFLP and heteroduplex mobility assays:** Portions of the AbNIK1 gene potentially carrying mutations were PCR-amplified using relevant primers and either directly digested with different restriction enzymes according to the manufacturer's instructions or mixed for duplex formation as described by Sutra et al. (2001). Digestion products or mixtures of heteroduplexes and homoduplexes were then separated on non-denaturing polyacrylamide gels (Avenot et al. 2005).

Table 9.1 Primer sets used for amplification of the AbNIK1 gene from *Alternaria brassicicola* genomic DNA. Fragment-numbering corresponds to nucleotide positions in the AbNIK1 gene, with the first nucleotide of the start codon considered as position +1 (Avenot et al. 2005)

Primer set	5¢ primer	3¢ primer	Amplified DNA fragment
1	5¢-TGCCAAAGGTTCTTCAGTCC-3¢	5¢-CAGCCCAAATACCAGCAACT-3¢	+154 to +995
2	5¢-TTGCCAGTCAAGTCACATTCC-3¢	5¢-CCATGGCGTTTACATTGACA-3¢	+900 to +1350
3	5¢-AGAGATGTCGGAACTGAAGGCAAGC-3¢	5¢-ATTTCTCGCACCTGTGTTGTG-3¢	+1254 to +1657
4	5¢-CGAAGTCACAAAGATTGCGCGAG-3¢	5¢-TACTTCGTACTTGCAGCGTGAG-3¢	+1511 to +1932
5	5¢-TCGCACAAGAAGTCAGCAAGGTC-3¢	5¢-TGAATTTGCCCTCCATTGCAGC-3¢	+1780 to +2247
6	5¢-CGACATGGTGATGAGACTGGAT-3¢	5¢-GTTGCGCATGCGTCAGATCGGTG-3¢	+2030 to +2487
7	5¢-CGCAACGAGAGATGTTGAC-3¢	5¢-GATGAGGTTAAGCTTATCG-3¢	+2482 to +2906
8	5¢-CCAGATCATTATGTCTTCCA-3¢	5¢-GCACCTCCTCAACCGAGTGA-3¢	+2832 to +3237
9	5¢-CGGATGAGATTACTGAGCAC-3¢	5¢-ACTTCTCCAGAATCCTAC-3¢	+3166 to +3573
10	5¢-CGTTCCAGATTCTTCTGGC-3¢	5¢-GAGAGCCGAATTAGTAGGG-3¢	+3505 to +3911
11	5¢-CGTCATTGCACTAACGGCGCAC-3¢	5¢-TCAGCTACTGGACTCCCAGC-3¢	+3794 to +4304

9.3.4 Biochemical and Molecular Analyses for Mechanism of Resistance in *Brassica* to *Alternaria*

1. **Plant growth, testing, and sampling:** Twenty seedlings per each tested cultigen (*S. alba*, "Kamienna Głowa," PGH05I, and PGH09K) were grown from seeds for 45 days, and then inoculated with suspension of mixed *A. brassicicola* isolates and tested as per the optimized protocol. Each day, samples were collected from two plants per cultigen, at the same time (~11 am). At 3 dpi, Kamienna Glowa' started showing mild Alternaria dark spot symptoms, which exacerbated at 4 dpi; hence, sampling was stopped at this time. Collected samples were snap frozen in liquid nitrogen and stored at −80 °C until extracted.
2. **Biochemical assays:** The homogenized samples were processed as in Nowicki et al. (2016). Extraction of soluble proteins (supernatant) was done with 5× (v: w) 50 mM potassium phosphate, pH 7.2, whereas the reactive oxygen species (ROS) and free phenols were assayed from the methanolic extracts (18,000 × g for 15 min at 4 °C each, with 10:1 (v: w) of 80% methanol in water), supernatants combined, and pellets were used for lignin assessment.
3. **Assessment of ROS generation (H_2O_2, HO*, and O_2^-):** Generation of either ROS species followed the published methods (Loreto et al. 2001; Loreto and Velikova 2001; Habu and Ibeh 2015; Moukette et al. 2015). Methanolic extracts were assayed at A390, A560, and A530 for H_2O_2, HO*, and O_2^-, respectively. The H_2O_2 as one of the standards; in the HO* and O_2^- assays, the absorbance readout was recalculated per mg of fresh tissue.
4. **Assessment of plant red-ox enzymic activities:** To analyze the dynamics of polyphenol oxidase (PPO), peroxidase (POX), superoxide dismutase (SOD), polygalacturonase (PGA), and catalase (CAT), standard methodology published elsewhere were used (Korgan et al. 2011; Moukette et al. 2015; Sheehy et al. 1988). Buffer extracted samples (as above) were assayed for protein using the BCA kit and analyzed for linear assay range on a randomly chosen sample (0 dpi). Proper assays were run in microtiter plates for 3–5 min, respectively, whereas absorbance readouts were recorded every minute in the Epoch2T reader. Protein activity was calculated as change in absorbance per minute and per mg of protein (PPO) or metabolized substrate/min/mg protein (POX, SOD, PGA, CAT).
5. **Accumulation of free phenols and deposition of lignins:** Levels of lignins deposition and free phenolics were assessed with published spectrophotometric methods (Campbell and Ellis 1992; Cvikrova et al. 1992). Methanolic extracts and pellets were used, and the assays used technical lignins and gallic acid as the respective standards (Nowakowska et al. 2019).

9.3.5 Evaluation of Partial Resistance to *Alternaria brassicicola*

Three inoculation methods (cotyledon, detached leaf, and seedling) were evaluated by Doullah et al. (2006). The detached-leaf inoculation test is the most suitable for screening *B. rapa* genotypes because clear symptoms are observed on the leaves in less than 24 h. There is a significant positive correlation between the results from the detached-leaf inoculation test and seedling inoculation test, an established method considered to yield reliable results. In addition, it is very easy to screen plants for resistance on a large scale and to maintain standard physical conditions using detached leaves. For successful infection, inoculum concentration should be adjusted to 5×10^4 conidia mL^{-1} and incubation temperature should be between 20 and 25 °C. The 3rd/4th true leaves from 30 days old plants are optimal for inoculation.

For this test, two leaves (3rd/4th true leaves) are collected from 30 days old plants having approximately six leaves. A replicate for the detached-leaf tests consists of three randomly selected leaves from a set of 3rd/4th true leaves collected from several plants that are pooled. *Alternaria brassicicola* inoculum (5×10^4 conidia mL^{-1}) is sprayed on the upper surface of the collected leaves to run-off and inoculated leaves placed in a plastic box (length × width × height = 32 × 23 × 5 cm) where strips of moist tissue paper are put on four vertical sides to maintain humidity at 98%. The plastic box is inoculated at 22 ± 1 °C for 3 days in the dark in an incubator or controlled temperature room (Doullah et al. 2006).

9.3.6 Assessment of Methods of Inoculation for Resistance to *Alternaria*

Out of four methods of inoculation (seed inoculation, cotyledonary leaf inoculation, detached-true leaf inoculation, and detached green pod inoculation), detached-true leaf inoculation is more efficient and reliable for screening germplasm for resistance to Alternaria blight of rapeseed and mustard.

9.3.6.1 Method of Inoculation

A conidial suspension of *A. brassicae* containing 1.5×10^4 conidia/mL is prepared, and then with the help of Eppendorf micropipette, three separate 10 μL drops, each containing 4–6 conidia, are placed on the upper surface of true leaves at three equidistant places. Similarly, distilled water drops (without conidia) are placed on other cotyledons, which serve as checks. All the treatments are replicated three times for each cultivar with each isolate. Observations on disease severity are recorded on 0–5 rating scale (Saharan et al. 2016).

9.3.6.2 Detached-True Leaf Inoculation

True leaves are collected from 30 days old plants, surface sterilized with sterilized distilled water with the help of sterilized cotton, and leaves inoculated following the method described above. Inoculated leaves are kept in Petri plate moist chamber, and incubated at 20 ± 2 °C for 7 days, and leaves scored for disease severity.

9.3.6.3 Infection Score and Disease Index

A 0–5 rating scale is followed, where 0 = no visible symptom; 1 = 1–10% leaf or pod area covered with small spot; 2 = 11–25% leaf or pod area covered with bigger (more than 3 mm) spot; 3 = 26–50% leaf or pod area covered with bigger (more than 3 mm) spots with initiation of coalescing on leaves and deep lesions on pods; 4 = 51–75% leaf or pod area covered with bigger commonly coalescing spot, and 5 = 76–100% leaf or pod area covered with very large spots giving blighting appearance (Vishwanath and Kolte 1999; Saharan et al. 2016).

9.4 *Brassica-Erysiphe* Techniques

9.4.1 Collection, Preservation, and Cultivation of Crucifer's Powdery Mildew

Symptoms of powdery mildew on the infected host parts (leaves, stem, branches, and siliquae) are conspicuous in the form of white floury patches consisting of mycelium, conidia, and conidiophores. To locate formation of chasmothecia careful search is needed especially at the maturity stage of the crop. Chasmothecia are visible in the form of brown to dark brown pinhead size bodies on the infected leaves, stem, and siliquae. Special care should be taken to properly identify the host plant. It is always good to collect flowering, and/or fruiting parts of the host along with infected leaves. For the systematic study of powdery mildew, dry material is very satisfactory. For further study, infected parts of plants are collected, and placed between driers for a few days. The material is then transferred to suitable envelops, properly labeled, and laid aside for future study. To maintain the culture, plants should be grown under growth room/green house for continuous availability of inoculum, and pathogen to use in subsequent studies. In general, a succession of crops of young host must be ready to maintain powdery mildew on the host.

9.4.1.1 Maintenance of *Erysiphe* Isolates

A culture of a powdery mildew pathogen, later identified as *Erysiphe cruciferarum*, was isolated from an infected *B. napus* cv. Cobra plant growing in greenhouses at the University of East Anglia. To obtain a pure culture, conidiospores from a single colony were subcultured five times on *B. napus* cv. Cobra plants. This pure isolate, designated UEA1, was thereafter propagated on *B. napus* cv. Cobra plants growing in the controlled environment. *A. thaliana* plants with 6–8 true leaves (about 3–4 weeks old) were inoculated with *E. cruciferarum* UEA1 in a wind-still chamber, using heavily infected young leaves of *B. napus* cv. Cobra as the source of inoculum. For inoculation, an artist's fine brush was used to gently dislodge conidia from infected leaves onto *A. thaliana* plants approximately 20 cm below. The inoculated plants were kept under transparent plastic lids for 24 h, to give high humidity (>90% RH), after which the lids were removed (Adam and Somerville 1996).

9.4.2 Plant Material and Inoculation Methods

This technique has been developed using the *Arabidopsis thaliana* Col-0 genotype and the *edr1*-1, *eds1*-2, *pmr4*-1, *mlo2*-6 single, and *mlo2*-5 *mlo6*-2 *mlo12*-1 triple mutants in the Col-0 genetic background. Approximately 100 seeds were sown per pot of soil substrate, and five pots were used per genotype. After stratification for 2 days at 4 °C in darkness, plants were grown for 18 days at a day/night cycle of 10/14 h in a light chamber with 22 °C/20 °C day/night temperature, and a relative humidity of 60%. The *G. orontii* isolate MPIPZ was propagated on 4-week-old *eds1*-2 plants, and conidia were used at 14–21 dpi. Inoculations were performed in a simple 80 cm high cardboard settling tower whose opening was covered with 80 μm nylon mesh (Reuber et al. 1998). The tower contained up to nine pots per inoculation. Three consecutive rounds of infection have been used. In each round, pots of all genotypes were included. A fine paint brush was used to harvest conidia from four heavily infected leaves and to separate the conidia by brushing them through the nylon mesh. Inoculation density was approximately 750 spores/cm^2. Newly inoculated seedlings were then returned to the growth chamber.

9.4.2.1 Artificial Inoculation of Pathogen

Conflicting ideas of the relation of water to infection with powdery mildews has led to a variety of contrasting methods of making artificial inoculations. Suspending of conidia in water, spraying of plants with water before inoculation or incubating the plants in a moist chamber after inoculation are probably all of no value in obtaining infection in most cases, though they may incidentally serve such useful purposes as confinement of strains, reduction of light, and reduction of temperature (Yarwood 1957). Dusting dry conidia onto dry leaves, and leaving the inoculated plants in a dry environment (Yarwood 1936) will give heavy infection with most powdery mildews. Use of detached leaves (Yarwood 1946) has facilitated certain experiments with powdery mildews, especially in securing infection under conditions unfavorable to infection of entire plants. Growing powdery mildews on tissue cultures will probably be useful for specific purposes. Single spore inoculations are commonly successful (Homma 1937).

9.4.3 Primer Design and Test Specimen Methods

Erysiphales-specific primers for the ITS region were designed from 18S and 28S sequences (Mori et al. 2000). The forward primer, PMITS1 (5′-TCGGACTGGCC (T/C)AGGGAGA-3′) was designed from 18S sequences of *Blumeria graminis* (Gen Bank Accession AB022361), *Erysiphe heraclei* (AB022390), *Golovinomyces cichoracearum* var. *cichoracearum* (AB022359), *Phyllactinia moricola* (AB0022400), *Podosphaera xanthii* (as *Sphaerotheca cucurbitae*, AB022409), and *Pleochaeta shiraiana* (AB022402), whereas the reverse primer, PMITS2 (5-′-TCACTCGCCGTTACTGAGGT-3′) was designed from 28S sequences of *Erysiphe heraclei* (AB022391), *Erysiphe* (*Uncinula*) *mori* (AB022418),

Phyllactinia kakicola (AB022372), *Golovinomyces cichoracearum* var. *cichoracearum* (AB022360), *Sawadaea polyfida* var. *japonica* (AB022364), and *Blumeria graminis* (AB022362). The primer sites were chosen by comparing these sequences with those from a range of other fungi available in Gen Bank. In particular, 26 ascomycete sequences from species in the Pezizales, Ophiostomatales, Microascales, Saccharomycetales, Chaetothyriales, Eurotiales, Hypocreales, Phyllachorales, Cyttariales, Rhytismatales, Dothidales, Sordariales, Onygenales, Heliotales, and Schizosaccharomycetales were used. To gain the specificity required, particular attention was paid to ensure that the 3′ end of both primers consisted of a sequence unique to the powdery mildews (Innis and Gelfand 1990; Kwok et al. 1990).

9.4.4 DNA Extraction and PCR

DNA was extracted using a protocol modified from Hirata and Takamatsu (1996). A small amount (<1 mm^3) of the fungus was scraped from the host plant and placed into 50 μL of 5% Chelex 100 (Bio-Rad) containing 0.01% Triton X-100. This was incubated at 56 °C for 2 h (vortexing briefly after 1 h), vortexed, incubated in boiling water for 8 min, vortexed, and spun in a bench-top centrifuge at 20,000 g for 5 min. The top 40 μL was removed, added to 40 μL isopropanol, mixed, left on ice for 10 min, and centrifuged for 10 min at 20,000 g. The supernatant was removed, and the sample was dried and resuspended in 40 μL of water.

The initial PCR was performed in 25 μL containing 5 μL DNA, 200 μM of each dNTP, 1.5 mM $MgCl_2$, 2.5 μL 10 × buffer, 4 ng each of primers PMITS1, and PMITS2, 6% Tween 20 (Lab chem.), 0.5 units of Amplitaq Gold, and 100 ng BSA (Bresatec). Reactions were overlaid with 30 μL of light paraffin, and subjected to 10 min at 94 °C, 35 cycles of: 1 min at 94 °C, 1 min at 65 °C, 2 min at 72 °C, and a final extension of 10 min at 72 °C. PCR products were detected by running 4 μL on a 1.4% agarose gel in TBE buffer. A nested PCR was performed in 50 μL as outlined above, but without Tween 20, using primers PMITS1, and ITS4 (White et al. 1990) at twice the previous concentration, and containing 1 μL of a 1:100 dilution of the first round product, or if no product was visible then an undiluted 1 μL. Cycling times were the same, but with an annealing temperature of 60 °C. Nested PCR products were purified using a QIA quick PCR Purification Kit. To ensure that only the powdery mildew ITS region had been amplified, the products were sequenced directly using primers ITS5 (White et al. 1990), and ITS4, with an ABI PRISM BIGDYE Terminator Cycle Sequencing Kit. These sequences were used in similarity searches on the Gen bank database using the computer program Blast (Altschul et al. 1990) to confirm that they did belong to the Erysiphales.

9.4.5 Use of qPCR, and Spore Count Assays to Quantify Powdery Mildew

Two protocols for the quantification of the powdery mildew disease on *Arabidopsis* with reproducible results over a broad range of infection, phenotypes have been developed (Weßling and Panstruga 2012) (Fig. 9.1; Table 9.2). The techniques are also useful for the quantification of more subtle differences although such instances may require a higher number of experimental replications. For the qPCR-based method, efficient and specific primer combinations for the amplification of both host and pathogen DNA have been used. The quantified amount of *G. orontii* relative to *Arabidopsis* genomic DNA correlated well with the amount of fungal biomass. Fungal development was reflected accurately across both time, and different genotypes, and could also be determined upon the occurrence of localized powdery mildew-induced host cell death. 5 dpi was determined to be the optimal time point for comparative analysis among genotypes, but the method can also be used to resolve kinetics of powdery mildew pathogenesis.

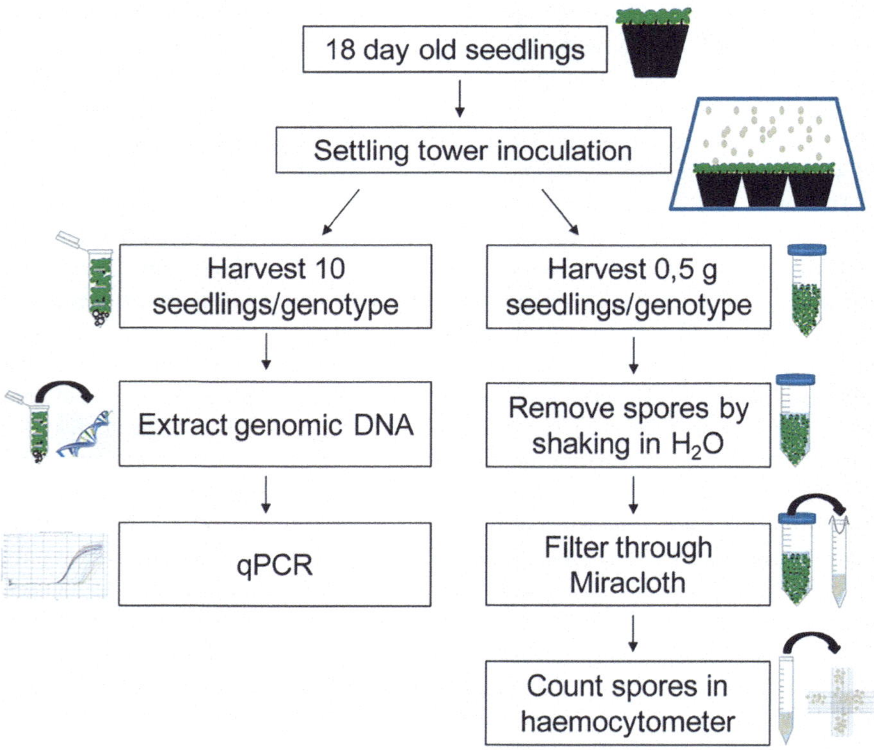

Fig. 9.1 Schematic overview of developed methods. Simplified view of the workflow of the qPCR- and spore count-based powdery mildew quantification procedures (Weßling and Panstruga 2012)

Table 9.2 Comparison of methods to assess powdery mildew infection (Weßling and Panstruga 2012)

Characters	Macroscopic categorization	Host cell entry counts	Conidiophore counts	Hyphal area quantification	qPCR	Conidia counts
Assay type	Semi-quantitative	Quantitative	Quantitative	Quantitative	Quantitative	Quantitative
Staining required	No	Yes	Yes	Yes	No	No
Microscopy required	No	Yes (hundreds of interaction sites)	Yes (multiple colonies)	Yes (multiple colonies)	No	Yes
Stage of pathogenesis scored	Late (Conidiation)	Early (host cell entry)	Late (Conidiation)	Early to late (host cell entry to conidiation)	Middle to late (hyphal expansion to conidiation)	Late (conidiation)
Importance of equal inoculation density	High (no normalization)	Low (internal normalization)	Low (internal normalization)	Low (internal normalization)	Low (internal normalization)	Medium (averaging effect of scoring multiple plants at once)
Suitable for high-throughput analysis	Yes	No	No	No	Yes	Yes

A spore count assay can also be useful for powdery mildew research. This assay has a wide dynamic range, accurately and reproducibly determined differences in conidiation by the pathogen. Both assays yielded comparable results for the susceptibility of the genotypes tested and can easily be adapted to other *Arabidopsis*-infecting powdery mildew species, and other plant-powdery mildew pathosystems. Both methods require similar amounts of time for data generation. The qPCR method is intrinsically more cost-intensive but allows simultaneous DNA extraction, and PCR analysis of many samples. Spore counts are more cost-effective, and owing to the lower number of practical steps involved, offer less possibilities for experimental error (Fig. 9.1). Currently, powdery mildew infection is either assessed semi-quantitatively by coarse categorization, and thus macroscopic symptoms or quantitatively by microscopic penetration rate counts at early, and/or conidiophore counts at late stages (Table 9.2). The latter methods require staining, mounting, and microscopy, and are therefore time-consuming, and labor-intensive. Both assays introduced here are quantitative and are applied at late stages of infection when differences between genotypes are more pronounced. They do not require staining, are easy to perform, and can generate quantitative data for larger numbers of genotypes (Table 9.2).

In addition, the spore count assay is based on many seedlings rather than few mature plants, which offers the advantage of an averaging effect, lowering the possibility of experimental outliers. These assays can overcome limitations of currently used methods of powdery mildew quantification. They might be particularly suitable for the assessment of larger numbers of different mutant lines, e.g., in the course of follow-up analyses of genes identified by omics approaches.

9.4.6 Characterization of the Disease Reaction Phenotypes

The disease reaction scores were evaluated by comparison with the level of infection observed on the fully susceptible Col-5. Two measures of infection were used. The level of fungal growth at 7 dpi was ranked on a scale from 0 to 3, with a score of 0 indicating that no fungal growth was apparent by eye, and no conidiophores developed. A score of 1 indicated that fungal growth was restricted, and no conidiation occurred. An intermediate density of fungal growth accompanied by the development of conidiophores was scored as a 2, and a score of 3 indicated dense fungal growth, and abundant conidiation (i.e., a fully compatible interaction). The host response to inoculation was scored as either a 1 or a 0 indicating the presence or absence, respectively, of chlorotic or necrotic flecks at the point of inoculation. For example, a disease reaction score of 0.1 (with the first number referring to the extent of fungal growth, and the second number referring to the presence of chlorotic/necrotic lesions) was given to highly resistant *Arabidopsis* accessions that supported no visible fungal growth, and that did produce chlorotic or necrotic lesions at the point of inoculation. A disease reaction score of 3.0 was assigned to accessions supporting dense fungal growth with no chlorotic or necrotic flecks (Adam and Somerville 1996).

9.4.7 Embryo Rescue Technique to Transfer Powdery Mildew Resistance

Fully developed flower buds were emasculated prior to anthesis and allowed to bloom. Fresh pollen grains from newly opened flowers were used to pollinate emasculated flowers. For embryo rescue, pistils were harvested 5–7 days after pollination, washed in 70% ethanol for 1 min, washed in 10% sodium hypochlorite solution for 10 min, and then rinsed twice with sterile distilled water. Ovaries were cultured for 7 days in hormone-free MS medium (Murashige and Skoog 1962) supplemented with 5% sucrose, 0.8% agar, and 500 mg/L casein hydrolysate (MS1) under continuous illumination with fluorescent lights (115 Lmol/m^2/s). Pistils were cut open aseptically, and ovules were transferred to MS medium with 3% sucrose, and 0.8% agar without casein hydrolysate (MS2). All ovules were incubated under fluorescent lights with 16 h photoperiod, and developed ovules were transferred to individual plant boxes. Well-established seedlings were transferred to plastic pots containing Cornell mix (Boodley and Sheldrake Jr 1972), covered with plastic bags, and placed in growth chambers for acclimatization with 6 h photoperiod day/night for two cycles per day with fluorescent lights. Fully developed plants were transferred to a greenhouse where they were grown to maturity (Tonguc and Griffiths 2004).

9.4.8 Identification of Molecular Markers Linked to Powdery Mildew R-Genes

1. **Collection of plant material:** The experimental material was comprised of four females, *viz.*, GM-1, GM-2, GM-3, and GDM-4 (*Brassica juncea* L.), and three males, *viz.*, NRCY-05-02 (*B. rapa* L.), Neelam (*B. napus L.*), and Pusa Swarnim (*B. carinata* L.).
2. **Generation of hybrid population:** Twelve F1s were developed by crossing four females with three males in an all possible crosses made (Excluding reciprocal). The seeds of all F1 hybrids were produced by hand emasculation and pollination while the parental seeds were obtained through selfing. Subsequently, in the next year, a part of the F1 seeds were utilized for raising the F1 plants. The F1s were screened for powdery mildew incidence and selfed seeds of all the 12 hybrids were collected as F2 seeds. However, only one cross, *i.e.*, GM-3 × Pusa Swarnim was advanced to F2 generation due to its lowest disease score.
3. **Development and identification of transgressive segregants (F2):** The F2 seeds of the cross GM-3 × Pusa Swarnim was preserved for raising the F2 generation. F2 evaluation trial for powdery mildew resistance was conducted along with the parents, *viz.*, GM-3 and Pusa Swarnim. The resistance plants were selected from the F2 population and the selected plants were used further for molecular analysis.
4. **Identification of DNA markers:** A set of 165 A-, B-, and C-genome-specific SSR primers including, functional SSR primer pairs were employed for the

parental screening to detect the polymorphism between the resistant parent Pusa Swarnim and susceptible parent, GM-3. The SSR primers were custom synthesized from Eurofins Bengaluru. The genomic DNA of the two parents was subjected to PCR amplification with individual DNA markers. The amplicons scored as codominant-differences in size of the amplicon corresponding to the target loci was noted. The polymorphic markers were noted and later used in Bulk Segregant Analysis (BSA).

5. **DNA extraction:** DNA from the F1 of the cross GM-3 × Pusa Swarnim along with the respective parents were extracted. The F1 was advanced to F2 generation and 210 plants were selected randomly for further molecular analysis (BSA). Leaves from 3- to 4-weeks-old plants were collected from field, surface sterilized with 70% alcohol and freshly used for DNA isolation. The genomic DNA was extracted following Doyle and Doyle (1990) method with a few minor modifications.
6. **Assessment of disease infection:** Observations of powdery mildew severity was recorded at 15 days before maturity of randomly selected ten plants from each plot/row of F2 population. In each F2 individual plants were randomly selected and powdery mildew severity was recorded by referring the 0–9 scale (Mayee and Datar 1986).
7. **Bulked segregant analysis (BSA):** F2 population of the cross GM-3 × Pusa Swarnim was raised which had individuals segregating for disease resistance and susceptibility. Percent Disease Index (PDI) was calculated with the above scales using the formula given by Wheeler (1969). The Percent Disease Index (PDI) for individual plants were plotted each from two extremes, *i.e.*, resistant and susceptible, was selected and their DNA was bulked. This was done with a view of gene interaction prevailing and thus resulting genotype-phenotype deviation expectable as a deviation from simple Mendelian situation (Govindaraj et al. 2005). The SSR marker found polymorphic between parents were screened across all the bulks and markers resulting in expected amplicon patterns were further shortlisted. Shortlisted markers were tested in the respective constituents of the bulks and selected ones were genotyped across all the F2 individual plants of a cross.
8. **Validation of molecular markers:** The 210 F2 individuals' plants of the cross GM-3 × Pusa Swarnim and parents were genotyped with shortlisted SSR markers through BSA. Marker segregation pattern was scored and recorded. The SSRs were scored as co-dominant marker type. After scoring the individual progenies, the data was entered in Microsoft Excel spread sheet and tested for goodness-of-fit through *Chi-square* analysis. The calculated χ^2 values were compared with table values at one degree of freedom for each marker locus.

$$X^2 = \sum_{i=n}^{k} \frac{(Oi - Ei)2}{Ei}$$

where $i = n = 1, 2, 3...210$ and $k = 1, 2, 3...$ (Class of frequency), $O =$ Observed frequency of kth class and E= Expected frequency of kth class. The data sheet was saved in a format suitable for Single Marker Analysis (SMA). A similar approach has also been adapted previously in mustard and other crops where gene interaction prevailed (Pushpa et al. 2016; Govindaraj et al. 2005; Zhang et al. 2009; Gemenet et al. 2010; Milad et al. 2011; Kapadia et al. 2019).

9.4.9 Molecular Identification of Anamorphic Powdery Mildews (Erysiphales)

The powdery mildew fungi (Erysiphales) are a common group of obligate plant pathogens that can be extremely difficult to identify, depending on the reproductive stage encountered. The ribosomal DNA Internal Transcribed Spacer (ITS) region was assessed for its usefulness to link anamorphic Erysiphales with their respective teleomorphs. PCR primers for the rDNA-ITS region were designed and found to have enhanced specificity for the 12 genera tested, even in the presence of contaminating fungi. Recent herbarium specimens of both anamorphic and teleomorphic material yielded sufficient DNA for amplification using a Chelex-based extraction method. The ITS regions from 25 anamorphic specimens were sequenced and compared with the ITS sequences of their suspected teleomorphs. In most cases, an ITS region sequence similarity of above 99% indicated that a correct match had been made. Although this technique will not always unambiguously identify an anamorphic specimen, it will provide valuable information to use in conjunction with morphological and host range data to aid in the final identification (Cunnington et al. 2003).

9.5 *Brassica-Hyaloperonospora* Techniques

9.5.1 Culturing of *Hyaloperonospora parasitica*

The biotrophic nature of this pathogen implies a sophistication of nutritional requirement which cannot be met upon the death of the host plant. *H. parasitica* was cultured on disinfected slices of swede root where aerial growth of mycelium was observed with conidiophores, conidia, antheridia, oogonia, and oospores (Von Guttenberg and Schmoller 1958). Ingram (1969) successfully established, and maintained cultures of the fungus on callus tissues derived from (a) a mature leaf of cabbage, (b) a mature root/hypocotyl of rape, and (c) a seedling hypocotyl of swede. The infected calluses were incubated either at 22 °C in the dark, or at 15 °C with 12 h fluorescent illumination photoperiod. To maintain the dual culture of the callus, and the pathogen, subculturing through transfer of explants of infected callus to fresh uninfected callus was necessary every 14–21 days. Calluses derived from the root/hypocotyl of rape grew faster than those from cabbage leaf and were transferred directly when subculturing. About 4 weeks after inoculation of the rape callus, small nodules of new tissue were developed on the infected callus. These nodules or the whole callus were successfully maintained when transferred to fresh culture medium. Conidia of *H. parasitica* were produced on infected callus tissue maintained at 15 °C and 12 h/photoperiod but production was much lower at 22 °C in the dark. Such conidia were used to infect detached cotyledons or leaf callus of cabbage or rape.

Attempts were made with limited success by Von Guttenberg and Schmoller (1958) to culture the fungus in the absence of living plant tissue. They obtained visible mycelial growth in filter, sterilized swede juice, but it ceased after 3 days when a yellow discoloration appeared suggesting that the medium was chemically unstable. Very limited mycelium was developed on swede seed-glucose agar, and maize decoctions-glucose agar, but more success was achieved with an agar medium containing 2% beer wort + 0.1% phosphate in which hyphae, and conidiophores were developed within, and outside the agar substrate. Similarly, less growth was achieved in oatmeal agar and rice starch agar.

Asada and Ohguchi (1981) studied the behavior of downy mildew fungus of Japanese radish on modified Knop's medium. An isolate of *H. parasitica* from naturally infected leaves was cultured on modified Knop's medium, and 0.1% streptomycin, using infected slices of radish root tissue as inoculum. Vigorous hyphal growth was observed spreading into the medium, and numerous haustoria like bodies were formed. The production of which was favored by low agar concentration, low pH (4), and high sucrose concentration (50 g/L). Growth was ceased 2 weeks after placing the tissue slices on the medium.

The response of *H. parasitica* to a liver medium was studied by McMeekin (1981). Washed, autoclaved 2 mm cubes of liver were placed in a 9 cm Petri dish, and covered with 20 mL of a mixture of 0.01% tryptone (Difco: pancreatic digest of casin), 0.04% K_2HPO_4, and 2% agar: a 1/10 dilution of the K_2HPO_4 and tryptone. Ten μg/mL streptomycin were added to this medium to control bacterial growth. When the agar was solidified, drops of a suspension of conidia in water were placed on the piece of fiver. The plates were incubated at 18 °C. After 4 days, the germ tubes in the control (without liver) disintegrate, but in a 5 mm diam. zone around the liver pieces, 90% of the germ tubes grow towards the liver. They grow from the surface towards the bottom of the medium and formed large swellings or lobes within the agar. The lobed germ tubes reached their maximum size after 4 days. When the plate was flooded with sterile distilled water, either 4 or 7 days after the beginning of conidial germination, the lobes on the germ tubes maintained the same size.

9.5.2 Maintenance of *Hyaloperonospora parasitica* Isolates and Production of Inoculum

Isolates of *H. parasitica* were maintained separately on cotyledons obtained from 6-days-old seedlings, raised in soil-less compost in a modified plant propagator (35.5 cm × 21.6 cm × 18 cm), sited in the glasshouse (Nashaat and Rawlinson 1994). The propagator was supplied with continuous filtered (spore-free) moist air at 18 ± 2 °C through a central flue conducting air from beneath the propagator to exhaust at two adjustable ventilators on the cover, and the junction between the cover, and the base (Jenkyn et al. 1973). Supplementary light was given to maintain a 16 h photoperiod. Cotyledons and a short length of hypocotyl were detached and transferred to folded filter paper (Whatman 12.5 cm, 113 v) supports in glass jars (8 cm diameter, 7 cm depth) containing 20 mL sterile distilled water. Cotyledons were then inoculated in a sterile air flow with 5 μL of conidial suspension on each half cotyledon with the aid of a micropipette. Conidial suspensions were prepared by tapping infected cotyledons to dislodge conidia into sterile distilled water; this minimized bacterial contamination. After inoculating the cotyledons, the glass jars were covered with clear plastic lids, sealed with parafilm, and incubated in a growth cabinet at 16 °C under 70 $\mu E/m^2\ s^1$ irradiance with a 16 h photoperiod for 7 days after which peak sporulation occurred.

Once a culture is actively growing, it must be maintained continuously on living host tissue to provide inoculum for experiment. Prepare inoculum as follows: Collect cotyledons and leaves with profuse sporulation from 15 to 20 seedlings placing the tissues in glass vial containing 3 mL double distilled water, shake the tube to wash conidiosporangia off, remove the mass of plant tissue with forceps, and transfer the suspension to two 1.5 mL Eppendorf centrifuge tubes. Wash the inoculum combining the pellets after the first centrifugation and replacing the final supernatant with 0.5 mL double distilled water. Quantify the suspension using a hemocytometer and adjust the concentration of inoculum to 5×10^6 conidiophores per mL. Inoculate cotyledons of 10–14-days-old seedlings by placing 1–2 μL of inoculum on each cotyledon, thereby delivering about 100–200 conidiosporangia per plant. Incubate in a clear plastic box to maintain high humidity conditions at 16–18 °C, 10 h photoperiod, 50–100 $\mu E\ m^2\ s^{-1}$. Spray plants daily with a fine mist of water. *H. parasitica* will begin to sporulate 4–5 days after inoculation. But peak sporulation will occur after 7–10 days and may continue for another week. Subculturing at weekly intervals is advisable (Dangl et al. 1992).

Percentage disease	Host–pathogen interaction
0.0	No infection
0.1	Traces of infection, generally confined to lower leaves, one plant in ten or fewer with lesions
1	Some plants infected, but one or two lesions per plant
5	Most plants infected with about 5% of the lower leaf area affected
10	Most plants with about 10% of the lower leaf area affected. Up to 5% infection on the upper stem leaves and bracts
25	About 25% of the lower leaf area affected. Leaf area may appear to be half affected and half unaffected. Infection frequent on upper stem leaves and bracts (up to 10% area affected)
50	About 50% of the leaf area affected. Affected area appears to be greater than unaffected
75	About 75% of the leaf area affected. Very little unaffected tissue observable
100	100% of leaf area affected

9.5.3 Germplasm Screening and Evaluation

Genotypes are grown in propagators, (as described under isolates maintenance in Sect. 9.5.2) except that two adjacent 5 cm "Jiffy-pots" for each line or cultivar are used as pots. The pots are placed on capillary matting to ensure a uniform water supply. Each propagator contains up to 13 accessions arranged as two randomized blocks (propagators) with each accession occurring only once in each propagator. Initially, 9–15 seedlings per accession are grown in each propagator, and these are thinned 6 days after sowing to 6–10 to decrease variability in growth. Sowing dates are staggered to produce seedlings at the required growth stage for inoculation at the same time. The average times required under these conditions to reach fully expanded cotyledons, first, and second true leaves are 7, 16, and 22 days, respectively (Nashaat and Rawlinson 1994). Seedlings are inoculated by spraying them to run-off with a suspension of conidia (2.5×10^6 conidia/mL). The propagators are sealed after inoculation to allow the relative humidity to rise to 100%, and then incubated in growth cabinets under the conditions described for isolates maintenance. Infection phenotypes are recorded 7–9 days after inoculation (on cotyledons, and leaves, respectively) using a 0–9 scale as mentioned below (Nashaat and Rawlinson 1994).

According to Williams and Leung (1981), single seeds are grown in 12-pack pots, and when the cotyledons have expanded, after 5–7 days, single 0.01–0.02 mL drops of a freshly collected conidial suspension containing approximately 1×10^5 conidia/mL are placed on each half of the two cotyledons using a finely tipped glass pipette. As plants of each 12-pack are inoculated, they are placed in glass or plastic boxes containing a 1–2 cm depth of warm water in the bottom. A tight fitting cover is placed on the box after the box is filled with plants. The box is then placed in a darkened incubator at 20 °C for 8–16 h. The atmosphere in the box will maintain the droplets on the cotyledons during which time germination, and penetration will occur. After incubation, plants are placed on a lighted greenhouse bench at 20–25 °C for 5 days then returned to humidity boxes at 20 °C for 16–24 h. Upon removal from the humid atmosphere, susceptible plants will have a profuse growth of *H. parasitica* conidiophores on the abaxial sides of the cotyledons, whereas resistant plants will exhibit varying degrees of sporulations and tissue necrosis which can be evaluated on a 0–9 scale (Fig. 9.2). A rating of 1 is given to rapidly occurring (24–48 h) hypersensitive necrotic flecking, without visible sporulation, found immediately under the droplets. Interaction phenotypes representing host–pathogen compatibility are expressed as increasing degree of sporulation on the abaxial side of the cotyledons, and decreasing degree of necrosis associated with tissue colonization. It is important to use freshly produced inoculum collected by washing off spores from leaves with distilled water. Older plants may be inoculated by atomizing a suspension of conidia on the foliage and holding them at 100% RH for 8–16 h.

Knight and Furber (1980) used descriptive key for the assessment of downy mildew disease prevalence and severity on winter oilseed rape varieties. A group of plants (10 plants/sample) chosen at random from a number of observation site were

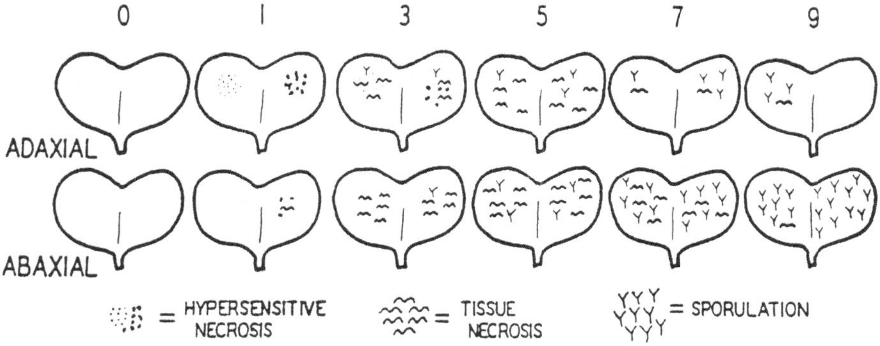

Fig. 9.2 Rating scale for downy mildew (*Hyaloperonospora parasitica*) interaction phenotypes on Chinese cabbage (Williams and Leung 1981)

assessed. However, different disease assessment scales have been used/suggested to classify downy mildew interaction phenotypes in cruciferous crops.

9.5.4 Preservation of *Hyaloperonospora parasitica*

Brassica leaves infected with the pathogen were collected, and conidia from such infected material were inoculated on cotyledons of a susceptible *Brassica* variety (Paul and Klodt-Bussmann 1993). Cotyledons are put into plastic boxes on moist filter paper and incubated at 15 °C and 70–80% RH. Under these environmental conditions, conidia, and conidiophores of the pathogens are ready to be harvested 6 days after inoculation. For each isolate, five cotyledons colonized with fresh conidia were collected in a glass vial in 10% (v/v) glycerine which serves as a cryo-protectant in the suspension medium. For each isolate, six glass vials are filled with the conidial suspension, and immediately stored in a freezer at −21 °C. After a storage interval, samples are taken out from the freezer, and thawed at room temperature (20 °C). After thawing for 10 min, 2 mL of the conidial suspension are added to each Petri dish which contains 15% (w/v) water-agar, and the percentage of conidial germination is assessed after 24 h at 15 °C. After a storage period of 8 days, it was found that the highest germination rate of 73% occurred using 10% (v/v) glycerine.

To culture and preserve Japanese radish downy mildew fungus, slices (1 cm) of Japanese radish root cv. Miyashige were inoculated with conidia of the fungus collected from naturally infected leaves and incubated at 20 °C for 6–8 days (Ohguchi and Asada 1981). Conidia produced on the infected slices were then collected to make a suspension and were used to inoculate other healthy slices. Numerous oospores are observed in these sliced tissues 6 days after inoculation. The conidial viability of *H. parasitica* derived from cabbage was longest, up to 130 days, when the spores were stored in air-dried soil at a constant temperature of 5 °C. Conidia kept at −25 °C, and relatively dry on leaf discs (air-dried at 20 °C) maintained a relatively high rate of germination after 1 year or longer (Krober 1970, 1981).

The pathogen is usually preserved by storing few sporulating cotyledons in small vials with tight lids inside a deep freezer (−25 to −30 °C) for up to 6 months. Temperature fluctuations should be avoided during this period. For reviving the pathogen, the vials containing the sporulating cotyledons are taken out of the freezer, and with their lids kept tightly on, transferred immediately to a container containing icy water. The temperature of the container and its content is then allowed to rise gradually, within l–2 h, to 15 °C. Thereafter, the vials are taken out of the container, and conidial suspension for inoculating fresh cotyledons is prepared in the normal way (Nashaat and Rawlinson 1994). A long-term preservation of *H. parasitica* conidia in freezer at −21 °C for a period of 1 year was possible using glycerine polyethylene glycol (PEG) 400 or 1000 as cryoprotective agents. Of these, 20% (V/V) glycerine/water solution was the most suitable, resulting in germination rates of the stored conidia of 21% after 1 year (Paul et al. 1998).

9.5.5 Artificial Inoculation of Excised Cotyledons

Cotyledons of radish, 9 days after sowing, were placed face downwards on a damp filter paper in a transparent plastic box (Bonnet and Blancard 1987). The box was then put in a growth chamber at 20 °C day, and 18 °C night, relative humidity of 90%, and illumination of 2000 lux for 12 h. With a micropipette, 50 μL of *H. parasitica* conidial suspension (16,000 sp/mL) were placed on the abaxial surface of the cotyledons. After 5 days, the conidia were collected by rubbing the cotyledons with a brush into 2 mL of water; the concentration of the suspension was then measured, and adjusted using a hemocytometer. When 15-day-old plants were inoculated, an excellent correlation was observed between the number of conidia and symptoms on leaves.

9.5.6 Methods of Breeding for Multiple Disease Resistance

To identify resistance to various Chinese cabbage pathogens, Williams and Leung (1981) developed methods for screening large populations of seedling plants. Screening of seedlings was preferred in the early stages of breeding programs because it takes less time and space. Plants which exhibited seedling resistance were later evaluated for mature plant resistance. Such procedures may involve simultaneous inoculation, and incubation of 1-week-old seedlings with *Plasmodiophora brassicae*, *Hyaloperonospora parasitica*, *Albugo candida*, *Phoma lingam,* and *Alternaria brassicicola* or *A. brassicae*. This can be followed by a sequential inoculation with turnip mosaic virus (TUMV), or/and *Erwinia carotovora* and/or *Xanthomonas campestris*. The interactions phenotypes of more than one pathogen on a single host can be relied on for evaluation (Fig. 9.3) by paying special attention to the following: (a) careful preparation, quantification, and

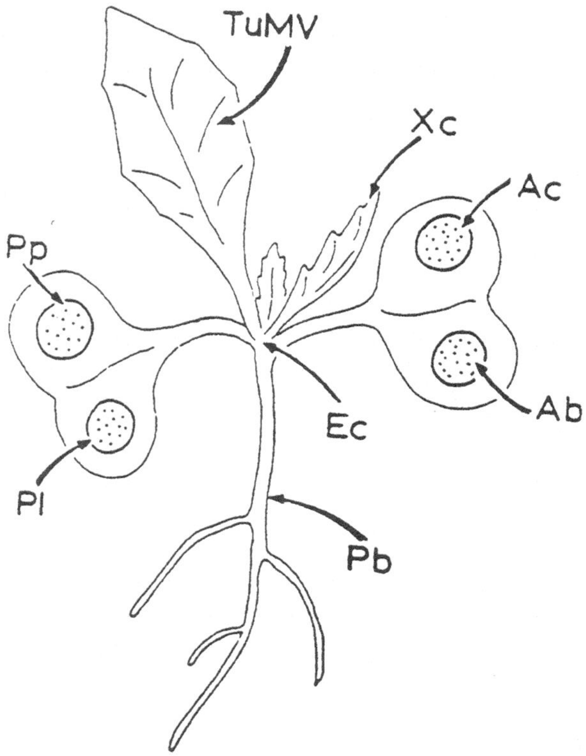

Fig. 9.3 Location of inoculum placement of eight pathogens in multiple disease screening of seedling Chinese cabbage. *Pb Plasmodiophora brassicae*, *Ec Erwinia carotovora*, *Ac Albugo candida*, *Pl Phoma lingam*, *Ab Alternaria brassicae*, *Hp Hyaloperonospora parasitica*, *Xc Xanthomonas campestris*, and *TUMV* turnip mosaic virus (Williams and Leung 1981)

delivery of precise amounts of virulent inoculum, (b) careful cultivation of host "target tissues" of known physiological age, and (c) optional incubation conditions for disease development.

Single seeds were sown in 12-pack pots, and when the cotyledons have expanded, after 5–7 days, single 0.01–0.02 mL drops of a freshly collected conidial suspension containing approximately 1×10^5 conidia/mL were placed on each of the two cotyledons with the aid of a finely tipped glass pipette (Williams and Leung 1981). As plants of each 12 pack were inoculated, they were placed in glass or plastic boxes containing a 1–2 cm depth of warm water. A tight fitting cover was placed on the box when it was filled with plants. The box was then placed in a darkened incubator set at 20 °C for 8–16 h. The atmosphere in the box maintained the droplets in the cotyledons during which time germination and penetration occurred. Plants were then transferred to the greenhouse bench at 20–25 °C for 5 days, then returned to the humidity boxes at 20 °C for 16–24 h. By then, the susceptible plants had a profuse growth of conidiophores on the lower sides of the cotyledons, whereas resistant plants exhibited varying degrees of sporulation, and tissue necrosis which was evaluated on a 0–9 scale as illustrated in Fig. 9.2. Williams and Leung (1981) also noted that it is important to use freshly produced inoculum collected by washing off spores from leaves with distilled water, and that older plants may be inoculated by atomizing the foliage with the suspension of conidia, and keeping them at 100% RH for 8–16 h.

Chinese cabbage was grown under the above conditions and sequentially inoculated with four pathogens (Williams and Leung 1981). Five days after sowing, seedlings were dipped in a spore suspension of *Plasmodiophora brassicae* spores, and transplanted, and then 1–2 days later inoculated with *Hyaloperonospora parasitica* conidia. Twelve days after sowing, the plants were evaluated for downy mildew resistance, and the susceptible plants removed. The remaining plants were then inoculated with turnip mosaic virus (TuMV) at 14 days, evaluated, re-inoculated, and rouged over the following 14 days. Surviving plants were then inoculated at 21 days after sowing for soft rot resistance. Two weeks later, 35 days after sowing, resistant plants could be removed from the pots and examined for club root. Plants withstanding all four diseases could then be potted, and vernalized or treated with benlate fungicide, and transplanted to the field. Further inoculations with TuMV, *Erwinia,* and *Hyaloperonospora* could be made in the field, and plants not treated with benomyl fungicide could be planted in *H. parasitica* infested field plots.

The procedures for the sequential inoculations were reported to be essentially the same as those for individual inoculations except when TuMV inoculation was to be followed by *E. carotovora* or *H. parasitica* (Williams and Leung 1981). The plants were maintained at 25 °C instead of returning them to cooler temperatures for enhancement of virus symptoms. Inoculation with a combination of any of the four pathogens was possible by following the appropriate portions of the total sequence in Fig. 9.3. Plants can also be inoculated with other pathogens such as *Albugo candida*, *Alternaria* spp., *Phoma lingam,* and *Xanthomonas campestris*. Inoculations with these pathogens were kept apart from those areas of the cotyledons, and leaves which were occupied by *Hyaloperonospora* (Fig. 9.4). It allowed large F_2, and backcross progenies to be efficiently screened. It was possible to screen approximately 600 plants per m^2 for the four diseases every 35 days. If resistance to each of the above pathogens were controlled by independent single recessive genes, a theoretical minimum population of 256 plants would be needed to recover the four recombinants. It is likely that far larger populations would be screened to accommodate the differing heritabilities for each form of resistance. An important consideration in selecting dominant forms of resistance in the production of F_1 hybrid Chinese cabbages was that resistance to different pathogens can be introduced into the hybrid from different inbred parents.

9.5.7 Identification of Heterothallism and Homothallism

Isolates of *H. parasitica* were collected from different host species and also from various geographical locations (Sherriff and Lucas 1989a, b). The isolates were maintained on seedlings of susceptible cultivars of respective hosts. Cotyledons with spores were excised, placed in 50 mL sterile distilled water (SDW), and shaken gently to dislodge the conidia. The conidial suspension was then filtered through

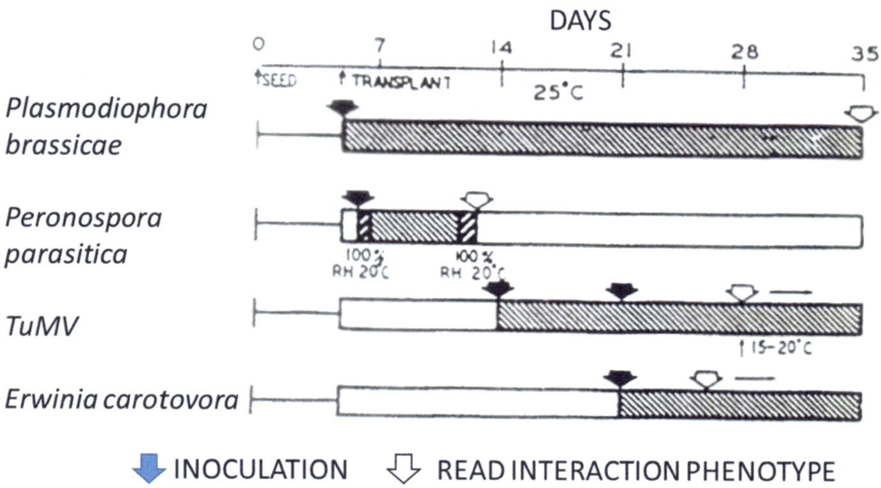

Fig. 9.4 Sequence for individual and multiple disease resistance screening in Chinese cabbage (Williams and Leung 1981)

three layers of cotton gauze and centrifuged at 1500 g. The conidial pellet was resuspended in SDW, centrifuged and finally resuspended in 1–2 mL SDW. Cotyledons of 7-day-old susceptible seedlings, raised in a 9 cm pot, were drop inoculated with the conidial suspension with the aid of a Pasteur pipette. Inoculated seedlings were then sealed in a 13 × 21 cm propagator and transferred to a growth room (19 + 1 °C; 16 h d, Osram cool-white fluorescent tubes, photon flux density 70 μmL^{-2} S^{-1}). Conidia were harvested 5–7 days after inoculation.

For microscopic examination, cotyledons and leaf pieces were cleared by boiling for 2 min in a lactophenol-ethanol solution containing 10 g phenol, 20 mL glycerol, 10 mL lactic acid and 20 mL 96% (V/V) ethanol (Sherriff and Lucas 1989b). Cleared cotyledons were rinsed in water and stored in 70% (V/V) glycerol and examined under a low powered microscope. Mature oospores were easily visible due to their brown pigmentation; young oospores tended to take up and retain green pigments from host tissues during cleaning.

9.5.8 Leaf Disc Test to Assess Resistance

To study the correlation between the interaction phenotypes obtained with the leaf disc test, and in the classical cotyledon test, leaf dicks are removed from sufficiently developed leaves (at least 100 cm^2). Different conditions are tested; they are either disinfected in a 3% calcium hypochloride solution or simply rinsed in sterile distilled water, then placed in agar plates (14 cm diameter) with the upper side on the agar surface. The agar medium is completed with nutrients (MS, Murashige and Skoog

1962) or no nutrient but contained either 50 or 100 ppm of benzimidazole. Five droplets of 20 μL of a spore suspension (5×10^4 spores mL^-) are deposited on each leaf dick. Working with five droplets is a way to spread the isolate on the disc (a spraying process, as used for infecting whole plants, is not possible on leaf disc placed in Petri dishes). After inoculation, Petri dishes are sealed and kept in a growth chamber in the dark for at least 8 h, then for 7 days under a 12 h photoperiod (16 °C night, and 20 °C day, 30 μE light intensity). At least six leaf discs per couple of accession/isolate are tested at least twice for each treatment. In another experiment on the effect of the leaf position on the disc resistance/susceptibility response, each leaf from the bottom (number 1) to the top of the plant is tested (the highest number is 17). These 17 leaf positions are obtained on 10–18-week-old plants, in order to cut leaf disc from sufficiently developed leaves (at least 100 cm^2). Ten discs per leaf position are tested for each genotype/isolate combination (Agnola et al. 2003).

9.5.9 Use of Rooted Leaves for Screening *Brassica* Germplasm

Fully developed leaves from the middle part of 20–30 days old were cut off tightly to the stem, and these whole petioles were immediately immersed into a solution of 10 mg/L of indole-butyric acid (IBA), and 5 mg/L of nicotinic acid (NOA) for 20 h. Then the leaves were thoroughly rinsed under tap water and transplanted into moist perlite in 250 mL plastic beakers closed for the following week by another transparent plates cover. Perlite was moistened with tap water but the leaf blades were weekly sprayed with diluted foliar fertilizer 0.3 g/L. Beakers were incubated in growth chamber (22 °C/18 °C day/night, 16 h day length) under batteries of fluorescent tubes, generating 140 $\mu mol/m^2$ of irradiance at the plant level. When rooting young cotyledons, the same procedure was used except the cotyledons with petioles were harvested from 10 to 11 days old seedlings. *Hyaloperonospora* isolate OLI from kohlrabi (*Brassica oleracea* var. *gongylodes* L.) was propagated on young detached cotyledons of cabbage cv. Kodanske on moist filter paper in plastic Petri dishes under fluorescent tubes at 16 °C. After 7–10 days, the cotyledons covered with sporulating mycelia were frozen and maintained at −80 °C. Water suspensions of conidia from this stock were used for the preparation of inocula diluted to 5×10^{-4} to 5×10^{-5} spores per mL. Surfaces of leaf discs rooted leaves or cotyledons were drop inoculated (7 μL droplets). Closed plastic containers with plant organs growing in perlite were incubated for the next 2 days in dark at 10 °C, and thereafter under fluorescence tubes at 16 °C/10 °C (day/night, day length 16 h) (Havranek et al. 2005).

9.5.10 Artificial Inoculation Technique Under Growth Chamber (Williams 1985b)

1. **Inoculation:** Place a 10 μL drop of conidial suspension on each ½ cotyledon at 24–48 h after expansion using micropipette. Prevent droplets from evaporating. Gently agitate inoculum to keep conidia from settling (Fig. 9.5).
2. **Incubation**: Infectional environment have to be maintained on inocula droplets on cotyledon 12–18 h, 16 °C, 100% RH. Place plants in dew chamber in dark (Fig. 9.6). The incubation environment needs to be maintained as in case of propagation of the plants. Remove plants from dew chamber and grow them for 5 days. Again keep the plants to dew chamber for 6 days post-inoculation, and maintained same condition as infectional environment, 16 °C, 100% RH for 24 h.
3. **Evaluation***:* Observe interaction phenotype on 7 days of inoculation. Interaction phenotypes (IP) rates for both symptoms ranging from a hypersensitive necrotic flecking at the site of inocula droplet to irregular narrow streaks of internal necrosis associated with more or less chlorosis. Signs of the pathogen range from no sporulation to very sparse, through increasingly heavy sporulation on both the top and bottom of the cotyledons. Most highly compatible IP's have little or no necrosis, none to light sporulation on the upper, adaxial, surface of cotyledon with heavy sporulation on the lower surface (Fig. 9.2).

Scale: Plants scored 0–9 at 7 days PI after having been 16 °C, 100% RH, for 24 h.

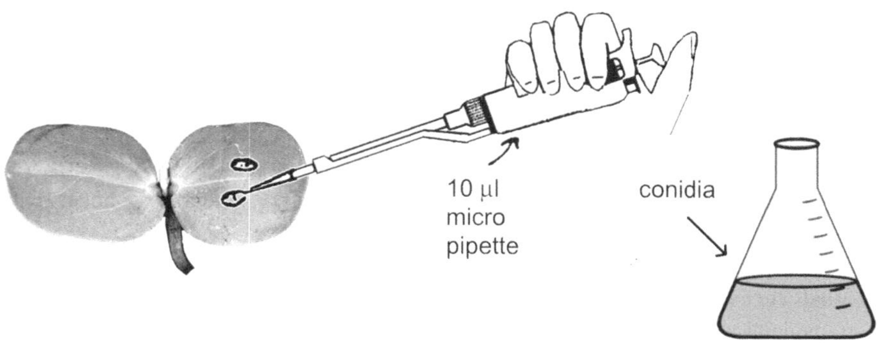

Fig. 9.5 Inoculation of leaves with conidial suspension

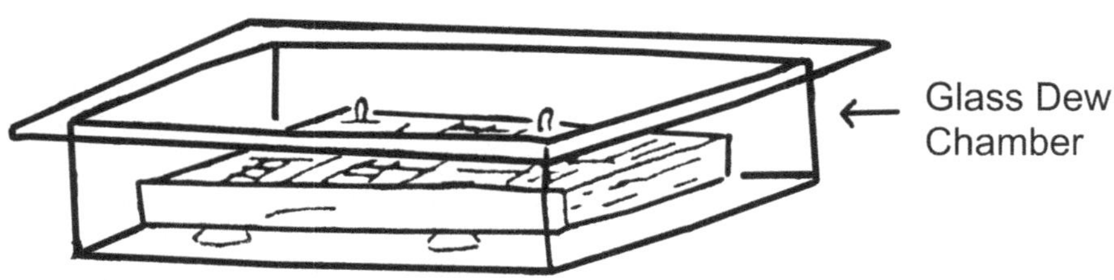

Fig. 9.6 Inoculated leaves in the dew chamber

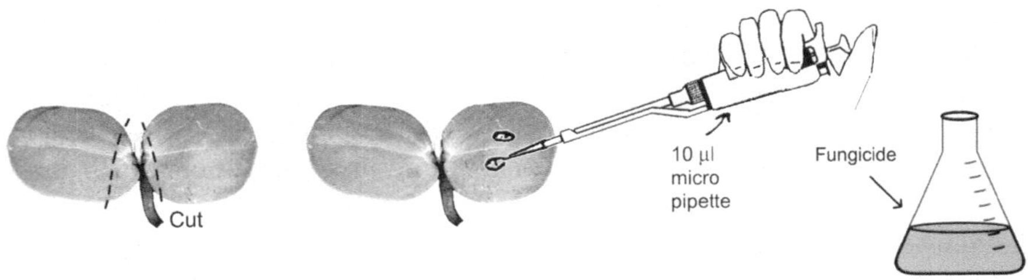

Fig. 9.7 Plants inoculation technique with fungicides

0 = No infection.
1 = Very minute to larger scattered necrotic flecks under inoculum drop, none to small amount of necrosis on lower cotyledon surface, no sporulation.
3 = Very sparse sporulation, 1 to few conidiophores on upper or lower surface. Necrotic flecking often present. Tissue necrosis present.
5 = Sparse scattered sporulation on either or both cotyledon surface, tissue necrosis.
7 = Abundant to heavy sporulation mainly on lower surface, light to scattered sporulation on upper surface, tissue necrosis, and chlorosis may be present.
9 = Heavy sporulation on lower surface, none to light sporulation on upper surface slight to no tissue necrosis. Chlorosis may be present.

Disease Index

$$\mathrm{DI} = \frac{\sum_{i=0}^{9} (i - j)}{n}$$

n = total plants, i = IP class, j = number of plants/class.

Selection*:* Susceptible controls, IP = 7–9 by 7 days, PI = partial resistance, IP = 3–5, resistant plants, IP = 0–1.

4. **Host recovery:** Selected plants may be grown in multipots until requiring transplanting, or vernalized, and flowered. Cotyledons of susceptible and partially resistant plants should be removed to prevent the spread of inoculum, and movement of the fungus, systemically into the plants. Plants may be rid of *H. parasitica* by spraying or dropping Ridomil systemic fungicide, 0.1 g/L on infection (Fig. 9.7).
5. **Resistance**: A wide range of host-specific variation exists in *H. parasitica* appears to be adaptable to numerous hosts other than on which it occurs naturally (Dickinson and Greenhalgh 1977).
   1. *B. oleracea* cc: Little effective R. exists, monogenic dominant race-specific R is in cc. I P.I.
      189029, and cc. c P.I. 245015 (Natti et al. 1967), cc. c cv. "January king," and wild cc, contains resistance (Greenhalgh and Dickinson 1975).

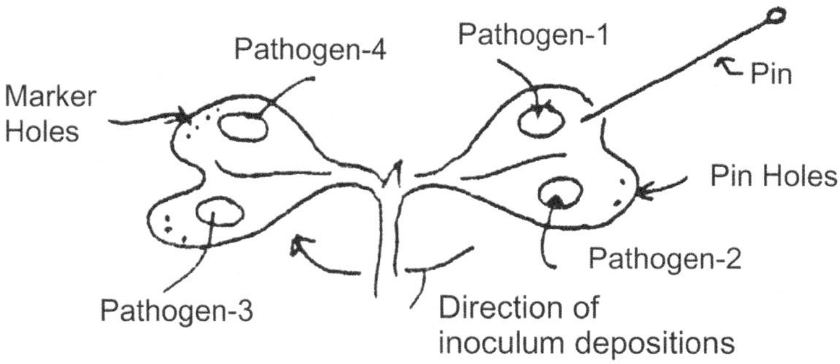

Fig. 9.8 Multiple inoculation methods

2. Resistance in cotyledons does not ensure resistance in other parts. Sources in cc. i. R. l.
   Gabrielson, CrGC.
3. aa: oligogenic r exists in aa. p (Leung 1981) source Cr GC.
4. Race-specific resistance can be found in a number of open-pollinated crucifers.
5. There is a great need to identify race nonspecific resistance to *H. parasitica.*

6. **Multiple pathogen inoculation:** Plants may be inoculated on one half of a cotyledon with one race of *H. parasitica,* and on other halves with other races. Mark initials inoculation with needle puncture on margin of cotyledon then proceed clockwise with other races or additional pathogens. Plants may be inoculated first on roots with *FOC, PB,* or *AR,* then on cotyledons with *HP, AC,* or *LM.* Multiple resistant survivors may later be inoculated with *TUM, XCC, EC,* and *EP. AC* and *HP* inoculum may be mixed, then applied. Interactions among host responses may occur depending on host and pathogen genotypes. For dual HP, and AC inoculations, infectional incubation temperature should be 16 °C (Figs. 9.8 and 9.9).

9.5.11 Measuring Systemic Infection of the Downy Mildew Pathogen

According to McMeekin (1971), seeds of *Brassicas* were first surface sterilized in sodium hypochlorite (10% Commercial Clorox) for 5 min, and then placed about 1.5 cm apart on either 1–2% agar or sterilized glass wool moistened with distilled water in the bottom of a moist chamber (McMeekin 1971). They were germinated in the dark at 20 °C. Seven days later most seedlings were about 2 cm tall. The cotyledons and roots were excised from these seedlings. The hypocotyl, relatively free of starch granules, and chloroplast, was placed with one end in 10 mL of test solution at the bottom of a Petri dish (5 cm diameter). Ten or more hypocotyls were kept upright in the dish by pushing them through a double layer of cheesecloth stretched over the dish and held in place by a rubber band.

Inoculum was applied either on the upper tip or on the side of the hypocotyl. A 1 mm square piece of *Brassicas* cotyledon covered with conidiophores was used as

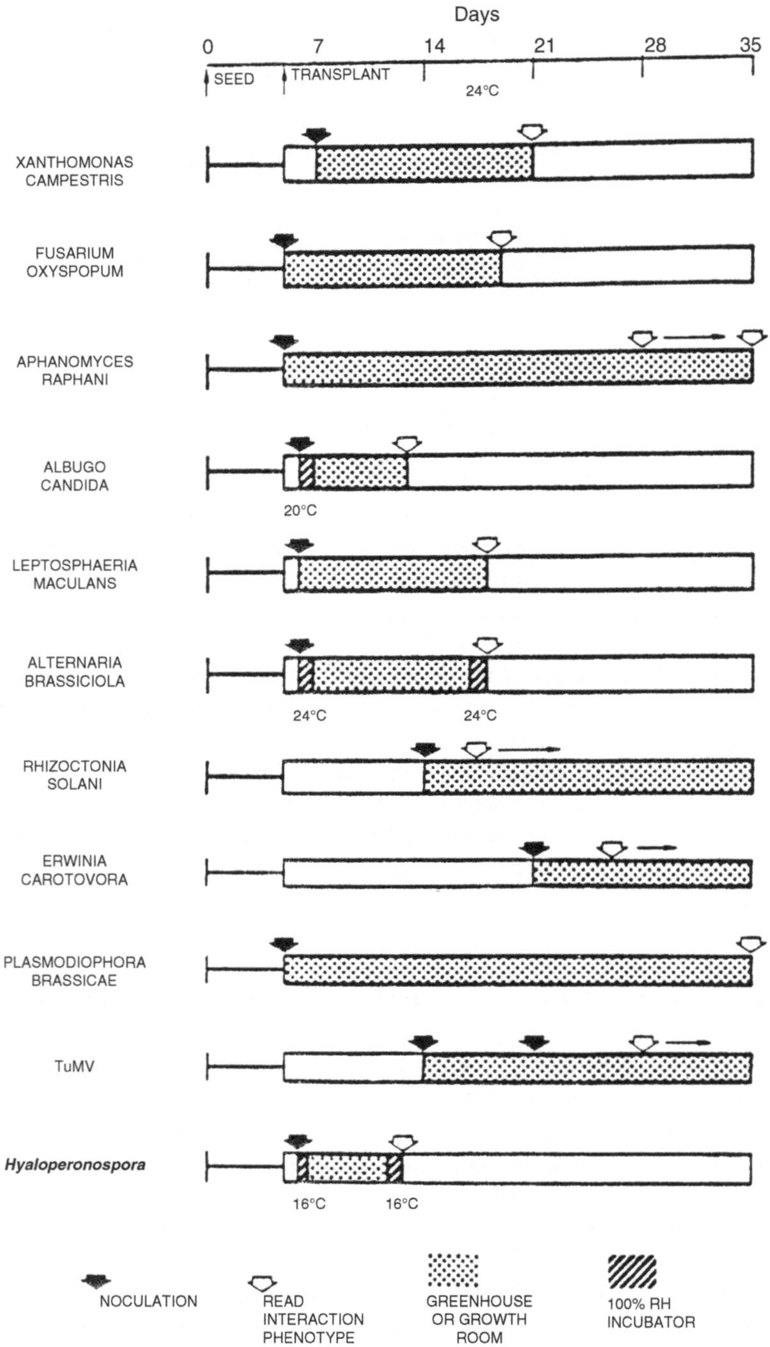

Fig. 9.9 Time course and environmental regime for multiple disease resistance screening of crucifer seedlings

inoculum. The Petri dish bottoms, containing the inoculated hypocotyls were placed in a moist chamber lined with wet paper toweling. The moist chamber was placed in a 15 °C incubator with a light intensity of 5 ft-c for about a week. The whole hypocotyl was removed from solution and fixed on a slide by 0.1% cotton or aniline blue in lactophenol (20% carbolic acid: 20% lactic acid: 40% glycerine: 20% distilled water). The hypocotyl was pressed evenly with another slide until it was flattened, and then a cover slip was applied. After a few hours, the mycelium was stained and could be seen within the host. The length of time between inoculation and fixation determined the intensity of the stain in the mycelium within the host tissue. The cotton blue stained the protoplasm of the fungus, but not the cell wall. Only the youngest fungal growth at the time of fixation was deeply stained in the final preparation. If the test solution favored or did not interfere with host or fungal growth, the fungus could grow from the point of inoculation to the lower tip of the hypocotyl that was immersed in the test solution for 4–5 days at 15 °C. At this time, the mycelium in the lower tip would stain dark blue, but the older mycelium at the point of inoculation took little stain. If the test solution was unfavorable, a "zone of inhibition" lacking fungal growth could be measured from the base of the hypocotyl to the point where fungal growth has stopped. Most of the mycelium was parallel to the stele. Solutions containing either antibiotic or sugars were tested with this method, and for a given concentration, the zone of inhibition was very consistent. It was possible to use this method without completely aseptic procedures and not have a serious problem with rotting. However, streptomycin sulfate (0.25 μg/mL) reduced the rotting without appearing to affect host or the pathogen (McMeekin 1971).

9.6 *Brassica-Leptosphaeria* Techniques

9.6.1 Quantitative Genes Mediated Resistance to *Leptosphaeria*

Three experiments were carried out to assess resistance to *L. macula*ns in the DYDH population following statistically valid designs under shade house and field conditions.

1. **Ascospore shower test:** The 2014 ascospore shower (tub) test using the mixed stubble collected after harvest of canola varieties was conducted as described previously (Marcroft et al. 2012) in a multi-phase experiment comprising a glasshouse, growth chamber, and shade house phase. Seedlings were grown in 20 cell (4 rows by 5 columns) seedling trays for approximately 2 weeks in a glasshouse. In each tray, a single DH line (or a parental line) was allocated to the 4 cells in each tray column, i.e., each tray contained 5 DH lines. Each DH line was replicated twice, i.e., was allocated to a tray column in two different trays. The two parental lines were replicated 30 times. The experimental unit at the glasshouse (and growth chamber) stage of the experiment is a tray column. Trays were then allocated to one of two growth chambers in such a way that each growth

chamber contains four seedlings, i.e., one experimental unit, of each DH line. The growth chamber phase of the experiment lasts 2 days in which time blackleg fungal spores have an opportunity to cause seedling infection. At the end of the growth chamber phase the four single DH line seedlings in a tray are transplanted into a pot which is then placed in a shade house. The shade house is divided into two bays (left and right) with pots arranged in a 4 column by 86 row arrays in one bay and a 4 column by 69 row arrays in the other bay, i.e., a total of 620 pots.

A restricted randomization was used to assign growth chamber tray experimental units to non-rectangular blocks of five pots in the shade house. Pots were kept in the shade house for approximately 3 weeks and then each plant within a pot was visually assessed for internal infection. Internal infection data was collected on 269 DH lines, the two parental lines and 6 Australian canola varieties: Crusher, CB-Telfer, Hyola444, ThumperTT, ATR-Marlin, and ATR-Stingray which were used as resistant and susceptible controls. The treatment and blocking structure for the 2014 tub test trial can be written using the symbolic notation of Wilkinson and Rogers (1973) as

Blocking structure: GC C GC. Tray C GC. GCCol. GCRow C SHBay C SHBay. SHBlock
Treatment structure: Geno

where GC indexes growth chambers and is a factor with two levels.
Tray is a unique identifier for trays within growth chambers, GGCol and GCRow indexes the spatial arrangement of tray experimental units within each growth chamber, SHBay is a factor which indexes the two sides of the shade house and SHBlock indexes non-rectangular blocks of five pots within each shade house bay.

2. **Field evaluation trials:** The design for field trial was a randomized complete block design where 600 plots were arranged in a 40 row by 15 column array with two replicate blocks comprising rows 1–20 and rows 21–40. Three hundred lines, comprising 276 DYDH lines, two parental lines and 22 other commercial lines were sown in each replicate block. The treatment and blocking structure for the 2015 field trial can be written as
Blocking structure: Block
Treatment structure: Geno

where Block is a factor with two levels and Geno is a factor with 300 levels.

The design for field trial was a randomized complete block design where 600 plots were arranged in a 50 row by 12 column arrays with two replicate blocks comprising columns 1–6 and columns 7–12. Two hundred and ninety eight lines, comprising 274 DH lines, two parental lines and 22 other commercial/advanced breeding lines (05-P71-11, ATR-Stingray, AV-Garnet, ATR-Marlin, BC78868-P2, BC78868-P3, BC 78868-P6, BLN3343-CO0402, CB-Telfer, Charlton-093, Crusher-TT, Hyola50, Monty-094, NMT370, Surpass501TT, Tarcoola-27,

Tarcoola-43, Tarcoola-22-094, Tarcoola-69-092, Tarcoola-141, Tarcoola-191, and TornadoTT) were sown once in each replicate block. A single commercial line was sown twice in each replicate block. The treatment and blocking structure for the 2016 field trial can be written as

Blocking structure: Block
Treatment structure: Geno

where Block is a factor with two levels and Geno is a factor with 299 levels. Each experimental unit (DH/parental/check cv.) was sown in 10 m rows with 50 cm row spacing. Two hundred seed from each line were sown in the blackleg nurseries under lateral move irrigation having mixed TT stubble as described previously. Expression of QR was assessed at the physiological maturity stage (BBCH 90). Disease severity was assessed as percent plant survival as well as percent internal infection for all field and shade house experiments (internal darkening, canker scores) as described previously (Marcroft et al. 2012; Raman et al. 2012a, b). Phenological development of DY population (maturity) was assessed in the field trial based on growth stages using the standard BBCH scale (an ordinal scale between 0 and 9 where 93% of the maturity scores were either: 7, 8, or 9).

3. **Genetic mapping:** After the assessment of disease severity of the parental lines, "Darmor-bzh" and "Yudal" and their DH progenies, fresh leaf samples were obtained in 96-well formatted tubes. Total genomic DNA was isolated following a high-throughput DNA protocol, essentially based on chloroform and phenol. A framework map of the DYDH population, based on 2094 DArTseq markers was constructed as described previously (Raman et al. 2014, 2017). PCR-based SCAR (sequence characterized amplified region) marker specific to *Bzh* gene for dwarfness was analyzed using the primer pairs: SCM07UP1 and LP2 as described by Barret et al. (1998). The quality of marker data was reassessed for its suitability for QTL identification by considering missing data, duplicate markers, and markers with excessive segregation distortion, and DH clones that may arise during the microspore culture process. A total of 249 markers with more than 70% missing data were removed. Two DH lines (DY043 and DY044) which had more than 70% of missing marker scores were also removed from the analysis. Thirteen DH individuals were removed from the marker data as the proportion of matching genotypes with another marker was more than 99%. The DH individual with the most missing data was the DH individual removed when considering clones. The final set of marker data containing 1842 markers scored as −1, or 1 or NA (missing) over 262 DH lines was used for QTL identification. Missing data for a DH line was imputed using the mean of the four nearest neighbors with non-missing data for that DH line using the ASMap (Taylor and Butler 2016) and pedicure (Butler 2016) software packages within the R (R Core Team 2014) computing environment.
4. **Statistical analysis of the phenotypic data and QTL mapping**: Reliability (broad-sense heritability, H2) and accuracy (Mrode and Thompson 2005) were

calculated from the extended baseline linear mixed for each trait within trials. The statistical analysis of each trait within a trial used linear mixed model technology and proceeded in six stages: (1) Fit a baseline linear mixed model. (2) Extend the baseline linear mixed model by considering other sources of variation, e.g., spatial variability in field trials. (3) Add markers to the (extended) baseline linear mixed model. (4) Consider each marker separately as a fixed effect in a genome scan model. (5) Based on the genome scan model select a subset of markers upon which backward elimination is performed. (6) A final set of markers is identified and declared as putative QTL. All traits considered, except maturity, were transformed using the logit transformation to better satisfy model assumptions of normality and constant variance.

The baseline statistical model for each trait in each trial can be written, using the symbolic notation described by Wilkinson and Rogers (1973), as trait ~1 + GnonDH + GDH + Blocking Structure + units where 1 is the overall mean, GnonDH and GDH are a partitioning of the treatment structure term Geno into non-doubled haploid lines and doubled haploid lines respectively, i.e., GnonDH indexes the parental, control, and commercial lines. The term Blocking Structure refers to factors which reflect trial design. Terms in bold font are considered random terms and associated with each of these terms is a variance parameter (often referred to as a variance component). The preferred method for estimating these parameters is residual (or restricted) maximum likelihood (REML) (Patterson and Thompson 1971). The term units refer to the residual and its associated variance parameter is the residual variance. Marker effects can be accommodated in the (extended) baseline linear mixed model by partitioning the effects of doubled haploid lines into marker effects plus random genetic variation. In the genome scan model, each marker is fitted as a fixed effect sequentially and for each marker a Wald type conditional F-statistic and an associated p-value is computed. LOD scores are calculated for each marker using the formula—log10 (pi) where pi is the p-value associated with the i-th marker. At the completion of fitting the genome scan model, a subset of markers where each marker had a *p*-value less than 0.01, i.e., an LOD score greater than 2, were chosen. Chosen markers that were within 30 cM of another chosen marker were culled to a single representative marker for that region based on LOD score. The chosen subset of markers were simultaneously fitted as fixed effects and backward elimination was conducted by dropping markers fitted as fixed effects one at a time until all remaining markers in the model had an LOD score of approximately 2. This final set of markers is referred to as the putative QTL marker set. The process of performing a genome scan, choosing a subset of markers for backward elimination and identifying a putative QTL marker set is performed a second time with the initial putative QTL marker set fitted as fixed effects in the genome scan and subsequent models. This approach to QTL identification is the approach considered by Nelson et al. (2014). All models were fitted using the statistical software package ASreml (Butler et al. 2009) within the R computing environment (Raman et al. 2018).

9.6.2 Cloning and Transformation of *Leptosphaeria* Avirulence Gene

1. **Fungal material, mapping population, and inoculum preparation:** The fungal F1 population was generated by crossing Lm isolate v23.1.3 [pathotype *AvrLm1*–4-5–6-7–8-(10)-11-J1-S, *AvrLep1*–3-4] to isolate 00–100 [*AvrLm2*–3-(5)-6-(8)-9-(10)-J1-S, *AvrLep1*–2-4] (Rouxel et al. 2011) by pairing single pycnidiospores isolates as described by Mengistu et al. (1995). Culturing and harvesting of the pycnidiospores were performed as described previously (Ghanbarnia et al. 2012). Isolates 00–100 and 2367 were obtained from the Canada. Isolate v23.1.3 (Rouxel et al. 2011) was provided by T. Rouxel, France.
2. **Plant material and virulence phenotyping:** Parental isolates, progenies of the mapping population and Lm transgenic lines were pathotypes for *AvrLm9* by pathology test on *B. napus* line Goeland, which carries the *Rlm9* resistance gene (Balesdent et al. 2005). Susceptible *B. napus* lines Topas DH16516 and Westar were used as controls. Pathology tests were performed using a 0–9 rating scale as described previously (Larkan et al. 2013). In addition, to show the specificity of response, additional *B. napus* differential lines, including Topas differential lines with single R-genes (Larkan et al. 2016b), *B. juncea* cv. Vulcan (*Rlm6*) and *B. napus* cv. *Roxet* (*Rlm7*), were included.
3. **DNA extraction and sequencing:** DNA extraction and resequencing of *L. maculans* isolates have been described elsewhere (Ghanbarnia et al. 2015). Sequence reads for each isolate were mapped to the reference genome v23.1.3 (Rouxel et al. 2011) and AAFC reference genome 00–100 using CLC.
4. **Mapping and marker development:** The genome sequences of the parental isolates v23.1.3 and 00–100 were utilized to generate SNP targets spanning the Lm genome for use in marker development. KASP assays (He et al. 2014) targeting individual SNPs were developed using Primer Picker software. Assays were first tested for clear detection of polymorphic alleles using DNA from the parental isolates before being applied to the entire F1 population. Assays were run and scored using a CFX96 Real-Time PCR System. Map distances (in cM) were calculated manually as described previously (Larkan et al. 2014).
5. **Cloning and transformation of candidate gene:** Genomic clones were generated by amplification of an 1813-bp fragment containing the avirulent (00–100) allele, including native promoter and 30 untranslated region (UTR), via PCR using Q5 High-Fidelity Master Mix. Cloning was conducted using Gateway technology by adding attB sites to *AvrLmJ1* primers. These were transferred to the fungal transformation vector pNL11 (Larkan et al. 2015) prior to *Agrobacterium*-medi (Lucas et al. 1988) and transformation (Utermark and Karlovsky 2008). SNP alleles (cDNAs only) were synthesized and cloned into the pLM4 vector, in which expression of the construct was driven by the *AvrLm1* promoter (Ghanbarnia et al. 2015). Positive transformants were tested on *B. napus* cvs. Goeland and Darmor which carry the *Rlm9* resistance gene. In addition, *B. napus* introgression lines (Larkan et al. 2016a, b) were used to further confirm that the interaction of *AvrLm9* was specific to *Rlm9*. These lines were

Topas-*Rlm1*, Topas-*Rlm2*, Topas-*Rlm3*, Topas-*Rlm4*, Topas-*LepR1*, Topas-*LepR2*, and Topas-*LepR3* (Larkan et al. 2016a, b).

6. **RNA extraction, reverse transcription, and quantitative reverse transcription-polymerase chain reaction (RT-PCR) analysis:** Total RNA was extracted from infected cotyledons of the susceptible cultivar *B. napus* Topas DH16516 using the TRIzol LS reagent, and subsequently purified with an RNAeasy Mini Kit according to the manufacturer's protocol. DNA was removed by on-column treatment with RNase-free DNase in solution. All samples were adjusted to 1 g l of RNA, and cDNA was synthesized using a Superscript III First-Strand Synthesis Super Mix Kit according to the manufacturer's protocol. qRT-PCR was performed using a CFX96 Touch Real-Time PCR System (*Bio-RAD*) and Sso Advanced Universal SYBR Green Super mix. RT-PCR was performed for the *AvrLm4*–7 and *AvrLm9* genes with three biological samples. Ct values were analyzed using the Q-gene program (Muller et al. 2002; Simon 2003). The *L. maculans* actin gene served as a reference.
7. **RNA sequencing analysis:** *L. maculans* genome-wide transcriptomic analyses were performed according to the methods described previously (Haddadi et al. 2016). Briefly, it was applied deep RNA sequencing using an Illumina HiSeq-2500 platform to assess genome-wide expression profiling of two natural *L. maculans* isolates harboring (v23.1.3/JN3) or lacking (00–100) *AvrLm4*–7 during cotyledon infection at different time points (0, 2, 4, 6, and 8 dpi). All reads were mapped to the genome sequence of *L. maculans* isolate v23.1.3 (Rouxel et al. 2011). Differentially expressed genes (DEGs) were detected using the edgeR (CLC) and DESeq2 packages (Love et al. 2014; Robinson et al. 2010). The RNA sequencing data corresponding to 00–100 were presented as log2 Reads Per Kilobase per Million mapped reads (RPKM) of the average of three biological replicates, whereas the data corresponding to v23.1.3/JN3 were presented as log2(RPKM) of the merged three technical replicates.
8. **Yeast two-hybrid assay:** For the yeast two-hybrid assay, the *AvrLm3* open reading frame (ORF) lacking signal peptide sequences was amplified using primers DspAvrLm3-NcoRI-F and DspAvrLm3-EcoRI-R from the *L. maculans*–*B. napus* cDNA library. Similarly, the *AvrLm4*–7 ORF lacking signal peptide sequences was amplified using the primer set DspAvrLm4–7-NdeI-F and DspAvrLm4–7-PstI-R. PCR products were cloned into pGBKT7 vector as bait constructs. *AvrLm9* lacking its signal peptide was amplified using primers DspAvrLm9-NdeI-F and DspAvrLm9-EcoRI-R and sub-cloned into the vector pGADT7 as a prey constructs (Clontech).

   The matchmaker GLA4 two-hybrid system and yeast strain Y2HGold (Clontech) was used. The yeast strain Y2HGold was co-transformed with bait and prey plasmid combinations using lithium acetate and polyethylene glycol 3350, as described previously (Gietz and Woods 2002). Transformants harboring both bait and prey plasmids were selected on plates containing minimal medium lacking leucine (Leu) and tryptophan (Trp) (MM-WL). For controls, empty prey vector pGBKT7 or pGADT7 was used as bait or prey. Next, one colony per combination was picked up from MM-WL plates to inoculate 1 mL of MM-WL culture. After

36 h of growth, the culture was centrifuged, the cells were resuspended in 25 μL of 0.9% NaCl from $OD_{600} = 1$ to $OD_{600} = 0.00001$, and then spotted on MM-WL and MM-AHWL plates supplemented with 40 μg/mL (X-α-D-Galactoside) (Clontech) and 200 ng/mL Aureobasidin A (Clontech). After 3 days of incubation, observations were recorded (Ghanbarnia et al. 2018).

9.6.3 Identification of QTLs in *Brassica* to *Leptosphaeria*

1. **Plant populations:** F1 seedlings, produced by crossing the Swedish spring type, blackleg-susceptible *B. napus* variety Topas with the Australian blackleg-resistant varieties AG-Castle and AV-Sapphire, were used to produce DH populations via microspore culture by the method of Ferreira et al. (1995) using modified Lichter medium (NLN) with 13% sucrose. The Topas/AG-Castle DH population (TC) utilized consisted of a total of 242 DH lines from two sub-populations of 139 and 103 DH lines, with each sub-population produced from a single F1 plant. The Topas/AV-Sapphire DH population (TS) contained a total of 109 lines, also consisting of two sub-populations of 66 and 43 DH lines. Seed of parental and DH lines used in the QTL trials was produced in tented field plots at Saskatoon, Canada during the 2007 (TS) and 2008 (TC) summer growing seasons, packaged into replicates of 100 seeds per line and sent to Australia for testing field resistance to blackleg. The Topas *B. napus* line was sourced from the AAFC Saskatoon collection. AG-Castle and AV-Sapphire lines were provided by DEDJTR Victoria.
2. **Field characterization of DH populations:** The parental lines and their respective DH populations, along with several *B. napus* variety checks were grown May through November in blackleg disease nurseries situated at Wagga Wagga (New South Wales) and two sites (Wonwondah and Green Lake) south of Horsham (Victoria). Rainfall differed by an average of 103.2 mm/year over the three trial years at each site, with the Wagga Wagga site receiving an average 38% more precipitation. Each trial was planted in a randomized block design with 3–4 replicates per line. The TC population was tested in three separate environments; both Wagga Wagga and Horsham (Wonwondah site) (3 replicates per trial) and once more at Wagga Wagga (4 replicates). The TS population was tested in four environments; Horsham (Wonwondah site), and Wagga Wagga, as 3-replicate trials and Wagga Wagga, and Horsham (Green Lake site), as 4-replicate trials. Emergence counts were made 4–5 weeks after sowing. For trials at the Horsham sites, local diseased canola stubble, collected in the previous year, was spread as the source of *L. maculans* inoculum, while at Wagga Wagga the trials were sown into in situ canola stubble from the preceding season as previously described. At the end of the growing season, trials were assessed for blackleg disease using both survival percent (S %) and internal infection (II) metrics. Survival percent was calculated for each row as standing plants at end of season/emergence count)/100, internal infection was determined by examining 5–20 random plants per row for internal infection after cutting the stem at the crown and rating fungal

discoloration on a 0–5 scale (0 = no infection, 1 = 1–25% infection, 2 = 26–50% infection, 3 = 51–75% infection, 4 = 76–99% infection, and 5 = 100% infection/missing plant), and median scores were calculated for each entry and metric using Microsoft Excel. Statistical analysis of populations and plotting of distribution histograms were performed using R software. Selected lines representative of the three major QTL identified from the TC DH population, both singularly and in combination were also entered into Saskatchewan disease nurseries at Melfort and Saskatoon in an attempt to test the efficacy of the QTL resistance against Canadian populations of *L. maculans*.

3. **Single-isolate characterization:** Each DH line in the TC and TS populations were characterized for the presence or absence of hypersensitive response during infection by the single spore-derived *L. maculans* isolates WA30 or v23.1.3, respectively, via cotyledon infection tests. Four plants for each line were scored at 14 days post-infection on a 0–9 scale as described by Larkan et al. (2013). Further differential pathology testing, using isolates of *L. maculans* varying in their avirulence gene profiles (00–100, 98–15, 99–53, 89–12, 2354-Rimmer Collection was performed with the parental lines and the R-gene control lines Topas DH16516 (a DH line of Topas with no effective blackleg resistance), Columbus (*Rlm1*, *Rlm3*), Quantum (*Rlm3*), Jet Neuf (*Rlm4*), Roxet (*Rlm7*), and Goeland (*Rlm9*) in order to produce a hypothesis for the likely R-gene content of each population. Inoculum for the tests was produced as described previously. Spores were diluted to 2×10^7 spores/mL and 10 μL was used to inoculate each of four small wounds on each plant (one wound at the center of each cotyledon lobe).
The transgenic isolate 3R11: AvrLm4–7 was produced by *Agrobacterium* transformation (Utermark and Karlovsy 2008). Briefly, a 2205 bp amplicon was produced from the isolate v23.1.3 (primers: GW-AvrLm47G-Fc; GGGGACAAGTTTGTA CAAAAAAGCAGGCTTCCTCTGGTAAGGAAGGTT TACCAATTATACACCT, GW-*AvrLm47G*-Rb; GGGG ACCACTTTGTACAAGAAAGCTGGGTCGGCGGTA GATTTGCTACTAAAAGGTAACTTT) which spanned 1213 bp upstream and 514 bp downstream of the *L maculans AvrLm4–7* CDS and inserted into the Gateway compatible fungal transformation vector pNL11 prior to transformation of the *L. maculans* isolate 3R11 as previously described (Parlange et al. 2009; Larkan et al. 2013, 2016a, b).
4. **Map construction and QTL identification:** The TC and TS populations were characterized using 225 and 212 simple sequence repeat (SSR) markers, respectively, as described previously. A subset of the TC population had previously been characterized using 279 DArT markers as part of a consensus map of *B. napus*, and this data was also incorporated into the TC map to provide additional resolution and to allow orientation to other maps (503 markers total, grouped into 360 bins). Linkage map construction and inclusive composite interval mapping (ICIM) was performed using QTL IciMapping version 3.3 (http://www.isbreeding.net). Prior to final map construction severely distorted markers (Chi-square Test, $p < 0.001$) were excluded. A threshold of LOD 6 was applied for initial grouping of markers. LGs consisting of >3 markers were

anchored to chromosomes based on BLAST-like alignment tool (BLAT) analysis of SSR marker sequences to the published *B. napus* genome before recalculating LGs at LOD 3.

Marker ordering was performed using nearest neighbor algorithm and rippling was performed once using the sum of adjacent recombination fractions (SARF) criterion. For QTL identification, the "ICIM-ADD" and "ICIM-EPI" functions of the software were utilized to investigate both additive and digenic epistatic QTL expressed in each environment and for each scoring metric (S and II). Permutation tests (1000 permutations, 95% confidence level, 1 cM interval) were performed for each scoring metric in each environment to determine significant LOD thresholds. The precision of QTL positions were then improved by re-analyzing the data using the determined LOD thresholds and a smaller (0.5 cM) scan interval. The "Multi-Environment Trials" (MET) function of the software was also utilized to determine the consensus positions for the major QTL and to assess the genotype by environment (G $\times$ E) interactions. QTL identified from individual environments were considered significant if they exceeded the LOD significance threshold and accounted for >5% of the variance. MET QTL were considered significant if they accounted for >5% of the variance and had a heritability (h2 = σ2 (Additive)/σ2 (Total)) > 0.5. QTLs positions were defined by both support and peak marker intervals, with support intervals representing the map interval in which the LOD exceeded the calculated LOD threshold (for single-environment QTL) or peak LOD-1 (for multi-environment analyses). The datasets for each population were analyzed both as total populations and as their separate component sub-populations. Final map figures were produced using Map chart 2.2 and Microsoft Power-Point software. To determine the location of the major QTL relative to the *B. napus* reference Darmor-bzh genome, the target sequences for the flanking SSR markers were obtained from the AAFC *Brassica* MAST Database (http://aafc-aac.usask.ca/BrassicaMAST/) and used to search the public *B. napus* genome database (http://www.genoscope.cns.fr) using default BLAT search Settings (Li et al. 2008; Zhang et al. 2012; Voorrips 2002; Larkan et al. 2016a, b).

9.7 *Brassica-Plasmodiophora* Techniques

9.7.1 Pathogen Inoculum Preparation for Artificial Inoculation

Plasmodiophora brassicae cannot be cultured axenically. Galls containing resting spores may be stored at −20 °C for several years with minimal loss of viability. When required, the galls are brought to room temperature and macerated either mechanically or with a pestle and mortar. The resultant suspensions are filtered through nylon or muslin sheets removing host debris and then centrifuged at approximately 3000 rpm for 10 min. This yields pellets of resting spores that are free from macromaterial. The supernatant liquid is decanted and discarded. Resting spore pellets are resuspended in sterile water and re-centrifuged until clean preparations, free from starch granules, are obtained. Inoculum may be applied directly to seedling roots by dipping them, it may be mixed with compost into which seed or seedlings are placed, or aliquots of standardized concentration of spores (normally 1×10^4–10^6 resting spores mL^{-1}) may be applied to the surface of compost in which seedlings will be grown (Dixon 1976). Inoculated test plants should be maintained at a basal temperature of 20–25 °C up to 6 weeks at which time galling symptoms will be apparent and may be assessed by the method of Dixon (1976). Worldwide physiologic specialization was analyzed by Toxopeus et al. (1986). On the microgeographic scale, there is substantial physiologic variation, whereas on the macroscale, it is more limited and reflects the dominant species of *Brassica* most commonly cultivated in a particular geographic region.

In another method to prepare resting spore's inoculum of *P. brassicae*, 5 g of air-dried clubroot galls are soaked in 200 mL distilled water for 2 h to soften the tissue before grinding in a mortar and pestle. After filtering through seven layers of cheesecloth, the spore concentration is estimated using a hemocytometer and adjusted to 5.0×10^7 spores/mL. *Brassica* seeds are sown in a soil-less potting mix, and plants are grown under 22 °C, a 16 h/8 h light/dark photoperiod in a growth chamber. Five-days-old seedlings of *Brassica* are inoculated by pipetting 5 mL of resting spore suspension (5.0×10^7 spores/mL) around the base of the seedlings shoots, where it meets the soil (Tu et al. 2018).

9.7.2 *In Vitro* Culture of *Plasmodiophora brassicae*

The Plasmodiophorales are generally considered to be obligate parasites (Webster 1980) although there have been several reports that plasmodial plant pathogens can be cultured in the laboratory. Diriwachter et al. (1979) reported that spores of the plasmodia *Spongospora subterranea* f. sp. *subterranea,* obtained from surface-sterilized diseased potato tubers, germinated readily on a simple agar medium to release amoebae which formed colonies, mixed with colonies of bacteria. After a few days, the amoebae encysted, which in turn released amoebae when transferred to fresh medium. These cultures can be maintained indefinitely when transferred to fresh medium in the presence of contaminating bacteria. They also described the development of disease symptoms on tomato plants inoculated with a mixture of amoebae and resting spores obtained *in vitro*. Jones (1928) described the isolation of eight cultures of *P. brassicae* which were maintained under laboratory conditions for 2 months. The cultures were obtained from surface-sterilized diseased roots which were placed in sterile distilled water. The organisms isolated were described as amoebae and were associated with bacteria. If the cultures were left without subculturing, the amoebae encysted. It was also shown that disease symptoms could be produced on turnip plants inoculated with the amoebae and cysts.

P. brassicae is capable of at least a short-lived period of saprotrophic amoebal growth. Spores extracted from surface-sterilized blocks of clubroot tissue are incubated at 25 °C on strength Ringer's agar, spread previously with a suspension of *Escherichia coli*. After a period of 5–10 days, outgrowth of amoebae commonly appears around the inoculum. Amoebae are subcultured on to fresh Ringer's-*E. coli* plates, where they continue to grow, and eventually encyst. An amoebal cyst suspension is prepared from these plates and used to inoculate Chinese cabbage (*Brassica campestris* var. *chinensis*) and Savoy cabbage (*B. oleracea*) seedlings. Out of 27 amoebal isolates tested, nine were found induced typical clubroot disease symptoms. However, when these putative positive *P. brassicae* amoebal isolates are subcultured, the subsequent generation of amoebal cysts failed to induce disease symptoms. A random genomic library is constructed in pUCl8 using *BamH* 1-digested DNA extracted from *P. brassicae* resting spores. The inserts from nine individual clones are then used to probe DNA extracted from 12 amoebal cultures, three showed positive hybridization to the *P. brassicae* probes but this hybridization signal was lost after subculture of the amoebae (Arnold et al. 1996).

9.7.3 Establishment of Callus Culture

The callus culture system can be established according to Bulman et al. (2011) with minor modifications. *Brassica* clubroot galls are harvested 5 weeks after inoculation. After washing thoroughly in tap water, galls are surface sterilized with 70% ethanol for 1 min, 30% bleach with 0.05% tween (1.4% sodium hypochlorite) for 20 min, followed by three washings in sterile distilled water. After washing, galls are cut into sections between 2 and 3 cm in length, which are sterilized with 70% ethanol for 1 min, 2% Chloramine-T for 20 min, followed by three washes in sterile distilled water. After this second sterilization, the sections are cut into 0.5 cm segments and are placed on a medium consisting of Gamborg B5 basal medium plus 3% sucrose, 300 mg/L timentin, and 1.5% agar. Lateral root segments of non-inoculated plants are used as control because mature main roots are inflexible and too difficult to cut into small segments. After 2 week incubation in the dark at 23 °C, gall segments showing callus growth on the cut edges are transferred to fresh medium minus timentin. At the same time, a thin section is removed from each cultured segment, divided into two pieces, with one piece placed into Luria–Bertani (LB) liquid medium and one onto Potato Dextrose Agar (PDA) medium to confirm that the gall/callus segment is free of any endophytic bacteria and fungi. Each medium is observed for 3 days to confirm no bacterial or fungal growth. Endophytic microbe-free calli are used to inoculate host seedlings (Tu et al. 2018).

9.7.4 Host Inoculation for Pathogenicity Test

To extract *P. brassicae* resting spores from dried rot tissue, approximately 1 g of dry root material is placed in a mortar and ground to a powder using a pestle. Twenty milli-liters of sterile distilled water is added, and the homogenate is then mixed and filtered through six layers of cheesecloth. To extract spores from frozen tissue, a modification of the procedure described by Williams (1966) can be employed (Tewari et al. 2005). Collections of *P. brassicae* from soil samples are recovered by planting 5-days-old Chinese cabbage (*B. rapa* var. *pekinensis*) "Granaat" seedlings (the universal susceptible) in the infested soil; plants are grown for 6 weeks, the galled roots are harvested, and the spores are extracted as described above. Resting spores are quantified with a hemocytometer, and used immediately or stored at 4 °C for a maximum of 2 days. Suspensions of *P. brassicae* resting spores, adjusted to a concentration of 1×10^7 spores/mL using sterile distilled water, are used to inoculate the host plants to be tested. Five-days-old seedlings, germinated on filter paper in Petri dishes, are inoculated by dipping the roots in the spore suspension. Controls are dipped in sterile distilled water. Seedlings are immediately planted in 4 × 4 cm plastic pots filled with Metro-Mix 290 soil at a density of one seedling per pot. The pots are thoroughly watered and transferred to a growth cabinet kept at 21:18 °C (day: night) with a 16 h photoperiod. The soil is kept saturated with water for the first week after inoculation, and then watered and fertilized as required (Strelkov et al. 2006).

Extraction of *P. brassicae* resting spores and host inoculation were conducted as described by Strelkov et al. (2006) for dried root tissues. The spore concentration was adjusted to 1×10^7 to 1×10^8 resting spores per milliliter with sterile distilled water and used for inoculation of the host roots. The roots of 7-days-old seedlings (pre-germinated on moistened filter paper in Petri dishes) were inoculated by dipping them in the resting spore suspension for 10 s. The inoculated seedlings were then planted immediately in 6 cm × 6 cm × 6 cm plastic pots filled with Metro-Mix 290 Growing Medium at a density of one seedling per pot. The pots were transferred to a green house kept at 20 ± 2 °C with a 16 h photoperiod (light intensity of 300 μmol-m^{-2}s^{-1}), with the soil kept saturated with water for the first week after inoculation. Subsequently, the seedlings were watered and fertilized as required. Twelve seedlings of each host genotype were inoculated with each population of *P. brassicae* and inoculation were repeated total of four times (Cao et al. 2009).

9.7.5 Germplasm Evaluation Technique

The various techniques employed to evaluate the *Brassica*/cruciferous germplasm, varieties, cultivars, lines/genotypes are explained as follows:

9.7.5.1 Clubroot Assessment Scale

The assessment of host resistance, a nine-grade scale has been used by Wesołowska (2014) at Poland to categories resistance against *P. brassicae*. The nine-grade scale based on root symptoms established by Williams (1985b), and modified by Reby and Michalik (2005) (Fig. 9.10) was used. The extent and severity of the clubroot symptoms were assessed visually according to a nine-grade scale: I—without visible disease symptoms; III—singular galls on lateral secondary roots; V—larger galls on lateral roots; VII—enlarging gall on the main and lateral roots; IX—large singular gall on root neck without root.

The disease index (DI) was calculated using the nine-grade scale according to the formula:

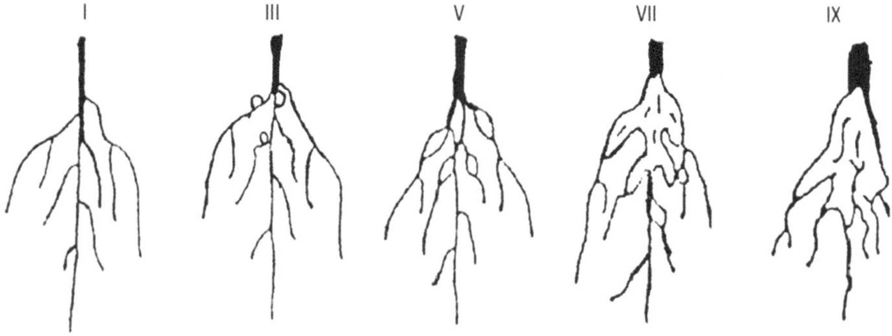

Fig. 9.10 Nine-grade scale based on clubroot symptoms (Wesołowska 2011)

$$\mathrm{DI} = \frac{\sum(n \times p)}{\sum n}$$

where n is the number of plants in the indicated class, p—class I–IX.

The disease indexes of the tested objects were calculated according to plant susceptibility at the 8-week seedling stage and evaluated once more during harvest, on the clubroot-infected field.

9.7.5.2 Development of Clubroot Sick Plot

The clubroot field was established at Poland by adding harvest remains and the soil from the field where heavily clubbed plants of white cabbage were observed by scattering on the surface of the area. The oilseed rape, which is susceptible to clubroot, was sown and plants of different cabbage cultivars artificially inoculated by *P. brassicae* were cultivated. Each year, clubbed plant roots were mixed with the soil for extra infection with the pathogen in the field, which was carried out after the autumn disease evaluation and harvests of tested plants. As a result, the field was contaminated by a mixture of different *P. brassicae* pathotypes. The procedures concerning the seed inoculation method and screening tests were the same as described in detail by Wesołowska (2011).

9.7.5.3 Germplasm Evaluation

A wide range of different methods to evaluate clubroot susceptibility screening have been developed. The soil inoculation method and dipping method seem to be the most popular. In such methods, plant materials are inoculated either by sowing seeds in an infected substrate or by dipping the roots of approximately 10-day-old seedlings in a resting spore suspension (Robak 1991). A double-stage evaluation system for the clubroot resistance was used by Wesołowska (2014). This procedure was established by Reby and Michalik (2005). Firstly, cabbage seed samples were inoculated by submerging them into the water pathogen spore suspension. This method is a modification of the one described by Lewis and Brokenshire (1978). The seed inoculation technique was found to be simple and not time-consuming. Additionally, after the plantlet stage screening, the resistant plants were transplanted into the clubroot-infested field for a second evaluation. This way of testing on adult plants in field conditions was described by Manzanares-Dauleux et al. (2000).

An evaluation of clubroot development in 8-week-old cabbage plants from the standard objects showed an influence of the pathogenicity level of the *P. brassicae* isolates used for seed inoculation on plant reaction. All of them were virulent against the susceptible HTM line plants. Heavily infested root systems were observed in plants from seeds inoculated by the *K* isolate from Krojczyn (DI-8.7), and the *P* isolate from Pobiednik (DI-8.0). All of the plants from the HTM line and "Badger Shipper" derived from the P isolate combinations were completely infested. The isolate from Pobiednik was found to be particularly aggressive not only to the susceptible HTM and "Badger Shipper" but also to the clubroot-resistant "Kilaxy" F1. The cultivar disease index calculated for that was 6.5, and only 9% of the plants were classified as healthy. Plants derived from Oregon123 were less susceptible to the P isolate (DI - 2.3) and characterized by 65% of the plants without clubroot galls.

9.7.5.4 The Germplasm Evaluation Technique Under Indian Conditions

The screening program can be started from the area where soil pH is at the level of 5.7 and severity of club root is very high. In every year, 50–60 genotypes can be tested in the sick plot and the same is repeated for two seasons. The genotypes which show moderately susceptible or susceptible reactions are discarded and the entries which show resistance are retained and further tested for 3–4 years to avoid any chance of escape. Entries of different *Brassica* species, viz., *B. campestris* var. *toria, B. campestris* var. *yellow sarson, B. campestris* var. *brown sarson, B. juncea, B. chinensis, B. napus, B. carinata, B. nigra, B. oleracea* var. *capitata, B. oleracea* var. *botrytis,* and *B. oleracea* var. *gongylodes* are tested in the sick plot. Each entry is sown in a single row of 4 m in length with 50 × 10 cm spacing. A susceptible check (YSB-9) is planted after every two test entries. Irrigation is given at 30, and 60 days after sowing. Fertilizer dose of N.P.K @ 80:40:0 kg/ha is applied or as per requirement.

After 70 days, plants are uprooted. Soil is carefully removed from the roots and washed thoroughly. The degree of severity is rated as described by Seaman et al. (1963). No clubs is 0; few small clubs on secondary roots is 1; considerable clubbing on the lateral roots is 2; and severe clubs on primary and secondary roots is 3. The number of plants in class 0 is multiplied by 0; in class 1 by 10; in class 2 by 60; in class 3 by 100, and the sum of the products is divided by the total number of plants to give the disease index. Initiation of formation of clubs in roots takes place between 25 and 32 days after sowing in the susceptible cultivar YSB-9. More than 90% plants of the susceptible check are affected, bearing large size clubs in primary and secondary roots throughout the plot confirming the uniform distribution of pathogen in the sick plot (Chatopadhyay et al. 2001).

9.7.5.5 Greenhouse Screening for Clubroot Resistance

Screening of *Brassica* germplasm with *P. brassicae* pathotypes has been described by Hasan et al. (2012). Details of how to classify pathotypes, methods of inoculum preparation, and procedures of inoculation, data collection, and statistical analysis have been described.

1. **Isolates**: Five single spore-derived isolates of *P. brassicae* (pathotypes 2, 3, 5, 6, and 8), collected from Canadian *Brassica* crops and infested soil (Xue et al. 2008), were used to evaluate the *Brassica* germplasm for resistance to clubroot. These pathotypes were classified based on Williams (1966) system. These single spore-derived isolates were reproduced on the universally susceptible Chinese cabbage cultivar "Graanat" (*B. rapa* L. var. *pekinensis*) and galls were preserved in −20 °C temperature.
2. **Preparation of Inoculum:** Resting spore suspensions (inocula) were prepared on the day of inoculation from frozen galls following the protocol of Williams (1966) as described by Strelkov et al. (2007). Approximately 2.0–2.5 g of clubbed roots were ground in a mortar with a pestle by adding 50 mL of sterile de-ionized water (sdH_2O). The homogenate was filtered through eight layers of cheesecloth to remove the plant debris. The resting spores in suspension were quantified with a hemocytometer and adjusted to minimum concentration of 1.0 $\times$ 10^7 resting spores mL^{-1}, to be the optimum concentration for successful inoculation (Voorips and Visser 1993).
3. **Inoculation:** Seeds were germinated under room temperature on moistened Whatman No. 1 filter paper in 90 × 15 mm size Petri dishes, and the plates were kept moist with tap water to provide adequate moisture for germination of the seeds. Seven-days-old seedlings were inoculated by dipping the entire root in the resting spore suspension for 10 s (Nieuwhof and Wiering 1961). The inoculated seedlings were immediately planted in 9 or 18 cell trays filled with sunshine professional growing mix. The seedlings were grown in a greenhouse at 21 ± 2 °C with a 16-h photoperiod. To ensure the optimum conditions for resting spore germination and root hair infection by the zoospores, the trays were kept saturated with water for the first week. From the second week, the trays were watered once a day. Quick release mixed fertilizer, viz., 15-30-15 (N-P-K) was applied once a week to provide nutrients for the seedlings.
4. **Data Collection:** Seedlings were evaluated for clubroot resistance 42–45 days after inoculation. Roots were dug out, washed in tap water and examined for gall formation. The severity of gall development was rated on a 0–3 scale as described by Kuginuki et al. (1999), where 0-no galling; 1-one or few small galls on the lateral roots; 2-moderate galling on the lateral roots; and 3-severe galling on the lateral roots or on the main root.
5. **Evaluation of germplasm**: Evaluation of the *Brassica* germplasm was conducted in two phases. In the first phase, all 275 *Brassica* lines were evaluated for resistance to pathotypes 2, 3, 5, 6, and 8. For this, nine seedlings of each line/cultivar were evaluated against each pathotype in one replication. A total of 59 cultivars/lines exhibiting resistance to multiple pathotypes in the first screening were selected for evaluation in a second, replicated screening against the pathotype 3 and 5. Eighteen seedlings of each line were tested against each

pathotype, with the experiment repeated three times. Chinese cabbage (*B. rapa* var. *pekinensis*) cultivar Granaat (ECD 05) was used as a susceptible control.

6. **Statistical analysis:** Percent resistance for each cultivar/line against each pathotype was calculated using the following formula:

$$\text{Percent resistance } (\%) = \frac{\text{Number of resistant seedlings}}{\text{Total no.of seedlings inoculated}} \times 100$$

where seedlings with no visible gall at 42–45 days after inoculation were considered as resistant.

The disease score was calculated for each cultivar/line using the following formula (Strelkov et al. 2007):

$$\text{Average score} = \sum (n \times 0 + n \times 1 + n \times 2 + n \times 3)/N$$

where *N* is the total number of plants, *n* is the number of plants in each class and 0, 1, 2, and 3 are the symptom severity classes, as described above. An index of disease (ID) was also calculated for each cultivar/line, using the method of Horiuchi and Hori (1980) as modified by Strelkov et al. (2006):

$$\text{ID } (\%) = \frac{\sum (n \times 0 + n \times 1 + n \times 2 + n \times 3)}{N \times 3} \times 100$$

where *n* is the number of plants in each class, *N* is the total number of plants, and 0, 1, 2, and 3 are the symptom severity classes. The ID value was calculated for each cultivar/line against each treatment from each replication.

The reaction of different cultivar/line against different pathotypes was analyzed with SAS software version 9.2 (SAS Institute Inc. 2008). Experiments were conducted in a split-plot design where cultivars/lines were main plots and pathotypes were the sub-plots. Residuals for ID were calculated using SAS PROC GLM. These residuals were used to estimate skewness and kurtosis with SAS PROC UNIVARIATE, and homogeneity of variance by PROC GLM. However, skewness and kurtosis of ID residual indicated that the data were not normally distributed and residual variances were not homogeneous. Therefore, data from replicated experiments were analyzed according to a generalized linear mixed model of split-plot design by using the SAS PROC GLIMMIX, where pathotypes and cultivars/lines were considered as fixed effects and replication as the random effect.

9.7.6 Artificial Pathogen Inoculation Techniques

Different species of *Brassica* germplasm have been used for inoculum extraction and conducive growth conditions after inoculation as per methods suggested by Buczacki et al. (1975). The different inoculation techniques followed are:

1. **The slurry method:** In which 400 mL of the spore suspension is mixed with 6000 g soil and the seeds are sown in the slurry (Toxopeus and Janssen 1975).
2. **The modified slurry methods:** Where after mixing of spore suspension and soil, 7-days-old seedlings are planted in slurry.
3. **A root dip method:** Proposed by Johnston (1968) in which the roots of 7-days-old seedlings are dipped in the spore suspension for 2 h.
4. **Seed Dipping:** Dipping of seeds in spore suspension for 6 h before sowing is also recommended.
5. **An injection method:** Injection method was also suggested by Schoeller and Grunewald (1987). In this method, 7-days-old seedlings are planted in the non-infected soil. Close to each plantlet, a hole is made in which 2 mL of the spore suspension is added so that roots of the plantlets are contaminated with the suspension.

After making a comparison of these inoculation techniques, Schoeller and Grunewald (1987) reported that only with the modified slurry method, the root dip method, and the injection method could infect 100% of tested plants. Rod (1988) described partly modified inoculation method which is very easy and rapid to use under glasshouse or phytotron conditions.

9.7.7 PCR Detection of Clubroot Pathogen

A simple, one-step polymerase chain reaction (PCR) protocol was developed to detect the pathogen in plant, and soil samples at Canada (Cao et al. 2007). The primers TC1F and TC1R, based on a *P. brassicae* partial 18S ribosomal RNA (rRNA) gene sequence from GenBank, yielded a 548-bp product in the optimized PCR. A second pair of primers, TC2F and TC2R, which amplified a fragment of the 18S and internal transcribed spacer (ITS) 1 regions of the rDNA repeat, also was tested and produced a 519-bp product. Neither set of primers amplified any DNA fragment from non-infected plant hosts, non-infested soil, or common soil fungi and bacteria tested in this study. Quantities of 100 fg or less of total *P. brassicae* DNA, or 1×10^3 resting spores per gram of soil, could be detected consistently using these primers and PCR protocol, corresponding to an index of disease of 11% or lower when the soil was bioassayed. The protocol also enabled detection of *P. brassicae* in symptomless root tissue 3 days after inoculation with the pathogen. Therefore, the PCR assay described by Cao et al. (2007) provide a reliable diagnosis for routine detection of *P. brassicae* in plant and soil materials in a specific and rapid manner.

9.7.7.1 PCR Technique to Identify Pathogen

PCR amplification was conducted to confirm the pathogen. DNA was extracted from root galls and healthy roots (control) of Shepherd's purse (*Capsella bursa-pastoris*) (SP). Two primers, TC2F (5′-AAACAACGAGTCAGCTTGAATGCTAGTGTG-3′) and TC2R (5′-CTTTAGTTGTGTTTCGGCTAGGATGGTTCG-3′), were used to detect *P. brassicae* (Cao et al. 2007). No PCR amplifications were observed with the control DNA as template. A fragment of the expected size (approximately 520 bp) was obtained when DNA was amplified from diseased roots of SP. These results suggest that the pathogen in the galled roots of SP is *P. brassicae*. Pathogenicity of *P. brassicae* in SP was tested on plants of both SP and Chinese cabbage (CC) (*B. campestris* ssp. *pekinensis*). A resting spore suspension prepared from naturally infected SP roots was mixed with a sterilized soil in two plastic pots, resulting in a final concentration of 5×10^6 spores/g soil. Soil treated with the same volume of sterile water was used as a control. Seeds of SP and CC were pre-germinated on moist filter paper for 2 days (20 °C) and seeded into the infested and control pots. One seed per pot was planted for CC and four seeds per pot for SP. The pots were placed in a chamber at 15–25 °C under 12 h light and 12 h dark. Plants in each pot were uprooted after 4 weeks and the roots of each plant were washed under tap water and rated for clubroot disease. No disease symptoms were observed in the control treatments of SP or CC. Plants of both species showed symptoms of clubroot, with the disease incidence of 62.5% and 100% on SP and CC, respectively. The pathogen was isolated from diseased roots of each plant and confirmed as *P. brassicae* based on morphological characteristics and PCR detection (Ren et al. 2014).

9.7.8 Isolation of Resting Spores

Plasmodiophora brassicae resting spores can be isolated according to the procedure of Castlebury et al. (1994). Approximately 2 g of fresh galled roots of canola cv. Westar or Chinese cabbage (*B. rapa* var. *pekinensis*) cv. Granaat were homogenized in a blender in 50 mL of sterile distilled water (sdH_2O) for 2 min at high speed, and the resultant homogenate was filtered through eight layers of cheesecloth. The filtrate was centrifuged at 4 °C in 50-mL centrifuge tubes at 4000 × *g* for 10 min in a swinging-bucket rotor. The supernatant was discarded and the pellet was resuspended in 5 mL of 50% sucrose and centrifuged again at 2000 × *g* for 10 min. The resulting supernatant was transferred to a new 50-mL tube and diluted with 30 mL of sdH_2O before centrifuging at 4000 × *g* for 10 min. The pellet obtained was resuspended in 5 mL of sdH_2O and centrifuged as above to remove any remaining sucrose. The supernatant was removed and the resting spore pellet resuspended in 5 mL of sdH_2O. The resting spores then were purified using a continuous gradient of LUDOX (HS-40 colloidal silica), 40 wt% suspensions in water, as described by Castlebury et al. (1994). The purified *P. brassicae* resting spore suspension normally was stored at 4 °C for less than 48 h before DNA extraction. Spore concentrations were quantified using a hemocytometer (Cao et al. 2007).

9.7.8.1 Inoculum Preparation of Spores

The resting spores for plant inoculation were prepared as described by Feng et al. (2013a). Galls were homogenized in 10% (w/v) sucrose in a blender. The slurry was passed through eight layers of cheese cloth and suspension was centrifuged at 50 × g for 5 min. After transfer into a new tube, the supernatant was centrifuged at 2000 × g for 5 min. The upper layer of the resultant pellet was suspended in 5 mL of water by gentle pipetting and transferred into new tube containing 40 mL of water. After centrifugation at 2000 × g for 5 min. the supernatant was removed. The resting spores in the pellet were resuspended in 2 mL of water and spore concentration was adjusted to 1×10^7 spores/mL. The resting spores were surface disinfected with colistin sulfate at 1 μg/mL and vancomycin hydrochloride at 1 μg/mL at 4 °C in the dark for 24 h. The spores were used as inoculum after removal of the antibiotics by washing twice with 40 mL of distilled water. For DNA extraction, the resting spores were prepared with the same protocol except that after overnight treatment with antibiotics at 4 °C, the spore suspension were further treated at 37 °C with 10 units/mL of DNase I for 2 h to remove genomic DNA from the host plant or other organisms released into the suspension during homogenization of galls (Zhang et al. 2015).

9.7.8.2 Resting Spore Purification and Single Spore Isolation

Currently, there is not a standard resting spore purification protocol that satisfies the requirement for molecular studies. The routinely used protocols as described by Bryngelsson et al. (1988) and Asano et al. (1999) are sufficient in quality for inoculation purposes, but not for DNA or RNA extraction when contamination from other organisms needs to be excluded. Similarly, although there are several reports on the isolation of single spores of *P. brassicae* (Xue et al. 2008), the techniques described are not capable of producing sufficient volumes of material. Therefore, while these techniques may be suitable for many purposes, further improvements are needed to develop protocols that would enable the purification and recovery of much larger numbers of single resting spores, which would increase efficiency and facilitate molecular studies (Feng et al. 2014).

1. **Isolation of zoospores**. A protocol to isolate secondary zoospores was reported by Feng et al. (2012). By washing the roots of infected canola seedlings 7 days after inoculation, the secondary zoospores were isolated without contamination of resting spores and demonstrated that the collected secondary zoospores could initiate infection. This technique allowed investigation of the infection processes of primary and secondary zoospores separately and will be useful in genetic studies in which resistance can be partitioned into different types based on the infection stage at which it is expressed. This study has stimulated several additional studies to investigate *P. brassicae* gene expression (Feng et al. 2013a) and the plant responses to primary and secondary infection (Deora et al. 2012; Feng et al. 2013c).
2. ***In vitro* dual culture:** Bulman et al. (2011) standardized an *in vitro* dual culture system to study the genomics of *P. brassicae* and related protist species. In the absence of exogenous plant growth regulators, stable, long-term *B. rapa* cell callus cultures infected by *P. brassicae* were produced and demonstrated to be an excellent starting material for gene discovery. By using this technique on a related plasmodiophorid species, *Spongospora subterranean* (Wallr.) Lagerh., Feng

et al. (2014) constructed a pilot scale DNA library and found that almost all of the DNA clones were from the protist rather than the plant host. This technique provides an avenue to remove plant contaminants from the pathogen materials, which is critical in genomics studies.

9.7.9 A Real-Time PCR Assay to Detect *Plasmodiophora brassicae* Under Field Conditions

A protocol using real-time polymerase chain reaction (PCR) for the direct detection and quantification of *Plasmodiophora brassicae* in soil samples has been developed and used on naturally and artificially infested soil samples containing different concentrations of *P. brassicae* under Swedish soil samples collected from seven fields across three locations. Species-specific primers and a TaqMan fluorogenic probe were designed to amplify a small region of *P. brassicae* ribosomal DNA (Table 9.3). Total genomic DNA was extracted and purified from soil samples using commercial kits. The amount of pathogen DNA was quantified using a standard curve generated by including reactions containing different amounts of a plasmid carrying the *P. brassicae* target sequence. The PCR assay was optimized to give high amplification efficiency and three to four copies of the target DNA sequence were detected. Regression analysis showed that the standard curve was linear over at least six orders of magnitude ($R^2 > 0.99$) and that the amplification efficiency was >92%. The detection limit in soil samples corresponded to 500 resting spores g^{-1} soil. The inter sample reproducibility was similar to, or higher than, that of assays for other pathogens quantified in soil samples. Bait plants were used to validate the real-time PCR assay. The protocol developed was used to investigate the spatial distribution of *P. brassicae* DNA in different fields and a significant difference was found between in-field sampling points. The reproducibility of soil sampling was evaluated and showed no significant differences for samples with low levels of inoculum, whereas at higher levels differences occurred. Indicator kriging was used for mapping the probability of detecting *P. brassicae* within a 2-ha area of a field. A threshold level of 5 fg plasmid DNA g^{-1} soil, corresponding to approximately 3×10^3 *P. brassicae* resting spores g^{-1} soil has been suggested for growing resistant cultivars. These results provide a robust and reliable technique for predicting the risk of disease development and for assessing the distribution of disease within fields (Wallenhammar et al. 2012).

Table 9.3 Primer and probe sequences used for the amplification and detection of the *Plasmodiophora brassicae* target sequence (Wallenhammar et al. 2012)

Primer/probe	Sequence[a]
Primer forward (PbF)	5¢-AAA CAA CGA GTC AGC TTG AAT GC-3¢
Primer reverse (PbR)	5¢-TTC GCG CAC AAG CAC TTG-3¢
Probe (PbP)	5¢-CGC GCC ATG CGA CAC TGT TAA ATT G-3¢

[a]Bases 4–23 of PbF are the same bases as in primer PbITS6 (Faggian et al. 1999)

In China: SYBR Green I real-time PCR assay has been developed to detect and quantify *Plasmodiophora brassicae* ribosomal DNA (rDNA) and internal transcribed spacer (ITS). A pair of primers PBF1/PBR1 was designed based on the conservative region of rDNA-ITS of *P. brassicae*. The positive plasmid pB12 was obtained and used as the template to create standard curve. The specificity, sensitivity, and reproducibility of real-time PCR were evaluated, respectively. Naturally and artificially infested soil samples containing different concentrations of *P. brassicae* were detected. The results demonstrated that standard curve established by recombinant plasmid had shown a fine linear relationship between threshold cycle and template concentration. The melting curve was specific with the correlation coefficient of 0.995, and the amplification efficiency was 93.8%. The detection limit of *P. brassicae* genomic DNA was approximately 40 copies per 25 μL. The sensitivity of the assay was at least 100-fold higher than conventional PCR. Only DNA from *P. brassicae* could be amplified and detected using this assay, suggesting the high specificity of this assay. The coefficient of variation was less than 3%, indicating that the PCR method revealed high reproducibility. The detection limit in soil samples corresponded to 1000 resting spores g^{-1} soil. Bait plants were used to validate the real-time PCR assay. This developed real-time PCR assay allows for fast and sensitive detection of *P. brassicae* in soil and should be useful in disease management and pathogen interception so as to prevent further spread of *P. brassicae* (Jin-ping et al. 2013).

9.7.9.1 Bioassay Technique of Wallenhammar et al. (2012)

Bait plants were used to validate the real-time PCR assay developed. The level of *P. brassicae* in naturally infested soil samples was assessed using a bioassay method previously developed by Wallenhammar (1996) and modified by Wallenhammar et al. (2000). The pots used were $90 \times 90 \times 95$ mm^3 (vefi). Infected plants were scored according to the following classes: 0 = no galls; 1 = enlarged lateral roots; 2 = enlarged tap root; 3 = enlarged napiform tap root; 4 = enlarged napiform tap root, lateral roots healthy; and 5 = enlarged napiform tap root, lateral roots infected. A disease severity index (DSI) was calculated according to the equation:

$$\text{DSI} = \frac{\sum(N1 \times 1)}{(N \times 4)} \times 100\%$$

where *N*, class no.1, no. of plants in each class; *N*, total no. of plants

Samples of approximately 500 g were withdrawn for further DNA detection and dried stored at room temperature until extraction.

1. **DNA extraction from infected roots, fungal material, and plant material:** A modified E.Z.N.A. SP Plant DNA Miniprep Kit extraction method was used to extract DNA from plant and fungal material. A total of 30–130 mg of fresh mycelia, apothecia, or infected root tissue was incubated with two to six stainless steel beads (2–3 mm in diameter) in 1.5-mL cetyltrimethyl ammonium bromide (CTAB) solution (20 g CTAB, 12.11 g Tris, 81.8 g NaCl and 7.4 g Na_2-ethylene

diamine tetra-acetic acid (EDTA) L^{-1}; pH = 8.0) and 20 μL proteinase K (20 mg mL^{-1}, GeneScan) at 60 °C for 2 h after 30 s shaking at speed setting 4.5–5.5 using a Hybaid Ribolyser (Hybaid). The suspension was then centrifuged at >10,000 g for 5 min and 1 mL of the supernatant was mixed with 5 μL RNase A in a 1.5-mL tube and incubated at 60 °C for 15 min. After centrifugation at 20,000 g for 1 min, 600 μL supernatant were mixed with 210 μL SP2 buffer and incubated in a refrigerator for 5 min. The suspension was centrifuged at >10,000 g for 10 min, and a 400-μL portion of the supernatant was transferred to a 1.5-mL tube and mixed with 600 μL SP3 buffer. The supernatant was then transferred to a HiBind DNA Mini Column, washed and eluted in 100 μL elution buffer according to the manufacturer's protocol. All DNA solutions were stored at −20 °C.

2. **DNA extraction from soil:** Soil samples were air-dried at room temperature (20 °C), plant material was removed by hand and the soil was pulverized using a mallet. DNA was extracted from 350-mg soil samples (two replicates per sample) using the Fast-DNA SPIN Kit for Soil according to the manufacturer's instructions with the following modifications. After cell lysis for 30 s at speed setting 5.5, samples were centrifuged at 14,000 g for 5 min. At the tenth step, 600 μL of the supernatant were discarded and 600 μL of the resuspended mixture transferred to a SPIN Filter. DNA was eluted in 100 μL DES. Further purification of the eluate was performed using the Wizard DNA Clean-Up System and Illustra Micro Spin S-300 HR Columns according to the manufacturer's instructions with the following modifications. In the Promega kit, 900 μL of the resin was used and incubated for 5 min at 60 °C together with the DNA solution. DNA was eluted in 100 μL. The Micro Spin S-300 HR Columns were centrifuged for 2 min before adding the DNA eluate.
3. **Primers and probes:** Primers and a probe with putative specificity to *P. brassicae* as mentioned below were designed using a ribosomal RNA gene sequence available in GenBank (accession number AF231027) and amplified a 10^3-bp fragment. The forward primer was based on the sequence of the forward primer used by Faggian et al. (1999). The 5′ terminal reporter dye of the probe was VIC and the 3′ quencher was TAMRA. Primer and probe sequences were checked for identity against other known sequences using the BLASTN program. (http://www.ncbi.nlm.nih.gov/blast/Blast.cgi). Primers were purchased from MWG-Biotech.
4. **Real-time PCR:** Real-time quantitative PCR was performed using the 7300 Real-Time PCR System in a total volume of 25 μL. The reaction mixture included 1 × Taq Man Universal PCR Master Mix No AmpErase UNG, 0.12 μM PbF, 0.6 μM PbR, 0.2 μM PbP, and 5 μL template DNA. Soil DNA extracts was diluted fivefold before analysis and was analyzed in duplicate. The thermal cycling conditions consisted of an initial denaturation for 10 min at 95 °C followed by 45 cycles at 95 °C for 15 s and finally 59 °C for 60 s.
5. **Quantification:** Since *P. brassicae* is an obligate biotroph, it is difficult and laborious to extract pure genomic DNA from spores to use as a standard. Therefore, the amount of pathogen DNA was quantified using a standard curve

generated by including reactions containing different amounts of a plasmid carrying the *P. brassicae* target sequence. The target DNA used to generate the calibration curves was a 607-bp sequence, amplified using a universal forward primer (NS7: 5′- to fg (10^{-15} g) plasmid DNA g^{-1} soil by taking into account that 350 mg soil were used for the DNA extraction and that 1 μL of the original DNA extract was analyzed. By analyzing the series of artificially infested soil samples containing different amounts of *P. brassicae* resting spores, it was possible to convert fg plasmid DNAg^{-1} soil to number of *P. brassicae* resting spores g^{-1} soil. Amplification efficiency (*E*) was calculated as a percentage using the equation:

$$E = 100 \times \left(10 - \frac{1}{k} - 1\right)$$

where *k* is the slope of the equation describing the plot of Ct versus the logarithm of the DNA amount. GAGGCAATAACAGGTCTGTGATGC-3′) (White et al. 1990), and a *P. brassicae*-specific reverse primer (Pb3R: 5-′-ACCATACCCAGGGCGATTG-3′), which was cloned into pGEM-T vectors according to the manufacturer's instructions. Plasmid DNA was prepared using a QI Aprep Spin Mini prep Kit, and the DNA concentration was determined spectrophotometrically. A tenfold dilution series of the plasmid standard was analyzed in triplicate using the real-time PCR assay and Ct values were then plotted against the amount of plasmid DNA to create a standard curve. Soil samples with low levels of *P. brassicae* DNA outside the range of the standard curve were quantified by extrapolating the standard curve to facilitate analysis of the results. The starting quantity of DNA in soil samples was transformed from ag (10^{-18} g) plasmid DNA.

6. **Evaluation of inhibition and repeatability:** Inhibition of the PCR reaction by contaminants in the DNA extracts was assessed for the soil samples analyzed in this study by analyzing a separate PCR reaction containing the soil DNA extract and 2 μL of the *P. brassicae* plasmid standard containing approximately 150 gene copies. The same amount of plasmid standard was added to a water control. The percentage inhibition effect (*I*) was calculated using the equation:

$$I = 100 \times \frac{\{(Q_{s+sp}) - (Q_s + Q_{sp})\}}{Q_s + Q_{sp}}$$

where Q_s is the quantity of the unknown sample, Q_{sp} is the quantity of the standard spiked in water, and Q_{s+sp} is the quantity of the unknown sample spiked with the standard. The inter sample repeatability was evaluated on separate extractions of ten soil subsamples quantified in a single real-time PCR assay.

7. **Distribution of *P. brassicae* inoculum and risk assessment:** Indicator kriging was used for mapping the probability of detecting *P. brassicae* within a 2-ha area of field D at farm 2 (Gotene). This method is a nonlinear form of ordinary kriging in which the original data are transformed to a binary scale (0–1) (Burrough and

McDonnell 1998). This approach, which has been applied frequently in soil mapping (Van Meirvenne and Goovaerts 2001; Lark and Ferguson 2004), is appropriate when the goal is to produce an estimate of the probability that a given threshold is exceeded. In this case, observations with a data value (z) of less than or equal to a specified threshold (T) (here T was set to 5.0 fg plasmid DNA g^{-1} soil) were assigned a value of 0. Other observations were given a value of 1. The transformed data were then used in variogram analysis and ordinary kriging (Webster and Oliver 1992; Burrough and McDonnell 1998). The map displays the probability of T being exceeded. The software GS+ (Gamma Design) was used for the geostatistical analysis and ARCGIS (ESRI) was used for mapping (Wallenhammar et al. 2012).

9.7.9.2 Bioassay Technique of Jin-ping et al. (2013)

Bait plants *Brassica pekinensis* (Lour.) Rupr. variety Zhongbai was used to validate the developed real-time PCR assay method. The level of *P. brassicae* in naturally infested soil samples was assessed using a bioassay method previously developed by Wallenhammar (1996) and modified by Wallenhammar et al. (2000). The pots used were 40 × 40 × 40 mm (vefi). Infected plants were scored according to the following classes: 0 = no galls; 1 = enlarged lateral roots; 2 = enlarged tap root; 3 = enlarged napiform tap root; 4 = enlarged napiform tap root, lateral roots healthy; and 5 = enlarged napiform tap root, lateral roots infected. A disease severity index (DSI) was calculated according to the equation:

$$\mathrm{DSI} = \frac{\sum(N1 \times 1)}{(N \times 4)} \times 100\%$$

where N, class no.1, no. of plants in each class; N, total no. of plants

1. **DNA extraction from infected roots, fungal material, bacterial material, and soil**: All the mycelia, apothecia were incubated in solid media, and then transferred into liquid media for 7 d at 28 °C. A modified CTAB method was used to extract DNA from infected root tissue, fungal and bacterial material. Frozen mycelium was placed in 1.5 mL microtubes, 600 μL of CTAB extraction buffer (2% CTAB, 0.3 mol L^{-1} NaCl, 1 mol L^{-1} Tris-HCl, 0.5 mol L^{-1} EDTA (pH 8.0)) was added, and the tubes were vortexed and then incubated at 65 °C for 30 min. Chloroform-isoamyl alcohol (24:1 mixture of chloroform and isoamyl alcohol) (1 volume) was added to each tube, and the tubes were mixed and centrifuged for 10 min at 13,000 × g. The aqueous phase was transferred to a new tube, and the chloroform extraction was repeated. DNA was precipitated overnight at −20 °C after the addition of 0.1 volume of 3 mol L^{-1} sodium acetate (pH 8.0) and two volumes of cold 100% ethanol. The supernatant was discarded, and the pellets were washed with 70% ethanol and then dried by vacuum centrifugation. DNA was resuspended in 80 μL of TE (10 mmol L^{-1} Tris-HCl, 0.1 mmol L^{-1} EDTA (pH 8.0)) and then diluted 1:100 for use in PCRs in TE. The internal transcribed spacer (ITS) region of *P. brassicae* was based on GenBank

(Y12831) and used for designing the primer PBF1 (5-′-GCTCCTGCGTGTCGCTGTTATT-3′) and PBR1 (5-′-CTGCGTTCTTCATCGTTGTGA-3′). Soil samples were air-dried at room temperature (20 ± 2 °C), plant material was removed by hand and the soil was pulverized using a mallet. DNA was extracted twice independently from 200 mg soil samples using the DNA SPIN Kit according to the manufacturer's instructions.

2. **Real-time PCR assay:** DNA concentration was determined spectrophotometrically. A tenfold dilution series of the plasmid standard was analyzed in triplicate using the real-time PCR assay. Each reaction of 25 μL contained 10 ng of plasmid pB12, 0.5 μL of each 15 μmol L^{-1} primer (PBF1 and PBR1), 12.5 μL of 2 × SYBR Green Master Mix, and 0.4 μL of a 1000 × diluted reference dye. The PCR reaction was set up in triplicate for each DNA sample. Prior to amplification, an initial denaturation step was performed (95 °C for 3 min) ensuring complete denaturation of the DNA and activation of the *Taq* polymerase. The reactions were performed in a iCyeler iQ5 Sequence Detection System. This was followed by 40 cycles of 30 s at 95 °C, 30 s at 60 °C, and 35 s at 72 °C. Fluorescence was detected after each cycle. Ct values were then plotted against the amount of plasmid DNA to create a standard curve.
3. **Sensitivity and reproducibility of real-time PCR:** The tenfold serial dilutions of the *P. brassicae* genomic DNA were made to measure the sensitivity of the assay. Every dilution repeated five times. Real-time PCR was performed using the optimal reaction system. Conventional PCR amplification was performed as control. The inter sample repeatability was evaluated on ten separate soil DNA extracts of *P. brassicae* from one soil samples analyzed in the same real-time PCR. Standard deviation (SD) of the Ct values was determined based on the ten replicates for each amount of target DNA.
4. **The detection of artificially and naturally infested soils:** The artificially and naturally infested soil samples were detected using the optimal real-time PCR reaction system. The artificially infested soils sample that contained 1.0×10^8, 1.0×10^7, 1.0×10^6, 1.0×10^5, 1.0×10^4, 1.0×10^3, 5.0×10^2, 10, and 0 resting spores g^{-1} soil (DW) were measured to validate the sensitivity of the real-time PCR and determine the threshold level of clubroot. The total of 160 soil samples of wide type and the sterilized soil samples from lab were detected using real-time PCR and conventional PCR. The thermal cycling parameters of conventional PCR were initial denaturation at 94 °C for 3 min followed by 40 cycles consisting of denaturation at 94 °C for 1 min, annealing at 56 °C for 1 min, and extension at 72 °C for 1.5 min. A final extension at 72 °C for 7 min was done at the end of the amplification. Negative controls (no DNA template) were used in every experiment to test for the presence of contamination in reagents (Jin-ping et al. 2013).

9.7.9.3 PCR and Quantitative PCR Assay

Primers for the identification of the pathotypes-specific gene(S) were derived from Feng et al. (2013b) with other primers used. The genomic locations of the primers are illustrated in Fig. 9.11. PCR analyses were conducted in an Eppendorf

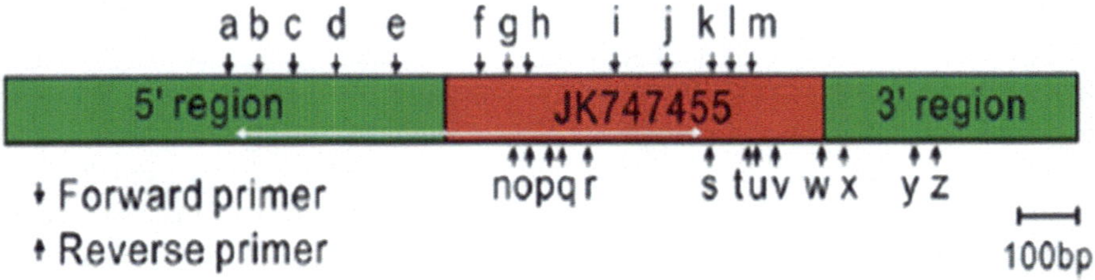

Fig. 9.11 Diagram of the genomic region of the *Cr 811* gene, JK 747455 GenBank accessions number of the expressed sequence tag that is a part of *Cr811* "5" region and "3" region. Sequences obtained by genome walking a to z primers whose sequences are linked. The white line denotes the sequenced cDNA (Zhang et al. 2015)

Mastercycler thermal cycler. Except in the genome walking analysis, all PCR analyses were run in a final volume of 20 μL in the Promega master mix with 0.25 μM of each primer and 100 ng in the template; for all primers, the annealing temperature was 52 °C, and/or 56 °C, all elongation time was 45 s, and the cycle number was 35. Quantitative PCR (qPCR) assays were conducted in a Step-One Real-Time PCR System using the GoTaq 2-step RT qPCR System (Promega) for RNA analysis or the power SYBR Green PCR Master Mix (Life Technologies) for DNA analysis. The *P. brassicae Actin 1* gene (GenBank accession number AAR88383) was used for normalization. Primer efficiency was verified using standard curves generated from the purified PCR product of the targeted fragments. The quantity of *Cr811* relative to *Actin1*, at the DNA or RNA level was calculated via the delta-delta method, with relative expression or amount $= 2^{-(\Delta CT\ \mathrm{sample} - \Delta CT\ \mathrm{control})}$. All qPCR amplification was performed in triplicate for each of three template preparations derived from different biological samples. Data from qPCR were analyzed for statistical significance using the general linear model (GLM) procedure of the Statistical Analysis System. All data were subjected to one-way analysis of variance and when appropriate, a multiple range test was conducted by Fisher's least significant difference (LSD) at $P \leq 0.01$ (Zhang et al. 2015).

9.7.9.4 Quantification and Control Assessment of Clubroot with qPCR

Conventionally, the detection of *P. brassicae* is carried out by gall observations and bait tests. With the advent of molecular techniques, PCR amplification of DNA has been used to aid detection and study the diversity of *P. brassicae* (Buhariwalla et al. 1995; Buhariwalla and Mithen 1995; Cao et al. 2007; Faggian et al. 1999; Faggian and Strelkov 2009; Ito et al. 1997, 1999; Wallenhammar and Arwidsson 2001). Methods of *P. brassicae* PCR detection are often based on ribosomal RNA (rRNA) gene internal transcribed spacer regions (ITS) (Ito et al. 1999). Detection and quantification is central to investigating the control of *P. brassicae*. Wakeham et al. (2017) quantified *P. brassicae* spores using real-time PCR (qPCR) and investigated the effect of Limex (LimeX70) and conventional lime-based products on clubroot control. Limex is a form of $CaCO_3$ produced as a by-product of the sugar beet refining process (Harling 2007). Limex contains P_2O_5, MgO, and SO_3. The addition of calcium carbonate to clubroot contaminated land has been shown to improve control. However, large quantities of calcium carbonate are required to produce consistent results. This can raise the pH of the soil to levels which are inconsistent with the other crop production in the rotation (potatoes). The number of *P. brassicae* resting spores in soil was monitored in different control treatments and the subsequent effect on crop yield was ascertained.

9.7.10 Genetics and Molecular Mapping of R-Genes to *Plasmodiophora* Pathotypes

1. **Plant material:** The parent materials included a clubroot-resistant rutabaga inbred line Rutabaga-BF and a susceptible spring *B. napus* canola cultivar A07-29NI (Rahman et al. 2011). The line Rutabaga-BF was developed from the cultivar Brookfield through self-pollination of a single clubroot-resistant plant. Seeds of Brookfield were received, and A07-29NI is a canola cultivar was developed. The line Rutabaga-BF showed complete resistance to Canadian *P. brassicae* pathotypes 2, 3, 5, 6, and 8.
2. **Crossing and population development:** The parent Rutabaga-BF was vernalized for 8 weeks at 4 °C with an 8/16 h (day/night) photoperiod and crossed to the susceptible parent A07-29NI. The F1 plants were vernalized for 8 weeks, and F2 seeds were produced through self-pollination of the F1 plants. Backcross (BC1) seeds were produced by crossing the F1 to A07-29NI. The F2 and BC1 seeds were harvested from the individual F1 plants and maintained as separate families.
3. **Doubled haploid (DH) lines production:** Doubled haploid lines were produced from five F1 plants through application of microspore culture technique (Coventry et al. 1988). For this, flower buds of 3.5–4.5 mm in length, where the size of the petals was about two-thirds of the length of the anther, were used for microspore isolation. The buds were collected in a falcon tube and were placed on a bed of ice to prevent microspore degeneration. Prior to microspore isolation, the buds were sterilized by treating with 6% calcium hypochlorite ($Ca\ (OCl)_2$) solution for 15 min, and then rinsed three times with cold sterile milli-Q Millipore water. About 15–20 sterilized flower buds were homogenized in 25 mL liquid B5 medium (Gamborg et al. 1968) containing 13% sucrose using a sterile mortar and pestle. The homogenate was poured through two layers of nested sterile filters (64 μm top (NTX64) and 41 μm bottom (NTX41)) into a sterile falcon tube. The falcon tube was centrifuged for 5 min at 1000 rpm, and the dark green supernatant was decanted leaving the pellet of yellow spores at the bottom of the tube. Microspores were washed three times with B5 liquid medium. After washing, the clean microspore pellet was resuspended in 30 mL NLN 13 medium (Lichter 1982) containing 50 mg/L colchicine and poured into sterile Petri plates. The Petri plates were sealed with parafilm and incubated in dark in an incubator at 30 °C for 24 h. After 24 h, the suspension was transferred to a falcon tube and centrifuged at 1000 rpm for 5 min. The supernatant was

decanted and 25 mL of NLN 13 medium (without colchicine) was added to the microspores and cultured in Petri plates for 14 days at 30 °C. After 14 days, the plates were transferred on a shaker rotating at 59 rpm to provide enough air to the developing embryos. The torpedo-shaped embryos were transferred to solid B5 medium (liquid B5 with 20% sucrose and 2% agar) at a density of 20 embryos per Petri plate. The plates were kept at 4 °C for 14 days, and then transferred to room temperature under light (16/8 h day/night) for germination of the embryos and development of plantlets. Seedlings with well-developed roots and true leaves were transferred to soil-free growth media and covered with transparent cups for about a week to maintain high humidity.

At flowering, the plants producing visible pollen were considered to be doubled haploids, while the sterile plants with smaller size flower/petal and lacking pollen grains were considered as haploids. The haploid plants were treated with 0.35% aqueous solution of colchicines for chromosome doubling. For this, roots were washed and immersed in colchicine solution for 1 h. After this, roots were washed thoroughly under running tap water and the shoots were clipped off leaving only a few auxiliary buds (lateral buds). The treated plants were transplanted to soil-free growth medium in pots. Fertile plants with visible pollen grains were self-pollinated by bag isolation and harvested as doubled haploids.

4. **Pathogen isolate:** Single spore-derived isolates of *P. brassicae*, classified as pathotypes 2, 3, 5, 6, or 8 based on Williams (1966) differentials, were used. These isolates were obtained from University of Alberta, in the form of galls. Resting spore suspensions (inoculum) from the preserved galls were prepared following the protocol described by Strelkov et al. (2007), and concentration of the suspension was adjusted to 1×10^7 to 1×10^8 resting spores/mL.
5. **Resistance test:** Four F2 and their corresponding BC1 families were evaluated for resistance to *P. brassicae* pathotype 3, and the DH population was evaluated for resistance to *P. brassicae* pathotypes 2, 3, 5, 6, and 8. Seedlings germinated on moistened Whatman filter paper No. 1, at the age of 7 days after seeding, were inoculated by dipping the roots in resting spore suspension (Nieuwhof and Wiering 1961). The inoculated seedlings were transplanted to 3 cm × 3 cm × 5 cm ($L \times W \times D$) cells filled with Sunshine Professional Natural and Organic Mixes #4, placed in a tray (2 m × 4 m), and grown in a greenhouse at 21 ± 2 °C with a 16 h photoperiod. After transplanting, 2 mL inoculum was pipetted to each cell to ensure successful inoculation. During the first 7 days, the cells were kept saturated with water to ensure sufficient moisture for development of the pathogen. To ensure acidic condition of the soil, HCl solution (10% *v*/*v*) at 20 mL/tray (2 m × 4 m) was added each day during this time. Starting the second week, watering was done once a day. Seedlings were evaluated for clubroot resistance 42–45 days after inoculation, and the severity of gall development was rated on a 0–3 scale, where 0 = no galling, 1 = one or few tiny galls on lateral roots, 2 = moderate galling on lateral roots, and 3 = severe galling on lateral or main root. The details of inoculum preparation, inoculation technique, and scoring are described (Rahman et al. 2011; Hasan et al. 2012). Based on

disease score, plants with scores of 0 or 1 were classified as resistant, while the plants with scores of 2 or 3 were classified as susceptible.

6. **DNA extraction:** Leaf samples of the parents and the DH population (3–5-weeks-old plants) grown in greenhouse were collected in aluminum foil and kept at −80 °C until use. DNA was extracted using a Wizard Genomic DNA purification kit following instructions of the manufacturer. The DNA pellet was air-dried for 60 min and rehydrated in 200 μL DNA rehydration solution (TE buffer) overnight at 4 °C. The quality and quantity of the DNA was estimated by spectrometry using 260/280 nm absorbance ratio, and the concentration of DNA was estimated at 260 nm with an ND-2000 Nano Drop spectrophotometer.
7. **Molecular markers:** Simple sequence repeat (SSR) primer sequences were obtained from Agriculture and Agri-Food Canada (AAFC), Saskatoon and the SSR primers published by different researchers, such as Lowe et al. (2002, 2004), Suwabe et al. (2002, 2003), Piquemal et al. (2005), Saito et al. (2006), Cheng et al. (2009), and Kato et al. (2012), and SSR markers designed based on *B. rapa* cultivar Chiifu-401 whole-genome sequence assembly version 1.5 (Cheng et al. 2011) were used. To design SSR primers from the *B. rapa* genome sequence, an open access software program, QDD version 3, was used to detect microsatellites (SSRs) in the whole genome and primers were designed from the targeted regions (Meglecz et al. 2010).
8. **Polymerase chain reaction (PCR):** Polymerase chain reaction was performed in a total volume of 14 μL containing 45 ng (15 ng/μL × 3 μL) of DNA, 2.5 μL of 1 × PCR buffer, 1.0 μL of 25 mmol/L $MgCl_2$, 0.35 μL of 0.4 mmol/L dNTPs, 0.5 μmol/L of each of the forward and reverse SSR primers, 0.25 pmol/L of labeling dye, 0.125 U of GoTaq DNA Polymerase enzyme, and 6.995 μL of ddH20. Amplification reactions were carried out using a Gene-Amp PCR system 9700 starting with 95 °C for 2 min, followed by 35 cycles of denaturation at 95 °C for 30 s, annealing at 56 °C for 30 s, and extension at 72 °C for 45 s, with a final extension at 72 °C for 15 min. The PCR products were labeled with an M13 tailing technique as described by Schuelke (2000). For this, each forward SSR primer was attached to a universal M13 primer sequence 5=-CACGACGTTGTAAAACGAC-3=fluorescently labeled with infrared FAM, VIC, NED, and PET dyes. Electrophoretic separation and detection of the amplification products was performed using a capillary ABI sequencer No. 3730 and was scored and analyzed using Genemapper software version 3.7.
9. **Map construction and identification of the locus controlling clubroot resistance:** All primer pairs that showed polymorphism between the two parents were used to genotype the DH population. Based on segregation in F2, BC1, and DH populations, it was evident that a major gene is involved in the control of resistance to *P. brassicae* pathotypes in Rutabaga-BF. Therefore, in the first step, single marker analysis was performed with the DH population using the polymorphic SSR markers to identify the chromosome(s) carrying the resistance gene(s). This analysis indicated that SSR markers from chromosome A8 were associated with resistance.

Therefore, a genetic linkage map of this chromosome was constructed to identify the genomic region associated with resistance as well as to identify markers from this region. The linkage map was constructed using the program Joinmap 4.0 (Van Ooijen 2006). For this, an LOD threshold of 3.0 and a recombination frequency of 0.40 were used to group the marker loci. The recombination frequencies were transformed into centi Morgans (cM) using the Kosambi mapping function (Kosambi 1943). Composite interval mapping analysis was performed by using the linkage map to identify the genomic region carrying the clubroot resistance gene as well as to identify the linked markers for use in marker-assisted selection.

For this, genotypic and phenotypic data were subjected to Inclusive Composite Interval Mapping (ICIM) analysis using the software program QTL IciMapping version 4.0 (http://www. isbreeding.net), and the linkage map was scanned for association with phenotypic variation for resistance to *P. brassicae* pathotypes 2, 3, 5, 6, and 8 at every 0.1 cM interval. Empirical threshold levels, which confirmed the existence of the locus at a significance level of 0.05, were obtained by analyzing 1000 permutations of the dataset following the method described by Churchill and Doerge (1994). The location of the resistance locus was declared at the position where the LOD score exceeded the threshold significance level of 3.0.

10. **Candidate gene prediction:** To identify the physical location of the clubroot resistance gene in the *Brassica* genome, SSR marker sequences flanking the resistance locus were mapped on *B. rapa* cultivar Chiifu-401 genome sequence assembly version 1.5 (Cheng et al. 2011; http://www.brassicadb.org) by using electronic-PCR (e-PCR) (Schuler 1997; Rotmistrovsky et al. 2004). e-PCR is a computational procedure that searches DNA sequences that closely match the PCR primers in context of correct order, orientation, and spacing, thereby identifying sequence(s) within the genome whose size is similar to the PCR product to be generated by use of these primer pairs. When the primer sequences matched with a fragment of DNA sequence from the target chromosome and the length of the deduced fragment to be amplified by the primers is similar to the actual size of the fragment amplified by these primers, the sequences were considered to be the marker sequence. The sequence of the A8 chromosome segment of *B. rapa* cultivar Chiifu-401 (genome sequence assembly version 1.5), flanked by the SSR markers associated with resistance, was scanned using Augustus, a server-based software program that predicts the ab initio gene in the eukaryotic genomic sequence (Stanke et al. 2004; http://www.augustus.gobics.de), for detection of the gene(s) in this genomic region to be associated with resistance. The protein sequences, predicted based on the gene sequences, were scanned for similarity with *Arabidopsis thaliana* protein(s) of known function (https://www.arabidopsis.org) using the software program WU-BLAST2 (Gish 2003; https://www.arabidopsis.org/wublast/index2.jsp) (Hasan and Rahman 2016).

9.7.11 Identification of Parents and F2 Lines for R-QTLs in *Brassica* to *Plasmodiophora*

1. **Phenotypic performance of parents and F2 lines:** According to the field identification for three successive years (E107.6459, N29.5709) where serious clubroot disease every year caused mainly by the fourth race of *P. brassicae* (identified according to Williams host system), "GZ87" exhibited complete resistance with a disease index of 0, while "263" which is a founder parent in cabbage breeding program showed stable susceptibility to *P. brassicae* (with average DIC of 72.6% and disease index of 33.7). Under the lab identification, similarly, "GZ87" exhibited higher resistance (DIC = 0) than "263" (DIC = 80.8). Wide variations were detected among the F2 lines for all the seven traits, of which FW, LR, and RFR showed normal distributions, while DIC, FR/R, NFR, and PCR exhibited skew distributions. Three traits, particularly PCR and NFR, were found to be significantly correlated with DIC ($R = -0.881$ and -0.640, respectively). Therefore, QTLs were subsequently screened for DIC, PCR, and NFR.
2. **SNP detection and genetic map construction:** Among 52,157 SNPs in the *Brassica* 60K SNP array, 21,646 (41.5%) could be aligned to the *B. oleracea* genome. These SNPs distributed unevenly on the nine chromosomes of *B. oleracea*, with the highest number on chromosome C03 (3558) and the fewest on chromosome C05 (1264), being in accordant with the chromosome length of *B. oleracea*. Among the C-genome SNPs, 3218 (14.9%) were polymorphic between the two parents, forming 615 bins. After removing the serious missing bins, the skewed bins ($P \leq 0.01$) and the redundant bins, 565 bins were allocated into nine linkage groups (LGs), with the fewest on C01 (16 bins) and the most number on C03 (130 bins). The lengths of LGs ranged from 17.2 to 187.7 cM, summed in a total length of 1027.98 cM, with an average distance of 1.82 cM between neighboring bins. The bins showed general coincidence between their genetic and physical positions for all LGs except C01 which has a limited number of bins.
3. **QTLs for CR-associated traits:** With significant LOD thresholds of 4.4, 4.5, and 4.4 which were set by 1000-permutation test in IciMapping, 2, 13, and 8 QTLs for DIC, NFR, and PCR, were detected, respectively, individually explaining 6.1–17.7% of the phenotypic variation. Overlaps were detected between two rounds, such as *DIC.I-1* and *DIC.II-1*, and *PCR.I-2* and *PCR.II-5*. Overlaps were further detected among traits, including one three-traits-overlapped QTL on LG C08 and four overlapped QTLs between NFR and PCR on LG C03, C04, and C06. The QTLs on LG C08 (*DIC.I-1, DIC.II-1, NFR.II-8*, and *PCR.II-6*) exhibited 10.2–17.8% of the total phenotypic variation, with an overlapped genetic range of 64.5–67.5 cM which corresponding to a 0.6 Mb interval (32.7–33.3 Mb) on chromosome C08. It is worth mentioning that two overlapped QTL regions on LG C06 between NFR and PCR also exhibited high R^2 values. In details, *NFR.I-5, PCR.I-2*, and *PCR.II-5* overlapped on 86.5–92.5 cM which corresponding to 1.98–3.54 Mb on chromosome C06, and

NFR.II-5 and *PCR.II-4* overlapped on 77.5–83.5 cM, being corresponded to 3.26–3.94 Mb of C06.

4. **Comparative analysis among CR loci:** According to Bolbase2 which provides information on syntenic regions between *B. oleracea* and *B. rapa* (Yu et al. 2013), a QTL interval on C03 (6.28–6.32 Mb, overlapped by *NFR.II-3* and *PCR. II-2*) was partially syntenic to 0.59–6.22 Mb on A02 of *B. rapa* where located the *CRc*- and *Pb(Anju)*2-linked marker "m6R" (from 2,112,653 to 2,113,153 bp) (Sakamoto et al. 2008; Nagaoka et al. 2010; Lee et al. 2016). A QTL on C07 (*NFR.II-7*) was found to be syntenic to 11,077,112–11,414,470 bp on A08 of *B. rapa*, being partially overlapped with a reported CR loci *Crr1* which was fine mapped to 10,692,602–11,617,700 bp on A08 in *B. rapa* (Hasan and Rahman 2016). The three important regions on C08 and C06 identified no accordance with reported CR loci. (Peng et al. 2018).

9.7.12 Transfer of Clubroot Resistance to *Brassica*

1. **Plant materials:** The ECD set was obtained from Warwick HRI's Vegetable Gene bank, United Kingdom. The ECD set includes five accessions of *B. rapa* (ECD1–5), five accessions of *B. napus* (ECD6–10), and five accessions of *B. oleracea* (ECD11–15). These 15 accessions were evaluated to identify clubroot disease-resistant materials against Canadian field isolates. After controlled environment evaluation, clubroot resistance in four turnips, ECD1, ECD2, ECD3, and ECD4 were identified and used as resistant male parents to cross with a susceptible female parent of *B. rapa* accession, yellow sarson (YS). The clubroot-resistant turnips were winter types, with brown seed coat color and self-incompatible. In contrast, the yellow sarson was susceptible to clubroot, with yellow seed, spring growth habit, and self-compatibility characteristics. Four crosses YS × ECD1, YS × ECD2, YS × ECD3, and YS × ECD4 were made in the greenhouse at the University of Manitoba after 6 weeks of vernalization for all four ECDs. Subsequently, five F1 individual plants from each cross were used to make backcrosses to the susceptible parent (YS) to produce a total 20 BC1 families. Since clubroot-resistant turnips were self-incompatible and open-pollinated accessions therefore genetic purity was unknown. All the 20 BC1 families were produced from individual F1 plants separately and further phenotypic evaluation was performed on the family basis in each of these four BC1 populations.
2. **Clubroot disease evaluation and statistical analysis:** Clubroot disease evaluation was performed in greenhouses using Canadian field isolates. The clubroot pathogen was collected from infected canola fields in AB, Canada. One part infected soil was thoroughly mixed with three parts Metro-Mix soil for even distribution of pathogen resting spores. Ten to twelve seeds were planted in 15 cm pots with three replicates per genotype. Resistant and susceptible parental genotypes were seeded as controls in each experiment. Six weeks after inoculation, seedlings were classified into four groups corresponding to four levels of

resistance based on the distribution of galls on roots. A score of zero was given for plants without galls on both primary and secondary roots, one for plants with a few small galls on the secondary roots, two for plants with a few small galls on both primary and secondary roots and three for plants with several big galls on both primary and secondary roots. Groups 0 and 1 were considered clubroot-resistant reactions, and groups 2 and 3 as clubroot susceptible reactions for phenotype evaluation of BC1 families. Segregation of clubroot disease-resistant and susceptible plants in each family of BC1 populations was analyzed using the Chi-square test for goodness-of-fit.

3. **DNA extraction and PCR for polymorphism detection:** Plant genomic DNA was extracted using a modified CTAB method from young leaves of parental genotypes and segregating populations (Li and Quiros 2001). Genomic DNA was used for PCR amplification and polymorphism detection. SRAP PCR amplification was performed as described by Li and Quiros (2001). Standard PCR was carried out for SSR and SCAR amplification with M13 tailed forward primers and M13 labeled primer with four different colors (Gao et al. 2014). SRAP and SSR PCR products were separated with an ABI 3130 x l Genetic Analyzer and five fluorescent dye set, 6-FAM (blue), VIC (green), NET (yellow), PET (red), and LIZ (orange) are used for signal detection. The LIZ orange color fluorescent dye was used as a size standard with the other four colors for PCR product separation.
4. **Analysis of the clubroot resistance gene-specific molecular markers on chromosome A03:** Fine mapping of clubroot resistance gene in Chinese cabbage was previously carried out and molecular markers linked to the gene on chromosome A03 were developed (Gao et al. 2014). For better clarification and consistency with *Brassica* gene nomenclature, previously fine mapped gene in Chinese cabbage was designated here as *BraA.CR.a.* These molecular markers were employed to determine haplotypes of resistance genes on chromosome A03. Four SSR markers, FSASS45c, FSASS79b, FSASS72b, and FSASS45b were used to select clubroot-resistant individual plants with clubroot resistance genes on chromosome A03. These SSR markers were analyzed in all BC1 segregating populations developed from each of ECD clubroot resistance parental genotypes. Data were analyzed by the ABI GeneScan software and visualized by Genographer software to score genotypes.
5. **Mapping of other clubroot resistance genes in ECDs:** Further analysis focused on the BC1 genotypes, which did not show co-segregation with gene-specific molecular markers on chromosome A03, or when the segregation ratio was not 1:1. From the selected BC1 genotypes, eight resistant plants and eight susceptible plants from each line were used for bulk segregant analysis using SRAP molecular markers. Bulk segregant analysis was performed using a total 150 SRAP primer pairs. SRAP PCR was performed using the standard protocol described by Li and Quiros (2001). In bulk segregant analysis, SRAP markers associated with clubroot phenotypes were identified and their PCR products were sequenced. The sequence information of these SRAP markers was compared against the *B. rapa* genome sequence1. Genomic regions between extreme SRAP marker sequences were extracted from the *B. rapa* genome sequence using BEDtools with Linux

commands (Quinlan and Hall 2010). Subsequently, blastx analysis was carried out for the extracted genomic regions on the NCBI database with least Blast Expected Value of 1.10^{-3} using the blast2go software v2.5. Predicted candidate genes were identified within 2 Mb region and new SSR markers were developed to analyze the candidate clubroot resistance loci (Hirani et al. 2018).

9.8 *Brassica-Sclerotinia* Techniques

9.8.1 A Rapid Screening Technique for Resistance

A method to inoculate detached bean leaves with *S. sclerotiorum* without using a carbon-nitrogen source has been described by Leone and Tonneijck (1990). Leaves are incubated in plastic trays to maintain air humidity above 95%. Spores of the pathogen inoculated in water alone are generally unable to cause infection. Additions of KH_2PO_4 (62.5 mM) or of mixtures of inorganic phosphate and glucose (1.1 or 5.5 mM) to the inoculum stimulate fungal infection. Also spore concentration (2×10^5 or 2×10^6 spore's mL^{-1}) influences the ability to produce lesions. The fungus concentration (2×10^6 spores mL^{-1}) is able to infect primary bean leaves when inoculated in an inorganic phosphate solution. Since the procedure is simple and does not require blooming plants, it can be rapidly adopted in germplasm screening and breeding programs.

Wu and Liu (1991) indicated the possibility of *in vitro* selection of rape (*B. napus* L.) callus cultures resistant to oxalic acid. However, Callahan and Rowe (1991) suggested that oxalic acid is not the sole inhibitory factor, yet other unidentified macromolecular components share a code terminate role in the inhibitory effect of lucerne *S. trifoliorum* system.

9.8.2 Germplasm Screening and Evaluation

One technique uses ascospores produced by apothecia under growth room conditions. The ascospores are collected and stored on microfilters, then suspended in water for inoculation. Flowering plants are sprayed with the ascospore suspension and kept in high humidity conditions. Symptoms development is recorded after 2–3 weeks. The second technique involves inoculation of stem of flowering plants with *S. sclerotiorum* growing on sterile medium. Inoculum is affixed to stem of plants using adhesive tape and the length of lesion developed around the inoculum is measured after 2–3 weeks (Thompson and Kondra 1983).

Similar techniques were developed by Pierre et al. (1992) where sclerotia of the pathogen are obtained from diseased plant tissue or produced *in vitro*. Sclerotia are put in pots between two layers of perlite, kept outside. Pans are watered once in a week and covered with black plastic cover. Three months later, sclerotia are sorted, cleaned, and put in pots filled with compost in a green house kept at 15–18 °C. Stipes appear 2–6 weeks later or else. The induction takes place in the field. Sclerotia put in sealed bags of plastic etamine are placed 10 cm deep in the soil. Sclerotia must be entirely surrounded by the soil. Four months later, sclerotia are washed and sorted out and put in small closed crystallizers on a cotton wool bed in the light, at a humidity approaching saturation and at a temperature of approximately 20 °C. Stipes develop 2–3 weeks later. As soon as stipes appear, sclerotia can be used directly in the field, or prepare ascospore inoculation for spray inoculation. Light is essential to obtain apothecia. Pots containing sclerotia with stipes are put in a greenhouse or in a long-day lit room sheltered from direct sunlight. Crystallizers can either be covered with a transparent lid, or be kept in the open air, if the atmosphere has a high relative humidity. In the case of pots, sclerotia and their stipes are covered with small, transparent plastic beakers. Inoculum can be collected in two ways: In the first case, when discs of apothecia are formed, spores are collected over several weeks twice to three times per week. When the beaker lid is removed, a cloud of ascospores is released and gathered on a millipore filter with meshes inferior to 5 μm using rapid aspiration. The principle of the device is simple, a funnel shaped container with a filter at the bottom supported by a thin grid. Aspiration is carried out using a household vacuum cleaner. In the second case, mature apothecia are sampled in the morning and put in a Petri dish. Spores are released by progressive drying. For sclerotial soil inoculation, as soon as stipes appear, two small boxes containing 5–10 sclerotia are put at soil level for each 15 m^2 plot at the beginning of rapeseed flowering. Symptoms of stem rot appear about a month later. It is recommended to cover the trial with a canvas sheet for a month to maintain temperature and humidity.

For ascospores spray inoculation. Ascospores collected on filters or in Petri dishes are put into suspension by scrapping. Most often spores appear in the form of clusters which can be separated by strong manual shaking followed by a few seconds in an ultrasonic vat. The volume of spore suspension is adjusted to obtain 10^4–10^5 viable spores/mL. Ascospores inoculum spraying is usually carried out when rapeseed petals fall, in the first period of flowering. A complementary supply of petals, naturally dried or heated for 1 h at 100 °C can be obtained. Scheibert-Bohm et al. (1981) used several inoculation methods, viz., 1—infected kernels of oats and rice, 2—mycelium suspension and/or cotton, 3—mycelium suspension, cotton and parafilm, 4—tooth prick and or tooth pick and cotton, and 5—infected pieces of agar. Best results are obtained with infected oat kernel and infected agar pieces under long day and low temperature conditions. In infected oat kernels, a mixture of 1:1 kernels and vermiculate is used as inoculum. It is spread around the plants or over the planted seeds. 50 g of infected oat kernels and 50 g of vermiculate is used per plastic dish. The infected agar pieces of 5 mm size are pressed to the hypocotyls of the test plant at three-leaf stage and wrapped with parafilm.

In case of cauliflower, the procedure involves inoculating fresh curd with a precolonized cruciferous petal in a drop of water and keeping it in a humid chamber at 25 ± 1 °C to serve as an inoculum source. To screen a crop in the field, 1 × 1 mm pieces of the colonized curd is then placed in the centers of the curds of the crop. Humidity is maintained by irrigation before and after inoculation and by mist spraying immediately after inoculation (Kapoor et al. 1986).

9.8.3 A Green House Spray and Drop-Mycelium Inoculation Method

In a spray-mycelium inoculation method, *S. sclerotiorum* mycelia are cultured in liquid potato dextrose broth and homogenized before spraying on the soybean leaves. In "drop-mycelium" methods, a drop of homogenized mycelium suspension is dropped on the tips of main stems. Inoculated plants are incubated in a greenhouse chamber with 60–80% relative humidity. Plant mortality and Area Under Wilt Progress Curve (AUWPC) are used to measure disease severity daily from 3 to 14 days after inoculation (DAI). The spray-mycelium method and the drop-mycelium method are compared with the cut-petiole method in the greenhouse. Significant differences in disease ratings of plant mortality and AUWPC to *Sclerotinia* stem rot are found among tested genotypes. The results obtained with the spray-mycelium and drop-mycelium inoculation methods are significantly ($R >$ 0.73, $P < 0.01$) correlated with the results obtained with the cut-petiole inoculation method for both of the plant mortality and AUWPC. Compared with the cut-petiole method, the spray-mycelium and the drop-mycelium methods uses less inoculation time and are less expensive in terms of materials. Both of these new methods are low cost, efficient, and reliable and can be valuable for large-scale evaluation of germplasm and breeding lines for resistance to *Sclerotinia* stem rot in a greenhouse or other similar facilities (Chen and Wang 2005).

9.8.4 Detection of *Sclerotinia* by ELISA

The ELISA technique used by Walcz et al. (1985) for detection of *Sclerotinia* in sunflower plants showed good specificity and sensitivity in detecting "*Sclerotinia* antigen" at a concentration as low as 10 ng mL^{-1}. The pathogen is incubated on PDA (pH 5.7) for 1 week at 25 °C and the mycelia are scraped with a scalpel from the surface. A modified Lilly-Barnett liquid medium (pH 5.7) is also used and the mycelia are collected by filtration. Growth of the fungus is scraped from three Petri dish cultures with 15 mL phosphate buffer saline (PBS) (pH 7.2) containing 0.5% formalin. The suspension is sonicated at 0 °C for 5 min. Rabbits are immunized with this suspension into the marginal ear vein. Doses of 0.2, 0.5, 1.0, and 2.0 mL are given at 4 days intervals. The 2.0 mL dose is then repeated twice. Four days later the ELISA titer of the serum is tested and the rabbit is bled when the titer reaches 1:32,000 or higher. In the case of lower titers, the 2.0 mL dose is repeated until satisfactory titer is reached. Sera are stored at −20 °C without preservative. Anti-*Sclerotinia* conjugate is prepared from an antiserum which gives an absorbance (A) value of 0.30 at dilution of 1:32,000. The anti-*Sclerotinia* rabbit serum is purified for 1 g G according to the method of Clark and Adams (1977). This purified 1 g G is coupled with the enzyme, horse-radish peroxidase as described by Avrameas (1969).

In ELISA procedure, polystyrene micro-test plates are coated with 100 μL per well of anti-*Sclerotinia* 1 g G diluted 1:5000 in coating buffer (0.79 g Na_2CO_3, 500 mL distilled water, pH 9.6). After overnight incubation at 4 °C, the wells are emptied and washed 3 × with PHS containing 0.05% Tween-20 (PBS-T). Then 100 μL plant extract (0.1 g of the plant is ground in 1 mL extracting buffer, (PBS-T with 2% polyvinyl-pyrrolidone), PVP is added to the wells. Each sample is replicated twice. After incubating for 1 h at 30 °C, the wells are washed again and 100 μL of the conjugate diluted 1:10,000 in PBS-T with 0.5% bovine serum albumin (Fluka) is added. Again the plates are incubated for 1 h and washed and then 150 μL of substrate solution (30 mL citric acid buffer, pH 5.6), 10 mg *o*-phenylandiamine-dihydrochloride, OPD and 5 μL H_2O_2 are added to the wells. The reaction is stopped after 5 min by adding 50 μL 4N H_2SO_4. Absorbances (A) values are recorded by a ELISA Reader at 492 nm on every plate positive (*Sclerotium homogenate*) and negative (extract of healthy plant) samples are replicated six times. 0.1 g of sclerotium is homogenized in 1 mL of extracting buffer. After 1 h of sedimentation, the supernatant is collected and used for the assay. When a sample gives an A value 3 × higher than the mean of the negative ones, it is considered to be positive. In practice, A values of negative samples remain below 0.20.The reactivity of rabbit antisera is tested by a modified ELISA in the following sequence: (1) wells coated with the antigen (*Sclerotium* extract); (2) serial dilutions of antisera (from rabbit immunized with *Sclerotinia*); (3) conjugate (anti-rabbit-l g G labeled with horse-radish peroxidase); and (4) substrate.

Seeds of oilseed rape are sown in pots and inoculated through wounds on the stems at 6–8 leaf stage of development with agar discs 0.5 mm diameter originating from *S. sclerotiorum* cultures maintained on PDA for a week. The sites of inoculation are covered with sterile wet cotton and aluminum foil. The infected plants are kept under alternating conditions of temperature and light (14–25 °C, dark-daylight, respectively), in a greenhouse and sprayed with water, 8–12 plants of each species are inoculated and un-inoculated ones are used as control. Stems and leaf pieces with early symptoms of disease and without symptoms are collected. Single pieces are put into nylon bags and coded. In the laboratory, the samples are divided aseptically into two parts, one for culturing on PDA and the other for ELISA test. ELISA differentiates well between different fungi causing simultaneous infection with similar symptoms. PCR-based assays for detecting inoculum of *Sclerotinia* spp. have been described by Freeman et al. (2002).

9.8.5 Medium for Production of Oxalic Acid

Three media used for production of oxalic acid were evaluated for their efficiency by Pierson and Rhodes (1992). The liquid media evaluated were:

(a) Potato dextrose broth
(b) PDB with addition of 56 mM sodium succinate (PDB+S)
(c) Maxwell and Lumsden medium (M&L). It is prepared using the inorganic salts and yeast extract medium described by Maxwell and Lumsden (1970) with 23.7 mM D-glucose and 56 mM sodium succinate. The pH of the medium is not adjusted before autoclaving.

Media (125 mL/250 mL Erlenmeyer flask) are autoclaved for 15 min at 121 °C. A single ascospore isolate of *S. trifoliorum* is collected from red clover (*Trifolium pratense* L) or alfalfa plot. Subcultures of this isolate are grown on potato dextrose agar acidified to pH 4 with 1 mL of 85% lactic acid per liter of medium (APDA). Plugs of agar plus mycelium or agar only 8 mm in diam. are cut from 3-days-old cultures or from un-inoculated APDA plates which serve as controls. Three mycelial or three control plugs are transferred to each flask of liquid medium. Cultures are incubated at 15 °C on a rotary shaker at 120 rpm with a 12 light/dark cycle provided by cool-white fluorescent lights (3000–4000 lux). Cultures are sampled 3, 5, 7, 10, 12, and 14 days after inoculation by removing 2 mL of medium per flask with a sterile Pasteur pipet. Samples are stored at −5 °C prior to oxalic acid determination. After 14 days of incubation, the final pH of the medium is measured.

Oxalic acid concentrations are measured calorimetrically, oxalate, zirconium, and quercetin solutions are prepared as described by Allan et al. (1986). An oxalic acid standard solution (1 mg/mL) is prepared after drying oxalic acid at 120 °C for 2 h. To prepare a standard curve 0.8 mL zirconium solution, 0.8 mL quercetin solution, 0.5 mL concentrated HCI, 0.1 mL of un-inoculated (control) culture medium and 0.1 mL, 20. 30, 40, or 50 μL oxalic acid standard solution are added to a series of 10 mL volumetric flasks. Flasks are filled with double distilled water and allowed to incubate at 21 °C for a minimum of 10 min. Absorbance of the solution at 430 nm is measured using a Spectronic-20. Linear regression is used to obtain standard curves. New standard curves are established each day and whenever new reagents are made. To determine concentrations of oxalic acid in inoculated culture media, 0.1 mL of inoculated medium is substituted for control medium and oxalic acid. Inoculated medium is diluted with control medium if necessary. Oxalic acid determinations are made on a minimum of three aliquots per flask per sampling date. After 14 days of incubation, *S. trifoliorum* in PDB+S produces approximately 9–12 times more oxalic acid than in M&L or PDB, respectively.

9.8.6 Identification and Pyramiding of R-QTLs in *Brassica* to *Sclerotinia*

1. **Plant materials and breeding process:** A schematic diagram of the breeding process is presented in Fig. 9.12. Interspecific hexaploid hybrid (AnAnCnCnCoCo) was generated between a Chinese semi-winter rapeseed variety "Zhongshuang 9" (female parent) which possesses partial resistance to *S. sclerotiorum* and a wild *B. oleracea* genotype "C01" (*B. incana*, male parent) which exhibited high resistance to *S. sclerotiorum* (Mei et al. 2015). The hexaploid progeny was backcrossed with Zhongshuang 9 for one round and then selfed until BC1F8. The selection in BC1F1 and BC1F2 was reported previously (Mei et al. 2015), and two elite BC1F2 individual plants, i.e., BC1F2-1 and BC1F2-3, were selected to develop the following generations.
2. **Marker-assisted selection:** According to the resistant QTLs identified from the wild *B. oleracea* "C01" (Mei et al. 2013), 28 simple sequence repeat (SSR)

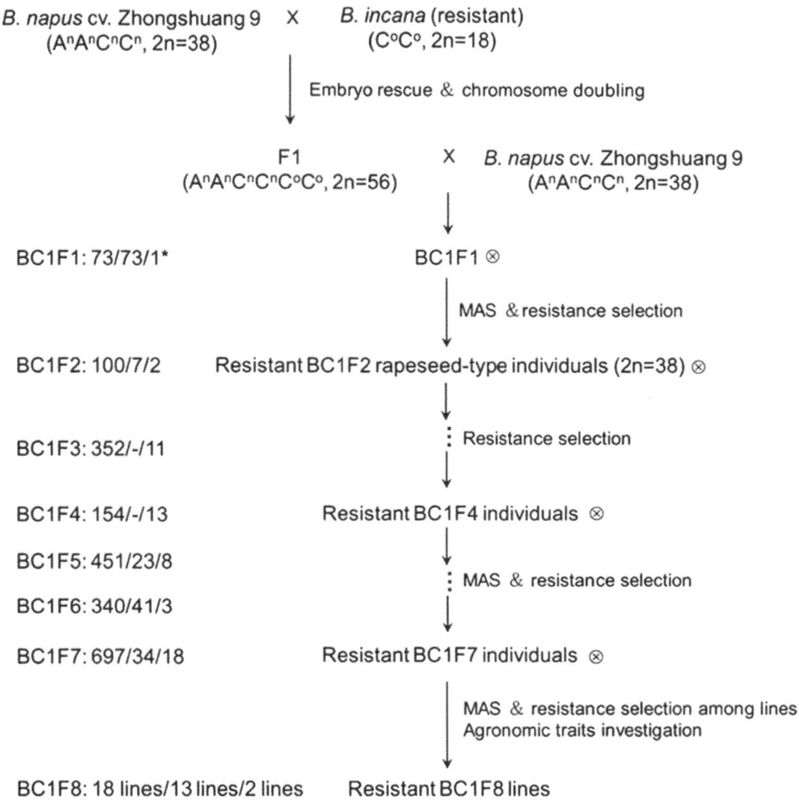

Fig. 9.12 A diagram of breeding process of *Sclerotinia*-resistant *Brassica napus* lines by using *Brassica incana* as the donor of resistance. Asterisk represents the number of plants for selection/the number of plants selected by MAS/the number of plants planted for the next generation (Mei et al. 2020)

markers located within and adjacent to all the QTL intervals were screened for polymorphism between wild *B. oleracea* and Zhongshuang 9. Seven polymorphic markers, which linked to one major QTL for leaf resistance on linkage group (LG) C01 and two major QTLs on LG C09 for both stem and leaf resistance, were finally used to trace the three resistance loci. For one locus, plants exhibited the same genotypes (bands) as the wild *B. oleracea* "C01" in all the markers linked to the QTL was considered to carry this resistant locus.

3. **Resistance evaluation:** The detached inoculation technique was conducted under controlled environment according to Mei et al. (2013) by using the same *S. sclerotiorum* isolate at ~80 days after sowing (DAS) (around nine-leaf stage) for leaf resistance and ~205 DAS (around 2-week post end flowering) for stem resistance. The infection temperature was maintained at 22 ± 1 °C. Lesion size/length on detached leaves/stems was recorded at 2/3 days after inoculation. In

order to evaluate the resistance level of offspring, the susceptibility (S) relative to the rapeseed parental line Zhongshuang 9 was calculated with the equation $S = V/V_{\text{Zhongshuang 9}}$, where V is the lesion size (leaf) or lesion length (stem) in the progenies, and $V_{\text{Zhongshuang 9}}$ is that of Zhongshuang 9.
Natural field resistance was investigated in BC1F8 generation. Lines were arranged in a randomized complete block design with three replications. Plot size was 2 by 2.4 m, with 60 plants in six rows. Plots were naturally infested with *S. sclerotiorum,* and disease incidence (percentage of infected plants in total plants) was recorded at 210 DAS (3-week post end flowering). Disease index (I) was calculated by the formula

$$I = \frac{\Sigma\,(di \times li)}{L \times 4} \times 100$$

where di, li, and L represent the disease grade (classified on a scale of 0–3) of each plant, the number of plants in each grade and the total number of plants, respectively (China 2011). Disease severity was classified on a scale of 1–5 based on disease incidence and disease index according to the agricultural trade standard on investigation of rape Sclerotiniose (China 2011).

4. **Seed yield and seed quality:** Seed setting (seeds per pod) was investigated in five plants per line, with 30 continuous siliques sampled from the main inflorescence of each plant. Seed yield per single plant was measured from ten open-pollinated plants in each plot. The 1000-seeds weight was measured from five open-pollinated plants. Seed quality was determined by near-infrared reflectance spectroscopy (NIRS) from 5 to 10 self-pollinated plants in each line, with three technical replicates.
5. **Data analysis:** Statistical analyses such as analysis of variance (ANOVA), *F,* and *t* tests were performed using the SAS software, version 6.07 (SAS Institute 1992) (Mei et al. 2020).

9.8.7 Identification of Genome-Wide Associated R-Loci in *Brassica* to *Sclerotinia*

1. **Plant materials:** The association panel comprised 448 rapeseed inbred lines, including 420 accessions from Asia, 18 accessions from Europe, 5 accessions from North America, and 5 accessions from Australia (Liu et al. 2016). Population structure and genetic relatedness analyses showed that most of the accessions included were distantly related (Liu et al. 2016).
2. **Field experiments:** All 448 accessions were grown in two consecutive years. All field experiments followed a randomized complete block design with two replications. Each accession was grown in a plot with 10–12 plants in each of two rows, with a distance of 21 cm between plants within each row and 30 cm between rows. The field management regime followed essentially regular breeding practices.

3. **Assessment of resistance to *S. sclerotiorum*:** The *S. sclerotiorum* isolate SS-1 was maintained and cultured on potato dextrose agar (PDA, 25% potato, 2.5% dextrose, and 1.5% agar, pH 5.8; Wu et al. 2013). The isolate was maintained at 4 °C in darkness and cultured twice prior to inoculation at 23 °C in darkness. Mycelial agar plugs (7 mm in diameter) punched from the margin of a 2-day-old culture of *S. sclerotiorum* grown on PDA were used as the inoculum. Sclerotinia stem rot resistance was assessed using the detached stem inoculation assay according to Mei et al. (2012) with some modifications. The stems of 3–6 individuals of each accession in each replicate with a length of approximately 30 cm were cut using a sharp knife at a height of 20 cm above the ground, and the cut ends were immediately wrapped with polyethylene to retain freshness. Subsequently, the detached stems were placed in plastic trays (56 × 38 × 15 cm) with wet gauze at the bottom of the trays. Each 30-cm stem was inoculated with a mycelial agar plug at two points at an interval of 10–13 cm. The inoculated stems were sprayed with a fine mist of water, and the plastic tray was covered with plastic film to maintain a high level of relative humidity. The plastic trays with inoculated stems were maintained at 22 ± 2 °C in darkness. The length of lesions along the stems was measured after 5 days post-inoculation (dpi). This assessment was conducted with two replications at the stage of flowering termination.
4. **Statistical Analysis** Broad-sense heritability was calculated as $h^2 = \alpha^2{}_g/(\alpha^2{}_g + \alpha^2{}_{ge}/n + \alpha^2{}_e/nr)$, where $s^2{}_g$ is the genetic variance; $s^2{}_{ge}$ is the interaction variance of the genotype with the environment; $s^2{}_e$ is the error variance; n is the number of environments; and r is the number of replications. The estimates of $s^2{}_g$, $s^2{}_{ge}$, and $s^2{}_e$ were obtained through ANOVA (analysis of variance) using SAS9.3. Best line a run biased predictions (BLUPs) for each line across the two environments were calculated using a MIXED procedure in SAS 9.3 as phenotypes for the association analysis.
5. **Genome-wide association analysis** The association panel was genotyped using *Brassica* 60K Illumina Infinium R HD Assay SNP arrays (Clarke et al. 2016) according to the manufacturer's protocol. Quality control and mapping of the SNPs to *B. napus* "Darmor-bzh" reference genome (Chalhoub et al. 2014) were performed and 25,573 SNPs retained for further analysis as described earlier (Liu et al. 2016). Principal component analysis (PCA) based on all the 25,573 SNPs was performed using the GCTA tool (Yang et al. 2011). The relative kinship matrix of 448 *B. napus* lines was generated using the SPAGeDi software package (Hardy and Vekemans 2002). Negative values between two accessions were set to 0. Trait-SNP association was performed using the PCACK model implemented in TASSEL 4.0 (Bradbury et al. 2007). The significance of the associations between SNPs and the trait was based on a single threshold of $P < 3.91 \times 10^{-5}$ ($P = 1/n$, where $n =$ the number of markers used; $-\log_{10}(1/25{,}573) = 4.4$). The linkage disequilibrium (LD) measurement parameter r^2 was employed to estimate the LD between SNPs using the software package TASSEL 4.0 (Bradbury et al. 2007). Only homozygous SNPs were used, and heterozygous SNPs were set to missing. Marker haplo types and LD blocks were defined for each associated locus when flanking markers showed strong LD ($r^2 > 0.4$) with a

lead SNP, as described by Hatzig et al. (2015). The interval of the LD block was considered for candidate gene disclosure.

6. **Alignment of SSR resistance QTLs to the *B. napus* genome** Quantitative trait locus for SSR resistance were collected from previous bi-parental QTL mapping studies (Zhao et al. 2006; Wu et al. 2013; Wei et al. 2014) and a GWAS (Wei et al. 2015). All QTLs were aligned to the *B. napus* "Darmor-bzh" reference genome (Chalhoub et al. 2014) based on the physical positions of markers in confidence intervals or LD blocks. The physical localization of the markers of RFLP (restricted fragment length polymorphism), SSR (single sequence repeat), and SNP (single nucleotide polymorphism) was determined using electronic PCR [e-PCR, (Schuler 1997)] or Blast N (Altschul et al. 1997) employing the marker primer or probe sequences. For SSR markers, e-PCR was performed with the primer sequences, using the genome sequences as templates (Cai et al. 2012). The parameters for e-PCR were set to allow three mismatches and one gap for a given primer pair. If a pair of primers had multiple amplicons, subsequently the amplicon for the primers were selected according the linkage group of the QTL (Cai et al. 2012). For RFLP markers, the probe sequences were used to perform BlastN1 http://www.illumina.com/searches against the reference genome sequences with an *E*-value of 1E − 20. If the probe aligned to multiple loci, subsequently hits were selected according to the linkage group of the QTL from the top several blast hits. The physical localization of the SNP markers used in the GWAS was performed (Wei et al. 2015; Liu et al. 2016; Wu et al. 2016).

9.8.8 Introgression of R-Genes from Wild Relatives in *Brassica* to *Sclerotinia*

1. **Plant materials and resistance screening.** Genetic materials and stem inoculation procedures were same as described earlier. Eighty eight ILs were raised in a randomized block design with two replications over three crop seasons against an aggressive and virulent strain (PAU-4) of *S. sclerotiorum*. Analysis of variance (ANOVA) was undertaken and standard deviation and standard errors were estimated through Multivariate General Linear Model (M-GLM) using software, SPSS17.0.
2. **DNA extraction for library preparation.** Genomic DNA of ILs and their parents were isolated from young leaf tissue using a CTAB (cetyltrimethyl ammonium bromide) extraction method, with the chloroform-isoamyl alcohol purification step repeated twice to assure good quality DNA. RNase was added between the two chloroforms: isoamyl alcohol solvent extractions by incubating the samples at 37 °C for 15 min to allow for a single DNA precipitation step at the end of the protocol. Quality of the extracted DNA was assessed using a NanoDrop^{R} 2000 spectrophotometer and 1% (w/v) agarose gel. Sample DNA having 260/280 absorbance ratio of 1.8–2.0, and with no evidence of substantial band shearing or contamination (either RNA or polysaccharide), was used for library construction.

3. **Resequencing and genotyping by sequencing.** Direct resequencing (21 xs) of four ILs and genotyping by sequencing of 88 ILs was outsourced. For this, high quality DNA of each sample was digested with appropriate combination of restriction enzymes based on in silico evaluation. This was followed by several rounds of PCR amplification. All samples were then individually pooled and then size-selected for the required fragments to complete the library construction. High quality libraries with appropriate insert sizes were subsequently used for pair-end sequencing on Illumina^R HiSeq platform, with the read length of 150 bp at each end. The sequences and corresponding sequencing quality information were stored in a FASTQ file format. The adapter sequences were removed from raw reads using the software Cutadapt. The available reference genome of *B. juncea* v1.5 was used for reference-based alignments of whole-genome sequences (21 xs) from the four most prominent ILs, using software bowtie2. Initially, one introgression line was aligned into reference genome and SNP called by NGSEP-GBS pipeline. Total SNPs with base quality more 30 were replaced in background genome reference (*B. juncea* v1.5) using a perl script, pseudomaker.pl implemented in SEG-Map to construct the first step of mock-up pseudomolecule (s), which were then used as a reference for subsequent construction of pseudomolecules of sequential replacement of SNPs for remaining three ILs in the same way as described above. Resultant mock-up reference was then used for aligning sequence tags from 88 ILs. SNPs identification was then carried out by using NGSEP-GBS pipeline. Resultant marker dataset comprised 31,896,811 SNPs. These were then filtered to include only quality SNPs for further analysis. Filtering parameters were: minimum mapping quality, minor allele frequency (0.08), bi-allelic SNVs, minimum number of sample genotyped, maximum observed heterozygosity (30%), and maximum missing calls (30%). After filtering, 88,624 SNPs remained. These were imputed using fcGENE and Beagle softwares.
4. **Genome-wide association analysis (GWAS) based on SNP genotyping.** *Sclerotinia* resistance lesion length (cm) data were first normalized through rank and log transformation. GWAS was carried out using transformed values of lesion lengths and 88,624 SNPs. Kinship matrix and Covariates data was generated through "MVP. Data" function of the software R (MVP-package). Discriminant Analysis of Principal Components (DAPC) was implemented in R software package "adegenet." PC and DAPC values were then used as covariates in different GWAS analysis algorithms to reduce false positives by accounting for the effects of population stratification. Manhattan plots were generated with multi-model plotting using MVP tools. GLM, MLM, and Farm CPU methods were implemented in the software R (MVP-package) (https://github.com/XiaoleiLiuBio/MVP), using default settings to identify marker trait associations. Identified markers were subsequently confirmed using alternate algorithms as implemented in TASSEL5.0 with DAPC (in adegenet) as covariate for GWAS with model; GLM + DAPC [GLM (T)], and MLM + kinship [MLM (T)]. Visualization of the significant QTLs and SNPs was done using Manhattan plots, generated using the R-package qqman75. Unadjusted *p*-values <0.001

were used to declare significance as the Bonferroni and FDR tests which are considered far too conservative for association studies involving thousands of candidate variants at very low minor allele frequencies. Bonferroni correction also assumes independence of every association test, an assumption which is rarely met due to linkage disequilibrium between SNP markers. The 50 kb flanking regions on each side of resistance was screened associated peak SNPs to predict candidate genes using Blast2GO v5.2.5 tool (Gotz et al. 2008). *Arabidopsis* protein database was probed for gene finding as well as blast search. The gene ontologies of *Arabidopsis* orthologs were used for this analysis as these are better curated than those of *B. juncea*. Positions of the predicted candidate genes w.r.t. the SNPs were projected through blast search of their sequences against *B. juncea* mock-up pseudomolecules. Predicted candidate genes were then verified for their function from literature for their relevance to the target trait (Atri et al. 2019).

9.8.9 Marker-Assisted Introgression of Resistance into *Brassica juncea* Against *Sclerotinia*

1. **Plant Materials:** A total of 96 *B. juncea–E. cardaminoides* ILs were developed previously through wide hybridization between *B. juncea* and *E. cardaminoides* as described.
2. **Chromosome preparation and in situ hybridization:** The method for preparation of chromosome spreads was used as described by Rana et al. (2017). To prepare species-specific genome in situ hybridization (GISH) was probes, purified DNA of *B. cardaminoides,* and *B. nigra,* and *B. juncea* were extracted using the DNeasy plant mini kit (Qiagen) as per manufacturer's specifications. Genomic DNA of *E. cardaminoides* was labeled with rhodamine-5-dUTP dye (red color). DNA of *B. nigra* was labeled with fluorescein-12-dUTP dye (green) using a nick translation kit. Two-step hybridization to perform GISH as described earlier (Rana et al. 2017). Visualization was carried out with Zeiss fluorescent microscope. Digital images were captured using IsisR software. Images were cropped and optimized with Image J using only functions affecting the whole image.
3. **Screening of ILs for resistance responses to *S. sclerotiorum* inoculation** For field evaluations, ILs were planted in a randomized complete block design, with 20 plants in paired rows of 2 m length with inter-row spacing 30 cm. The phenotypic screening was carried out in two replications. Foggers (with one fogger per 9 m^2) were used to maintain high humidity for disease development, operating 2–3 times a day for 15 min each. Phenotyping of ILs for their resistance responses was carried out. Stem inoculations were carried out by using isolate PAU-4, an aggressive and virulent strain of *S. sclerotiorum* collected from infested *Brassica* fields. PAU-4 is a known highly virulent isolate on *B. juncea* (Rana et al. 2017). Sclerotium was first surface-sterilized, cultured, and subcultured using standard procedures (Clarkson et al. 2003; Li et al. 2006).

The parents, viz., the recipient (*B. juncea* cv. RLC-1) as well as the donor (*E. cardaminoides*), were used as susceptible and resistant checks, respectively. The stem inoculation method (Buchwaldt et al. 2005; Bradley et al. 2006) was used for disease inoculation. The ILs was evaluated for disease incidence and development at 4–5 weeks post-inoculation. Ten plants per genotype per replication were scored for disease incidence by measuring stem lesion lengths according to the method described by Li et al. (2006, 2007b). The resistance responses of ILs were categorized into five classes, based on stem lesion length, viz., highly resistant (HR); (0 < 2.5 cm), resistant (R); (2.5 < 5.0 cm), moderately resistant (MR); (5.0 < 7.5 cm), susceptible (S); (7.5 < 10.0 cm), and highly susceptible (HS); (>10.0 cm) as categorized earlier by Garg et al. (2010). Analysis of variance (ANOVA) was undertaken and standard deviation and standard errors estimated using SAS software.

4. **DNA extraction:** A total of 96 ILs comprising all five disease expression categories (HR, R, MR, S, and HS) were used for molecular profiling. DNA was harvested from young leaves with minor modifications of CTAB (Cetyltrimethylammonium Bromide) standard procedure (Doyle and Doyle 1990). Isolated DNA was further digested with RNase at 37 °C for 1 h. DNA was quantified by NanoDropR 2000 spectrophotometer and for purity by assessing the OD260/OD280 ratio and Qubit^R 2.0 flurometer quantitation to measure DNA concentration. DNA samples with OD260/OD280 ratio of 1.8–2.0 and total amount >1.5 μg were qualified for library construction.
5. **SSR genotyping:** A total of 96 ILs, along with resistant and susceptible parents, were first genotyped by using 100 polymorphic SSR markers. These markers were identified as transferable based on studies involving over 650 A- and B-genome-specific markers (Kim et al. 2009; Lowe et al. 2004). Sequences of 48 B-genome-specific SSRs were obtained under MTA from Isobel Parkin. PCR reactions were performed in a 384-well Applied Biosystems thermocycler. Amplified DNA product was fractioned using an automated high-throughput electrophoresis system. Allelic polymorphism of all markers was recorded and mapping positions were inferred from published data (Lowe et al. 2002, 2004; Kim et al. 2009).
6. **Genotyping by sequencing:** A smaller set of 84 ILs was genotyped by sequencing (GBS) (Elshire et al. 2011). Genomic services were outsourced. For this, high quality DNA of each sample was digested with appropriate combination of restriction enzymes based on in silico evaluation. This step was followed by several rounds of PCR amplification. Samples were then individually pooled and size-selected for the required fragments to complete the library construction. High quality libraries with appropriate insert sizes were then used for pair-end sequencing on Illumina^R HiSeq platform, with the read length of 150 bp at each end. The sequences and corresponding sequencing quality information were stored in a FASTQ file. The adapter sequences were removed from raw reads using the software Cutadapt (Martin 2011). The available genome sequence of *B. juncea* v1.51 (Yang et al. 2016) was used for reference-based alignments of whole-genome sequences (25×) from the four

most prominent ILs (>25×), using software bowtie2 (Langmead and Salzberg 2012). Initially, one introgression line was aligned into the reference genome and SNP called using the NGSEP-GBS pipeline (Duitama et al. 2014). Total SNPs were replaced in genome reference using a perl script, pseudomaker.pl implemented in SEG-Map (Zhao et al. 2010) to construct the first step mock-up pseudomolecules, which were then used as a reference for next ILs. This process was repeated four times to construct final mock-up reference for alignment of sequence tags. Identification of SNPs was carried out by using NGSEP-GBS pipeline (Duitama et al. 2014) after aligning the paired end reads of 84 introgressed lines on final mock-up reference genome. The resulting marker dataset comprised 30,863,034 SNPs. These were then filtered to include only quality SNPs for further analysis. Filtering parameters were: minimum mapping quality, minor allele frequency (0.1), only bi-allelic SNPs, minimum number of samples genotyped, maximum observed heterozygosity, and maximum missing calls (30%) were used for finding putative SNPs. After filtering, 78,578 SNPs were identified and imputed using fcgene and Beagle (Browning and Browning 2016) software.

7. **Association mapping based on SSR genotyping:** The normalization of phenotypic data was done by using PBTools software2 in R-Package version 1.5 (R Core Team 2013). The marker trait associations (MTAs) were identified by using two models executed in TASSEL version 2.1 (Bradbury et al. 2007). GLM (generalized linear model) and MLM (mixed linear model) models were used. Bayesian model-based software, STRUCTURE version 2.2 (Pritchard et al. 2000), was used to determine the population structure by using multi-locus (SSR) genotypic data. Resultant Q-matrix was used as a covariate during association mapping analysis to reduce the bias from population structure. Association mapping was implemented in TASSEL software version 2.1 (Bradbury et al. 2007), measuring the non-random association between marker alleles from different loci (Zhu et al. 2008). Squared correlation coefficients between marker trait data (R2 values) and associated probabilities were calculated and converted into $-\log_{10}(P)$ values. The associated values are calculated with a false discovery rate (FDR) of 0.09 (Storey and Tibshirani 2003) to reduce the false marker trait associations. LD, also known as gametic phase disequilibrium, was established between markers (Flint-Garcia et al. 2003). Annotation, or gene prediction study, of significant markers was carried out using MEGANTE software (Numa and Itoh 2014).
8. **Genome-Wide Association Analysis (GWAS) based on SNP genotyping:** For genome-wide association analysis (GWAS), resistance responses (in terms of lesion length) were first converted into rank data and then transformed to log (x) for three crop seasons. These were also pooled over seasons. A principal component analysis (PCA) was performed across the introgressed lines to identify population stratification by MVP-GWAS tool. The imputed dataset of 78,578 SNP markers were used to calculate the PCs. First three components showed maximum variance. MVP tool5 was used for marker trait association with two different models, MLM and Farm CPU. GWAS were performed using

MLM association accounting for kinship, and GLM and Farm CPU were selected with PCs as covariate in MVP tool. R software package "adegenet" 6 was used for applying discriminant analysis of principal components (DAPC) in the association analysis. After DAPC correction, first three discriminant function was used as covariate in GLM model and kinship for MLM model for association analysis by software TASSELv5.2 (Bradbury et al. 2007). Manhattan plots were generated with multi-model plotting using MVP tools.

9. **SNP Validation:** Six peak SNPs associated with trait variation were identified for validation. Primers were designed from flanking region of SNP using Primer 3 software and their thermodynamic properties were confirmed by Vector NTI. Eight ILs were selected that differed for resistance responses. Genomic DNA from these test genotypes were amplified using designed primers. The PCR products were then purified and used for Sanger sequencing.
10. **In silico prediction of candidate genes:** The 25 kb flanking regions on each side of resistance associated peak SNP found to predict candidate genes using *B. juncea* pseudomolecules as a reference. The predicted genes and their orthologous sequences were then annotated by BLAST run against the *A. thaliana* database using Blast2GO v5.2.5 tool (Gotz et al. 2008). *Arabidopsis* protein database was used for gene finding as well as blast search. Protein IDs generated allowed annotation against all flowering plant databases (NCBI). These were further enriched by the biological functions inferred from the putative *Arabidopsis* orthologs. The gene ontologies of *Arabidopsis* orthologs was used for all analysis because they are far better curated than for *B. juncea*. Positions of the predicted candidate genes w.r.t. the SNPs were detected by blast searching sequences from the predicted genes against *B. juncea* mock-up pseudomolecules. Functions of the predicted candidate genes were verified from literature to determine their relevance for the trait in question (Rana et al. 2019).

9.8.10 Germplasm Screening Against *Sclerotinia* Under Controlled Conditions

1. **Leaf disc inoculation assay (LDIA):** Young, fully expanded leaves aged 33, 35, 50, 54, and 105 days were detached from plants grown in the greenhouse and transferred to the laboratory. Using a 20-mm diameter cork borer and a cutting board, leaf discs were punched out and laid on moist filter paper in 55-mm diameter Petri plates (one disc per plate). Using a sharp scalpel blade, six small injuries were made in the middle of half the discs of each cultivar. Mycelial plugs (4 mm in diameter) taken from the growing edge of 3-day-old cultures on ¼ strength potato dextrose agar were used for inoculation. Inoculum plugs were placed with the mycelium side down in the middle of each leaf disc. Control leaf discs were treated with uninfected ¼ PDA plugs. The Petri dishes were then immediately sealed with two layers of Parafilm to maintain high humidity and incubated at 25 °C in darkness. A completely randomized experimental design

with three replications was used. Leaf lesions were measured 24 h after inoculation. Lesion size was used as the measure of disease. Four treatments (including wounded with/without inoculum and non-wounded with/without inoculum) with three replications for every cultivar/species were used.

2. **Reaction of intact plants to the pathogen:** The bioassay was based on the inoculation of leaves on intact plants. For this purpose, a fully expanded leaf of normal color and shape was selected from each plant and marked for inoculation. The treatments were wounded and non-wounded intact leaves with inoculum, and wounded leaves with ¼ PDA plugs as a control. The inoculation site for each selected leaf was the first one-third from the leaf tip. Wounded and non-wounded leaves were inoculated with 4-mm diameter agar plugs containing hyphae of *S. sclerotiorum* which had been transferred from the margins of 3-day-old cultures. The mycelium-containing surface of the plugs was laid on the adaxial side of leaves. The purpose of wounding was to provide an entry point for the pathogen.
3. **Screening with oxalic acid:** Oxalic acid treatments were applied by attaching a small hydrophilic cotton wool ball soaked in 10 mM (pH = 4) solution of the acid to the leaf surface using a small stripe of adhesive tape. Plugs of ¼ PDA without the fungal mycelium were used on the control plants. All pots were immediately covered with transparent plastic bags to keep the humidity high. The experiment was established in the evening to provide dark conditions in the greenhouse. A completely randomized experimental design with three replications was used. Infection progress was measured as the diameter of necrotic areas at the inoculation points after 36 h and 60 h of incubation. In another experiment, 105-day-old plants were inoculated without the oxalic acid treatment. A completely randomized experimental design with eight replications was used. To study the regular incubation and measurement intervals, lesion diameter was measured after 24, 48, and 72 h. Data were analyzed using SPSS software. Newman-Keuls test was used for mean comparison to classify the reaction of genotypes (Rahmanpour et al. 2011).

9.9 *Brassica-Xanthomonas* Techniques

9.9.1 Plant Culture and Pathogen Inoculation

Surface-sterilized seeds of six oilseed rape (*B. napus*) cultivars were grown in pots (0.6 l). When the seedlings had grown up to the four-leaf stage, they were divided into two groups, i.e., the control and the one with pathogen inoculation. The pathogenic bacteria (*X. campestris* pv. *campestris*, Xcc) strain (KACC No-10377) was obtained from the Korean Agricultural Culture Collection. Bacterium inoculum was cultured in Yeast Dextrose Calcium Carbonate (YDC) agar plate for 48 h at 30 °C, and then the bacterial cells were scraped from plates and adjusted to a concentration of 108 CFU/mL (0.2 OD A600 nm) with 0.85% NaCl solution. The inoculation process was followed by clipping of the leaf edges near the veins using mouth–tooth forceps. For every inoculation, the forceps was dipped into the bacterial suspension; the four youngest fully expanded leaves were inoculated. Fourteen days after inoculation, the leaves of Xcc non-inoculated (control) or Xcc-inoculated plants were, respectively, collected for the evaluation of different biochemical defense markers; they were immediately frozen in liquid nitrogen (N) and stored in a deep freezer (−80 °C) for further analysis.

1. **Bacterial populations**: Two leaf discs from distinct infiltrated leaves of different inoculated plants were sampled at 14 days after inoculation and were homogenized in 200 mL sterile water. Serial dilutions of the homogenates were performed and 10 μL drops were spotted in triplicate for each dilution on plates supplemented with appropriate antibiotics. The plates were incubated at 30 °C for 48 h and colonies were counted in spots containing 1–30 colonies (Xu et al. 2008).
2. **Plant material.** The screening against Xcc, at least one of the parents used for each of the *B. oleracea* and *B. napus* crosses was a doubled haploid (DH) line. In cases where DH lines were not available, selections obtained previously by selfing single plants were used. Plants were raised from seeds sown in plastic pots with Levington M2 compost in a glasshouse with a minimum temperature of 20/15 °C (day/night), venting at 22/17 °C, and supplementary lighting from October to March to give 16-h days. Eight-week-old plants of cabbage lines BOH 85c (DH line derived from cv. Böhmerwaldkohl), PI 436 (cv. Heh Yeh da Ping Tou), Badger Inbred-16 (BI-16 derived from cv. Early Fuji), and rape lines N-o-9 and Cobra were vernalized for 10 weeks at 4 °C. Other lines flowered without need of vernalization.
   The plants were transferred to a glasshouse with a minimum temperature of 15/10 °C and venting at 18/12 °C (day/night); the crosses were made in this glasshouse. Parental lines were hybridized by bud pollination; cross pollination by insects was prevented by covering shoots with perforated polyethylene bags. All non-DH parental plants were self-pollinated to determine heterozygosity. Single plants of lines BOH 85c, PI 436606, and BI-16 were crossed to the susceptible DH line A12DHd as the female parent. Reciprocal crosses were made between a *B. carinata* line derived from PI 199947 and the susceptible line HRI 6987. Reciprocal crosses were also made between *B. napus* lines derived from Cobra, CrGC5, and N-o-1 and the susceptible DH line N-o-9. Crosses between Cobra and N-o-1 and CrGC5 and N-o-1 were also made. In most cases, F1 plants were vernalized by cold treatment, selfed to obtain an F2 population, and backcrossed to parental lines. Vernalization was not required for *B. carinata* and for *B. napus* lines derived from the cross between CrGC5 and N-o-1. A *B. napus* mapping population (N-o-72-8) of DH lines derived from a single F1 plant from an N-o-9 × N-o-1 cross was used (Sharpe et al. 1995). Thirty-four lines randomly selected from the population were used in the first assay. After analysis of the results, 13 additional lines were selected based on existing restriction fragment length polymorphism (RFLP) recombinant data to improve the accuracy of the mapping. Three plants were tested per line.

3. **Disease assays.** Plants were raised from seed sown in 9-cm plastic pots filled with Levington M2. Pots were placed in a glasshouse under the conditions described previously. Four-week-old plants were inoculated with mouse tooth forceps following a method of Vicente et al. (2001). Race type strains of races 1, 3, and 4 described by Vicente et al. (2001) were used in this study. The *B. oleracea* lines derived from the crosses A12DHd × BOH 85c, and A12DHd × PI 436606 were inoculated with the race 3 type strain, HRI 5212 (National Collection of Plant Pathogenic Bacteria 528). The *B. oleracea* lines derived from BI-16 were inoculated with the race 1 type strain, HRI 3811 (PHW 1205), and the race 3 type strain in separate leaves of the same plants. These plants were retested at 8 weeks with the race 1 type strain. Plants of the *B. carinata* lines were tested with the race 1 type strain and the race 4 type strain (HRI 1279A) in separate leaves of the same plants. The *B. napus* lines were inoculated with the race 4 type strain. Twenty-nine DH lines from the *B. napus* mapping population were tested with an additional race 4 isolate (HRI 6189). Inoculations with a single strain were made onto three leaves per plant; inoculations with two strains were made onto four leaves per plant, two with each strain. Approximately 10–12 points of inoculation are required per leaf.
4. **Scoring of resistance:** The results were recorded 2 and 3 weeks after inoculation. The total number of inoculated points and the number of points showing symptoms were recorded and the percentage of infected points was calculated. The severity of symptoms was assessed on a six-point scale of 0–9 based on the relative lesion size (0, no symptoms; 1, small necrosis or chlorosis surrounding the infection point; 3, typical small V-shaped lesion with black veins; 5, typical lesion half way to the middle vein; 7, typical lesion progressing to the middle vein; and 9, lesion reaching the middle vein).
5. **Segregation analysis:** Plants were grouped into various categories based on resistance scores and percentage of inoculated points showing symptoms. Generally, plants with a score of 0, 1, or 3 with less than 25% of points showing symptoms were considered resistant; plants with a score of 3 with more than 25% of points showing symptoms and with a score of 5 with less than 50% of points showing symptoms were considered partially resistant; plants with a score of 5 with more than 50% of points showing symptoms and with a score of 7 with less than 75% of points showing symptoms were considered susceptible; and plants with a score of 7 with more than 75% of points showing symptoms and with a score of 9 were considered very susceptible. The overall distinction between resistance and susceptibility in lines derived from the crosses was made by comparison with the reaction of the parental lines. In most cases, there was a clear distinction between resistant and susceptible plants, but in some cases (e.g., segregants from BI-16 that showed partial resistance) the response was more quantitative, with many plants showing intermediate scores. Chi-square (χ^2) goodness-of-fit was used to test the hypothesis that one locus controls resistance in the crosses, according to the segregation ratios expected for the F2 (3:1 or 1:3) backcross to the resistant or the susceptible parent (1:1) and F1-derived DH lines (1:1). The results obtained in inoculations of segregating populations (F2 and

backcrosses to the resistant or susceptible parent) with two different races and the results of inoculations at 4 and 8 weeks were compared through linear regression analysis using GenStat 5th edition, release 4.21.
The angular transformation of the percentage of points with symptoms and the resistance scores were analyzed separately. The independence of segregation and flower color in F2 and backcross to the resistant parent populations derived from crosses with BI-16 was tested with *t* tests (for transformed percentage of points with symptoms) and Mann-Whitney U tests (for resistance scores). Linkage analysis of qualitative segregation of resistance symptoms in *B. napus* was carried out in conjunction with segregation data for RFLP markers used previously to develop a linkage map for a DH population (Sharpe et al. 1995). The data were analyzed with the program MAPMAKER v.3.0 (Lander et al. 1987) with a minimum LOD (log of the odds ratio of linkage versus no linkage) score of 3.0. Distances were expressed in centi Morgans using Kosambi's map function (Kosambi 1944).

9.9.2 Introgression of Black Rot Resistance Through Embryo rescue

1. **Development of plant materials:** Inbred of black rot susceptible variety "Pusa Sharad" of cauliflower (*Brassica oleracea botrytis* group, 2n = 18, C) was crossed with a resistant accession "NPC-9" (*Brassica carinata A. Braun*; 2n = 4x = 34, BC genome) maintained by repeated selfing. The susceptible parent "Pusa Sharad" was planted sequentially from end of September to November. Seeds of "NPC-9" were sown, 2 months later than the recurrent parent Pusa Sharad to ensure synchronous flowering. Fully developed unopened flower buds which are expected to open within next 1–2 days were selected for pollination. These flower buds were emasculated and pollinated on the day of anthesis with fresh pollen grains collected from freshly opened flowers of donor parent. Nearly 10–15 flower buds were pollinated in each inflorescence to ensure better seed set. Simultaneously, the other flower buds which have not been pollinated were removed to avoid any unwanted seed set. All plants were covered with selfing bags to avoid any out crossing and pollen contamination. After pollination, the pollinated buds were covered with butter paper bags for 5–6 days to prevent out crossing. Development of plant materials is explained in flow diagram (Fig. 9.13).
2. ***In vitro* embryo rescue:** Effect of days after pollination on siliqua and ovule culture were studied under two temperature regime, viz., low temperature (Day/Night, 19.4 °C/4.3 °C) and high temperature (Day/Night, 26.4 °C/10.4 °C). The cross combination "Pusa Sharad × NPC-9" was advanced *in vitro* through immature siliqua and direct ovule culture. A total of 480 immature siliqua were harvested from 5 to 21 DAP at 2 days interval for the rescue of hybrid embryos. Harvested siliqua were placed in a 100 mL conical flask washed by tap water and then surface sterilized with 0.1% $HgCl_2$ solution for 1 min followed by 4–5 times thorough washings with double distilled sterile water. The siliqua were then sterilized with 70% alcohol for 5–7 min and followed by 2–3 wash with

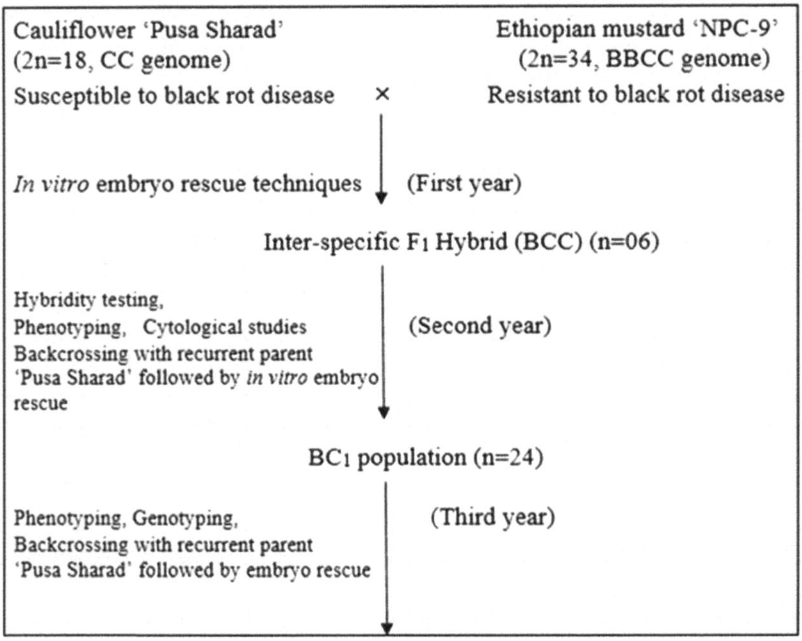

Fig. 9.13 Flow chart of development of plant materials using *in vitro* embryo rescue technique (Sharma et al. 2017)

doubled distilled sterile water. For siliqua culture, parts of the ovary stalks were transversely cut on autoclaved filter paper in Petri dishes and the ovaries were placed with the basal cut end on the media. Siliqua culture medium was Murashige and Skoog's (1962) supplemented with 3% (w/v) sucrose, 7% agar and casein hydrolysate (400 mg/L) at pH = 5.8, autoclaved for 22 min at 121 °C. The siliqua culture was maintained in culture room at 25 ± 2 °C having automatically set timings with white fluorescent lights under a 16/8 h (light/dark) photoperiod. The cultured siliqua were carefully dissected aseptically 5–15 days later and enlarged/turgid ovules were cultured on basal MS medium supplemented with 3% (w/v) sucrose and kept in the dark until germination. In direct ovule culture method, hybrid siliqua were harvested from 11 to 21 DAP at 2 days interval. Ovules from the siliquae were carefully dissected on filter paper in Petri dishes aseptically and cultured on basal MS medium with 6% (w/v) sucrose and were kept in dark until germination. Subculturing of the ovules was done on the same medium at 15 days interval. Putative hybrid plantlets were multiplied *in vitro* via shoot tip culture on MS medium with 1 mg BAP l^{-1}, 3% (w/v) sucrose. Well-developed plantlets were transferred on MS medium with 0.5 mg IBA L^{-1} for *in vitro* rooting. *In vitro* hardening of roots was done by keeping bare roots in test tubes filled with tap water and covered with polythene for 7–10 days. Well-established 3–4-week-old seedlings were transferred in the plastic pots filled with solarite and covered with polythene bags for acclimatization. All plastic pots were placed in shade house for acclimatization under natural photoperiod followed by transplanting in the earthen pots filled with pot mixture (Garden soil + FYM+ Sand) which were then shifted to open field conditions for advancing growing plants till maturity. The two embryo rescue methods (siliqua and ovule culture) employed less than two temperature regimes were statistically analyzed using t-test (Gupta and Kapoor 2002).

3. **Production of BC1 plants:** The F1 interspecific hybrid plants were grown under open field conditions. The highly resistant F1 individual plants were backcrossed with recipient recurrent parent "Pusa Sharad" of cauliflower and the BC1 plants were generated through direct ovule culture method of embryo rescue. Protocol for direct ovule culture and plantlets regeneration was same as described earlier for F1 production. BC1 plants were grown under natural open conditions.
4. **Hybridity confirmation:** Dominant morphological anther tip pigmentation marker was used to confirm hybridity of interspecific F1 plants. B-genome-specific 30 SSR markers were retrieved from publicly available data bank in the *Brassica* microsatellite information exchange (www.brassica.info/resource/markers/ssrexchange.php). *Brassica* B- and C-genome-specific primers DB and DC were used for interspecific hybridity testing as described earlier (Yan et al. 2014; Sharma et al. 2017).

9.9.3 Germplasm Screening for Black Rot Resistance

Screening of germplasm for resistance to Xcc races is a crucial step for resistance breeding and screening for black rot pathogen for the identification and phenotyping of resistant plants/sources which is carried out mostly by a leaf cut and dips method (Saha et al. 2014; Sharma et al. 2016). In this technique, the bacterial culture suspension is prepared at the concentration of 10^8–10^9 cfu/mL followed by clipping of secondary veins of leaf at margins with the help of scissor immersed in bacterial suspension. Sharma et al. (2016, 2017) performed the inoculation at 10 points per leaf on youngest leaves replicated thrice. The observations of disease incidence or resistance were recorded at weekly interval and final scoring was done 30 days after inoculation. The inoculated plants were assessed for pathogen reaction based on 6 point disease rating scale of 0–9 (Vicente et al. 2002), where 0: no symptoms, 1: small necrosis around inoculation point, 3: typical small V-shaped lesion and vein blackening, 5: lesion half way to middle vein, 7: lesion progressing towards middle vein, and 9: lesion reaching the middle vein. The other suggested method for inoculation is screening of germplasm by wound inoculation (Shae and Kado 1988) in which two of the true leaves per plant are pierced on either side of midrib with needles immersed in bacterial culture (Griffiths and Roe 2005) at two- to three true leaf stage. Wound inoculation is also practiced through injection, infiltration method, carborundum abrasion method, pressurized spraying, or sandpaper method (Maji and Nath 2015; Penazova et al. 2018). Plant evaluation for screening is carried out 14 days after inoculation using 1–5 rating scale, in which 1 being completely resistant and 5 being completely susceptible (Griffiths et al. 2009). Lema et al. (2012) used disease rating scale of 1–9 (1: no visible symptoms; 3: an isolated diffused chlorotic zoning around necrotic lesions and 25% vein blackening at inoculated area; 5: vein blackening at 50% inoculated area; 7: complete leaf chlorosis and vein blackening, and 9: severely diseased with V-shape chlorotic lesions). The disease score of 1–3 is considered as resistant, 4–6 is intermediate, and 7–9 is susceptible. Inoculation via hydathodes comprises spraying of suspension or immersion of undamaged healthy plants. In the spraying method, the trays of seedling plants are exposed to cold humid conditions for the opening of hydathodes and stomata followed by inoculums spraying with hand automizer. The application of bacterial suspension is done on lower leaf surface followed by covering of sprayed plants with polyethylene bag to increase humidity and are exposed to 28 °C with 16/8 h photoperiod. After 2 days, they are transferred to greenhouse conditions (Penazova et al. 2018). In the injection method of inoculation, the suspension is injected into midrib and veins of lower leaf surface via syringe about 0.1 mL per plant. In the carborundum abrasion method, 800-mesh powder is used followed by inoculums spreading through cotton swab (Penazova et al. 2018). Lee et al. (2013) suggested clipping with the help of mouse tooth forceps as better technique for inoculation of bacterial suspension as compared to piercing with pins or scissor cutting to study the disease reaction. Hence, more efficient germplasm screenings are essential to identify novel-resistant sources and to establish new combinations of resistant genes to accelerate resistance breeding for black rot and provide durable protection to *Brassica* crops in all the *Brassica* growing areas worldwide (Singh et al. 2018).

9.10 *Brassica-Turnip Mosaic Virus* Techniques

9.10.1 Collection of Samples in the Field

Leaf, petiole, and shoot (young stem) samples excluding inflorescences were taken from 24,689 dicotyledonous plants 664 M.W. (Schwinghamer et al. 2010) from commercial crops, pastures, and roadsides. There were 27 species including grains, forage crops, and weeds. All of them germinated in autumn and grew over winter and spring, except for a few turnip weed and wild radish plants that probably survived from the preceding summer. Samples were collected between July and October, i.e., late vegetative through podding/late flowering stages, except for forage turnip (*Brassica rapa* L. subsp. *rapa*) (one mid-vegetative crop in early June) and coriander (one near-mature crop in late November). The majority of plants were collected by sampling that was unbiased with respect to symptoms, e.g., 12 plants at each of 5 points on W-transect. Where symptoms suggestive of TuMV were evident such as leaf mosaic, or stem or shoot necrosis, representative plants were sampled as well as symptomless plants or unbiased samples. Crop plant species were identified by morphology and seed source. Weeds were identified by morphological keys, comparison with reference specimens and by ribosomal DNA sequences for ten vegetative stage turnip weed specimens (Schwinghamer et al. 2010).

9.10.2 Virus Detection, Preservation, and Identification

Initially, fresh field specimens were tested for viruses by enzyme-linked immunosorbent assays (ELISAs), using purified anti-virus immunoglobulins, alkaline phosphatase conjugates, and ELISA protocols from Agdia, Inc. Infection of fresh leaves of *B. napus* differential lines inoculated with TuMV isolates was identified by ELISA with TuMV monoclonal antibody EMA67 (Jenner et al. 1999). However, the majority of testing of field specimens and isolates was done on stored petiole and stem blots on nitrocellulose membranes (Schwinghamer et al. 2010) by tissue blot immunoassays (TBIAs). These TBIAs were indirect procedures that involved incubation with antigen-specific primary antibodies followed by alkaline phosphatase-conjugated secondary antibodies (anti-rabbit or anti-mouse), and finally chromogenic substrate (nitroblue tetrazolium plus 5-bromo-4-chloro-3-indolyl phosphate) to detect virus in blots or tissue prints on nitrocellulose membranes. Primary antibodies for TuMV TBIA and potyvirus TBIA were, respectively, TuMV polyclonal and general potyvirus monoclonal PTY 1, both from ELISA kits supplied by Bio-Rad Phyto-Diagnostics, INRA, France. Primary antibodies for CaMV and Pea seed-borne mosaic virus (PSbMV) TBIAs were polyclonals from DSMZ (Deutsche Sammlung von Mikroorganismen und Zellkulturen). The TBIA procedure of Lin et al. (1990) and Hsu and Lawson (1991) were modified as follows:

Tris-buffered saline (TBS) (0.02 M Tris–HCl, pH 7.5, 0.15 M NaCl) was used rather than phosphate-buffered saline as basic buffer for antibody dilution and membrane washing (Hammond and Jordan 1990); 1 μg/mL polyvinyl alcohol (Sigma MW 30,000–70,000) was used as pre-incubation blocking agent (Makkouk and Comeau 1994); 2% polyvinyl-pyrrolidone (Sigma PVP-40) and 0.2% egg albumin were added to basic buffer for dilution and incubation of monoclonal antibodies and alkaline phosphatase-conjugated antibodies.

For TBIAs that used polyclonals as primary antibodies, including TuMV, Cauliflower mosaic virus (CaMV), and other TBIAs, binding to host antigens was eliminated by dilution and pre-incubation at 37 °C for 2 h in strained leaf extract (1 g leaf per 50 mL TBS) of uninfected analogous host, usually wild radish, before adding the antibody mixture to membranes. Sensitivity of Potyvirus TBIA was maximized by heating membranes at 60 °C in 0.05 M Tris pH 9.5 (Jordan and Hammond 1991) for 5 min to expose antigenic determinants before blocking and incubating. Voucher specimens representing TuMV-positive hosts were lodged in the NSW Plant Pathology Herbarium (reference numbers prefaced by "DAR") and comprised one or more of the following: pressed leaves, diced leaves (desiccated over CaCl2 or frozen at −80 °C), blots on replicated nitrocellulose membranes, or isolates transmitted from field specimens and stored as desiccated or frozen diced leaves.

9.10.3 Isolates of *TuMV*

There were six isolates of *TuMV* transmitted first by aphids (*Lipaphis pseudobrassicae* or *Brevicoryne brassicae*) or rub inoculation and propagated by 1–12 serial rub inoculations of hosts including Chinese cabbage, turnip weed, mustard, and Buchan weed. All six TuMV isolates produced clear symptoms in these hosts and were positive in TuMV TBIA, positive in potyvirus TBIA, and negative in CaMV TBIA. Source plants and/or isolates except for 4297A were also tested for a range of other mechanically transmissible viruses by species-specific TBIAs (AMV, BYMV, CMV, TSWV, and PSbMV) and found to be negative. The attempt to isolate single variants of TuMV that might occur within isolates was done by means of single lesion isolations. For TuYV, used one isolate (3434B). This was maintained in canola by aphid (*M. persicae*) transmission and determined to be BWYV-like by reaction with polyclonal antiserum to BWYV. Identification as TuYV rather than BWYV was based on infectivity in canola (Graichen and Rabenstein 1996). The experiment was conducted in a glasshouse and included seven host genotypes (six mustard and one canola) and three inoculum classes: aphids (*M. persicae*) fed on TuMV-infected turnip weed, aphids fed on TuYV-infected canola, and non-acquisition fed aphids (mock inoculum) from a colony maintained on virus-free canola. There were seven replicates (49 individually potted plants) for TuMV and TuYV and six replicates (42 plants) for the mock inoculations. Field observations in previous years suggested that one mustard line (887–1) was particularly susceptible to TuMV, one line (J397) was relatively resistant, and the other four were intermediate.

9.10.4 Glasshouse Inoculations

Rub inoculations of *B. napus* differential lines S6, S1, R4, and 165 were conducted as described by Jenner and Walsh (1996). Five plants of each line were inoculated for each of the six NSW grain belt TuMV isolates. Inoculations of canola cv. Outback with TuMV isolates UK 1, CDN 1, and CZE 1 (Jenner and Walsh 1996) were done the same way. The inoculated plants were sown in pine bark/sand-based potting mixture in 14 cm diameter × 14.5 cm high black plastic pots and thinned to either one to three plants per pot. Freshly harvested virus infected leaves and uninfected leaves were used for inocula and mock inocula, respectively. Extracts for rub inoculations were prepared by crushing leaves with a mortar and pestle in 0.07 M sodium phosphate buffer, pH 7.4, with 0.02 M Na_2SO_3 added immediately before use. The pestles wetted with freshly prepared extracts were used to inoculate test plants by gently rubbing cotyledons or first or second true leaves after pre-dusting with carborundum, using the forefinger of the other hand as support. Aphid transmissions from field plant specimens were achieved by transferring colonizing aphids directly to test seedlings. Aphid transmission for the mustard pathogenicity experiment involved an acquisition access period of 48 h and inoculation access period (IAP) of 72 h, using six *Myzus persicae* per inoculated plant, inoculated turnip weed as source for TuMV, and inoculated canola as source for TuYV. Inoculations were done and plants grown year round in a glasshouse with evaporative cooling (10–28 °C) fitted with shade cloth in October to March. From flowering onward, nutrient status was maintained with inorganic fertilizer (Thrive®). Insects were eliminated by screens and regular monitoring. Occasional occurrences of fungal foliar diseases were controlled with application of commercially recommended foliar fungicides.

9.10.5 Plants, Chemical Treatments, and Virus Inoculation Method

Leafy mustard (*B. juncea* var. *foliosa Bailey*) plants were grown in a greenhouse at 23 °C and 20 °C (day and night air temperature, respectively) with a 16-h photoperiod. The four to five leaf-stage seedlings were subjected to chemical pre-treatment for four consecutive days, and then the lower fully developed leaves were used for inoculation with TuMV. Six days later, the upper third fully developed leaf of each plant was collected for further study. Stock solutions of AA and SHAM were dissolved in de-ionized water. Mustard plants were sprayed with water (control plants) or freshly diluted solution of either 0.1 mM AA or 3 mM SHAM on both the adaxial and abaxial surfaces. SHAM was sprayed daily for 10 days.

Isolate CHZH33 in B strain of TuMV was procured for inoculation. This strain infects a wide range of plant species, mostly of the family Brassicaceae and contains a single copy of the genome, which is a single-stranded positive sense RNA (Ma et al. 2010). The main visible symptoms of systemic infection are mosaic, chlorosis, and deformity. Viral inoculum was prepared in 5 mM potassium phosphate buffer with pH 7.2 according to the method described by Tan et al. (2004). For inoculation, the plants were first sprayed with carborundum and then inoculated by rubbing 100 μL of viral inoculum onto each leaf with cotton swabs. Mock-inoculation controls were prepared by following the same procedure using only buffer. The upper newly developed, un-inoculated leaves were collected at different times for determining leaf rate of respiration and quantitative reverse transcription polymerase chain reaction (qRT-PCR) measurements. Experiments were replicated three times.

Pot positions were randomized within each of seven replicate blocks on separate benches. Plants were inoculated (IAP initiated) 18 days after sowing, maintained and watered until maturity 105 days after inoculation, and pods and seed harvested after plants had desiccated. Infection status at 21 days post-inoculation was determined by TBIA and/or characteristic symptoms for TuMV, and solely by TBIA for TuYV which was symptomless. The TBIAs showed that there was no unintended spread of the viruses between inoculum classes, indicating that aphids were eliminated successfully after the IAP. Data including plant height, seed weight, pod count, and days to death were analyzed using generalized linear models (GLM) (McCullagh and Nelder 1989) to test the effects of infection (TuMV, TuYV, nil) and genotype on the mustard plants. All GLM analyses were conducted using ASReml, a statistical package that fits linear mixed models using Residual Maximum Likelihood

(Gilmour et al. 2009). Eighteen mustard plants that failed to become infected were combined with plants of their respective genotypes in the "non-viruliferous" inoculum class. Pairwise comparisons were conducted using least significant differences at $P < 0.05$. Canola was not included in the analysis as none of the seven replicate TuMV-inoculated canola plants became infected.

9.10.6 Serological Detection

The collected samples were tested for the presence of the most common oilseed rape viruses: *turnip yellows virus* (TuYV), *cauliflower mosaic virus* (CaMV), and *turnip mosaic virus* (TuMV), using commercial double antibody sandwich (DAS)-ELISA diagnostic kits according to the manufacturer's instructions. Plant tissues were prepared in an extraction buffer at a ratio of 1:10 (wt/vol). After incubation with p-nitrophenyl phosphate at room temperature for 2 h in the dark, absorbance at 405 nm was measured with an ELISA microplate reader (Multiscan Ascent, Finland). The samples were considered positive if their absorbance value was twice as high as the negative control. Commercial positive and negative controls were included in each ELISA plate.

1. **Aphid transmission method:** For aphid-borne TuYV inoculation, nymphs of *Myzus persicae* (Sulzer) were allowed to feed on the leaves of six selected serologically positive samples, i.e., one sample from each locality, over an acquisition access period (AAP) of 24 h. Groups of 5–7 aphids for each of the six isolates were then placed onto three plants of each *Capsella bursa-pastoris, Physalis floridana,* and *B. napus* cv. "Banaćanka" for a 4-day inoculation access period (IAP). The plants with aphids were placed into separate cages under controlled conditions, at 22 °C, for a 16 h photoperiod and watered regularly (Milošević et al. 2016).
2. **RT-PCR detection:** The presence of TuYV in rapeseed plants was further confirmed by the conventional reverse transcription (RT)-PCR. Total RNAs were extracted by the RNeasy Plant Mini Kit from 100 mg of leaf tissue of six selected samples originating from different localities. RT-PCR was performed using the One-Step RT-PCR Kit with primers TuYVorf0F and TuYVorf0R amplifying the P0 gene, the most useful for delineation of *Poleroviruses* (Schubert et al. 1998). The RT-PCR reaction mixture included 400 μM of each of the four dNTPs, 1 μL of RT-PCR enzyme mix, 0.6 μM of each primer, and 1 μL of extracted RNA in a final volume of 25 μL. Amplifications were performed in a thermal cycler with the following cycling parameters: reverse transcription at 50 °C for 30 min and an initial PCR denaturation step at 95 °C for 15 min, followed by 35 cycles consisting of a denaturation step for 30 s at 94 °C, primer annealing for 30 s at 55 °C, and extension for 60 s at 72 °C. The final extension was performed at 72 °C for 10 min. The amplified products were analyzed by 1% agarose gel electrophoresis and visualized under a UV trans-illuminator. A tissue

sample from the healthy oilseed rape leaf was used as a negative control in RT-PCR assays (Milošević et al. 2016).

3. **Sequencing and phylogenetic analysis:** After purification with a QIAquick PCR Purification Kit, the amplified products from one selected isolate (119-TuYV) were sequenced directly in both directions, using the same primers as in RT-PCR. Additionally, a previously identified oilseed rape TuYV isolate 114-TYuV (Milošević et al. 2015) was also included in the investigation. The sequence of the Serbian TuYV isolate was compared with the previously reported isolates available in GenBank (http://www.ncbi.nlm.nih.gov/BLAST/), using the ClustalW program (Thompson et al. 1994) and MEGA5 software (Tamura et al. 2011). A p-distance model was applied for nucleotide (nt) and deduced amino acid (aa) sequence analyses. A phylogenetic tree was constructed using the TuYV P0 gene sequence generated in this study, one generated in another study by Milošević et al. (2015) and 11 P0 sequences of TuYV, *Beet western yellows virus*, *Beet chlorosis virus*, and *Cucurbit aphid-borne yellows virus* isolates retrieved from GenBank, using the maximum parsimony method implemented in MEGA5. Intra- and inter-group diversity values were calculated as the average genetic distance using Kimura 2-parameter model Gamma distributed (K2+G), which was chosen as the best-fitting model of nt substitution (Milošević et al. 2016).

9.10.7 Respiration in Systemic Infected Leaves

Respiration of leaves was measured by the method of Millenaar et al. (2002) using a Clark-type oxygen electrode at different days post-inoculation (DPI). Plant leaves were detached and cut into pieces with a razor blade. Leaf samples (0.1 g) were transferred to airtight cuvette containing 2 mL of air saturated 20 mM potassium phosphate buffer (pH 6.8) and respiration was measured as the decrease in O_2 concentration at 25 °C. Cyanide (CN)-resistant respiration was measured in the presence of 1 mM KCN. For each virus–host combination examined, triplicate measurements were recorded, in order to have statistically valid comparisons of alternative pathway capacity.

9.10.8 RNA Extraction and Transcript Level Estimation by Real-Time Quantitative PCR

Total RNA was extracted from each third leaf using the TRIzol reagent according to the manufacturer's specifications. Genomic DNA was removed using the RNeasy mini kit. Complementary DNA was synthesized using 2 μg of RNA with the Revert Aid first-strand cDNA synthesis, using M-MLV Reverse Transcriptase and oligo (dT) 18 primers (Takara) (Qi et al. 2010). Primers for real-time PCR were designed using http://www.genescript.com/. Specific primers at 5′ end were designed for analyzing transcript levels of the AOX1a. TuMV-specific primers were also designed based on the sequence encoding the coat protein (CP). The primers used for the amplification of each cDNA as per list of Zhu et al. (2012). All amplified products were subjected to gel electrophoresis to confirm the presence of a single band (Qi et al. 2010). RT-PCR was performed with an iCycler iQ Multicolor Real-Time PCR Detection System. Each reaction (20 μL total volumes) consisted of 10 μL of iQ SYBR Green Supermix, 1 μL of diluted cDNA, and 0.1 μM each of forward and reverse primers. PCR cycling conditions were as follows: 95 °C for 3 min, followed by 40 cycles of 95 °C for 10 s and 59 °C for 45 s. Fluorescence data were collected during the 59 °C step. Mustard 25S was used as an internal control. Relative gene expression was calculated as described by Livak and Schmittgen (2001), Zhu et al. (2012).

9.11 *Brassica* Species Genomics Techniques

9.11.1 RNA Sequencing of *Brassica napus*

1. **RNA extraction:** Frozen roots were placed in 15 mL Falcon tubes and ground using 5–7 mm diameter steel beads at 1100–1200 rpm (3–4 rounds of 1 min each) in a Genogrinder, followed by additional grinding in a mortar with a pestle in the presence of liquid nitrogen. Total RNA was extracted with a combination of TRIzol reagent and the RNeasy kit to increase the quantity and purity of the RNA. DNase treatment was performed on-column with the RNase-Free DNase set. The quantity of RNA was assessed with a Nanodrop 2000c spectrophotometer and quality was checked in a 2200 Tape Station.
2. **RNA-Seq:** Three micrograms of RNA per pooled sample corresponding to each biological replicate with RNA Integrity Numbers (RIN) > 9 were sent for sequencing at Oklahoma State Genomics, the US libraries were prepared with a TruSeq Stranded mRNA library preparation kit for 48 samples and samples were sequenced using TG Nextseq 500/550 High Output Kit v2 (75 cycles), in a NextSeq500 equipment. The raw data reads were filtered using the bcl2fastq conversion software provided by Illumina. This software uses read indexes to assign each read to the corresponding sample and then removes any adapters, barcodes, or primers using a trimmomatic-like algorithm. Any reads with an average Q-score <30 were removed and trimming of read ends was performed for bins of 15 bp that had a Q-score <30. Sequencing reads were deposited in the NCBI Sequence Read Archive (SRA) under accession number: PRJNA597078. The reads from all treatments were mapped to the *B. napus* reference genome (AST_PRJEB5043_v1) using the genome and gene annotation deposited in Ensembl (Chalhoub et al. 2014; Ensembl 2018; Kersey et al. 2018). Genome indexes were generated with Bowtie2 v2.3.3.1 (Langmead and Salzberg 2012) and transcriptome indexes with TopHat v2.1.1 (Trapnell et al. 2009). Reads mapped with TopHat were used along gene models (obtained from the Ensembl website) as input for Cufflinks v.2.2.1 to generate all potential transcripts in each sample. Cufflinks was run using the GTF-guide option, multi-read correction and

fragment bias correction (multi-read correction and fragment bias correction were also used in the subprograms downstream). GTF files generated from all treatments and replicates were used to generate a consensus merged transcript gtf file with Cuffmerge. This consensus file was employed as a reference for re-mapping reads for quantification. Cuffquant was run with the merged gtf file to create ready-to-use files with quantified transcripts to perform the necessary comparisons of differential expression among treatments. Cuffdiff was used to compare cuffquant files, contrasting controls with inoculated plants at each time point and for each genotype separately. Expression levels were quantified as Fragments Per Kilobase of transcript per Million mapped reads (FPKM) and significant differential expression for each evaluated transcript was assessed after multiple comparison testing with the Benjamini-Hochberg correction (Benjamini and Hochberg 1995). The number of mapped reads from all treatments was compared using a one-way ANOVA implemented in R. FPKM values of all genes from treated and inoculated replicates were log-transformed to build PCA plots using the R-package ggfortify (https://github.com/sinhrks/ggfortify) (Son et al. 2018).

3. **cDNA Synthesis and Quantitative Reverse Transcription-PCR (qRT-PCR):** Five hundred nanograms of RNA per sample were subjected to DNAse treatment using RNase-free-DNase I. cDNA was synthesized using the Revert Aid H Minus Reverse transcriptase with oligo dT, following manufacturer's specifications. To test for residual DNA contamination, a 1:100 dilution of cDNA was used to perform PCR with primers corresponding to a constitutively expressed Clathrin Adaptor Complex (CAC) gene, producing 125 bp and 288 bp bands on cDNA and DNA, respectively. PCR analysis was conducted with 1× Buffer + KCl, 2.5 mM $MgCl_2$, 0.2 mM of each dNTP, 0.2 μM or each primer, and 1 unit of high-fidelity Taq polymerase. Cycling conditions were 94 °C for 3 min, followed by 35 cycles of 94 °C for 30 s, 60 °C for 30 s, and 72 °C for 1 min, finishing with an extension step of 72 °C for 10 min, and a 4 °C step until samples were recovered from the thermocycler. To validate the differential expression detected by RNA-seq analysis, 15 primer pairs from genes showing significant differential expression in at least one time point or host genotype were evaluated for efficiency (range of 90–110%) using a five-point dilution series (1:4 serial dilutions) of a mix of all cDNA samples. Ten target primers within the expected efficiency range were used for final RNA-seq validation. Five reference gene primers with valid efficiencies were tested for stability across all samples using Best Keeper (Pfaffl et al. 2004), and the three genes showing the best correlation and lower standard deviation were chosen for relative quantification. Primers were designed using Primer3 (NIH 2012; Untergasser et al. 2012) with the following parameters: Na^+ concentration 50 mM, Mg^{++} concentration 3 mM, dNTPs concentration 0.8 mM, oligo length: min 18, opt 22, max 30 bp, melting temperature: min 60, opt 62, max 64 °C, GC content: min 35, opt 50, max 65%, amplicon length: min 70, opt 100, max 250 bp. Relative quantification of expression was performed using the $2^{-\Delta\Delta}$CT method (Livak and Schmittgen 2001). All qRT-PCR experiments were performed with four biological replicates

and three technical replicates per biological replicate. The experiments were run on a Vii A7 Real-Time PCR system. Reactions were performed in 10 μL with 5 μL of an in-house made SYBR Green Reagent Mix, 2.5 μL of the pair of mixed primers (3.2 μM), and 2.5 μL of a 1:100 dilution of cDNA. Cycling conditions were 95 °C for 2 min followed by 40 cycles of 95 °C for 30 s and 60 °C for 1 min. A melting curve stage was added to verify product specificity: 95 °C for 15 s, 60 °C for 1 min, and 95 °C for 15 s. Log_2-fold change values from the qRT-PCR analysis were compared with the RNA-seq values of the respective genes to validate the RNA-seq data. Correlation coefficients were calculated using the CORREL function from Microsoft Excel for each day and host genotype using the ten target genes (Son et al. 2018).

9.11.2 Identification of *Brassica oleracea* Genes That Encode NBS Domain and NBS-Associated Conserved Domains

In the draft genome of *B. oleracea*, NBS-encoding genes were identified through Hidden Markov Model (HMM) profile corresponding to the Pfam NBS (NB-ARC) family PF00931 domain using HMMER V3.0 program with "trusted cutoff" as threshold. From the selected protein sequences screened through NBS domain, high quality sequences were aligned through CLUSTALW and used to construct *B. oleracea*-specific NBS profile using the "hmmbuild" module by HMMER V3.0 program. With this model final set of NBS-encoding proteins were identified and only 157 proteins were selected as NBS candidate genes with stringent parameters. The NBS R-gene family is subdivided into different groups based on the structure of the N-terminal and C-terminal domains of the protein. For the identification of N-terminal and C-terminal domains of NBS-encoding genes, HMM Pfam and HMM Smart was used for detection. Further, employed PAIRCOIL2 (P score cutoff of 0.025) and MARCOIL programs with a threshold probability of 90 to confirm Coiled-Coil (CC) motif. From the result generated by these programs, the selected overlapping sequences as candidate genes with CC motif. Used same procedures to identify genes that contain TIR domain only and excluded the NBS-encoding genes as TIR-X genes. NBS-encoding genes in *A. thaliana* and *B. rapa* have been reported earlier but in order to get the latest NBS-encoding genes in these two species for comparative analysis. It was followed by the same procedures to screen NBS candidate genes in *B. rapa* and *A. thaliana* for consistency (Finn et al. 2011; Thompson et al. 1994; McDonnell et al. 2006).

1. **Assigning the location of NBS-encoding genes to *B. oleracea* and *B. rapa* genome:** The physical position of NBS-encoding genes was mapped to the 9 and 10 pseudomolecular chromosomes of *B. oleracea* and *B. rapa* using GFF file which was downloaded from Bolbase and BRAD database, respectively. After that, in-house perl script to draw graphic potryl of NBS-encoding genes on pseudomolecular chromosomes with SVG module was used (Cheng et al. 2011; Yu et al. 2013; Altschul et al. 1997).

2. **Identification of tandem duplicated arrays:** To detect the generated mechanism of NBS-encoding genes, BLASTP program was employed to identify the tandem duplicated genes using protein sequences with *E*-value cutoff $\leq 1e-20$, and one unrelated gene was allowed within a tandem array (Altschul et al. 1997).
3. **Alignment and phylogenetic analysis of NBS-encoding genes:** According to location of conserved domains for NBS (Nucleotide-binding Site) in complete predicted NBS protein sequences, conserved domain sequences of NBS-encoding genes were extracted and aligned using the program Clustal W with default options for the phylogenetic analysis among three species. The poor alignment sequences were excluded by manually curation using Jalview. The resulting sequences were used to construct a phylogenetic tree using Maximum Likelihood (ML) method in MEGA 5.0 with 1000 replications (Thompson et al. 1994; Clamp et al. 2004).
4. **Orthologous gene pairs between *B. rapa, A. thaliana,* and *B. oleracea*:** Orthologous gene pairs provide information about the evolutionary relationship between different species. Two steps were used to detect gene pairs precisely. First, MCscan program was employed to identify orthologous regions with the parameters ($e = 1e - 20$, $u = 1$, and $s = 5$. Parameter of $s = 5$) between *B. rapa, A. thaliana,* and *B. oleracea* genomes. Second, after extracting orthologous regions that contained NBS-encoding genes, orthologous gene pairs of NBS-encoding genes were extracted (Tang et al. 2008).
5. **Non-synonymous/synonymous substitution (Ka/Ks) ratios of gene pairs between *B. rapa, A. thaliana,* and *B. oleracea*:** For the estimation of selection mode for the NBS-encoding genes among *B. oleracea, B. rapa,* and *A. thaliana,* the ratio of the rates of non-synonymous to synonymous substitutions (Ka/Ks) of all orthologous gene pairs were calculated for each branch of the phylogenetic tree using PAML software. For each sub-tree of NBS orthologous gene pairs among three species, model 1 with a free Ka/Ks ratio was calculated separately for each branch. The Ka/Ks values associated with terminal branches between modern species and their most recent reconstructed ancestors were employed in the subsequent analyses. In order to detect selection pressure, Ka/Ks ratio greater than 1, less than 1, and equal to 1 represents positive selection, negative or stabilizing selection, and neutral selection, respectively (Yang et al. 2007).
6. **RNA-seq data analysis of NBS-encoding genes:** For expression profiling of NBS-encoding genes, RNA-seq data that was generated earlier and submitted into GEO database was used. Transcript abundance is calculated by fragments per kilobase of exon model per million mapped reads (FPKM) and the FPKM values were log2 transformed. A hierarchical cluster was created using the Cluster 3.0 and heat map generated using Tree View Version 1.60 software (Eisen et al. 1998; Yu et al. 2014).

References

Adachi Y, Watanabe H, Tanabe K, Doke N, Nishimura S, Tsuge T (1993) Nuclear ribosomal DNA as a probe for genetic variability in the Japanese pear pathotype of *Alternaria alternata*. Appl Environ Microbiol 59:3197–3205

Adam L, Somerville SC (1996) Genetic characterization of five powdery mildew disease resistance loci in *Arabidopsis thaliana*. Plant J 9:341–356

Agnola B, Boury S, Monot C, Quillevere A, Herve I, Silue D (2003) Evidence that a leaf-disk allows assessment of isolatespecific resistance in *Brassica oleracea* crops against downy mildew (*Peronospora parasitica*). Eur J Plant Pathol 109:471–478

Alex LA, Borkovich KA, Simon MI (1996) Hyphal development in *Neurospora crassa*: involvement of a two-component histidine kinase. Proc Natl Acad Sci U S A 93(8):3416–3421

Allan AL, Fernandez B, Robio BS (1986) Spectrophotometer determination of oxalate in aqueous solution. Microchem J 34:51–55

Altschul SF, Gish W, Miller W, Myers E, Lipmann DJ (1990) Basic local alignment search tool. J Mol Biol 219:403

Altschul SF, Madden TL, Schäffer AA, Zhang J, Zhang Z, Miller W, Lipman DJ (1997) Gapped BLAST and PSI-BLAST: a new generation of protein database search programs. Nucleic Acids Res 25:3389–3402

Arnold DL, Blakesley D, Clarkson JM (1996) Evidence for the growth of *Plasmodiophora brassicae in vitro*. Mycol Res 100:535–540

Arora H, Padmaja KL, Paritosh K, Mukhi N, Tewari A, Mukhopadhyay A, Gupta V, Pradhan AK, Pental D (2019) BjuWRR1, a CC-NB-LRR gene identified in *Brassica juncea*, confers resistance to white rust caused by *Albugo candida*. Theor Appl Genet 132:2223–2236. https://doi.org/10.1007/s00122-019-03350-z

Arumugam N, Gupta V, Jagannath A, Mukhopadhyay A, Pradhan AK, Burma PK, Pental D (2007) A passage through in vitro culture leads to efficient production of marker-free transgenic plants in *Brassica juncea* using the Cr−*loxP* system. Transgenic Res 16:703–712

Asada Y, Ohguchi T (1981) Behaviour of downy mildew fungus of Japanese radish on modified knop's medium. Ann Phytopathol Soc Jpn 47:71–74

Asano T, Kageyama K, Hyakumachi M (1999) Surface disinfestations of resting spores of *Plasmodiophora brassicae* used to infect hairy roots of *Brassica* spp. Phytopathology 89:314–319

Atri C, Akhatar J, Gupta M, Gupta N, Goyal A, Rana K, Kaur R, Mittal M, Sharma A, Singh MP, Sandhu PS, Barbetti MJ, Banga SS (2019) Molecular and genetic analysis of defensive responses of *Brassica juncea—B. fruticulosa* introgression lines to Sclerotinia infection. Sci Rep 9:17089. https://doi.org/10.1038/s41598-019-53444-3

Avenot HF, Simoneau P, Iacomi B, Bataille-Simoneau N (2005) Characterization of mutations in the two-component histidine kinase gene AbNIK1 from *Alternaria brassicicola* that confer high dicarboximide and phenylpyrrole resistance. Curr Genet 47(4):234–243. https://doi.org/10.1007/s00294-005-0568-2

Avrameas S (1969) Coupling of enzymes to proteins with glutaraldehyde, use of conjugates for the detection of antigens and antibodies. Immunochemistry 5:43

Bains PS, Tewari JP (1987) Purification, chemical characterization and host-specificity of the toxin produced by *Alternaria brassicae*. Physiol Mol Plant Pathol 30:259–271

Balesdent MH, Barbetti M, Hua L, Sivasithamparam K, Gout L, Rouxel T (2005) Analysis of *Leptosphaeria maculans* race structure in a worldwide collection of isolates. Phytopathology 95:1061–1071

Barret P, Delourme R, Foisset N, Renard M (1998) Development of a SCAR (sequence characterized amplified region) marker for molecular tagging of the dwarf BREIZH (*Bzh*) gene in *Brassica napus* L. Theor Appl Genet 97:828–833. https://doi.org/10.1007/s001220050962

Bassey EO, Gabrielson RL (1983a) Factors affecting accuracy of 2, 4-D assays of crucifer seed for *Alternaria brassicicola* and relation of assays to seedling disease potential. Seed Sci Technol 11:411–420

Bassey EO, Gabrielson RL (1983b) The effects of humidity, seed infection level, temperature and nutrient stress on cabbage seedling disease caused by *Alternaria brassicicola*. Seed Sci Technol 11:403–410

Benjamini Y, Hochberg Y (1995) Controlling the false discovery rate: a practical and powerful approach to multiple testing. J R Stat Soc 57:289–300. https://doi.org/10.1111/j.2517-6161.1995.Tb02031.x

Beynon KI, Brown KF (1969) The identification of fungicide antagonists in leaf exudate. Phytopathol Z 64:213–220

Black MC, Dainello FJ (1986) Comparison of percent leaf area with white rust lesions and two other methods for evaluating partial resistance to Albugo occidentalis in spinach. Phytopathology 76:1087. (Abstr.)

Bonnet A, Blancard D (1987) Resistance of radish (*Raphanus sativa* L.) to downy mildew, *Peronospora parasitica*. Cruciferae Newslett 12:98–99

Boodley JW, Sheldrake R Jr (1972) Cornell peat-lite mixes for commercial plant growing. Cornell Univ Coop Exten Info Bull 43:1–8

Bradbury PJ, Zhang Z, Kroon DE, Casstevens TM, Ramdoss Y, Buckler ES (2007) TASSEL: software for association mapping of complex traits in diverse samples. Bioinformatics 23:2633–2635

Bradley CA, Henson RA, Porter PM, LeGare DG, Del Rio LE, Khot SD (2006) Response of canola cultivars to *Sclerotinia sclerotiorum* in controlled and field environments. Plant Dis 90:215–219

Browning BL, Browning SR (2016) Genotype imputation with millions of reference samples. Am J Hum Genet 98:116–126. https://doi.org/10.1016/j.ajhg.2015.11.020

Bryngelsson T, Gustafsson M, Gréen B, Lind-Halldén C (1988) Uptake of host DNA by the parasitic fungus *Plasmodiophora brassicae*. Physiol Mol Plant Pathol 33(2):163–171

Buchwaldt L, Li R, Hegedus DD, Rimmer S (2005) Pathogenesis of *Sclerotinia sclerotiorum* in relation to screening for resistance. In: Proc 13th intel Sclerotinia workshop, Monterey, CA, p 22

Buczacki ST, Toxopeus H, Johnston P, Dixon TD, Hobolt LA (1975) Study of physiological specialisation in *Plasmodiophora brassicae*: proposals for attempted rationalisation through an international approach. Trans Br Mycol Soc 65:295–303

Buhariwalla H, Mithen R (1995) Cloning of a *Brassica* repetitive DNA element from resting spores of *Plasmodiophora brassicae*. Physiol Mol Plant Pathol 47:95–101

Buhariwalla S, Greaves S, Magrath R, Mithen R (1995) Development of specific PCR primers for the amplification of polymorphic DNA from the obligate root pathogen *Plasmodiophora brassicae*. Physiol Mol Plant Pathol 47:83–94

Bulman S, Candy JM, Fiers M, Lister R, Conner AJ, Eady CC (2011) Genomics of biotrophic, plant-infecting plasmodiophorids using in vitro dual culture. Protist 162(3):449–461

Burrough PA, McDonnell RA (1998) Principles of geographical information systems. Oxford University Press, Oxford

Butler D (2016) Pedicure: pedigree tools. www.asreml.org

Butler DG, Cullis BR, Gilmour AR, Gogel BJ (2009) ASReml-R reference manual. Release 3.0. Technical report, Queensland Department of Primary Industries, Brisbane, QLD

Cai G, Yang Q, Yang Q, Zhao Z, Chen H, Wu J, Fan C, Zhou Y (2012) Identification of candidate genes of QTLs for seed weight in *Brassica napus* through comparative mapping among *Arabidopsis* and *Brassica* species. BMC Genet 13:105. https://doi.org/10.1186/1471-2156-13-105

Callahan FE, Rowe DE (1991) Use of the host-pathogen interaction system to test whether malic acid is the sole pathogenic determinant in the exudate of *Sclerotinia trifoliorum*. Phytopathology 81:1546–1550

Campbell MM, Ellis BE (1992) Fungal elicitor-mediated responses in pine cell cultures: cell wall-bound phenolics. Phytochemistry 31:737–742

Cao T, Tewari J, Strelkov SE (2007) Molecular detection of *Plasmodiophora brassicae*, causal agent of clubroot of crucifers, in plant and soil. Plant Dis 91:80–87

Cao T, Kim YM, Kav NNV, Strelkov SE (2009) A proteomic evaluation of Pyrenophora tritici-repentis, causal agent of tan spot of wheat, reveals major differences between virulent and avirulent isolates. Proteomics 9:1177–1196. https://doi.org/10.1002/pmic.200800475

Castlebury LA, Maddox JV, Glawe DA (1994) A technique for the extraction and purification of viable *Plasmodiophora brassicae* resting spores from host tissue. Mycologia 86:458–460

Chalhoub B, Denoeud F, Liu S, Parkin IA, Tang H, Wang X, Chiquet J, Belcram H, Tong C, Samans B, Corréa M, Da Silva C, Just J, Falentin C, Koh CS, Le Clainche I, Bernard M, Bento P, Noel B, Labadie K, Alberti A, Charles M, Arnaud D, Guo H, Daviaud C, Alamery S, Jabbari K, Zhao M, Edger PP, Chelaifa H, Tack D, Lassalle G, Mestiri I, Schnel N, Le Paslier MC, Fan G, Renault V, Bayer PE, Golicz AA, Manoli S, Lee TH, Thi VH, Chalabi S, Hu Q, Fan C, Tollenaere R, Lu Y, Battail C, Shen J, Sidebottom CH, Wang X, Canaguier A, Chauveau A, Bérard A, Deniot G, Guan M, Liu Z, Sun F, Lim YP, Lyons E, Town CD, Bancroft I, Wang X, Meng J, Ma J, Pires JC, King GJ, Brunel D, Delourme R, Renard M, Aury JM, Adams KL, Batley J, Snowdon RJ, Tost J, Edwards D, Zhou Y, Hua W, Sharpe AG, Paterson AH, Guan C, Wincker P (2014) Early allopolyploid evolution in the post-Neolithic *Brassica napus* oilseed genome. Science 345(6199):950–953

Chandna R, Augustine R, Bisht NC (2012) Evaluation of candidate reference genes for gene expression normalization in *Brassica juncea* using real time quantitative RT-PCR. PLoS One 7:e36918

Chatopadhyay AK, Moitra AK, Bhunia CK (2001) Evaluation of Brassica species for resistance to *Plasmodiophora brassicae* causing club root of rapeseed mustard. Indian Phytopathol 54 (1):131–132

Chen Y, Wang D (2005) Two convenient methods to evaluate soybean for resistance to *Sclerotinia sclerotiorum*. Plant Dis 89:1268–1272

Cheng X, Xu J, Xia S, Gu J, Yang Y, Fu J, Qian X, Zhang S, Wu J, Liu K (2009) Development and genetic mapping of microsatellite markers from genome survey sequences in *Brassica napus*. Theor Appl Genet 118(6):1121–1131

Cheng F, Liu S, Wu J, Fang L, Sun S, Liu B, Li P, Hua W, Wang X (2011) BRAD, the genetics and genomics database for *Brassica* plants. BMC Plant Biol 11:136. https://doi.org/10.1186/1471-2229-11-136

China MoAotPsRo (2011) Rules for investigation and forecast technology of rape sclerotiniose [*Sclerotinia sclerotiorum* (Lib.) de Bary]. Standards Press of China

Churchill GA, Doerge RW (1994) Empirical threshold values for quantitative trait mapping. Genetics 138:963–971

Clamp M, Cuff J, Searle SM, Barton GJ (2004) The Jalview Java alignment editor. Bioinformatics 20(3):426–427. https://doi.org/10.1093/bioinformatics/btg430

Clark MF, Adams AN (1977) Characteristics of the micro plate method of enzyme linked immuno-sorbent assay for viruses. J Gen Virol 34:475–483

Clarke WE, Higgins EE, Plieske J, Wieseke R, Sidebottom C, Khedikar Y, Batley J, Edwards D, Meng J, Li R, Lawley CT, Pauquet J, Laga B, Cheung W, Iniguez-Luy F, Dyrszka E, Rae S, Stich B, Snowdon RJ, Sharpe AG, Ganal MW, Parkin I (2016) A high-density SNP genotyping array for *Brassica napus* and its ancestral diploid species based on optimized selection of single-locus markers in the allotetraploid genome. Theor Appl Genet 129:1887–1899

Clarkson JP, Staveley J, Phelps K, Young CS, Whipps JM (2003) Ascospore release and survival in *Sclerotinia sclerotiorum*. Mycol Res 107:213–222. https://doi.org/10.1017/s0953756203007159

Corpet F (1988) Multiple sequence alignment with hierarchical clustering. Nucleic Acids Res 16:10881–10890

Coventry J, Kott L, Beversdorf WD (1988) Manual for microspore culture technique for *Brassica napus*. University of Guelph, Guelph

Cunnington JH, Takamatsu S, Lawrie AC, Pascoe IG (2003) Molecular identification of anamorphic powdery mildews (*Erysiphales*). Aust Plant Pathol 32:421–428

Cvikrova M, Binarova P, Eder J, Nedelnik J (1992) Accumulation of phenolic acids in filtrate-treated alfalfa cell cultures derived from genotypes with different susceptibility to *Fusariumoxysporum*. J Plant Physiol 140:21–27

Dangl JL, Holub EB, Bebener T, Lehnackers H, Ritter C, Crute IR (1992) Genetic definition of loci involved in *Arabidopsis*-pathogen interactions. In: Koncz C, Chua NH, Schell J (eds) Methods in *Arabidopsis* research. World Scientific Press, Singapore, pp 393–418

Deora A, Gossen BD, McDonald MR (2012) Infection and development of *Plasmodiophora brassicae* in resistant and susceptible canola cultivars. Can J Plant Pathol 34:239–247

Dhaka N, Mukhopadhyay A, Paritosh K, Gupta V, Pental D, Pradhan AK (2017) Identification of genic SSRs and construction of a SSR-based linkage map in *Brassica juncea*. Euphytica 213:15

Dickinson CH, Greenhalgh JR (1977) Host range and taxonomy of *Peronospora* on crucifers. Trans Br Mycol Soc 69:111–116

Diriwachter G, Herter G, Gindrat D (1979) Observations sur la culture de *Spongospora subterranea* (Plasmodiophoromycetes) en milieu gelose. Ber Schweiz Bot Ges 89:105–113

Dixon GR (1976) Methods used in Western Europe and U S A for testing *Brassica* seedling resistance to clubroot. Plant Pathol 25:129–134

Doullah M, Meah M, Okazaki K (2006) Development of an effective screening method for partial resistance to *Alternaria brassicicola* (dark leaf spot) in *Brassica rapa*. Eur J Plant Pathol 116:33–43

Doyle J, Doyle J (1990) Isolation of plant DNA from fresh tissue. Focus 12:13–15

Duitama J, Quintero JC, Cruz DF, Quintero C, Hubmann G, Foulquie-Moreno MR, Verstrepen KJ, Thevelein JM, Tohme J (2014) An integrated framework for discovery and genotyping of genomic variants from high-throughput sequencing experiments. Nucleic Acids Res 42:e44

Dunn CL, Brown KF, Montagne JTW (1969) Antagonism between fungicides and water soluble exudates from leaves of plants. Phytopathol Z 64:112–118

Eisen MB, Spellman PT, Brown PO, Botstein D (1998) Cluster analysis and display of genome-wide expression patterns. Proc Natl Acad Sci U S A 95(25):14863–14868. https://doi.org/10.1073/pnas.95.25.14863

Elshire RJ, Glaubitz JC, Sun Q, Poland JA, Kawamoto K, Buckler ES, Mitchell SE (2011) A Robust, simple genotyping-by-sequencing (GBS) approach for high diversity species. PLoS One 6:e19379. https://doi.org/10.1371/journal.pone.0019379

Ensembl (2018) Ensembl plants. *Brassica napus* (AST_PRJEB5043_v1). http://plants.ensembl.org/*Brassica*_napus/Info/Index

Faggian R, Strelkov SE (2009) Detection and measurement of *Plasmodiophora brassicae*. In: *Plasmodiophora brassicae* (clubroot)—a plant pathogen that alters host growth and productivity, Dixon RJ (ed). J Plant Growth Reg 28(Spl):282–287

Faggian R, Bulman SR, Lawrie AC, Porter IJ (1999) Specific PCR primers for the detection of *Plasmodiophora brassicae* in soil and water. Phytopathology 89:392–397

Fan Z, Rimmer SR, Stefansson BR (1983) Inheritance of resistance to *Albugo candida* in rape (*Brassica napus* L.). Can J Genet Cytol 25:420–424

Feng J, Xiao Q, Hwang SF, Strelkov SE, Gossen BD (2012) Infection of canola by secondary zoospores of *Plasmodiophora brassicae* produced on a non host. Eur J Plant Pathol 132:309–315

Feng J, Hwang SF, Strelkov SE (2013a) Assessment of gene expression profiles in primary and secondary zoospores of *Plasmodiophora brassicae* by dot-blot and real-time PCR. Microbiol Res 168:518–524

Feng J, Hwang SF, Strelkov SE (2013b) Studies into primary and secondary infection processes by *Plasmodiophora brassicae* on canola. Plant Pathol 62:177–183

Feng J, Hwang SF, Strelkov SE (2013c) Genetic transformation of the obligate parasite *Plasmodiophora brassicae*. Phytopathology 103:1052–1057

Feng J, Cao T, Rennie DC, Strelkov SE, Hwang SF (2014) Host-parasite interactions in clubroot of crucifers. Can J Plant Pathol 36(Suppl 1):113–321

Ferreira ME, Rimmer SR, Williams PH, Osborn TC (1995) Mapping loci controlling *Brassica napus* resistance to *Leptosphaeria maculans* under different screening conditions. Phytopathology 85:213–217

Finn RD, Clements J, Eddy SR (2011) HMMER web server: interactive sequence similarity searching. Nucleic Acids Res 39(Web Server issue):W29–W37. https://doi.org/10.1093/nar/gkr367

Finn RD, Attwood TK, Babbitt PC, Bateman A, Bork P, Bridge AJ, Chang HY, Dosztányi Z, El-Gebali S, Fraser M, Gough J, Haft D, Holliday GL, Huang H, Huang X, Letunic I, Lopez R, Lu S, Marchler-Bauer A, Mi H, Mistry J, Natale DA, Necci M, Nuka G, Orengo CA, Park Y, Pesseat S, Piovesan D, Potter SC, Rawlings ND, Redaschi N, Richardson L, Rivoire C, Sangrador-Vegas A, Sigrist C, Sillitoe I, Smithers B, Squizzato S, Sutton G, Thanki N, Thomas PD, Tosatto SC, Wu CH, Xenarios I, Yeh LS, Young SY, Mitchell AL (2017) InterPro in 2017-beyond protein family and domain annotations. Nucleic Acids Res 45:D190–D199

Flint-Garcia SA, Thornsberry JM, Buckler ES (2003) Structure of linkage disequilibrium in plants. Annu Rev Plant Biol 54:357–374

Fox DT, Williams PH (1984) Correlation of spore production by *A. candida* and a visual white rust rating scale. Can J Plant Pathol 6:175–178

Freeman J, Ward E, Calderon C, Mc Cartney AA (2002) Polymerase chain reaction (PCR) assay for the detection of inoculum of *Sclerotinia sclerotiorum*. Eur J Plant Pathol 108:877–886

Gamborg OL, Miller RA, Ojima K (1968) Nutrient requirements of suspension cultures of soybean root cells. Exp Cell Res 50:151–158

Gao F, Hirani AH, Liu J, Liu Z, Fu G, Wu C, McVetty PBE, Li G (2014) Fine mapping a clubroot resistance locus in Chinese cabbage. J Am Soc Hortic Sci 139:247–252

Garg H, Atri C, Sandhu PS, Kaur B, Renton M, Banga SK, Singh H, Singh C, Barbetti MJ, Banga SS (2010) High level of resistance to *Sclerotinia sclerotiorum* in introgression lines derived from hybridization between wild crucifers and the crop *Brassica* species *B. napus* and *B. juncea*. Field Crops Res 117(1):51–58

Gemenet DC, Wachira FN, Pathak RS, Munyiri SW (2010) Identification of molecular markers linked to drought tolerance using bulked segregant analysis in Kenyan maize (*Zea mays* L.) landraces. J Anim Plant Sci 9(1):1122–1134

Ghanbarnia K, Lydiate DJ, Rimmer SR, Li G, Kutcher HR, Larkan NJ, McVetty PBE, Fernando WGD (2012) Genetic mapping of the *Leptosphaeria maculans* avirulence gene corresponding to the *LepR1* resistance gene of *Brassica napus*. Theor Appl Genet 124:505–513

Ghanbarnia K, Fudal I, Larkan NJ, Links MG, Balesdent MH, Profotova B, Fernando DWG, Rouxel T, Borhan H (2015) Rapid identification of the *Leptosphaeria maculans* avirulence gene *AvrLm2* using an intraspecific comparative genomics approach. Mol Plant Pathol 16:699–709

Ghanbarnia K, Ma L, Larkan NJ, Haddadi P, Fernando WGD, Borhan MH (2018) *Leptosphaeria maculans* AvrLm9: a new player in the game of hide and seek with AvrLm4-7. Mol Plant Pathol 19(7):1754–1764

Gietz RD, Woods RA (2002) Transformation of yeast by lithium acetate/single-stranded carrier DNA/polyethylene glycol method. Methods Enzymol 350:87–96

Gilmour AR, Gogel BJ, Cullis BR, Thompson R (2009) ASReml user guide release 3.0 VSN International Ltd, Hemel Hempstead, HP1 1ES, UK. www.vsni.co.uk

Gish W (2003) WU BLAST. http://blast.wustl.edu

Gotz S, Garcia-Gomez JM, Terol J, Williams TD, Nagaraj SH, Nueda MJ, Robles M, Talon M, Dopazo J, Conesa A (2008) High–throughput functional annotation and data mining with the Blast2GO suite. Nucleic Acids Res 36:3420–3435. https://doi.org/10.1093/nar/gkn176

Govindaraj P, Arumugachamy S, Maheswaran M (2005) Bulked segregant analysis to detect main effect QTL associated with grain quality parameters in Basmati 370/ASD16 cross in rice (*Oryza sativa* L.) using SSR markers. Euphytica 144:61–68

Goyal BK, Verma PR, Spurr DT (1995) Temperature effects on oospore development of *Albugo candida* race 2V in detached *Brassica juncea* leaves. Indian J Mycol Plant Pathol 26:224–228

Goyal BK, Verma PR, Spurr DT, Reddy MS (1996a) *Albugo candida* staghead formation in *Brassica juncea* in relation to plant age, inoculation sites, and incubation conditions. Plant Pathol 45:787–794

Goyal BK, Verma PR, Seguin-Swartz G, Spurr DT (1996b) Growth of *Albugo candida* in leaf callus culturesof *Brassica rapa*. Can J Plant Pathol 18:225–232

Graichen K, Rabenstein F (1996) European isolates of Beet western yellows virus (BWYV) from oilseed rape (*Brassica napus* L. ssp. *napus*) are non-pathogenic on sugar beet (*Beta vulgaris* L. var. *altissima*) but represent isolates of Turnip yellows virus (TuYV). J Plant Dis Prot 103:233–245

Greenhalgh JR, Dickinson CH (1975) Differential reactions of three crucifers to infection by *Peronospora parasitica* Pers. (Ex Fr.) Fr. Phytopathol Z 84:131–141

Griffiths PD, Roe C (2005) Response of *Brassica oleracea* var. *capitata* to wound and spray inoculations with *Xanthomonas campestris* pv. *campestris*. Hortic Sci 40(1):47–49. https://doi.org/10.21273/HORTSCI.40.1.47

Griffiths PD, Marek LF, Robertson LD (2009) Identification of crucifer accessions from the NC-7 and NE-9 plant introduction collections that are resistant to black rot (*Xanthomonas campestris* pv. *campestris*) races 1 and 4. Hortic Sci 44:284–288

Grontoft M, O'Connor D (1990) Green house method of testing resistance of cotyledons to downy mildew in *Brassica napus* and *B. campestris*. Plant Breed 105:160–165

Guillemette T, Iacomi-Vasilescu B, Simoneau P (2004) Conventional and real-time PCR-based assay for detecting pathogenic *Alternaria brassicae* in cruciferous seed. Plant Dis 88:490–496

Gupta SC, Kapoor VK (2002) Fundamentals of mathematical statistics, 10th edn. Sultan Chand and Sons, New Delhi

Habu JB, Ibeh BO (2015) In vitro antioxidant capacity and free radical scavenging evaluation of active metabolite constituents of *Newbouldia laevis* ethanolic leaf extract. Biol Res 48:16

Haddadi P, Ma L, Wang H, Borhan MH (2016) Genome-wide transcriptome analyses provides insights into the lifestyle transition and effector repertoire of *Leptosphaeria maculans* during colonization of *Brassica napus* seedlings. Mol Plant Pathol 17(8):1196–1210. https://doi.org/10.1111/mpp.12356

Hammond M, Jordan RL (1990) Dot blots (viruses) and colony screening. In: Hamptom R, Ball S, De Boer S (eds) Serological methods for the detection and identification of viral and bacterial plant pathogens—a laboratory manual. APS Press, 389St Paul, MN

Hardy OJ, Vekemans X (2002) SPAGeDi: a versatile computer program to analyse spatial genetic structure at the individual or population levels. Mol Ecol Notes 2:618–620. https://doi.org/10.1046/j.1471-8286.2002.00305.x

Harling R (2007) Clubroot control using novel and sustainable methods. Final report DEFRA project HH3227TFV

Hasan MJ, Rahman H (2016) Genetics and molecular mapping of resistance to *Plasmodiophora brassicae* pathotypes 2, 3, 5, 6, and 8 in rutabaga (*Brassica napus* var. *napobrassica*). Genome 59:805–815. https://doi.org/10.1139/gen-2016-0034

Hasan M, Strelkov SE, Howard RJ, Rahman H (2012) Screening of *Brassica* germplasm for resistance to *Plasmodiophora brassicae* pathotypes prevalent in Canada for broadening diversity in clubroot resistance. Can J Plant Sci 92:501–515

Hatzig SV, Frisch M, Breuer F, Nesi N, Ducournau S, Wagner MH, Leckband G, Abbadi A, Snowdon RJ (2015) Genome-wide association mapping unravels the genetic control of seed germination and vigor in *Brassica napus*. Front Plant Sci 6:221. https://doi.org/10.3389/fpls.2015.00221

Havranek P, Navratilova B, Chytilova V (2005) Use of rooted leaves for screening of Brassica germplasm response to clubroot (*Plasmodiophora brassicae*) and downy mildew (*Hyaloperonospora parasitica*). Hortic Sci (Prague) 32:1–5

He C, Holme J, Anthony J (2014) SNP genotyping: the KASP assay. In: Fleury D, Whitford R (eds) Crop breeding. Methods in molecular biology (methods and protocols), vol 1145. Humana Press, New York

Hirani AH, Gao F, Liu J, Fu G, Wu C, McVetty PBE, Duncan RW, Li G (2018) Combinations of independent dominant loci conferring clubroot resistance in all four turnip accessions (*Brassica rapa*) from the European clubroot differential set. Front Plant Sci 9:1628. https://doi.org/10.3389/fpls.2018.01628

Hirata T, Takamatsu S (1996) Nucleotide sequence diversity of rDNA internal transcribed spacers extracted from conidia and cleistothecia from several powdery mildew fungi. Mycoscience 37:283–288

Hodgkin T (1990) *In vitro* pollen selection in *Brassica napus* L. for resistance to phytotoxic compounds from *Alternaria brassicicola* (Schw.) Wilts. Sex Plant Reprod 3:116–120

Hodgkin T, MacDonald MV (1986) The effect of a phytotoxin from *Alternaria brassicicola* on *Brassica* pollen. New Phytol 104:631–636

Homma Y (1937) Erysiphaceae of Japan. J Fac Agric Hokkaida Univ 38:183–461

Horiuchi S, Hori M (1980) A simple greenhouse technique for obtaining high levels of clubroot incidence. Bull Chugoku Natl Agril Expt Sta Ser E (Env Div) 17:33–55

Hsu HT, Lawson RH (1991) Direct tissue blotting for detection of Tomato spotted wilt virus. Plant Dis 75:292–295

Iacomi-Vasilescu B, Avenot H, Bataille-Simoneau N, Laurent E,Guenard M, Simoneau P (2004) *In vitro* fungicide sensitivity of *Alternaria* species pathogenic to crucifers and identification of *Alternaria brassicicola* field isolates highly resistant to both di-carboximides and phenylpyrroles. Crop Prot 23:481–488

Ingram DS (1969) The susceptibility of *Brassica* callus to infection by *Peronospora parasitica*. J Gen Microbiol 58:391–401

Innis MA, Gelfand DH (1990) Optimization of PCRs. In: Innis MA, Gelfand DH, Sninsky JJ, White TJ (eds) PCR protocols: a guide to methods and applications. Academic, San Diego, pp 3–12

Ito S, Maechara T, Tanaka S, Kameya-Iwaki M, Yano S, Kishi F (1997) Cloning of single-copy DNA sequence unique to *Plasmodiophora brassicae*. Physiol Mol Plant Pathol 50:289–300

Ito S, Maehara T, Maruno E, Tanaka S, Kameya-Iwaki M, Kishi F (1999) Development of a PCR-based assay for the detection of *Plasmodiophora brassicae* in soil. J Phytopathol 147:83–88

Jenkyn JF, Hirst JM, King G (1973) An apparatus for the isolated propagation of foliar pathogens and their hosts. Ann Appl Biol 73:9–13

Jenner CE, Walsh JA (1996) Pathotypic variation in turnip mosaic virus with special reference to European isolates. Plant Pathol 45:848–856

Jenner CE, Keane GJ, Jones JE, Walsh JA (1999) Serotypic variation in turnip mosaic virus. Plant Pathol 48:101–108

Jin-ping L, Yan L, Yan-xia S, Xue-wen X, A-li C, Bao-ju L (2013) Development of a real-time PCR assay for *Plasmodiophora brassicae* and its detection in soil samples. J Integr Agric 12 (10):1799–1806

Jobling SA, Gehrke L (1987) Enhanced translation of chimaeric messenger RNAs containing a plant viral untranslated leader sequence. Nature 325:622–625

Johnston TD (1968) Clubroot in *Brassica*: a standard inoculation technique and the specification of races. Plant Pathol 17:184–187

Jones PM (1928) Morphology and cultural history of *Plasmodiophora brassicae*. Arch Protistenkd 62:313–327

Jordan R, Hammond J (1991) Comparison and differentiation of potyvirus isolates and identification of strain-, virus-, subgroup-specific and potyvirus group-common epitopes using monoclonal antibodies. J Gen Virol 72:25–36

Kapadia VN, Sasidharan N, Patel DA, Kalyanrao (2019) Identification of the SSR marker linked to powdery mildew resistant gene(s) in *Brassica* spp. J Pharmacogn Phytochem 8(3):388–395

Kapoor KS, Gill HS, Sharma SR (1986) A technique for artificial inoculation of curd stage of cauliflower with *Sclerotinia sclerotiorum*. Cruciferae Newslett 11:13

Kato T, Hatakayema K, Fukino N, Matsumoto S (2012) Identification of a clubroot resistance locus conferring resistance to a *Plasmodiophora brassicae* classified into pathotype group 3 in Chinese cabbage (*Brassica rapa* L.). Breed Sci 62:282–287. https://doi.org/10.1270/jsbbs.62.282

Kersey PJ, Allen JE, Allot A, Barba M, Boddu S, Bolt BJ, Carvalho-Silva D, Christensen M, Davis P, Grabmueller C, Kumar N, Liu Z, Maurel T, Moore B, McDowall MD, Maheswari U, Naamati G, Newman V, Ong CK, Paulini M, Pedro H, Perry E, Russell M, Sparrow H, Tapanari E, Taylor K, Vullo A, Williams G, Zadissia A, Olson A, Stein J, Wei S, Tello-Ruiz M, Ware D, Luciani A, Potter S, Finn RD, Urban M, Hammond-Kosack KE, Bolser DM, De Silva N, Howe KL, Langridge N, Maslen G, Staines DM, Yates A (2018) Ensembl Genomes 2018: an integrated omics infrastructure for non-vertebrate species. Nucleic Acids Res 46: D802–D808. https://doi.org/10.1093/nar/gkx1011

Kim H, Choi SR, Bae J (2009) Sequenced BAC anchored reference genetic map that reconciles the ten individual chromosomes of *Brassica rapa*. BMC Genomics 10:432. https://doi.org/10.1186/1471-2164-10-432

Knight C, Furber MJ (1980) Winter oilseed rape disease variety reaction. Nat Inst Agric Bot 15:276–280

Kolte SJ, Sachan JN, Rimmer SR (1998) A new simple technique for development of seedling infection of white rust of rapeseed-mustard from oospores of *A. candida*. Indian Phytopathol 51:287–289

Korgan S, Wolski EA, Cicore P, Suarez P, Capezio S, Huarte MA, Andreu AB (2011) *Solanum tarijense* reaction to *Phytophthora infestans* and the role of plant defense molecules. Plant Breed 130:231–236

Kosambi DD (1943) The estimation of map distances from recombination values. Ann Eugenics 12:172–175

Kosambi DD (1944) The estimation of map distances from recombination values. Ann Eugenics 12:172–175

Krober H (1970) Longevity of conidia of several *Peronospora* species under different environmental conditions. Phytopathol Z 69:64–70

Krober H (1981) Germination and longevity of conidia of some Peronosporaceae kept under different conditions. Z Pflanzenkrankh Und Pflanzenschutz 88:510–517

Kuginuki Y, Hiroaki Y, Hirai M (1999) Variation in virulence of *Plasmodiophora brassicae* in Japan tested with clubroot-resistant cultivars of Chinese cabbage (*Brassica rapa* L. ssp. *pekinensis*). Eur J Plant Pathol 105:327–332

Kwok S, Kellogg DE, McKinnery N, Spasic D, Goda L, Levenson C, Sninsky JJ (1990) Effects of primer-template mismatches on the polymerase chain reaction: human immunodeficiency virus type 1 model studies. Nucleic Acids Res 18:999–1005

Lakra BS, Saharan GS (1988) Influence of host resistance on colonization and incubation period of *A. candida* in mustard. Cruciferae Newslett 13:108–109

Lakra BS, Saharan GS (1989a) Location and estimation of oospores of *Albugo candida* in infected plant parts of mustard. Indian Phytopathol 42:467

Lakra BS, Saharan GS (1989b) Sources of resistance and effective screening techniques in *Brassica-Albugo* system. Indian Phytopathol 42:293. (Abstr.)

Lakra BS, Saharan GS (1989c) Sporangial viability of *Albugo candida* infecting mustard under different storage conditions. Oil Crops Newslett 6:22–24

Lander ES, Green P, Abrahamson J, Barlow A, Daly MJ, Lincoln SE, Newburg L (1987) MAPMAKER: an interactive computer package of constructing primary genetic linkage maps of experimental and natural populations. Genomics 1:174–181

Langmead B, Salzberg SL (2012) Fast gapped-read alignment with Bowtie 2. Nat Methods 9:357–360. https://doi.org/10.1038/nmeth.1923

Lark RM, Ferguson RB (2004) Mapping risk of soil nutrient deficiency or excess by disjunctive and indicator kriging. Geoderma 118:39–53

Larkan NJ, Lydiate DJ, Parkin IA, Nelson MN, Epp DJ, Cowling WA, Rimmer SR, Borhan MH (2013) The *Brassica napus* blackleg resistance gene LepR3 encodes a receptor-like protein triggered by the *Leptosphaeria maculans* effector AVRLM1. New Phytol 197:595–605. https://doi.org/10.1111/nph.12043

Larkan NJ, Lydiate DJ, Yu F, Rimmer SR, Borhan MH (2014) Co-localization of the blackleg resistance genes *Rlm2* and *LepR3* on *Brassica napus* chromosome A10. BMC Plant Biol 14:387. https://doi.org/10.1186/s12870-014-0387-z

Larkan NJ, Ma L, Borhan H (2015) The *Brassica napus* receptor-like protein RLM2 is encoded by a second allele of the *LepR3/Rlm2* blackleg resistance locus. Plant Biotechnol J 13:983–992

Larkan NJ, Raman H, Lydiate DJ, Robinson SJ, Yu F, Barbulescu DM, Raman R, Luckett DJ, Burton W, Wratten N, Salisbury PA, Rimmer SR, Borhan MH (2016a) Multi-environment QTL studies suggest a role for cysteine-rich protein kinase genes in quantitative resistance to blackleg disease in *Brassica napus*. BMC Plant Biol 16:183. https://doi.org/10.1186/s12870-016-0877-2

Larkan NJ, Yu F, Lydiate DJ, Rimmer SR, Borhan MH (2016b) Single R gene introgression lines for accurate dissection of the *Brassica-Leptosphaeria* pathosystem. Front Plant Sci 7:1771. https://doi.org/10.3389/fpls.2016.01771

Leckie D, Astley D, Crute IR, Ellis PR, Pink DAC, Boukema I, Monteiro AA, Dias JS (1996) The location and exploitation of genes for pest and disease resistance in European gene bank collection of Horticultural *Brassicas*.In*:* Proc intl symp on brassicas, 9th crucifer genetics workshop, Dias JS, Crute IR, Monteiro AA (eds). Acta Hortic 407:95–101

Lee JH, Kim J-C, Jang KS, Choi YH, Ahn KG, Choi GJ (2013) Development of efficient screening method for resistance of cabbage cultivars to black rot disease caused by *Xanthomonas campestris* pv. *campestris*. Res Plant Dis 19:95–101

Lee J, Izzah NK, Choi BS, Joh HJ, Lee SC, Perumal S, Seo J, Ahn K, Jo EJ, Choi GJ, Nou IS, Yu Y, Yang TJ (2016) Genotyping-by-sequencing map permits identification of clubroot resistance QTLs and revision of the reference genome assembly in cabbage (*Brassica oleracea* L.). DNA Res 23:29–41. https://doi.org/10.1093/dnares/dsv034

Lema M, Velasco P, Soengas P, Francisco M, Cartea ML (2012) Screening for resistance to black rot in *Brassica oleracea* crops. Plant Breed 131:607–613

Leone G, Tonneijck AEG (1990) A rapid procedure for screening the resistance of bean cultivars (*Phaseolus valgaris* L) to *Botrysis cinerea* and *Sclerotinia sclerotiorum*. Euphytica 48:87–90

Leung H (1981) Breeding for multiple disease resistance in cytoplasmic male strile Brassica campestris L. MSc Thesis, Department of Plant Pathology, University of Wisconsis, Madison, USA, 73 p

Lewis SJ, Brokenshire T (1978) A comparison of plant inoculation techniques for the determination of *Plasmodiophora brassicae* populations. In: Proc. Woronin +100 conference, Madison, pp 54–67

Li G, Quiros C (2001) Sequence-related amplified polymorphism (SRAP), a new marker system based on a simple PCR reaction: its application to mapping and gene tagging in *Brassica*. Theor Appl Genet 103:455–461

Li CX, Li H, Sivasithamparam K, Fu TD, Li YC, Liu SY, Barbetti MJ (2006) Expression of field resistance under Western Australian conditions to *Sclerotinia sclerotiorum* in Chinese and Australian *Brassica napus* and *Brassica juncea* germplasm and its relation with stem diameter. Aust J Agric Res 57:1131–1135

Li CX, Sivasithamparam K, Walton G, Salisbury P, Burton W, Banga SS, Banga S, Chattopadhyay C, Kumar A, Singh R, Singh D, Agnohotri A, Liu SY, Li YC, Fu TD, Wang YF, Barbetti MJ (2007a) Expression and relationships of resistance to white rust (*Albugo candida*) at cotyledonary, seedling, and flowering stages in *Brassica juncea* germplasm from Australia, China, and India. Aust J Agric Res 58:259–264

Li CX, Li H, Siddique AB, Sivasithamparam K, Salisbury P, Banga SS, Banga S, Chattopadhyay C, Kumar A, Singh R, Singh D, Agnihotri A, Liu SY, Li YC, Tu J, Fu TD, Wang YF, Barbetti MJ (2007b) The importance of the type and time of inoculation and assessment in the determination of resistance in *Brassica napus* and *B. juncea* to *Sclerotinia sclerotiorum*. Aust J Agric Res 58:1198–1203. https://doi.org/10.1071/AR07094

Li CX, Sivasithamparam K, Walton G, Fels P, Barbetti MJ (2008) Both incidence and severity of white rust disease reflect host resistance in *Brassica juncea* germplasm from Australia, China and India. Field Crops Res 106:1–8

Lichter R (1982) Induction of haploid plants from isolated pollen of *Brassica napus*. Z Pflanzenphysiol 105(5):427–434. https://doi.org/10.1016/j.bbr.2011.03.031

Lin NS, Hsu YH, Hsu HT (1990) Immunological detection of plant viruses and a mycoplasmalike organism by direct tissue blotting on nitrocellulose membranes. Phytopathology 80:824–828. https://doi.org/10.1094/Phyto-80-824

Liu JQ, Rimmer SR, Scarth R (1989) Histopathology of compatibility and incompatibility between oilseed rape and *Albugo candida*. Plant Pathol 38:176–182

Liu Y, Huang X, Li M, He P, Zhang Y (2016) Loss-of-function of *Arabidopsis* receptor-like kinase BIR1 activates cell death and defense responses mediated by BAK1 and SOBIR1. New Phytol 212:637–645

Livak KJ, Schmittgen TD (2001) Analysis of relative gene expression data using real-time quantitative PCR and the $2^{-\Delta\Delta C_T}$ method. Methods 25:402–408

Loreto F, Velikova V (2001) Isoprene produced by leaves protects the photosynthetic apparatus against ozone damage, quenches ozone products, and reduces lipid peroxidation of cellular membranes. Plant Physiol 127:1781–1787

Loreto F, Mannozzi M, Maris C, Nascetti P, Ferranti F, Pasqualini S (2001) Ozone quenching properties of isoprene and its antioxidant role in leaves. Plant Physiol 126:993–1000

Love MI, Huber W, Anders S (2014) Moderated estimation of fold change and dispersion for RNA-Seq data with DESeq2. bioRxiv. https://doi.org/10.1101/002832

Lowe AJ, Jones AE, Raybould AF, Trick M, Moule CJ, Edwards KJ (2002) Transferability and genome specificity of a new set of microsatellite primers among *Brassica* species of the U triangle. Mol Ecol Notes 2:7–11. https://doi.org/10.1046/j.1471-8286.2002.00126.x

Lowe A, Moule C, Trick M, Edwards K (2004) Efficient large scale development of microsatellites for marker and mapping applications in *Brassica* crop species. Theor Appl Genet 108:1103–1112. https://doi.org/10.1007/s00122-003-1522-7

Lucas JA, Crute IR, Sherriff C, Gordon PL (1988) The identification of a gene for race-specific resistance to *Peronospora parasitica* (downy mildew) in *Brassica napus* var. *oleifera* (oilseed rape). Plant Pathol 37:538–545

Ma J, Hou X, Xiao D, Qi L, Wang F, Sun F, Wang Q (2010) Cloning and characterization of the *BcTuR3* gene related to resistance to *Turnip mosaic virus* (TuMV) from non-heading chinese cabbage. Plant Mol Biol Rep 28:588–596

MacDonald MV, Ingram DS (1985) *In vitro* selection for resistance to *Alternaria brassicicola* in *Brassica napus* spp. *oleifera* (winter oilseed rape) using partially purified culture filtrates. Cruciferae Newslett 10:97–100

MacDonald MV, Ingram DS (1986) Towards the selection *in vitro* for resistance to *Alternaria brassicicola* (Schw.) Wilts., in *Brassica napus* ssp. *oleifera* (Metzg.) Sinsk., winter oilseed rape. New Phytol 104:621–629

MacDonald MV, Ingram DS, Hodgkin T (1986) Selection *in vitro* for resistance to *Alternaria brassicicola* in *Brassica napus* spp. *oleifera*, winter oilseed rape. In: Proc crucifer genetic workshop 3, crucifer genetic co-operative, University of Wisconsin, Madison, Wisconsin, p 66

Maji A, Nath R (2015) Pathogenicity test by using different inoculation methods on *Xanthomonas campestris* pv. *campestris* caused of black rot of cabbage. IMPACT Int J Res Appl Nat Soc Sci 3 (2):53–58

Makkouk K, Comeau A (1994) Evaluation of various methods for the detection of barley yellow dwarf virus by the tissue-blot immunoassay and its use for virus detection in cereals inoculated at different growth stages. Eur J Plant Pathol 100:71–80

Manzanares-Dauleux MJ, Barret P, Thomas G (2000) Development of a pathotype specific SCAR marker in *Plasmodiophora brassicae*. Eur J Plant Pathol 106(8):781–787

Marchler-Bauer A, Bryant SH (2004) CD-Search: protein domain annotations on the fly. Nucleic Acids Res 32:W327–W331

Marchler-Bauer A, Lu S, Anderson JB, Chitsaz F, Derbyshire MK, DeWeese-Scott C, Fong JH, Geer LY, Geer RC, Gonzales NR, Gwadz M, Hurwitz DI, Jackson JD, Ke Z, Lanczycki CJ, Lu F, Marchler GH, Mullokandov M, Omelchenko MV, Robertson CL, Song JS, Thanki N, Yamashita RA, Zhang D, Zhang N, Zheng C, Bryant SH (2011) CDD: a conserved domain database for the functional annotation of proteins. Nucleic Acids Res 39:D225–D229

Marcroft SJ, Elliott VL, Cozijnsen AJ, Salisbury PA, Howlett BJ, Van De Wouw AP (2012) Identifying resistance genes to *Leptosphaeria maculans* in Australian *Brassica napus* cultivars based on reactions to isolates with known avirulence genotypes. Crop Pasture Sci 63:338–350. https://doi.org/10.1071/CP11341

Martin M (2011) Cutadapt removes adapter sequences from high-throughput sequencing reads. EMBnet J 17:10–12

Maxwell DP, Lumsden RD (1970) Oxalic acid production by *Sclerotinia sclerotiorum* in infected bean and in culture. Phytopathology 60:1395–1398

Mayee CD, Datar VV (1986) Phytopathometry Bulletin-1. Marathawada Agricultural University, Parbhani, p 46

McCullagh P, Nelder JA (1989) Generalized linear models, 2nd edn. Chapman & Hall, London

McDonnell AV, Jiang T, Keating AE, Berger B (2006) Paircoil2: improved prediction of coiled coils from sequence. Bioinformatics 22(3):356–358. https://doi.org/10.1093/bioinformatics/bti797

McMeekin D (1971) Measuring systemic infection of crucifers by downy mildew. Plant Dis Rep 55:877–878

McMeekin D (1981) The response of *Peronospora parasitica* to a liver medium. Mycologia 73:551–554

Meena PD (2007) Reoprt on Research Attachment. DFID INDO-UK Collaboration on Oilseeds (Rapeseed-Mustard) Phase II 1998-2007. Rothamsted Research, Harpenden, Herts, UK, 52 p

Meglecz E, Costedoat C, Dubut V, Gilles A, Malausa T, Pech N, Martin JF (2010) QDD: a user-friendly program to select microsatellite markers and design primers from large sequencing projects. Bioinformatics 26(3):403–404. https://doi.org/10.1093/bioinformatics/btp670. PMID: 20007741

Mehra S, Pareek A, Bandyopadhyay P, Sharma P, Burma P, Pental D (2000) Development of trans-genicsin Indian oilseed mustard (*Brassica juncea*) resistant to herbicide phosphinothricin. Curr Sci 78:1358–1364

Mei J, Wei D, Disi JO, Ding Y, Liu Y, Qian W (2012) Screening resistance against *Sclerotinia sclerotiorum* in *Brassica* crops with use of detached stem assay under controlled environment. Eur J Plant Pathol 134:599–604. https://doi.org/10.1007/s10658-012-0040-3

Mei J, Ding Y, Lu K, Wei D, Liu Y, Disi JO, Li J, Liu L, Liu S, McKay J, Qian W (2013) Identification of genomic regions involved in resistance against *Sclerotinia sclerotiorum* from wild *Brassica oleracea*. Theor Appl Genet 126:549–556. https://doi.org/10.1007/s00122-012-2000-x

Mei J, Liu Y, Wei D, Wittkop B, Ding Y, Li Q, Li J, Wan H, Li Z, Ge X, Frauen M, Snowdon RJ, Qian W, Friedt W (2015) Transfer of *Sclerotinia* resistance from wild relative of *Brassica oleracea* into *Brassica napus* using a hexaploidy step. Theor Appl Genet 128:639–644

Mei J, Shao C, Yang R, Feng Y, Gao Y, Ding Y, Li J, Qian W (2020) Introgression and pyramiding of genetic loci from wild *Brassica oleracea* into *B. napus* for improving *Sclerotinia* resistance of rapeseed. Theor Appl Genet 133:1313–1319. https://doi.org/10.1007/s00122-020-03552-w

Mengistu A, Hill CB, Williams PH (1995) A tooth pick method for mating *Leptosphaeria maculans*, the causal agent of blackleg of crucifers. Plant Dis 79:755–756

Meyers BC, Kozik A, Griego A, Kuang H, Michelmore RW (2003) Genome-wide analysis of NB-LRR-encoding genes in *Arabidopsis*. Plant Cell 15:809–834

Milad SI, Wahba LE, Barakat MN (2011) Identification of RAPD and ISSR markers associated with flag leaf senescence under water stressed conditions in wheat (*Triticum aestivum* L.). Aust J Crop Sci 5(3):337–343

Millenaar FF, Gonzalez-Meler MA, Siedow JN, Wagner AM, Lambers H (2002) Role of sugars and organic acids in regulating the concentration and activity of the alternative oxidase in *Poa annua* roots. J Exp Bot 53:1081–1088

Milošević D, Marjanović-Jeromela A, Ignjatov M, Jovičić D, Stanković I, Bulajić A, Krstić B (2015) First report of Turnip yellows virus on Oilseed rape in Serbia. Plant Dis 99(12):1869

Milošević D, Ignjatov M, Nikolić Z, Stanković I, Bulajić A, Marjanović-Jeromela A, Krstić B (2016) The presence of Turnip yellows virus in oilseed rape (*Brassica napus* L.) in Serbia. Pestic Phytomed (Belgrade) 31(1–2):37–44

Mori Y, Sato Y, Takamatsu S (2000) Evolutionary analysis of the powdery mildew fungi using nucleotide sequences of the nuclear ribosomal DNA. Mycologia 92:74–93

Moukette BM, Pieme CA, Njimou JR, Biapa CPN, Marco B, Ngogang JY (2015) *In vitro* antioxidant properties, free radicals scavenging activities of extracts and polyphenol composition of a non-timber forest product used as spice: *Monodora myristica*. Biol Res 48:15

Mrode R, Thompson R (2005) Linear models for the prediction of animal breeding values. https://books.google.com.au/books?id=bnewaF4Uq2wC; https://doi.org/10.1079/9780851990002.0000

Muller PY, Janovjak H, Miserez AR, Dobbie Z (2002) Processing of gene expression data generated by quantitative real-time RT–PCR. Biotechniques 32:1372–1379

Murashige T, Skoog F (1962) A Revised medium for rapid growth and bio assays with tobacco tissue cultures. Physiol Plant 15:473–497

Nagaoka T, Doullah MA, Matsumoto S, Kawasaki S, Ishikawa T, Hori H, Okazaki K (2010) Identification of QTLs that control clubroot resistance in *Brassica oleracea* and comparative analysis of clubroot resistance genes between *B. rapa* and *B. oleracea*. Theor Appl Genet 120:1335–1346. https://doi.org/10.1007/s00122-010-1259-z

Nashaat NI, Rawlinson CJ (1994) The response of oilseed rape (*Brassica napus* ssp. oleifera) accessions with different gluconsinolate and erucic acid contents to four isolates of *Peronospora parasitica* (downy mildew) and the identification of new sources of resistance. Plant Pathol 43:278–285

Natti JJ, Dickson MH, Atkin JD (1967) Resistance of *Brassica oleracea* varieties to downy mildew. Phytopathology 57:144–147

Nelson MN, Rajasekaran R, Smith A, Chen S, Beeck CP, Siddique KH, Cowling WA (2014) Quantitative trait loci for thermal time to flowering and photoperiod responsiveness discovered in summer annual-type *Brassica napus* L. PLoS One 9:e102611. https://doi.org/10.1371/journal.pone.0102611

Nieuwhof M, Wiering D (1961) Testing cabbage plants for clubroot resistance. Euphytica 10 (2):191–200. https://doi.org/10.1007/BF00022210

NIH (2012) Primer3. http://bioinfo.ut.ee/primer3-0.4.0/

Nowakowska M, Wrzesińska M, Kaminski P, Szczechura W, Lichocka M, Tartanus M, Kozik EU, Nowicki M (2019) *Alternaria brassicicola*—Brassicaceae pathosystem: insights into the infection process and resistance mechanisms under optimized artificial bio-assay. Eur J Plant Pathol 153:131–151. https://doi.org/10.1007/s10658-018-1548-y

Nowicki M, Klosinska U, Szczechura W, Markiewicz M, Sowik I, Michalczuk L, Kozik E U (2016) Cellular and biochemical mechanisms of cucumber resistance against *Pseudoperonospora cubensis*. In: Cucurbitaceae 2016, XIth Eucarpia meeting on cucurbit genetics & breeding, July 24–28, 2016, Warsaw, Poland, Cucurbitaceae 2016 Organizing Committee, pp 128–131

Numa H, Itoh T (2014) MEGANTE: a Web-based system for integrated plant genome annotation. Plant Cell Physiol 55:e2. https://doi.org/10.1093/pcp/pct157

O'Brien MJ, Webb RE (1958) Preservation of conidiaof *A. occidentalis* and *Peronospora effusa* obligate parasites of spinach. Plant Dis Rep 42:1312–1315

Ohguchi T, Asada Y (1981) Oospore formation of Japanese radish downy mildew fungus. Ann Phytopathol Soc Jpn 47:75–77

Panjabi P, Jagannath A, Bisht NC, Padmaja KL, Sharma S, Gupta V, Pradhan AK, Pental D (2008) Comparative mapping of *Brassica juncea* and *Arabidopsis thaliana* using Intron Polymorphism (IP) markers: homoeologous relationships, diversification and evolution of the A, B and C *Brassica* genomes. BMC Genomics 9:113. https://doi.org/10.1186/1471-2164-9-113

Panjabi-Massand P, Yadava SK, Sharma P, Kaur A, Kumar A, Arumugam N, Sodhi YS, Mukhopadhyay A, Gupta V, Pradhan AK, Pental D (2010) Molecular mapping reveals two independent loci conferring resistance to *Albugo candida* in the east European germplasm of oilseed mustard *Brassica juncea*. Theor Appl Genet 121:137–145. https://doi.org/10.1007/s00122-010-1297-6

Parlange F, Daverdin G, Fudal I, Kuhn ML, Balesdent MH, Blaise F, Grezes-Besset B, Rouxel T (2009) *Leptosphaeria maculans* avirulence gene AvrLm4-7 confers a dual recognition specificity by the Rlm4 and Rlm7 resistance genes of oilseed rape, and circumvents Rlm4-mediated recognition through a single amino acid change. Mol Microbiol 71:851–863. https://doi.org/10.1111/j.1365-2958.2008.06547.x

Patterson HD, Thompson R (1971) Recovery of inter-block information when block sizes are unequal. Biometrika 58:545–554. https://doi.org/10.1093/biomet/58.3.545

Paul VH, Klodt-Bussmann E (1993) First results from simple methods of preservation of *Peronospora parasitica*. IOBC/WPRS Bull 16:25–28

Paul VH, Klodt-Bussmann E, Dapprich PD, Capelli C, Tewari JP (1998) Results on preservation, epidemiology, and aggressiveness of *Peronospora parasitica* and results with regard to the disease resistance of the pathogen on *Brassica napus*. Bull OILB/SROP 2:49–56

Penazova E, Kopta T, Jurica M, Pecenka J, Eichmeier A, Pokluda R (2018) Testing of inoculation methods and susceptibility testing of perspective cabbage breeding lines (*Brassica oleracea* convar. *capitata*) to the black rot disease caused by *Xanthomonas campestris* pv. *campestris*. Acta Univ Silv Mendel Brun 66:139–148

Peng L, Zhou L, Li Q, Wei D, Ren X, Song H, Mei J, Si J, Qian W (2018) Identification of quantitative trait loci for clubroot resistance in *Brassica oleracea* with the use of *Brassica* SNP microarray. Front Plant Sci 9:822. https://doi.org/10.3389/fpls.2018.00822

Perry JB, Williams PH (1984) Controlled droplet applicators for inoculation with plant pathogens. Plant Dis 68:107–108

Pfaffl MW, Tichopad A, Prgomet C, Neuvians TP (2004) Determination of stable housekeeping genes, differentially regulated target genes and sample integrity: BestKeeper–Excel-based tool using pair-wise correlations. Biotechnol Lett 26:509–515. https://doi.org/10.1023/B:BILE.0000019559.84305.47

Pierre JG, Charlet CD, Brun H, Penaual A, Vaysse C, Lacombe JP, Martin D, Regnault Y, Labrouhe DT, Sarrazin JF (1992) Techniques for the artificial contamination of oilseed rape in trial plots. CETIOM 26–33

Pierson PE, Rhodes LH (1992) Effect of culture medium on the production of oxalic acid by *Sclerotinia trifoliorum*. Mycologia 84:467–469

Piquemal J, Cinquin E, Couton F, Rondeau C, Seignoret E, Doucet I, Perret D, Villeger M, Vincourt P, Blanchard P (2005) Construction of an oilseed rape (*Brassica napus* L.) genetic map with SSR markers. Theor Appl Genet 111:1514–1523

Pound GS, Williams PH (1963) Biologal races of *Albugo candida*. Phytopathology 53:1146–1149

Pritchard JK, Stephens M, Donnelly P (2000) Inference of population structure using multilocus genotype data. Genetics 155:945–959

Pushpa HD, Devendra K, Yadava, Naveen S, Vasudev S, Saini N, Muthusamy V, Prabhu K (2016) Validation of molecular markers linked to low glucosinolate QTLs for marker assisted selection in Indian mustard [*Brassica juncea* (L.) Czern & Coss.]. Indian J Genet 76(1):64–68. https://doi.org/10.5958/0975-6906.2016.00009.2

Qi J, Yu S, Zhang F, Shen X, Zhao X, Yu Y, Zhang D (2010) Reference gene selection for real-time quantitative polymerase chain reaction of mRNA transcript levels in chinese cabbage (*Brassica rapa* L. ssp. *pekinensis*). Plant Mol Biol Rep 28:597–604

Quinlan AR, Hall IM (2010) Bedtools: a flexible suite of utilities for comparing genomic features. Bioinformatics 26:841–842

R Core Team (2013) *R:* a language and environment for statistical computing. R Foundation for Statistical Computing, Vienna

R Core Team (2014) R: a language and environment for statistical computing. R Foundation for Statistical Computing, Vienna. http://www.R-project.org/

Rahman H, Shakir A, Jakir Hasan M (2011) Breeding for clubroot resistant spring canola (*Brassica napus* L.) for the Canadian prairies: can the European winter canola cv. mendel be used as a source of resistance? Can J Plant Sci 91:447–458. https://doi.org/10.1093/bioinformatics/btq033

Rahmanpour S, Backhouse D, Nonhebel HM (2011) Reaction of *Brassica* species to *Sclerotinia sclerotiorum* applying inoculation techniques under controlled conditions. Crop Breed J 1 (2):143–149

Raman R, Taylor B, Lindbeck K, Coombes N, Barbulescu D, Salisbury P, Raman H (2012a) Molecular mapping and validation of *Rlm1* genes for resistance to *Leptosphaeria maculans* in canola (*Brassica napus* L). Crop Pasture Sci 63:1007–1017. https://doi.org/10.1071/CP12255

Raman R, Taylor B, Marcroft S, Stiller J, Eckermann P, Coombes N, Rehman A, Lindbeck K, Luckett D, Wratten N, Batley J, Edwards D, Wang X, Raman H (2012b) Molecular mapping of qualitative and quantitative loci for resistance to *Leptosphaeria maculans* causing blackleg disease in Canola (*Brassica napus* L.). Theor Appl Genet 125:405–418. https://doi.org/10.1007/s00122-012-1842-6

Raman H, Raman R, Kilian A, Detering F, Carling J, Coombes N, Diffey S, Kadkol G, Edwards D, McCully M, Ruperao P, Parkin IA, Batley J, Luckett DJ, Wratten N (2014) Genome-wide delineation of natural variation for pod shatter resistance in *Brassica napus*. PLoS One 9: e101673. https://doi.org/10.1371/journal.pone.0101673

Raman H, Raman R, McVittie B, Orchard B, Qiu Y, Delourme R (2017) A major locus for manganese tolerance maps on chromosome A09 in a doubled haploid population of *Brassica napus* L. Front Plant Sci 8:1952. https://doi.org/10.3389/fpls.2017.01952

Raman H, Raman R, Diffey S, Qiu Y, McVittie B, Barbulescu DM, Salisbury PA, Marcroft S, Delourme R (2018) Stable quantitative resistance loci to blackleg disease in canola (*Brassica napus* L.) over continents. Front Plant Sci 9:9

Rana K, Atri C, Gupta M, Akhatar J, Sandhu PS, Kumar N, Jaswal R, Barbetti MJ, Banga SS (2017) Mapping resistance responses to *Sclerotinia* infestation in introgression lines of *Brassica juncea* carrying genomic segments from wild *Brassicaceae B. fruticulosa*. Sci Rep 7:5904

Rana K, Atri C, Akhatar J, Kaur R, Goyal A, Singh MP, Kumar N, Sharma A, Sandhu PS, Kaur G, Barbetti MJ, Banga SS (2019) Detection of first marker trait associations for resistance against *Sclerotinia sclerotiorum* in *Brassica juncea–Erucastrum cardaminoides* introgression lines. Front Plant Sci 10:1015. https://doi.org/10.3389/fpls.2019.01015

Reby E, Michalik B (2005) Ocena podatności odmian wzorcowych kapusty z genami odporności na krajowe izolaty kiły (*Plasmodiophora brassicae*). In: Michalik B, Żurawicz E (eds) Zmienność genetyczna i jej wykorzystanie w hodowli roślin ogrodniczych. ISiK, Skierniewice, pp 37–42

Ren L, Fang XP, Sun CC, Chen KR, Liu F, Li M, Xu L (2014) First report of clubroot on *Capsella bursa-pastoris* caused by *Plasmodiophora brassicae* in Sichuan Province of China. Plant Dis 98 (5):687

Reuber TL, Plotnikova JM, Dewdney J, Rogers EE, Wood W, Ausubel FM (1998) Correlation of defense gene induction defects with powdery mildew susceptibility in *Arabidopsis* enhanced disease susceptibility mutants. Plant J 16(4):473–485

Robak J (1991) Variability of the pathotypes of *Plasmodiophora brassicae* Wor.present in Poland and their pathogenicity to the cultivars and breeding lines of *Brassica oleracea* [in Polish]. *Habilitation monograph*. Research Institute of Vegetable Crops, Skierniewice

Robinson MD, McCarthy DJ, Smyth GK (2010) edgeR: a Bio-conductor package for differential expression analysis of digital gene expression data. Bioinformatics 26:139–140

Rod J (1988) New inoculation method for the testing of *Brassica* for resistance to *Plasmodiophora brassicae* Worn. Crucferae Newslett 13:102–103

Rogers SO, Bendich AJ (1994) Extraction of total cellular DNA from plants, algae and fungi. Plant Mol Biol Manual D1:1–8

Rotmistrovsky K, Jang W, Schuler GD (2004) A web server for performing electronic PCR. Nucleic Acids Res 32(Web server issue):108–112. https://doi.org/10.1093/nar/gkh450. PMID:15215361

Rout K, Yadav BG, Yadava SK, Mukhopadhyay A, Gupta V, Pental D, Pradhan AK (2018) QTL landscape for oil content in *Brassica juncea*: analysis in multiple bi-parental populations in high and "0" erucic background. Front Plant Sci 9:1448

Rouxel T, Grandaubert J, Hane JK, Hoede C, Van de Wouw AP, Couloux A, Dominguez V, Anthouard V, Bally P, Bourras S, Cozijnsen AJ, Ciuffetti LM, Degrave A, Dilmaghani A, Duret L, Fudal I, Goodwin SB, Gout L, Glaser N, Linglin J, Kema GH, Lapalu N, Lawrence CB, May K, Meyer M, Ollivier B, Poulain J, Schoch CL, Simon A, Spatafora JW, Stachowiak A, Turgeon BG, Tyler BM, Vincent D, Weissenbach J, Amselem J, Quesneville H, Oliver RP, Wincker P, Balesdent MH, Howlett BJ (2011) Effector diversification within compartments of the *Leptosphaeria maculans* genome affected by repeat-induced point mutations. Nat Commun 2:202

Saha P, Kalia P, Sharma P, Sharma TR (2014) Race-specific genetics of resistance to black rot disease [*Xanthomonas campestris* pv. *campestris* (Xcc) (Pammel) Dowson] and the development of three random amplified polymorphic DNA markers in cauliflower. J Hortic Sci Biotechnol 89:480–486

Saharan GS, Kaushik CD, Kaushik JC (1988) Sources of resistance and epidemiology of white rust of mustard. Indian Phytopathol 41:96–99

Saharan GS, Naresh M, Meena PD (2016) Alternaria blight of crucifers: biology, ecology and disease management. Springer, Singapore, p 326. ISBN 978-981-10-0019-5

Saito M, Kubo N, Matsumoto S, Suwabe K, Tsukada M, Hirai M (2006) Fine mapping of the clubroot resistance gene, Crr3, in *Brassica rapa*. Theor Appl Genet 114:81–91. https://doi.org/10.1007/s00122-006-0412-1

Sakamoto K, Saito A, Hayashida N, Taguchi G, Matsumoto E (2008) Mapping of isolate-specific QTLs for clubroot resistance in Chinese cabbage (*Brassica rapa* L. ssp. *pekinensis*). Theor Appl Genet 117:759–767. https://doi.org/10.1007/s00122-008-0817-0

SAS Institute (1992) SAS Technical Report P-229, SAS/STAT software: changes and enhancements. Author, Cary, NC

SAS Institute Inc. (2008, 2009, 2013) SAS^R 9.3 Software. SAS Institute Inc., Cary

Scheibert-Bohm F, Schlosser E, Schuster W (1981) Studies on the infection of rape by *Sclerotinia sclerotiorum*. Angew Bot 55:37–47

Schoeller M, Grunewald J (1987) Comparison of inoculum techniques for testing *Brassica* seedling resistance against *Plasmodiophora brassicae* Wor. Cruciferae Newslett 12:86–87

Schubert J, Rabenstein F, Graichen K, Richter K (1998) Comparison of the 5′-end nucleotide sequences of luteoviruses from oilseed rape and sugar beet. Arch Phytopathol Plant Protect 3:519–530

Schuelke M (2000) An economic method for the fluorescent labeling of PCR fragments. Nat Biotechnol 18(2):233–234. https://doi.org/10.1038/72708. PMID: 10657137

Schuler GD (1997) Sequence mapping by electronic PCR. Genome Res 7(5):541–550. https://doi.org/10.1101/gr.7.5.541. PMID: 9149949

Schwinghamer MW, Thomas JE, Schilg MA, Parry JN, Dann EK, Moore KJ, Kumari SG (2010) Mastreviruses in chickpea (*Cicer arietinum*) and other dicotyledonous crops and weeds in Queensland and northern New South Wales, Australia. Aust Plant Pathol 39:551–561

Seaman WL, Walker JC, Larson RH (1963) A new race of *Plasmodiophora brassicae* affecting Badger Shipper cabbage. Phytopathology 53:1426–1429

Shae JJ, Kado CI (1988) Whole plant wound inoculation for consistent reproduction of black rot of crucifers. Phytopathology 78:981–986

Sharma S, Padmaja KL, Gupta V, Paritosh K, Pradhan AK, Pental D (2014) Two plastid DNA lineages—*rapa/oleracea* and *nigra*—within the tribe Brassiceae can be best explained by reciprocal crosses at hexaploidy: evidence from divergence times of the plastid genomes and R-block Genes of the A and B genomes of *Brassica juncea*. PLoS One 9:e93260

Sharma BB, Kalia P, Yadava DK, Dingh D, Sharma TR (2016) Genetics and molecular mapping of black rot resistance locus *Xca1bc* on chromosome B-7 in Ethiopian mustard (*Brassica carinata* A. Braun). PLoS One 11:e0152290. https://doi.org/10.1371/journal.pone.0152290

Sharma BB, Kalia P, Dingh D, Sharma TR (2017) Introgression of black rot resistance from *Brassica carinata* to cauliflower (*Brassica oleraceabotrytis* group) through embryo rescue. Front Plant Sci 8:1255

Sharpe AG, Parkin IAP, Keith DJ, Lydiate DJ (1995) Frequent nonreciprocal translocations in the amphidiploid genome of oilseed rape (*Brassica napus*). Genome 38:1112–1121

Sheehy RE, Kramer M, Hiatt WR (1988) Reduction of polygalacturonase activity in tomato fruit by antisense RNA. Proc Natl Acad Sci 85:8805–8809

Sherriff C, Lucas JA (1989a) Cytogenetic study of heterothalic and homothallic isolates of *Peronospora parasitica*. Mycol Res 92:302–310

Sherriff C, Lucas JA (1989b) Heterothallism and homothallism in *Peronospora parasitica*. Mycol Res 92:311–316

Siebert PD, Chenchik A, Kellogg DE, Lukyanov KA, Lukyanov SA (1995) An improved PCR method for walking in unclonedgenomic DNA. Nucleic Acids Res 23:1087–1088

Simon P (2003) Q-Gene: processing quantitative real-time RT–PCR data. Bioinformatics 19:1439–1440

Singh DV, Singh J (1983) A technique for inoculating *Albugo candida* on Lahi (Indian mustard). Indian Phytopathol 36:139–140

Singh US, Doughty KJ, Nashaat NI, Bennett RN, Kolte SJ (1999) Induction of systemic resistance to *Albugo candida* in *Brassica juncea* by pre- or co-inoculation with an incompatible isolate. Phytopathology 89:1226–1232

Singh S, Dey SS, Bhatia R, Batley J, Kumar R (2018) Molecular breeding for resistance to black rot [*Xanthomonas campestris* pv. *campestris* (Pammel) Dowson] in *Brassicas*: recent advances. Euphytica 214:196

Son K, Yu S, Shin W, Han K, Kang K (2018) A simple guideline to assess the characteristics of RNA-Seq Data. BioMed Res Int 1–9. https://doi.org/10.1155/2018/2906292

Stanke M, Steinkamp R, Waack S, Morgenstern B (2004) AUGUSTUS: a web server for gene finding in eukaryotes. Nucleic Acids Res 32:W309–W312

Stone JR, Verma PR, Dueck J, Spurr DT (1987) Control of *Albugo candida* race 7 in *Brassica campestris* cv. Torch by foliar, seed and soil applications of metalaxyl. Can J Plant Pathol 9:137–145

Storey JD, Tibshirani R (2003) R: statistical significance for genome-wide experiments. Proc Natl Acad Sci U S A 100:9440–9445

Strelkov SE, Tewari JP, Smith-Degenhardt E (2006) Characterization of *Plasmodiophora brassicae* populations from Alberta, Canada. Can J Plant Pathol 28:467–474

Strelkov SE, Manolii VP, Cao T, Xue S, Hwang SF (2007) Pathotype classification of *Plasmodiophora brassicae* and its occurrence in *Brassica napus* in Alberta, Canada. J Phytopathol 155:706–712. https://doi.org/10.1111/j.1439-0434.2007.01303.x

Stringam GR (1971) Genetics of four hypocotyls mutants in *Brassica campestris* L. J Heridity 62:248–250

Sutra L, Simoneau P, Gardan L (2001) Genetic diversity of *Pseudomonas syringae* pathovars and related species assessed by heteroduplex mobility assay. In: De Boer SH (ed) Plant pathogenic bacteria. Kluwer, Dordrecht, pp 128–130

Suwabe K, Iketani H, Nunome T, Kage T, Hirai M (2002) Isolation and characterization of microsatellites in *Brassica rapa* L. Theor Appl Genet 104:1092–1098

Suwabe K, Tsukazaki H, Iketani H, Hatakeyama K, Fujimura M, Nunome T, Fukuoka H, Matsumoto S, Hirai M (2003) Identification of two loci for resistance to clubroot (*Plasmodiophora brassicae* Woronin) in *Brassica rapa* L. Theor Appl Genet 107:997–1002

Tamura K, Peterson D, Peterson N, Stecher G, Nei M, Kumar S (2011) MEGA5: molecular evolutionary genetics analysis using maximum likelihood, evolutionary distance, and maximum parsimony methods. Mol Biol Evol 28:2731–2739

Tan Z, Wada Y, Chen J, Ohshima K (2004) Inter- and intralineage recombinants are common in natural populations of Turnip mosaic virus. J Gen Virol 85:2683–2696. https://doi.org/10.1099/vir.0.80124-0

Tang H, Wang X, Bowers JE, Ming R, Alam M, Paterson AH (2008) Unraveling ancient hexaploidy through multiply-aligned angiosperm gene maps. Genome Res 18(12):1944–1954. https://doi.org/10.1101/gr.080978.108

Taylor J, Butler D (2016) ASMap: linkage map construction using the mstmap algorithm. https://CRAN.R-project.org/package=ASMap

Tewari JP, Bains PS (1988) Mode of action of the host-specific toxins produced by *Alternaria brassicae*. In: Proc 5th intel congress of plant pathology, Kyoto, 20–27 August 1988, p 234

Tewari JP, Strelkov SE, Orchard D, Hartman M, Lange RM, Turkington T (2005) Identification of clubroot of crucifers on canola (*Brassica napus*) in Alberta. Can J Plant Pathol 27:143–144

Thompson JR, Kondra ZP (1983) Techniques for artificially inoculating oilseed rape with *Sclerotinia sclerotiorum*. In: Proc 6th intl rapeseed conf, vol 2, p 185 (Abstr.)

Thompson JD, Higgins DG, Gibson TJ, Clustal W (1994) improving the sensitivity of progressive multiple sequence alignment through sequence weighting, position-specific gap penalties and weight matrix choice. Nucleic Acids Res 22(22):4673–4680. https://doi.org/10.1093/nar/22.22.4673

Tonguc M, Griffiths PD (2004) Transfer of powdery mildew resistance from *Brassica carinata* to *Brassica oleracea* through embryo rescue. Plant Breed 123:587–589

Toxopeus H, Janssen AMP (1975) Clubroot resistance in turnip.-II. the 'slurry' screening method and clubroot races in the Netherlands. Euphytica 751–755

Toxopeus H, Dixon GR, Mattusch P (1986) Physiological specialisation in *Plasmodiophora brassicae*—an analysis by international experimentation. Mycol Res 87:279–287

Trapnell C, Pachter L, Salzberg SL (2009) TopHat: discovering splice junctions with RNA-Seq. Bioinformatics 25:1105–1111. https://doi.org/10.1093/bioinformatics/btp120

Tu J, Bush J, Bonham-Smith P, Wei Y (2018) Live cell imaging of *Plasmodiophora brassicae*—host plant interactions based on a two step axenic culture system. Microbiologyopen 8(6):e00765

Untergasser A, Cutcutache I, Koressaar T, Ye J, Faircloth BC, Remm M, Rozen SG (2012) Primer3–new capabilities and interfaces. Nucleic Acids Res 40:e115. https://doi.org/10.1093/nar/gks596

Utermark J, Karlovsky P (2008) Genetic transformation of filamentous fungi by *Agrobacterium tumefaciens*. Nat Protoc Exch 427:1–18

Van Meirvenne M, Goovaerts P (2001) Evaluating the probability of exceeding a site-specific soil cadmium contamination threshold. Geoderma 102:75–100

Van Ooijen J (2006) JoinMap^R 4, Software for the calculation of genetic linkage maps in experimental populations. Kyazma BV, Wageningen, p 33

Vannacci G (1981) Seedborn *Alternaria brassicicola:* detection by means of symptoms on seedlings. Acta Hortic 111:123–129

Verma U, Bhowmik TP (1986) A simple method of inoculating white rust on rapeseed and mustard. Int J Trop Plant Dis 4:41–43

Verma PR, Petrie GA (1978) A detached-leaf culture technique for the study of white rust disease of *Brassica* species. Can J Plant Sci 58:69–73

Verma PR, Petrie GA (1979) Effect of fungicides on germination of *Albugo candida* oospores *in vitro* and on the foliar phase of the white rust disease. Can Plant Dis Surv 59:53–59

Verma PR, Petrie GA (1980) Effect of seed infestation and flower bud inoculation on systemic infection of turnip rape by *Albugo candida*. Can J Plant Sci 60:267–271

Verma PR, Harding H, Petrie GA, Williams PH (1975) Infection and temporal development of mycelium of *Albugo candida* in cotyledons of four *Brassica spp*. Can J Bot 53:1016–1020

Verma PR, Spurr DT, Petrie GA (1983) Influence of age, and time of detachment on development of white rust in detached *Brassica campestris* leaves at different temperatures. Can J Plant Pathol 5:154–157

Verma PR, Saharan GS, Bartaria AM, Shivpuri A (1999) Biological races of *Albugo candida* on *Brassica juncea* and *B. rapa* var. Toria in India. J Mycol Plant Pathol 29:75–82

Vicente JG, Conway J, Roberts SJ, Taylor JD (2001) Identification and origin of *Xanthomonas campestris* pv. *campestris* races and related pathovars. Phytopathology 91:492–499

Vicente JG, Taylor JD, Sharpe AG, Parkin IA, Lydiate PDJ, King GJ (2002) Inheritance of race-specific resistance to *Xanthomonas campestris pv. campestris* in *Brassica* genomes. Phytopathology 92:1134–1141. https://doi.org/10.1094/PHYTO.2002.92.10.1134

Vishwanath, Kolte SJ (1999) Methods of inoculation for resistance to Alternaria blight of rapeseed and mustard. J Mycol Plant Pathol 29:96–99

Von Guttenberg H, Schmoller H (1958) Culture tests with *Peronospora brassicae*. Arch Mikrobiol 30:268–279

Voorips RE, Visser DL (1993) Examination of resistance to clubroot in accessions of *Brassica oleracea* using a glasshouse seedling test. Eur J Plant Pathol 99:269–276

Voorrips RE (2002) MapChart: software for the graphical presentation of linkage maps and QTLs. J Heredity 93:77–78

Vyalykh AK, Lanetskii VP (1974) Luminescent microscopic method for determining viability of spores of the genus *Albugo*. Mikol Fitopatol 8:155–157 (Russian)

Wakeham AJ, Faggian R, Kennedy R (2017) Assessment of the response of *Plasmodiophora brassicae* in contaminated horticultural land, using lime based fertilizer concentrations. Food Energy Secur 6:122

Walcz I, Parsa AS, Eniody L, Szabo LG (1985) Detection of *Sclerotinia sclerotiorum* in sunflower by enzyme linked immunosorbent assay (ELISA). Trans Br Mycol Soc 85:485–488

Wallenhammar AC (1996) Prevalence of *Plasmodiophora brassicae* in a spring oil seed rape growing area in central Sweden and factors influencing soil infestation levels. Plant Pathol 45:710–719

Wallenhammar AC, Arwidsson O (2001) Detection of *Plasmodiophora brassicae* by PCR in naturally infested soils. Eur J Plant Pathol 107(3):313–321

Wallenhammar AC, Johnsson L, Gerhardson B (2000) Agronomic performance of partly clubroot-resistant spring oilseed turnip rape lines. J Phytopathol 148:495–499

Wallenhammar AC, Almquist C, Söderström M, Jonsson A (2012) In field distribution of *Plasmodiophora brassicae* measured using quantitative real-time PCR. Plant Pathol 61:16–28

Wang X, Wu J, Liang J, Cheng F, Wang X (2015) *Brassica* database (BRAD) version 2.0: integrating and mining Brassicaceae species genomic resources. Database 2015:bav093. https://doi.org/10.1093/database/bav093

Webster J (1980) Introduction to fungi, 2nd edn. Plasmodiophoromycetes. pp 42–53

Webster R, Oliver MA (1992) Sample adequately to estimate variograms of soil properties. J Soil Sci 43:177–192

Wei D, Mei J, Fu Y, Disi JO, Li J, Qian W (2014) Quantitative trait loci analyses for resistance to *Sclerotinia sclerotiorum* and flowering time in *Brassica napus*. Mol Breed 34:1797–1804

Wei L, Jian H, Lu K, Filardo F, Yin N, Liu L, Qu C, Li W, Du H, Li J (2015) Genome-wide association analysis and differential expression analysis of resistance to Sclerotinia stem rot in *Brassica napus*. Plant Biotechnol J 14(6):1368–1380. https://doi.org/10.1111/pbi.12501

Wesołowska M (2011) Isolate pathogenicity recognition of *Plasmodiophora brassicae* Wor. in different areas of Poland. Folia Hort 23(2):165–174

Wesołowska M (2014) Identification of *Plasmodiophora brassicae* Worn.isolate suppressing clubroot resistance in 'Kilaxy' F_1 white cabbage. Folia Hortic 26:57–62

Weßling R, Panstruga R (2012) Rapid quantification of plant-powdery mildew interactions by qPCR and conidiospore counts. Plant Methods 8:35

Wheeler BEJ (1969) An introduction to plant disease. Wiley, London, p 301

White TJ, Bruns T, Lee S, Taylor J (1990) Amplification and direct sequencing of fungal ribosomal RNA genes for phylogenetics. In: Innis MA, Gelfand DH, Sninsky JJ, White TJ (eds) PCR protocols: a guide to methods and applications. Academic, New York, pp 315–322

Wilkinson GN, Rogers CE (1973) Symbolic description of factorial models for analysis of variance. JR Stat Soc Ser C 22:392–399. https://doi.org/10.2307/2346786

Williams PH (1966) A system for the determination of races of *Plasmodiophora brassicae* that infect cabbage and rutabaga. Phytopathology 56:624–626

Williams PH (1985a) White rust [*Albugo candida* (Pers. ex. Hook.) Kuntze.]. In: Crucifer Genetics Cooperative (CRGC) Resource Book. University Wisconsin, USA, pp 1–7

Williams PH (1985b) Crucifer genetics co-operative resource book. Downy mildew. University of Wisconsin, Madison, pp D5–D7

Williams PH, Leung H (1981) Methods of breeding for multiple disease resistant Chinese cabbage. In: Talekar NS, Griggs TD (eds) Chinese cabbage. Proc. 1st intern. symposium. The Asian Vegetable Research and Development Center, Shanhua, pp 393–403

Wu CR, Liu HL (1991) Selection of rape (*Brassica napus* L) callus cultures resistant to oxalic acid. Cruciferae Newslett 16–15:80–81

Wu J, Cai G, Tu J, Li L, Liu S, Luo X, Zhou L, Fan C, Zhou Y (2013) Identification of QTLs for resistance to Sclerotinia stem rot and BnaC.IGMT5.a as a candidate gene of the major resistant QTL SRC6 in *Brassica napus*. PLoS One 8:e67740. https://doi.org/10.1371/journal.pone.0067740

Wu J, Zhao Q, Liu S, Shahid M, Lan L, Cai G, Zhang C, Fan C, Wang Y, Zhou Y (2016) Genome-wide association study identifies new loci for resistance to Sclerotinia stem rot in *Brassica napus*. Front Plant Sci 7:1418. https://doi.org/10.3389/fpls.2016.01418

Xu RQ, Blanvillain S, Feng JX, Jiang BL, Li XZ, Wei HY, Kroj T, Lauber E, Roby D, Chen B, He YQ, Lu GT, Tang DJ, Vasse J, Arlat M, Tang JL (2008) AvrAC (Xcc8004), a type III effector with a leucine-rich repeat domain from *Xanthomonas campestris* pathovar *campestris* confers avirulence in vascular tissues of *Arabidopsis thaliana* ecotype Col-0. J Bacteriol 190:343–355

Xue S, Cao T, Howard RJ, Hwang SF, Strelkov SE (2008) Isolation and variation in virulence of single-spore isolates of *Plasmodiophora brassicae* from Canada. Plant Dis 92(3):456–462

Yan M, Liu X, Chunyun G, Liu L, Xiang J, Ying L, Liu Z (2014) Cloning of *TTG1* gene and PCR identification of genomes A, B and C in *Brassica* species. Genetica 142:169–176. https://doi.org/10.1007/s10709-014-9764-7

Yang B, Srivastava S, Deyholos MK, Kav NNV (2007) Transcriptional profiling of canola (*Brassica napus* L.) responses to the fungal pathogen *Sclerotinia sclerotiorum*. Plant Sci 173:156–171

Yang J, Weedon MN, Purcell S, Lettre G, Estrada K, Willer CJ, Smith AV, Ingelsson E, O'Connell JR, Mangino M, Mägi R, Madden PA, Heath AC, Nyholt DR, Martin NG, Montgomery GW, Frayling TM, Hirschhorn JN, McCarthy MI, Goddard ME, Visscher PM, GIANT Consortium (2011) Genomic inflation factors under polygenic inheritance. Eur J Hum Genet 19(7):807–812

Yang J, Liu D, Wang X, Ji C, Cheng F, Liu B, Hu Z, Chen S, Pental D, Ju Y, Yao P, Li X, Xie K, Zhang J, Wang J, Liu F, Ma W, Shopan J, Zheng H, Mackenzie SA, Zhang M (2016) The genome sequence of allopolyploid *Brassica juncea* and analysis of differential homoeolog gene expression influencing selection. Nat Genet 48:1225–1232. https://doi.org/10.1038/ng.3657

Yarwood CE (1936) The diurnal cycle of the powdery mildew *Erysiphe polygoni*. J Agric Res 52:645–657

Yarwood CE (1946) Detached leaf culture. Bot Rev 12:1–56

Yarwood CE (1957) Powdery mildews. Bot Rev 23:235–301

Yu F, Gugel RK, Kutcher HR, Peng G, Rimmer SR (2013) Identification and mapping of a novel blackleg resistance locus LepR4 in the progenies from *Brassica napus* × *B. rapa* subsp. *sylvestris*. Theor Appl Genet 126:307–315. https://doi.org/10.1007/s00122-012-1919-2

Yu J, Tehrim S, Zhang F, Tong C, Huang J, Cheng X, Dong C, Zhou Y, Qin R, Hua W, Liu S (2014) Genome-wide comparative analysis of NBS-encoding genes between *Brassica* species and *Arabidopsis thaliana*. BMC Genomics 15:3. https://doi.org/10.1186/1471-2164-15-3

Zalewski A (1883) ZurKenntniss der Gattung *Cystopus*. Bot Zentralbl *15*:215–224

Zhang XW, Yuan YX, Wang XW (2009) QTL mapping for TuMV resistance in Chinese cabbage [*Brassica campestris* L. ssp. *pekinensis* (Lour.) Olssom]. Acta Hortic Sin 36:731–736

Zhang L, Li H, Wang J (2012) The statistical power of Inclusive Composite Interval Mapping in detecting digenic epistasis showing common F_2 segregation ratios. J Integr Plant Biol 54 (4):270–279

Zhang H, Feng J, Zhang S, Zhang S, Li F, Strelkov SE, Sun R, Hwang SF (2015) Resistance to *Plasmodiophora brassicae* in *Brassica rapa* and *Brassica juncea* genotypes from China. Plant Dis 99:776–779

Zhao J, Udall JA, Quijada PA, Grau CR, Meng J, Osborn TC (2006) Quantitative trait loci for resistance to *Sclerotinia sclerotiorum* and its association with a homeologous non-reciprocal transposition in *Brassica napus* L. Theor Appl Genet 112:509–516

Zhao Q, Huang X, Lin Z, Han B (2010) SEG-Map: a novel software for genotype calling and genetic map construction from next-generation sequencing. Rice 3:98–102. https://doi.org/10.1007/s12284-010-9051-x

Zhu C, Gore M, Buckler ES, Yu J (2008) Status and prospects of association mapping in plants. Plant Genome 1:5–20

Zhu X, Li X, Chen W, Chen J, Lu W, Chen L, Fu D (2012) Evaluation of new reference genes in papaya for accurate transcript normalization under different experimental conditions. PLoS One 7(8):e44405. https://doi.org/10.1371/journal.pone.0044405

Future Research Priorities

10

Keywords

Future research priorities · Next-generation sequencing · Disease resistance · Resistance genes · Major pathosystems · Marker-assisted selection

10.1 Introduction

The advances done in molecular technology and advent of next-generation sequencing technology has allowed to search genetics of host resistance in crucifers' host–pathogen interactions in more precise way. The host and pathogen have ever evolutionary biological system in the nature to overcome each other for survival and fitness, therefore, always use new arsenal. To combat the pathogen attack, updating of resistance mechanisms is inevitable. Although significant progress has been made in comprehension of genomics of host resistance in crucifers, still there is a need to have more data on the following aspects to improve disease resistance programs of crucifers breeding.

An international collaborative network is creating a world database for each pathogen of crucifers to resolve the nomenclature issue of R-genes and pathotypes. Since dynamic changes can happen with both host cultivars and pathogens in terms of cultivar resistance breakdowns and/or pathogens losing/gaining in terms of their pathogenicity/virulence over time, it is imperative that any universal database set up to be regularly updated and kept current.

Except for the four R-genes that have been cloned, the QTL for the major diseases in *Brassica* have been identified but the causal genes remain unknown. Cloning of R-genes has historically been undertaken using traditional methods, such as map-based cloning. However, the recent genomics revolution has provided significant insights into the presence and absence of expressed genes in the *Brassica* A- and C-genomes using a pan-transcriptome approach. This, along with the recent

G. S. Saharan et al., *Genomics of Crucifer's Host-Resistance*,
https://doi.org/10.1007/978-981-16-0862-9_10

pan-genome research readily allows cloning of R-genes in crops such as *Brassica* that have complex genomes. More R-gene cloning work in *Brassica* will help us better understand how the plant defends itself against these pathogens, and we can use this knowledge to exploit novel R-genes in a quicker and more effective manner for breeding purposes.

Effective deployment of R-genes in breeding *B. napus*-resistant crops for enduring sustainability demands detailed knowledge about the ecology and life history of both host and pathogen and how their interactions may shape their evolution over time. Such knowledge can allow prediction of the host resistance at different stages of the infection processes for the pathogen, potentially allowing identification and targeting of the resistance gene(s) effective at the earliest infection stage in order to maximize impedance and/or even prevent the establishment of the pathogen and also curtail subsequent pathogen inoculum production and consequent secondary spread of the disease. While this approach could be particularly applicable for a disease such as clubroot, there are still challenges as it is not yet possible to identify the co-relationship between root hair infection (primary) and root cortex infection (secondary). On the other hand, knowledge about controlling pathogen infection at different developmental stages of the host (e.g., seedling vs. adult) or even in different plant organs (e.g., cotyledons, leaves, stems, pods, and root) where the pathogen infection and colonization has the greatest adverse impact on the productivity of the plant (plant stems for *L. maculans* and *S. sclerotiorum*) have become more forthcoming for the major pathosystems in *B. napus*. Clearly, improving our understanding of the mechanism(s) at both life stages of the pathogen infection and colonization processes, the development stages of the host, and the host–pathogen interactions should open up new opportunities to better understand and exploit approaches that maximize control of pathogen infection and spread to minimize consequent yield losses.

Sequencing of *Brassica* genomes will allow new discoveries and understandings regarding the genetic relationships of different R-genes, within and between members of the Brassicaceae. Although hybridization between divergent groups in the Brassicaceae, which contains rich resources of R-genes, is already possible, particularly in India, this is an area of great future potential that is likely to lead to identification of many new R-genes as these methodologies become more widely adopted and as rates of success with wide hybridization events improve. While scientists are able to make informed decisions about utilizing race-specific and/or non-race-specific resistance genes, further understanding of the benefits of different patterns of deployment for each type individually or rotation of both types, will identify better approaches towards maximizing the durability and effectiveness of both types of resistances, particularly R-genes. Further, field deployment of race-specific and/or non-race-specific resistance genes will have a large impact on the dynamics within pathogen populations. Combining race-specific and non-race-specific genes together in the one cultivar will also potentially increase durability of R-genes deployed within *Brassica* species and has already been demonstrated for R-genes against *L. maculans*.

Improved resistance from gene pyramiding through marker-assisted selection was reported for the *P. brassicae–B. rapa* ssp. *pekinensis* pathosystem. However, this remains a controversial approach with some suggesting that such an approach could lead to an undesired outcome of breakdown of the multiple R-genes pyramided. As we gain more knowledge about the molecular mechanisms of the plant immunity signaling pathway and the role of R-genes, this will open up more effective and durable strategies to overcome pathogen attack.

For most of the *Brassica* species, the resistance resources for diseases are highly limited. In general, the A-genome is rich in Turnip mosaic virus, Black rot, Downy mildew, and clubroot resistance, whereas the B-genome possesses black rot and Blackleg resistance, and the C-genome harbors stem rot, Fusarium wilt, and downy mildew resistance. Fortunately, the six species in the *Brassica* genus and others, such as *B. incana*, *B. cretica* (C-genome), and *Brassica fruticulosa* (B-genome), as well as its close Brassicae relatives, could be used to facilitate resistance gene exchanges in breeding programs.

Interspecies crossing within the *Brassica* genus are widely adopted using embryo rescue, reciprocal crossing, and MAS. There are few black rot-resistant resources in the C-genome of B. oleracea, whereas high resistance is present in both the A- and B-genomes. Thus, interspecies hybridization has been used to transfer and utilize the resistance found in the A- and B-genomes. In addition, distant hybridization has been used by breeders to generate novel-resistant germplasms. In contrast, high-level resistance has been observed in a few wild species, such as *Erucastrum cardaminoides*, *Diplotaxis tenuisiliqua*, and *Sinapis arvensis*. High levels of stem rot resistance from crosses between *B. napus/B. juncea* and *E. cardaminoides/ D. tenuisiliqua/Erucastrum abyssinicum* have been obtained. The novel resistance germplasms generated in such studies provide valuable materials in future breeding programs for *Brassica* crops. However, there is still much work to be done for them to be practically applied in commercial cultivars. There are rich resources with clubroot resistance in *B. rapa* but very few in *B. oleracea*, and breeders have spent ~17 years introgressing resistance from Chinese cabbage (*B. rapa* subsp. *pekinensis*) cv. Parkin to different *B. oleracea* cultivars, which are now widely used.

In many cases, there is one main disease in a specific production region and application of cultivars with resistance to that main disease is appropriate to realize the balance between defense and growth. However, for some continuous or high-intensity cropping regions, various pathogens may accumulate; therefore, cultivars with multiple resistances to different diseases are in great need. Currently, MAS using abundant molecular markers enables us to realize the pyramiding of R alleles for different diseases and breed multi-resistant cultivars. MAS combined with other methods, such as hybridization and microspore culture, has greatly shortened the breeding circle.

A single resistance gene is easily overcome because of pathogen variations and global climate changes. More durable resistance is urgently needed to ensure *Brassica* crop production. Complete race-specific resistance genes are very effective in a short period but are easily overcome by the pathogens; polygene-inherited resistance is thought to be more durable, but its effects might be unstable owing to variable environmental conditions. Thus, pyramiding qualitative genes with major quantitative loci in cultivars represents an ideal means to ensure the effectiveness and durability of resistance. Thus, combining quantitative resistance with single R-genes is a promising strategy for resistance breeding. In addition, co-evolution between the host and the pathogen has been extensively studied; the mapped seven Blackleg resistance genes/alleles and their contrasting avirulence genes have indicated a rapid and fierce arms race between *Brassica* hosts and Lm, and the single use of one type of pyramiding could pose a high selection pressure on the pathogen and thus raise the risk of pathogen mutation and host resistance loss. Thus, the rationalized pyramiding of dissimilar sets of resistance genes/QTLs should be deployed in different cultivars to provide heterogeneity in the selection pressure on the pathogen population and result in more durable resistance. In addition, pyramided resistance could be integrated with agricultural, physical, chemical, and biological controls to realize IPM, further maximize durability and guarantee stability.

Current molecular and omics methods, including transcriptomics, proteomics, and metabolomics, provide new opportunities for mining genes in the resistance-regulating network, which could be either directly used in resistance breeding or used in indirect pre-breeding studies to promote understanding of *Brassica*–pathogen interactions. In addition, the sequenced *Brassica* accessions do not contain all R-genes due to variations between individuals, whereas the establishment of the pan-genomics could facilitate gene mining from a wider platform.

In *Brassica* crops, although over ten R-genes have been characterized to date, most avirulence or interaction genes in the pathogens have not yet been thoroughly characterized. Thus, the next emphasis should be on the establishment of global collections of pathogen isolates for the identification of important avirulence or interaction genes. In addition, for the host, the development of a series of ILs through recurrent backcrossing to "Mendalize" the quantitative loci would be beneficial.

The next efforts should be more accurate identification of the R-genes and clarification of their molecular mechanisms using emerging high-efficiency genomic, post-genomic, and omics methods, and more efficient application of the R resources through an integration of approaches, such as haploid culture, MAS, distant introgression, genome design, pyramiding, and transgenic breeding, to control the diseases and secure crucifers production.

On the host side, a number of disease resistance genes in Brassicas and *A. thaliana* have been postulated, some have been mapped, but none have been cloned. The effort to identify more disease resistance genes should be continued, and some of the most important crops could be improved with the incorporation of disease resistance genes from closely related species via interspecific crosses and embryo rescue, or via transformation. It is also possible that the engineering of the promoters of known R-genes to recognize multiple effectors could contribute to the development of broad spectrum disease-resistant crops. The control of the disease could be improved by pyramiding disease resistance genes and/or by using multiple lines with different gene combinations that could be mixed and balanced to match the avirulence genes present in the bacterial population.

The operating mode of signaling network in host to multiple stresses remains a challenging question that needs further investigation. Emerging imaging techniques at nanoscale resolution and new proteomics approaches dedicated to deciphering protein complexes are very promising and should lead to a better understanding of the mechanisms underlying how a single Ca^{2+} sensor acts at the crossroads of various signaling pathways to help plants to cope with various environmental challenges.

There is need to fully understand the complex stress-response system of the plants to minimize differences in the tolerance capacity of transgenic plants between laboratories and crop fields.

R-genes identification and inheritance governing seedling and adult plants resistance along with their sensitivity to temperature, light, and soil conditions may be explored during different *Brassica* host–pathogen system.

The role of WRKY genes and calmodulin-binding transcription activators (CAMTA) in providing resistance to Brassica species against different pathogens needs further characterization.

Fitness cost to multiple disease resistance in *Brassica* with linkage disequilibrium and confounded pleiotropy needs more data evidenced from ecological and physiological studies.

It is necessary to determine whether net costs of R-genes resistance with infection occur in natural plant populations or similar net costs occur in other plant stress system.

Integration among plant breeders, plant pathologist, and marker-assisted selection for the development of resistant varieties is required.

An International and National collaborative network among the institutes and scientists are needed may be in a phase wise so that the variety developed may be evaluated against all the pathogen races/pathotypes.

Index

A
Abiotic stresses, 275, 310, 313, 337, 360, 389–391, 398, 407–421, 424, 425, 430–435, 468, 470, 474, 482, 496, 551, 557, 558, 601, 602, 617, 618, 620, 634, 636
Accessions, 68–70, 72, 74–76, 79, 88, 90, 91, 93, 95, 97, 98, 114–117, 123, 124, 127–129, 133, 134, 137–140, 142, 144, 148–150, 152, 155, 156, 158, 198, 200, 215–220, 227–230, 232–234, 236, 237, 241, 249–251, 274, 279, 291, 292, 310, 311, 313, 317, 472, 490, 491, 495, 507, 509, 524, 533–535, 540, 541, 543, 550, 552, 565, 608, 613, 615, 634–636, 663, 668, 672, 678, 684, 710, 714, 720, 728, 729, 739, 748, 776
Accumulation of ROS, 470, 481, 513
Acquired resistance (AR), 366, 687
Africa, 310, 361, 363, 366, 370, 379, 384, 387, 400, 401, 403–406, 634
Albugo candida, 68–73, 77, 90, 93, 149, 158, 197–205, 277, 279, 316, 322, 327, 384, 472, 507–510, 559, 606, 608, 628, 652–655, 657, 659, 680–682
Albugo-Arabidopsis pathosystem, 24–27
Alternaria brassicae, 3, 4, 73–75, 77, 206–210, 267–272, 274–277, 279, 283, 286–288, 319, 321, 323, 329–331, 361, 474, 481, 484, 486–488, 492–497, 511, 512, 557–559, 611, 612, 623, 625–628, 660, 661, 666, 680, 681
Alternaria brassicicola, 73–76, 275–277, 279, 281, 287, 321, 324, 329, 337, 361, 413, 414, 487, 488, 496, 497, 512, 565, 566, 611, 612, 626, 627, 661, 662, 664–666, 680
Alternaria raphani, 321, 361, 512
Angola, 366, 370
Arabidopsis species, 6, 67, 274, 397, 469, 513, 604
Arabidopsis thaliana, 6, 10, 20, 23, 38, 67, 72, 78, 79, 87, 90, 91, 109, 116, 118, 119, 121–123, 125, 133, 138, 144, 148–150, 157, 158, 200–202, 217, 219, 220, 227–229, 231–235, 239, 296, 310, 323, 325, 327, 330, 397, 408–410, 412–414, 416, 418, 471–474, 476, 477, 482, 483, 512, 514–517, 531–534, 543, 549, 562, 563, 566, 570–572, 603, 604, 607, 608, 611, 613, 616, 633, 667, 668, 718, 735, 750, 751, 776
Area under disease progress curve (AUDPC), 199, 200, 210, 212, 213, 545
Argentina, 361, 362, 366, 369, 372, 373, 378, 383, 386, 387
Artificial inoculations, 76, 80, 277, 290, 490, 627, 668, 680, 685–687, 697–698
  bioassay techniques, 709, 712
  breeding for multiple disease resistance, 680
  designation and nomenclature of R-genes and pathotypes, 92
  evaluation of resistance, 131
  gene expression analysis, 410
  genome sequencing, 108, 144, 148, 150–151, 306, 315, 316, 471, 473, 608, 621, 637, 638
  germplasm evaluation, 280, 700–703
  germplasm screening, 512, 655, 660–661, 678, 722–723, 735–736, 741–742
  green house screening, 558
  host–pathogen interaction, 35, 41, 87, 92, 151, 288, 468, 470–472, 483, 488, 554, 609, 620, 773, 774
  identification of QTL's, 121, 154–156, 483, 695–697
  identification of R-genes, 76, 80

G. S. Saharan et al., *Genomics of Crucifer's Host-Resistance*,
https://doi.org/10.1007/978-981-16-0862-9

Artificial inoculations (*cont.*)
inoculation, 80, 277, 290, 490, 627, 668, 680, 685, 697
inoculum preparation, 652–653, 693, 697–698, 702, 706, 716
in-vitro culture, 698
molecular mapping, 483, 602, 715–718
molecular markers, 290, 489
Australia, 71, 73, 79, 93–95, 112, 113, 115–117, 127, 129–131, 133, 141, 142, 145, 146, 299, 316, 321, 361, 365, 369, 371, 373, 375, 377, 378, 382, 386, 387, 399–406, 517, 518, 546, 547, 569, 695, 728
Austria, 361, 364, 371, 373, 381, 385
Avirulence (AVR), 70, 87, 92, 93, 147, 220, 227, 232–234, 316, 325, 469, 470, 475, 483, 519, 520, 551, 569, 602, 606, 622, 693–696, 776
Avirulence genes
complementary, 228, 229, 231, 237, 569, 622
dominant genes, 197, 202, 229, 236, 242, 249
enhanced *R*-genes, 221
modifiers, 211
partial, 205, 220, 228, 231, 232, 238
polygenic, 235, 236, 247, 251
recessive genes, 238, 249

B

Bacillus subtilis, 524, 525, 565
Basal defense, 414, 415, 428, 481, 524
B. chinensis (Chinese cabbage), 70, 72, 89, 95–97, 121, 125, 126, 134, 144, 157, 226, 231, 235, 236, 238, 239, 249, 251, 289–291, 297–300, 315, 320, 329, 331, 336, 489, 490, 536, 537, 540, 541, 549–551, 679–683, 699, 700, 702–704, 706, 721, 744, 775
Belarus, 364, 371, 382
Belgium, 364, 371, 381, 385, 386
Biochemical basis of resistance, 611
Biochemical compounds
brassinosteroids, 325
camalexin, 23, 476, 480–482, 514, 612
carbohydrates, 319, 323, 324, 331, 558
enzymes, 23, 123, 127, 132, 154, 271, 288, 292, 330, 336, 337, 423, 470, 477, 480, 482, 488, 491, 513, 527, 528, 533, 563, 612, 619, 659, 663, 731, 733
lipids, 42
metabolites, 23, 42, 44, 310, 424, 476, 477, 480–482, 512, 523, 525, 526, 528, 535, 558, 562, 566, 612, 631
nucleic acid, 322
phenolic compounds, 305, 330, 470, 476, 513, 611, 629
protein, 78, 131, 286, 308, 482, 512, 620, 665, 737
starch, 523, 536, 676, 687, 697
sugar, 7, 123, 125, 132, 554, 556, 605, 714
Biological controls, 2, 294, 336, 492, 524, 525, 548, 602, 776
Biological control agents
bacteria, 525
fungi, 524
Biotic stresses, 125, 155, 319–338, 361, 408, 409, 416, 421, 424–426, 430–431, 470, 471, 512, 527, 556–560, 562, 615, 617
Biotrophic pathogens, 21, 24, 306, 430, 472, 479, 517, 638
B. oleracea var. *acephala* (Kale), 68, 75, 153, 155, 333, 545, 629
B. oleracea var. *alboglobra* (Chinese kale), 153
B. oleracea var. *botrytis* (Broccoli), 76, 77, 89, 95, 116, 117, 138–140, 144, 153, 155, 614, 615, 627, 634, 636, 702
B. oleracea var. *botrytis* (Cauliflower), 68, 75–77, 88, 95, 138, 139, 151–153, 155, 226, 227, 229, 231, 236, 250, 251, 277, 279, 283, 298, 299, 311–314, 319, 320, 323, 330, 333, 361, 383, 467, 552, 608, 615, 621, 624, 627, 628, 634–637, 723, 739, 741, 743
B. oleracea var. *capitata* (Cabbage), 68, 74, 76, 77, 86, 88, 89, 116–118, 129, 134, 138–140, 142, 153, 155, 156, 211, 224, 226, 229, 231, 233, 235–237, 249–251, 276, 277, 297–300, 302, 312, 316, 319–322, 327, 332–334, 377, 399–405, 467, 507, 514, 537, 545, 553, 615, 624, 627, 635, 661, 662, 675, 676, 679, 684, 701, 702, 719, 737
B. oleracea var. *gemmifera* (Brussel's sprouts), 76
B. oleracea var. *gongylodes* (Kohlrabi), 116, 117, 151, 153, 155, 684, 702
B. oleracea var. *italic* (Broccoli), 226, 231–233, 236, 277, 292, 297, 298, 311, 319, 320, 322, 545, 567
B. oleracea var. *sabauda* (Savoy cabbage), 155, 699

B. oleracea var. viridis (Kale), 236, 237, 301, 304, 319
Bolivia, 362, 369, 373, 383
Bosnia-Herzegovina, 368, 381
B. pekinensis (Chinese cabbage), 70, 235
B. rapa subsp. *oleifera* (Turnip rape), 75, 657
B. rapa subsp. *pekinensis* (Chinese cabbage), 116–118, 138, 332, 334, 335, 775
B. rapa subsp. *rapifera* (Turnip), 89, 116, 117, 138, 291, 297, 298, 304, 315, 319, 322, 490, 537, 629, 698, 742–745
B. rapa subsp. var. Brown Sarson (Brown sarson), 69, 71, 79, 88, 282, 702
B. rapa subsp. var. Toria (Toria), 69, 71, 75, 80, 384
B. rapa subsp. var. Yellow Sarson (Yellow sarson), 69, 71, 75, 79, 88, 133, 279, 282, 361, 661, 702, 720
Brassica tournefortii, 11
Brassica alba, 70, 72, 73, 75, 76, 78, 79, 88, 133, 276, 277, 626, 627
Brassica campestris, 87, 88, 144, 158, 226, 227, 268, 269, 284, 485, 494, 546, 562, 625, 699, 702, 706
Brassica carinata, 68, 70–80, 82, 88, 94, 116–118, 124, 128, 131–133, 137–140, 143, 144, 197–200, 202–204, 206–209, 214, 215, 249–251, 267, 268, 276–278, 281–283, 285, 287, 288, 292, 304, 310–314, 316, 320, 361, 476, 484, 485, 487, 488, 491, 512, 514, 545–547, 551, 552, 604, 611, 613, 614, 616, 621, 623, 625–629, 633–637, 673, 702, 737–739
Brassica fructiculosa, 77
Brassica genome complexity, 1, 32–34, 607
Brassica germplasm, 31, 67, 74, 154, 554, 607, 660–661, 684, 702–704
Brassica juncea, 68–80, 82, 87, 88, 90, 92, 94, 95, 124, 128–134, 136–140, 142–144, 150, 155, 158, 197–204, 206–211, 213–215, 226–228, 230, 231, 249, 267–269, 275–279, 282, 284–286, 288, 304, 308–311, 313–316, 320, 321, 323–327, 329–334, 336, 361, 373, 378, 379, 384, 472–476, 483–486, 488, 494, 496, 507–513, 518, 525, 526, 546, 547, 549, 551, 552, 557–559, 604, 609, 611–613, 620, 621, 625–629, 631–632, 634, 636, 637, 653–655, 657–660, 673, 693, 702, 731–735, 745, 775
Brassica napus, 68–73, 75–80, 82, 87, 89, 93–95, 98, 99, 109, 113–118, 120, 121, 124, 126–135, 137–148, 150–152, 155–158, 197–202, 204, 205, 210, 226–229, 231, 235–239, 241, 243, 248–251, 267–269, 275–277, 282–286, 289–291, 296–298, 300–303, 307–309, 311, 313–317, 320, 321, 323–337, 361, 383, 430, 473, 475, 476, 478, 480, 481, 484–486, 489, 490, 494, 496, 512, 513, 517, 518, 522, 523, 536, 539, 540, 543, 545–547, 550–552, 554, 557, 561, 566, 567, 569, 604, 609, 612–614, 616, 620, 621, 625–628, 632–634, 636, 661, 667, 673, 693–697, 702, 715, 720, 722, 727, 729, 730, 736–739, 743, 744, 746, 748, 774, 775
Brassica nigra, 67, 76, 77, 89, 94, 116, 118, 128, 130, 137, 139, 140, 142–144, 150–152, 158, 197, 202, 249, 267, 268, 277, 285, 310, 311, 313, 314, 316, 320, 325, 473, 475, 476, 484, 485, 604, 620, 621, 625, 627, 634, 635, 637, 660, 702, 732
Brassica rapa, 6, 67, 267, 361, 467, 506, 604, 654, 775
Brassica species, 2, 6, 67, 267, 467, 507, 602, 660, 774
  coenospecies, 73, 276, 606, 608, 626
  cytodemes, 606
  genetic diversity, 236, 313, 551, 606, 617, 621, 635
  wild relatives, 67, 129, 138, 507, 606–608, 616, 621, 622
Brassica vegetables, 137, 291, 313, 321, 490, 551, 635
Brazil, 361–363, 366, 369, 372, 373, 383, 386, 388
Broad spectrum resistance, 95, 138, 200, 202, 241, 513, 605
Bulgaria, 361, 364, 368, 370, 372, 373, 378, 381, 385

C

Camalexin, 476, 480–482, 514, 612
Canada, 71, 73, 88, 89, 93, 95, 98, 115–117, 121, 141, 142, 146, 155, 198, 199, 241, 243, 298, 300, 305, 306, 315, 361, 365, 368, 369, 372, 373, 378, 379, 382, 383, 386, 387, 389, 390, 393, 394, 399–401, 403–405, 517, 518, 520, 521, 536, 538, 545, 630, 631, 693, 695, 705, 717, 720
Capsella bursa-pastoris, 73, 76, 127, 132, 276, 397, 626, 627, 746
Chile, 361, 362, 373, 383, 386
China, 68, 71, 93–95, 98, 115–118, 129–131, 133, 134, 141, 142, 144, 146, 155, 197,

202, 321, 361–364, 366, 371, 373, 377, 378, 380, 384, 387, 401, 405, 524, 526, 546, 709, 728
Chitinase genes, 286, 474, 483, 486, 524, 557–559, 611
Clubroot
compatible interaction, 672
Colletotrichum, 78, 87, 362, 474, 481, 608
Columbia, 72, 91, 92, 215, 251, 365, 472, 572
Costa Rico, 372
CR genotypes, 301
Croatia, 372, 381, 385
Crucifers, 66, 67, 71, 72, 74–76, 79, 80, 87–91, 95–99, 117, 118, 127, 128, 132, 133, 137, 138, 141–148, 196–198, 202, 211, 228, 231, 235, 239, 266, 267, 276, 279, 281, 284, 288, 306, 313, 360, 361, 389, 396, 399, 400, 434, 468, 470–473, 479, 483, 488, 506, 507, 512–514, 547, 552, 570–573, 601, 602, 604, 606–613, 615–617, 620–625, 627, 636, 638, 652, 667, 687, 688, 773, 776
Crucifers' vegetables, 138, 547
Cultivars
resistant, 68, 79, 112, 114, 130, 147, 279–281, 289, 297, 298, 300, 303–305, 307, 308, 315, 467, 478, 489, 507, 520, 544–547, 550, 553, 554, 569, 606, 608–610, 612–614, 616, 621–625, 629, 630, 632, 638, 661
susceptible, 131, 145, 197, 199, 226, 233, 235, 238, 241, 250, 268, 297, 299, 303, 305, 307, 308, 315, 417, 478, 521, 538, 540, 544, 545, 553, 569, 610, 623, 625, 630, 632, 633, 661, 682, 701
tolerant, 130, 268, 616, 625
Cysteine-rich protein, 567
Czech Republic, 305, 372, 521, 630
Czechoslovakia (Former), 201, 240, 361, 368, 374, 381

D

Defense mechanism, 42, 293, 491, 524, 533
Defense related genes, 156, 287, 412, 417, 424, 562, 567
Deployment of R, 305, 521, 622, 623, 630, 774
Denmark, 95, 250, 361, 364, 368, 372, 374, 377, 382, 383, 385, 387
Differential gene expression, 245
Differential resistance, 226, 227, 229
Differential set, 95, 241
Disease assessment, 233, 614, 679
Disease cycle, 573, 612
Disease incidence, 302, 706, 728, 733, 741
Disease index, 210, 212, 213, 227, 228
Disease intensity, 80, 81, 206, 207, 286, 486
Disease management
host resistance, 226, 407–410
integrated, 422
Disease ratings, 724, 741, 742
Disease reaction phenotypes, 216, 218, 672
Disease resistance, 196–201, 206, 207, 209, 211, 213–215, 217, 219, 226–229, 231–233, 235, 238, 239, 243, 245, 246, 248–251, 413–415, 431, 432, 468–471, 475–488, 505–573, 602, 604–610, 612–617, 620, 622–628, 632–634, 637, 638, 674, 680–683, 688, 773, 776, 777
Disease resistances, 2, 3, 6–8, 11, 13–15, 19, 20, 22–24, 33, 35, 36, 38, 39, 66–68, 70–78, 87, 93, 116, 122, 128, 130, 131, 148–151, 153, 154, 156, 157, 265–338
Disease severity, 201, 214, 288, 488, 510, 548, 558, 563, 569, 656, 666, 691, 709, 712, 724, 728
Disease tolerance sources, 80
Distant hybridization, 267, 290–291, 313–314, 490, 551–552, 615, 616, 621, 635, 775
DNA sequence analysis, 150, 483

E

Ecuador, 372, 383
Effector proteins, 39, 234, 469, 470, 474, 562
Effectors' genes, 24, 40, 108, 315, 609, 620, 638
Egypt, 361, 363, 370, 374, 378, 379, 384
Embryo, 739–741
Embryo rescues, 267, 280, 281, 283–285, 290, 292, 298, 310–314, 468, 484–485, 490, 491, 514, 551–552, 615, 616, 621, 623, 625, 633–637, 673, 739–741, 775, 776
Enzymes
chitinases, 288, 336, 488
cytokinin oxidase, 124, 125
dehydrogenase, 125
Erosion of R genes, 536–539, 622, 623
Eruca sativa (Taramira), 70, 71, 73, 74, 137, 276, 277, 282, 361, 378, 626, 627
Erysiphe cruciferarum, 3, 4, 77, 79, 81, 214, 215, 217, 219, 220, 277, 279, 293, 294, 378, 473, 476, 491, 492, 514, 558, 559, 608, 621, 628, 667
Estonia, 364, 381
Europe, 93, 95, 98, 141, 146, 299, 321, 364, 368, 370, 371, 380, 383, 385, 387, 399–406, 472, 517, 728
Evolution of host-pathogen, 151
Expression of genes

metabolic pathways, 42, 44, 123, 305, 553, 560, 629
primary stage of infection, 123

F

Faroe Islands, 372
Finland, 361, 364, 368, 369, 372, 374, 378, 381, 385, 396, 746
France, 95, 112, 299, 305, 361, 364, 368, 372, 374, 377, 378, 381, 383, 385, 387, 518, 520, 630, 663, 693, 743
Functional
host resistance, 417, 423
Fusarium, 86, 153, 224, 325, 373, 387, 388, 425, 426, 475, 516, 524, 562, 563, 608, 613, 775

G

Gene expression, 123, 124, 126, 144, 145, 267, 280, 285, 328, 331, 408, 410, 412, 413, 418, 422, 431, 432, 468, 475, 478, 482, 483, 485
Gene expression in plants
healthy plant, 433, 540, 660, 702, 725
infected plant, 295, 516, 531, 653, 709, 712, 725, 728
Gene expressions, 516, 522, 523, 553, 556, 567, 615, 625, 659, 707, 748
Gene ontologies, 732, 735
Genes
characterization, 95, 119, 121, 122, 143, 147, 239, 288, 315, 483, 488, 553, 567, 601–607, 610, 620, 638, 777
cloning, 67, 78, 94, 119, 121, 122, 146, 148, 150, 151, 239, 289, 471, 483, 489, 512, 553, 554, 602, 607, 773
differential expression, 7, 478, 526, 604, 749
expressions, 68, 119, 122, 125, 131, 143, 147, 150, 200, 272, 275, 281, 286, 288, 301, 330, 409, 412–417, 419, 423, 424, 430, 432, 470–472, 475–477, 481, 483, 486, 488, 494, 496, 513, 523, 541, 543, 551, 554, 556, 559, 560, 562, 571, 604, 605, 619, 694
functions, 86, 87, 92, 150, 224, 415, 416, 424, 431, 432, 469, 473, 476, 478, 513, 517, 520, 541, 543, 549, 553, 556, 560, 562, 619, 731, 735, 750
inheritance, 86, 95, 124, 134, 196–200, 202, 204, 206, 218, 221, 227, 231, 233, 235–237, 242, 247–250, 266, 268, 271, 278, 296, 303, 314, 468, 493, 512, 552, 554, 606, 610, 611, 613, 615, 624, 625, 628, 636
interactions, 78, 87, 92, 95, 124, 131, 139, 143, 147, 156, 197–199, 234, 248, 281, 284, 286, 297, 315, 319, 360, 410, 411, 417, 424, 431, 467, 471, 472, 474, 477, 478, 482, 483, 486, 511, 512, 516, 523, 539, 541, 543, 549, 553, 556, 562, 570, 601–603, 608, 609, 611, 613, 615, 619, 620, 624, 774, 776
isolation, 130, 149, 329, 602
mapping, 78, 92, 95, 115, 116, 119, 199, 215–221, 224, 225, 231, 234, 238–246, 271, 289, 291, 296, 297, 306, 309, 316, 474, 475, 483, 489, 490, 493, 541, 549, 550, 553, 557, 602, 603, 616, 620, 631, 637, 693, 721
mutations, 92, 119, 221, 224, 274, 316, 479, 495, 541, 549, 551, 568, 569, 776
ontologies, 732, 735
regulations, 123, 139, 411–414, 418–420, 422–426, 432, 474, 477, 479, 482, 514, 516, 560, 605, 609, 619
sequencing, 99, 119, 120, 141, 143, 148, 158, 244, 245, 306, 308, 309, 315, 316, 411, 417, 473, 483, 512, 526, 553, 607, 608, 610, 619, 631, 632, 637, 638, 694, 774
structures, 119, 131, 158, 290, 307, 310, 330, 331, 474, 483, 490, 555, 634, 637, 750
Genetic, 86
Genetic mapping, 121, 141, 245, 473, 691
Genetic variations, 7, 10, 24, 114, 200, 206, 287, 318, 487, 567, 568, 572, 605, 692
Genetics of disease resistance, 610–615
Genetics of host-pathogen relationship, 36
Genome comparison, 14
Genome complexity, 32
Genome prediction, 112–114, 658, 718, 734
Genome wide association, 114–116, 141, 154, 308, 309, 317, 475, 483, 540, 550, 631, 729, 731, 734
Genomics
pathogenesis, 620
pathogens, 144, 151, 156, 319, 468, 495, 602, 604, 606, 616, 620, 624, 708, 773, 774
pathotypes, 236, 241, 243–246, 620
Genomics approaches
bioinformatics, 43, 67, 125, 156, 411, 471, 483, 602, 607, 621, 658
biometabolomics, 44, 478, 482, 601, 620–622
epigenetics, 28, 33, 43, 418–419, 471, 473

Genomics approaches (*cont.*)
genomics, 121, 122, 124, 471, 543
metabolomics, 42, 44, 416, 471, 480, 552, 554, 776
metagenomics, 45, 471
pangenomics, 19, 45, 150–153, 158, 471, 601, 608, 776
pathogenomics, 45, 471
pharmacogenetics, 45
phylogenomics, 45, 471
proteomics, 42, 43, 125, 126, 471, 473, 478, 482, 483, 553, 601, 619, 620, 776, 777
toxicogenomics, 44
transcriptomics, 2, 42, 43, 121, 123–127, 152, 471, 473, 478, 482, 483, 496, 522, 553, 601, 620, 776
Geographical distribution (GD), 3, 4, 388, 398, 406, 472, 602
Germany, 95, 116, 117, 250, 299–301, 305, 361, 362, 364, 368, 372–374, 377, 378, 381, 383, 385, 387, 520, 538, 630
Germplasm evaluation, 280, 700–704
Germplasm screenings, 512, 655–656, 660–661, 678–679, 722–723, 735–736, 741–742
Greece, 250, 362, 368, 370, 372, 374, 378, 381, 385
Global distribution
disease, 3, 4, 602
pathotypes, 608
Guatemala, 368, 372, 374, 382
Guyana, 373, 386

H

High efficiency integrated breeding (HIB), 607
Homeostasis, 124–126, 415, 420, 422, 482, 526
Homoeologous exchanges (HEs), 18, 151, 152
Homologus DNA fragments, 148–150
Hormonal responses
metabolism, 420, 481
salicylic acid, 522
signaling, 3, 420–421, 430, 477, 481, 522, 603
Hormones, 470
auxins, 125, 529
cytokinins, 125, 481
ethylene, 398, 408, 431, 470, 529
gibberellins, 409, 529
Host-defense to biotrophs
host defense to hemibiotrophs, 66, 415, 472, 474, 482, 561, 607, 613, 620
host defense to necrotrophs, 21, 66, 267, 322, 472, 474–476, 482, 606, 607, 613, 616, 620
mechanisms, 322, 469, 471, 474, 613, 620
Host gene expressions, 123
Host growth stages, 472
Host-pathogen interaction, 35, 41, 288
Host resistance
approaches, 125, 130, 468, 474, 513, 601, 620, 774
biochemical, 131, 482, 512, 620
durable, 131, 514, 602, 609, 775
functional, 232, 242, 245, 478, 482, 513, 601, 620
genetical, 124, 378
genomics, 132, 468, 474, 483, 599–638, 773
histological, 197
induced, 130, 198, 221, 287, 409, 413, 415–419, 423, 425, 430, 474, 480, 487
manipulation, 268, 280, 468, 562, 615, 625, 626
mechanisms, 288, 410–421, 426, 774
molecular, 143, 199, 232–235, 242, 288, 398, 407, 408, 411, 415, 424, 425, 431, 468, 471–476, 482, 483, 488, 512, 601, 609, 620, 773
multiple, 206, 215, 221, 249, 250, 359–436, 477, 550, 617–619
principles, 1–46
Hungary, 361, 362, 364, 368, 370, 372, 374, 381, 385, 387
Hyaloperonospora parasitica (*H. parasitica*), 4, 8, 11, 23, 87, 88, 90, 92, 93, 95, 109, 122, 144, 145, 149, 226–233, 278, 279, 373, 477, 479, 480, 508, 514, 515, 567, 568, 571, 572, 606, 614, 615, 623, 628, 675–682, 686, 687
Hyaloperonospora –*Arabidopsis* pathosystem, 23–24
Hybridization
interspecific, 137, 267, 284, 285, 311, 313, 552, 621, 634, 636

I

Iceland, 368, 372, 382
Identification of R-genes
Brassica crops, 3, 67
Brassica species, 67, 94, 121, 128–129
wild relatives, 67, 607
In vitro, 77, 127, 268, 284–285, 311, 312, 330, 430, 469, 484–485, 516, 555–558, 565, 625, 634, 635, 659, 661, 698–699, 707, 722, 739, 740
In vivo, 77, 425, 553, 555, 556, 660
India, 68–74, 79, 88, 89, 95, 116, 117, 130, 131, 133, 138, 139, 198, 228, 276, 299, 316,

361–364, 366, 367, 369–371, 373, 374, 377, 378, 380, 383, 384, 387, 389, 394, 396, 400–406, 546, 559, 626, 654, 774
Indonesia, 299, 367, 371
Induced resistance (IR), 508, 513, 570, 571, 610
Inheritance of disease resistance
  Alternaria blight, 613
  clubroots, 613
  downy mildew, 613
  Fusarium wilt, 224
  Sclerotinia Stem rot, 14–18, 34, 98, 142–144, 309, 315, 321, 379, 467, 547, 608, 624, 631, 724
  Turnip mosaic virus, 249–250
  powdery mildews, 217, 613
  White rust, 198, 202, 613
  Xanthomonas black rot, 551–554
Inheritance of resistance genes
  additive, 268, 512, 611, 613, 614, 625
Integrated disease management, 622
Interspecific hybridization, 12, 34, 70, 137, 157, 285, 289–290, 300, 313, 314, 489, 551, 552, 606, 616, 623, 635, 636
Intracellular receptors, 15, 25, 27–29, 471, 474, 606
Introgressions, 66, 108, 115, 120, 121, 128, 129, 132, 133, 137, 143, 144, 155, 156, 158, 267, 271–272, 280, 286, 291, 292, 298, 299, 303, 307–312, 314, 315, 317, 318
Invertase, 236, 468, 475, 484, 486, 490, 493, 523, 545, 552, 557, 607, 608, 615, 616, 621, 623–625, 631–637, 693, 730–735, 739–741, 776
Iran, 361, 362, 364, 367, 370, 371, 374, 378, 380, 384, 396
Iraq, 370, 371, 374, 384, 387
Ireland, 361, 364, 368, 372, 374, 377, 381, 385
Isolates, 68, 86–99, 108, 112, 114, 115, 119, 124, 126, 127, 130, 132, 134, 136, 139, 143–147, 149, 158, 198–200, 226–230, 232–235, 237–239, 241, 243, 249, 251, 268, 281, 288, 291, 297–303, 308, 315, 316, 472–474, 483, 488, 490, 507–510, 514, 515, 517–520, 522, 535–537, 540, 541, 543, 547, 548, 550, 551, 555, 559, 560, 569, 606, 608, 612, 614, 623, 626, 632, 638, 652–653, 656, 659, 662, 665–668, 676–679, 682, 684, 693, 694, 696, 699, 702, 703, 707, 716, 720, 726, 727, 729, 732, 738, 743–747, 776
Israel, 364, 367, 370, 371, 374, 380, 384, 387
Italy, 89, 361, 362, 364, 368, 370, 372, 374, 381, 385, 387

J
Japan, 89, 98, 116, 117, 138, 241, 303, 361–364, 366, 367, 371, 374, 377, 378, 380, 384, 387, 401, 403, 405, 536, 537

K
Korea, 95, 98, 362, 364, 367, 371, 374, 378, 380, 385
Kyrgyzstan, 364, 371

L
Latvia, 364, 372, 377, 382, 385
Leaf and stem rot, 247, 248
Leptosphaeria maculans, 3, 4, 11, 14, 93–95, 109, 111, 112, 114–116, 145–148, 267, 278, 286, 315, 316, 319, 321, 325, 330, 337, 363, 475, 481, 484, 486, 517–520, 560, 561, 566–569, 613, 628, 689, 693–696, 774
Linkage markers, 473
Lithuania, 364, 372, 382, 387
Luxembourg, 372

M
Major diseases
  distribution, 4
  global ranking, 4
  host range, 4
  yield losses, 4
Major gene
  mapping, 95, 301
  mode of segregating, 203, 215
  monogenic, 251
Malawi, 361, 366, 371, 375, 380, 384
Malaysia, 367, 371, 375, 384, 387
Management of disease resistance
  arginase gene, 533, 623
  biocontrol agents, 329, 336, 524–526, 561–566, 623
  deployment of R-genes cvs, 305, 306, 521–522, 623, 624, 630–631, 774
  integrated approach, 536, 623, 624
  non-host resistance, 8, 27, 29, 35, 470, 474, 480, 508, 513, 514, 523, 606, 617, 623
  omics approaches, 6, 67, 144, 158, 470, 475, 483, 552–554, 601, 604, 606, 607, 672
  pyramiding of R-genes, 554, 623
  slow white rust genotypes, 507
  systemic resistance, 15, 287, 477, 487–488, 507, 562, 566, 616, 621, 623

Management of disease resistance (*cont.*)
transgenic approaches, 483, 556–560
under field conditions, 12, 80, 82, 142, 298, 506, 537, 544, 545, 560, 610, 620, 622, 708–715
Mechanism, 2, 119, 269, 408, 467, 508–512, 601, 657, 773
Mechanism of resistance, 145, 147, 199, 304–305, 512, 524, 665
Metabolic pathways, 42, 44, 123, 305, 553, 560, 629
Metabolites
amino acids, 13, 43, 525, 534, 551
brassinosteroid, 325
camalexin, 23, 476, 480–482, 514, 612
fatty acids, 477
glucosinolates, 424, 476, 481, 526
induction, 408, 431, 476, 482, 620
lipids, 482
starch, 523, 536, 537, 676, 687, 698
sugars, 22, 132, 689
trehalase, 523, 524
Mexico, 95, 365, 368, 372, 375, 378, 382, 386
Microarray analysis, 43, 272, 412, 415, 493
MicroRNAs, 476
miRNAs
pathogenesis, 417
Molecular analysis, 144, 145, 556, 563, 665, 673, 674
Molecular approaches, 66, 139–141, 266, 468, 484–488, 607, 608, 615, 616, 621, 623, 624
Molecular host resistance
genetic modification, 419
Molecular markers, 86, 94, 144, 146, 148, 153, 154, 158, 196, 224, 241, 286–287, 289–292, 307, 314, 316–319, 473, 475, 483, 487, 489, 490, 541, 549, 552, 553, 602, 613, 616, 621, 633, 636, 673–675, 717, 721, 775
Molecular plant pathology, 472
Molecular techniques
pathogenic variability, 34, 66, 280, 512, 607
pathogenicity assay, 286, 487
PCR, 714
PCR (QRT-PCR), 526, 659, 749
pyramiding of QTL's, 267, 308, 615, 633
qPCR, 714–715
RT-PCR, 139, 141, 275, 496, 659, 694, 746–747
single spore isolates, 95
single spore isolation, 707–708
transfer of resistance, 268, 279, 511, 621, 624, 625
Montenegro, J.D., 372, 382, 385
Multiple disease resistance
multiple pathotypes, 121, 540, 608
mutational approach, 92
nomenclature, 65, 87, 90, 92, 94, 146, 147, 519, 608, 721, 773
Mustards, 68, 74–76, 81–85, 88, 89, 137, 198, 199, 209–211, 213, 276, 277, 286, 310, 312, 313, 319, 322, 383, 402, 403, 405, 467, 486, 511, 518, 520, 526, 547, 549, 552, 554, 557–559, 604, 611, 615, 624, 627, 634–636, 666, 675, 744–746, 748
Mutagenic resistance, 292–294, 477, 491–492, 616
Mutants, 81–83, 85, 92, 93, 119, 130, 221–225, 272, 274, 275, 279, 293, 412–414, 418, 424, 426, 430, 470, 472, 476–480, 492, 493, 495, 496, 513, 514, 527, 529, 531, 533–535, 543, 544, 549, 561, 563, 571, 605, 609, 610, 613, 663, 668, 672
Mutational approach, 92–93
Myrosinase, 126, 337

N
NBS encoding genes, 6, 7, 13–15, 17, 18, 604, 750, 751
Nepal, 361, 367, 370, 371, 375, 380, 384, 387
Netherlands, 361, 364, 368, 372, 375, 377, 381, 385, 387
New Zealand, 93, 95, 146, 235, 361, 362, 365, 369, 371, 375, 377, 378, 382, 386, 387, 405
Norway, 298, 361, 364, 368, 372, 375, 377, 381, 387

O
Oilseed rape, 75, 93, 98, 125, 127, 130, 131, 134, 145, 146, 226, 238, 241, 291, 300, 301, 304, 467, 490, 517, 518, 520, 536, 537, 539, 540, 551, 554–556, 615, 624, 629, 678, 701, 736, 746, 747
Omics, 67, 150–154, 157, 158, 468, 470, 474, 475, 483, 552–554, 601, 606–608, 672, 776
Over expression of
enzymes inhibitors, 470, 513
PR proteins, 470, 513
phytoalexins, 470, 480, 513, 612, 629

P
Pakistan, 361, 363, 364, 370, 371, 375, 380, 385, 547
Papua New Guinea, 365, 369, 371, 375, 386
Partial
  Avirulence genes, 433
Partial resistance, 3, 72, 95, 96, 98, 127, 129, 130, 147, 200, 201, 204, 211, 217, 226, 231–233, 280, 281, 475, 476, 543, 603, 611, 612, 614, 615, 655, 665–666, 686, 726, 738
Pathogen
  genomes, 115, 119, 122, 124, 125, 142, 144, 147, 196, 232, 249, 287, 314, 318, 468, 470, 471, 487, 506, 534, 538, 553, 608, 611, 613, 622, 637, 774
  perception, 470, 478, 620
  phylogeny, 287, 487
  quantification, 471, 670, 680, 708, 710
Pathogenesis
  effector genes, 620
  genomics, 414, 611, 620
  miRNAs, 417, 418, 482, 619
  Pathotype specific R, 118, 120, 134
Pathogenic variability
  identification, 512, 607
Pathotypes, 66, 73, 74, 87–92, 95–98, 108, 112, 116, 118–121, 124, 134, 136, 141–145, 198, 199, 236, 237, 241–243, 245, 276, 281, 297, 298, 300, 303–306, 315, 316, 473, 507, 512, 517, 518, 521, 522, 536, 538–541, 543, 545, 547, 554, 607, 608, 610, 613, 614, 616, 621–623, 626, 629–631, 638, 693, 701–704, 715–718, 773, 777
Phenylpropanoid, 153, 478, 482, 527
Philippines, 361, 364, 367, 371, 375, 377, 384, 387, 401
Physiologic specialization, 698
Phytoalexins, 271, 281, 288, 305, 470, 475–477, 480, 481, 488, 513, 527, 528, 612, 629
Phytohormones, 390, 398, 423, 472, 474, 477, 618, 619
Plasmodiophora brassicae
  genomes, 613
  molecular biology, 472, 483, 610
  recombination, 197, 223, 240
Poland, 299, 361, 362, 364, 372, 375, 378, 381, 383, 385, 700, 701
Polymorphism, 116, 136, 154, 156, 200, 221, 234, 244, 251, 287, 317, 471, 487, 571, 573, 658, 674, 717, 721, 727, 730, 733, 737
Portugal, 68, 72, 95, 364, 368, 372, 375, 377, 381, 385
Post-penetration, 513
Primers, 75, 148–150, 287, 312, 487, 635, 659, 663, 664, 668–670, 673–675, 691, 693, 694, 696, 705, 708–711, 713, 714, 717, 718, 721, 730, 735, 741, 746–750
Programmed cell death (PCD), 398, 425, 428, 477, 478, 513, 514, 522, 557
Protease inhibitors, 332, 336
Proteins, 78, 92, 93, 108, 116, 118–121, 125–127, 131, 132, 150, 151, 153, 155, 157, 200, 232–234, 239, 242, 243, 275, 276, 281, 286–288, 290, 308, 310, 315, 319, 323–337, 398, 408–412, 415–416, 418–425, 427–432, 468–475, 477–482, 486, 488, 490, 496, 497, 513, 522–526, 533, 534, 541, 551, 553, 555, 558–560, 563, 565, 567, 570, 603, 606, 607, 611, 612, 618–620, 631, 632, 638, 658, 665, 718, 732, 735, 748, 750, 751, 777
Proteomics, 125–127, 471, 473, 478, 482, 483, 552, 553, 601, 619, 620, 776, 777
Protocols, 272, 307, 320, 483, 494, 558, 616, 623–625, 637, 652, 659, 665, 669, 670, 691, 694, 703, 705, 707, 708, 710, 716, 721, 729, 730, 741, 743
Puerto Rico, 365, 368, 372, 375, 386, 388
Pythium, 373

Q
Quantitative trait loci (QTL), 14, 78, 86, 95, 96, 98, 99, 109–115, 119, 121, 122, 124, 128, 136, 141–143, 145, 147, 148, 153–155, 157, 158, 238–244, 247, 248, 250, 267, 271, 296–298, 302–304, 473–476, 480, 483, 493, 522, 539–540, 543, 553, 602, 606, 608, 613–616, 620, 621, 629, 691, 692, 695–697, 718–720, 727, 730, 773

R
Races, 66–70, 72, 77, 86–92, 94–97, 108, 114, 124, 125, 131, 137–140, 144, 146, 196–202, 204, 205, 224, 226, 235, 236, 250, 251, 277, 281, 296–298, 300, 302, 310–312, 314, 315, 321, 471, 472, 475–477, 541, 543, 552, 603, 605, 607, 608, 613,

614, 621, 628, 634–638, 654, 655, 657, 687, 719, 738, 739, 741, 776, 777
Rapeseed-mustard, 72, 74, 79, 133, 215, 282, 467, 624
Raphanus raphanistrum, 12, 133, 320, 397
Raphanus sativus, 68, 71, 76, 77, 87, 89, 116, 117, 133, 134, 138, 143, 197, 201, 202, 235, 239, 277, 296, 303–304, 313, 327, 329, 512, 518, 551, 554, 616, 627, 636
Rating scales, 655, 666, 679, 693, 741, 742
Receptor like kinase (RLK), 3, 111, 116, 151, 153, 155, 310, 411, 412, 420, 471, 473, 603, 619
Relationship between diseases, 278, 627–629
Research priorities, 773–777
Resistance
  induction, 288, 425, 470, 475, 477, 478, 487–488, 507–508, 513, 523, 570–572, 616, 621, 623
  mechanisms, 119, 124, 125, 139, 141, 145, 156, 196, 232, 246, 249, 271, 304, 306, 310, 318, 331, 408, 411, 419, 424, 426, 467–477, 480, 482, 483, 508–513, 522, 523, 526, 536, 549, 553, 555, 566, 568, 570, 573, 601–603, 606, 609–615, 618, 620, 623, 624, 629, 631, 773, 775
  modulation, 539, 543
  partial, 122, 127, 150, 613, 614
Resistance genes (R-genes)
  characterization, 67, 78, 95, 108, 119, 121, 122, 138, 143, 147
  cloning, 67, 78, 94, 119, 121, 122, 146, 148, 150, 151
  deletion, 119, 121, 136
  designation, 87–92
  expression, 68, 82, 119, 122, 123, 125, 126, 129, 131, 141, 143–145, 148
  genetics, 68, 73, 90, 92, 109, 112, 113, 119, 121, 125, 127, 128, 131, 138–140
  identification, 66–159
  interaction, 78, 87, 92, 95, 97, 124, 127, 131, 139, 143, 147, 151, 156, 157
  isolation, 130, 149
  novel sources, 94–112, 147, 268
  pathotype specific, 73, 116, 118, 120, 125, 134–136, 143
  polygenic, 86, 95, 124, 130, 140
  QTLs, 78, 86, 95, 109, 112, 114, 116, 119, 121, 122, 124, 128, 135, 138, 141, 143, 145, 148, 153–158
  qualitative, 66, 94, 95, 112, 124, 146, 148
  quantitative, 66, 78, 94, 112, 114, 115, 119, 124, 128, 130, 136, 139–141, 146, 148, 156
Resistant, 68–70, 72–74, 76–82, 86, 87, 89, 93, 95, 108, 118–121, 124–126, 128–132, 134, 139, 141–143, 145, 148, 150, 153, 155, 197–207, 209, 214, 215, 217–219, 221–223, 225–229, 232, 233, 235–237, 239–241, 243, 245, 247, 249, 250, 269, 271, 276, 277, 279–281, 285, 286, 288, 289, 291, 292, 297–301, 303–305, 307, 308, 312–316, 319, 329, 330, 411, 414, 417, 418, 424, 468, 470, 474–478, 481, 482, 484–486, 488–491, 493, 496, 507, 511–513, 515, 520–522, 524, 526, 534, 536, 538, 540, 541, 543–546, 549, 550, 552–555, 557, 567, 568, 570–573, 601, 603, 607–611, 613–617, 621–623, 625–633, 635–637, 653, 655, 657, 661–663, 672, 674, 678, 681, 682, 686, 687, 690, 701, 702, 704, 708, 717, 720–722, 726, 727, 733, 738, 739, 741, 742, 744, 774, 775, 777
Recognition
  R-genes cvs, 623
R-loci, 215, 474, 483, 728–730
Role of R-genes, 775
Romania, 361, 364, 368, 372, 376, 377, 381, 385, 387

S

Sanitization, 539
Sao Tome and Principe, 370
Sclerotinia sclerotiorum, 3, 5, 14, 22, 98, 109, 127–134, 142, 143, 248, 275, 286, 307–309, 315, 316, 319, 321, 326, 329–331, 379, 486, 496, 548, 561, 567, 612, 631–633, 722, 724–730, 732, 736, 774
Sclerotinia stem rot
  Inheritance of disease resistance, 14–16, 33
Serbia, 362, 368, 372, 382, 385
Signaling, 157, 287, 331, 360–433, 469, 470, 472, 474, 476–479, 481, 482, 488, 513, 523, 535, 562, 563, 565, 567, 605, 606, 611, 617–620, 777
Serology, 746
Signaling molecules
  ethylene, 476, 620
  jasmonic acid, 476, 620
  salicylic acid, 398, 418, 423, 476
Signaling network system
  abiotic stresses, 360, 389, 391, 398, 408, 411–413, 415, 417–421, 424, 429–432, 434
  biotic stresses, 408
  complexity, 411, 479, 607

epigenetics, 418–419
genes involved, 409, 410, 412
hormonal stresses, 420–421
micro RNA's analysis, 417–418
molecular mechanisms, 411–421
multiple stresses, 360–436, 560, 600, 601, 617–619, 777
receptor proteins, 408, 411–412
role of CMLS, 423–424
role of CPK's, 426
systemic signaling, 419–420
transcription factors, 408, 411, 413–417, 424, 425, 428, 431, 432
Signaling pathways, 21, 23, 398, 408–410, 413–416, 421–432, 469, 471, 472, 474, 476–478, 481, 522, 523, 526, 534, 603, 618, 619, 775, 777
Sinapis alba (*Sinapis arvensis*), 73, 76, 77, 79, 108, 128, 130, 133, 137, 284, 314, 485, 518, 554, 621, 627, 637, 775
Sinapis species (White Mustard), 76, 276, 554, 627
Singapore, 361, 371, 380
Sisymbrium species, 90
Slovakia, 364, 372, 381, 385
Sources of resistance to
Albugo, 621
Alternaria, 612, 621
Colletotrichum, 4, 87, 362, 474, 481, 608
Erysiphe, 621
Fusarium, 224, 608
Hyaloperonospora, 233, 234, 621
Leptosphaeria, 608, 621
Plasmodiophora, 243, 245, 608, 621
Sclerotinia, 247–249, 608, 626
Xanthomonas, 138–140, 250, 251, 608, 621
Sources of R-genes
rutabaga, 235, 241
Use of R-genes, 13–15, 18, 520, 629–631
transfer, 467, 606, 615, 621, 624
South Africa, 361, 363, 370, 373, 376, 379, 384, 387
Spain, 74, 361, 364, 372, 374, 376, 381, 385
Spongospora subterranean, 707
Sri Lanka, 361, 367, 371, 376, 380, 387, 401
Sweden, 95, 361, 364, 368, 372, 376, 379, 382, 383, 385, 387, 547
Switzerland, 361, 364, 368, 372, 376, 379, 381, 385
Systemic acquired resistance (SAR), 19, 93, 331, 430, 470, 471, 473, 476, 479, 513, 543–544, 570, 572

T
Taiwan, 74, 361, 364, 367, 371, 376, 380, 385, 387
Techniques, 86, 121, 130, 137, 139, 143, 154, 157, 206, 239, 247, 268, 272, 279–281, 283–285, 287, 307, 308, 310–315, 331, 468, 471, 474, 483–485, 487, 494, 512, 516, 522, 552, 553, 558–560, 609, 610, 612, 623–625, 632–636, 652–751, 777
Technology
modern, 474, 607
traditional, 286, 486, 556, 607
Thailand, 364, 367, 371, 376, 380, 387
Toria, 282, 653, 702
Transcription, 331, 408, 411, 413–417, 423–426, 428, 431, 432, 473, 476, 477, 479, 482, 514, 523, 526, 556, 565, 567, 618, 619, 659, 694, 746, 777
Transcriptomic analysis, 121, 274, 496, 522, 694
Transcriptomics, 123–127, 152, 157, 471, 473, 478, 482, 483, 523, 552, 601, 602, 620, 776
Transcriptional programming, 476
Transfer of disease resistance, 86, 94, 144, 146, 148, 153, 158, 286–287, 289–291, 307, 308, 314, 316, 317, 473, 483, 486, 489, 549, 602, 621, 674, 717, 721, 775
biotechnological approaches, 79, 279, 284–288, 486, 513, 608, 612, 616
development of R-cvs, 116
development of transgenics, 130, 275, 319–338, 496, 511, 623
distant hybridization, 267, 290, 313–314, 489, 551–552, 615, 616, 621, 635, 775
embryo rescue, 37, 267, 280, 281, 283, 284, 290, 292, 298, 310, 311, 468, 484–485, 489, 514, 551–552, 615, 616, 621, 623, 625, 633–637, 673, 741, 776
endo-chitinase gene, 275–276, 496, 616, 621
gene silencing, 67, 272, 307, 473, 494, 495, 516, 608, 616
germplasm sources, 267–268, 625–626
hexaploidy approach, 632
hybridization, 267, 268, 275, 279, 280, 283, 289, 290, 300, 310, 311, 313, 314, 318, 320
inter-or-intra specific, 271, 493, 512, 611, 616, 621
introgression breeding, 267, 484, 621
microarray based, 272, 493

Transfer of disease resistance (*cont.*)
  molecular markers, 2, 42, 473, 475, 489, 541, 613, 616, 633, 636
  mutagenic, 279, 292–294, 477, 489, 514, 616, 621
  non-host plants, 268–272, 274, 474, 492–495, 511, 512, 616, 621
  pyramiding of R QTL's, 726–728
  somatic hybridization, 268, 283, 285, 310, 311, 313–314, 485, 551–552, 616, 621, 623, 625, 634–637
Transfer of R-genes, 289–294, 306, 314–315, 489, 490, 616, 637–638
Transgenics, 130–132, 239, 267, 272, 275–276, 281, 286, 319–338, 412, 414, 415, 419, 468, 476, 483, 486, 493, 496–497, 511, 516, 541, 549, 555–561, 611, 612, 615, 623, 625, 659, 693, 696, 776, 777
Trinidad and Tobago, 362, 369, 376, 386
Turkey, 361, 363, 364, 367, 370, 371, 376, 379, 380, 385, 387
Turnip mosaic virus (TuMV), 3, 5, 40, 134, 135, 249, 250, 322, 323, 331, 383, 408, 416, 473, 474, 483, 549–551, 608, 614, 623, 680–682, 742–746, 775
Turnip rape
  vascular system, 321
  Zoospores, 657
Turnips, 68, 75, 95, 97, 118, 124, 125, 141, 149, 151, 153, 236, 238, 239, 243, 245, 250, 251, 289, 291, 384, 399–403, 489, 490, 720

U
Uganda, 361, 366, 371, 376
Ukraine, 364, 372, 382, 383, 385, 387
United Kingdom, 364, 368, 382, 386, 720
United States, 224, 363, 365, 369, 382, 386

V
Venezuela, 373, 376, 383
Verticillium, 327, 383, 412, 476, 482, 516, 560
Vietnam, 367, 371, 376, 379, 380, 387
Virulence
  patterns, 280, 468, 538, 539

W
Weed infestation, 396

Y
Yugoslavia (Serbia and Montenegro), 368, 372, 376, 381

Z
Zambia, 362, 363, 371

The manufacturer's authorised representative in the EU is Springer Nature Customer Service Centre GmbH, Europaplatz 3, 69115 Heidelberg, Germany. If you have any concerns regarding our products, please contact ProductSafety@springernature.com

Printed and bound by CPI Group (UK) Ltd, Croydon, CR0 4YY
23/09/2026
02221523-0008